NUTRITION AND ENHANCED SPORTS PERFORMANCE

NUTRITION AND ENHANCED SPORTS PERFORMANCE

MUSCLE BUILDING, ENDURANCE, AND STRENGTH

Edited by

DEBASIS BAGCHI, PhD, MACN, CNS, MAIChE

Department of Pharmacological and Pharmaceutical Sciences
University of Houston College of Pharmacy
Houston, TX, USA

SREEJAYAN NAIR, PhD

University of Wyoming
School of Pharmacy
Laramie, WY, USA

CHANDAN K. SEN, PhD, FACN, FACSM

Center for Regenerative Medicine & Cell-Based Therapies
The Ohio State University Wexner Medical Center
Columbus, OH, USA

AMSTERDAM • BOSTON • HEIDELBERG • LONDON
NEW YORK • OXFORD • PARIS • SAN DIEGO
SAN FRANCISCO • SINGAPORE • SYDNEY • TOKYO

Academic Press is an imprint of Elsevier

Academic Press is an imprint of Elsevier
32 Jamestown Road, London NW1 7BY, UK
225 Wyman Street, Waltham, MA 02451, USA
525 B Street, Suite 1800, San Diego, CA 92101-4495, USA

Medicine is an ever-changing field. Standard safety precautions must be followed, but as new research and clinical experience broaden our knowledge, changes in treatment and drug therapy may become necessary or appropriate. Readers are advised to check the most current product information provided by the manufacturer of each drug to be administered to verify the recommended dose, the method and duration of administrations, and contraindications. It is the responsibility of the treating physician, relying on experience and knowledge of the patient, to determine dosages and the best treatment for each individual patient. Neither the publisher nor the authors assume any liability for any injury and/or damage to persons or property arising from this publication.

Notice
No responsibility is assumed by the publisher for any injury and/or damage to persons or property as a matter of products liability, negligence or otherwise, or from any use or operation of any methods, products, instructions or ideas contained in the material herein.

Because of rapid advances in the medical sciences, in particular, independent verification of diagnoses and drug dosages should be made

British Library Cataloguing-in-Publication Data
A catalogue record for this book is available from the British Library

Library of Congress Cataloging-in-Publication Data
A catalog record for this book is available from the Library of Congress

ISBN: 978-0-12-396454-0

For information on all Academic Press publications
visit our website at elsevierdirect.com

Typeset by MPS Limited, Chennai, India
www.adi-mps.com

Printed and bound in United States of America

13 14 15 16 17 10 9 8 7 6 5 4 3 2 1

Dedication

Dedicated to my Beloved Nephew Krishnendu Bagchi

Contents

8. Immune Function, Nutrition, and Exercise

WATARU AOI, YUJI NAITO AND TOSHIKAZU YOSHIKAWA

9. The Immune Response to Exercise: Effects on Cellular Mobilization, Immune Function and Muscle Regeneration

DANIEL J. FREIDENREICH AND JEFF S. VOLEK

III
SPORTS AND NUTRITION

10. Vegetarian Athletes

JING ZHOU, JIA LI AND WAYNE W. CAMPBELL

11. Nutrition in Combat Sports

GUILHERME G. ARTIOLI, EMERSON FRANCHINI, MARINA Y. SOLIS, ALINE C. TRITTO AND ANTONIO H. LANCHA JR

12. Sumo Wrestling: An Overview

TAISHI MIDORIKAWA, SHIZUO SAKAMOTO AND MASAKATSU KONDO

13. Bioenergetics of Cyclic Sports Activities on Land: Walking, Running and Cycling

PAOLA ZAMPARO, CARLO CAPELLI AND SILVIA POGLIAGHI

14. Bioenergetics of Cyclic Sport Activities in Water: Swimming, Rowing and Kayaking

PAOLA ZAMPARO, AND MARCO BONIFAZI

15. Performance Enhancement Drugs and Sports Supplements: A Review of the Evidence

GARY GAFFNEY

16. Nutrition and Ultra-Endurance: An Overview

BEAT KNECHTLE

17. Exercise and Nutritional Benefits for Individuals with a Spinal Cord Injury or Amputation

JUSTIN W. KEOGH

18. An Overview of Doping in Sports

FABIAN SANCHIS-GOMAR, VLADIMIR MARTINEZ-BELLO, HELIOS PAREJA-GALEANO, THOMAS BRIOCHE AND MARI CARMEN GÓMEZ-CABRERA

19. Nutrition In Paralympics

AMITAVA DAS, DEBASIS BAGCHI AND CHANDAN K. SEN

20. An Overview on the History of Sports Nutrition Beverages

GUSTAVO A. GALAZ

IV MOLECULAR MECHANISMS

21. α-Hydroxy-Isocaproic Acid (HICA)—Effects on Body Composition, Muscle Soreness and Athletic Performance

TUOMO OJALA, JACOB M. WILSON, JUHA J. HULMI, TUOMO KARILA, TIMO A. SEPPÄLÄ AND ANTTI A. MERO

22. Role of Mammalian Target of Rapamycin (mTOR) in Muscle Growth

EVGENIY PANZHINSKIY, BRUCE CULVER, JUN REN, DEBASIS BAGCHI AND SREEJAYAN NAIR

23. Stress Proteins and Heat Shock Proteins: Role in Muscle Building and Sports Nutrition

MIKA VENOJÄRVI, NIKU OKSALA, SUSANNA KINNUNEN AND MUSTAFA ATALAY

24. Anabolic and Catabolic Signaling Pathways that Regulate Skeletal Muscle Mass

JOHN J. MCCARTHY

25. Muscle Growth, Repair and Preservation: A Mechanistic Approach

ROBERT M. ERSKINE AND HANS DEGENS

26. Nitric Oxide, Sports Nutrition and Muscle Building

LAWRENCE J. DRUHAN

35. Anabolic Training Response and Clinical Implications

JAN SUNDELL

36. Requirements of Energy, Carbohydrates, Proteins, and Fats for Athletes

CHAD M. KERKSICK AND MICHELLE KULOVITZ

37. An Overview of Branched-Chain Amino Acids in Exercise and Sports Nutrition

HUMBERTO NICASTRO, DANIELA FOJO SEIXAS CHAVES AND ANTONIO HERBERT LANCHA JR

38. Water, Hydration and Sports Drink

FLAVIA MEYER, BRIAN WELDON TIMMONS AND BOGUSLAW WILK

39. Physiological Basis for Creatine Supplementation in Skeletal Muscle

WILLIAM J. KRAEMER, HUI-YING LUK, JOEL R. LOMBARD, COURTENAY DUNN-LEWIS AND JEFF S. VOLEK

40. Oral Bioavailability of Creatine Supplements

DONALD W. MILLER, SAMUEL AUGUSTINE, DENNIS H. ROBINSON, JONATHAN L. VENNERSTROM AND JON C. WAGNER

41. An Overview of Carnitine

RICHARD J. BLOOMER, TYLER M. FARNEY AND MATTHEW J. MCALLISTER

42. An Overview of the Influence of Protein Ingestion on Muscle Hypertrophy

KOJI OKAMURA

43. An Overview of Ornithine, Arginine and Citrulline in Exercise and Sports Nutrition

KOHEI TAKEDA AND TOHRU TAKEMASA

Preface

Nutrition plays a pivotal role in sustaining human health, extending healthy life span, enhancing sports performance, and offsetting the challenges of advancing age. As current research improves our understanding of dietary requirements in different contexts of human function, a comprehensive appraisal of the role of nutrition in human health and performance is warranted. It is this need that has inspired the current volume.

This volume is divided into six thematic sections. The introductory section presents a general overview of the role of nutrition in human health. This section is a compilation of chapters reviewing nutritional prophylaxis in human health, including food exchange values, personalized nutrition and a critical assessment of antioxidants, vitamins, membrane stabilizers, minerals, herbal extracts and other nutritional supplements and their influence on human health and exercise performance. This section also addresses performance enhancement drugs and sports supplements, discussing concerns associated with the benefits and risks associated with the use of performance-enhancing supplements.

The second section is based on types of physical exercises. Cardiovascular training, resistance training, psychological aspects, bone health and bone mineral density, and immune function are discussed. Issues related to bone health and hormone replacement therapy, primarily affecting the elderly and post-menopausal women, are also covered in this section.

The third section addresses nutritional requirements in the context of sporting events. Special considerations for vegetarian athletes are discussed in the first chapter in this section. In addition to an overview of sports nutrition, this section covers combat sports, sumo wrestling, endurance training, high-altitude sports, winter sports, paralympic games, zero-gravity (space) fitness, and extreme sports such as bungee jumping, hang gliding and sky diving. Other related issues such as doping in sports and a history of sports nutrition beverages are also addressed.

Molecular mechanisms implicated in muscle building constitute the theme of the fourth section. Molecular and cellular underpinnings of muscle growth, repair and preservation are discussed with reference to how they may influence physical performance. Intracellular signaling mechanisms underlying tissue growth and adaptation are reviewed with the aim of identifying key hubs of regulation of exercise performance. For example, the emergent significance of nitric oxide in exercise performance and muscle building is discussed in this section.

Section five presents a collection of review articles addressing food products, minerals, supplements, phytochemicals, testosterone, amino acids, transition metals, small molecules and other ergogenic agents that have been implicated in muscle building and human performance. The scope of nutritional intervention to manage post-exercise immune suppression is discussed in this section. In addition, carbohydrate and glycogen metabolism and the roles of adaptogens are discussed. Clinical implications of anabolic training and protein, carbohydrate and fat requirements for athletes are also covered. The significance of amino acids, including branched-chain amino acids, and the importance of hydration are addressed. Creatine, carnitine, ornithine, arginine, citrulline, ursolic and maslinic acid, HMB and GAKIC are discussed at length. The role of chromium(III), vanadium, zinc and plant borates, as well as astaxanthin, caffeine, quercetin and tongkat ali are covered.

Finally, the sixth section draws together the aforementioned topics by reviewing nutrition and dietary recommendations for bodybuilders.

With emphasis on sports and nutrition in human health, physiological basis of muscle function, and functional food and nutraceuticals affecting human health and performance, this volume is an indispensable resource for all those who seek to be current in their understanding of nutritional advantages that may bolster human health and sports performance.

The editors extend their sincere thanks and gratitude to all the eminent contributors and especially to Ms. Nancy Maragioglio, Mrs. Mara Conner and Ms. Megan Wickline for their continued support, cooperation and fruitful suggestions.

Debasis Bagchi, PhD, MACN, CNS, MAIChE
University of Houston College of Pharmacy, Houston, TX, USA

Sreejayan Nair, MPharm, PhD
University of Wyoming School of Pharmacy, Laramie, WY, USA

Chandan K. Sen, PhD, FACN, FACSM
The Ohio State University Wexner Medical Center, Columbus, OH, USA

List of Contributors

Asif Ali, PhD Department of Biochemistry, Jawaharlal Nehru Medical College, AMU, Aligarh, India

Anthony L. Almada, MSc, FISSN IMAGINutrition, Inc., Dana Point, CA, USA

Ezra A. Amsterdam, MD, FACC Division of Cardiovascular Medicine, University of California, Davis, CA, USA

Wataru Aoi, PhD Laboratory of Health Science, Graduate School of Life and Environmental Sciences, Kyoto Prefectual University, Kyoto, Japan

Philip E. Apong R&D, Iovate Health Sciences International Inc., Oakville, ON, Canada

Guilherme G. Artioli, PhD Laboratory of Applied Nutrition and Metabolism, School of Physical Education and Sport, University of São Paulo, São Paulo, Brazil

Mustafa Atalay, MD, MPH, PhD Institute of Biomedicine, Physiology, University of Eastern Finland, Kuopio, Finland

Samuel Augustine, PharmD, FAPhA Department of Pharmacy Practice, Creighton University, Omaha, NE, USA

Debasis Bagchi, PhD, MACN, CNS, MAIChE University of Houston College of Pharmacy, Houston, TX, USA

Raza Bashir, MSc Iovate Health Sciences International Inc., Oakville, ON, Canada

John C. Blocher Biothera, Eagan, MN, USA

Richard J. Bloomer, PhD Cardiorespiratory/Metabolic Laboratory, The University of Memphis, Memphis, TN, USA

Marco Bonifazi Department of Neurological, Neurosurgical and Behavioral Sciences, Faculty of Medicine, University of Siena, Italy

Rachel Botchlett Faculty of Nutrition, Texas A&M University, College Station, TX, USA

Thomas Brioche Department of Physiology, Faculty of Medicine, University of Valencia, Valencia, Spain; Fundación Investigación Hospital Clínico Universitario/INCLIVA, Valencia, Spain; Laboratory M2S (Movement, Sport and Health Sciences), UFR-APS, Rennes Cedex, France

Wayne W. Campbell, PhD Department of Nutrition Science, Purdue University, West Lafayette, IN, USA

Bob Capelli Cyanotech Corporation, Kailua-Kona, HI, USA

Carlo Capelli, BA Department of Neurological and Movement Sciences, School of Exercise and Sport Sciences, University of Verona, Verona, Italy

Philippe Connes, PhD UMR Inserm 665, Centre Hospitalier Universitaire; Laboratoire ACTES EA3596, Département de Physiologie, Pointe à Pitre, Guadeloupe; Laboratory of Excellence GR-Ex, PRES Sorbonne Paris Cité, Paris, France

Don J. Cox Biothera, Eagan, MN, USA

Brent C. Creighton, MA Human Performance Laboratory, Department of Kinesiology, University of Connecticut, Storrs, CT, USA

Bruce Culver, PhD University of Wyoming, School of Pharmacy and the Center for Cardiovascular Research and Alternative Medicine, University College of Health Sciences, Laramie, WY, USA

Rui Curi Department of Physiology and Biophysics, Institute of Biomedical Sciences, University of São Paulo. São Paulo, Brazil

Gerald R. Cysewski, PhD Cyanotech Corporation, Kailua-Kona, HI, USA

Amitava Das Department of Surgery, The Ohio State University Wexner Medical Center, Columbus, OH, USA

Hans Degens, MSc, PhD Institute for Biomedical Research into Human Movement & Health, Faculty of Science & Engineering, Manchester Metropolitan University, Manchester, United Kingdom

Chariklia K. Deli, BSc, MSc Department of Physical Education and Sport Science, University of Thessaly, Trikala, Greece; Institute of Human Performance and Rehabilitation, Center for Research and Technology – Thessaly, Trikala, Greece

Zsolt Demetrovics, PhD Institute of Psychology, Eötvös Loránd University, Budapest, Hungary

Lawrence J. Druhan, PhD The Levine Cancer Institute, Carolinas Healthcare System, Charlotte, NC, USA

Stéphane Dufour, PhD Université de Strasbourg, Faculté des Sciences du Sport, Strasbourg, France; Faculté de Médecine, Strasbourg, France

Michael J. Duncan Human Performance Laboratory, Department of Biomolecular and Sports Sciences, Coventry University, Coventry, United Kingdom

Courtenay Dunn-Lewis Human Performance Laboratory, Department of Kinesiology, University of Connecticut, Storrs, CT, USA

Robert M. Erskine, BSc, PhD School of Sport and Exercise Sciences, Liverpool John Moores University, Liverpool, United Kingdom

Brad Evers VDF FutureCeuticals, Inc., Momence, IL, USA

Nir Eynon, PhD Institute of Sport, Exercise and Active Living (ISEAL); School of Sport and Exercise Science, Victoria University, Australia

Tyler M. Farney, MS Cardiorespiratory/Metabolic Laboratory, The University of Memphis, Memphis, TN, USA

Ioannis G. Fatouros, BSc, MSc, PhD Department of Physical Education and Sport Science, University of Thrace, Komotini, Greece; Institute of Human Performance and Rehabilitation, Center for Research and Technology – Thessaly, Trikala, Greece

Fabrice Favret, PhD Faculté de Médecine, Strasbourg, France; Université de Strasbourg, Faculté des Sciences du Sport, Strasbourg, France

Maria Lucia Fleiuss de Farias, PhD Endocrinology Service, Hospital Universitário Clementino Fraga Filho, Universidade Federal do Rio de Janeiro, Brazil

Emerson Franchini, PhD Combat Sports and Martial Arts Research Group, School of Physical Education and Sport, University of São Paulo, São Paulo, Brazil

Daniel J. Freidenreich Human Performance Laboratory, Department of Kinesiology, University of Connecticut, Storrs, CT, USA

Mari Carmen Gómez-Cabrera Department of Physiology, Faculty of Medicine, University of Valencia, Valencia, Spain

Gary Gaffney, MD Division of Child and Adolescent Psychiatry, University of Iowa, College of Medicine, Iowa City, IA, USA

Gustavo A. Galaz, MSc Iovate Health Sciences, Oakville, ON, Canada

Kalliopi Georgakouli, BSc, MSc Department of Physical Education and Sport Science, University of Thessaly, Trikala, Greece; Institute of Human Performance and Rehabilitation, Center for Research and Technology – Thessaly, Trikala, Greece

Frederico Gerlinger-Romero Department of Physiology and Biophysics, Institute of Biomedical Sciences, University of São Paulo. São Paulo, Brazil

Mark D. Griffiths Psychology Division, Nottingham Trent University, Nottingham, United Kingdom

Lucas Guimarães-Ferreira Exercise Metabolism Research Group, Center of Physical Education and Sports, Federal University of Espírito Santo. Vitória, Brazil

Safia Habib, PhD Department of Biochemistry, Jawaharlal Nehru Medical College, AMU, Aligarh, India

Erik D Hanson Institute of Sport, Exercise and Active Living (ISEAL); College of Health and Biomedicine, Victoria University, Melbourne, Australia

Hande Hofmann, ScD Institute of Preventive Pediatrics, Technische Universität München, Munich, Germany

Juha J. Hulmi Department of Biology of Physical Activity & Neuromuscular Research Center, University of Jyväskylä, Jyväskylä, Finland

John Hunter VDF FutureCeuticals, Inc., Momence, IL, USA

Athanasios Z. Jamurtas, BSc, MSc, PhD Department of Physical Education and Sport Science, University of Thessaly, Trikala, Greece; Institute of Human Performance and Rehabilitation, Center for Research and Technology – Thessaly, Trikala, Greece

Usha Jenkins, BS Cyanotech Corporation, Kailua-Kona, HI, USA

Asker Jeukendrup Gatorade Sports Science Institute, Barrington, IL, USA; Loughborough University, Loughborough, UK

C. Tissa Kappagoda, MD, PhD Division of Cardiovascular Medicine, University of California, Davis, CA, USA

Tuomo Karila Hospital Orton, and Dextra Sports and Injury Clinic, Helsinki, Finland

Justin W.L. Keogh Bond University, Robina, Queensland, Australia

Chad M. Kerksick, PhD Health, Exercise and Sports Sciences Department, University of New Mexico, Albuquerque, NM, USA

Susanna Kinnunen Institute of Biomedicine, Physiology, University of Eastern Finland, Kuopio, Finland

Erik P. Kirk, PhD, CSCS, CPT Department of Kinesiology and Health Education, Southern Illinois University Edwardsville, Edwardsville, IL, USA

Edeth K. Kitchens, PhD Kitsto Consulting LLC, Frisco, TX, USA

Beat Knechtle Institute of General Practice and for Health Services Research, University of Zurich, Zurich, and Gesundheitszentrum St Gallen, St Gallen, Switzerland

Masakatsu Kondo Department of Exercise Physiology, Nihon University, Tokyo, Japan

William J. Kraemer Human Performance Laboratory, Department of Kinesiology, University of Connecticut, Storrs, CT, USA

Michelle Kulovitz, PhD Department of Kinesiology, California State University – San Bernadino, San Bernadino, CA, USA

Antonio H. Lancha Jr, PhD Laboratory of Applied Nutrition and Metabolism, School of Physical Education and Sport, University of São Paulo, São Paulo, Brazil

John M. Lawler Department of Health and Kinesiology and Faculty of Nutrition, Texas A&M University, College Station, TX, USA

Jia Li, MS Department of Nutrition Science, Purdue University, West Lafayette, IN, USA

Jan Lingen, Dipl Sports Sci Institute of Preventive Pediatrics, Technische Universität München, Munich, Germany

Joel R. Lombard Department of Chemistry/Biology, Springfield College, Springfield, MA, USA

Hui-Ying Luk Human Performance Laboratory, Department of Kinesiology, University of Connecticut, Storrs, CT, USA

Vladimir Martinez-Bello Department of Teaching of Musical, Visual and Corporal Expression, Faculty of Teaching, University of Valencia, Spain

Matthew J. McAllister, MS Cardiorespiratory/Metabolic Laboratory, The University of Memphis, Memphis, TN, USA

John J. McCarthy, PhD Center for Muscle Biology, Department of Physiology, College of Medicine, University of Kentucky, Lexington, KY, USA

Brian K. McFarlin University of North Texas, Applied Physiology Laboratory, Denton, TX, USA

Antti A. Mero Department of Biology of Physical Activity & Neuromuscular Research Center, University of Jyväskylä, Jyväskylä, Finland

Flavia Meyer, MD, PhD Universidade Federal do Rio Grande do Sul (UFRGS), Porto Alegre, Brazil

Taishi Midorikawa College of Health and Welfare, J. F. Oberlin University, Tokyo, Japan; Faculty of Sport Sciences, Waseda University, Saitama, Japan

Donald W. Miller, PhD Department of Pharmacology and Therapeutics, University of Manitoba, Winnipeg, MB, Canada

Hiroyoshi Moriyama, PhD, FACN Ryusendo Co. Ltd, Tokyo, Japan

Igor Murai Laboratory of Applied Nutrition and Metabolism, Physical Education and Sports School, University of São Paulo, São Paulo, Brazil

Sreejayan Nair, PhD University of Wyoming, School of Pharmacy and the Center for Cardiovascular Research and Alternative Medicine, College of Health Sciences, Laramie, WY, USA

Yuji Naito Department of Molecular Gastroenterology and Hepatology, Graduate School of Medical Science, Kyoto Prefectural University of Medicine, Kyoto, Japan

Yasmin Neggers, PhD, RD Department of Human Nutrition and Hospitality Management, The University of Alabama, Tuscaloosa, AL, USA

Humberto Nicastro, PhD Laboratory of Nutrition and Metabolism, School of Physical Education and Sports, University of São Paulo, São Paulo, Brazil

Sonja E. Nodland, PhD Biothera, Eagan, MN, USA

Tuomo Ojala, MSc Department of Biology of Physical Activity & Neuromuscular Research Center, University of Jyväskylä, Jyväskylä, Finland

Koji Okamura, PhD Graduate School of Sport and Exercise Sciences, Osaka University of Health and Sport Sciences, Kumatori, Japan

Niku Oksala Department of Vascular Surgery, Tampere University Hospital, Tampere, Finland

Evgeniy Panzhinskiy, PhD University of Wyoming, School of Pharmacy and the Center for Cardiovascular Research and Alternative Medicine, College of Health Sciences, Laramie, WY, USA

Helios Pareja-Galeano Department of Physiology, Faculty of Medicine, University of Valencia, Valencia, Spain; Fundación Investigación Hospital Clínico Universitario/INCLIVA, Valencia, Spain

Aurélien Pichon, PhD Laboratory of Excellence GR-Ex, PRES Sorbonne Paris Cité, Paris, France; Université Paris, Bobigny, France

Zbigniew Pietrzkowski Applied BioClinical, Inc., Irvine, CA, USA

Carlos Hermano J. Pinheiro, PhD, MSc, PT Department of Physiology and Biophysics, Institute of Biomedical Sciences, University of São Paulo. São Paulo, Brazil

Silvia Pogliaghi Department of Neurological and Movement Sciences, School of Exercise and Sport Sciences, University of Verona, Verona, Italy

Hartley Pond VDF FutureCeuticals, Inc., Momence, IL, USA

Jun Ren, PhD University of Wyoming, School of Pharmacy and the Center for Cardiovascular Research and Alternative Medicine, College of Health Sciences, Laramie, WY, USA

Beatriz Gonçalves Ribeiro, PhD Josué de Castro Nutrition Institute, Universidade Federal do Rio de Janeiro, Rio de Janeiro, Brazil

Dennis H. Robinson, PhD Department of Pharmaceutical Sciences, University of Nebraska Medical Center, Omaha, NE, USA

Shizuo Sakamoto Faculty of Sport Sciences, Waseda University, Saitama, Japan

Fabian Sanchis-Gomar, MD Department of Physiology, Faculty of Medicine, University of Valencia, Valencia, Spain; Fundación Investigación Hospital Clínico Universitario/INCLIVA, Valencia, Spain

Martin Schönfelder, ScD Research Institute of Molecular Sport and Rehabilitation Medicine, Paracelsus Medical Private University of Salzburg, Austria

Annie Schtscherbyna, PhD Josué de Castro Nutrition Institute, Universidade Federal do Rio de Janeiro, Rio de Janeiro, Brazil

John Seifert, PhD Montana State University, Bozeman, MT, USA

Daniela Fojo Seixas Chaves Laboratory of Nutrition and Metabolism, School of Physical Education and Sports, University of São Paulo, São Paulo, Brazil

Chandan K. Sen, PhD Department of Surgery, The Ohio State University Wexner Medical Center, Columbus, OH, USA; The Ohio State University Wexner Medical Center, Columbus, OH, USA

Timo A. Seppälä Department of Drugs, Alcohol and Addiction, National Institute for Health and Welfare, Helsinki, Finland

Yoshiaki Shiojima Ryusendo Co. Ltd, Tokyo, Japan

Wagner Silva Dantas Laboratory of Applied Nutrition and Metabolism, Physical Education and Sports School, University of São Paulo, São Paulo, Brazil

Bryan K. Smith, PhD, CSCS Department of Kinesiology and Health Education, Southern Illinois University Edwardsville, Edwardsville, IL, USA

JohnEric W. Smith Gatorade Sports Science Institute, Bradenton, FL, USA

Marina Y. Solis, PhD Division of Rheumatology, Clinics Hospital, School of Medicine, University of São Paulo, São Paulo, Brazil

Bruce R. Stevens, PhD Department of Physiology and Functional Genomics, University of Florida College of Medicine, Gainesville, FL, USA

Sidney J. Stohs, PhD, FACN, CNS, ATS Creighton University Medical Center, Omaha, NB, USA; Kitsto Consulting LLC, Frisco, TX, USA

Jan Sundell Department of Medicine, University of Turku, and Turku University Hospital, Turku, Finland

Attila Szabo Institute of Psychology, Eötvös Loránd University, Budapest, Hungary

Tomohisa Takagi Department of Molecular Gastroenterology and Hepatology, Graduate School of Medical Science, Kyoto Prefectural University of Medicine, Kyoto, Japan

Tohru Takemasa, PhD, D. Med Sci Exercise Physiology, Health and Sport Sciences, University of Tsukuba, Japan

Shawn M. Talbott, PhD, CNS, LDN, FACN, FACSM, FAIS GLH Nutrition, LLC, Draper, UT, USA

Brian Weldon Timmons, PhD McMaster University, Hamilton, ON, Canada

Aline C. Tritto, BSc Laboratory of Applied Nutrition and Metabolism, School of Physical Education and Sport, University of São Paulo, São Paulo, Brazil

Jonathan L. Vennerstrom, PhD Department of Pharmaceutical Sciences, University of Nebraska Medical Center, Omaha, NE, USA

Mika Venojärvi Institute of Biomedicine, Physiology, University of Eastern Finland, Kuopio, Finland

John B. Vincent, PhD Department of Chemistry, The University of Alabama, Tuscaloosa, AL, USA

Jeff S. Volek, PhD, RD Human Performance Laboratory, Department of Kinesiology, University of Connecticut, Storrs, CT, USA

Brittanie M. Volk, MA, RD Human Performance Laboratory, Department of Kinesiology, University of Connecticut, Storrs, CT, USA

Jon C. Wagner, PharmD Vireo Resources LLC, Plattsmouth, NE, USA

Ankita Wal Pranveer Singh Institute of Technology, Kanpur, UP, India

Pranay Wal Pranveer Singh Institute of Technology, Kanpur, UP, India

Boguslaw Wilk Children's Exercise & Nutrition Centre, McMaster Children's Hospital, Hamilton, ON, Canada

Jacob M. Wilson Department of Health Sciences and Human Performance, University of Tampa, Tampa, FL, USA

Guoyao Wu Department of Animal Science and Faculty of Nutrition, Texas A&M University, College Station, TX, USA

Toshikazu Yoshikawa Department of Molecular Gastroenterology and Hepatology, Graduate School of Medical Science, Kyoto Prefectural University of Medicine, Kyoto, Japan

Paola Zamparo, PhD Department of Neurological and Movement Sciences, School of Exercise and Sport Sciences, University of Verona, Verona, Italy

Nelo Eidy Zanchi, PhD Department of Physiology and Biophysics, Institute of Biomedical Sciences, University of São Paulo, São Paulo, Brazil

Jing Zhou, BS Department of Nutrition Science, Purdue University, West Lafayette, IN, USA

SECTION 1

NUTRITION AND HUMAN HEALTH

CHAPTER

1

Nutritional Supplementation in Health and Sports Performance

Sidney J. Stohs[1,2] *and Edeth K. Kitchens*[2]

[1]Creighton University Medical Center, Omaha, NB, USA [2]Kitsto Consulting LLC, Frisco, TX, USA

INTRODUCTION

The nutritional status of an athlete is a major determinant of health, fitness and sports performance. Nutrition plays a central role in adaptation, rehydration, refueling, and repair as well as recovery from injury [1–6]. As a consequence, for optimal performance it is essential that athletes be in the best possible nutritional and metabolically balanced state.

Athletes as well as the general population may be overfed and still be deficient in a wide range of essential nutrients including vitamin A, vitamin B_6, vitamin B_{12}, vitamin C, vitamin D, vitamin E, vitamin K, folic acid, iodine, iron, zinc, calcium, magnesium, and selenium [7–13]. These nutritional deficiencies can be extrapolated to athletes, with some indications that the incidence of some deficiencies may be higher among athletes than in the general population. Examples of nutritional deficiencies that have been specifically reported among athletes include iron [1,14], magnesium [14,15], sodium [1,5,16], zinc [1], calcium and vitamin D [17], vitamin C, vitamin E, and vitamin A [18].

The primary reason for nutritional imbalance is consumption of refined foods and dietary supplements that are high in calories from sugars, starches and fats, and low in vitamins, minerals, trace elements and fiber as the result of the refining and manufacturing processes. The net effect is that athletes may consume an excess of calories in conjunction with lower levels of essential nutrients.

At least 85% of athletes and those engaged in regular exercise consume nutritional and dietary supplements on a daily basis [2]. Approximately 50–70% of the adult population consume dietary supplements daily, with women consuming supplements more regularly then men [19]. Other studies indicate that 72% of cardiologists, 59% of dermatologists, 91% of orthopedists and physicians, and 82% of nurses recommended dietary supplements to their patients, while almost 70% of physicians and 89% of nurses at least occasionally used dietary supplements themselves [20,21]. Thus, oral nutrition in the form of dietary supplements is accepted by a large percentage of athletes, the general population, and a wide range of healthcare providers.

DEFINITIONS

The US Congress defined the term dietary supplement with the passage of the Dietary Supplement Health and Education Act (DSHEA) of 1994. Therefore, a dietary supplement is defined as a product taken by mouth that contains a "dietary ingredient" intended to supplement the diet. Furthermore, dietary ingredients may include: vitamins, minerals, amino acids, herbs or other botanical products, and substances such as enzymes, glandulars, organ tissues, and metabolites.

Dietary supplements may also be extracts or concentrates, and can occur in forms such as capsules, tablets, powders, liquids, gelcaps, softgels, or bars. They must be labeled as dietary supplements since by law they are a special category of "foods" and not drugs. The definition of dietary supplements used in the USA differs from the definition used in Europe, where the term refers to vitamins and minerals, and herbal products are regulated separately as herbal medicines or herbal remedies [22].

Nutrition and Enhanced Sports Performance.
DOI: http://dx.doi.org/10.1016/B978-0-12-396454-0.00001-1

Based on DSHEA, structure /function claims can be made for dietary supplements which describe the role of a dietary ingredient or nutrient that is intended to affect normal structure or function in humans. Several examples of structure/function claims include statements such as "calcium builds strong bones", "fiber maintains bowel integrity", "omega-3-fatty acids support heart health", and "chromium helps maintain blood glucose levels in the normal range". When a dietary supplement includes a structure/function claim, it must state that the FDA has not evaluated the claim, and must further affirm that the product is not intended to "diagnose, treat, cure or prevent any disease", since legally only a drug can make these claims.

All nutrition facts panels on labels contain Daily Values (DVs) which represent the recommended daily intake (RDI) of each nutrient that is considered to be adequate to meet the requirements of 97–98% of normal, healthy individuals in all demographics in the United States, based on 2000 kcal per day. RDIs are a reflection of the older Recommended Dietary Allowances (RDAs) which were calculated based on Estimated Average Requirements (EARs) or the amount of a nutrient believed to satisfy the needs of 50% of the people in a demographic. RDA values are usually about 20% higher than EARs. Finally, a system of nutrition recommendations entitled the Dietary Reference Intake (DRI) values was introduced by the Institute of Medicine of the US National Academy of Sciences in 1997 to broaden the RDAs. DRIs have not been widely adopted.

Unfortunately, these multiple systems of recommended essential nutrient intake add much confusion and questionable clarity to the widely asked question "how much of each nutrient is needed for optimal physical performance and health?" In reality, these systems are based largely on the smallest amount of a nutrient needed to prevent a deficiency or disease state, and do not reflect the amount of each nutrient required to provide optimal health and peak physical performance. Furthermore, they project the minimal needs of healthy individuals, with little or no allowance for stressful situations such as intense exercise or disease.

The widely help misconception that only 100% of the DV amount of each essential nutrient is required for good health clearly is not true. Supplementation with a multivitamin/mineral product containing 100% of the DVs may decrease the prevalence of suboptimal levels of some nutrients in athletes, but it will not provide optimal nutritional requirements. Furthermore, providing 100% of the DV for vitamins and minerals does not enhance the levels of various markers of anti-inflammatory activity, antioxidant capacity, or immune response [8,23].

NUTRITIONAL SUPPLEMENT RECOMMENDATIONS FOR ATHLETES

Many factors are involved in determining how much of the various essential nutrients are required by an individual to meet daily needs and support optimal sports performance. These factors include: age, weight, gender, stress levels, physical condition, daily physical activity, gastrointestinal health, general health, metabolic rate, disease states, and recovery from injury or surgery. As a consequence, it is apparent that one size (amount) does not fit all, and as previously noted, supplementing with a product that contains 100% of DVs does not adequately meet the overall needs. Metabolism can be equated to a chain which is as strong as its weakest link.

With these considerations in mind, what should be the approximate level of daily intake of nutritional supplements to facilitate optimal performance for an athlete who may be consuming 3000–6000 kcal per day, keeping in mind that DVs are based on a 2000 kcal per day intake? The average athlete should consume dietary supplements daily that contain at least 200–300% of the DVs for vitamins and minerals, and may require 400–600% of the DVs, depending upon the intensity and duration of daily activities. For example, consuming a product with 100% of the DVs for vitamins and minerals two to six times daily or a product with 200–300% of the DVs twice a day with meals may be appropriate.

With respect to optimal nutritional needs, vitamin D is an excellent example of inter-individual variability. In addition to its role in calcium absorption and bone health, vitamin D has been shown to play a role in many other body functions, including immune system support, cognition, cardiovascular health, prevention of some forms of cancer, and blood sugar regulation [24]. The current DV for vitamin D is 400 IU, and many athletes do not meet this minimal requirement [17]. A recommendation by the Institute of Medicine to increase the DV for vitamin D to 600 IU for most adults, based on needs for bone health, has not been adopted by the FDA to date. However, for optimal overall health, blood levels of at least 40 ng/mL of the active intermediate 25-hydroxyvitamin D are recommended, and from 2000 to 4000 IU of vitamin D are needed daily to achieve this level [24].

In order to avoid vitamin A toxicity, vitamin A should be used in the form of beta-carotene or mixed carotenoids and not retinol or its esters retinyl palmitate and retinyl acetate. Beta-carotene is converted into vitamin A as it is needed by the body [25], and exhibits a much greater safety profile.

Vegetarian diets are becoming more popular, and if vitamin B_{12} supplements are not being used, deficiencies will occur. Low vitamin B_{12} levels are associated with anemia, lack of endurance, weakness, neurological abnormalities, acidosis, elevated homocysteine levels, decreased HDL levels, and possibly platelet aggregation [26], all of which may contribute to decrements in athletic performance.

Typical multivitamin/mineral supplements do not contain adequate amounts of calcium, magnesium, or vitamin D. Products that contain only vitamin D and calcium with no magnesium are inadequate, and should be avoided. High calcium intake in the absence of magnesium inhibits the absorption of magnesium, and may be one of the reasons for the high incidence of magnesium deficiency in the USA as well as the frequency of leg cramps among athletes and the general public [27,28].

Calcium and magnesium products that are available in chelated and absorbable forms as calcium citrate, fumarate, hydroxyapatite, aspartate or other amino acid chelates [29], and magnesium citrate, ascorbate, aspartate or other amino acid chelate should be used [30,31]. Products that contain magnesium oxide are poorly absorbed and should be avoided [31].

Omega-3 fatty acids are required for normal cell and organ function and are present in every cell in the body. As the result of widespread omega-3-fatty acid deficiency, an estimated 84,000 people die prematurely each year in the USA [32]. In the USA, vegetarians exhibit particularly low intake of this critical nutrient [26]. Omega-3 fatty acids exhibit numerous beneficial functions, including protective effects associated with muscles, joints, cardiovascular system, immune system, brain and nervous system, gastrointestinal system, as well as bones, lungs, liver, skin, eyes, hair, and other organs and tissues [33].

Docosahexaenoic acid (DHA) and eicosapentaenoic acid (EPA) are the primary omega-3 fatty acids responsible for these health effects. They are derived primarily from fish oils although they may also be obtained from krill (zooplankton), while DHA may be extracted from algae and EPA from yeast. α-Linolenic acid (ALA) is another omega-3 fatty acid that is derived from plant sources such as flax seed oil, canola oil, soybean oil, nuts, and some berries. Because very little ALA is converted into EPA and DHA, it possesses only a fraction of the health benefits of DHA and EPA [34].

The American Heart Association recommends 400–500 mg of DHA/EPA per week (two servings of oily fish) for general health and wellness. Athletes, however, should consider consuming 1 gram of DHA/EPA three to four times daily, and endurance athletes should consider taking 5–10 grams per day based on need [35,36]. A high quality product from a reputable manufacturer should be used. Concerns expressed regarding possible contamination of fish oils with heavy metals and pesticides are unfounded and do not present a health threat [37].

The development, repair, and preservation of muscles are of paramount concern for athletes. The Institute of Medicine has established 0.8 g/kg body weight as the DRI for protein for adults [38]. However, this amount of protein is too low to support and preserve muscle mass in athletes, and as a consequence the recommended daily protein intake for strength and speed athletes is in the range of 1.2–1.5 g/kg body weight [39], assuming normal kidney function.

The branched chain amino acids L-leucine, L-valine, and L-isoleucine inhibit breakdown of skeletal muscle and promote muscle repair. Of these three amino acids, L-leucine has been shown to stimulate muscle protein synthesis and is believed to play a primary regulatory role in muscle protein metabolism [40,41].

An alternative and/or supplementary method for promoting and preserving muscle is the use of nutritional supplements that contain a mixture of L-leucine, L-valine, and L-isoleucine or contain L-leucine as the primary amino acid [42,43]. A daily intake of 8–12 grams per day of L-leucine or 12–18 grams per day of a mixture of the branched chain amino acids may be appropriate, particularly in athletes with compromised renal function or who are not able to tolerate high levels of protein [44]. Use of beta-hydroxy beta-methyl butyrate, a metabolite of L-leucine, constitutes an alternative approach to supporting muscle development [45]. Ingestion of protein levels that constitute approximately 30% of the total daily caloric intake is appropriate when not involved in competition.

The conditionally essential amino acids L-arginine and L-glutamine should be considered also for nutritional supplementation of athletes. L-Glutamine plays important roles in protein repair and synthesis, wound healing, acid–base balance, immune system support, gut barrier function, and cellular differentiation, and serves as an energy source [46–48]. Supplementation can be provided with 8 to 12 grams of L-glutamine per day in divided doses.

L-Arginine plays important roles in cell division, wound healing, support of the immune system, removal of ammonia from the body, synthesis of nitric acid, and blood pressure regulation [49–52]. Supplementation can be provided with 10 to 14 grams of L-arginine per day in divided doses.

Finally, it should be noted that rehydration of athletes with water only following intensive exercise and dehydration is inadequate for recovery and subsequent performance [53]. Furthermore, rehydration with a product that contains primarily water, sugar and sodium chloride is superior to water alone, but does not support optimal performance. A more complex product containing water, sodium, potassium, calcium, magnesium, carbohydrates, vitamins and selected amino acids should be consumed [53].

SAFETY ISSUES

The US Poison Control Center has repeatedly reported no deaths associated with vitamins, minerals, herbal ingredients, or dietary supplements in general [54]. Furthermore, no deaths have been reported in association with these nutrients in recent years. In fact, deaths that were reported in the past were directly related to inappropriate use. All substances can be toxic if taken in sufficiently large doses.

At the same time, the scientific literature does not support the premise that taking vitamin and mineral products in amounts that exceed 100% of the daily value results in toxicity. Excess amounts of water-soluble vitamins are readily excreted and toxicities do not occur with consumption of vitamins in amounts up to 10–20 times the DVs. Greater caution is required for most minerals, which should not be used in amounts greater than approximately 6 times the DV unless specific deficiencies are documented. The recommendations presented above are designed to address optimum nutritional needs while maintaining a wide margin of safety.

SUMMARY AND CONCLUSIONS

Over 50% of the adult population of the United States and most athletes consume nutritional supplements on a daily basis. Yet, large percentages of the populace, including athletes, are deficient in multiple vitamins and minerals because of poor dietary habits and consumption of foods with inadequate amounts of essential nutrients.

Consuming multivitamin/mineral products that contain up to 100% of the DV is inadequate to provide blood and tissue levels needed for good nutrition, let alone appropriate nutrition necessary to support optimal performance. In general, consuming up to 300–600% of the DV for most common vitamins and minerals may be necessary in order to meet the nutritional needs of athletes. Intake will depend on overall daily caloric consumption and the rigor of the training and exertion involved in a particular physical activity.

Omega-3 fatty acids impact the functions of all organs, and deficiencies are common. To appropriately address omega-3 fatty acid needs, athletes are encouraged to consume a minimum of 3–4 grams of DHA/EPA daily, while larger amounts may be required by endurance athletes.

High protein intake and exercise are two factors that preserve and enhance muscle. A protein intake in the range of at least 1.2–1.5 g/kg/day is recommended. Where renal function is an issue or an alternative approach is necessary to preserve and enhance muscle health, high intake of L-leucine (8–12 grams/day) or a combination of the branched-chain amino acids L-leucine, L-valine and L-isoleucine (12–18 grams/day) may be appropriate. These branched-chain amino acids in general, and L-leucine in particular, promote protein synthesis.

L-Arginine and L-glutamine are two conditionally essential amino acids that support the immune system, cell division, and wound healing. Athletes can benefit from ingesting 8–10 grams of L-glutamine per day and 10–14 grams of L-arginine per day, particularly during periods of intense exercise and training.

Finally, hydration with a complex product that contains water, multiple minerals, vitamins, carbohydrates, and selected amino acids is superior to water alone or water plus sugar and sodium chloride.

In summary, nutritional status is a significant factor in determining overall performance and endurance of athletes. The above recommendations are designed to assist in providing optimal nutritional support for athletes, recognizing that specific needs and the timing of nutritional intake will vary depending upon the rigor of the physical activity and the goals that are involved.

References

[1] Kreider RB, Wilborn CD, Taylor L, Campbell B, Almada AL, Collins R, et al. ISSN exercise & sport nutrition review: research & recommendations. J Int Soc Sports Nutr 2010;7:1–35.

[2] Maughan RJ, Depiesse F, Geyer H. The use of dietary supplements by athletes. J Sports Sci 2007;25:S103–13.

[3] Tipton KD. Nutrition for acute exercise-induced injuries. Ann Nutr Metab 2010;57:S43–53.

[4] Maughan RJ, Burke LM. Practical nutritional recommendations for the athlete. Nestle Nutr Inst Workshop Ser 2011;69:131–49.

[5] Jeukendrup AE. Nutrition for endurance sports: marathon, triathlon, and road cycling. J Sports Sci 2011;29:S91–9.

[6] Stillingwerff T, Maughan RJ, Burke LM. Nutrition for power sports: middle-distance running, track cycling, rowing, canoeing/kayaking, and swimming. J Sports Sci 2011;29:S79–89.

[7] Ahuja JKC, Goldman JD, Moshfegh AJ. Current status of vitamin E nutriture. Ann NY Acad Sci 2004;1031:387–90.

[8] Sebastian RS, Cleveland LE, Goldman JD, Moshfegh AJ. Older adults who use vitamin/mineral supplements differ from nonusers in nutrition intake adequacy and dietary attitudes. J Am Diet Assoc 2007;107:1322–32.

[9] Ma J, Johns RA, Stafford RS. Americans are not meeting current calcium recommendations. Am J Clin Nutr 2007;85:1361–6.

[10] Park S, Johnson M, Fischer JG. Vitamin and mineral supplements: barriers and challenges for older adults. J Nutr Elder 2008;27:297–317.

[11] Schleicher RL, Carroll MD, Ford ES, Lacher DA. Serum vitamin C and the prevalence of vitamin C deficiency in the United States: 2003–2004 National Health and Nutrition Examination Survey (NHANES). Am J Clin Nutr 2009;90:1252–63.

[12] Nelson FH. Magnesium, inflammation, and obesity in chronic disease. Nutr Rev 2010;68:333–40.

[13] McCann JC, Ames BN. Adaptive dysfunction of selenoproteins from the perspective of the triage theory: why modest selenium

deficiency may increase risk of diseases of aging. FASEB J 2011;25:1793–814.

[14] Lundy B. Nutrition for synchronized swimming: a review. Int J Sport Nutr Exerc Metab 2011;21:436–45.

[15] Nielsen FH, Lukaski HL. Update on the relationship between magnesium and exercise. Magnes Res 2006;19:1801–89.

[16] Snell PG, Ward R, Kandaswami C, Stohs SJ. Comparative effects of selected non-caffeinated rehydration sports drinks on short-term performance following moderate dehydration. J Int Soc Sports Nutr 2010;7:28–36.

[17] Larson-Meyer DE, Willis KS. Vitamin D and athletes. Curr Sports Med Rep 2010;9:220–6.

[18] Machefer G, Groussard C, Zouhal H, Vincent S, Youssef H, Faure H, et al. Nutritional and plasmatic antioxidant vitamins status of ultra-endurance athletes. J Am Coll Nutr 2007;26:311–6.

[19] Bailey RL, Fulgoni III VL, Keast DR, Dwyer JT. Examination of vitamin intake among US adults using dietary supplements. J Acad Nutr Diet 2012;112:657–63.

[20] Dickinson A, Boyon N, Shao A. Physicians and nurses use and recommend dietary supplements: report of a survey. Nutr J 2009;8. doi:10.1186/1475-2891-8-29 epub.

[21] Dickinson A, Shao A, Boyon N, Franco JC. Use of dietary supplements by cardiologists, dermatologists and orthopedists: report of a survey. Nutr J 2011;10. doi:10.1186/1475-2891-10-20 epub.

[22] Goldstein LH, Elias M, Ron-Avraham G, Biniaurishvili BZ, Madjar M, Kamargash I, et al. Consumption of herbal remedies and dietary supplements amongst patients hospitalized in medical wards. Brit J Clin Pharmacol 2007;64:373–80.

[23] McKay DL, Perrone G, Rasmussen H, Dallal G, Hartman W, Cao G, et al. The effects of a multivitamin/mineral supplement on micronutrient status, antioxidant capacity and cytokine production in healthy older adults consuming a fortified diet. J Am Coll Nutr 2000;19:613–20.

[24] Holick MF. Vitamin D: evolutionary, physiological and health perspectives. Curr Drug Targets 2011;12:4–18.

[25] Grune T, Lietz G, Palou A, Ross AC, Stahl W, Tang G, et al. Beta-carotene is an important vitamin A source for humans. J Nutr 2010;140:S2268–85.

[26] Li D. Chemistry behind vegetarianism. J Agric Food Chem 2011;3:777–84.

[27] Gums JG. Magnesium in cardiovascular and other disorders. Am J Health-Syst Pharm 2004;61:1569–15676.

[28] Nayor D. Widespread deficiency with deadly consequences. Life Ext May 2008:77–83.

[29] Hanzlik RP, Fowler SC, Fisher DH. Relative bioavailability of calcium from calcium formate, calcium citrate, and calcium carbonate. J Pharmacol Exp Ther 2005;313:1217–1222.

[30] Ranade VV, Somberg JC. Bioavailability and pharmacokinetics of magnesium after administration of magnesium salts to humans. Am J Ther 2001;8:345–57.

[31] Firoz M, Graber M. Bioavailability of US commercial magnesium preparations. Magnes Res 2001;14:257–62.

[32] Danaei G, Ding EL, Mozaffarian D, Taylor B, Rehm J, Murray CJ, et al. The preventable causes of death in the United States: comparative risk assessment of dietary, lifestyle, and metabolic risk factors. PLoS Med 2009; 6:1–23.

[33] Calder PC, Deckelbaum RJ. Omega-3 fatty acids; time to get the message right! Curr Op Clin Nutr Metab Care 2008;11:91–3.

[34] Wang C, Harris WS, Chung W, Lichtenstein AH, Balk EM, Kupelnick B, et al. n-3 fatty acids from fish or fish-oil supplements, but not α-linolenic acid, benefit cardiovascular disease outcomes in primary- and secondary-prevention studies: a systematic review. Am J Clin Nutr 2006;84:5–18.

[35] Davis M. Broad spectrum cardiac protection with fish oil. Life Ext Sept 2006:37–48.

[36] Mazza M, Pomponi M, Janiri L, Brai P, Mazza S. Omega-3 fatty acids and antioxidants in neurological and psychiatric diseases: an overview. Prog Neuro-Psychopharm Biol Psych 2007;31:12–26.

[37] Anon. Fish oil and omega-3 fatty acid supplements. <https://www.consumerlab.com/reviews/fish_oil_supplements_review/omega3/>. Last updated August 22, 2012.

[38] Paddon-Jones D, Short KR, Campbell WW, Volpi E, Wolfe RR. Role of protein in the sarcopenia of aging. Am J Clin Nutr 2008;87:1562S–6S.

[39] Tipton KD, Wolfe RR. Protein and amino acids for athletes. J Sports Sci 2004;21:65–79.

[40] Garlick PJ. The role of leucine in the regulation of protein metabolism. J Nutr 2005;135:15535–65.

[41] Koopman R, Verdijk L, Manders RFJ, Gijsen AP, Gorselink M, Pijpers E, et al. Co-ingestion of protein and leucine stimulates muscle protein synthesis rates to the same extent in young and elderly lean men. Am J Clin Nutr 2006;84:623–32.

[42] Desikan V, Mileva I, Garlick J, Lane AH, Wilson TA, McNurlan MA. The effect of oral leucine on protein metabolism in adolescents with type 1 diabetes mellitus. Int J Ped Endocrinol 2010; doi:10.1155/2010/493258 epub.

[43] Baptista IL, Leal ML, Artioli GG, Aoki MS, Fiamoncini J, Turri AO, et al. Leucine attenuates skeletal muscle wasting via inhibition of ubiquitin ligases. Musc Nerve 2010;41:800–8.

[44] Laviano A, Muscaritoli M, Cascino A, Preziosa I, Inui A, Mantovani G, et al. Branched-chain amino acids: the best compromise to achieve anabolism? Curr Opin Clin Nutr Metab Care 2005;8:408–14.

[45] Zanchi NE, Gerlinger-Romero L, De Sigueira Filho MA, Felitti V, Lira FS, Seelaender M, et al. HMB supplementation: clinical and athletic performance-related effects and mechanisms of action. Amino Acids 2011;40:1015–25.

[46] Novak F, Heyland DK, Avenell A, Drover JW, Su X. Glutamine supplementation in serious illness: a systematic review of the evidence. Crit Care Med 2002;30:2022–9.

[47] De-Souza DA, Greene LJ. Intestinal permeability and systemic infections in critically ill patients: effects of glutamine. Crit Care Med 2005;33:1125–35.

[48] Burrin DG, Stoll B. Metabolic fate and function of dietary glutamate in the gut. Am J Clin Nutr 2009;90:850S–6S.

[49] Daly JM, Reynolds J, Thom A, Kinsley L, Dietrick-Gallagher M, Shou J, et al. Immune and metabolic effects of arginine in the surgical patient. Ann Surg 1998;208:512–23.

[50] Wilmore D. Enteral and parenteral arginine supplementation to improve medical outcomes in hospitalized patients. J Nutr 2004;134:28635–75.

[51] Zhou M, Martindale RG. Arginine in the critical care setting. J Nutr 2007;137:1687S–92S.

[52] Buijs N, van Brokhorst-de van der Schueren MAE, Languis JAE, Leemans CR, Kuik DJ, Vermeulen MA, et al. Perioperative arginine-supplemented nutrition in malnourished patients with head and neck cancer improves long-term survival. Am J Clin Nutr 2011;73:323–32.

[53] Snell PG, Ward R, Kandaswami C, Stohs SJ. Comparative effects of selected non-caffeinated rehydration sports drinks on short-term ;performance following moderate rehydration. J Int Soc Sports Nutr 2010;728. doi:10.1186/1550-2783-7-28 epub.

[54] Bronstein AC, Spyker DA, Cantilena LR, Green JR, Rumack BH, Dart RC. 2010 annual report of the American Association of Poison Control Centers' National Poison Data System (NPDS): 28th annual report. Clin Tox 2011;49:910–41.

CHAPTER

2

Glycemic Index, Food Exchange Values and Exercise Performance

Athanasios Z. Jamurtas[1,2], *Chariklia K. Deli*[1,2], *Kalliopi Georgakouli*[1,2] and *Ioannis G. Fatouros*[2,3]

[1]Department of Physical Education and Sport Science, University of Thessaly, Trikala, Greece [2]Institute of Human Performance and Rehabilitation, Center for Research and Technology – Thessaly, Trikala, Greece [3]Department of Physical Education and Sport Science, University of Thrace, Komotini, Greece

GLYCEMIC INDEX

Carbohydrates typically constitute the greatest percentage of all three energy-providing nutrients of a diet [1]. There is increasing evidence that the type of carbohydrates being consumed is an important factor which influences exercise performance [2] and the risk of developing obesity and chronic diseases associated with it [3].

The concept of the Glycemic Index (GI) was developed in 1981 by Jenkins and colleagues in order to define the type of carbohydrates ingested by estimating the responses of blood glucose and insulin levels after the ingestion. GI is defined as the incremental area under the blood glucose response curve of a test food that provides a fixed amount of carbohydrates (usually a 50 g carbohydrates portion), expressed as a percentage of the response to the same amount of carbohydrate from a reference food (glucose or white bread) consumed by the same subject [4,5]. Most often glucose is used as the reference food, which has been defined to have a GI value of 100 [6]. Table 2.1 illustrates the categorization of GI values.

Long-term compliance with a low-GI diet may induce favorable metabolic effects [8,9], but the planning of such a diet is complicated. Foods containing simple carbohydrates (monosaccharides, disaccharides, oligosaccharides) have a high GI. Conversely, foods containing complex carbohydrates (starch, fiber) usually have a low GI. However, the GI of a food cannot be predicted by its carbohydrate content alone, as it depends on factors such as pH, cooking, processing, and other food components (fiber, fat, protein) [5]. There are some published studies on GI values of several food items as they were tested in healthy [4,10] or diabetic individuals [11–14]. Despite efforts to systematically tabulate published and unpublished sources of reliable GI values [15,16], no standardized method is currently available to determine dietary GI in national food databases. Considering that the GI value of a food may vary significantly from place to place due to the aforementioned factors (i.e., pH, cooking, etc.), it has already been demonstrated that "the need for GI testing of local foods is critical to the practical application of GI to diets" [17]. Moreover, there are different glycemic responses among individuals after consumption of a carbohydrates meal. This may be attributed to the presence of diabetes [18,19] and different individual characteristics, such as age, sex, body weight, and race [20]. Therefore, it would be prudent for researchers to consider all these factors which can alter results, leading to erroneous conclusions, especially when investigating the association of GI in diseases such as Type 2 diabetes mellitus (T2DM), where glycemic and insulinemic responses vary significantly.

GLYCEMIC LOAD

Glycemic load represents a ranking system for carbohydrates ingested and reflects the total exposure to glycemia during a 2 hour period, as first proposed by

Nutrition and Enhanced Sports Performance.
DOI: http://dx.doi.org/10.1016/B978-0-12-396454-0.00002-3

TABLE 2.1 Methods of Assessing Glycemia Following Ingestion of a Meal

Method	Value
Glycemic Index[a]	
Low	<55
Medium	55–69
High	≥70
Glucose Load[b]	
Low	≤10
Medium	11–19
High	≥20

[a]*Aston et al., 2006 [6].*
[b]*Brand-Miller et al., 2003 [7].*

better predictor of glycemic response and insulin demand than GI [24]. However, the incorporation of GL to examine postprandial effects or during subsequent exercise must be used with caution, especially in populations where controlling the glycemia and/or insulinemia is a critical issue for their health. From the equation of the GL, a low-GI/high-carbohydrates food or a high-GI/low-carbohydrates food can have the same GL [25]. Although the effects on postprandial glycemia may be similar, there is evidence that the two approaches will have very different influences on β-cell function [26], triglyceride concentrations [26], free fatty acid levels [26], and effects on satiety [27]. These influences should also be considered in seeking to effectively manage and control postprandial response to carbohydrates with the use of the GL.

Salmeron and his colleagues [21]. It constitutes a mathematical product incorporating both the carbohydrates content in grams per serving, and the GI score of the food, and is calculated by the following equation:

$$GL = \text{carbohydrates content} \times GI/100$$

Similar to GI, foods are classified as having low, medium, or high GL. Owing to the limited experience with the use of GL values, Brand-Miller and colleagues [7] suggested, as a starting point, that the preliminary cutoffs be ≤10 for a low GL, between 11 and 19 for a medium GL, and ≥20 for a high GL.

As mentioned already, the GI concept is based on the ingestion of a fixed amount of available carbohydrates, usually 50 g [4]. However, it has been demonstrated that the amount of carbohydrates ingested accounts for the 65–66% of the variability in glucose response, and the GI of the carbohydrates explains a similar degree (60%) of the variance [22]. Together, the amount and the GI of carbohydrates account for approximately 90% of the total variability in the blood glucose response. Hence, considering only the GI to explain the different metabolic responses in the postprandial phase, and not the amount of the carbohydrates ingested, could lead to an incomplete estimation of these responses. Additionally, because of the broad ranges in the amounts of carbohydrates consumed per serving [23], the development of practical recommendations for everyday settings would be inadequate.

The GL concept overcomes the above-mentioned restrictions of the GI concept as it takes into account not only the GI but also the serving sizes [24]. Hence, GL is determined by the overall glycemic effect and insulin demands of the diet and not just by the amount of carbohydrates. Consequently, GL may serve as a

GLYCEMIC INDEX, GLYCEMIC LOAD AND METABOLIC RESPONSES

The GI is an indicator of the effect of food on postprandial blood glucose and insulin levels. Consumption of high-GI foods causes a sharp, large peak in postprandial blood glucose level and subsequently a large rise in blood insulin level [28] and inhibition of glucagon release [6]. In normal individuals, insulin induces the uptake of ingested glucose by the liver and peripheral tissues (skeletal muscle and adipose tissue) [29]. Also insulin affects metabolism in many ways: increases glycogen synthesis in the liver and skeletal muscle, decreases gluconeogenesis and glycogenolysis in the liver, increases lipogenesis and inhibits lipolysis in adipose tissue [6,30]. Elevated insulin levels due to consumption of high-GI foods result in continuous uptake of nutrients and their suppressed mobilization from tissues, often leading to hypoglycemia. On the other hand, consumption of a low-GI food causes slower and smaller glycemic and insulinemic responses [6].

All these metabolic changes attributable to the GI value of foods lead to the suggestion that low-GI diets reduce the risk of metabolic diseases. Low dietary GI may improve health by contributing to decrease in body weight [3–33] and/or favorable changes in lipid profile [34,35], while high dietary GI may have the opposite effect [8,36]. Some of the health benefits gained by avoiding high-GI diets include the prevention and control of obesity and chronic diseases associated with it [3], such as T2DM [37].

Similarly to GI, the GL of a diet is proposed to have an effect on metabolic responses in the postprandial state, and these responses have been associated with several chronic diseases as well as with altered metabolic responses during exercise.

Glycemic Index, Glycemic Load and Chronic Disease Risk

Despite knowledge of the immediate metabolic benefits from compliance with a low-GI diet, a long-term metabolic effect of such a diet has yet to be established [38]. The relationship of GI or GL to the risk of developing a chronic disease has been assessed through prospective cohort studies or meta-analyses, although the outcomes are inconsistent (Table 2.2). Several studies indicate that both GI and GL are associated with a risk of developing a chronic disease [39–41], whereas other studies support such association only for GI [8,42,43] or GL [21,44–48]. There are also studies that do not support any relationship between these two indices and disease risk [49–57].

A meta-analysis of observational studies by Barclay et al. (2008) suggests that low-GI and/or low-GL diets are independently associated with a reduced risk of T2DM, heart disease, gallbladder disease, breast cancer, and all diseases combined [37]. It is notable though that the overwhelming majority of the subjects in the above meta-analysis were women (90%). Thus these findings may not be generalizable to men [37]. Nutritional interventions where the concept of GI is used are essential for the prevention [43] and management of T2DM [7]. Compliance with a low-GI diet has been suggested to be of benefit in improving insulin sensitivity and reducing glycated hemoglobin concentrations in T2DM patients [31,58–60]. Additionally, low-GI diets as well as diets with high cereal fiber content are shown to be independently associated with a reduced risk for developing T2DM in men [21,61] and in women [43]. As for the GL, it has been demonstrated to be positively associated with increased risk of T2DM in men with low cereal fiber intake [21], but not in women [43].

The association of dietary GI and GL with the risk of CVD is not well established, and it may vary among different populations. Hardy et al. [62] has shown that high GI is associated with increased incidence of coronary heart disease (CHD) in African Americans, while high GL is associated with an increased incidence of CHD in Whites. Men with high GI and high GL are at greater risk for CVD than women [63]. Among women though, the risk for CHD is greater with high GL with BMI > 23 [47] or BMI > 25 [48]. Nevertheless, there are also reports that do not support any relation between high GI or high GL and the occurrence of CVD in men [64] or in women [56,57].

The occurrence of cancer has been linked with factors involving glucose metabolism [65]. Nevertheless, the evidence on the relationship between the GI and the risk of several types of cancer is inconsistent. Studies have shown that both high-GI and high-GL diets increase the risk of breast cancer [40], whereas others support the connection of high risk with high GI alone [42], or high GL alone [44–46]. Nevertheless, there are also studies that do not report an association between either high GI or high GL and increased risk for developing breast cancer [51,52]. Similarly, some investigators found an increased risk of endometrial cancer in women with the consumption of high-GI or high-GL foods [42], while others do not agree with these findings [53]. Likewise, some cohort studies do not support an association between either high GI or high GL and increased risk for developing colorectal cancer [49,66] despite the positive association that has been demonstrated [41]. Regarding pancreatic cancer, most of the data indicate that there is no relationship between GI or GL and increased risk for developing the disease in either men or women [54,55]; however, a positive relationship has also been reported for high GL in overweight and sedentary women [67]. It is clear that published results have failed so far to establish a clear association between GI and elevated cancer risk. Therefore, further long-term studies are needed to determine the role of GI in carcinogenesis.

Diets with high GI or GL are also reported to independently increase the risk of gallbladder disease [37,39]. In regard to age-related cataract, no association has been observed between disease risk and either the GI or GL of the diet [50].

Obesity is a very common public health problem that increases the risk of developing several of the aforementioned diseases, such as CVD, T2DM and certain types of cancer, and also osteoarthritis [68]. A restriction in saturated fat intake for weight loss has had modest success, while a restriction in refined carbohydrates is the most recent approach to weight loss and related reduction of disease risk [69]. Low dietary GI may contribute to decrease in body weight in a most effective way [31–33]. This could be due to different substrate oxidation postprandial as a result of a low-GI meal consumption compared with a high-GI meal [31,70,71], which leads to different fuel portioning [31]. In addition, low-GI diets may assist with weight control by effecting satiety [31,70]. At the same time, high-GI foods may promote excessive weight gain due to hormonal responses that seem to lower circulating levels of metabolic fuels, stimulate appetite, and favor storage of fat [72]. Indeed, body mass index has been positively associated with the GI (long term) in women [33].

An important finding regarding the GI influence on health, is that of the DECODE study (1999) [73]. According to this study, postprandial hyperglycemia is considered a risk factor for mortality not only for patients who suffer from chronic disease, but also for people with normal fasting blood glucose. Therefore,

TABLE 2.2 Glycemic Index, Glycemic Load, and Chronic Diseases

Study	Disease	Study Protocol	Estimation Method	Subjects	Average Follow Up (years)	Disease Incidences	Results
NEOPLASM							
Silvera et al. (2005)	Breast cancer	Prospective cohort; Data from the Canadian National Breast Screening Study	FFQ	49 111 women (40–59 y)	16.6	1450	GI: ↑ risk in post-menopausal women; GL: No association with the disease risk
Jonas et al. (2003)	Breast cancer	Cohort study; Data from the Cancer Prevention Study-II	SFFQ	63 307 postmenopausal women (40–87 y)	5	1442	GL, GI: no association with breast cancer
Sieri et al. (2007)	Breast cancer	Cohort study; Data from the Hormones and Diet in the Etiology of Breast Tumors Study	SFFQ	8959 women (34–70 y)	11.5	289	GI: ↑ risk with HGI in postmenopausal GL: ↑ risk with HGL in postmenopausal women; ↑ risk with HGL in women with BMI < 25
Larsson et al. (2009)	Breast cancer	Data from the Swedish Mammography Cohort, a population-based cohort	FFQ	61 433 women (39–76y)	17.4	2952	GI, CHO: No association with risk of overall breast cancer; GL: Positively association with risk of overall breast cancer
Wen et al. (2009)	Breast cancer	Data from the Shanghai Women's Health Study, a population-based cohort study	FFQ	73 328 Chinese women	7.35	616	GI: No association with breast cancer risk GL, CHO: Positive linear association with breast cancer risk in premenopausal women or women <50 y
Shikany et al. (2011)	Breast cancer	Data from the Women's Health Initiative	FFQ	148 767 postmenopausal women (50–79 y)	8	6115	GI, GL, CHO: No association with total breast cancer risk
Sieri et al. (2012)	Breast cancer	Prospective cohort study; Data from the Italian section of the European Prospective Investigation into Cancer and Nutrition	FFQ, SLC	26 066 women (age?)	11	879	GI, Total CHO: No association with the risk of breast cancer; GL: Positive association with the risk of breast cancer
Larsson et al. (2007)	Colorectal cancer	Prospective study; Data from the Swedish Mammography Cohort Study	FFQ	36 616 women born between 1914 and 1948	15.6	870	GI, GL, CHO: No association with colorectal cancer risk
Higginbotham et al. (2004)	Colorectal cancer	Cohort study; Data from the Women Health Study	FFQ, RFQ	38 451 women (≥45 y)	7.5	174	GI, GL: Positive relationship with colorectal cancer

Li et al. (2011)	Colorectal cancer	Prospective cohort; Data from cohort of Chinese women	FFQ	73 061 women (40–70 y)	9.1	475	GI, GL, CHO: No association with colorectal cancer risk
Patel et al. (2007)	Pancreatic cancer	Prospective cohort; Data from the Cancer Prevention Study-II	FFQ	124 907 men and women (62.7 ± 6.35 y)	9	401	GI, GL, CHO intake: No association with pancreatic cancer
Jiao et al. (2009)	Pancreatic cancer	Prospective cohort study; Data from the NIH-AARP Diet and Health Study	FFQ	482 362 (280 542 men and 201 820 women)	7.2	1151 (733 men and 418 women)	GI, GL: No association with the risk of pancreatic cancer
Michaud et al. (2002)	Pancreatic cancer	Prospective cohort study; Data from the Nurses' Health Study	FFQ	88 802 women	18	180	GI: No association with the risk of pancreatic cancer GL: Positive relationship with pancreatic cancer risk in sedentary and overweight women
Silvera et al. (2005)	Endometrial cancer	Prospective study; Data from the Canadian National Breast Screening Study	FFQ	34 391 women ((5.10) in DnV (2007))40–59 y)	16.4	426	GI, GL: Overall positive association with endometrial cancer, particularly among obese women, premenopausal women, and postmenopausal women who use hormone replacement therapy
Larsson et al. (2006)	Endometrial cancer	Prospective study; Data from the Swedish Mammography Cohort Study	FFQ	61 226 women born between 1914 and 1948	15.6	608	GI, GL: No overall association with endometrial cancer; a 1.9 non-significant increase in disease risk in overweight women with low physical activity
DIABETES							
Salmeron et al. (1997)	Type 2 diabetes	Prospective study; Data from the Health Professionals Follow up Study	SFFQ	42 759 men (40–75y)	6	523	GL: ↑risk with H-GL in men with low cereal fiber intake
Schulze et al. (2004)	Type 2 diabetes	Cohort study; Data from the Nurses' Health Study II	SFFQ	91 249 women (24–44 y)	8	741	GI: HGI↑risk of Type 2 diabetes; GL: No association
CVD							
Liu et al. (2000)	Cardiovascular Disease	Cohort study; Data from the Nurse's Health Study	SFFQ, MHQ	75 521 women (38–63 y)	10	761	GL: Positive association with CHD;↑risk with HGL in women with BMI > 23

(*Continued*)

TABLE 2.2 (*Continued*)

Study	Disease	Study Protocol	Estimation Method	Subjects	Average Follow Up (years)	Disease Incidences	Results
Oh et al. (2005)	Stroke	Prospective cohort study; Data from the Nurse's Health Study	SFFQ	78 799 women (30–55 y)	18	1020	HCHO intake: positive association; GL: Positive association with ↑ risk in women with BMI ≥ 25
Levitan et al. (2007)	CVD	The Cohort of Swedish Men	FFQ	36 246 men (46–79 y)	6 y follow-up for incidence of CVD, and mortality; 8 y for all-cause mortality	2181 incidence of CVD, and 785 CV deaths; 2959 deaths of all causes combined	GI: No association with ischemic CVD or mortality; GL: No association with ischemic cardiovascular disease or mortality; Positive association with the risk of hemorrhagic stroke
Hardy et al. (2010)	CHD with or without type 2 diabetes	Data from the Atherosclerosis Risk in Communities Study	SFFQ	13 051 Whites and African Americans (1378 with diabetes and 11 673 without diabetes) (45–64y)	17	1683 (371 with diabetes,1312 without diabetes)	GI: Positive association with the risk of CHD in African Americans; GL: Positive association with the risk of CHD in Whites; more pronounced association in Whites without diabetes
Levitan (2010a)	Myocardial infarction	Participants from the Swedish Mammography Cohort	FFQ, MHQ	36 234 women (48–83 y)	9	1138	GI, GL: No association with myocardial infarction in women
Levitan (2010b)	Heart failure	Prospective, observational study with participants from the Swedish Mammography Cohort	FFQ, MHQ	36 019 women (48–83)	9	639 (54 died and 585 hospitalized for heart failure for the first time)	GI, GL: No association with the risk of heart failure events in women
Burger et al. (2011)	CHD and stroke	Data from the EPIC-MORGEN Study, a large prospective cohort study	FFQ	10 753 women and 8855 men (21–64 y)	11.9	CHD: 300 women and 581 men. Stroke: 109 women and 120 men	GL: Positive association with CHD risk in men only, after adjustment for established CVD risk factors. Slightly ↑ CHD risk in men only, after inclusion of nutritional factors. No association with ↑ stroke risk in men or women. GI: No association with ↑ CHD risk; Positive association with stroke risk in men only, after adjustment for CVD risk factors and nutrients

OTHER DISEASES							
McKeown et al. (2004)	Insulin resistance, metabolic syndrome	Data from the fifth examination cycle of the Framingham Offspring Study	SFFQ, MHQ	2834 (1290 men and 1544 women) (26–82 y)	4	2834	GI: Positive association with prevalence of the metabolic syndrome; Positive association with insulin resistance. GL: No association with prevalence of the metabolic syndrome; Positive association with insulin resistance
Schaumberg et al. (2004)	Age-related cataract	Data from the Nurse's Health Study and the Health Professionals Follow-up Study	SFFQ	71 919 women and 39 926 men (>45 y)	10	3258 (women) 1607 (men)	GI, GL: No association with cataract extraction
Tsai et al. (2005)	Gallbladder disease	Prospective cohort study; Data from the Health Professionals Follow up Study	FFQ, MHQ	44 525 men (40–75 y)	12	1710	GL, GI, CHO intake: positive association with the disease risk

FFQ, Food Frequency Questionnaire; SFFQ, Semiquantitative Food Frequency Questionnaire; RFQ, Risk Frequency Questionnaire; MHQ, Medical History Questionnaire; CHD, coronary heart disease; H-CHO, high carbohydrate; T2DM, non-insulin dependent diabetes mellitus; BLC, blood lipid concentrations; RRR, reduced rank regression; UL, uterine leiomyomata; DHI, diet history interviews; GDM, gestational diabetes mellitus; SLC, Standardized Lifestyle Questionnaire the prevention of hyperglycemic situations should also be targeted in healthy people.

GLYCEMIC INDEX, METABOLIC RESPONSES AND EXERCISE PERFORMANCE

GI has recently gained interest in the field of sports nutrition. Different GIs have been shown to produce diverse metabolic responses during exercise [74,75], and some authors support the connection of pre-exercise low GI with enhanced physical performance [76–80], while others dispute such a connection [75,81,82].

The intensity of the glycemic and insulinemic responses postprandial and during exercise, are considered main contributors to achieved performance. The maintenance of euglycemia during exercise, and the avoidance of rebound hypoglycemia. have been shown to be of great importance in the delay of fatigue and improvement of performance, especially during prolonged exercise. The preservation of carbohydrates and the use of fat as the main substrate for energy are of prime pursuit when it comes to endurance exercise. Since the GI represents the blood glucose response of carbohydrate-containing foods, the perspective of controlling the glycemia and insulinemia and, as a consequence, substrate utilization during exercise by manipulating the GI of pre-exercise meals, would be of critical significance for athletic performance. Glucose utilization during exercise is mediated by insulin release by the β-cells of the pancreas, which is stimulated by high glucose concentrations, as it happens after the consumption of a rich-carbohydrate meal. Moreover, high-GI pre-exercise meals have been connected with increased carbohydrates oxidation and diminished fat oxidation [1,71,80]. Such a connection is further indicated by elevated serum levels of GLU during exercise and suppression of circulating free fatty acid and triglyceride [74,80].

Substrate utilization during exercise is also dependent on the intensity of physical activity, as carbohydrates are the preferred energy source in high-intensity exercise whereas, during low-to-moderate-intensity physical activity, the energy comes mainly from the catabolism of free fatty acids (FFA). Nevertheless, pre-exercise glucose ingestion suppresses lipolysis to a point at which it limits fat oxidation, even during low-intensity exercise [83]. It seems that a pre-exercise carbohydrates meal can affect the typical intensity-dependent substrate utilization, probably by the carbohydrates-induced rise in insulin that inhibits the mobilization and hence availability of circulating FFA. Additionally, the increased carbohydrates oxidation that has been reported after a high-GI pre-exercise meal and the subsequent inhibition of long-chain fatty acid entrance into the mitochondria [84] could also explain the reduced fat oxidation. However, such an inhibition of free fatty acid entry into the mitochondria and a restricted energy production from fat oxidation could also result in earlier fatigue and deterioration of physical performance.

The Effect of Different GI of Pre-Exercise Meals on Physical Performance

As has already been mentioned, different GI pre-exercise meals result in different metabolic responses. There is evidence that these responses affect, in turn, physical performance. However, despite those studies that report an enhancement in athletic performance, mainly due to low-GI pre-exercise meals [76–79], there are also reports that do not support any improvement in physical performance [75,81,82]. Physical performance is mainly being estimated through the Time Trial (TT), that is the time needed for covering a predefined distance, Time to Exhaustion (TE), or total work and power output, along with other physiological parameters such as heart rate (HR), rate of perceived exertion (RPE), VO_{2max}, and respiratory exchange ratio (RER). Table 2.3 summarizes all of the studies that have been reviewed in this chapter.

Improvement of Physical Performance

An improvement by an average of 3 min in the time needed to cycle a 40 km distance was observed after the ingestion of a low-GI carbohydrates meal 45 min before the onset of the exercise, when compared with the consumption of a high-GI meal in well-trained male cyclists [77]. Although improvement in exercise performance has also been reported in previous studies [80,85], the 3.2% improvement in performance in Moore's study appears somewhat lower than the 7.9–59% range of improvement in performance previously observed. The authors attribute this lower improvement to a possible overestimation of the beneficial effect that low-GI meals are actually having on performance. Additionally, lower RPE during the TT, and lower whole blood GLU and insulin concentration 45 min postprandial, were demonstrated in the low-GI meal compared with the high-GI meal. Another interesting finding in Moore's study was the lower carbohydrates oxidation during exercise in the high-GI trial, which is in contrast to the typical inhibition of FFA and increased carbohydrates oxidation following the consumption of high-GI pre-exercise meals [74,76,78,80].

In the study by Karamanolis et al. (2011) [76], carbohydrates oxidation during prolonged exercise till

TABLE 2.3 Glycemic Index and Metabolic Responses During Exercise

Studies	Study Protocol	Subjects	Estimated Indices	Results
Wee et al. (2005)	HGI (GI = 80), LGI (GI = 36) breakfast meals ingested 3 h prior to 30 min running at ~70% VO_{2max}.	7 male recreational runners (31 ± 4 y)	HR, RER, total CHO & fat oxidation, GLU, FFA, glycerol, glucagon, insulin, glucose AUC, insulin AUC	GLU: ↓ in HGI below baseline at 10 min of exercise, ↑ in HGI than LGI at the end of exercise; INS: 2-fold ↑ in HGI than in LGI; Glucagon: ↑ in LGI than HGI postprandial; FFA: ↓ in both meals postprandial, ↑ suppression in HGI than in LGI; LA: ↓ in LGI than in HGI in serum and in muscle after exercise; Muscle glycogen: 15% ↑ in the HGI, 46% ↑ net muscle glycogen utilization in the HGI than in the LGI
IMPROVEMENT OF PERFORMANCE				
Moore et al. (2010)	Two standardized meals with HGI (GI = 72) or LGI (GI = 30) ingested 45 min prior to exercise. Each trial separated by 7 days.	10 well-trained male cyclists (8 ± 6 y)	40 km TT, HR, RPE, fat & CHO oxidation, VO_{2max}, RER, GLU, INS, FFA, TGA, LA	TT: ↑ performance in LGI than in HGI meal; RPE, Fat oxidation, GLU, INS: ↓ in the LGI than in HGI; RER, CHO oxidation: ↓ in the HGI than in the LGI trial; HR, FFA, LA: ↑ in time than baseline for both GI trials; VO_{2max}, TGA: No differences between trials
Karamanolis et al. (2011)	HGI (GI = 83), LGI (GI = 29) and control meal ingested 15 min prior to exercise. Each trial separated by 7 days.	9 recreational runners (26 ± 3 y)	Treadmill TE, VO_{2max}, RER, HR, GLU, INS, glycerol, LA, fat & CHO oxidation	TE: 23% ↑ in LGI vs PL; VO_{2max}: ↑ in the HGI vs PL from 45 min till the TE; GLU: ↑ in the HGI vs PL and LGI 15 min postprandial; INS: ↑ in the HGI vs PL 15 min postprandial; Glycerol: ↑ in the PL vs LGI at 60 min and TTE; LA: ↑ in HGI vs LGI at TE; CHO oxidation: ↑ in the HGI vs LGI at 30 min & 45 min and vs PL at 30 min; RER, HR, fat oxidation: No differences between trials
Kirwan et al. (2001)	HGI (GI = 82), Moderate GI (GI = 61), & control (water) whole-food breakfast cereals, ingested 45 min prior to exercise. Each trial separated by 7 days.	6 healthy active men (22 ± 1 y)	Cycling TE, VO_{2max}, RER, HR, GLU, INS, Epinephrine, Norepinephrine, glycerol, FFA, fat & CHO oxidation	TE: 23% ↑ in the MGI meal; GLU: ↑ postprandial in both HGI and MGI, ↑ at 60 and 90 min of exercise in the MGI; INS: ↑ postprandial in both HGI and MGI, ↑ at the start of exercise and returned to pre-meal at 30 min of exercise; FFA: ↓ before exercise and remained ↓ at 30, 60, 120 min of exercise in both HGI and MGI; CHO oxidation: ↑ in the MGI than in control; Significant association between total CHO oxidation and TE; Epinephrine, Norepinephrine: Not affected
Wong et al. (2008)	LGI (GI = 37) or HGI (GI = 77) meal providing 1.5 g CHO/kg body mass, ingested 2 h prior to exercise. Each trial separated by 7 days.	8 endurance-trained male runners (33 ± 1.7y)	21 km TT, HR, RPE, GLU, LA, INS, FFA, cortisol, glycerol, CHO and fat oxidation, hematocrit, Hb, PV, RER, RPT, RAD, & GFS, GLU IAUC, SL	TT: ↑ performance in the LGI trial; GLU: higher in the LGI trial; LA, RPE: ↑ in both trials; INS: No differences between trials; HR: ↑ in both trials, higher in the LGI trial at the end of exercise; FFA: ↓ until 10 km of exercise in both trials, higher in the LGI trial at the end of exercise; Glycerol: ↑ in both trials, higher in the LGI trial at 10 km until the end of

(Continued)

TABLE 2.3 (*Continued*)

Studies	Study Protocol	Subjects	Estimated Indices	Results
				exercise; PV, SL, RPT: No significant differences between trials; RER, CHO oxidation: ↑ in both trials, higher in the HGI trial at 5–10 km of exercise; Fat oxidation: ↑ in both trials, higher in the LGI trial at 5–15km; RAD: ↑ at the end of the exercise in both trials; GFS: No differences between meals; GLU IAUC: higher in the HGI meal
Wu & Williams (2006)	LGI (GI = 37) or HGI (GI = 77) meal providing 2 g CHO/kg body mass, ingested 3 h prior to exercise. Each trial separated by 7 days.	8 male recreational runners (28.9 ± 1.5 y)	TE , RER, RPE, RPT, GFS, BM, LA, FFA, GLU, INS, glycerol, hemoglobin, PV, GLU IAUC, INS IAUC, energy expenditure, CHO oxidation, fat oxidation	TE: higher in the LGI trial; GLU: higher in the LGI trial at 15–30 min of exercise; INS, LA, HR, RPE, PV, RPT, BM: No significant differences between trial; GLU IAUC, INS IAUC: postprandial higher in the HGI meal; FFA: higher in the LGI trial; Glycerol: higher in the LGI trial at 45 min of exercise until exhaustion; CHO oxidation, RER: higher in the HGI trial; Fat oxidation: higher in the LGI trial at 15 min and 45–90 min of exercise
NO EFFECT ON PHYSICAL PERFORMANCE				
Jamurtas et al. (2011)	LGI (GI = 30), HGI (GI = 70) meal providing 1.5 g CHO/kg of body weight or placebo (control) meal, ingested 30 min before 1 hour of cycling (65% VO_{2max}) followed by cycling to exhaustion (90% VO_{2max}). Each trial separated by 7 days.	8 untrained healthy males (22.8 ± 3.6y)	TE, RPE, HR, ventilation, GLU, INS, LA, β-Endorphin, RER, CHO oxidation, Fat oxidation	TE, RPE, HR, RER, ventilation: No differences between trials; INS: higher in the HGI than control at 20 min of exercise; β-Endorphin, RER: ↑ at the point of exhaustion in all trials; CHO oxidation, Fat oxidation, GLU, LA: No differences between trials;
Febbraio et al. (2000)	HGI, LGI meal providing 1 g CHO/kg body wt, or placebo (control) meal, ingested 30 min before 120 min of submaximal exercise followed by a 30 min performance trial. Each trial separated by 7 days.	8 endurance-trained men (26 ± 6 y)	TW, VO2, HR, GLU, FFA, INS, Total glucose Ra and Rd, LA, glycogen, Total CHO oxidation, Glucose oxidation, Fat oxidation	TW: No differences between trials during the performance trial; VO_2, HR, INS, LA, Glycogen: No differences between trials during the submaximal exercise; FFA, GLU: lower in the HGI trial at some points of submaximal exercise; Glucose Ra, Glucose Rd: higher in the HGI trial throughout submaximal exercise, lower in the control vs LGI trial at some points of submaximal exercise; CHO oxidation: higher in the HGI trial; Glucose oxidation: higher in the GI trials vs control, higher in the HGI vs LGI trial
Kern et al. (2007)	HGI (117 ± 15) or MGI (88 ± 13) meal providing 1 g CHO/kg body weight, ingested 45 min prior to 45 min of submaximal exercise followed by a 15 min	8 endurance-trained male (n = 4) and female (n = 4) cyclists (30 ± 5 y)	GLU, INS, LA, TG, FFA, BHB, Power output	GLU, LA, TG, BHB: No differences between trials during the submaximal exercise; INS: lower in the MGI trial; FFA: higher in the MGI trial; Power output: No differences between trials during the performance trial

	performance trial. Each trial separated by at least 7 days.			
Little et al. (2010)	HGI (GI = 76), LGI (GI = 26) meal providing 1.5 g CHO/kg body weight or placebo (control), ingested 2 h before 90 min of high-intensity, intermittent exercise including 15 min of a repeated-sprint test at the end of exercise. Each trial separated by at least 7 days.	16 male athletes (22.8 ± 3.2 y)	15 min DC, VO_2, RER, RPE, Fat oxidation, CHO oxidation, GLU, LA, INS, FFA, CAT, Glycogen	15 min DC:↑performance in both GI trials; VO_2, RER, GLU, LA: No differences between trials; Fat oxidation: lower in the HGI vs control at 33–40 min of exercise (Collection Period 2), lower in the LGI vs control at 63–70 min of exercise (Collection Period 3);↑time for all trials; CHO oxidation: No differences between trials,↓time for all trials; INS: lower in the control vs GI trials; FFA: higher in the control vs GI trials; CAT: higher in the HGI vs control at the end of exercise; Glycogen: higher in the GI trials after 75 min of exercise; RPE: lower in the LGI vs control
CHO INGESTION DURING EXERCISE				
Little et al. (2009)	Fasted control, HGI (GI = 81) or LGI (GI = 29) meal providing 1.3 g CHO/kg body weight, ingested 3 h before high-intensity, intermittent exercise and halfway through. Performance estimated by DC on five 1-min sprints during last 15 min of exercise. Each trial separated by at least 7 days.	7 male athletes (23.3 ± 3.8 y)	DC, VO_{2max}, RER, Fat oxidation, CHO oxidation, RPE, HR, GLU	DC:↑performance in the GI trials; VO_{2max}, HR, RPE, GLU: No differences between trials; RER: higher in the HGI vs control; Fat oxidation: lower in the HGI vs control; CHO oxidation: higher in the HGI vs control, higher in the LGI vs control at some points of exercise
Wong et al. (2009)	HGI (GI = 83), LGI (GI = 36) meal providing 1.5 g CHO/kg body mass or Placebo (Control), ingested 2 h before exercise. 2 mL/kg body mass of a 6.6% CHO-electrolyte solution was provided immediately before exercise and every 2.5 km after the start of running. Each trial separated by at least 7 days.	9 male endurance runners (24 ± 2.4 y)	21 km TT, RS, VO_{2max}, GLU, GLU IAUC, LA, INS, FFA, Glycerol, CHO oxidation, Fat oxidation, Blood osmolality, Na^+, K^+, %BM change, SL	TT: No improvement in performance for all trials; RS, VO_{2max}: No differences between trials throughout the performance run; GLU, INS, %BM change, SL, LA, RER, Blood osmolality, Na^+, K^+, CHO oxidation, Fat oxidation: No differences between trials; FFA: No differences between trials,↑at the end of exercise in the three trials; Glycerol: higher in the control vs HGI trial, higher in the control vs LGI trial at some points of exercise
CHO LOADING				
Hamzah (2009)	High CHO/LGI, high CHO/HGI or habitual diet (control), 5 days prior to exercise. Each trial separated by at least 11 days.	9 healthy active males (23.9 ± 4.3 y)	TE, DC, VO_2, RPE, HR, GLU, INS, NEFA, Glycerol, Fat oxidation, CHO oxidation	TE, DC, GLU, INS, NEFA, VO_2, RPE, HR: No differences between trials; Glycerol, Fat oxidation: No differences between the GI trials, higher in the control trial at some points of exercise; CHO oxidation: No differences between the GI trials, lower in the control trial at some points of exercise

CHO, Carbohydrates; HGI, High Glycemic Index; LGI, Low Glycemic Index; TT, Time Trial; TE, Time to Exhaustion; DC, Distance Covered; HR, Heart Rate; RER, Respiratory Exchange Ratio; RPE, Rating of Perceived Exertion; RPT, Rating of Perceived Thirst; RAD, Rating of Abdominal Discomfort; GFS, Gut Fullness Scale; GLU, Glucose; INS, Insulin; FFA, Free Fatty Acids; TG, Triglycerides; LA, Lactic Acid/Lactate; TW, Total Work; BHB, β-hydroxybutyrate; CAT, Catecholamines; RS, Running Speed; %BM change, percent Body Mass change; SL, Sweet Loss.

exhaustion was lower in the low-GI (GI = 29) group compared with the high-GI (GI = 83) and placebo (PL) group, whereas fat oxidation was similar among groups. In this study, the three different GI pre-exercise meals were consumed 15 min prior to exercise. The lower carbohydrates oxidation was accompanied by an improvement in the performance, as the TE was 23% greater in the low-GI trial compared with the PL trial. This seems to be the result of the prevention of hyperinsulinemia and a better maintenance of blood GLU concentration throughout the exercise and mainly due to significantly higher responses of blood GLU at the time to exhaustion. Lactate was lower in the low-GI than in the high-GI trial only at the TE, when increased levels of anaerobic glycolysis were present. Glycerol mobilization was better in the low-GI, which probably suggests that a better preservation of carbohydrates was attained that led to less fatigue and improvement in exercise performance.

Wee et al. (2005) [74] investigated the effect of pre-exercise breakfast of either high or low GI on muscle glycogen metabolism 3 h postprandial and during 30 min submaximal running. Higher values of GLU and INS were reported postprandial for both the HGI and the LGI meal, with GLU falling below baseline after the first 10 min of exercise in HGI meal but been greater than those of the LGI meal at the end of exercise. Insulin appeared to be 2-fold higher in the HGI compared with the LGI but only at the onset of exercise. FFA and glycerol were suppressed throughout the postprandial period for both meals, but the suppression was higher in the high-GI compared with the L-GI, which demonstrated higher levels of FFA and glycerol than the high-GI during exercise. No differences in RER were observed between the two pre-exercise dietary approaches. However, higher levels of lactate were recorded both in serum and muscle for the high-GI meal. Carbohydrates oxidation during exercise was 12% lower in the LGI, with a compensatory increase in fat oxidation compared with the HGI meal, such that the overall energy expenditure was similar.

An attempt has also been made to ascertain whether a moderate- vs a high-GI pre-exercise meal could lead to different metabolic responses and performance. Kirwan et al. (2001) [80] used a moderate-GI (M-GI, GI = 61) vs a high-GI (GI = 82) breakfast cereal to test this hypothesis. Six healthy active men were tested at ~60% VO_{2peak} cycling till exhaustion, 45 min after the consumption of the meals. Performance was improved in the M-GI meal, with a 23% improvement in TE compared with a 5% improvement in the high-GI meal, which was no different than ingesting water. Elevated GLU and INS levels were observed postprandial in both meals, with GLU being higher at 60 min and 90 min of exercise in the M-GI meal, whereas INS was elevated only at the start of exercise and returned to pre-meal values within 30 min of exercise. Circulating FFA were suppressed before exercise and remained that way at 30, 60 and 120 min of exercise in both the high-GI and M-GI meals compared with control. Total carbohydrates oxidation was higher in the M-GI than in control trial, and additionally a significant association between total carbohydrates oxidation and time to exhaustion was observed.

In a study by Wu and Williams (2006) [79], at 70% VO_{2max} an improvement by approximately 8 minutes in running time to exhaustion was observed 3 hours after the consumption of a low-GI meal, when compared with an isocaloric high-GI meal trial, in eight healthy male recreational runners. Plasma glucose level was higher during the high-GI trial than in the low-GI trial only for the first 30 min of exercise. FFA, glycerol levels, and fat oxidation rate were significantly higher, while RER and carbohydrate oxidation rate were significantly lower in the low-GI trial compared with the high-GI trial. These findings are consistent with data from other studies suggesting that the consumption of a pre-exercise low-GI meal increases fat oxidation and availability of FFA, and suppresses carbohydrates oxidation, resulting in delayed onset of carbohydrates depletion and delay of fatigue; this may be a way to improve endurance exercise performance [86]. Lactate concentrations, heart rate and ratings of perceived exertion did not significantly differ between the two GI trials. The results suggest that consumption of a low-GI meal 3 h before exercise leads to greater endurance performance than after the consumption of a high-GI meal.

Moreover, in a study by Wong and his colleagues (2008) [78], the consumption of a low-GI meal 2 h before a 21 km performance run resulted in improved running performance time by 2.8% compared with an isocaloric high-GI meal trial, in endurance-trained male runners. During exercise, blood glucose concentration was higher in the low-GI trial while serum insulin concentration did not differ between trials. Substrate oxidation differed between the two GI trials during exercise: carbohydrate oxidation was 9.5% lower and fat oxidation was 17.9% higher in the low-GI trial compared with the high-GI trial. Moreover, heart rates were higher at 15 km and at the end of the exercise in the low-GI trial than in the high-GI trial, whereas RPE did not differ between the two GI trials. All these findings led to the suggestion that a low-GI meal 2 h before a 21 km performance run results in improved running performance time compared with a high-GI meal.

Even though there is a consensus in the aforementioned studies indicating an increase in exercise performance following a low-GI meal prior to exercise, the appropriate time of ingestion of the meal is unclear, since GI meals were ingested from 15 minutes to 3 hours prior to exercise.

No Effect of GI on Exercise Performance

Although an association between GI and exercise performance has been suggested, there are a number of studies that have found no influence of GI of the diet on exercise performance.

In a study by Jamurtas et al. (2011), the ingestion of different GI meals before cycling until exhaustion did not affect exercise performance and metabolic responses during exercise. Eight untrained healthy males ingested a rich-carbohydrates meal (1.5 g/kg of body weight) with low- or high-GI or a placebo 30 min before exercise. Subjects performed a submaximal cycling exercise (65% of VO_{2max}) for 60 min and then cycled until exhaustion (90% of VO_{2max}). No differences in VO_2 during the submaximal exercise were observed between trials. Also there were no differences in mean values of TE, RPE, ventilation, substrate oxidation, and metabolic responses between trials throughout the cycling exercise. In this study a significant increase in β-endorphin was found at the end of the exercise in all trials. According to the authors, β-endorphin levels after the ingestion of the meals did not differ at rest due to the amount of carbohydrates consumed (1.5 g/kg of body weight) nor during 1 hour of cycling due to exercise intensity [81].

Febbraio and colleagues (2000) conducted a study to investigate the effects of GI meals consumed 30 min before prolonged exercise on substrate oxidation and performance. Eight trained males consumed a low-GI or high-GI meal or placebo and 30 min later cycled at 70% peak oxygen uptake for 120 min (submaximal exercise) followed by a 30 min performance cycle. No difference in work output during the performance trial was observed among the three dietary groups. Substrate oxidation was somewhat different between groups: total carbohydrates and glucose oxidation were significantly increased in the high-GI group compared with the low-GI group. Also FFA and blood glucose concentration were lower in the high-GI group at some points of the submaximal exercise [75].

Kern et al. (2007) investigated the effects of pre-exercise consumption of carbohydrates sources with different GI on metabolism and cycling performance. Eight endurance-trained male and female cyclists ingested 1 g carbohydrates/kg body weight from either raisin with moderate GI (88 ± 13) or sports gel with high GI (117 ± 15). Forty-five minutes later they completed a submaximal exercise on a cycle ergometer at 70% VO_{2max} and then a 15 min performance trial. No differences in work output during the performance trial were observed between the moderate-GI (189.5 ± 69.8 kJ) and high-GI (187.9 ± 64.8 kJ) trial. There were minor differences in metabolic responses between trials. During submaximal exercise, serum glucose, blood lactate, serum triglycerides and serum beta-hydroxybutyrate levels did not differ between the two trials. However, insulin level was significantly higher in the high-GI trial, while FFA levels were significantly higher in the moderate-GI trial. The authors of this study concluded that raisins may be a more suitable source for carbohydrates utilization than a sports gel during short-term exercise bouts [82].

Little and his colleagues (2010) conducted a study to examine the effects of pre-exercise carbohydrates-rich meals with different GI on high-intensity, intermittent exercise performance when no exogenous carbohydrates are provided during exercise. Sixteen male athletes performed 90 min of high-intensity intermittent running trial (two 45 min sections separated by a 15 min break), either fasted (control), 2 h after ingesting a low-GI (GI = 26) pre-exercise meal, or 2 h after ingesting an isoenergetic (carbohydrates: 1.5 g/kg body mass) high-GI (GI = 76) pre-exercise meal. Total distance covered was significantly higher in the GI trials compared with the control trial, whereas no significant differences were evident between the two GI trials. Mean VO_{2max} (~63% VO_{2peak}), RER and RPE were not different between trials. Fat oxidation increased and carbohydrates oxidation decreased in both GI trials during exercise, and no significant differences between the GI trials were observed. There were no differences in metabolic responses between the GI trials. These results indicate that, although ingestion of a rich-carbohydrates meal 2 h before prolonged, high-intensity intermittent exercise improves performance when compared with fasted control, the GI of a pre-exercise meal does not affect exercise performance and metabolic responses during exercise [87].

Ingestion of Carbohydrates During Exercise Following a Pre-Exercise Meal

The ingestion of carbohydrates during exercise following a pre-exercise meal with different GI has not been adequately investigated.

Little and colleagues (2009) investigated the influence of GI meals consumed 3 h before and halfway through high-intensity, intermittent exercise that simulated a soccer match on exercise performance. Seven male athletes participated in three trials (ingested a high-GI or low-GI meal 3 h before exercise or fasted) and performed two 45 min of intermittent treadmill running separated by a 15 min break with the consumption of a small amount of the same pre-exercise meal. Performance was improved after the consumption of the GI meals as estimated by the total distance covered on five 1 min sprints during the last 15 min of exercise compared with fasted control. However, no significant differences were found in the total distance covered between the two GI meals. Oxygen uptake (~58% of VO_{2max}), RPE and HR (~70% of maximum) did not differ between the GI meal trials during the soccer match. Metabolic responses and substrate oxidation were not different between the GI trials during exercise although some differences in GI trials compared with fasted control were observed; the high-GI trial impaired fat oxidation during exercise when compared with a low-GI meal. Nevertheless, manipulation of the GI of a meal consumed 3 h before exercise does not seem to affect high-intensity, intermittent exercise performance [88].

Wong and O'Reilly (2009) investigated the effect of consuming a carbohydrates-electrolyte solution during a 21 km run on exercise performance, 2 h after the ingestion of meals with different GI. Nine male endurance runners consumed a high-GI (GI = 83) or a low-GI (GI = 36) meal (both meals provided 1.5 g carbohydrates/kg body mass) or a control meal (GI = 0), and 2 h later they completed a 21 km performance run on a treadmill (5 km at a fixed intensity of 70% VO_{2max} and 16 km as fast as possible). No differences in time to completion were observed between the trials. Running speeds, percentage VO_{2max}, metabolic responses and substrate oxidation did not significantly differ between the trials. Plasma Na^+ and K^+ levels, percent body-mass change, and sweat loss did not differ between trials. The results indicate that the GI of a pre-exercise meal does not affect running performance and metabolic responses when a carbohydrates-electrolyte solution is consumed during exercise [24].

Carbohydrates Loading

It has been suggested that a meal rich in carbohydrates eaten prior to exercise increases liver and muscle glycogen stores which in turn improve prolonged exercise performance. Several loading strategies have been tested for this reason and some of them include the concept of the GI.

In a study by Hamzah et al. (2009), the GI of a high-carbohydrates diet consumed 5 days before exercise had no impact on running capacity performed after 12 h overnight fast. Nine healthy males performed three treadmill runs to exhaustion at 65% $VO2_{max}$: after a habitual diet (control trial), after a high-GI/high carbohydrates diet, and after an isocaloric low-GI/high carbohydrates diet in randomized counterbalanced order. No significant differences in TE or running distance covered were observed between the high-GI and low-GI diet. Moreover glucose and insulin responses as well as carbohydrate and fat oxidation did not differ between the GI diets at any point of the trial. However, fat oxidation and glycerol level were higher, and carbohydrates oxidation was lower, in the control trial at some points of exercise compared with the GI trials. These results suggest that the GI of a high-carbohydrates diet consumed for 5 days prior to an intensive exercise may not influence running exercise capacity and substrate oxidation during exercise [89].

GLYCEMIC LOAD, METABOLIC RESPONSES AND EXERCISE PERFORMANCE

Glycemic Load has recently been integrated into sports nutrition because it may play an important role in the overall glycemic effect of a diet. It has been suggested that GL may have an effect on metabolic responses with regard to fat and carbohydrates oxidation or performance during exercise [22]. For example, a pre-exercise low-GL meal can reduce fluctuations in glycemic responses during the postprandial period as well as during subsequent exercise when compared with a high-GL meal [22].

Very few research attempts have been made to clarify the metabolic responses of different GL in postprandial period or during exercise. In fact, only three studies so far have examined whether metabolic responses during exercise, exercise performance [22,90], or the immune response [91] are affected by different GL of pre-exercise meals (Table 2.4).

As mentioned already, the GL value can be reduced either by decreasing the amount of carbohydrates consumed or by lowering the dietary GI. In the study of Chen and colleagues [22], three isoenergetic dietary approaches consumed 2 h before a preloaded 1 h run and 10 km time trial, were used to investigate the effect of different GL on metabolic responses and endurance running performance. The first approach consisted of high carbohydrates content, high GI and high GL (H-H trial); the second of high carbohydrates content, low GI and low GL (L-L trial); the third of reduced carbohydrates content replaced by the corresponding amount in fat, high GI and low GL (H-L trial). The major finding of this study was that pre-exercise low-GL (H-L and L-L) meals induced smaller metabolic changes during the postprandial period and during exercise than did the high-GL meals (H-H). Moreover, higher total carbohydrates oxidation during postprandial, exercise, and recovery period was observed in the high-GL trial compared with the other

TABLE 2.4 Glycemic Load and Metabolic Responses During Exercise

Study	Study Protocol	Subjects	Estimated Indices	Results
Chen et al. (2008)	Three different GI and GL isocaloric meals 2 h before exercise testing. H-H: high CHO, high GI & GL; L-L: low CHO, low GI & GL; H-L: low CHO, high GI, low GL. Each trial separated by 7 days.	8 male healthy runners (24.3 ± 2.2 y)	Preloaded 10 km TT, substrate oxidation rates, HR, RER, RPE, LA, GLU, INS, total CHO, FFA, thirst	10 km TT: No difference between trials. H-H trial vs L-L & H-L trials: ↑ Average RER, ↑ LA during exercise; ↑ total CHO oxidation during exercise; ↑ 2 h AUC, ↑ INS postprandial.
Chen et al. (2008)	Three different GI and GL isocaloric meals 2 h before exercise testing. H-H: high CHO, high GI & GL; L-L: low CHO, low GI & GL; H-L: low CHO, high GI, low GL. Each trial separated by 7 days.	8 male healthy runners (24.3 ± 2.2 y)	Preloaded 10 km TT, Hct, Hb, leucocytes, cytokines, cortisol	H-L trial vs L-L and H-H trials: ↑ neutrophils, lymphocytes and monocytes at 60 min of exercise and post TT; ↑ total leucocytes throughout the 2 h recovery; ↑ IL-6 and IL-10 throughout exercise, post-TT and after 2 h recovery; Cortisol: negative correlation with neutrocytes immediately after exercise and after 2 h recovery
Chen et al. (2008)	3-day CHO loading with different GL meals before exercise testing. H-H: high CHO, high GI & GL; L-L: low CHO, low GI & GL; H-L: low CHO, high GI, low GL. Each trial separated by 7 days.	9 male healthy runners (20.1 ± 0.8 y)	Preloaded 10 km TT, substrate oxidation rates, HR, RPE, RER, LA, GLU, INS, total CHO, FFA, thirst	10 km TT: ↑ in L-L than in the H-L trial. High CHO trials vs H-L trial: ↑ average RER; ↑ CHO oxidation postprandial and during exercise; ↑ GLU during exercise and recovery; ↓ NEFA & glycerol during exercise and recovery; ↑ LA during exercise. H-H vs L-L and H-L trials: ↑ 2 h AUC before exercise; ↑ INS before exercise than baseline

GI, Glycemic Index; GL, Glycemic Load; TT, Time-trial; RER, Respiratory Exchange Rate; RPE, Ratio of Perceived Exertion; LA, Lactic Acid; GLU, Glucose; INS, Insulin; CHO, Carbohydrates; FFA, Free Fatty Acids; NEFA, Non Esterified Fatty Acids; AUC, Area under the Blood Glucose Curve; Hct, Hematocrit; Hb, Hemoglobin; IL-6/10, Interleukin-6/10.

two low-GL trials. Regarding glucose and insulin responses, the postprandial 2 h incremental area under the blood glucose curve (AUC) was larger and serum insulin concentration at 60 min and 120 min in the postprandial period was greater in the high-GL trial compared with the other two. Serum insulin concentration, though, was not different between the three trials throughout the exercise or during the 2 h recovery period. No differences in time to complete the preloaded 10 km performance run, or heart rate, or rating of perceived exertion and perceived thirst were observed among the three dietary approaches. However, a higher average respiratory exchange rate and blood lactate throughout the exercise period was observed in the H-H trial. Another interesting finding in this study was that, despite the big difference in micronutrients in the two equally low-GL meals (L-L and H-L) achieved by changing either the GI or the amount of the carbohydrates by replacing carbohydrates with fat, they both produced similar fat oxidation during exercise. Galgani and colleagues (2006) also demonstrated that different carbohydrates amount and type, but similar GL meals, can result in similar postprandial glucose and insulin responses [92] with those reported during exercise in Chen's study [22], although the subjects and the protocol in the two studies were very different.

In a subsequent study [90] using the same data from the project cited above, the authors also investigated the influence of a GI and/or GL meal on immune response to prolonged exercise. The major finding was that the consumption of a high-carbohydrates meal (H-H and L-L), resulted in less perturbation of the circulating numbers of leucocytes, neutrophils and T lymphocyte subsets, and decreased the elevation of plasma IL-6 concentrations immediately after exercise and during the 2 h recovery period, compared with the low-carbohydrates meals (H-L). These responses were accompanied by an attenuated increase in plasma IL-10 concentrations at the end of the 2 hrecovery period. It seems that the amount of carbohydrates consumed in a pre-exercise meal plays a significant role in modifying the immunoendocrine response to prolonged exercise irrespective of its GI and GL values.

In a further study of the same laboratory [91], the researchers expanded their previous investigation by incorporating pre-exercise carbohydrates loading. It was shown that loading with an equally high carbohydrates amount may produce similar muscle glycogen supercompensation status and exercise metabolic responses during subsequent endurance running, regardless of their difference in GI and GL. It appears that in contrast with simple pre-exercise meals [22], it is the amount rather than the nature of the carbohydrates consumed during the 3 days regimen that may be the most dominant factor in metabolism and endurance running performance.

FOOD EXCHANGE VALUES IN HEALTH AND EXERCISE

Food exchanges are a method of meal planning for healthy eating designed by the American Dietetic Association and the American Diabetic Association, based on the National Nutrient Database for Standard Reference [93]. This system uses food exchange lists that each contains foods of approximately the same amount of carbohydrates, calories and fat per serving size. One serving in a group is called an "exchange". An exchange has about the same amount of carbohydrate, protein, fat and calories and the same effect on blood glucose as a serving of every other food in that same group.

Food exchange lists can be especially useful for health professionals, individuals affected with diabetes and those who are monitoring their blood glucose homeostasis for health reasons. The exchange system can simplify meal planning and ensure a consistent, nutritionally balanced diet, while it adds variety in one's diet because exchanges in the same group can be eaten interchangeably. Moreover, insulin-dependent individuals can easily find the ratio of carbohydrate to insulin doses.

According to the American Dietetic Association, the latest edition of food exchange lists [94] continues the principles of the previous lists [95], arranging the foods into eleven broad categories and more subcategories or listing them based on their nutrient content. All foods within a list have approximately the same carbohydrate, protein, fat and caloric value per specified serving. The 11 lists are:

1. **Starch list** (Each item on this list contains approximately 15 g of carbohydrate, 3 g of protein, a trace of fat, and 80 calories)
2. Sweets, desserts and other carbohydrates list
3. Fruit list (Each item on this list contains about 15 g of carbohydrate and 60 calories)
4. Non-starchy vegetables list (Each vegetable on this list contains about 5 g of carbohydrate, 2 g of protein, 0 g of fat, and 25 calories)
5. Meat and meat substitutes list (Each serving of meat and substitute on this list contains about 7 g of protein)
6. Milk list (Each serving of milk or milk product on this list contains about 12 g of carbohydrate and 8 g of protein)
7. Fat list (Each serving on the fat list contains 5 g of fat and 45 calories)
8. Fast foods list
9. Combination foods list (Many foods we eat are combinations of foods that do not fit into only one exchange list)
10. Free foods list (A free food is any food or drink that contains less than 20 calories or less than 5 g of carbohydrate per serving)
11. Alcohol list.

Unfortunately, currently there are not any existing food exchange lists that can be used in an international scale; though some scientific groups are making an effort to create national food lists that are more appropriate for specific populations [96,97]. Food exchange lists have not been used in the sports science field either. It would be interesting to investigate how useful it is for athletes to use these tables for menu planning and by extension to regulate properly their carbohydrate intake. Moreover, it is currently unknown whether food exchange lists could be part of dietetic manipulations in order to augment athletes' exercise capacity and performance.

Calculation of Food Exchanges

- *Number of starch exchanges:* Divide the number of carbohydrates per serving by 15 in order to get the number of starch exchanges the food has.
- *Number of meat exchanges:* Multiply the starch number by three and subtract it from the number of grams of protein in a serving. This total will then be divided by seven.
- *Number of fat exchanges:* Multiply the number of starches by 80 and the number of meats by 45. Subtract both of these numbers from the total number of calories then divide the answer by 45 to get the number of fat exchanges.

CONCLUSIONS

The concepts of GI and GL are relatively new, but there is a growing interest in their use in scientific fields like nutrition or sport science. It is not clear whether different GI intake is associated with higher prevalence of chronic disease risk. More research is needed to elucidate the long-term metabolic effects of different GI food intake and its association with carcinogenesis in men. Research has been shown that a low-GI/rich-carbohydrates meal prior to prolonged exercise positively affects metabolic responses in favor of exercise performance. However, there are a noticeable number of studies that have found no relation between GI manipulation and exercise performance, even if some of those have shown differences in metabolic responses during prolonged exercise. Glycemic load is a relatively new concept introduced in sports nutrition. A limited number of studies that examined the effects of different pre-exercise glycemic loads on metabolic effects and exercise performance have produced inconsistent results. Additional studies elucidating the time of ingestion of different-GI food, and the amount of carbohydrate intake, in order to enhance performance are warranted.

References

[1] Díaz EO, Galgani JE, Aguirre CA. Glycaemic index effects on fuel partitioning in humans. Obes Rev 2006;7(2):219–26.

[2] O'Reilly J, Wong SHS, Chen Y. Glycaemic index, glycaemic load and exercise performance. Sports Med 2010;40(1):27–39.

[3] daSilva LMV, deCàssia GAR. Effect of the glycemic index on lipid oxidation and body composition. Nutr Hosp 2011;26 (1):48–55.

[4] Jenkins DJ, Wolever TM, Taylor RH, Barker H, Fielden H, Baldwin JM, et al. Glycemic index of foods: a physiological basis for carbohydrate exchange. Am J Clin Nutr 1981;34(3):362–6.

[5] FAO/WHO. Carbohydrates in human nutrition. Report of a joint FAO/WHO expert consultation. FAO 1998;66:1–140.

[6] Aston LM. Glycaemic index and metabolic disease risk. Proc Nutr Soc 2006;65:125–34.

[7] Brand-Miller JC, Holt SHA, Petocz P. Reply to R Mendosa. Am J Clin Nutr 2003;77:994–5.

[8] McKeown NM, Meigs JB, Liu S, Rogers G, Yoshida M, Saltzman E, et al. Dietary carbohydrates and cardiovascular disease risk factors in the Framingham offspring cohort. J Am Coll Nutr 2009;28(2):150–8.

[9] Frost G, Leeds AA, Doré CJ, Madeiros S, Brading S, Dornhorst A. Glycaemic index as a determinant of serum HDL-cholesterol concentration. Lancet 1999;353(9158):1045–8.

[10] Lok KY, Chan R, Chan D, Li L, Leung G, Woo J, et al. Glycaemic index and glycaemic load values of a selection of popular foods consumed in Hong Kong. Br J Nutr 2010;103(4):556–60.

[11] Wolever TMS, Katzman-Rellea L, Jenkinsa AL, Vuksana V, Jossea RG, Jenkins DJA. Glycaemic index of 102 complex carbohydrate foods in patients with diabetes. Nutr Res 1994;14(5):651–69.

[12] Gannon MC, Nuttall FQ, Krezowski RA, Billington CJ, Parker S. The serum insulin and plasma glucose responses to milk and fruit products in type 2 (non-insulin-dependent) diabetic patients. Diabetologia 1986;29:784–91.

[13] Bornet FRJ, Costagliola D, Rizkalla SW, Blayo A, Fontvieille A-M, Haardt M-J, et al. Insulinemic and glycemic indexes of six starch-rich foods taken alone and in a mixed meal by type 2 diabetics. Am J Clin Nutr 1987;45:588–95.

[14] Jenkins DJA, Wolever TMS, Jenkins AL, Thorne MJ, Lee R, Kalmusky J, et al. The glycaemic index of foods tested in diabetic patients: a new basis for carbohydrate exchange favouring the use of legumes. Diabetologia 1983;24:257–64.

[15] Foster-Powell K, Holt SHA, Brand-Miller JC. International table of glycemic index and glycemic load values: 2002. Am J Clin Nutr 2002;76:5–56.

[16] Atkinson FS, Foster-Powell K, Brand-Miller JC. International tables of glycemic index and glycemic load values: 2008. Diabetes Care 2008;31:2281–3.

[17] Brand-Miller J, McMillan-Price J, Steinbeck K, Caterson I. Dietary glycemic index: health implications. J Am Coll Nutr 2009;28(4):446–9.

[18] Simpson RW, McDonald J, Wahlqvist ML, Atley L, Outch K. Food physical factors have different metabolic effects in non-diabetics and diabetics. Am J Clin Nutr 1985;42:462–9.

[19] Simpson RW, McDonald J, Wahlqvist ML, Atley L, Outch K. Macronutrients have different metabolic effects in nondiabetics and diabetics. Am J Clin Nutr 1985;42:449–53.

[20] Kolata G. Diabetics should lose weight, avoid diet fads. [News]. Science 1987;235(4785):163–4.

[21] Salmeron J, Ascherio A, Rimm EB, Colditz GA, Spiegelman D, Jenkins DJ, et al. Dietary fiber, glycemic load, and risk of NIDDM in men. Diabetes Care 1997;20(4):545–50.

[22] Chen YJ, Wong SH, Wong CK, Lam CW, Huang YJ, Siu PM. Effect of preexercise meals with different glycemic indices and loads on metabolic responses and endurance running. Int J Sport Nutr Exerc Metab 2008;18(3):281–300.

[23] Colombani PC. Glycemic index and load – dynamic dietary guidelines in the context of diseases. Physiol Behav 2004;83:603–10.

[24] Wong SH, O'Reilly J. Glycemic index and glycemic load. Their application in health and fitness. ACSM's Health Fit J 2009;14 (6):18–23.

[25] Barclay AW, Brand-Miller JC, Wolever TM. Glycemic index, glycemic load, and glycemic response are not the same. Diabetes Care 2005;28(7):1839–40.

[26] Wolever TMS, Mehlin C. High-carbohydrate–low-glycaemic index dietary advice improves glucose disposition index in subjects with impaired glucose tolerance. Br J Nutr 2002;87:477–87.

[27] Ball SD, Keller KR, Moyer-Mileur LJ, Ding YW, Donaldson D, Jackson WD. Prolongation of satiety after low versus moderately high glycemic index meals in obese adolescents. Pediatrics 2003;111:488–94.

[28] Wolever TMS, Bolognesi C. Source and amount of carbohydrate affect postprandial glucose and insulin in normal subjects. J Nutr 1996;126:2798–806.

[29] Abdul-Ghani MA, DeFronzo RA. Pathogenesis of insulin resistance in skeletal muscle. J Biomed Biotechnol 2010; doi: 10.1155/2010/476279.

[30] Ferrannini E, Galvan AQ, Gastaldelli A, Camastra S, Sironi AM, Toschi E, et al. Insulin: new roles for an ancient hormone. Eur J Clin Invest 1999;29:842–52.

[31] Marsh K, Barclay A, Colagiuri S, Brand-Miller J. Glycemic index and glycemic load of carbohydrates in the diabetes diet. Curr Diab Rep 2011;11:120–7.

[32] Ma Y, Olendzki B, Chiriboga D, Hebert JR, Li Y, Li W, et al. Association between dietary carbohydrates and body weight. Am J Epidemiol 2005;161:359–67.

[33] Hare-Bruun H, Flint A, Heitmann BL. Glycemic index and glycemic load in relation to changes in body weight, body fat distribution, and body composition in adult Danes. Am J Clin Nutr 2006;84:871–9.

[34] Thomas DE, Elliot EJ, Baur L. Low glycaemic index or low glycaemic diets for overweight and obesity. Cochrane Database Syst Rev 2007;18(3):CD005105. doi:10.1002/14651858.CD005105.

[35] Sloth B, Krog-Mikkelsen I, Flint A, Tetens I, Bjrck I, Vinoy S, et al. No difference in body weight decrease between a low-glycemic-index and a high-glycemic-index diet but reduced

LDL cholesterol after 10-wk ad libitum intake of the low-glycemic-index diet. Am J Clin Nutr 2004;80:337–47.

[36] Denova-Gutiérrez E, Huitrón-Bravo G, Talavera JO, Castañón S, Gallegos-Carrillo K, Flores Y, et al. Dietary glycemic index, dietary glycemic load, blood lipids, and coronary heart disease. [Article]. J Nutr Metab 2010;1–8. doi:10.1155/2010/170680.

[37] Barclay AW, Petocz P, McMillan-Price J, Flood VM, Prvan T, Mitchell P, et al. Glycemic index, glycemic load, and chronic disease risk – a meta-analysis of observational studies. Am J Clin Nutr 2008;87(3):627–37.

[38] Radulian G, Rusu E, Dragomir A, Posea M. Metabolic effects of low glycaemic index diets. [Research Support, Non-U.S. Gov't Review]. Nutr J 2009;8:5. doi:10.1186/1475-2891-8-5.

[39] Tsai C-J, Leitzmann MF, Willett WC, Giovannucci EL. Dietary carbohydrates and glycaemic load and the incidence of symptomatic gall stone disease in men. Gut 2005;54:823–8.

[40] Sieri S, Pala V, Brighenti F, Pellegrini N, Muti P, Micheli A, et al. Dietary glycemic index, glycemic load, and the risk of breast cancer in an Italian prospective cohort study. Am J Clin Nutr 2007;86:1160–6.

[41] Higginbotham S, Zhang Z-F, Lee I-M, Cook NR, Giovannucci E, Buring JE, et al. Dietary glycemic load and risk of colorectal cancer in the womens health study. J Natl Cancer Inst 2004;96 (3):229–33.

[42] Silvera NSA, Jain M, Howe GR, Miller AB, Rohan TE. Dietary carbohydrates and breast cancer risk: a prospective study of the roles of overall glycemic index and glycemic load. Int J Cancer 2005;114:653–8.

[43] Schulze MB, Liu S, Rimm EB, Manson JE, Willett WC, Hu FB. Glycemic index, glycemic load, and dietary fiber intake and incidence of type 2 diabetes in younger and middle-aged women. Am J Clin Nutr 2004;80:348–56.

[44] Larsson SC, Bergkvist L, Wolk A. Glycemic load, glycemic index and breast cancer risk in a prospective cohort of Swedish women. [Research Support, Non-U.S. Gov't]. Int J Cancer 2009;125(1):153–7. doi:10.1002/ijc.24310.

[45] Wen W, Shu XO, Li H, Yang G, Ji BT, Cai H, et al. Dietary carbohydrates, fiber, and breast cancer risk in Chinese women. [Research Support, N.I.H., Extramural]. Am J Clin Nutr 2009;89 (1):283–9. doi:10.3945/ajcn.2008.26356.

[46] Sieri S, Pala V, Brighenti F, Agnoli C, Grioni S, Berrino F, et al. High glycemic diet and breast cancer occurrence in the Italian EPIC cohort. Nutr Metab Cardiovasc Dis 2012;doi:10.1016/j.numecd.2012.01.001.

[47] Liu S, Willett WC, Stampfer MJ, Hu FB, Franz M, Sampson L, et al. A prospective study of dietary glycemic load, carbohydrate intake, and risk of coronary heart disease in US women. Am J Clin Nutr 2000;71:1455–61.

[48] Oh K, Hu FB, Cho E, Rexrode KM, Stampfer MJ, Manson JE, et al. Carbohydrate intake, glycemic index, glycemic load, and dietary fiber in relation to risk of stroke in women. Am J Epidemiol 2005;161:161–9.

[49] Larsson SC, Giovannucci E, Wolk A. Dietary carbohydrate, glycemic index, and glycemic load in relation to risk of colorectal cancer in women. [Research Support, Non-U.S. Gov't]. Am J Epidemiol 2007;165(3):256–61. doi:10.1093/aje/kwk012.

[50] Schaumberg DA, Liu S, Seddon JM, Willett WC, Hankinson SE. Dietary glycemic load and risk of age-related cataract. Am J Clin Nutr 2004;80:489–95.

[51] Shikany JM, Redden DT, Neuhouser ML, Chlebowski RT, Rohan TE, Simon MS, et al. Dietary glycemic load, glycemic index, and carbohydrate and risk of breast cancer in the women's health initiative. Nutr Cancer 2011;63(6):899–907.

[52] Jonas CR, McCullough ML, Teras LR, Walker-Thurmond KA, Thun MJ, Calle EE. Dietary glycemic index, glycemic load, and risk of incident breast cancer in postmenopausal women. Cancer Epidemiol Biomarkers Prev 2003;12:573–7.

[53] Larsson SC, Friberg E, Wolk A. Carbohydrate intake, glycemic index and glycemic load in relation to risk of endometrial cancer: a prospective study of Swedish women. Int J Cancer 2006;120:1103–7.

[54] Patel AV, McCullough ML, Pavluck AL, Jacobs EJ, Thun MJ, Calle EE. Glycemic load, glycemic index, and carbohydrate intake in relation to pancreatic cancer risk in a large US cohort. Cancer Causes Control 2007;18:287–94.

[55] Jiao L, Flood A, Subar AF, Hollenbeck AR, Schatzkin A, Stolzenberg-Solomon R. Glycemic index, carbohydrates, glycemic load, and the risk of pancreatic cancer in a prospective cohort study. [Research Support, N.I.H., Intramural]. Cancer Epidemiol Biomarkers Prev 2009;18(4):1144–51. doi:10.1158/1055-9965.EPI-08-1135.

[56] Levitan EB, Mittleman MA, Wolk A. Dietary glycemic index, dietary glycemic load, and incidence of heart failure events: a prospective study of middle-aged and elderly women. [Research Support, N.I.H., Extramural Research Support, Non-U.S. Gov't]. J Am Coll Nutr 2010;29(1):65–71.

[57] Levitan EB, Mittleman MA, Wolk A. Dietary glycaemic index, dietary glycaemic load and incidence of myocardial infarction in women. [Research Support, N.I.H., Extramural Research Support, Non-U.S. Gov't]. Br J Nutr 2010;103(7):1049–55. doi:10.1017/S0007114509992674.

[58] Thomas D, Elliott EJ. Low glycaemic index, or low glycaemic load, diets for diabetes mellitus. [Meta-Analysis Review]. Cochrane Database Syst Rev 2009;1:CD006296. doi:10.1002/14651858.CD006296.pub2.

[59] Psaltopoulou T, Ilias I, Alevizaki M. The role of diet and lifestyle in primary, secondary, and tertiary diabetes prevention: a review of meta-analyses. [Review]. Rev Diabet Stud 2010;7 (1):26–35. doi:10.1900/RDS.2010.7.26.

[60] Vega-López S, Mayol-Kreiser SN. Use of the glycemic index for weight loss and glycemic control: a review of recent evidence. Curr Diab Rep 2009;9(5):379–88.

[61] Dong J-Y, Zhang L, Zhang Y-H, Qin L-Q. Dietary glycaemic index and glycaemic load in relation to the risk of type 2 diabetes: a meta-analysis of prospective cohort studies. Br J Nutr 2011;106:1649–54.

[62] Hardy DS, Hoelscher DM, Aragaki C, Stevens J, Steffen LM, Pankow JS, et al. Association of glycemic index and glycemic load with risk of incident coronary heart disease among Whites and African Americans with and without type 2 diabetes: the atherosclerosis risk in communities study. [Comparative Study Research Support, N.I.H., Extramural]. Ann Epidemiol 2010;20 (8):610–6. doi:10.1016/j.annepidem.2010.05.008.

[63] Burger KN, Beulens JW, Boer JM, Spijkerman AM, van der AD. Dietary glycemic load and glycemic index and risk of coronary heart disease and stroke in Dutch men and women: the EPIC-MORGEN study. [Clinical Trial Research Support, Non-U.S. Gov't]. PLoS One 2011;6(10):e25955. doi:10.1371/journal.pone.0025955.

[64] Levitan EB, Mittleman MA, Hakansson N, Wolk A. Dietary glycemic index, dietary glycemic load, and cardiovascular disease in middle-aged and older Swedish men. [Research Support, N.I.H., Extramural Research Support, Non-U.S. Gov't]. Am J Clin Nutr 2007;85(6):1521–6.

[65] Augustin LS, Franceschi S, Jenkins DJ, Kendall CW, La Vecchia C. Glycemic index in chronic disease: a review. [Review]. Eur J Clin Nutr 2002;56(11):1049–71. doi:10.1038/sj.ejcn.1601454.

[66] Li HL, Yang G, Shu XO, Xiang YB, Chow WH, Ji BT, et al. Dietary glycemic load and risk of colorectal cancer in Chinese women. [Research Support, U.S. Gov't, P.H.S.]. Am J Clin Nutr 2011;93(1):101–7. doi:10.3945/ajcn.110.003053.

[67] Michaud DS, Liu S, Giovannucci E, Willett WC, Colditz GA, Fuchs CS. Dietary sugar, glycemic load, and pancreatic cancer risk in a prospective study. [Research Support, U.S. Gov't, P.H.S.]. J Natl Cancer Inst 2002;94(17):1293–300.

[68] Branca F, Nikogosian H, Lobstein T. The challenge of obesity in the WHO European Region and the strategies for response. Denmark: WHO Europe; 2007.

[69] Preuss HG. Bean amylase inhibitor and other carbohydrate absorption blockers: effects on diabesity and general health. J Am Coll Nutr 2009;28(3):266–76.

[70] Brand-Miller JC, Holt SHA, Pawlak DB, McMillan J. Glycemic index and obesity. Am J Clin Nutr 2002;76:281–5.

[71] Stevenson EJ, Williams C, Mash LE, Phillips B, Nute ML. Influence of high-carbohydrate mixed meals with different glycemic indexes on substrate utilization during subsequent exercise in women. Am J Clin Nutr 2006;84:354–60.

[72] Ludwig DS. Dietary glycemic index and obesity. J Nutr 2000;130:280–3.

[73] DECODE study group. Glucose tolerance and mortality: comparison of WHO and American diabetes association diagnostic criteria. The DECODE study group. European diabetes epidemiology group. diabetes epidemiology: collaborative analysis of diagnostic criteria in Europe. Lancet 1999;354(9179):617–21.

[74] Wee S, Williams C, Tsintzas K, Boobis L. Ingestion of a high-glycemic index meal increases muscle glycogen storage at rest but augments its utilization during subsequent exercise. J Appl Physiol 2005;99:707–14.

[75] Febbraio MA, Campbell SE, Keenan J, Angus DJ, Garnham AP. Preexercise carbohydrate ingestion, glucose kinetics, and muscle glycogen use: effect of the glycemic index. J Appl Physiol 2000;89:1845–51.

[76] Karamanolis IA, Laparidis KS, Volaklis KA, Douda HT, Tokmakidis SP. The effects of pre-exercise glycemic index food on running capacity. Int J Sports Med 2011;32:666–71.

[77] Moore LJS, Midgley AW, Thurlow S, Thomas G, McNaughton LR. Effect of the glycaemic index of a pre-exercise meal on metabolism and cycling time trial performance. J Sci Med Sport 2010;13(1):182–8. doi:10.1016/j.jsams.2008.11.006.

[78] Wong SHS, Siu PM, Lok A, Chen YJ, Morris J, Lam CW. Effect of the glycaemic index of pre-exercise carbohydrate meals on running performance. Eur J Sport Sci 2008;8(1):23–33.

[79] Wu CL, Williams C. A low glycemic index meal before exercise improves endurance running capacity in men. Int J Sport Nutr Exerc Metab 2006;16:510–27.

[80] Kirwan JP, O'Gorman DJ, Cyr-Campbell D, Campbell WW, Yarasheski KE, Evans WJ. Effects of a moderate glycemic meal on exercise duration and substrate utilization. [Research Support, Non-U.S. Gov't Research Support, U.S. Gov't, P.H.S.]. Med Sci Sports Exerc 2001;33(9):1517–23.

[81] Jamurtas AZ, Tofas T, Fatouros I, Nikolaidis MG, Paschalis V, Yfanti C, et al. The effects of low and high glycemic index foods on exercise performance and beta-endorphin responses. J Int Soc Sports Nutr 2011;8:15.

[82] Kern M, Heslin CJ, Rezende RS. Metabolic and performance effects of raisins versus sports gel as pre-exercise feedings in cyclists. [Comparative Study Randomized Controlled Trial Research Support, Non-U.S. Gov't]. J Strength Cond Res 2007;21(4):1204–7. doi:10.1519/R-21226.1a.

[83] Horowitz JF, Mora-Rodriguez R, Byerley LO, Coyle EF. Lipolytic suppression following carbohydrate ingestion limits fat oxidation during exercise. Am J Physiol 1997; 273(36):768–75.

[84] Sidossis LS, Horowitz JF, Coyle EF. Load and velocity of contraction influence gross and delta mechanical efficiency. [Research Support, Non-U.S. Gov't]. Int J Sports Med 1992;13 (5):407–11. doi:10.1055/s-2007-1021289.

[85] Thomas DE, Brotherhood JR, Brand JC. Carbohydrate feeding before exercise: effect of glycemic index. [Comparative Study Research Support, Non-U.S. Gov't]. Int J Sports Med 1991;12 (2):180–6. doi:10.1055/s-2007-1024664.

[86] Hargreaves M. Pre-exercise nutritional strategies: effects on metabolism and performance. [Review]. Can J Appl Physiol 2001;26(Suppl):S64–70.

[87] Little JP, Chilibeck PD, Ciona D, Forbes S, Rees H, Vandenberg A, et al. Effect of low- and high-glycemic-index meals on metabolism and performance during high-intensity, intermittent exercise. [Research Support, Non-U.S. Gov't]. Int J Sport Nutr Exerc Metab 2010;20(6):447–56.

[88] Little JP, Chilibeck PD, Ciona D, Vandenberg A, Zello GA. The effects of low- and high-glycemic index foods on high-intensity intermittent exercise. [Comparative Study Research Support, Non-U.S. Gov't]. Int J Sports Physiol Perform 2009; 4(3):367–80.

[89] Hamzah S, Higgins S, Abraham T, Taylor P, Vizbaraite D, Malkova D. The effect of glycaemic index of high carbohydrate diets consumed over 5 days on exercise energy metabolism and running capacity in males. [Clinical Trial]. J Sports Sci 2009;27 (14):1545–54. doi:10.1080/02640410903134115.

[90] Chen YJ, Wong SHS, Wong CK, Lam CW, Huang YJ, Siu PMF. The effect of a pre-exercise carbohydrate meal on immune responses to an endurance performance run. Br J Nutr 2008;100 (6):1260–8. doi:10.1017/S0007114508975619.

[91] Chen Y, Wong SHS, XU X, Hao X, Wong CK, Lam CW. Effect of CHO Loading Patterns on Running Performance. Int J Sports Med 2008;29:598–606.

[92] Galgani J, Aguirre C, Díaz E. Acute effect of meal glycemic index and glycemic load on blood glucose and insulin responses in humans. Nutr J 2006;5:22.

[93] US Department of Agriculture. (2006). Agricultural Research Service USDA National Nutrient Database for Standard Reference, Release 19.

[94] American Diabetes Association. Nutrition recommendations and interventions for diabetes: a position statement of the American Diabetes Association. Diabetes Care 2007;30:S48–65. doi:10.2337/dc07-S048.

[95] Daly A, Franz M, Holzmeister LA, Kulkari K, O'Connell B, Wheeler M, et al. Exchange Lists for Meal Planning. VA: American Diabetes Association and American Dietetic Association; 2003.

[96] Coulibaly A, O'Brien H, Galibois I. Development of a Malian food exchange system based on local foods and dishes for the assessment of nutrient and food intake in type 2 diabetic subjects. S Afr J Clin Nutr 2009;22(1):31–5.

[97] Bawadi HA, Al-Shwaiyat NM, Tayyem RF, Mekary R, Tuuri G. Developing a food exchange list for Middle Eastern appetizers and desserts commonly consumed in Jordan. Nutr Diet 2009;66 (1):20–6.

CHAPTER

3

Performance Enhancement Drugs and Sports Supplements for Resistance Training

Lucas Guimarães-Ferreira[1], *Wagner Silva Dantas*[2], *Igor Murai*[2], *Michael J. Duncan*[3] *and Nelo Eidy Zanchi*[4]

[1]Exercise Metabolism Research Group, Center of Physical Education and Sports, Federal University of Espirito Santo, Vitória, Brazil [2]Laboratory of Applied Nutrition and Metabolism, Physical Education and Sports School, University of São Paulo, São Paulo, Brazil [3]Human Performance Laboratory, Department of Biomolecular and Sports Sciences, Coventry University, Coventry, United Kingdom [4]Department of Physiology and Biophysics, Institute of Biomedical Sciences, University of São Paulo, São Paulo, Brazil

INTRODUCTION

The term ergogenic is derived from the Greek words *ergon* (work) and *gennan* (to produce). Hence ergogenic refers to any strategy that enhances work capacity. Individuals engaged in physical training have been using sports ergogenics in order to improve athletic performance or potentiate the physiological adaptations to training. Williams [1] listed five categories of sports ergogenics: (a) nutritional aids; (b) pharmacological aids; (c) physiological aids; (d) psychological aids; and (e) mechanical or biomechanical aids.

In general, nutritional sports ergogenics are designed to enhance energy production and/or improve body composition (promoting muscle growth and decreasing body fat). Nutritional supplements have been highly commercialized, and their utilization is widespread among athletes and non-athletes. It is estimated that 40–88% of athletes consume sports supplements [2], and in the USA more than 3 million use, or have used, these ergogenic aids [3]. Pharmacological sports ergogenics are a slightly different type of product and comprise drugs that enhance athletic performance, usually as hormones found naturally in the human body, but administered in higher doses. Although these drugs may be an effective sports ergogenic, their use also can present health risks. Doping is the use of drugs or even some natural substances in an attempt to improve sports performance during competitions. Many athletic governing bodies have developed anti-doping programs and policies. The World Anti-Doping Agency (WADA) was established in 1999 as an international independent agency that works on scientific research, education, development of anti-doping capacities, and monitoring of anti-doping programs around the world. The list of prohibited substances with some examples is presented on Table 3.1.

Despite a great number of scientific publications investigating the physiological effects of nutritional supplements and pharmacological substances, many of these products have been used without knowledge of the effects on human metabolism caused by their chronic administration. Before the utilization or prescription of any ergogenic aid, it is important to consider some questions about that substance: Is it effective? Is it safe? Is it legal and ethical? In this chapter we will discuss the most widely utilized drugs and supplements among individuals engaged in resistance training, focusing on their effects on strength and body composition, the safety of their utilization, and their mechanisms of action.

Nutrition and Enhanced Sports Performance.
DOI: http://dx.doi.org/10.1016/B978-0-12-396454-0.00003-5

TABLE 3.1 List of Prohibited Substances and Methods

SUBSTANCES AND METHODS PROHIBITED AT ALL TIMES (IN AND OUT OF COMPETITION)	
Prohibited substances	Examples
Anabolics agents	Anabolic androgenic steroids (AAS); Clenbuterol
Peptide hormones, growth factors, and related substances	Growth hormone (GH); insulin-like growth factor I (IGF-I); chorionic gonadotrophin
Beta-2 agonists	Salbutamol; formoterol
Hormone and metabolic modulators	Aromatase inhibitors; selective estrogen receptor modulators
Diuretics and other marking agents	Acetazolamide; bumetanide
Prohibited methods	Examples
Enhancement of oxygen transfer	Blood doping
Chemical and physical manipulation	Tampering of samples collected during doping control
Gene doping	Transfer of nucleic acids or nucleic acid sequences; use of normal or genetically modified cells
SUBSTANCES PROHIBITED IN COMPETITION	
Prohibited substances	Examples
Stimulants	Adrenaline; sibutramine
Narcotics	Diamorphine (heroin)
Cannabinoids	Marijuana
Glucocorticosteroids	Dexamethasone

The complete list can be found on the WADA website [http://www.wada-ama.org/en/World-Anti-Doping-Program/Sports-and-Anti-Doping-Organizations/International-Standards/Prohibited-List/].

TESTOSTERONE AND ANABOLIC STEROIDS

The use of testosterone and related anabolic steroids (AS) is a widespread phenomenon among top athletes, amateurs, and for a large part of the population who desire to improve their appearance. The popularity of AS is also related to their potency in increasing both strength and skeletal muscle mass. Although not cited in this chapter, AS are widely used in recovery from injury and catabolic process. Altogether, this text will focus on the effects and mechanisms of action of AS on strength and mass of untrained and trained subjects.

Physiological Production of Testosterone and Related Side Effects of Abuse

Testosterone is mainly produced in Leydig cells in the testes, with a small portion coming from the adrenal cortex and the peripheral conversion of androstenedione. Testosterone increases skeletal muscle mass, strength, decreases body fat, and importantly, enables periods of intensive training to be sustained. An important question to be addressed is: Is it necessary to perform resistance training to increase muscle mass? The literature shows that testosterone alone is capable of increasing muscle mass. Moreover, this increase seems to be dose-dependent, thus making its use of extreme importance in the rehabilitation process. However, through case reports, AS have been associated with increased risk factors for cardiovascular disease, alterations in liver function, and changes in behavior mostly related to increased aggressiveness.

Testosterone and AS Combined with Resistance Training

AS seems to be a vital component to increase muscle mass in the presence of resistance training. For example, in a double-blinded intervention, manipulations of testosterone levels through gonadotropin releasing hormone (GnRH) once every 4th week during 12 weeks of resistance training resulted in a decrease in muscle mass in the group receiving GnRH. There was, however, no difference in isometric knee extensor strength between groups. Surprisingly, suppression of testosterone was not correlated with reductions in mRNA expression of genes coding for the regulatory factors MyoD and myogenin (a family of transcription factors related to proliferation and differentiation, respectively), insulin like growth factor (IGF-1), myostatin, and androgen receptors (AR) [4]. Thus, improvements in the muscle mass and strength appear to be dependent on the initial dose of AS and the initial conditioning level of participants. For example, young healthy individuals receiving testosterone enanthate (300 mg/week) for 6 weeks demonstrated consistent strength gains. Similarly, a supraphysiological dose of testosterone (600 mg/week) for 10 weeks in trained men produced a significant increase in muscle strength and in the cross-sectional area of quadriceps [5]. Another important finding is that the use of AS in conjunction with heavy resistance training seems to be associated with changes in muscle pennation angle, which means that more muscle can be put inside a fascicle or a pack of muscles [6]. Tables 3.2 and 3.3 present the main results in the literature regarding these aspects.

Testosterone use in High-Level Powerlifters

High-level powerlifters who reported the use of testosterone for several years (100–500 mg/week for a period of 9 ± 3.3 years) present a higher degree of muscle hypertrophy than do high-level power lifters who have reported the use of anabolic steroids [24,25]. Testosterone induces the hypertrophy of both type I and type II fibers. However, there is evidence suggesting that the largest difference in muscle fiber size between steroid users and non-users is observed in slow type I muscle fibers [24–26]. In the trapezius fibers of steroid users, the area of type I muscle fibers is 58% larger than in non-users, whereas the area of type II muscle fibers is 33% larger than in non-users [24]. The same tendency is observed in the vastus lateralis and, overall, type I muscle fibers are more sensitive to anabolic agents than type II fibers [27]. Importantly, it has been established that 300 mg and 600 mg testosterone increase type I muscle fibers, whereas type II muscle fibers enlarge only after administration of 600 mg testosterone [28].

Protein Synthesis and Myonuclear Content

Increased protein synthesis is a vital mechanism by which testosterone increases skeletal muscle mass. For example, intramuscular injection of 200 mg testosterone enanthate in healthy individuals induced a twofold increase in protein synthesis compared with control subjects, without changes in protein breakdown [29].

Adult muscles contain a specific number of myonuclei, which are able to send their transcripts to a specific location. With this concept in mind, every myonucleus is responsible for the transcription of mRNA that can move throughout the cell to be transcripted in a muscle protein, but instead they can be restricted to move within a specific volume, a concept called "nuclear domain". In this respect, Kadi et al. [30,31] stated that a fiber that enlarges over 36% in its area also increases the number of myonuclei to maintain an adequate proportion between myonucleus and cytoplasm. On the other hand, they stated that skeletal muscle myonuclei do not increase in number unless skeletal muscle fiber exceeds 26%. In this way, one mechanism by which testosterone facilitates the hypertrophy of muscle fibers seen in drug users is to promote myonuclear accretion [6,24]. To illustrate this, in high-level powerlifters, the mean number of nuclei per cross section is significantly higher in steroid users compared with non-users, and myonuclear accretion is greater in type I fibers (+23%) compared with type II muscles (+14%) [24]. This is in accordance with the larger hypertrophy of type I muscle fibers seen in steroid users.

Androgen Receptor

After being produced or injected, testosterone or AS act in the cell through their ligands, the androgenic receptor (AR). AR belongs to a superfamily of transcription factors; when AS binds to AR, the AR is translocated to the nucleus and activates a series of steroid responsive elements inside the nucleus, increasing the rates of transcription. The degree of AR expression among human muscles varies substantially. For example, AR expression in myonuclei in trapezius muscle is about 60% higher than in the vastus lateralis [32]. Of note, steroid-using athletes show a higher percentage of AR in the trapezius muscle than do non-steroid users. Similarly, after a month of AS administration, the number of AR is enhanced, but it returns to the basal levels after 6 months [15].

Conclusions

AS are amongst the most powerful agents capable of increasing both muscle mass and strength. However, their usage can inhibit the natural process of testosterone production, which is a significant problem among heavy AS users. The literature on this topic has demonstrated several potential medical problems related to brain disorders (aggressiveness, bipolarity, etc.) and prostate cancer development. Conversely, elevated testosterone levels, especially in competitive sport, gives to the user a marked advantage. The solution to this problem is to enhance information about both the positive and negative effects and consequences of its use.

TABLE 3.2 Effects of Different Anabolic–Androgenic Regimens, Associated or Not with Strength Training, on Muscle Strength and Hypertophy in Young Men

Authors	Drug	Maximum Dosage	Duration	Route of Administration	Treatment	Lean Mass (Mean Changes)	Muscle Strength (Mean Changes)
Woodhouse et al. [7]	Testosterone enanthate	600 mg/week	20 weeks	IM	25 mg/day	↑ 22 ± 5 cm^3	NM
					50 mg/day	↑15 ± 15 cm^3	
					125 mg/day	↓ 21 ± 7 cm^3	
					300 mg/day	↓ 30 ± 8 cm^3	
					600 mg/day	↓ 34 ± 10 cm^3	
King et al. [8]	Androstenedione + ST	300 mg/day	12 weeks	PO	Androstenedione + ST	↑ 4.7%	↑ 29%
					Placebo + ST	↑ 4.5%	↑ 29%
Bhasin et al. [5]	Testosterone enanthate + GnRH	600 mg/week	20 weeks	IM	25 mg/day	↑ 0.4 ± 0.3 kg	↓1.2 ± 7.4 kg
					50 mg/day	↑ 1.1 ± 0.9 kg	↑ 22.7 ± 7.6 kg
					125 mg/day	↑ 2.9 ± 0.8 kg	↑ 18.4 ± 10 kg
					300 mg/day	↑ 5.5 ± 0.7 kg[a]	↑ 72.2 ± 12.4 kg[a]
					600 mg/day		↑ 76.5 ± 12.2 kg[a]
Rogerson et al. [9]	Testosterone enanthate + ST	3.5 mg/kg	6 weeks	IM	Testosterone (week 0)	~79 ± 9 kg (BMI)	~300 ± 80 kg
					Placebo (week 0)	~77 ± 5 kg	~305 ± 85 kg
					Testosterone (week 3)		~340 ± 70 kg[a]
					Placebo (week 3)		~340 ± 70 kg
					Testosterone (week 6)	~84 ± 9 kg[b]	~360 ± 50 kg[a]
					Placebo (week 6)	~77 ± 7 kg	~340 ± 60 kg
Bhasin et al. [10]	Testosterone enanthate + ST	600 mg/week	10 weeks	IM	Testosterone enanthate + ST	↑ 9.3%	↑ 37.2%
					Placebo + ST	↑ 2.7%	↑ 19.8%
					Testosterone enanthate	↑ 4.5%	↑ 12.6%
					Placebo	↑ 0.4%	↑ 2.9%

[a] $p < 0.001$;
[b] $p < 0.01$.
IM, intramuscular; PO, oral; ST, strength training.

CREATINE MONOHYDRATE

Creatine (Cr) is one of the most widely available nutritional supplements used by athletes and physical activity practitioners over the past two decades. Evidence supporting its ergogenic efficacy and safety profile make this nutritional supplement one of the most researched in sport science worldwide. In 2009, the annual spend on creatine-containing nutritional supplements was estimated to be approximately $2.7 billion in the USA alone [33]. In this respect, the aim of this chapter is to highlight the applications of Cr supplementation as an ergogenic aid, to elucidate briefly its mechanisms of action on skeletal muscle. and to present scientific research findings in order to establish trustworthy information about this ergogenic agent.

Creatine Synthesis and Mechanisms of Action

Cr is a compound synthesized predominately by the liver and to a lesser extent by the kidneys and pancreas in an amount of approximately 1–2 g/day and

TABLE 3.3 Effects of Different Anabolic–Androgenic Regimens, Associated or Not with Strength Training, on Muscle Strength and Hypertophy in Older Men

Authors	Drug	Maximum Dosage	Duration	Route of Administration	Treatment	Lean Mass (Mean Changes)	Muscle Strength (Mean Changes)
Sullivan et al. [11]	Testosterone enanthate + ST	100 mg	5 years	SC	Low intensity ST +P	↑ 3.5 ± 1.9%	↑ 9.7 ± 7.0 kg
					Low intensity ST +T	↑ 7.2 ± 1.8%	↑ 25.4 ± 6.8 kg
					High intensity ST +P	↑ 1.2 ± 1.9%	↑ 35.9 ± 7.0 kg[d]
					High intensity ST +T	↑ 8.6 ± 1.8%	↑ 39.4 ± 6.6 kg[d]
Broeder et al. [12]	Androstenediol + ST	200 mg/day	12 weeks	PO		↑ 3.4%	↑ 23.6%[h]
	Androstenedione + ST					↓ 2.4%	↑ 19.2%[h]
Lambert et al. [13]	Testosterone enanthate + ST	100 mg/week	12 weeks	PO	P + ST	Not significant	↑ 19.3%
					P	Not significant	Not significant
					ST +P	↑ thigh muscle cross-sectional area (1.6 cm^2)	↑ ~21%
					T	Not significant	Not significant
					ST +T	↑ thigh muscle cross-sectional area (~4 cm^2)[g]	↑ ~18%
					T	↑ 4.2 ± 0.6 kg[a]	
					P	↓ 2 ± 1.0 kg	
Ferrando et al. [14]	Testosterone enanthate	300 mg/week	6 months	SC	T enanthate	↑ 4.2 ± 0.4 kg (↑ 6.7%)[a]	1 month: ↑4.9 ± 1.7 kg 6 months: ↑10.4 ± 2.1 kg[a]
					P	↓ 2 ± 0.6 kg (↓ 3%)	1 month: ↑2.3 ± 1.6 kg 6 months: ↓0.9 kg ± 1.2 kg
Ferrando et al. [15]	Testosterone enanthate	U	6 months	SC	T enanthate	↑ 4.2 ± 0.4 kg (↑ 6.7%)[a]	NM
					P	↓ 2 ± 0.6 kg (↓ 3%)	
Vonk-Emmelot et al. [16]	Testosterone undecenoate	160 mg/day	6 months	PO	T	↑ 1.8%[a]	↓ 1.7%
					P	↔ 0.6%	↓ 3.9%
Snyder et al. [17]	Testosterone (testoderm®)	6 mg/day	3 years	T (scrotum)	T	↑ 1.9 ± 2.0 kg[a]	↓ 117.28 ± 32.37 N.m (↓13%)
					P	↔ 0.2 ± 1.5 kg	↓ 116.06 ± 34.13 N.m (↓12.74%)

(Continued)

TABLE 3.3 (*Continued*)

Authors	Drug	Maximum Dosage	Duration	Route of Administration	Treatment	Lean Mass (Mean Changes)	Muscle Strength (Mean Changes)
Sattler et al. [18]	Testosterone + GnRH	7.5 mg GnRH and 5 or 10 g/day of 1% T transdermal gel	16 weeks	To + IM	High dose of T only	↑ 1.7 ± 1.7 kg	↑ 30.2 ± 33%
					High dose of T + GnRH	↑ 2.7 ± 2.0 kg	NM
					Low dose of T only	↑ 0.8 ± 1.2 kg	NM
					Low dose of T + GnRH	↑ 1.3 ± 1.7 kg	NM
Bhasin et al. [19]	Testosterone enanthate	600 mg/week	20 weeks	SC	25 mg/day	↓ 0.3 ± 0.5 kg	↑ 0.8 ± 6.7 kg
					50 mg/day	↑ 1.7 ± 0.4 kg[d]	↑ 11.5 ± 4.7 kg
					125 mg/day	↑ 4.2 ± 0.6 kg[d]	↑ 28 ± 6.8 kg[d]
					300 mg/day	↑ 5.6 ± 0.5 kg[d]	↑ 51.7 ± 4.7 kg
					600 mg/day	↑ 7.3 ± 0.4 kg[d]	↑29.8 ± 6.4 kg[d]
Sheffield-Moore et al. [20]	Testosterone enanthate	100 mg	5 months	SC	Continuous T	↑ 3.12 ± 2.0 kg[c]	↑ 13.4 ± 7.2 kg[c]
					Monthly cycled T	↑ 2.67 ± 1.3 kg[c]	↑ 11.9 ± 10.9 kg[c]
					P	↓ 1.0 ± 1.6 kg	↔ 2.2 ± 9.5 kg
Schroeder et al. [21]	Oxymetholone	50 mg/day	12 weeks	PO		↑ 3.3 ± 1.2 kg[a]	↑ 8.2 ± 9.2%[f]
		100 mg/day				↑ 4.2 ± 2.4 kg[a]	↑ 13.9 ± 8.1%[g]
Schroeder et al. [22]	Oxandrolone	20 mg	12 weeks	PO	O	↑ 3.0 ± 1.5 kg[a]	↑ 6.3 ± 6.6% (leg press) and ↑ 6.3 ± 8.3% (leg extension)[b]
					P	↔ 0.1 ± 1.5 kg	↔ 0.1 ± 1.5 kg
Storer et al. [23]	Testosterone enanthate	125 mg/week	5 weeks	IM		↑ 2.4 ± 0.096 kg[c]	Not significant
		300 mg/week				↑ 3.1 ± 0.133 kg[a]	↑ 19%

[a] $p < 0.001$,
[b] $p < 0.003$,
[c] $p < 0.01$,
[d] $p < 0.05$,
[e] $p < 0.005$,
[f] $p < 0.04$,
[g] $p < 0.002$,
[h] $p \leq 0.03$.
IM, intramuscular; PO, oral; SC, subcutaneous; To, topical; ST, strength training; NM, not measured; ST, strength training; O, oxandrolone; T, testosterone; P, placebo.

ingested exogenously from animal source foods, especially meat and fish [34]. Cr synthesis involves three amino acids: glycine (Gly), arginine (Arg) and methionine (Met), and the action of three enzymes: L-arginine: glycine amidinotransferase (AGAT), methionine adenosyltransferase (MAT) and guanidinoacetate methyltransferase (GAMT) [34,35]. Cellular Cr uptake occurs in a process against its gradient concentration, which is sodium and chloride dependent and controlled by the specific transporter Creat-1. Approximately 95% of Cr stores are found in skeletal muscle, especially in fast-twitch fibers, and the remaining 5% is distributed in the brain, testes and bone [34,36]. Cr exists in a free and phosphorylated form (phosphorylcreatine, or PCr) inside the cell; both forms can be spontaneously and irreversibly degraded into creatinine and, at a rate of approximately 2 g/day, are excreted by the kidneys [37].

The Cr–PCr system plays an important role in providing rapid energy to tissues whose energetic demand is high, such as skeletal muscle and the brain [38]. PCr transfers a N-phosphoryl group to ADP to re-phosphorylate the ATP molecule through a reversible reaction catalyzed by the enzyme creatine kinase (CK). The existence of different CK isoforms enables a link between the sites of ATP generation (i.e., mitochondria; Mt-CK) to those of ATP consumption, such as skeletal muscle and brain (i.e., MM-CK and BB-CK, respectively) [38]. In 1992, Harris and his colleagues showed a significant increase in the total Cr content of skeletal muscle (e.g., quadriceps femoris) after two or more days of Cr supplementation at a dose of 5 g, four to six times a day [39]. This supplementation protocol was observed to increase the Cr content in skeletal muscle by more than 20%, of which approximately 20% is in a phosphorylated form. Once the plateau of concentration (around 150–160 mmol/kg dry muscle) is reached, Cr storage in human skeletal muscle cannot be exceeded, suggesting that Cr supplementation provides larger effects in subjects whose Cr intake in diet is low [39,40]. Thus, there are different mechanisms by which Cr plays an important role as an ergogenic nutritional supplement, including: increased PCr content, leading to faster ATP re-phosphorylation, higher training volume, and reduced damage and inflammation [35].

The effectiveness of Cr supplementation in increasing skeletal muscle size (i.e., hypertrophy), strength and lean body mass is well established. For example, Willoughby and Rosene [41] conducted a 12-week random, double-blind and placebo-controlled study to investigate the effects of Cr supplementation (6 g/day) and heavy resistance training (85–90% 1 repetition maximum) on myosin heavy chain (MHC) and myofibrilar protein content of young male untrained subjects. Their study has shown that heavy resistance training *per se* is a sufficient stimulus to increase the expression of MHC isoforms concomitantly with an increase of myofibrilar protein content; and, when associated with Cr supplementation, this increase seems to be augmented.

Creatine Supplementation and Performance

The effects of oral Cr supplementation as an ergogenic compound in sports settings have been supported by a large number of prior studies. These widespread supplementation loading protocols (i.e., short-term vs long-term) have demonstrated similar effectiveness in increasing skeletal muscle Cr content and exercise performance [35,36]. The traditional short-term (5 to 6 days) loading protocol consists of a high-dose of approximately 20 g/day which can rapidly elevate intramuscular Cr content by approximately 20% [39], whereas the long-term protocol is characterized by ingestion of 3 g/day for a period of at least 4 weeks [42]. An extensive number of studies have demonstrated improvements in short-duration anaerobic sports, such as cycling, intermittent running and sprints (for details, see [43,44]), or resistance training, following one of these Cr supplementation protocols. For example, in a random, double-blind placebo-controlled study by Volek et al. [45], the authors demonstrated for the first time significant improvements in peak power output during a high-intensity resistance training session on trained young subjects after one week of Cr supplementation at 25 g/day. The exercise protocol consisted of 10 maximum repetition (RM) distributed in 5 sets of bench press and 5 sets of jump squat performed in 10 repetitions with an intensity of 30% 1 RM. No significant improvements were observed in the placebo group. Some years later, Volek et al. [46] demonstrated a significant increase in muscle fiber cross-sectional areas (e.g., Type I, IIA and IIAB) following 12 weeks of resistance training in conjunction with Cr supplementation, which was much greater than found with resistance training alone, thus showing an important role of Cr supplementation on physiological adaptations of skeletal muscle. Furthermore, in 2003, Rawson and Volek [47] reviewed 22 studies in order to evaluate the response of Cr supplementation associated with resistance training on muscle strength. This review showed an average increase of 8% more than in the groups who performed resistance training alone (i.e., placebo associated with resistance training), therefore confirming the efficacy of Cr supplementation as an ergogenic nutritional supplement.

Conclusion

As highlighted above, studies using randomized double-blind placebo-controlled designs support the efficacy of Cr supplementation as one of the most beneficial nutritional supplements to improve sports performance, primarily in those sports in which the energetic demand is high, such as sprint cycling, intermittent high-intensity running, and resistance training. Once the storage of this compound in tissues reaches a plateau, the total content cannot be exceeded and the remaining proportion is degraded in an irreversible process into creatinine and excreted by the kidneys into urine. Thus, it seems that the effects of Cr supplementation will be greater in subjects whose dietary intake of this compound is low (e.g., vegetarians).

BETA-HYDROXY BETA-METHYLBUTYRATE (HMB)

HMB Metabolism

Beta-hydroxy beta-methylbutyrate (HMB) is a metabolite of the amino acid leucine. The first step in HMB metabolism is the reversible transamination of leucine to alpha-ketoisocaproate (KIC), which occurs mainly extrahepatically. Following this enzymatic reaction, KIC is converted to HMB by two routes: by cytosolic enzyme KIC dioxygenase; or through the formation of isovaleryl-CoA by branched-chain ketoacid dehydrogenase (BCKD) in liver, which after some steps results in HMB formation. Under normal conditions the majority of KIC is converted into isovaleryl-CoA, in which approximately 5% of leucine is metabolized into HMB [48]. Endogenous HMB production is about 0.2–0.4 g of HMB per day, depending on the content of leucine in the diet. Based on the fact that L-leucine (the precursor of HMB) is not synthesized by the human body, this quantity is reached via dietary protein intake. The vast majority of studies on this topic have employed a bolus of 3 g/day of HMB, based on evidence that this dose produces better results than 1.5 g/day and is equivalent to 6 g/day [49].

Effects of HMB Supplementation

Nissen et al. [50] conducted one of the first studies addressing the effects of oral supplementation with different doses of HMB. Individuals were supplemented with 0, 0.5 and 3.0 g/day of HMB in conjunction with a resistance training program for 3 weeks. In the first 2 weeks, urinary excretion of 3-methyl-histidine decreased, indicating attenuation of muscle proteolysis, and at the end of the protocol the muscle damage indicators – creatine kinase (CK) and lactate dehydrogenase (LDH) activities – were lower in the supplemented group. A significant increase in fat-free mass and strength was reported when 3.0 g/day of HMB was supplemented in association with resistance training for 7 weeks [51].

However, controversial results have been reported in human studies assessing the effects of oral supplementation of HMB in tandem with resistance training [49–55]. In previously untrained individuals, HMB supplementation (3.0 g HMB per day) during a resistance training program did not change body composition, muscular strength levels, and biochemical markers of protein turnover and muscle damage [52], increase muscle mass [51], or potentiate the strength gain and fat-free mass gain in elderly subjects [53]. In addition, in athletes highly conditioned to resistance training, HMB was unable to promote gains in strength and fat-free mass in water polo, rowing, or football athletes [52,54,56], and did not elicit attenuation of muscle damage markers (CPK and LDH) or gains in speed [56].

In untrained individuals, oral supplementation of HMB in association with resistance training may elicit gains in strength and muscle mass because these effects appear to be more prominent among those who are in the initial phase of training. Untrained individuals submitted to a resistance training program exhibit lower levels of muscle damage markers when supplemented with 3.0 g/day of HMB [49,57]. If HMB reduces the muscle protein catabolism associated with exercise, resistance-trained athletes may not respond to HMB supplementation in the same manner as untrained individuals, because of training-induced suppression of protein breakdown. To confirm the anti-catabolic properties of HMB, further research using more precise techniques is required, since most studies addressing this issue have used the urinary excretion of 3-methyl-histidine as an indicator of muscle catabolism, and this technique has been criticized.

It has been demonstrated that ingestion of 3.0 g/day of HMB increases its plasma levels and promotes gains in fat-free mass and peak isometric torque during a resistance-training program. Greater amounts of HMB (6.0 g/day) did not elicit the same effect [49]. Furthermore, 8 weeks of HMB supplementation (up to 76 mg/kg/day) appears to be safe and does not alter or adversely affect hematological parameters, hepatic and renal function in young male adults [58].

Mechanisms of Action

Based on studies evaluating the mechanisms of action of HMB, it is postulated that such supplementation could involve the following mechanisms: (1) increased sarcolemmal integrity, (2) increased metabolic efficiency, (3) upregulation of insulin-like growth factor I (IGF-I) expression in liver and skeletal muscle, (4) stimulation of protein synthesis by increasing the mTOR signaling pathway, and (5) suppression of proteolysis by the inhibition of the ubiquitin–proteasome system.

The protective effect of HMB against contractile activity-induced damage may be associated with increased stability of muscle plasma membrane. HMB is converted to b-methylglutaryl-CoA (HMG-CoA) for cholesterol synthesis, and inhibition of HMG-CoA reductase affects the electrical properties of cell membrane in skeletal muscle [48]. In addition, HMB supplementation may also promote an increase in acetyl-CoA content through the conversion of HMG-CoA into acetoacetyl-CoA by HMG-CoA synthase in mitochondria, increasing metabolic efficiency [59,60]. HMB supplementation also has been reported to stimulate lipolysis in adipose tissue and increase fatty acid oxidation capacity of skeletal muscles (see [61] for a review).

One other mechanism underlying the effects of HMB supplementation is the increased expression of IGF-I expression in liver and skeletal muscles. Kormasio et al. [62] demonstrated *in vitro* that HMB could stimulate IGF-I expression, as well as myogenic regulatory factors (MRFs) and thymidine incorporation (an indicator of DNA synthesis). Later, Gerlinger-Romero et al. [63] demonstrated that supplementation with HMB promoted increased growth hormone (GH) and IGF-I expression in pituitary and liver, respectively. *In vivo* and *in vitro* animal data also pointed to a possible role of HMB in stimulation of the mTOR signaling pathway and inhibition of the ubiquitin-proteasome system, a proteolytic system involved in skeletal muscle atrophy [64,65]. More studies are needed to determine whether the actions of HMB on protein synthesis and degradation signaling pathways are direct or mediated by an increased expression of IGF-I, as well to determine the molecular basis of HMB supplementation in humans.

Conclusions

In recent years, growing interest in HMB supplementation has arisen from previous demonstrations of its effects on fat-free mass and strength gains in combination with resistance exercise, its anti-catabolic properties, and speculations related to the mechanisms of action involved. Most studies have employed 3 g/day of HMB, based on evidence that this dose produces better results than 1.5 g/day and is equivalent to 6 g/day. Although, in untrained individuals, HMB supplementation appears to act as an effective ergogenic, in well-trained individuals and athletes the positive effects of HMB are less clear. The physiological mechanisms involved increased sarcolemmal integrity and metabolic efficiency, stimulation of GH-IGF-I axis, stimulation of protein synthesis, and suppression of proteolysis. Although some of these mechanisms were demonstrated in animal and *in vitro* studies, human studies are needed and could provide new insights into the mechanisms underlying the effects of supplementation.

CAFFEINE

Caffeine is one of the most widely used ergogenic aids in the world. Due to its widespread availability and presence in nutritional products and foodstuffs it is important to understand in what ways caffeine may be ergogenic for athletic performance. A considerable number of studies have consistently evidenced enhanced aerobic endurance performance following caffeine ingestion (see [66] for a review). This has been coupled with reduced ratings of perceived exertion (RPE) during submaximal, aerobically based exercise [67,68]. However, while the study of the effect of caffeine ingestion on aerobically based exercise has a long pedigree, with studies from Costill's Laboratory in the 1970s evidencing increased lipolysis and sparing of muscle glycogen during endurance exercise following caffeine ingestion [69], the research base pertaining to the ergogenic effect of caffeine for high-intensity anaerobic and strength-based performance is less well-developed (see [70] for a review).

Moreover, in the context of strength and power performance specifically, research studies documenting the ergogenic effects of acute caffeine ingestion are equivocal. This research base is still developing, and disparities in findings on this topic arise due to differences in methodologies and protocols employed, differences in the caffeine dose that is administered, and differences in the participant group being examined (e.g., trained vs untrained).

For example, Astorino et al. [71] reported a nonsignificant 11% and 12% increase in total mass lifted at 60% of 1 repetition maximum (1RM) in the bench press after caffeine ingestion (6 mg/kg) compared with placebo. Goldstein et al. [72] replicated this design in a sample of resistance-trained females and reported that caffeine ingestion (6 mg/kg) resulted in a significant increase in 1RM bench press performance but did not result in any ergogenic effect in repetitions to failure at 60% 1RM. Similarly, Duncan and Oxford [73] reported that (5 mg/kg) caffeine ingestion resulted in significant increases in repetitions to failure during the bench press exercise in a group of moderately trained men. Green et al. [74] also reported that acute caffeine ingestion (6 mg/kg) resulted in an increased number of repetitions and higher peak heart rate during leg press to failure at 10RM. Likewise, other studies have reported increases in total weight lifted during bench press performance [75] following caffeine ingestion (5 mg/kg), an increased number of repetitions during the first set of leg extension performance to failure during a multi-set protocol (6 mg/kg caffeine with 10 mg/kg aspirin) [76], and increases in peak torque during 3–5 repetitions of leg extension and flexion [77] following caffeine ingestion (7 mg/kg) in highly trained strength and power athletes.

Conversely, studies have reported that caffeine ingestion does not enhance strength and power performance during resistance exercise. Beck et al. [78]

reported that ingestion of a caffeine-based supplement (containing 201 mg caffeine) did not significantly enhance 1RM bench press strength in 31 men from different training status backgrounds. Likewise, Williams et al. [79] reported that ingestion of 300 mg of combined caffeine and ephedra did not influence 1RM bench press, lat pull down, and Wingate Anaerobic Test performance. More recently, Astorino et al. [80] used a multi-exercise protocol whereby 14 resistance trained men completed 4 sets of bench press, leg press, bilateral row and shoulder press exercises to failure following caffeine (6 mg/kg) ingestion. They reported that acute caffeine ingestion did not offer any practical benefit in enhancing strength performance over placebo. However, in their study, although as a whole there was no significant difference in total weight lifted in the caffeine and placebo conditions, Astorino et al. [80] identified that, within their sample, there were 9 participants who responded positively to caffeine ingestion. Consequently, they concluded that further research was needed to examine responder vs non-responder status in the context of resistance exercise performance.

Mechanisms of Action

Several possible explanations for the effect of caffeine in enhancing high-intensity exercise performance generally and strength and power performance specifically have been proposed. Early data suggested that increased lipolysis and sparing of muscle glycogen [69] and increased motor unit recruitment [81] resulted in potential ergogenic effects. Subsequent research revealed that caffeine does not spare glycogen [82]; and, as short-term, resistance exercise is not limited by carbohydrate availability, increased lipolysis does not explain the ergogenic effect of caffeine [70]. Research also shows no effect of caffeine ingestion on muscle activation during short-term, high-intensity exercise [83], limiting the hypothesis that increased motor unit recruitment may be responsible for caffeine's effects.

Increased intracellular calcium concentrations [68] and/or altered excitation-contraction coupling [84] have also been proposed as potential mechanisms for caffeine's action in strength and power exercise. However, these effects only occur at supraphysiological caffeine doses, impractical for humans, have only been demonstrated in animal-based research [85], and have subsequently been discounted in relation to anaerobically based exercise [70].

Other authors have suggested that the effect of caffeine lies within the central nervous system (CNS) as caffeine delays fatigue via stimulation of the CNS by acting as an adenosine antagonist [86]. Adenosine, an endogenous neuromodulator, has an inhibitory effect on central excitability. It preferentially inhibits the release of excitatory neurotransmitters, decreasing the firing rate of central neurons, and a reversal in the inhibitory effects of adenosine after caffeine administration has been reported [87]. However, the data supporting the effect of caffeine ingestion on adenosine antagonism is predominantly based on animal studies which, for example, document increased run time to fatigue (60% longer) in rats with caffeine compared with an adenosine antagonist [86].

One other, less explored, potential mechanism for the ergogenic effect of caffeine on high-intensity exercise performance is through the electrolyte potassium [88]. The net movement of potassium out of muscle cells during exercise, results in an increase in extracellular potassium concentration, and disturbance of the muscle resting membrane potential, which may cause muscle fatigue [89]. As caffeine decreases plasma potassium concentrations at rest and post exercise [68,89], caffeine may enhance performance due to an increase in potassium uptake and maintenance of resting membrane potential of muscle cells. However, although decreased plasma potassium concentration may contribute to ergogenic effects of caffeine, it is unlikely to be the sole mediator of its effects [88].

Psychophysiological Mechanisms

An alternative and emerging mechanism which has been posited to explain, at least in part, the effect of acute caffeine ingestion on strength and power performance, is that caffeine ingestion modifies the perceptual responses to exercise. For example, data reporting no differences in RPE during resistance exercise to failure, in spite of increased muscular performance, suggests that caffeine blunts RPE responses to resistance exercise [71]. Dampened perceptual responses have also been reported for RPE and muscle pain perception by Duncan and Oxford [90]. In their study, 18 moderately trained males performed bench press repetitions to failure at 60% 1RM following ingestion of caffeine (5 mg/kg) or a placebo. Caffeine ingestion resulted in increased repetitions to failure, lower RPE and lower muscle pain perception compared with placebo. There are comparatively fewer studies that have examined this issue in the context of strength and power performance. However, these results support assertions made by authors based on aerobic exercise that have suggested acute caffeine ingestion can dampen exercise-induced muscle pain [91]. In the case of strength and power performance, the hypoanalgesic effects of caffeine ingestion on muscle pain perception are important, as excess pain sensation may reduce individuals' performance and eventual participation in exercise [92].

Embryonic data are also emerging that suggest acute caffeine ingestion positively influences perceptual responses to strength and power exercise across the spectrum of psychophysiological responses to high-intensity exercise aside from RPE and pain perception. Recent data examining the efficacy of a caffeine-containing energy drink on resistance exercise performance acknowledged the limitations of unidimensional perceptual measures such as RPE and reported improved readiness to invest both mental and physical effort in the presence of the energy drink [93]. Thus, acute caffeine ingestion may favorably influence a range of psychophysiological variables which could augment the quality of resistance exercise training. However, data reporting on psychophysiological factors aside from RPE are scarce and in some cases have not yet examined the effect of caffeine, but rather have examined caffeine-containing energy drinks. It is therefore difficult to make robust conclusions regarding the impact of acute caffeine ingestion on the psychophysiological responses to resistance exercise, such as readiness to invest effort, until further research has been conducted on this topic.

Conclusions

The extant research base relating to the ergogenic effect of caffeine ingestion on strength and power performance is growing, with the majority of studies suggesting that acute ingestion in the range of 5–7 mg/kg will have a positive effect on resistance exercise performance. This may be due to physiological mechanisms such as adenosine antagonism, increases in cellular potassium uptake, dampening of perception of muscle pain and exertion, positive change to psychophysiological responses to exercise, or a combination of these mechanisms. Future studies are needed, however, to elucidate these mechanisms, alongside greater clarity as to what role training status and responder vs non responder status plays in any ergogenic effect of caffeine on strength and power performance.

References

[1] Williams M. Beyond training: How athletes enhance performance legally and illegally. Champaign IL: Leisure Press; 1989. p. 214.

[2] Silver MD. Use of ergogenic aids by athletes. J Am Acad Orthop Surg 2001;9:61–70.

[3] Palmer ME, Haller C, McKinney PE, Klein-Schwartz W, Tschirgi A, Smolinske SC, et al. Adverse events associated with dietary supplements: an observational study. Lancet 2003;361: 101–6.

[4] Kvorning T, Andersen M, Brixen K, Schjerling P, Suetta C, Madsen K. Suppression of testosterone does not blunt mRNA expression of myoD, myogenin, IGF, myostatin or androgen receptor post strength training in humans. J Physiol 2007;578:579–93.

[5] Bhasin S, Woodhouse L, Casaburi R, Singh AB, Bhasin D, Berman N, et al. Testosterone dose-response relationships in healthy young men. Am J Physiol Endocrinol Metab 2001;281: E1172–81.

[6] Blazevich AJ, Giorgi A. Effect of testosterone administration and weight training on muscle architecture. Med Sci Sports Exerc 2001;33:1688–93.

[7] Woodhouse LJ, Gupta N, Bhasin M, Singh AB, Ross R, Phillips J, et al. Dose-dependent effects of testosterone on regional adipose tissue distribution in healthy young men. J Clin Endocrinol Metab 2004;89:718–26.

[8] King DS, Sharp R l, Vukovich MD, Brown GA, Reifenrath TA, Uhl NL, et al. Effect of oral androstenedione on serum testosterone and adaptations to resistance training in young men: a randomized controlled trial. JAMA 1999;281:2020–8.

[9] Rogerson S, Weatherby RP, Deakin GB, Meir RA, Coutts RA, Zhou S, et al. The effect of short-term use of testosterone enanthate on muscular strength and power in healthy young men. J Strength Cond Res 2007;21:354–61.

[10] Bhasin S, Storer TW, Berman N, Callegari C, Clevenger B, Phillips J, et al. The effects of supraphysiologic doses of testosterone on muscle size and strength in normal men. N Engl J Med 1996;335:1–7.

[11] Sullivan DH, Roberson PK, Johnson LE, Bishara O, Evans WJ, Smith ES, et al. Effects of muscle strength training and testosterone in frail elderly males. Med Sci Sports 2005;37:1664–72.

[12] Broeder CE, Quindry J, Brittingham K, Panton L, Thomson J, Appakoundu S, et al. The Andro Project: physiological and hormonal influences of androstenedione supplementation in men 35 to 65 years old participating in a high-intensity resistance training program. Arch Intern Med 2000;160:3093–104.

[13] Lambert CP, Sullivan DH, Freeling AS, Lindquist DM, Evans WJ. Effects of testosterone replacement and/or resistance exercise on the composition of megestrol acetate stimulated weight gain in elderly men: a randomized controlled trial. J Clin Endocrinol Metab 2002;87:2100–6.

[14] Ferrando AA, Sheffield-Moore M, Paddon-Jones D, Wolfe RR, Urban RJ. Differential anabolic effects of testosterone and amino acid feeding in older men. J Clin Endocrinol Metab 2003;88:358–62.

[15] Ferrando AA, Sheffield-Moore M, Yeckel CW, Gilkison C, Jiang J, Achacosa A, et al. Testosterone administration to older men improves muscle function: molecular and physiological mechanisms. Am J Physiol Endocrinol Metab 2002;282:E601–7.

[16] Vonk-Emmelot MH, Verhaar HJ, Pour HR, Aleman A, Lock TM, Bosch JL, et al. Effect of testosterone supplementation on functional mobility, cognition, and other parameters in older men: a randomized controlled trial. JAMA 2008;299:39–52.

[17] Snyder PJ, Peachey H, Hannoush P, Berlin JA, Loh L, Lenrow DA, et al. Effect of testosterone treatment on body composition and muscle strength in men over 65 years of age. J Clin Endocrinol Metab 1999;84:2647–53.

[18] Sattler F, Bhasin S, He J, Chou CP, Castaneda-Sceppa C, Yarasheski K, et al. Testosterone threshold levels and lean tissue mass targets needed to enhance skeletal muscle strength and function: The HORMA Trial. J Gerontol A Biol Sci Med Sci 2011;66:122–9.

[19] Bhasin S, Woodhouse L, Casaburi R, Singh AB, Mac RP, Lee M, et al. Older men are as responsive as young men to the anabolic effects of graded doses of testosterone on the skeletal muscle. J Clin Endocrinol Metab 2005;90:678–88.

[20] Sheffield-Moore M, Dillon EL, Casperson SL, Gikinson CR, Paddon-Jones D, Durham WJ, et al. A randomized pilot study of monthly cycled testosterone replacement or continuous

testosterone replacement versus placebo in older men. J Clin Endocrinol Metab 2011;96:1831–7.
[21] Schroeder ET, Terk M, Sattler FR. Androgen therapy improves muscle mass and strength but not muscle quality: results from two studies. Am J Physiol Endocrinol Metab 2003;285:16–24.
[22] Schroeder ET, Singh A, Bhasin S, Storer TW, Azen C, Davidson T, et al. Effects of an oral androgen on muscle and metabolism in older, community-dwelling men. Am J Physiol Endocrinol Metab 2003;284:120–8.
[23] Storer TW, Woodhouse L, Magliano L, Singh AB, Dzekov C, Dzekov J, et al. Changes in muscle mass, muscle strength, and power but not physical function are related to testosterone dose in health older men. J Am Geriatr Soc 2008;56:1991–9.
[24] Kadi F, Eriksson A, Holmner S, Thornell LE. Effects of anabolic steroids on the muscle cells of strength-trained athletes. Med Sci Sports 1999;31:1528–34.
[25] Kadi F. Adaptation of human skeletal muscle to training and anabolic steroids. Acta Anaesthesiol Scand Suppl 2000;646:1–52.
[26] Eriksson A, Kadi F, Malm C, Thornell LE. Skeletal muscle morphology in power-lifters with and without anabolic steroids. Histochem Cell Biol 2005;124:167–75.
[27] Hartgens F, Kuipers H, Wijnen JA, Keizer HA. Body composition, cardiovascular risk factors and liver function in long-term androgenic-anabolic steroids using bodybuilders three months after drug withdrawal. Int J Sports Med 1996;17:429–33.
[28] Sinha-Hikim I, Artaza J, Woodhouse L, Gonzalez-Cadavid N, Singh AB, Lee MI, et al. Testosterone-induced increase in muscle size in healthy young men is associated with muscle fiber hypertrophy. Am J Physiol Endocrinol Metab 2002;283: E154–64.
[29] Ferrando AA, Tipton KD, Doyle D, Phillips SM, Cortiella J, Wolfe RR. Testosterone injection stimulates net protein synthesis but not tissue amino acid transport. Am J Physiol Endocrinol Metab 1998;275:E864–71.
[30] Kadi F, Schjerling P, Andersen LL, Charifi N, Madsen JL, Christensen LR, et al. The effects of heavy resistance training and detraining on satellite cells in human skeletal muscles. J Physiol 2004;558:1005–12.
[31] Kadi F, Charifi N, Denis C, Lexell J, Andersen JL, Schjerling P, et al. The behaviour of satellite cells in response to exercise: what have we learned from human studies? Pflugers Arch 2005;451:319–27.
[32] Kadi F, Bonnerud P, Eriksson A, Thornell LE. The expression of androgen receptors in human neck and limb muscles: effects of training and self-administration of androgenic-anabolic steroids. Histochem Cell Biol 2000;113:25–9.
[33] Nutrition Business Journal. Sports nutrition & weight loss report. Nutr Bus J 2009;14:1.
[34] Wyss M, Kaddurah-Daouk R. Creatine and creatinine metabolism. Physiol Rev 2000;80:1107–213.
[35] Gualano B, Roschel H, Lancha-Jr AH, Brightbill CE, Rawson ES. In sickness and in health: the widespread application of creatine supplementation. Amino Acids 2012;43:519–29.
[36] Gualano B, Artioli GG, Poortmans JR, Lancha Jr AH. Exploring the therapeutic role of creatine supplementation. Amino Acids 2010;38:31–44.
[37] Wallimann T, Tokarska-Schlattner M, Schlattner U. The creatine kinase system and pleiotropic effects of creatine. Amino Acids 2011;40:1271–96.
[38] Wallimann T, Wyss M, Brdiczka D, Nicolay K, Eppen-Berger HM. Intracellular compartmentation, structure and function of creatine kinase isoenzymes in tissues with high and fluctuating energy demands: the "phosphocreatine circuit" for cellular energy homeostasis. Biochem J 1992;281:21–40.
[39] Harris RC, Soderlund K, Hultman E. Elevation of creatine in resting and exercised muscle of normal subjects by creatine supplementation. Clin Sci (Lond) 1992;83:367–74.
[40] Greenhalff P. Creatine and its application as an ergogenic aid. Int J Sports Nutr 1995;5:S100–10.
[41] Willoughby DS, Rosene J. Effects of oral creatine and resistance training on myosin heavy chain expression. Med Sci Sports 2001;33:1674–81.
[42] Hultman E, Soderlund K, Timmons JA, Cederblad G, Greenhaff PL. Muscle creatine loading in men. J Appl Physiol 1996;81:232–7.
[43] Bemben MG, Lamont HS. Creatine supplementation and exercise performance: recent findings. Sports Med 2005;35:107–25.
[44] Cooper R, Naclerio F, Allgrove J, Jimenez A. Creatine supplementation with specific view to exercise/sports performance: an update. J Int Soc Sports Nutr 2012;9:33.
[45] Volek JS, Kraemer WJ, Bush JA, Boetes M, Incledon T, Clark KL, et al. Creatine supplementation enhances muscular performance during high-intensity resistance exercise. J Am Diet Assoc 1997;97:765–70.
[46] Volek JS, Duncan ND, Mazzetti SA, Staron RS, Putukian M, Gomez AL, et al. Performance and muscle fiber adaptations to creatine supplementation and heavy resistance training. Med Sci Sports 1999;31:1147–56.
[47] Rawson ES, Volek JS. Effects of creatine supplementation and resistance training on muscle strength and weightlifting performance. J Strength Cond Res 2003;17:822–31.
[48] Zanchi NE, Gerlinger-Romero F, Guimarães-Ferreira L, de Siqueira Filho MA, Felitti V, Lira FS, et al. HMB supplementation: clinical and athletic performance-related effects and mechanisms of action. Amino Acids 2011;40:1015–25.
[49] Gallagher PM, Carrithers JA, Godard MP, Schulze KE, Trappe SW. Beta-hydroxybeta-methylbutyrate ingestion, part I: effects on strength and fat free mass. Med Sci Sports 2000;32:2109–15.
[50] Nissen S, Sharp R, Rathmacher JA, Rice D, Fuller Jr JC, Connelly AS, et al. The effect of leucine metabolite b-hydroxy b-methylbutyrate (HMB) on muscle metabolism during resistance-exercise training. J Appl Physiol 1996;81:2095–104.
[51] Jowko E, Ostaszewski P, Jank M. Creatine and b-hydroxybmethylbutyrate (HMB) additively increase lean body mass and muscle strength during a weight-training program. Nutrition 2001;17:558–66.
[52] Slater GJ, Jenkins D, Longan P, Lee H, Vukovich M, Rathmacher JA, et al. Beta-hydroxy-betamethylbutyrate (HMB) supplementation does not affect changes in strength or body composition during resistance training in trained men. Int J Sport Nutr Exerc Metab 2001;11:384–96.
[53] Vukovich MD, Slater G, Macchi MB, Turner MJ, Fallon K, Boston T, et al. beta-Hydroxy-betamethylbutyrate (HMB) kinetics and the influence of glucose ingestion in humans. J Nutr Biochem 2001;12:631–9.
[54] Ransone J, Neighbors K, Lefavi R, Chromiak J. The effect of betahydroxy beta-methylbutyrate on muscular strength and body composition in collegiate football players. J Strength Cond Res 2003;17:34–9.
[55] Hoffman JR, Cooper J, Wendell M, Im J, Kang J. Effects of betahydroxy beta-methylbutyrate on power performance and indices of muscle damage and stress during high-intensity training. J Strength Cond Res 2004;18:747–52.
[56] Kreider RB, Ferreira M, Greenwod M, Wilson M, Grindstaff P, Plisk S, et al. Effects of calcium b-HMB supplementation during training on markers of catabolism, body composition, strength and sprint performance. J Exerc Physiol Online 2000;3:48–59.

[57] Panton LB, Rathmacher JA, Baier S, Nissen S. Nutritional supplementation of the leucine metabolite beta-hydroxy-betamethylbutyrate (HMB) during resistance training. Nutrition 2000;16:734–9.
[58] Gallagher PM, Carrithers JA, Godard MP, Schulze KE, Trappe SW. Beta-hydroxybeta- methylbutyrate ingestion, part II: effects on hematology, hepatic and renal function. Med Sci Sports 2000;32:2116–9.
[59] Nissen SL, Abumrad NN. Nutritional role of the leucine metabolite b-hydroxy-b-methylbutyrate (HMB). J Nutr Biochem 1997;8:300–11.
[60] Van Koverin M, Nissen SL. Oxidation of leucine and alphaketoisocaproate to b-hydroxy-b-methylbutyrate in vivo. Am J Physiol Endocrinol Metab 1992;262:27.
[61] Slater GJ, Jenkins D. Beta-hydroxy-beta-methylbutyrate (HMB) supplementation and the promotion of muscle growth and strength. Sports Med 2000;30:105–16.
[62] Kormasio R, Riederer I, Butler-Browne G, Mouly V, Uni Z, Halevy O. Beta-hydroxybeta- methylbutyrate (HMB) stimulates myogenic cell proliferation, differentiation and survival via the MAPK/ERK and PI3K/ Akt pathways. Biochim Biophys Acta 2009;1793:755–63.
[63] Gerlinger-Romero F, Guimarães-Ferreira L, Giannocco G, Nunes MT. Chronic supplementation of beta-hydroxy-beta methylbutyrate (HMβ) increases the activity of the GH/IGF-I axis and induces hyperinsulinemia in rats. Growth Horm IGF Res 2011;21:57–62.
[64] Eley HL, Russel ST, Baxter JH, Mukerji P, Tisdale MJ. Signaling pathways initiated by beta-hydroxy-beta-methylbutyrate to attenuate the depression of protein synthesis in skeletal muscle in response to cachectic stimuli. Am J Physiol Endocrinol Metab 2007;293:923–31.
[65] Smith HJ, Mukerji P, Tisdale MJ. Attenuation of proteasome induced proteolysis in skeletal muscle by {beta}-hydroxy-{beta}-methylbutyrate in cancer-induced muscle loss. Cancer Res 2005;65:277–83.
[66] Graham T. Caffeine and exercise: Metabolism, endurance and performance. Sports Med 2001;31:785–807.
[67] Doherty M, Smith P. Effects of caffeine ingestion on rating of perceived exertion during and after exercise: A meta-analysis. Scand J Med Sci Sports 2005;15:69–78.
[68] Doherty M, Smith PM, Hughes MG, Davison RC. Caffeine lowers perceptual response and increases power output during high-intensity cycling. J Sports Sci 2004;22:637–43.
[69] Costill D, Dlasky GP, Fink WJ. Effects of caffeine on metabolism and exercise performance. Med Sci Sports 1978;29:999–1012.
[70] Astorino TA, Roberson DW. Efficacy of acute caffeine ingestion for short-term, high-intensity exercise performance: A systematic review. J Strength Cond Res 2010;24:257–65.
[71] Astorino TA, Rohmann RL, Firth K. Effect of caffeine ingestion on one-repetition maximum muscular strength. Eur J Appl Physiol 2008;102:127–32.
[72] Goldstein E, Jacobs PL, Whitehurst M, Penhollow T, Antonio J. Caffeine enhances upper body strength in resistance trained women. J Int Soc Sports Nutr 2010;7:18.
[73] Duncan M, Oxford S. The effect of caffeine ingestion on mood state and bench press performance to failure. J Strength Cond Res 2012;25:178–85.
[74] Green J, Wickwire P, McLester J, Gendle S, Hudson G, Pritchett R, et al. Effects of caffeine on repetitions to failure and ratings of perceived exertion during resistance training. Int J Sports Phys Perform 2007;2:250–9.
[75] Woolf K, Bidwell WK, Carlson AG. The effect of caffeine as an ergogenic aid in anaerobic exercise. Int J Sport Nutr Exerc Metab 2008;18:412–29.
[76] Hudson GM, Green JM, Bishop PA, Richardson MT. Effects of caffeine and aspirin on light resistance training performance, perceived exertion, and pain perception. J Strength Cond Res 2008;22:1950–7.
[77] Jacobsen B, Weber M, Claypool I, Hunt L. Effect of caffeine on maximal strength and power in elite male athletes. Br J Sports Med 1992;26:276–80.
[78] Beck TW, Housh TJ, Malek MH, Mielke M, Hendrix R. The acute effects of a caffeine containing supplement on bench press strength and time to running exhaustion. J Strength Cond Res 2008;22:1654–8.
[79] Williams AD, Cribb PJ, Cooke MB, Hayes A. The effect of ephedra and caffeine on maximal strength and power in resistance trained athletes. J Strength Cond Res 2008;22:464–70.
[80] Astorino TA, Martin BJ, Schachtsiek L, Wong K, Ng K. Minimal effect of acute caffeine ingestion on intense resistance training performance. J Strength Cond Res 2011;25:1752–8.
[81] VanHandel P. Caffeine. In: Williams MH, editor. Ergogenic aids in sport. Champaign, Ill: Human Kinetics; 1983. p. 128–63.
[82] Jackman M, Wendling P, Friars D, Graham TE. Metabolic, catecholamine and endurance responses to caffeine during intense exercise. J Appl Physiol 1996;81:1658–63.
[83] Greer F, Morales J, Coles M. Wingate performance and surface EMG frequency variables are not affected by caffeine ingestion. Appl Physiol Nutr Metab 2006;31:597–603.
[84] Carr A, Dawson B, Schneiker K, Goodman C, Lay B. Effect of caffeine supplementation on repeated sprint running performance. J Sports Med Phys Fitness 2008;48:472–8.
[85] Rosser J, Walsh B, Hogan MC. Effect of physiological levels of caffeine on Ca2+ handling and fatigue development in Xenopus isolated single myofibers. Am J Physiol 2009;296: R1512–1517.
[86] Davis JM, Zhao Z, Stock HS, Mehl KA, Buggy J, Hand G. Central nervous system effects of caffeine and adenosine on fatigue. Am J Physiol 2003;284:R399–404.
[87] Kalmar JM, Cafarelli E. Caffeine: A valuable tool to study central fatigue in humans? Exerc Sport Sci Rev 2004;32:143–7.
[88] Crowe MJ, Leicht A, Spinks W. Physiological and cognitive responses to caffeine during repeated high-intensity exercise. Int J Sport Nutr Exerc Metab 2006;16:528–44.
[89] Lindinger MI, Graham TE, Spriet LL. Caffeine attenuates the exercise-induced increase in plasma [K+] in humans. J Appl Physiol 1993;74:1149–55.
[90] Duncan MJ, Oxford SW. Acute caffeine ingestion enhances performance and dampens muscle pain following resistance exercise to failure. J Sports Med Phys Fitness 2012;52:280–5.
[91] Gliottoni RC, Meyers JR, Arngrimsson SA, Broglio SP, Motl RW. Effect of caffeine on quadriceps muscle pain during acute cycling exercise in low versus high caffeine consumers. Int J Sport Nutr Exerc Metab 2009;19:150–61.
[92] Astorino TA, Terzi MN, Roberson DW, Burnett TR. Effect of caffeine intake on pain perception during high-intensity exercise. Int J Sport Nutr Exerc Metab 2011;21:27–32.
[93] Duncan MJ, Stanley M, Parkhouse N, Smith M, Cook K. Acute caffeine ingestion enhances strength performance, perceived exertion and muscle pain perception during resistance exercise. Eur J Sports Sci 2012; In Press; e-pub ahead of print.

SECTION 2

EXERCISE AND HUMAN HEALTH

CHAPTER

4

Exercise and Cardiovascular Disease

C. Tissa Kappagoda and Ezra A. Amsterdam

Division of Cardiovascular Medicine, University of California, Davis, CA, USA

INTRODUCTION

In 1999 the Surgeon General of the United States issued a landmark statement [1] drawing public attention to an issue which had engaged the attention of the healthcare professions for several years. The nation was alerted to the fact that physical activity had not only been declining progressively in the US population over the previous two decades but had also ushered in a near epidemic of non-communicable diseases. The decline in physical activity was evident in both leisure time [2] and employment-related activities [3]. This initiative by the Surgeon General resulted eventually in the development of a national plan for increasing physical activity in the population [4]. The plan stipulated that each state would endeavor to track and monitor the following five selected behavioral indicators relating to exercise:

1. The proportion of adults in the state who achieve at least 150 minutes a week of moderate-intensity aerobic physical activity or 75 minutes a week of vigorous-intensity aerobic physical activity or an equivalent combination of moderate- and vigorous-intensity aerobic activity.
2. The proportion of adults in the state who achieve more than 300 minutes a week of moderate-intensity aerobic physical activity or 150 minutes a week of vigorous-intensity aerobic physical activity or an equivalent combination of moderate- and vigorous-intensity aerobic physical activity.
3. The proportion of adults in the state who engage in no leisure-time physical activity.
4. The proportion of students in grades 9–12 in the state who achieve 1 hour or more of moderate and/or vigorous-intensity physical activity daily.
5. The proportion of students in grades 9–12 in the state who participate in daily physical education.

However, perusal of the annual reports of the states shows that there are significant differences in the data reported. For instance, while Missouri [5] approaches the national average for children in high school, California does not appear to gather this information [6].

Although the majority of these ideas have been incorporated into the guidelines issued by the US Department of Health and Human Services aimed at increasing the level of physical activity among Americans [7], the data emerging from the Centers of Disease Control and Prevention (CDC) in recent years, relating to obesity, diabetes and the metabolic syndrome do not indicate that these initiatives have yielded the anticipated benefits [8]. Aside from its link with diabetes mellitus, recent meta-analyses have shown that obesity *per se* is associated with cardiovascular diseases and several other conditions which adversely affect the overall quality of life (e.g., [9]). The findings of a recent population based study (2000 to 2005) Medical Expenditures Panel (self reported) Survey has shown that even in the short term (6 years) the coexistence of obesity-related problems such as hypertension and diabetes increase mortality in morbidly obese subjects [10].

PHYSIOLOGICAL RESPONSES OF THE CARDIOVASCULAR SYSTEM TO EXERCISE

Determinants of Functional Capacity (see [11] for review)

The Heart

The ability to exercise is dependent upon the delivery of an adequate blood supply to the muscles involved in the activity. The demand dictated by

Nutrition and Enhanced Sports Performance.
DOI: http://dx.doi.org/10.1016/B978-0-12-396454-0.00004-7

activity is met by changes in two fundamental physiological variables: heart rate (HR) and stroke volume (SV) operating against a background of an adequate venous return.

Thus,

$$\text{Cardiac Output [CO] } (L/\text{min}) = \{\text{Stroke Volume (SV) (mL/beat)} \times \text{Heart Rate (HR) (beat/min)}\}/1000 \quad (1)$$

Based on the Fick equation:

$$\text{CO (L/min)} = \text{oxygen consumption } [VO_2] \text{ (mL/min)} / a - v \text{ oxygen difference (mL/100 mL)}/10 \quad (2)$$

Combining equations 1 and 2:

$$[\text{SV (mL/beat)} \times \text{HR (beats/min)}]/1000 = \{VO_2(\text{mL/min})/a - v \text{ oxygen difference (mL/100 mL)}\}/10$$

By rearranging the terms:

$$VO_2 \propto SV \times HR \times a - v \text{ difference}$$

Therefore, oxygen consumption at maximum exercise, ($VO_{2\ max}$) is determined by the maximum HR, SV and the a-v difference (see ref. [12] and [11] for review). During upright aerobic exercise the HR and SV change in a predictable fashion in normal subjects increasing from a value at rest of ~70 mL/beat to approximately 150 mL/beat during maximum effort (Figure 4.1). In healthy young individuals, following exercise training this fundamental relationship persists with the modification that the stroke volume increases by approximately 3–5 mL/heartbeat at submaximal heart rates, resulting in an increase in CO of approximately 0.5 to 0.7 L/min. Numerous studies have shown that the functional capacity as measured by $VO_{2\ max}$ could be enhanced by aerobic and, to a lesser degree, strength training.

The increase in SV during exercise is the result of a combination of factors: (a) Starling mechanism resulting from an increase in venous return, (b) increase in myocardial contractility, and (c) an increase in the force of contraction resulting from an increase in rate *per se* (Bowditch phenomenon).

Lungs

The ability to perform physical exercise is not only dependent on the ability of the cardiovascular system to supply oxygen to the muscles but also on the ability of the lungs to clear carbon dioxide from the blood. This process can be separated into four components: (i) ventilation, which moves air from the atmosphere to the lungs, (ii) simultaneous transport of carbon dioxide to the lungs and oxygen to exercising muscles, (iii) diffusion which promotes the gas exchange between the lungs and the blood, (iv) capillary gas exchange, or the exchange of oxygen and carbon dioxide. The first two processes have been classified as external respiration and the last as internal respiration. The cardiovascular system links these two phases of respiration (see [15] for complete discussion).

An increase in cardiac output results in a corresponding increase in pulmonary blood flow. There is also a concurrent increase in ventilation, and these two factors facilitate both the increase in oxygen uptake and removal of carbon dioxide. During exercise, there is also a greater extraction of oxygen by exercising muscles from the blood as it perfuses the muscles, resulting in a widening of the arterio-venous oxygen difference.

Peripheral Circulation

Dynamic exercise in normal subjects is associated with an increase in CO that is facilitated by an increase in SV and HR as described above. There is also a preferential redistribution of the CO to exercising muscles, with a corresponding reduction in flow to "non-essential" organs such as the kidney and gut [16]. The blood flow to the brain is preserved while the flow to the skin is reduced until the need for thermoregulation results in vasodilatation in this region, usually at maximal exercise [17]. A concurrent increase in venous return enhances diastolic filling, which maintains the CO at a higher level during exercise. These changes in the peripheral circulation result in an increase in systolic and mean blood pressures with little or no change in diastolic blood pressure.

The relative increase in blood flow to exercising muscles is reflected as an overall fall in peripheral resistance, which is expressed as:

$$\text{Total Peripheral Resistance} = (\text{Mean Arterial Pressure} - \text{Mean Venous Pressure})/\text{CO}$$

where mean arterial pressure is given by diastolic + (systolic − diastolic)/3.

The resistance is usually expressed as mmHg/L per minute (also known as Wood Units). During graded dynamic exercise there is a disproportionate increase in CO compared with the changes in blood pressure. Total peripheral resistance is determined by the degree of vasodilatation resulting from an increased metabolic demand in exercising muscles and vasoconstriction in organs not involved with exercise such as the kidney and the gastrointestinal tract. These changes not only cause a redistribution of blood flow but also result in a reduction in overall peripheral resistance (despite a 4–5 fold increase in cardiac output). Thus, when a healthy individual transitions from rest to maximum

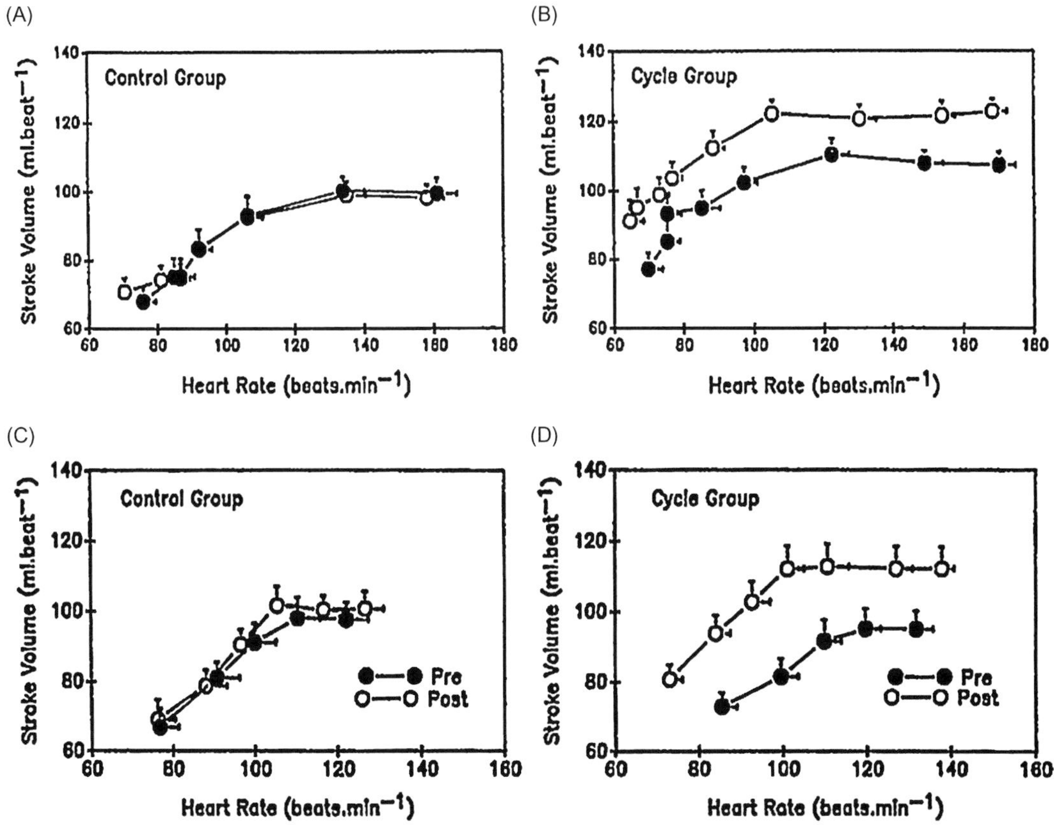

FIGURE 4.1. **Effect of training on the stroke volume/heart rate relationship.** Upper panel: Normal subjects [13]; (A) control subjects and (B) experimental subjects who underwent exercise training on a bicycle ergometer. Lower panel: subjects who had recovered from coronary artery bypass surgery [14]; (C) control subjects and (D) experimental subjects who underwent exercise training on a bicycle ergometer.

exercise, the resistance falls from nearly 20 Wood units to 5. The reduction in vascular resistance with respect to workload or oxygen consumption has a curvilinear profile (see Figure 4.2).

Assessment of Functional Capacity

The conventional measure of functional capacity in healthy individuals is VO_{2max} which is determined during progressively increasing workloads until a plateau is achieved (despite increases in load). However, in those with compromised left ventricular function, it is not possible to establish a meaningful maximal value because it is often difficult to achieve a plateau expressed in these terms. A pragmatic alternative is the measurement of a symptom-limited maximum value (VO_{2peak}).

Another approach to defining an index of exercise capacity is to use a respiratory parameter derived from a submaximal test such as the anaerobic (or ventilatory) threshold (VT) which is defined as the level of activity at which the expired gas volume increases at a rate that is faster than the oxygen uptake. VT is analogous to the anaerobic threshold, the physiological basis of which is believed to be the inability of the oxygen supply to match the requirement of the exercising muscle. This imbalance is compensated by anaerobic glycolysis in muscles, which in turn generates lactic acid and eventually lactate as the final product. During this phase there is an increase in ventilatory volume to eliminate the excess carbon dioxide produced during the conversion of lactic acid to lactate [18]. VT usually occurs at approximately 45% to 65% of measured peak or maximal VO_{2max} in healthy untrained subjects, and it generally occurs at a higher percentage of exercise capacity in endurance-trained individuals [19].

Several methods have been proposed for determination of VT, and the three used most frequently are: (i) the departure of VO_2 from a line of identity drawn through a plot of volume of expired CO_2 against VO_2, often called the V-slope method; (ii) the point at which a systematic increase in the ventilatory equivalent for oxygen (minute ventilation/VO_2) occurs without an increase in the ventilatory equivalent for carbon dioxide (minute ventilation/VCO_2); and (iii) the point at which a systematic rise in end-tidal oxygen partial

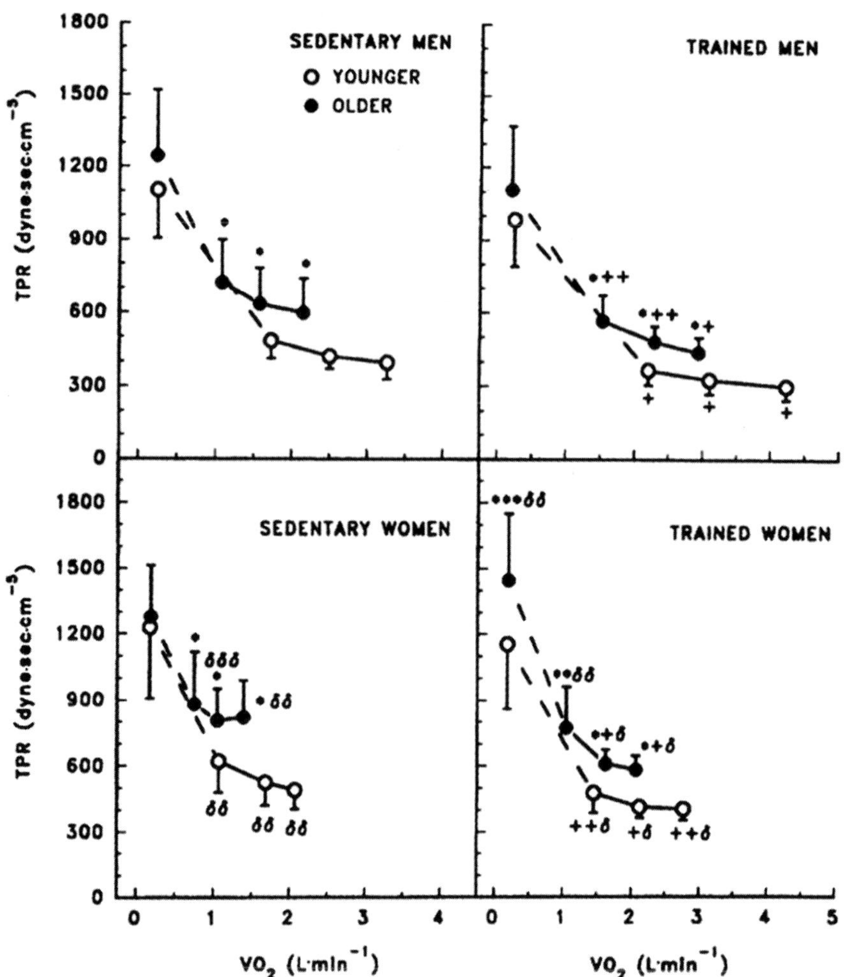

FIGURE 4.2 **Changes in peripheral resistance associated with dynamic exercise.** To convert from dyn s/cm^5 to Wood units divide the value by 80 [12]. Graphs showing total peripheral resistance (TPR) at rest and during exercise in younger and older sedentary men, trained men, sedentary women, and trained women. Statistically significant differences are designated for corresponding values at rest and at each level of submaximal or maximal exercise. *$p < 0.001$, **$p < 0.01$, and ***$p < 0.05$ vs younger subjects of the same sex and training status. + $p < 0.001$ and ++$p < 0.01$ vs sedentary subjects of similar age and the same sex. δ $p < 0.001$, δδ $p < 0.01$, and δδδ $p < 0.05$ vs men of similar age and training status. Training is associated with a reduction in peripheral vascular resistance during exercise in both men and women, regardless of age.

pressure occurs without a decrease in the end-tidal carbon dioxide pressure ($P_{ET}CO_2$). Despite the apparent simplicity of the concept of using ventilatory parameters to estimate exercise capacity, there is no general agreement regarding kinetics of oxygen transport during exercise [20].

EXERCISE AND HEALTH

All-cause Mortality

One of the earliest *observational* studies that described the effect of regular physical activity on all-cause mortality was that reported by Paffenbarger et al [21], who examined physical activity and other lifestyle characteristics of approximately 17,000 alumni of Harvard University aged 35–74 years in an attempt to link physical activity levels with all-cause mortality and length of life. The data were gathered from the alumni using a questionnaire which was mailed to the subjects. They found that the death rates declined steadily as energy expended increased from less than 500 to 3500 kcal per week, beyond which rates increased slightly. Rates were one quarter to one third lower among alumni expending 2000 or more kcal during exercise per week than among less active men. These differences were preserved even after adjustments for *hypertension, cigarette smoking, extremes or gains in body weight, or early parental death.*

The National Health Interview Survey, which is an ongoing multipurpose, in-person, health survey of the civilian, non-institutionalized US population conducted by the National Center for Health Statistics, has yielded information regarding the degree of adherence to the levels of physical activity recommended in the *2008 Physical Activity Guidelines for Adults* [22] by both men and women >18 years. The findings have suggested that adherence to these levels was associated with substantial survival benefits: the all-cause mortality risks dropped by 27% among people without existing chronic comorbidities, and by almost half in people with chronic comorbidities [22]. Such lower risks occurred regardless of age and obesity levels. These findings were corroborated by another survey based on the National Health and Nutrition

Examination Survey (NHANES I, 1971–1975) and the associated Epidemiologic Follow-up Study. It examined both leisure and non-leisure-time activities [23].

Coronary Artery Disease

Coronary heart disease (CHD) is recognized as one of the leading causes of death among men and women in the United States. The benefits derived from exercise are multiple (see Box 4.1) and can be viewed as occurring in two categories of people:

1. those with no previous history of CHD (i.e., associated with primary prevention programs); and
2. those who have recovered from a major cardiac event (i.e., associated with secondary prevention).

The benefits enjoyed by these two categories of patients are reflected in a reduced rate of major cardiac events and enhanced survival. It should be recognized that the studies relating to the first group are observational in nature whereas the latter is supported by multiple randomized clinical trials.

Exercise and Primary Prevention

The Health Professionals' Follow-up Study, tracked a cohort of 44,452 US men at 2-year intervals from 1986 through January 31, 1998, to assess potential CHD risk factors and to identify newly diagnosed cases of CHD (incidence of nonfatal myocardial infarction or fatal CHD) and their relationship to leisure-time physical activity [25]. The relative risk corresponding to moderate (4–6 METs) and high (6–12 METs) activity intensities were 0.94 and 0.83 compared with low activity intensity (<4 METs) ($p = 0.02$ for trend). A half-hour per day or more of brisk walking was associated with an 18% risk reduction (RR, 0.82; 95% CI, 0.67–1.00). Walking pace was associated with reduced CHD risk independent of the number of walking hours. Total physical activity, running, weight training, and walking were each associated with reduced CHD risk. Average exercise intensity was associated with reduced risk independent of the number of MET-hours spent in physical activity.

Similar conclusions were drawn from the findings of the Women's Health Study, which examined the relationship between physical activity and CHD among women, focusing on walking (a light-to-moderate activity depending on pace). The study cohort consisted of 39,372 healthy female health professionals aged 45 years or older, enrolled throughout the United States between September 1992 and May 1995, with follow-up to March 1999. Recreational activities, including walking and stair climbing, were reported at study entry. A total of 244 cases of CHD occurred during the follow

BOX 4.1

BENEFITS OF EXERCISE IN SECONDARY PREVENTION PROGRAMS

Exercise capacity	
Estimated METS	+35%
Peak VO_2	+15%
Peak anaerobic threshold	+11%
Reduction in obesity indices	
Body mass index	−1.5%
Percent fat	−5%
Metabolic syndrome	−37% (Prevalence)
Improvements in lipids	
Total cholesterol	−5%
Triglycerides	−15%
HDL-C	+6% (13–16% increase in subgroups with low HDL-C levels)
LDL-C	−2%
LDL-C/HDL-C	5%
Inflammatory markers	
High sensitivity – CRP	−40%

Changes in autonomic regulation of the circulation reflected in improved baroceptor function and increased heart rate variability. Both the heart rate and blood pressure at submaximal workloads decrease, thereby reducing myocardial oxygen demand.

Improvements in behavioral characteristics: e.g., depression, anxiety, somatization, hostility.

Improvements in overall quality of life and its components.

Improvement in blood rheology and viscosity.

Improvements in endothelial function.

Reduction in homocysteine levels.

Reduction in overall morbidity and mortality (especially associated with depression and psychological distress).

Reduction in hospitalization costs.

Adapted from [24].

up. Adjusting for potential confounders, the relative risks (RRs) of CHD for <200, 200–599, 600–1499, and ≥1500 kcal/week expended on all activities were 1.00 (referent), 0.79 , 0.55 and 0.75 respectively (p for linear trend = .03). Vigorous activities were associated with lower risk (RR, 0.63) comparing highest and lowest categories. The inverse association between physical activity and CHD risk did not differ by weight or cholesterol levels (p for interaction = 0.95 and 0.71, respectively), but there were significant interactions with smoking and hypertension status. Physical activity was inversely related to risk in current smokers but not in hypertensive women (p for interaction = 0.01 and 0.001, respectively) [26].

Exercise in Secondary Prevention of CHD

These benefits refer to patients with established CHD who participate in secondary prevention programs designed to prevent new cardiac events and cover a range of factors that have a bearing on the outcome of these patients. The major benefit that accrues from regular exercise is an increase in functional capacity, which is reflected best in the relationship between the heart rate and stroke volume during exercise (Figure 4.1). Exercise training in both normal subjects and those with CHD has the capacity to increase the stroke volume at comparable heart rates, resulting eventually in an increase in work capacity. However, much of the activity in life is undertaken at submaximal levels of exercise, and they would be carried out at a lower heart rate and blood pressure in trained individuals compared with those who are untrained. As discussed previously, it is clear that the myocardial oxygen demand during submaximal activity would be lower in trained subjects. It is likely that the reduction in perceived exertion observed in them during submaximal exercise tests is a reflection of this physiological adaptation.

A recent systematic review of 47 randomized clinical trials has addressed the issue of all-cause and cardiac mortality in patients who enroll in cardiac rehabilitation programs. The analysis was based on 10,794 patients who were randomized to exercise-based cardiac rehabilitation or usual care. In the medium to longer term (i.e., 12 or more months follow-up), exercise-based cardiac rehabilitation reduced overall and cardiovascular mortality [RR 0.87 (95% CI 0.75–0.99) and 0.74 (95% CI 0.63–0.87), respectively], and hospital admissions [RR 0.69 (95% CI 0.51–0.93)] in the shorter term (<12 months follow-up), with no evidence of heterogeneity of effect across trials. Neither intervention had any effect on the occurrence of non-fatal myocardial infarction or the need for revascularization [27]. The mortality benefits (both overall and cardiovascular) were evident in trials that extended for periods longer than 12 months.

Hypertension

Several studies have addressed the issue of blood pressure and peripheral vascular resistance on blood pressure and autonomic function following exercise training. Cornelissen et al. [28] studied the effect of exercise training in middle-aged sedentary men and women at two levels of training intensity. They found that both levels of training intensity reduced the systolic blood pressure to a similar degree. These changes were also accompanied by a reduction in heart rate at submaximal workloads. These findings on the effect of exercise on blood pressure are essentially similar to those reported by Ogawa [12] and Bonanno and Lies [29]. A meta-analysis of clinical trials which included exercise as a component also concluded that physical activity contributed to better control of blood pressure [30].

It has been suggested that in patients with CHD these changes in blood pressure following exercise are associated with changes in peripheral inflammatory biomarker concentrations [31]. The concentrations of C-reactive protein, interleukin 6, fibrinogen, and vascular cell adhesion molecule 1 were lower after exercise training. There was also a concurrent increased in high-density lipoprotein cholesterol. One of the possible mechanisms responsible for the improvement in blood pressure is that exercise attenuates endothelial dysfunction and inflammation and improves the nitric oxide bioavailability. It also increases the number of endothelial progenitor cells, with a concurrent reduction in the level of pro-inflammatory cytokines and C-reactive protein.

Body Weight

The Body Mass Index (BMI) is a common means of estimating overweight and obesity in both adults and children. It is a calculated number based on height and weight, and is used to compare and analyze the health effects in all people. The conventional formula for its calculation is the weight in kilograms divided by height in meters squared. For instance, if the weight is 68 kg and the height is 165 cm (1.65 m), the BMI is: $68 \div (1.65)^2 = 24$. Box 4.2 shows the interpretation of BMI values.

The BMI number and body fat are fairly closely related but there is some variation associated with gender, race and age. For instance, at the same BMI:

- women tend to have more body fat than men;
- older people, on average, tend to have more body fat than younger adults;

BOX 4.2

INTERPRETATION OF BMI VALUES

<18.5	Underweight
18.5–24.9	Normal
25.0–29.9	Overweight
≥30.0	Obese

- highly trained athletes may have a high BMI because of increased muscularity rather than increased body fatness.

The BMI is only one factor related to risk for disease. For assessing someone's likelihood of developing overweight- or obesity-related diseases, the National Heart, Lung, and Blood Institute guidelines recommend reviewing other predictors:

- the individual's waist circumference (because abdominal fat is a predictor of risk for obesity-related diseases);
- other risk factors the individual has for diseases and conditions associated with obesity (for example, high blood pressure or serum cholesterol).
- Ethnicity or race is also a factor. The World Health Organization has recommended that the normal/overweight threshold for South East Asian body types be lowered to a BMI of 23. The new cut-off BMI index for obesity in Asians is 27.5 compared with the traditional WHO figure of 30. An Asian adult with a BMI of 23 or greater is now considered overweight and the ideal normal range is 18.5–22.9.

Although weight management is beyond the scope of this chapter, it should be recognized that an effective exercise program should be an integral part of any weight management program, coupled with an appropriate regimen for reducing calorie intake. The recommendations of the CDC with respect to exercise are addressed later in this chapter.

Diabetes Mellitus

It has been recognized for many years that physical exercise is an integral part of the management of Type 2 diabetes. The Diabetes Prevention Program (DPP) clinical trial provided clear evidence that intensive lifestyle changes that included physical exercise were successful in reducing the incidence of diabetes [32]. The goals for the intensive lifestyle intervention were to achieve and maintain a weight reduction of at least 7% of initial body weight through healthy eating and physical activity, and to achieve and maintain a level of physical activity of at least 150 min/week (equivalent to 700 kcal/week) through moderate-intensity activity (such as walking or bicycling).

The study was undertaken on non-diabetic individuals with a high risk of progression to type 2 diabetes. It is very likely that the majority of these subjects met the criteria for the metabolic syndrome. Those individuals with conditions that might increase the risk of adverse effects from the interventions or severely reduce life-expectancy were excluded. The subjects were assigned at random to one of three intervention groups: an intensive lifestyle intervention focusing on a healthy diet and exercise, and two masked medication treatment groups (metformin or placebo) each combined with standard diet and exercise recommendations. Participants were recruited during a 32 month period, and were followed for an additional 39 to 60 months after the close of recruitment. At the end of the follow up (average 2.8 years) the incidence of diabetes was 11.0, 7.8, and 4.8 cases per 100 person-years in the placebo, metformin, and intensive lifestyle groups, respectively. The lifestyle intervention reduced the incidence by 58% and metformin by 31% as compared with placebo, the lifestyle intervention being significantly more effective than metformin [33].

Subsequently, all active DPP participants were eligible for continued follow-up. Approximately 900 people from each of the three original groups (88% of the total) enrolled for a median additional follow-up of 5.7 years. All three groups were offered group-implemented lifestyle intervention. Metformin treatment was continued in the original metformin group (850 mg twice daily as tolerated), with participants unmasked to assignment, and the original lifestyle intervention group was offered additional lifestyle support. During the 10.0-year (Interquartile range: 9.0–10.5) follow-up since randomization to DPP, it was found that the incidence of diabetes remained lowest in the intensive lifestyle group [34].

A systematic review of trials examined the effects of aerobic or resistance exercise training on clinical markers of CHD risk, including glycemic control, dyslipidemia, blood pressure, and body composition in patients with type 2 diabetes [35]. Aerobic exercise training alone or combined with resistance training (RT) significantly reduced HbA1c by 0.6% and by 0.67% (95% CI: −0.93 to −0.40), respectively. In addition, there were significant reductions in systolic blood pressure by 6.08 and 3.59 mmHg and triglycerides by 0.3 mmol/L (for both forms of exercise) respectively. Waist circumference was significantly reduced by 3.1 cm (95% CI −10.3 to −1.2) with combined aerobic and resistance exercise, although fewer studies and more heterogeneity of the responses were observed in the latter two markers.

BOX 4.3

CENTERS FOR DISEASE CONTROL AND PREVENTION RECOMMENDATIONS FOR EXERCISE

2 hours and 30 minutes (150 minutes) of *moderate-intensity aerobic activity* (i.e., brisk walking) every week **and** *muscle-strengthening activities* on 2 or more days a week that work all major muscle groups (legs, hips, back, abdomen, chest, shoulders, and arms).

OR

1 hour and 15 minutes (75 minutes) of *vigorous-intensity aerobic activity* (i.e., jogging or running) every week **and** *muscle-strengthening activities* on 2 or more days a week that work all major muscle groups (legs, hips, back, abdomen, chest, shoulders, and arms).

OR

An equivalent mix of *moderate- and vigorous-intensity aerobic activity* **and** *muscle-strengthening activities* on 2 or more days a week that work all major muscle groups (legs, hips, back, abdomen, chest, shoulders, and arms).

HOW MUCH EXERCISE IS ENOUGH? [36]

There are multiple recommendations from several national bodies stipulating the amount of exercise that should be undertaken daily to maintain health. The current consensus is that 30 minutes of moderate-intensity activity/day would provide substantial benefits across a broad range of health outcomes for sedentary adults. However, this amount of exercise may be inadequate and need to be revised upward, to prevent weight gain in those who appear to be consuming an appropriate number of calories in their diet. For persons who are exercising 30 min/day and consuming what appears to be an appropriate number of calories, but are still having trouble controlling their weight, additional exercise or caloric restriction is recommended to reach energy balance and minimize the likelihood of further weight gain. In addition to aerobic exercise, it is desirable that people engage in activities that build musculoskeletal fitness, such as resistance training and flexibility exercises, at least twice a week. The current recommendations of the CDC are shown in Box 4.3 [37].

ARE THE BENEFITS OF EXERCISE UNIVERSAL?

In a recent study Bouchard et al. [38] have suggested that certain individuals may be genetically predisposed to not benefit from exercise training. In support of this claim they have provided a systematic analysis of six short-term clinical trials (<6 months) in which certain factors pertaining to cardiovascular risk were measured. The factors were blood pressure, HDL cholesterol concentration, plasma insulin and plasma triglycerides. Using their definitions of "technical error" of the measurements, it was found that 8.4%, 12.2%, 10.4%, and 13.3% of subjects had adverse changes in plasma insulin, systolic blood pressure, triglycerides, and HDL cholesterol, respectively. However, the average changes in HDL cholesterol (~0.04 mmol/L) and systolic blood pressure (<4 mmHg) were probably too small to make an impression on conventional calculations of coronary heart disease risk (such as that based on the Framingham Score) in a population in which almost 40% of the subjects were in their early to mid-thirties.

References

[1] Centers for Disease Control and Prevention. Physical activity and health—a report of the surgeon general 1999. Available from: <http://www.cdc.gov/nccdphp/sgr/intro.htm>.

[2] Anderssen N, Jacobs DR, Sidney S, Bild DE, Stemfeld B, Slattery ML, et al. Change and secular trends in physical activity patterns in young adults: a seven-year longitudinal follow-up in the coronary artery risk development in young adults study (CARDIA). Am J Epidemiol 1996;143(4):351–62.

[3] Church TS, Thomas DM, Tudor-Locke C, Katzmarzyk PT, Earnest CP, Rodarte RQ, et al. Trends over 5 decades in U.S. occupation-related physical activity and their associations with obesity. PLoS ONE 2011;6(5): e19657.

[4] Centers for Disease Control and Prevention. State indicator report on physical activity, 2010. Atlanta, GA: U.S. Department of Health and Human Services, 2010. [cited 2012 May 29th, 2012]; Available from: <http://www.cdc.gov/physicalactivity/downloads/PA_State_Indicator_Report_2010.pdf>.

[5] Department of Health and Human Services Centers for Disease Control and Prevention. State indicator report on physical activity, 2010 Missouri Action Guide 2010 [cited 2012 April 27, 2012]; Available from: <http://www.nspapph.org/images/resources2/state_indicator_reports_2010/PA%20State%20Action%20Guide%20Missouri%204%2028%2010.pdf>.

[6] Department of Health and Human Services Centers for Disease Control and Prevention. State indicator report on physical activity, 2010 California Action Guide. 2010 [cited 2012 April 27th, 2012]; Available from: <http://www.californiahealthykids.org/articles/CA_PA.pdf>.

[7] Office of Disease Prevention & Health Promotion U.S. Department of Health and Safety. 2008 Physical Activity Guidelines for Americans. 2008 [cited; Available from: <http://health.gov/paguidelines/guidelines/default.aspx>].

[8] Prevention Centers for Disease Control and Prevention. Diabetes Data & Trends. 2012 March 22, 2012 [cited 2012 May 29th, 2012]; Available from: <http://apps.nccd.cdc.gov/DDTSTRS/default.aspx>.
[9] Guh D, Zhang W, Bansback N, Amarsi Z, Birmingham CL, Anis A. The incidence of co-morbidities related to obesity and overweight: a systematic review and meta-analysis. BMC Public Health 2009;9(1):88.
[10] Jerant A, Franks P. Body mass index, diabetes, hypertension, and short-term mortality: a population-based observational study, 2000–2006. J Am Board Fam Med 2012;25(4):422–31.
[11] Kappagoda T, Amsterdam E. Exercise and heart failure in the elderly. Heart Fail Rev 2012;17(4–5):635–62.
[12] Ogawa T, Spina RJ, Martin III WH, Kohrt WM, Schechtman KB, Holloszy JO, et al. Effects of aging, sex, and physical training on cardiovascular responses to exercise. Circulation 1992;86 (2):494–503.
[13] Haennel R, Teo K, Quinney A, Kappagoda T. Effects of hydraulic circuit training on cardiovascular function. Med Sci Sports Exerc 1989;21(5):605–12.
[14] Haennel RG, Quinney HA, Kappagoda CT. Effects of hydraulic circuit training following coronary artery bypass surgery. Med Sci Sports Exerc 1991;23(2):158–65.
[15] Balady GJ, Arena R, Sietsema K, Myers J, Coke L, Fletcher GF, et al. Clinician's guide to cardiopulmonary exercise testing in adults. Circulation 2010;122(2):191–225.
[16] Kenney WL, Ho CW. Age alters regional distribution of blood flow during moderate-intensity exercise. J Appl Physiol 1995;79 (4):1112–9.
[17] Rowell L. Human cardiovascular adjustments to exercise and thermal stress. Physiol Rev 1974;54:75–159.
[18] Wasserman K, Beaver W, Whipp B. Gas exchange theory and the lactic acidosis (anaerobic) threshold. Circulation 1990;81 (Supp II:14–II:30).
[19] Davis JA, Vodak P, Wilmore JH, Vodak J, Kurtz P. Anaerobic threshold and maximal aerobic power for three modes of exercise. J Appl Physiol 1976;41(4):544–50.
[20] Xu F, Rhodes EC. Oxygen uptake kinetics during exercise. Sports Med 1999;27(5):313–27.
[21] Paffenbarger RS, Hyde R, Wing AL, Hsieh CC. Physical activity, all-cause mortality, and longevity of college alumni. N Engl J Med 1986;314(10):605–13.
[22] Schoenborn CA, Stommel M. Adherence to the 2008 adult physical activity guidelines and mortality risk. Am J Prev Med 2011;40(5):514–21.
[23] Arrieta A, Russell LB. Effects of leisure and non-leisure physical activity on mortality in U.S. Adults over two decades. Ann Epidemiol 2008;18(12):889–95.
[24] Lavie CJ, Milani RV. Cardiac rehabilitation and exercise training in secondary coronary heart disease prevention. Prog Cardiovasc Dis 2011;53(6):397–403.
[25] Tanasescu M, Leitzmann MF, Rimm EB, Willett WC, Stampfer MJ, Hu FB. Exercise type and intensity in relation to coronary heart disease in men. JAMA 2002;288(16):1994–2000.
[26] Lee IM, Rexrode KM, Cook NR, Manson JE, Buring JE. Physical activity and coronary heart disease in women: is "no pain, no gain" passé? JAMA 2001;285(11):1447–54.
[27] Heran BS, Chen J, Ebrahim S, Moxham T, Oldridge N, Rees K, et al. Exercise-based cardiac rehabilitation for coronary heart disease. Cochrane Database Syst Rev 2011;(7): CD001800.
[28] Cornelissen VA, Verheyden B, Aubert AE, Fagard RH. Effects of aerobic training intensity on resting, exercise and post-exercise blood pressure, heart rate and heart-rate variability. J Hum Hypertens 2009;24(3):175–82.
[29] Bonanno JA, Lies JE. Effects of physical training on coronary risk factors. Am J Cardiol 1974;33:760–4.
[30] Dickinson HO, Mason JM, Nicolson DJ, Campbell F, Beyer FR, Cook JV, et al. Lifestyle interventions to reduce raised blood pressure: a systematic review of randomized controlled trials. J Hypertens 2006;24(2):215–33.
[31] Swardfager W, Herrmann N, Cornish S, Mazereeuw G, Marzolini S, Sham L, et al. Exercise intervention and inflammatory markers in coronary artery disease: a meta-analysis. Am Heart J 2012;163(4):666–676.e3.
[32] The Diabetes Prevention Program. Design and methods for a clinical trial in the prevention of type 2 diabetes. Diabetes Care 1999;22(4):623–34.
[33] Knowler WC, Barratt-Connor E, Fowler SE, Hamman RF, Lachin JM, Walker EA, Diabetes Prevention Program Research Group, et al. Reduction in the incidence of type 2 diabetes with lifestyle intervention or metformin. N Engl J Med 2002;346 (6):393–403.
[34] Diabetes Prevention Program Research Group, Knowler WC, Fowler SE, Hamman RF, Christophi CA, Hoffman HJ, et al. 10-year follow-up of diabetes incidence and weight loss in the Diabetes Prevention Program Outcomes Study. Lancet 2009;374 (9702):1677–86.
[35] Chudyk A, Petrella RJ. Effects of Exercise on cardiovascular risk factors in type 2 diabetes. Diabetes Care 2011;34 (5):1228–37.
[36] Blair SN, LaMonte MJ, Nichaman MZ. The evolution of physical activity recommendations: how much is enough? Am J Clin Nutr 2004;79(5):913S–20S.
[37] Prevention Ca. How much physical activity do adults need? December 1, 2011 [cited; Available from: <http://www.cdc.gov/physicalactivity/everyone/guidelines/adults.html>.
[38] Bouchard C, Blair S, Church T, Earnest C, Hagberg J, Häkkinen K, et al. Adverse metabolic response to regular exercise: is it a rare or common occurrence? PLoS One 2012;7(5): e37887.

CHAPTER

5

Resistance Training and Physical Exercise in Human Health

Bryan K. Smith and Erik Kirk

Department of Kinesiology and Health Education, Southern Illinois University Edwardsville, Edwardsville, IL, USA

RESISTANCE TRAINING IN HUMAN HEALTH

Evidence showing the relationship between aerobic exercise and improvements in human health is abundant. Aerobic exercise was the recommended mode of exercise in the original physical activity recommendations for developing and maintaining fitness in healthy adults published by the American College of Sports Medicine in 1978 [1]. Through subsequent updates, aerobic exercise continued to be the primary recommendation for improving human health [2,3]. Resistance training (RT) was briefly discussed in the original recommendation [1], but it was not a recommended mode of activity until the second update in 1990 [3]. The addition of RT to the second update stemmed from its effectiveness for the development and maintenance of fat-free mass (FFM). The rationale for the inclusion in the third update in 1998 [2] did not change. However, the current recommendations [4] move beyond the original rationale and emphasize the role of RT in improving cardiovascular risk factors and lowering all-cause mortality.

Resistance training involves the voluntary activation of specific skeletal muscles against some form of external resistance, which can be provided by a variety of modalities such as machines or free weights. Traditionally, the primary purpose of RT has been for increasing FFM and/or increasing muscular strength and endurance [5]. Although RT is the most effective way to increase FFM and muscular endurance, there is a growing body of evidence supporting the importance of RT in overall human health. There is now considerable evidence supporting improvements in cardiovascular risk factors [5–7] following properly designed RT programs.

There are direct relationships between RT and improvements in cardiovascular risk factors such as blood pressure (BP), blood lipids, glucose management, visceral adiposity, and weight management. Weight loss and the prevention of weight gain also have direct relationships with improvements in cardiovascular risk factors. Therefore, it is possible that RT can directly and indirectly (through weight management) improve many cardiovascular risk factors. This chapter will focus on the improvements in cardiovascular risk factors (BP, blood lipids, glucose management, visceral adiposity, and weight management) subsequent to RT. Weight management will be discussed in detail because weight loss and the prevention of weight regain are both associated with beneficial improvements in BP, blood lipids, and glucose management [6].

Blood Pressure

Several meta-analyses have shown the beneficial effect of RT on controlling BP [8–10]. In the most recent meta-analysis, Cornelissen et al. [9] examined randomized control trials that were ≥4 week in length. The final analysis included 1012 participants from 28 different trials. Overall, RT significantly reduced systolic and diastolic BP by 3.9 and 3.6 mmHg, respectively. When participants were grouped into prehypertensive, normohypertensive, and hypeternsive, RT significantly reduced systolic and diastolic BP by 3.9 and 3.9 mmHg in normotensive and prehypertensive study groups but did not significantly reduce systolic and diastolic BP in hypertensive groups (2.8 and 3.8 mmHg, respectively). It should be noted that although there were not significant

Nutrition and Enhanced Sports Performance.
DOI: http://dx.doi.org/10.1016/B978-0-12-396454-0.00005-9

reductions in BP in the hypertensive groups, there were only four studies in the analysis that had hypertensive subjects. Thus there is a need for further research with this particular population.

Although these reductions in BP are relatively moderate, they are clinically significant. Reductions in BP of only 3 mmHg have been associated with a 4% decrease in all-cause mortality, an 8–14% reduction in stroke, and a 5–9% reduction in cardiac morbidity [11]. Since there appears to be no lower threshold between BP and cardiovascular risk, even minor reductions can have clinical significance in individuals with optimal BP and especially for those that are prehypertensive [12].

Research has also shown that RT can prevent the development of hypertension. For example, a recent study by Maslow et al. [13] examined the role of muscular strength and the development of hypertension in men. Using data from the Aerobics Center Longitudinal Study, they compared muscular strength (1-repetition maximal bench press and leg press) to the incident rate of hypertension during the follow-up period (avg. 19 years). The results showed that men who were prehypertensive at baseline had a reduced risk of developing hypertension if they had middle or high levels of muscular strength. The authors reported that they were the first to demonstrate an inverse relationship between muscular strength and the development of hypertension.

Blood Lipids

Traditionally, aerobic exercise has been the recommended mode of exercise for reducing elevated blood lipid levels—total cholesterol (TC), low-density lipoproteins (LDL), and triglycerides (TG)—and for elevating low levels of high-density lipoproteins (HDL) [14]. The evidence supporting a positive role of RT on lipid modification has been questionable. One possible explanation for the lack of support for RT's role in lipid management may be that many recent reviews have been subjective reports rather than a more objective meta-analytic approach [15]. Several recent narrative reviews [7,16] reported no effect of RT on lipid management whereas a recent meta-analysis reported clinically significant changes in blood lipids following RT [8].

Kelley et al. [8] pooled data from 29 adult studies which included 1329 participants (~50% exercise/control). All of the included studies were randomized control trials that were a minimum of 4 weeks in duration. The results of this analysis showed significant reductions in TC (5.5 mg/dL, 2.7%), LDL (6.1 mg/dL, 4.6%), TC/HDL (0.5 mg/dL, 11.6%), non-HDL (8.7 mg/dL, 5.6%), and TG (8.1 mg/dL, 6.4%). There was an increase in HDL (0.7 mg/dL, 1.4%) but this increase was not statistically significant. The reduction in TC and non-HDL represent reductions in coronary heart disease risk of 5% [17,18]. The reduction in TG represents a 3% and 7% improvement in coronary heart disease risk in men and women, respectively [19]. The reduction in TC/HDL represents a 21% improvement in coronary heart disease risk in men [20].

It should be noted that all of the above changes were evident despite non-significant changes in body weight or body mass index (BMI). Despite no changes in body weight or BMI, there were significant improvements in percent body fat (−1.8%) and FFM (+1.0 kg), suggesting that a decrease in body fat and an increase in FFM may play an essential role in the management of blood lipids.

Glucose Management

Similar to aerobic exercise, RT has been reported to enhance insulin action [21]. RT has been shown to increase muscular strength and power [22] and FFM [23], which could improve glycemic control by augmenting the skeletal muscle storage of glucose and assist in the prevention of sarcopenia. These findings support the hypothesis that increases in skeletal muscle mass are related to decreases in hemoglobin A1C [24,25]. What is not clear, however, is whether the improvement in A1C is due to an increase in the qualitative aspects of muscle function and/or skeletal muscle size. Interestingly, research has indicated that improvements in insulin action can occur without a change in FFM [26], suggesting that qualitative changes in skeletal muscle function play a key role in the RT-induced improvements in insulin action. For example, Holten et al. [27] reported that, in individuals with type 2 diabetes who performed one-legged RT, insulin action improved independently of increased FFM. Even more interesting is that the insulin receptor, glycogen synthase, protein content of GLUT4, protein kinase B-α/β, and total glycogen synthase activity were up-regulated. Collectively, these findings suggest that RT improves the insulin signaling pathway, which may partly explain the improved insulin action with RT.

Visceral Adiposity

Obesity in general is associated with an increase in cardiovascular and metabolic disorders. However, evidence now supports the notion that location of excess adiposity plays a significant factor in the health risks associated with obesity. Specifically, visceral adipose tissue (VAT) has been shown to be an independent predictor of hypertension [28] and insulin resistance [29–31].

Both aerobic exercise and energy restriction resulting in a 4–10% decrease in body weight have been shown to significantly decrease VAT [32,33]. Weight loss of this magnitude is difficult to achieve and maintain [34,35] and requires very high exercise levels up to 420 min/wk [35–37]. Weight loss and exercise levels of these magnitudes are unrealistic for the vast majority of the population, thus there is a need for alternative strategies to decrease VAT.

Recent evidence suggests that exercise levels that are less than the current exercise recommendations and at lower energy expenditures can result in a change in VAT in the absence of weight loss [33,38]. When compared with aerobic exercise, RT exercise generally results in significantly lower energy expenditure and is not generally associated with weight loss. Thus, RT has the potential to reduce levels of VAT.

Several studies have examined changes in VAT following RT. In older adults (men and women), decreases in VAT of 10–11% have been reported [39,40]. There is evidence to suggest that RT may play a role in preventing the accumulation of VAT during weight regain. Hunter et al. [41] did a one year follow-up on pre-menopausal women following a diet-induced weight loss study. There was no significant increase in VAT in either the aerobic or RT group, but there was a 25% increase in the non-exercising control group. There was no significant difference between the aerobic or RT group despite the fact that the aerobic group had a much higher level of exercise energy expenditure. Schmitz et al. [42], using a similar design, did a 2 year follow-up on post-menopausal women. In the non-exercising control group, VAT increased by 21% versus only a 7% increase in VAT in the RT group. To date, there is no evidence in men that suggests RT can prevent the development of VAT during weight regain.

A recent meta-analysis by Ismail et al. [43] compared the role of aerobic and RT on VAT. A total of 35 studies were included and all were randomized control trials that included either aerobic exercise and/or RT program for a minimum of 4 weeks. When compared with a control group, RT did not result in any significant decrease in VAT. When compared with aerobic exercise, the results favored aerobic exercise, but there was no significant difference in the changes in VAT between the two modes of exercise.

As VAT also has been linked to insulin resistance [44], it is possible that RT-induced reduction in VAT improves insulin sensitivity. Research has shown that a diet-induced weight loss (15%) combined with RT resulted in a 37% reduction in VAT and increased insulin sensitivity by 49%. However, others have shown [39] that, although RT reduced VAT by 11.2% and improved insulin action by 46.3%, no significant relationship between the improvements in insulin sensitivity and the losses in VAT was found. Further research is warranted to determine the link between reduction in VAT and improvement in insulin action.

Weight Management

Research showing beneficial effects of RT on the musculoskeletal systems has led to recommendations that it be included in an overall fitness program for all adults [4]. However, what has not been emphasized by these recent public health guidelines is the research demonstrating the potential benefits of RT as a weight gain prevention strategy, a role generally ascribed to aerobic exercise training. RT may also play an important role in the prevention of weight gain. In fact, RT may elicit changes in fat mass (FM) of a magnitude similar to that induced by aerobic training while simultaneously increasing FFM. For example, in a meta-analysis of 53 papers published from the late 1960s through the mid-1980s on the effect of cycling, running/jogging, and RT programs on changes in body mass, FM and FFM, Ballor and Poehlman [45] noted that, in men, the reductions in FM and percent fat were not different between exercise modes. The report lacked statistical power to report any differences among the women. Resistance training may also play a role in promoting free-living physical activity [46,47]. Resistance training has been shown to improve strength, balance and locomotion [46]. If overweight and obese individuals increase strength in response to RT, increases in activities of daily living may occur. This in turn would contribute to total energy expenditure and may promote weight maintenance or weight loss.

Resistance training has been associated with increases in FFM and resting energy expenditure (REE). It has been demonstrated that RT of 12–16 weeks increases FFM on average by 1.7 kg with no changes in total body mass for both men and women [48,49]. This compares with no increases in FFM for either men or women when aerobic exercise is performed [50]. Thus, RT could potentially increase FFM, which normally does not occur with aerobic exercise. FFM is highly correlated with REE, which may lead to an increase in total energy expenditure. REE accounts for the largest component (60–75%) of total daily energy expenditure and therefore plays a significant role in the regulation of energy balance.

Resistance Training and Resting Energy Expenditure

An increase in REE can have a significant impact on total energy expenditure and the creation of a negative energy balance. Theoretically, even small changes in

REE may significantly affect the regulation of body weight and body composition. Results from intervention studies [51,52] have found increases in REE of 150 to 350 kcal/day after 8 to 16 weeks of RT in both younger and older individuals. The investigators attributed the increase in REE to a decrease in FM and increase in FFM. However, several investigators have reported increases in REE in the absence of increases in FFM in normal weight individuals. Pratley and colleagues [53] examined changes in REE following 16 weeks of RT in 13 normal weight men. REE increased by ~120 kcal/day. Although FFM increased during the training program (+1.6 kg), the increase in REE persisted after the authors controlled for changes in FFM. This finding suggests that REE may be increased even in the absence of increased FFM. The investigators speculated that the increase in REE may have been related to the increase in plasma levels of norepinephrine. Similar results have been found in overweight and obese individuals. For example, Byrne and Wilmore [54] reported that in 20 sedentary, moderately obese women (age 38 years), there was a significant increase in REE (44 kcal/day) following RT for 20 weeks, 4 days per week. In addition, most RT protocols are of high intensity and are composed of at least three sets of 6–8 RM, performed 3–4 days per week, involving 12–14 different exercises. The higher intensity may not induce a significantly greater difference than a more moderate RT protocol involving only one set and may lead to poor compliance in obese individuals [55]. Therefore, exercise modalities such as RT that increase compliance and adherence could be advantageous in preventing weight gain.

Resistance Training and 24-hour Energy Expenditure

The accelerated increase in the development of overweight and obesity, combined with the difficulty in treating these conditions, suggests that innovative strategies for obesity prevention need to be developed and evaluated. Resistance exercise training offers an innovative and time-efficient approach to obesity prevention that differs in concept from that of traditional aerobic exercise. Aerobic exercise results in significant increases in energy expenditure during, and for a short time following, cessation of the activity [56,57]. Although the energy expenditure of RT is relatively low during the activity [58,59], the accumulated energy expenditure across 24 hours may be substantial enough to aid in weight management [47,60,61]. Additionally, the increased energy expenditure associated with aerobic exercise may increase energy intake and decrease daily physical activity energy expenditure [62] whereas the minimal energy expenditure associated with RT is unlikely to alter energy intake. RT also has great potential to increase functional ability, which may increase daily physical activity levels, leading to an increase in 24-h EE beyond increases in REE alone [46,47].

Resistance Training and Daily Physical Activity

Unlike aerobic exercise, which may actually lead to a decrease in daily physical activity [62], RT has the potential to increase daily physical activity through an improved ability to complete activities of daily living (ADL). With aging comes a decrease in the ability to complete ADL [63–65]. Although improvements in general physical fitness can attenuate this decline, the decline will occur regardless of overall physical fitness levels [63–65]. One possible explanation for the decrease in ADL is a decrease in muscular strength, which also is a side effect of aging [66]. If muscular strength can be maintained or increased, it is possible that ADL can be maintained and possibly increased. Although ADL are generally very short in duration, when summed up over the course of a day, they have the potential to considerably increase 24-h EE, which can play a significant role in weight management.

The decrease in muscular strength with aging is accompanied by a decrease in cardiovascular fitness which also will contribute to a decrease in ADL. It could be argued that aerobic exercise and RT both have the potential to increase ADL. When one considers that most ADL are short in duration and of a very low intensity, it becomes apparent that muscular strength may play a greater role in completing ADL than does cardiovascular fitness. There is considerable evidence in older adults showing that ADL are improved following RT [46,67–70]. Currently, there is little if any evidence to support this relationship in young adults. Considering the decline in levels of physical fitness and the increasing rates of obesity, it is easy to speculate that an increase in muscular strength in physically unfit, overweight/obese young and middle age adults would result in less fatigue when doing ADL. As fatigue from doing ADL decreases, there is the potential to increase daily physical activity, which would lead to an increase in 24-h EE.

There is considerable evidence that shows an increase in physical function measures such as 6-meter walks, repeated chair stands, stair climbs, and walking speed following RT [71–73]. In theory, similar to ADL, as physical function improves and the fatigue associated with these simple tasks decreases, there is greater opportunity for individuals to be more active over the course of the day. Although these are simple tasks of very low intensity, the accumulated daily energy expenditure associated with these tasks can contribute to an increase in 24-h EE. To date, no research has been published that examines the relationship between increased ADL and/or physical function and changes in 24-h EE subsequent to RT.

Resistance Training and Fat Free Mass

Results of recent investigations on RT and body composition, ranging in duration from 9 to 26 weeks (mean duration = 15 weeks), have shown an average increase in FFM of approximately 1.7 kg (range 0.7 to 3.6 kg), with no change in total body mass [47,48,50,60,61,74]. It appears that these changes may occur in both younger [60,75–78] and older individuals [47,48,50,61,74,79], men [60,61,75,80] as well as women [54,61,78], and in both normal weight ($BMI \leq 25\ kg/m^2$) [47,48, 60,61,75,78] and overweight individuals [53,74,76,81]. The changes in FFM should be interpreted with caution, as FFM was assessed by imprecise methods. For example, most investigators report changes in FFM assessed by using skinfolds or hydrostatic weighing, which may not accurately assess FFM compared with other modalities such as Dual Energy X-ray Absorptiometry (DXA). DXA may reduce the error in measurement that is normally associated with hydrostatic weighing and could provide different results [82].

The importance of an increase in FFM is its positive association with REE ($r > 0.85$) [83]. For example, Bosselaers and colleagues examined REE in 10 RT men and women and 10 healthy controls matched for age, sex and percent body fat in a cross-sectional study [84]. They found a higher REE in the RT individuals compared with untrained controls (+354 kcal/day). This difference was eliminated after the researchers adjusted REE for differences in FFM assessed by hydrostatic weighing (RT, 69 ± 3.4 kg vs Controls, 60.0 ± 2.6 kg). This finding suggests that the elevated REE in RT subjects is attributable to their greater quantity of FFM. Increases in FFM as a result of RT may provide enough of an increase in REE to prevent weight gain in young adults.

Resistance Training Without Energy Restriction and Body Weight

RT without diet does not result in decreased body weight. Short-term studies of less than 6 months [85–88] and long-term studies of at least 6 months or greater [26,47,67,86,89–92] in women [48,78,87,91,93] and men [53,74,81,85,87,94–96], and younger [60,76,78,85,86,93] and older [53,74,81,94,96,97] participants generally show no significant changes in BMI or body weight, although they generally show changes for body composition, i.e., FFM increases and FM decreases. For example, Treuth et al. [81] compared two groups of older men. One group completed a 16 wk RT protocol (age 60 y) and the other group acted as a control group (age 62 y). Body weight and BMI did not change in either group. The RT resulted in significant decreases in percent body fat (1.8%) and FM (1.7 kg) and a significant increase in FFM (1.7 kg). In the control group, percent body fat increased by 1.4 %, FM increased by 1.4 kg, and FFM decreased by 0.4 kg, all of which were not significant. Cullinen and Caldwell [98] have reported similar results in young women (mean age = 26 y). After 12 weeks of RT, body weight and BMI had not changed in either the RT or the control group. In the RT group there were significant decreases in percent body fat (2.6%) and FM (1.6 kg), and a significant increase in FFM (2.0 kg). There were no significant changes in the control group. Results of the aforementioned studies suggest the potential for using RT for the prevention of weight gain.

Resistance Training During Energy Restriction

Resistance training may help maintain FFM and REE during diet-induced weight loss [99]. During a very low calorie diet (VLCD) of 500–800 kcal/day, significant weight loss occurs. Dieting, without exercise, results in the loss of both body fat and a substantial quantity of FFM. The result of a significant loss of FFM is a potential reduction in REE [45].

Participating in RT during caloric restriction may maintain or even increase FFM while having a similar reduction in body weight to that with diet alone. A number of studies support this claim [99–101]. Ballor et al. [99] examined the effects of diet alone (1340 kcal/d) or diet combined with RT on body weight in obese women. After 8 weeks there were no significant differences in the amount of weight lost between the diet-only group (−4.47 kg) and diet plus RT group (−3.89 kg). However, FFM increased in the diet plus RT group (1.07 kg) compared with the diet-only group (−0.91 kg). It was concluded that adding RT to a caloric restriction program results in comparable reductions in body weight and maintenance of FFM compared with diet only. In a similar study by Geliebter et al. [100] a moderate caloric deficit (1248 kJ/d) alone or combined with moderate RT resulted in no significant difference in the average amount of weight loss between the diet-only group (−9.5 kg) compared with a diet plus RT group (−7.9 kg) in obese men and women. However, the diet-only group lost a significantly greater amount of FFM (−2.7 kg) compared with the RT plus diet group (−1.1 kg), suggesting that RT may attenuate the loss in FFM on a moderate restricted diet compared with diet alone, with no significant difference in the amount of weight lost. These investigations are limited by the short duration (less than 6 months) and use of moderate restricted diet, whose effects may differ from those of a more severe caloric restricted diet.

Resistance Training Versus Aerobic Exercise During Energy Restriction

There is debate as to whether RT or aerobic exercise during a VLCD is more effective in attenuating the loss of FFM and promoting FM reduction. It would be desirable to attenuate the loss of FFM and promote FM reduction in order to minimize the loss of tissue with higher energy restriction and removing excess body fat that is associated with obesity and increased disease risk [102]. Bryner et al. [103] examined the effects on FFM in 20 participants randomized to an 800-calorie per day diet plus either resistance or aerobic training. The aerobic exercise group lost a significant amount of FFM, but the RT group maintained FFM following the 12 weeks of training. Consequently, REE decreased in the aerobic exercise group but increased in the RT group. These results suggest that RT may assist in preserving FFM during energy restriction, which could assist in the management of body weight. For example, Pronk et al. [104] compared the effects a VLCD (520 kcal/day) combined with either aerobic exercise or RT. Aerobic exercise was performed for 4 d/wk at 75% of heart rate reserve for 45 min each session, while RT individuals performed three sets of seven repetitions at 75% of 1-RM on eight exercises for 4 d/wk. After 12 weeks there was no difference in amounts of weight or FFM loss between groups, suggesting that RT is no more effective at retaining FFM and promoting weight loss than aerobic exercise in individuals on a VLCD. However, most investigations are of short duration, usually only 8–12 weeks, which may affect the outcomes of RT on body composition. In one of the few long-term studies (48 weeks) Wadden et al. [105] examined the effects of RT or aerobic exercise combined with diets ranging from 900 to 1250 kcal/d on body weight and FFM. They found that RT did not result in significantly better preservation of FFM than did diet alone. Similar findings were reported by Donnelly et al. [106] using a VLCD combined with either aerobic training performed for 4 d/wk, 45 min, at 70% of heart rate reserve versus RT 4 d/wk, three sets of seven repetitions at 75% of 1-RM in obese women. At the conclusion of 12 weeks of training there was no difference in the amount of weight loss (− 16.6 kg vs −16.1 kg), FFM (− 4.8 kg vs −4.7 kg) or FM (− 9.0 kg vs −9.3 kg) for the aerobic trained group compared with the RT group. Thus, the evidence suggests that RT may not moderate the declines in FFM or induce greater weight loss than aerobic exercise alone in investigations lasting between 12 and 48 weeks when combined with energy restriction.

The Role of Resistance Training Versus Aerobic Exercise and Weight Maintenance

There appear to be successful strategies for maintenance of weight loss that may be essential to long-term weight management success. In addition to behavioral predictors, such as cognitive control of overeating, increased physical activity has been associated with successful weight maintenance in an observational investigation [107]. Increased physical activity is often advocated as a solution to improve weight maintenance [108–110]. However, in controlled intervention trials with prescribed exercise training, the beneficial effects of physical activity on long-term weight maintenance have not been demonstrated unequivocally [107]. Nevertheless, high levels of physical activity were common among individuals in the National Weight Control Registry cohort, a group of individuals who maintained at least a 30 pound weight loss for at least 1 year. Twenty-four percent of men and 20% of women in this cohort reported regular weight-lifting exercises, a much higher percentage than expected in the general population (20% men, 9% women) [111]. The reality of the difficulty in the maintenance of body weight provides a strong rationale for the development and evaluation of strategies to prevent or slow the rate of weight gain.

Despite the potential benefits of RT on weight maintenance, few investigations have been performed to elucidate the role of RT on weight maintenance following weight loss. If RT is effective in maintaining body weight following weight loss it may also be effective as a weight maintenance strategy. For example, Ballor et al. evaluated the effects of 12 weeks of aerobic exercise versus RT in obese men and women following a 10% weight loss [50]. Individuals were randomized to either an aerobic exercise or RT group 3 days per week. Aerobic exercise was performed at 50% of VO_{2max} for 60 min each session, while RT individuals performed three sets of eight repetitions at 65% of 1-RM on seven exercises. After 12 weeks following the 10% reduction in initial body weight the aerobic training group lost more weight (− 2.5 ± 3 kg) than the RT group (+ 0.5 ± 2 kg). The increase in body weight for the RT group may be attributable to the increase in FFM (+ 1.5 kg) compared with a decrease in FFM in the aerobic trained group (− 0.6 kg). Interestingly, REE did not change significantly in the aerobic exercise group (− 12 kcal/day) but increased in the RT group (+ 79 kcal/day), which over the long-term may further attenuate the increase in body weight. In a longer study of 24 weeks following a 13% weight loss in obese men, Borg et al. evaluated the effects of aerobic exercise or RT combined with dietary counseling compared with a control group who received dietary counseling only [112]. Individuals performed either aerobic exercise or RT 3 days per week. Aerobic exercise was performed at 65% of VO_{2max} for 45 min each session, while the RT group performed three sets of eight repetitions at 70% of 1-RM. After 24 weeks, neither aerobic exercise nor RT improved weight maintenance when compared with the control group. However, both aerobic exercise and RT attenuated the regain in body weight. It was concluded that the poor adherence in the aerobic trained group may have contributed to the lack of weight maintenance compared with the control group. Thus, RT may be an important therapeutic method of weight maintenance by increasing exercise adherence compared with aerobic exercise. However, both of the aforementioned investigations did not supervise all of the exercise sessions, but relied on self-report, making it difficult to determine if the individuals performed all of the prescribed exercise. These results may further substantiate the significant role that RT may have on preventing weight gain. However, although these studies involve the effects of RT on weight maintenance following weight loss, collectively they suggest that RT could be potentially effective in the prevention of weight gain in young overweight adults, due to the success of preventing weight gain following weight loss.

SUMMARY

Traditionally, RT has been recommended for building and maintaining FFM and initially was not considered an important mode of physical activity for improving or maintaining human health. As new evidence emerged, it became apparent that RT may be more important to human health beyond building and maintaining FFM. Although aerobic exercise is still considered the most important mode of physical activity for improving human health, it is quite apparent that RT can be a valuable mode of physical activity for improving health as well. The latest physical activity recommendations emphasize the importance of RT in building strength and FFM as well as the role of RT in improving numerous cardiovascular risk factors. RT is a growing field of research, and one can speculate that RT's role in human health will continue to expand as it has since the original physical activity recommendations were published in 1978.

References

[1] American College of Sports Medicine. Position Statement on the recommended quantity and quality of exercise for developing and maintaining fitness in healthy adults. Med Sci Sports Exerc 1978;10(3):vii–x.

[2] American College of Sports Medicine. American college of sports medicine position stand. The recommended quantity and quality of exercise for developing and maintaining cardiorespiratory and muscular fitness, and flexibility in healthy adults. Med Sci Sports Exerc 1998;30(6):975–91.

[3] American College of Sports Medicine. American college of sports medicine position stand. The recommended quantity and quality of exercise for developing and maintaining cardiorespiratory and muscular fitness in healthy adults. Med Sci Sports Exerc 1990;22(2):265–74.

[4] Garber CE, Blissmer B, Deschenes MR, Franklin BA, Lamonte MJ, Lee IM, et al. American college of sports medicine position stand. Quantity and quality of exercise for developing and maintaining cardiorespiratory, musculoskeletal, and neuromotor fitness in apparently healthy adults: guidance for prescribing exercise. Med Sci Sports Exerc 2011;43(7):1334–59.

[5] Hass CJ, Feigenbaum MS, Franklin BA. Prescription of resistance training for healthy populations. Sports Med 2001;31 (14):953–64.

[6] National Heart, Lung, and Blood Institute. Clinical guidelines on the identification, evaluation, and treatment of overweight and obesity in adults; The evidence report. National Institutes of Health; 1998.

[7] Williams MA, Haskell WL, Ades PA, Amsterdam EA, Bittner V, Franklin BA, et al. Resistance exercise in individuals with and without cardiovascular disease: 2007 update: a scientific statement from the American heart association council on clinical cardiology and council on nutrition, physical activity, and metabolism. Circulation 2007;116(5):572–84.

[8] Kelley GA, Kelley KS. Impact of progressive resistance training on lipids and lipoproteins in adults: a meta-analysis of randomized controlled trials. Prev Med 2009;48(1):9–19.

[9] Cornelissen VA, Fagard RH, Coeckelberghs E, Vanhees L. Impact of resistance training on blood pressure and other cardiovascular risk factors: a meta-analysis of randomized, controlled trials. Hypertension 2011;58(5):950–8.

[10] Cornelissen VA, Fagard RH. Effect of resistance training on resting blood pressure: a meta-analysis of randomized controlled trials. J Hypertens 2005;23(2):251–9.

[11] Whelton SP, Chin A, Xin X, He J. Effect of aerobic exercise on blood pressure: a meta-analysis of randomized, controlled trials. Ann Intern Med 2002;136(7):493–503.

[12] Chobanian AV, Bakris GL, Black HR, Cushman WC, Green LA, Izzo Jr. JL, et al. Seventh report of the joint National committee on prevention, detection, evaluation, and treatment of high blood pressure. Hypertension 2003;42(6):1206–52.

[13] Maslow AL, Sui X, Colabianchi N, Hussey J, Blair SN. Muscular strength and incident hypertension in normotensive and prehypertensive men. Med Sci Sports Exerc 2010;42(2):288–95.

[14] Rosamond W, Flegal K, Furie K, Go A, Greenlund K, Haase N, et al. Heart disease and stroke statistics—2008 update: a report from the American heart association statistics committee and stroke statistics subcommittee. Circulation 2008;117(4):e25–146.

[15] Sacks HS, Berrier J, Reitman D, Ancona-Berk VA, Chalmers TC. Meta-analyses of randomized controlled trials. N Engl J Med 1987;316(8):450–5.

[16] Braith RW, Stewart KJ. Resistance exercise training: its role in the prevention of cardiovascular disease. Circulation 2006;113 (22):2642–50.

[17] Secondary prevention by raising HDL cholesterol and reducing triglycerides in patients with coronary artery disease: the Bezafibrate Infarction Prevention (BIP) study. Circulation 2000;102(1):21–7.

[18] Consensus conference. Lowering blood cholesterol to prevent heart disease. JAMA 1985;253(14):2080–6.

[19] Hokanson JE, Austin MA. Plasma triglyceride level is a risk factor for cardiovascular disease independent of high-density lipoprotein cholesterol level: a meta-analysis of population-based prospective studies. J Cardiovasc Risk 1996;3(2):213–9.

[20] Kinosian B, Glick H, Preiss L, Puder KL. Cholesterol and coronary heart disease: predicting risks in men by changes in levels and ratios. J Investig Med 1995;43(5):443–50.

[21] Andersen JL, Schjerling P, Andersen LL, Dela F. Resistance training and insulin action in humans: effects of de-training. J Physiol 2003;551(Pt 3):1049–58.
[22] Kirk EP, Washburn RA, Bailey BW, LeCheminant JD, Donnelly JE. Six months of supervised high-intensity low-volume resistance training improves strength independent of changes in muscle mass in young overweight men. J Strength Cond Res 2007;21(1):151–6.
[23] Washburn RA, Donnelly JE, Smith BK, Sullivan DK, Marquis J, Herrmann SD. Resistance training volume, energy balance and weight management: rationale and design of a 9 month trial. Contemp Clin Trials 2012;33(4):749–58.
[24] Fenicchia LM, Kanaley JA, Azevedo Jr. JL, Miller CS, Weinstock RS, Carhart RL, et al. Influence of resistance exercise training on glucose control in women with type 2 diabetes. Metabolism 2004;53(3):284–9.
[25] Baldi JC, Snowling N. Resistance training improves glycaemic control in obese type 2 diabetic men. Int J Sports Med 2003;24 (6):419–23.
[26] Hunter GR, Bryan DR, Wetzstein CJ, Zuckerman PA, Bamman MM. Resistance training and intra-abdominal adipose tissue in older men and women. Med Sci Sports Exerc 2002;34 (6):1023–8.
[27] Holten MK, Zacho M, Gaster M, Juel C, Wojtaszewski JF, Dela F. Strength training increases insulin-mediated glucose uptake, GLUT4 content, and insulin signaling in skeletal muscle in patients with type 2 diabetes. Diabetes 2004;53(2):294–305.
[28] Rheaume C, Arsenault BJ, Belanger S, Perusse L, Tremblay A, Bouchard C, et al. Low cardiorespiratory fitness levels and elevated blood pressure: what is the contribution of visceral adiposity? Hypertension 2009;54(1):91–7.
[29] Tulloch-Reid MK, Hanson RL, Sebring NG, Reynolds JC, Premkumar A, Genovese DJ, et al. Both subcutaneous and visceral adipose tissue correlate highly with insulin resistance in African Americans. Obes Res 2004;12(8):1352–9.
[30] Nakamura T, Tokunaga K, Shimomura I, Nishida M, Yoshida S, Kotani K, et al. Contribution of visceral fat accumulation to the development of coronary artery disease in non-obese men. Atherosclerosis 1994;107(2):239–46.
[31] Fujioka S, Matsuzawa Y, Tokunaga K, Tarui S. Contribution of intra-abdominal fat accumulation to the impairment of glucose and lipid metabolism in human obesity. Metabolism 1987;36 (1):54–9.
[32] Ross R, Rissanen J. Mobilization of visceral and subcutaneous adipose tissue in response to energy restriction and exercise. Am J Clin Nutr 1994;60(5):695–703.
[33] Kay SJ, Fiatarone Singh MA. The influence of physical activity on abdominal fat: a systematic review of the literature. Obes Rev 2006;7(2):183–200.
[34] Hansen D, Dendale P, Berger J, van Loon LJ, Meeusen R. The effects of exercise training on fat-mass loss in obese patients during energy intake restriction. Sports Med 2007;37(1):31–46.
[35] Franz MJ, VanWormer JJ, Crain AL, Boucher JL, Histon T, Caplan W, et al. Weight-loss outcomes: a systematic review and meta-analysis of weight-loss clinical trials with a minimum 1-year follow-up. J Am Diet Assoc 2007;107(10):1755–67.
[36] Donnelly JE, Blair SN, Jakicic JM, Manore MM, Rankin JW, Smith BK. American college of sports medicine position stand. Appropriate physical activity intervention strategies for weight loss and prevention of weight regain for adults. Med Sci Sports Exerc 2009;41(2):459–71.
[37] Knowler WC, Barrett-Connor E, Fowler SE, Hamman RF, Lachin JM, Walker EA, et al. Reduction in the incidence of type 2 diabetes with lifestyle intervention or metformin. N Engl J Med 2002;346(6):393–403.
[38] Johnson NA, Sachinwalla T, Walton DW, Smith K, Armstrong A, Thompson MW, et al. Aerobic exercise training reduces hepatic and visceral lipids in obese individuals without weight loss. Hepatology 2009;50(4):1105–12.
[39] Ibanez J, Izquierdo M, Arguelles I, Forga L, Larrion JL, Garcia-Unciti M, et al. Twice-weekly progressive resistance training decreases abdominal fat and improves insulin sensitivity in older men with type 2 diabetes. Diabetes Care 2005;28(3):662–7.
[40] Treuth MS, Hunter GR, Kekes-Szabo T, Weinsier RL, Goran MI, Berland L. Reduction in intra-abdominal adipose tissue after strength training in older women. J Appl Physiol 1995;78 (4):1425–31.
[41] Hunter GR, Brock DW, Byrne NM, Chandler-Laney PC, Del Corral P, Gower BA. Exercise training prevents regain of visceral fat for 1 year following weight loss. Obesity 2010;18 (4):690–5.
[42] Schmitz KH, Hannan PJ, Stovitz SD, Bryan CJ, Warren M, Jensen MD. Strength training and adiposity in premenopausal women: strong, healthy, and empowered study. Am J Clin Nutr 2007;86(3):566–72.
[43] Ismail I, Keating SE, Baker MK, Johnson NA. A systematic review and meta-analysis of the effect of aerobic vs. resistance exercise training on visceral fat. Obes Rev 2012;13(1):68–91.
[44] Virtanen KA, Iozzo P, Hallsten K, Huupponen R, Parkkola R, Janatuinen T, et al. Increased fat mass compensates for insulin resistance in abdominal obesity and type 2 diabetes: a positron-emitting tomography study. Diabetes 2005;54(9):2720–6.
[45] Ballor DL, Poehlman ET. A meta-analysis of the effects of exercise and/or dietary restriction on resting metabolic rate. Eur J Appl Physiol Occup Physiol 1995;71(6):535–42.
[46] Fiatarone MA, O'Neill EF, Ryan ND, Clements KM, Solares GR, Nelson ME, et al. Exercise training and nutritional supplementation for physical frailty in very elderly people. N Engl J Med 1994;330(25):1769–75.
[47] Hunter GR, Wetzstein CJ, Fields DA, Brown A, Bamman MM. Resistance training increases total energy expenditure and free-living physical activity in older adults. J Appl Physiol 2000;89 (3):977–84.
[48] Treuth MS, Hunter GR, Weinsier RL, Kell SH. Energy expenditure and substrate utilization in older women after strength training: 24-h calorimeter results. J Appl Physiol 1995;78 (6):2140–6.
[49] Katch FI, Drumm SS. Effects of different modes of strength training on body composition and anthropometry. Clin Sports Med 1986;5(3):413–59.
[50] Ballor DL, Harvey-Berino JR, Ades PA, Cryan J, Calles-Escandon J. Contrasting effects of resistance and aerobic training on body composition and metabolism after diet-induced weight loss. Metabolism 1996;45(2):179–83.
[51] Kraemer WJ, Volek JS, Clark KL, Gordon SE, Incledon T, Puhl SM, et al. Physiological adaptations to a weight-loss dietary regimen and exercise programs in women. J Appl Physiol 1997;83 (1):270–9.
[52] Roth SM, Martel GF, Ivey FM, Lemmer JT, Tracy BL, Hurlbut DE, et al. Ultrastructural muscle damage in young vs. older men after high-volume, heavy-resistance strength training. J Appl Physiol 1999;86(6):1833–40.
[53] Pratley R, Nicklas B, Rubin M, Miller J, Smith A, Smith M, et al. Strength training increases resting metabolic rate and norepinephrine levels in healthy 50- to 65-yr-old men. J Appl Physiol 1994;76(1):133–7.
[54] Byrne HK, Wilmore JH. The effects of a 20-week exercise training program on resting metabolic rate in previously sedentary, moderately obese women. Int J Sport Nutr Exerc Metab 2001;11 (1):15–31.

[55] Feigenbaum MS, Pollock ML. Prescription of resistance training for health and disease. Med Sci Sports Exerc 1999;31(1):38–45.
[56] Goran MI, Poehlman ET. Endurance training does not enhance total energy expenditure in healthy elderly persons. Am J Physiol 1992;263(5 Pt 1):E950–7.
[57] Ezell DM, Geiselman PJ, Anderson AM, Dowdy ML, Womble LG, Greenway FL, et al. Substrate oxidation and availability during acute exercise in non-obese, obese, and post-obese sedentary females. Int J Obes Relat Metab Disord 1999;23 (10):1047–56.
[58] Phillips WT, Ziuraitis JR. Energy cost of the ACSM single-set resistance training protocol. J Strength Cond Res 2003;17(2):350–5.
[59] Melby C, Scholl C, Edwards G, Bullough R. Effect of acute resistance exercise on postexercise energy expenditure and resting metabolic rate. J Appl Physiol 1993;75(4):1847–53.
[60] Dolezal BA, Potteiger JA. Concurrent resistance and endurance training influence basal metabolic rate in nondieting individuals. J Appl Physiol 1998;85(2):695–700.
[61] Lemmer JT, Ivey FM, Ryan AS, Martel GF, Hurlbut DE, Metter JE, et al. Effect of strength training on resting metabolic rate and physical activity: age and gender comparisons. Med Sci Sports Exerc 2001;33(4):532–41.
[62] Meijer EP, Westerterp KR, Verstappen FT. Effect of exercise training on physical activity and substrate utilization in the elderly. Int J Sports Med 2000;21(7):499–504.
[63] Schulz R, Curnow C. Peak performance and age among superathletes: track and field, swimming, baseball, tennis, and golf. J Gerontol. 1988;43(5):P113–20.
[64] Schultz AB. Muscle function and mobility biomechanics in the elderly: an overview of some recent research. J Gerontol A Biol Sci Med Sci 1995;(50 Spec No):60–3.
[65] Klitgaard H, Mantoni M, Schiaffino S, Ausoni S, Gorza L, Laurent-Winter C, et al. Function, morphology and protein expression of ageing skeletal muscle: a cross-sectional study of elderly men with different training backgrounds. Acta Physiologica Scandinavica 1990;140(1):41–54.
[66] Lexell J. Human aging, muscle mass, and fiber type composition. J Gerontol A Biol Sci Med Sci 1995;(50 Spec No):11–6.
[67] Hunter GR, Wetzstein CJ, McLafferty Jr. CL, Zuckerman PA, Landers KA, Bamman MM. High-resistance versus variable-resistance training in older adults. Med Sci Sports Exerc 2001;33 (10):1759–64.
[68] Sauvage Jr. LR, Myklebust BM, Crow-Pan J, Novak S, Millington P, Hoffman MD, et al. A clinical trial of strengthening and aerobic exercise to improve gait and balance in elderly male nursing home residents. Am J Phys Med Rehabil 1992;71 (6):333–42.
[69] Parker ND, Hunter GR, Treuth MS, Kekes-Szabo T, Kell SH, Weinsier R, et al. Effects of strength training on cardiovascular responses during a submaximal walk and a weight-loaded walking test in older females. J Cardiopulm Rehabil 1996;16 (1):56–62.
[70] Fiatarone MA, Marks EC, Ryan ND, Meredith CN, Lipsitz LA, Evans WJ. High-intensity strength training in nonagenarians. Effects on skeletal muscle. JAMA 1990;263(22):3029–34.
[71] Miszko TA, Cress ME, Slade JM, Covey CJ, Agrawal SK, Doerr CE. Effect of strength and power training on physical function in community-dwelling older adults. J Gerontol A Biol Sci Med Sci 2003;58(2):171–5.
[72] Hruda KV, Hicks AL, McCartney N. Training for muscle power in older adults: effects on functional abilities. Can J Appl Physiol 2003;28(2):178–89.
[73] Galvao DA, Taaffe DR. Resistance exercise dosage in older adults: single- versus multiset effects on physical performance and body composition. J Am Geriatr Soc 2005;53(12):2090–7.
[74] Miller JP, Pratley RE, Goldberg AP, Gordon P, Rubin M, Treuth MS, et al. Strength training increases insulin action in healthy 50- to 65-yr-old men. J Appl Physiol 1994;77(3):1122–7.
[75] Van Etten LM, Westerterp KR, Verstappen FT. Effect of weight-training on energy expenditure and substrate utilization during sleep. Med Sci Sports Exerc 1995;27(2):188–93.
[76] Broeder CE, Burrhus KA, Svanevik LS, Volpe J, Wilmore JH. Assessing body composition before and after resistance or endurance training. Med Sci Sports Exerc 1997;29(5):705–12.
[77] Kraemer WJ, Keuning M, Ratamess NA, Volek JS, McCormick M, Bush JA, et al. Resistance training combined with bench-step aerobics enhances women's health profile. Med Sci Sports Exerc 2001;33(2):259–69.
[78] LeMura LM, von Duvillard SP, Andreacci J, Klebez JM, Chelland SA, Russo J. Lipid and lipoprotein profiles, cardiovascular fitness, body composition, and diet during and after resistance, aerobic and combination training in young women. Eur J Appl Physiol 2000;82(5–6):451–8.
[79] Hunter GR, Weinsier RL, Bamman MM, Larson DE. A role for high intensity exercise on energy balance and weight control. Int J Obes Relat Metab Disord 1998;22(6):489–93.
[80] Broeder CE, Burrhus KA, Svanevik LS, Wilmore JH. The effects of either high-intensity resistance or endurance training on resting metabolic rate. Am J Clin Nutr 1992;55(4):802–10.
[81] Treuth MS, Ryan AS, Pratley RE, Rubin MA, Miller JP, Nicklas BJ, et al. Effects of strength training on total and regional body composition in older men. J Appl Physiol 1994;77(2):614–20.
[82] Tataranni PA, Ravussin E. Use of dual-energy X-ray absorptiometry in obese individuals. Am J Clin Nutr 1995;62(4):730–4.
[83] Sparti A, DeLany JP, de la Bretonne JA, Sander GE, Bray GA. Relationship between resting metabolic rate and the composition of the fat-free mass. Metabolism 1997;46(10):1225–30.
[84] Bosselaers I, Buemann B, Victor OJ, Astrup A. Twenty-four-hour energy expenditure and substrate utilization in body builders. Am J Clin Nutr 1994;59(1):10–2.
[85] Mazzetti SA, Kraemer WJ, Volek JS, Duncan ND, Ratamess NA, Gomez AL, et al. The influence of direct supervision of resistance training on strength performance. Med Sci Sports Exerc 2000;32(6):1175–84.
[86] Kokkinos PF, Hurley BF, Vaccaro P, Patterson JC, Gardner LB, Ostrove SM, et al. Effects of low- and high-repetition resistive training on lipoprotein-lipid profiles. Med Sci Sports Exerc 1988;20(1):50–4.
[87] Staron RS, Karapondo DL, Kraemer WJ, Fry AC, Gordon SE, Falkel JE, et al. Skeletal muscle adaptations during early phase of heavy-resistance training in men and women. J Appl Physiol 1994;76(3):1247–55.
[88] Hass CJ, Garzarella L, de Hoyos D, Pollock ML. Single versus multiple sets in long-term recreational weightlifters. Med Sci Sports Exerc 2000;32(1):235–42.
[89] Gettman LR, Culter LA, Strathman TA. Physiologic changes after 20 weeks of isotonic vs isokinetic circuit training. J Sports Med Phys Fitness 1980;20(3):265–74.
[90] Taaffe DR, Pruitt L, Pyka G, Guido D, Marcus R. Comparative effects of high- and low-intensity resistance training on thigh muscle strength, fiber area, and tissue composition in elderly women. Clin Physiol 1996;16(4):381–92.
[91] Nelson ME, Fiatarone MA, Layne JE, Trice I, Economos CD, Fielding RA, et al. Analysis of body-composition techniques and models for detecting change in soft tissue with strength training. Am J Clin Nutr 1996;63(5):678–86.
[92] Van Etten LM, Westerterp KR, Verstappen FT, Boon BJ, Saris WH. Effect of an 18-wk weight-training program on energy expenditure and physical activity. J Appl Physiol 1997;82 (1):298–304.

[93] Keeler LK, Finkelstein LH, Miller W, Fernhall B. Early-phase adaptations of traditional-speed vs. superslow resistance training on strength and aerobic capacity in sedentary individuals. J Strength Cond Res 2001;15(3):309–14.
[94] Gettman LR, Ayres JJ, Pollock ML, Durstine JL, Grantham W. Physiologic effects on adult men of circuit strength training and jogging. Arch Phys Med Rehabil 1979;60(3):115–20.
[95] Van Etten LM, Verstappen FT, Westerterp KR. Effect of body build on weight-training-induced adaptations in body composition and muscular strength. Med Sci Sports Exerc 1994;26 (4):515–21.
[96] Hagerman FC, Walsh SJ, Staron RS, Hikida RS, Gilders RM, Murray TF, et al. Effects of high-intensity resistance training on untrained older men. I. Strength, cardiovascular, and metabolic responses. J Gerontol A Biol Sci Med Sci 2000;55(7): B336–46.
[97] Hurley BF, Seals DR, Ehsani AA, Cartier LJ, Dalsky GP, Hagberg JM, et al. Effects of high-intensity strength training on cardiovascular function. Med Sci Sports Exerc 1984;16 (5):483–8.
[98] Cullinen K, Caldwell M. Weight training increases fat-free mass and strength in untrained young women. J Am Diet Assoc 1998;98(4):414–8.
[99] Ballor DL, Katch VL, Becque MD, Marks CR. Resistance weight training during caloric restriction enhances lean body weight maintenance. Am J Clin Nutr 1988;47(1):19–25.
[100] Geliebter A, Maher MM, Gerace L, Gutin B, Heymsfield SB, Hashim SA. Effects of strength or aerobic training on body composition, resting metabolic rate, and peak oxygen consumption in obese dieting subjects. Am J Clin Nutr 1997;66 (3):557–63.
[101] Ross R, Pedwell H, Rissanen J. Response of total and regional lean tissue and skeletal muscle to a program of energy restriction and resistance exercise. Int J Obes Relat Metab Disord 1995;19(11):781–7.
[102] Poehlman ET, Melby C. Resistance training and energy balance. Int J Sport Nutr 1998;8(2):143–59.
[103] Bryner RW, Ullrich IH, Sauers J, Donley D, Hornsby G, Kolar M, et al. Effects of resistance vs. aerobic training combined with an 800 calorie liquid diet on lean body mass and resting metabolic rate. J Am Coll Nutr 1999;18(2):115–21.
[104] Pronk NP, Donnelly JE, Pronk SJ. Strength changes induced by extreme dieting and exercise in severely obese females. J Am Coll Nutr 1992;11(2):152–8.
[105] Wadden TA, Vogt RA, Andersen RE, Bartlett SJ, Foster GD, Kuehnel RH, et al. Exercise in the treatment of obesity: effects of four interventions on body composition, resting energy expenditure, appetite, and mood. J Consult Clin Psychol 1997;65(2):269–77.
[106] Donnelly JE, Pronk NP, Jacobsen DJ, Pronk SJ, Jakicic JM. Effects of a very-low-calorie diet and physical-training regimens on body composition and resting metabolic rate in obese females. Am J Clin Nutr 1991;54(1):56–61.
[107] Fogelholm M, Kukkonen-Harjula K. Does physical activity prevent weight gain–a systematic review. Obes Rev 2000;1 (2):95–111.
[108] Jeffery RW, Drewnowski A, Epstein LH, Stunkard AJ, Wilson GT, Wing RR, et al. Long-term maintenance of weight loss: current status. Health Psychol 2000;19 (1 Suppl):5–16.
[109] McGuire MT, Wing RR, Klem ML, Seagle HM, Hill JO. Long-term maintenance of weight loss: do people who lose weight through various weight loss methods use different behaviors to maintain their weight? Int J Obes Relat Metab Disord 1998;22(6):572–7.
[110] Mustajoki P, Pekkarinen T. Maintenance programmes after weight reduction–how useful are they? Int J Obes Relat Metab Disord 1999;23(6):553–5.
[111] Wing RR, Hill JO. Successful weight loss maintenance. Annu Rev Nutr 2001;21:323–41.
[112] Borg P, Kukkonen-Harjula K, Fogelholm M, Pasanen M. Effects of walking or resistance training on weight loss maintenance in obese, middle-aged men: a randomized trial. Int J Obes Relat Metab Disord 2002;26(5):676–83.

CHAPTER

6

Psychology and Exercise

Attila Szabo[1,2], *Mark D. Griffiths*[3] *and Zsolt Demetrovics*[1]

[1]Institute of Psychology, Eötvös Loránd University, Budapest, Hungary [2]Institute for Health Promotion and Sport Sciences, Eötvös Loránd University, Budapest, Hungary [3]Psychology Division, Nottingham Trent University, Nottingham, United Kingdom

ACUTE AND CHRONIC PSYCHOLOGICAL EFFECTS OF EXERCISE

Research evidence reveals that physical activity yields numerous health benefits [1–4]. There is also scholastic evidence linking regular exercise and/or sport with positive mental wellbeing [5–9], as well as lower psychophysiological reactivity to mental stress [10–13]. The acute psychological benefits of exercise on various measures of affect and state anxiety are consistently demonstrated in the literature [7,14–22]. Since a single bout of acute exercise yields immediate psychological benefits, it may be seen as a suitable non-pharmaceutical antidote to stress and various mood disorders, in addition to its other health benefits. It is therefore not surprising then that the American College of Sports Medicine (ACSM) launched the 'Exercise is Medicine' program initiative [23] to make physical exercise part of both prevention and treatment of various morbidities.

Research has confirmed that different forms of exercise can trigger positive psychological changes [16,24–27]. The mechanisms by which acute exercise leads to improved wellbeing are primarily based on the volume and/or the duration and intensity of exercise (as a mediator of the psychological effect). The most popular theories are the endorphin hypothesis [28], the amine hypothesis [28], and the thermogenic hypothesis [29]. However, most of these theories have been challenged, because it is now evident that the intensity of exercise has little or no role in the acute psychological benefits of exercise on feelings states [22,30,31]. A placebo mechanism, that complements other mechanisms, has recently been proposed [32].

Ekkekakis [30] reviewed over one hundred research papers and concluded that exercise performed at self-selected intensity triggers effects in wellbeing and may be appropriate from a public health perspective. In considering the duration of exercise, research has shown that a number of positive psychological changes occur even after brief 10-minute bouts of physical exercise [14,33,34]. Therefore, brief exercise bouts are sufficient for experiencing psychological benefits, in contrast to physiological effects that require greater volumes [35]. However, using a cluster randomized cross-over design, Sjögren [36] found that an average of 5 minutes training per working day decreased the prevalence of headache, neck, shoulder and low back symptoms, and alleviated the intensity of headaches, neck and low back pain among symptomatic office workers. The intervention also improved subjective physical wellbeing. Therefore, physical benefits— despite the possibility that they may occur via placebo effects— also occur after short bouts of exercise.

Long-term regular exercise also benefits one's psychological health. A recent review by Gogulla, Lemke, and Hauer [37] showed that most research reports claim that physical exercise results in a significant reduction of depression and fear of falling in healthy elderly participants. However, the evidence was not convincing in elderly people with cognitive impairment. The reviewed studies also suggested that high-intensity aerobic or anaerobic exercise appears to be the most effective in reducing depression, while Tai-Chi and multimodal training are more effective in reducing the fear of falling. Another recent study showed that physical exercise training helps in reducing symptoms of worry among generalized anxiety disordered patients [38]. In contrast to a waiting list

Nutrition and Enhanced Sports Performance.
DOI: http://dx.doi.org/10.1016/B978-0-12-396454-0.00006-0

control group, the symptoms of worry decreased after 6 weeks of bi-weekly interventions in female participants in both aerobic exercise and resistance training exercise groups. Consequently, from a mental health perspective, both aerobic (endurance) and anaerobic (strength) exercises have beneficial long-term effects.

In an earlier review, Herring, Connor, and Dishman [39] concluded that exercise training significantly reduced anxiety symptoms when compared with no-treatment conditions. The authors noted that the exercise interventions that resulted in the largest anxiety improvements were those that (i) lasted not longer than 12 weeks, (ii) used exercise sessions lasting at least 30 minutes, and (iii) measured persistent anxiety lasting for more than 1 week. Milani and Lavie [40] showed that, apart from the anxiety-mediating effects of exercise, regular physical activity is also beneficial in the management of stress-related illnesses. The authors found that psychosocial stress is an independent risk factor for mortality in patients with coronary artery disease, and regular exercise training could effectively reduce its prevalence. In their study, the authors claimed that exercise training reduced mortality in patients with coronary artery disease, and that the observed effect may be mediated (at least in part) by the positive effects of exercise on psychosocial stress.

MOTIVATION FOR EXERCISE BEHAVIOR: WHY DO PEOPLE EXERCISE?

Motivation for exercise could be physically or psychologically oriented. Physical motives include (i) being in better physical condition, (ii) having a better looking and healthier body, (iii) having greater strength and endurance, and/or (iv) facilitating weight loss. However, the work-for and achievement of a physical goal also inherently triggers psychological rewards. Individuals participate in physical activity for one or more specific reasons. The reason is often an intangible social reward that itself stems from psychological needs of the person, like being with old buddies or making new friends. The personal experience of the anticipated reward strengthens the exercise behavior. The key point here is that there is always an anticipated reward, and the degree of fulfillment of that reward strongly predicts the continuance of the exercise behavior. Behaviorists, adhering to one of the most influential schools of thought in the field of psychology, postulate that most human behavior can be understood and explained through reinforcement and punishment. The gist of the theory is the operant conditioning-based governance of behavior, which involves positive reinforcement, negative reinforcement, and punishment [41]. Positive reinforcement is a motivational incentive for engaging in an activity to gain a reward that is subjectively pleasant or desirable (e.g., increased muscle tone). The reward then becomes a motivational incentive that increases the likelihood that the behavior will reoccur. In contrast, negative reinforcement is a motivational incentive for doing something to avoid a noxious or unpleasant event (e.g., gaining weight). The avoidance or reduction of the noxious stimulus is the reward, which then increases the probability that the behavior will reoccur. Here, the behavior is essentially used as a coping mechanism by the individual. It should also be noted that while both positive and negative reinforcers increase the likelihood of engaging in the behavior [41], their mechanisms are different because in positive reinforcement there is a 'gain' following the action (e.g., feeling revitalized), whereas in behaviors motivated by negative reinforcement one attempts— for whatever reason—to 'avoid' or prevent something bad, unpleasant, and/or simply undesirable (e.g., feeling guilty or fat if a planned exercise session is missed). Punishment, on the other hand, refers to situations in which the imposition of some noxious or unpleasant stimulus or event (or alternately the removal of a pleasant or desired stimulus or event) reduces the probability of a given behavior reoccurring. In contrast to reinforcers, punishers suppress the behavior and, therefore, exercise or physical activity, reading or other desirable behaviors should never be used (by teachers, parents, or coaches) as punishment.

Habitual exercisers may be motivated by positive reinforcement associated with muscle gain. However, numerous exercisers are motivated by negative reinforcement (e.g., to avoid gaining weight). Every time a person undertakes behavior to avoid something negative, bad, or unpleasant, the motive behind that behavior is classified as negative reinforcement. In these situations, the person involved *has to do it* in contrast to *wants to do it*. In the punishment situation, the person has to do it in a similar way to negative reinforcement, with the difference that (unless we talk about rare instances of self-punishment) the source of obligation (i.e., one has to do it) comes from an outside source (e.g., a parent, a teacher, the law, etc.) rather than from the inside. It is very important to differentiate between imposed punishment and self-selected negative reinforcement in exercise behavior.

There are many examples in other sport areas where a behavior initially driven by positive reinforcement may turn into negatively reinforced behavior. For example, an outstanding football player who starts playing the game *for fun*, after being discovered as a talent and being offered a service contract in a team, becomes a professional player who upon signing the contract is *expected* to perform. Although the player may still enjoy playing (especially when all goes well), the pressure or expectation to perform is the "has to do" new facet of football playing and the negatively reinforcing component of their sporting activity.

THEORIES AND MODELS ACCOUNTING FOR THE PSYCHOLOGICAL BENEFITS OF EXERCISE

The Sympathetic Arousal Hypothesis

Back in the 1980s, Thompson and Blanton [42] developed the Sympathetic Arousal Hypothesis on the basis of the factual information that regular exercise (especially aerobic exercise like running) if performed for a sustained period, resulted in decreased heart rate at rest. While heart rate is only a rough measure of the body's sympathetic activity (which is directed by the autonomic nervous system), it is, nevertheless, a sensitive measure and it is often used to mirror sympathetic activity. A lower resting heart rate after training results from the adaptation of the person to exercise. With repeated exercise, the person develops a more efficient cardiovascular system characterized by lower basal heart rate, lower sympathetic activity, and lesser arousal at rest. This new state of lowered arousal may induce relaxation, tranquility, and a positive engagement in the habituated exerciser [43].

The Cognitive Appraisal Hypothesis

A psychological explanation based on negative reinforced behavior stems from Szabo [44]. According to this model, some exercisers workout to escape from their psychological hardship [45]. They use exercise as a means of coping with stress. Once the person uses exercise for coping with hardship, the affected individual starts to depend on the adopted form of exercise, because every session brings the desired psychological effect. Therefore, the person experiences a form of psychological relief after exercise. When exercise is prevented for some reason, the exerciser loses the means of coping, and the lack of exercise triggers the opposite effect, that is negative psychological feeling states like irritability, guilt, anxiousness, sluggishness, etc. These feelings collectively are known as withdrawal symptoms experienced because of no- or reduced exercise. Avoidance of these symptoms is a negative reinforcer for exercise behavior.

The Affect Regulation Hypothesis

The affect regulation hypothesis posits that exercise has a dual effect on mood. First it increases the positive affect (defined as momentary psychological feeling states of somewhat longer duration than momentary emotions) and therefore contributes to an improved general mood state (defined as prolonged psychological feeling states lasting for several hours or even days). Second, exercise decreases the negative affect or the transient state of guilt, irritability, sluggishness, anxiety, etc. and therefore contributes to an improved general mood state [46].

The Thermogenic Regulation Hypothesis

This model is based on physiological evidence that physical exercise increases body temperature. A warm body temperature induces a relaxing state with concomitant reduction in anxiety (similar to sun-tanning, Turkish or warm bath, and sauna effects). Therefore, physical exercise reduces anxiety [47,48] via an increased state of physical relaxation. Lower levels of anxiety and states of relaxation are therefore positive reinforcers in exercise behavior. A relaxed body relaxes the mind and yields a positive subjective feeling state.

The Catecholamine Hypothesis

This hypothesis is driven by the observation that increased levels of catecholamines may be measured (in the peripheral blood circulation) after exercise [49]. Catecholamines, among other functions, are involved in the stress response and sympathetic responses to exercise. In light of the catecholamine hypothesis, it is speculated that central catecholaminergic activity is altered by exercise. Because central catecholamine levels are involved in regulating mood and affect and play an important role in mental dysfunctions like depression, the alteration of catecholamines by exercise may be an attractive explanation. However, to date, there is inconclusive evidence for this hypothesis. Indeed, it is unclear whether the peripheral changes in catecholamines have an effect on brain catecholamine levels or vice versa. Furthermore, the changes in brain catecholamine levels during exercise in humans are unknown, because direct measurement in the human brain is not possible.

The Endorphin Hypothesis

This model is attractive and popular in the literature because it is connected to the "runner's high" phenomenon (i.e., a pleasant feeling state associated with positive self-image, sense of vitality, control, and a sense of fulfillment reported by runners as well as by other exercisers after a certain amount and intensity of exercise). This feeling has been associated with increased levels of endogenous opioids and catecholamines observed after exercise. The theory behind this model is that exercise leads to increased levels of endorphins in the brain, which act as internal psychoactive agents yielding a sense of euphoria. In fact, this hypothesis is analogous to substance or recreational drug addiction (e.g., heroin, morphine, etc.) with the exception that the psychoactive agent (beta endorphin) is endogenously generated from within the body during exercise rather than being exogenously generated from a substance outside the body.

THE "RUNNERS' HIGH" PHENOMENON AND THE ACUTE PSYCHOLOGICAL EFFECTS OF EXERCISE

> "I believe in the runner's high, and I believe that those who are passionate about running are the ones who experience it to the fullest degree possible. To me, the runner's high is a sensational reaction to a great run! It's an exhilarating feeling of satisfaction and achievement. It's like being on top of the world, and truthfully ... there's nothing else quite like it!" —*Sasha Azevedo (http://www.runtheplanet.com/resources/historical/runquotes.asp)*

For many decades, marathon runners, long-distance joggers, and even regular joggers have reported a feeling state of strong euphoria masking the fatigue and pain of physical exertion caused by very long sessions of exercise. This euphoria triggers a sensation of "flying", effortless movement, and has become a legendary goal referred to as "the zone" [50]. The existence of runner's high is subject of heated debate in scholastic circles. The question is whether a biochemical explanation for the runner's high exists, or it is a purely subjectively (psychologically) conceptualized and popularized terminology. Runners (and most if not all habitual exercisers) experience withdrawal symptoms when their exercise is prevented. The symptoms include guilt, irritability, anxiety, and other unpleasant feelings [44]. Research has shown that the human body produces its own opiate-like peptides, called endorphins. Like morphine, these peptides can cause dependence [51] and consequently may be the route of withdrawal symptoms. In general, endorphins are known to be responsible for pain and pleasure responses in the central nervous system. Morphine and other exogenous opiates bind to the same receptors that the body intended for endogenous opioids or endorphins, and since morphine's analgesic and euphoric effects are well documented, comparable effects for endorphins can be anticipated [52].

Research has been conducted to examine the effects of fitness levels, gender, and exercise intensity on endogenous opioid—mainly beta-endorphin—production during cycling, running on a treadmill, participating in aerobic dance, and running marathons. Research by Biddle and Mutrie [53] reported that aerobic exercise can cause beta-endorphin levels to increase five-fold compared with baseline levels. Fitness level of the research participants appears to be irrelevant as both trained and untrained individuals experienced an increase in beta-endorphin levels, although the metabolism of beta-endorphins appeared to be more efficient in trained athletes [54].

Goldfarb et al. [55] examined gender differences in beta-endorphin production during exercise. Their results did not show any gender differences in beta-endorphin response to exercise. Other studies have demonstrated that both exercise intensity and duration are factors in increasing beta-endorphin concentrations. For example, the exercise needs to be performed at above 60% of the individual's maximal oxygen uptake (VO_{2max}) [54] and for at least 3 minutes [56] to detect changes in endogenous opioids.

Researchers have further examined the correlation between exercise-induced increase in beta-endorphin levels and mood changes, using the Profile of Mood States (POMS) inventory [51]. Here, the POMS was administered to all participants before and after their exercise session. The participants gave numerical ratings to five negative categories of mood (i.e., tension, depression, anger, fatigue, and confusion) and one positive category (vigor). Adding the five negative affect scores and then subtracting from the total, the vigor score yields a "total mood disturbance" (TMD) score. In Farrell's study the TMD scores improved by 15 and 16 raw score units from the baseline, after participants exercised at 60% and 80% VO_{2max}. Quantitatively, mood improved about 50%, which corresponds to clinical observations that people's moods are elevated after vigorous exercise workouts. Using radioimmunoassay techniques, Farrell et al. [51] also observed two- to five-fold increase in plasma beta-endorphin concentrations as measured before and after exercise.

However, Farrell et al.'s research is inconclusive. First, only six well-trained endurance athletes were studied, and the six showed large individual variations in beta-endorphin response to submaximal treadmill exercise. Second, the exercise-induced changes in mood scores were not statistically significantly different between pre- and post-exercise scores. Third, no significant relationship between mood measures obtained with the POMS inventory and plasma beta-endorphin levels was found. Therefore, the obtained results do not conclusively prove that beta-endorphins cause mood elevations. However, a more questionable issue—also recognized by Farrell et al.—is that the beta-endorphin measure in the experiment comes from plasma, which means that this type of beta-endorphin is located in the periphery. Because of its chemical makeup, beta-endorphin cannot cross the blood brain barrier (BBB). Hence, plasma beta-endorphin fluctuations do not reflect beta-endorphin fluctuations in the brain. Some researchers have speculated that endogenous opiates in the plasma may act centrally and therefore can be used to trace CNS activity [53]. At this time, such models concerning beta-endorphins only rely on circumstantial evidence that two opioids (i.e., met-enkephalin and dynorphin) show a modification mechanism that might possibly transport them across the BBB [52]. Unfortunately, direct measurement of changes in brain beta-endorphins involves cutting open the brain and employing radioimmunoassay techniques on brain slices. Animal studies, using rats, have been performed and they have shown an increase in opioid receptor binding after exercise [57].

In humans, to work around this problem, researchers proposed that naloxone could be useful in testing whether beta-endorphins play a role in CNS-mediated responses like euphoria and analgesia. Since it is a potent opioid receptor antagonist, it competes with beta-endorphin to bind to the same receptor. Thus, injection of naloxone into humans should negate the euphoric and analgesic effects produced by exercise, if indeed beta-endorphin facilitates such effects. Such research has found that naloxone decreases the analgesic effect reportedly caused by runner's high, but other researchers who have conducted similar experiments remain divided about these results. As for naloxone's effects on mood elevation, Markoff, Ryan, and Young [58] observed that naloxone did not reverse the positive mood changes induced by exercise.

Mounting evidence demonstrates that beta-endorphins are not necessary for the euphoria experienced by exercisers. Harte, Eifert, and Smith [59] noted that, although exercise produces both positive emotions and a rise in beta-endorphin levels, the two are not necessarily connected. Indeed, physically undemanding activities like watching comedy programs or listening to music produce elevations in mood identical to those resulting from exercise [60,61], although accompanying elevations in beta-endorphins were not be observed after watching comedy programs [62] or music [63]. Similarly, Harte et al. [59] found that both running and meditation resulted in significant positive changes in mood. In addition to taking mood measures, Harte et al. have also measured plasma beta-endorphin levels of the participants. As expected, those in the meditation group did not show a rise in beta-endorphin levels, despite reported elevations in mood. Such results seem to further question the link between mood improvement and changes in beta-endorphin levels following exercise.

Answering the improved mood and increased beta-endorphin levels connection question inversely, experiments were carried out in which beta-endorphin was directly injected into the bloodstream of healthy participants. The results failed to show any changes in mood [53]. On the other hand, beta-endorphin injections had a positive effect on clinically depressed patients [53]. Furthermore, electroconvulsive therapy, used to treat patients with depression, also increased plasma beta-endorphin levels.

The lack of beta-endorphin release during meditation, and the lack of mood alteration after beta-endorphin injection, call for attention on factors that influence beta-endorphin levels. In an effort to consolidate peripheral beta-endorphin data with the central nervous effects, researchers have realized that the peripheral opioid system requires further investigation. Taylor et al. [64] proposed that, during exercise, acidosis is the trigger of beta-endorphin secretion in the bloodstream. Their results showed that blood pH level strongly correlated with beta-endorphin level (i.e., acidic conditions raise the concentration of beta-endorphin; buffering the blood attenuates this response). The explanation behind such observations is that acidosis increases respiration and stimulates a feedback inhibition mechanism in the form of beta-endorphin. The latter interacts with neurons responsible for respiratory control, and beta-endorphin therefore serves the purpose of preventing hyperventilation [64]. How then is this physiological mechanism connected to CNS-mediated emotional responses? Sforzo [52] noted that, since opioids have inhibitory functions in the CNS, if a system is to be activated through opioids, at least one other neural pathway must be involved. Thus, instead of trying to establish how peripheral amounts of beta-endorphin act on the CNS, researchers could develop an alternate physiological model demonstrating how the emotional effects of opioids may be activated through the inhibition of peripheral sympathetic activity [52].

While the "runner's high" phenomenon has not been empirically established as a fact, and beta-endorphins' importance in this event is questionable, other studies have shown how peripheral beta-endorphins affect centrally-mediated behavior. Electro-acupuncture used to treat morphine addiction by diminishing cravings and relieving withdrawal symptoms, caused beta-endorphin levels to rise [65]. Since exercise also increases beta-endorphin levels in the plasma, McLachlan et al. [65] investigated whether exercise could lower exogenous opiate intake. Rats were fed morphine and methadone for several days and then randomly divided into two groups of exercisers and non-exercisers. At that time, voluntary exogenous opiate intake was recorded to see if the exercise would affect the consumption of opiate in exercising rats. The results showed that, while opiate consumption had increased in both groups, exercising rats did not consume as much as non-exercising animals, and the difference was statistically significant [65]. These findings suggest that exercise decreases craving.

In conclusion, the connection between beta-endorphins and runner's high is an elegant explanation but without sufficient empirical support. It is likely that the intense positive emotional experience, to which athletes, runners, and scientists refer as the runner's high, is evoked by several mechanisms acting jointly. Szabo [60] has shown that, while exercise and experiencing humor are equally effective in decreasing negative mood and increasing positive mood, the effects of exercise last longer than those of humor. These results are evidence for the involvement of more than one mechanism in mood alterations after physically active and relatively passive interventions.

THE DARK SIDE OF PHYSICAL ACTIVITY: EXERCISE ADDICTION

Beside the many advantageous effects of physical training, excessive exercise also has the potential to have adverse effects on both physical and mental health, and to lead to exercise addiction. Currently, exercise addiction is not cited within any officially recognized medical or psychological diagnostic frameworks. However, it is important, on the basis of the known and shared symptoms with related morbidities, that the dysfunction receives attention in a miscellaneous category of other or unclassified disorders. Based on symptoms with diagnostic values, exercise addiction could potentially be classified within the category of *behavioral addictions* [66–69]. Despite increased usage of the term 'exercise addiction', several incongruent terminologies are still in use for this phenomenon [70]. The most popular is arguably exercise dependence [71,72]. Others refer to the phenomenon as obligatory exercising [73] and exercise abuse [74], while in the media the condition is often described as compulsive exercise [75].

The Symptoms of Exercise Addiction

Regarding the symptoms, exercise addiction is characterized by six common symptoms of addiction: salience, mood modification, tolerance, withdrawal symptoms, personal conflict, and relapse [76–78]. However, it is important to clarify whether exaggerated exercise behavior is a primary problem in the affected person's life or emerges as a secondary problem in consequence of another psychological dysfunction. In the former case, the dysfunction is classified as primary exercise addiction because it manifests itself as a form of behavioral addiction. In the latter case it is termed as secondary exercise addiction because it co-occurs as a consequence of another dysfunction, typically with eating disorders such as anorexia nervosa or bulimia nervosa [79–81]. In the former, the motive for over-exercising is typically geared towards avoiding something negative [78], although the affected individual may be totally unaware of their motivation. It is a form of escape response to a source of disturbing, persistent, and uncontrollable stress. However, in the latter, excessive exercise is used as a means of achieving weight loss (in addition to very strict dieting). Thus, secondary exercise addiction can have a different etiology than primary exercise addiction. Nevertheless, it should be highlighted that many symptoms and consequences of exercise addiction are similar whether it is a primary or secondary exercise addiction. The distinguishing feature between the two is that in primary exercise addiction the exercise is the objective, whereas in secondary exercise addiction weight loss is the objective, while exaggerated exercise is one of the primary means in achieving the objective.

Measurement of Exercise Addiction

In measuring exercise addiction, two popular scales are worth noting. The Exercise Dependence Scale (EDS) [82–84] conceptualizes compulsive exercise on the basis of the DSM-IV criteria for substance abuse or addiction [85], and empirical research shows that it is able to differentiate between at-risk, dependent, and non-dependent athletes, and also between physiological and non-physiological addiction. The EDS has seven subscales: (i) tolerance, (ii) withdrawal, (iii) intention effect, (iv) lack of control, (v) time, (vi) reduction of other activities, and (vii) continuance. To generate a quick and easily administrable tool for surface screening of exercise addiction, Terry, Szabo and Griffiths [86] developed the 'Exercise Addiction Inventory' (EAI), a short six-item instrument aimed at identifying the risk of exercise addiction. The EAI assesses the six common symptoms of addictive behaviors mentioned above: (i) salience, (ii) mood modification, (iii) tolerance, (iv) withdrawal symptoms, (v) social conflict, and (vi) relapse. Both measures have been psychometrically investigated and proved to be reliable instruments [87].

Epidemiology of Exercise Addiction

Studies of exercise addiction prevalence have been carried out almost exclusively on American and British samples of regular exercisers. In five studies carried out among university students, Hausenblas and Downs [84] reported that between 3.4% and 13.4% of their samples were at high risk of exercise addiction. Griffiths, Szabo, and Terry [88], reported that 3.0% of a British sample of sport science and psychology students were identified as at-risk of exercise addiction. These research-based estimates are in concordance with the argument that exercise addiction is *relatively* rare [89,90] especially when compared with other addictions [91]. Nevertheless, given the severity of the problem, even a tenth of positive diagnoses among the high-risk cases may be large (i.e., 0.3% is 30/10,000 cases).

Among those who are also professionally connected to sport, the prevalence may be even higher. For example, Szabo and Griffiths [92] found that 6.9% of British sport science students were at risk of exercise addiction. However, in other studies where more-involved exercisers were studied, much higher estimates have generally been found. Blaydon and Lindner [93] reported that 30.4% of triathletes could be diagnosed with primary exercise addiction, and a further 21.6% with secondary exercise addiction. In another study, 26% of 240 male and 25% of 84 female runners were classified as "obligatory exercisers" [94]. Lejoyeux et al. [95] found that 42% of clients of a Parisian fitness room could be identified as exercise addicts. Recently, he reported lower rates of just under 30% [96]. However, one study that surveyed 95 'ultra-marathoners' (who typically run 100 km races) reported only three people (3.2%) as at-risk for exercise addiction [97]. Gender, however, can have a moderating effect on ideal-weight goals and exercise dependence symptoms [98]. It is evident that, besides differences in the applied measures and criteria, these appreciable differences in the estimates may be attributable to the sample selection, small sample size, and the sampling method. With the exception of the study by Lejoyeux et al. [95] that applied consecutive sampling, all the aforementioned studies used convenience sampling.

To date, the only national representative study is the one carried out by Mónok et al. [87] on a Hungarian adult population aged 18–64 years ($N = 2,710$), and assessed by both the EAI and the EDS. According to their results 6.2% (EDS) and 10.1% (EAI) of the population were characterized as nondependent-symptomatic exercisers, while the proportions of the at-risk exercisers were 0.3% and 0.5%, respectively.

CONCLUSIONS

This chapter reviewed physical exercise (both acute and chronic) and showed that it can have advantageous and disadvantageous effects. Long-term exercising at an optimal level can significantly contribute to physical and psychological health, whereas, in some cases, excessive exercisers can develop exercise addiction that can have various harmful effects. Similarly to other behaviors that can also become addictive, it was demonstrated that exercising also has the potential to develop over-engagement that might lead to negative consequences. The task and responsibility of researchers and healthcare promoters is to communicate clearly on these issues. More specifically, they should promote exercising as a behavior to improve health but also draw attention to the possible harms related to over-exercising and addiction.

Acknowledgements

This work was supported by the Hungarian Scientific Research Fund (grant numbers: 83884). Zsolt Demetrovics acknowledges financial support of the János Bolyai Research Fellowship awarded by the Hungarian Academy of Science.

References

[1] Bellocco R, Jia C, Ye W, Lagerros YT. Effects of physical activity, body mass index, waist-to-hip ratio and waist circumference on total mortality risk in the Swedish National March Cohort. Eur J Epidemiol 2010;25:777–88.

[2] Blair SN, Kohl HW, Barlow CE. Physical activity, physical fitness, and all-cause mortality in women: do women need to be active?. J Am Coll Nutr 1993;12:368–71.

[3] Lee DC, Sui X, Ortega FB, et al. Comparisons of leisure-time physical activity and cardiorespiratory fitness as predictors of all-cause mortality in men and women. Br J Sports Med 2011;45:504–10.

[4] Powell KE, Blair SN. The public health burdens of sedentary living habits: theoretical but realistic estimates. Med Sci Sports Exerc 1994;26:851–6.

[5] Biddle S. Exercise and psychosocial health. Res Q Exerc Sport 1995;66:292–7.

[6] Biddle SJH, Fox KR, Boutcher SH, editors. Physical activity and psychological well-being. London: Routledge; 2000.

[7] Biddle SJH, Mutrie N. Psychology of physical activity determinants, Well-being and interventions. London: Routledge; 2001.

[8] Brown WJ, Mishra G, Lee C, Bauman A. Leisure time physical activity in Australian women: relationship with well being and symptoms. Res Q Exerc Sport 2000;71:206–16.

[9] Tseng CN, Gau BS, Lou MF. The effectiveness of exercise on improving cognitive function in older people: a systematic review. J Nurs Res 2011;19:119–31.

[10] Norris R, Carroll D, Cochrane R. The effects of aerobic and anaerobic training on fitness, blood pressure, and psychological stress and well-being. J Psychosom Res 1990;34:367–75.

[11] Norris R, Carroll D, Cochrane R. The effects of physical activity and exercise training on psychological stress and well-being in an adolescent population. J Psychosom Res 1992;36:55–65.

[12] Rosenfeldt F, Braun L, Spitzer O, et al. Physical conditioning and mental stress reduction–a randomised trial in patients undergoing cardiac surgery. BMC Complement Altern Med 2011;11:20.

[13] Stein PK, Boutcher SH. The effect of participation in an exercise training program on cardiovascular reactivity in sedentary middle-aged males. Int J Psychophysiol 1992;13:215–23.
[14] Anderson RJ, Brice S. The mood-enhancing benefits of exercise: memory biases augment the effect. Psychol Sport Exerc 2011;12:79–82.
[15] Berger BG, Motl RW. A selective review and synthesis of research employing the profile of mood states. J Appl Sport Psychol 2000;12:69–92.
[16] Dasilva SG, Guidetti L, Buzzachera CF, et al. Psychophysiological responses to self-paced treadmill and overground exercise. Med Sci Sports Exerc 2011;43:1114–24.
[17] Fontaine KR. Physical activity improves mental health. Phys Sportsmed 2000;28:83–4.
[18] Hoffman MD, Hoffman DR. Exercisers achieve greater acute exercise-induced mood enhancement than nonexercisers. Arch Phys Med Rehabil 2008;89:358–63.
[19] O'Connor PJ, Raglin JS, Martinsen EW. Physical activity, anxiety and anxiety disorders. Int J Sport Psychol 2000;31:136–55.
[20] Paluska SA, Schwenk TL. Physical activity and mental health: current concepts. Sports Med 2000;29:167–80.
[21] Raglin JS. Exercise and mental health: beneficial and detrimental effects. Sports Med 1990;9:323–9.
[22] Szabo A. Acute psychological benefits of exercise performed at self-selected workloads: implications for theory and practice. J Sport Sci Med 2003;2:77–87.
[23] Jonas S, Phillips EM. ACSM's exercise is medicine™: a clinician's guide to exercise prescription. Philadelphia (PA): Wolters Kluwer, Lippincott Williams & Wilkins; 2009.
[24] Lavey R, Sherman T, Mueser KT, Osborne DD, Currier M, Wolfe R. The effects of yoga on mood in psychiatric inpatients. Psychiatr Rehabil J 2005;28:399–402.
[25] Petruzzello SJ, Snook EM, Gliottoni RC, Motl RW. Anxiety and mood changes associated with acute cycling in persons with multiple sclerosis. Anxiety Stress Coping 2009;22:297–307.
[26] Rokka S, Mavridis G, Kouli O. The impact of exercise intensity on mood state of participants in dance aerobics programs. Stud Phys Cult Tourism 2010;17:241–5.
[27] Valentine E, Evans C. The effects of solo singing, choral singing and swimming on mood and physiological indices. Br J Med Psychol 2001;74:115–20.
[28] Dunn AL, Dishman RK. Exercise and the neurobiology of depression. Exerc Sport Sci Rev 1991;19:41–98.
[29] Koltyn KF. The thermogenic hypothesis. In: Morgan WP, editor. Physical activity and mental health. Washington: Taylor and Francis; 1997. p. 213–26.
[30] Ekkekakis P. Let them roam free? Physiological and psychological evidence for the potential of self-selected exercise intensity in public health. Sports Med 2009;39:857–88.
[31] Stoll O. Endorphine, Laufsucht und runner's high. Aufstieg und niedergang eines mythos (Endogenous opiates, "runner's high" and "Exercise Addiction" – The rise and decline of a myth). Leipziger Sportwissenschaftliche Beitraege 1997;28:102–21.
[32] Szabo A, Abraham J. The psychological benefits of recreational running: a field study. Psychol Health Med 2012.
[33] Hansen CJ, Stevens LC, Coast JR. Exercise duration and mood state: how much is enough to feel better? Health Psychol 2001;20:267–75.
[34] Sullivan AB, Covington E, Scheman J. Immediate benefits of a brief 10-minute exercise protocol in a chronic pain population: a pilot study. Pain Med 2010;11:524–9.
[35] Garber CE, Blissmer B, Deschenes MR, et al. American College of Sports Medicine position stand. Quantity and quality of exercise for developing and maintaining cardiorespiratory, musculoskeletal, and neuromotor fitness in apparently healthy adults: guidance for prescribing exercise. Med Sci Sports Exerc 2011;43:1334–59.
[36] Sjögren T. Effectiveness of a workplace physical exercise intervention on the functioning, work ability, and subjective well-being of office workers – a cluster randomised controlled cross-over trial with a one-year follow-up. Jyväskylä: University of Jyväskylä; 2006.
[37] Gogulla S, Lemke N, Hauer K. Effects of physical activity and physical training on the psychological status of older persons with and without cognitive impairment. Z Gerontol Geriatr 2012;45:279–89.
[38] Herring MP, Puetz TW, O'Connor PJ, Dishman RK. Effect of exercise training on depressive symptoms among patients with a chronic illness: a systematic review and meta-analysis of randomized controlled trials. Arch Intern Med 2012;172:101–11.
[39] Herring MP, O'Connor PJ, Dishman RK. The effect of exercise training on anxiety symptoms among patients: a systematic review. Arch Intern Med 2010;170:321–31.
[40] Milani RV, Lavie CJ. Reducing psychosocial stress: a novel mechanism of improving survival from exercise training. Am J Med 2009;122:931–8.
[41] Bozarth MA. Pleasure systems in the brain. In: Warburton DM, editor. Pleasure: the politics and the reality. New York: John Wiley & Sons; 1994. p. 5–14.
[42] Thompson JK, Blanton P. Energy conservation and exercise dependence: a sympathetic arousal hypothesis. Med Sci Sports Exerc 1987;19:91–9.
[43] Gauvin L, Rejeski WJ. The Exercise-Induced Feeling inventory: development and initial validation. J Sport Exerc Psychol 1993;15:403–23.
[44] Szabo A. The impact of exercise deprivation on well-being of habitual exercisers. Aust J Sci Med Sport 1995;27:68–75.
[45] Morris M. Running round the clock. Running 1989;104:44–5.
[46] Hamer M, Karageorghis CI. Psychobiological mechanisms of exercise dependence. Sports Med 2007;37:477–84.
[47] De Vries HA. Tranquilizer effect of exercise: a critical review. Phys Sportsmed 1981;9:47–53.
[48] Morgan WP, O'Connor PJ. Exercise and mental health. In: Dishman RK, editor. Exercise adherence: its impact on public health. Champaign, IL: Human Kinetics; 1988. p. 91–121.
[49] Cousineau D, Ferguson RJ, de Champlain J, Gauthier P, Cote P, Bourassa M. Catecholamines in coronary sinus during exercise in man before and after training. J Appl Physiol 1977;43:801–6.
[50] Goldberg A. The sports mind: A workbook of mental skills for athletes. Northampton, MA: Competitive Advantage; 1988.
[51] Farrell PA, Gates WK, Maksud MG, Morgan WP. Increases in plasma beta-endorphin/beta-lipotropin immunoreactivity after treadmill running in humans. J Appl Physiol 1982;52:1245–9.
[52] Sforzo GA. Opioids and exercise: and update. Sports Med 1988;7:109–24.
[53] Biddle S, Mutrie N. Psychology of physical activity and exercise: a health-related persective. London: Springer Verlag; 1991.
[54] Goldfarb AH, Jamurtas AZ. Beta-endorphin response to exercise. An update. Sports Med 1997;24:8–16.
[55] Goldfarb AH, Jamurtas AZ, Kamimori GH, Hegde S, Otterstetter R, Brown DA. Gender effect on beta-endorphin response to exercise. Med Sci Sports Exerc 1998;30:1672–6.
[56] Kjaer M, Dela F. Endocrine response to exercise. In: Hoffman-Goetz L, editor. Exercise and immune function. Boca Raton, FL: CRC; 1996. p. 6–8.
[57] Sforzo GA, Seeger TF, Pert CB, Pert A, Dotson CO. In vivo opioid receptor occupation in the rat brain following exercise. Med Sci Sports Exerc 1986;18:380–4.
[58] Markoff RA, Ryan P, Young T. Endorphins and mood changes in long-distance running. Med Sci Sports Exerc 1982;14:11–5.

[59] Harte JL, Eifert GH, Smith R. The effects of running and meditation on beta-endorphin, corticotropin-releasing hormone and cortisol in plasma, and on mood. Biol Psychol 1995;40:251–65.
[60] Szabo A. Comparison of the psychological effects of exercise and humour. In: Lane AM, editor. Mood and human performance: conceptual, measurement, and applied issues. Hauppauge, NY: Nova Science Publishers, Inc.; 2006. p. 201–16.
[61] Szabo A, Ainsworth SE, Danks PK. Experimental comparison of the psychological benefits of aerobic exercise, humor, and music. Humor 2005;18:235–46.
[62] Berk LS, Tan SA, Fry WF, et al. Neuroendocrine and stress hormone changes during mirthful laughter. Am J Med Sci 1989;298:390–6.
[63] McKinney CH, Tims FC, Kumar AM, Kumar M. The effect of selected classical music and spontaneous imagery on plasma beta-endorphin. J Behav Med 1997;20:85–99.
[64] Taylor DV, Boyajian JG, James N, et al. Acidosis stimulates beta-endorphin release during exercise. J Appl Physiol 1994;77:1913–8.
[65] McLachlan CD, Hay M, Coleman GJ. The effects of exercise on the oral consumption of morphine and methadone in rats. Pharmacol Biochem Behav 1994;48:563–8.
[66] Albrecht U, Kirschner NE, Grusser SM. Diagnostic instruments for behavioural addiction: an overview. Psychosoc Med 2007;4: Doc11
[67] Demetrovics Z, Griffiths MD. Behavioral addictions: past, present and future. J Behav Addict 2012;1:1–2.
[68] Grant JE, Potenza MN, Weinstein A, Gorelick DA. Introduction to behavioral addictions. Am J Drug Alcohol Abuse 2010;36:233–41.
[69] Griffiths MD. Exercise addiction: a case study. Addict Res 1997;5:161–8.
[70] Allegre B, Souville M, Therme P, Griffiths MD. Definitions and measures of exercise dependence. Addict Res Theory 2006;14:631–46.
[71] Cockerill IM, Riddington ME. Exercise dependence and associated disorders: a review. Couns Psychol Q 1996;9:119–29.
[72] Hausenblas HA, Symons Downs D. How much is too much? The development and validation of the exercise dependence scale. Psychol Health 2002;17:387–404.
[73] Pasman LN, Thompson JK. Body image and eating disturbance in obligatory runners, obligatory weightlifters, and sedentary individuals. Int J Eat Disord 1988;7:759–69.
[74] Davis C. Exercise abuse. Int J Sport Psychol 2000;31:278–89.
[75] Dalle Grave R, Calugi S, Marchesini G. Compulsive exercise to control shape or weight in eating disorders: prevalence, associated features, and treatment outcome. Compr Psychiatry 2008;49:346–52.
[76] Brown RIF. Some contributions of the study of gambling to the study of other addictions. In: Eadington WR, Cornelius JA, editors. Gambling behavior and problem gambling. Reno: University of Nevada Press; 1993. p. 241–72.
[77] Griffiths MD. A 'components' model of addiction within a biopsychosocial framework. J Subst Use 2005;10:191–7.
[78] Szabo A. Addiction to exercise: a symptom or a disorder? New York: Nova Science Publishers Inc.; 2010.
[79] Bamber DJ, Cockerill IM, Carroll D. The pathological status of exercise dependence. Br J Sports Med 2000;34:125–32.
[80] Blaydon MJ, Lindner KJ, Kerr JH. Metamotivational characteristics of eating-disordered and exercise-dependent triathletes: an application of reversal theory. Psychol Sport Exerc 2002;3:223–36.
[81] de Coverley Veale DM. Exercise dependence. Br J Addict 1987;82:735–40.
[82] Costa S, Cuzzocrea F, Hausenblas HA, Larcan R, Oliva P. Psychometric examination and factorial validity of the exercise dependence scale-revised in Italian exercisers. J Behav Addict 2012;1:1–5.
[83] Symons Downs D, Hausenblas HA, Nigg CR. Factorial validity and psychometric examination of the exercise dependence scale-revised. Meas Phys Educ Exerc Sci 2004;8:183–201.
[84] Hausenblas HA, Symons Downs D. Exercise dependence: a systematic review. Psychol Sport Exerc 2002;3:89–123.
[85] American Psychiatric Association. text revision ed. Diagnostic and statistical manual for mental disorders. 4th ed. Washington: American Psychiatric Publishing; 2000
[86] Terry A, Szabo A, Griffiths MD. The exercise addiction inventory: a new brief screening tool. Addict Res Theory 2004;12:489–99.
[87] Mónok K, Berczik K, Urbán R, et al. Psychometric properties and concurrent validity of two exercise addiction measures: a population wide study. Psychol Sport Exerc 2013;6:739–46.
[88] Griffiths MD, Szabo A, Terry A. The exercise addiction inventory: a quick and easy screening tool for health practitioners. Br J Sports Med 2005;39:e30.
[89] De Coverley Veale DMW. Does primary exercise dependence really exist? In: Annett J, Cripps B, Steinberg H, editors. Exercise addiction. Motivation and participation in sport and exercise. Leicester: The British Psychological Society; 1995. p. 1–5.
[90] Szabo A. Physical activity as a source of psychological dysfunction. In: Biddle SJ, Fox KR, Boutcher SH, editors. Physical activity and psychological well-being. London: Routledge; 2000. p. 130–53.
[91] Sussman S, Lisha N, Griffiths M. Prevalence of the addictions: a problem of the majority or the minority? Eval Health Prof 2011;34:3–56.
[92] Szabo A, Griffiths MD. Exercise addiction in British sport science students. Int J Ment Health Addict 2007;5:25–8.
[93] Blaydon MJ, Lindner KJ. Eating disorders and exercise dependence in triathletes. Eat Disord 2002;10:49–60.
[94] Slay HA, Hayaki J, Napolitano MA, Brownell KD. Motivations for running and eating attitudes in obligatory versus nonobligatory runners. Int J Eat Disord 1998;23:267–75.
[95] Lejoyeux M, Avril M, Richoux C, Embouazza H, Nivoli F. Prevalence of exercise dependence and other behavioral addictions among clients of a Parisian fitness room. Compr Psychiatry 2008;49:353–8.
[96] Lejoyeux M, Guillot C, Chalvin F, Petit A, Lequen V. Exercise dependence among customers from a Parisian sport shop. J Behav Addict 2012;1:28–34.
[97] Allegre B, Therme P, Griffiths MD. Individual factors and the context of physical activity in exercise dependence: a prospective study of 'ultra-marathoners'. Int J Ment Health Addict 2007;5:233–43.
[98] Cook B, Hausenblas H, Rossi J. The moderating effect of gender on ideal-weight goals and exercise dependence symptoms. J Behav Addict 2013;2: (in press).

CHAPTER

7

Bone Health, Bone Mineral Density and Sports Performance

Annie Schtscherbyna[1], *Beatriz Gonçalves Ribeiro*[1] and *Maria Lucia Fleiuss de Farias*[2]

[1]Josué de Castro Nutrition Institute, Universidade Federal do Rio de Janeiro, Rio de Janeiro, Brazil [2]Endocrinology Service, Hospital Universitário Clementino Fraga Filho, Universidade Federal do Rio de Janeiro, Brazil

INTRODUCTION

Bone is a living tissue with two main functions: structural support and mineral storage. It must be simultaneously stiff for protection of internal organs, flexible to support deformations without breaking, and light to allow movements. Bone is also a reservoir of several ions, containing 99% of total body calcium [1].

Many factors can influence bone health: heredity, physical activity, food intake (especially related to calcium, vitamin D and energy) and serum vitamin D sufficiency. Physiological events such as pubertal age, menstrual cycles, parity, breast-feeding, menopause, and aging also interfere with the skeleton [2,3]. Incidence of bone diseases is growing worldwide, and the complex interaction of etiological factors requires greater research [1].

Physical activity seems to be the major influence in bone mineral density (BMD) [4] as bone mass increases in response to active and passive activities [2,5]. Exercise is often recommended for both prevention and treatment of low BMD, and athletes have higher BMD values than non-athletes [6,7]. However, extreme exercises may be detrimental to bone health, especially in the young [8].

Besides the progressive growing literature related to bone health and performance, several doubts remain and many topics must be addressed.

BONE HEALTH

Bone mass seems to be polygenic, and multiple genes may be involved in both the attainment of bone mass and in the control of bone turnover. Candidate genes include the vitamin D receptor (VDR) gene [9,10], the vitamin D promoter region of the osteocalcin gene [11], as well as genes for type 1 collagen (COL1A1), the estrogen receptor [12], and certain cytokines [1].

Food consumption is another important factor for bone mineral accrual, especially related to calcium, vitamin D and energy. The calcium requirement is elevated, especially in adolescence (Recommended Dietary Allowance 1300 mg/day) [13]. Studies in adolescents showed that the consumption of foods rich in calcium, such as milk and cheese, are usually insufficient in this period of life [14,15]. As an adaptive process, there is an increase in the capacity of calcium intestinal absorption during childhood and puberty [16], but this adjustment is not sufficient to avoid a negative balance when the long-term calcium intake is 400 mg/day or less [17].

During childhood and adolescence the skeleton grows and changes its shape by bone modeling, a process in which bone formation is not preceded by bone resorption. After peak bone mass is achieved, skeleton integrity depends on bone renewal, known as remodeling or turnover. This process, where bone formation occurs only at sites previously resorbed, guarantees the substitution of old packets of bone by new ones at a rate of 10–15% annually on average. Bone remodeling occurs at all bone surfaces and is more active within the trabecular compartment than in cortical bone [18].

Bone remodeling is also committed to calcium and phosphorus homeostasis. Approximately 99% of the

Nutrition and Enhanced Sports Performance.
DOI: http://dx.doi.org/10.1016/B978-0-12-396454-0.00007-2

total body calcium is found in bone and teeth, 1% in other tissues, and 0.1% in the extracellular space. When calcium intake or absorption is inadequate, parathyroid hormone (PTH) is secreted to avoid hypocalcemia by increasing the conversion of 25-hydroxyvitamin D into 1,25di-hydroxyvitamin D and intestinal absorption, increasing the renal tubular calcium reabsorption but also increasing bone resorption. If the stimulus persists, secondary hyperparathyroidism will cause a negative bone balance [1].

Another important nutrient to maintain mineral equilibrium in the body is vitamin D, an essential steroid hormone. The cholecalciferol vitamin D_3 (of animal origin) is synthesized in the skin through the action of ultraviolet light on 7-dehydrocholesterol, a cholesterol derivative distributed widely in the body [1]. The consumption of food rich in this vitamin, such as fish liver oil and saltwater fishes like sardine and herring, and others with a small quantity such as egg, meat, milk and butter, is extremely important to bone health [19]. As expected, unfortunately, vitamin D intake is insufficient in different regions of the world [20]. The earliest manifestations of vitamin D deficiency are muscle weakness and increased risk of infection. In children a nonspecific symptom of hypovitaminosis D is decreased appetite [14]. More severe and long-term deficiencies cause accelerated bone loss due to secondary hyperparathyroidism and impair skeleton mineralization, causing diffuse bone pain and osteomalacia (corresponding to rickets in childhood) [1,21].

Energy availability is also extremely important to bone and corresponds to the amount of energy that remains available to the body after, for example, expenditure in exercise training. It is the energy used to support all other bodily functions, including reproductive and endocrine system functions. Low energy availability can result from an insufficient energy diet, intentional restrictive or disordered eating (DE) behaviors, or it can be related to extreme weight control measures (fasting, diet pills, laxatives, diuretics, or enemas) [22]. When energy deficits reach a critical level, the functions of the reproductive system can be disrupted [23].

Energy deficiency can suppress gonadotropin-releasing hormone (GnRH) from the hypothalamus, inhibiting the release and pulsatility of luteinizing hormone (LH) [24] and follicle-stimulating hormone (FSH) from the pituitary gland [25]. The inhibited release of other hormones from the pituitary gland can cause decreased circulating levels of thyroid hormones (especially triiodothyronine or T_3), insulin, insulin-like growth-factor-1 (IGF-1) and leptin [26], and increased levels of other hormones, such as cortisol and growth hormone (GH). All these alterations are important to signal the body to increase food intake and return to its normal body composition [27].

The risk of developing a fragility fracture is dependent on the maximum amount and strength of bone achieved in any person's lifetime, and the rate at which the bone is subsequently lost. Evidence indicates that childhood and the adolescent years provide a 'window of opportunity' to maximize bone mass and strength [28]. Adolescence is characterized by an accelerated bone turnover, but formation must exceed resorption to guarantee a progressive bone gain until peak bone mass is reached. The amount of bone mass acquired in youth will serve as a reservoir for the rest of life and will be one of the major factors protecting against osteopenia and osteoporosis in future life [29]. Peak bone mass is attained by the third decade of life in both women and men [30,31]. Moreover, bone maturation in females is reported to be complete at the end of adolescence [32]. Although the general understanding is that peak bone mass occurs by the end of the second or early in the third decade of life [33], accumulation is drastically reduced by 16 years of age in both the lumbar spine and femoral neck [32]. During pubertal development, bone mass is directly correlated to age, height and weight [34]. Maximum BMD gain depends on a lower rate of remodeling [2] and is impaired in situations of irregular or insufficient production of sex steroids.

After puberty initiation, sex steroids and GH will act synergistically to increase IGF-1 production by the liver and bone cells. This will lead to linear growth and bone expansion in both genders [24,26,27].

In female adolescents, the rising level of estrogen slows bone remodeling by inhibiting local production of cytokines, favoring bone accretion and cortical thickness [35]. A chronic inadequate food consumption is relatively common in adolescents and can interfere in this process. As a consequence of decrease in GnRH and gonadotropins, the ovaries fail to secrete estrogen and also to ovulate, leading to irregular menstrual cycles (oligomenorrhea) or the cessation of menstruation (hypothalamic amenorrhea) [26,28]. Chronic estrogen deficiency will compromise peak bone mass and cause low bone mass [29]. BMD declines as the number of missed menstrual cycles accumulates [36], and the loss of BMD may not be fully reversible [37]. Girls and young women with low BMD may be susceptible to stress fractures, particularly if they participate in high-impact sports (e.g., gymnastics) or sports in which repetitive mechanical joint stresses are sustained (e.g., running) [38].

In men, the pathogenesis of low BMD has not been well investigated. As a consequence, prevention and treatment are not well understood. Estrogen seems to mediate sex steroid action in bone in males too. Therefore, a similar sequence of events is expected: the progressive rise in testosterone secretion by the testes will lead to a proportional increase in estrogen levels, stimulating IGF-1 production and decreasing bone remodeling, and consequently increasing bone mass. Bone mass acquired by healthy young men is higher that that of young women. However, there are data on a significant progressive loss of bone beginning in the third decade and continuing throughout life in male athletes [39].

Anthropometric characteristics have also been shown to influence bone mass [40,41]. Low weight and percent of body fat are critical for the maintenance of hormones production (already described) and are directly correlated with BMD [42]. Also fat-free mass seems to have a mechanical impact on cortical bone, whereas trabecular bone loss is more related to fat mass [43].

Because of this complexity, an increase in bone diseases and even fractures is expected among all groups not yet considered at risk, like children [2], adolescents [31], men [7], and also in athletes [44,45]. Osteoporosis is a systemic skeletal disease characterized by low bone mass and microarchitecture deterioration of bone tissue, with a consequent increase in bone fragility and susceptibility to fractures [1,46]. It is oligosymptomatic before fractures occur, and in elderly people half of the vertebral fractures are silent (nonclinical). In this setting, osteoporosis must be suspected, actively sought, and treated to reduce risk of complications.

Osteoporosis is not only caused by accelerated bone mineral loss in adulthood, but may also be associated with not accumulating optimal BMD during childhood and adolescence [47]. Chronic illnesses and prolonged immobilization in the young may contribute to a lower peak bone mass [48]. Also metabolic abnormalities (e.g., diabetes mellitus, hyperthyroidism), gastrointestinal disease (e.g., celiac disease), exposure to certain drugs (e.g., glucocorticoid), cigarette consumption, and excess of alcohol may contribute to bone loss [49].

BONE MINERAL DENSITY

Although BMD is only one aspect of bone strength, this chapter focuses on BMD because screening and diagnosis of osteoporosis are still based on bone densitometry. Early BMD testing in elite athletes has important clinical relevance, as it may detect the potential risk for osteoporosis in the future [39]. A number of techniques are available to evaluate bone density, and these vary in both clinical and research utility and in general availability. In order to be useful, such methods need to be accurate, precise, rapid, reliable, inexpensive, and expose patients to minimal radiation. Furthermore, they should rely on adequate reference data for the population studied [49]. These include dual-energy X-ray absorptiometry (DXA), spine and peripheral quantitative computed tomography, and quantitative ultrasound, but the gold standard for clinical evaluation of BMD is DXA. This exam has been shown to predict fracture risk [50].

Definitions: Osteopenia, Osteoporosis and Low Bone Mineral Density

Criteria for diagnosing osteopenia and osteoporosis apply to postmenopausal women and men 50 years and older. The difference between a patient's BMD and average peak BMD obtained in young adults of the same gender is measured in standard deviations and is called T-score. These criteria derive from epidemiological data relating BMD to subsequent fractures in Caucasian postmenopausal women. Current osteopenia and osteoporosis definitions according to DXA results are [51,52]:

1. *Osteoporosis:* BMD at or below −2.5 standard deviations (≤ − 2.5 SD) from peak bone mass;
2. *Osteopenia (sometimes called "Low bone mass"):* BMD between −1 and −2.5 SD (< − 1 and > − 2.5 SD);
3. *Normal:* BMD at or above −1 SD (≥ − 1 SD) from peak bone mass.

The current WHO definition of osteopenia and osteoporosis should not be applied to a younger population. In growing children and adolescents, for example, there is no agreement for adjusting BMD for bone size, pubertal stage, skeletal maturity, or body composition. Instead, the International Society for Clinical Densitometry (ISCD) recommends that BMD in these young populations should be compared with age- and sex-matched controls and expressed as Z-scores: values at or below −2.0 SD (≤ − 2.0 SD) identify "low bone mineral density for chronologic age" [52–54]. Only in the presence of high risk for fractures can a young patient be considered and treated as "osteoporotic". These conditions include chronic malnutrition, eating disorders, hypogonadism, glucocorticoid exposure, and previous fractures [52,54]. Thus, in the diagnosis and assessment of most disorders, the patient's history, physical and biochemical examinations are important features for diagnostic and therapeutic considerations [55].

Athletes with a BMD Z-score below −1.0 SD (< − 1.0 SD) warrant further investigation, even in the absence of a prior fracture. The American College of Sports Medicine (ACSM) defines the term "low bone mineral density" in the presence of nutritional deficiencies, estrogen deficiency, stress fractures, and/or other secondary clinical risk factors for fracture, together with a BMD Z-score between −1.0 and −2.0 SD (< − 1.0 SD and > − 2.0 SD) [41]. Bone densitometry may also be used to evaluate the response to treatment, and the current consensus is that bone mass should be measured at least twice following initiation of treatment at annual intervals or greater [55].

BONE AND PHYSICAL ACTIVITY

It is well known that mechanical loading, generated by physical activities, plays an important role in bone development [56–58]. The study of the human skeletal response to exercise offers a potential means of exploring the bone adaptive process. Exercise causes bone remodeling through cell mechano-transduction, leading to a simultaneous increase in both resorption and deposition of bone tissue [59,60]. In the long term, there is an increase in bone mass [61], confirmed by prospective studies [62]. This could also happen with a shorter training period (14–15 wk training) [63,64].

The geometry and the long-axis distribution of bone mass are key determinants of bone strength [65]. Changes in bone geometry with exercise have been poorly studied, due in part to a paucity of appropriate available technologies.

Studies investigating athletes and non-athletes and studies comparing exercise type demonstrate that osteogenic effects are greatest for activities associated with high-strength, high-frequency loading distributed unevenly over the skeleton [66]. Weight-bearing exercise (e.g., gymnastics and running) improves BMD whereas non-weight-bearing exercise (e.g., swimming and water polo) does not have an equivalent beneficial effect [67–70]. Athletes in weight-bearing sports usually have 5–15% higher BMD than non-athletes [71–73]. Nonspecific resistance exercise does not impact bone density, geometry, or microstructure in young men [74–77].

Low BMD is found in almost 22% of athletes depending on the modality studied [78,79]. The relatively high prevalence in low BMD in athletes, together with the high risk of trauma-related fractures caused by falls or non-traumatic stress fractures induced by overtraining (fatigue fractures), warrants a recommendation that BMD be monitored [43].

Although it is weight-bearing, endurance running exercise has been associated with deleterious effects on bone in some populations, including reduced spinal BMD in endurance runners [80], and stress fractures in runners [38] and military recruits [81]. Both endurance runners and recruits are regularly exposed to high-intensity exercise, while military recruits also frequently perform "common" physical training sessions during which individual relative exercise intensities range from 53% to 73% of heart rate reserve [82]. Recruits with the lowest aerobic fitness will experience the highest relative exercise intensities during these activities and have an increased risk of stress fractures [83]. These findings suggest that high cardiovascular intensity itself might, in part, contribute to a deleterious effect of exercise on bone. The ACSM recommends three to five sessions of impact exercises a week and two to three sessions of resistance exercises, with each session lasting 30–60 min [84].

Bone and the Young Athlete

Pre- and early puberty may be the most opportune time to strengthen the female skeleton [85,86]. Therefore, exercising during growth could represent a primary prevention against the effects of aging on bone. Although physical activity has been shown to improve the BMD throughout the growing period, there are few studies dealing with its effects on bone geometry. A training program for 1 year in prepubertal subjects seems not to influence hip structure [87,88].

Environmental and genetic factors influence an individual's ability to attain peak bone strength during adolescence and young adulthood. Beyond growth, the question of whether adolescent bone remains responsive to exercise warrants attention. Despite cross-sectional studies giving new information on the long-term adaptation of bone to physical activity [89], there are no relevant data on how bone geometry may respond to short-term constraints during adolescence. In the same way, whether such structural response can be linked to the period of growth or the duration of training stressors imposed on the skeleton is not clearly identified. Tenforde and Fredericson [90] in a review study conclude that high-impact exercises promote peak bone mass and may help maintain bone geometry.

Frost and Schönau [89] proposed the concept of the muscle-bone unit in children and adolescents, which suggests that the development of optimal bone strength relies primarily on muscle, because muscle generates the largest mechanical load and strain on bone. Thus sports that involve high acceleration and that produce larger loads on bone (e.g., soccer) will result in greater bone strength than do sports that require submaximal muscle forces (e.g., long-distance running). Duncan et al. [91] studied adolescent girls aged 15–18 years and reported that runners experienced greater BMD values in the femoral neck, legs, and total body than swimmers and cyclists; these results provide evidence supporting the hypothesis that weight-bearing exercises promote bone health. No differences in bone mass have been reported in male adolescent cyclists compared with sedentary control subjects [92]. Male athletes involved in martial arts (judo and karate) have greater BMD in the legs and total body than control subjects or those who participated in water polo [75]. These studies [75,91,92] demonstrate that high impact and weight-bearing activities enhance BMD, particularly in anatomic locations directly loaded by those sports.

Ferry et al. [93] evaluated regional BMD, body composition and hip geometry in elite female adolescent soccer players and swimmers. Lean body mass was higher in the lower extremities of the soccer players and in the upper extremities of the swimmers. However, when compared with the swimmers, the soccer players had higher BMD values at nearly every anatomic location. In the swimmers, no significant differences to reference values were noted.

Participation in sports during the age range in which growth and skeletal maturity occur may result in a higher peak bone mass. For example, participating in sports before menarche has been shown to produce the greatest changes in BMD in the dominant as opposed to the nondominant arm in squash and tennis players, but those differences diminished as age increased at the onset of sports participation [94]. Early puberty may be the most critical time to participate in sports that emphasize weight-bearing and high-impact exercises [86].

Delayed onset of menarche and menstrual dysfunction in young women has been associated with lower BMD in athletes and non-athletes [45]. In swimmers, younger ages of onset training were associated with the presence of menstrual dysfunction [95]. The delayed onset of amenorrhea does not cause osteoporosis immediately, but skeletal demineralization begins moving her BMD in that direction. Similarly, resuming regular menses does not immediately restore optimal bone health, but mineral accumulation favors the improvement of BMD [41]. The impact of exercise practice and bone formation in the maturity period must be studied to a greater extent.

An elevated prevalence of low BMD for age, as well as cross-sectional evidence of suppressed bone mineral accumulation, was recently recognized in a sample of adolescent endurance runners. Bone mass was lower among adolescent runners who exhibited a history of oligomenorrhea or amenorrhea, elevated dietary restraint, lower body mass index (BMI) or lean tissue mass levels, and in those with the longest history of participation in an endurance running sport [85]. Each of these factors may be associated with energy deficiency and is therefore consistent with the results of several controlled laboratory studies that demonstrate a direct negative effect of low energy availability on factors that promote bone formation [86,96]. It remains to be determined whether low bone mass for age among adolescents is irreversible or if bone can undergo "catch-up" mineralization at the end of or after the second decade of life. Several studies among girls with anorexia nervosa have evaluated this possibility, yet they have yielded inconclusive results [97,98]. However, some evidence points to the possibility that girls who recover weight and menstrual function, or who have late pubertal onset, may increase bone mass to near normal levels [99,100]. A study with highly trained female runners reinforced that, at least partially, low BMD is reversible before the age of 30 years, even when competitive running is continued. This seems to be related to restored menses and an increase in body fat [101].

It is important to emphasize that adolescent female athletes may be at a greater risk of having inadequate energy intake [102], and also the ones with DE may have a lower calcium intake adequacy [103].

Bone and the Female Athlete Triad

The rise of the practice of sports by women, and their predisposition to irregular food behaviors, are widely documented [40,41]. Within this context, the number of athletes who developed the Female Athlete Triad (FAT) or presented partial symptoms of it has grown. The term FAT refers to the simultaneous presence of DE, amenorrhea, and osteopenia or osteoporosis [41].

The presence of both an energy deficiency and estrogen deficiency exacerbate alterations of bone metabolism in exercising women [104]. Although low BMD has been associated with DE even in eumenorrheic athletes [105], BMD is lower in amenorrheic athletes than in eumenorrheic athletes [106].

Athletes with menstrual disorders may present skeletal abnormalities, including difficulty in reaching peak bone mass, a low BMD, scoliosis, and stress fractures [107]. An athlete's BMD reflects her cumulative history of energy availability and menstrual status as well as her genetic endowment and exposure to other nutritional, behavioral, and environmental factors. Therefore, it is important to consider both where her BMD is currently and how it is moving along the BMD spectrum [41].

The prevalence of FAT in different sports varies from 0% to 1.36%. In contrast, the isolated components, i.e., DE, menstrual dysfunctions, and bone dysfunctions, vary from 16.8% to 60%, 9.8% to 40%, and 0% to 21.8%, respectively [40,78,79]. Sports that emphasize leanness (e.g., gymnastics and running) increase the risk of FAT [41].

Bone and the Male Athlete

Data about BMD and fracture risk in males, in particular elite male athletes, are limited [108]. However, there is a growing literature on this group [39].

In a study with 63 healthy males (47 cyclists and 16 triathletes), 15 (23.8%) were classified as having low BMD

[39]. Another study, of 23 professional male cyclists, reported that 15 (65.2%) had low BMD values [43].

As in female athletes involved in weight-restricted sports, professional jockeys have an elevated rate of bone loss and reduced bone mass that appears to be associated with disrupted hormonal activity. This may have occurred in response to the chronic weight cycling habitually experienced in competition period as a response to low hormone levels [108].

CONCLUSIONS

Due to the complexity of factors influencing bone mass, athletes should be routinely screened for low BMD. It is important to monitor signs and symptoms and to study the causes of bone disorders. This will allow early diagnosis and interventions on skeletal alterations, thus preserving the athletes' health.

References

[1] Holick MF. Vitamin D deficiency. N Engl J Med 2007;357:266–81.
[2] Soyka LA, Fairfield WP, Klibanski A. Hormonal determinants and disorders of peak bone mass in children. J Clin Endocrinol Metab 2000;85:3951–63.
[3] Rizzoli R, Bonjour JP. Determinants of peak bone mass and mechanisms of bone loss. Osteoporosis Int 1999;9:S17–23.
[4] Wosje KS, Specker BL. Role of calcium in bone health during childhood. Nutr Rev 2000;58:253–68.
[5] Okano H, Mizunuma H, Soda MY, Matsui H, Aoki I, Honjo S, et al. Effects of exercise and amenorrhoea on bone mineral density in teenage runners. Endocr J 1995;42:271–6.
[6] Arasheben A, Barzee KA, Morley CP. A meta-analysis of bone mineral density in collegiate female athletes. J Am Board Fam Med 2011;24:728–34.
[7] Rector RS, Rogers R, Ruebel M, Widzer MO, Hinton PS. Lean body mass and weight-bearing activity in the prediction of bone mineral density in physically active men. J Strength Cond Res 2009;23:427–35.
[8] Heinonen A, Oja P, Kannus P, Sievänen H, Haapasalo H, Mänttäri A, et al. Bone mineral density in female athletes representing sports with different loading characteristics of the skeleton. Bone 1995;17:197–203.
[9] Nakamura O, Ishii T, Ando Y, Amagai H, Oto M, Imafuji T, et al. Potential role of vitamin D receptor gene polymorphism in determining bone phenotype in young male athletes. J Appl Physiol 2002;93:1973–9.
[10] Morrison NA, Qi JC, Tokita A, Kelly PJ, Crost L, Nguyen TV, et al. Prediction of bone density from vitamin D receptor alleles. Nature 1994;367:284–7.
[11] Morrison NA, Yeoman R, Kelly PJ, Eisman JA. Contribution of trans-acting factor alleles to normal physiological variability: vitamin D receptor gene polymorphisms and circulating osteocalcin. Proc Natl Acad Sci USA 1992;89:6665–9.
[12] Kobayashi S, Inoue S, Hoso T, Ouchi Y, Shiraki M, Orimo H. Association of bone mineral density with polymorphism of the estrogen receptor gene. J Bone Miner Res 1996;11:306–11.
[13] Rees JM, Christine MT. Nutritional influences on physical growth and behavior in adolescence. In: Adams G, editor. Biology of adolescent behaviour and development. California: Sage Publications; 1989. p. 195–222.
[14] Bueno AL, Czepielewski MA, Raimundo FV. Calcium and vitamin D intake and biochemical tests in short-stature children and adolescents. Eur J Clin Nutr 2010;64:1296–301.
[15] Key JD, Key JR LL. Calcium needs of adolescents. Curr Opin Pediatr 1994;6:379–82.
[16] Matkovic V. Calcium metabolism and calcium requirements during skeletal modeling and consolidation of bone mass. Am J Clin Nutr 1991;54: 2458–2260.
[17] Allen LH. Calcium bioavailability and absorption: a review. Am J Clin Nutr 1982;35:783–808.
[18] Uusi-Rasi K, Sievanen H, Pasanen M, Beck TJ, Kannus P. Influence of calcium intake and physical activity on proximal femur bone mass and structure among pre- and postmenopausal women. A 10-year prospective study. Calcif Tissue Int 2008;82:171–81.
[19] Calvo MS, Whiting SJ, Barton CN. Vitamin D intake: a global perspective of current status. J Nutr 2005;135:310–6.
[20] Weng FL, Shults J, Leonard MB, Stallings VA, Zemel BS. Risk factors for low serum 25-hydroxyvitamin D concentrations in otherwise healthy children and adolescents. Am J Clin Nutr 2007;86:150–8.
[21] Lips P, Van Schoor NM. The effect of vitamin D on bone and osteoporosis. Best Pract Res Clin Endocrinol Metab 2011;25:585–91.
[22] Loucks AB, Manore MM, Sanborn CF, Sundgot-Borgen J, Warren MP. The female athlete triad: position stand. Med Sci Sports Exerc 2007;39:1867–81.
[23] Loucks AB, Thuma JR. Lutenizing hormone pulsatility is disrupted at a threshold of energy availability in regularly menstruating women. J Clin Endocrinol Metab 2003;88:297–311.
[24] Loucks AB, Verdun M, Heath EM. Low energy availability, not stress of exercise, alters LH pulsatility in exercising women. J Appl Physiol 1998;84:37–46.
[25] De Souza MJ, Miller BE, Loucks AB, Luciano AA, Pescatello LS, Campbell CG, et al. High frequency of luteal phase deficiency and anovulation in recreational women runners: blunted elevation in follicle-stimulating hormone observed during luteal-follicular transition. J Clin Endocrinol Metab 1998;83:4220–32.
[26] Stafford DE. Altered hypothalamic–pituitary–ovarian axis function in young female athletes: implications and recommendations for management. Treat Endocrinol 2005;4:147–54.
[27] De Souza MJ, Leidy HJ, O'Donnell E, Lasley B, Williams NI. Fasting ghrelin levels in physically active women: relationship with menstrual disturbances and metabolic hormones. J Clin Endocrinol Metab 2004;89:3536–42.
[28] Greene DA, Naughton GA. Adaptive skeletal responses to mechanical loading during adolescence. Sports Med 2006; 36:723–32.
[29] Boot AM, Ridder MAJ, Pols HAP, Krenning EP, Muinck Keizer-Schrama SMPF. Bone mineral density in children and adolescents: relation to puberty, calcium intake, and physical activity. J Clin Endocrinol Metab 1997;82:57–62.
[30] Teegarden D, Proulx WR, Martin BR, Zhao J, McCabe GP, Lyle RM, et al. Peak bone mass in young women. J Bone Miner Res 1995;10:711–5.
[31] Specker BL, Wey HE, Smith EP. Rates of bone loss in young adult males. Int J Clin Rheumtol 2010;5:215–28.
[32] Theintz G, Buchs B, Rizzoli R, Slosman D, Clavien H, Sizonenko PC, et al. Longitudinal monitoring of bone mass accumulation in healthy adolescents: evidence for a marked reduction after 16 years of age at the levels of lumbar spine and femoral neck in female subjects. J Clin Endocrinol Metab 1992;75:1060–5.

[33] Baxter-Jones AD, Faulkner RA, Forwood MR, Mirwarld RL, Bailey DA. Bone mineral accrual from 8 to 30 years of age: an estimation of peak bone mass. J Bone Miner Res 2011;26:1729–39.
[34] Bianchi ML. Osteoporosis in children and adolescents. Bone 2007;41:486–95.
[35] Libanati C, Baylink DJ, Lois-Wenzel E, Srinidavan N, Mohan S. Studies on potential mediators of skeletal changes occurring during puberty in girls. J Clin Endocrinol Metab 1999;84:2807–14.
[36] Drinkwater BL, Bruemner B, Chesnut III CH. Menstrual history as a determinant of current bone density in young athletes. JAMA 1990;263:545–8.
[37] Keen AD, Drinkwater BL. Irreversible bone loss in former amenorrheic athletes. Osteoporos Int 1997;7:311–5.
[38] Bennell K, Matheson G, Meewisse G, Brukner P. Risk factors for stress fractures. Sports Med 1999;28:91–122.
[39] FitzGerald L, Carpenter C. Bone mineral density results influencing health-related behaviors in male athletes at risk for osteoporosis. J Clin Densitom 2010;13:256–62.
[40] Schtscherbyna A, Soares EA, Oliveira FP, Ribeiro BG. Female athlete triad in elite swimmers of the city of Rio de Janeiro, Brazil. Nutrition 2009;25:634–9.
[41] American College of Sports Medicine. Position stand: the female athlete triad. Med Sci Sports Exerc 2007;39:1–9.
[42] Reid IR. Relationships among body mass, its components, and bone. Bone 2002;31:547–55.
[43] Medelli J, Lounana J, Menuet J, Shabani M, Cordero-MacIntyre Z. Is osteopenia a health risk in professional cyclists? J Clin Densitom 2009;12:28–34.
[44] De Souza MJ, Williams NI. Beyond hypoestrogenism in amenorrheic athletes: energy deficiency as a contributing factor for bone loss. Curr Sports Med Rep 2005;4:38–44.
[45] Torstveit MK, Sundgot-Borgen J. The female athlete triad exists in both elite athletes and controls. Med Sci Sports Exerc 2005;37:1449–59.
[46] National Institutes of Health Consensus Development Panel. Osteoporosis prevention, diagnosis, and therapy. JAMA 2001;285:785–95.
[47] Ebeling PR. A 12-month prospective study of the relationship between stress fractures and bone turnover in athletes. Calcif Tissue Int 1998;63:80–5.
[48] Kanis JA. Causes of osteoporosis. Osteoporosis. Oxford: Blackwell Scientific; 1994. p. 81–113.
[49] National Osteoporosis Foundation, Clinical indications for bone mass measurements. The Scientific Advisory Board of the National Osteoporosis Foundation 1989;4:1–28.
[50] Kanis JA, McCloskey EV, Johansson H, Oden A, Melton III LJ, Khaltaev N. A reference standard for the description of osteoporosis. Bone 2008;42:467–75.
[51] World Health Organization Study Group. Assessment of fracture risk and its application to screening for postmenopausal osteoporosis. WHO Technical Report Series 843. Geneva 1994:2–25.
[52] Brandão CMA, Camargos BM, Zerbini CA, Plapler PG, Mendonça LMA, Albergaria BH, et al. Posições oficiais 2008 da Sociedade Brasileira de Densitometria Óssea. Arq Bras Endocrinol Metabol 2009;53:107–12.
[53] Gordon CM, Bachrach LK, Carpenter TO, Crabtree N, El-Hajj Fuleihan G, Kutilek S, et al. Dual energy X-ray absorptiometry interpretation and reporting in children and adolescents: the 2007 ISCD Pediatric Official Positions. J Clin Densitom 2008;11:43–58.
[54] World Health Organization. Assessment of osteoporosis at the primary health care level. Summary Report of a WHO Scientific Group. WHO, Geneva 2007.
[55] Consensus Development Statement. Who are candidates for prevention and treatment for osteoporosis? Osteoporosis Int 1997;7:1–6.
[56] Skerry TM. Mechanical loading and bone: what sort of exercise is beneficial to the skeleton? Bone 1997;20:179–81.
[57] Kannus P, Sievanen H, Vuori I. Physical loading, exercise and bone. Bone 1996;S18:1–3.
[58] Carter DR, Van Der Meulen MCH, Beaupre GS. Physical factors in bone growth and development. Bone 1996;S18:5–10.
[59] Skerry TM, Suva LJ. Investigation of the regulation of bone mass by mechanical loading: from quantitative cytochemistry to gene array. Cell Biochem Funct 2003;21:223–9.
[60] Wallace JD, Cuneo RC, Lundberg PA, Rosen T, Jorgensen JO, Longobardi S, et al. Responses of markers of bone and collagen turnover to exercise, growth hormone (GH) administration, and GH withdrawal in trained adult males. J Clin Endocrinol Metab 2000;85:124–33.
[61] Wilks DC, Winwood K, Gilliver SF, Kwiet A, Chatfield M, Michaelis I, et al. Bone mass and geometry of the tibia and the radius of master sprinters, middle and long distance runners, race-walkers and sedentary control participants: a pQCT study. Bone 2009;45:91–7.
[62] Bradney M, Pearce G, Naughton G, Sullivan C, Bass S, Beck T, et al. Moderate exercise during growth in prepubertal boys: changes in bone mass, size, volumetric density, and bone strength: a controlled prospective study. J Bone Miner Res 1998;13:1814–21.
[63] Casez JP, Fischer S, Stussi E, Stalder H, Gerber A, Delmas PD, et al. Bone mass at lumbar spine and tibia in young males—impact of physical fitness, exercise, and anthropometric parameters: a prospective study in a cohort of military recruits. Bone 1995;17:211–9.
[64] Etherington J, Keeling J, Bramley R, Swaminathan R, McCurdie I. The effects of 10 wk military training on heel ultrasound attenuation and bone turnover. Calcif Tissue Int 1999;64:389–93.
[65] Jergas M, Gluer CC. Assessment of fracture risk by bone density measurements. Semin Nucl Med 1997;27:261–75.
[66] Umemura Y, Ishiko T, Tsujimoto H. The effects of jump training on bone hypertrophy in young and old rats. Int J Sports Med 1995;16:364–7.
[67] Lehtonen-Veromaa M, Möttönen T, Irjala K, Nuotio I, Leino A, Viikari J. A 1-year prospective study on the relationship between physical activity, markers of bone metabolism, and bone acquisition in peripubertal girls. J Clin Endocrinol Metab 2000;85:3726–32.
[68] Pettersson U, Nordstrom P, Alfredson H, Henriksson-Larsén K, Lorentzon R. Effect of high impact activity on bone mass and size in adolescent females: a comparative study between two different types of sports. Calcif Tissue Int 2000;67:207–14.
[69] Nikander R, Sievanen H, Heinonen A, Kannus P. Femoral neck structure in adult female athletes subjected to different loading modalities. J Bone Miner Res 2005;20:520–8.
[70] Colletti LA, Edwards J, Gordon L, Shary J, Bell NH. The effects of muscle-building exercise on bone mineral density of the radius, spine, and hip in young men. Calcif Tissue Int 1989;45:12–4.
[71] Fehling P, Aleket L, Clasey J, Rector A, Stillman RJ. A comparison of bone mineral densities among female athletes in impact loading and active loading sports. Bone 1995;17:205–10.
[72] Risser WL, Lee EJ, Leblanc A, Poindexter HB, Risser JM, Schneider V. Bone density in eumenorrheic female college athletes. Med Sci Sports Exerc 1990;22:570–4.
[73] Robinson TL, Snow-Harter C, Taaffe DR, Gillis D, Shaw J, Marcus R. Gymnasts exhibit higher bone mass than runners despite similar prevalence of amenorrhea and oligomenorrhea. J Bone Miner Res 1995;10:26–35.
[74] Greene DA, Naughton GA, Bradshaw E, Moresi M, Ducher G. Mechanical loading with or without weight-bearing activity:

influence on bone strength index in elite female adolescents engaged in water polo, gymnastics, and track-and-field. J Bone Miner Metab 2012;30:580–7.

[75] Andreoli A, Monteleone M, Van Loan M, Promenzio L, Tarantino U, De Lorenzo A. Effects of different sports on bone density and muscle mass in highly trained athletes. Med Sci Sports Exerc 2001;33:507–11.

[76] MacKelvie KJ, McKay HA, Khan KM, Crocker PR. A school-based exercise intervention augments bone mineral accrual in pubertal girls. J Pediatr 2001;139:501–8.

[77] Nilsson M, Ohlsson C, Mellstrom D, Lorentzon M. Sport-specific association between exercise loading and density, geometry, and microstructure of weight-bearing bone in young adult men. Osteoporos Int 2012; (Epub ahead of print).

[78] Vardar SA, Vardar E, Altun GD, Kurt C, Ozturk L. Prevalence of the female athlete triad in Edirne, Turkey. J Sports Sci Med 2005;4:550–5.

[79] Nichols JF, Rauh MJ, Lawson MJ, Ji M, Barkai H. Prevalence of the female athlete triad syndrome among high school athletes. Arch Pediatr Adolesc Med 2006;160:137–42.

[80] Hind K, Truscott JG, Evans JA. Low lumbar spine bone mineral density in both male and female endurance runners. Bone 2006;39:880–5.

[81] Lappe J, Cullen D, Haynatzki G, Recker R, Ahlf R, Thompson K. Calcium and vitamin D supplementation decreases incidence of stress fractures in female navy recruits. J Bone Miner Res 2008;23:741–9.

[82] Rayson MP, Wilkinson DM, Blaker S. The physical demands of CMS(R): an ergonomic assessment. Bristol, UK: Optimal Performance Ltd; 2002.

[83] Välimäki VV, Alfthan H, Lehmuskallio E, Löyttyniemi E, Sahi T, Suominen H, et al. Risk factors for clinical stress fractures in male military recruits: a prospective cohort study. Bone 2005;37:267–73.

[84] American College of Sports Medicine. Physical activity and bone health. Med Sci Sports Exerc 2004; Available from: <http://www.acsm-msse.org> [accessed 23.10.12].

[85] Barrack MT, Rauh MJ, Nichols JF. Cross-sectional evidence of suppressed bone mineral accrual among female adolescent runners. J Bone Miner Res 2010;25:1850–7.

[86] MacKelvie KJ, Khan KM, McKay HA. Is there a critical period for bone response to weight-bearing exercise in children and adolescents? a systematic review. Br J Sports Med 2002;36:250–7.

[87] Alwis G, Linden C, Ahlborg HG, Dencker M, Gardsell P, Karlsson MK. A 2-year school-based exercise programme in pre-pubertal boys induces skeletal benefits in lumbar spine. Acta Paediatr 2008;97:1564–71.

[88] Alwis G, Linden C, Stenevi-Lundgren S, Ahlborg HG, Besjakov J, Gardsell P, et al. A one-year exercise intervention program in pre-pubertal girls does not influence hip structure. BMC Musculoskelet Disord 2008;9:9.

[89] Frost HM, Schönau E. The "muscle-bone unit" in children and adolescents: a 2000 overview. J Pediatr Endocrinol Metab 2000;13:571–90.

[90] Tenforde AS, Fredericson M. Influence of sports participation on bone health in the young athlete: a review of the literature. PM&R 2011;3:861–7.

[91] Duncan CS, Blimkie CJ, Cowell CT, Burke ST, Briody JN, Howman-Giles R. Bone mineral density in adolescent female athletes: Relationship to exercise type and muscle strength. Med Sci Sports Exerc 2002;34:286–94.

[92] Rico H, Revilla M, Hernandez ER, Gomez-Castresana F, Villa LF. Bone mineral content and body composition in postpubertal cyclist boys. Bone 1993;14:93–5.

[93] Ferry B, Duclos M, Burt L, Therre P, Le Gall F, Jaffré C, et al. Bone geometry and strength adaptations to physical constraints inherent in different sports: comparison between elite female soccer players and swimmers. J Bone Miner Metab 2011;29:342–51.

[94] Kannus P, Haapasalo H, Sankelo M, Sievänen H, Pasanen M, Heinonen A, et al. Effect of starting age of physical activity on bone mass in the dominant arm of tennis and squash players. Ann Intern Med 1995;123:27–31.

[95] Schtscherbyna A, Barreto TM, Oliveira FP, Luiz RR, Soares EA, Ribeiro BG. Age of onset training but not body composition is crucial in menstrual dysfunction in adolescent competitive swimmers. Rev Bras Med Esporte 2012;18:161–3.

[96] Chan JL, Mantzoros CS. Role of leptin in energy-deprivation states: normal human physiology and clinical implications for hypothalamic amenorrhoea and anorexia nervosa. Lancet 2005;366:74–85.

[97] Misra M, Prabhakaran R, Miller KK, Goldstein MA, Mickley D, Clauss L, et al. Weight gain and restoration of menses as predictors of bone mineral density change in adolescent girls with anorexia nervosa-1. J Clin Endocrinol Metab 2008;93:1231–7.

[98] Compston JE, McConachie. C, Stott C, Hannon RA, Kaptoge S, Debiram I, et al. Changes in bone mineral density, body composition and biochemical markers of bone turnover during weight gain in adolescents with severe anorexia nervosa: a 1-year prospective study. Osteoporos Int 2006;17:77–84.

[99] Hind K. Recovery of bone mineral density and fertility in a former amenorrheic athlete. J Sports Sci Med 2008;7:415–8.

[100] Fredericson M, Kent K. Normalization of bone density in a previously amenorrheic runner with osteoporosis. Med Sci Sports Exerc 2005;37:1481–6.

[101] Hind K, Zanker C, Truscott J. Five-year follow-up investigation of bone mineral density by age in premenopausal elite-level long-distance runners. Clin J Sport Med 2011;21:521–9.

[102] Bass S, Inge K. Nutrition for special populations: Children and young athletes. In: Burke LM, Deakin V, editors. Clinical sports nutrition. Sydney: McGraw-Hill; 2006. p. 589–632.

[103] Costa NF, Schtscherbyna A, Soares EA, Ribeiro BG. Disordered eating among adolescent female swimmers: dietary, biochemical, and body composition factors. Nutrition 2012; (Epub ahead of print).

[104] De Souza MJ, West SL, Jamal SA, Hawker GA, Gundberg CM, Williams NI. The presence of both an energy deficiency and estrogen deficiency exacerbate alterations of bone metabolism in exercising women. Bone 2008;43:140–8.

[105] Cobb KL, Bachrach LK, Greendale G, Marcus R, Neer RM, Nieves J, et al. Disordered eating, menstrual irregularity, and bone mineral density in female runners. Med Sci Sports Exerc 2003;35:711–9.

[106] Rencken ML, Chesnut III CH, Drinkwater BL. Bone density at multiple skeletal sites in amenorrheic athletes. JAMA 1996;276:238–40.

[107] Zanker CL, Cooke CB, Truscott JG, Oldroyd B, Jacobs H. Annual changes of bone density over 12 years in an amenorrheic athlete. Med Sci Sports Exerc 2004;36:137–42.

[108] Dolan E, McGoldrick A, Davenport C, Kelleher G, Byrne B, Tormey W, et al. An altered hormonal profile and elevated rate of bone loss are associated with low bone mass in professional horse-racing jockeys. J Bone Miner Metab 2012;30:534–42.

[109] Patel A, Coates PS, Nelson JB, Trump DL, Resnick NM, Greenspan SL. Does bone mineral density and knowledge influence health-related behaviors of elderly men at risk for osteoporosis? J Clin Densitom 2003;6:323–30.

CHAPTER

8

Immune Function, Nutrition, and Exercise

Wataru Aoi[1], Yuji Naito[2] and Toshikazu Yoshikawa[2]

[1]Laboratory of Health Science, Graduate School of Life and Environmental Sciences, Kyoto Prefectural University, Kyoto, Japan [2]Department of Molecular Gastroenterology and Hepatology, Graduate School of Medical Science, Kyoto Prefectural University of Medicine, Kyoto, Japan

INTRODUCTION

The immune system protects the body against not only ectogenetic factors (e.g., bacilli and viruses) but also internal factors (e.g., cancer) via non-specific (innate) and specific (acquired or adaptive) mechanisms. Physiological and psychological stresses can alter immune function. Growing evidence has shown that physical exercise modifies both innate and acquired immune systems (Figure 8.1). The effect is closely associated with changes in the number and function of circulating leukocytes, which are mediated via the neuro-immune-endocrine system. Stress hormones and inflammatory cytokines as well as oxidative stress, which can be induced by acute exercise, alter the number and activity of T lymphocytes, natural killer (NK) cells, neutrophils, and macrophages. Regular exercise decreases circulating levels of inflammatory cytokines and oxidative stress and also enhances the function of immune cells in the resting state. In contrast, strenuous exercise increases the production of inflammatory cytokines in muscle tissues and causes delayed-onset muscle damage. In addition, growing evidence suggests that exercise-induced mechanical stress induces the secretion of certain immunoregulatory proteins, including myokines. Myokines are secreted from skeletal muscle cells into the circulation without inducing inflammation. In contrast, the secretion of inflammatory adipokines is reduced by the reduction in body fat that accompanies exercise. These changes in the level of cytokine secretion from metabolic organs affect levels of circulating leukocytes and directly regulate immune function in other organs. Immune function changes in response to acute and chronic exercise and regulates physiological and pathological states such as fatigue, exercise performance, the etiology and development of common diseases, and infection risk (Figure 8.2). These various aspects of the immune response can be affected by the individual's dietary habits. In addition to major nutrients, phytochemicals may also attenuate immune suppression and excess inflammation after high-intensity exercise, thus some factors may have therapeutic efficacy.

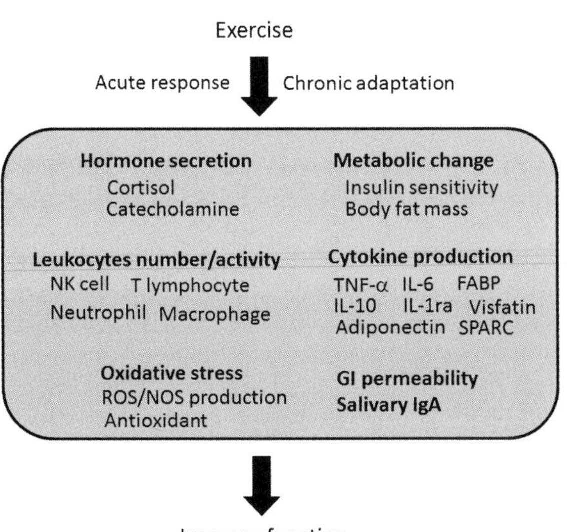

FIGURE 8.1 **Exercise modifies immune systems.** The immune function is regulated by various factors such as the number and activity of leukocytes, hormones, cytokines, oxidative stress, and metabolic factors. Acute exercise transiently changes the level of these immune-related factors. In addition, chronic exercise can adaptively change these factors in the resting state.

Nutrition and Enhanced Sports Performance.
DOI: http://dx.doi.org/10.1016/B978-0-12-396454-0.00008-4

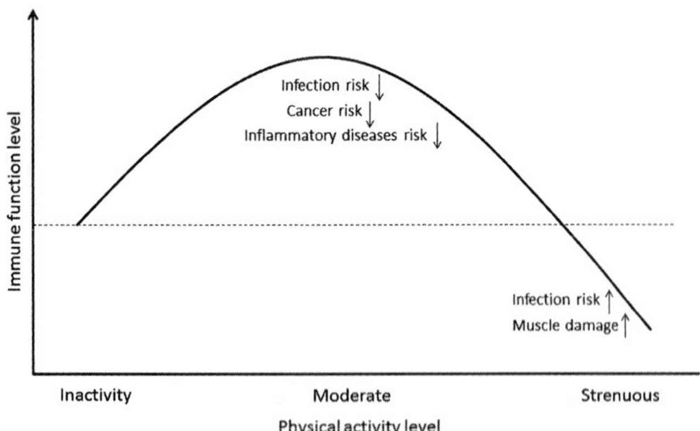

FIGURE 8.2 Physical activity level and immune function. Moderate-intensity exercise improves immune function compared with inactive state, which decreases the incidence of upper respiratory tract infections, cancer, and inflammation-related diseases. On the other hand, strenuous exercise transiently suppresses immune function and can increase the risk of upper respiratory infection and delayed-onset muscle damage.

EXERCISE AND UPPER RESPIRATORY TRACT INFECTION

From a clinical point of view, the most useful index of immune function is the incidence of upper respiratory tract infection (URTI). Several studies have reported that moderate-intensity exercise reduces the incidence of URTI. Matthews et al. [1] reported that healthy adults aged 20–70 years who performed regular moderate-intensity exercise (above 3 metabolic equivalents [METS] for <2 h per day) lowered the risk of URTI by 20% compared with that in inactive subjects. Recent larger scale studies [2,3] also showed that perceived physical fitness and the frequency of aerobic exercise are associated with a decrease in the number of days spent suffering from a URTI as well as the severity of symptoms during cold season. In addition, a study in over 20,000 adults showed that habitual exercise at a low to moderate frequency is beneficial in lowering influenza-associated mortality [4]. Even among elderly people, more active subjects develop fewer URTI symptoms [5]. A randomized control study showed that 8 weeks of moderate-intensity exercise were effective in reducing a given patient's illness burden [6].

Various mechanisms have been proposed to explain the immune activation triggered by moderate regular exercise. Measurements of salivary immunoglobulin A (IgA) levels are widely used as a noninvasive tool to evaluate immune function in athletes. IgA is secreted by plasma cells in the mucous membranes of the intestines and respiratory tract, saliva, urine, tears, and mother's milk. Salivary IgA plays an important role in intraoral specific immunity—it deactivates antigen infectiousness by: preventing adhesion to the mucous membrane epithelium in the upper respiratory tract; creating a barrier to antigen transcytosis; and extruding antigens that have invaded the lamina propria of the mucous membrane. Levels of salivary IgA and URTI prevalence are inversely correlated. A cross-sectional study [7] demonstrated that elderly people who took >7,000 steps per day and engaged in a free-living level of daily physical activity had higher levels of IgA in their saliva. Similarly, several intervention studies [8,9] showed that long-term, moderate regular exercise increases salivary IgA levels in both elderly and young subjects, which likely contributes to the reduced susceptibility to URTI. Moderate regular exercise affects not only salivary levels of IgA but also the circulating leukocyte population profile. Fairey et al. [10] reported that 15 weeks of moderate exercise training increased NK cell activity. Although NK cell activity is suppressed by dietary restriction and the associated weight loss, a combined program of light- to moderate-intensity aerobic and resistance exercise prevents the reduction in NK cell activity observed in obese women [11]. Other intervention studies [12,13] in elderly subjects revealed that long-term commitment to an exercise regimen increases the absolute number of T cells and Th cells; it also enhances the activity of CD28-expressing Th cells and Th1 cells, leading to up-regulated cytokine activity and Th-cell proliferation and differentiation. These changes in the profile of the leukocyte population would contribute to the prevention of infection.

A number of studies have suggested that acute, strenuous, and prolonged exercise increases URTI susceptibility. The symptoms of URTI are increased by 2–6-fold for several weeks following participation in marathon or ultramarathon events [14,15]. Susceptibility to infections following excessive physical activity is associated with an increase in the production of immunosuppressive factors such as adrenocortical hormones and anti-inflammatory cytokines, leading to a decrease in the number and activity of circulating NK cells and T cells as well as a lower IgA concentration in the saliva [16]. Therefore, individuals performing strenuous exercise may exhibit impaired immunocompetence. This period of impaired immunocompetence is referred to as an "open window" [17]—a period when reduced resistance to viral infections allows infectious microorganisms to gain access to the body more easily.

Cross-sectional studies have reported that regular, high-intensity exercise also leads to immunosuppression in the resting state. High-performance endurance athletes are more easily affected by URTI than are sedentary individuals. This typically occurs when athletes are over-training or training heavily prior to a competition. Longitudinal studies of high-level athletes also revealed that the incidence of URTI increases during intense training. These findings indicate an association between increased URTI risk and lower concentrations of salivary IgA as well as systemic immune parameters such as NK cell activity [18–21]. Furthermore, subjects who engaged in a more rigorous exercise program (>11 h/week) had approximately threefold higher levels of interleukin (IL)-2, IL-4, and IL-10 production (as determined by antigen-stimulated whole blood culture) than subjects who engaged in low (3–6 h/week) or moderate (7–10 h/week) levels of exercise [22]. These data suggest that an enhanced anti-inflammatory cytokine response to antigen challenge is associated with an increased risk of URTI.

EXERCISE AND CANCER

Cancer is a heterogeneous group of diseases with multiple causes, and immune dysfunction is closely associated with cancer progression. The tissues and blood of cancer patients exhibit increased levels of inflammatory cytokines, including IL-1, IL-6, and tumor necrosis factor (TNF)-α released from the macrophage or monocyte lineage; reduced levels of IL-2, interferon (IFN)-γ, and class-II MHC molecules, and reduced NK cell activity [23–25]. The treatment of cancer patients often requires the use of a therapeutic method to enhance immunity [26].

Numerous epidemiological investigations have reported on the average individual's level of physical activity and its relationship with the incidence of cancer in Europe, the United States, and Japan. The general consensus among the authors of these studies is that physical activity can prevent cancer in the colon, breast, uterus, pancreas, and lungs. Not only overall physical activity but also occupational and recreational physical activity levels have been shown to reduce the individual's cancer risk [27–30]. Davey Smith et al. [31] showed that men who walk at a slow pace have a greater cancer risk than those who walk at a fast pace, which suggests a connection between exercise intensity and cancer prevention. This relationship is also affected by the frequency of exercise as well as exercise conditions [32]. The World Cancer Research Fund/American Institute for Cancer Research (WCRF/AICR) funded a review of these epidemiological studies. The results were presented in a report entitled "Food, Nutrition, Physical Activity, and the Prevention of Cancer: a Global Perspective [33]". This report stated that physical activity was the only lifestyle change that was sure to reduce an individual's risk of colon cancer. There is no evidence that strenuous exercise increases cancer risk. On the contrary, the mortality of elite athletes is lower than that of the general population [34,35].

Although the exact mechanism underlying the beneficial results reported in epidemiological studies remains unclear, various potential mechanisms have been suggested. Because NK cells destroy carcinoma cells [36], exercise is considered to suppress tumorigenesis by regulating the activation and proliferation of circulating NK cells. NK cells recognize carcinoma cells by identifying mutations of the tumor histocompatibility antigen [37], which means that many tumor cells go undetected. Tumor suppression may also involve antioxidants. In addition, antioxidant action may be involved in antitumorigenesis because gene mutations result from oxidative damage [38], although it is not clear whether exercise leads to prevention of the tumorigenesis via activation of antioxidant enzymes. Other factors such as anti-inflammatory factors, improved insulin sensitivity, and exercise-induced increases in gastrointestinal transit speed have been suggested, but there is no related evidence. Research has suggested that exercise-induced anti-tumorigenesis may be mediated by muscle-secreted myokines. We have recently reported a myokine (secreted protein acidic and rich in cysteine [SPARC]) that is secreted from skeletal muscle into the circulation in response to a single bout of exercise and can prevent colon tumorigenesis [39]. Regular exercise prevented the formation of aberrant crypt foci, which are the precursor lesions of colon adenocarcinoma, on the mucosal surface of the colon in a colon cancer animal model. However, the inhibitory effect of exercise on foci formation was not observed in SPARC-null mice. Cell culture experiments revealed that SPARC secretion from myocytes was induced by cyclic stretch. SPARC inhibited proliferation in this model. In addition, Hojman et al. [40] reported that serum obtained from exercised mice inhibited caspase activation and proliferation in breast cancer cells. Further experiments will be necessary to elucidate the mechanism of exercise-induced cancer prevention.

EXERCISE AND INFLAMMATION

Metabolic disorders and cardiovascular diseases are associated with low-grade continuous inflammation [41,42]. When aging individuals lead a sedentary lifestyle, they increase chronic inflammation and oxidative stress in skeletal muscle, blood, and other tissues. The primary sources of cytokine production are not clear, but it is assumed that certain adipokines, such as TNF-α and IL-6, which are secreted from accumulated visceral adipose tissue, can induce inflammation. These pro-inflammatory cytokines impair glucose transport via the inhibition of insulin signal transduction, which involves activation of the insulin receptor, phosphatidylinositol 3-kinase (PI3-K), and Akt, followed by IκB kinase activation and degradation [43]. Growing evidence suggests that additional adipokines, including resistin, fatty acid binding protein (FABP), and visfatin, can also induce insulin resistance during inflammation [44–46]. In addition, a reduction of circulating adiponectin, an adipokine with anti-inflammatory properties, occurs with obesity and leads to insulin resistance in skeletal muscle and liver [47,48]. Indeed, insulin resistance is associated with elevated TNF-α expression in human skeletal muscle [49]. This indicates that the TNF-α generated by myocytes and other cell types disturbs insulin signaling.

Inflammatory cytokines induce protein degradation through activation of the ubiquitin-proteasome pathway. This is one of the major causes of protein degradation. *In vitro* studies have revealed that the addition of oxidants and TNF-α to myotubes increases protein degradation rates, the ubiquitination of proteins such as myosin, and expression of the main components of the ubiquitin-proteasome pathway [50,51]. Muscle ring finger 1 (MuRF1) and atrogin-1 have been identified as ubiquitin ligases with increased activation during atrophy [52,53]. NF-κB can regulate the ubiquitin-proteasome proteolytic pathway through the induction of MuRF1 and proteasome expression [54,55]. Furthermore, it has been shown that the 20S proteasome can selectively degrade oxidatively-modified proteins without the need for ubiquitination [56]. These observations suggest that protein degradation could be the link among oxidative stress, the inflammatory cascade, and muscle atrophy. Hyperactivity of NF-κB and ubiquitin-proteasome pathway has been identified as a major cause of age-related muscle atrophy [57].

Low- to moderate-intensity training in healthy elderly persons can reduce resting levels of pro-inflammatory markers such as monocytes, C-reactive protein, and IL-6 [58,59]. Regular exercise reduces circulating levels of adipokines such as resistin, visfatin, and FABP, which are involved in metabolic disorders and inflammation [60–62]. The effect of exercise on circulating adiponectin remains to be elucidated. Several studies have suggested that the improvement in insulin sensitivity induced by regular exercise is not mediated by changes in plasma adiponectin [63,64]. However, the ratio of high-molecular-weight proteins to total adiponectin was increased by regular exercise; there was a positive correlation between the increase in the adiponectin ratio and the improvement of insulin sensitivity in older insulin-resistant adults [65]. In addition, it has been shown that a receptor for adiponectin in muscle is elevated in response to physical exercise [66], which potentiates adiponectin's metabolic signal transduction and thus improves aerobic metabolism. IL-6 (a myokine) is transiently secreted from muscle cells in response to a single bout of exercise. IL-6 is considered to reduce inflammation [67]. The level of TNF-α is markedly elevated in anti-IL-6-treated mice and in IL-6-null mice. Recombinant IL-6 infusion inhibits the endotoxin-induced increase in circulating levels of TNF-α in healthy humans [68]. The exercise-induced elevation of IL-6 increases circulating levels of anti-inflammatory cytokines such as IL-1 receptor antagonist and IL-10 [69]. Therefore, the regulation of these adipokines and myokines likely contributes to the prevention of metabolic syndrome by daily exercise.

Appropriate daily exercise increases the antioxidant capacity of skeletal muscle and other tissues through the expression and activation of related enzymes (e.g., superoxide dismutase [SOD] and glutathione peroxidase) [70,71]. There is growing evidence that stimulation by a low concentration of ROS induces the expression of antioxidant enzymes such as SOD and glutathione peroxidase, and other defense systems [72]. The phenomenon is denoted as "hormesis". Namely, ROS generated by regular, moderate exercise can act as signaling factors for upregulation of the defense system. Regular exercise blunts TNF-α expression in muscle and blood [73,74], which results partly from the increase in antioxidant enzymes: TNF-α expression is induced by NF-κB signaling. Furthermore, the inhibition of TNF-α and oxidative stress would lead to an improvement in age-related muscle dysfunction, including protein degradation, muscle fiber apoptosis, and impaired glucose uptake.

DELAYED-ONSET MUSCLE DAMAGE

Unaccustomed and strenuous exercise causes muscle damage that presents clinically as muscular pain and involves protein degradation and ultrastructural changes (delayed-onset muscle damage). Previous studies have shown that delayed-onset muscle damage is mainly induced by mechanical stress, especially eccentric muscle contraction [75], and disturbances of calcium homeostasis [76]. Strenuous exercise leads to phagocyte infiltration into the damaged muscle; this inflammatory response induces delayed-onset muscle damage [77]. In response, certain redox-sensitive transcription factors are relocated to the nucleus; they regulate inflammatory mediators, such as cytokines, chemokines, and adhesion molecules. The infiltration of phagocytes into the tissues expressing these mediators results in proteolysis and ultrastructural damage.

TABLE 8.1 Potential Beneficial Nutrients for Exercise-Induced Immune Response

Nutrient	Suggested Immune Regulatory Function
Carbohydrates	Energy substrate of leukocyte; insulin signaling; regulation of stress hormones and pro-inflammatory cytokines
Antioxidant Vitamins	Antioxidant; regulation of stress hormones and pro-inflammatory cytokines
Protein/Amino Acids	
Glutamine	Energy substrate of leukocytes and intestine
BCAA	Protein metabolism; anti-inflammation; maintenance of glutamine level
Protein hydrolysate	Anti-inflammation
Fermented milk	Antioxidant; anti-inflammation
Other Phytochemicals	
Probiotics	Regulation of intestinal immunity
Astaxanthin	Antioxidant; anti-inflammation; metabolic improvement
Quercetin	Antioxidant; metabolic improvement
Curcumin	Antioxidant; anti-inflammation
β-glucan	Antioxidant; anti-inflammation

BCAA, branched-chain amino acids

Damaged muscle tissue also exhibits increased oxidative damage to cellular components such as lipids, proteins, and DNA [78].

It is well-known that a single bout of exercise improves glucose uptake into skeletal muscle via insulin-dependent and -independent signal transduction mechanisms. This effect is observed for several hours after exercise and often persists until the next day. Elevated glucose uptake requires translocation of glucose transporter 4 (GLUT4) to the plasma membrane after activation of insulin-sensitive signaling [79]. Insulin-stimulated glucose uptake in skeletal muscle decreases after strenuous exercise, partly due to elevated levels of inflammatory cytokines. In particular, TNF-α is well known to impair insulin transduction in muscle tissue. This cytokine blocks insulin-induced glucose uptake and GLUT4 translocation by blocking IR activation and PI3-K/Akt signaling [80]. Oxidative stress also blocks insulin signal transduction in damaged muscle [81]. Recently, we reported that 4-hydroxy-2-nonenal modification of insulin receptor substrate-1 was elevated in the muscle from exercised mice [82]. This finding led to the conclusion that insulin-sensitive glucose transport into muscle can be diminished in damaged muscle after exercise due to oxidative modification of insulin-signaling proteins. In addition to glucose metabolic dysfunction, elevated arterial stiffness and reduced force generation are also observed in subjects with delayed-onset muscle damage [83,84]. These observations suggest that muscle-damaging exercise may be inappropriate for health promotion in patients with metabolic diseases. In addition, the transient reduction in glucose metabolism may represent a disadvantage during pre-game conditioning among athletes.

NUTRITION AND EXERCISE-INDUCED IMMUNE CHANGES

It is important to maintain immune function in order to avoid deficiencies in the levels of nutrients that play an essential role in the activation, interaction, and differentiation of immune cells. Malnutrition decreases immune defenses against invading pathogens and makes the individual more susceptible to infection. Proper nutrition attenuates the inflammation induced by intense exercise. A number of studies have investigated the effect of macro- and micronutrients on exercise-induced immune suppression and excess inflammation (Table 8.1).

Carbohydrates

Prolonged heavy exercise after several days on a very low carbohydrate diet (<10% of dietary energy intake from carbohydrates) increases levels of stress hormones (e.g., adrenaline and cortisol) and cytokines (e.g., IL-6, IL-1 receptor antagonist, and IL-10) [85,86]. Athletes with a carbohydrate-deficient diet are vulnerable to the immunosuppressive effects of cortisol, including the suppression of antibody production, lymphocyte proliferation, and NK cell cytotoxic activity. In contrast, the consumption of carbohydrates (about 1 L/h of a 6% carbohydrate beverage) during exercise attenuates the elevation in plasma catecholamines, cortisol, and cytokines [87]. Carbohydrate ingestion also suppresses IFN-γ production by stimulated T lymphocytes, IL-6, IL-10, and an IL-1-receptor antagonist found in the plasma. Carbohydrate ingestion thus attenuates the pro-inflammatory cascade

[88,89]. Ingestion during exercise has little effect on salivary IgA levels. One study showed that triathletes with a diet high in carbohydrates (12 g-CHO $kgbm^{-1}$ day^{-1}) exhibited higher salivary IgA concentrations after training for 6 days than did the self-selected intake group [90]. Diet appears to have no effect on NK cell activity [91]. The immune changes induced by carbohydrate intake may contribute to the prevention of URTI following prolonged heavy exercise. The underlying mechanism remains unclear but likely involves the endocrine system (e.g., insulin or glucagon), glucose metabolism, and the energy substrate supply to leukocytes and metabolic organs.

Antioxidant Vitamins

Several vitamins are essential for normal immune function. Low levels of lipid-soluble vitamins A and E and the water-soluble vitamins, folic acid, vitamin B6, vitamin B_{12}, and vitamin C, impair immune function and decrease the resistance to infection. Oxidative stress can reduce levels of circulating leukocytes through apoptosis [92] and is therefore though to contribute to the immune suppression induced by exercise. Vitamins C and E are major antioxidants that are effective in scavenging reactive oxygen species in both intracellular and extracellular fluids, which may inhibit the leukocyte apoptosis induced by oxidative stress. Leukocytes contain high levels of vitamin C, which has been reported to have a variety of anti-infective functions, including the promotion of T-lymphocyte proliferation, the prevention of corticosteroid-induced suppression of neutrophil activity, and the inhibition of virus replication [93]. Several randomized, double-blind, placebo-controlled studies have demonstrated that daily supplementation of a high-dose (600–1500 mg) of vitamin C reduced the incidence of URTI symptoms among high-intensity athletes who competed in ultramarathons [94,95]. The combined intake of vitamins C and E significantly suppressed the elevation in cortisol concentration induced by prolonged exercise [96]. The influence of vitamin C and E supplementation on the URTI has not been reproducible. Vitamin E does not appear to contribute to the increases in plasma cytokines, perturbations in other measures of immunity, or oxidative stress.

The enhancement of antioxidant capacity is one way to attenuate delayed-onset muscle damage by regulating oxidative stress. It has been reported that a short- or long-term intake of antioxidant vitamins limits the accumulation of oxidative products and the expression of inflammatory factors in damaged muscle [78,97,98]. These factors may prevent the damage induced by muscle contraction that involves excess mechanical stress (e.g., resistance exercise). The ingestion of antioxidant vitamins may counteract the energy metabolism improvement and antioxidant action associated with regular exercise [99,100], which would counteract benefits such as improved performance and disease prevention. Antioxidants would counteract the oxidative stress that induces healthy adaptations in skeletal muscle. Levels of PPAR gamma coactivator-1 alpha, a key factor regulating metabolic pathways are tightly correlated with intracellular redox levels [101]. On the other hand, some antioxidants (e.g., astaxanthin), some polyphenols, and α-lipoic acid enhance the metabolic improvements induced by exercise [102–105]. Each antioxidant must be considered separately.

Amino Acids

Glutamine is the most abundant free amino acid in human muscle and plasma and is utilized by leukocytes, the gut mucosa, and bone marrow stem cells. Prolonged exercise is associated with a fall in the plasma concentration of glutamine, which can suppress immune function [106,107]. Thus, dietary glutamine supplements may be beneficial in maintaining plasma glutamine concentrations and preventing immune suppression following prolonged exercise. Castell et al. [108] showed that glutamine supplementation (5 g in 330 mL water) immediately after and 2 h after a marathon reduced the incidence of URTI (in the 7 days following the race). Furthermore, glutamine had a beneficial effect on gut function, morbidity and mortality, and on some aspects of immune cell function in clinical studies of diseased or traumatized patients [109,110]. Most studies, however, have not been able to demonstrate that exercise-induced reductions in plasma glutamine levels cause impaired immunity or reduce host protection against viruses in athletes.

Branched-chain amino acids (BCAA), leucine, isoleucine, and valine, account for 35–40% of the dietary essential amino acids in body protein and 14–18% of the total amino acids in muscle proteins [111]. It is well-known that BCAA supplementation before exercise attenuates the breakdown of muscle proteins during exercise and promotes protein synthesis in skeletal muscle. A few studies have suggested that BCAA may mediate immune regulation. Bassit et al. [112] reported that supplementation with BCAA (6 g/day for 15 days) prior to a race in triathletes and marathoners prevented the decline in mitogen-stimulated lymphocyte proliferation after the race as compared with treatment with a placebo. Although the related mechanism is unclear, BCAA may maintain the plasma glutamine concentration. Recent studies showed that a BCAA supplement can attenuate the muscle damage induced by exercise and promote subsequent recovery. Shimomura et al. [113] reported that ingestion of 5 g of BCAAs before exercise can reduce muscle soreness and muscle fatigue for several days after exercise. More recently, a randomized, double-blind, placebo-controlled study demonstrated that BCAA supplementation before and following resistance exercise reduces plasma creatine kinase levels and muscle soreness in male athletes [114].

Probiotics

Exercise also induces mechanical changes in gastrointestinal (GI) barrier function. Exercise increases GI permeability through several mechanisms related to reduced blood flow and hyperthermia in the gut. One study has reported that GI permeability increases after treadmill running at 80% VO_{2max}, which is associated with core temperature [115]. Many kinds of bacteria are used as probiotics, with the most common strains belonging to the *Lactobacilli* and *Bifidobacteria* genuses. A number of studies have shown the beneficial effects in humans, but few have focused on athletes. A double-blind, placebo-controlled cross-over trial investigated the use of *L. fermentum* in elite runners over the 4-month winter training season [116]. Athletes taking the probiotic supplement reported a 50% reduction in the days plagued by respiratory symptoms during the supplementation period (30 days) as compared with the placebo group (72 days). Illness severity was also lower for episodes occurring during the supplementation period. Another study also demonstrated beneficial effect of *L. fermentum* against gastrointestinal and respiratory-tract illness symptoms [117]. In contrast, a randomized double-blind intervention study in which runners took either a placebo or *L. rhamnosus* for 3 months leading up to a marathon reported no significant difference in either URTI or GI symptom episodes in the 2 weeks after the marathon [118], although a tendency toward shorter episodes of GI disturbance was reported in the probiotic group. Probiotic supplementation with *L. casei* DN-114001 by commando cadets during a 3-week training course, followed by a 5-day combat course, had little effect on the incidence of URTI [119]. In addition, a recent double-blind study showed that regular ingestion of *L. salivarius* does not protect against URTI or affect blood leukocyte counts or levels of salivary IgA among endurance athletes [120]. The benefits of probiotics for highly active individuals may vary according to the species of bacteria used. Further research will be necessary to elucidate the effect of probiotics on exercise-induced immune suppression.

Other Nutrients

Several phytochemicals are useful for promoting the health effects of exercise, maintaining homeostasis, and preventing muscle aging. Human or animal studies have shown that caffeine, quercetin, and β-glucan are effective against URTI and changes in cytokine and hormone levels after exercise [121–125]. A double-blind, placebo-controlled study with 40 cyclists showed that ingestion of quercetin (1000 mg) for 3 weeks significantly reduced URTI incidence during the 2-week period following 3 days of exhaustive exercise [122]. However, a single ingestion of quercetin increases plasma quercetin levels but did not suppress post-exercise inflammation or immune changes relative to treatment with a placebo [123]. Recently, it has been shown that β-glucan from mushroom pleurotus ostreatus reduced the incidence of URTI symptoms and increased the number of circulating NK cells during heavy physical training in athletes [125].

Several factors reduce muscle inflammation following exercise. The carotenoid astaxanthin can attenuate exercise-induced damage in mouse skeletal muscle and heart, including associated neutrophil infiltration that induces further damage [126]. In a randomized, double-blind study in soccer players, the astaxanthin group exhibited significantly lower post-exercise levels of creatine kinase and AST than the placebo group [127]. The treatment had no effect on resistance exercise-induced muscle damage [128]. In addition, we have shown that *L. helveticus*–fermented milk prevents the muscle damage induced by acute exercise through the activation of antioxidative enzymes including Mn-SOD and glutathione-S-transferase in skeletal muscle [129]. Fermented milk contains small peptides, which are more easily absorbed by the intestines than are amino acids or large oligopeptides, which may lead to physiological benefits. Peptide-rich wheat gluten hydrolysate also reduces the elevation of creatine kinase after races in athletes [130], which supports the anti-inflammatory properties of small peptides in muscle tissue.

CONCLUSION

Moderate habitual exercise improves immune function and thus reduces URTI risk, the severity of inflammation-related diseases, and the incidence of cancer. On the other hand, acute and habitual strenuous exercise increases URTI risk. Strenuous exercise elevates the inflammatory response in skeletal muscle, leading to delayed-onset muscle damage. Many nutritional programs have been designed to prevent URTI risk and the muscle damage induced by strenuous exercise. A general consensus remains elusive. The effects of exercise likely vary according to environmental conditions, patient age, and the individual's body characteristics. Further research should help in the elaboration of nutritional guidelines to improve immune function.

References

[1] Matthews CE, Ockene IS, Freedson PS, Rosal MC, Merriam PA, Hebert JR. Moderate to vigorous physical activity and risk of upper-respiratory tract infection. Med Sci Sports Exerc 2002;34:1242–8.

[2] Kostka T, Praczko K. Interrelationship between physical activity, symptomatology of upper respiratory tract infections, and depression in elderly people. Gerontology 2007;53:187–93.

[3] Nieman DC, Henson DA, Austin MD, Sha W. Upper respiratory tract infection is reduced in physically fit and active adults. Br J Sports Med 2011;45:987–92.

[4] Wong CM, Lai HK, Ou CQ, Ho SY, Chan KP, Thach TQ, et al. Is exercise protective against influenza-associated mortality? PLoS One 2008;3:e2108.

[5] Kostka T, Drygas W, Jegier A, Praczko K. Physical activity and upper respiratory tract infections. Int J Sports Med 2008;29:158–62.

[6] Barrett B, Hayney MS, Muller D, Rakel D, Ward A, Obasi CN, et al. Meditation or exercise for preventing acute respiratory infection: a randomized controlled trial. Ann Fam Med 2012;10:337–46.

[7] Shimizu K, Kimura F, Akimoto T, Akama T, Otsuki T, Nishijima T, et al. Effects of exercise, age and gender on salivary secretory immunoglobulin A in elderly individuals. Exerc Immunol Rev 2007;13:55–66.

[8] Akimoto T, Kumai Y, Akama T, Hayashi E, Murakami H, Soma R, et al. Effects of 12 months of exercise training on salivary secretory IgA levels in elderly subjects. Br J Sports Med 2003;37:76–9.

[9] Klentrou P, Cieslak T, MacNeil M, Vintinner A, Plyley M. Effect of moderate exercise on salivary immunoglobulin A and infection risk in humans. Eur J Appl Physiol 2002;87:153–8.

[10] Fairey AS, Courneya KS, Field CJ, Bell GJ, Jones LW, Mackey JR. Randomized controlled trial of exercise and blood immune function in postmenopausal breast cancer survivors. J Appl Physiol 2005;98:1534–40.

[11] Scanga CB, Verde TJ, Paolone AM, Andersen RE, Wadden TA. Effects of weight loss and exercise training on natural killer cell activity in obese women. Med Sci Sports Exerc 1998;30:1666–71.

[12] Shimizu K, Kimura F, Akimoto T, Akama T, Tanabe K, Nishijima T, et al. Effect of moderate exercise training on T-helper cell subpopulations in elderly people. Exerc Immunol Rev 2008;14:24–37.

[13] Zhao G, Zhou S, Davie A, Su Q. Effects of moderate and high intensity exercise on T1/T2 balance. Exerc Immunol Rev 2012;18:98–114.

[14] Peters EM, Bateman ED. Ultramarathon running and upper respiratory tract infections. An epidemiological survey. S Afr Med J 1983;64:582–4.

[15] Nieman DC, Johanssen LM, Lee JW, Arabatzis K. Infectious episodes in runners before and after the Los Angeles Marathon. J Sports Med Phys Fitness 1990;30:316–28.

[16] Walsh NP, Gleeson M, Shephard RJ, Gleeson M, Woods JA, Bishop NC, et al. Position statement. Part one: Immune function and exercise. Exerc Immunol Rev 2011;17:6–63.

[17] Pedersen BK, Bruunsgaard H. How physical exercise influences the establishment of infections. Sports Med 1995;19:393–400.

[18] Fahlman MM, Engels HJ. Mucosal IgA and URTI in American college football players: a year longitudinal study. Med Sci Sports Exerc 2005;37:374–80.

[19] Gleeson M, Hall ST, McDonald WA, Flanagan AJ, Clancy RL. Salivary IgA subclasses and infection risk in elite swimmers. Immunol Cell Biol 1999;77:351–5.

[20] Gleeson M, Pyne DB, Austin JP, Lynn FJ, Clancy RL, McDonald WA, et al. Epstein-Barr virus reactivation and upper-respiratory illness in elite swimmers. Med Sci Sports Exerc 2002;34:411–7.

[21] Gleeson M, McDonald WA, Cripps AW, Pyne DB, Clancy RL, Fricker PA. The effect on immunity of long-term intensive training in elite swimmers. Clin Exp Immunol 1995;102:210–6.

[22] Gleeson M, Bishop N, Oliveira M, Tauler P. Influence of training load on upper respiratory tract infection incidence and antigen-stimulated cytokine production. Scand J Med Sci Sports 2011;doi:10.1111/j.1600-0838.2011.01422.x.

[23] Monteleone G, Pallone F, Stolfi C. The dual role of inflammation in colon carcinogenesis. Int J Mol Sci 2012;13:11071–84.

[24] Fantini MC, Pallone F. Cytokines: from gut inflammation to colorectal cancer. Curr Drug Targets 2008;9:375–80.

[25] Szkaradkiewicz A, Karpiński TM, Drews M, Borejsza-Wysocki M, Majewski P, Andrzejewska E. Natural killer cell cytotoxicity and immunosuppressive cytokines (IL-10, TGF-beta1) in patients with gastric cancer. J Biomed Biotechnol 2010;2010:901564.

[26] Dougan M, Dranoff G. Immunotherapy of cancer. In: Wang R, editor. Innate immune regulation and cancer immunotherapy. New York: Springer; 2012. p. 391–414.

[27] Norat T, Bingham S, Ferrari P, Slimani N, Jenab M, Mazuir M, et al. Meat, fish, and colorectal cancer risk: the European Prospective Investigation into cancer and nutrition. J Natl Cancer Inst 2005;97:906–16.

[28] Wei EK, Giovannucci E, Wu K, Rosner B, Fuchs CS, Willett WC, et al. Comparison of risk factors for colon and rectal cancer. Int J Cancer 2004;108:433–42.

[29] Nilsen TI, Vatten LJ. Prospective study of colorectal cancer risk and physical activity, diabetes, blood glucose and BMI: exploring the hyperinsulinaemia hypothesis. Br J Cancer 2001;84:417–22.

[30] Lee KJ, Inoue M, Otani T, Iwasaki M, Sasazuki S, Tsugane S, , et al. JPHC Study Group Physical activity and risk of colorectal cancer in Japanese men and women: the Japan public health center-based prospective study. Cancer Causes Control 2007;18:199–209.

[31] Davey Smith G, Shipley MJ, Batty GD, Morris JN, Marmot M. Physical activity and cause-specific mortality in the Whitehall study. Public Health 2000;114:308–15.

[32] Wu AH, Paganini-Hill A, Ross RK, Henderson BE. Alcohol, physical activity and other risk factors for colorectal cancer: a prospective study. Br J Cancer 1987;55:687–94.

[33] World Cancer Research Fund, and American Institute for Cancer Research. Physical activity. Food, nutrition, physical activity, and the prevention of cancer: a global perspective. Washington: World Cancer Research Fund, American Institute for Cancer Research; 2007. p. 198–209.

[34] Teramoto M, Bungum TJ. Mortality and longevity of elite athletes. J Sci Med Sport 2010;13:410–6.

[35] Sarna S, Kaprio J, Kujala UM, Koskenvuo M. Health status of former elite athletes. The Finnish experience. Aging (Milano) 1997;9:35–41.

[36] Smyth MJ, Cretney E, Kelly JM, Westwood JA, Street SE, Yagita H, et al. Activation of NK cell cytotoxicity. Mol Immunol 2005;42:501–10.

[37] Lanier LL. Up on the tightrope: natural killer cell activation and inhibition. Nat Immunol 2008;9:495–502.

[38] Klaunig JE, Kamendulis LM, Hocevar BA. Oxidative stress and oxidative damage in carcinogenesis. Toxicol Pathol 2010;38:96–109.

[39] Aoi W, Naito Y, Takagi T, Tanimura Y, Takanami Y, Kawai Y, et al. A novel myokine, secreted protein acidic and rich in cysteine (SPARC), suppresses colon tumorigenesis via regular exercise. Gut 2013;62:882–9.
[40] Hojman P, Dethlefsen C, Brandt C, Hansen J, Pedersen L, Pedersen BK. Exercise-induced muscle-derived cytokines inhibit mammary cancer cell growth. Am J Physiol Endocrinol Metab 2011;301:E504–10.
[41] Monteiro R, Azevedo I. Chronic inflammation in obesity and the metabolic syndrome. Mediators Inflamm 2010;2010: pii: 289645.
[42] Wei Y, Chen K, Whaley-Connell AT, Stump CS, Ibdah JA, Sowers JR. Skeletal muscle insulin resistance: role of inflammatory cytokines and reactive oxygen species. Am J Physiol Regul Integr Comp Physiol 2008;294:R673–80.
[43] de Alvaro C, Teruel T, Hernandez R, Lorenzo M. Tumor necrosis factor alpha produces insulin resistance in skeletal muscle by activation of inhibitor kappaB kinase in a p38 MAPK-dependent manner. J Biol Chem 2004;279:17070–8.
[44] Steppan CM, Bailey ST, Bhat S, Brown EJ, Banerjee RR, Wright CM, et al. The hormone resistin links obesity to diabetes. Nature 2001;409:307–12.
[45] Fain JN. Release of inflammatory mediators by human adipose tissue is enhanced in obesity and primarily by the nonfat cells: a review. Mediators Inflamm 2010;2010:513948.
[46] Fukuhara A, Matsuda M, Nishizawa M, Segawa K, Tanaka M, Kishimoto K, et al. Visfatin: a protein secreted by visceral fat that mimics the effect of insulin. Science 2005;307:426–30.
[47] Okamoto Y, Arita Y, Nishida M, Muraguchi M, Ouchi N, Takahashi M, et al. An adipocyte-derived protein, adiponectin, adheres to injured vascular walls. Horm Metab Res 2001;32:47–50.
[48] Hotta K, Funahashi T, Arita Y, Takahashi M, Matsuda M, Okamoto Y, et al. Plasma concentrations of a novel, adipose-specific protein, adiponectin, in type 2 diabetic patients. Arterioscler Thromb Vasc Biol 2000;20:1595–9.
[49] Saghizadeh M, Ong JM, Garvey WT, Henry RR, Kern PA. The expression of TNF alpha by human muscle. Relationship to insulin resistance. J Clin Invest 1996;97:1111–6.
[50] Gomes-Marcondes MC, Tisdale MJ. Induction of protein catabolism and the ubiquitin-proteasome pathway by mild oxidative stress. Cancer Lett 2002;180:69–74.
[51] Li YP, Reid MB. NF-kappaB mediates the protein loss induced by TNF-alpha in differentiated skeletal muscle myotubes. Am J Physiol Regul Integr Comp Physiol 2000;279:R1165–70.
[52] Bodine SC, Latres E, Baumhueter S, Lai VK, Nunez L, Clarke BA, et al. Identification of ubiquitin ligases required for skeletal muscle atrophy. Science 2001;294:1704–8.
[53] Gomes MD, Lecker SH, Jagoe RT, Navon A, Goldberg AL. Atrogin-1, a muscle-specific F-box protein highly expressed during muscle atrophy. Proc Natl Acad Sci USA 2001;98:14440–5.
[54] Cai DS, Frantz JD, Tawa NE, Melendez PA, Oh BC, Lidov HG, et al. IKK beta/NF-kappa B activation causes severe muscle wasting in mice. Cell 2004;119:285–98.
[55] Wyke SM, Tisdale MJ. NF-kappaB mediates proteolysis-inducing factor induced protein degradation and expression of the ubiquitin-proteasome system in skeletal muscle. Br J Cancer 2005;92:711–21.
[56] Grune T, Merker. K, Sandig G, Davies KJ. Selective degradation of oxidatively modified protein substrates by the proteasome. Biochem Biophys Res Commun 2003;305:709–18.
[57] Bar-Shai M, Carmeli E, Ljubuncic P, Reznick AZ. Exercise and immobilization in aging animals: The involvement of oxidative stress and NF-κB activation. Free Radic Biol Med 2008;44:202–14.
[58] Borodulin K, Laatikainen T, Salomaa V, Jousilahti P. Associations of leisure time physical activity, self-rated physical fitness, and estimated aerobic fitness with serum C-reactive protein among 3,803 adults. Atherosclerosis 2006;185:381–7.
[59] Balducci S, Zanuso S, Nicolucci A, Fernando F, Cavallo S, Cardelli P, et al. Anti-inflammatory effect of exercise training in subjects with type 2 diabetes and the metabolic syndrome is dependent on exercise modalities and independent of weight loss. Nutr Metab Cardiovasc 2010;20:608–17.
[60] Kadoglou NP, Perrea D, Iliadis F, Angelopoulou N, Liapis C, Alevizos M. Exercise reduces resistin and inflammatory cytokines in patients with type 2 diabetes. Diabetes Care 2007;30:719–21.
[61] Haus JM, Solomon TP, Marchetti CM, O'Leary VB, Brooks LM, Gonzalez F, et al. Decreased visfatin after exercise training correlates with improved glucose tolerance. Med Sci Sports Exerc 2009;41:1255–60.
[62] Choi KM, Kim TN, Yoo HJ, Lee KW, Cho GJ, Hwang TG, et al. Effect of exercise training on A-FABP, lipocalin-2 and RBP4 levels in obese women. Clin Endocrinol (Oxf) 2009;70:569–74.
[63] Hulver MW, Zheng D, Tanner CJ, Houmard JA, Kraus WE, Slentz CA, et al. Adiponectin is not altered with exercise training despite enhanced insulin action. Am J Physiol Endocrinol Metab 2002;283:E861–5.
[64] Ryan AS, Nicklas BJ, Berman DM, Elahi D. Adiponectin levels do not change with moderate dietary induced weight loss and exercise in obese postmenopausal women. Int J Obes Relat Metab Disord 2003;27:1066–71.
[65] O'Leary VB, Jorett AE, Marchetti CM, Gonzalez F, Phillips SA, Ciaraldi TP, et al. Enhanced adiponectin multimer ratio and skeletal muscle adiponectin receptor expression following exercise training and diet in older insulin-resistant adults. Am J Physiol Endocrinol Metab 2007;293:E421–7.
[66] Blüher M, Bullen Jr. JW, Lee JH, Kralisch S, Fasshauer M, Klöting N, et al. Circulating adiponectin and expression of adiponectin receptors in human skeletal muscle: associations with metabolic parameters and insulin resistance and regulation by physical training. Clin Endocrinol Metab 2006;91:2310–6.
[67] Petersen AM, Pedersen BK. The role of IL-6 in mediating the anti-inflammatory effects of exercise. J Physiol Pharmacol 2006;57:43–51.
[68] Starkie R, Ostrowski SR, Jauffred S, Febbraio M, Pedersen BK. Exercise and IL-6 infusion inhibit endotoxin-induced TNF-alpha production in humans. FASEB J 2003;17:884–6.
[69] Steensberg A, Fischer CP, Keller C, Møller K, Pedersen BK. IL-6 enhances plasma IL-1ra, IL-10, and cortisol in humans. Am J Physiol Endocrinol Metab 2003;285:E433–7.
[70] Lambertucci RH, Levada-Pires AC, Rossoni LV, Curi R, Pithon-Curi TC. Effects of aerobic exercise training on antioxidant enzyme activities and mRNA levels in soleus muscle from young and aged rats. Mech Ageing Dev 2007;128:267–75.
[71] Nakao C, Ookawara T, Kizaki T, Oh-Ishi S, Miyazaki H, Haga S, et al. Effects of swimming training on three superoxide dismutase isoenzymes in mouse tissues. J Appl Physiol 2000;88:649–54.
[72] Limón-Pacheco J, Gonsebatt ME. The role of antioxidants and antioxidant-related enzymes in protective responses to environmentally induced oxidative stress. Mutat Res 2009;674:137–47.
[73] Greiwe JS, Cheng B, Rubin DC, Yarasheski KE, Semenkovich CF. Resistance exercise decreases skeletal muscle tumor necrosis factor a in frail elderly humans. FASEB J 2001;15:475–82.
[74] Gielen S, Adams V, Möbius-Winkler S, Linke A, Erbs S, Yu J, et al. Anti-inflammatory effects of exercise training in the skeletal muscle of patients with chronic heart failure. J Am Coll Cardiol 2003;42:861–8.
[75] Proske U, Morgan DL. Muscle damage from eccentric exercise: mechanism, mechanical signs, adaptation and clinical applications. J. Physiol 2001;537:333–45.

[76] Gissel H, Clausen T. Excitation-induced Ca^{2+} influx and skeletal muscle cell damage. Acta Physiol. Scand 2001;171:327–34.
[77] Tidball JG. Inflammatory cell response to acute muscle injury. Med Sci Sports Exerc 1995;27:1022–32.
[78] Aoi W, Naito Y, Takanami Y, Kawai Y, Sakuma K, Ichikawa H, et al. Oxidative stress and delayed-onset muscle damage after exercise. Free Radic Biol Med 2004;15:480–7.
[79] Bryant NJ, Govers R, James DE. Regulated transport of the glucose transporter GLUT4. Nat Rev Mol Cell Biol 2002;3:267–77.
[80] de Alvaro C, Teruel T, Hernandez R, Lorenzo M. Tumor necrosis factor alpha produces insulin resistance in skeletal muscle by activation of inhibitor kappaB kinase in a p38 MAPK-dependent manner. J Biol Chem 2004;279:17070–8.
[81] Singh I, Carey AL, Watson N, Febbraio MA, Hawley JA. Oxidative stress-induced insulin resistance in skeletal muscle cells is ameliorated by gamma-tocopherol treatment. Eur J Nutr 2008;47:387–92.
[82] Aoi W, Naito Y, Tokuda H, Tanimura Y, Oya-Ito T, Yoshikawa T. Exercise-induced muscle damage impairs insulin signaling pathway associated with IRS-1 oxidative modification. Physiol Res 2012;61:81–8.
[83] Powers SK, Jackson MJ. Exercise-induced oxidative stress: cellular mechanisms and impact on muscle force production. Physiol Rev 2008;88:1243–76.
[84] Barnes JN, Trombold JR, Dhindsa M, Lin HF, Tanaka H. Arterial stiffening following eccentric exercise-induced muscle damage. J Appl Physiol 2010;109:1102–8.
[85] Nieman DC, Henson DA, Smith LL, Utter AC, Vinci DM, Davis JM, et al. Cytokine changes after a marathon race. J Appl Physiol 2001;91:109–14.
[86] Nieman DC, Bishop NC. Nutritional strategies to counter stress to the immune system in athletes, with special reference to football. J Sports Sci 2006;24:763–72.
[87] Nieman DC, Davis JM, Henson DA, Gross SJ, Dumke CL, Utter AC, et al. Skeletal muscle cytokine mRNA and plasma cytokine changes after 2.5-h cycling: influence of carbohydrate. Med Sci Sports Exerc 2005;37:1283–90.
[88] Nieman DC, Davis JM, Henson DA, Walberg-Rankin J, Shute M, Dumke CL, et al. Carbohydrate ingestion influences skeletal muscle cytokine mRNA and plasma cytokine levels after a 3-h run. J Appl Physiol 2003;94:1917–25.
[89] Nieman DC, Henson DA, Davis JM, Dumke CL, Utter AC, Murphy EA, et al. Blood leukocyte mRNA expression for IL-10, IL-1ra, and IL-8, but not IL-6 increases after exercise. J Interferon Cytokine Res 2006;26:668–74.
[90] Costa RJ, Jones GE, Lamb KL, Coleman R, Williams JH. The effects of a high carbohydrate diet on cortisol and salivary immunoglobulin A (s-IgA) during a period of increase exercise workload amongst Olympic and Ironman triathletes. Int J Sports Med 2005;26:880–5.
[91] Nieman DC, Henson DA, Gojanovich G, Davis JM, Murphy EA, Mayer EP, et al. Influence of carbohydrate on immune function following 2-h cycling. Res Sports Med 2006;14:225–37.
[92] Tanimura Y, Shimizu K, Tanabe K, Otsuki T, Yamauchi R, Matsubara Y, et al. Exercise-induced oxidative DNA damage and lymphocytopenia in sedentary young males. Med Sci Sports Exerc 2008;40:1455–62.
[93] Maggini S, Wintergerst ES, Beveridge S, Hornig DH. Selected vitamins and trace elements support immune function by strengthening epithelial barriers and cellular and humoral immune responses. Br J Nutr 2007;98:S29–35.
[94] Peters EM, Goetzsche JM, Joseph LE, Noakes TD. Vitamin C as effective as combinations of anti-oxidant nutrients in reducing symptoms of upper respiratory tract infection in ultramarathon runners. S Afr J Sports Med 1996;11:23–7.
[95] Himmelstein SA, Robergs RA, Koehler KM, Lewis SL, Qualls CR. Vitamin C supplementation and upper respiratory tract infections in marathon runners. J Exerc Physiol 1998;1:1–17.
[96] Davison G, Gleeson M, Phillips S. Antioxidant supplementation and immunoendocrine responses to prolonged exercise. Med Sci Sports Exerc 2007;39:645–52.
[97] Rosa EF, Ribeiro RF, Pereira FM, Freymüller E, Aboulafia J, Nouailhetas VL. Vitamin C and E supplementation prevents mitochondrial damage of ileum myocytes caused by intense and exhaustive exercise training. J Appl Physiol 2009;107:1532–8.
[98] Bryer SC, Goldfarb AH. Effect of high dose vitamin C supplementation on muscle soreness, damage, function, and oxidative stress to eccentric exercise. Int J Sport Nutr Exerc Metab 2006;16:270–80.
[99] Ristow M, Zarse K, Oberbach A, Klöting N, Birringer M, Kiehntopf M, et al. Antioxidants prevent health-promoting effects of physical exercise in humans. Proc Natl Acad Sci USA 2009;106:8665–70.
[100] Gomez-Cabrera MC, Domenech E, Romagnoli M, Arduini A, Borras C, Pallardo FV, et al. Oral administration of vitamin C decreases muscle mitochondrial biogenesis and hampers training-induced adaptations in endurance performance. Am J Clin Nutr 2008;87:142–9.
[101] Powers SK, Talbert EE, Adhihetty PJ. Reactive oxygen and nitrogen species as intracellular signals in skeletal muscle. J Physiol 2011;589:2129–38.
[102] Aoi W, Naito Y, Takanami Y, Ishii T, Kawai Y, Akagiri S, et al. Astaxanthin improves muscle lipid metabolism in exercise via inhibitory effect of oxidative CPT I modification. Biochem Biophys Res Commun 2008;366:892–7.
[103] Murase T, Haramizu S, Shimotoyodome A, Tokimitsu I, Hase T. Green tea extract improves running endurance in mice by stimulating lipid utilization during exercise. Am J Physiol Regul Integr Comp Physiol 2006;290:R1550–6.
[104] Wu J, Gao W, Wei J, Yang J, Pu L, Guo C. Quercetin alters energy metabolism in swimming mice. Appl Physiol Nutr Metab 2012;37:912–22.
[105] Saengsirisuwan V, Perez FR, Sloniger JA, Maier T, Henriksen EJ. Interactions of exercise training and alpha-lipoic acid on insulin signaling in skeletal muscle of obese Zucker rats. Am J Physiol Endocrinol Metab 2004;287:E529–36.
[106] Mackinnon LT, Hooper SL. Plasma glutamine and upper respiratory tract infection during intensified training in swimmers. Med Sci Sports Exerc 1996;28:285–90.
[107] Kargotich S, Goodman C, Dawson B, Morton AR, Keast D, Joske DJ. Plasma glutamine responses to high-intensity exercise before and after endurance training. Res Sports Med 2005;13:287–300.
[108] Castell LM, Poortmans JR, Newsholme EA. Does glutamine have a role in reducing infections in athletes? Eur J Appl Physiol Occup Physiol 1996;73:488–90.
[109] Cavalcante AA, Campelo MW, de Vasconcelos MP, Ferreira CM, Guimarães SB, Garcia JH, et al. Enteral nutrition supplemented with L-glutamine in patients with systemic inflammatory response syndrome due to pulmonary infection. Nutrition 2012;28:397–402.
[110] Newsholme P. Why is L-glutamine metabolism important to cells of the immune system in health, postinjury, surgery or infection? J Nutr 2001;131:2515S–22S.
[111] Rennie MJ. Influence of exercise on protein and amino acid metabolism. In: Rowell LB, Shepherd JT, editors. Handbook of physiology, section 12: exercise: regulation and integration of multiple systems. New York: Oxford University Press; 1996. p. 995–1035.

[112] Bassit RA, Sawada LA, Bacurau RF, Navarro F, Martins Jr. E, Santos RV, et al. Branched-chain amino acid supplementation and the immune response of long-distance athletes. Nutrition 2002;18:376–9.

[113] Shimomura Y, Yamamoto Y, Bajotto G, Sato J, Murakami T, Shimomura N, et al. Nutraceutical effects of branched-chain amino acids on skeletal muscle. J Nutr 2006;136:529S–32S.

[114] Howatson G, Hoad M, Goodall S, Tallent J, Bell PG, French DN. Exercise-induced muscle damage is reduced in resistance-trained males by branched chain amino acids: a randomized, double-blind, placebo controlled study. J Int Soc Sports Nutr 2012;9:20.

[115] Lambert GP. Intestinal barrier dysfunction, endotoxemia, and gastrointestinal symptoms: the 'canary in the coal mine' during exercise-heat stress? Med Sport Sci 2008;53:61–73.

[116] Cox AJ, Pyne DB, Saunders PU, Fricker PA. Oral administration of the probiotic Lactobacillus fermentum VRI-003 and mucosal immunity in endurance athletes. Br J Sports Med 2010;44:222–6.

[117] West NP, Pyne DB, Cripps AW, Hopkins WG, Eskesen DC, Jairath A, et al. Lactobacillus fermentum (PCC®) supplementation and gastrointestinal and respiratory-tract illness symptoms: a randomised control trial in athletes. Nutr J 2011;10:30.

[118] Kekkonen RA, Vasankari TJ, Vuorimaa T, Haahtela T, Julkunen I, Korpela R. The effect of probiotics on respiratory infections and gastrointestinal symptoms during training in marathon runners. Int J Sport Nutr Exerc Metab 2007;17:352–63.

[119] Tiollier E, Chennaoui M, Gomez-Merino D, Drogou C, Filaire E, Guezennec CY. Effect of a probiotics supplementation on respiratory infections and immune and hormonal parameters during intense military training. Mil Med 2007;172:1006–11.

[120] Gleeson M, Bishop NC, Oliveira M, McCauley T, Tauler P, Lawrence C. Effects of a Lactobacillus salivarius probiotic intervention on infection, cold symptom duration and severity, and mucosal immunity in endurance athletes. Int J Sport Nutr Exerc Metab 2012;22:235–42.

[121] Fletcher DK, Bishop NC. Caffeine ingestion and antigen-stimulated human lymphocyte activation after prolonged cycling. Scand J Med Sci Sports 2012;22:249–58.

[122] Nieman DC, Henson DA, Gross SJ, Jenkins DP, Davis JM, Murphy EA, et al. Quercetin reduces illness but not immune perturbations after intensive exercise. Med Sci Sports Exerc 2007;39:1561–9.

[123] Konrad M, Nieman DC, Henson DA, Kennerly KM, Jin F, Wallner-Liebmann SJ. The acute effect of ingesting a quercetin-based supplement on exercise-induced inflammation and immune changes in runners. Int J Sport Nutr Exerc Metab 2011;21:338–46.

[124] Davis JM, Murphy EA, Brown AS, Carmichael MD, Ghaffar A, Mayer EP. Effects of oat beta-glucan on innate immunity and infection after exercise stress. Med Sci Sports Exerc 2004;36:1321–7.

[125] Bergendiova K, Tibenska E, Majtan J. Pleuran (β-glucan from Pleurotus ostreatus) supplementation, cellular immune response and respiratory tract infections in athletes. Eur J Appl Physiol 2011;111:2033–40.

[126] Aoi W, Naito Y, Sakuma K, Kuchide M, Tokuda H, Maoka T, et al. Astaxanthin limits exercise-induced skeletal and cardiac muscle damage in mice. Antioxid Redox Signal 2003;5:139–44.

[127] Djordjevic B, Baralic I, Kotur-Stevuljevic J, Stefanovic A, Ivanisevic J, Radivojevic N, et al. Effect of astaxanthin supplementation on muscle damage and oxidative stress markers in elite young soccer players. J Sports Med Phys Fitness 2012;52:382–92.

[128] Bloomer RJ, Fry A, Schilling B, Chiu L, Hori N, Weiss L. Astaxanthin supplementation does not attenuate muscle injury following eccentric exercise in resistance-trained men. Int J Sport Nutr Exerc Metab 2005;15:401–12.

[129] Aoi W, Naito Y, Nakamura T, Akagiri S, Masuyama A, Takano T, et al. Inhibitory effect of fermented milk on delayed-onset muscle damage after exercise. J Nutr Biochem 2007;18:140–5.

[130] Koikawa N, Nakamura A, Ngaoka I, Aoki K, Sawaki K, Suzuki Y. Delayed-onset muscle injury and its modification by wheat gluten hydrolysate. Nutrition 2009;25:493–8.

CHAPTER

9

The Immune Response to Exercise: Effects on Cellular Mobilization, Immune Function and Muscle Regeneration

Daniel J. Freidenreich and Jeff S. Volek

Human Performance Laboratory, Department of Kinesiology, University of Connecticut, Storrs, CT, USA

At any given time, millions of leukocytes travel through the circulation. However, this represents less than half of the total white blood cells in the body [1,2]. The rest of the leukocytes temporarily reside in what is known as the marginated pool of leukocytes. This "pool" acts like a storage depot, ready to release leukocytes when needed. For example, increases in catecholamines and cortisol in response to exercise can induce the release of leukocytes from the marginated pool. Indeed, after acute bouts of exercise, changes in the concentration and proportion of leukocyte subpopulations are observed. This chapter strives to answer the following questions: To what extent does endurance and resistance exercise affect circulating leukocytes? What functional changes in these cells, if any, occur after they are called from the marginated pool? Why are they called to the circulation in the first place?

THE EFFECTS OF ACUTE EXERCISE ON CIRCULATING LEUKOCYTE COUNTS

During the two decades from the late 1980s to the early 2000s several different research groups documented changes in circulating leukocyte counts after exercise. During the post exercise (PE) period, changes in leukocyte subpopulations follow a distinguishable temporal pattern. While it is acknowledged that various factors can impact the magnitude and time course of the exercise-induced leukocyte response, some factors have a greater impact (acute program variables and age) while others have a lesser influence (e.g., nutrition and sex) [1,3–5].

Endurance Exercise

Monocytes

Monocytes generally increase during exercise and remain elevated during the early PE period for up to 15 min [1,4–12]. From 30 min to about 4 h post exercise, monocytes appear to return and remain at baseline values [1,4,6,7,9,12]. Thereafter a second surge of monocytes in the circulation appears between 5 h and 11 h PE, peaking at 6 h before returning to baseline by 12 h [4,8,9,13].

Neutrophils

The predominant pattern for neutrophils is a continuous increase beginning during exercise and continuing up to 9 h PE with a peak occurring between 2 to 2.5 h [1,5,7,9,10,12,14]. Neutrophils then appear to return and remain at baseline for the next several days [9,12–14,19–21]. However, some studies show no change in neutrophils during the first 1 h post exercise [1,4–7,9,10,12,14–18].

CD4 T Helper Cells

CD4 T helper cells increase during exercise and remain elevated during the PE period for up to 15 min [5–7,12,13,19–21]. Thereafter CD4 T cells return and remain at baseline from 30 min to 24 h PE [5–7,10,12,14,22].

Nutrition and Enhanced Sports Performance.
DOI: http://dx.doi.org/10.1016/B978-0-12-396454-0.00009-6

CD8 Cytotoxic T Cells

CD8 T cells remain at pre-exercise values during exercise and reach peak values immediately PE [4–7,10,12,14,22]. Between 30 min and 2 h PE they then either return to baseline or drop below [4–7,10,12,14,22]. From 3.5 to 24 h PE they return to and remain at baseline [4,7,12]. Reports are evenly split regarding the period from 30 min to 2 h PE as to whether CD8 T cells return to baseline or decrease below it. The cause for this disparity may be due to the intensity and/or duration of exercise. CD8 T cells remain at baseline when the exercise is light to moderate (50–65% VO_{2max}) and of moderate duration (45–60 min) [4,7,14,22] but drop below baseline when the exercise intensity is moderate to high (85% to 150% individual anaerobic threshold (IAT) [5,6,10,12].

CD19 B Cells

CD19 B cells appear to increase immediately PE [4–6]. Between 5 min and 2.5 h PE, B cells either return to baseline or remain elevated [4–7,12]. Thereafter B cells remain at baseline for the duration of the recovery period up to 24 h [5,12].

Natural Killer Cells

Natural killer (NK) cells generally increase during exercise and remain elevated for 15 minutes PE [4,5,7,14,22]. NK cells then return to baseline by 30 min PE and generally remain at baseline for the remainder of the recovery period, except for a potential but transient decrease at 1 h PE [4,5,7,14,22].

Resistance Exercise

Monocytes

Monocytes increase during exercise and show a sustained increase for 2 h PE [3,8,23,24]. Peak levels occur either immediately PE or at 2 h PE [8,23,24].

Neutrophils

Neutrophils increase immediately PE and remain elevated for 2 h PE [3,8,23–26]. The exact peak time is unclear as studies have shown neutrophils peaking immediately PE or at the final measured time points (90 or 120 min) after exercise [8,23–26].

CD4 T Helper Cells

CD4 T Helper cells have shown either no change during and after exercise, or they increase between the immediate post and 15 min post time points prior to returning to baseline by 30 min PE [3,24,26,27].

CD8 Cytotoxic T Cells

CD8 cytotoxic T cells increase immediately PE, followed by a return to baseline 15–45 min after exercise [26,27].

CD19 B Cells

CD19 B cells increase immediately after exercise and return to baseline sometime before 2 h after exercise [3,8,28,29]. Unfortunately a more refined timeframe for their return to baseline has not yet been resolved.

Natural Killer Cells

Natural killer cells increase during exercise through 15 min PE [3,8,26,27]. NK cells then either return to baseline 30–45 min after exercise or decrease below baseline until 2 h after exercise [3,24,26,27]. NK cells may then return to baseline by 4 h PE [3,30].

EXERCISE AND IMMUNE FUNCTION

Chronic participation in moderate exercise as prescribed by the ACSM guidelines has shown decreases in the number of incidences of upper respiratory tract infections (URTI) [31,32]. In an attempt to explain the reduction in illness that appears to be associated with chronic moderate exercise, components of immune cell function have been examined after acute bouts of both endurance and resistance exercise. Although only acute benefits of exercise are covered here, there is a great deal of work concerning the effects of chronic exercise on immune function [33,34]. In this section we discuss immune functional changes after endurance and resistance exercise.

Endurance Exercise

An excellent review of endurance exercise and changes in immune function has been previously authored by Rowbottom [1]. This section contains key points summarized below with additional supporting information.

T Cell Proliferation

Early research indicated that exercise causes a temporary reduction in T cell proliferation after both high and moderate intensity as well as long and short duration [1]. However, this theory may not be entirely true. The proliferation assay used to determine changes in T cell proliferation is a whole blood assay which uses a constant number of cells. However, due to a large increase in the number of NK cells, the proportion of lymphocytes consisting of T cells can decrease, reducing the number of T cells in the cell culture for the assay [1,4,5,8,11]. In order to combat this dilemma, changes in T cell proliferation can be calculated on a per T cell basis [1]. When T cell proliferation is expressed on a per T cell basis, high intensity exercise of short duration does not affect T cell proliferation, while high intensity long duration exercise still results in a decrease [4,8,13].

B Cells and Ig

B cells secrete immunoglobulins, and *in vivo* assessments of Ig production show that Ig do not significantly change in the blood after exercise [1]. Even high intensity exercise of long duration does not increase circulating Igs, and those studies that have seen increases have shown them to be negligible [1,15–17].

Natural Killer Cell Cytotoxic Activity

It is commonly found that during and immediately after exercise, NK cell cytotoxic activity (NKCA) increases, with higher intensity endurance exercise resulting in greater increases in NKCA [13,19–21]. However during the recovery period after long duration endurance exercise, NKCA can be depressed [13,19–21]. Changes in NKCA may be an artifact attributed to changes in the number of NK cells. During and immediately after endurance exercise, NK cell numbers increase markedly. Thus, during whole blood assays, the proportion of NK cells is higher and so the NKCA at these times could appear to be elevated, due to a change in the proportion of NK cells. In a similar fashion to T cell proliferation, NKCA can be reported on a per NK cell basis. When this correction is performed, moderate exercise has been shown to not change NKCA whereas more intense exercise can increase NKCA [1,2,13,35].

Neutrophil Function

Neutrophil function can be assessed by measuring their phagocytic and degranulation capabilities. The phagocytic ability of neutrophils has been shown to remain unchanged or to increase after acute endurance exercise [3–5,36–40]. Degranulation assessed by an increase in circulating proteases shows increases in neutrophil function after long duration moderate intensity exercise and high intensity exercise until fatigue [6,7,9,10,12,41]. Chemotaxis has been shown to increase after both moderate and intense endurance exercise [4,6,7,9,12,36,39].

Resistance Exercise

Compared with endurance exercise, there is significantly less information on the function of immune cells after resistance exercise.

Natural Killer Cell Cytotoxic Activity

In one study NKCA was not affected immediately PE, but 2 h after exercise NKCA decreased below baseline levels [8,9]. NKCA was measured and corrected on a per NK cell basis. These changes in NKCA were observed when NK cells increased by 225% immediately after exercise and at 2 h when NK cells had decreased below baseline, demonstrating that changes in NKCA do not mirror changes in NK cell number [5,7–10,12,14].

T Cell Proliferation

T cell proliferation has demonstrated inconsistent findings during the PE period whereby they either increase, decrease, or do not change following resistance exercise [4–10,12,14,18,29,42,43]. These discrepancies could be due to methodological differences or differences in the resistance exercise protocol used.

MUSCLE DAMAGE AND LEUKOCYTE INFILTRATION

Muscle actions have two components: a shortening phase known as the concentric contraction and a muscle lengthening phase known as the eccentric contraction. The eccentric contraction can be thought of as a brake or a shock absorber [9,12,14,44]. It is the eccentric phase that is primarily responsible for the mechanical muscle damage that elicits an immune response during the post exercise period. Mechanically, muscle damage has been described by the popping sarcomere hypothesis [5–7,12,45]. During the eccentric contraction sarcomeres are not stretched uniformly, resulting in some sarcomeres that are stretched beyond their optimum length [5–7,10,12,14,22,44]. Sarcomeres that are excessively stretched become weak and stretch faster than other sarcomeres, weakening further and potentially reaching a point where thin and thick filaments no longer overlap [4–7,10,12,14,22,44]. This leads to shearing of myofibrils and potential architectural deformations in the muscle fiber, leading to further damage [4–7,10,12,14,22,44]. The myofiber itself can display z-disk streaming, loss of structural proteins, sarcolemmal rupture, widening and detachment of the endomysium and perimysium and even necrosis [4,7,12,46–48]. Muscle cells can also leak enzymes such as creatine kinase and lactate dehydrogenase and proteins such as myosin heavy chain and myoglobin [4,7,14,22,49–51]. In addition, if the muscle damage is significant, then leukocytes can infiltrate into the endomysium and perimysium and even into the muscle fiber itself [5,6,10,12,47,52]. Phagocytes such as macrophages/monocytes and neutrophils are the major subsets of leukocytes that infiltrate the muscle tissue [4–6,52].

Muscle Damage and Endurance Sports

Muscle damage with concomitant leukocyte infiltration into the skeletal muscle after endurance sports has been controversial [4–7,12,48,53–55]. The reason for this is most likely the mode of endurance exercise and the variable amount of muscle damage accrued after the exercise bout. The eccentric component of muscle movement is the primary cause of muscle damage during activity, but many endurance activities do not cause ample eccentric stress. Endurance activities such as cycling, rowing and swimming provide little eccentric strain, and even running may not provide a great deal of eccentric muscle disruption unless downhill segments are part of the running course [5,12,44,49,53]. As an example of the differences in muscle damage based on the incorporation of eccentric strain, a 230 km cycle race, a 67 km marathon (with a 30 km downhill component) and a 3.8 km run (entirely downhill) were compared for markers of muscle damage after exercise. Plasma concentrations of CK and Type I MHC in order from greatest to least in the circulation were 67 km marathon > 3.8 km run > 230 km cycling race [4,5,7,14,22,49]. Therefore when assessing leukocyte infiltration into the skeletal muscle, modes of exercise that stress the eccentric component, such as downhill running or eccentric cycling, are often used in the exercise protocol to determine if leukocyte infiltration occurs after endurance exercise.

The few studies that have looked specifically at endurance exercise have found mixed results [4,5,7,14,22,48,53–55]. When infiltrating leukocytes were observed, it was in the presence of ultrastructural muscle damage such as z-disk streaming, focal damage, and either elevated plasma CK or significant decreases in isokinetic strength of the quadriceps [3,8,23,24,48,53]. When infiltrating leukocytes were not observed, no markers of ultrastuctural damage were assessed, plasma CK increases were modest, and isokinetic force was not significantly decreased [8,23,24,54,55]. In order to classify the severity of muscle damage, Paulsen created three distinct levels of muscle damage as gauged by the decrease in maximal voluntary contraction force-generating capacity during an isokinetic movement [3,8,23–26,56]. According to this classification system *mild exercise-induced muscle damage* is characterized by a decrease in force-generating capacity of <20%, no necrosis, possible leukocyte accumulation, plasma/serum CK activity below ~1,000 IU/L and quick recovery from exercise. *Moderate exercise-induced muscle damage* is characterized by a decrease in force-generating capacity of 20–50%, possible necrosis, a great likelihood of leukocyte accumulation and a recovery time of 2–7 days. *Severe exercise-induced muscle damage* is characterized by a decrease in force-generating capacity of ≥50%, necrosis, leukocyte accumulation, plasma/serum CK activity >10,000 IU/L and a recovery time of >1 week [8,23–26,56].

According to this classification scheme and based on the reported CK values, the studies which did not observe a significant increase in leukocyte accumulation after eccentrically biased exercise fit into the mild exercise-induced muscle damage category. On the other hand, the studies which found a significant increase in infiltrating leukocytes fit into the moderate and mild categories. Muscle damage that falls into the mild category has a small chance of observing leukocyte infiltration, therefore the lack of observation of leukocyte infiltration in some studies may be because the protocols in these studies did not induce enough muscle damage to warrant the efflux of circulating leukocytes out of the circulation and into the muscle tissue to aid in the repair and regeneration process. Leukocytes such as neutrophils and monocytes/macrophages enter the muscle tissue to phagocytose particles and organelles (such as mitochondria) released from damaged muscle fibers and to release growth factors to aid in the repair of damaged tissue [3,24,26,27,48,57–59]. In addition, their presence in necrotic muscle fibers indicates they may mediate the process of muscle fiber death [26,27,52]. Therefore if muscle damage is mild, the signal to leukocytes to leave the circulation and to penetrate the muscle tissue may not be strong enough and so leukocyte accumulation in muscle tissue may not be observed. The timeline of leukocyte infiltration into damaged muscle tissue is not well documented in the endurance exercise literature, but is well documented in the resistance training and eccentrically biased exercise literature.

Muscle Damage Following Resistance Training and Eccentrically Biased Exercise

Although a few studies have determined the extent of muscle damage and leukocyte infiltration into the skeletal muscle after resistance exercise, most literature on damage and leukocyte accumulation in the skeletal muscle describes studies using eccentrically biased exercise. Eccentrically biased protocols such as 300 eccentric repetitions of the quadriceps on an isokinetic dynamometer are used to assess muscle damage and leukocyte infiltration much more commonly than are either endurance or resistance training protocols. The reason for this is that protocols which place more stress on the eccentric contraction will most likely produce the greatest muscle damage and so act as a better exercise stimulus to study muscle damage and leukocyte infiltration. The greater number of studies which use either resistance training or eccentrically biased protocols allows for the distinction of a timeline of muscle damage and leukocyte accumulation after exercise.

There is little to no increase in leukocyte infiltration in the first 3–8 h after exercise [3,8,28,29,60–62]. A significant increase in intramuscular (in the muscle tissue) leukocytes is observed beginning 20–24 h after exercise and this increase is maintained 48 h later [3,8,26,27,60–62]. In contrast to this commonly observed pattern, sedentary individuals have demonstrated an increase in intramuscular leukocytes as early as 30 min post exercise, and the increased leukocyte accumulation continued for 168 h after exercise, peaking at 4 and 7 days post exercise [3,24,26,27,52]. Not only did these sedentary individuals show an accelerated timeline of increase in the skeletal muscle, but the number of intramuscular leukocytes was also much greater [3,30,52]. Therefore, training status may impact the timing and magnitude of increase in intramuscular leukocytes after exercise. In addition, it has also been shown that the timing of intramuscular leukocyte infiltration may occur simultaneously, or close to the timing of ultrastructural muscle damage [31,32,62], although more research is required to show this definitively. Ordinarily intramuscular leukocytes such as macrophages surround damaged muscle fibers in the endomysium and perimysium, but when muscle fibers become necrotic, leukocytes can penetrate into the muscle fiber [33,34,47,52].

CHAPTER SUMMARY

Exercise can change the number of circulating cells during recovery, modify various functional attributes of these cells, and leukocytes can infiltrate the muscle tissue to aid in the repair and regeneration process. Although endurance and resistance exercise appear to display unique leukocyte temporal patterns and functional changes, the shortage of studies concerning resistance exercise limits direct comparisons between these exercise types. In general, monocytes, neutrophils and natural killer cells increase to the greatest extent in the circulation after exercise, but monocytes and neutrophils remain elevated for a longer duration than do natural killer cells. The reason for this may be that monocytes and neutrophils can infiltrate the muscle tissue if muscle fibers become damaged. Once in the muscle tissue monocytes can mature into macrophages and, along with neutrophils, aid in repair and regeneration. These phagocytic cells accomplish their tasks by moving through the endomysium and perimysium to find damaged muscle fibers, phagocytose particles and mitochondria released from the damaged muscle fibers and then release growth factors to aid in the regeneration of the damaged muscle fibers. In order to determine how exercise may impact immune health, changes in the functional capabilities of various leukocyte subpopulations have been studied. Changes have been seen in lymphocyte proliferation, cytotoxic activity, phagocytic and chemotactic capabilities. The relationship between these functional changes and immune health is a topic of research for future studies. In summary, exercise induces several acute changes in the cells of the immune system, and these changes can affect the recovery after exercise and potentially our health.

References

[1] Rowbottom DG, Green KJ. Acute exercise effects on the immune system. Med Sci Sports Exerc 2000;32:S396.

[2] Kuebler WM, Goetz AE. The marginated pool. Eur Surg Res 2002;34:92–100.

[3] Freidenreich DJ, Volek JS. Immune responses to resistance exercise. Exerc Immunol Rev 2012;18:8.

[4] Nieman DC, Miller AR, Henson DA, et al. Effect of high- versus moderate-intensity exercise on lymphocyte subpopulations and proliferative response. Int J Sports Med 1994;15:199–206.

[5] Gabriel H, Schwarz L, Steffens G, Kindermann W. Immunoregulatory hormones, circulating leucocyte and lymphocyte subpopulations before and after endurance exercise of different intensities. Int J Sports Med 1992;13:359–66.

[6] Gabriel H, Urhausen A, Kindermann W. Circulating leucocyte and lymphocyte subpopulations before and after intensive endurance exercise to exhaustion. Eur J Appl Physiol Occup Physiol 1991;63:449–57.

[7] Shinkai S, Watanabe S, Asai H, Shek PN. Cortisol response to exercise and post-exercise suppression of blood lymphocyte subset counts. Int J Sports Med 1996;17:597–603.

[8] Nieman DC, Henson DA, Sampson CS, et al. The acute immune response to exhaustive resistance exercise. Int J Sports Med 1995;16:322.

[9] Smith LL, Bond JA, Holbert D, et al. Differential white cell count after two bouts of downhill running. Int J Sports Med 1998;19:432–7.

[10] Nieman DC, Berk LS, Simpson-Westerberg M, et al. Effects of long-endurance running on immune system parameters and lymphocyte function in experienced marathoners. Int J Sports Med 1989;10:317–23.

[11] Hinton JR, Rowbottom DG, Keast D, Morton AR. Acute intensive interval training and in vitro t-lymphocyte function. Int J Sports Med 1997;18:130–5.

[12] Gabriel H, Urhausen A, Kindermann W. Mobilization of circulating leucocyte and lymphocyte subpopulations during and after short, anaerobic exercise. Eur J Appl Physiol Occup Physiol 1992;65:164–70.

[13] Nieman DC, Miller AR, Henson DA, et al. Effects of high- vs moderate-intensity exercise on natural killer cell activity. Med Sci Sports Exerc 1993;25:1126–34.

[14] Tvede N, Pedersen BK, Hansen FR, et al. Effect of physical exercise on blood mononuclear cell subpopulations and

in vitro proliferative responses. Scand J Immunol 1989;29: 383–9.
[15] Eliakim A, Wolach B, Kodesh E, et al. Cellular and humoral immune response to exercise among gymnasts and untrained girls. Int J Sports Med 1997;18:208–12.
[16] Hanson PG, Flaherty DK. Immunological responses to training in conditioned runners. Clin Sci 1981;60:225–8.
[17] Nehlsen-Cannarella SL, Nieman DC, Jessen J, et al. The effects of acute moderate exercise on lymphocyte function and serum immunoglobulin levels. Int J Sports Med 1991;12:391–8.
[18] Gabriel H, Brechtel L, Urhausen A, Kindermann W. Recruitment and recirculation of leukocytes after an ultramarathon run: preferential homing of cells expressing high levels of the adhesion molecule LFA-1. Int J Sports Med 1994;15(Suppl. 3):S148–53.
[19] Pedersen BK, Tvede N, Hansen FR, et al. Modulation of natural killer cell activity in peripheral blood by physical exercise. Scand J Immunol 1988;27:673–8.
[20] Pedersen BK, Tvede N, Klarlund K, et al. Indomethacin in vitro and in vivo abolishes post-exercise suppression of natural killer cell activity in peripheral blood. Int J Sports Med 1990;11: 127–31.
[21] Shek PN, Sabiston BH, Buguet A, Radomski MW. Strenuous exercise and immunological changes. Int J Sports Med 2007;16: 466–74.
[22] Fry RW, Morton AR, Crawford GP, Keast D. Cell numbers and in vitro responses of leucocytes and lymphocyte subpopulations following maximal exercise and interval training sessions of different intensities. Eur J Appl Physiol Occup Physiol 1992;64:218–27.
[23] Mayhew DL, Thyfault JP, Koch AJ. Rest-interval length affects leukocyte levels during heavy resistance exercise. J Strength. Cond. Res. 2005;19:16–22.
[24] Ramel A, Wagner KH, Elmadfa I. Acute impact of submaximal resistance exercise on immunological and hormonal parameters in young men. J Sports Sci 2003;21:1001–8.
[25] Ramel A, Wagner KH, Elmadfa I. Correlations between plasma noradrenaline concentrations, antioxidants, and neutrophil counts after submaximal resistance exercise in men. Br J Sports Med 2004;38:E22.
[26] Simonson SR, Jackson CG. Leukocytosis occurs in response to resistance exercise in men. J Strength Cond Res 2004;18: 266.
[27] Stock C, Schaller K, Baum M, Liesen H, Weiss M. Catecholamines, lymphocyte subsets, and cyclic adenosine monophosphate production in mononuclear cells and CD4 + cells in response to submaximal resistance exercise. Eur J Appl Physiol Occup Physiol 1995;71:166–72.
[28] Miles MP, Leach SK, Kraemer WJ, Dohi K, Bush JA, Mastro AM. Leukocyte adhesion molecule expression during intense resistance exercise. J Appl Physiol 1998;84:1604–9.
[29] Dohi K, Mastro AM, Miles MP, et al. Lymphocyte proliferation in response to acute heavy resistance exercise in women: influence of muscle strength and total work. Eur J Appl Physiol 2001;85:367–73.
[30] Koch AJ, Potteiger JA, Chan MA, Benedict SH, Frey BB. Minimal influence of carbohydrate ingestion on the immune response following acute resistance exercise. Int J Sport Nutr Exerc Metab 2001;11:149–61.
[31] Gleeson M. Immune function in sport and exercise. J Appl Physiol 2007;103:693–9.
[32] Nieman DC. Special feature for the Olympics: effects of exercise on the immune system: exercise effects on systemic immunity. Immunol Cell Biol 2000;78:496–501.
[33] MacKinnon LT. Chronic exercise training effects on immune function. Med Sci Sports Exerc 2000;32:S369.
[34] Miles M, Kraemer W, Grove D, et al. Effects of resistance training on resting immune parameters in women. Eur J Appl Physiol 2002;87:506–8.
[35] Nieman DC, Ahle JC, Henson DA, et al. Indomethacin does not alter natural killer cell response to 2.5 h of running. J Appl Physiol 1995;79:748–55.
[36] Ortega E, Collazos ME, Maynar M, Barriga C, la Fuente De M. Stimulation of the phagocytic function of neutrophils in sedentary men after acute moderate exercise. Eur J Appl Physiol Occup Physiol 1993;66:60–4.
[37] Gabriel H, Müller HJ, Kettler K, Brechtel L, Urhausen A, Kindermann W. Increased phagocytic capacity of the blood, but decreased phagocytic activity per individual circulating neutrophil after an ultradistance run. Eur J Appl Physiol Occup Physiol 1995;71:281–4.
[38] Gabriel H, Müller HJ, Urhausen A, Kindermann W. Suppressed PMA-induced oxidative burst and unimpaired phagocytosis of circulating granulocytes one week after a long endurance exercise. Int J Sports Med 1994;15:441–5.
[39] Syu G-D, Chen H-I, Jen CJ. Differential effects of acute and chronic exercise on human neutrophil functions. Med Sci Sports Exerc 2012;44:1021–7.
[40] Hack V, Strobel G, Rau JP, Weicker H. The effect of maximal exercise on the activity of neutrophil granulocytes in highly trained athletes in a moderate training period. Eur J Appl Physiol Occup Physiol 1992;65:520–4.
[41] Robson PJ, Blannin AK, Walsh NP, Castell LM, Gleeson M. Effects of exercise intensity, duration and recovery on in vitro neutrophil function in male athletes. Int J Sports Med 1999;20:128–35.
[42] Miles MP, Kraemer WJ, Nindl BC, et al. Strength, workload, anaerobic intensity and the immune response to resistance exercise in women. Acta Physiol Scand 2003;178:155–63.
[43] Potteiger JA, Chan MA, Haff GG, et al. Training status influences T-cell responses in women following acute resistance exercise. J Strength Cond Res 2001;15:185–91.
[44] Morgan DL, Proske U. Popping sarcomere hypothesis explains stretch-induced muscle damage. Clin Exp Pharmacol Physiol 2004;31:541–545.
[45] Morgan DL. New insights into the behavior of muscle during active lengthening. Biophys J 1990;57:209–21.
[46] Fridén J, Lieber RL. Segmental muscle fiber lesions after repetitive eccentric contractions. Cell Tissue Res 1998;293:165–71.
[47] Stauber WT, Clarkson PM, Fritz VK, Evans WJ. Extracellular matrix disruption and pain after eccentric muscle action. J Appl Physiol 1990;69:868–74.
[48] Hikida RS, Staron RS, Hagerman FC, Sherman WM, Costill DL. Muscle fiber necrosis associated with human marathon runners. J Neurol Sci 1983;59:185–203.
[49] Koller A, Mair J, Schobersberger W, et al. Effects of prolonged strenuous endurance exercise on plasma myosin heavy chain fragments and other muscular proteins. Cycling vs running. J. Sports Med. Phys. Fitness 1998;38:10–7.
[50] Overgaard K, Fredsted A, Hyldal A, Ingemann-Hansen T, Gissel H, Clausen T. Effects of running distance and training on Ca^{2+} content and damage in human muscle. Med Sci Sports Exerc 2004;:821–9.
[51] Goodman C, Henry G, Dawson B, et al. Biochemical and ultrastructural indices of muscle damage after a twenty-one kilometre run. Aust J Sci Med Sport 1997;29:95–8.
[52] Paulsen G, Crameri R, Benestad HB, et al. Time course of leukocyte accumulation in human muscle after eccentric exercise. Med Sci Sports Exerc 2010;42:75–85.
[53] Fielding RA, Manfredi TJ, Ding W, Fiatarone MA, Evans WJ, Cannon JG. Acute phase response in exercise. III. Neutrophil

and IL-1 beta accumulation in skeletal muscle. Am J Physiol Regul Integr Comp Physiol 1993;265:R166–72.

[54] Malm C, Nyberg P, Engstrom M, et al. Immunological changes in human skeletal muscle and blood after eccentric exercise and multiple biopsies. J Physiol 2000;529(**Pt 1**):243–62.

[55] Malm C, Sjödin TLB, Sjöberg B, et al. Leukocytes, cytokines, growth factors and hormones in human skeletal muscle and blood after uphill or downhill running. J Physiol 2004;556: 983–1000.

[56] Paulsen G, Mikkelsen UR, Raastad T, Peake JM. Leucocytes, cytokines and satellite cells: what role do they play in muscle damage and regeneration following eccentric exercise? Exerc Immunol Rev 2012;18:42–97.

[57] Lu H, Huang D, Saederup N, Charo IF, Ransohoff RM, Zhou L. Macrophages recruited via CCR2 produce insulin-like growth factor-1 to repair acute skeletal muscle injury. FASEB J 2011;25: 358–69.

[58] Tidball JG, Wehling-Henricks M. Macrophages promote muscle membrane repair and muscle fibre growth and regeneration during modified muscle loading in mice in vivo. J Physiol 2006;578:327–36.

[59] Broek Ten RW, Grefte S, den Hoff Von JW. Regulatory factors and cell populations involved in skeletal muscle regeneration. J Cell Physiol 2010;224:7–16.

[60] MacIntyre DL, Reid WD, Lyster DM, Szasz IJ, McKenzie DC. Presence of WBC, decreased strength, and delayed soreness in muscle after eccentric exercise. J Appl Physiol 1996;80: 1006–13.

[61] Raastad T, Risoy BA, Benestad HB, Fjeld JG, Hallén J. Temporal relation between leukocyte accumulation in muscles and halted recovery 10–20 h after strength exercise. J Appl Physiol 2003;95:2503–9.

[62] Mahoney DJ, Safdar A, Parise G, et al. Gene expression profiling in human skeletal muscle during recovery from eccentric exercise. Am J Physiol Regul Integr Comp Physiol 2008;294: R1901–10.

S E C T I O N 3

SPORTS AND NUTRITION

CHAPTER

10

Vegetarian Athletes

Jing Zhou, Jia Li and Wayne W. Campbell

Department of Nutrition Science, Purdue University, West Lafayette, IN, USA

Vegetarianism is the dietary practice of not consuming meat and possibly other animal-derived foods and beverages. Vegetarian diets can be divided into, but not limited to, four categories: lacto-ovo-vegetarian (with dairy and eggs), lacto-vegetarian (with dairy), ovo-vegetarian (with eggs), and vegan diets (devoid of all animal products). Based on a 2012 national poll, approximately 4% of the adult population in the United States is vegetarian [1]. Factors for choosing a vegetarian diet include environmental concerns, ethical issues, and avoidance of animal-borne diseases [2]. Vegetarian diets are rich in fiber, antioxidants, phytochemicals, fruits, vegetables, and carbohydrate, while lower in protein, saturated fat, cholesterol, and processed foods. The benefits of consuming vegetarian diets are implicated in the prevention of chronic diseases, such as type 2 diabetes, obesity, hypertension, and cancer [2].

The Position of the American Dietetic Association (ADA) and Dietitians of Canada (DC) on Vegetarian Diets states that well-planned vegetarian diets can meet the nutritional needs of competitive athletes [3]. The impact of vegetarian diets on athletic performance is not extensively studied, although several elite vegetarian athletes have risen to the top of the sports world [4]. This chapter will focus on the risks of developing certain nutrient inadequacies among vegetarian athletes and their potential impact on performance. We will also summarize and discuss current studies on vegetarian diets and athlete performance. Recommendations for successfully planning a vegetarian diet will also be discussed.

NUTRITIONAL CONSIDERATIONS FOR VEGETARIAN ATHLETES

Athletes consuming a vegetarian diet may be at a greater risk of developing insufficiencies for the following nutrients: proteins, essential fatty acids, iron, zinc, calcium, vitamin D, and vitamin B_{12} [5]. The higher risks may be the result of increased needs and losses during training, lowered absorption and digestion rates of vegetarian foods, the uneven distribution of nutrients in meat and plant products, and poorly planned meals. Thus, it is recommended for vegetarian athletes to consider both their exercise training and dietary practices for health and performance, and to regularly monitor their nutrition status. Major functions of these nutrients regarding athletic performance and potential consequences when insufficient or deficient are listed in Table 10.1. Related research articles are cited for further reading.

Energy

Inadequate energy intake is one of the major nutritional concerns among vegetarian athletes. Vegetarians tend to have lower energy intake compared with non-vegetarians [6,7]. Athletes have an increased energy need, and it is more challenging for vegetarian athletes to consume adequate energy from their diet, because vegetarian diets have lower energy density and higher fiber content. Inadequate energy intake may result in inferior performance, undesirable weight loss and body composition, as well as inadequacies of certain macro- and micronutrients. For example, more proteins and amino acids in the body will be metabolized for energy when energy intake is insufficient. This is especially detrimental when essential amino acids are metabolized with inadequate intake. However, well-planned vegetarian diets can meet the increased energy needs of athletes. Inclusion of a wide variety of food choices and frequent meals are two strategies for vegetarian athletes to increase energy intake [2]. Vegetarian athletes need to

Nutrition and Enhanced Sports Performance.
DOI: http://dx.doi.org/10.1016/B978-0-12-396454-0.00010-2

TABLE 10.1 Potential Risk of Nutrient Inadequacies and Impact on Performance for Vegetarian Athletes

Nutrients	Nutrient Functions Related to Performance	Impact on Performance when Inadequate	Primary Food Sources	Suggested Readings
Proteins & essential amino acids	Maintenance, repair, and synthesis of skeletal muscle	Reduced muscle mass	Soy products, beans and legumes, eggs, tofu, dairy products	a
Essential fatty acids (n-3 fatty acids)	Attenuate tissue inflammatory process and oxidative stress	Muscle fatigue, pain, and swelling as a result of inflammation	Fish, eggs, canola oil, flaxseed, nuts, soybeans	b, [17]
Iron	Oxygen-carrying capacity and energy production; synthesis of hemoglobin and myoglobin	Impaired muscle function and limited work capacity; lowered oxygen uptake; lactate buildup; muscle fatigue	Fortified foods, legumes, dried beans, soy foods, nuts, dried fruits, and green leafy vegetables	c–f
Zinc	Growth, building, and repair of muscle tissue; energy production; immune status	Decreases in cardiorespiratory function, muscle strength, and endurance	Legumes, whole grains, cereals, nuts and seeds, soy and dairy products	g,h
Vitamin B_{12}	Proper nervous system function; homocysteine metabolism; production of red blood cells; protein synthesis; tissue repair and maintenance	Anemia; reduced endurance and aerobic performance; neurological symptoms	Dairy products, eggs, fortified foods and beverages	i
Vitamin D	Calcium metabolism; bone health; development and homeostasis of the nervous system and skeletal muscle; cardiovascular fitness	Lower muscle strength and muscle mass; inflammatory disease; increased incidence of bone fracture	Dairy products, eggs, fortified foods and beverages	j,k, [59]
Calcium	Growth, maintenance, and repair of bone tissue; maintenance of blood calcium concentration; regulation of muscle contraction; normal blood clotting; nerve transmission	Increased risk of low bone-mineral density and stress fractures; menstrual dysfunction among female athletes	Dairy products, calcium-fortified tofu, calcium-fortified foods and beverages	[28,59]

[a] *Phillips SM. The science of muscle hypertrophy: making dietary protein count. Proc Nutr Soc. 2011; 70(1): 100–3.*
[b] *Mickleborough TD. Omega-3 Polyunsaturated Fatty Acids in Physical Performance Optimization. Int J Sport Nutr Exerc Metab. 2012.*
[c] *Lukaski HC, Hall CB, Siders WA. Altered metabolic response of iron-deficient women during graded, maximal exercise. Eur J Appl Physiol Occup Physiol. 1991; 63(2): 140–5.*
[d] *Friedmann B, Weller E, Mairbaurl H, Bartsch P. Effects of iron repletion on blood volume and performance capacity in young athletes. Med Sci Sports Exerc. 2001; 33(5): 741–6.*
[e] *Hinton PS, Giordano C, Brownlie T, Haas JD. Iron supplementation improves endurance after training in iron-depleted, nonanemic women. J Appl Physiol. 2000; 88(3): 1103–11.*
[f] *Brownlie Tt, Utermohlen V, Hinton PS, Giordano C, Haas JD. Marginal iron deficiency without anemia impairs aerobic adaptation among previously untrained women. Am J Clin Nutr. 2002; 75(4): 734–42.*
[g] *Lukaski HC. Low dietary zinc decreases erythrocyte carbonic anhydrase activities and impairs cardiorespiratory function in men during exercise. Am J Clin Nutr. 2005; 81(5): 1045–51.*
[h] *Lukaski HC, Bolonchuk WW, Klevay LM, Milne DB, Sandstead HH. Maximal oxygen consumption as related to magnesium, copper, and zinc nutriture. Am J Clin Nutr. 1983; 37(3): 407–15.*
[i] *Herrmann W, Obeid R, Schorr H, Hubner U, Geisel J, Sand-Hill M, et al. Enhanced bone metabolism in vegetarians—the role of vitamin B12 deficiency. Clin Chem Lab Med. 2009; 47(11): 1381–7.*
[j] *Vitamin d and athletic performance: the potential role of muscle. Asian J Sports Med. 2011; 2(4): 211–9.*
[k] *Vitamin D, muscle function, and exercise performance. Pediatr Clin North Am. 2010; 57(3): 849–61.*

routinely monitor their weight to ensure that they are consuming adequate energy.

Macronutrients

Protein and Essential Amino Acids

Protein is an essential macronutrient required for muscle synthesis and recovery. The Recommended Dietary Allowance (RDA) for protein intake is $0.8\ g\ kg^{-1}\ d^{-1}$ for adults older than 18 years. Generally, athletes, even vegetarian athletes, tend to consume more proteins than the RDA [8,9]. However, some plant protein sources are of poor quality and low digestibility, such as cereals. Vegetarian athletes consuming vegan diets or poorly planned meals may not be consuming enough proteins or certain essential amino acids, i.e., lysine, threonine, tryptophan, or sulfur-containing amino acids [9,10]. Thus, even with an adequate total protein intake, vegetarian athletes may not meet their essential amino acids needs [9]. This could lead to sub-optimal health, as well as inferior performance. The RDA of 0.8 g kg^{-1} d^{-1} for protein applies equally to healthy vegetarians and omnivores, provided the vegetarians consume a variety of complementary plant proteins [11].

The ADA, DC, and American College of Sports Medicine position recommends higher protein intakes for omnivorous endurance and strength athletes, at 1.2–1.4 g kg^{-1} d^{-1} and 1.2–1.7 g kg^{-1} d^{-1}, respectively [5]. A further 10% increase in protein intake is advised for vegetarians, to achieve approximately 1.3–1.8 g kg^{-1} d^{-1}[8]. Soy and bean products included in vegetarian diets can help provide essential amino acids which other plant protein sources may be lacking. Other good sources of plant proteins include beans, legumes, peas, nuts and seeds. Protein or amino acid supplementation is not necessary provided that adequate energy intake and a variety of plant protein sources are consumed when planning a vegetarian diet for athletes. Adequate energy intake, especially from carbohydrate, can spare amino acids from energy production versus protein synthesis [12].

Carbohydrate

The recommended carbohydrate intake for athletes ranges from 6 to 10 g kg^{-1} d^{-1}[5]. Adequate carbohydrate intake optimizes glycogen stores and provides readily available energy. It is well accepted that optimal carbohydrate intake helps sustain physical performance [13], especially endurance performance [14] Several cohort studies have confirmed that vegetarians consume a higher percentage of energy from carbohydrate than omnivores [6,15]. It is less likely for vegetarian athletes to have inadequate carbohydrate intake provided that adequate energy intake is achieved.

Fat and Essential Fatty Acids

Fat is the macronutrient that provides energy and aids the absorption of fat-soluble vitamins A, D, E, and K [5,16]. Hormone production and maintenance of cell membranes also utilize fatty acids. Research has shown that vegetarians tend to consume less fat—especially vegans, who exclude all animal products [6,15]. Intakes of n-3 fatty acids are marginal among vegetarians, including essential fatty acids eicosapentaenoic acid (EPA), docosahexaenoic acid (DHA) [6,17]. EPA and DHA are important for cardiovascular health, eye and brain development, as well as controlling exercise-induced inflammation. Good sources of fat in a vegetarian diet include walnuts, flaxseed, canola oil, and soy. Foods or drinks fortified with DHA or EPA are also available in the market now and can be consumed to ensure adequate intake.

Micronutrients

Iron

Iron functions in the delivery of oxygen to tissues and energy production at the cellular level. Athletes, especially female athletes, are at greater risk for developing depletion of iron stores, with or without anemia [5]. Hemolysis, increased iron loss, low iron intake, and poor absorption due to inflammation of the intestine resulting from training may compromise iron balance [18].

Vegetarians and omnivores may consume an equal amount of total iron from their diets [6]. However, vegetarian diets contain mainly non-heme iron that has lower availability than heme iron from animal products. Even though both inhibitors and enhancers of iron absorption are present in plant food sources, such as phytates and organic acids, respectively, iron digestion and absorption efficiency are lower for vegetarian diets. Studies found that iron stores in vegetarians are lower than in omnivores, while no significant difference in the incidence of iron-deficient anemia between the two populations [19]. Since it is a slow process (3–6 months) to reverse iron-deficient anemia, preventative measures should be taken. An approximately two-fold higher iron intake, at 14 mg d^{-1} and 32 mg d^{-1} for adult men and adult premenopausal women, respectively, is advised for nonathletic vegetarians [20]. Considering that athletes have a higher iron loss associated with their training regimen, vegetarian athletes may benefit from an iron intake greater than these amounts. Female vegetarians who do not have adequate energy intake are at a greater risk for developing iron-deficient anemia. It is strongly recommended that vegetarian athletes monitor iron status on a regular basis to prevent iron deficiency.

Vitamin B_{12}

Similar to iron, animal products are the major sources of vitamin B_{12}. Physiological processes including red blood cell production, protein synthesis, tissue repair, and maintenance of the central nervous system all depend on availability of vitamin B_{12} in the body. The RDA for Vitamin B_{12} is 2.4 µg d^{-1} for adults. Vegetarians tend to have inadequate intake and low serum vitamin B_{12} concentrations. Strict vegetarians (vegans) have lower serum B_{12} concentrations than those in lacto-ovo-vegetarians and omnivores [21] and are at increased risk for vitamin B_{12} depletion [9]. Deficiency of vitamin B_{12} may result in macrocytic anemia with lowered oxygen transport, which consequently leads to inferior athletic performance. Prolonged vitamin B_{12} deficiency may cause irreversible neurological damage. It is recommended that vegetarians regularly consume reliable vitamin B_{12} sources, such as fortified foods and supplements.

Zinc

Zinc is required for a broad range of cellular and physiological functions. Specific performance-related functions of zinc include building and repair of muscle tissue, energy production, glucose metabolism, regulation of thyroid hormone, and protein use [22]. Zinc deficiency will not only result in poor exercise performance, but also overall health problems. Animal products contribute 50–70% of dietary zinc in an omnivorous diet [23]. Legumes, whole grains, nuts, seeds, soy and dairy products, which are abundant in vegetarian diets, may provide adequate total zinc intake. But zinc bioavailability in these foods is lower, possibly due to the high phytate content [24]. Research has shown that zinc intakes are near or below the recommended levels among vegetarians [25]. With the lowered dietary zinc absorption and increased zinc loss in sweat and urine associated with training [26], vegetarian athletes are more prone to develop zinc insufficiency even though overt zinc deficiency among western vegetarians is rare [3].

Vitamin D

Vitamin D plays an important role in calcium absorption, regulation of serum calcium concentration, and maintenance of bone health. It is also involved in the development and homeostasis of the nervous system and skeletal muscle, reducing inflammation, and lowering the risks of chronic diseases [5,27]. Vitamin D status depends on cutaneous production of vitamin D_3 upon sunlight exposure and dietary intake. The RDA of vitamin D is 15 $\mu g\, d^{-1}$ for people aged 19–50 years old. Vegetarians have a lower vitamin D intake than non-vegetarians, even though both groups have an intake below the RDA [25]. Vegans who avoid dairy products have the lowest vitamin D intake and low serum 25-hydroxyvitamin D concentrations [15,25]. Reduced bone mass and increased incidences of bone fractures are reported in vegans [28]. It is recommended that vegetarians include vitamin D fortified foods in their diets or vitamin D supplements to ensure adequate intake and prevention of bone fractures [28,29]. Vitamin D-fortified juice, soymilk, and cereals are commercially available for vegans in place of milk products for increasing vitamin D intake.

Calcium

Calcium is especially important for athletes. It is involved in the growth, maintenance, and repair of bone tissue, regulation of muscle contraction, nerve conduction, and normal blood clotting [5]. Adequate calcium and vitamin D intakes improve bone mineral density and prevent fractures. Vegetarians, except for vegans, have similar or higher calcium intake than those of non-vegetarians [15,25]. Vegans who avoid all dairy products have the lowest calcium intake, which falls below the RDA [15,25]. A 30% higher risk of fracture is also observed among vegans, compared with lacto-ovo-vegetarians or omnivores [28]. Other dietary factors associated with vegetarian diets may affect calcium status, including sodium, phosphate, and potassium [30]. Athletes tend to consume sport drinks in order to maintain electrolyte balance. This is potentially problematic for vegan athletes with low calcium intake, because high sodium intake increases urinary calcium excretion. On the other hand, fruits and vegetables are high in potassium and magnesium, which produce a high renal alkaline load, reducing urinary calcium losses and bone resorption [31]. Calcium absorption may be inhibited by oxalates and phytates that are abundant in a vegetarian diet. Some good sources of calcium include green-leafy vegetables (e.g., bok choy, broccoli, Chinese cabbage, collards, and kale), calcium-set tofu, and calcium fortified fruit juices.

The Female Athlete Triad

Vegetarianism may be adopted by some athletes in order to achieve an ideal body composition or weight. With constant monitoring and nutrient assessment, an athlete can achieve their goal safely and effectively following a well-planned vegetarian diet. However, female vegetarians, especially vegans, may be at a greater risk for the female athlete triad, which encompasses disordered eating, amenorrhea, and osteoporosis [32]. There is no causal relationship shown between vegetarianism and disordered eating, but young women with anorexia nervosa more frequently adopt a vegetarian diet [9,33]. With the lower intake of iron, calcium, dietary fat, and energy associated with vegetarian diet, female vegetarian athletes may have a higher risk of developing the female athlete triad [5].

Recommended Practices when Consuming a Vegetarian Diet

Vegetarian diets can meet the needs of competitive athletes, provided the meals are well planned and wise food choices are made [3]. There are several strategies that can help ensure adequate nutrient intake, such as consumption of supplementation and fortified foods and beverages. Constant assessment of dietary adequacy and blood work analyses are recommended to monitor diet and health of vegetarian athletes [3].

VEGETARIAN DIET AND ATHLETIC PERFORMANCE

Vegetarian diets are considered by endurance athletes for higher carbohydrate intake, improvements in body composition, and weight control to enhance athletic performance.

High Carbohydrate Content of Vegetarian Diet

In 1988, Nieman [34] reported consistency among several studies performed in the 1900s that found endurance performance was enhanced in vegetarian athletes compared with their omnivorous counterparts. One study measured the maximum number of times that 25 students could lift a weight on a pulley by squeezing a handle [35]. The vegetarian group performed 69 contractions while the omnivore group performed 38. Another study engaged athletes trained on a full-flesh diet, athletes who abstained from meat, and sedentary vegetarians [36]. The maximum length of time that their arms could be held out horizontally was measured. Only 2 of the 15 omnivores were able to maintain longer than 15 minutes, while 22 of 32 vegetarians exceeded 15 minutes, with one surpassing 3 hours. Nieman suggested that the superior performance was partly due to the vegetarians' motivation to perform and partly due to higher carbohydrate intake [34]. He pointed out that the proportion of carbohydrate utilized for energy production increases with increasing exercise intensity. Since vegan or lacto-ovo-vegetarian diets typically contain a greater proportion of carbohydrate than non-vegetarian diets, the resulting increased glycogen storage may be beneficial during prolonged endurance exercise [2].

The 2003 consensus of the International Olympic Committee on nutrition for athletes [37] indicated that: "A high carbohydrate diet in the days before competition will help enhance performance, particularly when exercise lasts longer than about 60 minutes" and "Athletes should aim to achieve carbohydrate intakes that meet the fuel requirements of their training programs and also adequately replace their carbohydrate stores during recovery between training sessions and competition. This can be achieved when athletes consume carbohydrate-rich snacks and meals that also provide a good source of protein and other nutrients".

While most studies promoting higher carbohydrate intake for athletes have focused on the effect of carbohydrate intake during or at the period immediately before exercise, the long-term benefit of higher dietary carbohydrate intake has not been addressed. The concept of reduced carbohydrate intake (train low) during chronic endurance training was highlighted by recent findings demonstrating that the transcription of a number of metabolic genes involved in training adaptations are enhanced when exercise is undertaken with low muscle glycogen content [38,39]. There is no clear evidence that consuming a diet low in carbohydrate during training enhances exercise performance [40]. The effects of high carbohydrate intake on athletic performance are not likely due to consuming a vegetarian diet *per se*.

Vegetarian Diets and Athletic Performance

In contrast to the results of studies done in the 1900s that showed an improved performance of vegetarian athletes [34], Hanne et al. found no differences in aerobic or anaerobic capacities between habitual vegetarians and non-vegetarians [41]. In this study, researchers did a series of cross-sectional comparisons between 49 vegetarians (29 men and 20 women), all of whom consumed a vegetarian diet for at least 2 years, and 49 age, sex, body size and athletic activities matched non-vegetarian athletes. Anthropometric, metabolic, and fitness parameters tested include: pulmonary functions, heart rate, blood pressure, physical working capacity at a heart rate of 170, predicted maximum oxygen uptake (VO_{2max}), perceived rate of effort from a cycle ergometer stress test, and total power, peak power, and percent fatigue from a Wingate anaerobic test. The authors concluded that habitual consumption of vegetarian diets does not influence athletic performance. However, as pointed out by Venderley and Campbell [2], the variety of vegetarian diets the subjects followed, the different sports they performed, and the large age range of subjects are each potential confounders of the data.

To control for the macronutrient distribution of diets, two reports were published based on a study that investigated endurance performance and immune parameters in 8 male athletes after a lacto-ovo-vegetarian diet and a mixed western diet were consumed for 6 weeks [42,43]. Both diets consisted of 57% energy from carbohydrate, 14% from protein, and 29% from fat. No differences were observed between the intervention and control diet for VO_{2max}, maximal voluntary contraction, and isometric endurance at 35% of maximal voluntary contraction on quadriceps muscle and elbow flexors. Despite that, the researchers observed a decrease in serum testosterone concentration in the athletes only after they consumed the lacto-ovo-vegetarian diet for 6 weeks. This was inferred to be a result of increased fiber intake in the lacto-ovo-vegetarian diet because fiber binds to steroid hormones *in vitro*. Since testosterone augments muscle protein synthesis and muscle mass more quickly than performance, the short duration of this intervention study could be a reason why the authors did not see any changes in performance.

Immune parameters were also tested in the same study [42]. The concentration and proliferation of blood mononuclear cells, spontaneous or stimulated natural killer cell activity, were not different whether the subjects consumed the mixed western diet or the vegetarian diet. As the authors mentioned, another inconsistency between the two diets was fat composition. The polyunsaturated fatty acid/saturated fatty acid ratio was 1.24 for the vegetarian diet and 0.5 for the mixed western diet. Since diets rich in polyunsaturated fats may suppress lymphocyte blastogenesis, the potential beneficial effect of vegetarian diets might be masked by the high polyunsaturated fatty acid content.

Similarly, differences in dietary fiber and lipid contents existed in endurance runners who consumed vegetarian versus regular western diets [44]. During the study, both diets had the same ratio of carbohydrate: fat: protein, 60:30:10 percent of total energy intake. One hundred ten runners freely chose the amounts and types of food to consume daily within their dietary groups. Among the 55 athletes who finished a 1000 km run in 20 days, 30 were from the regular western diet group and 25 from the lacto-ovo-vegetarian group. The diets didn't influence the percentage of the subjects finishing the run in each group (50% in both groups). The researchers concluded that the lacto-ovo-vegetarian diet served to the runners "fulfilled the demands of sport nutrition". However, they also reported lower iron stores measured by serum ferritin level before the race, despite a higher iron intake, in the lacto-ovo-vegetarian group compared with the regular western diet group.

Reduced iron storage was also reported by Snyder et al. [45] when they cross-sectionally compared female runners who consumed a modified vegetarian diet (<100 g red meat per week) with those who consumed a diet which included red meat. Athletes in the modified vegetarian group had lower serum ferritin, but higher total iron binding capacity compared with athletes in the red meat group. Serum iron concentration and percentage transferrin saturation were not different between groups. These results suggest that the female runners who consumed a modified vegetarian diet had non-anemic iron deficiency, which could reduce endurance capacity. While female athletes, the group most vulnerable to iron deficiency, were investigated in this study, the iron status of male athletes has not been addressed, yet. The effect of iron status on endurance or resistance performance among vegetarian athletes remains to be clarified.

To summarize the studies discussed above, no apparent differences in athletic performance between vegetarian and non-vegetarian athletes were observed. In their 2006 review of the nutritional considerations for vegetarian athletes, Venderley and Campbell [2] proposed that "better research is needed to directly compare athletic performance of vegetarian and non-vegetarians since few well conducted studies exist". Thus, future research needs to focus on properly controlling energy and macronutrient contents of diets, standardizing the sports athletes are engaged in, and directly evaluating various endurance or resistance performance parameters of vegetarian and non-vegetarian athletes.

Vegetarian Diets and Physical Performance of the General Population

While data from studies conducted with vegetarian athletes are limited, research conducted with non-athletes can provide important information about how vegetarian diets influence physical performance of the general population. In 1970, Cotes et al. did a cross-sectional comparison between 14 healthy women who had consumed a vegan diet with a vitamin B_{12} supplement, 66 non-vegan housewives with comparable social background, and 20 office cleaners who had a comparable level of customary activity [46]. No differences were found among the three groups in forced expiratory volume and forced vital capacity by spirometry; cardiovascular response to submaximal exercise on a cycle ergometer; or estimates of thigh muscle width from circumference and skinfold measurement. The authors concluded that dietary deficiency of animal protein did not impair the physiological responses to submaximal exercise.

In 1999, Campbell et al. [47] reported that older men who consumed a lacto-ovo-vegetarian diet containing marginal total protein for 12 weeks experienced declines in whole-body density, fat-free mass, and whole-body muscle mass, despite performing resistance exercise thrice weekly. In contrast, older men who consumed an omnivorous diet with sufficient total protein experienced increases in these body composition parameters after performing the same resistance exercise program for 12 weeks. While these findings might suggest that an omnivorous diet is superior to a lacto-ovo-vegetarian diet to promote anabolic body composition responses to resistance training among older men, this conclusion is not appropriate because both the quantity and predominant sources of protein differed between groups. It is of interest to note that the only dietary advice the men in the lacto-ovo-vegetarian group were provided was to avoid consuming any flesh foods (meats, poultry, and fish). The resulting marginal total protein intake reinforces the need for careful consideration of food and nutrient intakes when adopting a vegetarian diet pattern. Simply omitting meats from the diet may inadvertently result in unintended nutrient insufficiencies and compromised physiological responses to exercise training.

Subsequent research by Haub et al. [48] indicates that resistance-training-induced improvements in muscle strength and size were comparable between groups of older men who consumed lacto-ovo-vegetarian vs omnivore diets with sufficient total protein for 12 weeks. Maximal dynamic strength of all the muscle groups trained and cross-sectional muscle area of the vastus lateralis improved with no significant difference between groups. Body composition, resting energy expenditure, and concentration of muscle creatine, phosphocreatine and total creatine were not different between groups or changed over time. The findings from these two studies [47,48] suggest that body composition and physical performance are not compromised (or enhanced) by consumption of a vegetarian diet with sufficient total protein.

The Australian Longitudinal Study on Women's Health cross-sectionally compared the health and well-being of a total of 9113 women defined as young vegetarian, semi-vegetarian, or non-vegetarian [49]. Vegetarians and semi-vegetarians had lower BMIs than non-vegetarians and tended to exercise more. Lord et al. investigated the relationship between predominant source of protein consumed and muscle mass index of older women (age 57–75 yrs) [50] and reported that animal protein intake was the only independent predictor of muscle mass index. The observational study by Aubertin-Leheudre et al. confirmed Lord et al.'s finding by examining muscle mass in participants classified as omnivorous (n = 20, age 43 ± 13 yrs) and vegetarians (n = 19, age 48 ± 12 yrs) [51]. Total protein ingested exceeded the RDA for both groups while the vegetarian group had significantly lower muscle mass index. These results indicated that vegetarian athletes need to be cautious about their body composition when choosing dietary protein sources. Since certain amino acids are lacking in most plant foods, decreased muscle hypertrophy is possible without careful diet planning. However, there's no clear evidence to prove a lowered muscle hypertrophy in vegetarian athletes.

Baguet et al. [52] compared carnosine content and buffering capacity in sprint-trained omnivorous athletes who were assigned to consume either vegetarian or omnivorous diets for 5 weeks. Soleus carnosine content non-significantly increased (+11%) with the omnivorous diet but non-significantly decreased (−9%) in the vegetarian diet group. Carnosine synthase mRNA expression decreased in the vegetarian group. Since muscle carnosine content is considered to be an indicator of muscle buffering capacity *in vitro* [53] and muscle buffering capacity was positively correlated with high-intensity exercises performance [12,54], vegetarian athletes may be at risk of decreased performance. The researchers pointed out that beta-alanine, the rate-limiting precursor of carnosine synthesis, is naturally present in meat, fish and poultry, as well as trace quantities in vegetable oils. Vegetarian athletes who are engaged in high-intensity exercises were recommended to take beta-alanine into consideration when planning their diets.

Finally, vegetarian diets typically contain higher amounts of antioxidant nutrients than omnivorous diets, which may help athletes reduce exercise-induced oxidative stress. However, Szeto et al. [55] upon investigating the long-term effects of vegetarian diets on biomarkers of antioxidant status, concluded that while vegetarians consume more antioxidants (including vitamin C), they did not have better antioxidant status. In their 2010 review, Trapp et al. [56] indicated that most of the literature on this topic focused on supplementations rather than antioxidant-rich diets. Therefore "further research is required to assess the antioxidant status of vegetarians, and whether vegetarian athletes have an advantage in overcoming exercise-induced oxidative stress".

TAKE-HOME MESSAGES

During the past 25 years, research on vegetarian diets and athletes has mainly focused on three areas: (i) nutrient adequacy; (ii) oxidative stress and health; and (iii) performance. The following statements summarize the predominant views and opinions from scientific reviews [2,9,34,56–59] and position statements from professional societies [3,8].

Nutrient Adequacy

- The energy and nutrient needs of athletes can be met by a vegetarian diet.
- Consuming adequate energy to meet requirements from a wide variety of predominantly or exclusively plant-based sources is central to helping vegetarian athletes obtain sufficient protein and micronutrients.
- While vegetarian diets typically have lower protein density, the dairy and eggs in lacto-ovo-vegetarian diets provide athletes with high-quality sources of complete proteins.

- Vegetarian diets generally provide ample carbohydrate to maximize body glycogen stores and support endurance performance.
- Athletes and non-athletes who consume vegetarian diets may be at higher risk of inadequate intake of iron for body stores, which could lead to iron deficiency. Omnivorous diets help maintain sufficient iron stores.
- Athletes who are strict vegetarians (vegans) are especially advised to carefully monitor their diet to ensure that their energy, protein, and micronutrient needs are met.

Oxidative Stress and Health

- Limited research indicates that immune function is comparable in athletes who consume vegetarian vs omnivorous diets.
- Consumption of a vegetarian diet may help reduce exercise-induced oxidative stress, due to higher antioxidant intake and status.
- Vegetarian diets are considered heart-healthy and improve metabolic and physiological coronary risk factors.

Performance

- Research consistently indicates that habitually consuming a vegetarian diet does not positively or negatively impact physical performance of athletes.

References

[1] Stahler C. How Often Do Americans Eat Vegetarian Meals? And How Many Adults in the U.S. Are Vegetarian? 2012. Available from: <http://www.vrg.org/blog/2012/05/18/how-often-do-americans-eat-vegetarian-meals-and-how-many-adults-in-the-u-s-are-vegetarian/> [cited 10.01.2013].

[2] Venderley AM, Campbell WW. Vegetarian diets: nutritional considerations for athletes. Sports Med 2006;36(4):293–305.

[3] Craig WJ, Mangels AR. Position of the American dietetic association: vegetarian diets. J Am Diet Assoc 2009;109(7): 1266–82.

[4] Fuhrman J, Ferreri DM. Fueling the vegetarian (vegan) athlete. Curr Sports Med Rep 2010;9(4):233–41.

[5] Rodriguez NR, DiMarco NM, Langley S. Position of the American dietetic association, dietitians of Canada, and the American college of sports medicine: nutrition and athletic performance. J Am Diet Assoc 2009;109(3):509–27.

[6] Haddad EH, Tanzman JS. What do vegetarians in the United States eat? Am J Clin Nutr 2003;78(3 Suppl.):626S–32S.

[7] Kennedy ET, Bowman SA, Spence JT, Freedman M, King J. Popular diets: correlation to health, nutrition, and obesity. J Am Diet Assoc 2001;101(4):411–20.

[8] Position of the American dietetic association and dietitians of canada: vegetarian diets. J Am Diet Assoc 2003;103(6): 748–765.

[9] Barr SI, Rideout CA. Nutritional considerations for vegetarian athletes. Nutrition 2004;20(7-8):696–703.

[10] Young VR, Pellett PL. Plant proteins in relation to human protein and amino acid nutrition. Am J Clin Nutr 1994;59(5 Suppl.):1203S–12S.

[11] Trumbo P, Schlicker S, Yates AA, Poos M. Dietary reference intakes for energy, carbohydrate, fiber, fat, fatty acids, cholesterol, protein and amino acids. J Am Diet Assoc 2002;102(11): 1621–30.

[12] Edge J, Bishop D, Hill-Haas S, Dawson B, Goodman C. Comparison of muscle buffer capacity and repeated-sprint ability of untrained, endurance-trained and team-sport athletes. Eur J Appl Physiol 2006;96(3):225–34.

[13] Burke LM, Loucks AB, Broad N. Energy and carbohydrate for training and recovery. J Sports Sci 2006;24(7):675–85.

[14] Jeukendrup AE. Nutrition for endurance sports. marathon, triathlon, and road cycling. J Sports Sci. 2011;29 (Suppl 1): S91–9.

[15] Davey GK, Spencer EA, Appleby PN, Allen NE, Knox KH, Key TJ. EPIC-Oxford: lifestyle characteristics and nutrient intakes in a cohort of 33 883 meat-eaters and 31 546 non meat-eaters in the UK. Public Health Nutr 2003;6(3):259–69.

[16] Joint position statement: nutrition and athletic performance. American college of sports medicine, American dietetic association, and dietitians of Canada. Med Sci Sports Exerc 2000; 32 (12):2130–145.

[17] Kornsteiner M, Singer I, Elmadfa I. Very low n-3 long-chain polyunsaturated fatty acid status in Austrian vegetarians and vegans. Ann Nutr Metab 2008;52(1):37–47.

[18] DellaValle DM, Haas JD. Impact of iron depletion without anemia on performance in trained endurance athletes at the beginning of a training season: a study of female collegiate rowers. Int J Sport Nutr Exerc Metab 2011;21(6):501–6.

[19] Ball MJ, Bartlett MA. Dietary intake and iron status of Australian vegetarian women. Am J Clin Nutr 1999;70(3): 353–8.

[20] Dietary Reference intakes: Vitamin A, Vitamin K, Arsenic, Boron, Chromium, Copper, Iodine, Iron, Manganese, Molybdenum, Nickel, Silicon, Vanadium, and Zinc. In: F. a. N. B. Institute of Medicine, editors. National Academy Press; Washington DC. 2001.

[21] Janelle KC, Barr SI. Nutrient intakes and eating behavior scores of vegetarian and nonvegetarian women. J Am Diet Assoc 1995;95(2):180–6 189, quiz 187-188

[22] Lukaski HC. Vitamin and mineral status: effects on physical performance. Nutrition 2004;20(7–8):632–44.

[23] de Bortoli MC, Cozzolino SM. Zinc and selenium nutritional status in vegetarians. Biol Trace Elem Res 2009;127(3):228–33.

[24] Hunt JR. Bioavailability of iron, zinc, and other trace minerals from vegetarian diets. Am J Clin Nutr 2003;78(3 Suppl.): 633S–9S.

[25] Craig WJ. Nutrition concerns and health effects of vegetarian diets. Nutr Clin Pract 2010;25(6):613–20.

[26] DeRuisseau KC, Cheuvront SN, Haymes EM, Sharp RG. Sweat iron and zinc losses during prolonged exercise. Int J Sport Nutr Exerc Metab 2002;12(4):428–37.

[27] Holick MF. The vitamin D deficiency pandemic and consequences for nonskeletal health: mechanisms of action. Mol Aspects Med 2008;29(6):361–8.

[28] Appleby P, Roddam A, Allen N, Key T. Comparative fracture risk in vegetarians and nonvegetarians in EPIC-Oxford. Eur J Clin Nutr 2007;61(12):1400–6.

[29] Tenforde AS, Sayres LC, Sainani KL, Fredericson M. Evaluating the relationship of calcium and vitamin D in the prevention of stress fracture injuries in the young athlete: a review of the literature. PM R 2010;2(10):945–9.

[30] Craig WJ. Health effects of vegan diets. Am J Clin Nutr 2009;89 (5):1627S–33S.
[31] Tucker KL, Hannan MT, Kiel DP. The acid-base hypothesis: diet and bone in the Framingham Osteoporosis Study. Eur J Nutr 2001;40(5):231–7.
[32] Golden NH. A review of the female athlete triad (amenorrhea, osteoporosis and disordered eating). Int J Adolesc Med Health 2002;14(1):9–17.
[33] O'Connor MA, Touyz SW, Dunn SM, Beumont PJ. Vegetarianism in anorexia nervosa? A review of 116 consecutive cases. Med J Aust 1987;147(11–12):540–2.
[34] Nieman DC. Vegetarian dietary practices and endurance performance. Am J Clin Nutr 1988;48(3 Suppl.):754–61.
[35] Berry E. The effects of a high and low protein diet on physical efficiency. Am Phys Ed Rev 1909;14:288–97.
[36] Fisher I. The influence of flesh-eating on endurance. Yale Med J 1908;13:205–21.
[37] Burke LM. The IOC consensus on sports nutrition 2003: new guidelines for nutrition for athletes. Int J Sport Nutr Exerc Metab 2003;13(4):549–52.
[38] Hawley JA, Tipton KD, Millard-Stafford ML. Promoting training adaptations through nutritional interventions. J Sport Sci 2006;24(7):709–21.
[39] Baar K, Mcgee S. Optimizing training adaptations by manipulating glycogen. Eur J Sport Sci 2008;8(2):97–106.
[40] Burke LM. Fueling strategies to optimize performance: training high or training low? Scand J Med Sci Sports 2010;20 (Suppl. 2):48–58.
[41] Hanne N, Dlin R, Rotstein A. Physical fitness, anthropometric and metabolic parameters in vegetarian athletes. J Sports Med Phys Fitness 1986;26(2):180–5.
[42] Richter EA, Kiens B, Raben A, Tvede N, Pedersen BK. Immune parameters in male atheletes after a lacto-ovo vegetarian diet and a mixed Western diet. Med Sci Sports Exerc 1991;23 (5):517–21.
[43] Raben A, Kiens B, Richter EA, Rasmussen LB, Svenstrup B, Micic S, et al. Serum sex hormones and endurance performance after a lacto-ovo vegetarian and a mixed diet. Med Sci Sports Exerc 1992;24(11):1290–7.
[44] Eisinger M, Plath M, Jung K, Leitzmann C. Nutrient intake of endurance runners with ovo-lacto-vegetarian diet and regular western diet. Z Ernahrungswiss 1994;33(3):217–29.
[45] Snyder AC, Dvorak LL, Roepke JB. Influence of dietary iron source on measures of iron status among female runners. Med Sci Sports Exerc 1989;21(1):7–10.
[46] Cotes JE, Dabbs JM, Hall AM, McDonald A, Miller DS, Mumford P, et al. Possible effect of a vegan diet upon lung function and the cardiorespiratory response to submaximal exercise in healthy women. J Physiol 1970;209(1): Suppl:30P+.
[47] Campbell WW, Barton Jr. ML, Cyr-Campbell D, Davey SL, Beard JL, Parise G, et al. Effects of an omnivorous diet compared with a lactoovovegetarian diet on resistance-training-induced changes in body composition and skeletal muscle in older men. Am J Clin Nutr 1999;70(6):1032–9.
[48] Haub MD, Wells AM, Tarnopolsky MA, Campbell WW. Effect of protein source on resistive-training-induced changes in body composition and muscle size in older men. Am J Clin Nutr 2002;76(3):511–7.
[49] Baines S, Powers J, Brown WJ. How does the health and well-being of young Australian vegetarian and semi-vegetarian women compare with non-vegetarians? Public Health Nutr 2007;10(5):436–42.
[50] Lord C, Chaput JP, Aubertin-Leheudre M, Labonte M, Dionne IJ. Dietary animal protein intake: association with muscle mass index in older women. J Nutr Health Aging 2007;11(5):383–7.
[51] Aubertin-Leheudre M, Adlercreutz H. Relationship between animal protein intake and muscle mass index in healthy women. Br J Nutr 2009;102(12):1803–10.
[52] Baguet A, Everaert I, De Naeyer H, Reyngoudt H, Stegen S, Beeckman S, et al. Effects of sprint training combined with vegetarian or mixed diet on muscle carnosine content and buffering capacity. Eur J Appl Physiol 2011;111 (10):2571–80.
[53] Parkhouse WS, McKenzie DC, Hochachka PW, Ovalle WK. Buffering capacity of deproteinized human vastus lateralis muscle. J Appl Physiol 1985;58(1):14–7.
[54] Bishop D, Edge J, Goodman C. Muscle buffer capacity and aerobic fitness are associated with repeated-sprint ability in women. Eur J Appl Physiol 2004;92(4–5):540–7.
[55] Szeto YT, Kwok TC, Benzie IF. Effects of a long-term vegetarian diet on biomarkers of antioxidant status and cardiovascular disease risk. Nutrition 2004;20(10):863–6.
[56] Trapp D, Knez W, Sinclair W. Could a vegetarian diet reduce exercise-induced oxidative stress? A review of the literature. J Sports Sci 2010;28(12):1261–8.
[57] Nieman DC. Physical fitness and vegetarian diets: is there a relation? Am J Clin Nutr 1999;70(3 Suppl.):570S–5S.
[58] Berning JR. The vegetarian athlete. In: Maughan RJ, editor. Nutrition in sport. Oxford: Blackwell Science Ltd; 2000. p. 442–56.
[59] Fogelholm M. Dairy products, meat and sports performance. Sports Med 2003;33(8):615–31.

CHAPTER

11

Nutrition in Combat Sports

Guilherme G. Artioli[1,2], *Emerson Franchini*[2], *Marina Y. Solis*[1,3], *Aline C. Tritto*[1] *and Antonio H. Lancha Jr*[1]

[1]Laboratory of Applied Nutrition and Metabolism, School of Physical Education and Sport, University of São Paulo, São Paulo, Brazil [2]Martial Arts and Combat Sports Research Group, School of Physical Education and Sport, University of São Paulo, São Paulo, Brazil [3]School of Medicine, University of São Paulo, São Paulo, Brazil

INTRODUCTION

Combat sports refer to a group of individual sports that are very relevant to the world sportive scenario. Altogether, the Olympic disciplines of combat sports (i.e., judo, wrestling, taekwon-do and boxing) account for almost 25% of the total medals disputed in the Olympic games. In addition, professional boxing and mixed martial arts (MMA) are two non-Olympic sports that gather millions of spectators from all over the world and constitute a billion dollar industry [1].

Each specific combat sport has a unique combination of rules that confers singular characteristics to each one (e.g., grappling-based techniques or striking-based techniques; scoring system; number of rounds; recovery time between rounds; time duration of each round). Despite those differences, studies have shown that most combat sports can be characterized as high-intensity, intermittent sports [2–6] and consequential adaptations have been consistently found in these athletes [7,8]. Nonetheless, it seems that grappling combat sports rely more on the anaerobic lactic metabolism [4,9] whilst the striking combat sports are more dependent on the alactic anaerobic metabolism [2], although in both types of combat sports the actions used to score are often mantained by the anaerobic alactic metabolism. For all combat disciplines, however, aerobic metabolism is predominant during low-intensity efforts and recovery periods, as it is responsible for ATP and PCr resynthesis. Another important characteristic of all combat sports is the long duration of a competitive event. While a professional MMA fight may last up to five 5-minute rounds or a professional boxing fight may last twelve 3-minute rounds, Olympic judo, wrestling and taekwondo athletes may perform up to seven matches in the same day.

A common characteristic of all combat sports is that competitions are disputed in weight divisions. Although the weight classes promote more even matches in terms of body size, strength, speed and agility, most athletes tend to reduce significant amounts of body weight in short periods of time in order to qualify for a lighter weight division [10–12]. By doing so, athletes believe they will gain competitive advantage as they will compete against lighter, smaller and weaker opponents. Indeed, rapid weight loss practices are harmful to health and have great potential to impair performance [13], hence, successful management of body weight and body composition is crucial for any nutritional program for combat athletes.

Based on the characteristics of combat sports (i.e., high-intensity, intermittent and long-duration events) the major factors causing fatigue can be identified. In light of results from both *in vitro* and *in vivo* studies, it is possible to affirm that fatigue is caused, among other factors, by muscle acidosis [14,15], increased extracellular K^+ concentration [16], depletion of phosphocreatine [17] and depletion of muscle glycogen [18]. Dehydration and consequential thermal and electrolyte imbalance are also important factors related to fatigue in combat sports [19–21]. Knowledge of the causes of fatigue during both training and competition situations is essential for designing nutritional

Nutrition and Enhanced Sports Performance.
DOI: http://dx.doi.org/10.1016/B978-0-12-396454-0.00011-4

strategies aiming to delay performance decrements and maximize training adaptations as well as competitive performance. In this chapter, the aspects of sports nutrition relevant to combat sports will be discussed, including micro- and macronutrients, hydration, supplements and weight management.

ROLE OF NUTRIENTS

The human body obtains nutrients from the digestion and absorption of food, and they are needed for virtually all bodily functions. While the macronutrients (i.e., carbohydrates, fats and proteins) provide energy, the micronutrients (i.e., vitamins and minerals) are essential for a number of specific metabolic functions. A healthy and balanced diet must supply all nutrients to fulfil the requirements for energy and the other elements that support metabolism, including water. The individual need for each nutrient varies depending on age, gender, presence of medical conditions and level of physical activity [22]. In comparison with non-athletes, athletes generally have greater needs in terms of macronutrients and, in some specific cases, micronutrients. The success in supplying these extra needs is a key factor for any nutritional plan to support intensive training regimens and maximize competitive performance. This will discussed in the present section and, whenever applicable, the concepts will be extended to the specificities of combat sports.

Macronutrients

Carbohydrates

The most important role played by carbohydrates (CHO) is energy supply for cell functions. CHO act as energy substrates and can be either oxidized via aerobic metabolism (i.e., glycolytic pathway coupled with Krebs cycle and respiratory chain) or converted into lactate via anaerobic metabolism (i.e., anaerobic glycolysis). In both cases, energy is transferred and ATP is synthesized. Whereas the aerobic metabolism is more efficient (i.e., more ATP is synthesized) and the energy transferred is less readily available (i.e., the rate of ATP synthesis is lower), the anaerobic metabolism produces lower amounts of ATP per glucose molecule, but at very high rates, which is crucial for high-intensity exercises. In the context of sports nutrition, the energetic role of carbohydrates is even more evident because they will provide energy for muscle contraction and for sustaining the exercise.

Muscle glycogen content has been classically related to the ability to sustain exercise [23]. Although glycogen depletion has been consistently related as a major cause of fatigue in endurance exercise [24], data from both human [25] and animal [18] studies suggest that glycogen depletion also occurs in high-intensity exercises. That means that glycogen availability plays, at least, a permissive role in high-intensity performance [26] and may even be a limiting factor in competitive performance in many combat sports, especially when multiple bouts are performed.

Studies have shown that the acute ingestion of high-glycemic-index CHO is beneficial for performance in high-intensity intermittent exercises [27]. The ingestion of ~40 g of dextrose (e.g., 600 mL of a 6.5% dextrose beverage) immediately before a long-term high-intensity intermittent exercise accompanied by the ingestion of 200 mL of a 6.5% dextrose beverage at every 15 minutes has proven effective in enhancing exercise tolerance [27]. Other studies have also shown similar ergogenic effects of CHO ingestion before and during high-intensity exercises in a broad variety of exercise models [28].

According to some authors, approximately 40–60 g of high-glycemic index CHO should be consumed every hour during a continuous exercise [29]. Similar quantities are very likely to be beneficial to combat athletes during their training routines or during competitive events. Other recommendations state that athletes engaged in high-intensity sports, such as combat sports, should consume 10–12 g/kg/day of CHO [26]. The timing of CHO intake has to be individually adjusted, taking into account food preferences, training schedule and other activities performed by each athlete. A well planned diet should comprise the intake of appropriate amounts of CHO well suited with the training times. This will allow the athlete to perform better in the training sessions, which will improve training quality and maximize training adaptations. The ingestion of CHO-rich meals containing 200–300 g of high-glycemic-index CHO 3–4 hours before the exercise is another effective strategy to improve performance [30]. Whether the CHO is consumed in liquid solutions (e.g., carbohydrates-electrolytes beverages or in the form of supplements) or in solid food is irrelevant for exercise performance, as both liquid and solid carbohydrates are equally beneficial to performance [28].

The ingestion of CHO increases the availability of exogenous glucose for the muscle cells, which diminishes the rate of muscle glycogen usage by active muscles, sparing muscle glycogen and delaying its depletion [31]. Such an effect likely explains the ergogenic effect of CHO ingested immediately and during exercise. In addition, the ability of CHO to increase glycemia and prevent the fall in blood glucose concentration may also play a role in delaying fatigue and, at least in part, explain the ergogenic effects of CHO [28,29].

Because muscle glycogen depletion is a major cause of fatigue [18,24] and exercise tolerance is directly related to pre-exercise muscle glycogen content [23], pre-competition nutritional strategies must ensure maximal glycogen accumulation before competitive events. A classical study by Bergstrom and colleagues in 1967 [23] was the first to describe a protocol capable of markedly increasing muscle glycogen content and improving exercise capacity, as compared with a regular mixed diet. The so-called supercompensation diet consists in a 3-day period of CHO deprivation combined with high-intensity/high-volume endurance exercise followed by a 3-day period of high-CHO intake. In the first 3-day period, muscle glycogen is severely depleted, which stimulates insulin action and glycogen accumulation [32] over the following 3-day period of high-CHO availability. Although effective, this glycogen supercompensation protocol is somewhat aggressive as it requires a relatively long period of CHO deprivation and exhaustive exercise sessions, which may result in low adherence by athletes. Some investigations have found similar results of increased muscle glycogen content after less extreme regimens. The study by Arnall et al. [33], for example, has shown that a single bout of exercise that depletes muscle glycogen, when followed by 3 days of an extremely high-CHO diet (i.e., 85% of the total calories from CHO), can successfully increase muscle glycogen levels above baseline values. Moreover, this elevated muscle glycogen content can be maintained for 5 days if the athlete keeps his/her exercise levels at a minimum and consumes a CHO-rich diet (i.e., ~60% of the total calories from CHO) [33]. Other studies have also shown that even simpler procedures can maximize muscle glycogen. Sherman et al. [34] have observed supercompensated muscle glycogen after a 3-day period of a high-CHO diet (i.e., 70% from total calories) following a glycogen-depleting protocol (i.e., 5 days of exercises and 3 days of diet containing 50% of CHO). Interestingly, high-intensity exercises, such as those usually performed in combat sports' training sessions, seem to elicit faster responses of glycogen accumulation during the recovery from exercise. In combination with a high-CHO diet, this may result in muscle glycogen above normal levels after several hours. In fact, data from Fairchild et al. [35] has confirmed that short bouts of high-intensity exercise and subsequent intake of a CHO-rich diet (~10 g/kg/day) is an effective stimulus for faster muscle glycogen supercompensation, which occurs in less than 24 hours [35].

After the exercise, CHO also play fundamental roles in the recovery process, especially in the restoration of the muscle glycogen depleted during the exercise [26]. Therefore, it is necessary to consume adequate amounts of CHO not only before and during, but also after the exercise. In fact, studies have demonstrated that there is a close relationship between the amount of CHO consumed in the 24 hours after an exercise session and the amount of muscle glycogen replenished [26]. However, it seems to have a saturation point from which further increases in CHO intake do not result in further glycogen accrual. Interestingly, this saturation point matches quite well with the daily amount of carbohydrates recommended for athletes involved with high-intensity activities (i.e., 10–12 g/kg/day). Importantly, the highest rates of glycogen synthesis seem to occur at the early stages of post-exercise recovery (i.e., the first hours after the training session) [36,37], which means that the earlier an athlete consumes a CHO-rich meal after the exercise, the earlier he/she will be recovered and ready to perform at his/her best in the next training session. Furthermore, some evidence indicates that ingesting smaller amounts of CHO at every ~30 min is more effective in restoring muscle glycogen during the early phase of recovery than taking a high amount of CHO in a single bolus [26], probably due to a more sustained glucose and insulin availability. Thus, it is important for any athlete who is training at high intensities, as is the case of most combat athletes, to keep consuming high-CHO diets because they will provide not only the energy necessary for the training sessions, but they will also provide the substrates needed for restoring the glycogen depleted during the previous training sessions.

To summarize, combat sports athletes are recommended to consume CHO-rich diets containing 10–12 g/kg/day of CHO. The timing of the intake must ensure appropriate supply of energy for all training sessions throughout the day, which can be achieved by consuming a meal containing 200–300 g of CHO 3–4 h before any training session and smaller amounts immediately before and during the exercise (i.e., ~40 g before and 10–15 g every 15 min during the exercise). Although the form of CHO (liquid beverages/supplements or solid meals) has no influence on the ergogenic effects of CHO, the gastric discomfort related to the ingestion of solid meals immediately before or during strenuous exercise may hamper the use of solid meals at some specific times. Thus, the use of liquid beverages/supplements may help athletes to consume the recommended amounts of CHO at the most appropriate times. After a training session, the immediate intake of a CHO-rich meal will maximize glycogen restoration rates. Most important, the total amount consumed in the 24 hours following a training session will determine the amount of glycogen that will be restored, until a saturation point of 12 g/kg/day is reached. However, for short recovery periods between sessions (i.e., 8 hours or shorter), it is important to speed up glycogen recovery and consume large amounts of CHO straight after the training and keep ingesting 20–30 g of CHO by snacking every 30 minutes. For the competition day, all these strategies can be summed with any other strategy that results in muscle glycogen supercompensation.

Proteins

Skeletal muscle is the major deposit of protein molecules (about 40% of body weight in young males [38]) and it is in a constant balance between anabolism and catabolism, known as "protein turnover". Skeletal muscle protein turnover is the ratio between protein synthesis and protein breakdown rates [39]. Thus, protein balance could be: (a) neutral – protein synthesis and breakdown are equal, which results in muscle mass maintenance; (b) positive – protein synthesis is higher than breakdown, which results in muscle mass gain; or (c) negative – protein breakdown rate is greater than the rate of protein synthesis, which results in muscle mass loss.

In sports, it is well established that muscle mass loss can significantly change an athlete's performance, strength and power capacity. In this context, it is important to avoid muscle mass losses, guaranteeing neutral or positive protein balance. This is of special importance for combat sports athletes, as competitive performance is strongly dependent on strength and power.

Negative protein balance and muscle loss occur under catabolic conditions such as malnutrition, fasting state, cancer, AIDS, sepsis, burns, disuse and aging [40]. Rapid weight loss, a commonly practiced strategy for weight adjustment in combat sports, is another condition that results in negative protein balance and muscle loss [41]. Conversely, positive protein balance leading to muscle hypertrophy could be increased by physical training, stimulating protein synthesis and controlling protein breakdown [42,43]. Likewise, protein and essential amino acids consumption via food intake have a direct influence on protein synthesis [44]. Both amino acids intake and physical training are indepent factors that contribute to positive protein balance [45,46]. Therefore, gain of muscle mass (or muscle hypertrophy) is the result of the accumulation of successive periods of positive protein balance after exercise and protein consumption. Based on it, a combat sport athlete aiming to maximize his/her body composition and preserve muscle mass must be aware that resistance training and adequate protein consumption (i.e., proper amounts of high-quality protein ingested at the best times throughout a day) are key aspects for success.

In order to investigate the ideal amount of protein intake, Tarnopolsky and co-workers [44] conducted studies of strength-trained athletes and sedentary subjects taking three protein regimens: low protein (0.86 g protein/kg/day), moderate protein (1.40 g protein/kg/day) or high protein (2.40 g protein/kg/day). According to the protein synthesis, strength athletes' protein requirements were higher than those of sedentary subjects. For the sedentary population, low protein (0.86 g protein/kg/day) is sufficient to maintain protein balance. However, for strength athletes the low-protein diet did not provide adequate protein, suggesting that the moderate- and high-protein diets fit better with their needs. In addition to these results, it is interesting to observe that current dietary reference intakes (DRIs) recommend a low-protein diet (0.8 g protein/kg/day) for all individuals, with no recommendation for consumption of extra protein when combined with exercise [22]. Indeed, combat athletes' daily requirements for proteins are very similar to those of strength athletes because training regimens are quite intense and also involve resistance/power training sessions. However, it is important to highlight that excessive intake of protein might result in decreased protein synthesis rate, as demonstrated by Bolster et al. [46a]. In this study, the ingestion of 3.6 g/kg/day of protein reduced protein synthesis when compared with 1.8 g/kg/day. Although this was done in endurance-trained athletes, this result supports the concept that excessively high protein diets do not provide any further benefit to athletes and, in fact, may even be detrimental.

Considering the specific amount of daily protein intake, it is suggested that the total amount should be divided in four or five single doses to be taken throughout the day, giving priority to the periods following exercise bouts [46a,46b]. The timing of protein intake is another important point to be considered for muscle hypertrophy. Several studies have shown that carbohydrate alone is not capable of increasing protein synthesis or diminishing protein breakdown after exercise. However, carbohydrates ingested together with protein before and/or after resistance exercise promote a better anabolic response and a positive protein balance [43,47]. Cribb and Hayes [48] demonstrated that protein (0.1 g protein/kg/day) combined with CHO ingested immediately before and/or after an exercise training was capable of significantly increasing cross-sectional area of the type II fibers, contractile protein content, strength and lean body mass, when compared with the same amount of protein and CHO taken 5 hours before or after the workout.

Additionally, the amount of protein intake just after an exercise bout seems to be relevant. A study using 0, 5, 10, 20, or 40 g of whole-egg protein after an intense bout of resistance exercise demonstrated that muscle protein synthesis exhibits a dose-dependent response to dietary protein until a plateau is reached at 20 g. Therefore, the maximal anabolic response after the exercise is achieved with 20 g of protein. Ingesting higher amounts of protein will not further improve the anabolic response of the muscle but only increase amino acid oxidation, which indicates an excess of protein intake [49].

In respect of quality, milk, whey, casein and soy protein have proven effective in stimulating muscle protein synthesis when consumed after an exercise session [50–52]. However, not all high-quality protein influences the protein turnover to the same extent, as each protein elicits different physiological responses and possesses different digestibility and muscle retention [53]. Casein, for instance, coagulates and precipitates in stomach acid, which results in low rates of digestion and absorption [54]. Hence, casein promotes slower but more sustained rise in plasma amino acid. On the other hand, soy, milk and whey protein are considered "fast" proteins because they have rapid digestion, which leads to a large but transient rise in aminoacidemia (amount of amino acids in the blood). Evidence indicates that milk and whey protein are able to promote superior muscle mass accretion than casein and soy protein, but all four strategies are superior to carbohydrates alone [48,55]. These responses are probably related to the hyper-aminoacidemia provoked by milk and whey protein.

Considering the aforementioned, it is important to keep in mind that protein consumption plays a key role in maintaining muscle mass, although excessive ingestion will be not translated into more muscle mass. Protein intake should be adjusted per athlete's body weight. For combat athletes, a range between 1.8 and 2.4 g/kg/day should be achieved. Also, daily protein must be ingested four to five times throughout a day. Straight after exercise bouts, the doses should be of approximately 20 g of high-quality protein. Larger quantities of protein can be ingested in meals if this is necessary to achieve the recommended daily amounts. Protein quality and digestibility should be considered. That means that, after the training sessions, milk or whey proteins are preferable as they result in rapidly increased aminoacidemia. On the other hand, evidence suggests that consuming casein before bedtime may attenuate muscle catabolism during overnight sleep because of its slower absorption rates [56].

Micronutrients

Micronutrients (i.e., vitamins and minerals) are so named because the daily intake requirements for these nutrients are low. As they cannot be endogenously produced in sufficient amounts, adequate intake through diet is of great importance. Micronutrients do not provide energy and, therefore, they do not play any role in fattening or weight-gaining processes. Vitamins exert a large number of different functions in the human body, such as acting as coenzymes in several metabolic reactions, hormonal function, calcium metabolism, antioxidant, coagulation, and structure of tissues, among others. Unlike vitamins, minerals are inorganic compounds. Minerals are also essential to several metabolic pathways, cell signaling, synthesis and maintenance of tissues.

In theory, athletes involved with high-volume intensive training regimens could have increased requirements for daily vitamins and minerals intake. This would be due to increased rates of synthesis, maintenance and repair of muscle tissue, as well as to losses of some micronutrients in sweat. In addition, exercise stimulates metabolic pathways in which micronutrients are involved and can produce biochemical adaptations in muscle tissue that would increase the requirements of vitamins and minerals [57].

However, studies have shown that a balanced diet that adequately supplies the needs for energy will also adequately supply the needs for micronutrients, which is true for both non-athletic and athletic populations [57,58]. Hence, athletes in general do not benefit from supplementation with vitamins and minerals, unless some specific micronutrient deficiency is present. In these cases, it is important to make all necessary changes in the diet in order to ensure that all micronutrients will be eaten in adequate amounts. Also, in the event of micronutrient deficiency, supplementation with vitamins or minerals may be indicated until that specific deficiency is circumvented [59]. The countermeasures for vitamin or mineral deficiency are especially important for athletes because physical performance is impaired by micronutrient deficiency [60,61].

It is worthy to note that some specific athletic groups are at increased risk of vitamins and minerals deficiency, such as those who constantly restrict food intake, exclude specific groups of foods from the diet, or constantly cycle their body weight. This is quite common in some sports, especially in combat sports because of the weight classes issue. Thus, many combat sports athletes are at high risk for inadequate intake of vitamins and minerals. This is of greater concern for women during menstrual periods as they may lose a significant amount of micronutrients, especially iron, in the menses, which may result in anemia and negatively affect performance [62,63]. Besides iron, calcium and zinc deficiency are also relatively common among athletes, especially among female athletes and vegetarians. Calcium deficiency can decrease bone mineral density, making the bone structures more fragile and susceptible to fractures. In these cases, there may be an indication for supplementation [57]. Another group at risk for micronutrient deficiency is vegetarian athletes, to whom special attention in this regard must be given.

Finally, it is important to emphasize that maintaining healthy eating habits, such as preventing severe food restriction and refraining from the exclusion of particular groups of foods, is the most appropriate way to avoid micronutrients-related problems. Supplementation would be necessary only to overcome any eventual deficiency caused by unbalanced diets.

ROLE OF HYDRATION

Combat sports' training routines are often prolonged and intensive. Frequently, training rooms are not adequately ventilated or cooled and, in many combat sports, athletes use thick and heavy clothes, such as the traditional martial arts' "Gi". Altogether, these conditions lead to elevated sweating rates, which may cause loss of large amounts of body fluids throughout a prolonged session [64]. If an athlete fails to properly replace fluid and electrolytes, dehydration and electrolyte unbalance will probably occur and physical performance is most likely to be hampered.

Thirst is the main mechanism that drives voluntary fluid intake during exercise. However, voluntary rehydration usually does not fully replace sweat loss, and the consequence is a phenomenon known as voluntary dehydration [65]. It is well established that hypohydration has relevant health consequences, such as hyperthermia, impaired cardiovascular [66] and cognitive functions [67,68], among other deleterious effects. Compelling evidence also indicates that hypohydration is detrimental to exercise performance [69–71] and that high-intensity exercise capacity may be reduced even when hypohydration levels are as low as 2% of body weight [20]. Although maximal strength and short-term sprint capacity seem to be minimally or not affected by hypohydration [72,73], prolonged repeated sprint ability is negatively affected by moderate and high levels of dehydration [73], which means that almost all combat sports training and competitive situations are limited by dehydration. That is especially true if one takes into account that performance in combat sports is complex and multifactorial, being not limited only by physical capacities such as strength, power and endurance. In fact, performance in combat sports relies on a variety of physical (e.g., strength and anaerobic capacity), motor (e.g., specific skills), cognitive (e.g., ability to make fast decisions and to keep focused), and psychological (e.g., mood state and motivation) factors, so "field performance" may be more severely compromised by dehydration than is suggested by laboratory tests that assess only isolated physical attributes [73]. Therefore, nutritional plans for training and competition days should never neglect fluid and electrolyte replacement.

During a combat, adequate rehydration strategies may not be feasible. Nonetheless, athletes should ensure that their pre-combat hydration status is normal, so that performance decrements due to fluid loss during the fight will not be extreme. In some combat sports in which multiple combats are performed in the same day, rehydration between combats is essential for replacing water and electrolytes lost in the previous match. In others, such as boxing and MMA, rehydration strategies between rounds may be important in preventing hypohydration-induced fatigue. Athletes are advised not to let thirst drive the amount of fluid replaced during exercise, as it usually does not compensate for the amount lost in sweat [73].

Although some guidelines recommend fixed amounts of fluid to be replaced during exercise, some authors argue that it is not possible to determine how much fluid every athlete must intake, since water and electrolyte losses are largely variable depending on environmental conditions (e.g., temperature and humidity) and individual characteristics (e.g., acclimatization and electrolyte content in sweat) [73].

According to the American College of Sports Medicine [74], water and electrolyte replacement strategies should aim to eliminate water losses greater than 2% of body weight. This general recommendation is useful regardless of environmental conditions and individual characteristics because it is automatically adjusted to the rate of water loss during all types of exercise. The same guideline also states that athletes should avoid excessive fluid ingestion capable of increasing body weight during exercise [74]. However, caution must be used when extrapolating this last recommendation to athletes who start the exercise in hypohydrated state [73]. As previously mentioned, it is fundamental to ensure that athletes start their training sessions or competitive events in euhydrated state.

Evidence indicates that beverage temperature and flavor influence the amount of fluid ingested during exercise. Thus, these characteristics may facilitate voluntary fluid replacement and contribute to performance [75]. In fact, it has been shown that cold beverages (i.e., with temperature below 22°C) are more palatable, resulting in increased voluntary consumption of fluid during exercise [75]. Cold beverages also seem to play a role in cooling the body and controlling the rise in body temperature [75]. Because beverages containing CHO plus electrolytes are more efficient in delaying fatigue than water alone [76] and because carbohydrates *per se* clearly have ergogenic effects, as previously discussed in this chapter, athletes are recommended to consume cold CHO plus electrolytes drinks during exercise as a strategy to replace fluid, electrolytes and carbohydrates. Again, the volume to be replaced has to be individualized according to the amount of water lost in sweat, so that losses are less than 2% of body weight.

RAPID WEIGHT LOSS

Because most combat sports and martial arts competitions are divided into weight classes, the great majority of competitors undergo a number of aggressive methods to significantly reduce body weight in a short period of time. It is assumed that athletes believe that, by competing in a lighter weight class, they will get some competitive advantage against lighter, smaller and weaker opponents. In fact, several surveys have shown that rapid weight loss is probably the most remarkable common feature of all combat sports, as 60–90% of athletes from different countries have been engaged in rapid weight loss for the last 45 years at least – a tradition that does not seem to have changed over these years [11,12,77–88].

With regards to the methods commonly used to quickly reduce weight, athletes report restricting fluid and food intake in the week preceding the competition [12,81]. As competition approaches, athletes tend to restrict even more the intake of food and fluid. Hence, many athletes weigh-in without having had any meal of drink for 24 to 48 hours. This restricted pattern is usually combined with methods that induce water loss and leads to dehydration, such as: saunas, exercise in hot environments, exercise with winter clothing, plastic or rubberized suits, and spitting, among others [11,12,81,89,90]. Of greater concern, a considerable percentage of athletes uses more extreme and harmful methods, such as vomiting, using diet pills, laxatives and diuretics [12,81,90]. Athletes start reducing weight before competitions very young, many of them younger than 15 years of age [81]. A disturbing reported was of a 5-year-old wrestler who was encouraged to cut 10% of his body weight, which suggests that some athletes may be reducing weight at very early ages.

The negative impact of rapid weight loss and weight cycling on health is undisputed. Among other effects, it has been shown that rapid weight loss affects the cardiovascular system [91], suppresses the immune response [92–94], impairs hydroelectrolytic balance and thermal regulation [95], increases bone loss [96], and induces hormonal imbalance such as decreased serum testosterone and increased cortisol and GH levels [97,98]. Moreover, weight cycling may be associated with some eating disorders [90,99] and, during childhood and adolescence, it can lead athletes to a borderline undernourished state, putting growth and development at risk [100–102]. Once the competitive career is finished, retired athletes who used to cycle their body weight have greater chance of becoming overweight or obese than do athletes who did not cycle weight [103], being subject to all health problems related to obesity.

In light of these negative effects for a range of health-related outcomes, it is reasonable to assume that athletes who are engaged in rapid weight loss procedures are at risk. Obviously, those athletes who cut more weight, in shorter periods of time, more times per season and through more aggressive and extreme methods are at higher risk. Not surprisingly, the deaths of a few wrestlers [104] and judo players [1] have been attributed to rapid weight loss and consequential severe hyperthermia, dehydration and electrolyte imbalance. Because of the widespread use of rapid weight loss strategies among athletes and due to its potential to harm athlete's health, the American National Collegiate Athletic Association (NCAA) has launched a program aiming to control the abusive weight management practices among collegiate wrestlers. Despite some criticisms, the minimum weight program has proven effective in improving weight loss management behaviors among collegiate wrestlers [87,105]. Interestingly, the same wrestlers who were moderate in managing their body weight for NCAA-regulated competitions presented much more aggressive behaviors in international-style wrestling competitions, which are not under the NCAA minimum weight regulations [78]. These data suggest that athletes will probably not voluntarily adhere to less harmful behaviors unless a set of rules compels them to do so. Therefore, institutional regulations are warranted for all weight-classed sports, like those undertaken by NCAA for collegiate wrestling and proposed for judo [13].

Rapid weight loss generally results in reduced fat mass as well reduced lean body mass [106]. It seems possible to maximize fat loss during weight reduction by supplementing with BCAA [107], which could be explained by the reduced muscle catabolism triggered by the anti-catabolic effect that leucine exerts on skeletal muscle under atrophic conditions [40,108].

Despite the reduction on lean body mass, it seems consensual that rapid weight loss has minimal detrimental effect on maximal strength, muscle power, aerobic and anaerobic capacities if athletes have at least 3 hours to rehydrate and recover after weight loss [106,109–111]. If athletes do not have a minimum of 3 hours to rehydrate and/or refeed after rapid weight loss, both aerobic and anaerobic performance will probably be impaired [71,112–114]. However, if the weight reduction is gradual rather than rapid and achieved by a high-CHO diet, performance is less likely to be reduced [112]. Thus, in any combat sport where the period between weigh-in and competition varies from a few hours to a few days, the impact of weight loss can be largely minimized if the weight loss regimen is gradual, a high-CHO diet is adopted during weight loss period, and if the recovery period after weigh-in is used to fully replenish fluid, electrolytes and carbohydrates [97,106,115].

Although scientific evidence indicates that rapid weight loss, if followed by 3 h or longer "reload" period, has negligible effect on performance, it is important to emphasize that almost all studies have assessed performance after a ~5% of body weight reduction. Thus, there is currently no information available on the effects of larger weight reductions on physical capacity. In fact, studies have shown that a considerable percentage of athletes frequently reduces more than 5% of body weight; some of them reduce more than 10% of body weight [12,81]! It is quite possible that these athletes are competing below their physical, psychological and cognitive capacity as a consequence of the severe weight reduction.

Despite the lack of effect of acute weight loss on strength, a longitudinal study has demonstrated reduced strength after a wrestling season, during which athletes cycled their body weight [101], suggesting that weight loss has relevant long-term effects on muscle strength capacity. Besides physical capacities, competitive performance in combat sports is dependent on psychological aspects, such as mood state and cognitive function. Studies have shown that rapid weight loss negatively affects the profile of mood state and cognitive function, decreasing short-term memory, vigor, concentration and self-esteem as well as increasing confusion, rage, fatigue, depression and isolation [12,116,117]. According to Franchini et al. [1], lack of concentration can affect the ability to deal with distractions during high-level competitions; low self-esteem may result in difficulty in envisioning winning a match, especially against high-level opponents; confusion can impair the capacity to make decisions during the combat; and rage may result in lack of self-control. Although aggressiveness is relevant for combat sports, excessive rage may increase the possibility of illegal actions. Depression and isolation, in turn, can result in low adherence to training sessions. Obviously, all these changes can be detrimental for training and competitive performance [1].

In short, athletes are not recommended to cut weight before competitions. If strictly necessary, rapid weight loss should never exceed 5% of the body weight. Preferably, weight adjustments have to be done in a gradual fashion (i.e., no more than 1 kg per week) and include body fat reduction and muscle mass maximization, rather than acute dehydration. During the weight reduction period, a high-CHO diet is highly recommended. After the weigh-in, a carbohydrate load (i.e., meal containing 200–300 g of CHO) is also recommended. Despite the slight impact on physical capacity, rapid weight loss has negative effects on several health parameters. In addition, competitive performance may be impaired, since other factors (e.g., mood profile and cognition) associated with competitive performance are impaired. Rapid weight loss should especially be avoided if the athlete will knowingly have less than 3 hours to refeed and rehydrate after the weigh-in.

SUPPLEMENTS FOR COMBAT ATHLETES

As previously discussed in this chapter, data from literature suggest that the major causes of fatigue in most combat sports are: (i) muscle acidosis; (ii) muscle glycogen depletion; (iii) muscle phosphocreatine depletion; (iv) increased extracellular K^+ concentration; (v) dehydration and hydroelectrolytic imbalance. Based on that, it is conceivable that some specific supplements may be especially beneficial for the combat sport athlete, as they could delay the onset of fatigue or allow the athlete to perform at higher exercise intensity, therefore maximizing performance in competitions or in training sessions. Because dehydration and strategies to maximize glycogen and to minimize glycogen depletion have already been comprehensively discussed in this chapter, this section will focus on supplements capable of acting on the other three major causes of fatigue listed above.

With regards to the decrease in intramuscular pH observed during high-intensity exercises, studies have shown that nutritional interventions able to increase extracellular buffering capacity (e.g., sodium bicarbonate or sodium citrate ingestion [118,119]) or intramuscular buffering capacity (e.g., increase in muscle carnosine content via beta-alanine supplementation [15,120]) possess ergogenic effects and therefore have great potential to benefit combat sport athletes. In fact, Artioli et al. [121] have demonstrated that the acute ingestion of 300 mg/kg of body mass of sodium bicarbonate 120 min prior to exercise significantly improves performance in judo-specific and judo-related tests. Other studies indicate that similar effects can be achieved if a chronic (i.e., ~500 mg/kg of body mass split in four or five smaller single doses) rather than acute ingestion protocol is used [122]. The chronic loading protocol seems to be preferable because it is less likely to cause gastrointestinal side effects and it promotes a more sustained and prolonged effect on anaerobic performance than does acute ingestion. Chronic ingestion of sodium bicarbonate can improve performance up to 2 days after the cessation of ingestion [122] whereas acute ingestion will improve performance for no longer than 3–4 hours after ingestion.

While sodium bicarbonate and sodium citrate augment extracellular buffering capacity, beta-alanine supplementation was consistently shown to increase the concentration of carnosine in muscle cells [123]. Of note, carnosine is an intracellular cytoplasmic dipeptide, abundantly found in skeletal muscle, which exerts a relevant acid-base regulation function [124]. Carnosine is not taken up into the muscle cells, but it is synthesized inside muscle fibers from the amino acid histidine and beta-alanine [125]. The rate-limiting step for intramuscular carnosine synthesis is the availability of beta-alanine [124], an amino acid poorly found in diet. Therefore, supplementation with 1.6–6.4 g/day is the best way to significantly increase muscle carnosine content (doses are to be taken for 4 weeks or longer, and the expected increase in carnosine is >40%) [123]. Beta-alanine supplementation can elicit significant performance improvements in continuous and high-intensity intermittent exercises [15,126]. Clearly, the effects of beta-alanine supplementation are most likely to benefit combat sport athletes as well, as indicated by a study by Tobias and colleagues [127]. In this study, highly trained judo and jiu-jitsu athletes were randomly assigned to one of four groups: beta-alanine, sodium bicarbonate, beta-alanine plus sodium bicarbonate, and placebo. Athletes were assessed for intermittent anaerobic performance before and after supplementation. Interestingly, beta-alanine and sodium bicarbonate resulted in almost identical performance enhancements, indicating that one supplement is not superior to the other. Moreover, they appear to have additive effects, as the combination of both supplements yielded a twofold greater increase in performance in comparison with beta-alanine or sodium bicarbonate alone [127]. Hence, the use of beta-alanine at least 1 month before a major competition in addition to the use of sodium bicarbonate at least 5 days before the same competition will probably be highly beneficial for physical performance in most combat sports.

Muscle phosphorylcreatine (PCr) depletion during intensive exercise is another relevant factor causing fatigue. Increasing resting intramuscular PCr emerges, therefore, as an appealing way to delay PCr depletion and improve anaerobic performance. In fact, since 1992 studies with humans have consistently shown that creatine supplementation (~20 g for 5 days or longer) augments muscle PCr content at rest [128]. This increased muscle PCr is related to improved anaerobic performance, especially in high-intensity intermittent exercises [129], which highlights the potential ergogenic effects of creatine supplementation in combat sports. However, not every athlete will respond positively to creatine supplementation since the increase in muscle PCr is dependent on the initial concentration of muscle PCr which, in turn, is dependent on dietary patterns [128]. More precisely, athletes who normally eat high amounts of creatine-rich foods (e.g., red meat and fish) present high muscle PCr concentration and do not respond to creatine supplementation, neither increasing muscle PCr content nor improving performance. On the other hand, athletes who don't eat creatine sources in their diets (e.g., vegetarians) present low muscle PCr concentration and respond quite well to supplementation.

In those athletes who respond to creatine supplementation, there is a notable retention of water in muscle, which is due to an osmotic effect of creatine and leads to increased total body water [130]. Although such effect is completely harmless, it results in a modest increase in body mass. Even though only modest, the increase in body mass may represent an enormous obstacle for most combat athletes, as they normally weigh more than their weight classes' limit [81,84]. If an athlete does not need to make weight, then creatine supplementation is probably an effective supplement. On the other hand, if an athlete is usually above his/her weight class, creatine supplementation will probably worsen the weight-cutting problem, and alternative supplementation strategies would be preferred. In these cases, creatine may be taken during training periods (e.g., preparation or competitive phases) in order to maximize training adaptations and ceased approximately 4 weeks before the weigh-in, as this is the average wash-out time for creatine in humans [131]. This procedure will ensure that muscle creatine and, consequently, total body water and body mass return to pre-supplementation levels before the weigh-in. Alternatively, if the athlete will compete in an event where the weigh-in occurs >48 hours prior to the matches, the use of creatine (20 g/day taken in four 5 g single doses) may help in performance, as the first 48 hours of supplementation are those with the highest increase in muscle PCr [128]. High doses of creatine ingested after the weigh-in may benefit performance even when the time between weigh-in and the first match is shorter (~15 h or longer) [132].

Caffeine supplementation seems to be another useful ergogenic aid in combat sports. Interestingly, acute ingestion of 6 mg/kg body mass of caffeine decreases extracellular potassium concentration [133], which is one of the underlying mechanisms that explains the ergogenic effects of caffeine on high-intensity performance. As a matter of fact, several studies have demonstrated the ergogenic potential of caffeine ingestion (3–6 mg/kg of body mass 1–3 h prior to exercise) on high-intensity intermittent performance [129]. Besides the local effects on skeletal muscle, caffeine also increases plasma catecholamines concentration [134], which helps to explain its ergogenic properties.

In addition to beta-alanine, sodium bicarbonate or citrate, creatine and caffeine, a few other supplements may help to support combat athletes' training regimens. High-quality proteins and carbohydrates, for example, may be useful if an athlete is unable to ingest the recommended amounts, as previously discussed in this chapter. Similarly, vitamins or minerals may be valuable if a specific deficiency is detected, and electrolyte-carbohydrate beverages may help to prevent performance decrements during training and competition. Based on current literature, other supplements are less likely to benefit combat athletes, although new promising supplements may emerge in the near future. Nonetheless, the conscientious athlete will make his/her diet as healthy and complete as possible, leaving supplement to a minimum. Finally, an important caveat about the purity of some supplements found over the counters should be made: a considerable percentage of supplements may be contaminated with illegal substances [135]. Although such contamination is usually very small, it may be just enough to cause an athlete to fail anti-doping tests [136].

References

[1] Franchini E, Brito CJ, Artioli GG. Weight loss in combat sports: physiological, psychological and performance effects. J Int Soc Sports Nutr 2012;9(1):52. http://dx.doi.org/10.1186/1550-2783-9-52.

[2] Campos FA, Bertuzzi R, Dourado AC, et al. Energy demands in taekwondo athletes during combat simulation. Eur J Appl Physiol 2012;112(4):1221–8.

[3] Artioli GG, Gualano B, Franchini E, et al. Physiological, performance, and nutritional profile of the Brazilian Olympic Wushu (kung-fu) team. J Strength Cond Res 2009;23(1):20–5.

[4] Franchini E, Del Vecchio FB, Matsushigue KA, et al. Physiological profiles of elite judo athletes. Sports Med 2011;41 (2):147–66.

[5] Terbizan DJ, Seljevold PJ. Physiological profile of age-group wrestlers. J Sports Med Phys Fitness 1996;36(3):178–85.

[6] del Vecchio FB, Hirata SM, Franchini E. A review of time-motion analysis and combat development in mixed martial arts matches at regional level tournaments. Percept Mot Skills 2011;112(2):639–48.

[7] Mandroukas A, Heller J, Metaxas TI, et al. Deltoid muscle characteristics in wrestlers. Int J Sports Med 2010;31(3):148–53.

[8] Horswill CA, Miller JE, Scott JR, et al. Anaerobic and aerobic power in arms and legs of elite senior wrestlers. Int J Sports Med 1992;13(8):558–61.

[9] Franchini E, Sterkowicz S, Szmatlan-Gabrys U, et al. Energy system contributions to the special judo fitness test. Int J Sports Physiol Perform 2011;6(3):334–43.

[10] Artioli G, Scagliusi F, Kashiwagura D, et al. Development, validity and reliability of a questionnaire designed to evaluate rapid weight loss patterns in judo players. Scand J Med Sci Sports 2009 Sep 28.

[11] Brito CJ, Roas AF, Brito IS, et al. Methods of body mass reduction by combat sport athletes. Int J Sport Nutr Exerc Metab 2012;22(2):89–97.

[12] Steen SN, Brownell KD. Patterns of weight loss and regain in wrestlers: has the tradition changed? Med Sci Sports Exerc 1990;22(6):762–8.

[13] Artioli GG, Franchini E, Nicastro H, et al. The need of a weight management control program in judo: a proposal based on the successful case of wrestling. J Int Soc Sports Nutr 2010;7(1):15.

[14] Knuth ST, Dave H, Peters JR, et al. Low cell pH depresses peak power in rat skeletal muscle fibres at both 30 degrees C and 15 degrees C: implications for muscle fatigue. J Physiol 2006;575(Pt 3):887–99.

[15] Artioli GG, Gualano B, Smith A, et al. Role of beta-alanine supplementation on muscle carnosine and exercise performance. Med Sci Sports Exerc 2010;42(6):1162–73.

[16] Allen DG, Lamb GD, Westerblad H. Skeletal muscle fatigue: cellular mechanisms. Physiol Rev. 2008;88(1):287–332.

[17] Lanza IR, Befroy DE, Kent-Braun JA. Age-related changes in ATP-producing pathways in human skeletal muscle in vivo. J Appl Physiol 2005;99(5):1736–44.

[18] Yoshimura A, Toyoda Y, Murakami T, et al. Glycogen depletion in intrafusal fibres in rats during short-duration high-intensity treadmill running. Acta Physiol Scand 2005;185(1):41–50.

[19] Saltin B. Aerobic and Anaerobic Work Capacity after Dehydration. J Appl Physiol 1964;19:1114–8.

[20] Walsh RM, Noakes TD, Hawley JA, et al. Impaired high-intensity cycling performance time at low levels of dehydration. Int J Sports Med 1994;15(7):392–8.

[21] Kraemer WJ, Fry AC, Rubin MR, et al. Physiological and performance responses to tournament wrestling. Med Sci Sports Exerc 2001;33(8):1367–78.

[22] NRC. Dietary Reference Intakes for energy, carbohydrate, fiber, fat, fatty acids, cholesterol, protein and amino acids. The National Academies Press; 2002.

[23] Bergstrom J, Hermansen L, Hultman E, et al. Diet, muscle glycogen and physical performance. Acta Physiol Scand 1967;71 (2):140–50.

[24] Robinson TM, Sewell DA, Hultman E, et al. Role of submaximal exercise in promoting creatine and glycogen accumulation in human skeletal muscle. J Appl Physiol 1999;87(2): 598–604.

[25] Gollnick PD, Piehl K, Saltin B. Selective glycogen depletion pattern in human muscle fibres after exercise of varying intensity and at varying pedalling rates. J Physiol 1974;241(1):45–57.

[26] Burke LM, Kiens B, Ivy JL. Carbohydrates and fat for training and recovery. J Sports Sci 2004;22(1):15–30.

[27] Foskett A, Williams C, Boobis L, et al. Carbohydrate availability and muscle energy metabolism during intermittent running. Med Sci Sports Exerc 2008;40(1):96–103.

[28] Phillips SM, Sproule J, Turner AP. Carbohydrate ingestion during team games exercise: current knowledge and areas for future investigation. Sports Med 2011;41(7):559–85.

[29] Jeukendrup AE. Carbohydrate intake during exercise and performance. Nutrition 2004;20(7-8):669–77.

[30] Hargreaves M, Hawley JA, Jeukendrup A. Pre-exercise carbohydrate and fat ingestion: effects on metabolism and performance. J Sports Sci 2004;22(1):31–8.

[31] Yaspelkis III BB, Patterson JG, Anderla PA, et al. Carbohydrate supplementation spares muscle glycogen during variable-intensity exercise. J Appl Physiol 1993;75(4):1477–85.

[32] Wojtaszewski JF, Nielsen P, Kiens B, et al. Regulation of glycogen synthase kinase-3 in human skeletal muscle: effects of food intake and bicycle exercise. Diabetes 2001;50(2):265–9.

[33] Arnall DA, Nelson AG, Quigley J, et al. Supercompensated glycogen loads persist 5 days in resting trained cyclists. Eur J Appl Physiol 2007;99(3):251–6.

[34] Sherman WM, Costill DL, Fink WJ, et al. Effect of exercise-diet manipulation on muscle glycogen and its subsequent utilization during performance. Int J Sports Med 1981;2(2):114–8.

[35] Fairchild TJ, Fletcher S, Steele P, et al. Rapid carbohydrate loading after a short bout of near maximal-intensity exercise. Med Sci Sports Exerc 2002;34(6):980–6.

[36] van Hall G, Shirreffs SM, Calbet JA. Muscle glycogen resynthesis during recovery from cycle exercise: no effect of additional protein ingestion. J Appl Physiol 2000;88(5):1631–6.

[37] Ivy JL, Katz AL, Cutler CL, et al. Muscle glycogen synthesis after exercise: effect of time of carbohydrate ingestion. J Appl Physiol 1988;64(4):1480–5.

[38] Poortmans JR, Carpentier A, Pereira-Lancha LO, et al. Protein turnover, amino acid requirements and recommendations for athletes and active populations. Braz J Med Biol Res 2012 Jun 6.

[39] Burd NA, Tang JE, Moore DR, et al. Exercise training and protein metabolism: influences of contraction, protein intake, and sex-based differences. J Appl Physiol 2009;106(5):1692–701.

[40] Nicastro H, Artioli GG, Costa Ados S, et al. An overview of the therapeutic effects of leucine supplementation on skeletal muscle under atrophic conditions. Amino Acids 2010;40 (2):287–300.

[41] Artioli GG, Franchini E, Solis MY, et al. Recovery time between weigh-in and first match in State level judo competitions. Brazilian Journal of Physical Education and Sport 2011;25 (3):371–6.

[42] Phillips SM, Tipton KD, Ferrando AA, et al. Resistance training reduces the acute exercise-induced increase in muscle protein turnover. Am J Physiol 1999;276(1 Pt 1):E118–24.

[43] Tipton KD, Borsheim E, Wolf SE, et al. Acute response of net muscle protein balance reflects 24-h balance after exercise and amino acid ingestion. Am J Physiol Endocrinol Metab 2003;284 (1):E76–89.

[44] Tarnopolsky MA, Atkinson SA, MacDougall JD, et al. Evaluation of protein requirements for trained strength athletes. J Appl Physiol 1992;73(5):1986–95.

[45] Bohe J, Low A, Wolfe RR, et al. Human muscle protein synthesis is modulated by extracellular, not intramuscular amino acid availability: a dose-response study. J Physiol 2003;552(Pt 1):315–24.

[46] Phillips SM, Tipton KD, Aarsland A, et al. Mixed muscle protein synthesis and breakdown after resistance exercise in humans. Am J Physiol 1997;273(1 Pt 1):E99–107.

[46a] Bolster DR, Pikosky MA, Gaine PC, Martin W, Wolfe RR, Tipton KD, et al. Dietary protein intake impacts human skeletal muscle protein fractional synthetic rates after endurance exercise. Am J Physiol Endocrinol Metab 2005;289(4):E678–83.

[46b] Phillips SM. Protein requirements and supplementation in strength sports. Nutrition 2004;20(7-8):689–95.

[47] Biolo G, Tipton KD, Klein S, et al. An abundant supply of amino acids enhances the metabolic effect of exercise on muscle protein. Am J Physiol 1997;273(1 Pt 1):E122–9.

[48] Cribb PJ, Hayes A. Effects of supplement timing and resistance exercise on skeletal muscle hypertrophy. Med Sci Sports Exerc 2006;38(11):1918–25.

[49] Moore DR, Tang JE, Burd NA, et al. Differential stimulation of myofibrillar and sarcoplasmic protein synthesis with protein ingestion at rest and after resistance exercise. J Physiol 2009;587 (Pt 4):897–904.

[50] Rennie MJ, Wackerhage H, Spangenburg EE, et al. Control of the size of the human muscle mass. Annu Rev Physiol 2004;66:799–828.

[51] Tipton KD, Elliott TA, Cree MG, et al. Ingestion of casein and whey proteins result in muscle anabolism after resistance exercise. Med Sci Sports Exerc 2004;36(12):2073–81.

[52] Tang JE, Phillips SM. Maximizing muscle protein anabolism: the role of protein quality. Curr Opin Clin Nutr Metab Care 2009;12(1):66–71.

[53] Dangin M, Boirie Y, Guillet C, et al. Influence of the protein digestion rate on protein turnover in young and elderly subjects. J Nutr 2002;132(10):3228S–3233SS.

[54] Churchward-Venne TA, Burd NA, Phillips SM, et al. Nutritional regulation of muscle protein synthesis with resistance exercise: strategies to enhance anabolism. Nutr Metab (Lond) 2012;9(1):40.

[55] Wilkinson SB, Tarnopolsky MA, Macdonald MJ, et al. Consumption of fluid skim milk promotes greater muscle protein accretion after resistance exercise than does consumption of an isonitrogenous and isoenergetic soy-protein beverage. Am J Clin Nutr 2007;85(4):1031–40.

[56] Res PT, Groen B, Pennings B, et al. Protein Ingestion before Sleep Improves Postexercise Overnight Recovery. Med Sci Sports Exerc 2012;44(8):1560–9.

[57] Rodriguez NR, DiMarco NM, Langley S. Position of the American Dietetic Association, Dietitians of Canada, and the American College of Sports Medicine: Nutrition and athletic performance. J Am Diet Assoc 2009;109(3):509–27.

[58] Williams MH. Dietary supplements and sports performance: introduction and vitamins. J Int Soc Sports Nutr 2004;1:1–6.

[59] Driskell J. Vitamins and trace elements in sports nutrition. In: Driskell J, Wolinsky I, editors. Sports Nutrition. New York: CRC/Taylor & Francis; 2006.

[60] Lukaski HC. Vitamin and mineral status: effects on physical performance. Nutrition 2004;20(7-8):632–44.

[61] Woolf K, Manore MM. B-vitamins and exercise: does exercise alter requirements? Int J Sport Nutr Exerc Metab 2006;16 (5):453–84.

[62] Cowell BS, Rosenbloom CA, Skinner R, et al. Policies on screening female athletes for iron deficiency in NCAA division I-A institutions. Int J Sport Nutr Exerc Metab 2003;13 (3):277–85.

[63] Brownlie T, Utermohlen V, Hinton PS, et al. Tissue iron deficiency without anemia impairs adaptation in endurance capacity after aerobic training in previously untrained women. Am J Clin Nutr 2004;79(3):437–43.

[64] Brito CJ, Gatti K, Lacerda Mendes E, et al. Carbohydrate intake and immunosuppression during judo training. Medicina dello Sport 2011;64(4):393–408.

[65] Sawka MN. Physiological consequences of hypohydration: exercise performance and thermoregulation. Med Sci Sports Exerc 1992;24(6):657–70.

[66] Montain SJ, Sawka MN, Latzka WA, et al. Thermal and cardiovascular strain from hypohydration: influence of exercise intensity. Int J Sports Med 1998;19(2):87–91.

[67] Nybo L. Cycling in the heat: performance perspectives and cerebral challenges. Scand J Med Sci Sports 2010;20(Suppl 3):71–9.

[68] Ganio MS, Armstrong LE, Casa DJ, et al. Mild dehydration impairs cognitive performance and mood of men. Br J Nutr 2011;106(10):1535–43.

[69] Montain SJ, Smith SA, Mattot RP, et al. Hypohydration effects on skeletal muscle performance and metabolism: a 31P-MRS study. J Appl Physiol 1998;84(6):1889–94.

[70] Burdon CA, Johnson NA, Chapman PG, et al. Influence of beverage temperature on palatability and fluid ingestion during endurance exercise: a systematic review. Int J Sport Nutr Exerc Metab 2012;22(3):199–211.

[71] Craig EN, Cummings EG. Dehydration and muscular work. J Appl Physiol 1966;21(2):670–4.
[72] Judelson DA, Maresh CM, Farrell MJ, et al. Effect of hydration state on strength, power, and resistance exercise performance. Med Sci Sports Exerc 2007;39(10):1817–24.
[73] Maughan RJ, Shirreffs SM. Development of hydration strategies to optimize performance for athletes in high-intensity sports and in sports with repeated intense efforts. Scand J Med Sci Sports 2010;20(Suppl 2):59–69.
[74] Sawka MN, Burke LM, Eichner ER, et al. American College of Sports Medicine position stand. Exercise and fluid replacement. Med Sci Sports Exerc 2007;39(2):377–90.
[75] Burdon CA, Johnson NA, Chapman PG, et al. Influence of Beverage Temperature on Palatability and Fluid Ingestion During Endurance Exercise: a systematic review. Int J Sport Nutr Exerc Metab 2012 Jun 15; [Epub ahead of print].
[76] Maughan RJ, Fenn CE, Leiper JB. Effects of fluid, electrolyte and substrate ingestion on endurance capacity. Eur J Appl Physiol Occup Physiol 1989;58(5):481–6.
[77] ACSM. American College of Sports Medicine position stand on weight loss in wrestlers. Med Sci Sports 1976;8(2):xi–xiii.
[78] Alderman BL, Landers DM, Carlson J, et al. Factors related to rapid weight loss practices among international-style wrestlers. Med Sci Sports Exerc 2004;36(2):249–52.
[79] AMA. Wrestling and weight control. JAMA 1967;201(7):131–3.
[80] Artioli GG, Franchini E, Lancha Junior AH. Rapid weight loss in grapling combat sports: review and applied recommendations [in Portuguese]. Brazilian Journal of Kinanthropometry and Human Performance 2006;8(2):92–101.
[81] Artioli GG, Gualano B, Franchini E, et al. Prevalence, magnitude, and methods of rapid weight loss among judo competitors. Med Sci Sports Exerc 2010;42(3):436–42.
[82] Brownell KD, Steen SN, Wilmore JH. Weight regulation practices in athletes: analysis of metabolic and health effects. Med Sci Sports Exerc 1987;19(6):546–56.
[83] Fogelholm M. Effects of bodyweight reduction on sports performance. Sports Med 1994;18(4):249–67.
[84] Kiningham RB, Gorenflo DW. Weight loss methods of high school wrestlers. Med Sci Sports Exerc 2001;33(5):810–3.
[85] Oppliger RA, Landry GL, Foster SW, et al. Wisconsin minimum weight program reduces weight-cutting practices of high school wrestlers. Clin J Sport Med 1998;8(1):26–31.
[86] Oppliger RA, Steen SA, Scott JR. Weight loss practices of college wrestlers. Int J Sport Nutr Exerc Metab 2003;13(1):29–46.
[87] Oppliger RA, Utter AC, Scott JR, et al. NCAA rule change improves weight loss among national championship wrestlers. Med Sci Sports Exerc 2006;38(5):963–70.
[88] Tipton CM, Tcheng TK. Iowa wrestling study. Weight loss in high school students. JAMA 1970;214(7):1269–74.
[89] Artioli GG, Scagliusi FB, Polacow VO, et al. Magnitude e métodos de perda rápida de peso em judocas de elite. Revista de Nutrição 2007;20(3):307–15.
[90] Oppliger RA, Landry GL, Foster SW, et al. Bulimic behaviors among interscholastic wrestlers: a statewide survey. Pediatrics 1993;91(4):826–31.
[91] Allen TE, Smith DP, Miller DK. Hemodynamic response to submaximal exercise after dehydration and rehydration in high school wrestlers. Med Sci Sports 1977;9(3):159–63.
[92] Imai T, Seki S, Dobashi H, et al. Effect of weight loss on T-cell receptor-mediated T-cell function in elite athletes. Med Sci Sports Exerc 2002;34(2):245–50.
[93] Kowatari K, Umeda T, Shimoyama T, et al. Exercise training and energy restriction decrease neutrophil phagocytic activity in judoists. Med Sci Sports Exerc 2001;33(4):519–24.
[94] Umeda T, Nakaji S, Shimoyama T, et al. Adverse effects of energy restriction on changes in immunoglobulins and complements during weight reduction in judoists. J Sports Med Phys Fitness 2004;44(3):328–34.
[95] Oppliger RA, Case HS, Horswill CA, et al. American College of Sports Medicine position stand. Weight loss in wrestlers. Med Sci Sports Exerc 1996;28(6):ix–xii.
[96] Prouteau S, Pelle A, Collomp K, et al. Bone density in elite judoists and effects of weight cycling on bone metabolic balance. Med Sci Sports Exerc 2006;38(4):694–700.
[97] McMurray RG, Proctor CR, Wilson WL. Effect of caloric deficit and dietary manipulation on aerobic and anaerobic exercise. Int J Sports Med 1991;12(2):167–72.
[98] Strauss RH, Lanese RR, Malarkey WB. Weight loss in amateur wrestlers and its effect on serum testosterone levels. JAMA 1985;254(23):3337–8.
[99] Filaire E, Rouveix M, Pannafieux C, et al. Eating attitudes, perfectionism and body-esteem of male judoists and cyclists. J Sports Sci Med 2007;6:50–7.
[100] Horswill CA, Park SH, Roemmich JN. Changes in the protein nutritional status of adolescent wrestlers. Med Sci Sports Exerc 1990;22(5):599–604.
[101] Roemmich JN, Sinning WE. Weight loss and wrestling training: effects on nutrition, growth, maturation, body composition, and strength. J Appl Physiol 1997;82(6):1751–9.
[102] Roemmich JN, Sinning WE. Weight loss and wrestling training: effects on growth-related hormones. J Appl Physiol 1997;82(6):1760–4.
[103] Saarni SE, Rissanen A, Sarna S, et al. Weight cycling of athletes and subsequent weight gain in middleage. Int J Obes (Lond) 2006;30(11):1639–44.
[104] Hyperthermia and dehydration-related deaths associated with intentional rapid weight loss in three collegiate wrestlers—North Carolina, Wisconsin, and Michigan, November-December 1997. MMWR Morb Mortal Wkly Rep. 1998 Feb 20;47(6):105–8.
[105] Davis SE, Dwyer GB, Reed K, et al. Preliminary investigation: the impact of the NCAA Wrestling Weight Certification Program on weight cutting. J Strength Cond Res 2002;16 (2):305–7.
[106] Artioli GG, Iglesias RT, Franchini E, et al. Rapid weight loss followed by recovery time does not affect judo-related performance. J Sports Sci 2010;23:1–12.
[107] Mourier A, Bigard AX, de Kerviler E, et al. Combined effects of caloric restriction and branched-chain amino acid supplementation on body composition and exercise performance in elite wrestlers. Int J Sports Med 1997;18(1): 47–55.
[108] Baptista IL, Leal ML, Artioli GG, et al. Leucine attenuates skeletal muscle wasting via inhibition of ubiquitin ligases. Muscle Nerve 2010;41(6):800–8.
[109] Klinzing JE, Karpowicz W. The effects of rapid weight loss and rehydratation on a wrestling performance test. J Sports Med Phys Fitness 1986;26(2):149–56.
[110] Webster S, Rutt R, Weltman A. Physiological effects of a weight loss regimen practiced by college wrestlers. Med Sci Sports Exerc 1990;22(2):229–34.
[111] Serfass RC, Stull GA, Alexander FF, et al. The effects of rapid weight loss and attempted rehydration on strength and endurance of the handgripping muscles in college wrestlers. Res Q Exerc Sport 1984;55(1):46–52.
[112] Horswill CA, Hickner RC, Scott JR, et al. Weight loss, dietary carbohydrate modifications, and high intensity, physical performance. Med Sci Sports Exerc 1990;22(4):470–6.

[113] Fogelholm GM, Koskinen R, Laakso J, et al. Gradual and rapid weight loss: effects on nutrition and performance in male athletes. Med Sci Sports Exerc 1993;25(3):371–7.

[114] Burge CM, Carey MF, Payne WR. Rowing performance, fluid balance, and metabolic function following dehydration and rehydration. Med Sci Sports Exerc 1993;25(12):1358–64.

[115] Rankin JW, Ocel JV, Craft LL. Effect of weight loss and refeeding diet composition on anaerobic performance in wrestlers. Med Sci Sports Exerc 1996;28(10):1292–9.

[116] Filaire E, Maso F, Degoutte F, et al. Food restriction, performance, psychological state and lipid values in judo athletes. Int J Sports Med 2001;22(6):454–9.

[117] Degoutte F, Jouanel P, Begue RJ, et al. Food restriction, performance, biochemical, psychological, and endocrine changes in judo athletes. Int J Sports Med 2006;27(1):9–18.

[118] McNaughton LR, Siegler J, Midgley A. Ergogenic effects of sodium bicarbonate. Curr Sports Med Rep 2008;7(4):230–6.

[119] Requena B, Zabala M, Padial P, et al. Sodium bicarbonate and sodium citrate: ergogenic aids? J Strength Cond Res 2005;19 (1):213–24.

[120] Derave W, Everaert I, Beeckman S, et al. Muscle carnosine metabolism and beta-alanine supplementation in relation to exercise and training. Sports Med 2010;40(3):247–63.

[121] Artioli GG, Gualano B, Coelho DF, et al. Does sodium-bicarbonate ingestion improve simulated judo performance?. Int J Sport Nutr Exerc Metab 2007;17(2):206–17.

[122] McNaughton L, Thompson D. Acute versus chronic sodium bicarbonate ingestion and anaerobic work and power output. J Sports Med Phys Fitness 2001;41(4):456–62.

[123] Stellingwerff T, Decombaz J, Harris RC, et al. Optimizing human in vivo dosing and delivery of beta-alanine supplements for muscle carnosine synthesis. Amino Acids 2012;43 (1):57–65.

[124] Harris RC, Tallon MJ, Dunnett M, et al. The absorption of orally supplied beta-alanine and its effect on muscle carnosine synthesis in human vastus lateralis. Amino Acids 2006;30 (3):279–89.

[125] Bauer K, Schulz M. Biosynthesis of carnosine and related peptides by skeletal muscle cells in primary culture. Eur J Biochem 1994;219(1-2):43–7.

[126] Hobson RM, Saunders B, Ball G, et al. Effects of beta-alanine supplementation on exercise performance: a meta-analysis. Amino Acids 2012;43(1):25–37.

[127] Tobias GC, Benatti FB, Roschel H., et al. The additive ergogenic effects of beta-alanine and sodium bicarbonate in high-intensity exercise performance. Amino Acids 2013. [Epub ahead of print].

[128] Harris RC, Soderlund K, Hultman E. Elevation of creatine in resting and exercised muscle of normal subjects by creatine supplementation. Clin Sci (Lond) 1992;83(3):367–74.

[129] Tarnopolsky MA. Caffeine and creatine use in sport. Ann Nutr Metab 2010;57(Suppl 2):1–8.

[130] Powers ME, Arnold BL, Weltman AL, et al. Creatine Supplementation Increases Total Body Water Without Altering Fluid Distribution. J Athl Train 2003;38(1):44–50.

[131] McKenna MJ, Morton J, Selig SE, et al. Creatine supplementation increases muscle total creatine but not maximal intermittent exercise performance. J Appl Physiol 1999;87(6):2244–52.

[132] Oopik V, Paasuke M, Timpmann S, et al. Effects of creatine supplementation during recovery from rapid body mass reduction on metabolism and muscle performance capacity in well-trained wrestlers. J Sports Med Phys Fitness 2002;42(3):330–9.

[133] Crowe MJ, Leicht AS, Spinks WL. Physiological and cognitive responses to caffeine during repeated, high-intensity exercise. Int J Sport Nutr Exerc Metab 2006;16(5):528–44.

[134] Stuart GR, Hopkins WG, Cook C, et al. Multiple effects of caffeine on simulated high-intensity team-sport performance. Med Sci Sports Exerc 2005;37(11):1998–2005.

[135] Geyer H, Parr MK, Koehler K, et al. Nutritional supplements cross-contaminated and faked with doping substances. J Mass Spectrom 2008;43(7):892–902.

[136] Watson P, Judkins C, Houghton E, et al. Urinary nandrolone metabolite detection after ingestion of a nandrolone precursor. Med Sci Sports Exerc 2009;41(4):766–72.

CHAPTER

12

Sumo Wrestling

An Overview

Taishi Midorikawa[1,2], *Shizuo Sakamoto*[2] *and Masakatsu Kondo*[3]

[1]College of Health and Welfare, J.F. Oberlin University, Tokyo, Japan [2]Faculty of Sport Sciences, Waseda University, Saitama, Japan [3]Department of Exercise Physiology, Nihon University, Tokyo, Japan

INTRODUCTION

The winner of Sumo wrestling is decided on the inner circle, the "Dohyo" (diameter 4.55 m). Two wrestlers fight to push or throw the opponent out of the Dohyo or make any part of his body other than feet touch the ground [1]. Hence, Sumo wrestlers have to acquire a mix of power, agility, balance ability and aerobic capacity—although Sumo wrestlers have a relatively low VO_{2max} /body weight kg [2]—from participating in regular training (termed "Kei-ko"), which normally consists of wrestling exercises (e.g., pushing and throwing other Sumo wrestlers) and additional technical drills [3]. In addition, since there are no weight limits as in boxing or western wrestling, Sumo wrestlers who have greater body mass possess one of the most effective ways to win. Therefore, Sumo wrestlers are a group of athletes who have high levels of fat mass and fat-free mass, not often observed in most other forms of competitive sports [1]. Research on Sumo wrestlers may provide insight into weight gain to enhance sports performance.

ENERGY BALANCE

The large amounts of fat mass and fat-free mass in Sumo wrestlers is greatly influenced by energy intake and energy expenditure or physical activity levels during exercise training. Sumo wrestlers basically have two meals per day at about 1230 h and 1730 h and take a nap between the two meals. They start regular training from around 0600 h to 1000 h [4]. During the remaining time periods, Sumo wrestlers are permitted to spend time freely, such as in reading, writing, viewing television, cleaning and household chores.

A Sumo wrestler's meals (termed "Chanko-ryori") are abundant in calories, protein and carbohydrate [4]. A study of the diets of Sumo wrestlers in the 1970s found that the estimated daily energy intake for Sumo wrestlers was 5122–5586 kcal/day [4]. Moreover, according to a dietary survey (2 days of two meals/day) of 10 professional sumo wrestlers (standing height 186.0 ± 8.3 cm, body weight 152.0 ± 39.5 kg), the average estimated daily energy intake was 3939 kcal/day, and the diet of Sumo wrestlers was well balanced (the PFC ratio was 16.0% for protein, 28.2% for fat and 55.8% for carbohydrate) except for slightly inadequate calcium intake [5]. Based on these previous studies, Sumo wrestlers get an energy intake of about 4000 kcal/day to more than 5000 kcal/day in two meals per day.

In contrast, no studies on total energy expenditure or physical activity level are available for Sumo wrestlers. However, there are a few reports about resting energy expenditure (which constitutes 60–70% of total energy expenditure) for Sumo wrestlers. According to relatively recent studies using the Douglas bag technique, the measured resting energy expenditure was 2952 ± 302 kcal/day for 15 male college Sumo wrestlers (standing height 176.8 ± 3.5 cm, body weight 125.1 ± 12.9 kg) [6] and 2286 ± 350 kcal/day for 10 male college Sumo wrestlers (standing height 172.9 ± 8.4 cm, body weight 109.1 ± 14.7 kg) [7]. Even if calculated from the resting energy expenditure (2500 kcal/day) and physical activity level (1.75; the lowest limit physical activity level by [6]), the predicted total energy expenditure for Sumo wrestlers would be 4375 kcal/day. On the energy balance of Sumo wrestlers, the energy intake

Nutrition and Enhanced Sports Performance.
DOI: http://dx.doi.org/10.1016/B978-0-12-396454-0.00012-6

exceeds energy expenditure over a considerable period, for gaining weight. Sumo wrestlers simply have an energy imbalance, but it is important to point out that the weight gain is directly related to an enhancement of Sumo performance.

Because there is presently no information on energy expenditure during Sumo wrestling, future research will measure total energy expenditure for Sumo wrestlers with high accuracy using a doubly labeled water method. Moreover, since information about energy intake for Sumo wrestlers has only been reported from the diet survey of two meals/day, future research will also focus on a more realistic energy intake by observing all of what is eaten and drunk (i.e., including eating between meals).

FAT MASS AND FAT-FREE MASS FOR TOP LEAGUE ("SEKITORI")

Sumo wrestlers were found to have body weights in excess of 100 kg, with fat mass in excess of 30 kg and fat-free mass greater than 80 kg, which was larger than fat-free mass for bodybuilders [3]. The large amounts of fat and fat-free mass coupled with regular exercise training allowed Sumo wrestlers to be considered "obese athletes" [8]. Can Sumo wrestlers be considered "obese" or "obese athletes"?

The first scientific report on the physique of Sumo wrestlers was published more than a century ago [9]. Since then, more than 10 studies have reported the body composition of Sumo wrestlers (Table 12.1). As the first major finding in Table 12.1, the average value of BMI both in professional and college Sumo wrestlers is categorized as obese using the conventional WHO criteria (i.e., BMI>30 kg/m^2), but the mean value of percent fat using the methods of underwater weighing, air displacement plethysmography, and dual-energy X-ray absorptiometry is less than 30%. Because of their high fat-free mass, Sumo wrestlers can be misclassified as obese based on the BMI [6]. In fact, most Sumo wrestlers are able to maintain normal serum glucose and triglyceride levels despite a very large visceral adipose tissue area (151 ± 58 cm^2); however, daily exercise training does not reduce all cardiovascular disease risk factors such as insulin resistance [14]. Another point is that Sumo wrestlers in most published data have more than 80 kg of fat-free mass. According to the first study for Sumo wrestlers using the underwater weighing method, the average fat-free mass of wrestlers in the top league ("Sekitori") was 109 kg, including the largest one of 121.3 kg (standing height 186 cm, body weight 181 kg, percent fat 33.0% [3]). Since some Sumo wrestlers have a body mass above 200 kg, the upper limit of fat-free mass might approach 150 kg or a fat-free mass/standing height ratio of 0.7 kg/cm [3].

Sumo wrestlers are, according to their abilities, divided into the upper leagues (Sekitori), including the "Makuuchi" headed by the grand champion and "Juryo" division, and the lower leagues, which include "Makushita", "Sandanme", "Jonidan", and "Jonokuchi" division [4]. According to a report about the hierarchical differences in body composition of professional Sumo wrestlers, Sekitori division Sumo wrestlers were found to have larger fat-free mass than those who belong to the lower leagues, although adiposity level for the Sekitori division was equal to that in the lower divisions [11]. The cut-off point of fat-free mass index (i.e., fat-free mass [kg] /height [m]^2) which separated Sekitori wrestlers from other wrestlers was approximately 30 [11]. Additionally, force generation capability was higher in the upper-leagues' wrestlers than those in the lower-leagues [10]. Based on these previous studies, Sumo wrestlers, especially Sekitori wrestlers, can be considered to be "athletes".

ORGAN-TISSUE LEVEL BODY COMPOSITION

Although body composition studies for Sumo wrestlers have been developed over about 20 years based on a two-compartment model which classifies body weight as fat mass and fat-free mass, there is limited information about body composition at the organ-tissue level. Recently, the method of magnetic resonance imaging (MRI) has provided precise, reliable, and safe measurements of whole-body skeletal muscle mass and internal organs in adults [15], and does so in a relatively short period of time (i.e., it takes about 30 min to scan from the top of the head to the ankle joints with 1.0-cm slice thickness and 0-cm interslice gap in adults). According to the only published research on organ-tissue level composition for Sumo wrestlers, they were found to have greater skeletal muscle (36.9 ± 5.9 kg; max value 43.4 kg) than controls (24.5 ± 3.4 kg) (Table 12.2) [7]. Even if calculated from the skeletal muscle mass equation using fat-free mass [15],

$$\text{skeletal muscle mass} = 0.56 \times \text{fat-free mass} - 9.1,$$

the predicted skeletal muscle mass for Sumo wrestlers with 150 kg fat-free mass would be 74.9 kg. The upper limit of skeletal muscle mass in humans might be about 75 kg.

In addition, it was reported that Sumo wrestlers had greater liver and kidney masses, but not brain mass, compared with controls (Table 12.2) [7].

TABLE 12.1 Fat and Fat-Free Mass of Professional and College Sumo Wrestlers

Subjects	n	Age (year)	Standing Height (cm)	Body Weight (kg)	BMI (kg/m²)	%Fat (%)	Method	Fat Mass (kg)	Fat-free Mass (kg)	Reference
Professional Sumo Wrestlers										
Top league (Sekitori)	12	20.6	182.2 ± 4.4	109.5 ± 20.9	33.0[a]	17.9 ± 5.1	Skinfold thickness	19.6[a]	89.9[a]	Nishizawa et al., [4]
Lower league	84		178.2 ± 5.1	99.1 ± 17.3	31.2[a]	20.7 ± 6.4	Skinfold thickness	20.5[a]	78.6[a]	Nishizawa et al., [4]
Mixed	37	21.1 ± 3.6	178.9 ± 5.2	115.9 ± 27.4	36.2 ± 8.1	26.1 ± 6.4	Underwater weighing	31.4 ± 13.5	84.6 ± 15.8	Kondo et al., [3]
Mixed	23	22.0 ± 1.2	178.7 ± 1.1	115.1 ± 4.2	36.0 ± 1.3	27.3 ± 1.1	Underwater weighing	31.9 ± 2.1	83.2 ± 2.8	Kanehisa et al., [10]
Top league (Sekitori)	7	25.6 ± 2.9	180.1 ± 6.1	154.2 ± 24.9	47.5 ± 6.7	28.6 ± 5.1	Underwater weighing	45.4 ± 14.6	109.0 ± 10.7	Hattori et al., [11]
Middle league (Makushita)	12	20.7 ± 2.2	178.8 ± 5.8	105.7 ± 17.5	32.9 ± 4.4	22.2 ± 5.4	Underwater weighing	24.0 ± 8.8	81.7 ± 11.3	Hattori et al., [11]
Low league (Sandanme)	12	19.8 ± 3.5	179.4 ± 4.4	109.3 ± 18.2	34.0 ± 5.8	28.2 ± 4.2	Underwater weighing	31.3 ± 9.6	78.0 ± 9.2	Hattori et al., [11]
Mixed	10	23.2 ± 3.0	186.0 ± 8.3	152.0 ± 39.5	43.5 ± 8.2	39.1 ± 9.5	Bioelectrical impedance analysis	59.3[a]	90.3 ± 18.8	Tsukahara et al., [5]
Mixed	331	21.6 ± 3.7	179.2 ± 5.3	117.9 ± 21.5	36.6 ± 6.2	29.6 ± 6.6	Air displacement plethysmography	35.9 ± 13.5	81.9 ± 10.2	Kinoshita et al., [12]
College Sumo Wrestlers										
	13	19.8 ± 0.3	178.5 ± 1.6	111.2 ± 3.8	35.0 ± 1.2	24.8 ± 1.0	Underwater weighing	27.9 ± 2.0	83.3 ± 2.0	Kanehisa et al., [1]
	24	19.7 ± 1.2	177.8 ± 5.3	111.2 ± 21.9	35.2 ± 6.4	24.1 ± 7.3	Underwater weighing	28.0 ± 13.3	83.1 ± 10.1	Saito et al., [13]
	15	20.5 ± 0.5	176.8 ± 3.5	125.1 ± 12.9	40.0 ± 20.8	25.6 ± 3.6	B-mode ultrasound	32.4 ± 7.9	92.7 ± 6.0	Yamauchi et al., [6]
	10	19.4 ± 1.5	172.9 ± 8.4	109.1 ± 14.7	36.5 ± 4.3	27.7 ± 4.5	Underwater weighing	30.5 ± 7.6	78.6 ± 9.7	Midorikawa et al., [7]
	18	19 ± 1	177.2 ± 6.6	125.4 ± 15.0	40.0 ± 4.8	29.6 ± 4.2	Dual-energy X-ray absorptiometry	37.6 ± 9.3	87.8 ± 7.5	Midorikawa et al., [14]

[a]*BMI, fat mass and fat-free mass were calculated from mean standing height, body weight and %fat, respectively.*
Values are the means and standard deviations except for Kanehisa et al [1,10], which show standard error.

Moreover, the ratios of kidney mass to fat-free mass (0.6% vs 0.6%) were similar between Sumo wrestlers and controls, and the ratio of liver mass to fat-free mass was higher in Sumo wrestlers (3.1%) than in controls (2.6%) (Table 12.2) [7]. Previous studies have found that a reduction in the proportion of internal organ-tissue mass to fat-free mass was coupled with an increase in that of skeletal muscle mass in untrained individuals [16,18]. Furthermore, in a recent study comparing the body composition of untrained obese subjects (body weight 105.4 ± 10.8 kg) with that of intermediate-weight subjects (body weight 70.9 ± 11.6 kg), there were no differences of liver and kidney masses between groups (obese subjects 1.64 and 0.32 kg, intermediate-weight subjects 1.64 and 0.36 kg, respectively) even though obese subjects have a greater fat-free mass (i.e., liver mass and kidney mass/FFM ratios: obese subjects 2.5% and 0.5%, intermediate weight subjects 3.0% and 0.7%, respectively) [19]. However, the Sumo wrestlers had greater absolute liver and kidney masses in comparison with untrained controls. Additionally, the present study found that the ratio of internal organ mass to fat-free mass for Sumo wrestlers does not decline with greater FFM, unlike untrained individuals. Although the cause of the phenomenon for Sumo wrestlers has not yet

TABLE 12.2 Organ-Tissue Level Body Composition [7]

	Sumo Wrestlers n = 10	Controls n = 11
Organ-Tissue Mass (kg)		
Skeletal muscle	36.9 ± 5.9**	24.5 ± 3.4
Adipose tissue[a]	35.9 ± 8.9**	10.2 ± 3.5
Liver	2.40 ± 0.52**	1.40 ± 0.20
Brain	1.44 ± 0.07	1.46 ± 0.10
Heart[b]	0.60 ± 0.08**	0.34 ± 0.03
Kidney	0.49 ± 0.08**	0.33 ± 0.04
Residual[c]	31.4 ± 5.1**	23.7 ± 2.6
Organ-Tissue Mass/FFM (%)		
Skeletal muscle	46.9 ± 3.9	45.9 ± 3.0
Brain	1.9 ± 0.2**	2.8 ± 0.3
Heart	0.8 ± 0.0**	0.6 ± 0.0
Liver	3.1 ± 0.6*	2.6 ± 0.3
Kidney	0.6 ± 0.1	0.6 ± 0.1

[a]*It was assumed that 85% of adipose tissue was fat and 15% of adipose tissue was the remaining calculated fat-free component [16].*

[b]*$0.006 \times Weight^{0.98}$[17].*

[c]*Residual mass was calculated as body mass minus sum of other measured mass components.*

FFM, fat-free mass.

Sumo wrestlers vs Controls: * $p < 0.05$, ** $p < 0.01$ on unpaired t-test using SPSS 10.0.

been clarified, the increase in liver and kidney mass may be attributed to an increase in protein intake and to high metabolic stress of exercise training. Sumo wrestlers have a large skeletal muscle mass and internal organ mass during fat-free mass accumulation [7].

CONCLUSION

When an athlete with body weight in excess of 100 kg tries to increase fat-free mass or skeletal muscle mass to enhance sports performance, there is a simultaneous increase in fat mass. Although this type of phenomenon was observed for Sumo wrestlers, percent fat of upper-leagues wrestlers remained at about 25%. Therefore, the balance between an effective increase of fat-free mass or skeletal muscle mass and decrease of fat mass is an important and key point for the sport performance of heavyweight athletes.

References

[1] Kanehisa H, Kondo M, Ikegawa S, Fukunaga T. Characteristics of body composition and muscle strength in college sumo wrestlers. Int J Sports Med 1997;18:510–5.

[2] Beekley MD, Abe T, Kondo M, Midorikawa T, Yamauchi T. Conparison of maximum aerobic capacity and body composition of sumo wrestlers to athletes in combat and other sports. J Sports Sci Med 2006;5:13–20.

[3] Kondo M, Abe T, Ikegawa S, Kakami Y, Fukunaga T. Upper limit of fat-free mass in humans: a study on Japanese sumo wrestlers. Am J Hum Biol 1994;6:613–8.

[4] Nishizawa T, Akaoka I, Nishida Y, Kawaguchi Y, Hayashi E, Yoshimura T. Some factors related to obesity in the Japanese sumo wrestler. Am J Clin Nutr 1976;29:1167–74.

[5] Tsukahara N, Omi N, Ezawa I. Effects of special physical characteristics and exercise on bone mineral density in sumo wrestlers. J Home Econ Japan 1999;50:673–82.

[6] Yamauchi T, Abe T, Midorikawa T, Kondo M. Body composition and resting metabolic rate of Japanese college sumo wrestlers and untrained students: are sumo wrestlers obese? Anthropol Sci 2004;112:179–85.

[7] Midorikawa T, Kondo M, Beekley MD, Koizumi K, Abe T. High REE in sumo wrestlers attributed to large organ-tissue mass. Med Sci Sports Exerc 2007;39:688–93.

[8] Chiba K, Tsuchiya M, Kato J, Ochi K, Kawa Z, Ishizaki T. Cefotiam disposition in markedly obese athlete patients, Japanese sumo wrestlers. Antimicrob Agents Chemother 1989;33:1188–92.

[9] Miwa T. Anthropometric measurement of Sumo wrestlers. Bull Tokyo Anthropol Soc 1889;4:120–2 [in Japanese].

[10] Kanehisa H, Kondo M, Ikegawa S, Fukunaga T. Body composition and isokinetic strength of professional sumo wrestlers. Eur J Appl Physiol Occup Physiol 1998;77:352–9.

[11] Hattori K, Kondo M, Abe T, Tanaka S, Fukunaga T. Hierarchical differences in body composition of professional Sumo wrestlers. Ann Hum Biol 1999;26:179–84.

[12] Kinoshita N, Onishi S, Yamamoto S, Yamada K, Oguma Y, Katsukawa F, et al. Unusual left ventricular dilatation without functional or biochemical impairment in normotensive extremely overweight Japanese professional sumo wrestlers. Am J Cardiol 2003;91:699–703.

[13] Saito K, Nakaji S, Umeda T, Shimoyama T, Sugawara K, Yamamoto Y. Development of predictive equations for body density of sumo wrestlers using B-mode ultrasound for the determination of subcutaneous fat thickness. Br J Sports Med 2003;37:144–8.

[14] Midorikawa T, Sakamoto S, Ohta M, Torii S, Konishi M, Takagi S, et al. Metabolic profiles and fat distribution in Japanese college Sumo wrestlers. Int J Body Compos Res 2010;8:57–60.

[15] Abe T, Kearns CF, Fukunaga T. Sex differences in whole body skeletal muscle mass measured by magnetic resonance imaging and its distribution in young Japanese adults. Br J Sports Med 2003;37:436–40.

[16] Heymsfield SB, Gallagher D, Kotler DP, Wang Z, Allison DB, Heshka S. Body-size dependence of resting energy expenditure can be attributed to nonenergetic homogeneity of fat-free mass. Am J Physiol Endocrinol Metab 2002;282:132–8.

[17] Calder WA. Size, function, and life history. New York: Dover; 1996. p. 49.

[18] Gallagher D, Belmonte D, Deurenberg P, Wang Z, Krasnow N, Pi-Sunyer FX, et al. Organ-tissue mass measurement allows modeling of REE and metabolically active tissue mass. Am J Physiol 1998;275:249–58.

[19] Bosy-Westphal A, Reinecke U, Schlorke T, Illner K, Kutzner D, Heller M, et al. Effect of organ and tissue masses on resting energy expenditure in underweight, normal weight and obese adults. Int J Obes Relat Metab Disord 2004;28:72–9.

CHAPTER

13

Bioenergetics of Cyclic Sports Activities on Land

Walking, Running and Cycling

Paola Zamparo, Carlo Capelli and Silvia Pogliaghi

Department of Neurological and Movement Sciences, School of Exercise and Sport Sciences, University of Verona, Verona, Italy

ENERGY EXPENDITURE OF HUMAN LOCOMOTION

To determine the physical activity energy expenditure (PAEE) of various forms of human locomotion has a practical relevance for maintaining or regaining optimal body mass and composition, for developing optimal nutrition strategies in competitive athletes, as well as for improving their performance.

For competitive athletes, the knowledge of the PAEE of their specific locomotion mode allows an accurate quantification of the energy requirements of their diet [1]; moreover, the knowledge of the exercise intensity (elicited by a specific form of locomotion at a specific speed) relative to the individual's maximal exercise capacity, determines the relative contribution of fat and carbohydrates to energy production during exercise [2]. In turn, this is important for the determination of the optimal diet composition (nutrients) to reduce or postpone fatigue, to optimize recovery and to sustain muscle repair and hypertrophy [1].

Regarding the general population, physical exercise is commonly prescribed in association with a weight-reducing diet to favor short-term weight loss and to reduce long-term weight regain [3]. The choice of the type of exercise is normally based on personal preferences and opportunities, within the limitations of individual functional capacity and possible medical conditions, but the knowledge of the PAEE of different activities is essential to select the appropriate exercise intensity and duration, to generate an adequate energy deficit and to favour the utilization of fat deposits for energy production during exercise [3].

The energy expenditure of physical activity can be "roughly" estimated based on predictive equations and on "activity factors" (which take into account the exercise intensity) or by using the metabolic equivalents (METs), e.g., according to the ACSM guidelines [4]. However, for cyclic sport activities (on land and in water), PAEE can be accurately calculated when the energy cost (*C*) of that form of locomotion is known.

THE ENERGY COST OF LOCOMOTION

The energy cost (C) of human locomotion can be calculated as:

$$C = E'v^{-1} \tag{13.1}$$

where E' is the metabolic power (the energy expenditure per unit of time) and v is the speed of progression [5]. If the speed is expressed in m s^{-1} and E' in kJ s^{-1}, C results in kJ m^{-1}; it thus represents the energy expended to cover one unit of distance (the cost of transport) in analogy with the liters of gasoline needed to cover a km for a car: the larger this value, the less economical the car.

The importance of C in determining performance (for cyclic sports activities, on land and in water), can

Nutrition and Enhanced Sports Performance.
DOI: http://dx.doi.org/10.1016/B978-0-12-396454-0.00013-8

be appreciated by rearranging Eq. 13.1 and applying it to maximal conditions [5]:

$$v_{\max} = E'_{\max} C^{-1} \quad (13.2)$$

Equation 13.2 indicates that maximal speed (v_{max}) depends on the ratio of maximal metabolic power (E'_{max}) to C; hence, for a given athlete (for a given value of E'_{max}), the maximal speed he/she can attain in different forms of land/water locomotion is set essentially by the value of C of that form of locomotion.

It must be pointed out that this analysis can be applied only to those forms of locomotion (on land or in water) in which the speed of progression is the only determinant of performance, that are sufficiently standardized so as to make the use of a single value of C meaningful, and in which the driving energy is metabolic (e.g., it cannot be applied to alpine skiing, where one major driving force is that of gravity, or to sailing, where the driving force is that of wind). The forms of locomotion to which the analysis may be applied are the so-called "cyclic sports activities": e.g., walking, running and cycling (land locomotion), swimming, rowing and kayaking (water locomotion).

It goes without saying that an athlete can improve performance (can increase his/her $v_{\max}$) by increasing his/her maximal metabolic power ($E'_{\max}$, "physiological" parameters, the numerator of Eq. 13.2) and/or by decreasing his/her energy cost per unit distance (C, "technical" parameters, the denominator of Eq. 13.2). In the next section the physiological determinants of $v_{\max}$ (the contribution of the different energy sources to $E'_{\max}$) will be briefly described, whereas the following sections will be devoted to an analysis of the determinants of C in land locomotion. In Chapter 14 this analysis will be extended to water locomotion.

ENERGY SOURCES

The energy expenditure of locomotion at a constant, submaximal speed is based on "aerobic energy sources"; in these conditions the energy required to resynthesize ATP is completely derived from the oxidation of a mixture of carbohydrates (CHO) and fat substrates in the Krebs cycle. In these conditions, in Eq. 13.1,

$$E' = E'_{aer} = V'O_2$$

where $V'O_2$ is oxygen uptake.

The moles of ATP obtained per mole of oxygen consumed (the P/O_2 ratio) ranges from 5.6 (for lipids) to 6.2 (for CHO) and the substrates selection is a function of the relative exercise intensity [2]: the relative contribution of CHO to total energy production ranges from 25% at rest to 80–100% at high exercise intensities (100% for a respiratory exchange rate [RER] = 1). Especially in short-term, high-intensity bouts, CHO constitutes the main fuel for ATP production, and this causes rapid CHO depletion and consequent fatigue.

The exercise intensity that maximizes fat oxidation during exercise is called "$\text{Fat}_{\max}$" and corresponds to a value of roughly 50–60% of $V'O_{2max}$. $\text{Fat}_{\max}$ is used as landmark intensity for exercise prescription for overweight subjects and in cases of obesity [6].

At a given submaximal exercise intensity, substrates selection is also influenced by substrates availability: the relative contribution of CHO to oxidative ATP production can be reduced after prolonged exercise sessions, during fasting, in low CHO diets, or as a consequence of a suboptimal replenishment of CHO stores following training or competition. For the above reasons, fat oxidation during a constant-intensity exercise will progressively increase with exercise durations exceeding 30 min; this is the reason to prescribe training sessions above 30 min, at least 5 days per week, when weight loss is desired [3].

At maximal speeds (all-out tests) the contribution of anaerobic energy sources to $E'_{\max}$ (see Eq. 13.2) could play an important role, the more so the shorter the distance. In these conditions $E'_{\max}$ can be calculated as the sum of three terms, as originally proposed by [7] and later applied to running [8], cycling [9,10], kayaking [11] and swimming [12,13] (see also Chapter 14):

$$E'_{\max} = E'_{\text{AnL}} + E'_{\text{AnAl}} + E'_{\text{Aer}} \quad (13.3)$$

where the term E'_{AnL} depends on the anaerobic lactic energy sources (which can be calculated based on measured blood lactate concentration at the end of exercise) and the term E'_{AnAl} depends on the anaerobic alactic energy sources (which depend on the concentration of high energy substrates in the working muscles); for further details see the papers cited above. This approach allows estimation of the energy demands of human locomotion ($E'_{\max}$) in "square wave" exercises of intensity close or above maximal aerobic power where a true steady state of oxygen uptake cannot be attained and where energy contributions other than the aerobic one cannot be neglected. As also indicated in Chapter 14, in these conditions, the percentage contribution of the aerobic and anaerobic energy sources is independent of the mode of locomotion (e.g., running, skating or cycling) and depends essentially upon the duration of the exercise (see Figure 13.1).

AERODYNAMIC AND NON-AERODYNAMIC COST OF LOCOMOTION

Values of C as a function of v are reported in the literature for several forms of human locomotion (on land

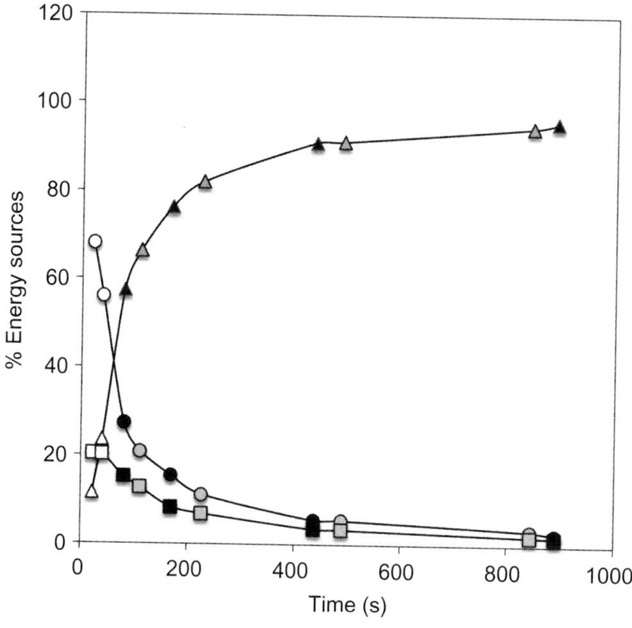

FIGURE 13.1 Percentage contribution of aerobic (Aer, triangles), anaerobic lactic (Al, squares) and anaerobic alactic (AnAl, circles) energy sources to overall energy expenditure during maximal trials in running (black symbols) and cycling (grey symbols, track bicycle; open symbols, four-wheels recumbent bicycle). *Data referring to a four-wheel recumbent bike are taken from [10]; data referring to cycling and running were calculated based on data reported by [9] and [14], respectively; since in the latter study [14] blood lactate concentration (La_b) at the end of the maximal trials was not assessed, it was assumed to amount to 11 mM, which is the average Lab value reported by [9] in maximal cycling trials over similar distances.*

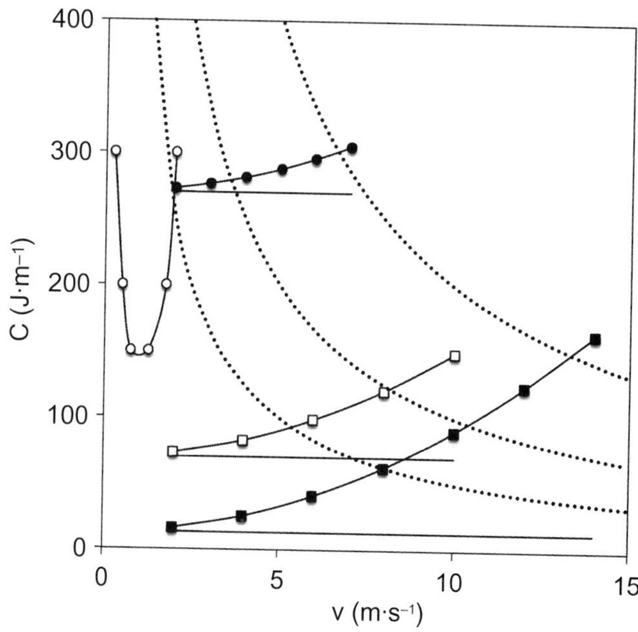

FIGURE 13.2 The energy cost C (J m^{-1}) to cover a given distance as a function of speed v (m s^{-1}) in land locomotion. Open dots, walking; filled dots, running; open squares, speed skating; filled squares, cycling. Upper curves represent the total energy cost and continuous lines the "non-aerodynamic" cost for each of these forms of locomotion (these two curves coincide for walking). Thin dotted descending lines represent iso-metabolic power hyperbolae of 2, 1 and 0.5 kW (from top to bottom). With a metabolic power input of 1 kW (a PAEE of 860 kcal/h, central dotted line), a subject can run at 4 m/s, skate at 8 m/s and cycle at 10 m/s. *Adapted from [5].*

and in water). In walking, the relationship between C and v has a characteristic "U shaped" curve: C attains a minimum at a speed of about 1–1.3 m s^{-1} (depending on gender and age), which is very close to the speed of spontaneous walking (*sws*); at this speed the energy cost of walking is about half that of running (see, e.g., [15,16]). At speeds lower and/or larger than the *sws* the C of walking increases; it even exceeds that of running at high walking speeds. The energy cost of running is almost constant and independent of the speed when assessed indoors (on a treadmill), but outdoors (on track) air resistance affects C, the more so the higher the speed; this can be generalized for all "outdoor cyclic sport activities".

In his review of the energetics of locomotion on land and in water, di Prampero [5] reports the following relationships between C (J m^{-1}) and v (m s^{-1}):

- for running, $C = 270 + 0.72\,v^2$;
- for speed skating on ice ("dropped posture"), $C = 70 + 0.79\,v^2$;
- for cycling (standard racing bike, "dropped posture"), $C = 13 + 0.77\,v^2$.

These data refer to a 70 kg body mass and 1.75 m stature subject, to flat and firm terrain, to sea level and in the absence of wind (see Figure 13.2).

In these forms of locomotion, it is thus sufficient to measure v in order to obtain an accurate estimate of the physical activity energy expenditure (PAEE) using the appropriate C vs v relationship. This allows us to prescribe physical activity as scientifically based "therapy", as we know the metabolic power elicited by the exercise at stake. As an example, when walking at the *sws*, net energy cost is about 2 J m^{-1} kg^{-1} whereas, in running, C (which is almost independent of the speed) amounts to about 4 J m^{-1} kg^{-1} (see, e.g., [15,16]). Thus, a subject of 70 kg body mass who covers 1 km by walking at his/her *sws* (about 1–1.3 m s^{-1} in healthy adults) will consume 140 kJ of energy, which corresponds to a PAEE of 33.5 kcal (1 kJ = 0.239 kcal); if he were covering the same distance by running, he would consume 67 kcal (see Table 13.1).

As an example, this approach, in connection with the use of GPS tracking of speed and altitude, has been successfully utilized to estimate the daily energy expenditure of a runner during an ultra-marathon event (LANY footrace 2011) [17].

TABLE 13.1 Values of C for Different Modes of Land Locomotion and in Different Conditions

Condition/speed		C (J m^{-1} kg^{-1})	PAEE (kcal)	Reference
Walking (C_{na}) at the *sws*	Flat, firm terrain	2.0	33	16
	Flat, soft sand	5.4	90	16
	Incline (+10%)	4.9	82	41
	Incline (−10%)	1.1	19	41
	Children (3–4 yr)	3.0	50	31
	Elderly (70 yr)	3.0	50	32
Running (C_{na}) at speeds of 2–4 m s^{-1}	Flat, firm terrain	4.0	67	16
	Flat, soft sand	6.4	107	16
	Incline (+10%)	6.0	100	41
	Incline (−10%)	2.2	36	41
Skipping (C_{na}) flat, firm terrain	2–3 m s^{-1}	5.5	92	27
Track cycling (C_{na} + C_a) flat, firm terrain	3 m s^{-1}	0.4	7	9
	6 m s^{-1}	0.7	11	9
	9 m s^{-1}	1.1	19	9
	12 m s^{-1}	1.8	30	9
	15 m s^{-1}	2.6	44	9
Cycling, 1820 Hobby Horse (C_{na})	2 m s^{-1}	1.7	29	52
Cycling, 1860 Bone Shaker (C_{na})	3 m s^{-1}	1.6	27	52
Ice skating, modern skates (C_{na})	4–10 m s^{-1}	1.4	24	51
Ice skating, 1800 BC (C_{na})	1.13 m s^{-1}	4.6	77	51
Cross country skiing, skating technique (C_{na})	4–6 m s^{-1}	1.9	32	50
Cross country skiing, 542 AD (C_{na})	1–2 m s^{-1}	4.8	81	50

Physical activity energy expenditure (PAEE) was calculated over a distance of 1 km for a 70 kg subject (in the absence of wind).
sws, self-selected walking speed; C_{na}, non-aerodynamic energy cost; C_a, aerodynamic energy cost.

The first term of the above-reported C vs v relationships represents the so-called "non-aerodynamic" energy cost of locomotion (C_{na}); it does not change as a function of speed and differs greatly across different forms of locomotion (it amounts to about 90% of C at top speeds in race walking and running and to less than 10% at top speeds in speed skating and cycling).

The last term of the above-reported C vs v relationships represents the so-called "aerodynamic" energy cost of locomotion (C_a), i.e., the energy that is expended to overcome air resistance ($W_a = k\,v^2$); C_a increases with the square of speed and, for a given speed, is similar for all forms of land locomotion. This component of C is negligible in walking and running (less than 10% at top running speeds); its importance increases with the speed of locomotion (90% or greater at top cycling speeds). On land locomotion, therefore:

$$C = C_{na} + C_a = C_{na} + k'\,v^2 \tag{13.4}$$

Equation 13.4 can be considered a "general equation" for the C vs v relationship in land locomotion and indicates that C is given by the sum of a term that is dependent on the speed and a term that is independent on it, the latter being the "specific determinant" of C for each locomotion mode. For instance, in aquatic locomotion the "general equation" for the C vs v relationship is: $C = k\,v^n$ (with $n \approx 2$), indicating that the energy needed to overcome water resistance is the most important component of C, in that environment (see Chapter 14).

Even if these considerations could not be applied to all forms of locomotion (e.g., they do not hold for walking), this generalization could help in understanding the importance of resistive forces (i.e., air and/or water resistance) in determining C in the two environments.

THE DETERMINANTS OF C IN LAND LOCOMOTION

The determinants of the cost of transport, in human locomotion, are the total work per unit distance (W_{tot}, $\mathrm{kJ\,m^{-1}}$) and the "locomotion efficiency" (η_L):

$$C = W_{tot}/\eta_L \tag{13.5}$$

C will then be larger the larger W_{tot} and the smaller η_L. A detailed description of the determinants of η_L is given in the section on efficiencies in land locomotion at the end of this Chapter.

In analogy with Eq. 13.4, W_{tot} can also be considered as the sum of two terms: the work to overcome aerodynamic forces ($W_a = k\,v^2$) and the work needed to overcome "non-aerodynamic" forces (W_{na}) (see, e.g., [9,10,18,19]); hence:

$$W_{tot} = W_{na} + W_a = (C_{na}\,\eta_L) + k\,v^2 \tag{13.6}$$

(therefore, in Eq. 13.4, $k' = k/\eta$).

The "aerodynamic" component of W_{tot} (W_a) depends, besides the speed, on frontal area (A), on the coefficient of air resistance (C_d) and on air density (ρ). As indicated above, W_a is negligible in walking and running whereas its importance is pivotal in cycling. The determinants of W_a will thus be described in detail in the section about cycling.

The "non-aerodynamic" component of W_{tot} (W_{na}) can be considered as the sum of two terms (see, e.g., [15,20–22]):

$$W_{na} = W_{ext} + W_{int} \tag{13.7}$$

where W_{ext} (the external work) is the work needed to raise and accelerate the body center of mass (BCoM) within the environment and W_{int} the work associated with the acceleration/deceleration of the limbs relative to the BCoM. W_{ext} can be calculated based on the changes in potential and kinetic energy of BCoM during a step/stride/cycle whereas W_{int} can be calculated based on the linear and rotational kinetic energy of the body segments.

While there is still a debate about considering these two components as two separate entities, most of the studies reported so far in the literature about human locomotion utilize this partitioning, and the reader is referred to specific papers for further details about this topic (e.g., [15,22–24]).

The Determinants of the External Work in Walking and Running

Legged locomotion is the result of the coordination of several muscles, exerting forces via tendons and producing the movement of several bones and body segments; however, the complex movements of the two basic gaits of land locomotion (walking and running) can be described by using two simple models: an inverted pendulum for walking and a spring (or a pogo stick) for running (see, e.g., [15]). These models (paradigms) explain the interplay among the three fundamental energies associated with BCoM: in walking the work done by the muscles to sustain locomotion is in part relieved by the exchange of potential ($E_p = m\,g\,h$) and kinetic ($E_k = ½\,m\,v^2$) energy, whereas in running these two forms of energy change in phase during the stride, and the work to sustain locomotion is in part relieved by the recoil of elastic energy (E_{el}).

In walking, E_p and E_k change in opposition of phase, as in a (inverted) pendulum, but losses are associated with the deviation from an ideal system (e.g., through friction) and with the transition from one "inverted swing" to the other [15,25]. The percent energy recovery (a parameter introduced by [26] to quantify the ability to save mechanical energy by using a "pendulum like motion") can be as high as 60–70% at the self-preferred walking speed (where the exchange between potential and kinetic energy is maximized), thus explaining why C is minimal at a this speed. At larger and smaller speeds the percent energy recovery decreases so that the work to sustain locomotion is bound to increase, as well as C.

In running, E_p and E_k change in phase so that no recovery between these two forms of energy is possible. In this gait, elastic energy has a crucial role since part of the energy of the system ($E_p + E_k$, in the flight phase) is transformed into elastic energy during the first half of the contact phase, via tendon stretch; a consistent part of this energy is then given back to the system in the second half of the contact phase via tendon recoil (see, e.g., [15]). This explains why C_{na} does not change as a function of speed in running: with increasing speed the ground reaction forces increase and, in proportion, the stretch of tendons and the recoil of E_{el} increase, thus reducing the need of muscle contraction and hence the increase in C_{na} that would otherwise occur.

Skipping is a gait that children display when they are about 4–5 years old, and it is also the gait of choice in low-gravity conditions. Skipping is a combination of walking and running in a single stride; it differs from walking because it has a flight phase (the duration of which is longer the higher the speed) and from running because a double support phase often occurs. In this form of locomotion (similar to horse gallop) E_p, E_k and E_{el} exchange during the stride in a combination of a pendulum-like and an elastic energy saving mechanism. The energy expenditure of this "mode of locomotion" is about 5–6 J m^{-1} kg^{-1} at a speed of 1–3 m s^{-1}[27].

To summarize: in walking and running, W_{ext} (and thus C) will be larger:

- the larger E_p (the larger the subject's body mass and the vertical displacement of the BCoM),
- the larger E_k (the larger the subject's body mass and the vertical and horizontal speeds of the BCoM),
- the lower the percent recovery between E_p and E_k (in walking)
- and the lower the recoil of elastic energy (in running).

Since some of these factors depend on environmental constraints (e.g., opposing winds, track stiffness and slope) these will be considered in the following paragraphs.

Body Mass, Gender and Age

Since the cost of transport depends on the mass of the subject (and on his/her resting metabolic rate), in land locomotion C is generally expressed per kg of body mass (e.g., in J m^{-1} kg^{-1}) and above resting values. In young children (3–4 years old) the thus calculated cost of transport (walking) could be as much as 70% larger than in adults [28]; the differences in C are lower the older the children, and C becomes similar to that of adults at an age of 9–10 years [28]. The mechanical determinants of these differences are discussed by [29–31]. The energy cost of walking is about 30% larger in the elderly (70 years of age) compared with young adults (see, e.g., [32]); the mechanical determinants of these differences are discussed in [32]. Moreover, the C of walking is affected by several pathological conditions (neurological or orthopedic disabilities), as reviewed by [33].

In "healthy young adults", when C is expressed per unit body mass and above resting values, inter-individual differences in the energy cost of walking and running (between genders and across ages and fitness levels) are negligible (see, e.g., [5,34]); this allows for a proper prescription of PAEE (at variance with water locomotion, see Chapter 14) in "standardized conditions" (e.g., on flat and firm terrain, at sea level and in the absence of wind).

Aerodynamic Resistance

In "non-standard conditions" several factors could affect C and this should be taken into account for a proper prescription of PAEE. As indicated by Eq. 13.6, one of these factors is air resistance. Even if the aerodynamic component of C is rather small in running (and altogether negligible in walking), it has been shown that shielding (running on a treadmill 1 m behind another runner with induced winds of 10–14 m s^{-1}) can reduce the C of running alone by about 6% [35]. This example indicates:

1. that indoor values of C (e.g., assessed on a treadmill) could be lower than those assessed in "ecological" conditions (outdoor); and
2. that C increases in the presence of wind.

These considerations, of course, hold for all forms of land locomotion.

Track Stiffness

The type of surface may affect the energy cost of walking and running. As an example, C is about 2.0–2.7 times larger when walking on soft sand than on firm terrain and 1.2–1.6 times larger in running [16,36]. The cost of walking on even more compliant surfaces—e.g., on soft snow—depends on the depth at which the feet sink and could be as high as 6 J m^{-1} kg^{-1} at a speed of 0.67 m s^{-1} and for a footprint depth of about 10 cm [37].

These differences in C could be attributed to the combined effect of the stiffness of the leg and that of track. The former is kept constant by a reflex control [38,39] so that the contact time (and hence the recoil of E_{el}) depends only on the latter, decreasing as the track stiffness increases. On soft terrains the efficiency of running is thus decreased [16] because the muscles must replace the energy that could not be recoiled by tendon and muscle elasticity. In walking the external mechanical work increases, on soft terrains, because the foot moves (sinks) on the surface: the recoil between E_k and E_p is thus bound to decrease, and this leads to an increase of W_{ext} and thus of C [16,36].

Gradient Locomotion

The energy cost of walking and running on a upward slope is larger than on flat terrain; however, the energy required to cover one unit of distance when going downhill is not a decreasing function of the slope. As shown by several authors (e.g., [15,40,41]) C increases when going uphill and decreases when going downhill, up to a slope of about −10–15%, to further increase at larger downhill slopes. As indicated by Minetti et al. [41] this state of affairs can be explained by differences in positive and negative work (these two quantities are equal on the level) and in efficiency (up to 25% uphill and −125% downhill) across gradients, the internal work of walking and running being almost unaffected by the slope. The relationship between C (J m^{-1} kg^{-1}) and gradient (*i*, from −0.45 downhill to

0.45 uphill) for walking (at *sws*) and running is well described by the following equations (41):

in walking: $C = 280.5i^5 - 58.7i^4 - 76.8i^3 + 51.9i^2 + 19.6i + 2.5$

in running: $C = 155.4i^5 - 30.4i^4 - 43.3i^3 + 46.3i^2 + 19.5i + 3.6.$

The Determinants of the Internal Work in Walking and Running

In walking and running, the internal work (W_{int}, $J\,m^{-1}\,kg^{-1}$) can be estimated by using the following equation (42):

$$W_{int} = q\,f\,v(1 + \{d/(1-d)\}^2) \qquad (13.8)$$

where f is the stride frequency (Hz), v is the forward speed ($m\,s^{-1}$) and q is a parameter which takes into account the inertial proprieties of the limbs and the mass partitioning between the limbs and the rest of the body; q is about 0.08 in walking and running on flat terrain and about 0.10 on gradient locomotion [43]. The term d is the duty factor, i.e., the fraction of the stride duration at which a single limb is in contact with the ground [44]; d is larger than 0.5 in walking (from 0.6 at $1\,m\,s^{-1}$ to 0.5 at $2\,m\,s^{-1}$ on flat terrain) and lower than 0.5 in running (from 0.45 at $2\,m\,s^{-1}$ to 0.3 at $3\,m\,s^{-1}$ on flat terrain). In walking and running, W_{int} constitutes 25–40% of W_{na} and ranges from 0.2 to $0.6\,J\,m^{-1}\,kg^{-1}$ [43].

The Effects of Fatigue on the Energy Cost of Running

The energy cost of running increases in fatiguing conditions, the more so the longer the duration of the run and the higher the exercise intensity (see, e.g., [45–48]). The "deterioration" of C varies among individuals (i.e., augmenters and non-augmenters) and has been attributed, among others, to changes in neuromuscular coordination, in biomechanical variables (such as stride length and frequency), to a reduced ability to store and release mechanical energy (e.g., to changes in muscle stiffness), as well as to metabolic factors (such as muscle glycogen depletion, thermal stress and dehydration).

PASSIVE LOCOMOTORY TOOLS ON LAND

The relationship between C, W_{tot} and η_L (Eq. 13.5) is also useful in understanding how locomotory passive tools "work". As indicated in Figure 13.2, compared with walking and running, locomotory tools such as skates and bicycles allow for a reduction in C; in land locomotion, these tools mainly reduce W_{tot} (whereas in water locomotion they mainly allow for an increase in η_L) compared with walking and running (i.e., land locomotion without passive aids).

Passive locomotory tools do not supply any additional energy to the body but provide effective compensation for the limitations of our biological actuators (muscles and tendons) [49]. As an example, all skating techniques (ice skating, cross country skiing, roller skating) allow for an increase in speed (the 1 hour record in roller skating is about twice that for running) because they allow for a reduction in W_{na} (and hence of C_{na}): this occurs because of a decrease in friction with the medium, because the contraction speed can be reduced (due to the presence of a gliding phase), because the vertical excursion of the body center of mass can be reduced, and so on [49]. Passive locomotory tools are often the result of a long evolution in their design, lasting hundreds and hundreds of years: for skates and skis, see [50,51]; for bicycles, see [52–54].

The Determinants of C in Cycling (the External Work)

The external mechanical work per unit distance in cycling can be considered as the sum of three terms [5]:

$$W_{ext} = W_a + W_r + W_i \qquad (13.9)$$

W_a is the work needed to overcome air resistance:

$$W_a = k\,v^2 = \tfrac{1}{2}\,A\,C_d\,\rho\,v^2 \qquad (13.10)$$

where, A is frontal area (of cyclist + bike), C_d is the coefficient of air resistance, ρ is air density and v is the speed. As shown by Wilson [54], A can vary from $0.55\,m^2$ (traditional bike, upright posture) to $0.36\,m^2$ (racing bike, dropped posture) whereas C_d can range from 1.15 to 0.88 (in these two cases). Lower values of A and especially of C_d (down to 0.10) can be observed for aerodynamic bikes [18] and for human powered vehicles (HPV) such as faired recumbent bicycles [54].

W_a can be reduced by up to about 30% by moving in the wake of someone else, depending on the distance from the leading cyclist and on the cycling speed (see, e.g., [55]), and significant reductions in W_a can also be obtained by reducing the air resistance of the equipment (clothing, helmets and shoes): as an example, loose clothing can increase W_a by 30% at speeds higher than $10\,m\,s^{-1}$ (see, e.g., [54,55]).

Finally W_a can be reduced by moving to altitude since air density (ρ) depends on pressure and temperature; considering that maximal metabolic power decreases with altitude, and that this effect opposes the decrease in W_a, it has been calculated that the 1 hour world record could be attained at an altitude of about 2500 m (see, e.g., [5,56]). This altitude corresponds to that of Mexico City (2300 m) where indeed several cycling records were attained between 1968 and 1984 on an outdoor track. The following, most recent, records were attained on indoor tracks, at near sea level, by using "aerodynamic bicycles" (with aero bars and disk wheels) now banned by the UCI (Union Cycliste Internationale) [57].

W_r is the work needed to overcome rolling resistance:

$$W_r = M\, g\, C_r \tag{13.11}$$

where M is the mass (of cyclist + bike), g is the acceleration of gravity and C_r is the coefficient of rolling resistance. C_r ranges from 0.002 to 0.010, depending on the roughness of the surface and on the characteristics of the wheels: the larger their diameter and the inflation pressure of the tires, the lower C_r (see, e.g., [54,55]). W_r can be reduced by reducing the mass of the bicycle (e.g., by using materials such as carbon fiber): the mass of a traditional bike (15–18 kg) is twice that of a racing bike (about 6–8 kg) and thus its W_r is double [54].

In this component of W_{na} are generally considered all friction forces (not only rolling resistance), such as bearing friction, dynamic tire deformations and gear energy losses; the contribution of these resistive forces (at least in modern bikes) is, however, negligible when compared with the friction at the wheel-to-ground contact.

W_i is the work against gravity:

$$W_i = M\, g \sin\theta \tag{13.12}$$

where M is the mass (of cyclist + bike), g is the acceleration of gravity and θ is the road incline. Thus, when a cyclist is climbing uphill, gravity slows the rate of ascent, whereas downhill, gravity speeds up the bike; since high downhill speeds do not compensate for the slower climbing rate the average speed on hills is less than that on the level. Lowering M will increase the climbing rate, improving the average speed on a hill course [55].

The internal work in cycling

This component of total mechanical work can be considered as the metabolic equivalent of additional work due to pedaling frequency [24,58]; this additional work seems to be attributable to "viscous internal work" (due to dissipation of internal energy that reduces the efficiency of movement) rather than to "kinematic work" due to the changes in linear and rotational kinetic energy of the limbs in respect to the center of mass. According to Minetti [24,52] this work (W_{int}, J m^{-1} kg^{-1}) can be estimated based on the following equation:

$$W_{int} = q\, f^3\, v^{-1} \tag{13.13}$$

where f is the pedaling frequency (Hz), v is the speed (m s^{-1}) and q is a term that accounts for the inertial parameters of the moving limbs. Therefore in cycling, as in walking and running, the internal work is larger the larger the forward speed and the larger the frequency of the moving limbs (Eqs. 13.8 and 13.13). On a stationary bike, the internal mechanical work rate per unit body mass (W kg^{-1}) can be calculated according to this general equation (52):

$$W'_{int} = 0.153\, f^3$$

Thus, for a 70 kg cyclist, W'_{int} corresponds to about 11 W at 60 rpm (1 Hz) and 86 W at 120 rpm (2 Hz).

EFFICIENCY IN LAND LOCOMOTION

The general term "efficiency" refers to the ratio between "energy output and energy input". It can thus be defined as the ratio of mechanical work (W) to metabolic energy needed to cover a unit distance (W/C, where both terms are expressed in J m^{-1} kg^{-1}) or as the ratio of mechanical work and metabolic energy per unit of time (W'/E' where both terms are expressed in J s^{-1} kg^{-1}, i.e., W kg^{-1}, or in W).

When the whole mechanical energy flux is known and when the metabolism is aerobic, this efficiency cannot exceed a value of 0.25–0.30 [59]. This limit is set by the product of the efficiency related to the phosphorylation of metabolic substrates to ATP molecules (0.60) and of the efficiency related to muscle contraction itself (from ATP to force/displacement generation: 0.50); this efficiency is defined as muscular efficiency (η_m) and depends on the fiber's type composition, on contraction length and speed and on the type of contraction (concentric or eccentric) (see, e.g., [49]).

The "locomotion" efficiency of a given form of locomotion is defined as the ability to transform metabolic energy into the "minimum" external work necessary to move [60]. As an example, in cycling on flat terrain, one has at least to overcome air and rolling resistance. Indeed, at constant speed, the sum of the resistive forces has to be equal to the sum of the propulsive forces. Since the cyclic motion implies a certain movement frequency, W_{int} has also to be accounted for, and thus, for this form of locomotion, η_L is well described by Eq. 13.5: $\eta_L = W_{tot}/C$.

From another point of view, η_L can be considered as the product of two main components: muscle efficiency (η_m) and the so-called "transmission efficiency" (η_T):

$$\eta_L = \eta_m \, \eta_T \tag{13.14}$$

Transmission efficiency η_T (in aquatic locomotion this is known as propelling efficiency, η_P) is the ratio between the "minimum" work for locomotion and W_{tot} and accounts for all the energy dissipation "outside" the involved muscles. Transmission efficiency could range from 0 (no "external work" production, such as in isometric contractions) to 1 (when all the work generated by the muscles is utilized for producing "external work"). Thus, η_L will range from 0 to about 0.3. In cycling, η_L is close to 0.25 (provided that all components of W_{tot} are taken in due consideration), and this indicates that in this form of locomotion η_m is nearly maximal and η_T is close to 1 (see Chapter 14 for a further discussion on this point).

Passive locomotory tools, such as bicycles, do indeed optimize both η_m (e.g., by allowing the muscles to work in the optimal region of the force/velocity relationship by using gears) and η_T (e.g., by reducing the work needed for locomotion, such as the vertical excursion of the BCoM) [49,52].

The values of η_L range from 0.2 to 0.4 in walking and from 0.4 to 0.7 in running (see, e.g., [16,20,22]); this indicates that, for these forms of locomotion, the nominator of Eq. 13.5 is "overestimated" (η_L could be at most 0.30). Indeed, in these forms of locomotion, W_{ext} is calculated based on the changes in kinetic and potential energy of the BCoM (see above) but these changes cannot be attributable to "muscular work" only, since they also depend on the recoil of elastic energy; the latter is difficult to measure, and hence to be taken into account; it increases with increasing speed and is larger in running than in walking. This state of affairs indicates that is difficult to estimate the contribution of η_m and η_T for these forms of locomotion and that more research is needed to better define the values of W_{ext} (and W_{tot}) in "legged" locomotion.

CONCLUSIONS

Walking and running are cyclic movements requiring limited skills and low-cost equipment. Locomotion with passive tools such as skis, skates or bicycles requires specific skills and adequate equipment; the cost of transport for these forms of locomotion is lower than for running or walking, and so is the impact load on the lower limbs.

Compared with walking, running is a high-impact activity and it has the highest cost of transport of all forms of land locomotion; this, for unfit subjects, could prevent the continuation of exercise for more than a few minutes. For this reason, walking is the first-choice activity prescribed for weight loss programs, especially so in relatively unfit and/or obese individuals.

In "healthy young adults" inter-individual differences in the energy cost of walking, running and cycling are negligible. This allows for a proper prescription of physical activity energy expenditure (by contrast with water locomotion, see Chapter 14) in "standardized conditions" (e.g., on flat and firm terrain, at sea level and in the absence of wind).

References

[1] American College of Sport Medicine Position Stand. Nutrition and Athletic performance. Med Sci Sport Exerc 2009;41:709–31.

[2] di Prampero PE. Energetics of muscular exercise. Rev Physiol Biochem Pharmacol 1981;89:143–222.

[3] American College of Sport Medicine Position Stand. Appropriate physical activity intervention strategies for weight loss and prevention of weight regain for adults. Med Sci Sport Exerc 2009;41:459–71.

[4] American College of Sport Medicine. Guidelines for exercise testing and prescription. 8th ed. Baltimore, MA: Lippincott Williams & Wilkins; 2010.

[5] di Prampero PE. The energy cost of human locomotion on land and in water. Int J Sports Med 1986;7:55–72.

[6] Achten J, Jeukendrup AE. Optimizing fat oxidation through exercise and diet. Nutrition 2004;20:716–27.

[7] Wilkie DR. Equations describing power input by humans as a function of duration of exercise. In: Cerretelli P, Whipp BJ, editors. Exercise bioenergetics and gas exchange. Amsterdam: Elsevier; 1980. p. 75–80.

[8] di Prampero PE, Capelli C, Pagliaro P, Antonutto G, Girardis M, Zamparo P, et al. Energetics of best performance in middle distance running. J Appl Phisiol 1993;74:2318–24.

[9] Capelli C, Schena F, Zamparo P, Dal Monte A, Faina M, di Prampero PE. Energetics of best performances in track cycling. Med Sci Sports Exerc 1998;30:614–24.

[10] Zamparo P, Capelli C, Cencigh P. Energy cost and mechanical efficiency of riding a four-wheeled human powered, recumbent vehicle. Eur J Appl Physiol 2000;83:499–505.

[11] Zamparo P, Capelli C, Guerrini G. Energetics of kayaking at sub-maximal and maximal speeds. Eur J Appl Physiol 1999;80:542–8.

[12] Capelli C, Termin B, Pendergast DR. Energetics of swimming at maximal speed in humans. Eur J Appl Physiol 1998;78:385–93.

[13] Figueiredo P, Zamparo P, Sousa A, Vilas-Boas JP, Fernandes RJ. An energy balance of the 200 m front crawl race. Eur J Appl Physiol 2011;111:767–77.

[14] Lacour JC, Padilla-Magnuacelaya S, Barthelemy JC, Dormois D. The energetics of middle distance running. Eur J Appl Physiol 1990;60:38–43.

[15] Saibene F, Minetti AE. Biomechanical and physiological aspects of legged locomotion. *Eur J Appl Physiol* 2003;88:297–316.

[16] Lejeune TM, Willems PA, Heglund NC. Mechanics and energetics of human locomotion on sand. J Exp Biol 1998;201:2071–80.

[17] Ardigò LP, Capelli C. Energy expenditure during the LANY footrace 2011-a case study. Appl Physiol Nutr Met 2012;37:1–4.

[18] Capelli C, Rosa G, Butti F, Ferretti G, Veicsteinas A, di Prampero PE. Energy cost and efficiency of riding aerodynamic bicycles. Eur J Appl Physiol 1993;67:144–9.

[19] di Prampero PE, Cortili G, Mognoni P, Saibene F. Equation of motion of a cyclist. J Appl Physiol 1979;47:201–6.

[20] Cavagna GA, Kaneko M. Mechanical work and efficiency in level walking and running. J Physiol 1977;268:467–81.

[21] Winter D. A new definition of mechanical work done in human movement. J Appl Physiol 1979;64:79–83.

[22] Willems PA, Cavagna GA, Heglund NC. External, internal and total work of human locomotion. J Exp Biol 1995;198:379–93.

[23] Aleshinsky SY. An energy sources and fractions approach to the mechanical energy expenditure. J Biomech 1986;19:287–93.

[24] Minetti AE. Bioenergetics and biomechanics of cycling: the role of the internal work. Eur J Appl Physiol 2010;111:323–9.

[25] Alexander RMcN. Energy saving mechanisms in walking and running. J Exp Biol 1991;160:55–69.

[26] Cavagna GA, Thys H, Zamboni A. The sources of external work in level walking and running. J Physiol 1976;262:639–57.

[27] Minetti AE. The biomechanics of skipping gaits: a third locomotion paradigm? Proc R Soc B 1988;265:1227–35.

[28] DeJager D, Willems PA, Heglund NC. The energy cost of walking in children. Eur J Appl Physiol 2001;441:538–43.

[29] Cavagna GA, Franzetti P, Fuchimoto T. The mechanics of walking in children. J Physiol 1983;343:323–39.

[30] Schepens B, Willems PA, Cavagna GA. The mechanics of running in children. J Physiol 1998;509:927–40.

[31] Schepens B, Willems PA, Cavagna GA, Heglund NC. Mechanical power and efficiency in running children. Eur J Appl Physiol 2001;442:107–16.

[32] Mian OS, Thom JM, Ardigò LP, Narici MV, Minetti AE. Metanolic cost, mechanical work and efficiency during walking in young and older men. Acta Physiol 2006;186:127–39.

[33] Waters RL, Mulroy S. Energy expenditure of normal and pathological gait. Gait Posture 1999;9:207–31.

[34] Åstrand PO, Rodhal K, Dahl HA, Strømme SB. Applied sport physiology. Textbook of work physiology. 4th ed. Champaign, IL: Human Kinetics; 2003. pp. 479–502.

[35] Pugh LG. The influence of wind resistance in running and walking and the mechanical efficiency of work against horizontal or vertical forces. J Physiol 1971;213:255–76.

[36] Zamparo P, Perini R, Orizio C, Sacher M, Ferretti G. Energy cost of walking and running on sand. Eur J Appl Physiol 1992;65:183–7.

[37] Pandolf KB, Haisman MF, Goldman RF. Metabolic energy expenditure and terrain coefficients for walking on snow. Ergonomics 1976;19:683–90.

[38] Green PR, McMahon TA. Reflex stiffness of man's antigravity muscles during knee-bends while carrying extra weights. J Biomech 1979;12:881–91.

[39] Mc Mhaon TA. Muscle reflexes and locomotion. Princeton, NJ: Princeton University Press; 1984.

[40] Margaria R. Sulla fisiologia e specialmente sul consume energetico della marcia e dalla corsa a varia velocità ed inclinazione del terreno. Atti Accademia Nazionale dei Lincei 1938;7:299–368.

[41] Minetti AE, Moia C, Roi G, Susta D, Ferretti G. Energy cost of walking and running at extreme uphill and downhill slopes. J Appl Physiol 2002;93:1039–46.

[42] Minetti AE. A model equation for the prediction of mechanical internal work of terrestrial locomotion. J Biomech 1988;31:463–8.

[43] Nardello F, Ardigò LP, Minetti AE. Measured and predicted mechanical internal work in human locomotion. Human Movement Sciences 2011;30:90–104.

[44] Alexander RMcN. Optimization and gaits in the locomotion of vertebrates. Physiol Rev 1980;69:1199–227.

[45] Brueckner JC, Atchou G, Capelli C, Duvallet A, Barrault D, Jousselin E, et al. The energy cost of running increases with the distance covered. Eur J Appl Physiol 1991;62:385–9.

[46] Guezennec CY, Vallier JM, Bigard AX, Durey A. Increase in the energy cost of running at the end of a triathlon. Eur J Appl Physiol 1996;73:440–5.

[47] Sproule J. Running economy deteriorates following 60 min of exercise at 80% $V'O_{2max}$. Eur J Appl Physiol 1998;7:366–71.

[48] Candau R, Belli A, Millet GY, Georges D, Barbier B, Rouillon JD. Energy cost and running mechanics during a treadmill run to voluntary exhaustion in humans. Eur J Appl Physiol 1998;77:479–85.

[49] Minetti AE. Passive tools for enhancing muscle-driven motion and locomotion. J Exp Biol 2004;207:1265–72.

[50] Formenti F, Ardigò LP, Minetti AE. Human locomotion on snow: determinants of economy and speed of skiing across the ages. Proc R Soc B 2005;272:1561–9.

[51] Formenti F, Minetti A. Human locomotion on ice: evolution of ice skating energetics through history. J Exp Biol 2007;201:1825–33.

[52] Minetti AE, Pinkerton J, Zamparo P. From bipedalism to bicyclism: evolution in energetics and biomechanics of historic bicycles. Proc R Soc Lond B 2001;268:1351–60.

[53] Sharp A. Bicycles and tricycles. Mineola, NY: Dover Publications; 2003.

[54] Wilson DG. Bicycling science. Cambridge, MA: The MIT press; 2004.

[55] Kyle CR. Selecting cycling equipment. In: Burke ER, editor. High tech cycling. Champaign, IL: Human Kinetics; 2003. p. 1–48.

[56] Bassett DR, Kyle CR, Passfiled L, Broker JP, Burke ER. Comparing cycling world records, 1967–1996: modelling with empirical data. Med Sci Sports Exerc 1999;31:1665–76.

[57] Kyle CR, Basset DR. The cycling world record. In: Burke ER, editor. High tech cycling. Champaign, IL: Human Kinetics; 2003. p. 175–96.

[58] Francescato MP, Girardis M, di Prampero PE. Oxygen cost of internal work during cycling. Eur J Appl Physiol 1995;72:51–7.

[59] Woledge RC, Curtin NA, Homsher E. Energetic aspects of muscle contraction. London: Academic Press; 1985.

[60] Cavagna GA. Muscolo e Locomozione. Milano: Raffello Cortina; 1988.

CHAPTER

14

Bioenergetics of Cyclic Sport Activities in Water

Swimming, Rowing and Kayaking

Paola Zamparo[1] *and Marco Bonifazi*[2]

[1]Department of Neurological and Movement Sciences, School of Exercise and Sport Sciences, University of Verona, Verona, Italy [2]Department of Neurological, Neurosurgical and Behavioral Sciences, Faculty of Medicine, University of Siena, Italy

ENERGETICS AND BIOMECHANICS OF AQUATIC LOCOMOTION

To determine the physical activity energy expenditure (PAEE) of various forms of human locomotion has a practical relevance for maintaining or regaining optimal body mass and composition, for developing optimal nutrition strategies for competitive athletes, as well as for improving their performance (see Chapter 13). As is the case for land locomotion, in the water environment PAEE can be accurately calculated when the energy cost (C) of that form of locomotion is known.

The energy cost per unit distance (C) is defined as:

$$C = E' v^{-1} \tag{14.1}$$

where E' is the metabolic power (the energy expenditure per unit of time) and v is the speed of progression [1]. If the speed is expressed in m s^{-1} and E' in kJ s^{-1}, C results in kJ m^{-1}; it thus represents the energy expended to cover one unit of distance while moving in water at a given speed (see Chapter 13 for references and more detailed information).

The contribution of the aerobic and anaerobic energy sources to total metabolic energy expenditure (E') depends on the intensity and the duration of the exercise (see Chapter 13) and, for all-out efforts, is independent of the mode of aquatic locomotion. As an example, for all-out efforts of the same duration (about 1 min, e.g., over a distance of 100 m in swimming and 250 m in kayaking), the aerobic and anaerobic metabolism respectively contribute about 40% and 20% of the total metabolic energy expenditure in both activities (Table 14.1); for more details see [1–3].

In analogy with land locomotion (see Chapter 13), the total mechanical power of aquatic locomotion (W'_{tot}) can be considered as the sum of two terms: the power needed to accelerate and decelerate the limbs with respect to the centre of mass (the internal power, W'_{int}) and the power needed to overcome external forces (the external power, W'_{ext}):

$$W'_{tot} = W'_{ext} + W'_{int} \tag{14.2}$$

The internal power in aquatic locomotion (swimming) was investigated in only a few studies [4–6]; in analogy with land locomotion W'_{int} (W) was found to depend on the frequency of the limb's motion (f, Hz):

$$W_{int} = q\, f^3 \tag{14.3}$$

where q is a term that accounts for the inertial parameters of the moving limbs; $q = 6.9$ for the leg kick and 38.2 for the arm stroke [4,5]. W'_{int} represents a larger fraction of W'_{tot} in the leg kick, compared to the arm stroke, due to the higher frequency of movement in the former compared to the latter. For the whole stroke (front crawl) W'_{int} ranges from 10 to 40 W at speeds of 1.0–1.4 m s^{-1}[5]. Passive locomotory tools that can decrease f have thus a direct influence on W'_{int}; as an example, the use of fins reduces W'_{int} compared with

Nutrition and Enhanced Sports Performance.
DOI: http://dx.doi.org/10.1016/B978-0-12-396454-0.00014-X

TABLE 14.1 Percentage Contribution of Aerobic (Aer), Anaerobic Lactic (Anl) and Anaerobic Alactic (Anal) Energy Sources to Overall Energy Expenditure During Maximal Swimming Trials (Average for All Strokes) And Maximal Trials On The Olympic Kayak

	Distance (m)	Time (s)	Speed (m s^{-1})	E_{Aer} (%)	E_{Al} (%)	E_{AnAl} (%)
Swimming	50	25.9	1.76	19.2	54.8	26.0
Swimming	100	57.5	1.59	37.4	44.2	19.4
Swimming	200	126.6	1.45	62.4	25.3	12.3
Kayaking	250	61.9	4.04	40.5	37.3	22.2
Kayaking	500	134.8	3.71	60.4	26.9	13.4
Kayaking	1000	289.0	3.46	83.3	8.8	7.9
Kayaking	2000	568.2	3.52	89.5	6.1	4.4

Adapted from [2,3].

barefoot leg kicking by about 60–90%, depending on the type of fins utilized (see, e.g., 4, 6).

The external power in aquatic locomotion can be further partitioned into: the power to overcome drag that contributes to useful thrust (W'_d) and the power that does not contribute to thrust (W'_k):

$$W'_{ext} = W'_d + W'_k \tag{14.4}$$

Both W'_d and W'_k give water kinetic energy, but only W'_d effectively contributes to propulsion (see, e.g., [7,8]). The efficiency with which the overall mechanical power produced by the swimmer/kayaker/rower is transformed into useful mechanical power (i.e., the "minimum" power needed for propulsion, see Chapter 13) is termed propelling efficiency (η_P) and it is calculated as:

$$\eta_P = W'_d / W'_{tot} \tag{14.5}$$

This parameter is of utmost importance in water locomotion since it indicates the capability of the swimmer/kayaker/rower to best transform his/her muscular power into power useful for propulsion. Propelling efficiency corresponds to the transmission efficiency (η_T) of land locomotion; it accounts for all the energy degradation "outside" the involved muscles (see Chapter 13); η_P could range from 0 (none of the power provided by the muscles is useful for propulsion) to 1 (all power is utilized for propulsion). In elite swimmers η_P could be as high as 0.35–0.40 (see, e.g., [5,9]) but this value is reduced in children, masters and unskilled swimmers (e.g., as low as 0.10–0.20; see [10,11]). These data indicate that, in swimming, far more than 50% of the mechanical power output is wasted in giving water kinetic energy that is not useful for propulsion. In rowing and kayaking η_P is larger: it could be as high as 0.65–0.75, depending on the speed and on the level of skill (see, e.g., [12]).

The efficiency with which the total mechanical power produced by the swimmer is transformed into external power is termed hydraulic efficiency (η_H) and is given by W'_{ext}/W'_{tot}. The efficiency with which the external mechanical power is transformed into useful mechanical power is termed Froude (theoretical) efficiency (η_F) and is given by W'_d/W'_{ext}. It follows that $\eta_P = \eta_F \eta_H$. Hence, if the internal power is nil or negligible (and if the hydraulic efficiency is close to 1) $\eta_P = \eta_F$. Thus, propelling efficiency will be lower than Froude efficiency the higher the internal mechanical power and the lower the hydraulic efficiency (see, e.g., [4,7,8]). A similar description of the power partitioning and of the efficiencies in human swimming is reported in the literature by other authors (e.g., [13,14]); these authors, however, do not take into account the contribution of internal power to total power production; therefore, in their calculations the implicit assumption is also made that hydraulic efficiency is 100% and hence that $\eta_F = \eta_P$.

Whereas the propelling, hydraulic and Froude efficiencies refer to the mechanical partitioning only, the performance (drag) efficiency (η_D) takes into account also the metabolic expenditure: it is the efficiency with which the metabolic power input (E') is transformed into useful mechanical power output (W'_d):

$$\eta_D = W'_d / E' \tag{14.6}$$

Drag efficiency corresponds to the locomotion efficiency (η_L) of land locomotion (see Chapter 13) since the numerator of (Eq. 14.6) takes into account only the "minimum" (useful) work (rate) to move in water (e.g., the power to overcome drag). This efficiency is of about 0.03–0.09 (see, e.g., [5,15,16]).

Since $\eta_L = \eta_m \eta_P$ (see Chapter 13), in water locomotion it is possible to estimate η_m based on measured η_D (η_L) and η_P. As shown by Zamparo [17] the efficiency so calculated turns out to be of about 0.20–0.25 and corresponds to the efficiency (sometimes indicated as gross, overall, or mechanical efficiency, η_O) that can be calculated by measuring all components of W'_{tot} by means of land-based swimming ergometers (see, e.g., [18]) or by direct calculation of W'_{int}, W'_k and W'_d (see, e.g., [5]):

$$\eta_O = W'_{tot}/E' \quad (14.7)$$

Combining Equations (14.5), (14.6) and (14.7) one obtains:

$$E' = W'_d/(\eta_P\ \eta_O) \quad (14.8)$$

which, at any given speed, relates the metabolic power input (E') with the power needed to overcome hydrodynamic resistance (W'_d) and with propelling (η_P) and overall (η_O) efficiency; since, for any given speed, $C = E'/v$ and $W'_d = W_d\ v$, it follows that:

$$C = W_d/(\eta_P\ \eta_O) \quad (14.9)$$

which, at any given speed, relates C with hydrodynamic resistance (W_d) and with propelling (η_P) and overall (η_O) efficiency. For further details see [1,19].

Equation 14.9 indicates that at any given speed and for a specific η_O, an increase in η_P and/or a decrease in W_d lead to a decrease in C (e.g., allowing the swimmer/rower/kayaker to spend less energy to cover a given distance, or to cover the same distance at a higher speed).

ENERGETICS OF SWIMMING

Effects of Speed and Stroke on C

In swimming, C increases (nonlinearly) as a function of the speed (see Figure 14.1).

The energy cost of swimming, for a given speed, is lowest for the front crawl, followed by the backstroke and the butterfly; the breaststroke being the most demanding stroke [2,15]. In young elite swimmers the relationships between C and v for the four strokes are well described by the following equations (adapted from [2]):

- front crawl: $C = 0.670\ v^{1.614}$
- backstroke: $C = 0.799\ v^{1.624}$
- breaststroke: $C = 1.275\ v^{0.878}$
- butterfly: $C = 0.784\ v^{1.809}$.

Therefore, a swimmer who covers 1 km by swimming at $1\ m\ s^{-1}$ will consume from 670 (front crawl) to 1275 (breaststroke) kJ of energy, which corresponds to a PAEE (the energy expenditure related to physical activity, see Chapter 13) of 160–305 kcal (1 kJ = 0.239 kcal).

When compared with data of C for land locomotion at similar speeds of progression (as an example with walking at the self-selected speed where v is about $1–1.3\ m\ s^{-1}$, see Chapter 13), it is apparent that locomotion in water is more energy demanding (about ten times larger at this speed in swimming compared with walking). This is essentially because hydrodynamic resistance (W_d) is much larger than air resistance (water density is about 800 times higher than air density) and to the fact that for land locomotion (e.g., cycling) $\eta_T \approx 1$ whereas, in water locomotion, propelling efficiency is much lower than that (see above); thus, as indicated by Eq. 14.9, in water locomotion, C is larger than on land because W_d is larger and η_P is lower.

Differences in C, in swimming, could be expected not only based on differences in speed and stroke, but also based on differences in skill level, hydrodynamic position, gender and age, as discussed below.

Effects of Training and Skill Level on C

According to Eq. 14.9, we can expect differences in C based on differences in W_d, η_P and η_O among swimmers. This is very much the case, and the ACSM guideline tables [20] indicate this by warning the readers that PAEE "can vary substantially from person to person during swimming as a result of different strokes and skill levels". Indeed, technical skill in swimming deeply influences C because of its effects on η_P (and, to a lesser extent, also on W_d). As indicated by

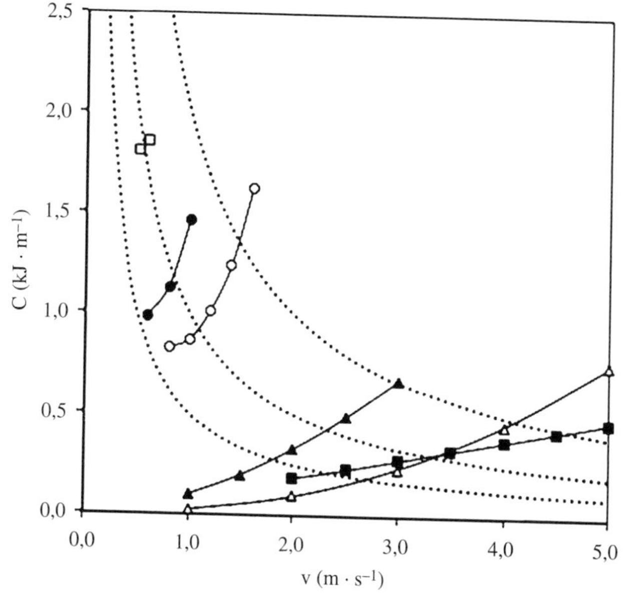

FIGURE 14.1 The energy cost (C) to cover a given distance as a function of the speed (v) in the front crawl (for subjects with different technical skills) and in boat locomotion. Open squares, recreational swimmers; full circles, good swimmers; open circles, elite swimmers; full triangles, slalom kayak; full squares, rowing shell; open triangles, flat water K1 kayak. Thin descending lines represent iso-metabolic power hyperbolae of 2, 1 and 0.5 kW (from top to bottom). With a metabolic power input of 1 kW (a PAEE of 860 kcal h^{-1}, central dotted line), a good swimmer can reach $0.85\ m\ s^{-1}$, and a competitive swimmer about $1.1\ m\ s^{-1}$, whereas a recreational swimmer can sustain a speed of only $0.6\ m\ s^{-1}$. At the same PAEE the speed attained with boat locomotion is much larger: about $2.3\ m\ s^{-1}$ with a slalom canoe, and about $3.5\ m\ s^{-1}$ with a rowing shell or a flat water K1 kayak. *Adapted from [3,15,41,42].*

Holmer [15], a recreational swimmer can spend as much as twice the energy that a good swimmer uses to proceed in water at the same speed, and the latter consumes about 20–30% more energy than an elite swimmer at comparable speeds (see Figure 14.1).

Training is expected to improve technical skills. As shown by Termin and Pendergast [21], 4 years of "high-velocity training" resulted in a 20% decrease in C and in a proportional (16%) improvement in the distance travelled per stroke (DS, an index of propelling efficiency). On the other hand, training could lead to a 15–35% decrease in water resistance (W_d, as reported by [22]); according to Eq. 14.9 this reduction contributes to the decrease in the energy cost of swimming observed after a training period.

Effects of Frontal Area and Underwater Torque on C

In analogy with aerodynamic resistance, hydrodynamic resistance (pressure drag) is defined as:

$$W_d = \tfrac{1}{2}\ \rho\ A\ C_x\ v^2 \qquad (14.10)$$

where ρ is water density, A is wetted frontal area and C_x is the coefficient of drag. C_x is generally calculated when all other terms of Eq. 14.10 are known; in front crawl swimming it amounts to about 0.3, with no differences as a function of speed or gender (see, e.g., [23]). The wetted frontal area (A) is thus, besides the speed, the major determinant of pressure drag (which, in turn, is the major determinant of drag, at least up to speeds of 1.2–1.4 m s^{-1}) (see, e.g., [23]). In turn, A depends on several factors:

1. on the overall surface area of the body (e.g., it is larger in heavier and taller swimmers);
2. on the fraction of the body that is submerged (which is larger the larger the "underwater weight" of the swimmer);
3. on the body incline (see, e.g., [23–26]).

The body incline can be measured/estimated in static (by means of an underwater balance) or dynamic (by kinematic analysis during swimming) conditions.

The effect of the "static position in water" on the energy cost of swimming can be described quantitatively by measuring the underwater torque (T), which is a quantitative measure of "the tendency of the legs to sink": the greater T the higher the energy required to cover a given distance at a given speed. While the function relating C and T is the same for both genders, men spend more energy per unit distance than women, as they have greater values of T[25–29]. At low swimming speeds (0.9–1.1 m s^{-1}), about 70% of the variability of C is explained by the variability of T, regardless of the gender, age and technical level of a swimmer. At high swimming speeds (>1.2–1.4 m s^{-1}) the hydrodynamic lift counteracts the tendency of the legs to sink and, at least in the front crawl, the body remains horizontal, whatever the static position would be; at these speeds no relationship between T and C could be observed any more [26].

The relationship between the static and dynamic position in water (e.g., assessed by measuring the inclination of the body relative to the waterline) depends thus on the speed; at high swimming speeds, A is essentially determined by body surface and by the fraction of the body that is submerged.

The Effect of Age and Gender on C

Both η_P and W_d depend on the anthropometric characteristics of the swimmer, and both change during growth (along with body development and training). Therefore C is expected to differ between children and adults and between males and females (see Table 14.2).

Females have indeed a lower C than males (C is about 20–30% lower in females than in males at submaximal swimming speeds) (see, e.g., [16,26,30,31]); the higher economy of female swimmers is traditionally attributed to a smaller hydrodynamic resistance (W_d) due to their smaller size, larger percentage of body fat, and more horizontal position (lower A) in comparison with male swimmers (see, e.g., [16,27,29]). Only a few studies have so far investigated the determinants of swimming economy in children and adolescents. These studies indicate that, at comparable speed, C is indeed lower in children than in adults (see, e.g., [32–34]), the more so the younger the subjects [29]. Also in this case the higher economy of children is attributed to a smaller hydrodynamic resistance due to their smaller size and more horizontal position in water [29,32]. No differences in C are observed between male and female swimmers before puberty [29].

TABLE 14.2 Variation in the Major Parameters that Affect Swimming Performance

Parameter	Changes with Increasing Speed	Females vs Males (at a given speed)	Children vs Adults (at a given speed)	Masters vs Young (at a given speed)
C	↑	↓	↑	↑
W_d	↑	↓	↓	↑
A	↓	↓	↓	↑
η_P	↓	=	↓	↓
DS	↓	↓	↓	↓

C, energy cost of locomotion; W_d, hydrodynamic resistance; A, wetted frontal area; η_P, propelling efficiency; DS, distance covered per stroke.

Regarding propelling efficiency, no differences are observed between male and female swimmers of the same age and technical skill even if differences can be observed in the distance covered per stroke (an index of propelling efficiency) [10]; young (pre-pubertal) swimmers are instead characterized by lower η_P values than adults, the more so the younger the age [10,29].

Master swimmers are characterized by a lower economy (higher energy cost of swimming) in comparison with young swimmers, the more so the older the age [11]; the decrease in maximal metabolic power (E') that occurs with age forces these swimmers to reduce their speed [36–38]; this, in turn, is associated with an unfavorable alignment of the body (an increase in A, and thus in W_d). More importantly, though, with age a deterioration in propelling efficiency occurs so that C increases due both to an increase in W_d and to a decrease in η_P [10,11].

These data indicate that, at variance with locomotion on land, the energy cost of swimming is dependent on so many factors (speed, stroke, age, gender and technical skill) that it is rather difficult to precisely quantify PAEE in swimming unless all these factors are taken into due consideration. In Table 14.3 the considerations reported above are summarized in an attempt to provide a rough estimation of C (and PAEE) in different populations of swimmers (and in boat locomotion, see below).

Nutrition Requirements and Substrate Utilization in Swimming

Dietary surveys of elite male swimmers (who undertake high-volume training programs) typically find self-reported daily energy intakes of 4000–6000 kcal day^{-1}[26]. These large energy requirements and the tight schedules can pose exceptional challenges to adequate nutrition in swimmers, especially at a young age when the dietary challenges of adolescence add to their training nutrition needs [39].

In the recent Olympic Games in London, the male winner of the endurance swimming race covered the 10-km distance in about 110 minutes, i.e., at an average speed of 1.52 m s^{-1}. This speed corresponds to an average value of C of 1.31 kJ m^{-1} (which can be calculated based on the C vs v relationship reported by [2] in elite front crawl male swimmers) and hence to an overall energy expenditure of about 13 106 kJ (3132 kcal). Assuming that the totality of ATP is resynthesized via oxidative metabolism and that an average RER of about 0.85 can be maintained in swimming at "self-selected speed" for 2 hours, the contributions of carbohydrates (F_G) and fats (F_L) used as fuel during a 10-km swimming race are about 49% (6396 kJ) and 51% (6709 kJ), respectively ($F_G = \{RER-0.707\}/0.293$; $F_L = 1-F_G$). The absolute amounts of carbohydrates

TABLE 14.3 Ratio of C in Different Populations/Modes of Aquatic Locomotion to C in Young Elite Male Swimmers

	C ratio	*C* (kJ m^{-1})	PAEE (kcal)	Reference
YEMS front crawl	1.00	0.67	160	[2]
YEMS backstroke	1.19	0.80	291	[2]
YEMS breaststroke	1.90	1.28	305	[2]
YEMS butterfly	1.17	0.78	187	[2]
GTSMS front crawl	1.62	1.09	260	[16]
GTSFS front crawl	1.15	0.77	185	[16]
Children (F, 12 yr) front crawl	1.22	0.82	196	[29]
Children (M, 13 yr) front crawl	1.46	0.98	234	[29]
Master (M, 30–40 yr) front crawl)	1.85	1.24	296	[11]
Master (M, 40–50 yr) front crawl	2.22	1.49	359	[11]
YEMS front crawl with fins	0.9	0.60	151	[5]
Kayaking (flat water K1 scull) at 1 (3) m s^{-1}	0.03	0.02 (0.24)	5 (57)	[3]
Kayaking (slalom canoe) at 1 (3) m s^{-1}	0.15	0.10 (0.68)	23 (162)	[41]
Rowing (two-oared shell) at 1 (3) m s^{-1}	0.13	0.09 (0.28)	21 (67)	[42]
Waterbike (double hull catamaran) at 1 (3) m s^{-1}	0.09	0.06 (0.35)	15 (84)	[43]
Paddle wheel boat at 1 (3) m s^{-1}	0.27	0.18 (1.13)	43 (270)	[43]

Data were calculated for a speed of 1 m s^{-1} (and for an additional speed of 3 m s^{-1} for boat locomotion). Physical activity energy expenditure (PAEE) is calculated over a distance of 1 km.

YEMS, young elite male swimmers; GTSMS, good technical skill male swimmers; GTSFS, good technical skill female swimmers.

(CHO) and fats used as fuel during a 10-km swimming race are hence of about 370 g and 170 g, respectively (38.9 kJ g^{-1} for fat and 17.2 kJ g^{-1} for CHO). Even if these calculations are based on a simplified approach, they indicate quantitative guidelines for re-feeding strategies and caloric supplementation during actual competitions for elite long-distance swimmers.

Of course this method can be utilized to estimate the amounts of CHO and fat used as fuel for any form of locomotion (on land and in water) for which the C vs v relationship is known and for which the RER can be estimated with sufficient accuracy.

PASSIVE LOCOMOTORY TOOLS IN WATER

Changes in the energy cost of aquatic locomotion can be observed when passive locomotory tools, such as fins and hand paddles, are used for moving in water. As an example, as shown by [5], at any given speed, fins reduce C of front crawl swimming by about 10% due to an increase in η_P of about 20% (with no changes in W_d). When swimming with the leg kick only, the decrease in C is about 50–55% when "double fins" are used and up to 60–70% when using a monofin [4,6], the reduction in C being associated with an increase in η_P that is larger the larger the surface area of the fin.

Besides these "small tools", aimed at improving the energy expenditure of swimming freely at the surface (or underwater), humans have learnt to use a variety of passive locomotory tools (such as human powered boats and watercrafts) in the attempt to improve the economy and/or the speed of locomotion in water. In analogy with locomotion on land, these tools do not supply any additional mechanical energy to the body but provide effective compensation for limitations in the anatomical design and for inadequacy in muscle performance (see, e.g., [40]); in other words these tools can improve either η_m or η_T (see Chapter 13).

ROWING AND KAYAKING (BOAT LOCOMOTION)

Propulsion, in boat locomotion, can be sustained by upper or lower limbs action. In the first case oars/paddles are used to propel the boat (e.g., with synchronous movements for rowing, or with alternate movements for canoeing or kayaking); in the second case propulsion is sustained by leg cycling (water-bikes and paddle wheel boats).

Compared with swimming, boat locomotion is characterized by far lower values of drag and far larger values of propelling efficiency; this means that, according to Eq. 14.3, C is much lower than in swimming (see below and Figure 14.1).

As is the case for swimming, however, the energy cost of boat locomotion increases (nonlinearly) as a function of the speed and depends on the type of boat (and on the skill level of the athlete). For kayaking with a flat-water sprint K1 scull, $C = 0.020\ v^{2.26}$ [3] and with a slalom canoe $C = 0.098\ v^{1.76}$ [41]; for rowing (two-oared shell with coxswain, for one rower) $C = 0.088\ v^{1.05}$ [42]; for pedalling with a water bike (a double hulled catamaran) $C = 0.063\ v^{1.57}$ [43], or with a paddle wheel boat (used on lakes and beaches for recreational purposes) $C = 0.179\ v^{2.66}$ [43]; in all these examples C is in kJ m^{-1} and v in m s^{-1}.

Data reported in Figure 14.1 indicate that, for a given metabolic power input (E') of 1 kW (corresponding to a PAEE of 860 kcal h^{-1}), the speed attained in boat locomotion is much larger than that attained in swimming: about 2.3 m s^{-1} with a slalom canoe and about 3.5 m s^{-1} with a rowing shell or a flat water K1 kayak compared with 0.6–1 m s^{-1} in swimming. Of course, these differences in speed (at a given E') have to be attributed to differences in C (see Eq. 14.1) that, in turn, have to be attributed to differences in W_d, η_P and η_O (see Eq. 14.9).

Propelling efficiency in boat locomotion ranges from 0.4 (for paddle wheel boats, a value similar to that of front crawl swimming), to 0.70 (in rowing and kayaking) and is proportional to the distance covered by the hull for each cycle (D/c: m $cycle^{-1}$, a parameter that corresponds to the "distance covered per stroke" in swimming) (see, e.g., [12,22]). D/c is larger the longer the propulsive tool (arms, oars, paddles): in the "aerobic speed range" D/c is about 8 m $cycle^{-1}$ in rowing, 4 m $cycle^{-1}$ in kayaking and 2.5 m $cycle^{-1}$ in front crawl swimming [12]. As shown by [22], the maximal speeds of different forms of locomotion in water are related to their maximal D/c ($v_{max} = 0.89\ D/c_{max} + 0.31$, $R = 0.98$) and inversely related to cycle frequency (CF, cycles min^{-1}) ($v_{max} = -0.06\ CF + 0.78$, $R = 0.89$).

Hydrodynamic resistance in boat locomotion is largely reduced, compared with swimming, because the boats/shells float on the surface; this reduces wetted area (A, see Eq. 14.10) and hence pressure and wave drag. Since wetted area depends on weight (e.g., of rower + shell), changes in drag could be expected when weight increases: e.g., in rowing, W_d increases by about 10% for a 20% increase in weight [44]. As previously indicated for swimming, training decreases the energy cost of boat locomotion (by about 25–33% in rowing and 6–10% in kayaking) due to both an increase in η_P and a decrease in active drag (e.g., after 4 years of training, active drag can be reduced by 18–50% in kayaking) [22].

CONCLUSIONS

Swimming is characterized by a high energy cost, requires confidence with water, and it is a specific skill that is preferably acquired at a young age. In the general population, the specific skill requirements of swimming may limit the utilization of this type of activity for weight management. Yet, the high PAEE of swimming makes this activity a valuable option compared with low-impact (walking) and high-impact (running) weight-bearing and non-weight-bearing (cycling, kayaking and rowing) activities. Indeed, even at a speed of 1 m s^{-1} (3.6 km h^{-1}) front crawl generates a PAEE of 260 kcal km^{-1} in good-level male swimmers. The above implies that with a 30 min exercise session, a 470 kcal expenditure can be generated (kcal km^{-1} × km h^{-1} = kcal h^{-1}). To generate a similar energy expenditure would require about 40 min of running or rowing at 3 m s^{-1} (10.5 km h^{-1}), 45 min of cycling at 9 m s^{-1} (32 km h^{-1}), 50 min of cross country skiing at 5 m s^{-1} (18 km h^{-1}), or 50 min of kayaking at 3 m s^{-1} (10.5 km h^{-1}).

References

[1] di Prampero PE, Pendergast DR, Zamparo P. Swimming economy (energy cost) and efficiency. In: Seifert L, Chollet D, Mujika I, editors. World book of swimming: from science to performance. NY: Nova Science Publishers; 2011. p. 297–312.

[2] Capelli C, Termin B, Pendergast DR. Energetics of swimming at maximal speed in humans. Eur J Appl Physiol 1998;78:385–93.

[3] Zamparo P, Capelli C, Guerrini G. Energetics of kayaking at sub-maximal and maximal speeds. Eur J Appl Physiol 1999;80:542–8.

[4] Zamparo P, Pendergast DR, Termin AB, Minetti AE. How fins affect the economy and efficiency of human swimming. J Exp Biol 2002;205:2665–76.

[5] Zamparo P, Pendergast DR, Mollendorf J, Termin A, Minetti AE. An energy balance of front crawl. Eur J Appl Physiol 2005;94:134–44.

[6] Zamparo P, Pendergast DR, Termin A, Minetti AE. Economy and efficiency of swimming at the surface with fins of different size and stiffness. Eur J Appl Physiol 2006;96:459–70.

[7] Alexander RMcN. Swimming. In: Alexander RMcN, Goldspink G, editors. Mechanics and energetics of animal locomotion. London: Chapman et al.; 1977. p. 222–48.

[8] Daniel TL. Efficiency in aquatic locomotion: limitations from single cells to animals. In: Blake RW, editor. Efficiency and economy in animal physiology. Cambridge: Cambridge University Press; 1991. p. 83–96.

[9] Figueiredo P, Zamparo P, Sousa A, Vilas-Boas JP, Fernandes RJ. An energy balance of the 200 m front crawl race. Eur J Appl Physiol 2011;111:767–77.

[10] Zamparo P. Effects of age and gender on the propelling efficiency of the arm stroke. Eur J Appl Physiol 2006;97:52–8.

[11] Zamparo P, Dall'Ora A, Toneatto A, Cortesi M, Gatta G. The determinants of performance in master swimmers: age related changes in propelling efficiency, hydrodynamic position and energy cost of front crawl. Eur J Appl Physiol 2012;112(12):3949–57.

[12] Pendergast DR, Zamparo P, di Prampero PE, Mollendorf J, Capelli C, Cerretelli P, et al. Energy balance of human locomotion in water. Eur J Appl Physiol 2003;90:377–86.

[13] Toussaint HM, Beleen A, Rodenburg A, Sargeant AK, De Groot G, Hollander AP, et al. Propelling efficiency of front crawl swimming. J Appl Physiol 1988;65:2506–12.

[14] Toussaint HM, Hollander AP, van den Berg C, Vorontsov AR. Biomechanics of swimming. In: Garrett W, Kirkendall DT, editors. Exercise and sport science. Philadelphia: Lippincott Williams and Wilkins; 2000. p. 639–59.

[15] Holmer I. Oxygen uptake during swimming in man. J Appl Physiol 1972;33:502–9.

[16] Pendergast DR, di Prampero PE, Craig Jr. AB, Wilson DR, Rennie DW. Quantitative analysis of the front crawl in men and women. J Appl Physiol 1977;43:475–9.

[17] Zamparo P. Assessing gross efficiency and propelling efficiency in swimming. Port J Sport Sci 2011;11:65–8.

[18] Zamparo P, Swaine IL. Mechanical and propelling efficiency in swimming derived from exercise using a laboratory-based whole-body swimming ergometer. J Appl Physiol 2012;113:584–94.

[19] Zamparo P, Capelli C, Pendergast DR. Energetics of swimming: a historical perspective. Eur J Appl Physiol 2011;111:367–78.

[20] American College of Sport Medicine. Guidelines for exercise testing and prescription. 8th ed. Baltimore, MA: Lippincott Williams & Wilkins; 2010.

[21] Termin B, Pendergast DR. Training using the stroke frequency-velocity relationship to combine biomechanical and metabolic paradigms. J Swim Res 2001;14:9–17.

[22] Pendergast DR, Mollendorf J, Zamparo P, Termin A, Bushnell D, Paschke D. The influence of drag on human locomotion in water. J Undersea Hyperb Med 2005;32:45–57.

[23] Zamparo P, Gatta G, Capelli C, Pendergast DR. Active and passive drag, the role of trunk incline. Eur J Appl Physiol 2009;106:195–205.

[24] Mollendorf JC, Termin A, Oppenheim E, Pendergast DR. Effect of swimsuit design on passive drag. Med Sci Sports Exerc 2004;36:1029–35.

[25] Zamparo P, Antonutto G, Capelli C, Francescato MP, Girardis M, Sangoi R, et al. Effects of body size, body density, gender and growth on underwater torque. Scand J Med Sci Sports 1996;6:273–80.

[26] Zamparo P, Capelli C, Cautero M, di Nino A. Energy cost of front crawl swimming at supra-maximal speeds and underwater torque in young swimmers. Eur J Appl Physiol 2000;83:487–91.

[27] di Prampero PE. The energy cost of human locomotion on land and in water. Int J Sports Med 1986;7:55–72.

[28] Capelli C, Zamparo P, Cigalotto A, Francescato MP, Soule RG, Termin B, et al. Bioenergetics and Biomechanics of front crawl swimming. J Appl Physiol 1995;78:674–9.

[29] Zamparo P, Lazzer S, Antoniazzi C, Cedolin S, Avon R, Lesa C. The interplay between propelling efficiency, hydrodynamic position and energy cost of front crawl in 8 to19-year-old swimmers. Eur J Appl Physiol 2008;104:689–99.

[30] Chatard JC, Lavoie JM, Lacour JR. Energy cost of front crawl swimming in women. Eur J Appl Physiol 1991;63:12–6.

[31] Montpetit R, Cazorla G, Lavoie JM. Energy expenditure during front crawl swimming: a comparison between males and females. In: Ungherechts BE, Wilke K, Reischle K, editors. Swimming science V. IL: Human Kinetics, Champaign; 1988. p. 229–36.

[32] Kjendlie PL, Ingjer F, Madsen O, Stallman RK, Gunderson JS. Differences in the energy cost between children and adults during front crawl swimming. Eur J Appl Physiol 2004;91:473–80.

[33] Poujade B, Hautier CA, Rouard A. Determinants of the energy cost of front crawl swimming in children. Eur J Appl Physiol 2002;87:1–6.

[34] Ratel S, Poujade B. Comparative analysis of the energy cost during front crawl swimming in children and adults. Eur J Appl Physiol 2009;105:543–9.

[35] Donato AJ, Tench K, Glueck DH, Seals DR, Eskurza I, Tanaka H. Declines in physiological functional capacity with age: a longitudinal study in peak swimming performance. J Appl Physiol 2003;94:764–9.

[36] Tanaka H, Seals DR. Age and gender interactions in physiological functional capacity: insight from swimming performance. J Appl Physiol 1997;83:846–51.

[37] Tanaka H, Seals DR. Endurance exercise performance in Master athletes: age-associated changes and underlying physiological mechanisms. J Physiol 2008;586:55–63.

[38] Zamparo P, Gatta G, di Prampero PE. The determinants of performance in master swimmers: an analysis of master world records. Eur J Appl Physiol 2012;112:3511–8.

[39] Burke LM. Nutrition for Swimming. In: Seifert L, Chollet D, Mujika I, editors. World book of swimming: from science to performance. NY: Nova Science Publishers; 2011. p. 513–26.

[40] Minetti AE. Passive tools for enhancing muscle-driven motion and locomotion. J Exp Biol 2004;207:1265–72.

[41] Pendergast DR, Bushnell D, Wilson DR, Cerretelli P. Energetics of kayaking. Eur J Appl Physiol 1989;59:342–50.

[42] di Prampero PE, Cortili G, Celentano F, Cerretelli P. Physiological aspects of rowing. J Appl Physiol 1971;31:853–7.

[43] Zamparo P, Carignani G, Plaino L, Sgalmuzzo B, Capelli C. Energetics of locomotion with pedal driven watercrafts. J Sport Sci 2008;26:75–81.

[44] Secher NH. Physiological and biomechanical aspects of rowing. Sports Med 1993;15:24–42.

CHAPTER

15

Performance Enhancement Drugs and Sports Supplements: A Review of the Evidence

Gary Gaffney

Department of Psychiatry, University of Iowa College of Medicine, Iowa City, IA, USA

PERFORMANCE-ENHANCING DRUGS

Introduction

For the past 25 years, the abuse of performance-enhancing drugs (PEDs), followed by the cat and mouse chase between athletes using illicit PEDs and anti-doping agencies responsible for drug monitoring, has captivated the sports-going public around the world [1]. No sport or competition escaped the scandals, controversies, and heated debates generated by PED use, nor the anti-doping laboratory sleuthing of the PED abusers [2]. Although sports leagues and sporting events almost universally implemented anti-doping rules and regulations over time, the abuse of the past reverberates almost daily. The era is referred to as 'Baseball's Steroid Era' in Cooperstown Hall of Fame voting. Tour de France winners over about ten years have been stripped of their championship titles. The Lance Armstrong doping saga proved the most vitriolic saga of PED abuse, playing out over years. Armstrong's duplicitous deceptions resulted in libel and defamation legal actions against accusers, who turned out to be correct. Armstrong may yet face serious fraud charges from the United States government, as the US Postal Service sponsored his doped-up cycling team. Olympic gold medals were recalled as competitors at the most elite level admitted they doped and thereby cheated the competition.

So what drugs or substances constitute PED abuse or sport doping when used by athletes to enhance performance? And how do such doping agents enhance performance? Why would the theoretical edge given to competitors drive them to elaborate schemes to cheat their competitors and deceive the fans who support their sport? How powerful are the ergogenic effects of these drugs?

Answers to these questions must always be considered putative, because the rigorous burden of proof for medical drug efficacy does not exist for PEDs and may never exist [3–6]. There are no double blind, placebo-controlled, cross-over clinical trials of a PED in actual sports competition using elite level or professional athletes who make up the high-profile doping cases. Therefore the rigorous studies and analyses considered necessary for medical grade clinical trials are lacking to prove the ergogenic benefits of PED use. Read reports of PED effectiveness with this in mind: no completely universally accepted rigorous scientific evidence exists for the efficacy of a drug enhancing performance at an Olympic or elite level. Nonetheless, a body of evidence exists for robust effects when ergogenic drugs target fundamental athletic dimensions: strength, power, endurance, focus, recovery.

PEDs may be grouped into several broad categories (Table 15.1):

1. improving concentration and alertness;
2. enhancing strength, power and explosiveness;
3. improving the oxygen-carrying capacity of blood;
4. augmenting recovery, recuperation, and reconstitution of the athlete;
5. ameliorating inflammation and pain.

Improving Concentration and Focus

Most elite athletes develop incredible powers of concentration over years of training and competition, as well as unparalleled motivation and drive. Skill,

Nutrition and Enhanced Sports Performance.
DOI: http://dx.doi.org/10.1016/B978-0-12-396454-0.00015-1

TABLE 15.1 Performance Enhancing Drugs—Benefits Estimated on a 1(+) to 4(++++) Scale

Class of Drug	Examples	Actions	Benefits	Effectiveness	Side Effects
Anabolic androgenic steroids	Testosterone, nandrolone	Enhance anabolic and androgenic characteristic	In medical uses, benefits include increased muscle mass, red blood cell mass	++++	Multitude of side effects including virilizing and feminizing of tissues, edema, mood changes, and even tumor generation
Peptide growth hormones	HGH, insulin, IGFs	Part of physiological growth stimulation for bone, muscle, and organs	Obvious physiological benefit; pharmacological use remains controversial in medicine and illicit use	Exogenous HGH +; exogenous insulin ++	Serious side effects depending on drug: from cardiac enlargement to death
Blood-stimulating drugs and procedures	EPO, blood doping, blood transfusions	Increase red blood cell mass thus increasing oxygen delivery to tissues	Improvement of endurance and overall enhancement of performance	++++	Serious side effects including heart attack, stroke, and death
Stimulants and 'cognitive enhancers'	Dexedrine, methylphenidate	Improve alertness, concentration, motor coordination and reduce fatigue	Enhanced cognitive and motor performance	++++	Anorexia, irritability, insomnia; in high doses, mood changes and psychosis
Anti-inflammatory drugs and analgesics	Acetaminophen, ibuprofen, narcotics, cortisone	Reduce pain and inflammation	Pain and inflammation management	Reduce pain, help recovery and rehabilitation from injury	Depending on drug, from slight (edema, ulcers) to serious (liver poisoning, kidney damage)
Diuretics	Lasix	Enhance kidney function	Reduced weight, may mask PED use	As a PED, none	Electrolyte imbalance depending on drug; fatigue
Hormone-masking agents	Tamoxifen	Block estrogen receptors	Reduce side effects of testosterone, or mask illicit PEDs	May reduce annoying side effects of AASs	Edema, feminizing

strength and power athletes may improve performance with the use of pharmacologic agents to heighten the mental concentration involved in their sport. The stimulants, or analeptics, such as amphetamine compounds (Dexedrine, Adderall), or methylphenidate (Ritalin or Concerta), are powerful, rapid-onset agents that enhance mental focus [7,8]. These drugs improve concentration when prescribed for patients with ADHD; however the effect is nonspecific, also enhancing concentration of non-ADHD people. Studies find that adequately dosed stimulants also improve motor coordination, and reduce fatigue [7–9]. Side effects include anorexia, insomnia, rebound moodiness and dysphoria/irritability, and in high doses even psychosis. Without a Therapeutic Use Exemption (TUE) allowing an athlete to legally use the drugs, the stimulants are universally banned form sports competition.

Black-market modafinil—a nonstimulant cognitive enhancement drug—has also been employed by unscrupulous trainers to take advantage of concentration-enhancing effects [1]. This drug is approved for the treatment of narcolepsy; however there are preliminary results that strongly suggest it may improve concentration. Side effects appear to be minor. Modafinil is banned in sport without a doctor's prescription and TUE.

Legal supplements may enhance an athlete's focus on competition too. Caffeine in moderate doses appears to enhance concentration somewhat; in moderate doses the compound is legal in sports competition. Caffeine is often the psychoactive ingredient that matters in many of the sports energy drinks. Side effects can include insomnia, and possibly brief diuresis. In large super-pharmacologic doses, caffeine can result in a positive doping test.

A contemporary label for these drugs is 'cognitive-enhancers' or 'nootropic' [9]. Just as the majority of anabolic drugs are most likely used for 'appearance enhancement' and not performance enhancement, drugs like modafinil may find their way into the black market for those wishing to tip the academic scales in their favor, rather than increase cognitive performance during sports competition.

Enhancing Strength, Power, and Explosiveness: Anabolic Steroids and Peptide Hormones

Muscle mass and performance strength impress fans, intimidate opponents, and offer huge advantages to athletes with natural well-trained strength and power. Heavily muscled aggressive athletes dominate competition, although guile, skill, and motivation obviously remain components of a winning athletic performance [2]. Therefore, it was only a matter of time before strength athletes discovered the benefits of pharmacologic anabolic drugs that appear to enhance strength and power [2]. Dr. John Zeigler the team physician for Olympic Weightlifting, on a tip from a Russian physician, introduced first testosterone, then Dianabol (methandrostenolone). Zeigler, it is said, later regretted his pharmacological tinkering with performance enhancement.

Anabolic/androgenic steroids (AASs) are the most well-known of the anabolic agents (anabolic refers to the building of muscle tissue; androgenic refers to the development of male characteristics such as beard, male genitals, deep voice) [4,5]. Anabolic steroids fall into one of three categories, depending on side chains of the drug, off the main steroid ring—the classic four-ring structure of steroids found in corticosteroids (cortisol) and sex steroids (testosterone, estrogen, progesterone). AASs can be classified as:

- Class A: 17-beta ester AASs (nandrolone, stenbolone);
- Class B: 17-alpha alkyl AASs (stanozolol);
- Class C: testosterone alterations alkylated in A, B, or C steroids rings.

Some of the compounds are oral, most are parenteral (injected). Oral active anabolic steroids tend to be more toxic to organs such as liver.

These compounds exhibit differential effects on muscle building (anabolic effects) and masculinity (androgenic effects) [6]. The drugs have differences in pharmacokinetics, androgen receptor activity, and metabolic pathways in the body [10]. AAS compounds are often used in conjunction with each other (stacked) or rotated (cycled) according to street knowledge. Anabolic steroids most often are combined with other reputed PED classes to enhance the anabolic effects. To summarize, the evidence is strong that AASs in high pharmacological doses enhance strength, power, muscle size, and competition endurance [11]. The drugs also account for aggressiveness in competitors both on and off the field [1,2].

Side effects of the anabolic steroids include a host of deleterious effects depending on the age and sex of the user [11–16]. Males might expect muscle gain, as well as acne, water retention, gynecomastia, testicular atrophy and decreased sperm count, irritability, insomnia, and increased aggressiveness. Males may become infertile [17]. Females experience even more risks, including masculinity of the female genitalia, breast tissue loss, deepening voice, blood clotting irregularities, menstrual dysfunction, and (there is some evidence) birth defects in future children [10,18–20]. Lastly there can be liver tumors resulting from AAS use, although these tumors are reversible with cessation of the exogenous androgens [2]. Side effects likely result from high doses during long exposure times [14].

AAS use in adolescents presents particularly disconcerting problems; there is evidence that exposing the juvenile central nervous system to anabolic-androgenic drugs leads to abnormal brain development and subsequent increased aggressiveness [21,22].

As mentioned, there appear to be ergogenic benefits from anabolic steroids to enhance muscle size, performance, power and recovery. Recent reviews discuss the problems with elucidating conclusions on the research literature of the drugs. It is almost impossible to replicate the actual street use of these drugs because of excess doses, polypharmacy practices, and quirky exercise regiments of various disciplines of athletes and appearance-enhancing users. AASs increase the red blood cell mass. There is also strong evidence that aggressiveness increases, leading to focus and assertiveness in competition.

AAS users appear to use multiple drugs in enhancing athlete performance. Furthermore recent reports suggest that anabolic steroids can lead to dependence [5,16,23,24]. Although abuse of AASs is of great concern in athletes and strength performers, there is evidence that the highest percentage of use is for appearance-enhancing properties and that those who do so resort to polypharmacy with multiple PEDs [5].

Peptide Hormone Anabolic Drugs

The natural homeostatic system for muscle development in the body consists of a balance between many interacting neural-endocrine systems. The endocrine systems consist of steroid (testosterone, cortisone) and protein hormones. The anabolic steroids like testosterone stimulate muscle growth and repair as well as stimulating blood volume and aggressiveness. However, the protein hormones exert profound anabolic and metabolic effects too. Protein hormones include human growth hormone (HGH), insulin, and insulin-like growth hormones (IGFs), and have now been reported to be widely abused as illicit ergogenic PEDs [12,24–27].

Human growth hormone is a basic anabolic/metabolic protein hormone (as opposed to the steroid structure of the AASs). HGH is needed for proper growth and development of all organs. Lack of endogenous HGH causes dwarfism; physiological excess produces acromegaly or giantism.

In athletes, high doses of HGH may enhance muscle development, improve recovery and repair from competition, although the data are not totally clear [12].

Most users of HGH will stack the drug with other PEDs such as thyroid hormones or AASs. Side effects of HGH use are serious, and appear to be long term; the most serious is hypertrophy of myocardial (heart) musculature, leading to cardiac ischemia. It is suspected that many athletes who experience cardiac problems such as myocardial infarction or angina suffer from past abuse of HGH [26].

Insulin and the insulin-like compounds (IGF-1) are powerful anabolic hormones in biology [27,28]. Insulin insufficiency or insulin unresponsiveness leads to one of many forms of diabetes. Excess insulin leads to blood sugar irregularities. When administered to athletes (mostly body builders, but also used by other athletes, most famously by Victor Comte who recommended insulin to baseball, track and American football athletes), the hormone can increase muscle mass by its potent anabolic effects, in concert with other growth factors. Insulin is one of the most deadly drugs an athlete can use. Overdose can induce seizures, stupor, coma and can be rapidly fatal.

Often HGH users continue with significant use of AASs, as well as other street drugs, compounding problems with side effects [24]. Substance abuse treatment centers must now deal with AAS abuse and withdrawal.

PEDs to Enhance Oxygen Delivery

Various procedures improve the effectiveness and capacity of blood to deliver oxygen to the tissues [1,29–31]. Obviously, endurance athletes will be focused on improving the effectiveness of nourishment, recovery and regeneration of tissues, which is much dependent on oxygen delivery. However, strength athletes also know that recovery and regeneration in competition may be enhanced by increasing the oxygen-carrying capacity of blood.

Increasing the number of red blood cells (RBCs) by storing fresh blood then re-infusing into athletes later is the most straightforward method of blood doping. Variations of this method include using only one's own (autologous) packed RBCs, or even another person's (heterologous) blood transfusion. Anabolic steroids increase RBC mass. However the most successful method of enhancing RBCs and oxygen delivery was found with the use of erythropoietin, or EPO.

EPO is a natural hormone regulating RBC mass. Natural EPO is produced in the kidney. Because patients with kidney failure fail in EPO production, which produces dangerous anemia, EPO has been synthesized by recombinant DNA for medical uses. As with other medical advances, EPO was diverted into doping by athletes seeking a competitive edge [29]. There is a concern that a significant proportion of the EPO produced by drug companies is diverted to PED use.

It now appears almost universal that champion cyclists abused EPO over the past twenty years. The drug appears to be abused in marathons and in distance events to give the athlete an edge in endurance and recovery. However, even strength and skill athletes abuse EPO, as part of the anabolic chain.

EPO side effects are particularly deadly: increased blood mass may cause the blood to clot, producing heart attacks, stroke, or pulmonary embolism [32]. Many young trained endurance athletes who suffered early deaths likely abused EPO.

Drugs that Enhance Recovery from Training and Competition

Most PEDs will directly or indirectly lead to enhanced recovery from training or competition; this allows the athlete to train harder and longer without experiencing symptoms of over-training. Anabolic steroids, HGH, EPO and others are obvious doping agents; however, other drugs, including antidepressants, have been used to enhance recovery from stress.

Hormones such as cortisone and syndactin (ACTH) are used by some endurance athletes to enhance recovery and repair [33].

Anti-inflammatory Drugs and Analgesics

Athletes push their bodies to extremes in training and competition. Competitors suffer injuries. Rapid recovery from injury using rest, nutrition, analgesic drugs, and anti-inflammatory agents constitutes a part of the athlete's preparation for competition. Conditioning and proper techniques help to prevent injuries; however, sooner or later all athletes will suffer minor or major injuries.

Most analgesics and anti-inflammatory drugs are legal, often over the counter. However, the most powerful painkillers are highly regulated narcotics and steroids. These powerful drugs have been notoriously controversial and will remain that way because the drugs lead to abuse, dependence, addiction, and street value. Therefore, as with any highly regulated entity, there will be uses of the drug outside the accepted avenues of medical practice.

There are a multitude of analgesic drugs and narcotics. Effective and easily available agents include acetaminophen and non-steroid anti-inflammatory drugs (NSAIDs) such as ibuprofen. When used properly, these medications will reduce swelling and pain and enhance recovery from injury. However, even these widely available drugs carry serious side effects: for acetaminophen, severe liver damage, and for NSAIDs kidney and cardiac effects.

The narcotic drugs, when used properly in recovery from injury, are safe and beneficial. The athlete needs a TUE to compete while on these medications; otherwise s/he will test positive for a banned substance. Although the pain relief is dramatic, the side effects—sedation, constipation, GI upset, tolerance (requiring higher doses of the drug for the same effect), and withdrawal—are serious. Overuse of narcotic painkillers may mask the pain of a serious injury, leading to complications [34].

The corticosteroids (cortisol and cortisone and derivatives) act as powerful anti-inflammatory agents. These drugs can produce almost miraculous short-term effects which may be of benefit in short-term injury. Unfortunately, long-term use of the drugs causes problems with the immune system, water retention, mood changes and even psychosis. Even physicians underestimate the serious side effects of these powerful drugs.

When used correctly and judiciously, anti-inflammatory drugs will benefit the athlete during injury and recovery; when abused, the side effects can cause serious permanent injury and even death.

Masking Agents and Fluid Retention Drugs

To ether mask the anabolic drugs or treat the side effects of PEDs, other drugs have been used to defeat anti-doping measures. Diuretics increase urine volume; probenecid reduces concentration of acidic PED compounds; plasma expanders also dilute PED excretion [4,35].

To deal with side effects, diuretics reduce edema caused by AASs. Anti-estrogens, such as tamoxifen [36] block the estrogen effects of AASs such as gynecomastia.

Novel Drugs

There are several newer drugs that may offer benefits for athletes in some capacity. The Selective androgen receptor modulators (SARMs) may be orally active [36]. Therapeutically, such agents could be useful for conditions such as osteoporosis or severe androgen deficiency in older men. Undoubtedly these agents will be used by the black market if any advantage in building muscle mass or power is demonstrated.

Drug Testing

Drug testing at schools, workplaces, and in professional sports remains controversial. The issues differ between drug-free environments at work, in relationship to individual rights, and the issue of ensuring fair, illicit-drug-free sports competition. The apex of PED use appears to have occurred in sports in the late 1980s and 90s, depending on the sport. Baseball, professional cycling and the Olympics all struggled with PED use during the home run days of Mark McGwire and Barry Bonds, the Tour de France years of Lance Armstrong and Floyd Landis, and the Olympic taint of the East German doping machine, especially among women's swimming. Because the unethical use of these drugs appeared at the highest level (i.e., Ben Johnson at the 1988 Olympics), anti-doping agencies such as the World Anti-doping Agency (WADA) developed to combat sports doping [36].

The use of PEDs appears to be highly dependent on the sports culture. For instance, PED use is rampant in weightlifting, where power is of paramount importance, or sprinting where explosiveness separates winners from losers. Baseball, football, and other sports lie somewhere in between. Other sports such as professional basketball appear to be avoiding major contamination with PEDs, despite no paucity of recreational drug use.

Drug testing is unsavory, humiliating at times, expensive, and obviously challenged in courts. However, it is necessary to ensure fairness in competition, and to avoid a nuclear PED race among competitors which would expose all athletes to unwanted deleterious side effects. Considering the horrible side effects the German swimmers appear to have suffered (depression, gender ambiguity, birth defects), preventing illicit PED use is worth the efforts [10].

PERFORMANCE-ENHANCING SUPPLEMENTS

A supplement is defined by the 1994 Dietary Supplement Health and Education Act as a special category of 'food': "product taken by mouth that contains a dietary ingredient intended to supplement the diet" [37]. These ingredients may include vitamins, minerals, herbs, amino acids and proteins, and other substances such as enzymes, glands, and organ tissues. The product labeling must clearly indicate that the product is sold to supplement the diet.

Dietary supplements are not considered drugs, and therefore do not fall under the full regulatory responsibilities of the Food and Drug Administration (FDA). The FDA concentrates on good manufacturing practices, and ensures accuracy of supplement labeling. Supplement Manufacturers and marketers do not submit the large quantity of data that pharmaceutics companies submit to ensure the effectiveness and safety of legal drugs. However, if the supplement is new without a history of safe use over years of routine use then the FDA may regulate the supplement safety.

The FDA is particularly concerned about the toxic effects of supplements, and therefore their safe use in humans. The Federal Trade Commission (FTC) may require that the product demonstrate or substantiate claims of effectiveness, i.e., the product is not marketed in a misleading manner. However, there is a required disclaimer on supplement labels: "this statement has not been evaluated by the FDA. This product is not intended to diagnose, treat, cure, or prevent disease".

Despite the precautions and regulations noted above, reports indicate that many new supplement products have been introduced into the marketplace ignoring the regulations put in place by the Dietary Supplement Health and Education Act (DSHEA). Furthermore, the FDA has reviewed only a few of the numerous dietary supplements on the market. Although supplement claims of benefits should be backed up by scientific study, that goal remains distant.

Quality and Purity

Safeguards are employed by the most reputable of the supplement companies to ensure the safety of their products. The high-quality companies often employ research directors who either direct an active research effort for the company or review the available scientific literature and databases. Supplement companies can involve the FDA and FTC to review manufacturing processes for good manufacturing practices. Companies may also employ third-party testers to ensure their products are free of banned substances that may expose a user to rule-violations of his or her professional organization (e.g., the NFL). Nonetheless, the supplement user is responsible for dietary supplements that do not violate PED regulations of the sports regulatory agency.

When evaluating a dietary supplement, the user should consider the safety track record of the substance, the evidence for the supplement's intended ergogenic effect, and the quality control and professional practices of the manufacturer. Even with prescription medications, there is usually no definitive evidence of efficacy in all aspects of the drug's use; therefore the supplement user should always be vigilant for new safety warnings and new scientific developments on supplement efficacy.

The following categories of dietary ergogenic supplements will be considered in this chapter (Table 15.2):

1. Vitamins
2. Minerals
3. Pro-hormones
4. Dietary supplementation with protein compounds
5. Ergogenic aids
6. Weight control supplements
7. Endurance supplements.

Vitamins and Minerals

Long a staple of dietary supplements, the market for vitamin and mineral supplementation may have reached 11 billion dollars as 150 million Americans supplement their diet with vitamins and minerals [38]. For what benefit at this expense?

Vitamins can be classified into fat- and nonfat-soluble compounds. The fat-soluble vitamins include A, D, E and K. These vitamins are stored in body lipid tissues, which means in extreme cases toxicity can occur over time, notably for vitamin D.

Water-soluble vitamins include B and C. These vitamins are excreted in urine, thus preventing their build-up in the body.

For athletes with a balanced adequate diet, a once a day vitamin is an expensive supplementation strategy. There is scant evidence for the performance-enhancing effects of large doses of vitamins, with some exceptions. Vitamins E and C may serve as antioxidants; evidence suggests that increased doses of these vitamins may increase athletes' tolerance of heavy exercise, thus possibly improving performance [39].

On the other hand there is more evidence that mineral supplementation enhances athletic performance, with qualifications [39]. Calcium appears to benefit female athletes who are prone to osteoporosis. Calcium supplementation may manage body composition. Iron likewise appears to maintain a noncontroversial role in supplementation in part due to the role of iron in hemoglobin, the oxygen-carrying molecule in blood. Heavy exercise stresses the ability of the body to replenish the red blood cells needed for optimal oxygen-carrying capacity. For athletes, especially those prone to anemia, iron supplementation is necessary.

Maintenance of body electrolytes (salt or sodium chloride) improves oxygen uptake, endurance and maximal oxygen uptake. Zinc appears to reinforce the immune system in athletes. The role of electrolytes depends on adequate hydration; often overlooked, but water itself is an extremely important ergogenic aid for athletes who must maintain proper hydration for optimal performance [39].

Ergogenic Performance-Enhancing Supplements: Proteins, Pro-hormones and Creatine

Several classes of supplements are marketed as performance enhancing; scientific study data endorse the

TABLE 15.2 Effective Performance-Enhancing Supplements—Effectiveness Estimated on a 1(+) to 4(+ + + +) Scale

Class of Supplement	Actions	Benefits	Effectiveness	Side Effects
Vitamins	Supplementation with E and C may enhance antioxidant effects	Improve recovery from training	++	In excess may be toxic. Water-soluble vitamins rapidly excreted by kidneys
Minerals	Calcium in females	Involved in bone development and maintenance	+++ (important in females prone to osteoporosis)	
Electrolytes	Maintain physiological homeostasis	Critical for body physiological homeostasis	++++	
Amino acid and protein supplementation	Basic building blocks of muscle fibers	Ensure adequate foundation for exercise adaptation and repair	++++ (for basic muscle hypertrophy)	
Creatine	Enhances metabolism	Modestly increases muscle strength and recovery	++	Bloating, GI upset
Androgen pro-hormones	Putatively stimulate androgen (testosterone) receptors	Little	Little	Positive in Anti-doping testing

effectiveness of a few of these supplements. The protein and amino acid supplements are generally safe for athletes with normally functioning kidneys, and when used in appropriate training regimens promote a significant measure of enhanced performance. Although none of these supplements will provide ergogenic enhancement as powerful as do pharmaceutical grade PEDs, the compounds do offer some benefits to athletes.

Protein supplements, protein powders, and amino acids increase the amount of protein precursors of muscle tissue. Under rigorous training demands, athletes utilize a high level of protein to repair and build muscle fibers. Protein supplementation supplies the basic demands of developing new muscle structure or repairing stressed or injured muscle fibers. Recommended daily protein requirements of athletes currently are considered to be 1.4–2.0 g/kg of body weight. Although a quality diet is foremost in developing muscle growth following training, protein supplementation augments the diet. Protein supplementation appears to be most optimal when taken before exercise.

Essential amino acids also have been found to enhance protein synthesis in athletes. The branched-chain amino acids may be the most potent ergogenic enhancers.

Creatine monohydrate is a nitrous organic acid product of amino acid metabolism. Creatine remains the most effective non-drug ergonomic supplement available to athletes [40]. Evidence suggests that an athlete can engage in higher intensity exercise while taking the supplement, thus allowing more intense stimulation of the musculature, and thus greater muscle adaptions and hypertrophy; the gain is small, but significant. The most commonly reported side effect of creatine is weight gain. Further, there is evidence that creatine supplementation may lesson injury to the athlete.

Beta-hydroxy beta-methyl butyrate (HMB) once was a hot topic among athletes. Supplementation with HMB may prevent muscle breakdown and produce small gains in muscle mass; however, most studies do not present impressive positive results [41]. Theoretically, HMB slows muscle degradation and catabolic processes through a leucine mechanism. However, the final word on its effectiveness has yet to be spoken.

Several other amino acid supplements are marketed as performance enhancing: alpha-ketoglutarate, alpha-ketoisocaproate, and GH-releasing peptide analogs. Companies adopt a number of theoretical strategies to market these supplements; unfortunately the empirical evidence is thus far unimpressive.

There are several amino acid / protein supplements marketed as ergogenic, without proven scientific results backing claims: glutamine, similax, isoflavones, and myostatin inhibitors (sulfa-polysaccharides). Again, the marketing rhetoric exceeds the empirical evidence for effectiveness of these compounds.

Minerals such as chromate and borate have been marketed as effective ergogenic aids, or effective in weight loss; there is scant evidence of the effectiveness of these substances.

Testosterone pro-hormones and *Tribulus terrestris* are taken to either increase androgen levels in the athlete or to stimulate the release of testosterone-stimulating proteins from the pituitary. Dehydroepiandrosterone (DHEA) is the best known, although there are several other androgen precursors on the market. Although the story is long and involved with testosterone pro-hormones, it appears that these supplements offer little advantage to young athletes [42], who are releasing testosterone anyway in reaction to the stress of exercise. The amount of testosterone boost from pro-hormones appears to be trivial compared with physiological androgen. *Tribulus* supplementation is ineffective as a PED.

Weight Loss Supplements

In the past, weight loss supplements contained three major ingredients: ephedra, caffeine and aspirin/salicin. Ephedra is now banned by the FDA because of the risk of cardiac and psychiatric complications. The ECA supplements (or EC alone) did demonstrate effective weight loss in studies.

Green tea, and conjugated linoleic acids show promise in weight loss regimens [43]. Effects of most other supplements to increase metabolism or reduce appetite are unproven.

When looking at enhancing performance in athletic competition, athletes should understand that basic talent, hard work, and smart training and diet form the foundation of success. If there are medical conditions, such as anemia, asthma or ADHD, schedule the appropriate medical evaluation, and then ascertain the TUE to use medical treatments for legitimate medical problems. PEDs offer a quick pathway to 'success', but at the cost of ethical integrity, serious medical complications, and discipline from sports governing bodies.

References

[1] Fainaru-Wada M, Williams L. Game of shadows. barry bonds, BALCO, and the steroids scandal that rocked professional sports. New York, NY, USA: Gotham Books; 2006.

[2] Yesalis CE, Bahrke MS. History of doping in sport. Int Sports Stud 2001;24(1):42–76.

[3] Tokish JM, Kocher MS, Hawkins RJ. Ergogenic aids: a review of basic science, performance, side effects, and status in sports. Am J Sports Med 2004;32(6):1543–53.

[4] Ventura R, Segura J. Doping in sports. Handb Exp Pharmacol 2010;195:327–54.

[5] Kanayama G, Pope HG. Illicit use of androgens and other hormones: recent advances. Curr Opin Endocrinol Diabetes Obes 2012;19(3):211–9.

[6] Wood RI, Stanton SJ. Testosterone and sport: current perspectives. Horm Behav 2012;61(1):147–55.

[7] Deventer K, Roels K, Delbeke FT, Van Eenoo P. Prevalence of legal and illegal stimulating agents in sports. Anal Bioanal Chem 2011;401(2):421–32.

[8] Docherty JR. Pharmacology of stimulants prohibited by the World Anti-Doping Agency (WADA). Br J Pharmacol 2008;154 (3):606–22.

[9] Cakic V. Smart drugs for cognitive enhancement: ethical and pragmatic considerations in the era of cosmetic neurology. J Med Ethics 2009;35(10):611–5.

[10] Nikolopoulos DD, Spiliopoulou C, Theocharis SE. Doping and musculoskeletal system: short-term and long-lasting effects of doping agents. Fundam Clin Pharmacol 2011;25(5):535–63.

[11] Hartgens F, Kuipers H. Effects of androgenic-anabolic steroids in athletes. Sports Med 2004;34(8):513–54.

[12] Urban RJ. Growth hormone and testosterone: anabolic effects on muscle. Horm Res Pædiatr 2011;76(Suppl. 1):81–3.

[13] Dodge T, Hoagland MF. The use of anabolic androgenic steroids and polypharmacy: a review of the literature. Drug Alcohol Depend 2011;114(2–3):100–9.

[14] Van Amsterdam J, Opperhuizen A, Hartgens F. Adverse health effects of anabolic-androgenic steroids. Regul Toxicol Pharmacol 2010;57(1):117–23.

[15] Maravelias C, Dona A, Stefanidou M, Spiliopoulou C. Adverse effects of anabolic steroids in athletes. A constant threat. Toxicol Lett 2005;158(3):167–75.

[16] Welder A, Melchert RB. Cardiotoxic effects of cocaine and anabolic-androgenic steroids in the athlete. J Pharmacol Toxicol Methods 1993;29(2):61–8.

[17] Fronczak CM, Kim ED, Barqawi AB. The insults of illicit drug use on male fertility. J Androl 2012;33(4):515–28.

[18] Harmer PA. Anabolic-androgenic steroid use among young male and female athletes: is the game to blame? Br J Sports Med 2010;44(1):26–31.

[19] Cleve A, Fritzemeier K, Haendler B, Heinrich N, Möller C, Schwede W et al., Pharmacology and clinical use of sex steroid hormone receptor modulators. Handb Exp Pharmacol 2012;214:543–87.

[20] Ungerleider S. Faust's gold: inside the east german doping machine. New York, NY, USA: Thomas Dunne Books; 2001.

[21] Cunningham RL, Lumia AR, McGinnis MY. Androgenic anabolic steroid exposure during adolescence: ramifications for brain development and behavior. Horm Behav 2012; [In-press].

[22] Calfee R, Fadale P. Popular ergogenic drugs and supplements in young athletes. Pediatrics 2006;117(3):e577–89.

[23] Nyberg F, Hallberg M. Interactions between opioids and anabolic androgenic steroids: implications for the development of addictive behavior. Int Rev Neurobiol 2012;102:189–206.

[24] Brennan BP, Kanayama G, Hudson JI, Pope HG. Human growth hormone abuse in male weightlifters. Am J Addict 2011;20(1):9–13.

[25] Davis JK, Green JM. Ergogenic value and mechanisms of action. Sports Med 2009;39(10):813–32.

[26] Huckins DS, Lemons MF. Myocardial ischemia associated with clenbuterol abuse: report of two cases. J Emerg Med 2012; [In-press].

[27] Baumann GP. Growth hormone doping in sports: a critical review of use and detection strategies. Endocr Rev 2012;33 (2):155–86.

[28] Harridge SDR, Velloso CP. IGF-I and GH: potential use in gene doping. Growth Horm IGF Res 2009;19(4):378–82.

[29] Gaudard A, Varlet-marie E, Bressolle F, Audran M. Drugs for increasing oxygen transport and their potential use in doping 1 A review. Sports Med 2003;33(3):187–212.

[30] Thevis M, Kohler M, Schänzer W. New drugs and methods of doping and manipulation. Drug Discov Today 2008;13 (1–2):59–66.

[31] Wilber RL. Detection of DNA-recombinant human epoetin-alfa as a pharmacological ergogenic aid. Sports Med 2002;32(2): 125–42.

[32] Smith KJ, Bleyer AJ, Little WC, Sane DC. The cardiovascular effects of erythropoietin. Cardiovasc Res 2003; 59(3):538–48.

[33] Synacthen, ACTH, And Intrigue. Steroid Nation, <http://grg51.typepad.com/steroid_nation/2006/08/synacthen_acth_.html>; 13.08.06.

[34] Hemmersbach P, de la Torre P. Stimulants, narcotics and β-blockers: 25 years of development in analytical techniques for doping control. J Chromatogr B: Biomed Sci Appl 1996;687 (1):221–38.

[35] Cadwallader AB, De la Torre X, Tieri A, Botrè F. The abuse of diuretics as performance-enhancing drugs and masking agents in sport doping: pharmacology, toxicology and analysis. Br J Pharmacol 2010;161(1):1–16.

[36] Barroso O, Mazzoni I, Rabin O. Hormone abuse in sports: the antidoping perspective. Asian J Androl 2008;10(3):391–402.

[37] Dietary Supplement Health and Education Act of 1994. <http://www.fda.gov/RegulatoryInformation/Legislation/FederalFoodDrugandCosmeticActFDCAct/SignificantAmendmentstotheFDCAct/ucm148003.htm>.

[38] Nutritional Supplements in the U.S., 5th ed. <http://www.packagedfacts.com/Nutritional-Supplements-Edition-7131106/>.

[39] Kreider RB, Wilborn CD, Taylor L, Campbell B, Almada AL, Collins R, et al. ISSN exercise and sport nutrition review: research and recommendations. J Int Soc Sports Nutr 2010;7:1–43.

[40] King DS, Baskerville R, Hellsten Y, Senchina DS, Burke LM, Stear SJ, et al. A-Z of nutritional supplements: dietary supplements, sports nutrition foods and ergogenic aids for health and performance–Part 34. Br J Sports Med 2012;46(9):689–90.

[41] Stratton RJ, Elia M. A review of reviews: a new look at the evidence for oral nutritional supplements in clinical practice. Clin Nutr Suppl 2007;2(1):5–23.

[42] Brown GA, Vukovich M, King DS. Testosterone prohormone supplements. Med Sci Sports Exerc 2006;38:1451–61.

[43] Timcheh-Hariri A, Balali-Mood M, Aryan E, Sadeghi M, Riahi-Zanjani B. Toxic hepatitis in a group of 20 male body-builders taking dietary supplements. Food Chem Toxicol 2012;50(10): 3826–32.

CHAPTER

16

Nutrition and Ultra-Endurance

An Overview

Beat Knechtle

Institute of General Practice and for Health Services Research, University of Zurich, Zurich, and Gesundheitszentrum St Gallen, St Gallen, Switzerland

INTRODUCTION

Ultra-endurance performance is defined as an endurance performance lasting for 6 hours or longer [1]. Ultra-endurance athletes compete for hours, days, or even weeks, and face different nutritional problems which may occur singly or in combination. The continuous physical stress consumes energy, and an energy deficit occurs. Furthermore, ultra-endurance performances may lead to dehydration due to sweating.

We may separate these two problems into (i) energy deficit with corresponding loss in solid body masses such as fat mass and skeletal muscle mass; and (ii) dysregulation of fluid metabolism through dehydration or fluid overload, with the risk of exercise-associated hyponatremia (EAH).

Before considering potential aspects of nutrition during ultra-endurance races, we need to review the existing literature regarding the above-mentioned problems. The findings may help us to give recommendations or prescriptions for nutrition in ultra-endurance performances.

PROBLEMS ASSOCIATED WITH ULTRA-ENDURANCE PERFORMANCE

Energy Turnover and Energy Deficit in Ultra-Endurance

An ultra-endurance athlete competing for hours or days with or without breaks expends energy [2–22]. Meeting the energy demands of ultra-endurance athletes requires careful planning and monitoring of food and fluid intake [10,23]. Numerous controlled-case reports [2,10,13–19,24] and field studies [4,9,25–28] in ultra-endurance performances showed, however, that ultra-endurance athletes were unable to self-regulate diet or exercise intensity to prevent negative energy. Furthermore, the insufficient energy intake is also associated with malnutrition such as a low intake of antioxidant vitamins [29].

Generally, an adequate food and fluid intake is related to a successful finish in an ultra-endurance race [9,30,31]. An important key to a successful finish in an ultra-endurance race seems to be an appropriate nutrition strategy during the race [31]. An energy deficit impairs ultra-endurance performance. In ultra-cyclists, a significant negative relationship between energy intake and finish time in a 384-km cycle race has been demonstrated [28]. An ultra-endurance performance leads to an energy deficit [2,4–16,19,21–24,32–39]. In Table 16.1, results from literature are summarized and grouped by discipline (i.e., swimming, cycling, running, and the combination as triathlon). Regarding the single disciplines, the energy deficit seems higher in swimming than in cycling or running. This might be explained by the different environment (water) compared to cycling and running. For events lasting 24 hours or longer, the energy deficit is highest in multi-sports disciplines and cycling. In running, the energy deficit is around three times lower than in both triathlon and cycling.

Change in Body Mass During an Ultra-Endurance Performance

An ultra-endurance performance leads to a loss in body mass (Table 16.2) [2,6–8,10,12,13,16,18,20–22,32,33,36,38–51]. The loss in body mass occurs preferentially in

Nutrition and Enhanced Sports Performance.
DOI: http://dx.doi.org/10.1016/B978-0-12-396454-0.00016-3

TABLE 16.1 Energy Balance in Ultra-Endurance Athletes in Swimming, Cycling, Running and Triathlon

Distance and/or Time	Subjects	Total Energy Intake (kcal)	Total Energy Expenditure (kcal)	Total Energy Deficit (kcal)	Energy Deficit in 24 hours (kcal)	Energy Deficit per hour (kcal)	Reference
Swimming							
26.6 km	1 male	2105	5540	−3435	–	−429	[15]
26.6 km	1 male	–	–	–	–	−500	[32]
24-h swim	1 male	3900	11 460	−7480	−7480	−311	[33]
Mean ± SD						−413 ± 95	
Cycling							
12 hours indoor-cycling	1 male	2750	5400	−2647	–	−220	[19]
557 km in 24 h	1 male	5571	15 533	−9915	−9915	−413	[24]
617 km in 24 h	1 male	10 000	13 800	−3800	−3800	−158	[13]
694 km in 24 h	1 male	10 576	19 748	−9172	−9172	−382	[10]
24 h cycling	6 males	8450	18 000	−9590	−9590	−399	[11]
1,000 km in 48 h	1 male	12 120	16 772	−4650	−2325	−96	[21]
1,126 km in 48 h	1 male	11 098	14 486	−3290	−1645	−65	[34]
2,272 km in 5 d 7 h	1 male	51 246	80 800	−29 554	−5585	−232	[2]
4,701 km in 9 d 16 h	1 male	96 124	179 650	−83 526	−8352	−360	[7]
Mean ± SD					−6298 ± 3392	−258 ± 134	
Running							
160 km in 20 h	1 male	9600	8480	−1120	–	−56	[35]
320 km in 54 h	1 male	14 760	18 120	−3360	−1493	−62	[8]
501 km in 6 days	1 male	39 666	54 078	−14 412	−2402	−100	[6]
Atacama crossing	1 male	37 191	101 157	−63 966	−3046	−127	[22]
100 km	11 female	570	6310	−5750	–	−452	[36]
100 km	27 male	760	7420	−6660	–	−580	[37]
Mean ± SD					−2313 ± 780	−229 ± 227	
Triathlon							
Triple Iron ultra-triathlon	1 male	15 750	27 485	−11 735	−6869	−286	[38]
Triple Iron ultra-triathlon	1 male	22 500	28 600	−6100	−3404	−141	[16]
Gigathlon multi-stage triathlon	1 male	38 676	59 622	−20 646	−9937	−414	[14]
10 × Ironman triathlon	1 male	77 640	89 112	−11 480	−7544	−314	[39]
Mean ± SD					−6938 ± 2699	−288 ± 112	

the lower trunk [6,22,43]. Depending upon the length of an endurance performance and the discipline, the decrease in body mass corresponds to a decrease in fat mass [2,8,11,18,19,39,40,45–50] and/or skeletal muscle mass [2,8,17,39–42,44,45,49]. It seems that a concentric performance such as cycling rather leads to a decrease in fat mass [19,48] whereas an eccentric performance such as running rather leads to a decrease in muscle mass [42]. In runners, a decrease in both fat mass and skeletal muscle mass has been observed [41,42]. For swimmers, no change in body mass, fat mass, or skeletal muscle mass has been reported for 12-hour indoor pool swimmers [52]. In male open-water ultra-swimmers, however, a decrease in skeletal muscle mass was observed [53].

In some instances, an increase in body mass has been reported during ultra-endurance performances

TABLE 16.2 Change in Body Composition in Ultra-Endurance Athletes Competing in Swimming, Cycling, Running and Triathlon

Distance and/or Time	Subjects	Change in Body Mass (kg)	Change in Fat Mass (kg)	Change in Muscle Mass (kg)	Change in Body Water (L)	Reference
Swimming						
24-h swim	1 male	−1.6	−2.4	−1.5	−3.9	[33]
12-h swim	1 male	−1.1	–	−1.1	–	[32]
Cycling						
12-h indoor cycling	1 male	−0.4	−0.9	+0.2	–	[19]
617 km in 24 h	1 male	+4.0	+0.9	+2.9	–	[13]
1,000 km within 48 h	1 male	+2.5	−1	+0.4	+1.8	[21]
2,272 km in 5 d 7 h	1 male	−2.0	−0.79	−1.21	–	[2]
4,701 km in 9 d 16 h	1 male	−5	–	–	–	[7]
Running						
12-h run	1 male	+1.5	−4.4	+1.0	+4.9	[18]
320 km in 54 h	1 male	−0.4	−0.3	−1.0	–	[8]
501 km in 6 days	1 male	−3.0	−6.8	–	–	[6]
100 km in 762 min	11 females	−1.5	–	–	+2.2	[36]
100 km in 11 h 49 min	39 males	−1.6	−0.4	−0.7	+0.8	[40]
338 km in 5 days	21 males	–	–	−0.6	–	[42]
1,200 km in 17 days	10 males	–	−3.9	−2.0	+2.3	[41]
Triathlon						
Triple Iron ultra-triathlon	1 male	−1.1	−0.4	+1.4	+2.0	[38]
Triple Iron ultra-triathlon	1 male	+2.1	+0.4	+4.4	–	[16]
Deca Iron ultra-triathlon	1 male	+3.2	+2.4	+2.4	–	[43]
Quintuple Iron ultra-triathlon	1 male	−0.3	−1.9	–	+1.5	[20]
10 × Ironman triathlon	1 male	−1.0	−0.8	−0.9	+2.8	[39]
Ironman triathlon	27 males	−1.8	–	−1.0	–	[44]
Triple Iron ultra-triathlon	31 males	−1.7	−0.6	−1.0	–	[45]
10 × Ironman triathlon	8 males	–	−3	–	–	[46]
Mean ± SD		−0.45 ± 2.5	−1.41 ± 2.31	+0.08 ± 1.94	+1.51 ± 1.30	

[13,16,18,21,43] (Table 16.2) where also an increase in skeletal muscle mass was found [13,16,18,19,21,38,43] (Table 16.2). The increase in body mass was most probably due to fluid overload, which will be discussed in the next section. An increase in skeletal muscle mass might occur in cases where anthropometric methods were used and an increase in skinfold thicknesses and limb circumferences might occur. This will also be discussed in the next section. Overall, ultra-endurance athletes seem to lose ~0.5 kg in body mass and ~1.4 kg in fat mass where skeletal muscle mass seems to remain unchanged. However, total body water seems to increase by ~1.5 L [18,20,21,36,38–41] (Table 16.2).

Dehydration, Fluid Intake and Fluid Overload

Most endurance athletes are concerned with dehydration during an ultra-endurance performance. Body mass was found to decrease in a 24-hour ultra-marathon [54]. However, body mass reduction in ultra-endurance athletes seems rather to be due to a decrease in sold mass and not due to dehydration [45,47,55].

Dehydration refers both to hypohydration (i.e., dehydration induced prior to exercise) and to exercise-induced dehydration (i.e., dehydration that develops during exercise). The latter reduces aerobic endurance performance and results in increased body temperature, heart rate, perceived exertion, and possibly increased reliance on carbohydrate as a fuel source [56]. Fluid replacement is considered to prevent dehydration, and hypohydration has been shown to impair endurance performance [57]. Adequate fluid intake helps to prevent loss in body mass [26,58]. However, fluid overload may lead to an increase in body mass [59] and a decrease in plasma sodium [59], with the risk of developing exercise-associated hyponatremia [59–61].

Fluid overload may lead to a considerable increase in body mass [59]. For example, one athlete competing in a Deca Iron ultra-triathlon covering 38 km swimming, 1800 km cycling and 422 km running within 12 d 20 h showed an increase in body mass of 8 kg within the first 3 days [43]. In athletes with a post-race increase in body mass, an increase in skin-fold thicknesses and limb circumferences of the lower limb has been recorded [21,43]. In another athlete with an increase in body mass, an increase in skin-fold thicknesses at four skin-fold sites has been shown [13]. Both these races were held in rather hot environments where most probably fluid intake was rather high. However, in athletes with a decrease in body mass, an increase in skin-fold thicknesses at the lower limb has also been reported [2,39,50]. In one athlete with a decrease in body mass after a Triple Iron ultra-triathlon, a considerable swelling of the feet was described [38].

Most probably, the increase in body mass, skin-fold thicknesses and limb circumferences was due to an increase in body water [21,39,62] (Table 16.2). In several studies, an increase in total body water in ultra-endurance athletes has been reported [18,20,21,36,38–41, 46,63,64]. One might now argue about the potential reasons for the increase in both the skin-fold thicknesses and total body water. The increase in total body water might result from an increase in plasma volume [20,63–66], which might be due to sodium retention [63,65] caused by increased activity of aldosterone [20,67]. An association between an increase in plasma volume and an increase in the potassium-to-sodium ratio in urine might suggest that increased activity of aldosterone [68] may lead to retention of both sodium and fluid during an ultra-endurance performance [37]. In a multi-stage race over 7 days, total mean plasma sodium content increased and was the major factor in the increase in plasma volume [63].

Apart from these pathophysiological aspects, fluid overload might also lead to an increase in limb volume. A recent study showed an association between changes in limb volumes and fluid intake [69]. Since neither renal function nor fluid-regulating hormones were associated with the changes in limb volumes, fluid overload is the most likely reason for increase in both body mass and limb volumes. A study showed an association between increased fluid intake and swelling of the feet in ultra-marathoners [70].

Fluid Overload and Exercise-Associated Hyponatremia

Fluid overload might lead to exercise-associated hyponatremia (EAH), defined as a serum sodium concentration ($[Na^+]$) < 135 mmol/L during or within 24 h of exercise [71]. EAH was first described in the scientific literature in 1985 by Noakes et al. [72] in ultra-marathoners in South Africa as being due to 'water intoxication'.

Three main factors are responsible for the occurrence of EAH in endurance athletes: (i) overdrinking due to biological or psychological factors; (ii) inappropriate secretion of the antidiuretic hormone (ADH), in particular, the failure to suppress ADH-secretion in the face of an increase in total body water; and (iii) a failure to mobilize Na^+ from the osmotically inactive sodium stores or, alternatively, inappropriate osmotic inactivation of circulating Na^+ [71]. Because the mechanisms causing factors (i) and (iii) are unknown, it follows that the prevention of EAH requires that athletes be encouraged to avoid overdrinking during exercise.

EAH is the most common medical complication of ultra-distance exercise and is usually caused by excessive intake of hypotonic fluids [73,74]. The main reason for developing EAH is the behavior of overdrinking during an endurance performance by excessive fluid consumption [60] and/or inadequate sodium intake [75]. Subjects suffering EAH during an ultra-endurance performance consumed double the fluids compared with subjects without EAH [60]. Generally, fluid overload is reported for slower athletes [76]. However, in ultra-endurance athletes, faster athletes drink more than slower athletes but seem not to develop EAH [77,78].

The environmental conditions seem to influence the prevalence of EAH. Often, EAH is a common finding in ultra-endurance races held in extreme cold [75,79] or extreme heat [59,80]. In temperate climates, EAH is relatively uncommon [67,81–94]. There seems to be a gender difference whereby females seem to be at higher risk of developing EAH [79]. Compared with marathoners [76,95–97], the prevalence of EAH in ultra-marathoners is, however, not higher [85,94,98].

The prevalence of EAH seems also to be dependent upon the discipline (Table 16.3). While EAH was highly prevalent in ultra-swimming [79] and ultra-running [80], the prevalence of EAH was low [83,99] or even absent [82,84] in ultra-cycling. An explanation could be that

TABLE 16.3 Prevalence of Exercise-Associated Hyponatremia in Ultra-Endurance Athletes Competing in Swimming, Cycling, Running and Multi-Sports Disciplines

Distance and/or Time	Conditions	Subjects	Prevalence of EAH	Reference
Swimming				
26-km open-water ultra-swim	Moderate	25 males and 11 females	8% in males and 36% in females	[79]
Cycling				
665-km mountain bike race	Moderate	25 cyclists	0%	[82]
109-km cycle race	Moderate	196 cyclists	0.5%	[83]
720-km ultra-cycling race	Moderate	65 males	0%	[84]
Running				
161-km mountain trail run	Hot	45 runners	51%	[59]
161-km mountain trail run	Hot	47 runners	30%	[80]
60-km mountain run	Moderate	123 runners	4%	[85]
100-km ultra-marathon	Moderate	50 male runners	0%	[67]
100-km ultra-marathon	Moderate	145 male runners	4.8%	[78]
24-hour ultra-run	Moderate	15 males	0%	[86]
90-km ultra-marathon	Moderate	626 runners	0.3%	[72]
160-km trail race	Hot	13 runners	0%	[26]
Multi-disciplines				
100-mile Iditasport ultra-marathon	Cold	8 cyclists and 8 runners	44%	[75]
161-km race	Cold	20 athletes	0%	[87]
Kayak, cycling and running	Moderate	48 triathletes	2%	[88]
Ironman triathlon	Moderate	330 triathletes	1.8%	[89]
Ironman triathlon	Moderate	330 triathletes	18%	[90]
Ironman triathlon	Moderate	95 triathletes	9%	[91]
Ironman triathlon	Moderate	18 triathletes	28%	[92]
Triple Iron ultra-triathlon	Moderate	31 triathletes	26%	[93]

cyclists can individually drink by using their drink bottles on the bicycle. In addition, the length of an ultra-endurance race seems to increase the risk for EAH. The highest prevalence of EAH has been found in Ironman triathlons [90,92], Triple Iron ultra-triathlons [93] and ultra-marathons covering 161 km [59,80].

NUTRITIONAL ASPECTS IN ULTRA-ENDURANCE ATHLETES

Adequate energy and fluid intake is needed to successfully compete in an ultra-endurance race [100–108]. Most studies are descriptive in nature, reporting the distribution of carbohydrates, fat and protein the athletes ingested [2,6,7,13,14,16,100,103,104] (Table 16.4). Some studies report the kind of food [105–107]. Also, some studies investigated the aspect of supplements [109–112].

Intake of Carbohydrates

Carbohydrates are the main source of energy intake in ultra-endurance athletes [2,23,39,102]. When the intake of carbohydrates, fat and protein was analysed for ultra-endurance athletes, the highest percentage was found for carbohydrates. Ultra-endurance athletes consume ~68% of ingested energy as carbohydrates (Table 16.4).

Intake of Fat

An increased pre-race fat intake leads to an increase in intramyocellular lipids in ultra-endurance athletes

TABLE 16.4 Intake of Energy in Ultra-Endurance Athletes in Different Disciplines

Distance and/or Time	Subjects	Intake of Carbohydrates (%)	Intake of Fat (%)	Intake of Protein (%)	Reference
Cyclists					
617 km in 24 h	1 male	64.2	27	8.8	[13]
2272 km in 5 d 7 h	1 male	75.4	14.6	10.0	[2]
4,701 km in 9 d 16 h	1 male	75.2	16.2	8.6	[7]
Runners					
100 km	7 males	88.6	6.7	4.7	[103]
501 km in 6 days	1 male	40.0	34.6	25.4	[6]
1005 km in 9 days	1 male	62	27	11	[106]
Triathletes					
Deca Iron ultra-triathlon	1 male	67.4	15.6	17.0	[104]
Gigathlon	1 male	72	14	13	[14]
Triple Iron ultra-triathlon	1 male	72	20	8	[16]
Mean ± SD		68.5 ± 13.2	19.5 ± 8.5	11.8 ± 6.1	

[16]. Increased intramyocellular lipids might improve ultra-endurance performance; however, there are no controlled data in field studies of whether fat loading improves ultra-endurance performance. In a case report, ultra-endurance performance in a rower was enhanced following a high-fat diet for 14 days [113]. An increased fat intake during an ultra-endurance competition might improve performance. However, also for this aspect, no controlled data of field studies exist. In a case report on an ultra-marathoner competing in a 6-day ultra-marathon, the athlete consumed 34.6% of fat in his daily food intake [6]. Nonetheless, body fat decreased within the first 2 days and remained unchanged until the end of the race. In addition, performance slowed down after the first 2 days. Ultra-endurance athletes consume ~19% of ingested energy as fat, which is higher than energy consumed in the form of protein (Table 16.4).

Intake of Protein

Regarding protein intake, athletes consume ~19% of ingested energy as protein during racing. An observational field study at the 'Race across America' showed that ultra-endurance cyclists ingest rather large amounts of protein [105]. One might assume that athletes experienced a loss in skeletal muscle mass and try to prevent this loss by the use of amino acids. A recent study tried to investigate whether an increase in amino acids during an ultra-marathon may prevent skeletal muscle damage [114]. However, the intake of amino acids showed no effect on parameters related to skeletal muscle damage.

Intake of Ergogenic Supplements, Vitamins and Minerals

Vitamin and mineral supplements are frequently used by competitive and recreational ultra-endurance athletes during training [106,107,110,111] and competition [104–107]. In some studies, the intake of ergogenic supplements, vitamins and minerals in ultra-endurance athletes and its effect on performance have been investigated [109,110,112]. In long-distance triathletes, over 60% of the athletes reported using vitamin supplements, of which vitamin C (97.5%), vitamin E (78.3%), and multivitamins (52.2%) were the most commonly used supplements during training. Almost half (47.8%) the athletes who used supplements did so to prevent or reduce cold symptoms [112]. The regular intake of vitamins and minerals seems, however, not to enhance ultra-endurance performance [109,110]. In the 'Deutschlandlauf 2006' of over 1200 km within 17 consecutive stages, athletes with a regular intake of vitamin and mineral supplements in the 4 weeks before the race finished the competition no faster than athletes without an intake of vitamins and minerals [108]. Also in a Triple Iron ultra-triathlon, athletes with a regular intake of vitamin and mineral supplements prior to the race were not faster [110].

Fluid Intake During Endurance Performance

Ad libitum fluid intake seems to be the best strategy to prevent EAH and to maintain plasma sodium concentration [36,67,78,115–118]. A rather low fluid intake of 300–400 mL/h seems to prevent EAH [36,90,115]. A mean *ad libitum* fluid intake of ~400 mL/h maintained serum sodium concentration in a 4-h march [115], and fluid consumption of ~400 mL/h prevented EAH in a 161-km race in the cold [87].

Sodium Supplementation During Endurance Performance

One might argue that the supplementation with sodium during an endurance race might prevent EAH. However, two studies on Ironman triathletes showed that *ad libitum* sodium supplementation was not necessary to preserve serum sodium concentrations in athletes competing for about 12 h in an Ironman [119,120].

CONCLUSIONS AND IMPLICATIONS FOR FUTURE RESEARCH

Regarding these findings we see that ultra-endurance athletes face a decrease in body mass most probably due to a decrease in both fat mass and skeletal muscle mass. During racing, the athletes are not able to recover the energy deficit. Athletes tend with increasing length of an ultra-endurance performance to increase fluid intake, which seems to lead to both an increased risk for exercise-associated hyponatremia and limb swelling. In summary, an energy deficit seems to be unavoidable in ultra-endurance performances. Potential strategies might be to increase body mass by a pre-race diet to increase fat mass, and strength training to increase skeletal muscle mass. Another possibility could be to increase energy intake during racing by consuming a fat-rich diet. However, future studies are needed to investigate these aspects. Regarding fluid metabolism, the best strategy to prevent both exercise-associated hyponatremia and limb swelling is to minimize fluid intake to ~300–400 mL per hour.

References

[1] Zaryski C, Smith DJ. Training principles and issues for ultra-endurance athletes. Curr Sports Med Rep 2005;4:165–70.

[2] Bircher S, Enggist A, Jehle T, Knechtle B. Effects of an extreme endurance race on energy balance and body composition – a case report. J Sports Sci Med 2006;5:154–62.

[3] Hill RJ, Davies PS. Energy expenditure during 2 wk of an ultra-endurance run around Australia. Med Sci Sports Exerc 2001;33:148–51.

[4] Hulton AT, Lahart I, Williams KL, Godfrey R, Charlesworth S, Wilson M, et al. Energy expenditure in the Race Across America (RAAM). Int J Sports Med 2010;31:463–7.

[5] Kimber NE, Ross JJ, Mason SL, Speedy DB. Energy balance during an ironman triathlon in male and female triathletes. Int J Sport Nutr Exerc Metab 2002;12:47–62.

[6] Knechtle B, Bircher S. Changes in body composition during an extreme endurance run. Praxis (Bern 1994) 2005;94:371–7.

[7] Knechtle B, Enggist A, Jehle T. Energy turnover at the Race Across America (RAAM)—a case report. Int J Sports Med 2005;26:499–503.

[8] Knechtle B, Knechtle P. Run across Switzerland–effect on body fat and muscle mass. Praxis (Bern 1994) 2007;96:281–6.

[9] Kruseman M, Bucher S, Bovard M, Kayser B, Bovier PA. Nutrient intake and performance during a mountain marathon: an observational study. Eur J Appl Physiol 2005;94:151–7.

[10] White JA, Ward C, Nelson H. Ergogenic demands of a 24 hour cycling event. Br J Sports Med 1984;18:165–71.

[11] Peters EM. Nutritional aspects in ultra-endurance exercise. Curr Opin Clin Nutr Metab Care 2003;6:427–34.

[12] Bourrilhon C, Philippe M, Chennaoui M, Van Beers P, Lepers R, Dussault C, et al. Energy expenditure during an ultraendurance alpine climbing race. Wilderness Environ Med 2009;20:225–33.

[13] Knechtle B, Knechtle P, Müller G, Zwyssig D. Energieumsatz an einem 24 Stunden Radrennen: Verhalten von Körpergewicht und Subkutanfett. Österreichisches Journal für Sportmedizin 2003;33:11–8.

[14] Knechtle B, Bisig A, Schläpfer F, Zwyssig D. Energy metabolism in long-term endurance sports: a case study. Praxis (Bern 1994) 2003;92:859–64.

[15] Knechtle B, Knechtle P, Heusser D. Energieumsatz bei Langstreckenschwimmen – eine Fallbeschreibung. Österreichisches Journal für Sportmedizin 2004;33:18–23.

[16] Knechtle B, Zapf J, Zwyssig D, Lippuner K, Hoppeler H. Energieumsatz und Muskelstruktur bei Langzeitbelastung: eine Fallstudie. Schweizerische Zeitschrift für Sportmedizin und Sporttraumatologie 2003;51:180–7.

[17] Knechtle B, Knechtle P, Kaul R, Kohler G. Swimming for 12 hours leads to no reduction of adipose subcutaneous tissue–a case study. Praxis (Bern 1994) 2007;96:1805–10.

[18] Knechtle B, Zimmermann K, Wirth A, Knechtle P, Kohler G. 12 hours running results in a decrease of the subcutaneous adipose tissue. Praxis (Bern 1994) 2007;96:1423–9.

[19] Knechtle B, Früh HR, Knechtle P, Schück R, Kohler G. A 12 hour indoor cycling marathon leads to a measurable decrease of adipose subcutaneous tissue. Praxis (Bern 1994) 2007;96:1071–7.

[20] Knechtle B, Knechtle P, Andonie JL, Kohler G. Body composition, energy, and fluid turnover in a five-day multistage ultra-triathlon: a case study. Res Sports Med 2009;17:104–20.

[21] Knechtle B, Knechtle P, Kohler G. The effect of 1,000 km nonstop cycling on fat mass and skeletal muscle mass. Res Sports Med 2011;19:170–85.

[22] Koehler K, Huelsemann F, de Marees M, Braunstein B, Braun H, Schaenzer W. Case study: simulated and real-life energy expenditure during a 3-week expedition. Int J Sport Nutr Exerc Metab 2011;21:520–6.

[23] Lindeman AK. Nutrient intake of an ultraendurance cyclist. Int J Sport Nutr 1991;1:79–85.

[24] Bescós R, Rodríguez FA, Iglesias X, Benítez A, Marina M, Padullés JM, et al. High energy deficit in an ultraendurance athlete in a 24-hour ultracycling race. Proc (Bayl Univ Med Cent) 2012;25:124–8.

[25] Francescato MP, Di Prampero PE. Energy expenditure during an ultra-endurance cycling race. J Sports Med Phys Fitness 2002;42:1–7.

[26] Glace BW, Murphy CA, McHugh MP. Food intake and electrolyte status of ultramarathoners competing in extreme heat. J Am Coll Nutr 2002;21:553–9.

[27] Armstrong LE, Casa DJ, Emmanuel H, Ganio MS, Klau JF, Lee EC, et al. Nutritional, physiological, and perceptual responses during a summer ultraendurance cycling event. J Strength Cond Res 2012;26:307–18.

[28] Black KE, Skidmore PM, Brown RC. Energy intakes of ultraendurance cyclists during competition, an observational study. Int J Sport Nutr Exerc Metab 2012;22:19–23.

[29] Machefer G, Groussard C, Zouhal H, Vincent S, Youssef H, Faure H, et al. Nutritional and plasmatic antioxidant vitamins status of ultra endurance athletes. J Am Coll Nutr 2007;26:311–6.

[30] Stuempfle KJ, Hoffman MD, Weschler LB, Rogers IR, Hew-Butler T. Race diet of finishers and non-finishers in a 100 mile (161 km) mountain footrace. J Am Coll Nutr 2011;30:529–35.

[31] Knechtle B, Knechtle P, Rüst CA, Rosemann T, Lepers R. Finishers and nonfinishers in the 'Swiss Cycling Marathon' to qualify for the 'Race Across America'. J Strength Cond Res 2011;25:3257–63.

[32] Knechtle B, Baumann B, Knechtle P. Effect of ultra-endurance swimming on body composition–marathon swim 2006 from Rapperswil to Zurich. Praxis (Bern 1994) 2007;96:585–9.

[33] Knechtle B, Knechtle P, Kohler G, Rosemann T. Does a 24-hour ultra-swimming lead to dehydration?. J Hum Sport Exerc 2011;6:68–79.

[34] Stewart IB, Stewart KL. Energy balance during two days of continuous stationary cycling. J Int Soc Sports Nutr 2007;4:15.

[35] O'Hara WJ, Allen C, Shephard RJ, Gill JW. LaTulippe—a case study of a one hundred and sixty kilometer runner. Br J Sports Med 1977;11:83–7.

[36] Knechtle B, Senn O, Imoberdorf R, Joleska I, Wirth A, Knechtle P, et al. Maintained total body water content and serum sodium concentrations despite body mass loss in female ultra-runners drinking ad libitum during a 100 km race. Asia Pac J Clin Nutr 2010;19:83–90.

[37] Knechtle B, Senn O, Imoberdorf R, Joleska I, Wirth A, Knechtle P, et al. No fluid overload in male ultra-runners during a 100 km ultra-run. Res Sports Med 2011;19:14–27.

[38] Knechtle B, Vinzent T, Kirby S, Knechtle P, Rosemann T. The recovery phase following a Triple Iron triathlon. J Hum Kinet 2009;21:65–74.

[39] Knechtle B, Knechtle P, Schück R, Andonie JL, Kohler G. Effects of a Deca Iron Triathlon on body composition: a case study. Int J Sports Med 2008;29:343–51.

[40] Knechtle B, Wirth A, Knechtle P, Rosemann T. Increase of total body water with decrease of body mass while running 100 km nonstop–formation of edema? Res Q Exerc Sport 2009; 80:593–603.

[41] Knechtle B, Duff B, Schulze I, Kohler G. A multi-stage ultra-endurance run over 1,200 km leads to a continuous accumulation of total body water. J Sports Sci Med 2008;7:357–64.

[42] Knechtle B, Kohler G. Running 338 km within 5 days has no effect on body mass and body fat but reduces skeletal muscle mass – the Isarrun 2006. J Sports Sci Med 2007;6:401–7.

[43] Knechtle B, Marchand Y. Schwankungen des Körpergewichts und der Hautfaltendicke bei einem Athleten während eines Extremausdauerwettkampfes. Schweizerische Zeitschrift für Sportmedizin und Sporttraumatologie 2003;51:174–8.

[44] Knechtle B, Baumann B, Wirth A, Knechtle P, Rosemann T. Male ironman triathletes lose skeletal muscle mass. Asia Pac J Clin Nutr 2010;19:91–7.

[45] Knechtle B, Knechtle P, Rosemann T, Oliver S. A triple Iron triathlon leads to a decrease in total body mass but not to dehydration. Res Q Exerc Sport 2010;81:319–27.

[46] Knechtle B, Salas Fraire O, Andonie JL, Kohler G. Effect of a multistage ultra-endurance triathlon on body composition: World Challenge Deca Iron Triathlon 2006. Br J Sports Med 2008;42:121–5.

[47] Knechtle B, Wirth A, Knechtle P, Rosemann T, Senn O. Do ultra-runners in a 24-h run really dehydrate?. Ir J Med Sci 2011;180:129–34.

[48] Knechtle B, Wirth A, Knechtle P, Rosemann T. An ultra-cycling race leads to no decrease in skeletal muscle mass. Int J Sports Med 2009;30:163–7.

[49] Knechtle B, Duff B, Amtmann G, Kohler G. An ultratriathlon leads to a decrease of body fat and skeletal muscle mass–the Triple Iron Triathlon Austria 2006. Res Sports Med 2008;16:97–110.

[50] Knechtle B, Schwanke M, Knechtle P, Kohler G. Decrease in body fat during an ultra-endurance triathlon is associated with race intensity. Br J Sports Med 2008;42:609–13.

[51] Rogers G, Goodman C, Rosen C. Water budget during ultra-endurance exercise. Med Sci Sports Exerc 1997;29:1477–81.

[52] Knechtle B, Knechtle P, Kaul R, Kohler G. No change of body mass, fat mass, and skeletal muscle mass in ultraendurance swimmers after 12 hours of swimming. Res Q Exerc Sport 2009;80:62–70.

[53] Weitkunat T, Knechtle B, Knechtle P, Rüst CA, Rosemann T. Body composition and hydration status changes in male and female open-water swimmers during an ultra-endurance event. J Sports Sci 2012; doi:10.1080/02640414.2012.682083.

[54] Kao WF, Shyu CL, Yang XW, Hsu TF, Chen JJ, Kao WC, et al. Athletic performance and serial weight changes during 12- and 24-hour ultra-marathons. Clin J Sport Med 2008;18:155–8.

[55] Knechtle B, Knechtle P, Rosemann T, Senn O. No dehydration in mountain bike ultra-marathoners. Clin J Sport Med 2009; 19:415–20.

[56] Barr SI. Effects of dehydration on exercise performance. Can J Appl Physiol 1999;24:164–72.

[57] Cheuvront SN, Carter III R, Castellani JW, Sawka MN. Hypohydration impairs endurance exercise performance in temperate but not cold air. J Appl Physiol 2005;99:1972–6.

[58] Glace B, Murphy C, McHugh M. Food and fluid intake and disturbances in gastrointestinal and mental function during an ultramarathon. Int J Sport Nutr Exerc Metab 2002;12:414–27.

[59] Lebus DK, Casazza GA, Hoffman MD, Van Loan MD. Can changes in body mass and total body water accurately predict hyponatremia after a 161-km running race?. Clin J Sport Med 2010;20:193–9.

[60] Rothwell SP, Rosengren DJ, Rojek AM, Williams JM, Lukin WG, Greenslade J. Exercise-associated hyponatraemia on the Kokoda Trail. Emerg Med Australas 2011;23:712–6.

[61] Twerenbold R, Knechtle B, Kakebeeke TH, Eser P, Müller G, von Arx P, et al. Effects of different sodium concentrations in replacement fluids during prolonged exercise in women. Br J Sports Med 2003;37:300–3.

[62] Knechtle B, Morales NP, González ER, Gutierrez AA, Sevilla JN, Gómez RA, et al. Effect of a multistage ultraendurance triathlon on aldosterone, vasopressin, extracellular water and urine electrolytes. Scott Med J 2012;57:26–32.

[63] Fellmann N, Ritz P, Ribeyre J, Beaufrère B, Delaître M, Coudert J. Intracellular hyperhydration induced by a 7-day endurance race. Eur J Appl Physiol Occup Physiol 1999;80:353–9.

[64] Mischler I, Boirie Y, Gachon P, Pialoux V, Mounier R, Rousset P, et al. Human albumin synthesis is increased by an ultra-endurance trial. Med Sci Sports Exerc 2003;35:75–81.

[65] Leiper JB, McCormick K, Robertson JD, Whiting PH, Maughan RJ. Fluid homoeostasis during prolonged low-intensity walking on consecutive days. Clin Sci (Lond) 1988;75:63–70.

[66] Neumayr G, Pfister R, Hoertnagl H, Mitterbauer G, Prokop W, Joannidis M. Renal function and plasma volume following ultramarathon cycling. Int J Sports Med 2005;26:2–8.

[67] Bürge J, Knechtle B, Knechtle P, Gnädinger M, Rüst AC, Rosemann T. Maintained serum sodium in male ultra-marathoners—the role of fluid intake, vasopressin, and aldosterone in fluid and electrolyte regulation. Horm Metab Res 2011;43:646–52.

[68] Keul J, Kohler B, von Glutz G, Lüthi U, Berg A, Howald H. Biochemical changes in a 100 km run: carbohydrates, lipids, and hormones in serum. Eur J Appl Physiol Occup Physiol 1981;47:181–9.

[69] Bracher A, Knechtle B, Gnädinger M, Bürge J, Rüst CA, Knechtle P, et al. Fluid intake and changes in limb volumes in male ultra-marathoners: does fluid overload lead to peripheral oedema? Eur J Appl Physiol 2012;112:991–1003.

[70] Cejka C, Knechtle B, Knechtle P, Rüst CA, Rosemann T. An increased fluid intake leads to feet swelling in 100-km ultra-marathoners—an observational field study. J Int Soc Sports Nutr 2012;9:11.

[71] Noakes TD, Sharwood K, Speedy D, Hew T, Reid S, Dugas J, et al. Three independent biological mechanisms cause exercise-associated hyponatremia: evidence from 2,135 weighed competitive athletic performances. Proc Natl Acad Sci USA 2005;102:18550–5.

[72] Noakes TD, Goodwin N, Rayner BL, Branken T, Taylor RK. Water intoxication: a possible complication during endurance exercise. Med Sci Sports Exerc 1985;17:370–5.

[73] Rosner MH. Exercise-associated hyponatremia. Phys Sportsmed 2008;36:55–61.

[74] Rothwell SP, Rosengren DJ. Severe exercise-associated hyponatremia on the Kokoda Trail, Papua New Guinea. Wilderness Environ Med 2008;19:42–4.

[75] Stuempfle KJ, Lehmann DR, Case HS, Bailey S, Hughes SL, McKenzie J, et al. Hyponatremia in a cold weather ultraendurance race. Alaska Med 2002;44:51–5.

[76] Chorley J, Cianca J, Divine J. Risk factors for exercise-associated hyponatremia in non-elite marathon runners. Clin J Sport Med 2007;17:471–7.

[77] Knechtle B, Knechtle P, Rosemann T. No case of exercise-associated hyponatremia in male ultra-endurance mountain bikers in the 'Swiss Bike Masters'. Chin J Physiol 2011;54:379–84.

[78] Knechtle B, Knechtle P, Rosemann T. Do male 100-km ultra-marathoners overdrink?. Int J Sports Physiol Perform 2011; 6:195–207.

[79] Wagner S, Knechtle B, Knechtle P, Rüst CA, Rosemann T. Higher prevalence of exercise-associated hyponatremia in female than in male open-water ultra-endurance swimmers: the 'Marathon-Swim' in Lake Zurich. Eur J Appl Physiol 2012;112:1095–106.

[80] Hoffman MD, Stuempfle KJ, Rogers IR, Weschler LB, Hew-Butler T. Hyponatremia in the 2009 161-km Western States Endurance Run. Int J Sports Physiol Perform 2012;7:6–10.

[81] Cuthill JA, Ellis C, Inglis A. Hazards of ultra-marathon running in the Scottish Highlands: exercise-associated hyponatraemia. Emerg Med J 2009;26:906–7.

[82] Schenk K, Gatterer H, Ferrari M, Ferrari P, Cascio VL, Burtscher M. Bike Transalp 2008: liquid intake and its effect on the body's fluid homeostasis in the course of a multistage, cross-country, MTB marathon race in the central Alps. Clin J Sport Med 2010;20:47–52.

[83] Dugas JP, Noakes TD. Hyponatraemic encephalopathy despite a modest rate of fluid intake during a 109 km cycle race. Br J Sports Med 2005;39:e38.

[84] Rüst CA, Knechtle B, Knechtle P, Rosemann T. No case of exercise-associated hyponatraemia in top male ultra-endurance cyclists: the 'Swiss Cycling Marathon'. Eur J Appl Physiol 2012;112:689–97.

[85] Page AJ, Reid SA, Speedy DB, Mulligan GP, Thompson J. Exercise-associated hyponatremia, renal function, and nonsteroidal anti-inflammatory drug use in an ultraendurance mountain run. Clin J Sport Med 2007;17:43–8.

[86] Knechtle B, Knechtle P, Rosemann T. No exercise-associated hyponatremia found in an observational field study of male ultra-marathoners participating in a 24-hour ultra-run. Phys Sportsmed 2010;38:94–100.

[87] Stuempfle KJ, Lehmann DR, Case HS, Hughes SL, Evans D. Change in serum sodium concentration during a cold weather ultradistance race. Clin J Sport Med 2003;13:171–5.

[88] Speedy DB, Campbell R, Mulligan G, Robinson DJ, Walker C, Gallagher P, et al. Weight changes and serum sodium concentrations after an ultradistance multisport triathlon. Clin J Sport Med 1997;7:100–3.

[89] Wharam PC, Speedy DB, Noakes TD, Thompson JM, Reid SA, Holtzhausen LM. NSAID use increases the risk of developing hyponatremia during an Ironman triathlon. Med Sci Sports Exerc 2006;38:618–22.

[90] Speedy DB, Noakes TD, Rogers IR, Thompson JM, Campbell RG, Kuttner JA, et al. Hyponatremia in ultradistance triathletes. Med Sci Sports Exerc 1999;31:809–15.

[91] Speedy DB, Faris JG, Hamlin M, Gallagher PG, Campbell RG. Hyponatremia and weight changes in an ultradistance triathlon. Clin J Sport Med 1997;7:180–4.

[92] Speedy DB, Noakes TD, Kimber NE, Rogers IR, Thompson JM, Boswell DR, et al. Fluid balance during and after an iron-man triathlon. Clin J Sport Med 2001;11:44–50.

[93] Rüst CA, Knechtle B, Knechtle P, Rosemann T. Higher prevalence of exercise-associated hyponatremia in Triple Iron ultra-triathletes than reported for Ironman triathletes. Chin J Physiol 2012; in press.

[94] Knechtle B, Gnädinger M, Knechtle P, Imoberdorf R, Kohler G, Ballmer P, et al. Prevalence of exercise-associated hyponatremia in male ultraendurance athletes. Clin J Sport Med 2011;21:226–32.

[95] Hsieh M, Roth R, Davis DL, Larrabee H, Callaway CW. Hyponatremia in runners requiring on-site medical treatment at a single marathon. Med Sci Sports Exerc 2002;34:185–9.

[96] Kipps C, Sharma S, Pedoe DT. The incidence of exercise-associated hyponatraemia in the London marathon. Br J Sports Med 2011;45:14–9.

[97] Mettler S, Rusch C, Frey WO, Bestmann L, Wenk C, Colombani PC. Hyponatremia among runners in the Zurich Marathon. Clin J Sport Med 2008;18:344–9.

[98] Knechtle B, Knechtle P, Rosemann T. Low prevalence of exercise-associated hyponatremia in male 100 km ultra-marathon runners in Switzerland. Eur J Appl Physiol 2011;111:1007–16.

[99] Hew-Butler T, Dugas JP, Noakes TD, Verbalis JG. Changes in plasma arginine vasopressin concentrations in cyclists participating in a 109-km cycle race. Br J Sports Med 2010;44:594–7.

[100] Eden BD, Abernethy PJ. Nutritional intake during an ultraendurance running race. Int J Sport Nutr 1994;4:166–74.

[101] Bescós R, Rodríguez FA, Iglesias X, Knechtle B, Benítez A, Marina M, et al. Nutritional behavior of cyclists during a 24-hour team relay race: a field study report. J Int Soc Sports Nutr 2012;9:3.

[102] García-Rovés PM, Terrados N, Fernández SF, Patterson AM. Macronutrients intake of top level cyclists during continuous competition—change in the feeding pattern. Int J Sports Med 1998;19:61–7.

[103] Fallon KE, Broad E, Thompson MW, Reull PA. Nutritional and fluid intake in a 100-km ultramarathon. Int J Sport Nutr 1998;8:24–35.

[104] Knechtle B, Müller G. Ernährung bei einem Extremausdauerwettkampf. Deutsche Zeitschrift für Sportmedizin 2002;53:54–7.

[105] Knechtle B, Pitre J, Chandler C. Food habits and use of supplements in extreme endurance cyclists – the Race Across

America (RAAM) 2006. Schweizerische Zeitschrift für Sportmedizin und Sporttraumatologie 2007;55:102–6.
[106] Knechtle B, Knechtle P, Kaul R. Nutritional practices of extreme endurance swimmers – the Marathon swim in the lake of Zurich 2006. Pak J Nutr 2007;6:188–93.
[107] Knechtle B, Schulze I. Nutritional behaviours in ultra-endurance runners—Deutschlandlauf 2006. Praxis (Bern 1994) 2008;97:243–51.
[108] Enqvist JK, Mattsson CM, Johansson PH, Brink-Elfegoun T, Bakkman L, Ekblom BT. Energy turnover during 24 hours and 6 days of adventure racing. J Sports Sci 2010;28:947–55.
[109] Knechtle B, Knechtle P, Schulze I, Kohler G. Vitamins, minerals and race performance in ultra-endurance runners—Deutschlandlauf 2006. Asia Pac J Clin Nutr 2008;17:194–8.
[110] Frohnauer A, Schwanke M, Kulow W, Kohler G, Knechtle B. Vitamins, minerals and race performance in ultra-endurance triathletes – Triple Iron Triathlon Lensahn 2006. Pak J Nutr 2008;7:283–6.
[111] Haymes EM. Vitamin and mineral supplementation to athletes. Int J Sport Nutr 1991;1:146–69.
[112] Knez WL, Peake JM. The prevalence of vitamin supplementation in ultraendurance triathletes. Int J Sport Nutr Exerc Metab 2010;20:507–14.
[113] Robins AL, Davies DM, Jones GE. The effect of nutritional manipulation on ultra-endurance performance: a case study. Res Sports Med 2005;13:199–215.
[114] Knechtle B, Knechtle P, Mrazek C, Senn O, Rosemann T, Imoberdorf R, et al. No effect of short-term amino acid supplementation on variables related to skeletal muscle damage in 100 km ultra-runners – a randomized controlled trial. J Int Soc Sports Nutr 2011; 7; 8:6.
[115] Nolte H, Noakes TD, Van Vuuren B. Ad libitum fluid replacement in military personnel during a 4-h route march. Med Sci Sports Exerc 2010;42:1675–80.
[116] Nolte HW, Noakes TD, van Vuuren B. Protection of total body water content and absence of hyperthermia despite 2% body mass loss ('voluntary dehydration') in soldiers drinking ad libitum during prolonged exercise in cool environmental conditions. Br J Sports Med 2011;45: 1106–1012.
[117] Nolte HW, Noakes TD, Van Vuuren B. Trained humans can exercise safely in extreme dry heat when drinking water ad libitum. J Sports Sci 2011;29:1233–41.
[118] Tam N, Nolte HW, Noakes TD. Changes in total body water content during running races of 21.1 km and 56 km in athletes drinking ad libitum. Clin J Sport Med 2011;21:218–25.
[119] Hew-Butler TD, Sharwood K, Collins M, Speedy D, Noakes T. Sodium supplementation is not required to maintain serum sodium concentrations during an Ironman triathlon. Br J Sports Med 2006;40:255–9.
[120] Speedy DB, Thompson JM, Rodgers I, Collins M, Sharwood K, Noakes TD. Oral salt supplementation during ultradistance exercise. Clin J Sport Med 2002;12:279–84.

CHAPTER

17

Exercise and Nutritional Benefits for Individuals with a Spinal Cord Injury or Amputation

Justin W. L. Keogh

Bond University, Robina, Queensland, Australia

PARALYMPIC SPORT AND CLASSIFICATION SYSTEMS

Individuals with a disability (IWD) have often been viewed as 'patients' by medical practitioners and members of the public. However, many more IWD are now engaging in organized sports and other recreational activities, with a number of these individuals striving to become elite athletes and compete in the Paralympic Games.

A wide variety of athletes with a disability (AWD) currently compete in the Paralympic Games. At the 2008 Beijing Paralympic Games, AWD were grouped into five general classifications, these being spinal injury, amputee, visually impaired, cerebral palsy, and les autres, which literally means "and others" [1]. From 2010, athletes with an intellectual disability also became eligible to compete in Paralympic events [2]. In an attempt to ensure fairness within all Paralympic sports, AWD are classified according to: (i) the anatomical extent of their disability, e.g., degree of visual or intellectual impairment or location of amputation; or (ii) the functional consequence of their disability, e.g., physiological effects resulting from a spinal cord lesion.

It is acknowledged that each class of athletes who can participate in Paralympic competitions, as well as those IWD who are sedentary or recreationally active, have their own unique challenges in performing activities of daily living and in engaging in exercise, sporting and recreational activities. However, this chapter will focus on those with spinal cord injuries (SCI) and amputations because: (i) these individuals make up a large proportion of those who compete in Paralympic competitions; (ii) they may have significantly different body composition and reduced energy expenditure than able bodied individuals; and (iii) considerable research (especially for those with SCI) has been conducted in this area.

The general purpose of this chapter is to provide a framework for the benefits of exercise and nutrition in improving physical function and sporting performance for IWD and specifically for those with a SCI or amputation. To do this, some background information on body composition, energy expenditure and nutritional profile of these populations will also be given to better place in context the benefits of exercise and nutrition.

REVIEW METHODOLOGY

In order to systematically review the literature for the effects of exercise and nutrition as well as the nutritional profiles and knowledge of IWD, the following keywords and their derivatives were used: strength or resistance training; nutrition or diet or supplement; spinal cord injured or amputee or Paralympic. Database searches were performed in PubMed, SportDiscus, CINAHL, Health Source: Nursing/Academic, along with Google Scholar via the "cited by" function. The reference list of all articles found in the primary search was also searched in order to identify other relevant studies. To be eligible to be included in the review, all the studies had to be in English (or have results reported in English) and to be peer-reviewed.

ENERGY EXPENDITURE AND BODY COMPOSITION

Spinal cord injury and amputation can result in reductions in body, fat-free and muscle mass and also

Nutrition and Enhanced Sports Performance.
DOI: http://dx.doi.org/10.1016/B978-0-12-396454-0.00017-5

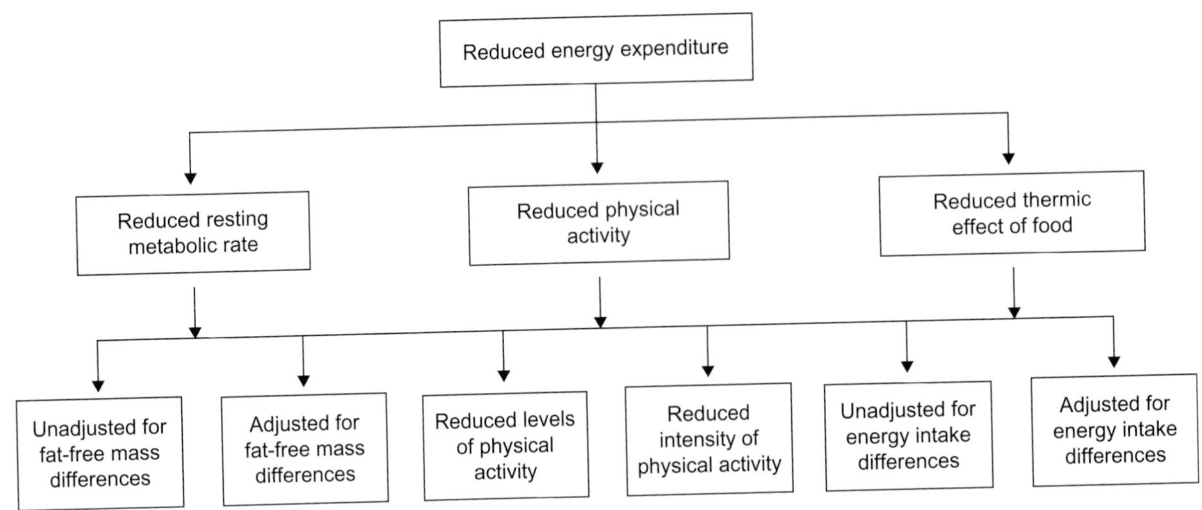

FIGURE 17.1 Potential factors contributing to the reduced energy expenditure of individuals with a disability compared with able-bodied individuals. *Adapted from Bucholz and Pencharz (2004) [9].*

indirectly increase body fat percentage. For individuals with a SCI, these effects are typically more pronounced the higher the spinal cord lesion. For those with an amputation, these effects are more pronounced on the side of the body with the amputation (residual) rather than the intact side [3,4] and for proximal vs distal amputations [5].

There are, however, many issues with assessing the body composition of Paralympic athletes and comparing these values to those of able-bodied individuals [6]. As body composition cannot be directly determined in a non-invasive manner, a variety of indirect (Level 2) and doubly indirect (Level 3) methods are used in research and practice [6]. The Level 2 methods, which include hydrodensitometry (underwater weighing), air displacement plethysmography and dual X-ray absorptiometry (DEXA), are the most accurate methods for able-bodied individuals. The doubly indirect (Level 3) methods, which include surface anthropometry and bioelectrical impedance (BIA), are much less expensive and more practical for use in the field and the clinic but are somewhat less accurate than the Level 2 methods.

Beyond the issues that all of these indirect and doubly indirect methods have in determining body composition in able-bodied individuals, there appear many more issues when applied to those with a disability. This reflects the fact that the assumptions on which many of these methods are based may be either invalid or of unknown validity in this population. For example, Ven de Vliet et al. [6] stated that differences between able-bodied individuals and IWD in bone density may affect the validity of two of the three Level 2 techniques (underwater weighing and air displacement plethysmography). The issues for the Level 3 methods may be even greater, as surface anthropometry and BIA may require the development of specific prediction equations for a variety of disability groups, including athletes [6]. Empirical support for this view is provided by Mojtahedi et al. [7] who assessed the body composition of 16 SCI athletes using DEXA, sum of 3 and 7 skinfolds, and BIA using generalized SCI and athletic-specific equations. Results indicated that although the Level 3 methods were moderately to strongly correlated to the DEXA values, errors of body fat percentage were relatively large, being 5–15 percentage points [7]. Ven de Vliet et al. [6] therefore currently recommends that the sum of skinfolds from body sites not affected by the disability might be used as a monitoring tool to detect changes in body composition, even if such values may not be easily converted to a body fat percentage.

The negative changes in body composition often seen in those with a SCI or amputation reflect an imbalance between energy expenditure and energy intake. Any reductions in energy expenditure may indirectly result in a cascade of effects that may further reduce physical fitness and function, physical independence and increase the risk of metabolic syndrome [8]. This sequence of events and their potential interactions is highlighted in Figure 17.1, which is adapted from Buchholz and Pencharz [9].

The greatest contributor to energy expenditure is resting metabolic rate (RMR), accounting for up to 65% [9]. The reductions in RMR for IWD was initially thought to reflect the loss of fat-free mass alone, but reduced sympathetic nervous system drive resulting from the lesion may also reduce RMR, with such an effect even observed after adjusting for differences in body composition between SCI and able-bodied individuals [9]. Individuals with a disability have also been shown to be less physically active than their able-bodied peers [10]. As physical activity may account for 25–30% of total energy expenditure, any significant reduction in the level or intensity of physical activity could further reduce the energy expenditure of IWD. The thermic effect of food (TEF), which reflects the energy expenditure of eating, swallowing and digestion, is the smallest contributor to energy expenditure, accounting for only 5–10% [9]. A reduced TEF could also contribute to the reduced energy expenditure of IWD [11], although only one of the three studies conducted to date has demonstrated a significantly reduced TEF in IWD when normalized to energy intake [9].

Although the review of Buchholz and Pencharz [9] adds much to our understanding, the exact contribution of differences in RMR, physical activity levels and TEF to the reduced energy expenditure and functional performance of individuals with SCI (and perhaps amputation) is not completely understood. Nevertheless, it appears important to develop a greater evidence basis for strategies that can safely and effectively increase fat-free, muscle and bone mass and reduce fat mass and improve physical fitness and function in these individuals. Two such interventions are exercise and dietary modifications.

EXERCISE ADAPTATIONS

Many studies over the last few decades have examined the physiological and functional adaptations that people with disabilities, especially spinal cord injuries, can obtain from physical exercise, with a number of recent reviews published in this area [12,13]. The vast majority of the studies in this area have had a rehabilitative focus and/or investigated the benefits of aerobic exercise, generally arm-cranking on physiological function. In a systematic review, Valent et al. [13] reported that of 14 studies of acceptable quality, the effects of training were quite variable, with ~18 ± 11% increases in VO_2 peak and 26 ± 16% increases in peak aerobic power output during maximal exercise. The variability in these responses would appear to reflect a variety of factors relating to the exercise prescription as well as the characteristics of the participants.

More recently, a greater number of studies have examined the benefits of higher intensity exercise for IWD. These studies have examined combined resistance (strength), endurance and/or flexibility training [14–16], resistance training [17–20] and combined technical and strength and conditioning training for AWD [21]. Two studies have also examined the benefits of wheelchair rugby training, which could be considered a form of resistance training for this population [22,23]. Resistance training would appear ideal for this population as the mechanical stimuli provided by this form of exercise constitute a potent stimulus for improving muscle mass, strength, power and improving functional performance [24,25]. A summary of the resistance training studies involving a gym-based training program is provided in Table 17.1.

Consistent with the review of Hicks et al. [12] for those with SCI, the results of the studies summarized in Table 17.1 reveal that resistance training offers many benefits for inactive and athletic IWD. Specifically, 8–16 weeks of resistance or combined resistance and aerobic training performed 2–3 times per week resulted in significant increases in strength (6–167%), mean power during 30 s Wingate tests (8–67%), peak power during 30 s Wingate tests (6–77%), maximal rate of force development (MRFD) (70%), and even aerobic power (10–30%). Although few studies measured changes in functional performance, these increases in strength, power and MRFD could improve mobility in transfer tasks (e.g., getting in and out of a wheelchair) and pushing up hills in a wheelchair. The potential for these proposed improvements in muscular function to improve athletic wheelchair performance was partially supported by the non-significant ($p = 0.058$) 6.2% increase in 10 m wheelchair sprinting speed after 8 weeks of bench press training [17]. Similarly, in a Paralympic swimmer with a SCI, 15 months of swimming and strength and conditioning training resulted in an increase in her mean and peak 30 s Wingate test power output by 67% and 77%, respectively [21]. Although it is impossible to determine the contribution of the strength vs swimming training, such improvements coincided with her setting Polish records in the 50 m, 100 m and 200 m freestyle events and winning four gold medals at the European Swimming Championships [21]. However, too few of these exercise studies have assessed changes in functional performance for a clear understanding of the magnitude of performance response that Paralympic athletes may gain, and whether such responses differ in magnitude to those of able-bodied athletes [12].

The significant increases in aerobic fitness in the three studies was also of substantial interest, with the magnitude of these changes lying within the ranges reported in the review of the literature for similar length aerobic training studies [13]. Additionally, Jacobs et al. [18],who used a matched pair design to compare the benefits of resistance and aerobic training for non-athletic SCI individuals, found significant increases in aerobic power for both groups, with the change in aerobic power for the resistance training (15%) comparable to that in the aerobic training group (12%). As aerobic training does not produce significant increases in muscular strength and power performance [18], it would appear that resistance training is the most effective exercise mode for this population, in that it

TABLE 17.1 Training-Related Changes in Body Composition and Physical Function of Individuals with a Disability

		Training Program Descriptors				Training-Related Changes			
Study	Subjects	Duration	Frequency	Training	Loading	Body Composition	Strength	Anaerobic Power	Aerobic Power
Inactive/Non-athletes									
Jacobs et al. [14]	10 SCI ♂ non-athletes T5–T12 39 ± 6 yr	12 wk	3/wk	RT + AT circuit	1 × 10 @ 50–60% 1RM		+12–30%*		+30% VO_2 peak*
Jacobs [18]	6 SCI ♂ & 3 ♀ non-athletes T6–T10 34 ± 8 yr	12 wk	3/wk	RT 6 exercises	3 × 6–10 @ 60–70% 1RM		+34–55%*	+8% MP-30 s* + 16% PP-30 s*	+15% VO_2 peak*
Klingenstierna et al. [19]	8 BKA ♂ non-athletes 61 yr	8–12 wk	2–3/wk	Isokinetic RT exercises	2 × 10 at each of 2 speeds	+4–18% Quad mean fiber area (amputated leg)* +19–40% Quad mean fiber area (intact leg)*	+40–79% (amputated leg)* +6–27% (intact leg)*		
Nash et al. [15]	7 SCI ♂ non-athletes T5–T12 39–58 yr	16 wk	3/wk	RT + AT circuit	1 × 10 at 50–60% 1RM		+39–60%*	+9% MP-30 s* + 6% PP-30 s*	+10% VO_2 peak*
Nolan [20]	8 amputees (4 TT, 3 TF, 1 BL) 6 ♂ & 2 ♀ non-athletes 41 ± 8 yr	10 wk	2/wk	RT + AT + balance	2 × 10–15 at each of 2 speeds		+17–42% intact limb*,[a] + 22–36% amputated limb*,[a]		
Rodgers et al. [16]	15 SCI T3–L5 & 3 others 44 ± 11 yr	6 wk	3/wk	RT + AT + stretching	5 × 75% 1RM 30 min @ 60% HRR		+45–167%*		+7% VO_2 peak +11% TTE
Athletes									
Tabecki et al. [21]	4 SCI C4–C6	15 month	3/wk	RT	Not stated			+67% MP-30 s[a] + 77% 1PP-30 s[a]	
Turbanski & Schmidtbleicher [17]	8 SCI ♂ athletes (2 tetra- & 6 paraplegic) 33 ± 11 yrs	8 wk	2/wk	RT Bench press	5 × 10–12 @ ~80% 1RM		+39%*	+70% MRFD*	

[a]*Percent change estimated from graphs and not including data for the BL amputee.*
**Significant increase.*
AT, aerobic training; BKA, below knee amputee; BL, bilateral; HRR, heart rate reserve; MP-30 s, mean power in 30 s Wingate test; MRFD, maximum rate of force development; PP-30 s, peak power in 30 s Wingate test; RT, resistance training; SCI, spinal cord injured; TF, transfemoral; TT, transtibial; TTE, time to exhaustion; VO_2 peak, peak volume of oxygen consumption during exercise test; wk, week(s); yr, year(s).

can produce both anaerobic and aerobic benefits. Such a finding would appear consistent with recent reviews for other populations with reduced muscle mass, and reduced strength and functional performance such as cancer patients [26,27] and older adults [28,29]. Resistance training is therefore now being acknowledged as a key form of exercise in these populations.

Consistent with the review of Hicks et al. [12], very few of these high-intensity exercise studies examined the potential benefits of such training on body composition. This was surprising for two reasons. The first was the negative changes in body composition observed in many IWD and their increased risk of chronic conditions such as osteoporosis and metabolic syndrome [3,8,30]. The second reason was that these forms of exercise have been shown to significantly improve many aspects of body composition, particularly increasing fat-free or muscle mass in young adults and at least maintaining this in able-bodied older individuals or patients with cancer. The lack of such studies involving IWD may, however, reflect the previously described difficulties in accurately measuring body composition in these populations, since many of the assumptions underlying common body composition assessment techniques may not apply to these individuals.

Even though IWD now have more opportunities to participate in exercise, sport and recreation activities [31], most of the published exercise studies, including the resistance training studies summarized in Table 17.1, still have involved inactive, non-athletic individuals, many of whom were middle aged or older. Extrapolating the results of this literature to Paralympic athletes might be challenging, as it is well known in the able-bodied literature that only very small increases (if any) in strength, power and body composition occur over the course of a year for elite athletes [32]. The results of Dallmeijer et al. [22] support this contention, since they found that, whereas novice players significantly increased muscular strength by 26% and tended to increase aerobic power (10%) from one session of wheelchair rugby per week for 12 weeks, experienced AWD had no significant change in strength (6%) or aerobic power (−8%).

NUTRITIONAL PRACTICES

Although nutrition and supplementation play important roles in elite able-bodied and Paralympic sports performance, relatively little is known regarding the nutrition practices and knowledge of non-athletic IWD or AWD compared with their able-bodied peers. Adequate nutrition intake is very important for IWD, especially those with a SCI injury, as they may be at higher risk of developing chronic metabolic syndrome and osteoporosis because of the reduced energy expenditure and muscle mass [3,8,30]. For AWD, nutrition and supplementation (especially electrolyte and fluid replacement) may be particularly important as they have many challenges in regulating their core temperature during exercising in hot humid conditions.

A number of studies have assessed the nutritional intakes of inactive individuals with SCI [33–37]. These studies, as well as those involving AWD [38–45], have utilized a variety of approaches to obtain nutritional data. The most common method appears to be a 3–7 day food recall, with many of these studies allowing the participants to use a simple method of approximating the weight of foods eaten via the use of common household measures which the researchers then converted to an approximate mass in grams. Results of these studies involving non-athletic IWD suggest that mean daily energy intakes may be greater in males than females with SCI [33,35] and that no significant differences occur in those with tetraplegia compared with paraplegia [33,35]. Proportional macronutrient intake indicates that non-athletic individuals with SCI consume sufficient carbohydrates (although this often consisted of too many simple sugars) and protein, but too much fat and too little fiber. Inadequate intakes of many vitamins and minerals were also reported [33,35].

A number of studies have also examined the nutritional intake of AWD. A summary of these studies is provided in Table 17.2. Most of these studies involved wheelchair athletes with a SCI or polio, although one study examined soccer players with an amputation [39]. These soccer players had substantially greater energy intakes (~3800 calories per day) than did the subjects of all the other studies involving wheelchair athletes, who consumed ~1500–2200 calories per day [39]. Such large differences in daily energy intake appear reasonable as the soccer players with amputations had greater levels of muscle mass and walked and ran during everyday activities, training and competition compared with wheelchair athletes in the other studies.

Inspection of the proportional macronutrient intakes revealed that the carbohydrate (48–54%), protein (17–20%) and fat (25–32%) intakes for AWD typically fell within the recommended ranges and had very little between-study group differences. However, when macronutrient intake was expressed in absolute amounts, some anomalies were observed. For example, soccer players with an amputation were found to have an average intake of 3.1 g protein per kg body mass per day, which is considerably higher than recommended [39]. Innocencio da Silva Gomes et al. [39] suggested that a reduction in the absolute protein intake of the soccer players would be beneficial, as it would allow greater carbohydrate intake (to allow

TABLE 17.2 Nutritional Intake of Athletes with a Disability

Study	Subjects	Data Collection Method	Daily Energy Intake	Daily CHO Intake	Daily Fat Intake	Daily Protein Intake	Deficiencies in Fiber, Vitamins & Minerals
Goosey-Tolfrey & Crosland [38]	14 ♀ mixed AWD 25 ± 8 yr	7 days weighed food intake	1520 cal 24 cal/kg	218 g 3.4 g/kg 54%	49 g 1 g/kg 29%	64 g 1 g/kg 17%	Fiber and iron
Goosey-Tolfrey & Crosland [38]	9 ♂ mixed AWD 28 ± 7 yr	7 days weighed food intake	2060 cal 32 cal/kg	283 g 4.3 g/kg 53%	60 g 1 g/kg 27%	90 g 1.4 g/kg 19%	Fiber
Innocencio da Silva Gomes et al. [39]	15 ♂ amputee soccer players 32 ± 6 yr	6 day food recalls at training camp	3830 cal 58 cal/kg	482 g 7.3 g/kg 50%	Not stated Not stated 29%	203 g 3.1 g/kg 21%	Vitamin E and calcium (forwards only)
Krempien & Barr [40]	24 ♂ & 8 ♀ SCI (12 para & 20 tetra) 31 ± 6 yr	3 day food recalls at home and training camps	2115 cal 33 cal/kg	285 g 4.4 g/kg 54%	68 g 1 g/kg 29%	93 g 1.4 g/kg 18%	Fiber, vitamin D, folate, calcium, magnesium and zinc
Krempien & Barr [41]	24 ♂ & 8 ♀ SCI (12 para & 20 tetra) 31 ± 6 yr	3 day food recalls at home and training camps	2115 cal 33 cal/kg	285 g 4.4 g/kg 54%	68 g 1 g/kg 29%	93 g 1.4 g/kg 18%	
Potvin et al. [42]	10 ♂ SCI 31 ± 8 yr	3 days weighed food intake	2139 cal 35 cal/kg	256 g 48%	76 g 32%	105 g 20%	Vitamin E and zinc
Rastmanesh et al. [43]	13 amputees and 59 SCI gender not stated 30 ± 8 yr	3 day food recalls at home	1690 cal				Fiber, vitamin C, D, E and calcium
Ribeiro et al. [44]	28 ♂ wheelchair rugby players with SCI 18–40 yr	3 day food recall at home	25 cal/kg	50%	31%	20%	Calcium
Ribeiro et al. [44]	32 ♂ wheelchair rugby players with polio 18–40 yr	3 day food recall at home	26 cal/kg	50%	29%	20%	Calcium
Wang et al. [45]	15 ♂ & 1 ♀ American wheelchair marathon athletes 36 ± 2 yr	Typical training day food recall	30 cal/kg	54%	26%	20%	Fiber
Wang et al. [45]	12 ♂ & 1 ♀ Japanese wheelchair marathon athletes 34 ± 3 yr	Typical training day food recall	28 cal/kg	52%	25%	22%	

cal, calories; SCI, spinal cord injured; yr, year(s).

better training and recovery at the training camp) and reduced fat intake (which was borderline high).

Although not all of these studies conducted a comprehensive assessment of fiber and micronutrient intakes, most reported only a few group mean intakes that were less than recommendations, with such deficiencies displaying no real patterns of difference between the types of AWD. The most commonly reported deficiencies for these athletes were seen for fiber, vitamin E and calcium. Further, some studies also reported the prevalence of athletes who were deficient in fiber and selected micronutrients [38,40,43].

To date, only one study has appeared to assess the nutritional strategies of AWD during competition [46]. This study used a 31-item questionnaire to obtain data from a group of 14 female and 12 male swimmers at the 2004 Paralympics Games. Pelly et al. [46] reported that 81% of all athletes stated receiving dietary advice prior to the competition, that most athletes (57–76%) made some adjustments to the composition of their diets—e.g., increased carbohydrate and fluid intake and reduced fat intake within the final 72 hours before competition—and that 88.5% had a high carbohydrate snack within the hour after competing. Although such dietary habits appeared quite consistent with nutritional guidelines, 73% of the swimmers felt that they lacked energy on competition day, with 69% feeling flat [46].

NUTRITIONAL KNOWLEDGE AND EDUCATION PROGRAMS

It has been hypothesized that the inadequacies of the nutritional intakes of AWD found in the literature may be a result of their relative lack of nutritional knowledge. As a result, Rastmanesh et al. [43] examined the nutritional knowledge of 72 AWD and assessed the effects of a nutrition education program on their nutritional knowledge and habits. Prior to the education program, results indicated a poor understanding of nutrition, with mean scores of ~40% for the AWD. Sub-scales relating to nutrition for AWD (13%), protein (30–31%), vitamins and minerals (33%) and calcium (35%) were particularly poorly answered across the education and control groups.

After completing the baseline knowledge questionnaire, the 42 AWD in the education group and their coaches were given a nutrition education booklet and attended four nutrition courses led by a dietician, with each course lasting 3 hours and separated by a period of a week. Over this time, the 30 control subjects were asked to maintain their regular nutritional habits. After completing the education program, the 42 AWD significantly improved their mean nutrition knowledge score from 42% to 81% correct, with the greatest improvements seen in sub-scales of nutrition for AWD (+61%), calcium (+57%) and carbohydrate (+54%). Over this period, no significant change in the nutrition knowledge occurred in the control group. More importantly, some improvements were observed in the dietary habits of the education group, with the 42 athletes significantly increasing their daily calcium intake from 715 mg to 840 mg post program, translating to a reduction from 70% to 47% who did not meet the recommended daily allowance of 800 mg. Similar significant reductions in the prevalence of those with inadequate intakes of vitamin C, vitamin D and fiber were also reported, although no numerical data were provided. There was also a non-significant trend for a reduction in adiposity for the education group as assessed via BMI, with mean group scores falling from 31.5 to 29.9 over the 4-week program.

SUPPLEMENT USAGE

Like many able-bodied athletes, AWD may use a variety of nutritional supplements, even if such use may not achieve the desired result or even be counterproductive. However, very little research has examined the supplementation usage by AWD. A small-scale study of 32 elite AWD found that 44% used a vitamin supplement when training at home, although this decreased to 34% when in a training camp [40]. Krempien and Barr [40] observed that vitamin supplementation increased the mean male athletes' intake of most micronutrients in which a deficiency was initially reported, although it had no effect on female athletes.

Tsitsimpikou et al. [47] examined the supplementation and medication practices of AWD competing in the 2004 Paralympic Games via self-reported declarations from the athletes (n = 680) as well as their application forms for the use of therapeutic good exemptions (n = 493). Primary results indicated that 64% of the AWD reporting using medications or supplements within the final 3 days before competition, with over 80% reporting the use of fewer than four such items. The primary nutritional supplements used were vitamins (44%), minerals/electrolytes (16%), amino acids/proteins (11%) and creatine (9%). Unsurprisingly, athletics and powerlifting accounted for 62% of total amino acid/protein consumption.

EFFECTS OF NUTRITION ON BODY COMPOSITION AND PERFORMANCE

A small number of studies have assessed the effect of nutritional interventions on aspects of performance in IWD. These studies have assessed the effect of

TABLE 17.3 Effect of Acute Glucose Supplementation on Physical Function in Individuals with a Disability

Study	Subjects	Design	Dosage	Performance Changes
Nash et al. [51]	2 ♂ & 1 ♀ incomplete SCI T4–C6 inactive 34–43 yr	Double-blind, placebo-controlled, cross-over design with both conditions consuming the same volume of fluid 20 minutes prior to exercise	48 g whey + 1 g/kg bodyweight maltodextrin vs 48 g soy protein drink + artificial sweetener	+18% time to fatigue and +38% distance walked
Spendiff & Campbell [48]	5 ♂ spina bifida & 3 ♂ SCI T5–T12 athletes 32 ± 4 yr	Double-blind, placebo-controlled, cross-over design with both conditions consuming 600 mL fluid 20 minutes prior to exercise	48 g dextrose monohydrate vs flavored water	+6% distance in 20 min arm cranking time trial*
Spendiff & Campbell [49]	5 ♂ spina bifida & 3 ♂ SCI T5–T12 athletes 31 ± 5 yr	Double-blind, placebo-controlled, cross-over design with both conditions consuming 600 mL fluid 20 minutes prior to exercise	72 g dextrose monohydrate vs 24 g dextrose monohydrate	+2% distance in 20 min wheelchair time trial
Temesi et al. [50]	5 ♂ C6–T3 SCI & 1 ♀ T7 SCI athletic/inactive mixed group 26–39 yr	Randomized, counter-balanced order with both groups consuming 500 mL fluid in 125-mL doses from 15 minutes prior to exercise to 30 minutes into exercise	0.5 g/kg body mass maltodextrin vs flavored water	+5% total work in 20 min time trial

**significant increase for intervention group.*
SCI, spinal cord injured; yr = year(s).

carbohydrate [48–50], carbohydrate and protein [51] and creatine [52,53] supplementation.

Carbohydrate supplementation is a relatively common practice used by endurance athletes in the lead-up to competition. This supplementation/loading may be performed in the last few days prior to competition, immediately prior (<30 minutes) and during exercise. The general premise of this approach is to cause a temporary increase in glycogen stores and/or blood glucose concentration so that the athlete can maintain a higher average power output/velocity during the event, thus improving their time to complete the distance. To date, four studies have examined the effect of carbohydrate supplementation during or just prior to performing aerobic exercise in AWD, with the results of this literature somewhat mixed (see Table 17.3).

These studies typically utilized a cross-over design whereby each IWD completed the performance protocol with and without the carbohydrate-supplemented fluid in a placebo-controlled, randomized order. Spendiff and Campbell [48] and Temesi et al. [50] utilized a protocol whereby they compared the benefits of carbohydrate supplementation to a zero carbohydrate fluid. In these studies, improvements in performance of 5–6% were observed across a 20-minute time trial, although this only reached statistical significance in one study [48]. Spendiff and Campbell [49] also compared the effect of a low (24 g) and high (72 g) carbohydrate supplement consumed in 600 mL of fluid 30 minutes prior to the 20-minute time trial, finding a 2% non-significantly greater improvement in performance for the high-carbohydrate supplement. The benefit of acute carbohydrate supplementation was also demonstrated by Nash et al. [51],who compared the effect of a rapidly digested protein and carbohydrate vs an artificially sweetened soy protein drink on the walking ability of three individuals with an incomplete SCI. Substantially greater time to exhaustion (+18%) and walking distance (+38%) were found when the participants consumed whey and carbohydrate compared with those who consumed the soy protein blend drink.

Creatine loading is used mainly by power athletes to improve repeated high- to supramaximal-intensity or maximal-strength performance. By supplementing creatine, the levels of creatine phosphate in the muscle increase, allowing an increased alactic anaerobic energy production and an improved ability to maintain creatine phosphate levels for repeated high-intensity efforts.

To date, two studies have examined the potential benefits of creatine supplementation for IWD (see Table 17.4). Both these studies utilized a cross-over design in which each AWD underwent performance testing with and without the creatine supplement in a placebo-controlled, randomized order. While the cross-over design and the creatine supplementation dosages in these studies were very similar, there was considerable between-study variation in the IWD groups and the performance measures assessed. Jacobs et al. [53] reported a significant 19% increase in aerobic power and a non-significant 7% increase in time to exhaustion for 16 SCI athletes during the VO_2 peak test. In contrast, Perret et al. [52] found no significant

TABLE 17.4 Effect of Chronic Creatine Supplementation on Physical Function in Individuals with a Disability

Study	Subjects	Design	Dosage	Aerobic Power Changes	Functional Performance Changes
Jacobs et al. [53]	16 ♂ SCI C5–C7 non-athletes 35 ± 9 yr	Double-blind, placebo-controlled, cross-over design, 7-day intervention & control phases with 21-day washout	20 g/day	+19% *	+7% TTE in VO_2 peak test
Perret et al. [52]	4 ♂ & 2 ♀ mixed AWD 33 ± 9 yr	Double-blind, placebo-controlled, cross-over design, 6-day intervention & control phases with 28-day washout	20 g/day		0% change in 800 m wheelchair race time

significant increase for intervention group.
TTE, time to exhaustion; VO_2 peak, peak volume of oxygen consumption during exercise test; yr, year(s).

change in the time to complete a 800 m wheelchair race for six mixed AWD.

Somewhat similar to the literature for carbohydrate supplementation, the mixed results and lack of literature for the effect of short-term creatine loading on athletic performance make it hard to offer any definitive conclusions on its effectiveness in this regard for AWD. Nevertheless, as the stronger of the two studies showed a significant improvement of 19% in aerobic power during a VO_2 peak test, more research appears warranted in this area. Interestingly, none of these studies have examined changes in body mass, composition, strength or power. This is surprising for two reasons. The first is that Tsitsimpikou et al. [47] found that creatine supplementation was more prevalent in power than endurance AWD. The second reason is that the physical qualities sought by power athletes, such as strength, power, speed and repeated sprint ability, are more commonly assessed variables in studies involving creatine supplementation in able-bodied athletes [54] and other chronic condition groups [55]. However, significant increases in strength have been reported to occur from creatine supplementation in humans with neuromuscular conditions in one study [56].

CONCLUSIONS AND AREAS FOR FUTURE RESEARCH

An examination of the literature has revealed a number of ways in which individuals with a SCI or amputation may improve body composition, physical fitness and function and athletic performance. While aerobic exercise has been traditionally recommended for IWD and older adults because of its ability to increase energy expenditure, reduce body fat and lessen the risk of cardio-metabolic disease, more recent research for both of these groups has indicated that increasing or at least maintaining muscle mass is extremely important.

Studies involving inactive and to a lesser extent athletic IWD have demonstrated the profound benefits of resistance training with regard to increasing muscular strength as well as anaerobic and aerobic power. Further efforts should therefore be made to educate these individuals and their clinicians about the benefits of resistance training and to ensure that appropriate resistance training programs are accessible and affordable.

Nutritional profiles of inactive and athletic individuals with SCI have revealed some alarming trends. While AWD appear to have relatively good nutritional intakes (at least when considered as a group), inactive individuals have a much poorer diet, consume too much fat and processed carbohydrates and too little fiber, vitamins and minerals. Therefore, inactive IWD should become a target for nutritional counseling, especially as it is known that such a group may have limited physical independence, poor quality of life and be at high risk for many chronic conditions. One challenge for nutritionists and dieticians who wish to work with these individuals would be the issue we face in accurately and reliably measuring body composition and energy expenditure for these individuals. Until the accuracy of Level 3 body composition assessment methods for IWD improves, practice in this area will involve a degree of trial and error with regards to the optimal energy, macro- and micronutrient intake that inactive and athletic IWD require on a daily basis.

Athletic individuals with a SCI or amputation have been shown to consume a variety of nutritional supplements, although the evidence for the effectiveness of this is rather limited. This may reflect: (i) the heterogeneity of the study samples; (ii) the wide variety of supplements available (each of which may have varying effects on different aspects of muscular function); (iii) the between-study variation in supplement dosage and duration; and (iv) the variety of outcome tests used to assess their effectiveness.

It may therefore be recommended that considerably more research be conducted across a range of disability groups in order to improve their body composition, physical fitness and function as well as athletic ability. Such research may concentrate on: (i) improving the validity and reliability of body composition and energy expenditure assessments in a variety of disability groups; (ii) examining the functional benefits of resistance and mixed training approaches; (iii) determining the primary barriers to good nutrition in inactive IWD so as to inform interventions and public health policy; and (iv) assessing the benefits of a variety of nutrition interventions and supplements. It is anticipated that such research and a heightened interest in the challenges that IWD face in their everyday activities should translate into improvements in health, independence, quality of life and athletic ability of these individuals by health professionals.

References

[1] International Paralympic Committee. Paralympic games, Available from: <http://www.paralympic.org/Paralympic_Games/>; 2009 [cited 02.03.2010].

[2] International Paralympic Committee. Classification, Available from: <http://www.paralympic.org/Sport/Classification/>; 2010 [cited 12.10.2010].

[3] Leclercq MM, Bonidan O, Haaby E, Pierrejean C, Sengler J. Study of bone mass with dual energy x-ray absorptiometry in a population of 99 lower limb amputees. Ann Readapt Med Phys 2003;46(1):24–30.

[4] Nolan L. Lower limb strength in sports-active transtibial amputees. Prosthet Orthot Int 2009;33(3):230–41.

[5] Sherk VD, Bemben MG, Bemben DA. Interlimb muscle and fat comparisons in persons with lower-limb amputation. Arch Phys Med Rehabil 2010;91(7):1077–81.

[6] Van de Vliet P, Broad E, Strupler M. Nutrition, body composition and pharmacology. The Paralympic Athlete. Chichester: Wiley-Blackwell; 2011. p. 172–97.

[7] Mojtahedi MC, Valentine RJ, Evans EM. Body composition assessment in athletes with spinal cord injury: comparison of field methods with dual-energy X-ray absorptiometry. Spinal Cord 2009;47(9):698–704.

[8] Mojtahedi MC, Valentine RJ, Arngrimsson SA, Wilund KR, Evans EM. The association between regional body composition and metabolic outcomes in athletes with spinal cord injury. Spinal Cord 2008;46(3):192–7.

[9] Buchholz AC, Pencharz PB. Energy expenditure in chronic spinal cord injury. Curr Opin Clin Nutr Metab Care 2004;7(6):635–9.

[10] Buchholz AC, McGillivray CF, Pencharz PB. Physical activity levels are low in free-living adults with chronic paraplegia. Obes Res 2003;11(4):563–70.

[11] Monroe M, Tataranni P, Pratley R, Manore M, Skinner J, Ravussin E. Lower daily energy expenditure as measured by a respiratory chamber in subjects with spinal cord injury compared with control subjects. Am J Clin Nutr 1998;68(6):1223–7.

[12] Hicks AL, Martin Ginis KA, Pelletier CA, Ditor DS, Foulon B, Wolfe DL. The effects of exercise training on physical capacity, strength, body composition and functional performance among adults with spinal cord injury: a systematic review. Spinal Cord 2011;49(11):1103–27.

[13] Valent L, Dallmeijer A, Houdijk H, Talsma E, van der Woude L. The effects of upper body exercise on the physical capacity of people with a spinal cord injury: a systematic review. Clin Rehabil 2007;21(4):315–30.

[14] Jacobs PL, Nash MS, Rusinowski JW. Circuit training provides cardiorespiratory and strength benefits in persons with paraplegia. Med Sci Sports Exerc 2001;33(5):711–7.

[15] Nash MS, van de Ven I, van Elk N, Johnson BM. Effects of circuit resistance training on fitness attributes and upper-extremity pain in middle-aged men with paraplegia. Arch Phys Med Rehabil 2007;88(1):70–5.

[16] Rodgers MM, Keyser RE, Rasch EK, Gorman PH, Russell PJ. Influence of training on biomechanics of wheelchair propulsion. J Rehabil Res Dev 2001;38(5):505–11.

[17] Turbanski S, Schmidtbleicher D. Effects of heavy resistance training on strength and power in upper extremities in wheelchair athletes. J Strength Cond Res 2010;24(1):8–16.

[18] Jacobs PL. Effects of resistance and endurance training in persons with paraplegia. Med Sci Sports Exerc 2009;41(5):992–7.

[19] Klingenstierna U, Renstrom P, Grimby G, Morelli B. Isokinetic strength training in below-knee amputees. Scand J Rehabil Med 1990;22(1):39–43.

[20] Nolan L. A training programme to improve hip strength in persons with lower limb amputation. J Rehabil Med 2012;44(3):241–8.

[21] Tabęcki R, Kosmol A, Mastalerz A. Effects of strength training on physical capacities of the disabled with cervical spine injuries. Hum Mov 2009;10(2):126–9.

[22] Dallmeijer AJ, Hopman MT, Angenot EL, van der Woude LH. Effect of training on physical capacity and physical strain in persons with tetraplegia. Scand J Rehabil Med 1997;29(3):181–6.

[23] Furmaniuk L, Cywi ska-Wasilewska G, Kaczmarek D. Influence of long-term wheelchair rugby training on the functional abilities of persons with tetraplegia over a two-year period post-spinal cord injury. J Rehabil Med 2010;42(7):688–90.

[24] Harris NK, Cronin J, Keogh J. Contraction force specificity and its relationship to functional performance. J Sports Sci 2007;25(2):201–12.

[25] Crewther B, Cronin J, Keogh J. Possible stimuli for strength and power adaptation: acute mechanical responses. Sports Med 2005;35(11):967–89.

[26] Keogh JWL, MacLeod RD. Body composition, physical fitness, functional performance, quality of life and fatigue benefits of exercise for prostate cancer patients: a systematic review. J Pain Symptom Manage 2012;43(1):96–110.

[27] Galvao DA, Newton RU, Taaffe DR, Spry N. Can exercise ameliorate the increased risk of cardiovascular disease and diabetes associated with ADT? Nat Clin Pract 2008;5:306–7.

[28] Granacher U, Zahner L, Gollhofer A. Strength, power, and postural control in seniors: considerations for functional adaptations and for fall prevention. Eur J Sport Sci 2008;8(6):325–40.

[29] Nelson ME, Rejeski WJ, Blair SN, Duncan PW, Judge JO, King AC, et al. Physical activity and public health in older adults: recommendation from the American college of sports medicine and the American heart association. Med Sci Sports Exerc 2007;39(8):1435–45.

[30] Tugcu I, Safaz I, Yilmaz B, Göktepe AS, Taskaynatan MA, Yazicioglu K. Muscle strength and bone mineral density in mine victims with transtibial amputation. Prosthet Orthot Int 2009;33(4):299–306.

[31] Keogh JWL. Paralympic sport: an emerging area for research and consultancy in sports biomechanics. Sports Biomech 2011;10(3):234–53.

[32] Argus CK, Gill ND, Keogh JWL, Hopkins WG, Beaven CM. Changes in strength, power and steroid hormones during a professional rugby union competition. J Strength Cond Res 2009;23(5):1583–92.

[33] Sabour H, Javidan AN, Vafa MR, Shidfar F, Nazari M, Saberi H, et al. Calorie and macronutrients intake in people with spinal cord injuries: an analysis by sex and injury-related variables. Nutrition 2012;28(2):143–7.

[34] Knight KH, Buchholz AC, Martin Ginis KA, Goy RE. Leisure-time physical activity and diet quality are not associated in people with chronic spinal cord injury. Spinal Cord 2011;49(3):381–5.
[35] Groah SL, Nash MS, Ljungberg IH, Libin A, Hamm LF, Ward E, et al. Nutrient intake and body habitus after spinal cord injury: an analysis by sex and level of injury. J Spinal Cord Med 2009; 32(1):25–33.
[36] Tomey KM, Chen DM, Wang X, Braunschweig CL. Dietary intake and nutritional status of urban community-dwelling men with paraplegia. Arch Phys Med Rehabil 2005;86(4):664–71.
[37] Walters JL, Buchholz AC, Martin Ginis KA. Evidence of dietary inadequacy in adults with chronic spinal cord injury. Spinal Cord 2009;47(4):318–22.
[38] Goosey-Tolfrey VL, Crosland J. Nutritional practices of competitive British wheelchair games players. Adapt Phys Activ Q 2010;27(1):47–59.
[39] Innocencio da Silva Gomes A, Goncalves Ribeiro B, de Abreu Soares E. Nutritional profile of the Brazilian Amputee Soccer Team during the precompetition period for the world championship. Nutrition 2006;22(10):989–95.
[40] Krempien JL, Barr SI. Risk of nutrient inadequacies in elite Canadian athletes with spinal cord injury. Int J Sport Nutr Exerc Metab 2011;21(5):417–25.
[41] Krempien JL, Barr SI. Eating attitudes and behaviours in elite Canadian athletes with a spinal cord injury. Eat Behav 2012; 13(1):36–41.
[42] Potvin A, Nadon R, Royer D, Farrar D. The diet of the disabled athlete. Sci Sports 1996;11(3):152–6.
[43] Rastmanesh R, Taleban FA, Kimiagar M, Mehrabi Y, Salehi M. Nutritional knowledge and attitudes in athletes with physical disabilities. J Athl Train 2007;42(1):99–105.
[44] Ribeiro SML, da Silva RC, de Castro IA, Tirapegui J. Assessment of nutritional status of active handicapped individuals. Nutr Res 2005;25(3):239–49.
[45] Wang JH, Goebert DA, Hartung GH, Quigley RD. Honolulu wheelchair marathon: a comparative study between American and Japanese participants. Sports Med Train Rehabil 1992;3(2):95–104.
[46] Pelly F, Holmes MA, Traven F, Neller AH, Burkett BJ. Competition nutrition strategies of the 2004 Australian Paralympic swimming team. J Sci Med Sport 2007;10(Suppl. 1):S50.
[47] Tsitsimpikou C, Jamurtas A, Fitch K, Papalexis P, Tsarouhas K. Medication use by athletes during the Athens 2004 Paralympic Games. Br J Sports Med 2009;43(13):1062–6.
[48] Spendiff O, Campbell IG. Influence of glucose ingestion prior to prolonged exercise on selected responses of wheelchair athletes. Adapt Phys Activ Q 2003;20(1):80–90.
[49] Spendiff O, Campbell IG. Influence of pre-exercise glucose ingestion of two concentrations on paraplegic athletes. J Sports Sci 2005;23(1):21–30.
[50] Temesi J, Rooney K, Raymond J, O'Connor H. Effect of carbohydrate ingestion on exercise performance and carbohydrate metabolism in persons with spinal cord injury. Eur J Appl Physiol 2010;108(1):131–40.
[51] Nash MS, Meltzer NM, Martins SC, Burns PA, Lindley SD, Field-Fote EC. Nutrient supplementation post ambulation in persons with incomplete spinal cord injuries: a randomized, double-blinded, placebo-controlled case series. Arch Phys Med Rehabil 2007;88(2):228–33.
[52] Perret C, Mueller G, Knecht H. Influence of creatine supplementation on 800 m wheelchair performance: a pilot study. Spinal Cord 2006;44(5):275–9.
[53] Jacobs PL, Mahoney ET, Cohn KA, Sheradsky LF, Green BA. Oral creatine supplementation enhances upper extremity work capacity in persons with cervical-level spinal cord injury. Arch Phys Med Rehabil 2002;83(1):19–23.
[54] Tarnopolsky MA. Building muscle: nutrition to maximize bulk and strength adaptations to resistance exercise training. Eur J Sport Sci 2008;8(2):67–76.
[55] Gualano B, Artioli G, Poortmans J, Lancha Jr A. Exploring the therapeutic role of creatine supplementation. Amino Acids 2010;38(1):31–44.
[56] Tarnopolsky M, Martin J. Creatine monohydrate increases strength in patients with neuromuscular disease. Neurology 1999;52(4):854–7.

CHAPTER

18

An Overview of Doping in Sports

Fabian Sanchis-Gomar[1,2], *Vladimir Martinez-Bello*[3], *Helios Pareja-Galeano*[1,2], *Thomas Brioche*[1,2,4] *and Mari Carmen Gómez-Cabrera*[1,2]

[1]Department of Physiology, Faculty of Medicine, University of Valencia, Valencia, Spain [2]Fundación Investigación Hospital Clínico Universitario/INCLIVA, Valencia, Spain [3]Department of Teaching of Musical, Visual and Corporal Expression, Faculty of Teaching, University of Valencia, Spain [4]Laboratory M2S (Movement, Sport and Health Sciences), UFR-APS, Rennes Cedex, France

INTRODUCTION

The word "doping" was first mentioned in 1889 in an English dictionary, initially describing a potion containing opium and used to "dope" horses. "Dope" was a spirit prepared from grapes, which Zulu warriors used as a "stimulant" at fights and religious procedures, also known as "doop" in Afrikaans or Dutch. Afterward, the concept of "dope" extended to other beverages with stimulating properties. Finally, the expression was introduced into English turf sports in about 1900 for illegal drugging of racehorses [1].

Thus, performance-enhancing substances and methods misuse is not exclusive to modern times. Throughout history, man has sought to improve his physical performance by artificial means. The first notion of doping dates back to antiquity. In brief, according to Milo of Croton (6th century BC), Greek athletes ingested different types of meat depending on the sport practiced: jumpers ate goat meat; boxers and wrestlers, bull meat; and wrestlers, pork because of its fatty meat. Pliny the Younger (1st century BC) indicated that endurance runners of ancient Greece used decoctions of horsetail to contract the spleen and prevent dropping out of long-term races. Also Galen and Philostratus reported that athletes tried to increase their performance by swallowing all kinds of substances. Natives of South America chew coca leaves and African natives, kola nut; the stimulant properties of ginseng, known for over three thousand years by the Chinese people, should also be remembered.

However, in modern times, drug abuse detection in sports was first tackled by the International Olympic Committee (IOC) in the 1960s when a Danish cyclist died during the Rome Olympic Games. Twelve years later, at the Summer Olympic Games in Munich, the first official anti-doping tests were performed systematically, whose pilot project was carried out at the Mexico City Olympic Games (1968). At those early stages, the only prohibited substances were those capable of producing a significant effect on performance only if administrated, in sufficient amounts, right before or during the sport competition [2].

From this period, the IOC, and later the World Anti-Doping Agency (WADA), controlled the "prohibited list" in accordance with new research in the doping field [3]. The revolutionary range of available drugs and the sophistication with which they have been applied in human doping has resulted in the introduction of evermore complex and sensitive methods designed to detect such misuse [4].

Doping is nowadays defined in the WADA Code as *"the occurrence of one or more of the anti-doping rule violations set forth in Article 2.1 through Article 2.8 of the Code"* [5]. Violations of the anti-doping rule include not only the use or attempted use of prohibited substances, but also: the presence of a prohibited substance, or its metabolites or markers, in an athlete's urine or blood sample; violation of the athlete's obligation to inform about his/her 'whereabouts'; tampering or attempted tampering with doping control procedures; possession of prohibited substances or the

Nutrition and Enhanced Sports Performance.
DOI: http://dx.doi.org/10.1016/B978-0-12-396454-0.00018-7

means for performing prohibited methods; and trafficking or attempted trafficking in a prohibited substance or the means for performing a prohibited method [5,6].

Following on from Botrè [7], the historical periods can be described in detail and partitioned into three main stages: (i) an early stage in which the abuse of drugs took place during competition and its detection was based on gas chromatography; (ii) the androgenic anabolic steroids age, around the 1970s, when the drugs were administered in both training periods and competition; and (iii) the most recent era in which protein chemistry, molecular biology, biomarkers, genetic engineering, and immunological techniques are applied routinely. A future period is feared by many as the possible "gene doping age" of "omics" technologies (see Table 18.1).

Therefore, it is now clear that there are several hundred forms of known and unknown doping substances and techniques used in the sports world. This chapter will provide a summary of the most commonly abused substances, their effects and adverse effects, and the alleged benefits together with current standards in doping. For this purpose, we mainly focus on five groups: (i) substances misused mostly in strength sports; (ii) hormones and metabolic modulators; (iii) blood doping in aerobic sport disciplines; (iv) masking agents; and (v) gene doping.

MAIN TENDENCIES OF DOPING IN THE STRENGTH FIELD

We can divide the prohibited substances into four major categories in strength sports: (i) androgenic-anabolic agents (AAA), which are certainly the most popular substances and the most used; (ii) Beta2-agonists such as clenbuterol; (iii) peptide hormones such as growth hormone (GH), somatomedins (IGF-I), and GH secretagogues; (iv) myostatin inhibition, a strategy which is definitely the future of doping in the field of muscle mass and strength.

Androgenic-Anabolic Agents (AAA)

Two major doping strategies are found in this category: direct and indirect androgen doping. Direct androgen doping could be defined as the exogenous administration of both endogenous (substance that can be produced naturally by the human body) and exogenous (substance that cannot usually be produced naturally by the human body) androgens [8]. Indirect androgen doping refers to the use of compounds that stimulate the production of endogenous testosterone [8].

Direct Androgen Doping and Detection Methods

ENDOGENOUS ANDROGENS

Athletes have used testosterone doping to avoid detection of synthetic androgens. Testosterone and epitestoterone are secreted endogenously by Leydig cells in the testes. Therefore, detection of illegal use depends on the distinction between endogenous and exogenous testosterone. Initially, anti-doping laboratories used the testosterone/epitestosterone ratio (T/E) to detect its abuse. Epitestosterone (17α-hydroxy-4-androstene-3-one) is a 17-epimer of testosterone and is biologically inactive. In normal times, the T/E ratio in urine is around 1, but because there is no interconversion between the two compounds, administration of exogenous testosterone results in an increase in the

TABLE 18.1 Evolution of the Main Challenges in Doping and in Control Analysis

Period	Challenge	Period	Challenge
Origin early 1970s	Stimulants, narcotics, drugs of abuse	Mid-1990–2000	Erythropoietin and analogs
Mid 1970s	Synthetic anabolic androgenic steroids (AAS)	2000–2005	Designer steroids, hormone and hormone receptor modulators
Late 1970s	Naturally occurring AAS (testosterone), synthetic AAS cocktails	2005–present	Peptide hormones
1980–1990	Beta-blockers, diuretics, cannabinoids, glucocorticoids	2003–present	Blood doping
Early to mid-1990s	Low concentration of AAS	2004	Tetrahydrogestrinone (THG)
Mid- to late 1990s	Human chorionic gonadotropin, endogenous testosterone and/or precursors	2008–present	Gene doping

Modified from Botrè [7].

T/E ratio [9]. Initially, the limit was fixed at 6 by the International Olympic committee (IOC). But variations occur in the ratio, due to genetic polymorphism [10,11], and for this reason the value is currently fixed at 4. However, it is possible to maintain this ratio below 4 [12] through the co-injection of testosterone and epitestosterone, but detection of high urinary concentrations of metabolites of the latter reveals the use of this method [13]. When the T/E value is suspicious, it is necessary to use gas chromatography combustion IRMS (isotope ratio mass spectrometry), which is based on the measurement of the 13C/12C isotope ratio in testosterone [14,15], to confirm if doping has taken place and, as synthetic testosterone has a lower 13C/12C ratio than the endogenous form, a lower 13C/12C ratio suggests exogenous testosterone use.

EXOGENOUS ANDROGENS (AAS)

Exogenous androgens are a synthetic derivative of the male hormone testosterone. The first AAS was synthesized during the 1960s and was named norbolethone. Today, among the most well-known AAS are nandrolone, stanazolol, tetra hydrogestrinone (THG), desoxymethyltestosterone, and dihydrotestosterone (DHT). Athletes use these substances because detecting them requires knowledge of the individuals' chemical signatures, which at present are in many cases unknown. Isotope ratio mass spectrometry can be used to detect some AAS, such as nandrolone and DHT, but others go undetected by this method. Currently, anti-doping laboratories are being employed to carry out bioassays to detect the use of other AAS.

Indirect Androgen Doping and Detection Methods

This strategy refers to using compounds that stimulate the production of endogenous testosterone. At least three methods are used: (i) androgen precursors; (ii) estrogen blockers; (iii) gonadotropins.

An androgen precursor is a substance that is metabolized in one or more steps to testosterone. For example, androstenedione is finally converted to testosterone. Use of androgen precursors can be detected because it leads to an increase in serum testosterone levels [16,17] and finally to an increase in the T/E ratio [18,19].

Another strategy is the use of estrogen antagonists and aromatase inhibitors, which is based on the fact that estrogen is a negative regulator of the hypothalamic-pituitary-gonadal axis which is more potent than testosterone itself [20,21]. Hence, the use of estrogen antagonists such as tamoxifen or, more recently, fulvestrant, or aromatase inhibitors like letrozole or vorozole, which leads to a rise in endogenous testosterone production [22–24].

Furthermore, in order to increase the endogenous production of testosterone, some athletes resort to gonadotropins. We can cite the luteinizing hormone (LH), and even the human chorionic gonadotropin (hCG) whose alpha subunit has a structure similar to LH. They are only banned for men because there is no evidence of their effectiveness in the development of muscle mass in women. Male athletes generally use the hCG in order to mask the use of exogenous testosterone, because its use does not affect the T/E ratio [25]. Another strategy for enhancing the endogenous levels of testosterone sometimes used by athletes is the pulsatile administration of gonadotropin releasing hormone (GnRH), which stimulates LH and follicle-stimulating hormone (FSH) production. Continuous administration desensitizes the gonadotrophs and leads to down-regulation of GnRH receptors and finally suppresses testosterone production.

Some of these substances can be detected by mass spectrometric methods [26,27]. Measurement of some metabolites of those products makes it possible to detect the use of androgen precursors (for example 4-hydroxyandrostenedione, an androstenedione metabolite). Use of LH and hCG can both be detected by immunoassays [28,29], but a positive immunoassay hCG test needs to be confirmed by immunoextraction and mass spectrometry [29].

Effect on Athletes and Adverse Effects

The desired effect of AAA could be divided into four categories: effects on body composition (weight, body dimension, lean body mass, fat and muscle mass); strength and muscle mass; hematology and endurance performance; and finally, recovery. Bhasin and co-workers [30,31] showed that testosterone produces a significant gain in muscle size and strength compared with placebo. In addition, they reported that testosterone plus training brings about a larger gain in muscle mass and strength compared with only training, placebo, or testosterone alone. Moreover, a dose–response effect between doses of testosterone received, serum testosterone levels, muscle mass, muscle volume, muscle strength and power, and decrease in fat mass [31] has been demonstrated. Although testosterone increases muscle power, it does not improve muscle fatigability or the quality of contractile muscle (the force generated by each unit of muscle volume). These results suggest that the effects of testosterone are specific to certain characteristics of muscle performance, but not to all. It is unclear if such a dose–response relationship exists in women, but the results reported in the records of East German athletes leave no doubt about the effect on strength [32]. The increase in muscle mass and strength induced by testosterone is associated with a hypertrophy of muscle fibers types I and II [33], which could be due to an increase in muscle protein synthesis and inhibition of protein degradation [34,35].

Pärssinen et al. [36] reported that there is an increased premature mortality (five-fold higher) of competitive powerlifters suspected to have used anabolic agents when compared with controls. Androgenic-anabolic agents use is linked to myocardial infarction and sudden death [37,38], septal and left ventricular hypertrophy, and cardiac arrhythmias [39–41], systolic and diastolic dysfunction (directly related to the dose and duration of AAA use) [42], impaired endothelial reactivity, and so on. One of the most commonly seen undesirable effects of androgens, particularly nonaromatizable androgens, is dyslipidemia due to a significant decrease in high-density lipoprotein cholesterol [43], a decrease in plasma high-density lipoprotein by more than 30% [44], and also an increase in hepatic lipase activity. Use of AAA has been associated with a wide range of psychiatric effects, but these effects are still not clear. Studies report cases of aggression, dysthymia, psychosis, and criminal behavior [45–47]. Endocrine side effects associated with taking anabolic steroids include gonadal function, with induction of hypogonadism hypogonatrope, leading: in men to testicular atrophy, abnormalities of spermatogenesis, impotence and changes in libido; and in women to hirsutism, menstruation cycle disorders, breast atrophy and a male pattern of baldness, hoarse voice and clitoromegaly [48]. Some cases of ruptures of biceps and quadriceps tendons have been reported in athletes using AAA [49,50]. Liver toxicity (hepatitis, hepatic adenomas) is a common effect reported in athletes using AAA. Acne is a common adverse effect of AAA therapy. The two most common forms of acne associated with AAA are acne fulminans and acne conglobata [51].

Beta2-Agonists: Clenbuterol

Beta2-agonists, also known as β2-adrenergic agonists or β2-adrenergic receptor agonists, act by binding to β2 adrenergic receptors present in smooth skeletal muscle and cardiac muscle. When used in aerosol, their action is relatively selective of bronchial muscles, and the indication of their medical use is asthma (as bronchodilator). Thus, β2-adrenergic agonists are usually used in oral forms, which make it possible to take doses from 10 to 100 times higher than those obtained by inhalation. They were initially used in endurance sport but they are also used with the objective of increasing muscle mass, since clenbuterol showed an anabolic effect in 1984 [52]. Clenbuterol is the only β2-adrenergic agonist classified in the anabolic category of the WADA prohibited list.

Effects on Athletes

Two studies in humans, using salbutamol, show that β2 agonists orally could induce muscle strength gain (+10–20%), but with great inter-individual variability and with no evidence of muscle mass gain [53,54]. Their effects are better demonstrated in the patient, to fight against muscular atrophy associated with spinal cord lesions, or in patients with scapulohumeral muscular dystrophy. Studies in animals allow doses closer to those used in athletes. Animal studies consistently show the anabolic effect of clenbuterol at a dose of 1–5 mg/kg body weight (+ 12–20% of muscle mass) [55–57]. However, it remains to be demonstrated that this increase in muscle mass in animals can be transferred to humans.

Adverse Effects

High doses of β2-adrenergic agonists can decrease performance. Duncan et al. [58], using doses known to promote anabolism (1–5 mg/kg) and studying the combined effects of clenbuterol and exercise, reported that after 4 weeks of administration of clenbuterol, and despite a significant gain in muscle mass (11%), treated rats were unable to maintain the same running speed as untreated rats. The same results were observed for swimming performance. Moreover, clenbuterol treatment can lead to heart failure. Studies in rats by imaging showed areas of necrosis with infiltration of collagen in the myocardium after treatment with clenbuterol [59]. Other studies also report a deleterious effect of clenbuterol in muscle cells, with myocyte necrosis in soleus of rats [59]. Finally, these results may explain the reported cases of myocardial infarction with normal coronary arteries, reported in some bodybuilders who used a cocktail of anabolic agents including clenbuterol.

Detection Method

In the case of clenbuterol, just the presence in the sample is considered as a doping situation. β2-adrenergic agonists can be detected in urine, hair, and serum by gas [60–62] and/or liquid chromatography, and after purification of the analytes by means of traditional isolation procedures, such as liquid–liquid extraction (LLE) and/or solid-phase extraction (SPE) [60,62], and immunoaffinity (IAC) based techniques, alone or in combination, or, very recently, by means of three-phase solvent bar microextraction by solid-phase microextraction coupled with liquid chromatography [63].

Peptide Hormones: the Growth Hormone Paradigm

Growth hormone (GH) is a pituitary polypeptide hormone with anabolic, growth-promoting activity and a lipolytic effect. It is a 191-amino-acid, 22-kDa protein. Growth hormone is believed to promote muscle growth and it also increases the rate of lipolysis leading to a decrease in fat mass; still today its use remains very difficult to detect.

Effects on Athletes and Adverse Effects

Growth hormone treatment induces persistent increases in isometric knee flexor strength, concentric knee flexor strength, and right-hand peak grip strength [64]. In the elderly, Vittone et al. [65] reported an increase in muscle mass and a decreased fat mass/muscle tissue ratio after GH treatment, leading to GH levels close to those observed in younger patients. Effects on muscle strength are conflicting: no or moderate gain (3–20%) [66,67]. The addition of GH to an intense fitness program, in young untrained subjects or in subjects performing regular weight training, does not induce additional gain of strength or muscle size, compared with the effects of training alone [68,69]. Meinhardt et al. [70] showed that GH (2 mg/d for 8 wk) had no effect on muscle strength (dead lift), power (jump height), or endurance (VO_2max) but did improve sprint capacity by 5.5% in men but not women (by 2.5%; not significant). Moreover, GH administration does not typically result in a net weight change, because the loss of fat is compensated for by a gain in lean body mass (of which a substantial part, 50–80%, represents retained fluid). This is true in the GH-deficient patient on GH replacement therapy as well as in normal subjects, including athletes, taking GH [70]. Another reason sometimes given for the use of GH by athletes is the belief that it accelerates recovery from injury. However, it is unknown whether after an injury GH plays a role in this response or whether local factors operating at the injury site are responsible. One study examined collagen synthesis in patellar tendon and quadriceps muscle in response to 14 days of high-dose GH treatment (33–50 μg/kg/day in noninjured young male volunteers); GH treatment increased collagen protein synthesis 1.3-fold over placebo (significant) in tendon and 5.8-fold (not significant) in muscle [71]. Another study examined the effect of various doses of GH (15, 30, and 60 μg/kg/day) on tibial fracture healing [72]. Just the highest dose of GH accelerated fracture healing by 29% in patients. These findings need to be corroborated and expanded before firm conclusions can be drawn about the effect of GH.

Hence, the evidence for GH as an ergogenic substance in healthy humans is weak and even more so in trained people. Yet athletes continue GH abuse in the belief that it improves their performance. Numerous reasons can be given why the scientific literature does not reflect GH use in the sports arena: GH doses are too low; duration of treatment is not long enough; GH in conjunction with anabolic steroids, insulin, and other doping agents may have greater ergogenicity than when given alone; GH in combination with exercise is particularly potent; athletes react to GH in a different manner than nonathletes, and so forth.

The adverse effects of GH use are summarized in Table 18.2. Adverse effects are dose-dependent, treatment duration dependent, and age dependent. Susceptibility varies among individuals; older people are more prone to side effects, even at low doses.

Detection Methods

The fact that exogenous GH is identical to the main isoform of endogenous GH (22-kDa isoform) renders its detection challenging. There is the GH isoform test based on the fact that there are various endogenous isoforms of GH whereas there is just one isoform of synthetic GH. Then, there is the GH biomarkers approach (only valid in blood, although there are methods for urine and saliva), and finally the genomic and proteomic approaches that are still being developed. Hence, the use of synthetic GH can be detected by the GH isoform test, first proposed by Wu et al. and Momomura et al. [73,74] which was developed and tested at the Athens Olympics (2004), then in Turin (2006) and in Beijing (2008). Now it is one of the current methods used by the WADA. The test consists of two GH immunoassays: one that is relatively specific for 22K-GH and another that is "permissive", that is it recognizes a number of pituitary isoforms in addition to 22K-GH. A dose of exogenous GH suppresses the endogenous forms. Thus, the ratio between 22K-GH and pituitary GH increases because most of the measurable GH is of exogenous origin [73,75]. For validation purposes, the WADA requires two independent assays, and thus two separate pairs of 22K-GH-specific (named "rec" for recombinant) and permissive (named "pit" for pituitary) assays [75]. Using these assays, the normal rec/pit ratio has a median value of approximately 0.8 and ranges from 0.1 to 1.2. The median value of less than 1 reflects the fact that 22K-GH accounts for only 75–80% of the GH isoforms. The current rec/pit ratio cutoffs ("decision limits") used by the WADA for evidence of doping is 1.81 for men and 1.46 for women (assay kit 1) and/or 1.68 for men and 1.55 for women (assay kit 2) [75a]. The isoform detection test performs well, but has a limited window of opportunity (12–24 h after the last GH injection). The isoform test can be circumvented by using cadaveric GH (with attendant risk of acquiring Creutzfeld-Jakob

TABLE 18.2 The Endocrine Side-Effects of Anabolic Androgenic Steroid Preparations and Recombinant Growth Hormone

Adverse Effect	Consequences
Sodium and fluid retention	Soft tissue swelling Paresthesias Nerve entrapment, carpal tunnel syndrome Joint stiffness Hypertension Peripheral edema
Insulin resistance	Carbohydrate intolerance Type II diabetes mellitus
Myalgias	
Arthralgias	
Gynecomastia in males	
Secondary hypogonadism	
Amenorrhea	
Ovarian cysts	
Testicular atrophy	
Virilization in females	
Clitoral hypertrophy	
Hoarse voice (women)	
Hirsutism, acne	
Hair loss	
Acromegalic changes expected with prolonged, high-dose GH	Acral enlargement Bone remodeling Arthritis Bone spurs Frontal bossing Dental malocclusion Spinal stenosis Disfigurement Cardiovascular changes Cardiac dysfunction

disease) or using GH secretagogues (resulting in only mild GH stimulation). To counter the two previously mentioned techniques, scientists fighting against doping have been creating indirect tests based on the appearance of biomarkers naturally produced in response to an increase in blood GH levels, whether injected exogenously or produced endogenously. Well-known effects of GH are the induction of IGF-I expression and promotion of collagen turnover in bone and connective tissues [76]. Thus, IGF-I and procollagen type III amino-terminal propeptide (P-III-NP) have been selected as relatively specific GH-responsive biomarkers suitable for an anti-doping test (there are many other biomarkers, such as IGF-binding protein (IGFBP) 2 and IGFBP3, osteocalcin, and type I collagen carboxyterminal cross-linked telopeptide (ICTP) but they were not selected). Serum levels of IGF-I and P-III-NP increase after GH administration and remain elevated for several days to weeks after a single GH dose. They are not completely specific for GH, but extensive validation studies have resulted in a discriminant formula that allows distinction of GH-induced elevation from most if not all nonspecific stimuli. The biomarker test has a window of opportunity of several days—realistically, probably 5–7 d. It was scheduled to be implemented by the WADA in time for the London 2012 Olympiad. Reports on new GH-responsive biochemical markers, such as mannan-binding lectin, will continue to appear in the literature. The specificity and sensitivity of such novel markers will have to be rigorously demonstrated before they are considered as an anti-doping strategy.

Research is continuing to identify additional indicators for GH use that may be useful for anti-doping purposes. In particular, genomic and proteomic approaches are being explored in an attempt to identify a "signature" that would be indicative of exogenous GH use [77]. A lot of work will be required before proteomic approaches become a realistic tool for anti-doping purposes.

Myostatin Inhibition

Myostatin, or growth differentiation factor 8 (GDF-8), is a protein that acts as a negative regulator of muscle mass. It is produced by the muscle itself and acts in an autocrine or paracrine fashion. The myostatin gene is expressed almost exclusively in cells of skeletal-muscle lineage [78].

Mice in which the myostatin gene has been inactivated show marked muscle hypertrophy, and it has been shown to be responsible for the "double-muscling" phenotype in cattle [79,80]. Furthermore, these mice showed less adipose tissue. By contrast, Reisz-Porszasz et al. [81], who specifically directed myostatin over-expression to striated muscle of mice, found a weight reduction of 20–25% in individual skeletal muscles within 42 days of postnatal life, and this also was solely due to a reduction in the fiber size. Moreover, the over-expression of myostatin additionally led to a reduction in heart weight (by 17%) and led to an increase in fat deposition. The function of myostatin appears to be present in some species of animals (mice and cattle) and humans. In fact, the identification of a myostatin mutation in a child with muscle hypertrophy and increased strength without any motor or cardiovascular anomalies has been reported [82], thereby providing strong evidence that myostatin does play an important role in regulating muscle mass in humans. The phenotypes of mice and cattle and this child lacking myostatin (muscle hypertrophy, increased strength, less adipose tissue) have posed the possibility that myostatin would be a very good solution in doping.

Two strategies could be considered. The first would be a genetic manipulation specific to striated muscle, like the mouse model of Reisz-Porszasz et al. [81]. The next more probable strategy would be the use of an inhibitor of myostatin. This inhibition could be achieved at a variety of levels, such as increasing the expression of follistatin, a natural antagonist of myostatin, or by blocking myostatin activity using a humanized monoclonal antibody, such as Wyeth's MYO-029 (its development was stopped). Over-expression of follistatin in skeletal muscle increased the weight of individual muscles of transgenic mice by 327% [83]. Muscle enlargement resulted through a combination of hyperplasia (66% increases) and hypertrophy (27%). Whittemore et al. [84] injected monoclonal antibodies directed against mouse myostatin into the peritoneal cavity of wild type mice on a daily basis and found a 20% increase in muscle mass within 4 weeks and an increase of muscle force (measured in grip strength) of 10%. The muscle enlarged solely by hypertrophy and, apart from the increased size, fiber morphology appeared normal. Importantly, blockade of myostatin did not affect the weight of other organs such as heart, and serum parameters for liver, kidney, muscle, bone, and glucose metabolism appeared normal.

METABOLIC MODULATORS AND RELATED SUBSTANCES

Another interesting and recent group of prohibited substances included in the 2012 WADA list of prohibited substances is the so-called "Hormone and Metabolic Modulators". A few years ago, several compounds that might increase athletic performance emerged. Exercise-mimetic drugs such as AICAR (5-aminoimidazole-4-carboxamide-1-β-D-ribofuranoside) and GW1516 (a peroxisome proliferator-activated receptor δ modulator) were flagged to mimic exercise and increase performance [85,86]. Questionable announcements about these drugs, such as "exercise in a pill", ensued in the media. The WADA has introduced both AICAR and GW1516 in the prohibited list as metabolic modulators in the class "Hormone and metabolic modulators" [87], reflecting a raising concern about the potential use of these compounds by cheating athletes. The main reason supporting this decision was that AICAR and GW1516, in combination with exercise, induce synergistically fatigue resistant type I fiber specification, mitochondrial biogenesis, angiogenesis, and improved insulin sensitivity, ultimately increasing physical performance [85,88]. Thus, since 2009, the prohibited list as established by the WADA included therapeutics such as the peroxisome-proliferator-activated receptor PPARδ-agonist GW1516 and PPARδ-AMPK agonist AICAR. Reliable methods for assessing both substances have been thereby included within anti-doping testing [89,90]. In addition, it should be mentioned that, although we included myostatin inhibitors under the heading "main tendencies of doping in the strength field", the agents modifying myostatin function(s) including, but not limited to, myostatin inhibitors are included in this group of prohibited substances [87]. Additionally, insulins have been recently included in this group [91].

Another group of drugs included in this section of prohibited substances are the selective androgen receptor modulators (SARMs). The first SARM was developed by Dalton et al. in 1998 [92]. Selective androgen receptor modulators have the advantage of dissociating the anabolic and androgenic effect of androgen. Hence, SARMs present just the anabolic properties in musculoskeletal tissues of androgen and for this reason have been included in the list of prohibited substances. Finally, almost all available SARMs can be detected either by liquid or gas chromatography tandem mass spectrometry [93,94].

Furthermore, we have recently proposed telmisartan, an angiotensin II receptor blocker (ARB), as a metabolic modulator in the context of performance-enhancing drugs in sport [95]. Feng et al. [96] have recently shown that telmisartan is effective in up-regulating the concentrations of both PPARδ and phospho-AMPKα in cultured myotubes. Chronic administration of telmisartan consistently prevented weight gain, increased slow-twitch skeletal muscle fibers in wild-type mice, and enhanced running endurance and post-exercise oxygen consumption, although these effects were absent in PPARδ-deficient mice. The mechanism is also involved in PPARδ-mediated stimulation of the AMPK pathway (e.g., the phospho-AMPKα level in skeletal muscle was up-regulated in mice treated with telmisartan as compared with control animals). It was very recently suggested that telmisartan may induce rather similar biochemical, biological, and metabolic changes (e.g., mitochondrial biogenesis and changes in skeletal muscle fibers type), such as those occurring with AICAR and GW1516 [95]. Thus, consideration should be given to including it among "metabolic modulators" in the prohibited list before it becomes available.

BLOOD DOPING AND ANTI-DOPING APPROACHES IN ENDURANCE SPORTS

The WADA defines blood doping as "the misuse of techniques and/or substances to increase red blood cells (RBCs) count, which allows the body to transport more O_2 to muscles, increasing performance" [97]. Prohibited procedures include the use of synthetic O_2 carriers, the transfusion of red blood cells, the infusion of hemoglobin (Hb), and the artificial stimulation of erythropoiesis. In this chapter we have not considered the synthetic O_2 carriers, such as Hb-based O_2 carriers, perfluorocarbons, or efaproxiral (RSR13).

Broadly, we can actually divide blood doping into two big groups [98–101]:

1. Substances that stimulate erythropoiesis (ESAs) or that mimic the action of erythropoietin (Epo): rhEpo (alfa, beta, delta or omega), darbepoetin-alfa, hypoxia-inducible factor (HIF) stabilizers, methoxy polyethylene glycol-epoetin β (CERA), peginesatide (Hematide; Affymax), prolyl hydroxylase inhibitor PHD-I FG-2216 (FibroGen).
2. Blood transfusions: including the use of autologous, homologous, or heterologous blood or red blood cell products of any origin.

Effects on Athletes and Adverse Effects

The most desirable effect for performance in endurance sports is the improvement of oxygen transport. The blood O_2-carrying capacity is maintained by the O_2-regulated production of Epo, which stimulates the proliferation and survival of red blood cell progenitors. The glycoprotein hormone Epo is known to be the major stimulator of erythropoiesis [102]. It has also been demonstrated that Epo also increases arterial O_2 content by decreasing plasma volume [103]. Epo gene expression is mainly induced under hypoxic conditions [104]. A primary function of Hb residing within RBCs is to bind O_2 under conditions of high O_2 concentration and transport and release it to tissues where O_2 is being consumed [100]. Thus, Hb concentration is a major determinant of oxygen delivery and exercise capacity, an effect that is accentuated in moderate hypoxia [105,106]. Blood transfusions or treatments with ESAs increase Hb and hence the oxygen-carrying capacity of blood; therefore, they increase performance.

The main risks and adverse effects of ESA abuse in healthy individuals are increased red cell mass, reduced blood flow due to increased viscosity, and increased likelihood of thrombosis and stroke [77].

Detection Methods

The International Olympic Committee officially prohibited the use of recombinant human erythropoietin (rhEpo) in 1990. Its detection is complicated because Epo synthesis occurs naturally in the body. Currently, the International Cyclist Union (UCI), the WADA and some International Sport Federations use the direct method to detect rhEpo abuse and the indirect method (based on the longitudinal follow-up of the athlete's hemogram) for targeting the athlete [107].

Direct Method of Detecting ESA Abuse

One of the greatest advances in doping control was rhEpo detection, which is based on a reliable test to differentiate between exogenous and endogenous Epo. The French National Anti-Doping Laboratory came up with a urine test based on isoelectrofocusing (IEF) combined with immunochemiluminescence [108]. Thus, the direct method is fundamentally based on the detection of rhEpo, darbepoetin, or CERA in a urine or serum sample because the migration differences observed when the different types of Epo are submitted to IEF analysis demonstrate that ESA misuse has taken place in spite of the presence of hEpo [108–113]. Endogenous Epo and the other ESAs are composed of protein isoforms that, depending on the synthesis procedure, differ in number and their relative abundance. Each Epo isoform has a different isoelectric point—pH value at which the isoform is uncharged—and therefore, when an electric field is applied in a pH range gel where the protein is loaded, there is no mobility; such isoelectric patterns are used to differentiate between the types of Epo. The rhEpo forms differ from natural purified urinary Epo, which has more acidic bands, probably due to post-translational modifications such as glycosylation. Such differences in urine analysis allowed us to ascribe excreted Epo to a natural or recombinant origin: interpretation of the results must be in accordance with WADA Technical Document EPO2009 [114].

Indirect Method of Detecting Blood Doping and ESA Abuse

The Australian Institute of Sport discovered a blood test that identifies markers that indicate recent use of rhEpo [115]. Biomarkers are, in fact, essential and a very useful tool for anti-doping laboratories. However, advances in metabolomics and proteomics will extend the potential of biomarkers approaches.

The indirect method is represented nowadays by the so-called Athletes Blood Passport (ABP). The fundamental principle of the ABP is based on the monitoring of an athlete's biological variables over time to facilitate indirect detection of doping on a longitudinal basis [116], and more particularly in Bayesian networks through a mathematical formalism based on probabilities and a distributed and flexible graphical representation [117], rather than on the traditional direct detection of doping. The hematological module of the ABP collects information on blood markers to detect the different methods of blood doping (autologous, homologous, or heterologous blood transfusions) and/or ESA abuse. The blood variables used to determine the ABP are hematocrit, hemoglobin, red blood cells count, mean corpuscular volume, mean corpuscular hemoglobin, mean corpuscular hemoglobin concentration, reticulocytes count, and percentage of reticulocytes. In addition, the multiparametric markers OFF-Hr score (index of stimulation) and ABPS (Abnormal Blood Profile Score) are calculated from this set of parameters [118].

The development of blood tests to identify athletes using the previously undetectable drug rhEpo has been one of the major aims in anti-doping research. Since 2000, a series of papers have been published showing the accuracy of a novel method for the indirect detection of rhEpo abuse [119–122]. Therefore, the indirect method to detect blood doping is based on the determination of several blood parameters: hemoglobin, hematocrit, reticulocytes (%) and on the calculation of the stimulation index (OFF-Hr Score). This index consists of applying a statistical model using two blood parameters: percentage of reticulocytes and hemoglobin (Hb in g/L) on the OFF-model score: $Hb - 60\sqrt{R}$. Both parameters are substantially altered after a period of Epo administration [119,121–123]. Reticulocytes are a crucial parameter in the ABP; their values increase significantly after rhEpo injection [121,124] or phlebotomy [125,126], and decrease when rhEpo treatment ceases [119] or blood is re-infused [127]. The accuracy of this method for the indirect detection of Epo abuse has been very well documented [119,121,122] which is why the OFF-Hr Score has been adopted by the anti-doping agencies [119]. However, it has been recently demonstrated that abusive rhEpo administration is possible without triggering ABP thresholds [128]. Thus, additional and more sensitive markers are needed to detect rhEpo abuse by means of the ABP model. For this purpose, several markers, such as total hemoglobin mass (Hbmass), bilirubin, ferritin, Hbmr, RBCHb/RetHb ratio, and/or Serum Epo concentration have been suggested [117,129–132].

MASKING AGENTS

According to the WADA, a substance or a method is considered doping when (among other criteria) its masking effect is recognized [133]. A masking method is a drug and/or a system used by athletes with the purpose of altering the presence of illegal substances controlled by the anti-doping authorities. The masking agents have the potential of masking the prohibited substance in the athlete's urine samples. According to the WADA prohibited list, group S5 "diuretics and other masking agents" are prohibited both in and out of competition in sports. In the 2012 list of prohibited substances they include diuretics, desmopressin, plasma expanders, probenecid, and other substances with similar biological effects [87]. Diuretics are therapeutic agents that are used to increase the rate of urine flow and sodium excretion in order to adjust the volume and composition of body fluids or to eliminate excess of fluids from tissues [134]. They are used in clinical therapy for the treatment of various diseases such as hypertension, heart failure, renal failure, and so on. Diuretics were first prohibited in sport in 1988 because they can be used by athletes as masking agents or for loss of body weight. Their potent ability to remove water from the body can cause a rapid weight loss that can be used to meet a weight category in sporting events such as Judo, Boxing, etc. Moreover they can be used to mask the administration of other doping agents by reducing their concentration in urine primarily because of an increase in urine volume [134]. Some diuretics also cause a masking effect by altering the urinary pH and inhibiting the excretion of acidic and basic drugs in urine [135]. In 2008, diuretics represented ~8% of all adverse analytical findings reported by the WADA's laboratories, with a total number of 436 cases [136], hydrochlorothiazide being the most common diuretic detected [134]. Over the years, the total number of positives for diuretics has been increasing. For instance, in 2012, Frank Schleck tested positive in the Tour de France for the banned diuretic xipamide and the IAAF suspended the Moroccan runner Mariem Alaoui Selsouli from the London Olympics after she tested positive for the diuretic furosemide.

Regarding the analytical methods for the identification of these prohibited compounds, initially the detection of diuretics was achieved using high-performance liquid chromatography with ultraviolet-diode array detection. However, gas chromatography / mass spectrometry has been the most widely used analytical technique for their detection [134].

In our laboratory we have studied whether normobaric intermittent hypoxia (NIH) and desmopressin (1-desamino-8-D-arginine-vasopressin), a modified form of the hormone vasopressin that works by limiting the amount of water that is eliminated in the urine, should be considered masking methods in sports. Regarding our experiments on intermittent hypoxia we have shown, in an animal study, that the hematological modifications (hemoglobin, hematocrit, and reticulocytes) achieved with an NIH protocol (12 h 21% O_2–12 h 12% O_2 for 15 days) are comparable with those that imply a treatment with 300 IU (3 times a week subcutaneously for 15 days) of rhEpo. Moreover we have also reported that NIH, after rhEpo administration, can significantly modify the main hematological parameters tested by the anti-doping authorities (hemoglobin, hematocrit, reticulocytes, and the OFF-Hr Score), making the misuse of rhEpo more difficult to detect using the indirect methods [133,137].

Regarding desmopressin, we have shown in humans that this drug decreases significantly the hematological values which are measured by the anti-doping authorities to detect blood doping [138]. Thus, and since 2011, it has been included in the WADA's list of prohibited substances.

GENE DOPING

Although there are no known cases of gene doping, the full publication of the human genome in 2004 [139] has prompted scientists' concern about a handful of the 25 000-odd genes detected which could become improvers of athletic performance [140]. Transferring selected genes into harmless viruses and then injecting them into the body improves specific traits in athletes [141]. Some candidates proposed by Lainscak et al. include the insulin growth factor (IGF-I) gene, which increases muscle hypertrophy and hyperplasia; the gene for mechano-growth factor (MGF), which has several roles in limiting fatigue; the AMPK gene, which makes it possible to accumulate glycogen in the muscle; and the angiotensin-converting enzyme gene, which if deleted can increase strength and if overexpressed can improve endurance. While Epo is the conventional method to increase red blood cell count, it can now be detected. However, similar effects can be gained by inhibiting the hematopoietic cell phosphatase (HCP) gene [140]. Also myostatin can be considered as a potential doping gene. Its removal or blockade results in a significant increase in muscle mass [142].

References

[1] Muller RK. History of doping and doping control. Handb Exp Pharmacol 2009;195:1–23.

[2] Fraser AD. Doping control from a global and national perspective. Ther Drug Monit 2004;26(2):171–4.

[3] Bowers LD. Anti-dope testing in sport: the history and the science. FASEB J 2012;26(10):3933–6.

[4] Teale P, Barton C, Driver PM, Kay RG. Biomarkers: unrealized potential in sports doping analysis. Bioanalysis 2009;1(6): 1103–18.

[5] WADA. World anti-doping code. 2009. Retrieved from: <http://www.wada-ama.org/en/World-Anti-Doping-Program/Sports-and-Anti-Doping-Organizations/The-Code/>.

[6] Kayser B, Broers B. The Olympics and harm reduction? Harm Reduct J 2012;9(1):33.

[7] Botrè F. New and old challenges of sports drug testing. J Mass Spectrom 2008;43(7):903–7.

[8] Basaria S. Androgen abuse in athletes: detection and consequences. J Clin Endocrinol Metab 2010;95(4):1533–43.

[9] Kicman AT, Brooks RV, Collyer SC, Cowan DA, Nanjee MN, Southan GJ, et al. Criteria to indicate testosterone administration. Br J Sports Med. 1990;24(4):253–64.

[10] Aguilera R, Chapman TE, Starcevic B, Hatton CK, Catlin DH. Performance characteristics of a carbon isotope ratio method for detecting doping with testosterone based on urine diols: controls and athletes with elevated testosterone/epitestosterone ratios. Clin Chem 2001;47(2):292–300.

[11] Jakobsson J, Ekstrom L, Inotsume N, Garle M, Lorentzon M, Ohlsson C, et al. Large differences in testosterone excretion in Korean and Swedish men are strongly associated with a UDP-glucuronosyl transferase 2B17 polymorphism. J Clin Endocrinol Metab 2006;91(2):687–93.

[12] Dehennin L. Detection of simultaneous self-administration of testosterone and epitestosterone in healthy men. Clin Chem 1994;40(1):106–9.

[13] Aguilera R, Hatton CK, Catlin DH. Detection of epitestosterone doping by isotope ratio mass spectrometry. Clin Chem 2002;48 (4):629–36.

[14] Catlin DH, Hatton CK, Starcevic SH. Issues in detecting abuse of xenobiotic anabolic steroids and testosterone by analysis of athletes' urine. Clin Chem 1997;43(7):1280–8.
[15] Shackleton CH, Phillips A, Chang T, Li Y. Confirming testosterone administration by isotope ratio mass spectrometric analysis of urinary androstanediols. Steroids 1997;62(4):379–87.
[16] Leder BZ, Longcope C, Catlin DH, Ahrens B, Schoenfeld DA, Finkelstein JS. Oral androstenedione administration and serum testosterone concentrations in young men. JAMA 2000;283 (6):779–82.
[17] Morales AJ, Haubrich RH, Hwang JY, Asakura H, Yen SS. The effect of six months treatment with a 100 mg daily dose of dehydroepiandrosterone (DHEA) on circulating sex steroids, body composition and muscle strength in age-advanced men and women. Clin Endocrinol (Oxf) 1998;49(4):421–32.
[18] Bowers LD. Oral dehydroepiandrosterone supplementation can increase the testosterone/epitestosterone ratio. Clin Chem 1999;45(2):295–7.
[19] Dehennin L, Ferry M, Lafarge P, Peres G, Lafarge JP. Oral administration of dehydroepiandrosterone to healthy men: alteration of the urinary androgen profile and consequences for the detection of abuse in sport by gas chromatography-mass spectrometry. Steroids 1998;63(2):80–7.
[20] Winters SJ, Janick JJ, Loriaux DL, Sherins RJ. Studies on the role of sex steroids in the feedback control of gonadotropin concentrations in men. II. Use of the estrogen antagonist, clomiphene citrate. J Clin Endocrinol Metab 1979; 48(2):222–7.
[21] Hayes FJ, Seminara SB, Decruz S, Boepple PA, Crowley Jr. WF. Aromatase inhibition in the human male reveals a hypothalamic site of estrogen feedback. J Clin Endocrinol Metab 2000;85(9):3027–35.
[22] Leder BZ, Rohrer JL, Rubin SD, Gallo J, Longcope C. Effects of aromatase inhibition in elderly men with low or borderline-low serum testosterone levels. J Clin Endocrinol Metab 2004;89 (3):1174–80.
[23] Taxel P, Kennedy DG, Fall PM, Willard AK, Clive JM, Raisz LG. The effect of aromatase inhibition on sex steroids, gonadotropins, and markers of bone turnover in older men. J Clin Endocrinol Metab 2001;86(6):2869–74.
[24] Brodie A, Lu Q, Long B. Aromatase and its inhibitors. J Steroid Biochem Mol Biol 1999;69(1-6):205–10.
[25] Cowan DA, Kicman AT, Walker CJ, Wheeler MJ. Effect of administration of human chorionic gonadotrophin on criteria used to assess testosterone administration in athletes. J Endocrinol 1991;131(1):147–54.
[26] Mareck U, Geyer H, Guddat S, Haenelt N, Koch A, Kohler M, et al. Identification of the aromatase inhibitors anastrozole and exemestane in human urine using liquid chromatography/tandem mass spectrometry. Rapid Commun Mass Spectrom 2006;20(12):1954–62.
[27] Mareck U, Sigmund G, Opfermann G, Geyer H, Thevis M, Schanzer W. Identification of the aromatase inhibitor letrozole in urine by gas chromatography/mass spectrometry. Rapid Commun Mass Spectrom 2005;19(24):3689–93.
[28] Kicman AT, Parkin MC, Iles RK. An introduction to mass spectrometry based proteomics-detection and characterization of gonadotropins and related molecules. Mol Cell Endocrinol 2007;260–262:212–27.
[29] Robinson N, Saudan C, Sottas PE, Mangin P, Saugy M. Performance characteristics of two immunoassays for the measurement of urinary luteinizing hormone. J Pharm Biomed Anal 2007;43(1):270–6.
[30] Bhasin S, Storer TW, Berman N, Callegari C, Clevenger B, Phillips J, et al. The effects of supraphysiologic doses of testosterone on muscle size and strength in normal men. N Engl J Med 1996;335(1):1–7.
[31] Bhasin S, Woodhouse L, Casaburi R, Singh AB, Bhasin D, Berman N, et al. Testosterone dose-response relationships in healthy young men. Am J Physiol Endocrinol Metab 2001;281 (6):E1172–81.
[32] Franke WW, Berendonk B. Hormonal doping and androgenization of athletes: a secret program of the German democratic republic government. Clin Chem 1997;43(7):1262–79.
[33] Sinha-Hikim I, Artaza J, Woodhouse L, Gonzalez-Cadavid N, Singh AB, Lee MI, et al. Testosterone-induced increase in muscle size in healthy young men is associated with muscle fiber hypertrophy. Am J Physiol Endocrinol Metab 2002;283(1): E154–64.
[34] Brodsky IG, Balagopal P, Nair KS. Effects of testosterone replacement on muscle mass and muscle protein synthesis in hypogonadal men—a clinical research center study. J Clin Endocrinol Metab 1996;81(10):3469–75.
[35] Ferrando AA, Sheffield-Moore M, Paddon-Jones D, Wolfe RR, Urban RJ. Differential anabolic effects of testosterone and amino acid feeding in older men. J Clin Endocrinol Metab 2003;88(1):358–62.
[36] Parssinen M, Kujala U, Vartiainen E, Sarna S, Seppala T. Increased premature mortality of competitive powerlifters suspected to have used anabolic agents. Int J Sports Med 2000;21 (3):225–7.
[37] Fineschi V, Baroldi G, Monciotti F, Paglicci Reattelli L, Turillazzi E. Anabolic steroid abuse and cardiac sudden death: a pathologic study. Arch Pathol Lab Med 2001;125(2):253–5.
[38] Fineschi V, Riezzo I, Centini F, Silingardi E, Licata M, Beduschi G, et al. Sudden cardiac death during anabolic steroid abuse: morphologic and toxicologic findings in two fatal cases of bodybuilders. Int J Legal Med 2007;121(1):48–53.
[39] Payne JR, Kotwinski PJ, Montgomery HE. Cardiac effects of anabolic steroids. Heart 2004;90(5):473–5.
[40] Sullivan ML, Martinez CM, Gallagher EJ. Atrial fibrillation and anabolic steroids. J Emerg Med 1999;17(5):851–7.
[41] Lau DH, Stiles MK, John B, Shashidhar, Young GD, Sanders P. Atrial fibrillation and anabolic steroid abuse. Int J Cardiol 2007;117(2):e86–7.
[42] D'Andrea A, Caso P, Salerno G, Scarafile R, De Corato G, Mita C, et al. Left ventricular early myocardial dysfunction after chronic misuse of anabolic androgenic steroids: a Doppler myocardial and strain imaging analysis. Br J Sports Med 2007;41 (3):149–55.
[43] Glazer G. Atherogenic effects of anabolic steroids on serum lipid levels. A literature review. Arch Intern Med 1991;151 (10):1925–33.
[44] Thompson PD, Cullinane EM, Sady SP, Chenevert C, Saritelli AL, Sady MA, et al. Contrasting effects of testosterone and stanozolol on serum lipoprotein levels. JAMA 1989; 261(8):1165–8.
[45] Pope Jr. HG, Kouri EM, Hudson JI. Effects of supraphysiologic doses of testosterone on mood and aggression in normal men: a randomized controlled trial. Arch Gen Psychiatry 2000;57 (2):133–40 discussion 55–6.
[46] Klotz F, Garle M, Granath F, Thiblin I. Criminality among individuals testing positive for the presence of anabolic androgenic steroids. Arch Gen Psychiatry 2006;63(11):1274–9.
[47] Clark AS, Henderson LP. Behavioral and physiological responses to anabolic-androgenic steroids. Neurosci Biobehav Rev 2003;27(5):413–36.
[48] Basaria S, Dobs AS. Safety and adverse effects of androgens: how to counsel patients. Mayo Clin Proc 2004;79(4 Suppl.): S25–32.

[49] Visuri T, Lindholm H. Bilateral distal biceps tendon avulsions with use of anabolic steroids. Med Sci Sports Exerc 1994;26 (8):941–4.

[50] David HG, Green JT, Grant AJ, Wilson CA. Simultaneous bilateral quadriceps rupture: a complication of anabolic steroid abuse. J Bone Joint Surg Br 1995;77(1):159–60.

[51] Melnik B, Jansen T, Grabbe S. Abuse of anabolic-androgenic steroids and bodybuilding acne: an underestimated health problem. J Dtsch Dermatol Ges 2007;5(2):110–7.

[52] Weppelman RM. Effects of gonadal steroids and adrenergic agonists on avian growth and feed efficiency. J Exp Zool 1984;232(3):461–4.

[53] Martineau L, Horan MA, Rothwell NJ, Little RA. Salbutamol, a beta 2-adrenoceptor agonist, increases skeletal muscle strength in young men. Clin Sci (Lond) 1992;83(5):615–21.

[54] Caruso JF, Signorile JF, Perry AC, Leblanc B, Williams R, Clark M, et al. The effects of albuterol and isokinetic exercise on the quadriceps muscle group. Med Sci Sports Exerc 1995;27 (11):1471–6.

[55] Choo JJ, Horan MA, Little RA, Rothwell NJ. Anabolic effects of clenbuterol on skeletal muscle are mediated by beta 2-adrenoceptor activation. Am J Physiol 1992;263(1 Pt 1):E50–6.

[56] Frances H, Diquet B, Goldschmidt P, Simon P. Tolerance to or facilitation of pharmacological effects induced by chronic treatment with the beta-adrenergic stimulant clenbuterol. J Neural Transm 1985;62(1–2):65–76.

[57] Rehfeldt C, Schadereit R, Weikard R, Reichel K. Effect of clenbuterol on growth, carcase and skeletal muscle characteristics in broiler chickens. Br Poult Sci 1997;38(4):366–73.

[58] Duncan ND, Williams DA, Lynch GS. Deleterious effects of chronic clenbuterol treatment on endurance and sprint exercise performance in rats. Clin Sci (Lond) 2000;98(3):339–47.

[59] Burniston JG, Ng Y, Clark WA, Colyer J, Tan LB, Goldspink DF. Myotoxic effects of clenbuterol in the rat heart and soleus muscle. J Appl Physiol 2002;93(5):1824–32.

[60] Keskin S, Ozer D, Temizer A. Gas chromatography-mass spectrometric analysis of clenbuterol from urine. J Pharm Biomed Anal 1998;18(4–5):639–44.

[61] Amendola L, Colamonici C, Rossi F, Botre F. Determination of clenbuterol in human urine by GC-MS-MS-MS: confirmation analysis in antidoping control. J Chromatogr B Analyt Technol Biomed Life Sci 2002;773(1):7–16.

[62] Koole A, Bosman J, Franke JP, de Zeeuw RA. Multiresidue analysis of beta2-agonist in human and calf urine using multimodal solid-phase extraction and high-performance liquid chromatography with electrochemical detection. J Chromatogr B Biomed Sci Appl 1999;726(1–2):149–56.

[63] Aresta A, Calvano CD, Palmisano F, Zambonin CG. Determination of clenbuterol in human urine and serum by solid-phase microextraction coupled to liquid chromatography. J Pharm Biomed Anal 2008;47(3):641–5.

[64] Svensson J, Sunnerhagen KS, Johannsson G. Five years of growth hormone replacement therapy in adults: age- and gender-related changes in isometric and isokinetic muscle strength. J Clin Endocrinol Metab 2003;88(5):2061–9.

[65] Vittone J, Blackman MR, Busby-Whitehead J, Tsiao C, Stewart KJ, Tobin J, et al. Effects of single nightly injections of growth hormone-releasing hormone (GHRH 1-29) in healthy elderly men. Metabolism 1997;46(1):89–96.

[66] Widdowson WM, Gibney J. The effect of growth hormone (GH) replacement on muscle strength in patients with GH-deficiency: a meta-analysis. Clin Endocrinol (Oxf) 2010;72(6):787–92.

[67] Widdowson WM, Gibney J. The effect of growth hormone replacement on exercise capacity in patients with GH deficiency: a metaanalysis. J Clin Endocrinol Metab 2008;93 (11):4413–7.

[68] Crist DM, Peake GT, Egan PA, Waters DL. Body composition response to exogenous GH during training in highly conditioned adults. J Appl Physiol 1988;65(2):579–84.

[69] Yarasheski KE, Zachweija JJ, Angelopoulos TJ, Bier DM. Short-term growth hormone treatment does not increase muscle protein synthesis in experienced weight lifters. J Appl Physiol 1993;74(6):3073–6.

[70] Meinhardt U, Nelson AE, Hansen JL, Birzniece V, Clifford D, Leung KC, et al. The effects of growth hormone on body composition and physical performance in recreational athletes: a randomized trial. Ann Intern Med 2010;152(9):568–77.

[71] Doessing S, Heinemeier KM, Holm L, Mackey AL, Schjerling P, Rennie M, et al. Growth hormone stimulates the collagen synthesis in human tendon and skeletal muscle without affecting myofibrillar protein synthesis. J Physiol 2010;588(Pt 2):341–51.

[72] Raschke M, Rasmussen MH, Govender S, Segal D, Suntum M, Christiansen JS. Effects of growth hormone in patients with tibial fracture: a randomised, double-blind, placebo-controlled clinical trial. Eur J Endocrinol 2007;156(3):341–51.

[73] Wu Z, Bidlingmaier M, Dall R, Strasburger CJ. Detection of doping with human growth hormone. Lancet 1999;353 (9156):895.

[74] Momomura S, Hashimoto Y, Shimazaki Y, Irie M. Detection of exogenous growth hormone (GH) administration by monitoring ratio of 20kDa- and 22kDa-GH in serum and urine. Endocr J 2000;47(1):97–101.

[75] Bidlingmaier M, Suhr J, Ernst A, Wu Z, Keller A, Strasburger CJ, et al. High-sensitivity chemiluminescence immunoassays for detection of growth hormone doping in sports. Clin Chem 2009;55(3):445–53.

[75a] World Anti-Doping Agency Guidelines: hGH isoform differential immunoassays for anti-doping analyses, version 1.0, <http://www.wada-ama.org>; June 2010.

[76] Kassem M, Brixen K, Blum WF, Mosekilde L, Eriksen EF. Normal osteoclastic and osteoblastic responses to exogenous growth hormone in patients with postmenopausal spinal osteoporosis. J Bone Miner Res 1994;9(9):1365–70.

[77] Duntas LH, Popovic V. Hormones as doping in sports. Endocrine 2012.

[78] Patel K, Amthor H. The function of Myostatin and strategies of Myostatin blockade—new hope for therapies aimed at promoting growth of skeletal muscle. Neuromuscul Disord 2005;15 (2):117–26.

[79] McPherron AC, Lee SJ. Double muscling in cattle due to mutations in the myostatin gene. Proc Natl Acad Sci U S A 1997;94 (23):12457–61.

[80] McPherron AC, Lawler AM, Lee SJ. Regulation of skeletal muscle mass in mice by a new TGF-beta superfamily member. Nature 1997;387(6628):83–90.

[81] Reisz-Porszasz S, Bhasin S, Artaza JN, Shen R, Sinha-Hikim I, Hogue A, et al. Lower skeletal muscle mass in male transgenic mice with muscle-specific overexpression of myostatin. Am J Physiol Endocrinol Metab 2003;285(4):E876–88.

[82] Schuelke M, Wagner KR, Stolz LE, Hubner C, Riebel T, Komen W, et al. Myostatin mutation associated with gross muscle hypertrophy in a child. N Engl J Med 2004;350(26):2682–8.

[83] Lee SJ, McPherron AC. Regulation of myostatin activity and muscle growth. Proc Natl Acad Sci U S A 2001;98(16):9306–11.

[84] Whittemore LA, Song K, Li X, Aghajanian J, Davies M, Girgenrath S, et al. Inhibition of myostatin in adult mice increases skeletal muscle mass and strength. Biochem Biophys Res Commun 2003;300(4):965–71.

[85] Narkar VA, Downes M, Yu RT, Embler E, Wang YX, Banayo E, et al. AMPK and PPARdelta agonists are exercise mimetics. Cell 2008;134(3):405–15.

[86] Lira VA, Benton CR, Yan Z, Bonen A. PGC-1alpha regulation by exercise training and its influences on muscle function and insulin sensitivity. Am J Physiol Endocrinol Metab 2010;299(2): E145–61.

[87] WADA. The 2012 prohibited list of Substances international standard, <wwwwada-amaorg.1-9>; 2012.

[88] Bostrom P, Wu J, Jedrychowski MP, Korde A, Ye L, Lo JC, et al. A PGC1-alpha-dependent myokine that drives brown-fat-like development of white fat and thermogenesis. Nature 2012.

[89] Thevis M, Moller I, Thomas A, Beuck S, Rodchenkov G, Bornatsch W, et al. Characterization of two major urinary metabolites of the PPARdelta-agonist GW1516 and implementation of the drug in routine doping controls. Anal Bioanal Chem 2009;396(7):2479–91.

[90] Thomas A, Beuck S, Eickhoff JC, Guddat S, Krug O, Kamber M, et al. Quantification of urinary AICAR concentrations as a matter of doping controls. Anal Bioanal Chem 2010;396 (8):2899–908.

[91] WADA. The 2013 prohibited list of Substances international standard, <www.wada-ama.org.1-9>; 2013.

[92] Dalton JT, Mukherjee A, Zhu Z, Kirkovsky L, Miller DD. Discovery of nonsteroidal androgens. Biochem Biophys Res Commun 1998;244(1):1–4.

[93] Thevis M, Kohler M, Maurer J, Schlorer N, Kamber M, Schanzer W. Screening for 2-quinolinone-derived selective androgen receptor agonists in doping control analysis. Rapid Commun Mass Spectrom 2007;21(21):3477–86.

[94] Thevis M, Kohler M, Thomas A, Maurer J, Schlorer N, Kamber M, et al. Determination of benzimidazole- and bicyclic hydantoin-derived selective androgen receptor antagonists and agonists in human urine using LC-MS/MS. Anal Bioanal Chem 2008;391(1):251–61.

[95] Sanchis-Gomar F, Lippi G. Telmisartan as metabolic modulator: a new perspective in sports doping? J Strength Cond Res 2012;26(3):608–10.

[96] Feng X, Luo Z, Ma L, Ma S, Yang D, Zhao Z, et al. Angiotensin II receptor blocker telmisartan enhances running endurance of skeletal muscle through activation of the PPAR delta/ AMPK pathway. J Cell Mol Med 2010.

[97] World Anti-Doping Agency. Available from: <www.wada-ama.org>; [cited 28.08.2012].

[98] Jelkmann W, Lundby C. Blood doping and its detection. Blood 2012;118(9):2395–404.

[99] Lundby C, Robach P, Saltin B. The evolving science of detection of 'blood doping'. Br J Pharmacol 2012;165(5):1306–15.

[100] Elliott S. Erythropoiesis-stimulating agents and other methods to enhance oxygen transport. Br J Pharmacol 2008;154(3): 529–41.

[101] Leuenberger N, Reichel C, Lasne F. Detection of erythropoiesis-stimulating agents in human anti-doping control: past, present and future. Bioanalysis 2012;4(13):1565–75.

[102] Ebert BL, Bunn HF. Regulation of the erythropoietin gene. Blood 1999;94(6):1864–77.

[103] Lundby C, Thomsen JJ, Boushel R, Koskolou M, Warberg J, Calbet JA, et al. Erythropoietin treatment elevates haemoglobin concentration by increasing red cell volume and depressing plasma volume. J Physiol 2007;578(Pt 1):309–14.

[104] Semenza GL, Wang GL. A nuclear factor induced by hypoxia via de novo protein synthesis binds to the human erythropoietin gene enhancer at a site required for transcriptional activation. Mol Cell Biol 1992;12(12):5447–54.

[105] Calbet JA, Lundby C, Koskolou M, Boushel R. Importance of hemoglobin concentration to exercise: acute manipulations. Respir Physiol Neurobiol 2006;151(2-3):132–40.

[106] Robach P, Calbet JA, Thomsen JJ, Boushel R, Mollard P, Rasmussen P, et al. The ergogenic effect of recombinant human erythropoietin on VO_2max depends on the severity of arterial hypoxemia. PLoS One 2008;3(8):e2996.

[107] Reichel C. Sports drug testing for erythropoiesis-stimulating agents and autologous blood transfusion. Drug Test Anal 2012;4(11):803–4.

[108] Lasne F, Martin L, Crepin N, de Ceaurriz J. Detection of isoelectric profiles of erythropoietin in urine: differentiation of natural and administered recombinant hormones. Anal Biochem 2002;311(2):119–26.

[109] Wide L, Bengtsson C. Molecular charge heterogeneity of human serum erythropoietin. Br J Haematol 1990;76(1): 121–7.

[110] Lasne F, de Ceaurriz J. Recombinant erythropoietin in urine. Nature 2000;405(6787):635.

[111] Lasne F. Double-blotting: a solution to the problem of non-specific binding of secondary antibodies in immunoblotting procedures. J Immunol Methods 2001;253(1-2):125–31.

[112] Reichel C, Abzieher F, Geisendorfer T. SARCOSYL-PAGE: a new method for the detection of MIRCERA- and EPO-doping in blood. Drug Test Anal 2009;1(11–12):494–504.

[113] Reichel C. SARCOSYL-PAGE: a new electrophoretic method for the separation and immunological detection of PEGylated proteins. Methods Mol Biol 2012;869:65–79.

[114] WADA. WADA Technical Document EPO2009. Harmonization of The Method for the Recombinant Erythropoietins (i.e. epoetins) and analogues (e.g. Darbopoietin and methoxypolyethylene glycol-epoitin beta), <wwwwada-amaorg.1-9>; 2009.

[115] Birchard K. Past, present, and future of drug abuse at the Olympics. Lancet 2000;356(9234):1008.

[116] Gilbert S. The biological passport. Hastings Cent Rep 2010;40 (2):18–9.

[117] Sottas PE, Robinson N, Saugy M. The athlete's biological passport and indirect markers of blood doping. Handb Exp Pharmacol 2010;195:305–26.

[118] WADA. The World Anti-Doping Code. Athlete biological passport operating guidelines and compilation of required elements, <wwwwada-amaorg.1-31>; 2010.

[119] Gore CJ, Parisotto R, Ashenden MJ, Stray-Gundersen J, Sharpe K, Hopkins W, et al. Second-generation blood tests to detect erythropoietin abuse by athletes. Haematologica 2003;88 (3):333–44.

[120] Parisotto R, Gore CJ, Emslie KR, Ashenden MJ, Brugnara C, Howe C, et al. A novel method utilising markers of altered erythropoiesis for the detection of recombinant human erythropoietin abuse in athletes. Haematologica 2000;85(6): 564–72.

[121] Parisotto R, Wu M, Ashenden MJ, Emslie KR, Gore CJ, Howe C, et al. Detection of recombinant human erythropoietin abuse in athletes utilizing markers of altered erythropoiesis. Haematologica 2001;86(2):128–37.

[122] Sharpe K, Ashenden MJ, Schumacher YO. A third generation approach to detect erythropoietin abuse in athletes. Haematologica 2006;91(3):356–63.

[123] Parisotto R, Ashenden MJ, Gore CJ, Sharpe K, Hopkins W, Hahn AG. The effect of common hematologic abnormalities on the ability of blood models to detect erythropoietin abuse by athletes. Haematologica 2003;88(8):931–40.

[124] Audran M, Gareau R, Matecki S, Durand F, Chenard C, Sicart MT, et al. Effects of erythropoietin administration in training

athletes and possible indirect detection in doping control. Med Sci Sports Exerc 1999;31(5):639–45.

[125] Morkeberg J, Belhage B, Ashenden M, Borno A, Sharpe K, Dziegiel MH, et al. Screening for autologous blood transfusions. Int J Sports Med 2009;30(4):285–92.

[126] Damsgaard R, Munch T, Morkeberg J, Mortensen SP, Gonzalez-Alonso J. Effects of blood withdrawal and reinfusion on biomarkers of erythropoiesis in humans: Implications for anti-doping strategies. Haematologica 2006; 91(7):1006–8.

[127] Morkeberg J, Sharpe K, Belhage B, Damsgaard R, Schmidt W, Prommer N, et al. Detecting autologous blood transfusions: a comparison of three passport approaches and four blood markers. Scand J Med Sci Sports 2009.

[128] Ashenden M, Gough CE, Garnham A, Gore CJ, Sharpe K. Current markers of the Athlete blood passport do not flag microdose EPO doping. Eur J Appl Physiol 2011.

[129] Morkeberg J, Sharpe K, Belhage B, Damsgaard R, Schmidt W, Prommer N, et al. Detecting autologous blood transfusions: a comparison of three passport approaches and four blood markers. Scand J Med Sci Sports 2009;21(2):235–43.

[130] Jelkmann W, Lundby C. Blood doping and its detection. Blood 2011;118(9):2395–404.

[131] Pottgiesser T, Echteler T, Sottas PE, Umhau M, Schumacher YO. Hemoglobin mass and biological passport for the detection of autologous blood doping. Med Sci Sports Exerc 2011.

[132] Morkeberg J. Detection of autologous blood transfusions in athletes: a historical perspective. Transfus Med Rev 2011;26 (3):199–208.

[133] Sanchis-Gomar F, Martinez-Bello VE, Gomez-Cabrera MC, Vina J. It is not hypoxia itself, but how you use it. Eur J Appl Physiol 2010;109(2):355–6.

[134] Cadwallader AB, de la Torre X, Tieri A, Botre F. The abuse of diuretics as performance-enhancing drugs and masking agents in sport doping: pharmacology, toxicology and analysis. Br J Pharmacol 2010;161(1):1–16.

[135] Ventura R, Segura J. Detection of diuretic agents in doping control. J Chromatogr B Biomed Appl 1996;687(1):127–44.

[136] WADA. The 2008 WADA Laboratory Statistics. Available at: <http://www.wada-ama.org.1-9>; 2009 [accessed 14.04. 2012].

[137] Sanchis-Gomar F, Martinez-Bello VE, Domenech E, Nascimento AL, Pallardo FV, Gomez-Cabrera MC, et al. Effect of intermittent hypoxia on hematological parameters after recombinant human erythropoietin administration. Eur J Appl Physiol 2009;107(4):429–36.

[138] Sanchis-Gomar F, Martinez-Bello VE, Nascimento AL, Perez-Quilis C, Garcia-Gimenez JL, Vina J, et al. Desmopresssin and hemodilution: implications in doping. Int J Sports Med 2010;31 (1):5–9.

[139] Consortium I. H. G. S. Finishing the euchromatic sequence of the human genome. Nature 2004;431(7011):931–45.

[140] Lainscak M, Osredkar J. Doping and the Olympic games: the good, the bad, and the ugly. Wien Klin Wochenschr 2009;121 (1–2):13–4.

[141] Adam D. Gene therapy may be up to speed for cheats at 2008 Olympics. Nature 2001;414(6864):569–70.

[142] Wells DJ. Gene doping: the hype and the reality. Br J Pharmacol 2008;154(3):623–31.

CHAPTER

19

Nutrition In Paralympics

Amitava Das[1], *Debasis Bagchi*[2] *and Chandan K.Sen* [1]

[1]Department of Surgery, The Ohio State University Wexner Medical Center, Columbus, OH, USA [2]University of Houston College of Pharmacy, Houston, TX, USA

INTRODUCTION

"... What I learned was that these athletes were not disabled, they were superabled. The Olympics is where heroes are made. The Paralympics is where heroes come." *Joey Reiman (author and purpose visionary)*

These words [1] capture the feelings of billions of people all over the globe who look upon these athletes as epitomes of courage, dedication and determination. The significance of the Paralympic Games has been highlighted by Giles Duley, the photographer who lost both his legs and an arm in an explosion in Afghanistan: *"since I lost my limbs I can see the true heroism of the competitors"* [2]. The discrimination of sports based on disability is in opposition to its fundamental principles of Olympism. The Olympics Charter [3] provides equal opportunities to athletics for all people by stating that: *"The practice of sport is a human right. Every individual must have the possibility of practicing sport, without discrimination of any kind and in the Olympic spirit, which requires mutual understanding with a spirit of friendship, solidarity and fair play .... Any form of discrimination with regard to a country or a person on grounds of race, religion, politics, gender or otherwise is incompatible with belonging to the Olympic Movement."* The Paralympic and Olympic movements are indivisible components of one noble mission. Emphasizing this fact, the chairman of the London Organizing Committee stated [4]: *"We want to change public attitudes towards disability, celebrate the excellence of Paralympic sport and to enshrine from the very outset that the two Games are an integrated whole."* The opening ceremony of the London 2012 Paralympics featured the renowned British theoretical physicist, cosmologist and recipient of the Albert Einstein Award, Professor Stephen Hawking, who suffers from motor neuron disease [5].

SPORTS NUTRITION AND ENHANCED PERFORMANCE

The nutritional strategy of competitive athletes substantially influences their sports performances [6]. Today, a well-developed nutritional plan is an integral component of the overall preparation to win. With the advancement in research in sports nutrition, the concept of the best diet for performance enhancement has undergone a paradigm shift. The guiding principles of sports nutrition have evolved, challenging earlier practices and notions [7]. Until 1968, protein-rich food was considered to be the chief source of fuel required for exercise. In the 1970s, that emphasis refocused on carbohydrate-rich foods [8]. Improving performance through dietary means often involves commercial or customized formulations. While opting for correct nutritional performance enhancers, regulatory and safety issues are of prime importance [10].

THE PARALYMPIC GAMES

Though the Ancient Olympics were first held in Greece in 776 BC, and the modern Olympics date back to the 17th century AD, the Paralympic games were initiated much later in the 20th century. Governed by the International Paralympic Committee, the games immediately follow the respective Olympic Games in the same city and involve athletes with an array of physical and intellectual disabilities. The International Stoke Mandeville Games, which later came to be known as The International Wheelchair and Amputee Sports (IWAS) World Games, included British veterans of the 2nd World War and paved the way to the 1988 Paralympic Games in Seoul. The name "Paralympic" is

Nutrition and Enhanced Sports Performance.
DOI: http://dx.doi.org/10.1016/B978-0-12-396454-0.00019-9

derived from the Greek preposition *pará* (meaning "beside" or "alongside"), suggesting that the games are held alongside the Olympic Games [11], which are organized biennially with alternating Summer and Winter Olympic Games.

CLASSIFICATION AND CATEGORIES AT THE PARALYMPIC GAMES

Because of the wide range of disabilities of Paralympic athletes, the Paralympics competitions are divided into several categories. The International Paralympic Committee (IPC) has recognized ten disability categories, applicable to both Summer and Winter Paralympics under which athletes with these disabilities can compete [12] (Figure 19.1).

The following eight different types of *physical impairment* are recognized:

- Impaired muscle power
- Impaired passive range of movement
- Loss of limb or limb deficiency
- Leg-length difference
- Short stature
- Hypertonia
- Ataxia
- Athetosis.

Visual impairment ranges from partial vision, sufficient to be evaluated as legally blind, to total blindness.

Intellectual disability is defined as significantly retarded intellectual functioning and associated inadequacies in adaptive behavior.

Though the classification determines the ability of an individual to be eligible to compete in a particular Paralympic sport and it is based on their activity limitation in a certain sport, the level of impairment will vary across individuals. To ensure fair competition amongst athletes with similar levels of ability, a classification system has been developed [13]. Until the 1980s, the Paralympic system for classifying athletes involved medical evaluation and diagnosis of impairment. The sole determinant of the class in which an athlete would compete was their medical condition, or more specifically the medical diagnosis. Only the cause of impairment contributed to the classification. The fact that different causes can lead to the same impairment did not attribute to the classification. In the 1980s, the classification system shifted from medical diagnosis to functional abilities of the athlete because of a change in views on disabled athletics from just a form of rehabilitation to an end in itself [9]. In this classification system, the effect of the athlete's impairment on his or her athletic performance is considered. As of 2012, the Paralympic sports comprise all of the sports contested in the Summer and Winter Paralympic Games— about 500 events. While some sports are open to multiple disability categories, others are limited to one.

NUTRITIONAL CONSIDERATIONS IN THE DISABLED

Physical Impairments

Nutrition represents a key component of the comprehensive health care strategy for persons with physical disability [14]. Since this category covers a wide

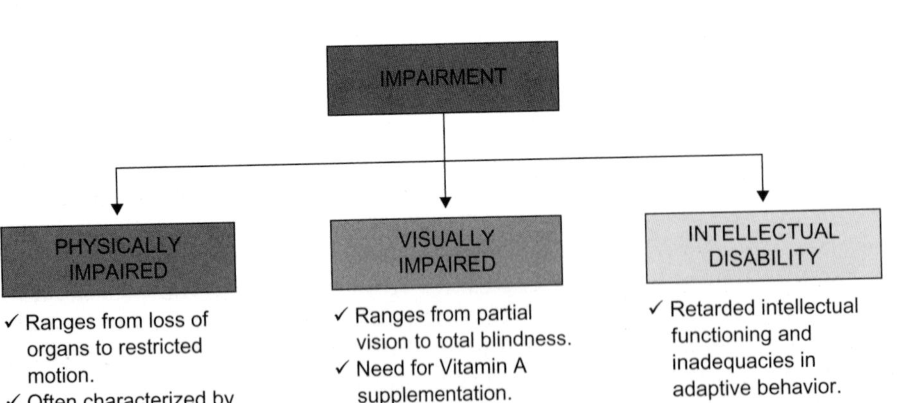

FIGURE 19.1 Summary of different forms of impairments along with their nutritional requirements. (This figure is reproduced in color plate section.)

range of disabilities ranging from loss of one or more organs to restricted motion, commenting on the nutritional requirements for this group as a whole becomes challenging. While some physically impaired individuals require excessive calories to compensate for exercise-associated excessive weight loss and tissue depletion, those restricted to wheelchairs require a diet restricted in calories [14]. Nutritional limitations like hypocalcemia and negative nitrogen balance may favor immobility [15]. Nutritional mineral requirements should be addressed in a balanced manner such that on one hand it meets the requirements of the sporting event such that appropriate calcium dosing for events involving excessive locomotor activities, while on the other hand, detrimental effects of excessive intake of dietary minerals should be matched with appropriate supplemental fluid intake [14]. Emphasis on protein nutrition is necessary for restitution of body tissues, to prevent infections and to maintain skin integrity [14]. People suffering from ataxia (lack of muscle coordination which adversely affects speech, eye movements, and other voluntary movements) may benefit from multivitamin supplementation (containing vitamin E—tocopherols and tocotrienols) with a special diet with minimized gluten content [16,17].

One of the most important physiological issues associated with physical impairment is the substantially increased cardiac burden for the same workload for amputees compared with non-amputees. The chronic overloading of the heart may result in heart failure. Increased mortality rates were reported in proximal limb amputees, due to cardiovascular disease [18]. People with limb amputation are reported to have higher mortality than those with deformity without loss of a body part and those with loss of a portion of hand or foot, and this increase was significant in the case of patients suffering from various heart diseases [18]. While the mortality in patients with "proximal" amputation of the lower limb was around 1.4-fold higher than that in disfigured veterans, "distal" amputation and amputation of the arm did not pose any death threat [19]. For bilateral leg amputation the threat of cardiovascular overload is substantially higher [19]. The increased mortality due to cardiovascular diseases in traumatic amputees is attributed to risk factors like hyperinsulinemia, increased coagulability, and increased autonomic responses [20]. Hemodynamic complications arising from proximity to the occluded femoral artery, and related factors such as shear stress, circumferential strain and reflected waves, are also suggested to explain why traumatic leg amputees are at increased risk for cardiovascular failure [21]. These cardiovascular challenges faced by amputees call for appropriate nutritional countermeasures [22].

Visual Impairments

The role of vitamins, especially vitamin A, in the visual cycle is well established. The promise of nutritional intervention in the development of cataracts and age-related macular degeneration (AMD) is well developed [23]. Vitamins C, E and carotenoids like lutein and zeaxanthin are predicted to derail the development and progression of cataracts, due to their antioxidant properties [24]. Visually challenged athletes must consider these options in developing their nutritional strategy.

Intellectual Disability

Diets have been reported to be effective in managing secondary complications like fatigue of people suffering from intellectual and developmental disabilities [25]. Previous reports suggest that, compared with the general population, intellectually disabled people are more prone to nutrition-related health problems, especially obesity and chronic constipation [26]. Nutritional strategy for such challenged people should therefore consider these special needs, with the aim to lower the risk of common chronic diseases and conditions associated with this population [25].

SPORTS NUTRITION OF PARALYMPIC ATHLETES

Best performance in a given sporting event depends on the optimal function of a number of body functions, including skeletal, respiratory, cardiovascular, and the nervous system [27]. While these considerations may be enough for able-bodied athletes, paralympians need additional nutritional support. Formulation of nutritional strategies for paralympians requires that several secondary factors be carefully considered. These depend on the specific nature of the individual's disability, their psychosocial state and their lifestyle. For paralympians on medication for their condition, interaction of nutritional factors with prescribed medication must be considered [28]. From a nutritional standpoint, the most critical considerations include energy utilization, muscle fuel, hydration status and the appropriate use of ergogenic aids [28] (Figure 19.2).

Energy Utilization

Energy utilization is substantially increased during exercise, giving rise to energy requirements which in turn are compensated though increased nutrient intake, especially in the form of additional carbohydrates [29]. The daily energy expenditure is calculated as the sum

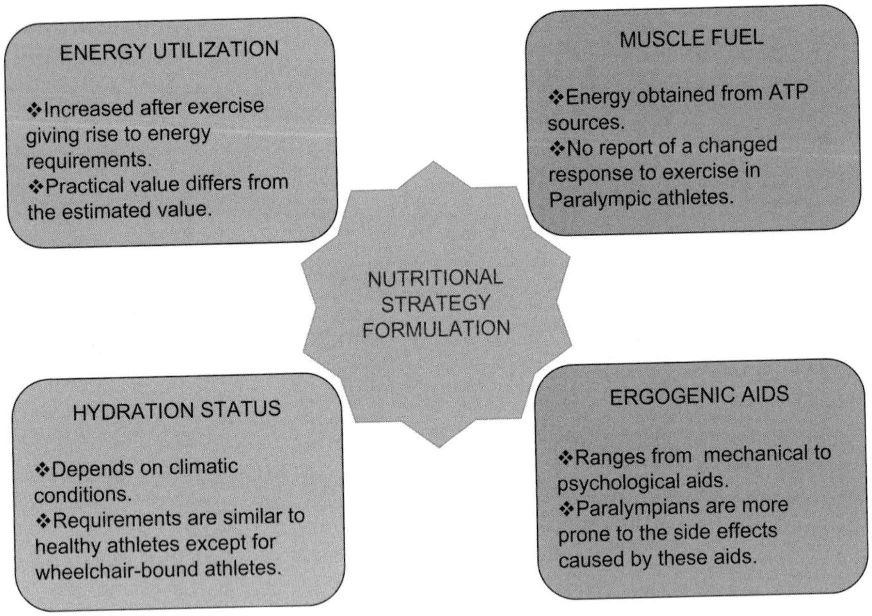

ENERGY UTILIZATION

❖Increased after exercise giving rise to energy requirements.
❖Practical value differs from the estimated value.

MUSCLE FUEL

❖Energy obtained from ATP sources.
❖No report of a changed response to exercise in Paralympic athletes.

NUTRITIONAL STRATEGY FORMULATION

HYDRATION STATUS

❖Depends on climatic conditions.
❖Requirements are similar to healthy athletes except for wheelchair-bound athletes.

ERGOGENIC AIDS

❖Ranges from mechanical to psychological aids.
❖Paralympians are more prone to the side effects caused by these aids.

FIGURE 19.2 **Factors to be considered while formulating a nutritional strategy. (This figure is reproduced in color plate section.)**

of four different energy outputs: resting metabolic rate (RMR), thermic effect of feeding (TFE), adaptive thermogenesis (AT), and thermic effect of activity (TEA). These indices for paralympic athletes differ from estimates calculated for the general population [28]. For example, RMR of paraplegic tennis and basketball players is 9–12% lower than the values estimated by the Harris-Benedict equation, while for quadriplegics the value is 15% less than that expected for the able-bodied population [30]. As a result of this, energy expenditure cannot be estimated accurately using conventional strategies. Thus, assessment of the nutritional strategy for paralympians is not straightforward [28].

Muscle Fuel

The energy required for muscle contraction is readily derived from the chemical energy stored in the form of ATP in the muscles, which later is resynthesized from carbohydrate, fat and phosphocreatinine [31]. The restricted capacity of the muscle and liver to store glycogen calls for adequate carbohydrate consumption to fuel the metabolic needs of training. The abundance of fat stores in the body of a healthy individual makes it possible to support energy requirements once the protein and carbohydrates are utilized. In paralympic athletes, expectations are comparable. Though the utilization of fat and carbohydrates may vary, the importance of the latter in disabled athletes as a fuel source remains the same as that for healthy athletes [28].

Hydration Status

Hydration status of an athlete plays a major role in determining sports performance. It is dependent on climatic conditions like temperature and humidity. Increase in the environmental temperature tends to increase the loss of water from the body in the form of sweat, which ultimately creates an imbalance of fluids and electrolytes that needs to be re-instated [32,33]. The rate of perspiration depends on environmental conditions, gender and the intensity of exercise. The hydration requirements of paralympians are similar to those of healthy athletes. However, in cases where the athlete uses a wheelchair, the fluid intake seems to decrease to avoid toilet-hygiene complexity [28].

Ergogenic Aids

Ergogenic aids play a vital role in enhancing performance both in healthy and disabled athletes. Ergogenic aids include mechanical aids such as ergogenic fabrics, pharmacological aids, physiological aids, nutritional aids such as dietary sports supplements, and psychological aids. Most ergogenic aids act as supplement to boost a particular physiological process during performance. However, lack of proper scientific knowledge about the use of these aids in performance enhancement calls for more research aimed at understanding their underlying mechanisms of action. For paralympians the need is even higher because the duration and intensity of their exercise differ considerably from those of healthy athletes, and they are more susceptible to potential side-effects of these aids [28].

Sport-Specific Nutrition

Research on able-bodied athletes provides valuable insights into the role of the nutritional regime in sports physiology, which enables us to predict the complications in disabled athletes and thus to formulate a strategy to counter those complications which would otherwise eventually have a negative impact on the health and performance of paralympians. Discussion in this section is broadly categorized into the events listed below:

1. Athletics
2. Cycling
3. Team sports
4. Power sports
5. Racquet events
6. Swimming
7. Skating and skiing.

Athletics

Literature addressing the role of nutrition in paralympic athletics is scanty. There is a dearth of knowledge about the nutritional requirements of athletes with disabilities, which gives rise to a crucial need to counsel them [34]. The literature on athletics in healthy individuals is inclined towards sprinting and long-distance running, which obviously would not be applicable for paralympians. Work examining the role of specific nutrients and nutritional supplements in paralympic sports performance is required.

Cycling

Cycling is a sport with one of the highest reported energy turnovers [35]. Early reports have demonstrated the positive effect of carbohydrates on improving and maintaining cycling performance [36]. The positive role of carbohydrates on cycling has been reported by Smith et al., who reported that the performance is improved in a dose-dependent manner [37]. Feeding carbohydrate during cycling did not affect the muscle glycogen breakdown [38]. Intake of sufficient fluids and high carbohydrate diet with minimal high dietary fiber content is recommended before a race [35].

Team Sports

Team sports in Paralympic games include football and goal ball. Depending on the role of the individual in a team, nutritional requirements are expected to vary. Because the exercise performed is intermittent, the team effort reduces the total energy burden on a single athlete and so the nutritional requirements for this type of sport will markedly differ from those of individual games.

Power Sports

The central factors that contribute to power sports are related to the muscles, both in composition and function. Important characteristics of muscle include the size of the muscle, the orientation and proportion of its constituent fibers, and the amount and structure of the related connective tissue [39]. Energy requirements of the athlete are governed by the total training load related to preparing for the final event [40]. Power-sport athletes are more inclined towards increasing power relative to body weight, for which they undertake some form of resistance training [41]. Judo and powerlifting are two major sports included in the current format of the paralympics. Held to be critical for this type of sport is protein intake, consisting of an appropriate balance of amino acids and dietary fats, which provide the essential fatty acids, along with vitamins and minerals which activate and regulate intercellular metabolic processes [42].

Racquet Events

In paralympics, major racket sports include table tennis and tennis [43]. The nutritional requirements for these sports depend on several factors such as the level of energy expenditure, game duration, and game format such as singles or doubles [44]. Both during training and within play, carbohydrates and fluids are known to influence the outcome of the game. Intake of appropriate vitamins and minerals is also recommended [44].

Swimming

Akin to other sports, the nutritional requirements associated with swimming also depend upon several factors, the most important among them being the type of event in question. Supplemental carbohydrates improve endurance performance during training sessions in exercise [45]. However, in swimming, the significance of carbohydrate supplementation is limited [46]. During swim training, depletion of glucose leads to accelerated protein catabolism [47], eventually leading to enhanced protein need. In swimming events, iron supplementation is recognized to be of significance.

Skating and Skiing

Both skating and skiing are part of the Winter Paralympic Games. The toughest problems for winter sport athletes in the extreme environmental conditions are increased energy expenditure, accelerated muscle and liver glycogen utilization, exacerbated fluid loss, and increased iron turnover [48]. Nutritional requirements may vary because of differences in physiological and physique characteristics, energy and substrate demands, and environmental training and competition conditions [48]. Ski jumpers aim to lower their body weight, and depend on carbohydrate diet throughout the competition [48].

CONCLUSION

Nutritional strategies play a key role in determining sports performance, including in paralympics. Nutritional considerations of the disabled are often not the same as those applicable for able-bodied humans. The nature of the disability and the specific sporting event determines the nutritional solutions required for best sporting performance. Consideration of chronic medication is of central importance in deciding nutritional solutions. Modified systemic responses to sports, such as an overworking cardiovascular system, as well as mind–body interaction demand special attention. At present, literature on specific nutritional needs for best sporting performance of disabled individuals is scanty. While general guidance may be obtained from data on able-bodied humans, caution should be exercised in directly translating such findings to paralympic athletes. Special needs of such athletes must be considered in light of current information related to the physiology of the specific disability [28].

References

[1] Available from: http://blog.gaiam.com/quotes/authors/joey-reiman/44838.

[2] Available from: http://www.guardian.co.uk/sport/2012/aug/26/paralympics-photographer-giles-duley.

[3] Available from: http://www.olympic.org/Documents/Olympic%20Charter/Charter_en_2010.pdf.

[4] Available from: http://tribune.com.pk/story/424810/london-prepares-for-welcome-again/.

[5] Available from: http://www.dailymail.co.uk/news/article-2195366/London-Paralympics-2012-Disabled-genius-Stephen-Hawking-star-turn-London-2012-Paralympics-Opening-Ceremony.html.

[6] Available from: http://sportsmedicine.osu.edu/patientcare/sports_nutrition/.

[7] Available from: http://www.topendsports.com/nutrition/olympic-nutrition.htm.

[8] Pelly FE, et al. Evolution of food provision to athletes at the summer Olympic Games. Nutr Rev 2011;69(6):321–32.

[9] Available from: http://en.wikipedia.org/wiki/Paralympic_Games.

[10] Available from: http://sportsmedicine.osu.edu/patientcare/sports_nutrition/performanceenhancers/.

[11] Available from: http://www.dailymail.co.uk/sport/other-sports/article-2193714/Paralympics-2012-Everything-need-know.html.

[12] Available from: http://www.paralympic.org/sites/default/files/document/120716152047682_ClassificationGuide_1.pdf.

[13] Available from: http://web.archive.org/web/20100217162608/http://www.paralympic.org.au/Sport/Classification/Athlete Classification.aspx.

[14] Available from: http://www.carryfitness.com/diet-for-physically-handicapped/.

[15] Available from: http://mynurse.weebly.com/impaired-physical-mobility.html.

[16] http://www.medicalnewstoday.com/articles/162368.php.

[17] Available from: http://www.ataxia.org/pdf/ataxia_diet_faq.pdf.

[18] Hrubec Z, Ryder RA. Traumatic limb amputations and subsequent mortality from cardiovascular disease and other causes. J Chronic Dis 1980;33(4):239–50.

[19] Hobson LB. Amputation as cause of cardiovascular disorders. Bull Prosthet Res 1979;(10-31):1–2.

[20] Modan M, et al. Increased cardiovascular disease mortality rates in traumatic lower limb amputees. Am J Cardiol 1998;82(10):1242–7.

[21] Naschitz JE, Lenger R. Why traumatic leg amputees are at increased risk for cardiovascular diseases. QJM 2008;101(4):251–9.

[22] Available from: http://www.nutritioncaremanual.org/vault/editor/Docs/AmputationsNutritionTherapy_FINAL.pdf.

[23] Available from: http://www.aoa.org/x4728.xml.

[24] Available from: http://www.aoa.org/x4734.xml.

[25] Available from: http://mtdh.ruralinstitute.umt.edu/Publications/StandardsStaff.htm.

[26] Stewart L, Beange H, Mackerras D. A Survey of Dietary Problems of Adults with Learning Disabilities in the Community. Mental Handicap Research 1994;7:1.

[27] Knuttgen HG. Basic Exercise Physiology. Nutrition. In: Maughan RJ, editor. Sport. Oxford: Blackwell Science; 2000.

[28] Vliet PV, Broad E, Strupler M. Nutrition, body composition and pharmacology. In: Thompson YVaW, editor. The Paralympic Athlete: Handbook of Sports Medicine and Science. Chichester: Wiley-Blackwell; 2011.

[29] Hardman AE. Exercise, Nutrition and Health. Nutrition. In: Sport, editor. R.J. Maughan. Oxford: Blackwell Science; 2000.

[30] Abel T, et al. Energy expenditure in ball games for wheelchair users. Spinal Cord 2008;46(12):785–90.

[31] Hultman E, Greenhaff PL. Carohydrate Metabolism in Exercise. Nutrition. In: Maughan RJ, editor. Sport. Blackwell Science; 2000.

[32] Mariott BM, editor. Nutrition Needs in Hot Environments: Applications for Military Personnel in Field Operations. National Academies Press; 1993.

[33] Mariott BM, editor. Fluid replacement and heat stress. Washinton DC: National Academies Press; 1994.

[34] Tsitsimpikou C, et al. Medication use by athletes during the Athens 2004 Paralympic Games. Br J Sports Med 2009;43(13):1062–6.

[35] Jeuendruo AE. Cycling. Nutrition. In: Maughan RJ, editor. Sport. Oxford: Blackwell Science; 2000.

[36] Christensen EH, Hansen O. Arbeitsfähigkeit und Ernährung. Skandinavisches Archiv Für Physiologie 1939;81(1):160–71.

[37] Smith JW, Zachwieja JJ, Horswill CA, Pascoe DD, Passe D, Ruby BC, Stewart LK. Evidence of a carbohydrate dose and prolonged exercise performance relationship. Medicine and Science in Sports and Exercise 2010;42:5.

[38] Jeukendrup AE. Nutrition for endurance sports: marathon, triathlon, and road cycling. J Sports Sci 2011;29(Suppl 1):S91–9.

[39] Rogozkin VA. Weight lifting and power events. Nutrition. In: Maughan RJ, editor. Sport. Oxford: Blackwell Science; 2000.

[40] Rogozkin VA. Some aspects of athletics nutrition. In: Parizkova VAR, editor. Nutrition , Physical Fitness and Health. International Series on Sport Science. Baltimore: University Park Press; 1978.

[41] Slater G, Phillips SM. Nutrition guidelines for strength sports: sprinting, weightlifting, throwing events, and bodybuilding. J Sports Sci 2011;29(Suppl 1):S67–77.

[42] Rogozkin VA. Principles of athletes' nutrition in the Russian Federation. World Rev Nutr Diet 1993;71:154–62.
[43] Lees A. Science and the major racket sports: a review. J Sports Sci 2003;21(9):707–32.
[44] Hargreaves M. Racquet sports. Nutrition. In: Maughan RJ, editor. Sport. Blackwell Science; 2000.
[45] Coyle EF, et al. Carbohydrate feeding during prolonged strenuous exercise can delay fatigue. J Appl Physiol 1983;55(1 Pt 1):230–5.
[46] O'Sullivan S, Sharp RL, King DS. Carbohydrate ingestion during competitive swim training. J Swimming Res 1994;10.
[47] Lemon PW, Mullin JP. Effect of initial muscle glycogen levels on protein catabolism during exercise. J Appl Physiol 1980;48(4):624–9.
[48] Meyer NL, Manore MM, Helle C. Nutrition for winter sports. J Sports Sci 2011;29(Suppl 1):S127–36.

CHAPTER

20

An Overview on the History of Sports Nutrition Beverages

Gustavo A. Galaz

Iovate Health Sciences, Oakville, ON, Canada

INTRODUCTION

Athletes have always been advised about what to eat, but the academic field now known as sport nutrition began in the exercise physiology laboratories. Historians consider the first studies of sport nutrition to be those of carbohydrate and fat metabolism conducted in Sweden in the late 1930s. In the late 1960s Scandinavian scientists began to study muscle glycogen storage, use, and resynthesis associated with prolonged exercise. Technology was also developed to help those scientists measure human tissue responses to exercise. In 1965 something else was born in the laboratory. At the University of Florida a team of researchers led by Dr. Robert Cade developed the first scientifically formulated beverage designed to replace fluids and salts lost through sweat during intense exercise [1].

The commercial success of this first beverage led to more development of sports nutrition supplements and diversification to functions beyond fluid replenishment, such as endurance, strength development and muscle growth.

BACKGROUND ON SPORTS BEVERAGES

A sports drink is any drink consumed in association with sport or exercise, either in preparation for exercise, during exercise itself, or as a recovery drink after exercise. The main role of a sports beverage is to stimulate rapid fluid absorption, to supply carbohydrate as substrate for use during exercise, to speed rehydration, and to promote overall recovery after exercise [2].

Basic sport drinks refer to those drinks formulated for quick replacement of fluids and electrolytes lost during exercise and that provide carbohydrate fuel to the muscles. Sports beverages usually contain a source of carbohydrates, various salts to provide electrolytes, and water. Secondary components of sports beverages include vitamins, minerals, choline and carbonation.

The hydration effect of sports beverages is not immediate since the fluid must be absorbed in the proximal small intestines and 50–60% of any given fluid ingested orally is absorbed here [1].

Sports drinks are hypertonic, isotonic, or hypotonic. Most sports beverages tend to be moderately isotonic, meaning their concentrations of salts and carbohydrates are similar to those found in the human body. Most sports drinks have a carbohydrate content of 6–9% weight/volume and contain small amounts of electrolytes in the form of salts, most commonly sodium [2].

Origins of Sports Beverages

The development of nutritional beverages specifically geared towards improving athletic performance started with studies on carbohydrate and fat metabolism conducted in Sweden in the 1930s and continued into the late 1960s. The team of scientists, led by Bjorn Ahlborg and Jonas Bergström, studied the relationship between muscle glycogen storage, use and re-synthesis during prolonged exercise to exhaustion in a group of volunteers. The research by the Swedish team demonstrated a performance-enhancing role for carbohydrates during endurance exercise and showed that glycogen content and long-term exercise capacity could be varied by instituting different diets after glycogen depletion [1,3].

The development of the very first sports beverage product, Gatorade® Thirst Quencher, was based on

Nutrition and Enhanced Sports Performance.
DOI: http://dx.doi.org/10.1016/B978-0-12-396454-0.00020-5

the early Scandinavian research that demonstrated a performance-enhancing function of carbohydrates during endurance exercise. In the summer of 1965, a team of researchers at the University of Florida, Gainesville, led by Dr Robert Cade, developed and formulated a beverage comprising glucose and electrolytes, with the goal of enhancing the performance of the school's football team [1].

At the time, it was common for football players to be admitted to the hospital because of heat exhaustion and dehydration, and as many as 25 football players in the USA died each year from heat-related illnesses [4]. Dr. Cade and his group formulated a new, precisely balanced carbohydrate–electrolyte beverage that would replace the key components lost by the Gators during games through sweating and exercise.

The effects of the new concoction on the football team were outstanding. Soon after the introduction of Gatorade, the team began to win against a number of more-favored rivals, finishing the season at 7–4. The progress of the team did not stop there, as the Gators went on to finish the 1966 season at 9–2 and won the Orange Bowl for the first time ever, defeating the Yellow Jackets of Georgia Tech. The success of the Gator football team in the late 1960s was attributed in large part to the use of Gatorade by the players. The use of Dr Cade's invention soon spread to other teams and football divisions as well. The Kansas City Chiefs began to use Gatorade, at the suggestion of Coach Ray Graves of the Florida Gators during a training exercise. The Chiefs continued taking Gatorade throughout the 1969 season and successfully concluded it with a Super Bowl victory against the Minnesota Vikings [5]. This gave birth to the multimillion dollar sports beverage industry, with sales at the end of the 1990s in excess of $2 billion [1].

The fast rise of Gatorade attracted the attention of the large multinational beverage companies, and they didn't take long to come up with their own versions of electrolyte sports drinks. The Coca Cola Company launched PowerAde, a beverage where the carbohydrates were supplied in the form of high-fructose corn syrup and maltodextrin. PepsiCo also launched its own version in the form of Allsport; in addition to the electrolyte component it provided 10% of the US daily value for thiamin, niacin, vitamins B6 and B12 and pantothenic acid [1].

In December 2000, Pepsi-Cola acquired Gatorade from the Quaker Oats company, expanding its sponsorship connection to the sport industry, thanks to Gatorade being already a major sponsor.

In 2008, the next generation of electrolyte sports beverages emerged as Pepsi launched a low-calorie version of Gatorade called simply "G2" in an effort to offset stagnant sales in the fizzy drink category in the USA. The new G2 represented the first true brand extension in the sports drinks category and was designed to keep athletes hydrated when not in training or competing [6].

Present State of Sports Beverages

Consumer interest in "natural" foods and beverages is driving demand for natural ingredients as well. New alternative sweeteners in the market are both low calorie and derived from "natural" sources, presenting opportunities for beverage manufacturers to offer products that can make both claims [7].

One such "natural", low-calorie alternative sweetener is Rebaudioside-A (or more commonly known as reb-A) (Stevia). Widely used in Asia, the high-intensity sweetener, which is an extract of the stevia leaf, was granted Generally Recognized as Safe (GRAS) status by the US Food and Drug Administration in 2008 [7].

The most recent generation of sports beverages includes offerings made entirely without artificial ingredients, such as G2 Natural and Code Blue, both of which feature stevia sweeteners and natural sea salt as a source of electrolytes. New "hybrid" beverages like FRS Healthy Energy are a combination of fruit juices with energy drinks that also claim to reduce exercise fatigue. Sport beverages are also becoming more specialized and sport specific. Drinks like Golazo, a Washington-based sports beverage targeted to soccer players, feature an all-natural ingredient claim as well as a reduced calorie claim. The Golazo line of beverages features coconut water as a natural source of electrolytes and organic agave syrup as a source of carbohydrate. The LifeAID beverage offers GolfAid and FitAid, two new energy and liquid supplements beverages targeted to golfers and cross-fit athletes respectively [8,9].

By 2001, sports and energy drinks tallied $2.92 billion [10]. Growth of hardcore drinks other than Gatorade also increased significantly, with sales more than doubling during the first decade of the 21st century. It is estimated that hardcore drinks will continue to grow and outpace the other two categories in sports supplements: powders and pills.

Glanbia's American Body Building brand (ABB) was one of the first innovators in ready-to-drink sports beverages for energy, recovery and weight control. At present, the brand has gone beyond its initial three core offerings and ventured into hybrid beverages such as Diet Turbo Tea, a zero calorie ice tea supplement featuring caffeine, ginseng, guarana and electrolytes [22].

As part of the new G-series, PepsiCo-Gatorade launched ready-to-drink protein beverages. The company's protein offerings are recovery type supplements intended to be taken post-game. One of them is a clear, fruit-flavored beverage containing protein derived from hydrolyzed whey and collagen. The other product is a protein shake, fortified with 20 g of protein derived from milk and whey protein concentrates [11].

Table 20.1 Illustrates the current state of the US sports beverage landscape.

HISTORY OF PROTEIN DRINKS

Background

Historically, strength athletes and bodybuilders have always consumed protein in levels well above the US Recommended Daily Allowance (RDA). Ample research has shown that, to maintain nitrogen balance and attenuate amino acid oxidation, both endurance and strength exercise increase protein requirements above the current RDA of 0.8 grams of protein per kilogram of body weight per day. Endurance athletes require 1.0–1.4 g/kg/d, while strength athletes need 1.4–1.8 g/kg/d to achieve nitrogen balance. Protein beverages are consumed by athletes as a means to facilitate protein supplementation in a convenient, cost effective and simple way [12,16].

Dairy proteins and soy proteins constitute the majority of protein sources used in sports nutrition products. Dairy proteins and soy proteins accounted for 77% and 23% of US protein sales respectively in 2004. Whey protein concentrate at 80% protein (WPC 80) is the most widely used protein ingredient in the sports nutrition industry. Nearly 37% of WPC 80 produced in 2007 was used in the sports nutrition sector. Regarding soy, soy protein isolate at 90% is the preferred ingredient due to its high protein content and higher nutritional quality among the different vegetable proteins [16].

Overview of Soy Protein Beverages

Soybeans had been used as an important food source in Eastern Asia for centuries prior to their introduction to North America. Early US interest in the soybean was for its oil, which was extracted by hydraulic and screw presses and later by direct solvent. The protein-rich byproduct, soy meal, was used primarily as cattle feed or fertilizer. Commercial processing of the soybean into food products like defatted soy flour, soy protein concentrate, and isolated soy protein started in the 1950s [13].

The new ingredients derived from soy were first used in food products such as infant formulas and replacements of whole-milk powder. One of the first entrants in the field of protein supplements for body builders was Bob Hoffman; although not a body builder himself, he was associated with the York Barbell Company and was the US Olympic weightlifting coach for many years. In 1951 he began to market a product called Hi-Protein Food, later rebranded as

TABLE 20.1 US Sports Beverage Landscape

Category	Subcategory	Key Ingredients	Function	Product Examples
Pre-workout	Energy	Caffeine, ginseng, electrolytes, guarana	Energy boost, focus	ABB Diet Turbo Tea, Golazo Sports Energy, Monster Rehab
	Energy + supplements	Botanicals, vitamins, glucosamine, electrolytes, amino acids, green tea extract	Sport-specific performance, joint health, anti-inflamatory	FitAid, GolferAid, Nawgan, Karbolic, FRS Healthy Energy, Xenergy, Speed Stack
Intra-workout	Electrolyte/hydration	Isomaltulose, glucose, glucose polymers, fructose, amino acids, electrolytes, sea salt, coconut water, vitamins	Endurance, hydration, glycogen replenishment	Gatorade Perform, G2, ABB Hydro Durance, PowerBar Perform, Golazo, Powerade Ion4, AllSport, Cytomax
Post-workout	Recovery	Maltodextrin, whey, electrolytes	Glycogen replenishment, muscle recovery	Gatorade G series Recover Drink, ABB Max Recovery, Powerbar Recovery, Code Blue
	Protein	Whey isolate, BCAAs, HWP, MPC	Muscle growth	Gatorade G Series Recover Shake, Muscle Milk, Pure Protein Shake, ABB Pure Pro 50, Designer Whey Shake, Isopure, OhYeah! Shake, FRS Healthy Protein
	Meal replacements	Casein, MPC, sunflower oil	Satiety, weight gain	Boost High Protein, Ensure, Special K protein shake, Met-Rx Original Meal Replacement, Carnation Instant Breakfast

BCAA, branched-chain amino acid; HWP, hydrolyzed whey protein; MPC, milk protein concentrate.

Hoffman's HI-Proteen, an early protein powder consisting of soy flour. The product contained 42% protein and was meant to be dissolved in milk. The beverage mix proved to be a success among the weightlifter community, and soon the brand expanded into fudge, tablets and bars [14,15].

Another pioneer in soy protein sports nutrition beverage powders was Joe Weider. Weider first developed a carbohydrate-based weight-grain powder where the consumer mixed an 8-ounce can of the product with a quart of milk. In 1952, Weider introduced "Hi-Protein Muscle Building Supplement". The supplement was marketed in body-building magazines as a carbohydrate-, fat- and sugar-free offering. The Weider company later expanded his line of products to include vitamin-minerals, weight reduction agents, and weight gain powders [14].

Initially, the low protein purity of soy-derived protein ingredients led to consumer dissatisfaction, mainly due to digestive issues as well as poor taste and poor miscibility. At present, soy protein is separated from soybeans through water extraction, followed by precipitation, washing and drying processes that yield protein concentrates of up to 70% protein or isolates with up to 90% protein. These procedures have greatly improved taste and successfully reduced miscibility issues. However, the image of poor tolerance of soy still remains with many consumers [16].

Despite soy protein having unique properties such as antioxidant isoflavones, and with solid research suggesting reduction of cardiovascular disease risk factors, the general perception that soy is inferior in quality to animal proteins still exists, limiting marketability of soy protein to strength athletes [16].

Alternative Vegetable Sources for Protein Beverages

There is a wide variety of semi-purified or purified proteins from other vegetable or animal sources. High costs, and the perceived advantages of milk, egg or soy proteins, prevent other proteins sources from being commercially important as protein supplements. Often these alternative proteins are mixed with amino acids or hydrolysates to improve their nutritional profiles, resulting in higher prices. Protein supplements using rice, pea, beet, wheat or other grain proteins have not been as successful with consumers as have milk- or soy-derived products [16].

Whey Protein Beverages

Whey accounts for all the watery material that remains after milk coagulation in cheese manufacturing. Liquid whey is 93% water and 0.6% protein, the rest being lactose, minerals and milk fat. The protein fraction of whey is equivalent to approximately 20% of the original protein in milk [16].

For many years and until the end of the twentieth century, whey products, which have a high protein efficiency ratio, constituted an under-utilized source of human food. Whey had usually been considered a waste product and was disposed in the most cost-effective manners or processed into whey powder and whey protein concentrate [17].

The use of dairy whey in commercial beverages started in Switzerland in 1952, with the introduction of Rivella Red, a carbonated, milk serum-based beverage. Originally created by Dr. Robert Barth and designed to be a commercial channel for a byproduct of Switzerland's vast rural dairy operations, Rivella was marketed as a healthy soft drink; soon after its launch, the beverage became a sponsor of the Swiss Olympic Team and the Swiss National Ski team. Although Rivella Red contained 35% whey by volume, it was fully deproteinized, so it was not branded as a protein beverage, but rather as a healthy natural source of minerals like calcium and magnesium as well as carbohydrates in the form of lactose [18].

Several significant technological developments at the end of the twentieth century prompted the rise of whey as an important functional ingredient. Ultrafiltration and ion exchange chromatography allowed the production of demineralized whey powders, and a number of new value-added products derived from whey reached the market for the first time. These included products such as whey protein isolates (WPI), a-lactalbumin, b-lactoglobulin, and immunoglobulin among others [19].

The development of techniques in ultrafiltration gave rise to the possibility of developing 'tailor-made' milk protein mixes. Companies like Davisco and Parmalat developed tailor-made whey-based ingredients with specific functional characteristics: such as enzyme-hydrolyzed whey with increased protein efficiency ratio, and whey protein concentrate with a standardized beta-lactoglobulin content. The availability of this type of product increased as more dairy companies turned to more efficient ways to reduce their waste [19].

By the 1990s, the newly emerged functional food market offered many opportunities to develop whey protein-based products. In 1998, the global market size for the functional food sector reached US$ 80 billion per year, and whey protein was poised to change its status from waste product to a major ingredient in this sector [17].

Whey protein gained popularity in sports nutrition based on essential amino acid composition, branched chain amino acid (BCAA) content, and sulfur amino acid content, as well as taste, acceptance, ease of mixing, and fast digestion. Whey protein, despite being costlier than other supplement proteins, became the protein of choice for weightlifters thanks to perceived advantages and good taste. The perception of whey as the ultimate protein was helped in part by articles and advertisements in popular magazines [16].

One of the first companies to successfully commercialize a whey protein drink mix was Next Proteins International from Carlsbad, California. Taking advantage of the mentioned technical improvements in whey protein production, Next launched the Designer Whey brand in 1993 [10].

By the early 2000s protein powder sales had reached $500 million, with companies like Next Protein and Optimum Nutrition leading the market [10].

Present Status of Protein Beverages

In addition to increased growth of whey protein drink mixes, the 2000s saw the arrival of ready-to-drink offerings such as Cytosport's Muscle Milk and Next Protein's Designer Whey high-protein Shakes [20].

Increased mainstream acceptance of whey protein-based protein drink mixes and ready-to-drink (RTDs) has resulted in their moving beyond the niche market of bodybuilders and hardcore athletes and into mass markets in the form of weight loss supplements and meal replacements, with companies like Walmart generating the most sales in the category. [20] In addition to Cytosport, mainstream food companies such as Nestlé and Kellogg's further helped launch protein beverages into the mainstream market with their Boost and Special K ready-to-drink protein shakes, respectively. These products not only provided liquid nutrition but also served an appetite-curving function.

Performance protein powder mixes have evolved from being regular protein concentrates and isolates into more complex products featuring protein hydrolysates and complex carbohydrates for sustained release energy [20].

Future of Sports Beverages

As of 2012, the US sales of non-aseptic sports drinks totaled $4.1 billion a year according to the beverage industry. Sales of all other hardcore sports drinks reached $300 million in 2010 (Figure 20.1) and are projected to reach $415 million by 2017 [20]. Over the last ten years, sales of hardcore drinks grew on average by 10.5% a year, outpacing the growth of sports powders and pill-form supplements since 2000. The subcategory has grown thanks to mass-market penetration by RTD brands. Powders and formulas will continue to be the largest subcategory in the sports nutrition sector, however. Overall, protein products in powdered or RTD format have become best sellers in the mass-market retail channel [20].

The rising popularity of protein in different forms of supplements and the fluctuation of prices of whey on the spot market will see alternative sources of protein like hemp, brown rice and pea becoming more interesting for manufacturers and formulators [20].

The next generation of sports drinks will continue to see more choice, with more flavor blends and better fidelity to standard, meaning the flavors will be closer to fresh and natural fruit. The sports drink of the future will be formulated for different target markets including age groups, gender, occasion use, and physical requirement. There will be more development of sport and occasion-specific drinks. We are already seeing examples of sport specific beverages such as Golazo and GolferAid. Hybridization among different

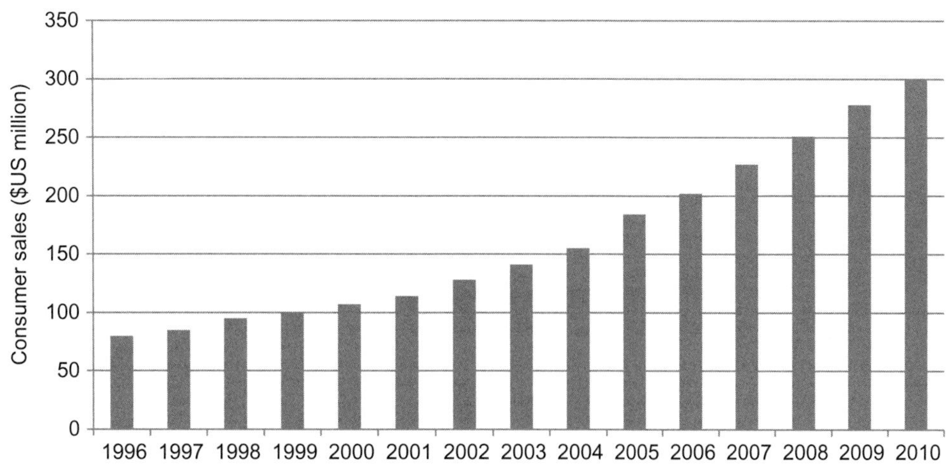

FIGURE 20.1 Sales of Hardcore Sports Drinks in the USA 1996–2010 [20]. (This figure is reproduced in color plate section.)

beverage categories will also continue, as supplements, phytochemicals and other botanical ingredients once found only at specialty store products become more widely used in mainstream mass-market beverages. The recent introductions of products like Coca Cola's Monster Rehab with whey protein hydrolysate and FRS Energy with quercetin point in this direction [21].

Another key driver will be packaging innovation with environmentally friendly sourcing of materials and more use of recyclables, minimizing waste [20].

CONCLUSION

Over the last 45 years the sports beverage industry has evolved from a single, locally distributed electrolyte drink into a multibillion dollar category within the larger sports supplement and functional foods markets. The growth of that industry has not been without sound scientific principles that back the formulations. Many research studies show clear evidence that drinking a properly formulated beverage during exercise provides better performance than drinking water alone [2,21].

As consumers' tastes have changed and become more sophisticated, the spectrum of sports drinks has become more diverse, and formulators have responded with beverages that cater to specific functions and stages of the physical activity. The market penetration of these beverages will continue to grow, thanks to more people becoming aware of the nutritional benefits of sports drinks and the marketing efforts of companies to create more robust and higher value-added products.

References

[1] Driskell J, Wolinsky I. Macroelements, water, and electrolytes in sports nutrition. Boca Raton: CRC Press; 1999.

[2] Shirreffs S. The optimal sports drink. "Sportmedizin und Sporttraumatologie" 2003;51:25–9.

[3] Bergstrom J, Jeramsen L, Hultman E, Saltin B. Diet, muscle glycogen and physical performance. Acta Physiol Scand 1967;71:140.

[4] Rovell D. First in thirst: how gatorade turned the science of sweat into a cultural phenomenon. New York: Amacom Books; 2005.

[5] Available from: http://www.gatorade.com/history.

[6] Pepsi unveils low-calorie Gatorade 'G2'. CNNMoney, <http://money.cnn.com/2007/09/07/news/companies/pepsi_newproducts/> 07.09.2007.

[7] Evani S. Trends in the US functional foods, beverages and ingredients market. Agriculture and Agri-Food Canada: Agri-Food Trade Service; 2009<http://www.agrireseau.qc.ca/Marketing-Agroalimentaire/documents/Tendances%20aliments%20fonctionnels%20-%20USA%20-%20AAC%2007-2009.pdf>

[8] Available from: http://vivagolazo.com/.

[9] Available from: http://www.frs.com, <http://www.lifeaidbevco.com/>

[10] Nutrition Business Journal. NBJ's supplement business report 2002. San Diego: Penton Media; 2002.

[11] Available from: http://www.gatorade.com.

[12] Paul G. The rationale for consuming protein blends in sports nutrition. J Am Coll Nutr 2009;28:464S–72S.

[13] Wilcke H. Soy protein and human nutrition: proceedings of the keystone conference on soy protein and human nutrition, held in keystone, Colorado, May 22–25, 1978. San Diego: Academic Press; 1979.

[14] Shurtleff W, Aoyagi A. History of soybeans and soyfoods in Canada 1831–2010: extensively annotated bibliography and sourcebook. Lafayette: Soyinfo Center; 2010.

[15] Fair JD. Muscletown USA: Bob hoffman and the manly culture of york barbell. Pennsylvania: Penn State Press; 1999.

[16] Driskell J, Wolinsky I. Energy yielding macronutrients and energy metabolism in sports nutrition. Boca Raton: CRC Press; 2000.

[17] McIntosh G, Royle P, Leu R. Whey proteins as functional food ingredients? Int Dairy J 1998;8:425–34.

[18] Available from: http://www.rivella.com.

[19] Nakai S, Modler H. Food proteins: processing applications. New York: John Wiley & Sons, Inc.; 2000.

[20] Nutrition Business Journal. NBJ's supplement business report 2011. San Diego: Penton Media; 2011.

[21] Maughan R, Murray R. Sports drinks: basic science and practical aspects. Boca Raton: CRC Press; 2001.

[22] Available from: http://www.abbperformance.com/.

SECTION 4

MOLECULAR MECHANISMS

CHAPTER

21

α-Hydroxy-Isocaproic Acid (HICA)—Effects on Body Composition, Muscle Soreness and Athletic Performance

Tuomo Ojala[1], Jacob M. Wilson[2], Juha J. Hulmi[1], Tuomo Karila[3], Timo A. Seppälä[4] and Antti A. Mero[1]

[1]Department of Biology of Physical Activity & Neuromuscular Research Center, University of Jyväskylä, Jyväskylä, Finland [2]Department of Health Sciences and Human Performance, University of Tampa, Tampa, FL, USA [3]Hospital Orton, and Dextra Sports and Injury Clinic, Helsinki, Finland [4]Department of Drugs, Alcohol and Addiction, National Institute for Health and Welfare, Helsinki, Finland

BACKGROUND

DL-α-hydroxy-isocaproic acid (HICA), also known as leucic acid or DL-2-hydroxy-4-methylvaleric acid, is an α-hydroxyl acid metabolite of leucine. HICA (molecular weight 132.16) is an end-product of leucine metabolism in human tissues such as muscle and connective tissue [1–3]. The part of HICA which is not bound to plasma proteins has a plasma concentration in healthy adults of about 0.25 ± 0.02 mmol/L whereas the plasma concentration of α-keto-isocaproic acid (KIC), the corresponding keto acid of leucine, is about 21.6 ± 2.1 mmol/L [1]. HICA can be measured in human plasma, urine and amniotic fluid [4–6]. Foodstuff produced by fermentation—e.g., certain cheeses, wines and soy sauce—contain HICA [7–11]. According to clinical and experimental studies, HICA can be considered as an anti-catabolic substance [12–19]. Although leucine has a unique role as a promoter of protein synthesis [20], metabolites of leucine may be more effective in preventing breakdown of proteins, particularly muscle proteins [15].

The roles and mechanisms of action of leucine and its metabolites are not yet thoroughly understood and often confusing. For instance, transaminated leucine metabolite KIC is anti-catabolic and reduces muscle protein degradation but, due to extensive first-pass metabolism, only when given as an intravenous infusion. On the other hand, it is a potent inhibitor of branched-chain α-keto acid dehydrogenase and may lead to increased catabolism of branched-chain amino acids (BCAAs) [21]. β-Hydroxy β-methylbutyric acid (HMB) or β-hydroxy β-methylbutyrate is another metabolite of leucine which also plays a role in protein synthesis and breakdown [22]. Specifically, HMB has been demonstrated to affect protein balance through an IGF-1-AKT-mTOR signaling cascade [23]. As such, HMB is able to speed repair of skeletal muscle damage and augment strength, power, and hypertrophy gains following chronic resistance exercise training, at least in situations where muscle damage occurs (for review, see [22]).

There are separate mechanisms to control protein synthesis and proteolysis [24]. In the absence of amino acids, the primary regulators of protein degradation rapidly and strongly increase. Regulatory amino acids (leucine, tyrosine, glutamine, proline, methionine, histidine, and tryptophan) as a group, as well as leucine alone, inhibit food deprivation-induced protein degradation [6]. Tischler et al. [15] suggested that the first step in muscle proteolysis is the oxidation of leucine, catalyzed by aminotransferase enzyme. The end product of the reaction is keto leucine (α-ketoisocaproate, KIC). However, in certain situations, the end product

Nutrition and Enhanced Sports Performance.
DOI: http://dx.doi.org/10.1016/B978-0-12-396454-0.00021-7

can be HICA as well. It is suggested that the aminotransferase enzyme is responsible for oxidizing leucine both to its keto and hydroxyl forms and both reactions are reversible [25]. In humans, the reaction between keto and hydroxyl leucine is an equilibrium reaction with a oxidoreduction equilibrium constant (thermodynamic constant) $K_{eq} = 3.1 \pm 0.2 \times 10^{-12}$ mol/L and reaction half-time is 230 min towards oxygenation. Keto acid is irreversibly oxidized by mitochondrial ketoacid dehydrogenase [26]. Irreversible degradation of keto acids is higher in liver than in muscle [26]. The branched-chain α-keto acid dehydrogenase complex is deemed to be the most important regulatory enzyme in the catabolism of leucine [27]. Skeletal muscle is considered to be the initial site of BCAA catabolism because of elevated BCAA aminotransferase [2]. According to Mortimore et al. [6], leucine alone accounts for about 60% of the total effectiveness of the group of regulatory amino acids, and the same effect is achieved with HICA alone, whereas KICA does not produce the same effect at similar concentrations. HICA has also a broad antibacterial effect, e.g., against MRSA (methicillin-resistant *Staphylococcus aureus*), which is resistant to several systemic antimicrobials [3]. No human studies, however, have been conducted to investigate that phenomenon.

EFFECTS OF ALFA-HYDROXY-ISOCAPROIC ACID ON BODY COMPOSITION, DELAYED-ONSET MUSCLE SORENESS AND PHYSICAL PERFORMANCE IN ATHLETES

The effects of HICA on body composition have been reported in several studies [14,19,28–32]. All the studies, with the exception of an open pilot study in wrestlers [19], a case report by Hietala et al. [19], and a double blind, placebo-controlled study by Mero et al. [28], have been conducted in animals. Mero et al. [28] used soccer players to assess the effects of HICA on body composition, exercise-induced delayed-onset muscle soreness (DOMS) and athletic performance.

Studies in rodents have shown that substitution of methionine, leucine, phenylalanine, or valine by their α-hydroxy analogs reduced growth of body weight [29,32]. More specifically, Chow & Walser [29] reported that leucine and its α-hydroxy analog (HICA) were equally effective at increasing growth. However, replacement of leucine with HICA reduced food intake and increased the volume of urine and its nitrogen concentration. Woods & Goldman [30] reported that, in rats, HICA supports growth of the rats using a leucine free diet without reducing the food intake. On the other hand Boebel & Baker [14] showed that leucine, valine or isoleucine analogs do not support the growth without the presence of their corresponding amino acids. Both α-keto and hydroxy analogs were equally effective in supporting the growth of body weight [14]. Thus, in summary, substitution of natural amino acids by their alpha-hydroxy analogs decreases growth in animals.

Flakoll et al. [31] reported that parenterally given KIC increased muscle mass and reduced fat mass in lambs. Mero et al. [28] found similar results using national-level soccer players when assessing the effects of oral HICA on body composition. According to their study, a 4-week supplementation period with HICA (1.5 g daily) increased whole-body lean mass in soccer players. This increase (~0.5 kg) was more pronounced in lower extremities. These findings suggest an interaction between the supplement and specificity of training of the athlete: specifically, the lower extremities of soccer players receive most of their training load. Athletes in this study already had an average protein intake of 1.6–1.7 g/kg/day. Thus, it can be concluded that this extra "amino acid" metabolite, even in addition to sufficient daily protein and leucine intake, increases lean muscle mass. It is likely that these changes are at least in part mediated through minimizing catabolic processes induced by exercise [28].

Delayed-Onset Muscle Soreness (DOMS)

DOMS is a sensation of muscular discomfort and pain during palpitation or active contractions that occurs in a delayed fashion after strenuous exercise. Subjects with DOMS have reduced range of motion of adjacent joints especially after unaccustomed exercise [33,34]. In addition to muscle tenderness with palpation, prolonged strength loss and a reduced range of motion are observed. These symptoms develop 24 to 48 hours after exercise, and they disappear within 5 to 7 days [33,34]. Although the pathophysiology of DOMS remains unknown, it has been reported that, after strenuous exercise, cell damage and inflammatory cells are observed in muscle [33,34].

Mero et al. [28] reported that soccer players had milder DOMS symptoms after using HICA compared with placebo. This was particularly evident during the 4th week of supplementation. Training alertness was also increased. These beneficial effects were significant after the 2nd week of HICA treatment and appeared to continue for weeks. Previously, similar results have been described with the combination of HMB and KIC, which reduced plasma creatine kinase activity, DOMS symptoms, muscle swelling and one-repetition maximum reduction [24]. In contrast, HMB alone had no effect on DOMS [35]. The mechanism by which HICA alleviates DOMS symptoms is unclear. It is known, however, that hydroxy metabolites of BCAAs (especially HICA), are potent inhibitors of catalytic protease enzymes. HICA has been shown to strongly inhibit several matrix metalloproteinases [18]—a family of peptidases, including collagenases for example, responsible for the degradation of the extracellular matrix during tissue remodeling. Inhibition of matrix metalloproteinases may be the most important mechanism related to HICA's anti-catabolic action, although other effects of HICA may have a contributory role. In severe catabolic states, HICA may act as an energy substrate and is oxidized instead of spare leucine [13,16].

Physical Performance

Mero et al. [28] reported no changes in physical performance after using HICA during the 4-week period even though the muscle mass of lower extremities increased significantly. They speculated that the training period was probably too short to observe strong training responses. Similar results have been reported using KIC [36]. It appears that even though the muscle mass increases and DOMS decreases, HICA has no effects on physical performance in short training periods.

In the future, studies should investigate whether Alfa-HICA is also effective when combined with whey protein, a typical recovery drink protein for athletes that has a high leucine content and is capable of increasing muscle size and enhancing recovery from resistance exercise [37,38].

CONCLUSION

Recent findings suggest that HICA relieves DOMS symptoms and protects muscle from catabolism. HICA may be beneficial for high-intensity training athletes who often experience stiff and sore muscles, which limit effective training.

References

[1] Hoffer LJ, Taveroff A, Robitaille L, Mamer OA, Reimer ML. Alpha-keto and alpha-hydroxy branched-chain acid interrelationships in normal humans. J Nutr 1993;123:1513–21.

[2] Holecek M. Relation between glutamine, branched-chain amino acids, and protein metabolism. Nutrition 2002;18(2):130–3.

[3] Sakko M, Tjäderhane L, Sorsa T, Hietala P, Järvinen A, Bowyer P, et al. α-Hydroxyisocaproic acid (HICA): a new potential topical antibacterial agent. Int J Antimicrob Agents 2012;39(6): 539–40.

[4] Jakobs C, Sweetman L, Nyhan WL. Hydroxy acid metabolites of branched-chain amino acids in amniotic fluid. Clin Chim Acta 1984;140(2):157–66.

[5] Mamer OA, Laschic NS, Scriver CR. Stable isotope dilution assay for branched chain alpha-hydroxy-and alpha-ketoacids: serum concentrations for normal children. Biomed Environ Mass Spectrom 1986;13(10):553–8.

[6] Mortimore GE, Pösö AR, Kadowaki M, Wert Jr JJ. Multiphasic control of hepatic protein degradation by regulatory amino acids. General features and hormonal modulation. J Biol Chem 1987;262(34):16322–7.

[7] Yamamoto A. Flavors of sake. II. Separation and identification of a hydroxyl carboxylic acid. Nippon Nogeikagaku Kaishi 1961;35:619.

[8] Van Wyk CJ, Kepner RE, Webb AD. Some volatile components of vitis vinifera variety white riesling. 2. Organic acids extracted from wine. J Food Sci 1967;32(6):664–8.

[9] Begemann WJ, Harkes PD. 1974a. Enhancing a fresh cheese flavor in foods. U. Lever Brothers Co. U.S.

[10] Begemann WJ, Harkes PD. 1974b. Process for enhancing a fresh cheese flavor in foods. U. Lever Brothers Co. U.S.

[11] Smit BA, Engels WJ, Wouters JT, Smit G. Diversity of L-leucine catabolism in various microorganisms involved in dairy fermentations, and identification of the rate-controlling step in the formation of the potent flavor component 3-methylbutanal. Appl Microbiol Biotechnol 2004;64(3): 396–402.

[12] Walser M. 1973. Composition for promotion of protein synthesis and suppression of urea formation in the body utilizing alpha-hydroxy-acid analogs of amino acids. US Patent 1 444 621.

[13] Walser M. 1978. Therapeutic compositions comprising alpha-hydroxy analogs of essential amino acids and their administration to humans for promotion of protein synthesis and suppression of urea formation. U.S., The Johns Hopkins University, Baltimore, Md. US Patent 4100161.

[14] Boebel K, Baker D. Comparative utilization of the alpha-keto and D- and L-alpha-hydroxy analogs of leucine, isoleucine and valine by chicks and rats. J Nutr 1982;112(10):1929–39.

[15] Tischler M, Desautels M, Goldberg A. Does leucine, leucyl-tRNA, or some metabolite of leucine regulate protein synthesis and degradation in skeletal and cardiac muscle? J Biol Chem 1982;257(4):1613–21.

[16] Lindgren S, Sandberg G, Enekull U, Werner T 1990a. Energy substrate containing hydroxycarboxylic acid and a glycerol ester. Kabivitrum Ab. Patent number EP 367734 A1.

[17] Lindgren S, Sandberg, G, Enekull U, Werner, T. 1990b. Energy substrate containing hydroxycarboxylic acid. Kabivitrum Ab. Patent number EP 363337 A1.

[18] Westermarck HW, Hietala P, et al. 1997. Use of alpha-hydroxy acids in the manufacture of a medicament for the treatment of inflammation, Exracta Oy. Patent Number WO 97/00676.

[19] Hietala P, Karila T, Seppälä T, Tähtivuori K, 2005. Nutrient supplement and use of the same. Oy Extracta ltd. Patent Number PCT/FI2005/050365.

[20] Kiebzak GM, Leamy LJ, Pierson LM, Nord RH, Zhang ZY. Measurement precision of body composition variables using the lunar DPX-L densitometer. J Clin Densitom 2000;3(1): 35–41.

[21] Shimomura Y, Murakami T, Nakai N, Nagasaki M, Harris RA. Exercise promotes BCAA catabolism: effects of BCAA supplementation on skeletal muscle during exercise. J Nutr 2004;134:1583–7.

[22] Wilson GJ, Wilson JM, Manninen AH. Effects on beta-hydroxy-beta-methylbutyrate (HMB) on exercise levels of age, sex and training experience: a review. Nutr Metab 2008;5(1)1. doi:10.1186/1743-7075-5-1.

[23] Kornasio R, Riederer I, Butler-Browne G, Mouly V, Uni Z, Halevy O. Beta-hydroxy-beta-methylbutyrate (HMB) stimulates myogenic cell proliferation, differentiation and survival via the MAPK/ERK and PI3K/Akt pathways. Biochim Biophys Acta 2009;1793(5):755–63.

[24] Van Someren K, Edwards A, Howatson G. Supplementation with beta-hydroxy-beta-methylbutyrate (HMB) and alpha-ketoisocaproic acid (KIC) reduces signs and symptoms of exercise-induced muscle damage in man. Int J Sport Nutr Exerc Metab 2005;15:413–24.

[25] Blanchard M, Green DE, Nocito-Carroll V, Ratner S. l-Hydroxy acid oxidase. J Biol Chem 1946;163:137–44.

[26] Suryawan A, Hawes JW, Harris RA, Shimomura Y, Jenkins AE, Hutson SM. A molecular model of human branched-chain amino acid metabolism. Am J Clin Nutr 1998;68 (1):72–81.

[27] Kobayashi R, Murakami T, Obayashi M, Nakai N, Jaskiewicz J, Fujiwara Y, et al. Clofibric acid stimulates branched-chain amino acid catabolism by three mechanisms. Arch Biochem Biophys 2002;407(2):231–40.

[28] Mero A, Ojala T, Hulmi J, Puurtinen R, Karila T, Seppälä T. Effects of alfa-hydroxy-isocaproic acid on body composition, doms and performance in athletes. J Int Soc Sports Nutr 2010;7 (1):1–8.

[29] Chow K, Walser M. Effects of substitution of methionine, leucine, phenylalanine, or valine by their alpha-hydroxy analogs in the diet of rats. J Nutr 1975;105(3):372–8.

[30] Woods M, Goldman P. Replacement of L-Phenylalanine and L-Leucine by a-Hydroxy analogues in the diets of germfree rats. J Nutr 1979;709:738–43.

[31] Flakoll P, VandeHaar M, Kuhlman G, Nissen S. Influence of α-ketoisocaproate on lamb growth, feed conversion and carcass composition. J Anim Sci 1991;69:1461–7.

[32] Pond W, Breuer L, Loosli K, Warner R. Effects of the a-Hydroxy analogues of isoleucine, lysine, threonine and tryptophan and the a-Keto analogue of tryptophan and the level of the corresponding amino acids on growth of rats. J Nutr 1964;83:45–55.

[33] Barlas P, Craig JA, Robinson J, Walsh DM, Baxter GD, Allen JM. Managing delayed-onset muscle soreness: lack of effect of selected oral systemic analgesics. Arch Phys Med Rehabil 2000;81:966–72.

[34] Lieber L, Friden J. Morphologic and mechanical basis of delayed-onset muscle soreness. J Am Acad Orthop Surg 2002;10 (1):67–73.

[35] Paddon-Jones D, Keech A, Jenkins D. Short-term β-hydroxy-β-methylbutyrate supplementation does not reduce symptoms of eccentric muscle damage. Int J Sport Nutr Exerc Metab 2001;11:442–50.

[36] Hulmi JJ, Lockwood CM, Stout JR. Effect of protein/essential amino acids and resistance training on skeletal muscle hypertrophy: a case for whey protein. Nutr Metab (Lond) 2010;7:51.

[37] Hulmi JJ, Kovanen V, Selänne H, Kraemer WJ, Häkkinen K, Mero AA. Acute and long-term effects of resistance exercise with or without protein ingestion on muscle hypertrophy and gene expression. Amino Acids 2009;37:297–308.

[38] Yarrow J, Parr J, White L, Borsa P, Stevens B. The effects of short-term alpha-ketoisocaproic acid supplementation on exercise performance: a randomized controlled trial. J Int Soc Sports Nutr 2007;4(2):10–5.

CHAPTER

22

Role of Mammalian Target of Rapamycin (mTOR) in Muscle Growth

Evgeniy Panzhinskiy[1], *Bruce Culver*[1], *Jun Ren*[1], *Debasis Bagchi*[2] *and Sreejayan Nair*[1]

[1]University of Wyoming, School of Pharmacy and the Center for Cardiovascular Research and Alternative Medicine, College of Health Sciences, Laramie, WY, USA [2]University of Houston College of Pharmacy, Houston, TX, USA

INTRODUCTION

The understanding of the growth and development of skeletal muscle is one of the most important steps in designing successful programs for muscle building during physical training. It also may provide potential therapeutic targets for the prevention and treatment of muscle wasting in metabolic and neuromuscular diseases. Skeletal muscles consist of fibers, so muscle mass is determined by number of muscle fibers and their size. During myogenesis, the extent of muscle cell multiplication largely determines how many muscle fibers are formed. Therefore, the number of muscle fibers is mainly determined by genetic and environmental factors which can influence prenatal myogenesis. However, in postnatal life, muscle size is determined by the size of individual fibers and external stimuli. Cell size in turn is determined by a balance between accumulation of new proteins and degradation of existing proteins [1]. The kinase mTOR (mammalian target of rapamycin) has been implicated as a key regulator of cell growth, which mediates the effect of growth factors, nutrients and other external stimuli on protein synthesis and degradation [2,3]. Recent studies provide an emerging concept which considers mTOR as a key convergence point for the signaling pathways regulating skeletal muscle growth. This review focuses on the role of mTOR in skeletal muscle growth.

MUSCLE GROWTH

The growth of skeletal muscle, like any other tissue, depends on increase in cell number (hyperplasia) and cell size (hypertrophy). Cell proliferation plays a major role during muscle development in embryo. In contrast, induction of protein synthesis associated with hypertrophy is the main characteristic of muscle growth in the adult state [1]. During embryonic development, muscular tissue is formed through the process called myogenesis. The skeletal muscles are derived from the somites, segmental derivatives of the paraxial mesoderm [4]. Initially, myogenic precursor cells of mesodermal origin enter the myogenic lineage, proliferate and divide to form a pool of myoblasts [5]. Subsequent signaling events cause the myoblasts to exit the cell cycle, to cease dividing and to differentiate. Cells begin to secrete fibronectin onto their extracellular matrix and express muscle cell-specific proteins. Finally myoblasts align and fuse to form multinucleated myotubes [6]. During postnatal muscle growth, satellite cell incorporation into the growing fibers occurs together with increased protein synthesis [7]. Satellite cells are mononuclear myoblast progenitor cells, which reside between the basement membrane and sarcolemma of associated muscle fibers (see [4] for review). However the total muscle fiber number remains unchanged after birth in mammals [5]. Unlike young muscle, in adults the increase in skeletal muscle

Nutrition and Enhanced Sports Performance.
DOI: http://dx.doi.org/10.1016/B978-0-12-396454-0.00022-9

mass happens mainly due to an increase in muscle fiber size, as opposed to the number, and doesn't require satellite cell incorporation [8]. This process of hypertrophy is associated with elevated synthesis of contractile proteins. Natural hypertrophy normally stops at full growth after sexual maturation, but resistance training or intake of anabolic steroids, such as testosterone, promote a hypertrophic response in both human and animal models [1]. Although myogenesis normally does not occur in adult skeletal muscle, it can be activated in satellite cells following injury or disease (reviewed in [9,10]). The satellite cells in adults are mitotically quiescent, but are capable of differentiation and incorporation of new nuclei into the growing muscle fibers or of forming new myofibers, leading to hypertrophy. This process also contributes to muscle hypertrophy induced as a compensatory response to myotrauma consequent to resistance training [11]. Rate of the muscle growth is determined by homeostasis between hypertrophy and atrophy, or muscle wasting. Atrophy is a decrease in cell size mainly caused by loss of organelles, cytoplasm and proteins and is characterized by increased protein degradation [1,12]. In summary, there are three main mechanisms which contribute to maintenance of muscle growth: myogenesis, hypertrophy and atrophy, and the mTOR pathway plays a pivotal role in regulating each of them.

mTOR SIGNALING PATHWAY

TORs are evolutionarily conserved proteins, which belong to a group of serine threonine kinases known as phosphatidylinositol kinase-related kinase (PIKK) family [13]. TOR was first identified in *Saccharomyces cerevisiae* mutants, TOR1-1 and TOR2-1, resistant to the growth inhibitory properties of rapamycin [14]. Rapamycin is macrocyclic lactone, isolated from *Streptomyces hygroscopius*, which inhibits proliferation of mammalian cells and possesses immunosuppressive properties. Rapamycin forms a complex with peptidyl-prolyl *cis/trans* isomerase FKBP12, and this complex then binds and inhibits TOR [13].

Biochemical studies with mammalian cells led to the identification and cloning of mTOR (reviewed in [15]). Insulin-like growth factor (IGF-1) stimulates phosphorylation of the insulin receptor substrate (IRS) and subsequent recruitment of phosphatidylinositol-3 kinase (PI3K), resulting in downstream activation of protein synthesis through the Akt pathway [16–18]. Transgenic mice with constitutively active Akt exhibit increase in the average cross-sectional area of individual muscles fibers, consequent to elevated protein synthesis [19]. These studies demonstrated that Akt by itself is sufficient to induce hypertrophy *in vivo*. Genetic experiments in *Drosophila* helped identify a RAFT-1 (homolog of mTOR) and p70S6 kinase as the downstream effectors of PI3K and Akt, which help to regulate cell size [20,21]. Genetic support of a linear Akt/mTOR pathway came from studies which demonstrated that tuberous sclerosis complex-1 and -2 proteins (TSC1 and TSC2) can inhibit mTOR [22]. Akt directly phosphorylates TSC2 and functionally inactivates TSC1-TSC2 heterodimer [23] (Figure 22.1). TSC2 acts as GTPase-activating protein for small GTPase Rheb (Ras homologue enriched in brain), which, when loaded with GTP, directly binds and activates mTOR kinase activity [24].

Purification of TOR1 and TOR2 from yeast led to the identification of two distinct TOR protein complexes, TORC1 and TORC2, which have varying sensitivity to rapamycin [25]. Mammalian TOR complex 1 (mTORC1) consists of the mTOR, raptor and mLST8 and is inhibited by binding of rapamycin [26]. mTORC1 has been shown to mediate the rapamycin-sensitive temporal control of cell growth [25]. It has been proposed that raptor functions as an adaptor to recruit substrates to mTOR [27]. Other studies show that upstream signals, including rapamycin-FKBP12 complex, regulate mTORC1 activity by facilitating raptor–mTOR interaction [28]. Skeletal muscle-specific ablation of raptor causes metabolic changes and results in muscle dystrophy [29]. mLST8 binds to the kinase domain of mTOR and is required for its full catalytic activity [25]. Mammalian TORC2 (mTORC2) contains mTOR, rictor and mLST8, and controls spatial aspects of cells growth [30]. mTORC2 is neither bound nor inhibited by FKB12-rapamycin and can be directly activated by PIP3K [26]. Rictor doesn't contain catalytic motifs, but knockdown of mTOR or rictor results in loss of actin polymerization and cell spreading [30]. Skeletal muscle-specific deletion of rictor in mice leads to impaired insulin-stimulated glucose transport, due to attenuation of glucose transporter-mediated exocytosis [31] (Figure 22.1).

The effect of mTOR on the translational machinery and protein synthesis is mediated by TORC1-dependent phosphorylation of the ribosomal protein S6 kinases (S6K1 and 2) and of 4E-binding protein 1 (4E-BP1) [2]. Activated S6K1 phosphorylates the 40S ribosomal protein S6, which leads to increased translation of a subset of mRNAs that contain the 5′ tract of oligo pyrimidine, although this mechanism has been recently debated (reviewed in [13]). S6K1 knockout mice have smaller muscle fibers, and their hypertrophic response to IGF-1 stimulation and Akt activation are significantly blunted [32]. S6K1 negatively regulates IRS1 both at the transcriptional level and through direct phosphorylation, which also contributes to development of obesity and diabetes [33]. Thus,

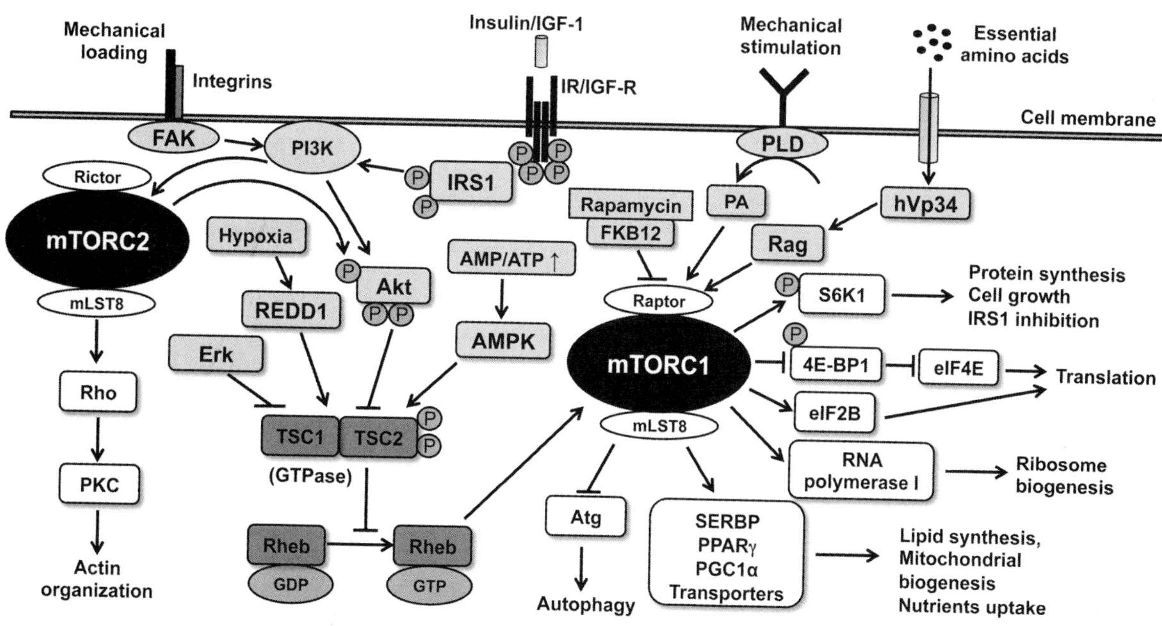

FIGURE 22.1 Model of mTOR signaling pathways regulating muscle growth. Growth factors (such as insulin and IGF1) stimulation activates IRS1/PI3K/Akt pathway, which inactivates TSC2 via phosphorylation, leading to the conversion of Rheb-GDP to Rheb-GTP and preventing inhibition of rapamycin-sensitive mTORC1. On the other hand, AMPK, induced by low levels of ATP, or REDD1, induced by hypoxia, enhances TSC2 activity, which inhibits mTORC1. mTORC1 can be also activated by mechanical loading or increase in essential amino acids concentration. mTORC1 can induce translation of mRNA and protein synthesis via the phosphorylation of 4E-BP1 and S6K1. mTORC1 also suppresses autophagy, regulates ribosome and mitochondrial biogenesis, and induces nutrients uptake and lipid synthesis. Rapamycin-insensitive mTORC2, which is activated by PIP3K, controls Akt activity and the actin cytoskeleton organization. Arrows represent activation, whereas bars represent inhibition. *Adapted with permission from [23].*

constitutive activation of mTOR signaling attenuates PI3K through a negative feedback loop via inhibition of IRS. S6K1 can also phosphorylate mTOR at Thr^{2446} and Ser^{2448}, although the consequence of this modification is unknown [34]. Another target of mTORC1 kinase activity is 4E-BP1, which on phosphorylation releases repressed cap-binding protein eIF4E, which subsequently binds to eIF4G to stimulate translation initiation [35]. 4E-BP1 and S6K1 are regulated by mTORC1 in a parallel, independent manner [36] (Figure 22.1). mTOR also enhances protein synthesis by activating eIF2B, a translation initiation factor [37]. Furthermore, exercise-induced expression of eIF2Bε is blocked by rapamycin.

The effects of mTORC1 are not limited to regulation of protein synthesis. In mammals TOR regulates ribosome biogenesis through control of RNA polymerase I, which is required for transcription of ribosomal RNAs [38]. Inhibition of mTOR signaling by rapamycin inactivates transcription initiation factor 1A and impairs transcription of RNA polymerase I. Another important function of mTOR is the regulation of the autophagy process. Autophagy is a catabolic process involving the degradation of a cell's own components, which involves the enclosure of organelles by double-membrane autophagosome and its subsequent delivery to lysosome for degradation and recycling. It is activated during starvation and serves as a conserved cytoprotective, rather than destructive, process [39]. TOR regulates autophagy in yeast and higher eukaryotes via inhibition of the autophagy-related protein kinase (ATG1), that mediates an early activation step in the autophagic process [40]. mTORC1 also promotes uptake of glucose, amino acids and iron via regulation of membrane trafficking of nutrient transporters [41]. In addition, mTOR plays a central role in transcription of nutrient-sensitive pathways involved in metabolic and biosynthetic pathways, for example PPARγ [42]. Early studies by Loewith and colleagues showed that mTORC2 plays an important role in cytoskeleton organization [25]. TORC2 signals to Rho1 GTPase, which regulates actin cytoskeleton assembly via the protein kinase C1 (PKC1) kinase pathway. mTORC2 signaling also affects actin cytoskeleton organization through a rapamycin-insensitive mechanism, although the direct targets of mTORC2 are unknown [30]. mTORC2 is required for the terminal differentiation of myoblasts, because disruption of mTORC2 by rictor knockdown blocks fusion of myoblasts [43]. Phosphorylation of Akt on Ser^{473} by mTORC2 is necessary for inhibition of

Rho-associated kinase 1 (ROCK1) required for myogenic differentiation. However, unlike mTORC1, detailed mechanisms of mTORC2 signaling in relation to muscle growth have not been fully characterized.

Although the classic mTOR activation route includes the IGF-1/PI3K/Akt pathway, mTOR can also be directly activated by a variety other stimuli. Nutrients, especially amino acids, have been shown to activate mTORC1 via inhibition of TSC1-TSC2 or via activation of Rheb [35]. Phosphatidic acid (PA) can activate mTOR signaling by binding directly to mTOR complex. In cultured cells, mitogen-mediated activation of phospholipase D (PLD) is required for mTOR-dependent phosphorylation of S6K1 and 4E-BP1 [44]. AMP-activated protein kinase (AMPK), which is activated in response to low energy levels into the cell (high AMP/ATP ratio), inhibits mTORC1-dependent phosphorylation of S6K1 and 4E-BP1 via enhancing TSC2 GAP activity through phosphorylation [45]. Environmental stresses, such as hypoxia, downregulate energy-demanding processes mediated by TORC1 via transcription of REDD1 (Regulated in development and DNA damage responses 1) [46]. Taken together, mTOR integrates a variety of signals, affecting muscle protein synthesis (MPS) and muscle growth (Figure 22.1). Detailed mechanisms of mTOR signaling and its relation to different stages of muscle growth are described below.

mTOR IN MYOGENESIS

Skeletal myogenesis is a highly coordinated cascade of events regulated by multiple signaling pathways. Little is known about the role of mTOR during muscle embryonic development. However, the recent studies by Hidalgo and colleagues shed some light on the importance of mTOR in early stages of myogenesis [47]. These authors showed that hypoxia causes growth retardation and motility defects in *Xenopus* embryos by affecting mitotic cells and inhibiting accumulation of muscle-specific proteins in somites. These changes were a result of inhibition of the Akt/mTOR pathway, which led to increased 4E-BP1 phosphorylation and translation repression. This observation is also consistent with early studies on mice embryos which show that mTOR knockout embryos die shortly after implantation, owing to impairment of proliferation of embryonic stem cells [48]. Consistent with these observations, rapamycin arrested cell proliferation in early mouse embryos [49]. Collectively, these studies suggest that rapamycin-sensitive mTOR function is required during embryonic muscle development.

Growth or regeneration of adult skeletal muscle requires satellite cell activation, proliferation and differentiation [50]. Studies using myogenic satellite cells isolated from 6-month-old pigs, revealed that both IGF-1 and the amino acid leucine stimulate activation and proliferation of quiescent satellite cells by increasing mTOR-mediated S6K1 and 4E-BP1 phosphorylation [51]. However, the role of mTOR in myogenic differentiation remains controversial. Rapamycin is known to inhibit differentiation of myoblasts in cell culture [52]. Shu and colleagues showed that expression of rapamycin-resistant mTOR mutant allowed C2C12 myotube differentiation even in the presence of rapamycin [53]. Interestingly, Erbay and coworkers showed that the kinase activity of mTOR is not required for the initiation of myoblast differentiation [54]. Subsequent studies showed that mTOR regulates the initial myoblast–myoblast fusion by controlling the expression of insulin-like growth factor-II [55]. However, the formation of mature myotubes by myoblast–myotube fusion requires the catalytic activity of mTOR [56]. Overexpression of angiotensin I-converting enzyme (ACE) suppresses myogenesis through inhibition of mTOR-mediated myotube maturation [57]. Furthermore, ACE inhibitor accelerated myogenesis through activation of mTOR signaling, leading to increased expression of myosin heavy chain and phosphorylation of S6K1 kinase [57]. These studies show that mTORC1 signaling is critical in myogenesis. Contrastingly however, other studies revealed that that expression of constitutively active, rapamycin-insensitive S6K1 or downregulation of 4E-BP1 failed to support myotubes differentiation in the presence of rapamycin [43]. Rapamycin also induced a significant and prolonged decrease in mTORC2-dependent phosphorylation of Akt on Ser^{473}. These results are consistent with the idea that Akt signaling is essential for myoblast differentiation [58]. Prolonged inhibition of mTORC1 by rapamycin may cause redistribution of mTOR from the mTORC2 complex, leading to decreased phosphorylation of Akt (Ser^{473}) and impaired myogenesis. Rapamycin also appears to destabilize the mTORC2 complex by inhibiting the interaction of mTOR and rictor in cytosolic and nuclear complexes [43].

mTOR IN MUSCLE HYPERTROPHY

For over a decade, increase in local production of IGF-1 was considered to be responsible for the induction of skeletal muscle hypertrophy in response to mechanical loading during exercise [18]. It has been assumed that this event leads to the increased activation of the PI3K/Akt signaling cascade, which in turns stimulates mTORC1-mediated induction of protein synthesis [12]. However, recent studies revealed that mTORC1-dependent phosphorylation of S6K1 is induced within a few hours following mechanical stimulation [59,60], whereas overload-induced IGF-1 production in skeletal muscle occurs only after 24 hours following the initiation of the workload [61]. Further studies using mice with muscle-specific dominant-negative mutation of the IGF-1 receptor showed that a functional IGF-1 receptor was not necessary for mTORC1 activation and the induction of skeletal muscle hypertrophy [62,63]. The aforementioned findings that growth factors are not required for the activation of mTOR during muscle hypertrophy, paved the way to studies which attempted to identify how the mechanical signal is transduced to mTOR. First, it was demonstrated that mechanical stimuli can activate mTOR signaling through a PI3K/Akt-independent pathway that requires phospholipase D-mediated phosphatidic production in both cultured myotubes [64] and in an *in vivo* model of resistance exercise [65]. On the other hand, over-expression of Rheb was sufficient to induce skeletal muscle hypertrophy independent of PI3K/Akt signaling [66]. More recently, Miazaki and coworkers have shown that the initial activation of mTORC1 in response to mechanical overload during the first 24 hours does not involve IGF-1/PI3K/Akt-dependent signaling. These authors demonstrated that extracellular signal-regulated kinases (Erk) signaling mediates early activation of mTORC1 signaling after workload, through phosphorylation of TSC2 at the Ser^{664} site, thereby inhibiting its GAP activity (Figure 22.1). This allows Rheb to accumulate in its active GTP-bound form [67]. In another study, treatment with rapamycin did not prevent compensatory hypertrophy in the plantaris muscle following 7 days of mechanical overload. Long-term activation of mTOR-dependent signaling following mechanical overload was independent of Erk-mediated activation, which occurs during the first 24 hours [67].

There are two main types of exercise training: endurance and resistance, and these appear to be differently influenced by mTOR signaling. Endurance training induces a partial fast-to-slow muscle phenotype switch and stimulates mitochondrial biogenesis, but not growth [68]. In contrast, resistance training mainly promotes hypertrophy by inducing increased MPS [60]. Isolated rat muscles electrically stimulated with high frequency, mimicking resistance training, showed acute increases in phosphorylation of Akt/mTOR signaling molecules [69]. However, electrical stimulation with low frequency, mimicking endurance training, activated AMPK signaling, which resulted in phosphorylation of TSC2 and inhibition of the mTOR pathway [69]. These findings partially explain the difference in adaptations in skeletal muscle caused by two different types of exercise. In contrast, *in vivo* models showed increased mTOR and S6K1 phosphorylation immediately after both endurance and resistance exercise, although resistance exercise resulted in a more prolonged activation of mTOR and S6K1. In addition, both types of exercise lead to AMPK activation in humans [70,71]. AMPK is a well-known upstream negative regulator of mTOR signaling in skeletal muscle [69]; in this scenario resistance training should have a negative effect on mTOR signaling. This paradox could be partially explained by the possibility of feeding-induced hyperaminoacidemia during or after the exercise, which overrides the inhibitory effect of AMPK activation on mTOR signaling. Indeed, feeding carbohydrate supplement enriched with essential amino acids overrules the inhibitory effect of AMPK on mTOR during resistance exercise in human muscle [72,73]. It appears that skeletal muscle hypertrophy in response to resistance training is delayed with aging [74]. Drummond and colleagues showed that despite lack of post-exercise differences in mTOR signaling, MPS was decreased in older men 1 hour after resistance exercise, due to lower activation of the Erk1/2 pathway and elevated AMPK phosphorylation [75].

Despite these advances in our understanding of the role of mTOR in increased skeletal muscle growth after the exercise, the upstream regulators that have the potential to sense tension in muscle and convert mechanical signals into molecular events leading to mTOR phosphorylation are not well defined. It has been shown that the rise in intracellular Ca^{2+} levels, caused by excess levels of amino acid, increases the direct binding of Ca^{2+}/calmodulin complex to mTORC1, leading to its activation [76]. Recently it has been demonstrated that $\alpha_7\beta_1$-integrin sensitizes skeletal muscle to mechanical strain and increases Akt-mTOR-S6K1 signaling during muscle hypertrophy following exercise training in α7BX2-integrin transgenic mice [77]. It is also possible that mTORC is sensitive to influx of Ca^{2+} across the sarcolemma into the muscle cell through stretch-activated channels [78]. Another intriguing hypothesis is that conversion of mechanical loading to cell signaling events in skeletal muscle may be mediated by changes in the phosphorylation status of focal adhesion kinase (FAK) [79]. It has been shown *in vitro* that FAK activation can affect mTOR signaling by phosphorylation of TSC2 [80]. Xia and coworkers showed that FAK can also affect activation of the Akt/mTOR pathway by increasing PI3K activity in fibroblasts [81]. However, there are limited data to suggest a definitive role of FAK-mediated activation of protein synthesis in skeletal muscle following exercise training.

Cytokines can also possibly contribute to induction of mTOR signaling in hypertrophic processes following mechanical overload. Specifically, the knockout of leukemia inhibitory factor (LIF) results in a failed hypertrophic response in plantaris muscle in response to mechanical loading, suggesting that LIF is a crucial mediator of the hypertrophic process [82]. Interestingly, LIF induces activation of the Akt/mTOR pathway in cardiac myocyte hypertrophy [83], and can possibly mediate exercise-induced activation of mTOR in skeletal muscle.

Taken together, it is clear that mTOR is a key player in the process of initiation of the protein synthesis during mechanical loading in skeletal muscle. However, it is unclear as to how exactly physical loading of the cell membrane is translated into a signal to induce mTOR-mediated phosphorylation of S6K1 and 4E-BP1, leading to increased initiation of protein translation in skeletal muscle.

mTOR IN MUSCLE ATROPHY

Muscle atrophy is defined as a decrease in the mass of the muscle and can be a partial or complete wasting away of a muscle [1]. Muscle atrophy is caused by several common diseases, such as cancer, diabetes and renal failure, and by severe burns, starvation and disuse of the muscles [12,84–86]. At the molecular level, the Akt/mTOR hypertrophy-inducing pathway is inhibited in an *in vivo* burn model of skeletal muscle atrophy [86]. In addition, sepsis causes a decrease in protein synthesis and an increase in atrophy, which can be attenuated by IGF-1 injections [87]. Muscle atrophy can be caused by injury to spinal cord (paraplegia), leading to impairment in motor or sensory function of the lower extremities [88,89]. In a rat model of paraplegia-induced muscle atrophy, authors found reduction in phosphorylation and expression levels of mTOR and S6K1, with no changes in AMPK activity 10 weeks after spinal cord transsection [90]. However, other reports using a hind limb suspension model of atrophy, showed an increase in AMPK activity 4–8 weeks after immobilization [85]. Most of the muscle mass loss occurs within the first 2 weeks of paraplegia [91], which coincides with increased AMPK activity in the early stages [85]. It is likely that AMPK negatively regulates mTOR signaling and protein synthesis in skeletal muscle during the initial stages after spinal cord transaction, and the further stages driven by AMPK-independent mTOR inhibition [90]. In addition to inhibition of protein synthesis, other distinct mechanisms are also involved in skeletal muscle atrophy [92]. The most important is the stimulation of proteolysis, due to activation of the ubiquitin-proteasome degradation system through the forkhead box protein O (FOXO) pathway [12]. Two genes encoding E3 ubiquitin ligases—the muscle ring finger 1 (MuRF1) and muscle atrophy F-box (MAFbx)—were shown to be significantly increased during skeletal muscle atrophy. These two proteins appear to play a key role in protein degradation during hypertrophy, because MuRF1$^{-/-}$ or MAFbx$^{-/-}$ mice have significantly less muscle mass loss during atrophy [93]. mTOR is required for activation of MuRF1 and MAFbx ubiquitin ligases, which does not involve the signaling mediated by the canonical phosphorylation of FOXO transcription factors [94]. Apoptosis, programmed cell death, and autophagy, cell self-degradation, might also contribute to atrophy. Ferreira and colleagues showed elevated lysosomal autophagic activity and increased apoptosis in soleus muscle of mice during the first 4 h after hindlimb suspension [95]. This supports the notion that, during immobilization atrophy, muscle wasting is initiated by an autophagic reaction followed by an inflammatory response [96]. mTOR is a well-known negative regulator of autophagy, so inhibition of mTOR signaling might be a key step in initiation of autophagy during muscle atrophy.

Upregulation of atrophy signaling during long-term muscle wasting occurs parallel to downregulation of muscle cell growth-associated signaling, and both processes are mediated by mTOR pathway. However, despite the important role of mTOR in muscle atrophy, it appears that the signaling events involved in decreased muscle protein synthesis are dictated by external stimuli of atrophy. For example, hypoxia-induced muscle atrophy, associated with chronic obstructive pulmonary disease (COPD) and obstructive sleep apnea (OSA), is mediated by unfolded protein response (UPR) rather than mTOR signaling [97]. Furthermore, according to the study of Nedergaard and coworkers, changes in muscle mass after limb immobilization and subsequent rehabilitation are not associated with increase in the expression or phosphorylation levels of components in mTOR/Akt pathway [98]. However, recent studies by Lang and colleagues demonstrated a significant delay of muscle regrowth in mTOR$^{+/-}$ mice following 7 days of hind limb immobilization compared with wild type (WT) mice [84]. Nevertheless, immobilization-induced decrease in the phosphorylation of 4E-BP1 did not differ between WT and mTOR$^{+/-}$ mice, suggesting that this model of muscle atrophy could be mTOR independent [84]. Accumulating evidence suggests that atrophy of skeletal muscle is associated with upregulation of myostatin— a TGFβ family member that acts as a negative regulator of muscle mass [99]. Studies by Amirouche and colleagues show that, although myostatin overexpression fails to alter the ubiquitin-proteasome pathway, myostatin negatively regulates the mTOR pathway [100].

Aging is also associated with muscle wasting, but it is a rather gradual degenerative loss of the ability to maintain skeletal muscle function and mass, a condition known as sarcopenia [101]. The exact biological mechanism underlying sarcopenia is unknown, but mTOR appears to play an important role in the process. A significant reduction in muscle sarcomere volume and phosphorylation levels of mTOR were detected in sedentary aged rats relative to young controls, but it was reversed by treadmill training [102]. These observations may have important clinical implications, as they suggest that exercise training in advanced age can directly activate mTOR signaling, making it a potential target for sarcopenia treatment. Other forms of sarcopenia, like sarcopenia of cirrhosis, also have been recently shown to be associated with impaired mTOR signaling [103].

NUTRITION AND mTOR-DEPENDENT MUSCLE GROWTH

Skeletal muscle has a high content of contractility proteins, so muscle building is largely dependent on a nutritional status. Under-nutrition decreases skeletal MPS in physically active adults by attenuation of mTOR-mediated 4E-BP1 phosphorylation. The availability of essential amino acids (EAAs) is the primary stimulator of MPS [104]. Among EAAs, leucine alone is able to stimulate MPS [105]. During pregnancy the fetal growth and development of skeletal muscle depend entirely on maternal nutrition. Maternal low protein diet attenuates fetal growth, but it can be reversed by leucine supplementation, which activates the mTOR pathway [105]. During the neonatal period, muscle growth is rapid due to an ability to increase protein synthesis in response to feeding, which significantly declines with development [106]. Leucine has been shown to stimulate protein synthesis by inducing the activation of mTORC1 and its downstream pathway in neonatal pigs, but this effect decreases with development [107]. Skeletal muscle protein turnover in adults is also largely affected by nutrients availability [108]. Rapamycin administration to humans blocks EAA-induced stimulation of MPS, indicating an mTORC1-dependent mechanism [109]. Changes in amino acids levels are sensed by nutrient sensor human vacuolar protein sorting 34 (hVps34) and transduced to mTORC1 in a Ca^{2+}-dependent manner [76]. Increased EAAs levels also lead to activation of mitogen-activated protein kinase kinase kinase kinase 3 (MAP4K3) and Rag GTPase, which results in mTORC1-dependent phosphorylation of S6K1 [110,111]. Stimulation of mTORC1 by essential amino acids was shown to not only elevate protein synthesis, but also increase expression of EAA transporters, which sensitizes cells to increased EAAs availability [112] (Figure 22.1). Consumption of a high-protein meal does not further stimulate MPS compared with a moderately sized protein meal [113]. This suggests that consumption of several moderately sized protein meals during the day will better support muscle growth than a single meal with high protein content. With aging, EAAs elicit smaller muscle protein anabolic response compared with young subjects [114]. EAAs supplementation also increases MPS induced by resistance training via the activation of the Akt/mTORC1/S6K1 pathway by leucine [73], and this response is not affected by aging [115]. Similar results were obtained by Morrison and colleagues using an endurance exercise model in rats, wherein phosphorylation of mTOR and S6K1 was observed following 3 hours of swimming only in rats that received carbohydrate-protein supplement, but not in fasted animals [116]. However, in elderly people, protein supplementation before and after exercise does not seem to affect muscle hypertrophy following resistance training [117]. Other factors, such as glycogen level in skeletal muscle, do not affect protein synthesis after resistance exercise, despite decreased mTOR phosphorylation in glycogen-depleted muscles [118]. Taken together, mTOR represents a nutrient sensitive pathway, regulating muscle growth in the fed state.

CONCLUSIONS

Over the past years, the mechanisms controlling muscle growth have attracted the attention of many scientists in various fields of study, including aging, prognosis of many diseases, and sports medicine. The mTOR complex appears to represent a common pathway on which several signals regulating protein turnover in muscle converge. mTOR is a key regulator of prenatal muscle development, and its activity is crucial for hypertrophy induced by resistance exercise in adults. Inhibition of mTOR appears to be critical in development of muscle atrophy. Recent findings offer a newer molecular explanation for training-induced muscle building. During mechanical overload, mTOR is activated independent of hormones and growth factors, implying that maximal muscle growth in athletes can be brought about by extensive training alone in the absence of external growth factors or supplements. In addition, recent advances in identifying the mechanisms of muscle loss may help develop novel drugs targeting mTOR to combat skeletal muscle atrophy in a variety of clinical conditions: from spinal cord trauma and cancer, to the loss of muscle mass during disuse or normal aging.

References

[1] Sandri M. Signaling in muscle atrophy and hypertrophy. Physiology (Bethesda) 2008;23:160–70.

[2] Hay N, Sonenberg N. Upstream and downstream of mTOR. Genes Dev 2004;18:1926–45.

[3] Teleman AA, Hietakangas V, Sayadian AC, Cohen SM. Nutritional control of protein biosynthetic capacity by insulin via myc in drosophila. Cell Metab 2008;7:21–32.

[4] Le Grand F, Rudnicki MA. Skeletal muscle satellite cells and adult myogenesis. Curr Opin Cell Biol 2007;19:628–33.

[5] Rehfeldt C, Fiedler I, Dietl G, Ender K. Myogenesis and postnatal skeletal muscle cell growth as influenced by selection. Livest Prod Sci 2000;66:177–88.

[6] Yaffe D, Feldman M. The formation of hybrid multinucleated muscle fibers from myoblasts of different genetic origin. Dev Biol 1965;11:300–17.

[7] Moss FP, Leblond CP. Satellite cells as the source of nuclei in muscles of growing rats. Anat Rec 1971;170:421–35.

[8] McCarthy JJ, Esser KA. Counterpoint: satellite cell addition is not obligatory for skeletal muscle hypertrophy. J Appl Physiol 2007;103:1100–2.

[9] Charge SBP, Rudnicki MA. Cellular and molecular regulation of muscle regeneration. Physiol Rev 2004;84:209–38.

[10] Hawke TJ, Garry DJ. Myogenic satellite cells: physiology to molecular biology. J Appl Physiol 2001;91:534–51.

[11] Schiaffino S, Pierobon Bormioli S, Aloisi M. The fate of newly formed satellite cells during compensatory muscle hypertrophy. Virchows Arch B Cell Pathol Zell-Pathol 1976;21:113–8.

[12] Glass DJ. Skeletal muscle hypertrophy and atrophy signaling pathways. Int J Biochem Cell Biol 2005;37:1974–84.

[13] Wullschleger S, Loewith R, Hall MN. TOR signaling in growth and metabolism. Cell 2006;124:471–84.

[14] Heitman J, Movva NR, Hall MN. Targets for cell cycle arrest by the immunosuppressant rapamycin in yeast. Science 1991;253:905–9.

[15] Fingar DC, Blenis J. Target of rapamycin (TOR): an integrator of nutrient and growth factor signals and coordinator of cell growth and cell cycle progression. Oncogene 2004;23:3151–71.

[16] Bodine SC, Stitt TN, Gonzalez M, Kline WO, Stover GL, Bauerlein R, et al. Akt/mTOR pathway is a crucial regulator of skeletal muscle hypertrophy and can prevent muscle atrophy in vivo. Nat Cell Biol 2001;3:1014–9.

[17] Rommel C, Bodine SC, Clarke BA, Rossman R, Nunez L, Stitt TN, et al. Mediation of IGF-1-induced skeletal myotube hypertrophy by PI(3)K/Akt/mTOR and PI(3)K/Akt/GSK3 pathways. Nat Cell Biol 2001;3:1009–13.

[18] DeVol DL, Rotwein P, Sadow JL, Novakofski J, Bechtel PJ. Activation of insulin-like growth factor gene expression during work-induced skeletal muscle growth. Am J Physiol Endocrinol Metab 1990;259:E89–95.

[19] Lai K-MV, Gonzalez M, Poueymirou WT, Kline WO, Na E, Zlotchenko E, et al. Conditional activation of akt in adult skeletal muscle induces rapid hypertrophy. Mol Cell Biol 2004;24:9295–304.

[20] Zhang H, Stallock JP, Ng JC, Reinhard C, Neufeld TP. Regulation of cellular growth by the Drosophila target of rapamycin dTOR. Genes Dev 2000;14:2712–24.

[21] Montagne J, Stewart MJ, Stocker H, Hafen E, Kozma SC, Thomas G. Drosophila S6 kinase: a regulator of cell size. Science 1999;285:2126–9.

[22] Tee AR, Fingar DC, Manning BD, Kwiatkowski DJ, Cantley LC, Blenis J. Tuberous sclerosis complex-1 and -2 gene products function together to inhibit mammalian target of rapamycin (mTOR)-mediated downstream signaling. PNAS 2002;99: 13571–6.

[23] Nair S, Ren J. Autophagy and cardiovascular aging: lesson learned from rapamycin. Cell Cycle 2012;11:2092–9.

[24] Long X, Lin Y, Ortiz-Vega S, Yonezawa K, Avruch J. Rheb binds and regulates the mTOR Kinase. Curr Biol 2005;15: 702–13.

[25] Loewith R, Jacinto E, Wullschleger S, Lorberg A, Crespo JL, Bonenfant D b, et al. Two TOR complexes, only one of which is rapamycin sensitive, have distinct roles in cell growth control. Mol Cell 2002;10:457–68.

[26] Oh WJ, Jacinto E. mTOR complex 2 signaling and functions. Cell Cycle 2011;10:2305–16.

[27] Hara K, Maruki Y, Long X, Yoshino K-i, Oshiro N, Hidayat S, et al. Raptor, a binding partner of target of rapamycin (TOR), mediates TOR action. Cell 2002;110:177–89.

[28] Kim D-H, Sarbassov DD, Ali SM, King JE, Latek RR, Erdjument-Bromage H, et al. mTOR interacts with raptor to form a nutrient-sensitive complex that signals to the cell growth machinery. Cell 2002;110:163–75.

[29] Bentzinger CF, Romanino K, Cloetta D, Lin S, Mascarenhas JB, Oliveri F, et al. Skeletal muscle-specific ablation of raptor, but not of rictor, causes metabolic changes and results in muscle dystrophy. Cell Metab 2008;8:411–24.

[30] Jacinto E, Loewith R, Schmidt A, Lin S, Ruegg MA, Hall A, et al. Mammalian TOR complex 2 controls the actin cytoskeleton and is rapamycin insensitive. Nat Cell Biol 2004;6:1122–8.

[31] Kumar A, Harris TE, Keller SR, Choi KM, Magnuson MA, Lawrence JC. Muscle-specific deletion of rictor impairs insulin-stimulated glucose transport and enhances basal glycogen synthase activity. Mol Cell Biol 2008;28:61–70.

[32] Ohanna M, Sobering AK, Lapointe T, Lorenzo L, Praud C, Petroulakis E, et al. Atrophy of S6K1-/- skeletal muscle cells reveals distinct mTOR effectors for cell cycle and size control. Nat Cell Biol 2005;7:286–94.

[33] Um SH, Frigerio F, Watanabe M, Picard F, Joaquin M, Sticker M, et al. Absence of S6K1 protects against age- and diet-induced obesity while enhancing insulin sensitivity. Nature 2004;431:200–5.

[34] Holz MK, Blenis J. Identification of S6 kinase 1 as a novel mammalian target of rapamycin (mTOR)-phosphorylating kinase. J Biol Chem 2005;280:26089–93.

[35] Burnett PE, Barrow RK, Cohen NA, Snyder SH, Sabatini DM. RAFT1 phosphorylation of the translational regulators p70 S6 kinase and 4E-BP1. PNAS 1998;95:1432–7.

[36] Hara K, Yonezawa K, Kozlowski MT, Sugimoto T, Andrabi K, Weng Q-P, et al. Regulation of eIF-4E BP1 phosphorylation by mTOR. J Biol Chem 1997;272:26457–63.

[37] Kubica N, Bolster DR, Farrell PA, Kimball SR, Jefferson LS. Resistance exercise increases muscle protein synthesis and translation of eukaryotic initiation factor 2Bε mRNA in a mammalian target of rapamycin-dependent manner. J Biol Chem 2005;280:7570–80.

[38] Mayer C, Zhao J, Yuan X, Grummt I. mTOR-dependent activation of the transcription factor TIF-IA links rRNA synthesis to nutrient availability. Genes Dev 2004;18:423–34.

[39] Klionsky DJ, Emr SD. Autophagy as a regulated pathway of cellular degradation. Science 2000;290:1717–21.

[40] Kamada Y, Funakoshi T, Shintani T, Nagano K, Ohsumi M, Ohsumi Y. Tor-mediated induction of autophagy via an Apg1 protein kinase complex. J Cell Biol 2000;150:1507–13.

[41] Edinger AL, Thompson CB. Akt maintains cell size and survival by increasing mTOR-dependent nutrient uptake. Mol Biol Cell 2002;13:2276–88.

[42] Kim JE, Chen J. Regulation of peroxisome proliferator-activated receptor-gamma activity by mammalian target of rapamycin and amino acids in adipogenesis. Diabetes 2004;53:2748–56.

[43] Shu L, Houghton PJ. The mTORC2 complex regulates terminal differentiation of C2C12 myoblasts. Mol Cell Biol 2009;29:4691–700.

[44] Fang Y, Vilella-Bach M, Bachmann R, Flanigan A, Chen J. Phosphatidic acid-mediated mitogenic activation of mTOR signaling. Science 2001;294:1942–5.

[45] Inoki K, Zhu T, Guan K-L. TSC2 mediates cellular energy response to control cell growth and survival. Cell 2003;115:577–90.

[46] Brugarolas J, Lei K, Hurley RL, Manning BD, Reiling JH, Hafen E, et al. Regulation of mTOR function in response to hypoxia by REDD1 and the TSC1/TSC2 tumor suppressor complex. Genes Dev 2004;18:2893–904.

[47] Hidalgo M, Le Bouffant R, Bello V, Buisson N, Cormier P, Beaudry M l, et al. The translational repressor 4E-BP mediates hypoxia-induced defects in myotome cells. J Cell Sci 2012.

[48] Murakami M, Ichisaka T, Maeda M, Oshiro N, Hara K, Edenhofer F, et al. mTOR is essential for growth and proliferation in early mouse embryos and embryonic stem cells. Mol Cell Biol 2004;24:6710–8.

[49] Martin PM, Sutherland AE. Exogenous amino acids regulate trophectoderm differentiation in the mouse blastocyst through an mTOR-dependent pathway. Dev Biol 2001;240:182–93.

[50] Rosenblatt JD, Yong D, Parry DJ. Satellite cell activity is required for hypertrophy of overloaded adult rat muscle. Muscle Nerve 1994;17:608–13.

[51] Han B, Tong J, Zhu MJ, Ma C, Du M. Insulin-like growth factor-1 (IGF-1) and leucine activate pig myogenic satellite cells through mammalian target of rapamycin (mTOR) pathway. Mol Reprod Dev 2008;75:810–7.

[52] Cuenda A, Cohen P. Stress-activated protein kinase-2/p38 and a rapamycin-sensitive pathway are required for C2C12 myogenesis. J Biol Chem 1999;274:4341–6.

[53] Shu L, Zhang X, Houghton PJ. Myogenic differentiation is dependent on both the kinase function and the N-terminal sequence of mammalian target of rapamycin. J Biol Chem 2002;277:16726–32.

[54] Erbay E, Chen J. The mammalian target of rapamycin regulates C2C12 myogenesis via a kinase-independent mechanism. J Biol Chem 2001;276:36079–82.

[55] Erbay E, Park I-H, Nuzzi PD, Schoenherr CJ, Chen J. IGF-II transcription in skeletal myogenesis is controlled by mTOR and nutrients. J Cell Biol 2003;163:931–6.

[56] Park IH, Chen J. Mammalian target of rapamycin (mTOR) signaling is required for a late-stage fusion process during skeletal myotube maturation. J Biol Chem 2005;280: 32009–17.

[57] Mori S, Tokuyama K. ACE activity affects myogenic differentiation via mTOR signaling. Biochem Biophys Res Commun 2007;363:597–602.

[58] Wilson EM, Rotwein P. Selective control of skeletal muscle differentiation by Akt1. J Biol Chem 2007;282:5106–10.

[59] Baar K, Esser K. Phosphorylation of p70S6k correlates with increased skeletal muscle mass following resistance exercise. Am J Physiol – Cell Physiol 1999;276:C120–7.

[60] Nader GA, Esser KA. Intracellular signaling specificity in skeletal muscle in response to different modes of exercise. J Appl Physiol 2001;90:1936–42.

[61] Adams GR, Haddad F, Baldwin KM. Time course of changes in markers of myogenesis in overloaded rat skeletal muscles. J Appl Physiol 1999;87:1705–12.

[62] Spangenburg EE, Le Roith D, Ward CW, Bodine SC. A functional insulin-like growth factor receptor is not necessary for load-induced skeletal muscle hypertrophy. J Physiol 2008;586: 283–91.

[63] Witkowski S, Lovering RM, Spangenburg EE. High-frequency electrically stimulated skeletal muscle contractions increase p70s6k phosphorylation independent of known IGF-I sensitive signaling pathways. FEBS Lett 2010;584:2891–5.

[64] Hornberger TA, Sukhija KB, Wang X-R, Chien S. mTOR is the rapamycin-sensitive kinase that confers mechanically-induced phosphorylation of the hydrophobic motif site Thr(389) in p70S6k. FEBS Lett 2007;581:4562–6.

[65] O'Neil TK, Duffy LR, Frey JW, Hornberger TA. The role of phosphoinositide 3-kinase and phosphatidic acid in the regulation of mammalian target of rapamycin following eccentric contractions. J Physiol 2009;587:3691–701.

[66] Goodman CA, Miu MH, Frey JW, Mabrey DM, Lincoln HC, Ge Y, et al. A phosphatidylinositol 3-kinase/protein kinase B-independent activation of mammalian target of rapamycin signaling is sufficient to induce skeletal muscle hypertrophy. Mol Biol Cell 2010;21:3258–68.

[67] Miyazaki M, McCarthy JJ, Fedele MJ, Esser KA. Early activation of mTORC1 signalling in response to mechanical overload is independent of phosphoinositide 3-kinase/Akt signalling. J Physiol 2011;589:1831–46.

[68] Salmons S, Henriksson J. The adaptive response of skeletal muscle to increased use. Muscle Nerve 1981;4:94–105.

[69] Atherton PJ, Babraj JA, Smith K, Singh J, Rennie MJ, Wackerhage H. Selective activation of AMPK-PGC-1alpha or PKB-TSC2-mTOR signaling can explain specific adaptive responses to endurance or resistance training-like electrical muscle stimulation. FASEB J 2005;19:786–8.

[70] Wilkinson SB, Phillips SM, Atherton PJ, Patel R, Yarasheski KE, Tarnopolsky MA, et al. Differential effects of resistance and endurance exercise in the fed state on signalling molecule phosphorylation and protein synthesis in human muscle. J Physiol 2008;586:3701–17.

[71] Dreyer HC, Fujita S, Cadenas JG, Chinkes DL, Volpi E, Rasmussen BB. Resistance exercise increases AMPK activity and reduces 4E-BP1 phosphorylation and protein synthesis in human skeletal muscle. J Physiol 2006;576:613–24.

[72] Moore DR, Atherton PJ, Rennie MJ, Tarnopolsky MA, Phillips SM. Resistance exercise enhances mTOR and MAPK signalling in human muscle over that seen at rest after bolus protein ingestion. Acta Physiol 2011;201:365–72.

[73] Dreyer HC, Drummond MJ, Pennings B, Fujita S, Glynn EL, Chinkes DL, et al. Leucine-enriched essential amino acid and carbohydrate ingestion following resistance exercise enhances mTOR signaling and protein synthesis in human muscle. Am J Physiol Endocrinol Metab 2008;294:E392–400.

[74] Raue U, Slivka D, Jemiolo B, Hollon C, Trappe S. Myogenic gene expression at rest and after a bout of resistance exercise in young (18-30 yr) and old (80-89 yr) women. J Appl Physiol 2006;101:53–9.

[75] Drummond MJ, Dreyer HC, Pennings B, Fry CS, Dhanani S, Dillon EL, et al. Skeletal muscle protein anabolic response to resistance exercise and essential amino acids is delayed with aging. J Appl Physiol 2008;104:1452–61.

[76] Gulati P, Gaspers LD, Dann SG, Joaquin M, Nobukuni T, Natt F, et al. Amino acids activate mTOR complex 1 via Ca^{2+}/CaM signaling to hVps34. Cell Metab 2008;7:456–65.

[77] Zou K, Meador BM, Johnson B, Huntsman HD, Mahmassani Z, Valero MC, Huey KA, Boppart MD. The $\alpha_7\beta_1$-integrin increases muscle hypertrophy following multiple bouts of eccentric exercise. J Appl Physiol 2011;111:1134–41.

[78] Yeung EW, Whitehead NP, Suchyna TM, Gottlieb PA, Sachs F, Allen DG. Effects of stretch-activated channel blockers on [Ca2 +]i and muscle damage in the mdx mouse. J Physiol 2005;562:367–80.

[79] Gordon SE, Fluck M, Booth FW. Selected contribution: skeletal muscle focal adhesion kinase, paxillin, and serum response factor are loading dependent. J Appl Physiol 2001; 90:1174–83.
[80] Gan B, Yoo Y, Guan J-L. Association of focal adhesion kinase with tuberous sclerosis complex 2 in the regulation of S6 kinase activation and cell growth. J Biol Chem 2006;281: 37321–9.
[81] Xia H, Nho RS, Kahm J, Kleidon J, Henke CA. Focal adhesion kinase is upstream of phosphatidylinositol 3-Kinase/Akt in regulating fibroblast survival in response to contraction of type I collagen matrices via a beta 1 integrin viability signaling pathway. J Biol Chem 2004;279:33024–34.
[82] Spangenburg EE, Booth FW. Leukemia inhibitory factor restores the hypertrophic response to increased loading in the LIF(-/-) mouse. Cytokine 2006;34:125–30.
[83] Oh H, Fujio Y, Kunisada K, Hirota H, Matsui H, Kishimoto T, et al. Activation of phosphatidylinositol 3-kinase through glycoprotein 130 induces protein kinase B and p70 S6 kinase phosphorylation in cardiac myocytes. J Biol Chem 1998;273: 9703–10.
[84] Lang SM, Kazi AA, Hong-Brown L, Lang CH. Delayed recovery of skeletal muscle mass following hindlimb immobilization in mTOR heterozygous mice. PLoS One 2012;7:e38910.
[85] Hilder TL, Baer LA, Fuller PM, Fuller CA, Grindeland RE, Wade CE, et al. Insulin-independent pathways mediating glucose uptake in hindlimb-suspended skeletal muscle. J Appl Physiol 2005;99:2181–8.
[86] Sugita H, Kaneki M, Sugita M, Yasukawa T, Yasuhara S, Martyn JAJ. Burn injury impairs insulin-stimulated Akt/PKB activation in skeletal muscle. Am J Physiol Endocrinol Metab 2005;288:E585–91.
[87] Svanberg E, Frost RA, Lang CH, Isgaard J, Jefferson LS, Kimball SR, et al. IGF-I/IGFBP-3 binary complex modulates sepsis-induced inhibition of protein synthesis in skeletal muscle. Am J Physiol Endocrinol Metab 2000;279:E1145–58.
[88] Taylor-Schroeder S, LaBarbera J, McDowell S, Zanca JM, Natale A, Mumma S, et al. Physical therapy treatment time during inpatient spinal cord injury rehabilitation. J Spinal Cord Med 2011;34:149–61.
[89] Malisoux L, Jamart C, Delplace K, Nielens H, Francaux M, Theisen D. Effect of long-term muscle paralysis on human single fiber mechanics. J Appl Physiol 2007;102:340–9.
[90] Dreyer HC, Glynn EL, Lujan HL, Fry CS, DiCarlo SE, Rasmussen BB. Chronic paraplegia-induced muscle atrophy downregulates the mTOR/S6K1 signaling pathway. J Appl Physiol 2008;104:27–33.
[91] Dupont-Versteegden EE, Houle JD, Gurley CM, Peterson CA. Early changes in muscle fiber size and gene expression in response to spinal cord transection and exercise. Am J Physiol – Cell Physiol 1998;275:C1124–33.
[92] Lecker SH, Jagoe RT, Gilbert A, Gomes M, Baracos V, Bailey J, et al. Multiple types of skeletal muscle atrophy involve a common program of changes in gene expression. FASEB J 2004;18:39–51.
[93] Bodine SC, Latres E, Baumhueter S, Lai VKM, Nunez L, Clarke BA, et al. Identification of ubiquitin ligases required for skeletal muscle atrophy. Science 2001;294:1704–8.
[94] Latres E, Amini AR, Amini AA, Griffiths J, Martin FJ, Wei Y, et al. Insulin-like Growth Factor-1 (IGF-1) inversely regulates atrophy-induced genes via the phosphatidylinositol 3-Kinase/Akt/Mammalian target of rapamycin (PI3K/Akt/mTOR) pathway. J Biol Chem 2005;280:2737–44.
[95] Ferreira R, Neuparth MJ, Vitorino R, Appell HJ, Amado F, Duarte JA. Evidences of apoptosis during the early phases of soleus muscle atrophy in hindlimb suspended mice. Physiol Res (Prague, Czech Repub) 2008;57:601–11.
[96] Appell HJ, Ascensao A, Natsis K, Michael J, Duarte JA. Signs of necrosis and inflammation do not support the concept of apoptosis as the predominant mechanism during early atrophy in immobilized muscle. Basic Appl Myol 2004;14:191–6.
[97] Etheridge T, Atherton PJ, Wilkinson D, Selby A, Rankin D, Webborn N, et al. Effects of hypoxia on muscle protein synthesis and anabolic signaling at rest and in response to acute resistance exercise. Am J Physiol Endocrinol Metab 2011;301: E697–702.
[98] Nedergaard A, Jespersen J, Pingel J, Christensen B, Sroczynski N, Langberg H, et al. Effects of 2 weeks lower limb immobilization and two separate rehabilitation regimens on gastrocnemius muscle protein turnover signaling and normalization genes. BMC Res Notes 2012;5:166.
[99] Reardon KA, Davis J, Kapsa RMI, Choong P, Byrne E. Myostatin, insulin-like growth factor-1, and leukemia inhibitory factor mRNAs are upregulated in chronic human disuse muscle atrophy. Muscle Nerve 2001;24:893–9.
[100] Amirouche A, Durieux A-C, Banzet S, Koulmann N, Bonnefoy R, Mouret C, et al. Down-regulation of Akt/Mammalian target of rapamycin signaling pathway in response to myostatin overexpression in skeletal muscle. Endocrinology 2009;150: 286–94.
[101] Rolland Y, Czerwinski S, van Kan G, Morley J, Cesari M, Onder G, et al. Sarcopenia: its assessment, etiology, pathogenesis, consequences and future perspectives. J Nutr Health Aging 2008;12:433–50.
[102] Pasini E, Le Douairon Lahaye S n, Flati V, Assanelli D, Corsetti G, Speca S, et al. Effects of treadmill exercise and training frequency on anabolic signaling pathways in the skeletal muscle of aged rats. Exp Gerontol 2012; 47:23–8.
[103] Dasarathy S. Consilience in sarcopenia of cirrhosis. J Cachexia, Sarcopenia Muscle 2012;3:225–37.
[104] Volpi E, Kobayashi H, Sheffield-Moore M, Mittendorfer B, Wolfe RR. Essential amino acids are primarily responsible for the amino acid stimulation of muscle protein anabolism in healthy elderly adults. Am J Clin Nutr 2003; 78:250–8.
[105] Teodoro GFR, Vianna D, Torres-Leal FL, Pantaleão LC, Matos-Neto EM, Donato J, et al. Leucine is essential for attenuating fetal growth restriction caused by a protein-restricted diet in rats. J Nutr 2012;142:924–30.
[106] Davis TA, Burrin DG, Fiorotto ML, Nguyen HV. Protein synthesis in skeletal muscle and jejunum is more responsive to feeding in 7- than in 26-day-old pigs. Am J Physiol Endocrinol Metab 1996;270:E802–9.
[107] Suryawan A, Nguyen H, Almonaci R, Davis T. Differential regulation of protein synthesis in skeletal muscle and liver of neonatal pigs by leucine through an mTORC1-dependent pathway. J Anim Sci Biotechnol 2012;3:3.
[108] Young VR, Yu YM, Fukagawa NK. Protein and energy interactions throughout life. Metabolic basis and nutritional implications. Acta Paediatr Scand Suppl 1991;373:5–24.
[109] Dickinson JM, Fry CS, Drummond MJ, Gundermann DM, Walker DK, Glynn EL, et al. Mammalian target of rapamycin complex 1 activation is required for the stimulation of human skeletal muscle protein synthesis by essential amino acids. J Nutr 2011;141:856–62.
[110] Findlay GM, Yan L, Procter J, Mieulet V, Lamb RF. A MAP4 kinase related to Ste20 is a nutrient-sensitive regulator of mTOR signalling. Biochem J 2007;403:13–20.

[111] Sancak Y, Peterson TR, Shaul YD, Lindquist RA, Thoreen CC, Bar-Peled L, et al. The rag GTPases bind raptor and mediate amino acid signaling to mTORC1. Science 2008;320:1496–501.
[112] Drummond MJ, Glynn EL, Fry CS, Timmerman KL, Volpi E, Rasmussen BB. An increase in essential amino acid availability upregulates amino acid transporter expression in human skeletal muscle. Am J Physiol Endocrinol Metab 2010;298:E1011–8.
[113] Symons TB, Sheffield-Moore M, Wolfe RR, Paddon-Jones D. A moderate serving of high-quality protein maximally stimulates skeletal muscle protein synthesis in young and elderly subjects. J Am Diet Assoc 2009;109:1582–6.
[114] Katsanos CS, Kobayashi H, Sheffield-Moore M, Aarsland A, Wolfe RR. Aging is associated with diminished accretion of muscle proteins after the ingestion of a small bolus of essential amino acids. Am J Clin Nutr 2005;82:1065–73.
[115] Pennings B, Koopman R, Beelen M, Senden JMG, Saris WHM, van Loon LJC. Exercising before protein intake allows for greater use of dietary protein-derived amino acids for de novo muscle protein synthesis in both young and elderly men. Am J Clin Nutr 2011;93:322–31.
[116] Morrison PJ, Hara D, Ding Z, Ivy JL. Adding protein to a carbohydrate supplement provided after endurance exercise enhances 4E-BP1 and RPS6 signaling in skeletal muscle. J Appl Physiol 2008;104:1029–36.
[117] Verdijk LB, Jonkers RAM, Gleeson BG, Beelen M, Meijer K, Savelberg HHCM, et al. Protein supplementation before and after exercise does not further augment skeletal muscle hypertrophy after resistance training in elderly men. Am J Clin Nutr 2009;89:608–16.
[118] Camera DM, West DWD, Burd NA, Phillips SM, Garnham AP, Hawley JA, et al. Low muscle glycogen concentration does not suppress the anabolic response to resistance exercise. J Appl Physiol 2012;113:206–14.

CHAPTER

23

Stress Proteins and Heat Shock Proteins

Role in Muscle Building and Sports Nutrition

Mika Venojärvi[1], *Niku Oksala*[2], *Susanna Kinnunen*[1] *and Mustafa Atalay*[1]

[1]Institute of Biomedicine, Physiology, University of Eastern Finland, Kuopio, Finland [2]Department of Vascular Surgery, Tampere University Hospital, Tampere, Finland

INTRODUCTION

Heat shock proteins (HSPs), also called stress proteins, are a group of highly conserved proteins that are an integral part of the cell's protective and antioxidant systems against stress and cell damage [1,2]. HSPs function in both physiological and stress-induced states. They aid in *de novo* synthesis of proteins by ensuring correct folding of nascent polypeptides and their translocation [3]. In stressed cells, they facilitate repair of denatured proteins and dissociation of initial loose protein aggregates [4]. HSPs also participate in the degradation of stress-damaged proteins in the proteasome system [5–7].

HEAT SHOCK PROTEIN FAMILY

HSPs are classified into families according to their molecular weight and homology and include HSP100, HSP90, HSP70, HSP60, HSP40 and small HSP families. Major isoforms of the HSP70 family includes HSP72, HSPA2, glucose regulated protein-78 (GRP78), HSP70B, HSP73 and GRP75 [8,9]. Isoforms of HSP are located mainly in the cytosol and nucleus, but are also found in lysosomes, the endoplasmic reticulum and the mitochondria [9]. HSP72 is the major inducible HSP found in the nucleus and cytosol [10], requires ATP for its chaperone activity, [3] and minimizes aggregation of newly synthesized proteins. Stress-induced HSP72 protects against aggregation of denatured proteins and inhibits stress-induced apoptosis even after activation of initiator and effector caspases [11,12]. HSP72 stabilizes the lysosomal membrane, inhibiting the release of hydrolases that lead to cell death [13]. Alterations in HSP72 synthesis may result in disturbances in cell proliferation [14]. HSP73 acts as a chaperone in non-stressed cells [15], but also has a role in the formation of clathrin and factors involved in intracellular transportation [16].

The HSP90 family, which includes HSP90α and HSP90β, participate in steroid hormone maturation and signaling [17]. HSP90 catalyzes the interaction with several substrate proteins and co-chaperones [18]. HSP90 is a potent autoantigen and is therefore presumed to have a role in inflammatory diseases and atherosclerosis [19]. This function is interesting, because HSP90 is involved in the activation of endothelial nitric oxide synthase (eNOS) causing an increase in the synthesis of NO [20].

The HSP60 family consists of stress-inducible HSP60 [10]. HSP60 is present in both the mitochondria and cytosol. HSP60 has a role in the innate immune response [21] and in mitochondrial protein synthesis [22,23]. HSP60 either inhibits caspase-3 [24] or facilitates the maturation of pro-caspase-3 to its active form [25]. Therefore, HSP60 has an important role in modulating the balance between cell death and survival [26]. The HSP40 family comprises HSP47, which is involved in pro-collagen synthesis [27]. HSP47 interacts closely with HSP70 [28]. HSP47 has the capability to bind to collagen types I–III [29]. HSP32 (HO-1) does not function as a chaperone, but it catalyzes the conversion of heme to iron, biliverdin and carbon monoxide, thus regulating inflammatory and immune responses and transplant rejection [30,31]. HO-1 induction is a sensitive marker of oxidative stress and also protects against

Nutrition and Enhanced Sports Performance.
DOI: http://dx.doi.org/10.1016/B978-0-12-396454-0.00023-0

oxidative damage [15,32,33]. HSP27 and αB crystalline belong to a family of small heat shock proteins. They stabilize actin microfilaments [34] and inhibit stress-induced apoptosis [35]. HSP27 is subject to modulation by phosphorylation [36]. Over-expression of HSP27 promotes endothelial cell migration [37].

REGULATION OF STRESS PROTEINS IN SKELETAL MUSCLE

The major transcription factor heat shock factor-1 (HSF-1) controls the expression of heat shock genes. It binds to the heat shock elements (HSE) which are present in multiple copies upstream of the heat shock genes. In addition to this classical regulatory mechanism, post-transcriptional mechanisms also modulate HSP synthesis [38]. The post-transcriptional mechanism acts via stabilization of HSP70 mRNA [39]. HSF-1 monomers are co-localized with HSP70 under physiological, non-stressed conditions. When the cells are exposed to stress, HSF-1 is rapidly activated [38]. This results in trimerization of HSF-1 monomers, translocation, hyperphosphorylation and binding of the active transcription factor to the promoter region [40–42]. This process is subject to negative feedback regulation because HSP70 binds to activated HSF-1, resulting in an inhibition of the heat shock response [43]. It has also been hypothesized that the lipid architecture of membranes enables them to act as sensors and modulate activation of HSF-1 [44]. Under certain conditions, HSF-1 inhibits activation of NF-kB, a major transcriptional regulator of proinflammatory genes [45].

Several mechanisms and strategies are suggested to induce an HSP response in skeletal muscle (Figure 23.1). Nevertheless, an increased intracellular pool of aggregating denatured proteins is the major inducer of HSP70 accumulation [46,47]. HSP synthesis is induced by a variety of stressful conditions, such as elevated core temperature, ischemia, increased intracellular calcium, electro-mechanical coupling, stress on intermediate filaments, glycogen and ATP depletion,

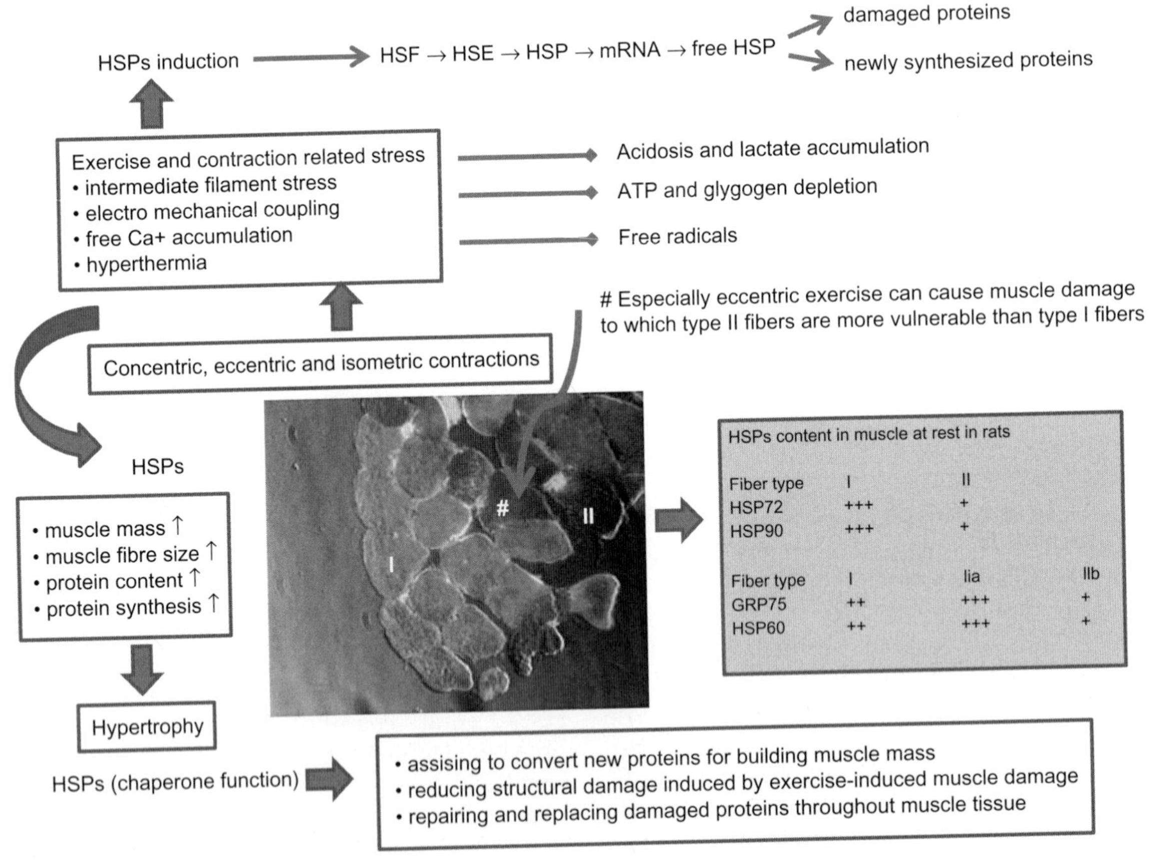

HSPs content in muscle at rest in rats

Fiber type	I	II	
HSP72	+++	+	
HSP90	+++	+	
Fiber type	**I**	**lia**	**IIb**
GRP75	++	+++	+
HSP60	++	+++	+

FIGURE 23.1 Exercise-related factors that can induce HSPs in the skeletal muscle and HSP protein contents at rest. A differential interference contrast microscopy (DIC) image of frozen section from vastus lateralis muscle of human subjects was double-stained with slow myosin heavy chain antibody (type I fiber) with red fluorochrome and anti-nitrotyrosine antibody (green) using Zeiss Axiovert 200 M (20× magnification). Type II (fast myosin heavy chain) fibers are unstained. Ca+, calcium; GRP75, glucose regulated protein; HSE, heat shock elements; HSF, heat shock factor; HSP, heat shock protein. (This figure is reproduced in color plate section.)

acidosis and oxidative stress, which all exist during exhaustive and prolonged physical exercise [2,48–50]. Therefore it is not surprising that it was shown by the late 1990s that exercise up-regulates the expression of HSPs [49,51,52]. It has been well documented that both acute and chronic exercise induce HSP expression [1,32,53]. Therefore, physical exercise can be considered a safe, physiological strategy to enhance skeletal muscle HSP levels [54]. The HSP70 family is the most studied HSP family in response to exercise in both humans and in different animal models [55,56]. Most of the human studies related to HSP70 response to exercise are focused on skeletal muscle or on lymphocytes [55,56].

STRESS PROTEINS AND EXERCISE

The continuous interplay between mechanisms that drive protein synthesis and those that boost protein degradation are important regulating factors of skeletal muscle mass [57]. Muscle protein synthesis is the driving force behind adaptive responses to exercise and is dependent upon both workload and intensity of training [58]. Mechanical tension, muscle damage, and metabolic stress are primary factors for initiating the hypertrophic response to resistance exercise. Stress proteins may have many important tasks for muscle building, including assisting the conversion of new proteins for building muscle mass, reducing structural damage consequent to exercise-induced muscle damage, and repairing and replacing damaged proteins throughout muscle tissue [2]. In cell culture and animal models, HSPs have previously been shown to increase muscle protein synthesis and content and increase muscle mass [59–62] in response to heat stress [59,62]. A single bout of heat stress increases muscle mass and protein synthesis [59,60]. In addition, heat stress may aid maintenance of muscle fiber size [63]. These muscle adaptations are mediated in part by HSPs, and therefore facilitate muscle protein synthesis [52,64,65].

HSP induction during physical exercise is suggested to occur in an intensity-and duration-dependent manner [66]. In humans, HSP70 induction after acute exercise may require subjects to exercise at lactate threshold [56]. Vogt et al. [67] showed that a high-intensity training group had significantly higher post-training HSP72 levels than a low-intensity training group. Liu et al.'s study [68] showed that an intensive 4-week rowing training significantly increased skeletal muscle HSP72. Nevertheless, it was difficult to conclude whether the HSP induction is due to metabolic stress or muscle damage. Fewer studies have addressed the HSP response in a sedentary or non-athletic population. In a study by Willoughby et al. [69] a 12-week training program involving seven subjects with complete motor spinal cord injury increased HSP72 in skeletal muscle. Induction of HSP60 expression in skeletal muscle of healthy humans has been reported after a short-term exercise program [70,71].

We previously showed that in streptozotocin-induced diabetic (SID) and non-diabetic control rats, 8 weeks of endurance training on a treadmill increased the levels of HSP72 both in red gastrocnemius (RG) and in superficial white portion of vastus lateralis (VL) muscles in control rats, but to a lesser extent in diabetic rats [1]. In this study, RG muscle was chosen to represent the response of mainly oxidative muscle fibers, and the superficial white portion of VL represented glycolytic muscle fibers. Similarly, in the same study endurance training increased GRP75 expression in rat skeletal muscle of non-diabetic rats, but not in diabetic rats [1]. We also observed that endurance training induced HSF-1 activation in control rats, but not in SID rats. This finding may explain the mechanisms involved in the impaired HSP response in SID. In addition, histological examinations revealed that tissues of non-diabetic animals with higher HSP protection were less vulnerable to inflammation after chronic exercise compared with diabetic animals, which have impaired HSP response to exercise. Furthermore, along with enhanced skeletal muscle HSP response, endurance training remarkably increased citrate synthase activity and triglyceride deposition in skeletal muscle, which represents an increased capacity for mitochondrial oxidation and fat oxidation, which are early adaptations of the skeletal muscle to endurance training [72,73]. Consistent with our results, another study has shown that up-regulation of GRP75 and HSP60 in response to endurance training were associated with increased skeletal muscle citrate synthase activity [74]. Collectively, these studies suggest a cross-talk between enhanced HSP response and increased oxidative capacity of skeletal muscle.

Induction of HSPs following "non-damaging" endurance type activities is thought to be mediated by redox signaling [2]. Non-damaging exercise protocols, where no inflammatory response has been present [75], may provide a better methodological approach to study exercise-induced regulation of HSPs. Damaging exercise models such as resistance training and downhill running have also been used to study the stress response of human muscle [71]. Interpretation of data from damaging exercise is more complicated due to inflammatory response, which may affect the HSP's response [76].

HSP mRNA increase during, immediately after, and several hours after exercise [2], and this response has been suggested to be muscle fiber-specific: HSP70 content was increased by 80% in human m. vastus lateralis, and this response was specific to type I fibers only [77]. In rats, the expression of HSP72 was higher in type I fibers compared with type II fibers at the basal level [78]. Similarly, in healthy and streptozotocin-induced diabetic rats, HSP72 protein levels were highest in type I red gastrocnemius muscle, but induction after 8 weeks of treadmill training was more apparent in type II white vastus lateralis muscle [1]. This pattern has not been observed for other HSPs, including HSP90, mitochondrial GRP75 and oxidative stress responsive HO-1 [1]. Nevertheless, Ornatsky et al. [79] have shown HSP60 content was highly expressed in heart followed by type IIa fibers of m. gastrocnemius, type I fibers of m. soleus and type IIb fibers of m. gastrocnemius. Also GRP75 expression has a similar pattern to that of HSP60 in rat muscles [79]. Therefore, a fiber-specific stress response may be limited to particular HSPs.

In addition to intensity, duration, quality (damaging/non-damaging) of exercise and fiber recruitment, study population should be taken into consideration before time-course of synthesis and degradation and fiber specificity of HSPs can be accurately defined [2]. High individual variations of the stress responses (magnitude and time-course) occur in human skeletal muscles [2]. Training status [80], recent activity levels [81], thermal history [82], energy availability [83], sex [84] and age [75] are possible determinants that affect the baseline levels of HSPs and the extent of HSP response [2].

STRESS PROTEINS AND DIETARY SUPPLEMENTS

In addition to physical exercise, a heat shock response can be induced or modulated by pharmacological agents or dietary supplements. These strategies can be utilized to induce protection against a potentially irreversible injury [15]. Several dietary antioxidant supplementation studies have aimed to protect against exercise-induced oxidative stress and muscle damage [85]. Alpha lipoic acid (α-LA) is a nutritional supplement and naturally occurring thiol group antioxidant precursor [86]. Protein thiolation regulates the function of some proteins and has been reported to be a mechanism that protects against oxidative stress [87]. It has also been suggested that LA might function as an HSP inducer [88]. The role of LA as an enhancer of HSP induction is further supported by previous study results of LA supplementation on HSP60 response in rat heart [86]. Furthermore, there is evidence that LA and its reduced form dihydrolipoic acid (DHLA) may have effects on regulatory proteins and on genes involved in normal growth and metabolism. At high doses, α-LA has been suggested to induce HSF-1 via increasing disulfide formation in certain target proteins [89]. In treadmill-exercised horses, lower-dose LA supplementation (25 mg kg^{-1} d^{-1}) for 5 weeks enhanced skeletal muscle HSP response along with an increased citrate synthase activity, a marker of oxidative metabolism [90]. In the same study, in non-supplemented horses, a negative correlation between the resting levels of muscle HSP60 and HSP25 and lower recovery level of plasma aspartate aminotransferase (ASAT), a lysosomal enzyme used as an indicator of exercise-induced muscle damage, support the cytoprotective role of HSPs in skeletal muscle. This is in line with the negative correlations between the plasma ASAT levels and HSP25 and HSP90 during recovery (24 and 48 h, respectively) [90]. Consistent with these observations, Williams et al. [91] have also demonstrated lower creatine kinase (CK) levels, another lysosomal enzyme marker of tissue damage, with LA-supplementation following endurance exercise.

However, other antioxidant and dietary supplementation strategies have given contradictory results. In physically active healthy non-athlete men, 3 h of knee extensor exercise at 50% of the maximal power output increased skeletal muscle HSP72 mRNA content 2.5-fold, and serum HSP72 protein increased 4-fold [92]. In the other experimental group of the same study, antioxidant co-supplementation with ascorbic acid and α-tocopherol (an isoform of vitamin E) gave a similar pattern of HSP72 mRNA expression and protein content both in skeletal muscle and plasma, although these changes were not statistically significant [92]. Surprisingly, supplementation with ascorbic acid, α- and γ-tocopherol attenuated the exercise-induced increase of HSP72 both in the skeletal muscle and in the circulation, suggesting that γ-tocopherol is a potent inhibitor of exercise-induced increase of HSP72 [92]. Consistent with this study, in an early investigation, exhaustive treadmill running induced leukocyte HSP72 mRNA expression and significantly increased granulocyte HSP72 protein levels only in the placebo group, but not in a supplementation group which received 500 IU daily of α-tocopherol for 8 days [93]. Nevertheless, in the same experimental settings, α-tocopherol did not affect HO-1 levels in lymphocytes and granulocytes [94]. In another antioxidant supplementation study in healthy, untrained males, oral supplementation of vitamin C 0.5 g/day for 8 weeks increased baseline HSP70 content in vastus lateralis muscle [76]. In the same experiment, cycling at 70% of VO_{2peak} for 45 min increased muscle HSP72 levels significantly and tended to increase HSP60 levels in controls groups, whereas no exercise-induced HSP response was seen in the vitamin C-supplemented group [76]. The authors concluded that, although vitamin C-supplementation up-regulated baseline expression of HSPs, it also attenuated exercise-induced HSP and antioxidant defenses. In a more recent study, interaction of antioxidant supplementation with aerobic training on systemic oxidative stress and HSP72 expression was investigated in the elderly at mean age of 74.6 yr. All subjects were supplemented with vitamin C (500 mg/day) and vitamin E (100 mg/day) for 8 weeks and randomly divided either to sedentary or training group [95]. Training consisted of an individualized aerobic training program of a supervised 1 h walking session 3 days per week for 8 weeks. Antioxidant supplementation decreased resting and post-exercise oxidative stress markers similarly in both groups, and lowered oxidative stress was parallel with decreased expression of HSP72 levels in monocytes and granulocytes in both groups [95]. However, 8 weeks of aerobic training had no effect on oxidative stress or leukocyte HSP defense in elderly people who received antioxidant supplement [95]. Because this study did not contain any appropriate control group of un-supplemented subjects for the training groups, we cannot obtain any evidence from it on the impact of antioxidant supplementation on exercise-induced HSP adaptations in the elderly. In addition to antioxidant supplementations, in an early study, estrogen administration to male rats attenuated post-exercise HSP72 over two-fold in the red and white vastus muscles compared with their vehicle-injected controls [96]. In the same study, female rats with no estrogen administration had significantly lower post-exercise HSP72 levels than their male counterparts [96]. These results suggest a role of estrogen for gender-specific HSP response to physical exercise, which may further give clues for a possible connection of HSPs to muscle building and body composition.

CONCLUSIONS

In conclusion, HSPs, essential components of tissue protection and protein homeostasis, are implicated in muscle protein synthesis, repair of damaged proteins, and muscle hypertrophy. Physical exercise is a safe tool to enhance muscle stress protein levels. Other strategies, including antioxidant nutrient supplementation studies, have recently gained interest. In spite of promising effects of LA on HSP response, further studies are required to allow us to conclude whether antioxidant supplementations can boost muscle stress protein responses.

References

[1] Atalay M, Oksala NK, Laaksonen DE, et al. Exercise training modulates heat shock protein response in diabetic rats. J Appl Physiol 2004;97(2):605–11.

[2] Morton JP, Kayani AC, McArdle A, Drust B. The exercise-induced stress response of skeletal muscle, with specific emphasis on humans. Sports Med 2009;39(8):643–62.

[3] McKay DB. Structure and mechanism of 70-kDa heat-shock-related proteins. Adv Protein Chem 1993;44:67–98.

[4] Buchner J. Supervising the fold: functional principles of molecular chaperones. Faseb J 1996;10(1):10–9.

[5] Li GC, Petersen NS, Mitchell HK. Induced thermal tolerance and heat shock protein synthesis in Chinese hamster ovary cells. Br J Cancer Suppl 1982;5:132–6.

[6] Morimoto RI. Regulation of the heat shock transcriptional response: cross talk between a family of heat shock factors, molecular chaperones, and negative regulators. Genes Dev 1998;12(24):3788–96.

[7] Riabowol KT, Mizzen LA, Welch WJ. Heat shock is lethal to fibroblasts microinjected with antibodies against hsp70. Science 1988;242(4877):433–6.

[8] Daugaard M, Rohde M, Jaattela M. The heat shock protein 70 family: highly homologous proteins with overlapping and distinct functions. FEBS Lett 2007;581(19):3702–10.

[9] Tavaria M, Gabriele T, Kola I, Anderson RL. A hitchhiker's guide to the human Hsp70 family. Cell Stress Chaperones 1996;1(1):23–8.

[10] Fink AL. Chaperone-mediated protein folding. Physiol Rev 1999;79(2):425–49.

[11] Jäättelä M, Wissing D, Kokholm K, Kallunki T, Egeblad M. Hsp70 exerts its anti-apoptotic function downstream of caspase-3-like proteases. Embo J 1998;17(21):6124–34.

[12] Mosser DD, Caron AW, Bourget L, et al. The chaperone function of hsp70 is required for protection against stress-induced apoptosis. Mol Cell Biol 2000;20(19):7146–59.

[13] Nylandsted J, Gyrd-Hansen M, Danielewicz A, et al. Heat shock protein 70 promotes cell survival by inhibiting lysosomal membrane permeabilization. J Exp Med 2004;200(4):425–35.

[14] Nollen EA, Morimoto RI. Chaperoning signaling pathways: molecular chaperones as stress-sensing 'heat shock' proteins. J Cell Sci 2002;115(Pt 14):2809–16.

[15] Benjamin IJ, McMillan DR. Stress (heat shock) proteins: molecular chaperones in cardiovascular biology and disease. Circ Res 1998;83(2):117–32.

[16] Rapoport I, Boll W, Yu A, Bocking T, Kirchhausen T. A motif in the clathrin heavy chain required for the Hsc70/auxilin uncoating reaction. Mol Biol Cell 2008;19(1):405–13.

[17] Srinivasan G, Post JF, Thompson EB. Optimal ligand binding by the recombinant human glucocorticoid receptor and assembly of the receptor complex with heat shock protein 90 correlate with high intracellular ATP levels in Spodoptera frugiperda cells. J Steroid Biochem Mol Biol 1997;60(1–2):1–9.

[18] Whitesell L, Lindquist SL. HSP90 and the chaperoning of cancer. Nat Rev Cancer 2005;5(10):761–72.

[19] Rigano R, Profumo E, Buttari B, et al. Heat shock proteins and autoimmunity in patients with carotid atherosclerosis. Ann N Y Acad Sci 2007;1107:1–10.

[20] Pritchard Jr KA, Ackerman AW, Gross ER, et al. Heat shock protein 90 mediates the balance of nitric oxide and superoxide anion from endothelial nitric-oxide synthase. J Biol Chem 2001;276(21):17621–4.

[21] Vabulas RM, Wagner H, Schild H. Heat shock proteins as ligands of toll-like receptors. Curr Top Microbiol Immunol 2002;270:169–84.

[22] Moseley P. Stress proteins and the immune response. Immunopharmacology 2000;48(3):299–302.

[23] Voos W, Rottgers K. Molecular chaperones as essential mediators of mitochondrial biogenesis. Biochim Biophys Acta 2002; 1592(1):51–62.

[24] Gupta S, Knowlton AA. Cytosolic heat shock protein 60, hypoxia, and apoptosis. Circulation 2002;106(21):2727–33.

[25] Xanthoudakis S, Roy S, Rasper D, et al. Hsp60 accelerates the maturation of pro-caspase-3 by upstream activator proteases during apoptosis. Embo J 1999;18(8):2049–56.

[26] Gupta S, Knowlton AA. HSP60, Bax, apoptosis and the heart. J Cell Mol Med 2005;9(1):51–8.

[27] Ohba S, Wang ZL, Baba TT, Nemoto TK, Inokuchi T. Antisense oligonucleotide against 47-kDa heat shock protein (Hsp47) inhibits wound-induced enhancement of collagen production. Arch Oral Biol 2003;48(9):627–33.

[28] Fan CY, Lee S, Cyr DM. Mechanisms for regulation of Hsp70 function by Hsp40. Cell Stress Chaperones 2003;8(4):309–16.

[29] Macdonald JR, Bachinger HP. HSP47 binds cooperatively to triple helical type I collagen but has little effect on the thermal stability or rate of refolding. J Biol Chem 2001;276(27):25399–403.

[30] Camara NO, Soares MP. Heme oxygenase-1 (HO-1), a protective gene that prevents chronic graft dysfunction. Free Radic Biol Med 2005;38(4):426–35.

[31] Srisook K, Kim C, Cha YN. Molecular mechanisms involved in enhancing HO-1 expression: de-repression by heme and activation by Nrf2, the "one-two" punch. Antioxid Redox Signal 2005;7(11–12):1674–87.

[32] Moran M, Delgado J, Gonzalez B, Manso R, Megias A. Responses of rat myocardial antioxidant defences and heat shock protein HSP72 induced by 12 and 24-week treadmill training. Acta Physiol Scand 2004;180(2):157–66.

[33] Motterlini R, Foresti R, Intaglietta M, Winslow RM. NO-mediated activation of heme oxygenase: endogenous cytoprotection against oxidative stress to endothelium. Am J Physiol 1996;270(1 Pt 2):H107–14.

[34] Lavoie JN, Gingras-Breton G, Tanguay RM, Landry J. Induction of Chinese hamster HSP27 gene expression in mouse cells confers resistance to heat shock. HSP27 stabilization of the microfilament organization. J Biol Chem 1993;268(5):3420–9.

[35] Welsh N, Margulis B, Borg LA, et al. Differences in the expression of heat-shock proteins and antioxidant enzymes between human and rodent pancreatic islets: implications for the pathogenesis of insulin-dependent diabetes mellitus. Mol Med 1995;1(7):806–20.

[36] Hirano S, Rees RS, Gilmont RR. MAP kinase pathways involving hsp27 regulate fibroblast-mediated wound contraction. J Surg Res 2002;102(2):77–84.

[37] Rousseau S, Houle F, Landry J, Huot J. p38 MAP kinase activation by vascular endothelial growth factor mediates actin reorganization and cell migration in human endothelial cells. Oncogene 1997;15(18):2169–77.

[38] Anckar J, Sistonen L. Heat shock factor 1 as a coordinator of stress and developmental pathways. Adv Exp Med Biol 2007;594:78–88.

[39] Kaarniranta K, Elo M, Sironen R, et al. Hsp70 accumulation in chondrocytic cells exposed to high continuous hydrostatic pressure coincides with mRNA stabilization rather than transcriptional activation. Proc Natl Acad Sci U S A 1998;95(5):2319–24.

[40] Baler R, Dahl G, Voellmy R. Activation of human heat shock genes is accompanied by oligomerization, modification, and rapid translocation of heat shock transcription factor HSF1. Mol Cell Biol 1993;13(4):2486–96.

[41] Sarge KD. Regulation of HSF1 activation and Hsp expression in mouse tissues under physiological stress conditions. Ann N Y Acad Sci 1998;851:112–6.

[42] Sarge KD, Murphy SP, Morimoto RI. Activation of heat shock gene transcription by heat shock factor 1 involves oligomerization, acquisition of DNA-binding activity, and nuclear localization and can occur in the absence of stress. Mol Cell Biol 1993;13(3):1392–407.

[43] Abravaya K, Myers MP, Murphy SP, Morimoto RI. The human heat shock protein hsp70 interacts with HSF, the transcription factor that regulates heat shock gene expression. Genes Dev 1992;6(7):1153–64.

[44] Vigh L, Horvath I, Maresca B, Harwood JL. Can the stress protein response be controlled by 'membrane-lipid therapy'? Trends Biochem Sci 2007;32(8):357–63.

[45] Wirth D, Bureau F, Melotte D, Christians E, Gustin P. Evidence for a role of heat shock factor 1 in inhibition of NF-kappaB pathway during heat shock response-mediated lung protection. Am J Physiol Lung Cell Mol Physiol 2004;287(5):L953–61.

[46] Dobson CM. Protein folding and misfolding. Nature 2003;426 (6968):884–90.

[47] Hartl FU, Hayer-Hartl M. Molecular chaperones in the cytosol: from nascent chain to folded protein. Science 2002;295 (5561):1852–8.

[48] Anckar J, Sistonen L. Regulation of HSF1 function in the heat stress response: implications in aging and disease. Annu Rev Biochem 2010.

[49] Essig DA, Nosek TM. Muscle fatigue and induction of stress protein genes: a dual function of reactive oxygen species? Can J Appl Physiol 1997;22(5):409–28.

[50] Liu Y, Steinacker JM. Changes in skeletal muscle heat shock proteins: pathological significance. Front Biosci 2001;6: D12–25.

[51] Kelly DA, Tiidus PM, Houston ME, Noble EG. Effect of vitamin E deprivation and exercise training on induction of HSP70. J Appl Physiol 1996;81(6):2379–85.

[52] Locke M. The cellular response to exercise: role of stress proteins. Exerc Sport Sci Rev 1997;25:105–36.

[53] Thompson HS, Clarkson PM, Scordilis SP. The repeated bout effect and heat shock proteins: intramuscular HSP27 and HSP70 expression following two bouts of eccentric exercise in humans. Acta Physiol Scand 2002;174(1):47–56.

[54] Atalay M, Oksala N, Lappalainen J, Laaksonen DE, Sen CK, Roy S. Heat shock proteins in diabetes and wound healing. Curr Protein Pept Sci 2009;10(1):85–95.

[55] Liu Y, Gampert L, Nething K, Steinacker JM. Response and function of skeletal muscle heat shock protein 70. Front Biosci 2006;11:2802–27.

[56] Yamada P, Amorim F, Moseley P, Schneider S. Heat shock protein 72 response to exercise in humans. Sports Med 2008;38 (9):715–33.

[57] Judge AR, Powers SK, Ferreira LF, Bamman MM. Meeting synopsis: advances in skeletal muscle biology in health and disease (Gainesville, Florida, February 22nd to 24th 2012)-Day 1: cell signaling mechanisms mediating muscle atrophy and hypertrophy and muscle force, calcium handling, and stress response. Front Physiol 2012;3:200.

[58] Atherton PJ, Smith K. Muscle protein synthesis in response to nutrition and exercise. J Physiol 2012;590(Pt 5):1049–57.

[59] Goto K, Okuyama R, Sugiyama H, et al. Effects of heat stress and mechanical stretch on protein expression in cultured skeletal muscle cells. Pflugers Arch 2003;447(2):247–53.

[60] Kobayashi T, Goto K, Kojima A, et al. Possible role of calcineurin in heating-related increase of rat muscle mass. Biochem Biophys Res Commun 2005;331(4):1301–9.

[61] Kojima A, Goto K, Morioka S, et al. Heat stress facilitates the regeneration of injured skeletal muscle in rats. J Orthop Sci 2007;12(1):74–82.

[62] Ohno Y, Yamada S, Sugiura T, Ohira Y, Yoshioka T, Goto K. A possible role of NF-kappaB and HSP72 in skeletal muscle hypertrophy induced by heat stress in rats. Gen Physiol Biophys 2010;29(3):234–42.

[63] Oishi Y, Hayashida M, Tsukiashi S, et al. Heat stress increases myonuclear number and fiber size via satellite cell activation in rat regenerating soleus fibers. J Appl Physiol 2009;107 (5):1612–21.

[64] Beck SC, De Maio A. Stabilization of protein synthesis in thermotolerant cells during heat shock. Association of heat shock protein-72 with ribosomal subunits of polysomes. J Biol Chem 1994;269(34):21803–11.

[65] Beckmann RP, Mizzen LE, Welch WJ. Interaction of Hsp 70 with newly synthesized proteins: implications for protein folding and assembly. Science 1990;248(4957):850–4.

[66] Ogawa K, Seta R, Shimizu T, et al. Plasma adenosine triphosphate and heat shock protein 72 concentrations after aerobic and eccentric exercise. Exerc Immunol Rev 2011;17:136–49.

[67] Vogt M, Puntschart A, Geiser J, Zuleger C, Billeter R, Hoppeler H. Molecular adaptations in human skeletal muscle to endurance training under simulated hypoxic conditions. J Appl Physiol 2001;91(1):173–82.

[68] Liu Y, Mayr S, Opitz-Gress A, et al. Human skeletal muscle HSP70 response to training in highly trained rowers. J Appl Physiol 1999;86(1):101–4.

[69] Willoughby DS, Priest JW, Nelson M. Expression of the stress proteins, ubiquitin, heat shock protein 72, and myofibrillar protein content after 12 weeks of leg cycling in persons with spinal cord injury. Arch Phys Med Rehabil 2002;83(5):649–54.

[70] Khassaf M, Child RB, McArdle A, Brodie DA, Esanu C, Jackson MJ. Time course of responses of human skeletal muscle to oxidative stress induced by nondamaging exercise. J Appl Physiol 2001;90(3):1031–5.

[71] Morton JP, MacLaren DP, Cable NT, et al. Time course and differential responses of the major heat shock protein families in human skeletal muscle following acute nondamaging treadmill exercise. J Appl Physiol 2006;101(1):176–82.

[72] Hoppeler H, Luthi P, Claassen H, Weibel ER, Howald H. The ultrastructure of the normal human skeletal muscle. A morphometric analysis on untrained men, women and well-trained orienteers. Pflugers Arch 1973;344(3):217–32.

[73] Schrauwen-Hinderling VB, Schrauwen P, Hesselink MK, et al. The increase in intramyocellular lipid content is a very early response to training. J Clin Endocrinol Metab 2003;88(4): 1610–6.

[74] Mattson JP, Ross CR, Kilgore JL, Musch TI. Induction of mitochondrial stress proteins following treadmill running. Med Sci Sports Exerc 2000;32(2):365–9.

[75] Vasilaki A, Jackson MJ, McArdle A. Attenuated HSP70 response in skeletal muscle of aged rats following contractile activity. Muscle Nerve 2002;25(6):902–5.

[76] Khassaf M, McArdle A, Esanu C, et al. Effect of vitamin C supplements on antioxidant defence and stress proteins in human lymphocytes and skeletal muscle. J Physiol 2003;549 (Pt 2):645–52.

[77] Tupling AR, Bombardier E, Stewart RD, Vigna C, Aqui AE. Muscle fiber type-specific response of Hsp70 expression in human quadriceps following acute isometric exercise. J Appl Physiol 2007;103(6):2105–11.

[78] Locke M, Noble EG, Atkinson BG. Inducible isoform of HSP70 is constitutively expressed in a muscle fiber type specific pattern. Am J Physiol 1991;261(5 Pt 1):C774–9.

[79] Ornatsky OI, Connor MK, Hood DA. Expression of stress proteins and mitochondrial chaperonins in chronically stimulated skeletal muscle. Biochem J 1995;311(Pt 1):119–23.

[80] Gonzalez B, Hernando R, Manso R. Stress proteins of 70 kDa in chronically exercised skeletal muscle. Pflugers Arch 2000;440 (1):42–9.

[81] Campisi J, Leem TH, Greenwood BN, et al. Habitual physical activity facilitates stress-induced HSP72 induction in brain, peripheral, and immune tissues. Am J Physiol Regul Integr Comp Physiol 2003;284(2):R520–30.

[82] Kregel KC. Heat shock proteins: modifying factors in physiological stress responses and acquired thermotolerance. J Appl Physiol 2002;92(5):2177–86.

[83] Febbraio MA, Steensberg A, Walsh R, et al. Reduced glycogen availability is associated with an elevation in HSP72 in contracting human skeletal muscle. J Physiol 2002;538(Pt 3):911–7.

[84] Paroo Z, Haist JV, Karmazyn M, Noble EG. Exercise improves postischemic cardiac function in males but not females: consequences of a novel sex-specific heat shock protein 70 response. Circ Res 2002;90(8):911–7.

[85] Atalay M, Lappalainen J, Sen CK. Dietary antioxidants for the athlete. Curr Sports Med Rep 2006;5:182–6.

[86] Oksala N, Atalay M, Laaksonen DE, et al. Heat shock protein 60 response to exercise in diabetes: effects ofalpha-lipoic acid supplementation. J Diabetes Complications 2006;20(4): 257–61.

[87] Packer L, Witt EH, Tritschler HJ. alpha-Lipoic acid as a biological antioxidant. Free Radic Biol Med 1995;19(2):227–50.

[88] Gupte AA, Bomhoff GL, Morris JK, Gorres BK, Geiger PC. Lipoic acid increases heat shock protein expression and inhibits stress kinase activation to improve insulin signaling in skeletal muscle from high-fat-fed rats. J Appl Physiol 2009;106 (4):1425–34.

[89] McCarty MF. Versatile cytoprotective activity of lipoic acid may reflect its ability to activate signalling intermediates that trigger the heat-shock and phase II reactions. Med Hypotheses 2001;57(3):313–7.

[90] Kinnunen S, Hyyppä S, Oksala N, et al. alpha-Lipoic acid supplementation enhances heat shock protein production and decreases post-exercise lactic acid concentrations in exercised standardbred trotters. Res Vet Sci 2009;87(3):462–7.

[91] Williams CA, Kronfeld, DS, Hess TM, Saker KE, Harris PA. Lipoic acid and vitamin E supplementation to horses diminishes endurance exercise induced oxidative stress, muscle enzyme leakage, and apoptosis. The Elite Race and Endurance Horse. CESMAS 2004, Oslo, Norway; p. 105–19.

[92] Fischer CP, Hiscock NJ, Basu S, et al. Vitamin E isoform-specific inhibition of the exercise-induced heat shock protein 72 expression in humans. J Appl Physiol 2006;100(5): 1679–87.

[93] Niess AM, Fehrenbach E, Schlotz E, et al. Effects of RRR-alpha-tocopherol on leukocyte expression of HSP72 in response to exhaustive treadmill exercise. Int J Sports Med 2002;23 (6):445–52.

[94] Niess AM, Sommer M, Schneider M, et al. Physical exercise-induced expression of inducible nitric oxide synthase and heme oxygenase-1 in human leukocytes: effects of RRR-alpha-tocopherol supplementation. Antioxid Redox Signal 2000;2 (1):113–26.

[95] Simar D, Malatesta D, Mas E, Delage M, Caillaud C. Effect of an 8-weeks aerobic training program in elderly on oxidative stress and HSP72 expression in leukocytes during antioxidant supplementation. J Nutr Health Aging 2012;16(2): 155–61.

[96] Paroo Z, Tiidus PM, Noble EG. Estrogen attenuates HSP 72 expression in acutely exercised male rodents. Eur J Appl Physiol Occup Physiol 1999;80(3):180–4.

CHAPTER

24

Anabolic and Catabolic Signaling Pathways that Regulate Skeletal Muscle Mass

John J. McCarthy

Center for Muscle Biology, Department of Physiology, College of Medicine, University of Kentucky, Lexington, KY, USA

INTRODUCTION

Skeletal muscle is one of the abundant tissues of the human body, accounting for roughly 40% of the body mass in men [1]. In addition to its primary function in locomotion, there is a growing recognition that skeletal muscle has an important role in whole-body metabolism and protein homeostasis during aging [2,3]. The clinical importance of skeletal muscle health is highlighted by the diverse patient population reported to show significant losses in skeletal muscle mass, including those suffering from systemic diseases (cancer, sepsis or HIV-AIDS), organ failure (cirrhosis, chronic kidney disease and chronic heart failure) or inactivity (as a result of obesity, rheumatoid arthritis or prolonged bed rest) as well as aging (sarcopenia). Currently, the most effective therapy for maintaining or restoring muscle mass is resistance exercise, which is often not a realistic option for the aforementioned patient populations. Given this state of affairs, there is great interest in defining the anabolic and catabolic signaling pathways that regulate skeletal muscle mass, as the basis for developing more effective therapies to prevent or restore the loss of muscle mass associated with systemic disease, bed rest and aging. Moreover, a better understanding of the signaling pathways that regulate skeletal muscle mass will allow for the design of more effective training and nutritional programs aimed at increasing muscle mass in athletes, with the goal of improving performance.

HISTORY

One of the most remarkable qualities of skeletal muscle, that is often taken for granted, is its ability to specifically alter its physical characteristics (phenotype) in response to a particular type of contractile activity. The most obvious example of this phenotypic plasticity is the significant increase in skeletal muscle mass following a progressive, high-resistance exercise training program. The observation that resistance exercise can increase muscle mass dates back to the to the ancient Greeks, when it was reputed that Milo of Crotona achieved his great strength by carrying a calf on his back every day until it was a bull. Current resistance exercise training programs have replaced the bull with barbells and dumbbells, prescribing three sets of 8–12 repetitions for each exercise, performed on alternating days, three times a week [4]. The dramatic increase in muscle size following resistance exercise is primarily the result of an increase in muscle fiber size (hypertrophy) with the contribution from an increase in muscle fiber number (hyperplasia) minor, if at all [5–9].

Skeletal muscle mass is primarily dictated by the balance between the rates of protein synthesis and degradation, with a net increase in the rate of protein synthesis leading to muscle hypertrophy [10]. Accordingly, resistance exercise has been shown in both humans and rodents to cause a net increase in the rate of protein synthesis [10–16]. While early animal

Nutrition and Enhanced Sports Performance.
DOI: http://dx.doi.org/10.1016/B978-0-12-396454-0.00024-2

studies reported changes in mitogen-activated protein kinase (MAPK) signaling in response to high-force contractions, Baar and Esser [17] provided the first mechanistic data linking high-force contractions to muscle hypertrophy via the prolonged activation of mTOR (mechanistic or mammalian target of rapamycin) signaling, the cell's master regulator of protein synthesis [17–20]. Bodine and colleagues [21] followed up with the seminal study demonstrating mTOR function was absolutely necessary for skeletal muscle hypertrophy [21]. The importance of mTOR activation was more recently confirmed in humans by Mayhew and co-workers [22], showing that changes in translational signaling following a bout of unaccustomed resistance exercise was predictive of the hypertrophic response after 16 weeks of resistance exercise training [22]. While mTOR signaling is considered to be the primary pathway regulating skeletal muscle mass, other signaling pathways have been shown to be capable of regulating skeletal muscle hypertrophy. The focus of this chapter will be to briefly discuss the role of mTOR signaling in muscle hypertrophy (see Chapter 22 by Panzhinskiy et al., for an in-depth review of mTOR signaling) followed by a review of other anabolic signaling pathways involved in the regulation of skeletal muscle mass. The chapter will conclude with a review of catabolic signaling pathways that promote muscle atrophy by inhibiting protein synthesis and/or increasing the rate of protein degradation.

ANABOLIC SIGNALING

mTOR Signaling

mTOR (mammalian target of rapamycin) is a member of the phosphoinositide 3-kinase (PI3K)-related kinase family that has been shown to form two separate multi-protein complexes designated mTOR complex 1 (TORC1) and 2 (mTORC2) [23]. mTORC1 is a master regulator of protein synthesis by integrating a number of upstream signals including growth factors (insulin, IGF-1), nutrients (amino acids) and mechanical strain, whereas TORC2 is primarily involved in the regulation of cytoskeleton dynamics and cell proliferation and survival [23,24].

Although it had been known for some time that insulin-like growth factor-1 is capable of inducing muscle fiber hypertrophy, the underlying mechanism remained unknown until Rommel and colleagues [25] provided evidence showing that IGF-1 activated TORC1 via the PI3K/AKT pathway [25,26]. Activation of the PI3K/AKT pathway by IGF-1 increases protein synthesis through AKT-mediated phosphorylation of tuberous sclerosis protein 2 (TSC2) protein, the primary inhibitor of TORC1 activity [27–30]. The phosphorylation of TSC2 by AKT inhibits its activity, resulting in the accumulation of the active form of Rheb (Ras homolog enriched in brain), a potent activator of TORC1 [31,32]. Once activated by Rheb, TORC1 stimulates protein synthesis by phosphorylating ribosomal S6 kinase ($p70^{S6k1}$) and eukaryotic initiation factor 4E binding protein (4E-BP1), two factors that enhance the initiation of mRNA translation. In addition to these factors, TORC1 activation has been shown to increase the expression of the ε-subunit of eIF2B holoenzyme thereby promoting the recruitment of the initiator methionine tRNA to the start codon and increasing protein synthesis [33]. The expression of eIF2Bε was shown to increase following resistance exercise training, with its abundance appearing to be an important determinant of the hypertrophic response [22,34].

In a search for proteins that were targeted for degradation by the muscle-specific ubiquitin ligase Fbxo32 (commonly known as MAFbx or atrogin), Lagirand-Cantaloube and co-workers [35] identified the eukaryotic initiation factor 3, subunit 5 (eIF3f) [35]. As the largest of the initiation factors, evidence indicates that the eIF3 complex is involved in nearly all aspects of translation initiation; most notable for the current discussion is the finding that eIF3f can serve as a docking site for TORC1 and $p70^{S6k1}$[36–38]. Upon activation, TORC1 is recruited to the eIF3f-$p70^{S6K1}$ complex, leading to the phosphorylation and subsequent release of active $p70^{S6K1}$ which is now capable of increasing protein synthesis by phosphorylating ribosomal protein S6 (rpS6) and eIF4B [37,38]. In a series of studies, it was demonstrated that the *in vivo* and *in vitro* over-expression of eIF3f induced myotube and muscle fiber hypertrophy, respectively, by stimulating protein synthesis through the phosphorylation of $p70^{S6K1}$, rpS6 and 4E-BP1 [35,39]. It will be important to determine if the regulation of muscle hypertrophy by eIF3f is conserved in humans and what role it has in the hypertrophic response following resistance exercise.

Wnt/β-Catenin Signaling

In their original description of IGF-1-mediated muscle hypertrophy, Rommel and colleagues [25] also identified glycogen synthase kinase 3β (GSK3β) as a downstream target of PI3K/AKT signaling [25]. In contrast to TORC1, phosphorylation by AKT leads to GSK3β inactivation which prevents further inhibition, via phosphorylation, of the translation initiation factor eIF2Bε [25]. The inhibition of GSK3β activity by IGF-1 or lithium (a known inhibitor of GSK3β) treatment, or over-expression of a dominant-negative form of

GSK3β, all resulted in a significant myotube hypertrophy, thus confirming the importance of GSK3β inhibition by IGF-1/PI3K/AKT signaling in the regulation of muscle mass [25,40].

In addition to eIF2Bε, the inactivation of GSK3β by IGF-1 via AKT phosphorylation also results in the stabilization of β-catenin protein [41]. β-catenin is known to exist in two separate intracellular pools: an actin cytoskeleton pool involved in the formation of adherens junctions and a cytoplasmic pool that is the downstream mediator of the Wnt signaling pathway involved in the regulation of gene expression [42]. Under resting conditions, cytoplasmic β-catenin is phosphorylated by GSK3β, targeting it for degradation by the proteosome [43]. Upon GSK3β inhibition, via phosphorylation of Ser9 by AKT or sequestration by Dishelved following Wnt activation, β-catenin accumulates in the cytoplasm and then ultimately translocates to the nucleus where it regulates the expression of target genes such as c-Myc and cyclin D [44,45]. These studies show that β-catenin levels can be regulated by a Wnt-dependent and -independent mechanism.

Following 7 days of mechanical overload induced by synergist ablation, nuclear β-catenin levels increased by over four-fold in the mouse plantaris muscle through a Wnt-independent mechanism [46]. In a follow-up study, Armstrong and colleagues [47] reported that the skeletal muscle-specific inactivation of β-catenin completely prevented muscle fiber hypertrophy, demonstrating the necessity of β-catenin in the regulation of muscle hypertrophy [47]. Associated with the increase in nuclear β-catenin levels during hypertrophy were increased myonuclear c-Myc expression (an established target gene of β-catenin) and a three-fold increase in total RNA [46]. The increase in the total RNA pool is indicative of ribosome biogenesis given that ~85% of total RNA consists of ribosomal RNA [48]. A greater ribosomal content of the cell results in an increase in protein synthesis as a consequence of greater translational capacity. An increase in the translational capacity of the cell represents an additional mechanism for increasing the rate of protein synthesis in response to a hypertrophic stimulus. While the necessity of enhanced translational efficiency through TORC1 activation is well-established, the importance of increased translational capacity to skeletal muscle hypertrophy will require further investigation [49]. Interestingly, in a review of the literature, Hannan et al. [50] concluded that increased translational capacity was required for cardiac hypertrophy and that increased translational efficiency alone was not sufficient to promote muscle hypertrophy [50].

Although the exact mechanism involved in the regulation of ribosome biogenesis during muscle hypertrophy is currently unknown, evidence from a number of different studies supports a model in which increased β-catenin expression up-regulates c-Myc expression, which subsequently drives transcription of ribosomal DNA through the regulation of RNA polymerase I and components of the preinitiation complex [46,51,52]. Future studies will need to determine the contribution of ribosome biogenesis, and the resulting greater translational capacity, to the increase in protein synthesis underlying muscle growth and if such a mechanism is operative in humans in response to resistance exercise training.

A non-canonical Wnt signaling pathway has been described that is capable of inducing muscle hypertrophy independent of GSK3β or β-catenin [53]. Experiments in myotubes revealed that Wnt7a binding to the frizzled homolog 7 (Fzd7) receptor activated AKT via a PI3K mechanism that did not involve IGF-1 signaling [53]. As expected, AKT activation by Wnt7a resulted in an increase in TORC1 activity, as assessed by rpS6 phosphorylation, and an increase in myotube diameter [53]. Consistent with these findings, over-expression of Wnt7a in skeletal muscle fibers increased cross-sectional area by 40–55%; however, whether or not this hypertrophy was caused by TORC1 activation and/or the increase in satellite cells associated with Wnt7a over-expression, remains to determined [53,54].

β-Adrenergic Receptor Signaling

The β-adrenergic receptors (β-AR) are members of the guanine nucleotide-binding G-protein-coupled receptor family, with the β_2 subtype being the most abundant in skeletal muscle [55–57]. In skeletal muscle, signaling through the β_2-AR occurs primarily via coupling with the G protein $G\alpha_s$, though there are reports describing signaling events involving $G\alpha_i$ and, more recently, $G\beta\gamma$ [58,59]. β_2-AR coupling to $G\alpha_s$ activates adenylyl cyclase production of cyclic-AMP (cAMP) which in turn activates downstream signaling through protein kinase A (PKA) [60]. Alternatively, β_2-AR signaling involving the $G\beta\gamma$ dimer initiates signaling through a PKA-independent pathway involving PI3K activation of AKT [61–63].

Historically, β-adrenergic receptor agonists (β-agonists) have been used to treat patients suffering from bronchial ailments such as asthma; however, Emery and co-workers [64] made the fortuitous discovery that administration of the β-agonist clenbuterol caused skeletal muscle hypertrophy in rats and was associated with a 34% increase in muscle protein synthesis [64]. The ability of β-agonists to increase skeletal muscle mass has since been confirmed by numerous studies and shown to be effective in humans as well (reviewed in Lynch GS, 2008) [60]. The increase in muscle mass following β-agonist administration is the result of hypertrophy (an increase in cell size) and not hyperplasia (an increase in cell number) or satellite cell activity [65–67].

The mechanism through which β-agonists exert their anabolic effect has been attributed to enhanced protein synthesis, though studies have reported a decrease in protein degradation as a contributing factor [64,68–73]. The increase in skeletal muscle mass following 14 days of clenbuterol treatment was completely blocked by co-administration of rapamycin, a potent inhibitor of TORC1 signaling, indicating that the anabolic action of β-agonists is the result of an increase in protein synthesis via TORC1 activation [74]. In support of this mechanism, the same authors reported that clenbuterol activated TORC1 signaling, as evidenced by increased phosphorylation of AKT, $p70^{S6K1}$ and 4E-BP1 [74]. Moreover, these findings suggest that, in skeletal muscle, β-agonists function through a β_2-AR/Gβγ complex that activates PI3K/AKT signaling. Interestingly, rapamycin or triciribine (an AKT inhibitor) was unable to block the muscle-sparing effects of clenbuterol following muscle denervation, suggesting a different mode of action, possibly involving the cAMP/PKA pathway downstream of β_2-AR/ $G\alpha_s$ [74,75].

In addition to $G\alpha_s$ and Gβγ, a lysophosphotidic acid (LPA)-induced muscle hypertrophy was uncovered that involved the $G\alpha_{i2}$ isoform [76]. In a series of loss- and gain-of-function experiments, Minetti and colleagues [76] showed that LPA-induced myotube hypertrophy increased protein synthesis by 40% as a result of TORC1 activation, as indicated by increased phosphorylation of its downstream targets $p70^{S6K1}$ and rpS6 [76]. TORC1 activation by LPA, however, was independent of AKT but required protein kinase C (PKC) inhibition of GSK3β, though the mechanism linking PKC to TORC1 activation remains to be identified [76].

Emerging Pathways

Unlike PI3K/AKT/TORC1 and β_2-AR signaling pathways, there are less well-established signaling pathways that have been shown to be capable of regulating muscle hypertrophy. Although it is still too early to know the importance of these "new" pathways in human skeletal muscle hypertrophy, an understanding of how each of these pathways function will provide a more comprehensive knowledge of the signaling pathways that regulate skeletal muscle hypertrophy.

Nitric Oxide (NO) Signaling

Inhibition of nitric oxide synthase (NOS) by N^G-nitro-L-arginine methyl ester (L-NAME) administration significantly blunted muscle hypertrophy induced by mechanical overload of the rat plantaris muscle [77,78]. Building on these early studies, Ito and colleagues [79] uncovered a signaling pathway by which nNOS production of nitric oxide mediates muscle hypertrophy [79]. Upon mechanical loading, nNOS-produced nitric oxide reacts with superoxide to generate peroxynitrite. Peroxynitrite then activates the TRPV1 (transient receptor potential cation channel, subfamily V, member 1) channel, causing an increase in intracellular Ca^{2+} levels via release from the sarcoplasmic reticulum. Elevation of intracellular Ca^{2+} causes an increase in protein synthesis through TORC1 activation by an unknown mechanism; though speculative at this time, the authors proposed a similar Ca^{2+}/calmodulin mechanism as that used by amino acids to stimulate TORC1 activity [79,80]. In agreement with these findings, the administration of the TRPV1 agonist capsaicin activated TORC1 signaling to a level comparable to that observed with mechanical overload [79].

PGC-α4 Signaling

PGC-1α (peroxisome proliferative activated receptor, gamma, coactivator 1 alpha) was originally identified as a coactivator of PPARγ in brown adipose tissue but has since been shown to have an important role in skeletal muscle adaptation to exercise [81,82]. Although the over-expression of PGC-1α in skeletal muscle was found to ameliorate the loss of muscle mass caused by denervation, there was no evidence to suggest PGC-1α might have a role in regulating skeletal muscle hypertrophy [83]. Recently, Ruas et al., [84] identified a splice variant of PGC-1α, PGC-1α4, capable of inducing muscle fiber hypertrophy when over-expressed both *in vitro* and *in vivo*[84]. The mechanism through which PGC-1α4 promotes hypertrophic growth appears to be through the down-regulation of myostatin expression while simultaneously increasing IGF-1 expression [84]. Importantly, in humans, PGC-1α4 expression was found to be dramatically increased in response to an 8-week training program consisting of both resistance and endurance exercises [84]. Moreover, leg press performance (number of repetitions) following the training program correlated with the increase in PGC-1α4 expression, suggesting PGC-1α4 expression might serve as a readout for optimizing a strength-training program [84].

MicroRNAs

MicroRNAs (miRs) are small (~22 nucleotides), noncoding RNAs that regulate gene expression through a post-transcriptional mechanism [85]. A small family of muscle-specific miRs, referred to as myomiRs, have been shown to have an important role in muscle development and disease [86]. The expression of these myomiRs has been found to change in response to both resistance and endurance exercise following a single bout or with training [87]. Davidsen and co-workers [88] identified a small group of miRs that were differentially expressed between low- and high-responders following a 12-week resistance exercise program [88]. In particular, the change in miR-378 expression in response to training was positively correlated with gains in skeletal muscle mass [88]. To have a better understanding of the molecular mechanisms through which myomiRs regulate skeletal muscle hypertrophy, future studies will need to focus on identifying target genes and their connection to anabolic signaling.

CATABOLIC SIGNALING

As dramatic as the increase in skeletal muscle mass can be as the result of resistance exercise training, so the loss of skeletal muscle mass can be following periods of disuse. Muscle atrophy can also occur as the consequence of certain systemic diseases (cancer, sepsis or HIV-AIDS), organ failure (cirrhosis, chronic kidney disease and chronic heart failure), inactivity (as a result of obesity, rheumatoid arthritis or prolonged bed rest) and aging. In contrast to muscle hypertrophy, muscle atrophy comes about when the rate of protein degradation exceeds the rate of protein synthesis, either through an increase in the rate of protein degradation and/or a decrease in the rate of protein synthesis.

In an effort to determine if a common mechanism might underlie muscle atrophy brought about by differing catabolic conditions, Bodine and colleagues [89] searched for genes that shared a similar pattern of expression in response to denervation, immobilization and hind limb unloading [89]. Two genes, MAFbx (Muscle Atrophy F-box) and MuRF-1 (Muscle RING Finger 1), were identified that were significantly up-regulated in all three models of muscle atrophy [89]. Concurrently, Gomes and co-workers reported the identification of atrogin-1 (also known as MAFbx) in atrophying muscle caused by food deprivation, diabetes, cancer and renal failure [90]. Further analysis revealed that both MAFbx and MuRF-1 skeletal muscle-specific ubiquitin ligases are involved in protein degradation through the ubiquitin-proteosome system [89,90]. The central importance of these two genes to the skeletal muscle atrophy was confirmed by experiments showing that inactivation of either MAFbx or MuRF-1 resulted in a blunted atrophic response under catabolic conditions [89].

AKT/Foxo Signaling

The ability of IGF-1 to block MAFbx expression under catabolic conditions induced by the synthetic glucocorticoid dexamethasone indicated, paradoxically, that the PI3K/AKT signaling pathway was involved in the regulation of muscle atrophy [91]. A downstream target of PI3K/AKT signaling through which IGF-1 could inhibit MAFbx expression was the Forkhead box O (FoxO) class of transcription factors; phosphorylation by AKT causes the Foxo protein to remain sequestered in the cytoplasm and unable to activate transcription of target genes such as MAFbx and MuRF-1[92,93]. Subsequent studies confirmed such a mechanism by providing evidence demonstrating that IGF-1 activation of AKT leads to repression of MAFbx and MuRF-1expression through FoxO phosphorylation [94,95]. Further, over-expression of a constitutively active form of FoxO3 isoform caused a significant increase in MAFbx expression and reduced myotube diameter by 50% [94]. Conversely, Southgate and colleagues [96] reported that FoxO1 was also able to promote muscle atrophy by inhibiting TORC1 signaling through the up-regulation of 4E-BP1 expression [96]. In addition to describing a key signaling pathway regulating muscle atrophy, these studies reveal the degree of crosstalk between anabolic and catabolic pathways in skeletal muscle.

Given that both MAFbx and MuRF-1 are E3 ubiquitin ligases, identifying the proteins each one targets for degradation is important for understanding their respective catabolic action and how it promotes muscle atrophy. MAFbx had been shown to target eIF3f and MyoD for degradation whereas thick myofilament proteins, such as myosin-binding protein C, myosin light chains 1 and 2 and myosin heavy chain, are targeted for degradation by MuRF-1 [35,97,98]. These findings suggest MAFbx and MuRF-1 have distinct functions in muscle atrophy; MAFbx modulates the expression level of genes involved in regulating protein synthesis while MuRF-1 influences the level of protein degradation [99].

Myostatin Signaling

Myostatin is a member of the transforming growth factor-β (TGF-β) superfamily that acts as a potent negative regulator of skeletal muscle mass [100]. Myostatin binds the activin type IIB receptor which in turn activates activin receptor-like kinase 4 (ALK4) or ALK5 receptor and initiates downstream signaling through Smad 2/Smad 3 transcription factor complex [101]. While the evidence is clear that inactivation of myostatin expression or activity promotes skeletal muscle hypertrophy, the mechanism by which myostatin induces muscle atrophy remains to be fully defined [100,102,103]. Studies have shown that myostatin is capable of inhibiting AKT/TORC1 signaling through a Smad2/3-dependent mechanism, though it is not clear if this results in Foxo activation of MAFbx and MuRF-1 expression [102,104–106]. Regardless, these findings provide another example of the crosstalk between the signaling pathways that regulate skeletal muscle mass.

AMPK Signaling

The AMP-activated protein kinase (AMPK) is a multi-protein kinase composed of a catalytic subunit (α) and two regulatory subunits (βγ) [107]. In response to cellular stress that causes a decrease in cellular energy levels (as sensed by an increase in the AMP:ATP ratio), AMPK is activated and functions to restore the energy status of the cell by turning on catabolic pathways and turning off ATP-consuming anabolic pathways such as protein synthesis [107]. The central role that AMPK plays in the regulation of cell metabolism, protein synthesis in particular, suggests that it could also be involved in regulating skeletal muscle mass.

Inoki and colleagues [108] provided evidence that AMPK was able to inhibit protein synthesis through phosphorylation of TSC2, a potent inhibitor of TORC1 activity [108]. In agreement with this finding, AMPK activity was reported to be increased during and immediately following a bout of high-resistance exercise

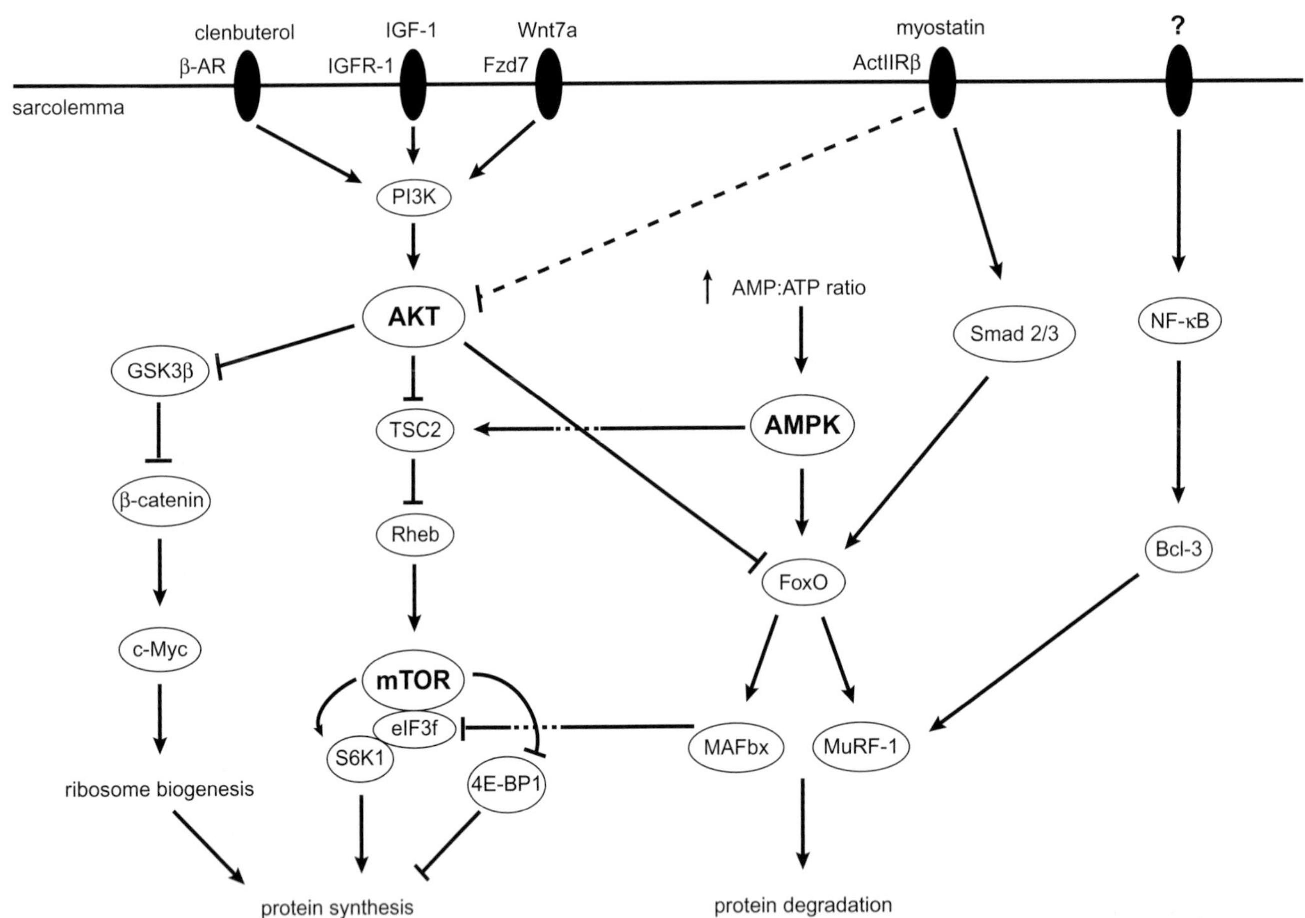

FIGURE 24.1 Anabolic and catabolic signaling pathways that regulate skeletal muscle mass. Clenbuterol, IGF-1 and Wnt 7a are able to activate PI3K/AKT signaling through binding to β-AR, IGFR-1 and Fzd7 receptor, respectively. AKT inhibits TSC2 activity, which results in Rheb activation of TORC1. TORC1 increases protein synthesis through phosphorylation of S6K1 ($p70S^{6K1}$) and 4E-BP1. AKT is also able to increase protein synthesis through inhibition of GSK3β. Inactivation of GSK3β leads to stabilization of β-catenin, resulting in c-Myc induction of ribosome biogenesis. Protein synthesis can be inhibited through AMPK activation of TSC2, a potent inhibitor of TORC1 signaling. AMPK can also increase protein degradation through FoxO regulation of MAFbx and MuRF-1 expression. MAFbx and MuRF-1 are muscle-specific ubiquitin ligase that target proteins for degradation such as eIF3f and myosin, respectively. FoxO up-regulation of MAFbx and MuRF-1 expression can also be mediated through myostatin binding to the ActIIRβ receptor with the downstream activation of Smad 2 and Smad 3 transcription factors. Activation of NF-κB through an unknown mechanism, that is independent of inflammatory cytokines such as TNF-α, increases MuRF-1 expression through activation of Bcl-3. An arrow indicates activation whereas a crossbar indicates inhibition.

which was associated with reduced phosphorylation of 4E-BP1 [109]. In AMPK knockout mice, muscle mass and fiber size were increased as a result of TORC1 activation [110]. Further, AMPK inactivation accelerated the rate of muscle hypertrophy induced by mechanical overload, thus confirming the idea that AMPK acts to limit skeletal muscle growth [111].

In addition to inhibiting TORC1 through TSC2 activation, AMPK has also been shown to restrict muscle growth by inducing expression of Foxo transcription factors and expression of target genes MAFbx and MuRF-1 [112]. In myotubes, activation of AMPK by the AMP analog AICAR (5-Aminoimidazole-4-carboxamide ribonucleotide) caused enhanced myofibrillar protein degradation that was likely the result of Foxo mediated up-regulation of MAFbx and MuRF-1 expression [112]. Collectively, these studies highlight the importance of AMPK regulation of catabolic pathways in limiting skeletal muscle growth.

NF-κB Signaling

NF-κB is a dimeric transcription factor involved in a range of biological processes including inflammatory and immune responses and is rapidly activated by inflammatory cytokines such as TNF-α [113]. In skeletal muscle, however, NF-κB expression was reported to dramatically increase during muscle atrophy that was independent of TNF-α [114]. Subsequent studies showed that NF-κB signaling was in fact necessary and sufficient for mediating the hypertrophic response [115,116]. The mechanism through which NF-κB signaling promoted atrophy remained unknown until recently when Jackman and colleagues (2012) showed that NF-κB directly regulates the expression of MuRF-1 through a Bcl-3 dependent mechanism [117]. Alternatively, a member of the TNF family of inflammatory cytokines, TWEAK (TNF-like weak inducer of apoptosis), has been shown to induce skeletal muscle atrophy through up-regulation of MuRF-1 via NF-κB activation [118].

SUMMARY

There have been significant advancements during the last decade in our understanding of the anabolic and catabolic signaling pathways involved in the regulation of skeletal muscle mass. These advances have uncovered the mechanistic details of such signaling pathways and demonstrated the complex crosstalk that occurs between pathways (Figure 24.1). Future studies will certainly continue to enhance our understanding of the aforementioned pathways as well as better define the importance of emerging pathways little considered in the field of skeletal muscle plasticity. While there still remains much to be learned, one conclusion that is clear is that the regulation of skeletal muscle mass represents the orchestrated output of multiple anabolic and catabolic signaling pathways.

References

[1] Clarys JP, Martin AD, Marfell-Jones MJ, Janssens V, Caboor D, Drinkwater DT. Human body composition: a review of adult dissection data. Am J Hum Biol 1999;11(2):167–74.

[2] Peppa M, Koliaki C, Nikolopoulos P, Raptis SA. Skeletal muscle insulin resistance in endocrine disease. J Biomed Biotechnol 2010;527850.

[3] Demontis F, Perrimon N. FOXO/4E-BP signaling in Drosophila muscles regulates organism-wide proteostasis during aging. Cell 2010;143(5):813–25.

[4] Hass CJ, Feigenbaum MS, Franklin BA. Prescription of resistance training for healthy populations. Sports Med 2001;31 (14):953–64.

[5] Snow MH, Chortkoff BS. Frequency of bifurcated muscle fibers in hypertrophic rat soleus muscle. Muscle Nerve 1987;10 (4):312–7.

[6] MacDougall JD, Sale DG, Alway SE, Sutton JR. Muscle fiber number in biceps brachii in bodybuilders and control subjects. J Appl Physiol 1984;57(5):1399–403.

[7] Larsson L, Tesch PA. Motor unit fibre density in extremely hypertrophied skeletal muscles in man. Electrophysiological signs of muscle fibre hyperplasia. Eur J Appl Physiol Occup Physiol 1986;55(2):130–6.

[8] Giddings CJ, Gonyea WJ. Morphological observations supporting muscle fiber hyperplasia following weight-lifting exercise in cats. Anat Rec 1992;233(2):178–95.

[9] Kelley G. Mechanical overload and skeletal muscle fiber hyperplasia: a meta-analysis. J Appl Physiol 1996;81(4):1584–8.

[10] Goldberg AL. Protein synthesis during work-induced growth of skeletal muscle. J Cell Biol 1968;36(3):653–8.

[11] Chesley A, MacDougall JD, Tarnopolsky MA, Atkinson SA, Smith K. Changes in human muscle protein synthesis after resistance exercise. J Appl Physiol 1992;73(4):1383–8.

[12] Phillips SM, Tipton KD, Aarsland A, Wolf SE, Wolfe RR. Mixed muscle protein synthesis and breakdown after resistance exercise in humans. Am J Physiol 1997;273(1 Pt 1):E99–107.

[13] Noble EG, Tang Q, Taylor PB. Protein synthesis in compensatory hypertrophy of rat plantaris. Can J Physiol Pharmacol 1984;62(9):1178–82.

[14] MacDougall JD, Tarnopolsky MA, Chesley A, Atkinson SA. Changes in muscle protein synthesis following heavy resistance exercise in humans: a pilot study. Acta Physiol Scand 1992;146 (3):403–4.

[15] MacDougall JD, Gibala MJ, Tarnopolsky MA, MacDonald JR, Interisano SA, Yarasheski KE. The time course for elevated muscle protein synthesis following heavy resistance exercise. Can J Appl Physiol 1995;20(4):480–6.

[16] Biolo G, Maggi SP, Williams BD, Tipton KD, Wolfe RR. Increased rates of muscle protein turnover and amino acid transport after resistance exercise in humans. Am J Physiol 1995;268(3 Pt 1):E514–20.

[17] Baar K, Esser K. Phosphorylation of p70(S6k) correlates with increased skeletal muscle mass following resistance exercise. Am J Physiol 1999;276(1 Pt 1):C120–7.

[18] Aronson D, Violan MA, Dufresne SD, Zangen D, Fielding RA, Goodyear LJ. Exercise stimulates the mitogen-activated protein

kinase pathway in human skeletal muscle. J Clin Invest 1997;99 (6):1251–7.

[19] Sherwood DJ, Dufresne SD, Markuns JF, Cheatham B, Moller DE, Aronson D, et al. Differential regulation of MAP kinase, p70(S6K), and Akt by contraction and insulin in rat skeletal muscle. Am J Physiol 1999;276(5 Pt 1):E870–8.

[20] Hayashi T, Hirshman MF, Dufresne SD, Goodyear LJ. Skeletal muscle contractile activity in vitro stimulates mitogen-activated protein kinase signaling. Am J Physiol 1999;277(4 Pt 1):C701–7.

[21] Bodine SC, Stitt TN, Gonzalez M, Kline WO, Stover GL, Bauerlein R, et al. Akt/mTOR pathway is a crucial regulator of skeletal muscle hypertrophy and can prevent muscle atrophy in vivo. Nat Cell Biol 2001;3(11):1014–9.

[22] Mayhew DL, Kim JS, Cross JM, Ferrando AA, Bamman MM. Translational signaling responses preceding resistance training-mediated myofiber hypertrophy in young and old humans. J Appl Physiol 2009;107(5):1655–62.

[23] Laplante M, Sabatini DM. mTOR signaling in growth control and disease. Cell 2012;149(2):274–93.

[24] Hornberger TA, Sukhija KB, Chien S. Regulation of mTOR by mechanically induced signaling events in skeletal muscle. Cell Cycle 2006;5(13):1391–6.

[25] Rommel C, Bodine SC, Clarke BA, Rossman R, Nunez L, Stitt TN, et al. Mediation of IGF-1-induced skeletal myotube hypertrophy by PI(3)K/Akt/mTOR and PI(3)K/Akt/GSK3 pathways. Nat Cell Biol 2001;3(11):1009–13.

[26] Vandenburgh HH, Hatfaludy S, Karlisch P, Shansky J. Mechanically induced alterations in cultured skeletal muscle growth. J Biomech 1991;24(Suppl 1):91–9.

[27] Dan HC, Sun M, Yang L, Feldman RI, Sui XM, Ou CC, et al. Phosphatidylinositol 3-kinase/Akt pathway regulates tuberous sclerosis tumor suppressor complex by phosphorylation of tuberin. J Biol Chem 2002;277(38):35364–70.

[28] Inoki K, Li Y, Zhu T, Wu J, Guan KL. TSC2 is phosphorylated and inhibited by Akt and suppresses mTOR signalling. Nat Cell Biol 2002;4(9):648–57.

[29] Manning BD, Tee AR, Logsdon MN, Blenis J, Cantley LC. Identification of the tuberous sclerosis complex-2 tumor suppressor gene product tuberin as a target of the phosphoinositide 3-kinase/akt pathway. Mol Cell 2002;10 (1):151–62.

[30] Potter CJ, Pedraza LG, Xu T. Akt regulates growth by directly phosphorylating Tsc2. Nat Cell Biol 2002;4(9):658–65.

[31] Garami A, Zwartkruis FJ, Nobukuni T, Joaquin M, Roccio M, Stocker H, et al. Insulin activation of Rheb, a mediator of mTOR/S6K/4E-BP signaling, is inhibited by TSC1 and 2. Mol Cell 2003;11(6):1457–66.

[32] Inoki K, Li Y, Xu T, Guan KL. Rheb GTPase is a direct target of TSC2 GAP activity and regulates mTOR signaling. Genes Dev 2003;17(15):1829–34.

[33] Webb BL, Proud CG. Eukaryotic initiation factor 2B (eIF2B). Int J Biochem Cell Biol 1997;29(10):1127–31.

[34] Kubica N, Bolster DR, Farrell PA, Kimball SR, Jefferson LS. Resistance exercise increases muscle protein synthesis and translation of eukaryotic initiation factor 2Bepsilon mRNA in a mammalian target of rapamycin-dependent manner. J Biol Chem 2005;280(9):7570–80.

[35] Lagirand-Cantaloube J, Offner N, Csibi A, Leibovitch MP, Batonnet-Pichon S, Tintignac LA, et al. The initiation factor eIF3-f is a major target for atrogin1/MAFbx function in skeletal muscle atrophy. EMBO J 2008;27(8):1266–76.

[36] Hinnebusch AG. eIF3: a versatile scaffold for translation initiation complexes. Trends Biochem Sci 2006;31(10):553–62.

[37] Holz MK, Ballif BA, Gygi SP, Blenis J. mTOR and S6K1 mediate assembly of the translation preinitiation complex through dynamic protein interchange and ordered phosphorylation events. Cell 2005;123(4):569–80.

[38] Harris TE, Chi A, Shabanowitz J, Hunt DF, Rhoads RE, Lawrence Jr. JC. mTOR-dependent stimulation of the association of eIF4G and eIF3 by insulin. EMBO J 2006;25(8):1659–68.

[39] Csibi A, Cornille K, Leibovitch MP, Poupon A, Tintignac LA, Sanchez AM, et al. The translation regulatory subunit eIF3f controls the kinase-dependent mTOR signaling required for muscle differentiation and hypertrophy in mouse. PLoS One 2010;5(2): e8994.

[40] Vyas DR, Spangenburg EE, Abraha TW, Childs TE, Booth FW. GSK-3beta negatively regulates skeletal myotube hypertrophy. Am J Physiol Cell Physiol 2002;283(2):C545–51.

[41] Satyamoorthy K, Li G, Vaidya B, Patel D, Herlyn M. Insulin-like growth factor-1 induces survival and growth of biologically early melanoma cells through both the mitogen-activated protein kinase and beta-catenin pathways. Cancer Res 2001;61 (19):7318–24.

[42] Valenta T, Hausmann G, Basler K. The many faces and functions of beta-catenin. EMBO J 2012;31(12):2714–36.

[43] Ikeda S, Kishida S, Yamamoto H, Murai H, Koyama S, Kikuchi A. Axin, a negative regulator of the Wnt signaling pathway, forms a complex with GSK-3beta and beta-catenin and promotes GSK-3beta-dependent phosphorylation of beta-catenin. EMBO J 1998;17(5):1371–84.

[44] He TC, Sparks AB, Rago C, Hermeking H, Zawel L, da Costa LT, et al. Identification of c-MYC as a target of the APC pathway. Science 1998;281(5382):1509–12.

[45] Shtutman M, Zhurinsky J, Simcha I, Albanese C, D'Amico M, Pestell R, et al. The cyclin D1 gene is a target of the beta-catenin/LEF-1 pathway. Proc Natl Acad Sci USA. 1999;96 (10):5522–7.

[46] Armstrong DD, Esser KA. Wnt/beta-catenin signaling activates growth-control genes during overload-induced skeletal muscle hypertrophy. Am J Physiol Cell Physiol 2005;289(4):C853–9.

[47] Armstrong DD, Wong VL, Esser KA. Expression of beta-catenin is necessary for physiological growth of adult skeletal muscle. Am J Physiol Cell Physiol 2006;291(1):C185–8.

[48] Ecker RE, Kokaisl G. Synthesis of protein, ribonucleic acid, and ribosomes by individual bacterial cells in balanced growth. J Bacteriol 1969;98(3):1219–26.

[49] Miyazaki M, Esser KA. Cellular mechanisms regulating protein synthesis and skeletal muscle hypertrophy in animals. J Appl Physiol 2009;106(4):1367–73.

[50] Hannan RD, Jenkins A, Jenkins AK, Brandenburger Y. Cardiac hypertrophy: a matter of translation. Clin Exp Pharmacol Physiol 2003;30(8):517–27.

[51] von Walden F, Casagrande V, Ostlund Farrants AK, Nader GA. Mechanical loading induces the expression of a Pol I regulon at the onset of skeletal muscle hypertrophy. Am J Physiol Cell Physiol 2012;302(10):C1523–30.

[52] Nader GA, McLoughlin TJ, Esser KA. mTOR function in skeletal muscle hypertrophy: increased ribosomal RNA via cell cycle regulators. Am J Physiol Cell Physiol 2005;289(6): C1457–65.

[53] von Maltzahn J, Bentzinger CF, Rudnicki MA. Wnt7a-Fzd7 signalling directly activates the Akt/mTOR anabolic growth pathway in skeletal muscle. Nat Cell Biol 2011;14(2):186–91.

[54] Le Grand F, Jones AE, Seale V, Scime A, Rudnicki MA. Wnt7a activates the planar cell polarity pathway to drive the symmetric expansion of satellite stem cells. Cell Stem Cell 2009;4(6): 535–47.

[55] Kim YS, Sainz RD, Molenaar P, Summers RJ. Characterization of beta 1- and beta 2-adrenoceptors in rat skeletal muscles. Biochem Pharmacol 1991;42(9):1783–9.

[56] Rattigan S, Appleby GJ, Edwards SJ, McKinstry WJ, Colquhoun EQ, Clark MG, et al. Alpha-adrenergic receptors in rat skeletal muscle. Biochem Biophys Res Commun 1986;136 (3):1071–7.

[57] Williams RS, Caron MG, Daniel K. Skeletal muscle beta-adrenergic receptors: variations due to fiber type and training. Am J Physiol 1984;246(2 Pt 1):E160–7.

[58] Gosmanov AR, Wong JA, Thomason DB. Duality of G protein-coupled mechanisms for beta-adrenergic activation of NKCC activity in skeletal muscle. Am J Physiol Cell Physiol 2002;283 (4):C1025–32.

[59] Yamamoto DL, Hutchinson DS, Bengtsson T. Beta(2)-Adrenergic activation increases glycogen synthesis in L6 skeletal muscle cells through a signalling pathway independent of cyclic AMP. Diabetologia 2007;50(1):158–67.

[60] Lynch GS, Ryall JG. Role of beta-adrenoceptor signaling in skeletal muscle: implications for muscle wasting and disease. Physiol Rev 2008;88(2):729–67.

[61] Murga C, Laguinge L, Wetzker R, Cuadrado A, Gutkind JS. Activation of Akt/protein kinase B by G protein-coupled receptors. A role for alpha and beta gamma subunits of heterotrimeric G proteins acting through phosphatidylinositol-3-OH kinasegamma. J Biol Chem 1998;273(30):19080–5.

[62] Murga C, Fukuhara S, Gutkind JS. A novel role for phosphatidylinositol 3-kinase beta in signaling from G protein-coupled receptors to Akt. J Biol Chem 2000;275(16):12069–73.

[63] Schmidt P, Holsboer F, Spengler D. Beta(2)-adrenergic receptors potentiate glucocorticoid receptor transactivation via G protein beta gamma-subunits and the phosphoinositide 3-kinase pathway. Mol Endocrinol 2001;15(4):553–64.

[64] Emery PW, Rothwell NJ, Stock MJ, Winter PD. Chronic effects of beta 2-adrenergic agonists on body composition and protein synthesis in the rat. Biosci Rep 1984;4(1):83–91.

[65] Maltin CA, Delday MI, Reeds PJ. The effect of a growth promoting drug, clenbuterol, on fibre frequency and area in hind limb muscles from young male rats. Biosci Rep 1986;6(3): 293–9.

[66] Maltin CA, Delday MI. Satellite cells in innervated and denervated muscles treated with clenbuterol. Muscle Nerve 1992;15 (8):919–25.

[67] Roberts P, McGeachie JK. The effects of clenbuterol on satellite cell activation and the regeneration of skeletal muscle: an autoradiographic and morphometric study of whole muscle transplants in mice. J Anat 1992;180(Pt 1):57–65.

[68] Maltin CA, Hay SM, McMillan DN, Delday MI. Tissue specific responses to clenbuterol; temporal changes in protein metabolism of striated muscle and visceral tissues from rats. Growth Regul 1992;2(4):161–6.

[69] Claeys MC, Mulvaney DR, McCarthy FD, Gore MT, Marple DN, Sartin JL. Skeletal muscle protein synthesis and growth hormone secretion in young lambs treated with clenbuterol. J Anim Sci 1989;67(9):2245–54.

[70] Reeds PJ, Hay SM, Dorwood PM, Palmer RM. Stimulation of muscle growth by clenbuterol: lack of effect on muscle protein biosynthesis. Br J Nutr 1986;56(1):249–58.

[71] Benson DW, Foley-Nelson T, Chance WT, Zhang FS, James JH, Fischer JE. Decreased myofibrillar protein breakdown following treatment with clenbuterol. J Surg Res 1991;50(1):1–5.

[72] Forsberg NE, Ilian MA, Ali-Bar A, Cheeke PR, Wehr NB. Effects of cimaterol on rabbit growth and myofibrillar protein degradation and on calcium-dependent proteinase and calpastatin activities in skeletal muscle. J Anim Sci 1989;67(12): 3313–21.

[73] Hesketh JE, Campbell GP, Lobley GE, Maltin CA, Acamovic F, Palmer RM. Stimulation of actin and myosin synthesis in rat gastrocnemius muscle by clenbuterol; evidence for translational control. Comp Biochem Physiol C. 1992;102(1):23–7.

[74] Kline WO, Panaro FJ, Yang H, Bodine SC. Rapamycin inhibits the growth and muscle-sparing effects of clenbuterol. J Appl Physiol 2007;102(2):740–7.

[75] Goncalves DA, Silveira WA, Lira EC, Graca FA, Paula-Gomes S, Zanon NM, et al. Clenbuterol suppresses proteasomal and lysosomal proteolysis and atrophy-related genes in denervated rat soleus muscles independently of Akt. Am J Physiol Endocrinol Metab 2012;302(1):E123–33.

[76] Minetti GC, Feige JN, Rosenstiel A, Bombard F, Meier V, Werner A, et al. Galphai2 signaling promotes skeletal muscle hypertrophy, myoblast differentiation, and muscle regeneration. Sci Signal 2011;4(201):ra80.

[77] Sellman JE, DeRuisseau KC, Betters JL, Lira VA, Soltow QA, Selsby JT, et al. In vivo inhibition of nitric oxide synthase impairs upregulation of contractile protein mRNA in overloaded plantaris muscle. J Appl Physiol 2006;100(1): 258–65.

[78] Smith LW, Smith JD, Criswell DS. Involvement of nitric oxide synthase in skeletal muscle adaptation to chronic overload. J Appl Physiol 2002;92(5):2005–11.

[79] Ito N, Ruegg UT, Kudo A, Miyagoe-Suzuki Y, Takeda S. Activation of calcium signaling through Trpv1 by nNOS and peroxynitrite as a key trigger of skeletal muscle hypertrophy. Nat Med 2013;19(1):101–6.

[80] Gulati P, Gaspers LD, Dann SG, Joaquin M, Nobukuni T, Natt F, et al. Amino acids activate mTOR complex 1 via Ca2+/CaM signaling to hVps34. Cell Metab 2008;7(5):456–65.

[81] Puigserver P, Wu Z, Park CW, Graves R, Wright M, Spiegelman BM. A cold-inducible coactivator of nuclear receptors linked to adaptive thermogenesis. Cell 1998;92(6): 829–39.

[82] Olesen J, Kiilerich K, Pilegaard H. PGC-1alpha-mediated adaptations in skeletal muscle. Pflugers Arch 2010;460(1): 153–62.

[83] Sandri M, Lin J, Handschin C, Yang W, Arany ZP, Lecker SH, et al. PGC-1alpha protects skeletal muscle from atrophy by suppressing FoxO3 action and atrophy-specific gene transcription. Proc Natl Acad Sci USA. 2006;103(44):16260–5.

[84] Ruas JL, White JP, Rao RR, Kleiner S, Brannan KT, Harrison BC, et al. A PGC-1alpha isoform induced by resistance training regulates skeletal muscle hypertrophy. Cell 2012;151(6): 1319–31.

[85] Pasquinelli AE. MicroRNAs and their targets: recognition, regulation and an emerging reciprocal relationship. Nat Rev Genet 2012;13(4):271–82.

[86] Williams AH, Liu N, van Rooij E, Olson EN. MicroRNA control of muscle development and disease. Curr Opin Cell Biol 2009;21(3):461–9.

[87] McCarthy JJ. The MyomiR network in skeletal muscle plasticity. Exerc Sport Sci Rev 2011;39(3):150–4.

[88] Davidsen PK, Gallagher IJ, Hartman JW, Tarnopolsky MA, Dela F, Helge JW, et al. High responders to resistance exercise training demonstrate differential regulation of skeletal muscle microRNA expression. J Appl Physiol 2011;110(2): 309–17.

[89] Bodine SC, Latres E, Baumhueter S, Lai VK, Nunez L, Clarke BA, et al. Identification of ubiquitin ligases required for skeletal muscle atrophy. Science 2001;294(5547):1704–8.

[90] Gomes MD, Lecker SH, Jagoe RT, Navon A, Goldberg AL. Atrogin-1, a muscle-specific F-box protein highly expressed during muscle atrophy. Proc Natl Acad Sci USA. 2001;98(25): 14440–5.

[91] Sacheck JM, Ohtsuka A, McLary SC, Goldberg AL. IGF-I stimulates muscle growth by suppressing protein breakdown and

expression of atrophy-related ubiquitin ligases, atrogin-1 and MuRF1. Am J Physiol Endocrinol Metab 2004;287(4):E591–601.
[92] Tang ED, Nunez G, Barr FG, Guan KL. Negative regulation of the forkhead transcription factor FKHR by Akt. J Biol Chem 1999;274(24):16741–6.
[93] Brunet A, Bonni A, Zigmond MJ, Lin MZ, Juo P, Hu LS, et al. Akt promotes cell survival by phosphorylating and inhibiting a Forkhead transcription factor. Cell 1999;96(6):857–68.
[94] Sandri M, Sandri C, Gilbert A, Skurk C, Calabria E, Picard A, et al. Foxo transcription factors induce the atrophy-related ubiquitin ligase atrogin-1 and cause skeletal muscle atrophy. Cell 2004;117(3):399–412.
[95] Stitt TN, Drujan D, Clarke BA, Panaro F, Timofeyva Y, Kline WO, et al. The IGF-1/PI3K/Akt pathway prevents expression of muscle atrophy-induced ubiquitin ligases by inhibiting FOXO transcription factors. Mol Cell 2004;14(3): 395–403.
[96] Southgate RJ, Neill B, Prelovsek O, El-Osta A, Kamei Y, Miura S, et al. FOXO1 regulates the expression of 4E-BP1 and inhibits mTOR signaling in mammalian skeletal muscle. J Biol Chem 2007;282(29):21176–86.
[97] Cohen S, Brault JJ, Gygi SP, Glass DJ, Valenzuela DM, Gartner C, et al. During muscle atrophy, thick, but not thin, filament components are degraded by MuRF1-dependent ubiquitylation. J Cell Biol 2009;185(6):1083–95.
[98] Lagirand-Cantaloube J, Cornille K, Csibi A, Batonnet-Pichon S, Leibovitch MP, Leibovitch SA. Inhibition of atrogin-1/MAFbx mediated MyoD proteolysis prevents skeletal muscle atrophy in vivo. PLoS One 2009;4(3):e4973.
[99] Foletta VC, White LJ, Larsen AE, Leger B, Russell AP. The role and regulation of MAFbx/atrogin-1 and MuRF1 in skeletal muscle atrophy. Pflugers Arch 2011;461(3):325–35.
[100] McPherron AC, Lawler AM, Lee SJ. Regulation of skeletal muscle mass in mice by a new TGF-beta superfamily member. Nature 1997;387(6628):83–90.
[101] Lee SJ. Regulation of muscle mass by myostatin. Annu Rev Cell Dev Biol 2004;20:61–86.
[102] Welle S, Burgess K, Mehta S. Stimulation of skeletal muscle myofibrillar protein synthesis, p70 S6 kinase phosphorylation, and ribosomal protein S6 phosphorylation by inhibition of myostatin in mature mice. Am J Physiol Endocrinol Metab 2009;296(3):E567–72.
[103] Zimmers TA, Davies MV, Koniaris LG, Haynes P, Esquela AF, Tomkinson KN, et al. Induction of cachexia in mice by systemically administered myostatin. Science 2002;296(5572):1486–8.
[104] Sartori R, Milan G, Patron M, Mammucari C, Blaauw B, Abraham R, et al. Smad2 and 3 transcription factors control muscle mass in adulthood. Am J Physiol Cell Physiol 2009;296 (6):C1248–57.
[105] Trendelenburg AU, Meyer A, Rohner D, Boyle J, Hatakeyama S, Glass DJ. Myostatin reduces Akt/TORC1/p70S6K signaling, inhibiting myoblast differentiation and myotube size. Am J Physiol Cell Physiol 2009;296(6):C1258–70.
[106] Lokireddy S, McFarlane C, Ge X, Zhang H, Sze SK, Sharma M, et al. Myostatin induces degradation of sarcomeric proteins through a Smad3 signaling mechanism during skeletal muscle wasting. Mol Endocrinol 2011;25(11):1936–49.
[107] Steinberg GR, Kemp BE. AMPK in Health and Disease. Physiol Rev 2009;89(3):1025–78.
[108] Inoki K, Zhu T, Guan KL. TSC2 mediates cellular energy response to control cell growth and survival. Cell 2003;115(5): 577–90.
[109] Dreyer HC, Fujita S, Cadenas JG, Chinkes DL, Volpi E, Rasmussen BB. Resistance exercise increases AMPK activity and reduces 4E-BP1 phosphorylation and protein synthesis in human skeletal muscle. J Physiol 2006;576(Pt 2):613–24.
[110] Lantier L, Mounier R, Leclerc J, Pende M, Foretz M, Viollet B. Coordinated maintenance of muscle cell size control by AMP-activated protein kinase. FASEB J 2010;24(9):3555–61.
[111] Mounier R, Lantier L, Leclerc J, Sotiropoulos A, Pende M, Daegelen D, et al. Important role for AMPKalpha1 in limiting skeletal muscle cell hypertrophy. FASEB J 2009;23(7): 2264–73.
[112] Nakashima K, Yakabe Y. AMPK activation stimulates myofibrillar protein degradation and expression of atrophy-related ubiquitin ligases by increasing FOXO transcription factors in C2C12 myotubes. Biosci Biotechnol Biochem 2007;71(7): 1650–6.
[113] Hayden MS, Ghosh S. NF-kappaB, the first quarter-century: remarkable progress and outstanding questions. Genes Dev 2012;26(3):203–34.
[114] Hunter RB, Stevenson E, Koncarevic A, Mitchell-Felton H, Essig DA, Kandarian SC. Activation of an alternative NF-kappaB pathway in skeletal muscle during disuse atrophy. FASEB J 2002;16(6):529–38.
[115] Hunter RB, Kandarian SC. Disruption of either the Nfkb1 or the Bcl3 gene inhibits skeletal muscle atrophy. J Clin Invest 2004;114(10):1504–11.
[116] Van Gammeren D, Damrauer JS, Jackman RW, Kandarian SC. The IkappaB kinases IKKalpha and IKKbeta are necessary and sufficient for skeletal muscle atrophy. FASEB J 2009;23(2): 362–70.
[117] Jackman RW, Wu CL, Kandarian SC. The ChIP-seq-Defined Networks of Bcl-3 Gene Binding Support Its Required Role in Skeletal Muscle Atrophy. PLoS One 2013;7(12):e51478.
[118] Bhatnagar S, Mittal A, Gupta SK, Kumar A. TWEAK causes myotube atrophy through coordinated activation of ubiquitin-proteasome system, autophagy, and caspases. J Cell Physiol 2012;227(3):1042–51.

CHAPTER

25

Muscle Growth, Repair and Preservation

A Mechanistic Approach

Robert M. Erskine[1] *and Hans Degens*[2]

[1]School of Sport and Exercise Sciences, Liverpool John Moores University, Liverpool, United Kingdom [2]Institute for Biomedical Research into Human Movement & Health, Faculty of Science & Engineering, Manchester Metropolitan University, Manchester, United Kingdom

INTRODUCTION

Skeletal muscle comprises numerous bundles of long, thin, multinucleated cells, or muscle fibers, each containing a multitude of myofibrils. Each myofibril is composed of myofilaments (comprising the contractile proteins actin and myosin) and a variety of structural proteins, all arranged in a regular configuration throughout the length of the myofibril, so as to form a series of contractile components, or sarcomeres. The maximum force that can be generated by a muscle fiber is proportional to the number of sarcomeres arranged in parallel, or fiber cross-sectional area (CSA), and ultimately the CSA of the whole muscle [1]. Therefore, there is a strong relationship between whole-muscle CSA and maximum isometric force measured *in vivo* [2–4].

Based on the correlation between muscle CSA and force-generating capacity, it is not surprising that an increase in muscle size following resistance training (RT) is accompanied by an increase in maximal muscle force [5–7]. Not only can this enhance the athletic performance of an individual but it can also reduce the elevated risk of falling and bone fracture in older people that is among other factors attributable to sarcopenia (the age-related loss of muscle mass). The question thus arises as to the mechanisms underlying overload-induced muscle hypertrophy.

A multitude of signaling molecules within the muscle fiber are thought to play an integral role in stimulating muscle protein synthesis (MPS) and degradation (MPD). If there is a positive net protein balance (NPB), i.e., when the rate of MPS exceeds that of MPD, the amount of contractile material will increase, enabling the muscle to hypertrophy and generate more force. Conversely, when NPB is negative, the muscle will decrease in size, or atrophy, and become weaker. This chapter will explore the specific signaling pathways involved in MPS and MPD, which help to explain how skeletal muscle adapts to overload, disuse, aging and muscle-wasting diseases. Furthermore, strategies used to preserve or maintain muscle mass during periods of disuse and wasting, such as RT and nutritional interventions, will be discussed.

In addition to the mechanisms underlying muscle growth and atrophy, there is still more to be learned about the systems associated with repair following exercise-induced muscle damage. Several studies have reported that disruption of the cytoskeletal structure of muscle fibers is accompanied by impairment of muscular function following damage-inducing exercise [8,9]. As well as structural damage to the sarcomere, eccentric exercise can cause raised intracellular calcium ion (Ca^{2+}) levels [10], decreased muscle force production [8,11], an increase in serum levels of muscle-specific proteins [12], an increase in muscle specific inflammation [13], an increase in proteolytic enzyme activity [14], and a delayed onset of muscle soreness [13]. The ultimate repair of the muscle requires the activation of satellite cells, and in this chapter we will consider the various MPD systems and the role of satellite cells in muscle damage and repair following exercise.

Nutrition and Enhanced Sports Performance.
DOI: http://dx.doi.org/10.1016/B978-0-12-396454-0.00025-4

MUSCLE GROWTH

While prenatal muscle growth is largely the result of muscle fiber formation, postnatal maturational muscle growth and that in response to RT is almost entirely attributable to fiber hypertrophy. Prenatal myogenesis, i.e., the formation of muscle fibers during embryonic development, involves the proliferation, migration, differentiation and fusion of muscle precursor cells to form post-mitotic multinucleated myotubes. Postnatal skeletal muscle growth is accompanied by an increase in the number of myonuclei per muscle fiber [15] that requires the activation of muscle stem cells, or satellite cells (located at the basal lamina that surrounds the muscle fiber), which proliferate and fuse with existing muscle fibers [16]. Once fully mature, skeletal muscle growth, or hypertrophy, is dependent upon a positive NPB, i.e., MPS must be greater than MPD [17,18], a process that is driven by an increase in the rate of MPS [19]. This leads to an accretion of myofibrillar proteins and an increase in muscle fiber CSA, which in turn leads to an increase in the overall CSA of the muscle, thus enabling more force to be produced. Resistance exercise, i.e., overloading the muscle, has been shown to increase MPS [17,18], and chronic resistance exercise, i.e., RT performed over many weeks, is a potent stimulus for skeletal muscle hypertrophy and strength gains [5–7]. However, exactly how overloading the muscle leads to a positive NPB and therefore an increase in muscle size has yet to be fully elucidated. It is thought that the process necessary for inducing muscle hypertrophy involves a myriad of molecules within the muscle fiber that form signaling cascades, eventually culminating in increased MPS and/or decreased MPD. Here we will discuss how insulin-like growth factor-I (IGF-I), mechanosensors, and amino acids might activate these specific signaling pathways that lead to MPS and ultimately to muscle growth, or hypertrophy.

The Role of IGF-I in Muscle Growth

IGF-I is produced by the liver and skeletal muscle and thus acts on muscle fibers in an endocrine and autocrine/paracrine manner [20,21]. This growth factor appears to play an integral role in activating a specific signaling pathway within the muscle fiber that stimulates MPS [22,23]. The local production and release of IGF-I during muscle contraction [24] activates this signaling cascade by binding to its receptor, located in the sarcolemma. This causes autophosphorylation of the insulin receptor substrate (IRS1) and subsequent phosphorylation of downstream molecules within this signaling pathway, which includes phosphatidylinositol-3 kinase (PI3K), protein kinase-B (PKB or Akt), the mammalian target of rapamycin complex 1 (mTORC1), 70-kDa ribosomal S6 protein kinase ($p70^{S6K}$), and eukaryotic initiation factor 4E binding protein (4E-BP) (Figure 25.1). In fact, $p70^{S6K}$ activation is related to gains in skeletal muscle mass following RT, both in rats [25] and humans [26], with the increase occurring mainly in type II fibers [27]. Together, these studies implicate mTORC1 and $p70^{S6K}$ as principal downstream mediators of IGF-I stimulation of skeletal muscle growth.

There is evidence that IGF-I produced in skeletal muscle is more important for developmental and exercise-induced muscle growth than IGF-I produced by the liver. This is indicated by the greater muscle mass in transgenic mice over-expressing IGF-I in skeletal muscle compared with wild-type mice [28,29] despite normal serum IGF-I levels [29]. Furthermore, low systemic IGF-I levels in liver-specific IGF-I knockout mice does not affect muscle size [30,31]. Also in young adult men, elevated levels of circulating IGF-I do not influence MPS following an acute bout of resistance exercise [32] or muscle hypertrophy in response to RT [33]. Although in rat skeletal muscle, local IGF-I gene expression increases proportionately to the progressive increase in external load [24], it is equivocal whether this occurs in human muscle [34–39]. Some of this controversy might be explained by the elevated expression of two isoforms of the *IGF-1* gene in animal skeletal muscle in response to mechanical stimulation [40,41]. Thus, at least two IGF-I isoforms exist: (i) IGF-IEa, which is similar to the hepatic endocrine isoform, and (ii) the less abundant IGF-IEb (in rats) or IGF-IEc (in humans), otherwise known as mechanical growth factor (MGF). However, it is not always clear which IGF-I isoform has been measured in the muscle [35]. Furthermore, the age of the participants also influences the findings, with MGF increasing in young but not old people following an exercise bout. Interestingly, IGF-IEa does not appear to change in young or older people following resistance exercise [36] despite its apparent hypertrophic effect [28].

Muscle-derived IGF-I does not appear to be the only regulator of adult muscle mass and function, as unloading induces atrophy even in mice that overexpress IGF-I in skeletal muscle [42], and overload induces hypertrophy even in transgenic mice that express a dominant negative IGF-I receptor in skeletal muscle [43]. Also in older people, enhanced muscle strength can be attained following RT without a significant change in muscle IGF-I gene expression [44]. Thus, it appears that for the development of hypertrophy in adult muscles, loading is more important than alterations in local and systemic IGF-I levels. This fits the notion that activation of mTORC1 and $p70^{S6K}$ can

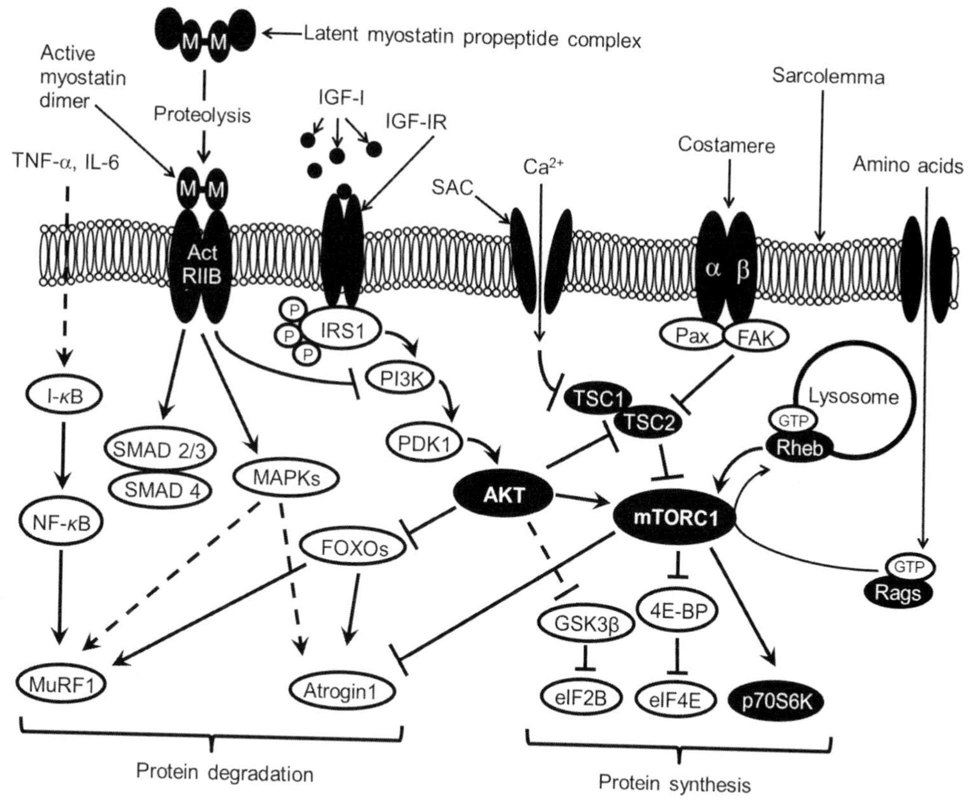

FIGURE 25.1 **The molecular signaling pathways associated with muscle hypertrophy and atrophy.** The binding of IGF-I to its receptor (IGF-IR) causes autophosphorylation of insulin receptor substrate (IRS1). Phosphatidylinositol 3-kinase (PI3K) is a lipid kinase that phosphorylates phosphatidylinositol (4,5)-bisphosphate, producing phosphatidylinositol (3,4,5)-trisphosphate, which is a membrane-binding site for phosphoinositide-dependent protein kinase (PDK1). Upon translocation to the sarcolemma, AKT (or protein kinase B, PKB) is phosphorylated by PDK1. Once activated, AKT phosphorylates mammalian target of rapamycin complex 1 (mTORC1) directly and by phosphorylating and inactivating the tuberous sclerosis complexes 1 and 2 (TSC1/2), which otherwise inhibit mTORC1 activation. Following resistance exercise, an influx of calcium ions (Ca^{2+}) via stretch-activated channels (SACs) and the activation of FAK in the costamere can inactivate TSC1/2, thus activating mTORC1. Amino acids entering the muscle fiber cause RagGTPase-dependent translocation of mTORC1 to the lysosome, where it is activated by ras homologous protein enriched in brain (Rheb). mTORC1 subsequently activates 70-kDa ribosomal S6 protein kinase ($p70^{S6K}$), and inhibits 4E-BP (also known as PHAS-1), which is a negative regulator of the eukaryotic translation initiation factor 4E (eIF-4E). Phosphorylated AKT also inhibits glycogen-synthase kinase 3β (GSK3β), a substrate of AKT that blocks protein translation initiated by the eIF-2B protein. All of these actions lead to increased protein synthesis. However, protein degradation can be induced by pro-inflammatory cytokines, such as tumor necrosis factor-α (TNF-α) and interleukin-6 (IL-6), which activate NF-κB via degradation of I-κB, leading to increased transcription of the E3 ubiquitin-ligase, muscle RING-finger protein-1 (MuRF1). Another ligase, Atrogin1 (also known as MAFbx), is up-regulated by mitogen-activated protein kinase (MAPK) p38, while both ligases are up-regulated by forkhead box (FOXO) transcription factors. However, phosphorylated AKT blocks the transcriptional up-regulation of Atrogin1 and MuRF1 by inhibiting FOXO1, while phosphorylated mTORC1 inhibits the up-regulation of Atrogin1 directly. Myostatin also increases protein degradation and decreases protein synthesis by activating MAPKs and the SMAD complex, and by inhibiting PI3K. In addition, myostatin inhibits the myogenic program, thus resulting in a decrease of myoblast proliferation.

also occur independently of PI3K activation following muscle overload [45].

The Role of Mechanosensors in Muscle Growth

The process that couples the mechanical forces during a muscle contraction with cell signaling and ultimately protein synthesis is called mechanotransduction. Stretch activated channels (SACs) are calcium (Ca^{2+}) and sodium permeable channels that increase their open probability (the fraction of time spent in the open state) in response to mechanical loading of the sarcolemma [46–48]. It has been proposed that SACs function as mechanosensors by allowing an influx of Ca^{2+} into the muscle fiber [49] following mechanical changes in the sarcolemma, which activate mTORC1 [50], leading to an increase in MPS [51]. Correspondingly, inhibition of SACs by streptomycin reduces skeletal muscle hypertrophy in response to mechanical overload [52,53] via attenuation of mTORC1 and $p70^{S6K}$ activation [52]. However, increasing the intracellular [Ca^{2+}] in combination with stretch increases MPS more than Ca^{2+} administration alone [51], thus suggesting that SACs are not the only mechanosensors in skeletal muscle.

Other mechanosensors might come in the form of costameres, intra-sarcolemmal protein complexes that are circumferentially aligned along the length of the muscle fiber [54]. Costameres mechanically link peripheral myofibrils via the Z-disks to the sarcolemma (Figure 25.2), thus maintaining the integrity of the muscle fiber during contraction and relaxation [54]. An individual costamere contains many proteins arranged in a complex structure [56,57], which comprises two different laminin receptors, a dystrophin/glycoprotein complex and an integrin-associated complex, that are localized in the sarcolemma and bound to intra- and extra-cellular structural proteins (Figure 25.2). In this way the force-producing contractile material is connected to the basal membrane and ultimately to adjacent muscle fibers [56,58,59].

Costameres are receptive to mechanical, electrical and chemical stimuli [57]. Indeed, mechanical tension is essential in regulating costameric protein expression, stability and organization, with talin and vinculin, for instance, being up-regulated in response to muscle contraction [60]. Regular contractions, as experienced during RT, increase the expression of costameric proteins, such as desmin [61], alpha-1-syntrophin and dystrophin [62] in humans, while focal adhesion kinase (FAK) and paxillin activity are increased in stretch-induced hypertrophied avian skeletal muscle [63]. The forces exerted on both the intracellular contractile proteins and the basal membrane during periods of loading are required to cause binding of basal membrane laminin to the receptors on the α and β integrins and on the dystrophin/glycoprotein complex [55].

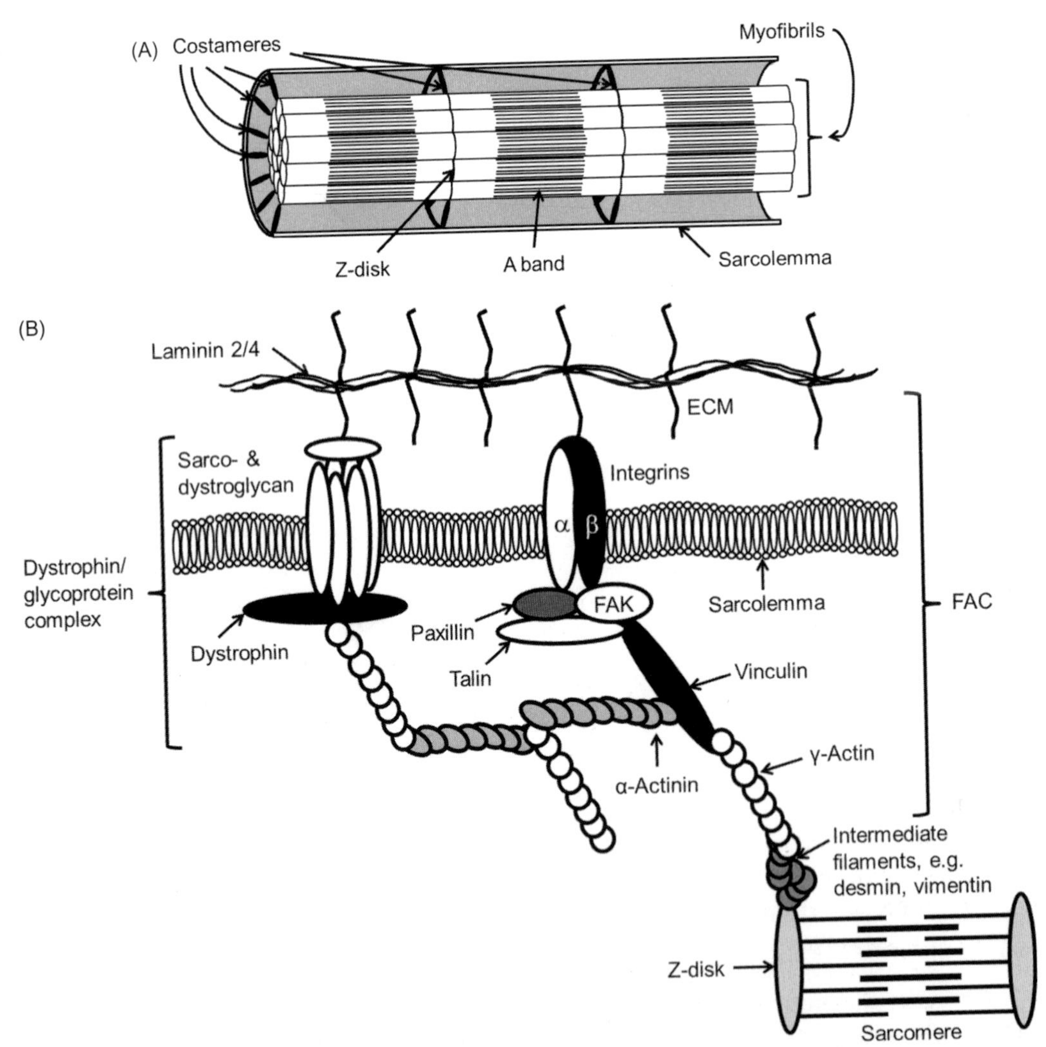

FIGURE 25.2 Schematic representation of (A) the location of costameres within a skeletal muscle fiber and (B) the proteins that constitute the costamere. Costameres are protein complexes circumferentially aligned along the length of the muscle fiber that connect peripheral myofibrils at the Z-disks to the sarcolemma and beyond to the extracellular matrix (ECM). The costamere comprises a dystrophin/glycoprotein complex and a focal adhesion complex (FAC), which includes the integrin-associated tyrosine kinase focal adhesion kinase (FAK). *Modified from [55] with permission.*

Interaction between integrins and the extra-cellular matrix causes rapid phosphorylation of FAK [64], which subsequently activates $p70^{S6K}$ independently of Akt [65,66]. This probably occurs via the phosphorylation and thus inactivation of tuberous sclerosis complex 2 [67,68], thus activating mTORC1 as shown in Figure 25.1.

The Role of Amino Acids in Muscle Growth

Both resistance exercise [18,69] and amino acid/protein ingestion [70,71] stimulate MPS independently, while a combination of the two augments MPS even further [72,73]. Both stimuli cause an increase in mTORC1 activation [74,75], but it is unclear whether they stimulate MPS via different signaling pathways, or whether the combination of the two stimulates the same pathway more than either stimulant on its own. It is thought that amino acids cause Rag GTPases to interact with raptor (a regulatory protein associated with mTORC1), leading to the translocation of mTORC1 to the lysosomal membrane, where Rheb (a Ras GTPase) activates mTORC1 [76–78], as shown in Figure 25.1. Of the essential amino acids (EAAs), the branched-chain amino acids (BCAAs: isoleucine, leucine, and valine), particularly leucine, are the most potent stimulators of the mTORC1 signaling pathway [79]. Leucine supplementation stimulates muscle protein accretion in cultured cells [80] and can also reduce MPD in healthy men [81], and it is possible that the action of leucine occurs via its metabolite, β-hydroxy-β-methylbutyrate (HMB) [82,83].

The timing of amino acid ingestion appears crucial for an optimal anabolic response to a single bout of resistance exercise: ingesting amino acids immediately before an exercise bout promotes a greater increase in MPS compared with ingestion immediately after the bout [84]. This effect was attributed to an increased blood flow during exercise, and therefore an increased delivery of amino acids to the active muscle when they were ingested prior to exercise [84]. As well as the timing, the amount of protein ingested is integral in producing an optimal anabolic environment following resistance exercise [85]. For example, the MPS dose response to ingested protein after a single bout of resistance exercise in healthy young men is saturated at 20 g protein, and any additional ingested protein is simply oxidized [85]. This suggests that if the rate of protein ingestion after resistance exercise exceeds the rate at which it can be incorporated into the muscle, the excess protein is not used for MPS. This dose-response relationship seems to be altered with age, as in older men increased rates of MPS were found when participants ingested 40 g protein following a resistance exercise bout [86]. Therefore, older muscle appears to be less sensitive to amino acids, which has been termed 'anabolic resistance'. Finally, the type and quality of the ingested protein appears to be important when it comes to MPS [70,71]. Following a single bout of resistance exercise and the ingestion of whey, soy or casein, each containing 10 g EAA, larger increases in blood EAA, BCAA, and leucine concentrations were found following the ingestion of whey compared with either soy or casein [70], suggesting a greater availability of these amino acids for protein synthesis following whey protein ingestion. This may be a reflection of the different rate of protein digestion and absorption of amino acids between the protein types [87–89] and explain why MPS was greater following ingestion of whey compared with casein, both at rest and after exercise [70]. In older muscle, the rate of MPS appears to be greater with whey than soy protein ingestion following resistance exercise [71], which could be due to the ~28% greater leucine content in whey versus soy protein [90] as well as differences in digestion and absorption rate.

There is therefore striking evidence to support the acute effects of amino acid ingestion and resistance exercise on MPS via their independent and complementary effects on mTORC1 activation. It is also well known that repeated bouts of resistance exercise over a prolonged period of time, i.e., a RT program, leads to gains in both muscle size and strength [5–7]. Therefore, the amplification of the anabolic environment within the muscle seen with the combination of both amino acid/protein ingestion and resistance exercise [72,73] suggests that RT with protein supplementation should confer greater gains in skeletal muscle size and strength than RT alone. However, the evidence for protein supplementation enhancing the increases in muscle size and strength following longer-term RT programs in young [91,92] and older [93,94] individuals is equivocal. The controversy surrounding the longer-term RT studies could be due to methodological differences/limitations between studies. For example, considerable inter-individual variability exists in the response to RT [95,96] and yet many studies have used small sample sizes [97–99] that may have significantly affected the statistical power required to detect an influence of protein supplementation. Different measures of muscle hypertrophy may also compound this discrepancy. For example, some studies have determined muscle thickness using ultrasonography [94,100] or whole-body fat-free mass assessed via dual-energy X-ray absorptiometry [91], while others have used magnetic resonance imaging to provide a more accurate assessment of muscle size, but still found no effect of protein supplementation on muscle hypertrophy following RT [92,99,101,102]. There are, however, circumstances where protein supplementation may have a beneficial effect on muscle hypertrophy and strength gains. For example, whole-body RT (incorporating multiple muscle groups rather than an individual muscle) could create a requirement for an increase in exogenous protein (due to a greater absolute MPD) that might not be satisfied by habitual protein intake alone. This might be particularly beneficial in the early phase of a RT program when MPS and MPD are likely to be higher than towards the end [103,104]. In addition, older individuals need to ingest at least twice as much protein (40 g) to maximally stimulate MPS [86] compared with the 20 g dose in younger people [85], which may reflect an 'anabolic resistance' at old age. Therefore, previous studies that have not shown a beneficial effect of protein supplementation on muscle size and strength gains in older people may have been administering inadequate amounts of protein per RT session.

MUSCLE ATROPHY

Skeletal muscle atrophy is known to occur in response to disuse and numerous chronic conditions, such as cancer, AIDS, and senile sarcopenia (the age-related loss of muscle mass). Here we will focus on the mechanisms underlying sarcopenia, which is the major cause of muscle weakness in older individuals [105]. Although the cause(s) of sarcopenia is unknown, disuse, chronic systemic inflammation and neuropathic changes leading to motoneuron death are thought to play an integral role [106]. Motoneuron death results in denervation of muscle fibers, and ultimately the loss of muscle fibers (hypoplasia). Selective atrophy of type II fibers [107] and a decrease in the proportion of type II fibers [108–110] are thought to be caused by denervation accompanied by reinnervation of these fibers by axonal sprouting from adjacent slow-twitch motor units [111,112].

Chronic Low-Grade Inflammation and Sarcopenia

Physiological aging is associated with chronic low-grade inflammation, a condition that has been termed 'inflammaging' [113]. Inflammaging is characterized by elevated serum levels of pro-inflammatory cytokines such as interleukin 1 (IL-1), IL-6 and tumor necrosis factor-α (TNF-α), as well as acute phase proteins such as C-reactive protein, or CRP [113,114], and increased circulating levels are associated with lower muscle mass and weakness in old age [115]. Furthermore, the levels of cytokines that counteract the inflammatory state, such as IL-10, are reduced with age [114,116]. The pro-inflammatory cytokines IL-1, IL-6 and TNF-α are produced by both skeletal muscle fibers and adipose cells, and are therefore also members of the adipokine family. As we age we accumulate more adipose tissue, which is deposited in the subcutaneous, visceral and intramuscular regions, and it is particularly the visceral fat that appears to contribute to the inflammatory environment [117].

TNF-α induces the production of reactive oxygen species (ROS), altering vascular permeability, which leads to leukocyte infiltration of the muscle fiber [14] and further ROS generation by the leucocytes. This release of ROS also activates NF-κB via degradation of I-κB (Figure 25.1), which results in the increased expression of key enzymes of the ubiquitin–proteasome MPD system [118]. In addition, TNF-α interferes with satellite cell differentiation and therefore muscle growth and regeneration in old age by reducing the expression of myogenic regulatory factors (MRFs) [106]. The MRFs are a family of muscle-specific transcription factors (MyoD, myogenin, MRF4 and myf-5) that regulate the transition from proliferation to differentiation of the satellite cell [119,120]. The TNF-α-induced activation of NF-κB results in a loss of MyoD mRNA [121] and, via activation of the ubiquitin–proteasome pathway (UPP) [122,123], breakdown of MyoD and myogenin. Thus, part of the attenuated hypertrophic response in elderly versus younger muscle [124] could be due to a decrease in MRFs as seen in overload-induced hypertrophy in older rats [125]. TNF-α also stimulates the release of proteolytic enzymes, such as lysozymes (e.g., cathepsin-B) from neutrophils [126], which are thought to contribute to the MPD process [127,128], but the main action of the pro-inflammatory cytokines in muscle atrophy is thought to occur via the UPP.

It has been suggested that the UPP cannot break down intact sarcomeres, so additional mechanisms are proposed to be involved [129]. For example, the activation of caspase-3 is thought to lead to cleavage of the myofilaments, actin and myosin, which are then degraded by the UPP [130–132]. The UPP involves numerous enzymes, or ligases, that regulate the ubiquitination (the coupling of ubiquitin to protein substrates), so that the 'tagged' protein fragments can be identified by the proteasome for the final step of MPD [133]. These ubiquitin ligases may also degrade MyoD and inhibit subsequent satellite cell activation and differentiation (see above), thus exacerbating the effects of MPD on muscle size by impairing satellite cell-associated muscle growth. Expression of two genes that encode the E3 ubiquitin ligases, muscle-specific atrophy F box (MAFbx, also known as Atrogin-1) and muscle RING Finger 1 (MuRF1), has been shown to increase during different types of muscle atrophy [134,135]. The structure and function of the UPP will be discussed in more detail, below, with regard to muscle damage and repair following exercise.

The Role of Myostatin in Muscle Atrophy

Myostatin, otherwise known as growth differentiation factor-8 (GDF-8), is part of the transforming growth factor β superfamily and is produced in skeletal muscle. The role of myostatin as a negative regulator of muscle mass has been demonstrated by knocking out the *gdf-8* gene in mice, which leads to a 2–3-fold increase in skeletal muscle mass [136]. Conversely, administering myostatin to wild-type mice induces substantial muscle wasting [137]. Further examples of myostatin's regulatory effect can be seen in bovine [138] and human [139] cases, where mutation of the *gdf-8* gene leads to a reduction in myostatin production and considerably enlarged skeletal muscles.

Once bound to the activin IIB receptor (ActRIIB), a signaling cascade is activated that leads to MPD. Key signaling proteins in this pathway include SMAD 2 and 3 [140], which form a complex with SMAD 4 that then translocates to the nucleus where it targets genes encoding MRFs [141], and inhibits differentiation via the reduction of MyoD expression [142] (Figure 25.1). In addition, myostatin reduces Akt/mTORC1/$p70^{S6K}$ signaling [143,144] (Figure 25.1) and is associated with smaller myotube size [143]. Accordingly, the inhibition of myostatin in mature mice leads to increased activation of $p70^{S6K}$, ribosomal protein S6 and skeletal muscle MPS [145]. Myostatin appears to not only inhibit satellite cell differentiation and MPS, but also to induce the expression of atrogenes (genes associated with muscle atrophy) via activation of the p38 mitogen-activated protein (MAP) kinase, Erk1/2, Wnt and c-Jun N-terminal kinase (JNK) signaling pathways [146–148] (Figure 25.1). This is in accord with the observation that myostatin induces cachexia by activating the UPP, i.e., phosphorylating the ubiquitin E3 ligases MAFbx and MuRF1, via FOXO1 activation rather than via the NF-κB pathway [149]. However, myostatin-induced atrophy persists despite inhibiting the expression of the two E3 ligases [143]. Therefore, it is likely that myostatin negatively regulates muscle growth via multiple pathways (Figure 25.1). A lower expression of myostatin may therefore help to maintain muscle mass at old age, a situation reflected by the attenuated loss of muscle mass and regenerative capacity in old myostatin-null mice compared with age-matched wild-type mice [150].

Combating Sarcopenia

RT has been shown to increase muscle size and strength in old [151], very old [152] and frail [153] individuals. This beneficial effect of RT is attributable to an increase in MPS in the atrophied muscle [154] and a reduction in MPD as a result of a reduced MAFbx and MuRF-1 gene expression [155,156]. A reduction in atrogene expression can be realized by the ability of phosphorylated Akt to block FoxO1, which would suppress the transcription of MuRF1 and MAFbx [157], as shown in Figure 25.1. In addition to the Akt/FoxO-1 pathway, mTORC1 also blocks MuRF1 and MAFbx transcription [157]. Therefore, while a degree of MPD is required for muscle remodeling, RT appears to reverse atrophy via the inhibiting effect of Akt/mTORC1 on MPD and the positive effect of mTORC1 activation on MPS [74,75], thus resulting in net protein synthesis (albeit to a lesser extent in elderly than in younger muscle).

The apparent 'anabolic resistance' to RT in older compared with young muscle [124,154] may not be due to an age-related reduction in the mechanosensitivity of the mTORC1 signaling pathway [158], although others do see a reduction in the translational signaling during overload [159]. It could therefore be that the rates of transcription and translation are reduced during overload. Another factor that might be considered is the proposed requirement of satellite cell recruitment for the development of hypertrophy. An impaired satellite cell recruitment would then result in impaired hypertrophy and part of the problem might be a decline in satellite cell number [160], particularly in type II muscle fibers of older people [161]. Yet, paradoxically, older muscle appears to have an increased regenerative drive and protein synthesis that is more pronounced the more severe the sarcopenia. For instance, in old rats, muscle mass decreases in spite of an increase in both $p70^{S6K}$ activation and MPS compared with young adult muscle [154]. IGF-I appears to play a key role in the activation and proliferation of satellite cells [162], while differentiation is regulated by MRFs [119,120]. However, despite an increased regenerative drive as reflected by elevated IGF-I expression [163] and MRF mRNA expression [164] in old rat muscle, MRF protein levels are reduced [164]. This reduction might be caused by a concomitant increase in Id protein expression [164], which inhibits MRF expression and DNA binding capacity. The result is that during a hypertrophic stimulus, satellite cell activation and proliferation can occur in older rat skeletal muscle but with limited differentiation [163]. In elderly human skeletal muscle, however, the capacity for satellite cells to proliferate and differentiate in response to RT does not appear to be diminished

[165,166]. It should be noted, however, that the relative age in the human studies was less than that in the rat studies and it may be that, beyond a given age in humans, the differentiation of satellite cells may also be diminished, particularly when associated with chronic low-grade systemic inflammation.

Extra stimulation of the mTORC1 pathway may overcome the anabolic blunting. As discussed above, older people need to ingest more protein than younger individuals to stimulate maximal MPS [85,86,167]. The greater activation of the mTORC1 pathway when combining RT with protein/amino acid ingestion may inhibit MPD and augment MPS, thereby improving the hypertrophic response. The observation that BCAA administration attenuates the loss of body mass in mice bearing a cachexia-inducing tumor [168] is promising and suggests that it may also enhance the hypertrophic response in this condition. Furthermore, HMB has been shown to attenuate the reduction in MPS in rodents following the administration of a cachectic stimulant [82,169].

In addition to RT and amino acid supplementation, various pharmaceutical therapies have been proposed to combat sarcopenia. Supplementation of the anabolic steroid testosterone augments muscle mass in older men, healthy hypogonadal men, older men with low testosterone levels, and men with chronic illness and low testosterone levels [170]. It is thought that testosterone can reverse sarcopenia by suppressing skeletal muscle myostatin expression, while simultaneously stimulating the Akt pathway [171] to increase MPS and decrease MPD. Furthermore, administration of a myostatin antagonist has led to satellite cell activation, increased MyoD protein expression, and greater muscle regeneration after injury in old murine skeletal muscle [172].

MUSCLE DAMAGE AND REPAIR

In normal skeletal muscle, cytoskeletal proteins act as a framework that keeps the myofibrils aligned in a lateral position by connecting the Z-disks to one another and to the sarcolemma [8,128]. Following eccentric exercise, Z-disk streaming (disturbance of Z-disk configuration) and misalignment of the myofibrils is a common characteristic [128]. Eccentric contractions are defined as contractions where muscles lengthen as they exert force and generally result in more muscle damage than concentric contractions [173]. It has been suggested that this is due to fewer motor units being recruited during eccentric exercise, leading to a smaller CSA of muscle being activated than during a concentric contraction at the same load [174]. It has also been demonstrated that the extent of muscle damage is due to strain (the change in length) rather than the amount of force generated by the muscle [175,176].

Eccentric exercise not only causes alterations in the cytoskeletal structure, but also increases in the activity of proteolytic enzymes [127,177,178]. The positive correlation between the proteolytic enzyme activity and a rise in serum levels of muscle-specific proteins, e.g., creatine kinase (CK), post exercise [127,177,178] suggests that the degree of activation of the proteolytic machinery is related to the degree of muscle damage. Therefore, it is feasible that the activity of these proteolytic enzymes may be required for the remodeling of skeletal muscle in response to exercise where the regulated degradation of cellular proteins [179] may be a prerequisite for subsequent adaptive repair and growth, in analogy to the restructuring of a damaged building that often is preceded by a certain amount of controlled deconstruction.

There are three main systems that contribute to the controlled MPD following muscle damage: (i) the release of calpain, a non-lysosomal, Ca^{2+}-dependent neutral protease that mediates the dismantling of myofibrils [180], (ii) the inflammatory response, which includes lysosomal proteolysis [126], and (iii) the ATP-dependent UPP, which coordinates the demolition of protein fragments liberated by the aforementioned degradation systems [133]. Recent findings suggest that myostatin is also implicated in the MPD process following damaging muscle contractions [181].

The Calpain Protein Degradation System

Calpain is a multidomain protein composed of two subunits: a catalytic 80-kDa subunit and a regulatory 30-kDa subunit [182]. In skeletal muscle, three homologous isozymes of calpain with different Ca^{2+} sensitivities have been identified [133]: μ-calpain (active at micromolar Ca^{2+} concentrations), m-calpain (active at millimolar Ca^{2+} concentrations), and n-calpain (requiring very high Ca^{2+} concentrations). It appears that, although the μ- and m-calpain 80-kDa subunits are quite different, both have similar binding domains: the proteolytic site of a cysteine proteinase, the calpastatin (an endogenous inhibitor of calpain activity) binding domain, and the Ca^{2+}-binding domain [10]. The 30-kDa subunit is extremely hydrophobic, which may help to act as an anchor to the membrane proteins [10]. While there is evidence to suggest that calpain is localized and activated at or around the sarcolemma, thus targeting the membrane-associated proteins [180], others have demonstrated that calpain also targets Z-disk proteins, such as desmin and α-actinin [183]. The action of other proteolytic complexes, including lysosomal enzymes and the UPP, may have a part to play in MPD immediately after damaging eccentric muscle contractions, but as their activity does not peak until later in the muscle damage/repair process [127,180], it is more likely that calpain and/or mechanical stress is the initial effector of cytoskeletal protein breakdown.

Calpain is activated by raised intracellular [Ca^{2+}] [180]. Initial mechanical damage to the sarcoplasmic reticulum (SR) and muscle plasma membrane caused by eccentric muscle contractions could lead to SR vacuolization and an increase in intracellular [Ca^{2+}] [173,184]. The intracellular [Ca^{2+}] could rise further following an increased open probability of SACs as a consequence of increased strain on the skeletal muscle fibers during forced lengthening [175]. It is thought that the activation of calpain is pivotal in the breakdown of cytoskeletal proteins, including desmin and α-actinin, rather than mechanical stress applied to the "over-stretched" sarcomeres during eccentric contractions *per se* [180]. The activity of calpain is not only dependent on the intracellular [Ca^{2+}] but also on the concentration of its inhibitor, calpastatin, and condition of degradable substrates, i.e., the ultrastructural proteins. To become fully active, calpain undergoes autolysis into its subunits. It is likely that the influx of excess Ca^{2+} into the muscle fiber (via the SACs, SR calcium channels and sarcolemmal lesions) binds to the specific domain on the 80-kDa calpain subunit, thereby inhibiting calpastatin. Once calpain is free of calpastatin, it may begin autolysis and/or bind to its substrate (with the help of Ca^{2+}) and begin the process of MPD [10]. A positive relationship between calpain activity and neutrophil accumulation within skeletal muscle after exercise suggests that the calpain-degraded protein fragments act as chemoattractants, thus localizing leukocytes to the site of muscle damage [180,185].

The Inflammatory Response

Exercise-induced damage to muscle fibers elicits an inflammatory response that results in movement of fluid, plasma proteins and leukocytes to the site of injury [173]. Leukocytes have the ability to break down intracellular proteins with the aid of lysosomal enzymes [128], but exactly how the inflammatory response regulates MPD and muscle repair following eccentric exercise is not entirely clear. However, the purpose of the post-exercise-induced inflammatory response is to promote clearance of damaged muscle tissue and prepare the muscle for repair [13], a process that is sub-classified into acute and secondary inflammation.

The acute phase response in skeletal muscle begins with the 'complement system' when fragments from the damaged fiber(s) serve as chemoattractants, luring leukocytes to the injured area [14,180]. As a consequence there is an accumulation of neutrophils, the histological hallmark of acute inflammation [13], in and around the site of injury, which peaks around 4 hours after exercise-induced damage has occurred [14]. The accumulation of neutrophils has been reported to be more significant after eccentric than concentric exercise, and is most likely related to the degree of damage incurred [14]. Pro-inflammatory cytokines such as IL-1, IL-6 and TNF-α, and acute phase proteins, e.g., CRP, act as mediators of inflammatory reactions. TNF-α induces the production of ROS, altering vascular permeability, which leads to leukocyte infiltration into the muscle fiber [14]. TNF-α also stimulates the release of cytotoxic factors from neutrophils, such as lysozymes and ROS [13,128], which are responsible for at least a part of the MPD process following exercise-induced damage [127,128].

It may take up to 7 days to see a significant infiltration of monocytes (precursors to macrophages) within the damaged muscle fiber [14], which carry out further phagocytic activity inside the muscle fiber. Furthermore, considerable increases in the quantity of lipofuscin granules (generally considered to be the indigestible residue of lysosomal degradation) in sore muscles 3 days after exercise suggests that lysozyme activity plays a major part in the secondary inflammatory process [126,186]. The role of the inflammatory response in muscle regeneration is therefore thought to be the further breakdown of damaged muscle proteins via lysozymes, the engulfing of protein fragments by macrophages, and the activation of the UPP by the pro-inflammatory cytokines released from the neutrophils. These cytokines simultaneously stimulate the proliferation of satellite cells, crucial for the regeneration of the damaged area [187].

The Ubiquitin–Proteasome Pathway

This pathway is recognized as the major non-lysosomal complex responsible for the degradation of cellular proteins. The UPP has received much attention due to its involvement in cellular processes where protein degradation is a key regulatory or adaptive event [188–190]. There are two types of ubiquitin in human skeletal muscle: free and conjugated [191]. In its free state, ubiquitin is a normal component of the non-stressed muscle fiber, but it also forms complexes, or conjugations, with abnormal proteins and then returns to its free state (Figure 25.3). The conjugation of ubiquitin with denatured proteins within the muscle fiber "tags" these proteins for recognition by a non-lysosomal protease to be subsequently degraded in a process that

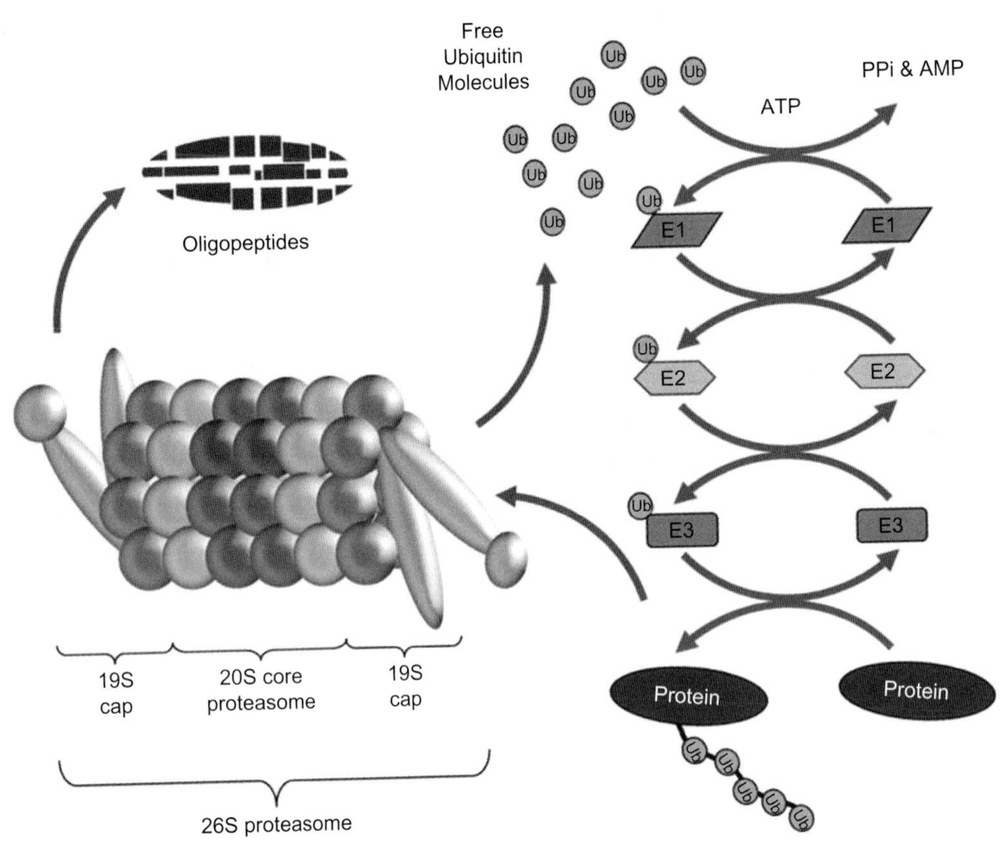

FIGURE 25.3 **Breakdown of protein fragment via the ubiquitin–proteasome pathway.** Multiple ubiquitin (Ub) molecules form a polyubiquitin chain in a process involving the Ub-activating (E1), Ub-conjugating (E2), and Ub-ligating (E3) enzymes. The targeted protein fragment is then selected for degradation, or "tagged", via the covalent attachment of the polyubiquitin chain. The tagging enables the 19S (PA700) module to recognize the protein fragment, so that it can be further degraded by the 20S core into oligopeptides after it has been de-ubiquitinated and the Ub molecules released recycled. *Adapted from [192] with permission.*

requires ATP [188]. The UPP, therefore, consists of two major components that represent the system's functionally distinct parts. Ubiquitin is the element that covalently binds to the protein due to be broken down, while the 26S proteasome, a large protease complex, catalyses the degradation of the ubiquitin-tagged proteins [133].

The cellular proteins are selected for degradation by the attachment of multiple molecules of ubiquitin, or a polyubiquitin chain, which is built by repeated cycles of conjugation via the action of E1, E2, and E3 conjugating enzymes (Figure 25.3). Ubiquitin is initially activated in the presence of ATP by the ubiquitin-activating enzyme, E1, which then transfers ubiquitin to E2, one of the ubiquitin-conjugating enzymes. E2 then binds the ubiquitin molecule to the protein substrate, which is selected for tagging by E3 [189]. The 26S proteasome is able to discriminate between ubiquitinated and non-ubiquitinated proteins, and rapidly degrades the polyubiquitinated proteins, deriving the energy for this process from ATP hydrolysis [193].

The 26S proteasome is composed of a 20S proteasome and two 19S (PA700) regulatory modules [133,189,190]. The 20S proteasome is the proteolytic core, containing multiple catalytic sites. PA700 binds to each end of the proteasome cylinder and elicits ATPase activity in order to unfold and/or translocate the ubiquitinated proteins to the catalytic sites within the 20S proteasome (Figure 25.3). PA700 is also thought to be responsible for disassembling the polyubiquitinated chain, a process requiring its isopeptidase activity.

The total amount of ubiquitin found in skeletal muscle is muscle fiber-type specific, with a greater abundance of ubiquitin found in type I fibers [194]. Furthermore, a three- to seven-fold higher density of conjugated ubiquitin was found at the Z-discs than anywhere else in the muscle fiber, which suggests that, like calpain, ubiquitin targets the cytoskeletal proteins of muscle fibers. One main difference between the two systems, however, is that calpain is activated a lot earlier than the ubiquitin-proteasome pathway [178,195]. Furthermore, the action of the UPP may be prolonged post-exercise in order to increase the intracellular concentration of amino acids [196], which would stimulate MPS via mTORC1 activation.

Repair Following Exercise-Induced Muscle Damage

Following the orderly demolition of damaged/cleaved muscle proteins (via the aforementioned MPD systems) in response to eccentric contractions, the damaged muscle fibers need to undergo repair. The activation, proliferation and fusion of satellite cells with damaged muscle fibers, and the subsequent differentiation into myoblasts, are crucial for this repair process [34,197]. In fact, it has been shown that while the hypertrophic response may be maintained in the absence of satellite cell recruitment, recovery was severely impaired under these conditions [197].

As previously discussed, IGF-I and MRFs are integral in the activation and differentiation of satellite cells [119,120,162], which probably explains why skeletal muscle IGF-I and MRF expression are increased following stretch-induced damage [34,37,198]. To repair the muscle, satellite cells fuse with damaged fibers and differentiate into myonuclei, but may even form new fibers in the case of complete fiber necrosis. The orderly proliferation and subsequent differentiation is crucial for optimal repair. For the initial proliferation of satellite cells the inflammatory environment is beneficial, but this inflammation must be transient to allow the cells to differentiate [199]. Therefore, it may be that chronic low-grade systemic inflammation, e.g., during aging, may underlie the delay in muscle regeneration [200].

SUMMARY

Skeletal muscle is able to grow, or hypertrophy, in response to a variety of anabolic stimuli, which include resistance exercise, amino acid ingestion, and an increase in IGF-I expression. All these stimuli are able to activate the mTORC1 signaling pathway, which stimulates MPS and inhibits MPD. When the rate of MPS exceeds MPD, there is a positive net protein balance (NPB) and an accretion of contractile material occurs, leading to muscle hypertrophy and an increase in strength. Inducing muscle hypertrophy can have beneficial effects on individuals suffering from cachectic conditions, such as cancer, AIDS, and sarcopenia, where muscle atrophy can have devastating effects on an individual's quality of life. Muscle atrophy occurs when there is a negative NPB, i.e., when the rate of MPD is greater than that of MPS. There are a number of stimuli that have been associated with muscle atrophy, including chronically elevated levels of pro-inflammatory cytokines (e.g., IL-1, IL-6 and TNF-α), a reduction in IGF-I and increased expression of myostatin. Furthermore, strenuous unaccustomed exercise can cause mechanical damage to the muscle, which activates MPD systems, including calpain, inflammation and the ubiquitin–proteasome protein degradation pathway. Damaged proteins within the muscle fiber are broken down, resulting in an increased intracellular amino acid concentration, which in turn activates mTORC1 and increases MPS, thus helping to repair the muscle. Elevated local IGF-I and MRF expression facilitates the repair process by activating satellite cells and enabling fusion with existing fibers. Many of the molecular signaling pathways associated with muscle hypertrophy, atrophy and repair have been identified. However, there is still much to be learned about these pathways, and understanding them may help us to prevent or reverse muscle atrophy associated with a host of muscle wasting conditions.

References

[1] Jones DA, Rutherford OM, Parker DF. Physiological changes in skeletal muscle as a result of strength training. Q J Exp Physiol 1989;74(3):233–56.

[2] Kanehisa H, Ikegawa S, Fukunaga T. Comparison of muscle cross-sectional area and strength between untrained women and men. Eur J Appl Physiol Occup Physiol 1994;68(2):148–54.

[3] Bamman MM, Newcomer BR, Larson-Meyer DE, Weinsier RL, Hunter GR. Evaluation of the strength-size relationship in vivo using various muscle size indices. Med Sci Sports Exerc 2000;32(7):1307–13.

[4] Fukunaga T, Miyatani M, Tachi M, Kouzaki M, Kawakami Y, Kanehisa H. Muscle volume is a major determinant of joint torque in humans. Acta Physiol Scand 2001;172(4):249–55.

[5] Erskine RM, Jones DA, Williams AG, Stewart CE, Degens H. Resistance training increases in vivo quadriceps femoris muscle specific tension in young men. Acta Physiol (Oxf) 2010;199(1):83–9.

[6] Jones DA, Rutherford OM. Human muscle strength training: the effects of three different regimens and the nature of the resultant changes. J Physiol 1987;391(1):1–11.

[7] Narici MV, Hoppeler H, Kayser B, Landoni L, Claassen H, Gavardi C, et al. Human quadriceps cross-sectional area, torque and neural activation during 6 months strength training. Acta Physiol Scand 1996;157(2):175–86.

[8] Friden J, Sjostrom M, Ekblom B. Myofibrillar damage following intense eccentric exercise in man. Int J Sports Med 1983;4(3):170–6.

[9] Lieber RL, Shah S, Friden J. Cytoskeletal disruption after eccentric contraction-induced muscle injury. Clin Orthop Relat Res 2002;(403 Suppl):S90–9.

[10] Belcastro AN, Albisser TA, Littlejohn B. Role of calcium-activated neutral protease (calpain) with diet and exercise. Can J Appl Physiol 1996;21(5):328–46.

[11] Clarkson PM, Sayers SP. Etiology of exercise-induced muscle damage. Can J Appl Physiol 1999;24(3):234–48.

[12] Ebbeling CB, Clarkson PM. Exercise-induced muscle damage and adaptation. Sports Med 1989;7(4):207–34.

[13] MacIntyre DL, Reid WD, McKenzie DC. Delayed muscle soreness. The inflammatory response to muscle injury and its clinical implications. Sports Med 1995;20(1):24–40.

[14] Evans WJ, Cannon JG. The metabolic effects of exercise-induced muscle damage. Exerc Sport Sci Rev 1991;19:99–125.

[15] Delhaas T, van der Meer SFT, Schaart G, Degens H, Drost MR. Steep increase in myonuclear domain size during infancy. Anat Rec 2013;296(2):192–97.

[16] Jacquemin V, Furling D, Bigot A, Butler-Browne GS, Mouly V. IGF-1 induces human myotube hypertrophy by increasing cell recruitment. Exp Cell Res 2004;299(1):148–58.

[17] Chesley A, MacDougall JD, Tarnopolsky MA, Atkinson SA, Smith K. Changes in human muscle protein synthesis after resistance exercise. J Appl Physiol 1992;73(4):1383–8.

[18] Phillips SM, Tipton KD, Aarsland A, Wolf SE, Wolfe RR. Mixed muscle protein synthesis and breakdown after resistance exercise in humans. Am J Physiol 1997;273(1 Pt 1):E99–107.

[19] Kumar V, Atherton P, Smith K, Rennie MJ. Human muscle protein synthesis and breakdown during and after exercise. J Appl Physiol 2009;106(6):2026–39.

[20] Goldspink G. Changes in muscle mass and phenotype and the expression of autocrine and systemic growth factors by muscle in response to stretch and overload. J Anat 1999;194 (Pt 3):323–34.

[21] Stewart CE, Rotwein P. Growth, differentiation, and survival: multiple physiological functions for insulin-like growth factors. Physiol Rev 1996;76(4):1005–26.

[22] Rommel C, Bodine SC, Clarke BA, Rossman R, Nunez L, Stitt TN, et al. Mediation of IGF-1-induced skeletal myotube hypertrophy by PI(3)K/Akt/mTOR and PI(3)K/Akt/GSK3 pathways. Nat Cell Biol 2001;3(11):1009–13.

[23] Bodine SC, Stitt TN, Gonzalez M, Kline WO, Stover GL, Bauerlein R, et al. Akt/mTOR pathway is a crucial regulator of skeletal muscle hypertrophy and can prevent muscle atrophy in vivo. Nat Cell Biol 2001;3(11):1014–9.

[24] DeVol DL, Rotwein P, Sadow JL, Novakofski J, Bechtel PJ. Activation of insulin-like growth factor gene expression during work-induced skeletal muscle growth. Am J Physiol 1990;259(1 Pt 1):E89–95.

[25] Baar K, Esser K. Phosphorylation of p70(S6k) correlates with increased skeletal muscle mass following resistance exercise. Am J Physiol 1999;276(1 Pt 1):C120–7.

[26] Terzis G, Georgiadis G, Stratakos G, Vogiatzis I, Kavouras S, Manta P, et al. Resistance exercise-induced increase in muscle mass correlates with p70S6 kinase phosphorylation in human subjects. Eur J Appl Physiol 2008;102(2):145–52.

[27] Koopman R, Zorenc AH, Gransier RJ, Cameron-Smith D, van Loon LJ. Increase in S6K1 phosphorylation in human skeletal muscle following resistance exercise occurs mainly in type II muscle fibers. Am J Physiol Endocrinol Metab 2006;290(6): E1245–52.

[28] Musaro A, McCullagh K, Paul A, Houghton L, Dobrowolny G, Molinaro M, et al. Localized Igf-1 transgene expression sustains hypertrophy and regeneration in senescent skeletal muscle. Nat Genet 2001;27(2):195–200.

[29] Coleman ME, DeMayo F, Yin KC, Lee HM, Geske R, Montgomery C, et al. Myogenic vector expression of insulin-like growth factor I stimulates muscle cell differentiation and myofiber hypertrophy in transgenic mice. J Biol Chem 1995;270 (20):12109–16.

[30] Ohlsson C, Mohan S, Sjogren K, Tivesten A, Isgaard J, Isaksson O, et al. The role of liver-derived insulin-like growth factor-I. Endocr Rev 2009;30(5):494–535.

[31] Yakar S, Liu JL, Stannard B, Butler A, Accili D, Sauer B, et al. Normal growth and development in the absence of hepatic insulin-like growth factor I. Proc Natl Acad Sci U S A 1999;96 (13):7324–9.

[32] West DW, Kujbida GW, Moore DR, Atherton P, Burd NA, Padzik JP, et al. Resistance exercise-induced increases in putative anabolic hormones do not enhance muscle protein synthesis or intracellular signalling in young men. J Physiol 2009;587(Pt 21):5239–47.

[33] West DW, Burd NA, Tang JE, Moore DR, Staples AW, Holwerda AM, et al. Elevations in ostensibly anabolic hormones with resistance exercise enhance neither training-induced muscle hypertrophy nor strength of the elbow flexors. J Appl Physiol 2010;108(1):60–7.

[34] Petrella JK, Kim J-S, Cross JM, Kosek DJ, Bamman MM. Efficacy of myonuclear addition may explain differential myofiber growth among resistance-trained young and older men and women. Am J Physiol Endocrinol Metab 2006;291(5): E937–46.

[35] Bamman MM, Shipp JR, Jiang J, Gower BA, Hunter GR, Goodman A, et al. Mechanical load increases muscle IGF-I and androgen receptor mRNA concentrations in humans. Am J Physiol Endocrinol Metab 2001;280(3):E383–90.

[36] Hameed M, Orrell RW, Cobbold M, Goldspink G, Harridge SD. Expression of IGF-I splice variants in young and old human skeletal muscle after high resistance exercise. J Physiol 2003;547 (Pt 1):247–54.

[37] Bickel CS, Slade J, Mahoney E, Haddad F, Dudley GA, Adams GR. Time course of molecular responses of human skeletal muscle to acute bouts of resistance exercise. J Appl Physiol 2005;98(2):482–8.

[38] Bickel CS, Slade JM, Haddad F, Adams GR, Dudley GA. Acute molecular responses of skeletal muscle to resistance exercise in able-bodied and spinal cord-injured subjects. J Appl Physiol 2003;94(6):2255–62.

[39] Psilander N, Damsgaard R, Pilegaard H. Resistance exercise alters MRF and IGF-I mRNA content in human skeletal muscle. J Appl Physiol 2003;95(3):1038–44.

[40] Yang S, Alnaqeeb M, Simpson H, Goldspink G. Cloning and characterization of an IGF-1 isoform expressed in skeletal muscle subjected to stretch. J Muscle Res Cell Motil 1996;17 (4):487–95.

[41] McKoy G, Ashley W, Mander J, Yang SY, Williams N, Russell B, et al. Expression of insulin growth factor-1 splice variants and structural genes in rabbit skeletal muscle induced by stretch and stimulation. J Physiol 1999;516(Pt 2):583–92.

[42] Criswell DS, Booth FW, DeMayo F, Schwartz RJ, Gordon SE, Fiorotto ML. Overexpression of IGF-I in skeletal muscle of transgenic mice does not prevent unloading-induced atrophy. Am J Physiol 1998;275(3 Pt 1):E373–9.

[43] Spangenburg EE, Le Roith D, Ward CW, Bodine SC. A functional insulin-like growth factor receptor is not necessary for load-induced skeletal muscle hypertrophy. J Physiol 2008;586 (1):283–91.

[44] Taaffe DR, Jin IH, Vu TH, Hoffman AR, Marcus R. Lack of effect of recombinant human growth hormone (GH) on muscle morphology and GH-insulin-like growth factor expression in resistance-trained elderly men. J Clin Endocrinol Metab 1996;81 (1):421–5.

[45] Hornberger TA, Sukhija KB, Wang XR, Chien S. mTOR is the rapamycin-sensitive kinase that confers mechanically-induced phosphorylation of the hydrophobic motif site Thr(389) in p70 (S6k). FEBS Lett 2007;581(24):4562–6.

[46] Franco Jr. A, Lansman JB. Calcium entry through stretch-inactivated ion channels in mdx myotubes. Nature 1990;344 (6267):670–3.

[47] Franco Jr. A, Lansman JB. Stretch-sensitive channels in developing muscle cells from a mouse cell line. J Physiol 1990;427:361–80.

[48] Guharay F, Sachs F. Stretch-activated single ion channel currents in tissue-cultured embryonic chick skeletal muscle. J Physiol 1984;352:685–701.

[49] Yeung EW, Whitehead NP, Suchyna TM, Gottlieb PA, Sachs F, Allen DG. Effects of stretch-activated channel blockers on [Ca2 +]i and muscle damage in the mdx mouse. J Physiol 2005;562(Pt 2):367–80.

[50] Gulati P, Gaspers LD, Dann SG, Joaquin M, Nobukuni T, Natt F, et al. Amino acids activate mTOR complex 1 via Ca2+/CaM signaling to hVps34. Cell Metab 2008;7(5):456–65.

[51] Kameyama T, Etlinger JD. Calcium-dependent regulation of protein synthesis and degradation in muscle. Nature 1979;279 (5711):344–6.

[52] Spangenburg EE, McBride TA. Inhibition of stretch-activated channels during eccentric muscle contraction attenuates p70S6K activation. J Appl Physiol 2006;100(1):129–35.

[53] Butterfield TA, Best TM. Stretch-activated ion channel blockade attenuates adaptations to eccentric exercise. Med Sci Sports Exerc 2009;41(2):351–6.

[54] Pardo JV, Siliciano JD, Craig SW. Vinculin is a component of an extensive network of myofibril-sarcolemma attachment regions in cardiac muscle fibers. J Cell Biol 1983;97(4):1081–8.

[55] Fluck M, Ziemiecki A, Billeter R, Muntener M. Fibre-type specific concentration of focal adhesion kinase at the sarcolemma: influence of fibre innervation and regeneration. J Exp Biol 2002;205(Pt 16):2337–48.

[56] Patel TJ, Lieber RL. Force transmission in skeletal muscle: from actomyosin to external tendons. Exerc Sport Sci Rev 1997;25:321–63.

[57] Ervasti JM. Costameres: the Achilles' heel of Herculean muscle. J Biol Chem 2003;278(16):13591–4.

[58] Morris EJ, Fulton AB. Rearrangement of mRNAs for costamere proteins during costamere development in cultured skeletal muscle from chicken. J Cell Sci 1994;107(Pt 3):377–86.

[59] Rybakova IN, Patel JR, Ervasti JM. The dystrophin complex forms a mechanically strong link between the sarcolemma and costameric actin. J Cell Biol 2000;150(5):1209–14.

[60] Tidball JG, Spencer MJ, Wehling M, Lavergne E. Nitric-oxide synthase is a mechanical signal transducer that modulates talin and vinculin expression. J Biol Chem 1999;274(46):33155–60.

[61] Woolstenhulme MT, Conlee RK, Drummond MJ, Stites AW, Parcell AC. Temporal response of desmin and dystrophin proteins to progressive resistance exercise in human skeletal muscle. J Appl Physiol 2006;100(6):1876–82.

[62] Kosek DJ, Bamman MM. Modulation of the dystrophin-associated protein complex in response to resistance training in young and older men. J Appl Physiol 2008;104(5):1476–84.

[63] Fluck M, Carson JA, Gordon SE, Ziemiecki A, Booth FW. Focal adhesion proteins FAK and paxillin increase in hypertrophied skeletal muscle. Am J Physiol 1999;277(1 Pt 1):C152–62.

[64] Cary LA, Guan JL. Focal adhesion kinase in integrin-mediated signaling. Front Biosci 1999;4:D102–13.

[65] Durieux AC, D'Antona G, Desplanches D, Freyssenet D, Klossner S, Bottinelli R, et al. Focal adhesion kinase is a load-dependent governor of the slow contractile and oxidative muscle phenotype. J Physiol 2009;587(Pt 14):3703–17.

[66] Klossner S, Durieux AC, Freyssenet D, Flueck M. Mechanotransduction to muscle protein synthesis is modulated by FAK. Eur J Appl Physiol 2009;106(3):389–98.

[67] Gan B, Yoo Y, Guan JL. Association of focal adhesion kinase with tuberous sclerosis complex 2 in the regulation of s6 kinase activation and cell growth. J Biol Chem 2006;281 (49):37321–9.

[68] Malik RK, Parsons JT. Integrin-dependent activation of the p70 ribosomal S6 kinase signaling pathway. J Biol Chem 1996;271 (47):29785–91.

[69] Biolo G, Maggi SP, Williams BD, Tipton KD, Wolfe RR. Increased rates of muscle protein turnover and amino acid transport after resistance exercise in humans. Am J Physiol 1995;268(3 Pt 1):E514–20.

[70] Tang JE, Moore DR, Kujbida GW, Tarnopolsky MA, Phillips SM. Ingestion of whey hydrolysate, casein, or soy protein isolate: effects on mixed muscle protein synthesis at rest and following resistance exercise in young men. J Appl Physiol 2009;107(3):987–92.

[71] Yang Y, Churchward-Venne TA, Burd NA, Breen L, Tarnopolsky MA, Phillips SM. Myofibrillar protein synthesis following ingestion of soy protein isolate at rest and after resistance exercise in elderly men. Nutr Metab (Lond) 2012;9(1):57.

[72] Tipton KD, Ferrando AA, Phillips SM, Doyle Jr. D, Wolfe RR. Postexercise net protein synthesis in human muscle from orally administered amino acids. Am J Physiol 1999;276(4 Pt 1): E628–34.

[73] Biolo G, Tipton KD, Klein S, Wolfe RR. An abundant supply of amino acids enhances the metabolic effect of exercise on muscle protein. Am J Physiol 1997;273(1 Pt 1):E122–9.

[74] Moore DR, Atherton PJ, Rennie MJ, Tarnopolsky MA, Phillips SM. Resistance exercise enhances mTOR and MAPK signalling in human muscle over that seen at rest after bolus protein ingestion. Acta Physiol (Oxf) 2011;201(3):365–72.

[75] Apro W, Blomstrand E. Influence of supplementation with branched-chain amino acids in combination with resistance exercise on p70S6 kinase phosphorylation in resting and exercising human skeletal muscle. Acta Physiol (Oxf) 2010;200 (3):237–48.

[76] Sancak Y, Peterson TR, Shaul YD, Lindquist RA, Thoreen CC, Bar-Peled L, et al. The Rag GTPases bind raptor and mediate amino acid signaling to mTORC1. Science 2008;320 (5882):1496–501.

[77] Sancak Y, Bar-Peled L, Zoncu R, Markhard AL, Nada S, Sabatini DM. Ragulator-Rag complex targets mTORC1 to the lysosomal surface and is necessary for its activation by amino acids. Cell 2010;141(2):290–303.

[78] Kim E, Guan KL. RAG GTPases in nutrient-mediated TOR signaling pathway. Cell Cycle 2009;8(7):1014–8.

[79] Anthony JC, Yoshizawa F, Anthony TG, Vary TC, Jefferson LS, Kimball SR. Leucine stimulates translation initiation in skeletal muscle of postabsorptive rats via a rapamycin-sensitive pathway. J Nutr 2000;130(10):2413–9.

[80] Haegens A, Schols AM, van Essen AL, van Loon LJ, Langen RC. Leucine induces myofibrillar protein accretion in cultured skeletal muscle through mTOR dependent and -independent control of myosin heavy chain mRNA levels. Mol Nutr Food Res 2012;56(5):741–52.

[81] Nair KS, Schwartz RG, Welle S. Leucine as a regulator of whole body and skeletal muscle protein metabolism in humans. Am J Physiol 1992;263(5 Pt 1):E928–34.

[82] Aversa Z, Bonetto A, Costelli P, Minero VG, Penna F, Baccino FM, et al. beta-hydroxy-beta-methylbutyrate (HMB) attenuates muscle and body weight loss in experimental cancer cachexia. Int J Oncol 2011;38(3):713–20.

[83] Pimentel GD, Rosa JC, Lira FS, Zanchi NE, Ropelle ER, Oyama LM, et al. beta-Hydroxy-beta-methylbutyrate (HMbeta) supplementation stimulates skeletal muscle hypertrophy in rats via the mTOR pathway. Nutr Metab (Lond) 2011;8(1):11.

[84] Tipton KD, Rasmussen BB, Miller SL, Wolf SE, Owens-Stovall SK, Petrini BE, et al. Timing of amino acid-carbohydrate ingestion alters anabolic response of muscle to resistance exercise. Am J Physiol Endocrinol Metab 2001;281(2):E197–206.

[85] Moore DR, Robinson MJ, Fry JL, Tang JE, Glover EI, Wilkinson SB, et al. Ingested protein dose response of muscle and albumin protein synthesis after resistance exercise in young men. Am J Clin Nutr 2009;89(1):161–8.

[86] Yang Y, Breen L, Burd NA, Hector AJ, Churchward-Venne TA, Josse AR, et al. Resistance exercise enhances myofibrillar protein synthesis with graded intakes of whey protein in older men. Br J Nutr 2012;:1–9.
[87] Dangin M, Boirie Y, Garcia-Rodenas C, Gachon P, Fauquant J, Callier P, et al. The digestion rate of protein is an independent regulating factor of postprandial protein retention. Am J Physiol Endocrinol Metab 2001;280(2):E340–8.
[88] Dangin M, Guillet C, Garcia-Rodenas C, Gachon P, Bouteloup-Demange C, Reiffers-Magnani K, et al. The rate of protein digestion affects protein gain differently during aging in humans. J Physiol 2003;549(Pt 2):635–44.
[89] Boirie Y, Dangin M, Gachon P, Vasson MP, Maubois JL, Beaufrere B. Slow and fast dietary proteins differently modulate postprandial protein accretion. Proc Natl Acad Sci USA 1997;94(26):14930–5.
[90] Drummond MJ, Rasmussen BB. Leucine-enriched nutrients and the regulation of mammalian target of rapamycin signalling and human skeletal muscle protein synthesis. Curr Opin Clin Nutr Metab Care 2008;11(3):222–6.
[91] Hartman JW, Tang JE, Wilkinson SB, Tarnopolsky MA, Lawrence RL, Fullerton AV, et al. Consumption of fat-free fluid milk after resistance exercise promotes greater lean mass accretion than does consumption of soy or carbohydrate in young, novice, male weightlifters. Am J Clin Nutr 2007;86 (2):373–81.
[92] Erskine RM, Fletcher G, Hanson B, Folland JP. Whey protein does not enhance the adaptations to elbow flexor resistance training. Med Sci Sports Exerc 2012;44(9):1791–800.
[93] Verdijk LB, Jonkers RA, Gleeson BG, Beelen M, Meijer K, Savelberg HH, et al. Protein supplementation before and after exercise does not further augment skeletal muscle hypertrophy after resistance training in elderly men. Am J Clin Nutr 2009;89(2):608–16.
[94] Candow DG, Little JP, Chilibeck PD, Abeysekara S, Zello GA, Kazachkov M, et al. Low-dose creatine combined with protein during resistance training in older men. Med Sci Sports Exerc 2008;40(9):1645–52.
[95] Erskine RM, Jones DA, Williams AG, Stewart CE, Degens H. Inter-individual variability in the adaptation of human muscle specific tension to progressive resistance training. Eur J Appl Physiol 2010;110(6):1117–25.
[96] Hubal MJ, Gordish-Dressman H, Thompson PD, Price TB, Hoffman EP, Angelopoulos TJ, et al. Variability in muscle size and strength gain after unilateral resistance training. Med Sci Sports Exerc 2005;37(6):964–72.
[97] Godard MP, Williamson DL, Trappe SW. Oral amino-acid provision does not affect muscle strength or size gains in older men. Med Sci Sports Exerc 2002;34(7):1126–31.
[98] Willoughby DS, Stout JR, Wilborn CD. Effects of resistance training and protein plus amino acid supplementation on muscle anabolism, mass, and strength. Amino Acids 2007;32 (4):467–77.
[99] Hulmi JJ, Kovanen V, Selanne H, Kraemer WJ, Hakkinen K, Mero AA. Acute and long-term effects of resistance exercise with or without protein ingestion on muscle hypertrophy and gene expression. Amino Acids 2009;37(2):297–308.
[100] Vieillevoye S, Poortmans JR, Duchateau J, Carpentier A. Effects of a combined essential amino acids/carbohydrate supplementation on muscle mass, architecture and maximal strength following heavy-load training. Eur J Appl Physiol 2010;110(3):479–88.
[101] Holm L, Olesen JL, Matsumoto K, Doi T, Mizuno M, Alsted TJ, et al. Protein-containing nutrient supplementation following strength training enhances the effect on muscle mass, strength, and bone formation in postmenopausal women. J Appl Physiol 2008;105(1):274–81.
[102] Coburn JW, Housh DJ, Housh TJ, Malek MH, Beck TW, Cramer JT, et al. Effects of leucine and whey protein supplementation during eight weeks of unilateral resistance training. J Strength Cond Res 2006;20(2):284–91.
[103] Hartman JW, Moore DR, Phillips SM. Resistance training reduces whole-body protein turnover and improves net protein retention in untrained young males. Appl Physiol Nutr Metab 2006;31(5):557–64.
[104] Phillips SM, Tipton KD, Ferrando AA, Wolfe RR. Resistance training reduces the acute exercise-induced increase in muscle protein turnover. Am J Physiol 1999;276(1 Pt 1):E118–24.
[105] Evans WJ. What is sarcopenia? J Gerontol A Biol Sci Med Sci 1995;50(Spec No):5–8.
[106] Degens H. The role of systemic inflammation in age-related muscle weakness and wasting. Scand J Med Sci Sports 2010;20 (1):28–38.
[107] Lexell J, Taylor CC, Sjostrom M. What is the cause of the ageing atrophy? Total number, size and proportion of different fiber types studied in whole vastus lateralis muscle from 15- to 83-year-old men. J Neurol Sci 1988;84(2–3):275–94.
[108] Larsson L, Sjodin B, Karlsson J. Histochemical and biochemical changes in human skeletal muscle with age in sedentary males, age 22–65 years. Acta Physiol Scand 1978;103 (1):31–9.
[109] Larsson L. Histochemical characteristics of human skeletal muscle during aging. Acta Physiol Scand 1983;117(3):469–71.
[110] Jakobsson F, Borg K, Edstrom L, Grimby L. Use of motor units in relation to muscle fiber type and size in man. Muscle Nerve 1988;11(12):1211–8.
[111] Brooks SV, Faulkner JA. Skeletal muscle weakness in old age: underlying mechanisms. Med Sci Sports Exerc 1994;26 (4):432–9.
[112] Faulkner JA, Larkin LM, Claflin DR, Brooks SV. Age-related changes in the structure and function of skeletal muscles. Clin Exp Pharmacol Physiol 2007;34(11):1091–6.
[113] Franceschi C, Bonafe M, Valensin S, Olivieri F, De Luca M, Ottaviani E, et al. Inflamm-aging. An evolutionary perspective on immunosenescence. Ann N Y Acad Sci 2000;908:244–54.
[114] Bartlett DB, Firth CM, Phillips AC, Moss P, Baylis D, Syddall H, et al. The age-related increase in low-grade systemic inflammation (Inflammaging) is not driven by cytomegalovirus infection. Aging Cell 2012;11(5):912–5.
[115] Visser M, Pahor M, Taaffe DR, Goodpaster BH, Simonsick EM, Newman AB, et al. Relationship of interleukin-6 and tumor necrosis factor-alpha with muscle mass and muscle strength in elderly men and women: the Health ABC Study. J Gerontol A Biol Sci Med Sci 2002;57(5):M326–32.
[116] Lio D, Scola L, Crivello A, Colonna-Romano G, Candore G, Bonafe M, et al. Gender-specific association between -1082 IL-10 promoter polymorphism and longevity. Genes Immun 2002;3(1):30–3.
[117] Pedersen BK. The diseasome of physical inactivity–and the role of myokines in muscle–fat cross talk. J Physiol 2009;587 (Pt 23):5559–68.
[118] Li YP, Schwartz RJ, Waddell ID, Holloway BR, Reid MB. Skeletal muscle myocytes undergo protein loss and reactive oxygen-mediated NF-kappaB activation in response to tumor necrosis factor alpha. Faseb J 1998;12(10):871–80.
[119] Szalay K, Razga Z, Duda E. TNF inhibits myogenesis and downregulates the expression of myogenic regulatory factors myoD and myogenin. Eur J Cell Biol 1997;74(4):391–8.
[120] Langen RC, Van Der Velden JL, Schols AM, Kelders MC, Wouters EF, Janssen-Heininger YM. Tumor necrosis factor-

alpha inhibits myogenic differentiation through MyoD protein destabilization. Faseb J 2004;18(2):227–37.

[121] Guttridge DC, Mayo MW, Madrid LV, Wang CY, Baldwin Jr. AS. NF-kappaB-induced loss of MyoD messenger RNA: possible role in muscle decay and cachexia. Science 2000;289 (5488):2363–6.

[122] Reid MB, Li YP. Tumor necrosis factor-alpha and muscle wasting: a cellular perspective. Respir Res 2001;2(5):269–72.

[123] Saini A, Al-Shanti N, Faulkner SH, Stewart CE. Pro- and anti-apoptotic roles for IGF-I in TNF-alpha-induced apoptosis: a MAP kinase mediated mechanism. Growth Factors 2008;26 (5):239–53.

[124] Welle S, Totterman S, Thornton C. Effect of age on muscle hypertrophy induced by resistance training. J Gerontol A Biol Sci Med Sci 1996;51(6):M270–5.

[125] Alway SE, Degens H, Krishnamurthy G, Smith CA. Potential role for Id myogenic repressors in apoptosis and attenuation of hypertrophy in muscles of aged rats. Am J Physiol Cell Physiol 2002;283(1):C66–76.

[126] Farges MC, Balcerzak D, Fisher BD, Attaix D, Bechet D, Ferrara M, et al. Increased muscle proteolysis after local trauma mainly reflects macrophage-associated lysosomal proteolysis. Am J Physiol Endocrinol Metab 2002;282(2):E326–35.

[127] Kasperek GJ, Snider RD. Increased protein degradation after eccentric exercise. Eur J Appl Physiol Occup Physiol 1985;54 (1):30–4.

[128] Friden J, Kjorell U, Thornell LE. Delayed muscle soreness and cytoskeletal alterations: an immunocytological study in man. Int J Sports Med 1984;5(1):15–8.

[129] Solomon V, Baracos V, Sarraf P, Goldberg AL. Rates of ubiquitin conjugation increase when muscles atrophy, largely through activation of the N-end rule pathway. Proc Natl Acad Sci U S A 1998;95(21):12602–7.

[130] Du J, Wang X, Miereles C, Bailey JL, Debigare R, Zheng B, et al. Activation of caspase-3 is an initial step triggering accelerated muscle proteolysis in catabolic conditions. J Clin Invest 2004;113(1):115–23.

[131] Lee SW, Dai G, Hu Z, Wang X, Du J, Mitch WE. Regulation of muscle protein degradation: coordinated control of apoptotic and ubiquitin-proteasome systems by phosphatidylinositol 3 kinase. J Am Soc Nephrol 2004;15(6):1537–45.

[132] Ottenheijm CA, Heunks LM, Li YP, Jin B, Minnaard R, van Hees HW, et al. Activation of the ubiquitin-proteasome pathway in the diaphragm in chronic obstructive pulmonary disease. Am J Respir Crit Care Med 2006;174(9):997–1002.

[133] DeMartino GN, Ordway GA. Ubiquitin-proteasome pathway of intracellular protein degradation: implications for muscle atrophy during unloading. Exerc Sport Sci Rev 1998;26:219–52.

[134] Gomes MD, Lecker SH, Jagoe RT, Navon A, Goldberg AL. Atrogin-1, a muscle-specific F-box protein highly expressed during muscle atrophy. Proc Natl Acad Sci U S A 2001;98 (25):14440–5.

[135] Lecker SH, Jagoe RT, Gilbert A, Gomes M, Baracos V, Bailey J, et al. Multiple types of skeletal muscle atrophy involve a common program of changes in gene expression. Faseb J 2004;18 (1):39–51.

[136] McPherron AC, Lawler AM, Lee SJ. Regulation of skeletal muscle mass in mice by a new TGF-beta superfamily member. Nature 1997;387(6628):83–90.

[137] Zimmers TA, Davies MV, Koniaris LG, Haynes P, Esquela AF, Tomkinson KN, et al. Induction of cachexia in mice by systemically administered myostatin. Science 2002;296(5572):1486–8.

[138] McPherron AC, Lee SJ. Double muscling in cattle due to mutations in the myostatin gene. Proc Natl Acad Sci U S A 1997;94 (23):12457–61.

[139] Schuelke M, Wagner KR, Stolz LE, Hubner C, Riebel T, Komen W, et al. Myostatin mutation associated with gross muscle hypertrophy in a child. N Engl J Med 2004;350 (26):2682–8.

[140] Sartori R, Milan G, Patron M, Mammucari C, Blaauw B, Abraham R, et al. Smad2 and 3 transcription factors control muscle mass in adulthood. Am J Physiol Cell Physiol 2009;296 (6):C1248–57.

[141] Rodino-Klapac LR, Haidet AM, Kota J, Handy C, Kaspar BK, Mendell JR. Inhibition of myostatin with emphasis on follistatin as a therapy for muscle disease. Muscle Nerve 2009;39 (3):283–96.

[142] Langley B, Thomas M, Bishop A, Sharma M, Gilmour S, Kambadur R. Myostatin inhibits myoblast differentiation by down-regulating MyoD expression. J Biol Chem 2002;277 (51):49831–40.

[143] Trendelenburg AU, Meyer A, Rohner D, Boyle J, Hatakeyama S, Glass DJ. Myostatin reduces Akt/TORC1/p70S6K signaling, inhibiting myoblast differentiation and myotube size. Am J Physiol Cell Physiol 2009;296(6):C1258–70.

[144] Amirouche A, Durieux AC, Banzet S, Koulmann N, Bonnefoy R, Mouret C, et al. Down-regulation of Akt/mammalian target of rapamycin signaling pathway in response to myostatin overexpression in skeletal muscle. Endocrinology 2009;150 (1):286–94.

[145] Welle S, Burgess K, Mehta S. Stimulation of skeletal muscle myofibrillar protein synthesis, p70 S6 kinase phosphorylation, and ribosomal protein S6 phosphorylation by inhibition of myostatin in mature mice. Am J Physiol Endocrinol Metab 2009;296(3):E567–72.

[146] Huang Z, Chen D, Zhang K, Yu B, Chen X, Meng J. Regulation of myostatin signaling by c-Jun N-terminal kinase in C2C12 cells. Cell Signal 2007;19(11):2286–95.

[147] Philip B, Lu Z, Gao Y. Regulation of GDF-8 signaling by the p38 MAPK. Cell Signal 2005;17(3):365–75.

[148] Yang W, Chen Y, Zhang Y, Wang X, Yang N, Zhu D. Extracellular signal-regulated kinase 1/2 mitogen-activated protein kinase pathway is involved in myostatin-regulated differentiation repression. Cancer Res 2006;66(3):1320–6.

[149] McFarlane C, Plummer E, Thomas M, Hennebry A, Ashby M, Ling N, et al. Myostatin induces cachexia by activating the ubiquitin proteolytic system through an NF-kappaB-independent, FoxO1-dependent mechanism. J Cell Physiol 2006;209 (2):501–14.

[150] Siriett V, Platt L, Salerno MS, Ling N, Kambadur R, Sharma M. Prolonged absence of myostatin reduces sarcopenia. J Cell Physiol 2006;209(3):866–73.

[151] Reeves ND, Narici MV, Maganaris CN. Effect of resistance training on skeletal muscle-specific force in elderly humans. J Appl Physiol 2004;96(3):885–92.

[152] Harridge SD, Kryger A, Stensgaard A. Knee extensor strength, activation, and size in very elderly people following strength training. Muscle Nerve 1999;22(7):831–9.

[153] Fiatarone MA, O'Neill EF, Ryan ND, Clements KM, Solares GR, Nelson ME, et al. Exercise training and nutritional supplementation for physical frailty in very elderly people. N Engl J Med 1994;330(25):1769–75.

[154] Kimball SR, O'Malley JP, Anthony JC, Crozier SJ, Jefferson LS. Assessment of biomarkers of protein anabolism in skeletal muscle during the life span of the rat: sarcopenia despite elevated protein synthesis. Am J Physiol Endocrinol Metab 2004;287(4):E772–80.

[155] Mascher H, Tannerstedt J, Brink-Elfegoun T, Ekblom B, Gustafsson T, Blomstrand E. Repeated resistance exercise training induces different changes in mRNA expression of

MAFbx and MuRF-1 in human skeletal muscle. Am J Physiol Endocrinol Metab 2008;294(1):E43–51.

[156] Jones SW, Hill RJ, Krasney PA, O'Conner B, Peirce N, Greenhaff PL. Disuse atrophy and exercise rehabilitation in humans profoundly affects the expression of genes associated with the regulation of skeletal muscle mass. Faseb J 2004;18 (9):1025–7.

[157] Sandri M, Sandri C, Gilbert A, Skurk C, Calabria E, Picard A, et al. Foxo transcription factors induce the atrophy-related ubiquitin ligase atrogin-1 and cause skeletal muscle atrophy. Cell 2004;117(3):399–412.

[158] Hornberger TA, Mateja RD, Chin ER, Andrews JL, Esser KA. Aging does not alter the mechanosensitivity of the p38, p70S6k, and JNK2 signaling pathways in skeletal muscle. J Appl Physiol 2005;98(4):1562–6.

[159] Thomson DM, Gordon SE. Impaired overload-induced muscle growth is associated with diminished translational signalling in aged rat fast-twitch skeletal muscle. J Physiol 2006;574 (Pt 1):291–305.

[160] Shefer G, Van de Mark DP, Richardson JB, Yablonka-Reuveni Z. Satellite-cell pool size does matter: defining the myogenic potency of aging skeletal muscle. Dev Biol 2006;294(1):50–66.

[161] Verdijk LB, Koopman R, Schaart G, Meijer K, Savelberg HH, van Loon LJ. Satellite cell content is specifically reduced in type II skeletal muscle fibers in the elderly. Am J Physiol Endocrinol Metab 2007;292(1):E151–7.

[162] Scime A, Rudnicki MA. Anabolic potential and regulation of the skeletal muscle satellite cell populations. Curr Opin Clin Nutr Metab Care 2006;9(3):214–9.

[163] Edstrom E, Ulfhake B. Sarcopenia is not due to lack of regenerative drive in senescent skeletal muscle. Aging Cell 2005;4 (2):65–77.

[164] Alway SE, Degens H, Lowe DA, Krishnamurthy G. Increased myogenic repressor Id mRNA and protein levels in hindlimb muscles of aged rats. Am J Physiol Regul Integr Comp Physiol 2002;282(2):R411–22.

[165] Verdijk LB, Gleeson BG, Jonkers RA, Meijer K, Savelberg HH, Dendale P, et al. Skeletal muscle hypertrophy following resistance training is accompanied by a fiber type-specific increase in satellite cell content in elderly men. J Gerontol A Biol Sci Med Sci 2009;64(3):332–9.

[166] Mackey AL, Esmarck B, Kadi F, Koskinen SO, Kongsgaard M, Sylvestersen A, et al. Enhanced satellite cell proliferation with resistance training in elderly men and women. Scand J Med Sci Sports 2007;17(1):34–42.

[167] Volpi E, Mittendorfer B, Rasmussen BB, Wolfe RR. The response of muscle protein anabolism to combined hyperaminoacidemia and glucose-induced hyperinsulinemia is impaired in the elderly. J Clin Endocrinol Metab 2000;85 (12):4481–90.

[168] Eley HL, Russell ST, Tisdale MJ. Effect of branched-chain amino acids on muscle atrophy in cancer cachexia. Biochem J 2007;407(1):113–20.

[169] Eley HL, Russell ST, Baxter JH, Mukerji P, Tisdale MJ. Signaling pathways initiated by beta-hydroxy-beta-methylbutyrate to attenuate the depression of protein synthesis in skeletal muscle in response to cachectic stimuli. Am J Physiol Endocrinol Metab 2007;293(4):E923–31.

[170] Bhasin S, Calof OM, Storer TW, Lee ML, Mazer NA, Jasuja R, et al. Drug insight: Testosterone and selective androgen receptor modulators as anabolic therapies for chronic illness and aging. Nat Clin Pract Endocrinol Metab 2006;2(3):146–59.

[171] Kovacheva EL, Hikim AP, Shen R, Sinha I, Sinha-Hikim I. Testosterone supplementation reverses sarcopenia in aging through regulation of myostatin, c-Jun NH2-terminal kinase, Notch, and Akt signaling pathways. Endocrinology 2010;151 (2):628–38.

[172] Siriett V, Salerno MS, Berry C, Nicholas G, Bower R, Kambadur R, et al. Antagonism of myostatin enhances muscle regeneration during sarcopenia. Mol Ther 2007;15(8):1463–70.

[173] Clarkson PM, Hubal MJ. Exercise-induced muscle damage in humans. Am J Phys Med Rehabil 2002;81(11 Suppl):S52–69.

[174] Enoka RM. Eccentric contractions require unique activation strategies by the nervous system. J Appl Physiol 1996;81 (6):2339–46.

[175] Lieber RL, Friden J. Muscle damage is not a function of muscle force but active muscle strain. J Appl Physiol 1993;74 (2):520–6.

[176] Lieber RL, Friden J. Mechanisms of muscle injury after eccentric contraction. J Sci Med Sport 1999;2(3):253–65.

[177] Arthur GD, Booker TS, Belcastro AN. Exercise promotes a subcellular redistribution of calcium-stimulated protease activity in striated muscle. Can J Physiol Pharmacol 1999;77 (1):42–7.

[178] Stupka N, Tarnopolsky MA, Yardley NJ, Phillips SM. Cellular adaptation to repeated eccentric exercise-induced muscle damage. J Appl Physiol 2001;91(4):1669–78.

[179] Ordway GA, Neufer PD, Chin ER, DeMartino GN. Chronic contractile activity upregulates the proteasome system in rabbit skeletal muscle. J Appl Physiol 2000;88(3):1134–41.

[180] Belcastro AN, Shewchuk LD, Raj DA. Exercise-induced muscle injury: a calpain hypothesis. Mol Cell Biochem 1998;179 (1–2):135–45.

[181] Ochi E, Hirose T, Hiranuma K, Min SK, Ishii N, Nakazato K. Elevation of myostatin and FOXOs in prolonged muscular impairment induced by eccentric contractions in rat medial gastrocnemius muscle. J Appl Physiol 2010;108(2):306–13.

[182] Suzuki K, Sorimachi H, Yoshizawa T, Kinbara K, Ishiura S. Calpain: novel family members, activation, and physiologic function. Biol Chem Hoppe Seyler 1995;376(9):523–9.

[183] Goll DE, Dayton WR, Singh I, Robson RM. Studies of the alpha-actinin/actin interaction in the Z-disk by using calpain. J Biol Chem 1991;266(13):8501–10.

[184] Warren GL, Hayes DA, Lowe DA, Armstrong RB. Mechanical factors in the initiation of eccentric contraction-induced injury in rat soleus muscle. J Physiol 1993;464:457–75.

[185] Raj DA, Booker TS, Belcastro AN. Striated muscle calcium-stimulated cysteine protease (calpain-like) activity promotes myeloperoxidase activity with exercise. Pflugers Arch 1998;435 (6):804–9.

[186] Friden J. Muscle soreness after exercise: implications of morphological changes. Int J Sports Med 1984;5(2):57–66.

[187] Chen SE, Jin B, Li YP. TNF-alpha regulates myogenesis and muscle regeneration by activating p38 MAPK. Am J Physiol Cell Physiol 2007;292(5):C1660–71.

[188] Attaix D, Aurousseau E, Combaret L, Kee A, Larbaud D, Ralliere C, et al. Ubiquitin-proteasome-dependent proteolysis in skeletal muscle. Reprod Nutr Dev 1998;38(2):153–65.

[189] Attaix D, Combaret L, Pouch MN, Taillandier D. Regulation of proteolysis. Curr Opin Clin Nutr Metab Care 2001;4(1):45–9.

[190] Ciechanover A. The ubiquitin-proteasome proteolytic pathway. Cell 1994;79(1):13–21.

[191] Thompson HS, Scordilis SP. Ubiquitin changes in human biceps muscle following exercise-induced damage. Biochem Biophys Res Commun 1994;204(3):1193–8.

[192] Milacic V, Dou QP. The tumor proteasome as a novel target for gold(III) complexes: implications for breast cancer therapy. Coord Chem Rev 2009;253(11–12):1649–60.

[193] Hershko A, Ciechanover A. The ubiquitin system. Annu Rev Biochem 1998;67:425–79.

[194] Riley DA, Ellis S, Giometti CS, Hoh JF, Ilyina-Kakueva EI, Oganov VS, et al. Muscle sarcomere lesions and thrombosis after spaceflight and suspension unloading. J Appl Physiol 1992;73(2 Suppl):33S–43S.

[195] Feasson L, Stockholm D, Freyssenet D, Richard I, Duguez S, Beckmann JS, et al. Molecular adaptations of neuromuscular disease-associated proteins in response to eccentric exercise in human skeletal muscle. J Physiol 2002;543(Pt 1):297–306.

[196] Tipton KD, Wolfe RR. Exercise-induced changes in protein metabolism. Acta Physiol Scand 1998;162(3):377–87.

[197] McCarthy JJ, Mula J, Miyazaki M, Erfani R, Garrison K, Farooqui AB, et al. Effective fiber hypertrophy in satellite cell-depleted skeletal muscle. Development 2011;138(17):3657–66.

[198] Yang H, Alnaqeeb M, Simpson H, Goldspink G. Changes in muscle fibre type, muscle mass and IGF-I gene expression in rabbit skeletal muscle subjected to stretch. J Anat 1997;190(Pt 4):613–22.

[199] Pelosi L, Giacinti C, Nardis C, Borsellino G, Rizzuto E, Nicoletti C, et al. Local expression of IGF-1 accelerates muscle regeneration by rapidly modulating inflammatory cytokines and chemokines. Faseb J. 2007;21(7):1393–402.

[200] Langen RC, Schols AM, Kelders MC, van der Velden JL, Wouters EF, Janssen-Heininger YM. Muscle wasting and impaired muscle regeneration in a murine model of chronic pulmonary inflammation. Am J Respir Cell Mol Biol 2006;35 (6):689–96.

CHAPTER

26

Nitric Oxide, Sports Nutrition and Muscle Building

Lawrence J. Druhan

The Levine Cancer Institute, Carolinas Healthcare System, Charlotte, NC, USA

INTRODUCTION

Nitric oxide (NO) is a naturally occurring diatomous free radical, consisting of one atom of nitrogen and one atom of oxygen. In the atmosphere, NO is formed by combustion, and is typically thought of as a noxious substance, being involved in ozone layer depletion, the formation of acid rain, and air pollution. Conversely, organic synthesis of NO is a critical component in many physiological processes. Prokaryotic synthesis of NO is part of the nitrogen cycle, during which atmospheric nitrogen is reduced to ammonium which can then be used for the production of the nitrogen-containing molecules, DNA, RNA and proteins, necessary for life. During the denitrification step of the nitrogen cycle, NO is generated from the nitrogen oxides produced by living organisms as a step in the completion of the nitrogen cycle.

The physiological function of NO was unknown prior to the landmark discoveries leading to the 1998 Nobel Prize in medicine. While investigating the mechanism of action of organic nitrates, which were known to induce relaxation of blood vessels, it was found that these drugs functioned by generating NO and that the generated NO increased the level of cyclic GMP in the vascular smooth muscle cells [1,2]. Subsequently, it was shown that a substance produced within the endothelium, termed endothelial derived relaxation factor (EDRF), activated the generation of the cGMP within the smooth muscle cells [3], and it was speculated that EDRF was NO. Subsequent experiments demonstrated that EDRF was indeed NO [4,5] and that an enzyme within the endothelium generated the NO from arginine [6,7]. These ground-breaking studies ushered in the era of NO signaling.

As a biological signaling molecule NO has many advantages: it is freely diffusible through biologic membranes and it has a short half life. Since the initial observation that NO is responsible for signaling between the endothelium and the vascular smooth muscle to induce vessel relaxation, NO signaling has been demonstrated in an extremely large range of physiological process; indeed every major organ system in the human body is either directly or indirectly affected by NO signaling. As such, it is not surprising that NO signaling has also been found to be involved in the physiology of skeletal muscle: in function, growth and repair. The following discussion will be directed toward the molecular mechanisms governed by NO in skeletal muscle, with an emphasis on potential nutritional modifications to augment the natural signaling processes and with an eye toward how these dietary modifications could and/or do affect muscle growth.

THE NITRIC OXIDE SYNTHASES

Enzymatic production of nitric oxide in humans is accomplished by the nitric oxide synthase family of enzymes. The three members of this family are neuronal NOS (NOS1 or nNOS), inducible NOS (NOS2 or iNOS) and endothelial NOS (NOS3 or eNOS). The conventional naming of the NOS enzymes implies cellular specificity; however, these enzymes were named according to the tissues/cells from which they were originally identified, and each of these enzymes has since been found in a much wider range of cells/tissues. The NOS enzymes all catalyze the same reaction: the conversion of arginine and oxygen to citrulline and NO using NADPH as the source for reducing equivalents. Additionally, the NOS

Nutrition and Enhanced Sports Performance.
DOI: http://dx.doi.org/10.1016/B978-0-12-396454-0.00026-6

enzymes share the same general structure, and utilize the same cofactors; however, they each have been found to elicit unique physiological functions. The production of varied function from the enzymatic generation of an identical product is possible because each of the three NOS enzymes differ in cellular and subcellular location and in regulation of activity [8]. Additionally, the activities of both nNOS and eNOS are positively regulated by increases in cellular calcium, whereas iNOS activity is calcium independent. Moreover, these enzymes are regulated by posttranslational modifications, including phosphorylation, acylation, and glutathionylation [9–12].

All three NOS isoforms have been identified in skeletal muscle. The principal NOS in skeletal muscle is nNOS [13]. Skeletal muscle contains an alternatively spliced variant of nNOS, nNOSμ. This splice variant is found in the cytoplasm and found localized to the membrane of the skeletal muscle via an association with the dystrophin protein complex [14], and mutations that cause human muscular dystrophy have been found to down-regulate nNOS activity and reduce muscle blood flow [15]. nNOSμ has been found to regulate skeletal muscle blood flow, and its activity can attenuate the activation of vasoconstriction to ensure adequate blood flow to the muscle during exercise [16,17]. Another nNOS splice variant, nNOSβ, has been found to be localized to the Golgi [18]. nNOSβ was shown to affect skeletal muscle force production. Expression of eNOS has been identified in skeletal muscle; however, the expression level is low. eNOS has been found to be associated with the vascular endothelium in skeletal muscle and with skeletal muscle mitochondria [19,20]. Although iNOS is generally not expressed in healthy human adult skeletal muscle, iNOS expression can be increased under some conditions [21].

Exercise has been found to both induce NO production and alter the expression of the various NOS isoforms. In humans the expression of nNOSμ has been shown to be increased by exercise [22]. In this study the authors demonstrated that nNOSμ was higher in endurance trained athletes compared with sedentary controls. Moreover, they showed that in sedentary subjects that underwent an exercise regimen, nNOSμ protein level was significantly increased. There was no detectable change in either eNOS or iNOS. Therefore, it is clear that the NOS family of enzymes is critical to the function of skeletal muscle, both at rest and during exercise.

THE NITRATE–NITRITE–NO PATHWAY

Nitrite and nitrate are formed in the body via the oxidation of NOS-derived NO. Additionally, these inorganic nitrogenous compounds are natural components of many healthy foods as well as being used as preservatives in many processed foods. Excess exposure to these compounds has been identified as carcinogenic [23,24]; however, there is increasing evidence that nitrite and nitrate are not just toxic byproducts but rather part of an intrinsic NO-recycling system that has physiologically relevant activity [25]. In addition to the recycling of NOS-derived NO, this pathway can directly convert inorganic nitrite and nitrate to NO in a NOS-independent fashion.

The NOS-independent generation of NO (and indeed the recycling of NOS-derived NO) functions through the action of an entero-salivary system. In this system, dietary nitrate is absorbed in the gastrointestinal tract and transported to the saliva glands via the circulatory system. Bacteria in the saliva glands convert the nitrate to nitrite via the action of nitrate reductase enzymes. The nitrite rich saliva is then swallowed and either converted to NO in the stomach or absorbed in the GI tract and passed into the circulation. The nitrite in circulation can be converted to NO via enzymatic nitrite reduction in a number of different tissues, including skeletal muscle [26,27]. Thus, once in circulation, nitrite can act as an endocrine-like carrier of NO activity.

The tissue metabolism of nitrite has been shown to be oxygen and pH dependent [26,28]. When the tissue oxygen concentration (or pH) is high, nitrite is preferentially oxidized to nitrate; however, at low oxygen tension, nitrite is converted to NO. Thus the nitrite-dependent generation of NO in skeletal muscle could be dependent upon the metabolic state of the tissue: under resting conditions the nitrite would be converted to nitrate and recycled whereas under strenuous exercise, when oxygen is limited (and pH is decreased), nitrite would be converted to NO. Somewhat surprisingly, however, dietary nitrate has been shown to result in lower oxygen demand during submaximal exercise [29]. Thus it seems that, even when tissue oxygen levels are not limiting, and the tissue pH is not significantly altered, nitrite can still affect skeletal muscle function.

SKELETAL MUSCLE FUNCTIONS MEDIATED BY NO

The cardiovascular function of NO in humans has been well characterized [30,31]. However, NO has also been found to affect a wide range of functions within skeletal muscle, including the regulation of muscle function, metabolism, and growth. It has been shown that NO regulates muscle force production [32]. In activated rat diaphragm, inhibition of NOS decreased maximum velocity of shortening, loaded shortening velocity, and power production. These affects were reversed by addition of an inorganic NO donor; however the NO donor had no affect alone. Thus, the endogenous generation of NO from NOS regulates muscle contraction. In regard to NO regulation of skeletal muscle metabolism, both oxygen consumption and glucose uptake are known to be regulated by NO [33–35], although, during exercise, it seems that the prevailing mechanism by which NO regulates skeletal muscle metabolism is via glucose homeostasis [35]. While these mechanisms could indirectly affect muscle growth, NO is also involved in the regulation of processes that are directly involved. These processes include transcription of skeletal muscle proteins, activation/proliferation of muscle satellite cells, and skeletal muscle blood flow.

Transcription of Skeletal Muscle Proteins and Activation of Satellite Cells

Muscle growth during exercise has been mainly attributed to increased protein transcription in the myofibrils and to the activity of satellite cells. Nitric oxide has been shown to be directly involved in muscle hypertrophy and fiber type transition observed in chronically loaded muscle in a rodent model [36]. Although this work did not examine the involvement of satellite cells, it was shown that load-induced protein synthesis (as indicated by muscle mass) was inhibited in animals treated with a NOS inhibitor, indicating that NO plays a role in muscle protein synthesis. These authors went on to demonstrate that this NOS-dependence of the observed muscle hypertrophy was not due to an alteration of growth factor induction or activation/proliferation of satellite cells [37]. Rather, the NOS-dependent component of the observed load-induced hypertrophy was due to an increase in mRNA for both actin and type 1 myosin heavy chain, indicating that endogenous production of NO was involved in the increase in synthesis of contractile proteins. Atrophy associated with functional unloading of muscle has been shown to be attenuated by the administration of arginine [38]. This study demonstrated that the decrease in myosin heavy chain type 1 associated with muscle unloading was diminished by the addition of arginine; however, this is likely a combination of alterations in protein synthesis and protein degradation found in unload-induced muscle atrophy [38,39].

Satellite cells are myocyte stem cells with the ability to generate new muscle fibers and to provide new myonuclei during muscle growth [40]. Moreover, increases in satellite cells have been observed in humans undergoing resistance training [41]. Thus, both myofibril protein synthesis and satellite cell number/activation are important for muscle growth not only in pathological settings (such as muscular dystrophy) but also in healthy subjects [42,43]. Nitric oxide has been shown to both activate satellite cells and to regulate protein synthesis. NO has been shown to stimulate the proliferation of satellite cells and to maintain the reserve pool of them [44]. Additionally, NO has been shown to activate satellite cells to enter the cell cycle for the production of new muscle [45,46]. These results (among others) have led to the hypothesis that therapies to increase muscle NO would be beneficial in pathophysiological settings that would benefit from increased muscle repair/growth, such as the muscular dystrophies [42].

Increase in muscle mass during resistance training has been shown to involve a number of mechanisms [47], including the NO-dependent mechanisms described above. Moreover, NO is known to be produced in working muscle [48]. Taken together, and with the above studies, these data indicate that it is plausible that increasing NO during resistance training by nutritional manipulation would increase the muscle growth response via stimulation of satellite cell proliferation/activation and/or by increasing protein synthesis in the myofibril. However, there are no studies to date directly examining this possibility in humans.

Mitochondrial Biogenesis

Mitochondrial biogenesis is necessary both during mitosis and to respond to the energy needs of developed cells. As such, this process is necessary for muscle growth. The biogenesis of mitochondria is a complex process involving the synthesis of mitochondrial proteins and DNA, the synthesis and import of nuclear encoded proteins, and the synthesis and import of lipids. This process is governed by a large and complex system involving many cellular proteins and transcription factors [49]. In skeletal muscle the process of mitochondrial biogenesis is initiated when the energy demand of the myocyte exceeds the capacity of the resident mitochondria. This energy imbalance can occur under a number of conditions, including exercise [50]. The energy imbalance triggers a signaling cascade that initiates mitochondrial biogenesis.

This signaling cascade is thought to be initiated by the activation of protein kinases, including AMP kinase (AMPK), which is activated in response to the increase in ATP consumption, and calcium-dependent kinases which are activated in response to changes in intracellular calcium. Activation of these kinase enzymes induces the expression of a family of transcription factors termed the peroxisome proliferator-activated receptor gamma coactivator (PGC-1) family. This family of transcription factors includes PCG-1α, PGC-1β, and PGC-1α-related coactivator (PRC). All of these transcription factors have been implicated in mitochondrial biosynthesis; however, it is PCG-1α that is thought to be the most significant member of this family in regard to the metabolic regulation of this process. Over-expression of PCG-1α demonstrates a significant increase in mitochondrial biogenesis accompanied by increased exercise capacity [51], while deletion of PGC-1α blunts the observed exercise-induced increase in mitochondrial enzymes [52]. Thus, PGC-1α has been referred to as the master regulator of mitochondrial biogenesis.

Nitric oxide has been found to stimulate mitochondrial biogenesis in many tissues, including skeletal muscle [53]. *In vitro* experiments in myoblasts and myotubes have demonstrated that NO donors increase mitochondrial biogenesis [54–56]. Studies have demonstrated that the NO-induced increase in mitochondria was via a cGMP-dependent mechanism, which led to the activation of PCG-1α and its downstream targets [57], and that this process resulted in functional mitochondria [56]. Moreover, it was demonstrated that AMPK, which activates PCG1-α, is regulated by NO [58] and that NO and AMPK act synergistically to regulate PCG-1α. The NOS involvement in this NO-driven process is likely tied to the known calcium dependence of both nNOS and eNOS. Indeed, NOS has been implicated in mitochondrial biogenesis [59]. It was demonstrated that mice deficient in eNOS had decreased PCG-1α mRNA and decreased mitochondria [56], while mice deficient in nNOS seem to have decreased mitochondria at basal conditions [60]. Thus it was concluded that, while eNOS plays a positive role in the regulation of mitochondrial biogenesis, nNOS plays a negative role, at least in mice. However, together pharmacological and genetic evidence indicates that neither eNOS nor nNOS seem to play a role in exercise-induced increase in mitochondrial biogenesis [59,61]. Thus it was concluded that, although NO does play a role in mitochondrial biogenesis under basal conditions, the observed increase in NO produced during exercise is not necessary for the observed exercise-induced increase in mitochondria [35].

Arginine supplementation as a means to induce mitochondrial biogenesis in skeletal muscle has not been tested in a healthy human population. However, in patients with type II diabetes, mitochondrial density in skeletal muscle is decreased [62], and long-term treatment of type II diabetics with arginine supplementation has demonstrated increased benefit of hypocaloric diet and exercise [63]. Thus it is possible, given the discussion above, that the long-term treatment of humans with arginine, at least in the diabetic setting, increases mitochondrial biogenesis in the skeletal muscle [64], and this increase in skeletal muscle mitochondria would be beneficial in regard to increasing muscle mass.

Skeletal Muscle Blood Flow

Given the known vasodilatory action of NO, it is not surprising that nitric oxide has been found to be involved in the regulation of skeletal muscle blood flow [65]. It has been demonstrated that infusion of the NOS inhibitor l-NMMA into resting human subjects decreased forearm blood flow. Moreover, the response to acetylcholine, which is known to activate endothelial NO production, was blunted whereas the response to an NO-donating drug was not inhibited [66]. This NO-dependent regulation of skeletal muscle blood flow has also been observed during recovery after exercise [67]. Some authors have also observed that muscle blood flow during exercise is also regulated by NO [68,69]. However, there are issues with when and how these reports were measuring blood flow, and the accumulating evidence indicates that NOS activity is not the driving factor regulating skeletal blood flow during exercise [67,70–72]. Indeed, a recent study using positron emission tomography indicated that NOS inhibition decreased resting muscle blood flow but did not decrease blood flow during exercise [73]. These authors did, however, demonstrate inhibition of skeletal muscle blood flow during exercise by a combination of NOS and COX2 inhibition. Thus, while NOS-dependent NO alone does not regulate exercise-induced changes in skeletal muscle blood flow, it can do so in synergy with other factors.

The stimulation of NOS-derived NO has also been tested in regard to the regulation of skeletal blood flow. Bolus oral consumption of 10 g of arginine failed to alter skeletal blood flow, either at rest or during exercise [74]. Additionally, although plasma arginine level was increased, there was no increase in plasma markers for increased NO synthesis. This observation could be unique to oral administration of arginine or dose dependent, as 30 g bolus intravenous arginine was found to decrease blood pressure and peripheral resistance, while 6 g of arginine (either IV or PO) did not induce a significant change [75]. However, in a recent study, acute administration of 6 g of arginine did increase muscle blood volume during recovery after exercise, but failed to increase markers of increased NO production [76]. Thus it seems clear that oral administration of arginine does not regulate muscle blood flow, but this lack of arginine-induced alteration cannot *per se* be attributed to any change in NOS-dependent NO generation. Although NOS-independent NO has not specifically been tested in regard to skeletal muscle blood flow in humans, administration of nitrate (which is converted to nitrite and then NO) has been shown to increase skeletal muscle blood volume [77], and dietary nitrate has been shown to regulate skeletal blood flow in rodents [78].

In summary, skeletal muscle blood flow is regulated directly by NO at rest and after exercise. However, during exercise, the contribution of NO to this regulation is less pronounced, but does have synergistic affects. Attempts to alter skeletal muscle blood flow via the consumption of potential NO-generating substances have not shown significant changes [79,80].

NUTRITIONAL MODIFICATION OF SKELETAL MUSCLE HYPERTROPHY

Since the identification of NO as a critical regulator of many physiological processes, there has been much interest in using nutritional modification to augment NO production for both the improvement of healthy humans and in the treatment of several pathophysiologic conditions [64,81–84], and these studies have shown varied results, both positive and negative. Moreover, it is clear that NO-directed supplementation can affect skeletal muscle hypertrophy and ameliorate some pathophysiologic conditions involving skeletal muscle dysfunction in animals [42]. Additionally, administration of an NO donor in combination with an anti-inflammatory drug has been demonstrated to be safe and have some efficacy in the treatment of muscular dystrophies [85]. However, while it is clear that NO signaling has the potential to positively regulate muscle growth, there have not been any peer-reviewed studies that directly examine if augmentation of NO signaling by dietary supplementation regulates muscle hypertrophy in humans.

Nutritional Supplementation for Enhanced Performance in Humans

There have been many studies examining the effect of NO-directed nutritional supplements on exercise performance in humans [86,87]. However, these studies use a wide range of supplements, have significantly variable duration of treatment, use variable exercise protocols, measure different outcomes, and have a variable subject population. Thus, it is not surprising that, whereas some studies indicated positive effects, an approximately equal number showed no effect. Attempts have been made to stratify these studies in order to draw some conclusions. Alvares et al. examined the differing ergogenic effects of acute versus chronic supplementation of arginine [86]. Of the five acute (treatment of less than 3 days prior to testing) studies reviewed, three demonstrated a significant increase in performance; of the eight chronic (treatment of 10 days to 6 months) studies, four demonstrated significant increases. It should be noted that in some studies the supplements used to increase circulating arginine levels included other components: glycine, α-ketoisocaproic acid, aspartate, α-ketoglutarate, and creatine. Moreover, none of the studies tested for markers of increased NO activity. Alvares et al. concluded that, while the supplements are well tolerated, there is not enough evidence to recommend the use of nutritional supplementation of arginine as an aid in exercise performance in healthy individuals.

A more recent and general review of the effects of NO-supplementation on exercise performance has come to similar conclusions [87]. In this review, Bescos et al. cite 42 studies that examined how various NO-related substances affect performance, including not only supplements aimed at increasing circulating arginine levels, but also those aimed at increasing circulating nitrite levels. Again, only approximately half of the studies reviewed that reported performance data indicated an increase in performance in response to supplementation. Bescos et al. draw several conclusions. The potential benefit of arginine supplementation is variable based on the training status of the individual, with benefits being seen in moderately and untrained subjects but little effect in well trained athletes, and this benefit in general relies on the combination of arginine with other substances. Citrulline supplementation alone does not enhance performance. Nitrate supplementation effectively enhances exercise performance in untrained or moderately trained subjects, but the limited data indicate that this affect is not seen in endurance-trained athletes. Thus it is clear that while attempts to augment NO signaling can improve exercise performance in general, this improvement is not seen in all settings. Moreover, it is impossible in many of the published studies to dissect out the effects of the supplementation on muscle hypertrophy (which would increase muscle strength and thus exercise performance) from the other physiological effects of augmented NO signaling. Indeed, neither of these reviews addressed the potential for the enhancement of skeletal muscle hypertrophy by NO-supplementation in detail.

Nutritional Supplementation for Enhanced Muscle Hypertrophy in Humans

As previously mentioned, there are no studies specifically examining the influences of NO-related nutritional supplements on muscle hypertrophy in humans.

However, there are studies that examine how muscle strength is affected in response to these supplements. Given that muscle strength correlates with muscle cross-sectional area [88], increases in muscle strength can thus be used as a surrogate marker for muscle hypertrophy. Additionally, it is clear that the initiation of muscle hypertrophy requires continuous muscle stimulation. Indeed, muscle hypertrophy in response to resistance exercise is virtually nonexistent during the first weeks of training [47]. Therefore, studies that examine the effects of the chronic administration of NO-related nutritional supplements on muscle strength can provide some data regarding whether skeletal muscle hypertrophy can be affected by nutritional supplementation of NO signaling.

Campbell et al. examined how chronic supplementation with l-arginine α-ketoglutarate (AAG) affected strength gain during an 8-week training protocol [89]. Administration of AAG significantly increased plasma arginine levels. The addition of AAG supplementation to a standardized exercise protocol over 8 weeks significantly increased muscle strength as assessed by a 1RM bench press test. From these data, it can be concluded that this supplement has augmented the exercise-induced muscle hypertrophy. Interestingly, there was no difference in body composition, muscle endurance, or aerobic capacity. It has been argued, rightly, that AAG can have affects outside its potential augmentation of NO signaling, because α-ketoglutarate could increase TCA cycle flux and initiate glutamate sparing [90]. As such, it cannot be conclusively concluded that it is solely the potential increase in NO signaling produced by AAG that elicits the effects observed. However, acute administration of AAG failed to alter muscle strength as measured by 1RM [90]. Thus the strength gain observed in the chronic supplementation model is not due to acute NO effects on muscle contractile function, bolstering the supposition that AAG supplementation did augment exercise-induced muscle hypertrophy.

Another study indicates that nutritional supplementation of NO signaling cannot in and of itself induce skeletal muscle hypertrophy [91]. Fricke et al. demonstrated that, in postmenopausal women, oral administration of l-arginine (18 g per day for 6 months) increased peak jump force, and they speculated that this supplementation could prevent the decline in muscle force observed in this population. However, they found no statistically significant changes in jump power, grip force, muscle cross-sectional area, or fat area. Thus, in this population, arginine supplementation does not affect muscle hypertrophy. Perhaps arginine supplementation alone cannot induce muscle hypertrophy, but it can augment exercise-induced hypertrophy.

CONCLUSION

It is clear that NO regulates many processes in skeletal muscle and that alterations in NO signaling, both pharmacologic and nutritional, can alter the response of skeletal muscle. These observations can be used to make a compelling argument for the use of nutritional substances in the augmentation of NO signaling to elicit beneficial effects in skeletal muscle strength, growth and performance. However, there is still not a clear consensus on whether any of the nutritional substances tested in rigorous controlled settings are truly eliciting an effect and whether any observed effect is truly due to an alteration in NO signaling, especially in regard to muscle growth. That notwithstanding, many studies have provided evidence that nutritional supplements aimed at altering NO signaling can be beneficial. More work is needed to definitively prove not only if NO supplementation can be beneficial to specific skeletal muscle processes, but also to better define optimal substance type, dosage, and treatment protocols.

References

[1] Schultz K, Schultz G. Sodium nitroprusside and other smooth muscle-relaxants increase cyclic GMP levels in rat ductus deferens. Nature 1977;265(5596):750–1.

[2] Katsuki S, Arnold W, Mittal C, Murad F. Stimulation of guanylate cyclase by sodium nitroprusside, nitroglycerin and nitric oxide in various tissue preparations and comparison to the effects of sodium azide and hydroxylamine. J Cyclic Nucleotide Res 1977;3(1):23–35.

[3] Furchgott RF, Zawadzki JV. The obligatory role of endothelial cells in the relaxation of arterial smooth muscle by acetylcholine. Nature 1980;288(5789):373–6.

[4] Palmer RM, Ferrige AG, Moncada S. Nitric oxide release accounts for the biological activity of endothelium-derived relaxing factor. Nature 1987;327(6122):524–6.

[5] Ignarro LJ, Buga GM, Wood KS, Byrns RE, Chaudhuri G. Endothelium-derived relaxing factor produced and released from artery and vein is nitric oxide. Proc Natl Acad Sci USA 1987;84(24):9265–9.

[6] Palmer RM, Ashton DS, Moncada S. Vascular endothelial cells synthesize nitric oxide from L-arginine. Nature 1988;333 (6174):664–6.

[7] Palmer RM, Moncada S. A novel citrulline-forming enzyme implicated in the formation of nitric oxide by vascular endothelial cells. Biochem Biophys Res Commun 1989;158(1):348–52.

[8] Zhang YH, Casadei B. Sub-cellular targeting of constitutive NOS in health and disease. J Mol Cell Cardiol 2012;52 (2):341–50.

[9] Robinson LJ, Ghanouni P, Michel T. Posttranslational modifications of endothelial nitric oxide synthase. Methods Enzymol 1996;268:436–48.

[10] Watanabe M, Itoh K. Characterization of a novel posttranslational modification in neuronal nitric oxide synthase by small ubiquitin-related modifier-1. Biochim Biophys Acta 2011;1814 (7):900–7.

[11] Zhou L, Zhu DY. Neuronal nitric oxide synthase: structure, subcellular localization, regulation, and clinical implications. Nitric Oxide 2009;20(4):223–30.

[12] Zweier JL, Chen CA, Druhan LJ. S-glutathionylation reshapes our understanding of endothelial nitric oxide synthase uncoupling and nitric oxide/reactive oxygen species-mediated signaling. Antioxid Redox Signal 2011;14(10):1769–75.

[13] Mungrue IN, Bredt DS. nNOS at a glance: implications for brain and brawn. J Cell Sci 2004;117(Pt 13):2627–9.

[14] Brenman JE, Chao DS, Gee SH, McGee AW, Craven SE, Santillano DR, et al. Interaction of nitric oxide synthase with the postsynaptic density protein PSD-95 and alpha1-syntrophin mediated by PDZ domains. Cell 1996;84(5):757–67.

[15] Grange RW, Isotani E, Lau KS, Kamm KE, Huang PL, Stull JT. Nitric oxide contributes to vascular smooth muscle relaxation in contracting fast-twitch muscles. Physiol Genomics 2001;5 (1):35–44.

[16] Thomas GD, Shaul PW, Yuhanna IS, Froehner SC, Adams ME. Vasomodulation by skeletal muscle-derived nitric oxide requires alpha-syntrophin-mediated sarcolemmal localization of neuronal Nitric oxide synthase. Circ Res 2003;92(5):554–60.

[17] Thomas GD, Sander M, Lau KS, Huang PL, Stull JT, Victor RG. Impaired metabolic modulation of alpha-adrenergic vasoconstriction in dystrophin-deficient skeletal muscle. Proc Natl Acad Sci U S A 1998;95(25):15090–5.

[18] Percival JM, Anderson KN, Huang P, Adams ME, Froehner SC. Golgi and sarcolemmal neuronal NOS differentially regulate contraction-induced fatigue and vasoconstriction in exercising mouse skeletal muscle. J Clin Invest 2010;120(3):816–26.

[19] Frandsen U, Lopez-Figueroa M, Hellsten Y. Localization of nitric oxide synthase in human skeletal muscle. Biochem Biophys Res Commun 1996;227(1):88–93.

[20] Bates TE, Loesch A, Burnstock G, Clark JB. Mitochondrial nitric oxide synthase: a ubiquitous regulator of oxidative phosphorylation? Biochem Biophys Res Commun 1996;218(1):40–4.

[21] Torres SH, De Sanctis JB, de LBM, Hernandez N, Finol HJ. Inflammation and nitric oxide production in skeletal muscle of type 2 diabetic patients. J Endocrinol 2004;181(3):419–27.

[22] McConell GK, Bradley SJ, Stephens TJ, Canny BJ, Kingwell BA, Lee-Young RS. Skeletal muscle nNOS mu protein content is increased by exercise training in humans. Am J Physiol Regul Integr Comp Physiol 2007;293(2):R821–8.

[23] Tannenbaum SR, Correa P. Nitrate and gastric cancer risks. Nature 1985;317(6039):675–6.

[24] Mensinga TT, Speijers GJ, Meulenbelt J. Health implications of exposure to environmental nitrogenous compounds. Toxicol Rev 2003;22(1):41–51.

[25] Lundberg JO, Weitzberg E, Gladwin MT. The nitrate-nitrite-nitric oxide pathway in physiology and therapeutics. Nat Rev Drug Discov 2008;7(2):156–67.

[26] Curtis E, Hsu LL, Noguchi AC, Geary L, Shiva S. Oxygen regulates tissue nitrite metabolism. Antioxid Redox Signal 2012;17 (7):951–61.

[27] Woolford G, Casselden RJ, Walters CL. Gaseous products of the interaction of sodium nitrite with procine skeletal muscle. Biochem J 1972;130(2):82P–3P.

[28] Lundberg JO, Weitzberg E. Nitrite reduction to nitric oxide in the vasculature. Am J Physiol Heart Circ Physiol 2008;295(2): H477–8.

[29] Larsen FJ, Weitzberg E, Lundberg JO, Ekblom B. Effects of dietary nitrate on oxygen cost during exercise. Acta Physiol (Oxf) 2007;191(1):59–66.

[30] Massion PB, Feron O, Dessy C, Balligand JL. Nitric oxide and cardiac function: ten years after, and continuing. Circ Res 2003;93(5):388–98.

[31] Bian K, Doursout MF, Murad F. Vascular system: role of nitric oxide in cardiovascular diseases. J Clin Hypertens (Greenwich) 2008;10(4):304–10.

[32] Morrison RJ, Miller III CC, Reid MB. Nitric oxide effects on shortening velocity and power production in the rat diaphragm. J Appl Physiol 1996;80(3):1065–9.

[33] Shen W, Xu X, Ochoa M, Zhao G, Wolin MS, Hintze TH. Role of nitric oxide in the regulation of oxygen consumption in conscious dogs. Circ Res 1994;75(6):1086–95.

[34] Balon TW, Nadler JL. Evidence that nitric oxide increases glucose transport in skeletal muscle. J Appl Physiol 1997;82 (1):359–63.

[35] McConell GK, Rattigan S, Lee-Young RS, Wadley GD, Merry TL. Skeletal muscle nitric oxide signaling and exercise: a focus on glucose metabolism. Am J Physiol Endocrinol Metab 2012;303(3):E301–7.

[36] Smith LW, Smith JD, Criswell DS. Involvement of nitric oxide synthase in skeletal muscle adaptation to chronic overload. J Appl Physiol 2002;92(5):2005–11.

[37] Sellman JE, DeRuisseau KC, Betters JL, Lira VA, Soltow QA, Selsby JT, et al. In vivo inhibition of nitric oxide synthase impairs upregulation of contractile protein mRNA in overloaded plantaris muscle. J Appl Physiol 2006;100(1):258–65.

[38] Lomonosova YN, Kalamkarov GR, Bugrova AE, Shevchenko TF, Kartashkina NL, Lysenko EA, et al. Protective effect of L-Arginine administration on proteins of unloaded m. soleus. Biochemistry (Mosc) 2011;76(5):571–80.

[39] Booth FW, Criswell DS. Molecular events underlying skeletal muscle atrophy and the development of effective countermeasures. Int J Sports Med 1997;18(Suppl. 4):S265–9.

[40] Kadi F, Charifi N, Denis C, Lexell J, Andersen JL, Schjerling P, et al. The behaviour of satellite cells in response to exercise: what have we learned from human studies? Pflugers Arch 2005;451(2):319–27.

[41] Kadi F, Schjerling P, Andersen LL, Charifi N, Madsen JL, Christensen LR, et al. The effects of heavy resistance training and detraining on satellite cells in human skeletal muscles. J Physiol 2004;558(Pt 3):1005–12.

[42] De Palma C, Clementi E. Nitric oxide in myogenesis and therapeutic muscle repair. Mol Neurobiol 2012;46(3):682–92.

[43] Babcock L, Escano M, D'Lugos A, Todd K, Murach K, Luden N. Concurrent aerobic exercise interferes with the satellite cell response to acute resistance exercise. Am J Physiol Regul Integr Comp Physiol 2012;302(12):R1458–65.

[44] Buono R, Vantaggiato C, Pisa V, Azzoni E, Bassi MT, Brunelli S, et al. Nitric oxide sustains long-term skeletal muscle regeneration by regulating fate of satellite cells via signaling pathways requiring Vangl2 and cyclic GMP. Stem Cells 2012;30 (2):197–209.

[45] Anderson JE. A role for nitric oxide in muscle repair: nitric oxide-mediated activation of muscle satellite cells. Mol Biol Cell 2000;11(5):1859–74.

[46] Wozniak AC, Anderson JE. Nitric oxide-dependence of satellite stem cell activation and quiescence on normal skeletal muscle fibers. Dev Dyn 2007;236(1):240–50.

[47] Schoenfeld BJ. The mechanisms of muscle hypertrophy and their application to resistance training. J Strength Cond Res 2010;24(10):2857–72.

[48] Kobzik L, Reid MB, Bredt DS, Stamler JS. Nitric oxide in skeletal muscle. Nature 1994;372(6506):546–8.

[49] Attardi G, Schatz G. Biogenesis of mitochondria. Annu Rev Cell Biol 1988;4:289–333.

[50] Hood DA, Irrcher I, Ljubicic V, Joseph AM. Coordination of metabolic plasticity in skeletal muscle. J Exp Biol 2006;209 (Pt 12):2265–75.

[51] Calvo JA, Daniels TG, Wang X, Paul A, Lin J, Spiegelman BM, et al. Muscle-specific expression of PPARgamma coactivator-1alpha improves exercise performance and increases peak oxygen uptake. J Appl Physiol 2008;104(5):1304–12.

[52] Geng T, Li P, Okutsu M, Yin X, Kwek J, Zhang M, et al. PGC-1alpha plays a functional role in exercise-induced mitochondrial biogenesis and angiogenesis but not fiber-type transformation in mouse skeletal muscle. Am J Physiol Cell Physiol 2010;298(3):C572–9.

[53] Nisoli E, Carruba MO. Nitric oxide and mitochondrial biogenesis. J Cell Sci 2006;119(Pt 14):2855–62.

[54] McConell GK, Ng GP, Phillips M, Ruan Z, Macaulay SL, Wadley GD. Central role of nitric oxide synthase in AICAR and caffeine-induced mitochondrial biogenesis in L6 myocytes. J Appl Physiol 2010;108(3):589–95.

[55] Lira VA, Brown DL, Lira AK, Kavazis AN, Soltow QA, Zeanah EH, et al. Nitric oxide and AMPK cooperatively regulate PGC-1 in skeletal muscle cells. J Physiol 2010;588(Pt 18):3551–66.

[56] Nisoli E, Falcone S, Tonello C, Cozzi V, Palomba L, Fiorani M, et al. Mitochondrial biogenesis by NO yields functionally active mitochondria in mammals. Proc Natl Acad Sci U S A 2004;101 (47):16507–12.

[57] Nisoli E, Clementi E, Paolucci C, Cozzi V, Tonello C, Sciorati C, et al. Mitochondrial biogenesis in mammals: the role of endogenous nitric oxide. Science 2003;299(5608):896–9.

[58] Lira VA, Soltow QA, Long JH, Betters JL, Sellman JE, Criswell DS. Nitric oxide increases GLUT4 expression and regulates AMPK signaling in skeletal muscle. Am J Physiol Endocrinol Metab 2007;293(4):E1062–8.

[59] Wadley GD, Choate J, McConell GK. NOS isoform-specific regulation of basal but not exercise-induced mitochondrial biogenesis in mouse skeletal muscle. J Physiol 2007;585(Pt 1):253–62.

[60] Schild L, Jaroscakova I, Lendeckel U, Wolf G, Keilhoff G. Neuronal nitric oxide synthase controls enzyme activity pattern of mitochondria and lipid metabolism. FASEB J 2006;20 (1):145–7.

[61] Wadley GD, McConell GK. Effect of nitric oxide synthase inhibition on mitochondrial biogenesis in rat skeletal muscle. J Appl Physiol 2007;102(1):314–20.

[62] Kelley DE, He J, Menshikova EV, Ritov VB. Dysfunction of mitochondria in human skeletal muscle in type 2 diabetes. Diabetes 2002;51(10):2944–50.

[63] Lucotti P, Setola E, Monti LD, Galluccio E, Costa S, Sandoli EP, et al. Beneficial effects of a long-term oral L-arginine treatment added to a hypocaloric diet and exercise training program in obese, insulin-resistant type 2 diabetic patients. Am J Physiol Endocrinol Metab 2006;291(5):E906–12.

[64] McConell GK. Effects of L-arginine supplementation on exercise metabolism. Curr Opin Clin Nutr Metab Care 2007;10 (1):46–51.

[65] Persson MG, Gustafsson LE, Wiklund NP, Hedqvist P, Moncada S. Endogenous nitric oxide as a modulator of rabbit skeletal muscle microcirculation in vivo. Br J Pharmacol 1990;100(3):463–6.

[66] Vallance P, Collier J, Moncada S. Effects of endothelium-derived nitric oxide on peripheral arteriolar tone in man. Lancet 1989;2(8670):997–1000.

[67] Radegran G, Saltin B. Nitric oxide in the regulation of vasomotor tone in human skeletal muscle. Am J Physiol 1999;276(6 Pt 2): H1951–60.

[68] Dyke CK, Proctor DN, Dietz NM, Joyner MJ. Role of nitric oxide in exercise hyperaemia during prolonged rhythmic handgripping in humans. J Physiol 1995;488(Pt 1):259–65.

[69] Gilligan DM, Panza JA, Kilcoyne CM, Waclawiw MA, Casino PR, Quyyumi AA. Contribution of endothelium-derived nitric oxide to exercise-induced vasodilation. Circulation 1994;90 (6):2853–8.

[70] Wilson JR, Kapoor S. Contribution of endothelium-derived relaxing factor to exercise-induced vasodilation in humans. J Appl Physiol 1993;75(6):2740–4.

[71] Endo T, Imaizumi T, Tagawa T, Shiramoto M, Ando S, Takeshita A. Role of nitric oxide in exercise-induced vasodilation of the forearm. Circulation 1994;90(6):2886–90.

[72] Frandsenn U, Bangsbo J, Sander M, Hoffner L, Betak A, Saltin B, et al. Exercise-induced hyperaemia and leg oxygen uptake are not altered during effective inhibition of nitric oxide synthase with N(G)-nitro-L-arginine methyl ester in humans. J Physiol 2001;531(Pt 1):257–64.

[73] Heinonen I, Saltin B, Kemppainen J, Sipila HT, Oikonen V, Nuutila P, et al. Skeletal muscle blood flow and oxygen uptake at rest and during exercise in humans: a pet study with nitric oxide and cyclooxygenase inhibition. Am J Physiol Heart Circ Physiol 2010;300(4):H1510–7.

[74] Tang JE, Lysecki PJ, Manolakos JJ, MacDonald MJ, Tarnopolsky MA, Phillips SM. Bolus arginine supplementation affects neither muscle blood flow nor muscle protein synthesis in young men at rest or after resistance exercise. J Nutr 2011;141 (2):195–200.

[75] Bode-Boger SM, Boger RH, Galland A, Tsikas D, Frolich JC. L-arginine-induced vasodilation in healthy humans: pharmacokinetic-pharmacodynamic relationship. Br J Clin Pharmacol 1998;46(5):489–97.

[76] Alvares TS, Conte CA, Paschoalin VM, Silva JT, Meirelles Cde M, Bhambhani YN, et al. Acute l-arginine supplementation increases muscle blood volume but not strength performance. Appl Physiol Nutr Metab 2012;37(1):115–26.

[77] Bailey SJ, Winyard P, Vanhatalo A, Blackwell JR, Dimenna FJ, Wilkerson DP, et al. Dietary nitrate supplementation reduces the O2 cost of low-intensity exercise and enhances tolerance to high-intensity exercise in humans. J Appl Physiol 2009;107 (4):1144–55.

[78] Ferguson SK, Hirai DM, Copp SW, Holdsworth CT, Allen JD, Jones AM, et al. Impact of dietary nitrate supplementation via beetroot juice on exercising muscle vascular control in rats. J Physiol 2012.

[79] Robinson TM, Sewell DA, Greenhaff PL. L-arginine ingestion after rest and exercise: effects on glucose disposal. Med Sci Sports Exerc 2003;35(8):1309–15.

[80] Fahs CA, Heffernan KS, Fernhall B. Hemodynamic and vascular response to resistance exercise with L-arginine. Med Sci Sports Exerc 2009;41(4):773–9.

[81] Stechmiller JK, Childress B, Cowan L. Arginine supplementation and wound healing. Nutr Clin Pract 2005;20(1):52–61.

[82] Kalil AC, Danner RL. L-Arginine supplementation in sepsis: beneficial or harmful? Curr Opin Crit Care 2006;12(4):303–8.

[83] Dong JY, Qin LQ, Zhang Z, Zhao Y, Wang J, Arigoni F, et al. Effect of oral L-arginine supplementation on blood pressure: a meta-analysis of randomized, double-blind, placebo-controlled trials. Am Heart J 2011;162(6):959–65.

[84] Machha A, Schechter AN. Inorganic nitrate: a major player in the cardiovascular health benefits of vegetables? Nutr Rev 2012;70(6):367–72.

[85] D'Angelo MG, Gandossini S, Martinelli Boneschi F, Sciorati C, Bonato S, Brighina E, et al. Nitric oxide donor and non steroidal anti inflammatory drugs as a therapy for muscular dystrophies: evidence from a safety study with pilot efficacy measures in adult dystrophic patients. Pharmacol Res 2012;65(4):472–9.

[86] Alvares TS, Meirelles CM, Bhambhani YN, Paschoalin VM, Gomes PS. L-Arginine as a potential ergogenic aid in healthy subjects. Sports Med 2011;41(3):233–48.

[87] Bescos R, Sureda A, Tur JA, Pons A. The effect of nitric-oxide-related supplements on human performance. Sports Med 2012;42(2):99–117.

[88] Maughan RJ, Watson JS, Weir J. Strength and cross-sectional area of human skeletal muscle. J Physiol 1983;338:37–49.

[89] Campbell B, Roberts M, Kerksick C, Wilborn C, Marcello B, Taylor L, et al. Pharmacokinetics, safety, and effects on exercise performance of L-arginine alpha-ketoglutarate in trained adult men. Nutrition 2006;22(9):872–81.

[90] Wax B, Kavazis AN, Webb HE, Brown SP. Acute L-arginine alpha ketoglutarate supplementation fails to improve muscular performance in resistance trained and untrained men. J Int Soc Sports Nutr 2006;9(1):17.

[91] Fricke O, Baecker N, Heer M, Tutlewski B, Schoenau E. The effect of L-arginine administration on muscle force and power in postmenopausal women. Clin Physiol Funct Imaging 2008;28 (5):307–11.

CHAPTER

27

Role of Nitric Oxide in Sports Nutrition

Safia Habib and Asif Ali

Department of Biochemistry, Jawaharlal Nehru Medical College, AMU, Aligarh, India

INTRODUCTION

Nitric oxide (NO) is a free radical that can function both as a cytoprotective and as a tumor-promoting agent. The basic reactions of NO can be divided into (a) a direct effect of the radical, where it alone plays a role in either damaging or protecting the cell milieu, and (b) an indirect effect in which byproducts of NO formed by convergence of two independent radical-generating pathways play a role in biological reactions which mainly involve oxidative and nitrosative stress. NO is known to interact with various biomolecules such as amino acids, nucleic acids, fatty acids, metal-containing compounds, etc. It is formed from L-arginine which is converted to L-citrulline by nitric oxide synthase (NOS) enzymes (Figure 27.1). These enzymes exist in three isoforms: NOS1 or neuronal nitric oxide synthase (nNOS), NOS2 or inducible nitric oxide synthase (iNOS), and NOS3 or endothelial nitric oxide synthase (eNOS). Each isoform is the product of a distinct gene [1]. Neuronal NOS and endothelial NOS are constitutively expressed and require elevated levels of Ca^{+2} along with activation of calmodulin to produce NO for a brief period of time [2]. Inducible NOS is induced by cytokines in almost every cell and generates a locally high level of NO for prolonged periods of time [3]. Inducible NOS is mainly expressed in macrophages, neutrophils and epithelial cells but its expression can also be induced in glial cells, liver and cardiac muscles. Endothelial NOS is constitutively expressed in the endothelial lining of blood vessels and depends on Ca^{+2} for cGMP-dependent smooth muscle relaxation thereby increasing blood flow; nNOS is also constitutively expressed in postsynaptic terminals of neurons and is Ca^{+2} dependent (Table 27.1). It is activated by Ca^{+2} influxes caused by the binding of the neurotransmitter glutamate to its receptor in the cell membrane. Neuronal NOS is also activated by membrane depolarization through the opening of voltage-gated channels [4]. The endogenously produced NO as well as exogenous nitrate and nitrite (which are a part of human diet as nutrients from cured meat, vegetables and preservatives) are responsible for direct reaction of NO with cellular components and its much pronounced potential role of nitrosative stress induction by its secondary intermediates termed reactive nitrogen species (RNS). Some examples of RNS are peroxynitrite ($ONOO^-$) formed by reaction of NO with superoxide anion ($O_2^{\bar{\cdot}}$) along with N_2O_2, nitroso-peroxocarbonate ($ONOOCO_2^-$), lipid peroxyradicals (LOO$^{\cdot}$), etc. [5]. The aim of this chapter is to highlight the significant role of RNS in muscle activity, sports nutrition and in general health and disease.

Nitric oxide and RNS, when produced within physiological limits, are known to play an important role in cell signaling and macrophage- and neutrophil-mediated immune responses which inhibit viruses, pathogens and tumor proliferation [6]. NO also acts on reactive oxygen species (ROS) as a detoxifier or ROS scavenger [7]. It is also known to act on vessels to regulate blood flow in muscles adapted to prolonged exercise. RNS can be detrimental when produced in high amounts in the intracellular compartments, leading to insulin sensitivity [8], cancers, and neurologic and cardiovascular pathologies. The cells respond by up-regulating antioxidants such as glutathione peroxidase, superoxide dismutase, catalase, and glutathione which have a protective role in scavenging free radicals [9].

According to the American Dietetic Association, Dietitians of Canada, and the American College of Sports Medicine, the selection of food, fluids, timing of intake and supplement choices affect exercise

Nutrition and Enhanced Sports Performance.
DOI: http://dx.doi.org/10.1016/B978-0-12-396454-0.00027-8

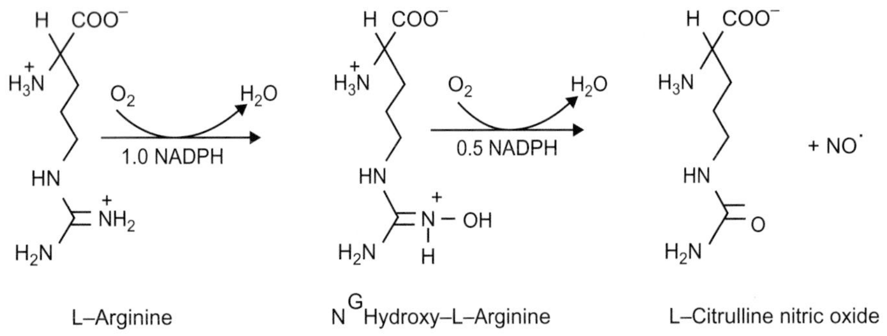

FIGURE 27.1 **Synthesis of nitric oxide from L-arginine.**

TABLE 27.1 Location and Activation of Nitric Oxide Isozymes

Type of Isozyme	Gene Loci	Availability	Location	Mode of Action
Neuronal nitric oxide synthase (NOS1)	On chromosome 12	Central and peripheral neurons, gastrointestinal cells	Cytoplasmic	Activated by calcium
Inducible nitric oxide synthase (NOS2)	On chromosome 17	Macrophages, neutrophils, hepatocytes	Cytoplasmic	Induced by cytokines (interleukin and tumor necrosis factor)
Endothelial nitric oxide synthase (NOS3)	On chromosome 7	Endocardium, myocardium, platelets	Plasma membrane	Activated by calcium

performance [10]. All the nutritional recommendations should be based on current scientific data and the needs of athletes as individuals. Athletes who follow a severe weight loss regime or restrict any of the food groups from their diet should be advised to judiciously use nutritional ergogenic aids to enhance their performance, keeping in mind their efficacy, potency and the effect of other endogenous substances on them which may either reduce their assumed effect or may induce toxicity. An endogenous agent that affects sports nutrition and its efficacy is NO (or RNS) which act on various macromolecules (food groups) leading to modifications of their structure and functions. The nitrosative stress induced in an athlete may affect overall performance. We have focused here on the role of NO in muscle activity.

RECOMMENDED NUTRITION CRITERIA FOR BETTER SPORTS PERFORMANCE

Diet is usually in the form of a complex matrix of food whence the nutrients have to be released before they can be absorbed and utilized. In general, a diet must fulfill the basic requirements of providing energy and body building through metabolic fuels, where mainly carbohydrates and lipids serve the function; whereas proteins must be supplied in the diet for the purpose of overall growth and turnover of tissue proteins. Vitamins, minerals and essential fatty acids are required for specific metabolic and physiological functions, along with water.

Adequate energy intake during times of high-intensity training is known to maintain body weight and maximize the training effect. According to the American College of Sports Nutrition, during times of high physical activity, energy (mainly carbohydrates) and macronutrient needs should be provided at a rate of 30 to 60 g per hour to maintain body weight and to replenish glycogen stores, along with proteins to build and repair tissues. Fat intake should also be sufficient to provide the essential fatty acids and fat-soluble vitamins to contribute energy for weight maintenance. Overall diet should provide moderate amounts of energy from fat, i.e., 20% to 25%, since body weight and composition can affect exercise performance [11,12]. Athletes should be well hydrated before their event and the beginning of exercise, in order to maximize exercise performance and improve recovery time. Since consuming adequate food and fluid before, during and after exercise in the form of electrolytes or sport drinks containing carbohydrates will help in maintaining blood glucose and the thirst mechanism, this will also provide fuel for the muscles along with the decreased risk of dehydration or hyponatremia. Exercise in hot conditions increases the rate of body heat storage and reduces the time required to reach a critical hypothalamic temperature that results in fatigue. This critical temperature appears to be associated with the dysfunction of brain's motor control centers. This may also lead to temperature-induced alteration in skeletal muscle function with increased requirement for anaerobic ATP provision. The duration of exercise that can be performed before this critical temperature is reached can be increased by ingesting fluids of a volume at least equal to that lost in sweat within 60 min prior to and during exercise. The heat dissipative mechanism involves heat-induced enhancement of muscle sympathetic nerve activity, producing stimulation of C III and C IV afferent nerves and causing heat induced release of NO, resulting in muscle and skin vasodilatation [13].

ROLE OF NITRIC OXIDE IN NUTRITIONAL SUPPLEMENTS

In addition to the required recommended dietary regime, nutritional supplements enhance exercise capacity beyond that afforded by regular food ingestion alone. Arginine and its metabolite creatine are the nutritional supplements that enhance exercise capacity, and therefore they are said to have an ergogenic effect. L-arginine after ingestion exerts its overall action in two modes:

1. acute effects which result in enhanced exercise capacity after ingestion of arginine;
2. chronic effects which are associated with the anabolism of muscle protein resulting from the stimulation of muscle protein synthesis.

Supplementation with arginine alone does not stimulate muscle protein synthesis, although exogenous arginine as a dietary supplement facilitates increase in body mass and functional capacity. Ingestion of arginine leads to the formation of ornithine, but one of its metabolic fates is conversion to NO via nitric oxide synthases. NO is known to have acute vasodilatory properties, and chronic exposure has been reported to slow down the process associated with the onset of atherosclerosis [14–16]. Supplementation with arginine, the precursor of NO, has a positive effect on muscle activity because of its indirect effect via the stimulation of NO production, which increases muscle blood flow. When sufficient arginine is ingested along with other amino acids, mainly essential amino acids, there is increased production of NO and an increase in the muscle blood flow which channels the absorbed amino acids to the muscles, increasing their uptake and incorporating them into muscle protein. Therefore, it could be expected that, along with the indirect effect of arginine, supplementing essential amino acids at concentrations above that present within the muscles would lead to an anabolic effect resulting in muscle building and better athletic performance. The vasodilatory property of NO has made athletes more exercise tolerant by increasing O_2 delivery to meet increased myocardial O_2 demands of strenuous physical activities [17,18]. It has also been reported that the activity of skeletal muscle is increased significantly by NO through S-nitrosylation of Ryanodine receptor (RYR) responsible for exporting calcium through cysteine rich calcium channels. NO is also known to inhibit sarcoplasmic/endoplasmic reticulum Ca^{+2}ATPase via tyrosine nitration, responsible for importing calcium.

Arginine is known to have anti-atherogenic and growth-promoting properties. It has also been reported that arginine (250 mg/kg/day) increases growth hormone secretion by inhibiting somatostatin secretion [19]. Besides arginine, creatine (a derivative of arginine, glycine and methionine) is also given as an ergogenic supplement. However, the side effects of excessive ingestion must be kept in mind since creatine may increase intracellular water and dilute electrolytes, thereby affecting fluid balance. The users should, therefore, be advised to pay attention to fluid need in hot climates. Large intake of creatine can reduce its endogenous synthesis, probably via feedback regulation. But the enzymes involved in creatine synthesis are reactivated when supplementation is discontinued. On average a person consumes 1 g/day of creatine from a regular diet. Creatine is degraded into creatinine and excreted in the urine at a rate of about 2 g/day. Potential side effects of creatine in a normal system, which is not under oxidative or nitrosative stress, are not known since most of the creatine ingested is removed from the plasma by the kidneys and excreted in urine.

The possible ergogenic effect of creatine has led to its widespread use as a supplement in sports. Many athletes ingest creatine over long periods of time for sports events and during training periods to increase strength and body mass. The normal concentration of total creatine in skeletal muscle is about 120 mmol/kg (dry mass), but muscle concentration of total creatine can indeed be increased by oral supplementation which involves a loading phase of 20 g creatine monohydrate for 4–6 days followed by a maintenance dose of 5 g daily for 2–3 weeks to ensure optimal uptake in muscle [20,21]. About 90–95% of the body's creatine is found is skeletal muscle and, of this, one third is free creatine and two thirds is phosphocreatine, which is responsible for the acute ergogenic effect of creatine making a phosphagen pool available for rapid resynthesis of ATP during periods of maximal ATP turnover. Creatine transport is increased when creatine is ingested together with carbohydrates or carbohydrate protein mixture [22,23]. Ingestion of creatine and 1 g glucose per kg body mass twice a day increases total muscle creatine by 9% more than with creatine intake alone.

Heavy sports training schedules and competitions are often associated with immune-suppression; therefore, athletes have to be provided with nutrients that display immunoregulatory properties. Among such immune-nutrients considerable attention has been paid to two amino acids, arginine and glutamine, where the dosage and administration schedules have to be decided according to the requirements.

All these ergogenic benefits of diet and dietary supplements are responsible for enhanced sports performance in a healthy individual, but if a system's homeostasis of pro-oxidants and antioxidants is disturbed then it leads to oxidative and nitrosative stress. Enhanced formation of ROS and RNS affect the oxidation/nitration of biomolecules such as lipids, protein and DNA. RNS are produced not only under pathological but also during physiological situations such as cellular metabolism, where they have beneficial effects. However, they can be detrimental when produced in high amounts in the intracellular compartments (Table 27.2). Cells generally respond to nitrosative stress by up-regulating antioxidants. Macrophage- and neutrophil-mediated immune responses involve the production and release of NO, which inhibits viruses, pathogens and tumor proliferation. NO reacts with ROS and acts as a detoxifier (ROS scavenger). It also acts on vessels to regulate blood flow, which is important for the adaptation of muscle for prolonged exercise.

Besides these beneficial effects, RNS are known to induce nitrosative stress in skeletal muscles, leading to increased proteolysis of muscle-specific structural proteins and their modifications. NO is also known to modify: insulin signaling, mitochondrial energy metabolism, endoplasmic reticulum stress, structure and function of protein, carbohydrates and lipids species, and creatine phosphokinase, an enzyme responsible for maintaining the phosphagen pool.

Peroxynitrite ($ONOO^-$) is a potent oxidant and nitrating species formed by rapid reaction of two free radicals: NO and $O_2^{\cdot-}$. It can modify a variety of biomolecules and can alter the functioning of intracellular organelles. $ONOO^-$ can modify different proteins but possesses high affinity for tyrosine residues in protein molecules; hence, 3-nitrotyrosine is a relatively specific marker of $ONOO^-$-mediated damage to proteins. Other markers of $ONOO^-$-induced protein modification are cysteine oxidations, oxidation/nitration of tryptophan, and protein carbonyls; dityrosine formation has been observed in atherosclerotic lesions, lung injury, hypertension, autoimmune myocarditis and rheumatoid arthritis (RA), etc. Furthermore, self proteins become immunologically active if their structure is altered. Accumulation of a variety of chemically modified proteins has been reported in inflamed tissues or apoptotic cells [24].

Besides, NO and its products have been reported to modify the structure of histones, collagen, and the structure and function of lipid molecules; effects on other macromolecules include altering the structure of DNA and rendering it immunogenic [25,26]. The potential role of nitrosative modifications affects the overall regulation of cellular functions (Figure 27.2).

TABLE 27.2 Physiopathological Properties of Nitric Oxide

S. No.	Properties
1	Potent antimicrobial activity of macrophages
2	Potent inhibitor of platelet aggregation, activation and adhesion through cGMP pathway
3	An important endogenous vasodilator and regulator of blood pressure
4	Acts as a neurotransmitter in brain and peripheral autonomic nervous system
5	Has a role in relaxation of skeletal muscles
6	Has a role in neurotoxicity
7	Has an important role in insulin signal transduction, through nitrosative modifications
8	Enzymes involved in cellular energy metabolism, including glycolysis, TCA cycle, and fatty acid oxidation, are susceptible to nitrosative modifications
9	Nitric oxide and its derivatives are also responsible for critically regulating gene transcription
10	Nitric oxide and its derivatives also influence key enzymes of lipid biosynthesis, e.g., COX-2 and cytochrome p-450.

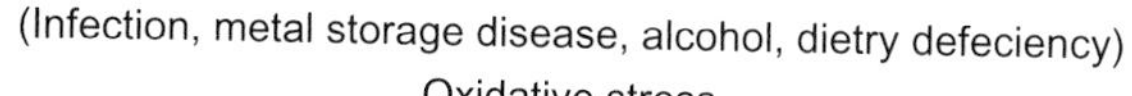

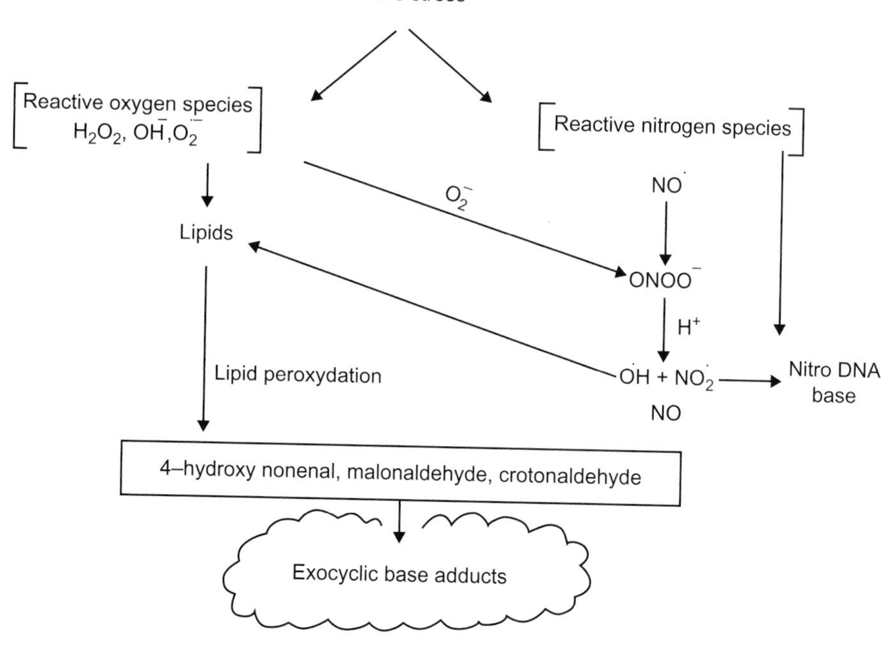

FIGURE 27.2 **Sources of oxidative and nitrosative damage.**

BIOCHEMISTRY OF NITROSATIVE PROTEIN MODIFICATIONS IN MUSCLES

Peroxynitrite is a powerful oxidant, far more reactive than its precursors, NO and $O_2^{\cdot-}$. Its reaction with proteins, inducing nitrosative modifications, is a selective process that targets precise molecular sites in proteins, leading to changes in their structure and function. In a biological system these modifications manifest mainly in two forms: either through S-nitrosylation of cysteine thiols or as nitration of tyrosine residues. $ONOO^-$-mediated tyrosine nitration proceeds through a radical mechanism, where it first reacts with carbon dioxide or metal centers and generates secondary nitrating species (Figure 27.3). Nitration is promoted by exposure of tyrosine, its location on a loop, its association with neighboring negative charge, and absence of proximal cysteine. Nitration is also enhanced in a hydrophobic environment because of the longer half-life of the nitrogen dioxide radical. $ONOO^-$ nitrates free or protein-bound tyrosine to form a stable product, 3-nitrotyrosine. S-nitrosylation occurs through the covalent attachment of a diatomic nitroso group to a reactive thiol sulfhydryl in a redox-dependent fashion.

Oxidative stress leads to the modification of collagen II (CII), an important component of human articular cartilage. In the inflamed joint, abnormally high fluxes of RNS give rise to chemical reactions that modify CII. It has been suggested that a breakdown of tolerance occurs because antibodies against modified self proteins are promiscuous and bind both the modified and unmodified self-antigen, leading to epitope spreading. Modifications of CII may contribute to the vicious cycle of chronicity by providing additional epitopes to which the immune system is intolerant, resulting in stimulation of the immune response against self antigens [27].

In terms of structure of skeletal muscles, elevated nitrite/nitrate concentrations modify skeletal muscle proteins by nitration leading to muscle wasting [28]. This is indicated by the presence of total ubiquitin conjugates and degradation fragments of myosin heavy chain and reduction in glutathione levels. Nitrosative stress is also responsible for promoting protein-S-glutathionylation, which plays an essential role in the control of cell signaling pathways of induced apoptosis and the mechanism involving messenger RNA stability. Nitrosative stress-induced protein tyrosine and tryptophan nitrations are also responsible for altered cytoskeletal protein function; tropomyosin, actin and myosin are reported to undergo structural and functional changes [29].

NO exerts control over vascular tone, platelet functions and prevents adhesion of leukocytes, but at an

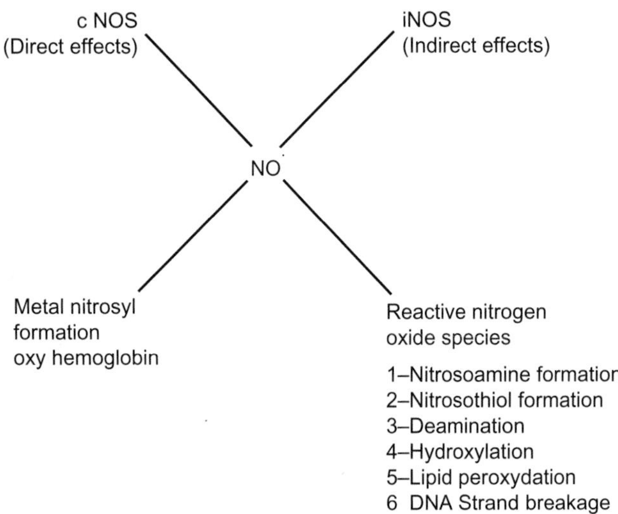

FIGURE 27.3 Nitrosative and oxidative reactions of nitric oxide.

increased concentration of ROS there is a decline in the amount of bioactive NO since it is converted to toxic $ONOO^-$. This can uncouple eNOS to become a dysfunctional RNS-generating enzyme that contributes to vascular oxidative stress and endothelial dysfunction, promoting atherogenesis.

Besides these effects on cellular proteins, nitrosative/nitrative stress is also known to inhibit creatine kinase activity [30] which is responsible for the maintenance of the phosphagen pool by forming phosphocreatine. The inhibition of this enzyme will down-regulate the ATP-phosphocreatine system that can provide energy at high rates, but only for a few seconds before the phosphocreatine (PCr) store is depleted. Increased NO concentration would result in a decline in creatine-involved temporal buffering, proton buffering and glycolysis regulation, because PCr is a limiting factor in maintaining ATP resynthesis during maximal short-term exercise. The amount of ATP in muscles is sufficient to sustain contractile activity for less than one second. The reservoir of high-energy phosphate in skeletal muscle is creatine phosphate. The ΔG^o of creatine phosphate is −10.3 kcal per mole, whereas that for ATP is only −7.5 kcal per mole. Resting muscle has a high concentration of creatine phosphate (25 mM) compared with ATP (4 mM). The creatine phosphate provides a high ATP concentration which is the major source of energy in athletes during the first 4 seconds of a short sprint. During muscle contraction, the ATP level remains high as long as creatine phosphate is present. But following contractile activity the levels of ADP and Pi rise. The reduced energy charge of active muscle stimulates glycogen breakdown, glycolysis, TCA cycle and oxidative phosphorylation.

NO and its derivative would result in reduced performance during short-term maximal exercise which is analogous to glycogen loading before endurance exercise. Increased level of RNS is known to affect training intensity, training stimulus and physiological adaptations to training.

NO also induces skeletal muscle insulin resistance through higher iNOS expression in some individuals, which would ultimately lead to diabetic complications. Induction of iNOS leads to NO production and formation of RNS which would result in the nitration of tyrosine at the insulin receptor and insulin receptor substrate, leading to improper insulin signal transduction and insulin resistance in skeletal muscle and liver. NO and RNS also interfere with the energy metabolism through S-nitrosylation and tyrosine nitration of enzymes involved in cellular energy metabolism, e.g., enzymes involved in glycolysis, TCA cycle, β oxidation, mitochondrial complex I and cytochrome c of the electron transport chain. Nitrosation of mitochondrial proteins using mitochondrial targeted S-nitrosothiol alters the activity of key TCA cycle enzymes in a reversible fashion, reducing the activity of important TCA cycle and electron transport proteins. This leads to the slow down of substrate oxidation and build-up of metabolic intermediates, affecting the overall potential of an athlete [31].

NO and RNS are also known to induce endoplasmic/sarcoplasmic stress through the inhibition of sarcoplasmic/endoplasmic reticulum Ca^{+2} (SERCA) and Ryanodine (RyR) receptors. This results in a negative effect on calcium homeostasis for maintaining cellular calcium stores responsible for activities of endoplasmic chaperones: e.g., protein disulfide isomerase, calreticulin and calnexin that are essential for protein folding and disulfide bond formation of newly synthesized proteins and their transport. Therefore, disruption of calcium homeostasis may lead to accumulation of unfolded or misfolded proteins in neurons, endothelial cells and foam cells, thus affecting overall protein maturation and apoptosis.

References

[1] Nathan C, Xie Q. Regulation of biosynthesis of nitric oxide. J Biol Chem 1994;926:13725–8.
[2] Bredt DS, Snyder SH. Nitric oxide a novel neuronal messenger. Neuron 1992;8:3–11.
[3] Wink DA, Vodototz Y, Laval J, Laval F, Dewhirst MW, Mitchell JB. The multifacet roles of nitric oxide in cancer. Carcinogenesis 1998;19:711–21.
[4] Uppu RM, Squadrito GL, Pryor WA. Acceleration of peroxynitrite oxidation by carbon dioxide. Arch Biochem Biophys 1996;327:335–43.
[5] Usmar VD, Wiseman H, Halliwell B. Nitric oxide and oxygen radicals: a question of balance. FEBS lett 1995;369:131–5.

[6] Xio L, Eneroth PHE, Qureshi GA. Nitric oxide synthase pathway may mediate human natural killer cell cytotoxicity. Scand J Immunol 1995;42:505–11.
[7] Gorbunov NV, Osipov AN, Day BW, Zayas RB, Kagan VE, Sayed NM. Reduction of ferryl myoglobin and ferryl hemoglobin by nitric oxide: A protective mechanism against ferryl hemoprotein induced oxidations. Biochemistry 1995;34:6689–99.
[8] Wiseman DA, Kalwat MA, Thurmond DC. Stimulus-induced S-nitrosylation of syntaxin 4 impacts insulin granule exocytosis. J Biol Chem 2011;286:16344–54.
[9] Wu D, Yotnda P. Production and detection of reactive oxygen species (ROS) in cancers. J Vis Exp 2011;57:3357–60.
[10] Rodrigue NR, Di Marco NM, Langley S. American college of sports medicine position stand. Nutrition and athletic performance. Med Sci Sports Exer 2009;41:709–31.
[11] Paquot N. Sports nutrition. Rev Med Liege 2001;56:200–3.
[12] Borsheim E, Tipton KD, Wolf SE, Wolfe RR. Essential amino acids and muscle protein recovery from resistance exercise. Am J Physiol Endocrinol Metab 2002;283:E648–57.
[13] Lindinger MI. Exercise in the heat: thermo regulatory limitations to performance in humans and horses. Can J Appl Physiol 1991;24:152–63.
[14] Tangphao O, Chalon S, Moreno H, Hoffman BB, Blaschke TF. Pharmacokineties of L-arginine during chronic administration to patients with hypercholesterolaemia. Clin Sci (Lond) 1999;96:199–207.
[15] Tsao PS, Theilmeier G, Singer AH, Leung LL, Cooke JP. L- arginine attenuates platelet reactivity in hypercholesterolemic rabbits. Arterioscler Thromb 1994;14:1529–33.
[16] Cooke JP, Tsao PS. Is NO an endogenous antiatherogenic molecule? Arterioscler Thromb 1994;14:653–5.
[17] Fisker S, Nielsen S, Ebdrup L, Bech JN, Christiansen JS, Pedersen EB, et al. The role of nitric oxide in L-arginine-stimulated growth hormone release. J Endocrinol Invest 1999;22:89–93.
[18] Fister S, Nielsen S, Ebdrup L, Bech JN, Christiansen JS, Pedersen EB, et al. L-arginine-induced growth hormone secretion is not influenced by co-infusion of the nitric oxide synthase inhibitor N-monomethyl-L-arginine in healthy men. Growth horm IGF Res 1999;9:69–73.
[19] Merimee TJ, Rabinowitz D, Riggs L, Burgess JA, Rimoin Dl, McKusick VA. Plasma growth hormone after arginine infusion clinical experiences. N Engl J Med 1967;276:434–9.
[20] Greenhaff PL. The creatine-phosphocreatine system: there's more than one song in its repertoire. J Physiol 2001;537:657–8.
[21] Hultman E, Soderlund K, Timmons JA, Cederblad G, Greenhaff PL. Muscle creatine loading in men. J Appl Physiol 1996;81:232–7.
[22] Kosjalka TR, Andrew CL, Brent RL. Effect of insulin on the uptake of creatine-1-14C by skeletal muscle in normal and X-irradiated rats. Proc Soc Exp Biol Med 1972;139:1265–71.
[23] Haugland RB, Chang DT. Insulin effect on creatine transport in skeletal muscle (38464). Proc Soc Exp Biol Med 1975;148:1–4.
[24] Khan MA, Dixit K, Jabeen S, Moinuddin, Alam K. Impact of peroxynitrite modification on structure and immunogenicity of H2A histone. Scand J Immunol 2008;69:99–109.
[25] Habib S, Moinuddin, Ali R. Acquired antigenicity of DNA after modification with peroxynitrite. Int J Biol Macromol 2005;35:221–5.
[26] Habib S, Moinuddin, Ali R. Peroxynitrite modified DNA: a better antigen for systemic lupus erythematosis anti-DNA autoantibodies. Biotechnol Appl Biochem 2006;43:65–70.
[27] Shahab U, Ahmad S, Moinuddin, Dixit K, Habib S, Alam K, et al. Hydroxy radical modification of collagen type II increases its arthritogenicity and immunogenicity. PLOS One 2012;7:e 31199.
[28] Wang YY, Lin SY, Chuang YH, Mao CH, Jung KC, Sheu WH. Protein nitration is associated with increased proteolysis in skeletal muscle of bile duct ligation-induced cirrhotic rats. Metabolism 2010;59:468–72.
[29] Starr ME, Ueda J, Yamamoto S, Evers BM, Saito H. The effects of aging on pulmonary oxidative damage, protein nitration, and extracellular superoxide dismutase down-regulation during systemic inflammation. Free Rad Biol Med 2011;50:371–80.
[30] Gross WL, Bak MI, Ingwall JS, Arstall MA, Smith TW, Balligand JL, et al. Nitric oxide inhibits creatine kinase and regulates heart contractive reserve. Proc Natl Acad Sci USA 1996;93:5604–9.
[31] Pacher P, Beckman SJ, Lucas L. Nitric oxide and peroxynitrite in health and disease. Physiol Rev 2007;87:315–424.

CHAPTER

28

Blood Rheology, Blood Flow and Human Health

Philippe Connes[1,2,3], *Stéphane Dufour*[4,5], *Aurélien Pichon*[3,6] *and Fabrice Favret*[4,5]

[1]UMR Inserm 665, Centre Hospitalier Universitaire [2]Laboratoire ACTES EA3596, Département de Physiologie, Pointe à Pitre, Guadeloupe [3]Laboratory of Excellence GR-Ex, PRES Sorbonne Paris Cité, Paris, France [4]Université de Strasbourg, Faculté des Sciences du Sport, Strasbourg, France, [5]Faculté de Médecine, Strasbourg, France [6]Université Paris, Paris Sorbonne Cité, Bobigny, France

BLOOD FLOW: CHARACTERISTICS, EXERCISE AND TRAINING

Current Knowledge on Skeletal Muscle Blood Flow Control

The amount of blood flowing into the large blood vessels feeding human skeletal muscles and subsequently perfusing the microvascular network is a tightly controlled variable based on the interplay between central (i.e., cardiac function) and peripheral (i.e., local skeletal muscle vasodilation) regulatory inputs [1–3]. In healthy subjects, skeletal muscle blood flow is tremendously well preserved to ensure that blood and oxygen (O_2) supply match O_2 demand over a wide range of exercise intensities [4]. According to the Poiseuille's Law [5], the increase in blood flow Q is governed by perfusion pressure ΔP, structural properties of the vascular network (length L and radius r), and blood viscosity η:

$$\dot{Q} = \frac{\Delta P \cdot \pi \cdot r^4}{8 \cdot L \cdot \eta} \qquad (28.1)$$

Both central and peripheral regulatory inputs arise from the integration of multiple signals mainly of neural, metabolic and mechanical origins [6], with the most prominent control being probably exerted by local skeletal muscle vasodilation [7], despite the exercise-induced elevation in sympathetic activity and its associated vasoconstrictor effect [8].

Signals of neural and/or mechanical origins have been proposed as likely mediators of the early increase (within 1 s) in blood flow observed at the beginning of exercise [9,10]. Most of the accumulating evidence currently suggests that neural signals together with the muscle pump are not mandatory in contraction-induced vasodilation and rather indicates an indirect mechanical control of early vasodilation [6,11]. The exact processes involved in the mechanically induced vasodilation are not clear but could be mediated by both endothelium-dependent and -independent pathways [12].

Once the early vasodilation is initiated, continuous muscle perfusion and delivery of O_2 requires the vasodilation to be sustained to ensure the continuation of exercise. Blood flow remarkably follows the energetic requirements of the contracting muscle tissue [13], making it appealing that metabolic signals produced by the contracting muscles and/or the bloodstream might contribute to the dynamic control of vascular tone either directly or through their stimulating effects on the endothelium [14]. Many vasoactive agents produced by the active muscles have been studied, including pCO_2, lactate, K^+ and adenosine [15]. However, their exact contribution to the exercise-induced vasodilation and hyperemia is still confusing [16] and might be combined with the potential vasoactive effect of other identified muscle-derived compounds such as interleukin 6 and 8 as well as nitric oxide (NO), prostacyclin and endothelium-derived

Nutrition and Enhanced Sports Performance.
DOI: http://dx.doi.org/10.1016/B978-0-12-396454-0.00028-X

hyperpolarizing factor (EDHF) [16] with or without mechanical stress [17,18]. Moreover, circulating erythrocytes can also release ATP and NO in plasma when they undergo de-oxygenation and/or mechanical deformation [19]. ATP binds to P2Y purinergic receptors on the endothelial muscle cells, where it triggers NO production [20]. This increase in NO activates soluble guanylyl cyclase in the vascular smooth muscle, increasing cGMP, which activates protein kinase G (PKG), leading to vasodilation [21].

The vasodilation initiated at the level of the terminal arterioles in the skeletal muscle vascular network needs to be conducted upstream, a phenomenon called "conducted vasodilation", in order to increase blood flow in the main feeding arteries. This vasodilation "spreading" would also facilitate a coordinated and homogenous distribution of blood flow within the active muscle [22,23].

As discussed in the section below, the mechanisms involved in the control of exercise-induced vasodilation and increased blood flow are well integrated and result in appropriate vascular functional responses in various conditions of submaximal but not maximal or supramaximal exercise intensities [24].

Blood Flow Responses to Acute Aerobic Exercise

The femoral artery blood flow during a rest-to-work transition exhibits a very large augmentation with a ~20 fold increase at the most extreme intensities of knee-extension exercises [25]. Many studies report that, under normal environmental conditions (normobaric and thermoneutral conditions), skeletal muscle blood flow increases linearly with the elevation of exercise intensity to match oxygen supply to oxygen demand [26,27]. This observation holds true for small muscle mass exercises, where skeletal muscle blood flow is mostly supported by profound vasodilation of the resistance vessels feeding the active muscle tissues (i.e., peripheral components) in combination with the increased cardiac output, perfusion pressure and blood flow redistribution (i.e., central components) [7]. At a given constant exercise intensity, skeletal muscle blood flow can eventually stabilize or progressively increase depending on the development of the slow component of oxygen uptake [28]. This slowly developing elevation of skeletal muscle blood flow and oxygen uptake is especially observed at exercise intensities greater than 75% of knee-extension peak power [28].

If the exercise involves a larger muscle mass (i.e., cycling or running), the response of skeletal muscle blood flow to increasing intensity is not linear. Indeed, the extent of vasodilation observed during knee-extension exercise applied to the whole active muscle mass would require enormous values of cardiac output, approaching 100 L/min, by far exceeding the pumping capacity of the human heart and thereby compromising the maintenance of an adequate arterial blood pressure [29]. Therefore, muscle vasodilation during whole-body exercises is largely restrained at high exercise intensities such that skeletal muscle blood flow typically increases linearly with increasing exercise intensity up to ~80% peak power and subsequently levels off or even declines when the subjects approach or surpass maximal aerobic power [4,13].

The skeletal muscle blood flow response to acute exercise can be profoundly altered by environmental conditions such as varying ambient O_2 availability and/or temperature. Hypoxia exposure causes a decrease of O_2 content in the arterial blood [30]. At rest and during submaximal knee-extension exercise intensities, the diminished arterial O_2 content triggers a compensatory vasodilation, increasing skeletal muscle blood flow such that bulk O_2 delivery and tissue O_2 uptake are maintained [14,31]. However, at peak knee-extension exercise, the hypoxic compensatory vasodilation is blunted such that peak muscle blood flow is reduced to normoxic value, leading to attenuated bulk O_2 delivery and maximal muscle O_2 uptake [32]. During whole-body exercise, similar observations pertained, with the exception that hypoxic environments markedly reduce peak muscle blood flow and exacerbate the hypoxia-induced reduction in bulk O_2 delivery and muscle O_2 uptake compared with knee-extension exercise [32].

Local muscle perfusion can also be tremendously altered by the consequences of changes in ambient temperature. The exacerbated thermoregulatory demand induced by hot environments leads to a competition between skeletal muscle and skin tissues for blood flow [33] to dissipate heat. However, with acute exposure to hot environments, muscles usually win and blood is not taken away from muscle tissue to feed the skin territories [34]. Indeed, heat stress *per se* has been reported to increase muscle blood flow both at rest and during submaximal small muscle mass exercises [35,36]. Nevertheless, a frequent consequence of heat stress is the development of progressive dehydration and its associated hemoconcentration [34]. During constant-load moderate-intensity exercise, the dehydration-induced hemoconcentration and parallel increase in arterial O_2 content combine with a reduced muscle blood flow to preserve bulk O_2 delivery such that muscle O_2 uptake does not decline until the subjects get very close to exhaustion [37]. However, when the intensity of exercise is higher (typically >80% peak power output), exercise tolerance and maximal oxygen uptake are markedly lowered with heat stress despite hemoconcentration, mainly as a result of the dramatic reduction in muscle blood flow and bulk O_2 delivery due to skin and peripheral vasodilation [24].

Training-Induced Improvement in Skeletal Muscle Blood Flow

Aerobic exercise training is known to induce structural and functional changes in the cardiovascular system, improve peak muscle blood flow during maximal knee-extension exercise or cycling exercise [38–41] to approximately the same extent (+ ~20–30%), and enhance maximal muscle oxygen uptake [25,27,42]. Conversely, peak muscle blood flows are markedly reduced in various pathologic states [43–46], but normalization may be achieved by exercise training [47].

The training-induced improvement in skeletal muscle blood flow is thought to be principally mediated by increased functional vasodilation of the resistance vessels and/or enlarged anatomical vascular cross-sectional area [26,39]. At the microvascular level, enlarged muscle capillary network has been documented after exercise training [48], presumably contributing to the improvement of the blood/myocyte surface exchange area, O_2 diffusion and muscle O_2 uptake [49]. How exercise training ultimately triggers structural adaptations (i.e., angiogenesis) in the vascular network might involve mechanical, intravascular and blood rheology-based signals as well as metabolic sensors. Both hypoxia-Inducible Factor-1α stabilization and Vascular Endothelial Growth Factor expression are involved in training-induced angiogenesis [50,51]. This angiogenesis allows the improvement of skeletal muscle blood flow as well as O_2 diffusion capacity, increasing aerobic performance [50,52,53].

HEMORHEOLOGY: INTERACTIONS WITH BLOOD FLOW, EXERCISE AND TRAINING

Classical Concepts in Hemorheology

Hemorheology deals with the study of the biophysical properties and flow properties of blood. Classically, scientists summarize the blood rheology field as being the study of blood viscosity. According to the Poiseuille law (Eq. 28.1), vascular geometry is the most important factor of blood flow resistance and, consequently, the impact of blood viscosity is often ignored, except in extreme cases such as in polycythemia where blood hyper-viscosity may promote thrombotic complications and increase cardiac work. However, as discussed below, blood rheology is a key regulator of vascular function and vascular resistance.

Blood Viscosity Regulates Vasodilation

While an increase of blood viscosity is usually considered as a risk for an increase in blood flow resistance, several studies demonstrated that a moderate rise in blood viscosity (if not supra-physiologic) may be very positive for the endothelial function and vasomotor tone [54,55] or even endurance performance [56]. A rise in blood viscosity causes an increase of the shear stress applied on the endothelial cells, leading to increased nitric oxide (NO) production by the endothelial NO synthase and facilitating vasodilation, tissue perfusion and O_2 delivery [54,55,57]. Patients with chronic vascular disease and endothelial dysfunction may be unable to cope as well as the healthy population with an increase of blood viscosity.

Acute physical exercise usually causes a rise of blood viscosity (+10–12%), which is mainly due to the decrease in plasma volume resulting in higher hematocrit (hemoconcentration), whereas training decreases hematocrit (auto-hemodilution phenomenon) resulting in a decrease of blood viscosity [58–60]. The physiological consequences of these acute changes during exercise have been poorly studied, but a recent work [60] studied the relationships between the changes in blood viscosity, vascular resistance, NO production and oxygen consumption induced by a submaximal exercise in a healthy population. Their findings suggested that a marked rise in blood viscosity during exercise may be necessary for NO production and adequate vasodilation, to reach the highest aerobic performance [60]. Nevertheless, an inadequate increase in hematocrit could also increase blood viscosity to very high level, then affecting cardiac and aerobic performance despite the increase in arterial O_2 content. Schuler et al. demonstrated that there is an optimal level of hematocrit and/or blood viscosity to reach the highest aerobic performance [56]. This optimal level could be perhaps obtained after an exposition to moderate hypoxia (live-high, train-low model), even if obvious positive results on aerobic performance remain to be assessed [61].

Other Hemorheological Factors, Blood Flow and Oxygen Delivery

Blood viscosity is influenced by several other hemorheological/hematological parameters such as plasma viscosity, red blood cell (RBC) deformability and aggregation, depending on the vascular compartment and the flow conditions. Blood is a shear-thinning fluid, meaning that increasing shear rate, which depends on vessel radius and blood flow, causes a decrease of the viscosity. RBC deformability and aggregation may affect blood flow in the macro- and micro-circulation independently of their effects on blood viscosity [62].

RBC Deformability

At high shear rate (arteries, large arterioles), any decrease in RBC deformability will increase blood viscosity because of the distortion of flow streamlines. In the microcirculation, such as in capillaries, RBC must extremely deform in order to pass through the vessels and reach the tissues to deliver oxygen. RBC deformability is determined by the geometric properties, cytoplasmic composition and viscosity, and membrane properties of RBC [62]. Exercise training improves baseline RBC deformability [63,64]. Having pliable RBCs is beneficial for microcirculatory blood flow, tissue oxygen delivery and aerobic performance [63]. During acute exercise, RBC deformability often decreases, probably because of the increased lactic acid production [60] and enhanced oxidative stress [65] caused by the physical effort. Lactic acid rapidly dissociates when it accumulates into blood, and both hydrogen ions and lactate anion may enter into RBCs, through the monocarboxylate transporter 1, accumulate and then decrease RBC deformability [66]. The production of reactive oxygen species during exercise may also affect the RBC membrane integrity by proteins and lipids oxidation [65,67]. However, some studies also reported a lack of change or a surprising improvement of RBC deformability in exercising people, and these specific responses seem to depend on the training status of subjects and the kind of exercise performed [59,68]. The greater ability of RBCs to be deformed under flow in some subjects could be positive for vasodilation since RBC stretching increases the release of ATP [69] and NO [70], two factors involved in vasomotor tone regulation. Several diseases are marked by RBC deformability alterations, and the best known is probably sickle cell anemia. The loss of RBC deformability in this disease is responsible for vaso-occlusion at the microcirculatory level and tissue ischemia. Waltz et al. [71] recently reported a positive association between RBC deformability and cerebral oxygenation in these sickle cell patients, emphasizing the key role of this RBC property in tissue oxygenation, even at the brain level [71].

RBC Aggregation

At low shear rate (veins, venules and small arterioles), blood viscosity is greatly influenced by RBC aggregation. RBC aggregation is a physiological and reversible process dominating under low shear or stasis conditions, characterized by the formation of regular structures where RBCs form face-to-face contacts resulting in structures looking like stacks of coins. It depends on cellular factors (i.e., RBC aggregability) and plasmatic factors such as fibrinogen level [72]. Inflammation and oxidative stress may promote RBC aggregation [73,74]. It appears that RBC aggregation that is too low or too high may adversely affect blood flow [75]. Increased RBC aggregation has been observed in various pathological situations such as diabetes [76] and cardiovascular disease [62,77].

Very few studies have investigated RBC aggregation responses to exercise/training [77], and most of them found that RBC aggregation increases with exercise, mainly because of the rise in plasma fibrinogen concentration. Simmonds et al. recently reported that increased RBC aggregation was associated with a decrease in oxygen consumption during submaximal exercise, suggesting a role of RBC aggregation in aerobic exercise capacity [78].

NUTRITION, BLOOD FLOW AND BLOOD RHEOLOGY

This section focuses on a selection of nutritional supplements that are thought or proven to impact on the cardiovascular system, skeletal muscle blood flow properties, and exercise performance. Other supplements than those discussed can be found in the literature, but we believe that those we discuss below represent the main ones experimentally tested.

NO Metabolism

NO is synthetized from L-arginine by NO synthase. Indeed, L-arginine has been considered as a potential ergogenic molecule that may improve skeletal muscle blood flow during exercise and then oxygen and nutrient delivery to active skeletal muscle [79]. Although L-arginine supplementation has beneficial effects on exercise capacity in various diseases, through an improvement of coronary and/or peripheral blood flow [80,81], its effect in healthy subject is still debated [80]. Studies involving untrained or moderately trained healthy subjects showed that NO donors could improve exercise tolerance, but no effect was demonstrated in well-trained subjects [79,82]. Moreover, it remains difficult to attribute the improvement of performance in healthy sedentary subjects to an increase in skeletal muscle blood flow since some studies reported that L-arginine supplementation does not improve skeletal blood flow [83,84].

The increase in plasma nitrite/nitrate (NOx) content, through nitrate rich diet (i.e. beetroot juice), might improve endurance performance capacity in peripheral artery disease patients [85] and moderately trained subjects [86,87] but not in high-level endurance-trained athletes [88,89]. When present, the beneficial effect of increased plasma NOx is thought to be mediated by subsequent increases in NO, which is known for its multiple physiological effects including stimulation of vasodilation and increase in blood flow but also modulation of muscle contraction and glucose uptake [90].

Recent evidence suggests that exercise capacity in endurance athletes is correlated with basal levels of plasma nitrite, albeit not with brachial endothelial function [91]. Therefore, the beneficial effect of elevated plasma nitrite/nitrate levels observed in endurance athletes might not be mechanistically linked to improved muscle blood flow and O_2 delivery but may likely be the result of NO-dependent improvement of mitochondrial function and optimized O_2 utilization leading to reduced O_2 cost of exercise [92,93]. However, a relationship has been well described between plasma NOx and basal endothelial function in trained [91] and untrained subjects [94]. Moreover, the plasma nitrite level observed at exhaustion during a progressive exercise test, taken as an index of NO synthase activity, is associated with endothelial function and exercise capacity in healthy subjects [95]. How much of the effect of plasma NOx can be ascribed to vascular effects is currently unclear, but plasma NOx have strong documented evidence towards reduced mean arterial pressure [96], increased vasodilation *in vitro* [97] and *in vivo* [98,99], presumably via the reduction of nitrite to NO by deoxyhemoglobin [98]. This mechanism is attracting substantial attention as it may be involved in hypoxic hyperemia [100] and hypoxia tolerance [101,102].

Hydration

The level of body hydration is crucial for human exercise performance [103–105], and a decrease of more than 2% of body weight alters endurance exercise capacity [106]. In the ecological field, it seems that the dehydration level affecting exercise performance is higher (4%) than in laboratory and standardized conditions [105]. Thermal loading is also a key factor of performance and, while it may improve short-duration exercise performance, it may affect endurance performance, particularly when associated with significant dehydration [107,108]. Under heat stress, dehydration causes large reductions in cardiac output and blood flow to the exercising musculature, impairing endurance performance [109] despite no significant or limited effect on blood rheology [110]. At a given submaximal intensity, dehydration and not heat stress alone causes a significant decrease in stroke volume and an increase in heart rate because of the reduction in blood volume [111,112]. However, when dehydration occurs in a hot environment, the decrease in stroke volume is amplified and the significant decrease in cardiac output leads to inadequate cardiovascular function with a rise in systemic vascular resistance and a fall of the mean arterial pressure [111]. This decrease in cardiac output could even lead to a decline in femoral and exercising muscle blood flow as a result of reduction in perfusion pressure [37].

Fluid replacement reduces the risk of heat illness, limits the cardiovascular stress and improves exercise performance by preventing or reducing dehydration [113]. Optimal performance is possible only when dehydration is minimized and sweat loss partially or totally compensated. Hydration also normalizes blood rheology during exercise, despite the stress caused by heat exposure [110,114]. Moreover, the inclusion of sodium in the rehydration solution restores plasma volume when ingested during exercise, and expands plasma volume if ingested pre-exercise [115].

Antioxidants

Exercise is characterized by an increase in oxygen reactive species production [116], which may either favor mitochondrial biogenesis or have deleterious effects by triggering apoptosis. Oxidative stress may impair NO metabolism, leading to alterations in vasodilation processes and, ultimately (if chronic or exaggerated), endothelial function [117]. Moreover, *in vitro*, superoxide anions decrease RBC deformability and increase the strength of RBC aggregates [118]. On the whole, exaggerated oxidative stress is widely suspected to cause cardiovascular impairment, but some of the alterations reported in the literature depend on the training status of the subjects/animals. For example, during exercise, oxidative stress decreases RBC deformability and increases osmotic fragility in untrained but not in trained rats [119]. Moreover, the exercise-induced oxidative stress is responsible for hemolysis in sedentary subjects, but not in trained humans [67].

Several studies have focused on the potential effects of antioxidants to reduce oxidative stress and improve aerobic exercise performance in both untrained and trained subjects [120,121]. The antioxidant vitamin C (ascorbic acid) and vitamin E (α, β, δ, and γ tocopherols and tocotrienols) have been commonly used in order to improve performance [122]. Kayatekin et al. demonstrated that the effects of exercise on RBC adhesion to the endothelium and RBC aggregation were not different between rats receiving vitamin C, E and zinc compared with non-supplemented untrained rats [123], suggesting limited action of these antioxidants on the RBC changes induced by exercise. Nevertheless, in another study, the oxidative stress induced by exhausting exercise was offset in sedentary subjects after a 2-month regimen in vitamins A, C and E. The various populations, the various mixtures of vitamins (at different doses) and the different exercises (type, duration, intensity) used in the studies may explain these controversies.

Polyphenols could also impact on the cardiovascular system and performance. A cumulative positive effect of exercise training and antioxidant supplementation (red wine!) has been observed on systolic blood pressure and high-density lipoprotein in hypertensive rats [124]. Red wine polyphenols could also act positively on aging-related endothelial dysfunction and decline in physical performance by normalizing oxidative stress and the expression of proteins involved in the formation of NO and the angiotensin II pathway [125]. Finally, slight consumption of red wine for 3 weeks decreased blood viscosity and RBC aggregation and improved RBC deformability: a finding which contributes to the "French" paradox [126].

In summary, the effects of antioxidant supplementation on the cardiovascular function and exercise performance differ according to the antioxidant characteristic, as described previously, but some kinds of polyphenols could significantly improve performance depending on the training status of the athletes.

Mineral and Trace Elements

Zinc plays a critical role in maintaining normal vascular function through its antioxidant and anti-inflammatory effects [127,128]. Athletes with low serum zinc level have elevated blood viscosity and reduced RBC deformability [59], and have a decrease in exercise performance [59,129]. Esen et al. demonstrated that zinc supplementation in humans increases skeletal muscle blood flow but this is not related to the changes in endothelial activity or vascular function [130]. Instead, exogenous zinc improves RBC deformability and RBC aggregation, leading to a decrease of blood viscosity, which in turn increases blood flow [130]. One could speculate that the greater RBC deformability and lower blood viscosity achieved under zinc supplementation [130,131] played a role in the lower anaerobic contribution to a 30 min endurance exercise in rats [132]: improved blood rheology may facilitate perfusion and oxygen tissue delivery. Experimentally, a double blind randomized trial of oral zinc supplementation in healthy volunteers demonstrated beneficial effect of zinc on blood viscosity and RBC aggregation, but no measurable effect was detected on exercise performance [133]. But, interestingly, in another study, Kaya et al. demonstrated a positive effect of melatonin supplementation on energetic metabolism and exercise endurance in swimming rats, and these effects seemed to be mediated by zinc, which increased under melatonin supplementation [134].

Another mineral, iron, is frequently lacking in athletes. In elite athletes, plasma ferritin has been observed to be negatively correlated with blood viscosity [135]. Mild iron deficiency commonly seen in athletes, before anemia occurs, is associated with an increase in blood viscosity and RBC aggregation, together with an increased subjective feeling of exercise overload [59]. Even without anemia, a low iron level is likely to impair exercise performance [136], but there are still some controversies concerning the benefit of iron supplementation in athletes [131].

Magnesium influences vascular tone by regulating endothelial function and vascular smooth muscle cell contraction [137]. Magnesium may stimulate the production of vasodilators, such as NO and prostacyclin, and promote endothelium-dependent and endothelium-independent vasodilation [138,139]. Animals with magnesium deficiency have high levels of endothelin-1 (a powerful vasoconstrictor), whose values are reduced after supplementation of this mineral [140]. Oral magnesium supplementation has been found to be useful in several diseases (except hypertension where the results are controversial [141]), with specific effects on platelet-dependent thrombosis [142], endothelial and vascular function [143] and RBC dehydration [144]. Magnesium deprivation also reduces endurance performance [136]. The main explanation is that magnesium is required for the activation of ATPases, which are involved in ATP synthesis. In addition, RBC deformability seems to vary with magnesium levels in athletes [145], suggesting that its level should be equilibrated to normal in athletes. All these effects have the capacity to improve vascular function and exercise performance, although the effects seem to depend on the athletes' training status [146].

CONCLUSION

Blood flow is modulated during exercise by the changes in perfusion pressure, vasodilation and blood rheology in normal conditions. Training-induced angiogenesis, vasodilation and hemorheological adaptations improve vascular function and blood flow. Nutritional supplements have been proposed in the literature for their ability to modulate skeletal muscle blood flow during acute or chronic exercise training. The present chapter mainly shows that:

1. Blood plasma nitrite/nitrate enrichment might improve endurance performance capacity in

patients and, potentially, in healthy subjects. However, the positive role of NO donors on blood flow or exercise performance is not apparent in well trained subjects.

2. Although antioxidants offer protection from the oxidative damage caused by exercise, few have demonstrated efficiency to improve exercise performance, with the most interesting, from our point of view, being polyphenols.
3. Minerals such as zinc and magnesium, by their positive effects on vascular function and blood rheology, seem to play a significant role in exercise performance, and body deficiency in one of them has deleterious effects.

In general, this chapter shows that nutritional supplementation has only small positive effects on blood flow or rheology, and positive effects are mainly reported in cases of deficit or illness.

References

[1] Gonzalez-Alonso J, Mortensen SP, Jeppesen TD, Ali L, Barker H, Damsgaard R, et al. Haemodynamic responses to exercise, ATP infusion and thigh compression in humans: insight into the role of muscle mechanisms on cardiovascular function. J Physiol 2008;586(9):2405–17.

[2] Dufour SP, Dawson EA, Stohr EJ. Central versus peripheral control of cardiac output in humans: insight from atrial pacing. J Physiol 2012;590(Pt 20):4977–8.

[3] Johnson BD, Harvey RE, Barnes JN. Exercise: where the body leads and the heart must follow. J Physiol 2012;590 (Pt 17):4127–8.

[4] Mortensen SP, Dawson EA, Yoshiga CC, Dalsgaard MK, Damsgaard R, Secher NH, et al. Limitations to systemic and locomotor limb muscle oxygen delivery and uptake during maximal exercise in humans. J Physiol 2005;566(Pt 1):273–85.

[5] Poiseuille JLM. Recherches sur les causes du mouvement du sang dans les vaiseaux capillaires. C R Acad Sci 1835;6:554–60.

[6] Clifford PS. Skeletal muscle vasodilatation at the onset of exercise. J Physiol 2007;583(Pt 3):825–33.

[7] Bada AA, Svendsen JH, Secher NH, Saltin B, Mortensen SP. Peripheral vasodilatation determines cardiac output in exercising humans: insight from atrial pacing. J Physiol 2012;590(Pt 8):2051–60.

[8] Thomas GD, Segal SS. Neural control of muscle blood flow during exercise. J Appl Physiol 2004;97(2):731–8.

[9] Clifford PS, Hellsten Y. Vasodilatory mechanisms in contracting skeletal muscle. J Appl Physiol 2004;97(1):393–403.

[10] Kirby BS, Carlson RE, Markwald RR, Voyles WF, Dinenno FA. Mechanical influences on skeletal muscle vascular tone in humans: insight into contraction-induced rapid vasodilatation. J Physiol 2007;583(Pt 3):861–74.

[11] Joyner MJ, Wilkins BW. Exercise hyperaemia: is anything obligatory but the hyperaemia? J Physiol 2007;583(Pt 3):855–60.

[12] Clifford PS, Kluess HA, Hamann JJ, Buckwalter JB, Jasperse JL. Mechanical compression elicits vasodilatation in rat skeletal muscle feed arteries. J Physiol 2006;572(Pt 2):561–7.

[13] Mortensen SP, Damsgaard R, Dawson EA, Secher NH, Gonzalez-Alonso J. Restrictions in systemic and locomotor skeletal muscle perfusion, oxygen supply and VO2 during high-intensity whole-body exercise in humans. J Physiol 2008;586 (10):2621–35.

[14] Casey DP, Joyner MJ. Compensatory vasodilatation during hypoxic exercise: mechanisms responsible for matching oxygen supply to demand. J Physiol. In Press Oct 22.

[15] Radegran G, Calbet JA. Role of adenosine in exercise-induced human skeletal muscle vasodilatation. Acta Physiol Scand 2001;171(2):177–85.

[16] Sarelius I, Pohl U. Control of muscle blood flow during exercise: local factors and integrative mechanisms. Acta Physiol (Oxf) 2010;199(4):349–65.

[17] Loot AE, Popp R, Fisslthaler B, Vriens J, Nilius B, Fleming I. Role of cytochrome P450-dependent transient receptor potential V4 activation in flow-induced vasodilatation. Cardiovasc Res 2008;80(3):445–52.

[18] Sun D, Huang A, Kaley G. Mechanical compression elicits NO-dependent increases in coronary flow. Am J Physiol Heart Circ Physiol 2004;287(6):H2454–60.

[19] Ellsworth ML, Sprague RS. Regulation of blood flow distribution in skeletal muscle: role of erythrocyte-released ATP. J Physiol 2012;590(Pt 20):4985–91.

[20] Burnstock G. Purinergic regulation of vascular tone and remodelling. Auton Autacoid Pharmacol 2009;29(3):63–72.

[21] Pandolfi A, De Filippis EA. Chronic hyperglicemia and nitric oxide bioavailability play a pivotal role in pro-atherogenic vascular modifications. Genes Nutr 2007;2(2):195–208.

[22] de Wit C. Different pathways with distinct properties conduct dilations in the microcirculation in vivo. Cardiovasc Res 2010;85(3):604–13.

[23] Twynstra J, Ruiz DA, Murrant CL. Functional coordination of the spread of vasodilations through skeletal muscle microvasculature: implications for blood flow control. Acta Physiol (Oxf) 2012;206(4):229–41.

[24] Gonzalez-Alonso J, Calbet JA. Reductions in systemic and skeletal muscle blood flow and oxygen delivery limit maximal aerobic capacity in humans. Circulation 2003;107(6):824–30.

[25] Richardson RS, Poole DC, Knight DR, Kurdak SS, Hogan MC, Grassi B, et al. High muscle blood flow in man: is maximal O2 extraction compromised? J Appl Physiol 1993;75(4):1911–6.

[26] Calbet JA, Lundby C. Skeletal muscle vasodilatation during maximal exercise in health and disease. J Physiol 2012;590 (Pt 24):6285–96.

[27] Andersen P, Saltin B. Maximal perfusion of skeletal muscle in man. J Physiol 1985;366:233–49.

[28] Jones AM, Krustrup P, Wilkerson DP, Berger NJ, Calbet JA, Bangsbo J. Influence of exercise intensity on skeletal muscle blood flow, O2 extraction and O2 uptake on-kinetics. J Physiol 2012;590(Pt 17):4363–76.

[29] Saltin B. Exercise hyperaemia: magnitude and aspects on regulation in humans. J Physiol 2007;583(Pt 3):819–23.

[30] Grocott MP, Martin DS, Levett DZ, McMorrow R, Windsor J, Montgomery HE. Arterial blood gases and oxygen content in climbers on Mount Everest. N Engl J Med 2009;360(2):140–9.

[31] Gonzalez-Alonso J, Richardson RS, Saltin B. Exercising skeletal muscle blood flow in humans responds to reduction in arterial oxyhaemoglobin, but not to altered free oxygen. J Physiol 2001;530(Pt 2):331–41.

[32] Calbet JA, Radegran G, Boushel R, Saltin B. On the mechanisms that limit oxygen uptake during exercise in acute and chronic hypoxia: role of muscle mass. J Physiol 2009;587(Pt 2):477–90.

[33] Wendt D, van Loon LJ, Lichtenbelt WD. Thermoregulation during exercise in the heat: strategies for maintaining health and performance. Sports Med 2007;37(8):669–82.

[34] Gonzalez-Alonso J, Crandall CG, Johnson JM. The cardiovascular challenge of exercising in the heat. J Physiol 2008;586 (1):45–53.

[35] Rowell LB. Human cardiovascular adjustments to exercise and thermal stress. Physiol Rev 1974;54(1):75–159.

[36] Smolander J, Louhevaara V. Effect of heat stress on muscle blood flow during dynamic handgrip exercise. Eur J Appl Physiol Occup Physiol 1992;65(3):215–20.

[37] Gonzalez-Alonso J, Calbet JA, Nielsen B. Muscle blood flow is reduced with dehydration during prolonged exercise in humans. J Physiol 1998;513(Pt 3):895–905.

[38] Juel C, Klarskov C, Nielsen JJ, Krustrup P, Mohr M, Bangsbo J. Effect of high-intensity intermittent training on lactate and H+ release from human skeletal muscle. Am J Physiol Endocrinol Metab 2004;286(2):E245–51.

[39] Blomstrand E, Krustrup P, Sondergaard H, Radegran G, Calbet JA, Saltin B. Exercise training induces similar elevations in the activity of oxoglutarate dehydrogenase and peak oxygen uptake in the human quadriceps muscle. Pflugers Arch 2011;462(2):257–65.

[40] Rud B, Foss O, Krustrup P, Secher NH, Hallen J. One-legged endurance training: leg blood flow and oxygen extraction during cycling exercise. Acta Physiol (Oxf) 2011;205(1):177–85.

[41] Roca J, Agusti AG, Alonso A, Poole DC, Viegas C, Barbera JA, et al. Effects of training on muscle O2 transport at VO_2max. J Appl Physiol 1992;73(3):1067–76.

[42] Snell PG, Martin WH, Buckey JC, Blomqvist CG. Maximal vascular leg conductance in trained and untrained men. J Appl Physiol 1987;62(2):606–10.

[43] Simon M, LeBlanc P, Jobin J, Desmeules M, Sullivan MJ, Maltais F. Limitation of lower limb VO(2) during cycling exercise in COPD patients. J Appl Physiol 2001;90(3):1013–9.

[44] Proctor DN, Parker BA. Vasodilation and vascular control in contracting muscle of the aging human. Microcirculation 2006;13(4):315–27.

[45] Kirby BS, Crecelius AR, Voyles WF, Dinenno FA. Impaired skeletal muscle blood flow control with advancing age in humans: attenuated ATP release and local vasodilation during erythrocyte deoxygenation. Circ Res 2012;111(2):220–30.

[46] Kingwell BA. Formosa M, Muhlmann M, Bradley SJ, McConell GK. Type 2 diabetic individuals have impaired leg blood flow responses to exercise: role of endothelium-dependent vasodilation. Diabetes Care 2003;26(3):899–904.

[47] Esposito F, Reese V, Shabetai R, Wagner PD, Richardson RS. Isolated quadriceps training increases maximal exercise capacity in chronic heart failure: the role of skeletal muscle convective and diffusive oxygen transport. J Am Coll Cardiol 2011;58 (13):1353–62.

[48] Hepple RT. Skeletal muscle: microcirculatory adaptation to metabolic demand. Med Sci Sports Exerc 2000;32(1):117–23.

[49] Mathieu-Costello O, Hepple RT. Muscle structural capacity for oxygen flux from capillary to fiber mitochondria. Exerc Sport Sci Rev 2002;30(2):80–4.

[50] Wagner PD. The critical role of VEGF in skeletal muscle angiogenesis and blood flow. Biochem Soc Trans 2011;39(6):1556–9.

[51] Egginton S. Physiological factors influencing capillary growth. Acta Physiol (Oxf) 2010;202(3):225–39.

[52] Gustafsson T, Rundqvist H, Norrbom J, Rullman E, Jansson E, Sundberg CJ. The influence of physical training on the angiopoietin and VEGF-A systems in human skeletal muscle. J Appl Physiol 2007;103(3):1012–20.

[53] Egginton S. Physiological factors influencing capillary growth. Acta Physiol (Oxf) 2011;202(3):225–39.

[54] Salazar Vazquez BY, Cabrales P, Tsai AG, Intaglietta M. Nonlinear cardiovascular regulation consequent to changes in blood viscosity. Clin Hemorheol Microcirc 2011;49 (1-4):29–36.

[55] Vazquez BY. Blood pressure and blood viscosity are not correlated in normal healthy subjects. Vasc Health Risk Manag 2012;8:1–6.

[56] Schuler B, Arras M, Keller S, Rettich A, Lundby C, Vogel J, et al. Optimal hematocrit for maximal exercise performance in acute and chronic erythropoietin-treated mice. Proc Natl Acad Sci U S A 2010;107(1):419–23.

[57] Vogel J, Kiessling I, Heinicke K, Stallmach T, Ossent P, Vogel O, et al. Transgenic mice overexpressing erythropoietin adapt to excessive erythrocytosis by regulating blood viscosity. Blood 2003;102(6):2278–84.

[58] Brun JF, Khaled S, Raynaud E, Bouix D, Micallef JP, Orsetti A. The triphasic effects of exercise on blood rheology: which relevance to physiology and pathophysiology? Clin Hemorheol Microcirc 1998;19(2):89–104.

[59] Brun JF, Varlet-Marie E, Connes P, Aloulou I. Hemorheological alterations related to training and overtraining. Biorheology 2010;47(2):95–115.

[60] Connes P. Hemorheology and exercise: effects of warm environments and potential consequences for sickle cell trait carriers. Scand J Med Sci Sports 2010;20(Suppl 3):48–52.

[61] Robach P, Siebenmann C, Jacobs RA, Rasmussen P, Nordsborg N, Pesta D, et al. The role of haemoglobin mass on VO(2)max following normobaric 'live high-train low' in endurance-trained athletes. Br J Sports Med 2012;46(11):822–7.

[62] Baskurt OK, Meiselman HJ. Blood rheology and hemodynamics. Semin Thromb Hemost 2003;29(5):435–50.

[63] Smith JA, Martin DT, Telford RD, Ballas SK. Greater erythrocyte deformability in world-class endurance athletes. Am J Physiol 1999;276(6 Pt 2):H2188–93.

[64] Zhao J, Tian Y, Cao J, Jin L, Ji L. Mechanism of endurance training-induced erythrocyte deformability in rats involves erythropoiesis. Clin Hemorheol Microcirc 2012 Apr 10.

[65] Yalcin O, Erman A, Muratli S, Bor-Kucukatay M, Baskurt OK. Time course of hemorheological alterations after heavy anaerobic exercise in untrained human subjects. J Appl Physiol 2003;94(3):997–1002.

[66] Connes P, Sara F, Hardy-Dessources MD, Etienne-Julan M, Hue O. Does higher red blood cell (RBC) lactate transporter activity explain impaired RBC deformability in sickle cell trait? Jpn J Physiol 2005;55(6):385–7.

[67] Senturk UK, Gunduz F, Kuru O, Kocer G, Ozkaya YG, Yesilkaya A, et al. Exercise-induced oxidative stress leads hemolysis in sedentary but not trained humans. J Appl Physiol 2005;99(4):1434–41.

[68] Connes P, Frank S, Martin C, Shin S, Aufradet E, Sunoo S, et al. New fundamental and applied mechanisms in exercise hemorheology. Clin Hemorheol Microcirc 2010;45(2-4):131–41.

[69] Gonzalez-Alonso J. ATP as a mediator of erythrocyte-dependent regulation of skeletal muscle blood flow and oxygen delivery in humans. J Physiol 2012;590(Pt 20):5001–13.

[70] Baskurt OK, Ulker P, Meiselman HJ. Nitric oxide, erythrocytes and exercise. Clin Hemorheol Microcirc 2011;49(1-4):175–81.

[71] Waltz X, Pichon A, Mougenel D, Lemonne N, Lalanne-Mistrih ML, Sinnapah S, et al. Hemorheological alterations, decreased cerebral microvascular oxygenation and cerebral vasomotion compensation in sickle cell patients. Am J Hematol 2012;87:1070–3.

[72] Baskurt OK, Meiselman HJ. Red blood cell "aggregability". Clin Hemorheol Microcirc 2009;43(4):353–4.

[73] Baskurt OK, Temiz A, Meiselman HJ. Red blood cell aggregation in experimental sepsis. J Lab Clin Med 1997;130(2):183–90.

[74] Baskurt OK, Temiz A, Meiselman HJ. Effect of superoxide anions on red blood cell rheologic properties. Free Radic Biol Med 1998;24(1):102–10.

[75] Yalcin O, Meiselman HJ, Armstrong JK, Baskurt OK. Effect of enhanced red blood cell aggregation on blood flow resistance in an isolated-perfused guinea pig heart preparation. Biorheology 2005;42(6):511–20.

[76] Simmonds MJ, Sabapathy S, Gass GC, Marshall-Gradisnik SM, Haseler LJ, Christy RM, et al. Heart rate variability is related to impaired haemorheology in older women with type 2 diabetes. Clin Hemorheol Microcirc 2010;46(1):57–68.

[77] Connes P, Simmonds MJ, Brun JF, Baskurt OK. Exercise hemorheology: classical data, recent findings and unresolved issue. Clin Hemorheol Microcirc 2013;53(1–2):187–99.

[78] Simmonds MJ, Minahan CL, Serre KR, Gass GC, Marshall-Gradisnik SM, Haseler LJ, et al. Preliminary findings in the heart rate variability and haemorheology response to varied frequency and duration of walking in women 65-74 yr with type 2 diabetes. Clin Hemorheol Microcirc 2012;51(2):87–99.

[79] Bescos R, Sureda A, Tur JA, Pons A. The effect of nitric-oxide-related supplements on human performance. Sports Med 2012;42(2):99–117.

[80] Paddon-Jones D, Borsheim E, Wolfe RR. Potential ergogenic effects of arginine and creatine supplementation. J Nutr 2004;134(10 Suppl):2888S–2894SS discussion 95S.

[81] Rector TS, Bank AJ, Tschumperlin LK, Mullen KA, Lin KA, Kubo SH. Abnormal desmopressin-induced forearm vasodilatation in patients with heart failure: dependence on nitric oxide synthase activity. Clin Pharmacol Ther 1996;60 (6):667–74.

[82] Sunderland KL, Greer F, Morales J. VO_2max and ventilatory threshold of trained cyclists are not affected by 28-day L-arginine supplementation. J Strength Cond Res 2011;25(3):833–7.

[83] Chin-Dusting JP, Alexander CT, Arnold PJ, Hodgson WC, Lux AS, Jennings GL. Effects of in vivo and in vitro L-arginine supplementation on healthy human vessels. J Cardiovasc Pharmacol 1996;28(1):158–66.

[84] Adams MR, Forsyth CJ, Jessup W, Robinson J, Celermajer DS. Oral L-arginine inhibits platelet aggregation but does not enhance endothelium-dependent dilation in healthy young men. J Am Coll Cardiol 1995;26(4):1054–61.

[85] Kenjale AA, Ham KL, Stabler T, Robbins JL, Johnson JL, Vanbruggen M, et al. Dietary nitrate supplementation enhances exercise performance in peripheral arterial disease. J Appl Physiol 2011;110(6):1582–91.

[86] Cermak NM, Gibala MJ, van Loon LJ. Nitrate supplementation's improvement of 10-km time-trial performance in trained cyclists. Int J Sport Nutr Exerc Metab 2012;22(1):64–71.

[87] Murphy M, Eliot K, Heuertz RM, Weiss E. Whole beetroot consumption acutely improves running performance. J Acad Nutr Diet 2012;112(4):548–52.

[88] Christensen PM, Nyberg M, Bangsbo J. Influence of nitrate supplementation on VO(2) kinetics and endurance of elite cyclists. Scand J Med Sci Sports 2013;23(1):e21-31.

[89] Peacock O, Tjonna AE, James P, Wisloff U, Welde B, Bohlke N, et al. Dietary Nitrate Does Not Enhance Running Performance in Elite Cross-Country Skiers. Med Sci Sports Exerc 2012;44 (11):2213–9.

[90] Stamler JS, Meissner G. Physiology of nitric oxide in skeletal muscle. Physiol Rev 2001;81(1):209–37.

[91] Totzeck M, Hendgen-Cotta UB, Rammos C, Frommke LM, Knackstedt C, Predel HG, et al. Higher endogenous nitrite levels are associated with superior exercise capacity in highly trained athletes. Nitric Oxide 2012;27(2):75–81.

[92] Larsen FJ, Weitzberg E, Lundberg JO, Ekblom B. Dietary nitrate reduces maximal oxygen consumption while maintaining work performance in maximal exercise. Free Radic Biol Med 2010;48(2):342–7.

[93] Bescos R, Rodriguez FA, Iglesias X, Ferrer MD, Iborra E, Pons A. Acute administration of inorganic nitrate reduces VO (2peak) in endurance athletes. Med Sci Sports Exerc 2011;43 (10):1979–86.

[94] Casey DP, Beck DT, Braith RW. Systemic plasma levels of nitrite/nitrate (NOx) reflect brachial flow-mediated dilation responses in young men and women. Clin Exp Pharmacol Physiol 2007;34(12):1291–3.

[95] Rassaf T, Lauer T, Heiss C, Balzer J, Mangold S, Leyendecker T, et al. Nitric oxide synthase-derived plasma nitrite predicts exercise capacity. Br J Sports Med 2007;41(10):669–73 discussion 73.

[96] Lidder S, Webb AJ. Vascular effects of dietary nitrate (as found in green leafy vegetables & beetroot) via the Nitrate-Nitrite-Nitric Oxide pathway. Br J Clin Pharmacol 2013;75 (3):677–96.

[97] Demoncheaux EA, Higenbottam TW, Foster PJ, Borland CD, Smith AP, Marriott HM, et al. Circulating nitrite anions are a directly acting vasodilator and are donors for nitric oxide. Clin Sci (Lond) 2002;102(1):77–83.

[98] Cosby K, Partovi KS, Crawford JH, Patel RP, Reiter CD, Martyr S, et al. Nitrite reduction to nitric oxide by deoxyhemoglobin vasodilates the human circulatio. Nat Med 2003;9 (12):1498–505.

[99] Ferguson SK, Hirai DM, Copp SW, Holdsworth CT, Allen JD, Jones AM, et al. Impact of dietary nitrate supplementation via beetroot juice on exercising muscle vascular control in rats. J Physiol 2013 15;591(Pt 2):547–57.

[100] Maher AR, Milsom AB, Gunaruwan P, Abozguia K, Ahmed I, Weaver RA, et al. Hypoxic modulation of exogenous nitrite-induced vasodilation in humans. Circulation 2008;117(5):670–7.

[101] Erzurum SC, Ghosh S, Janocha AJ, Xu W, Bauer S, Bryan NS, et al. Higher blood flow and circulating NO products offset high-altitude hypoxia among Tibetans. Proc Natl Acad Sci U S A 2007;104(45):17593–8.

[102] Vanhatalo A, Fulford J, Bailey SJ, Blackwell JR, Winyard PG, Jones AM. Dietary nitrate reduces muscle metabolic perturbation and improves exercise tolerance in hypoxia. J Physiol 2011;589(Pt 22):5517–28.

[103] Kraft JA, Green JM, Bishop PA, Richardson MT, Neggers YH, Leeper JD. Impact of dehydration on a full body resistance exercise protocol. Eur J Appl Physiol 2010;109(2):259–67.

[104] Jones LC, Cleary MA, Lopez RM, Zuri RE, Lopez R. Active dehydration impairs upper and lower body anaerobic muscular power. J Strength Cond Res 2008;22(2):455–63.

[105] Goulet ED. Effect of exercise-induced dehydration on endurance performance: evaluating the impact of exercise protocols on outcomes using a meta-analytic procedure. Br J Sports Med 2012 Jul 4.

[106] Kraft JA, Green JM, Bishop PA, Richardson MT, Neggers YH, Leeper JD. The influence of hydration on anaerobic performance: a review. Res Q Exerc Sport 2012;83(2):282–92.

[107] Lacerda AC, Gripp F, Rodrigues LO, Silami-Garcia E, Coimbra CC, Prado LS. Acute heat exposure increases high-intensity performance during sprint cycle exercise. Eur J Appl Physiol 2007;99(1):87–93.
[108] Linnane DM, Bracken RM, Brooks S, Cox VM, Ball D. Effects of hyperthermia on the metabolic responses to repeated high-intensity exercise. Eur J Appl Physiol 2004;93(1-2):159–66.
[109] Coyle EF. Physiological determinants of endurance exercise performance. J Sci Med Sport 1999;2(3):181–9.
[110] Tripette J, Loko G, Samb A, Gogh BD, Sewade E, Seck D, et al. Effects of hydration and dehydration on blood rheology in sickle cell trait carriers during exercise. Am J Physiol Heart Circ Physiol 2010;299(3):H908–14.
[111] Gonzalez-Alonso J, Mora-Rodriguez R, Below PR, Coyle EF. Dehydration markedly impairs cardiovascular function in hyperthermic endurance athletes during exercise. J Appl Physiol 1997;82(4):1229–36.
[112] Gonzalez-Alonso J, Mora-Rodriguez R, Coyle EF. Stroke volume during exercise: interaction of environment and hydration. Am J Physiol Heart Circ Physiol 2000;278(2):H321–30.
[113] Murray R. Rehydration strategies–balancing substrate, fluid, and electrolyte provision. Int J Sports Med 1998;19(Suppl 2): S133–5.
[114] Tripette J, Hardy-Dessources MD, Beltan E, Sanouiller A, Bangou J, Chalabi T, et al. Endurance running trial in tropical environment: a blood rheological study. Clin Hemorheol Microcirc 2011;47(4):261–8.
[115] Mora-Rodriguez R, Hamouti N. Salt and fluid loading: effects on blood volume and exercise performance. Med Sport Sci 2013;59:113–9.
[116] Powers SK, Jackson MJ. Exercise-induced oxidative stress: cellular mechanisms and impact on muscle force production. Physiol Rev 2008;88(4):1243–76.
[117] Kato GJ, Gladwin MT, Steinberg MH. Deconstructing sickle cell disease: reappraisal of the role of hemolysis in the development of clinical subphenotypes. Blood Rev 2007;21 (1):37–47.
[118] Baskurt OK, Meiselman HJ. Hemodynamic effects of red blood cell aggregation. Indian J Exp Biol 2007;45(1):25–31.
[119] Senturk UK, Gunduz F, Kuru O, Aktekin MR, Kipmen D, Yalcin O, et al. Exercise-induced oxidative stress affects erythrocytes in sedentary rats but not exercise-trained rats. J Appl Physiol 2001;91(5):1999–2004.
[120] Ristow M, Zarse K, Oberbach A, Kloting N, Birringer M, Kiehntopf M, et al. Antioxidants prevent health-promoting effects of physical exercise in humans. Proc Natl Acad Sci USA 2009;106(21):8665–70.
[121] Nikolaidis MG, Kerksick CM, Lamprecht M, McAnulty SR. Does vitamin C and E supplementation impair the favorable adaptations of regular exercise? Oxid Med Cell Longev 2012;707941 In Press.
[122] Sobal J, Marquart LF. Vitamin/mineral supplement use among athletes: a review of the literature. Int J Sport Nutr 1994;4 (4):320–34.
[123] Kayatekin BM, Uysal N, Resmi H, Bediz CS, Temiz-Artmann A, Genc S, et al. Does antioxidant supplementation alter the effects of acute exercise on erythrocyte aggregation, deformability and endothelium adhesion in untrained rats? Int J Vitam Nutr Res 2005;75(4):243–50.
[124] Soares Filho PR, Castro I, Stahlschmidt A. Effect of red wine associated with physical exercise in the cardiovascular system of spontaneously hipertensive rats. Arq Bras Cardiol 2011;96 (4):277–83.
[125] Dal-Ros S, Zoll J, Lang AL, Auger C, Keller N, Bronner C, et al. Chronic intake of red wine polyphenols by young rats prevents aging-induced endothelial dysfunction and decline in physical performance: role of NADPH oxidase. Biochem Biophys Res Commun 2011;404(2):743–9.
[126] Toth A, Sandor B, Papp J, Rabai M, Botor D, Horvath Z, et al. Moderate red wine consumption improves hemorheological parameters in healthy volunteers. Clin Hemorheol Microcirc; 2012 [Epub ahead of print].
[127] Prasad AS, Bao B, Beck FW, Kucuk O, Sarkar FH. Antioxidant effect of zinc in humans. Free Radic Biol Med 2004;37 (8):1182–90.
[128] Hennig B, Meerarani P, Toborek M, McClain CJ. Antioxidant-like properties of zinc in activated endothelial cells. J Am Coll Nutr 1999;18(2):152–8.
[129] Fanjiang G, Kleinman RE. Nutrition and performance in children. Curr Opin Clin Nutr Metab Care 2007;10 (3):342–7.
[130] Esen F, Gulec S, Esen H. Exogenous zinc improves blood fluidity but has no effect on the mechanisms of vascular response to acetylcholine iontophoresis in humans. Biol Trace Elem Res 2006;113(2):139–53.
[131] Brun JF. Hormones, metabolism and body composition as major determinants of blood rheology: potential pathophysiological meaning. Clin Hemorheol Microcirc 2002;26 (2):63–79.
[132] Kaya O, Gokdemir K, Kilic M, Baltaci AK. Zinc supplementation in rats subjected to acute swimming exercise: Its effect on testosterone levels and relation with lactate. Neuro Endocrinol Lett 2006;27(1-2):267–70.
[133] Khaled S, Brun JF, Micallel JP, Bardet L, Cassanas G, Monnier JF, et al. Serum zinc and blood rheology in sportsmen (football players). Clin Hemorheol Microcirc 1997;17 (1):47–58.
[134] Kaya O, Gokdemir K, Kilic M, Baltaci AK. Melatonin supplementation to rats subjected to acute swimming exercise: Its effect on plasma lactate levels and relation with zinc. Neuro Endocrinol Lett 2006;27(1-2):263–6.
[135] Khaled S, Brun JF, Wagner A, Mercier J, Bringer J, Prefaut C. Increased blood viscosity in iron-depleted elite athletes. Clin Hemorheol Microcirc 1998;18(4):309–18.
[136] Lukaski HC. Vitamin and mineral status: effects on physical performance. Nutrition 2004;20(7-8):632–44.
[137] Teragawa H, Matsuura H, Chayama K, Oshima T. Mechanisms responsible for vasodilation upon magnesium infusion in vivo: clinical evidence. Magnes Res 2002;15(3-4):241–6.
[138] Gold ME, Buga GM, Wood KS, Byrns RE, Chaudhuri G, Ignarro LJ. Antagonistic modulatory roles of magnesium and calcium on release of endothelium-derived relaxing factor and smooth muscle tone. Circ Res 1990;66(2):355–66.
[139] Pearson PJ, Evora PR, Seccombe JF, Schaff HV. Hypomagnesemia inhibits nitric oxide release from coronary endothelium: protective role of magnesium infusion after cardiac operations. Ann Thorac Surg 1998;65(4):967–72.
[140] Laurant P, Berthelot A. Endothelin-1-induced contraction in isolated aortae from normotensive and DOCA-salt hypertensive rats: effect of magnesium. Br J Pharmacol 1996;119 (7):1367–74.
[141] Cunha AR, Umbelino B, Correia ML, Neves MF. Magnesium and vascular changes in hypertension. Int J Hypertens 2012;754250 In Press.
[142] Shechter M, Merz CN, Paul-Labrador M, Meisel SR, Rude RK, Molloy MD, et al. Oral magnesium supplementation inhibits

platelet-dependent thrombosis in patients with coronary artery disease. Am J Cardiol 1999;84(2):152–6.

[143] Barbagallo M, Dominguez LJ, Galioto A, Pineo A, Belvedere M. Oral magnesium supplementation improves vascular function in elderly diabetic patients. Magnes Res 2010; 23(3):131–7.

[144] De Franceschi L, Bachir D, Galacteros F, Tchernia G, Cynober T, Alper S, et al. Oral magnesium supplements reduce erythrocyte dehydration in patients with sickle cell disease. J Clin Invest 1997;100(7):1847–52.

[145] Mel'nikov AA, Vikulov AD. Disorders in magnesium balance decrease erythrocyte deformability in athletes. Fiziol Cheloveka 2005;31(3):133–6.

[146] Newhouse IJ, Finstad EW. The effects of magnesium supplementation on exercise performance. Clin J Sport Med 2000;10 (3):195–200.

CHAPTER

29

Genetic Aspects of Sprint, Strength and Power Performance

Erik D Hanson[1,2] *and Nir Eynon*[1]

[1]Institute of Sport, Exercise and Active Living (ISEAL) [2]College of Health and Biomedicine, Victoria University, Australia

INTRODUCTION

When considering sprint, strength, and power performance within athletes of similar age, body composition, and training history, there is considerable variability across the group. One explanation for this discrepancy is the genetic makeup of the individual [1]. In the field of exercise and genetics, genotype is usually associated with either athletic performance or the response to training. The populations for these gene association studies range from untrained individuals up through elite athletes. Elite athletic cohorts are ideal targets for genetic research because of the relatively homogenous nature of the population and are often defined as having qualified for and completed in national or international competitions, such as the Olympic Games [1]. The similar training and dietary programs, access to coaches, physiotherapists, athletic trainers, etc., and training facilities make differences due to genetic factors more readily detectable in this group. Because elite athletes provide the best model for studying the influence of genetics on sprint, strength and power performance, summarizing this research and several key genes showing both associations with performance and biological mechanisms is the focus of this chapter. Additional examples in untrained individuals and also results from training studies are used where available.

Genetic variants (i.e., polymorphisms) are present throughout the human genome and are key to our understanding of the potential influence that genes may have on athletic performance. Identification of relevant variants and their relative importance has been an area of interest over the last several decades [1]. Along with environmental factors (training and diet), it is possible that elite athletes possess a 'blueprint' of genetic variants that permit them to succeed at the highest levels of competition. However, the direction and magnitude of the influence of those variants on athletic performance still needs clarification.

EXERCISE PERFORMANCE AND HERITABILITY

Research into genetics and athletic performance has been founded on familial and twin studies, which implicated genetics as a significant factor in athletic performance, even after adjusting for environmental factors. The first genome-wide linkage scan for athletic status reported the heritability of roughly 66% of athletic status in 700 British female dizygotic twin pairs [2]. More recent data from the HERITAGE family study have suggested that the heritability of changes in maximal oxygen uptake (VO_{2max}) with exercise training reaches 47% in sedentary subjects [3].

Muscular strength and mass are crucial components in sprint and power performance and are reported to be influenced by genetic factors. Heritability for muscle strength and mass assessed by twin and family studies is estimated at 31–78% [4], with the large range being attributed to differences between the muscle groups, contraction velocities and muscle lengths that were studied. Of particular interest are studies that have examined the heritability of maximal power and total explosive power, which have been estimated at 74% and 84%, respectively [5]. The ultimate question

Nutrition and Enhanced Sports Performance.
DOI: http://dx.doi.org/10.1016/B978-0-12-396454-0.00029-1

appears not to be whether performance-traits are heritable, but which particular genetic variants contribute substantially to sprint, strength and power performance and to classify and make use of these variants to identify exceptional athletes.

The genetic influence on sprint, strength and power performance has received less attention than that on endurance performance, with relatively few studies having identified associations between gene variants and power performance. Most studies to date have small sample sizes, and with a few notable exceptions, most variants have been tested in only one or two elite athletic cohorts. When non-elite populations are included, the results are similar, as significant variability within the participant population makes it difficult to isolate the small effects of each variant. Finally, the vast majority of work has only identified associations rather than determining potential mechanisms that might further explain these findings. Consequently, the genetic profile of sprint, strength and power athletes remains to be elucidated.

Despite the limitations, it is well accepted that genetics markedly influence sprint, strength and power performance. Several studies have provided evidence that skeletal muscle phenotypes are genetically influenced [5–7]. Without the appropriate genetic make-up, the chance of becoming an exceptional power athlete or sprinter is probably reduced, although exceptions do exist [8]. In this regard, the specific knowledge in the field of genetics and sprint, strength and power performance is summarized in the following sections, focusing primarily on elite athletes, to provide future perspectives and directions for research in this field.

Alpha Actinin 3 (ACTN3 R577X)

The α-actinins are a family of actin-binding proteins that help anchor the actin filaments. Of the four isoforms of α-actinin expressed in humans, α-actinin-3 (encoded by the *ACTN3* gene) is exclusively expressed in the fast, glycolytic muscle fibers that produce the 'explosive', powerful contractions [9]. A common variant (C→T, rs1815739) in the *ACTN3* gene results in a shift from arginine (R) to a premature stop codon (X) [10] and a complete deficiency of α-actinin-3. The frequency of the 577X null allele varies across ethnic groups, ranging from 10% to 50% depending on ethnicity, generally occurring in ~20% of the general population [9]. This variant has not been linked to any known disease. However, the specialized expression pattern and strong sequence conservation of α-actinin-3 over 300 million years suggest that it has a specific role in fast (type IIX) muscle fibers [9].

The association between *ACTN3* R577X and elite power performance has been quite consistent, with the initial findings in Australian athletes showing that the percentage of elite power athletes with the XX genotype was significantly lower than in healthy controls [11]. These findings have been replicated in many other studies, with the frequency distribution of the *ACTN3* XX genotype being significantly lower in track and field sprinters and throwers and in weight lifters than in controls and endurance athletes [12–16]. Moreover, in a large group of European athletes, sprint and power athletes were ~50% less likely to harbor the XX genotype than were sedentary controls [17]. *ACTN3* XX genotype is also associated with the level of athletic competition achieved, as no cases of *ACTN3* XX genotype were found in any Australian or Israeli elite sprinters [11,12], while elite Taiwanese short-distance swimmers have significantly lower *ACTN3* XX genotype frequency compared with their non-elite counterparts [18].

The effects of *ACTN3* genotype in non-athletic populations have been examined using more specific phenotypes, such as muscle strength, size and fiber type, sprint performance, and response to training. Studies have shown the RR genotype and R allele carriers may have greater muscle size and strength [19,20], faster sprint times [21], and higher percentages of fast-twitch muscle fibers [20,22] than individuals with the XX genotype. However, not all studies confirm these findings [23–25]. When considering the response to exercise training, women with the XX genotype had lower maximal arm strength but had greater gains with strength training [26]. Delmonico and colleagues [27] found that women with the RR genotype had lower muscle leg power initially but had greater increases following strength training than those with the XX genotype. The conflicting findings may be the result of heterogeneity amongst research participants coupled with small sample sizes, the use of men and/or women, and different muscle groups and tests.

The actn3 knock-out (KO) mouse model has been developed to examine the physiological and metabolic implications of *ACTN3* (XX genotype) deficiency [28]. Actn3 KO mice have reduced grip strength and muscle mass but have 33% greater running endurance than their wild-type (WT) littermates. This has been postulated to be the result of a decreased diameter of fast muscle fibers where α-actinin-3 is expressed and alters contractile and functional properties of the fast muscle fibers towards that of slow (type I) muscle fibers. Independent of a shift in fiber type proportions, the KO mouse muscle has decreased anaerobic enzymatic activity, increased oxidative metabolism [29], and slower contractile properties and enhanced recovery from fatigue [29,30]. Data from untrained humans supports the KO mouse model by demonstrating greater cross-sectional area, and higher percentage of type IIX fibers [20]. Collectively, these studies provide a plausible explanation for reduced power and sprint and improved endurance capacity in humans with the *ACTN3* XX genotype.

Angiotensin Converting Enzyme Insertion/ Deletion (ACE I/D)

Angiotensin Converting Enzyme (ACE) is a key component in the renin–angiotensin system, generating the vasoconstrictor angiotensin II (Ang II) and degrading vasodilator kinins [31]. It has been suggested that ACE plays an important role in both muscle and cardiovascular function during exercise [32].

The *ACE* I/D is the next most studied variant with respect to sprint, strength and power performance. The insertion (I allele, rs4646994) of a 287 base pair fragment, as opposed to its absence (deletion, D allele), is associated with lower ACE activity in both circulation and in heart tissue [33,34]. Both alleles have been associated with improved athletic performance, as the I allele has been linked with endurance phenotypes and the D allele with sprint and power phenotypes. Because this chapter focuses on sprint, strength and power performance, only studies including these phenotypes will be discussed.

Unlike *ACTN3*, elite power performance associations with the *ACE* I/D variant present more conflicting results, with an excess of DD genotypes and D allele frequencies [35–39], no genotype distribution differences [40–42], or an excess of II genotype and I allele frequencies [43,44] observed in sprint/power athletes vs endurance athletes and sedentary controls. Dividing athlete cohorts into elite-level and national-level generally does not show an association with the level of performance [36,38,39]. However, other studies have shown linear trends such that the D and I allele frequencies are overrepresented in elite power and elite endurance athletes, respectively, compared with ethnicity-matched controls [35,38]. Possible explanations for the conflicting results may include differences in level of athletic competition, gender, small sample sizes, and ethnicity, as the studies by Scott et al. [41], and Kim et al. [44] were performed in African-American and Asian populations, respectively while Amir et al. [43] did not account for the multi-ethnic nature of Israeli Caucasians.

Findings from the general population reveal similar results, as the D allele has been associated with muscle mass [45,46] but not typically strength or power [47,48]. The response to strength training also presents inconsistent findings [45,49,50]. In short, the relatively high number of studies that have shown *ACE* I/D variant to be associated with sprint, strength and power performance is suggestive of an underlying effect, but the number of discordant findings casts doubt on the overall effect of this variant on performance.

A possible mechanism for the *ACE* I/D genotype association that may be present in elite power and sprint athletes is through the renin–angiotensin system. ACE has been implicated in left ventricular, smooth, and also skeletal muscle hypertrophy [51–56], as inhibition of ACE attenuates overload-induced hypertrophy [55,56]. D allele carriers with higher ACE activity may have higher angiotensin II levels, which appear to translate into higher muscle mass at baseline in untrained men and women but did not influence the response to strength training [45,57]. The *ACE* I/D variant may also influence fiber type [58], with the DD genotype having a higher proportion of fast, type II (specifically IIX) and lower slow (type I) muscle fibers in untrained volunteers. The combination of increased muscle mass and fast-twitch fiber expression would favor power and sprint performance, but this is not supported by the ACE partial-KO mouse [59], as fiber type proportions were similar in the soleus muscle. Presently, the exact mechanism by which ACE may contribute to sprint, strength and power performance remains unknown.

Myostatin (*MSTN* K153R)

Myostatin (MSTN) is a negative regulator of skeletal muscle mass by controlling myoblast differentiation [60], and the loss or absence of MSTN leads to the double-muscling effect, which has been observed in animals [60,61] and humans [62]. Of the several genetic variants located within the *MSTN* gene, the K153R (rs1805086, also known as Lys153Arg) variant is the most commonly studied. The frequency distribution of the mutant *MSTN* KK genotype approaches 0% in Europeans but is higher (~9%) in African Americans, whereas heterozygosity (KR genotype) ranges between 3% in Europeans and 30% in African Americans. To the best of the authors' knowledge, no study to date has demonstrated association between *MSTN* K153R variant and sprint, strength and power performance using elite athletic cohorts. However, in non-athletic men, heterozygotes participants had lower vertical jump performance but not sprint performance compared with the KK participants [63]. There were no homozygotes for the rare R allele. Muscle strength and size were also shown to be lower in the R allele carriers, but only within African Americans [64]. The effect of genotype with strength training appears marginal, as there was a trend for women with the R allele to have a 68% greater hypertrophy response [65] in a very small sample size that was not confirmed in a larger cohort [64].

Beyond *ACTN3*, *ACE* and *MSTN*, other genetic variants have been associated with sprint, strength and power performance but are lacking replication in independent cohorts, present inconsistent results, or do not yet have biological mechanisms to support the associations.

Other Variants Associated with Power, Sprint and Strength Performance

Angiotensinogen (AGT *Met235Thr*)

A polymorphism (T→C, rs699) in the angiotensinogen gene (*AGT*) resulting in a replacement of methionine with threonine (*AGT* Met235Thr) has been associated with sprint and power performance in elite power athletes. AGT sits upstream of ACE in the renin–angiotensin pathway and may (i) indirectly influence performance via this pathway or (ii) have other direct effects, still to be identified. A study of Spanish elite power athletes had higher CC genotype frequency than either controls or endurance athletes [66], which is consistent with studies examining *ACE* DD genotype [35–39]. A possible mechanism for this association may be the production of higher levels of angiotensin II, which can be a skeletal muscle growth factor and therefore beneficial to muscle size, strength and power [45,55–57]. However, replication studies are required before definitive conclusions regarding this genotype on athletic performance can be drawn.

Adenosine Mono Phosphate Deaminase 1 (AMPD1 *C34T*)

The breakdown of ATP results in formation of adenosine monophosphate (AMP) and then inosine monophosphate (IMP) by AMPD during high-intensity exercise when ATP utilization exceeds resynthesis [67]. Similar to *ACTN3* R577X, a variant in the *AMPD1* gene results in a premature stop codon (*AMPD1* C34T, rs17602729) and complete deficiency of the AMPD protein and diminished AMP metabolism that produces muscle fatigue, weakness and cramping [68]. The null allele (TT genotype) is rare (~2% of population) [69], but heterozygotes display intermediate levels of AMPD activity [70], which may impair sprint and power exercise capacity [71]. As such, the CT genotype of *AMPD1* was observed in higher frequency in power vs endurance athletes [37]. In slight contrast, the CC genotype and C allele were consistently higher in different types of power athletes vs controls, and no power athlete had the TT genotype [72]. While power performance requires rapid ATP breakdown and resynthesis, the practical relevance of this variant needs to be clarified. Reduced anaerobic performance in physically active individuals with the TT genotype has been observed [73], but specific phenotype testing has not been assessed in elite athletes. The C allele may be beneficial for power performance but, given the low frequency of the TT genotype and the limited number of studies, the implications remain unclear.

Interleukin 6 (IL6 *C-174G*)

Skeletal muscle produces interleukin 6 (IL6) to increase substrate delivery and possibly reduce inflammation following exercise [74]. A functional C→G polymorphism in the 5′ flanking region of the *IL6* gene (rs1800795) shows that the mutant G allele is associated with increased transcriptional response *in vitro* [75] and *in vivo* [76]. The C allele and the CC genotype have been associated with a higher creatine kinase activity following eccentric exercise [77], suggesting that the G allele may protect skeletal muscle during powerful contractions and assist in repair, promoting beneficial adaptations following exercise training. Consequently, the GG genotype and G allele frequencies were observed to be higher in power vs endurance athletes and controls [78]. However, in a replication study with a larger sample size [79] that was also pooled with data from Ruiz et al. [78], no association with genotype was observed in the larger sample size, the pooled data, or when the athletes were stratified into elite vs national competitors. As such, the *IL6* variant does not appear to be a prominent factor in elite power performance. However, given the possible influence of *IL6* -174G/C in damage prevention and inflammation allows for speculation in this variant's role during recovery following exercise performance. Furthermore, these contradictory findings highlight the need to replicate studies in large cohorts containing different ethnic backgrounds.

Nitric Oxide Synthase 3 (NOS3*-786 T/C and Glu298Asp*)

Nitric oxide (NO) has effects on vascular tone [80,81] and blood supply to the working muscles [82,83], and may influence skeletal muscle glucose update during exercise [84], which is the preferred substrate during high-intensity, anaerobic sprint activities. There are two *NOS3* polymorphisms that have been associated with power performance: the *NOS3*-786 T/C polymorphism (rs2070744), which results in decreased gene promoter activity and nitric oxide synthesis [85], while the other is a missense glutamine/aspartame (*NOS3* Glu298Asp, rs1799983) that is associated with reduced endothelial activity, nitric oxide production [86,87], and better response to exercise training in non-athletic populations [88,89]. Both polymorphisms were hypothesized to be associated with endurance performance. Instead, the opposite was observed, as the TT genotype frequency of the *NOS3*-786 T/C polymorphism was higher in elite Spanish power athletes vs endurance athletes and sedentary controls [90], while the T allele frequency was higher in both Spanish and Italian power athletes compared with controls [42,90]. Looking at the Glu298Asp variant, the Glu298 allele frequency was higher in Italian power athletes than in controls [42]. Greater nitric oxide abundance, which may influence muscle hypertrophy [91,92], is suggested as the potential link between genotypes and power performance.

Polygenic Profiles

A common theme amongst all studies reviewed is the low percentage of the variation explained by each individual variant. To address this, Williams & Folland determined the probability for the existence of humans with a theoretically optimal polygenic profile for endurance sports by examining gene × gene interactions [93]. The optimal profile using the total genotype score (TGS, ranging from 0 to 100, with a higher score being better) was derived using an algorithm combining 23 different genotypes to explain individual variations in endurance performance. Two recent studies have examined the role of a polygenic profile on elite power performance. In the first study, TGS was determined using six polymorphisms that have been previously associated with power-related phenotypes, including *ACE, ACTN3, AGT, NOS3, IL6* and *MSTN* [94]. Elite power athletes had a significantly higher total genotype score than both endurance and controls, which had similar scores. Interestingly, five (9.4%) of the power athletes had perfect scores, yet none of these athletes were amongst the best sprinters. Using a similar approach, the same group attempted to predict elite sports performance based on genotype. Microarray analysis of 20 genes (36 variants within the selected genes) revealed that *IL6, NOS3* and *NAT2* (N-acetyltransferase) best predicted power and endurance performance and explained 21.4% of the variation [95]. However, the probability of an individual's possessing optimal polygenic profile (i.e., TGS = 100) for the six studied power-related polymorphisms is small at ~0.2% [94]. Assuming that other polymorphisms influencing sprint and power performance have yet to be identified, this further decreases the probability of a single person possessing the ideal polygenic profile.

CONCLUSIONS

The current literature supports the notion that genetics influence sprint, strength and power performance and that a favorable genetic endowment is advantageous in attaining elite athletic status. However, the specific variants across the genome that influences this phenomenon, and whether they act individually or in combination with other variants or environmental factors, requires further investigation. Presently, the *ACTN3* R577X variant provides the most consistent results and is supported by data from the actn3 KO mouse model. Both the KO mouse model and preliminary data in humans suggest a plausible biological mechanism behind the genotype:phenotype association, with a shift in the metabolic phenotype of type II (fast) muscle fibers away from anaerobic pathways towards the slower aerobic pathway normally associated with slow muscle fibers. The other variants provide less consistent results (i.e., *ACE* I/D) or have not been tested in multiple cohorts (e.g., *IL6, AMPD1, NOS3*), making it difficult to form firm conclusions.

Despite advances in our understanding of the genetic basis of sprint, strength and power performance, three important limitations have hampered the progression of genetics-based athletic research and need to be addressed: (i) Elite athletic populations are rare; thus, recruiting large numbers of these individuals is difficult. Large multi-site collaborations and data sharing between researchers and universities will be necessary. (ii) Some of the contradictory findings may be due to between-study differences, which reflects the absence of a universally accepted definition of what is considered 'elite-level'. Rather, the athletes' results in major competitions may be better indicators and could provide a more robust phenotype (i.e., speed, power, or strength) in itself. (iii) The analysis of single variants using low-throughput techniques that are based on poorly justified candidate genes was commonly (mis-)used. The use of the innovative Genome-Wide Association technique enables the detection of more than a million genetic variations across the human genome and may help resolve the current ambiguity within gene association studies by validating previous findings and identifying novel variants related to sprint/ power performance and muscle strength.

GLOSSARY

Polygenic trait: A trait or condition influenced by several genes, each contributing small effect size to the phenotypic variance.

Allele: The different forms that a particular polymorphism may take are called *alleles*; e.g., the angiotensin-converting enzyme (*ACE*) gene has two common alleles, *Insertion* (I) and *Deletion* (D), with three possible allele combinations or *genotypes*: II, ID, or DD.

Genotype: the genetic information specified by the maternal and paternal alleles at a given locus. The word "genotype" can also be used to represent the genetic information content at several relevant loci.

Mutation and polymorphism: Rare variations in gene structure (<1% of population) are known as *mutations*; whereas more common ones (≥1%) are called *polymorphisms*. The different types of polymorphisms include: (i) the presence/absence of an entire stretch of DNA (*insertion/deletion* polymorphisms); (ii) DNA duplication (*copy number variation*); (iii) *repeating patterns* of DNA that vary in the number of repeats (200–300 base pair (bp) stretches repeated a few to hundreds of times); and (iv) a single-bp change (*single-nucleotide polymorphism* (SNP)), which are the most common type of polymorphism.

Gene–environment interaction: An interaction effect is present when the response to an environmental factor varies depending on the genotype of the individual. In the context of elite athletes, an environmental challenge is any behavior or lifestyle factor that has a bearing on training, nutrition and socio-economic elements.

Candidate gene association studies: A method to assess the association of one or more specific genetic variants with outcomes or phenotypes of interest; genetic variants to be tested are selected according to their features, e.g., known or postulated biology or function. Case:control designs compare allele/genotype frequencies between cases (e.g., elite athletes) and controls (non-athletic people representative of the general population). Some studies also study a single cohort (e.g., athletes or non-athletes) and assess genotype effects on selected phenotypes (e.g., muscle power).

Genome-Wide Association Studies (GWAS): A more 'agnostic' approach that examines the association of genetic variation with outcomes or phenotypes of interest by analyzing 100,000 to millions of SNPs across the entire genome without any previous hypotheses about potential mechanisms.

Linkage studies: Both approaches (candidate gene association studies and GWAS) can be used on relatives in *genetic linkage* studies, in which the presence or absence of certain variant alleles in family members with or without a disease or phenotype is analyzed.

References

[1] Eynon N, Ruiz JR, Oliveira J, Duarte JA, Birk R, Lucia A. Genes and elite athletes: a roadmap for future research. J Physiol 2011;589:3063–70.

[2] De Moor MH, Spector TD, Cherkas LF, Falchi M, Hottenga JJ, Boomsma DI, et al. Genome-wide linkage scan for athlete status in 700 British female DZ twin pairs. Twins Res Hum Genet 2007;10:812–20.

[3] Bouchard C, Sarzynski MA, Rice TK, Kraus WE, Church TS, Sung YJ, et al. Genomic predictors of the maximal O uptake response to standardized exercise training programs. J Appl Physiol 2011;110:1160–70.

[4] Peeters MW, Thomis MA, Beunen GP, Malina RM. Genetics and sports: an overview of the pre-molecular biology era. Med Sport Sci 2009;54:28–42.

[5] Calvo M, Rodas G, Vallejo M, Estruch A, Arcas A, Javierre C, et al. Heritability of explosive power and anaerobic capacity in humans. Eur J Appl Physiol 2002;86:218–25.

[6] Seeman E, Hopper JL, Young NR, Formica C, Goss P, Tsalamandris C. Do genetic factors explain associations between muscle strength, lean mass, and bone density? a twin study. Am J Physiol 1996;270:E320–7.

[7] Thomis MA, Beunen GP, Maes HH, Blimkie CJ, Van Leemputte M, Claessens AL, et al. Strength training: importance of genetic factors. Med Sci Sports Exerc 1998;30:724–31.

[8] Lucia A, Olivan J, Gomez-Gallego F, Santiago C, Montil M, Foster C. Citius and longius (faster and longer) with no alpha-actinin-3 in skeletal muscles? Br J Sports Med 2007;41:616–7.

[9] Mills M, Yang N, Weinberger R, Vander Woude DL, Beggs AH, Easteal S, et al. Differential expression of the actin-binding proteins, alpha-actinin-2 and -3, in different species: implications for the evolution of functional redundancy. Hum Mol Genet 2001;10:1335–46.

[10] North KN, Yang N, Wattanasirichaigoon D, Mills M, Easteal S, Beggs AH. A common nonsense mutation results in alpha-actinin-3 deficiency in the general population. Nat Genet 1999;21:353–4.

[11] Yang N, MacArthur DG, Gulbin JP, Hahn AG, Beggs AH, Easteal S, et al. ACTN3 genotype is associated with human elite athletic performance. Am J Hum Genet 2003;73:627–31.

[12] Eynon N, Duarte JA, Oliveira J, Sagiv M, Yamin C, Meckel Y, et al. ACTN3 R577X polymorphism and Israeli top-level athletes. Int J Sports Med 2009;30:695–8.

[13] Niemi AK, Majamaa K. Mitochondrial DNA and ACTN3 genotypes in finnish elite endurance and sprint athletes. Eur J Hum Genet 2005;13:965–9.

[14] Papadimitriou ID, Papadopoulos C, Kouvatsi A, Triantaphyllidis C. The ACTN3 gene in elite Greek track and field athletes. Int J Sports Med 2008;29:352–5.

[15] Gineviciene V, Pranculis A, Jakaitiene A, Milasius K, Kucinskas V. Genetic variation of the human ACE and ACTN3 genes and their association with functional muscle properties in Lithuanian elite athletes. Medicina (B Aires) 2011;47:284–90.

[16] Roth SM, Walsh S, Liu D, Metter EJ, Ferrucci L, Hurley BF. The ACTN3 R577X nonsense allele is under-represented in elite-level strength athletes. Eur J Hum Genet 2008;16:391–4.

[17] Eynon N, Ruiz JR, Femia P, Pushkarev VP, Cieszczyk P, Maciejewska-Karlowska A, et al. The ACTN3 R577X polymorphism across three groups of elite male European athletes. PloS One 2012;7:e43132.

[18] Chiu LL, Wu YF, Tang MT, Yu HC, Hsieh LL, Hsieh SS. ACTN3 genotype and swimming performance in Taiwan. Int J Sports Med 2011;32:476–80.

[19] Walsh S, Liu D, Metter EJ, Ferrucci L, Roth SM. ACTN3 genotype is associated with muscle phenotypes in women across the adult age span. J Appl Physiol 2008;105:1486–91.

[20] Vincent B, De Bock K, Ramaekers M, Van den Eede E, Van Leemputte M, Hespel P, et al. ACTN3 (R577X) genotype is associated with fiber type distribution. Physiol Genomics 2007;32:58–63.

[21] Moran CN, Yang N, Bailey ME, Tsiokanos A, Jamurtas A, MacArthur DG, et al. Association analysis of the ACTN3 R577X polymorphism and complex quantitative body composition and performance phenotypes in adolescent Greeks. Eur J Hum Genet 2007;15:88–93.

[22] Ahmetov II, Druzhevskaya AM, Lyubaeva EV, Popov DV, Vinogradova OL, Williams AG. The dependence of preferred competitive racing distance on muscle fibre type composition and ACTN3 genotype in speed skaters. Exp Physiol 2011.

[23] Hanson ED, Ludlow AT, Sheaff AK, Park J, Roth SM. ACTN3 genotype does not influence muscle power. Int J Sports Med 2010;31:834–8.

[24] Norman B, Esbjornsson M, Rundqvist H, Osterlund T, von Walden F, Tesch PA. Strength, power, fiber types, and mRNA expression in trained men and women with different ACTN3 R577X genotypes. J Appl Physiol 2009;106:959–65.

[25] Santiago C, Rodriguez-Romo G, Gomez-Gallego F, Gonzalez-Freire M, Yvert T, Verde Z, et al. Is there an association between ACTN3 R577X polymorphism and muscle power phenotypes in young, non-athletic adults?. Scand J Med Sci Sports 2010;20:771–8.

[26] Clarkson PM, Devaney JM, Gordish-Dressman H, Thompson PD, Hubal MJ, Urso M, et al. ACTN3 genotype is associated with increases in muscle strength in response to resistance training in women. J Appl Physiol 2005;99:154–63.

[27] Delmonico MJ, Kostek MC, Doldo NA, Hand BD, Walsh S, Conway JM, et al. Alpha-actinin-3 (ACTN3) R577X polymorphism influences knee extensor peak power response to strength training in older men and women. J Gerontol A Biol Sci Med Sci 2007;62:206–12.

[28] MacArthur DG, Seto JT, Raftery JM, Quinlan KG, Huttley GA, Hook JW, et al. Loss of ACTN3 gene function alters mouse muscle metabolism and shows evidence of positive selection in humans. Nat Genet 2007;39:1261–5.

[29] MacArthur DG, Seto JT, Chan S, Quinlan KG, Raftery JM, Turner N, et al. An Actn3 knockout mouse provides mechanistic insights into the association between alpha-actinin-3 deficiency and human athletic performance. Hum Mol Genet 2008;17:1076–86.

[30] Chan S, Seto JT, MacArthur DG, Yang N, North KN, Head SI. A gene for speed: contractile properties of isolated whole EDL muscle from an alpha-actinin-3 knockout mouse. Am J Physiol Cell Physiol 2008;295:C897–904.

[31] Coates D. The angiotensin converting enzyme (ACE). Int J Biochem Cell Biol 2003;35:769–73.

[32] Jones A, Woods DR. Skeletal muscle RAS and exercise performance. Int J Biochem Cell Biol 2003;35:855–66.

[33] Rigat B, Hubert C, Alhenc-Gelas F, Cambien F, Corvol P, Soubrier F. An insertion/deletion polymorphism in the angiotensin I-converting enzyme gene accounting for half the variance of serum enzyme levels. J Clin Investig 1990;86:1343–6.

[34] Danser AH, Schalekamp MA, Bax WA, van den Brink AM, Saxena PR, Riegger GA, et al. Angiotensin-converting enzyme in the human heart. Effect of the deletion/insertion polymorphism. Circulation 1995;92:1387–8.

[35] Myerson S, Hemingway H, Budget R, Martin J, Humphries S, Montgomery H. Human angiotensin I-converting enzyme gene and endurance performance. J Appl Physiol 1999;87:1313–6.

[36] Costa AM, Silva AJ, Garrido ND, Louro H, de Oliveira RJ, Breitenfeld L. Association between ACE D allele and elite short distance swimming. Eur J Appl Physiol 2009;106:785–90.

[37] Juffer P, Furrer R, Gonzalez-Freire M, Santiago C, Verde Z, Serratosa L, et al. Genotype distributions in top-level soccer players: a role for ACE? Int J Sports Med 2009;30:387–92.

[38] Nazarov IB, Woods DR, Montgomery HE, Shneider OV, Kazakov VI, Tomilin NV, et al. The angiotensin converting enzyme I/D polymorphism in Russian athletes. Eur J Hum Genet 2001;9:797–801.

[39] Woods D, Hickman M, Jamshidi Y, Brull D, Vassiliou V, Jones A, et al. Elite swimmers and the D allele of the ACE I/D polymorphism. Hum Genet 2001;108:230–2.

[40] Papadimitriou ID, Papadopoulos C, Kouvatsi A, Triantaphyllidis C. The ACE I/D polymorphism in elite Greek track and field athletes. J Sports Med Phys Fitness 2009;49:459–63.

[41] Scott RA, Irving R, Irwin L, Morrison E, Charlton V, Austin K, et al. ACTN3 and ACE genotypes in elite Jamaican and US sprinters. Med Sci Sports Exerc 2010;42:107–12.

[42] Sessa F, Chetta M, Petito A, Franzetti M, Bafunno V, Pisanelli D, et al. Gene polymorphisms and sport attitude in Italian athletes. Genet Test Mol Biomark 2011;15:285–90.

[43] Amir O, Amir R, Yamin C, Attias E, Eynon N, Sagiv M, et al. The ACE deletion allele is associated with Israeli elite endurance athletes. Exp Physiol 2007;92:881–6.

[44] Kim CH, Cho JY, Jeon JY, Koh YG, Kim YM, Kim HJ, et al. ACE DD genotype is unfavorable to Korean short-term muscle power athletes. Int J Sports Med 2010;31:65–71.

[45] Charbonneau DE, Hanson ED, Ludlow AT, Delmonico MJ, Hurley BF, Roth SM. ACE genotype and the muscle hypertrophic and strength responses to strength training. Med Sci Sports Exerc 2008;40:677–83.

[46] Lima RM, Leite TK, Pereira RW, Rabelo HT, Roth SM, Oliveira RJ. ACE and ACTN3 genotypes in older women: muscular phenotypes. Int J Sports Med 2011;32:66–72.

[47] Thomis MA, Huygens W, Heuninckx S, Chagnon M, Maes HH, Claessens AL, et al. Exploration of myostatin polymorphisms and the angiotensin-converting enzyme insertion/deletion genotype in responses of human muscle to strength training. Eur J Appl Physiol 2004;92:267–74.

[48] McCauley T, Mastana SS, Folland JP. ACE I/D and ACTN3 R/X polymorphisms and muscle function and muscularity of older Caucasian men. Eur J Appl Physiol 2010;109:269–77.

[49] Giaccaglia V, Nicklas B, Kritchevsky S, Mychalecky J, Messier S, Bleecker E, et al. Interaction between angiotensin converting enzyme insertion/deletion genotype and exercise training on knee extensor strength in older individuals. Int J Sports Med 2008;29:40–4.

[50] Pescatello LS, Kostek MA, Gordish-Dressman H, Thompson PD, Seip RL, Price TB, et al. ACE ID genotype and the muscle strength and size response to unilateral resistance training. Med Sci Sports Exerc 2006;38:1074–81.

[51] Montgomery HE, Clarkson P, Dollery CM, Prasad K, Losi MA, Hemingway H, et al. Association of angiotensin-converting enzyme gene I/D polymorphism with change in left ventricular mass in response to physical training. Circulation 1997;96:741–7.

[52] Silva GJ, Moreira ED, Pereira AC, Mill JG, Krieger EM, Krieger JE. ACE gene dosage modulates pressure-induced cardiac hypertrophy in mice and men. Physiol Genomics 2006;27:237–44.

[53] Berk BC, Vekshtein V, Gordon HM, Tsuda T. Angiotensin II-stimulated protein synthesis in cultured vascular smooth muscle cells. Hypertension 1989;13:305–14.

[54] Geisterfer AA, Peach MJ, Owens GK. Angiotensin II induces hypertrophy, not hyperplasia, of cultured rat aortic smooth muscle cells. Circ Res 1988;62:749–56.

[55] Westerkamp CM, Gordon SE. Angiotensin-converting enzyme inhibition attenuates myonuclear addition in overloaded slow-twitch skeletal muscle. Am J Physiol Regul Integr Comp Physiol 2005;289:R1223–31.

[56] Gordon SE, Davis BS, Carlson CJ, Booth FW. ANG II is required for optimal overload-induced skeletal muscle hypertrophy. Am J Physiol Endocrinol Metab 2001;280:E150–9.

[57] Folland J, Leach B, Little T, Hawker K, Myerson S, Montgomery H, et al. Angiotensin-converting enzyme genotype affects the response of human skeletal muscle to functional overload. Exp Physiol 2000;85:575–9.

[58] Zhang B, Tanaka H, Shono N, Miura S, Kiyonaga A, Shindo M, et al. The I allele of the angiotensin-converting enzyme gene is associated with an increased percentage of slow-twitch type I fibers in human skeletal muscle. Clin Genet 2003;63:139–44.

[59] Zhang B, Shono N, Fan P, Ando S, Xu H, Jimi S, et al. Histochemical characteristics of soleus muscle in angiotensin-converting enzyme gene knockout mice. Hypertens Res 2005;28:681–8.

[60] McPherron AC, Lawler AM, Lee SJ. Regulation of skeletal muscle mass in mice by a new TGF-beta superfamily member. Nature 1997;387:83–90.

[61] McPherron AC, Lee SJ. Double muscling in cattle due to mutations in the myostatin gene. Proc Natl Acad Sci 1997;94:12457–61.

[62] Schuelke M, Wagner KR, Stolz LE, Hubner C, Riebel T, Komen W, et al. Myostatin mutation associated with gross muscle hypertrophy in a child. N Engl J Med 2004;350:2682–8.

[63] Santiago C, Ruiz JR, Rodriguez-Romo G, Fiuza-Luces C, Yvert T, Gonzalez-Freire M, et al. The K153R polymorphism in the myostatin gene and muscle power phenotypes in young, non-athletic men. PLoS One 2011;6:e16323.

[64] Kostek MA, Angelopoulos TJ, Clarkson PM, Gordon PM, Moyna NM, Visich PS, et al. Myostatin and follistatin polymorphisms interact with muscle phenotypes and ethnicity. Med Sci Sports Exerc 2009;41:1063–71.

[65] Ivey FM, Roth SM, Ferrell RE, Tracy BL, Lemmer JT, Hurlbut DE, et al. Effects of age, gender, and myostatin genotype on the hypertrophic response to heavy resistance strength training. J Gerontol A Biol Sci Med Sci 2000;55:M641–8.

[66] Gomez-Gallego F, Santiago C, Gonzalez-Freire M, Yvert T, Muniesa CA, Serratosa L, et al. The C allele of the AGT Met235Thr polymorphism is associated with power sports performance. Appl Physiol Nutr Metab 2009;34:1108–11.

[67] Norman B, Sabina RL, Jansson E. Regulation of skeletal muscle ATP catabolism by AMPD1 genotype during sprint exercise in asymptomatic subjects. J Appl Physiol 2001;91:258–64.

[68] Morisaki T, Gross M, Morisaki H, Pongratz D, Zollner N, Holmes EW. Molecular basis of AMP deaminase deficiency in skeletal muscle. Proc Natl Acad Sci 1992;89:6457–61.

[69] Norman B, Glenmark B, Jansson E. Muscle AMP deaminase deficiency in 2% of a healthy population. Muscle Nerve 1995;18:239–41.

[70] Norman B, Mahnke-Zizelman DK, Vallis A, Sabina RL. Genetic and other determinants of AMP deaminase activity in healthy adult skeletal muscle. J Appl Physiol 1998;85:1273–8.

[71] Lucia A, Martin MA, Esteve-Lanao J, San Juan AF, Rubio JC, Olivan J, et al. C34T mutation of the AMPD1 gene in an elite white runner. Br J Sports Med 2006;40:e7.

[72] Cieszczyk P, Ostanek M, Leonska-Duniec A, Sawczuk M, Maciejewska A, Eider J, et al. Distribution of the AMPD1 C34T polymorphism in Polish power-oriented athletes. J Sports Sci 2012;30:31–5.

[73] Fischer H, Esbjornsson M, Sabina RL, Stromberg A, Peyrard-Janvid M, Norman B. AMP deaminase deficiency is associated with lower sprint cycling performance in healthy subjects. J Appl Physiol 2007;103:315–22.

[74] Petersen AM, Pedersen BK. The anti-inflammatory effect of exercise. J Appl Physiol 2005;98:1154–62.

[75] Fishman D, Faulds G, Jeffery R, Mohamed-Ali V, Yudkin JS, Humphries S, et al. The effect of novel polymorphisms in the interleukin-6 (IL-6) gene on IL-6 transcription and plasma IL-6 levels, and an association with systemic-onset juvenile chronic arthritis. J Clin Investig 1998;102:1369–76.

[76] Bennermo M, Held C, Stemme S, Ericsson CG, Silveira A, Green F, et al. Genetic predisposition of the interleukin-6 response to inflammation: implications for a variety of major diseases?. Clin Chem 2004;50:2136–40.

[77] Yamin C, Duarte JA, Oliveira JM, Amir O, Sagiv M, Eynon N, et al. IL6 (-174) and TNFA (-308) promoter polymorphisms are associated with systemic creatine kinase response to eccentric exercise. Eur J Appl Physiol 2008;104:579–86.

[78] Ruiz JR, Buxens A, Artieda M, Arteta D, Santiago C, Rodriguez-Romo G, et al. The -174 G/C polymorphism of the IL6 gene is associated with elite power performance. J Sci Med Sport 2010;13:549–53.

[79] Eynon N, Ruiz JR, Meckel Y, Santiago C, Fiuza-Luces C, Gomez-Gallego F, et al. Is the -174 C/G polymorphism of the IL6 gene associated with elite power performance? a replication study with two different Caucasian cohorts. Exp Physiol 2011;96:156–62.

[80] Cooke JP, Rossitch Jr E, Andon NA, Loscalzo J, Dzau VJ. Flow activates an endothelial potassium channel to release an endogenous nitrovasodilator. J Clin Investig 1991;88:1663–71.

[81] Quyyumi AA, Dakak N, Andrews NP, Gilligan DM, Panza JA, Cannon III RO. Contribution of nitric oxide to metabolic coronary vasodilation in the human heart. Circulation 1995;92:320–6.

[82] Heydemann A, McNally E. NO more muscle fatigue. J Clin Investig 2009;119:448–50.

[83] Hickner RC, Fisher JS, Ehsani AA, Kohrt WM. Role of nitric oxide in skeletal muscle blood flow at rest and during dynamic exercise in humans. Am J Physiol 1997;273:H405–10.

[84] McConell GK, Kingwell BA. Does nitric oxide regulate skeletal muscle glucose uptake during exercise? Exerc Sport Sci Rev 2006;34:36–41.

[85] Nakayama M, Yasue H, Yoshimura M, Shimasaki Y, Kugiyama K, Ogawa H, et al. T-786-->C mutation in the 5′-flanking region of the endothelial nitric oxide synthase gene is associated with coronary spasm. Circulation 1999;99:2864–70.

[86] Wang XL, Sim AS, Wang MX, Murrell GA, Trudinger B, Wang J. Genotype dependent and cigarette specific effects on endothelial nitric oxide synthase gene expression and enzyme activity. FEBS Lett 2000;471:45–50.

[87] Yoshimura M, Yasue H, Nakayama M, Shimasaki Y, Sumida H, Sugiyama S, et al. A missense Glu298Asp variant in the endothelial nitric oxide synthase gene is associated with coronary spasm in the Japanese. Hum Genet 1998;103:65–9.

[88] Hand BD, McCole SD, Brown MD, Park JJ, Ferrell RE, Huberty A, et al. NOS3 gene polymorphisms and exercise hemodynamics in postmenopausal women. Int J Sports Med 2006;27:951–8.

[89] Rankinen T, Rice T, Perusse L, Chagnon YC, Gagnon J, Leon AS, et al. NOS3 Glu298Asp genotype and blood pressure response to endurance training: the HERITAGE family study. Hypertension 2000;36:885–9.

[90] Gomez-Gallego F, Ruiz JR, Buxens A, Artieda M, Arteta D, Santiago C, et al. The -786 T/C polymorphism of the NOS3 gene is associated with elite performance in power sports. Eur J Appl Physiol 2009;107:565–9.

[91] Kawada S, Ishii N. Skeletal muscle hypertrophy after chronic restriction of venous blood flow in rats. Med Sci Sports Exerc 2005;37:1144–50.

[92] Smith LW, Smith JD, Criswell DS. Involvement of nitric oxide synthase in skeletal muscle adaptation to chronic overload. J Appl Physiol 2002;92:2005–11.

[93] Williams AG, Folland JP. Similarity of polygenic profiles limits the potential for elite human physical performance. J Physiol 2008;586:113–21.

[94] Ruiz JR, Arteta D, Buxens A, Artieda M, Gomez-Gallego F, Santiago C, et al. Can we identify a power-oriented polygenic profile? J Appl Physiol 2010;108:561–6.

[95] Buxens A, Ruiz JR, Arteta D, Artieda M, Santiago C, Gonzalez-Freire M, et al. Can we predict top-level sports performance in power vs endurance events? a genetic approach. Scand J Med Sci Sports 2011;21:570–9.

CHAPTER

30

Unraveling the Function of Skeletal Muscle as a Secretory Organ

Role of Myokines on Muscle Regulation

Wataru Aoi[1], *Yuji Naito*[2], *Tomohisa Takagi*[2] *and Toshikazu Yoshikawa*[2]

[1]Laboratory of Health Science, Graduate School of Life and Environmental Sciences, Kyoto Prefectural University, Kyoto, Japan [2]Department of Molecular Gastroenterology and Hepatology, Graduate School of Medical Science, Kyoto Prefectural University of Medicine, Kyoto, Japan

INTRODUCTION

Adequate regular exercise has numerous health benefits. In the last few decades, epidemiological studies have shown that a dietary–exercise regimen reduces the risk of various common diseases such as type 2 diabetes, cardiovascular disease, and carcinogenesis. In addition, regular exercise improves the prognosis of existing diseases, including diabetes, ischemic heart disease, heart failure, and chronic obstructive pulmonary disease. Accumulating evidence has demonstrated the mechanisms underlying the benefits of acute and regular exercise. A single bout of exercise drastically changes various physiological parameters such as hormone production, blood flow, and the activity of the nervous and immune system, in addition to altering the expression/activity of certain genes and proteins in the skeletal muscle. For example, improvement of glucose metabolism is observed not only during exercise but also several hours after exercise and often persists until the next day. Further, regular exercise adaptively improves normal bodily functions including energy metabolism, brain-nervous system, endocrine system, and immune function, even in resting state, and the expression/activity of several key proteins in the skeletal muscle is involved in the development of this adaptation. Growing evidence indicates that bioactive proteins secreted by muscle cells are elevated in response to exercise and can regulate muscle itself, in addition to other organs, in an endocrine, autocrine, or paracrine manner, which is referred to as the myokine theory [1]. These bioactive proteins are suggested to mediate the acute and chronic effects of exercise, which would contribute in promoting health benefits along with maintaining physiological homeostasis and sports performance during exercise.

BIOACTIVE PROTEINS SECRETED FROM SKELETAL MUSCLE CELLS IN RESPONSE TO EXERCISE

Previously, several proteins that are secreted from muscle cells into the extracellular environment in response to exercise have been reported (Table 30.1). Many of them were suggested to be involved in the regulation of metabolic function in skeletal muscle itself and also in other metabolic organs. Interleukin (IL)-6 is well known as a representative secretory protein that is transiently elevated in muscle following a single bout of exercise [23]. IL-6 may act locally within the contracting skeletal muscle in a paracrine manner or be released into the circulation and may increase up to 100-fold, thus inducing systemic effects [2,3]. While this is controversial, IL-6 elevated by exercise in skeletal muscle can lead to additional improvement of insulin sensitivity in response to exercise [4]. Previous studies also showed that infusion of recombinant-IL-6 at the normal physiological level selectively stimulates

Nutrition and Enhanced Sports Performance.
DOI: http://dx.doi.org/10.1016/B978-0-12-396454-0.00030-8

TABLE 30.1 Bioactive Proteins Secreted from Skeletal Muscle in Response to Exercise.

Protein	Function	Target Organs	References
IL-6	Glucose metabolism, Lipid metabolism, Insulin secretion, Anti-inflammation	Skeletal muscle, Adipose tissue, Liver, Intestine, Neutophils	[2–10]
IL-15	Glucose metabolism, Fat metabolism, Muscle hypertrophy	Skeletal muscle	[12–14]
BDNF	Glucose metabolism	Skeletal muscle	[15]
FGF-21	Glucose metabolism	Skeletal muscle, Liver, Adipose tissue	[16]
Irisin	Lipid metabolism	Adipose tissue	[17]
LIF	Muscle hypertrophy	Skeletal muscle	[18]
IGF-1	Muscle hypertrophy,Osteogenesis	Skeletal muscle, Bone	[19,20]
Myostatin	Muscle anti-hypertrophy	Skeletal muscle	[21]
Fst/Fstl-1	Muscle hypertrophy, Endothelial function	Skeletal muscle, Endothelium	[11]
SPARC	Anti-tumorigenesis	Colon	[22]

IL-6, interleukin 6; IL-15, interleukin 15; BDNF, brain-derived neurotrophic factor; FGF-21, fibroblast growth factor; LIF, leukemia inhibitor factor; IGF-1, insulin like growth factor 1; Fst, follistatin; Fstl-1, follistatin-like 1; SPARC, secreted protein acidic and rich in cysteine.

lipid metabolism in skeletal muscle in healthy subjects [5] and in subjects with type 2 diabetes [6]. In addition, muscle-derived IL-6 has been suggested to play a role in increased lipolysis in adipose tissue through an endocrine mechanism [3]. In fact, recombinant IL-6 intra-lipid infusion elevates plasma fatty acid levels due to lipolysis of adipose tissue in healthy humans [7]. Furthermore, injection of IL-6 to rats catabolizes hepatic glycogen and accelerates glucose output into circulation [8], which may contribute to the maintenance of blood glucose and supply the required energy substrate during exercise. In addition to IL-6, other muscle-secreted proteins such as brain-derived neurotrophic factor, fibroblast growth factor 21, and IL-15 have been shown to be produced in skeletal muscle in response to acute or chronic exercise, and have been suggested to increase fat oxidation or glucose uptake in skeletal muscle [12,13,15,16]. A more recent study showed that peroxisome proliferator-activated receptor gamma coactivator-1 alpha (PGC-1α) expression in muscle stimulates an increase in the expression of FNDC5, a membrane protein that is cleaved and secreted as Irisin [17]. PGC-1α has been shown to play a central role in a family of transcriptional co-activators involved in aerobic metabolism; thus, a considerable amount of attention has been focused on it as a target for the prevention or treatment of metabolic syndrome through activation of lipid metabolism. Acute and regular exercise elevates PGC-1α expression in skeletal muscle [24,25] and, consequently, the secretion of Irisin from the muscle into circulation. Secreted Irisin acts on white adipose cells and facilitates brown-fat-like development, which may account for metabolic elevation and body fat reduction induced by exercise.

One of the other suggested functions of muscle-secreted proteins is anti-inflammation, and muscle-derived IL-6 likely contributes to reduction of inflammation when in circulation [23]. IL-6 can increase the levels if anti-inflammatory factors such as IL-10, IL-1 receptor agonist, and C-reactive protein in neutrophils and the liver [9,26]. Indeed, recombinant IL-6 infusion inhibits the endotoxin-induced increase in circulating levels of tumor necrosis factor (TNF-α), a representative pro-inflammatory cytokine [10]. On the other hand, IL-6 is recognized as a proinflammatory cytokine. In severe systemic infection, circulating IL-6 is drastically elevated and may reach over 10 000-fold the level in resting healthy state. In contrast, chronic low-grade elevation of IL-6 (below 10-fold of that in resting healthy state) is induced by sedentary life, obesity, and dietary habits, which are associated with the development of metabolic diseases, although regular physical activity reduces the elevation of circulating IL-6 in resting state along with metabolic improvement [27,28]. Therefore, it is necessary to consider as separate the exercise-induced secretion of IL-6, which is a transient/moderate elevation, and the pathological states, which are transient/high or chronic/low elevations.

It has been suggested that muscle-secreted proteins have further functions, and some proteins such as leukemia inhibitory factor, follistatin-like 1, and insulin growth factor-1 (IGF-1) contribute to muscle hypertrophy via autocrine or paracrine effects [11,14,18,19,21]. The secreted IGF-1 may also function as an osteogenic factor by stimulating differentiation and mineralization [20]. In addition, physiological elevation of IL-6 levels stimulates an insulin secretory hormone glucagon-like peptide-1, from intestinal L cells and pancreatic α cells, which ultimately improves insulin secretion from pancreas β cells [29].

APPROACH FOR IDENTIFICATION OF NEW MUSCLE-SECRETED PROTEINS

Many studies have suggested that several other proteins secreted from muscle have not been identified. For example, a bioinformatics study showed that the secretome of human muscle cells includes more than 300 proteins [30]. In addition, an *in vitro* study demonstrated that myocytes secrete many proteins into the medium during differentiation [31,32]. Furthermore, transcriptome and proteome studies of human and rodent muscle tissue have demonstrated that the expression of many genes and proteins increases in response to exercise [33–36]. Therefore, we recently tried to identify novel muscle-derived proteins that are secreted into the general circulation. The transcriptome of muscle tissue in sedentary and exercised young and old mice was compared. In total, 381 genes in gastrocnemius muscle were up-regulated in mice that exercised for 4 weeks compared with sedentary mice; on the other hand, 100 genes were down-regulated in 24-month-old sedentary mice compared with 3-month-old sedentary mice [22]. Among these genes, there were 24 common genes, including the secretory protein SPARC, a secreted matricellular glycoprotein.

The level of SPARC protein in gastrocnemius muscle was significantly elevated, and this elevation of muscle SPARC was found to be specifically pronounced around the plasma membrane in exercised muscle. In a human study, a time-course analysis of the serum levels of SPARC showed that the SPARC level was elevated in young healthy men immediately after a single bout of exercise at 70% VO_{2max} for 30 min, and then gradually decreased until it returned to the baseline level 6 h after exercise [22] (Figure 30.1). This exercise-induced increase in SPARC level appeared to be muscle specific, because no increase was observed in other organs where SPARC is abundant. Furthermore, 60 min cyclic stretching of C2C12 myotubes stimulated SPARC secretion into the extracellular medium. These findings suggest that a single bout of exercise accelerates SPARC secretion from contracting muscle into blood.

Regular exercise does not result in adaptive elevation of circulating SPARC in the resting state. The plasma level of SPARC between sedentary controls and mice that performed 4-week regular exercise did not differ significantly, although SPARC expression in the muscle was increased in the exercised mice [22]. In a human experiment, 4 weeks of exercise at 70% VO_{2max} did not change plasma SPARC between baseline and post-training in young healthy subjects. This finding suggests

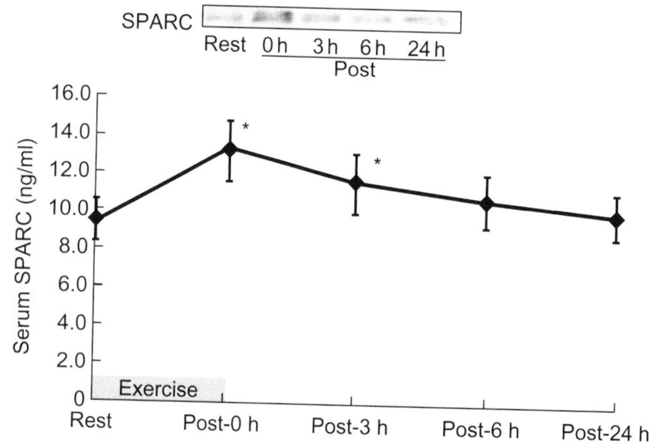

FIGURE 30.1 **A single bout of exercise increases circulating levels of SPARC in humans.** Time course of serum SPARC level after steady-state cycling at 70% maximal oxygen uptake (VO_{2max}) for 30 min (n = 10). *$P < 0.05$ versus resting state (Rest). Results are shown as mean ± standard error. *Data from Aoi et al. [22]. (This figure is reproduced in color plate section.)*

that the increase in SPARC level in muscle tissues due to regular exercise does not contribute substantially to the circulating concentration while at rest. However, regular exercise significantly promoted the acute exercise-induced increase in the serum level of SPARC.

SPARC IS A CANCER PREVENTIVE PROTEIN SECRETED BY SKELETAL MUSCLE

A number of epidemiological studies have been carried out on the average individual's level of physical activity and its relationship to the incidence of cancer in Europe, the United States, and Japan. The general consensus among the authors of these studies is that physical activity can prevent cancer in the colon, breast, uterus, pancreas, and lungs [37–43]. In particular, almost all investigation clearly demonstrated that physical activity significantly reduces the incidence of colon cancer. A review of these epidemiological studies by The World Cancer Research Fund / American Institute for Cancer Research (WCRF/AICR) showed that physical activity was the only lifestyle change that would certainly reduce an individual's risk of colon cancer [44]. Although the exact mechanism underlying the beneficial results obtained in epidemiological studies remains unclear, various potential mechanisms such as activation of the immune system and antioxidant status, anti-inflammation, improved insulin sensitivity and proportion of bile acids, and exercise-induced increases in gastrointestinal transit have been suggested [45–50]. Previously, we reported that regular exercise prevents the formation of aberrant crypt foci (ACF), which are the precursor lesions of colon adenocarcinoma, associated with anti-inflammation on the mucosal surface of the mouse colon [51]. However, the endogenous defense system, such as antioxidant and chaperone proteins, was unchanged [51], which suggested that the anti-tumorigenesis effect of regular exercise is affected by the levels of circulating factors rather than endogenous proteins in the colon.

SPARC is a matricellular protein that is primarily involved in development, remodeling, and tissue repair through modulation of cell–cell and cell–matrix interactions [52–54]. In addition, SPARC has been reported to have functions such as regulating angiogenesis and collagen production/fibrillogenesis, chaperoning, inhibiting adipogenesis, and further exerting anti-tumorigenetic effects [55–61]. Previous studies have revealed that a lack of SPARC increases pancreatic and ovarian tumorigenesis *in vivo* [60,61]. In addition, the presence of exogenous SPARC in cancer cell lines reduces cell proliferation *in vitro* [61,62]. Furthermore, epigenetic silencing of the *SPARC* gene via hypermethylation of its promoter is frequent in colon cancers, which leads to rapid progression of the tumor [63,64]. Moreover, modulation of SPARC expression affects the sensitivity of colorectal tumors to radiation and chemotherapy [65–67]. Interestingly, a clinical study showed that the 5-year survival of patients with tumors that expressed high levels of SPARC was significantly better than that of those with tumors that did not express SPARC [68]. Therefore, we examined the effect of the myokine SPARC on the onset of colon tumors by using SPARC-null mice. In a mouse model for colon cancer generated azoxymethane (AOM), regular low-intensity exercise, which consisted of treadmill running at 18 m/min and 3 times/week for 6 weeks, significantly reduced the formation of ACF in the colons of wild-type mice [22] (Figure 30.2). In contrast, more ACF were found in AOM-treated SPARC-null mice than in wild-type mice, and exercise did not have an inhibitory effect. In addition, we examined the effect of exogenous SPARC on ACF formation in the colon by injection of recombinant SPARC in the AOM-treated mice. Injection of SPARC, which is equivalent to the elevation in response to exercise, suppressed ACF formation. Furthermore, in a cell culture experiment, addition of recombinant SPARC to colon carcinoma cells inhibited cell proliferation in a dose-dependent manner. In contrast, addition of conditioned medium, from short interfering RNA-treated muscle cells, to the carcinoma cells accelerated the proliferation. These results suggested that secreted SPARC suppresses colon tumorigenesis, which is consistent with the findings of many previous studies [60–62,65] demonstrating that SPARC is a tumor suppressor.

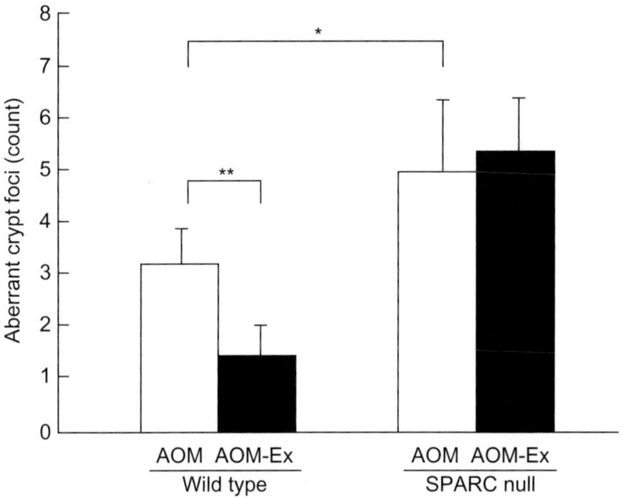

FIGURE 30.2 SPARC prevents tumorigenesis in colon. The numbers of aberrant crypt foci (ACF) on the mucosal surface of the colon were counted under a light microscope. In wild-type mice, regular low-intensity exercise significantly reduced the number of ACF in the colons of AOM-treated mice compared with sedentary mice. In contrast, more ACF were formed in AOM-treated SPARC-null mice than in wild-type mice, and exercise did not have an inhibitory effect. Results are shown as mean ± standard error (n = 10–12). AOM, AOM-treated sedentary mice; AOM-Ex, AOM-treated exercised mice. * $P < 0.05$; ** $P < 0.01$. *Data from Aoi et al. [22].*

A cause of ACF formation is dysregulation of apoptosis [66]. The terminal deoxyribonucleotidyl transferase dUTP nick end labeling (TUNEL) assay showed that regular exercise increased the number of apoptotic colon cells in wild-type mice; however, the number did not differ between sedentary and exercised SPARC-null mice [22]. Furthermore, the levels of cleaved caspase-3 and -8 were higher in wild-type mice than in SPARC-null mice, and regular exercise further increased the levels of these apoptosis markers in wild-type mice but not in SPARC-null mice. However, regular exercise did not affect the levels of B-cell lymphoma 2 (Bcl-2) or Bcl-2-associated X protein (Bax) in either wild-type or SPARC-null mice. These findings suggested that SPARC mediates exercise-induced colon reduction via caspase-3- and caspase-8-dependent apoptosis. In addition, we found the effect of exogenous SPARC on colon tumor by using colon carcinoma cells, and found that apoptosis of these cells was elevated by addition of recombinant SPARC in a dose-dependent manner. This *in vitro* result supported the hypothesis that SPARC prevents proliferation of colon tumor cells via increased apoptosis.

PERSPECTIVE

As was well-known from previous studies, exercise releases various metabolic factors from skeletal muscle into circulation. For example, lactate is generated from carbohydrates via glycolytic metabolism, and the amount depends on the intensity of exercise. After its release into blood, lactate is carried to other tissues and is utilized as a substrate of aerobic metabolism or gluconeogenesis. Recently, studies into further functions of such muscle-mediated metabolites, such as mitochondria biogenesis and energy substrate in brain [69,70], have suggested that lactate and others such as amino acids, ions, and ammonium, should be reconsidered as endocrine bioactive factors. In addition, microRNAs (miRNAs) may be secreted from muscle into circulation and function in an endocrine manner. Some miRNAs are taken into intracellular vesicles (e.g., exosomes) and released into circulation without being degraded by RNase [71]. In addition, the circulating miRNAs (c-miRNAs) can move from circulation into other cells and regulate their functions via regulation of gene expression at the post-transcriptional level through translational inhibition or mRNA degradation. Several miRNAs are highly enriched in skeletal muscle [72–75] and may be secreted from muscle into circulation. In the future, many other muscle-secreted bioactive factors including metabolites and microRNA could be identified, which may accelerate the understanding of the effect of exercise on improvement of physical performance and prevention of diseases.

References

[1] Pedersen BK, Steensberg A, Fischer C, Keller C, Keller P, Plomgaard P, et al. Searching for the exercise factor: is IL-6 a candidate. J Muscle Res Cell Motil 2003;24:113–9.

[2] Penkowa M, Keller C, Keller P, Jauffred S, Pedersen BK. Immunohistochemical detection of interleukin-6 in human skeletal muscle fibers following exercise. FASEB J 2003;17:2166–8.

[3] Fischer CP. Interleukin-6 in acute exercise and training: what is the biological relevance? Exerc Immunol Rev 2006;12:6–33.

[4] Benrick A, Wallenius V, Asterholm IW. Interleukin-6 mediates exercise-induced increase in insulin sensitivity in mice. Exp Physiol 2012;97:1224–35.

[5] van Hall G, Steensberg A, Sacchetti M, Fischer C, Keller C, Schjerling P, et al. Interleukin-6 stimulates lipolysis and fat oxidation in humans. J Clin Endocrinol Metab 2003;88:3005–10.

[6] Petersen EW, Carey AL, Sacchetti M, Steinberg GR, Macaulay SL, Febbraio MA, et al. Acute IL-6 treatment increases fatty acid turnover in elderly humans *in vivo* and in tissue culture *in vitro*: evidence that IL-6 acts independently of lipolytic hormones. Am J Physiol Endocrinol Metab 2005;288:E155–62.

[7] Lyngso D, Simonsen L, Bulow J. Metabolic effects of interleukin-6 in human splanchnic and adipose tissue. J Physiol 2002;543:379–86.

[8] Lienenlüke B, Christ B. Impact of interleukin-6 on the glucose metabolic capacity in rat liver. Histochem Cell Biol 2007;128:371–7.

[9] Steensberg A, Fischer CP, Keller C, Moller K, Pedersen BK. IL-6 enhances plasma IL-1ra, IL-10, and cortisol in humans. Am J Physiol Endocrinol Metab 2003;285:E433–7.

[10] Starkie R, Ostrowski SR, Jauffred S, Febbraio M, Pedersen BK. Exercise and IL-6 infusion inhibit endotoxin-induced TNF-alpha production in humans. FASEB J 2003;17:884–6.

[11] Ouchi N, Oshima Y, Ohashi K, Higuchi A, Ikegami C, Izumiya Y, et al. Follistatin-like 1, a secreted muscle protein, promotes endothelial cell function and revascularization in ischemic tissue through a nitric-oxide synthase-dependent mechanism. J Biol Chem 2008;283:32802–11.

[12] Tamura Y, Watanabe K, Kantani T, Hayashi J, Ishida N, Kaneki M. Upregulation of circulating IL-15 by treadmill running in healthy individuals: Is IL-15 an endocrine mediator of the beneficial effects of endurance exercise?. Endocr J 2011;58:211–5.

[13] Busquets S, Figueras M, Almendro V, López-Soriano FJ, Argilés JM. Interleukin-15 increases glucose uptake in skeletal muscle. An antidiabetogenic effect of the cytokine. Biochim Biophys Acta 2006;1760:1613–7.

[14] Quinn LS, Anderson BG, Drivdah RH, Alvarez B, Argilés JM. Overexpression of interleukin-15 induces skeletal muscle hypertrophy *in vitro:* implications for treatment of muscle wasting disorders. Exp Cell Res 2002;280:55–63.

[15] Matthews VB, Aström MB, Chan MH, Bruce CR, Krabbe KS, Prelovsek O, et al. Brain-derived neurotrophic factor is produced by skeletal muscle cells in response to contraction and enhances fat oxidation via activation of AMP-activated protein kinase. Diabetologia 2009;52:1409–18.

[16] Mashili FL, Austin RL, Deshmukh AS, Fritz T, Caidahl K, Bergdahl K, et al. Direct effects of FGF21 on glucose uptake in human skeletal muscle: implications for type 2 diabetes and obesity. Diabetes Metab Res Rev 2011;27:286–97.

[17] Boström P, Wu J, Jedrychowski MP, Korde A, Ye L, Lo JC, et al. A PGC1-α-dependent myokine that drives brown-fat-like development of white fat and thermogenesis. Nature 2012;481:463–8.

[18] Broholm C, Pedersen BK. Leukaemia inhibitory factor—an exercise-induced myokine. Exerc Immunol Rev 2010;16:77–85.

[19] Adams GR. Autocrine/paracrine IGF-I and skeletal muscle adaptation. J Appl Physiol 2002;93:1159–67.

[20] Yu Y, Mu J, Fan Z, Lei G, Yan M, Wang S, et al. Insulin-like growth factor 1 enhances the proliferation and osteogenic differentiation of human periodontal ligament stem cells via ERK and JNK MAPK pathways. Histochem Cell Biol 2012;137:513–25.

[21] Hittel DS, Berggren JR, Shearer J, Boyle K, Houmard JA. Increased secretion and expression of myostatin in skeletal muscle from extremely obese women. Diabetes 2009;58:30–8.

[22] Aoi W, Naito Y, Takagi T, Tanimura Y, Takanami Y, Kawai Y, et al. A novel myokine, secreted protein acidic and rich in cysteine (SPARC), suppresses colon tumorigenesis via regular exercise. Gut 2013;62:882–9.

[23] Pedersen BK, Fischer CP. Beneficial health effects of exercise—the role of IL-6 as a myokine. Trends Pharmacol Sci 2007;28:152–6.

[24] Baar K, Wende AR, Jones TE, Marison M, Nolte LA, Chen M, et al. Adaptations of skeletal muscle to exercise: rapid increase in the transcriptional coactivator PGC-1. FASEB J 2002;16:1879–86.

[25] Russell AP, Feilchenfeldt J, Schreiber S, Praz M, Crettenand A, Gobelet C, et al. Endurance training in humans leads to fiber type-specific increases in levels of peroxisome proliferator-activated receptor-gamma coactivator-1 and peroxisome proliferator-activated receptor-alpha in skeletal muscle. Diabetes 2003;52:2874–81.

[26] Heinrich PC, Castell JV, Andus T. Interleukin-6 and the acute phase response. Biochem J 1990;265:621–36.

[27] Kadoglou NP, Perrea D, Iliadis F, Angelopoulou N, Liapis C, Alevizos M. Exercise reduces resistin and inflammatory cytokines in patients with type 2 diabetes. Diabetes Care 2007;30:719–21.

[28] Nicklas BJ, Hsu FC, Brinkley TJ, Church T, Goodpaster BH, Kritchevsky SB, et al. Exercise training and plasma C-reactive protein and interleukin-6 in elderly people. J Am Geriatr Soc 2008;56:2045–52.

[29] Ellingsgaard H, Hauselmann I, Schuler B, Habib AM, Baggio LL, Meier DT, et al. Interleukin-6 enhances insulin secretion by increasing glucagon-like peptide-1 secretion from L cells and alpha cells. Nat Med 2011;17:1481–9.

[30] Bortoluzzi S, Scannapieco P, Cestaro A, Danieli GA, Schiaffino S. Computational reconstruction of the human skeletal muscle secretome. Proteins 2006;62:776–92.

[31] Chan XC, McDermott JC, Siu KW. Identification of secreted proteins during skeletal muscle development. J Proteome Res 2007;6:698–710.

[32] Henningsen J, Rigbolt KT, Blagoev B, Pedersen BK, Kratchmarova I. Dynamics of the skeletal muscle secretome during myoblast differentiation. Mol Cell Proteomics 2010;9:2482–96.

[33] Choi S, Liu X, Li P, Akimoto T, Lee SY, Zhang M, et al. Transcriptional profiling in mouse skeletal muscle following a single bout of voluntary running: evidence of increased cell proliferation. J Appl Physiol 2005;99:2406–15.

[34] Mahoney DJ, Parise G, Melov S, Safdar A, Tarnopolsky MA. Analysis of global mRNA expression in human skeletal muscle during recovery from endurance exercise. FASEB J 2005;19:1498–500.

[35] Guelfi KJ, Casey TM, Giles JJ, Fournier PA, Arthur PG. A proteomic analysis of the acute effects of high-intensity exercise on skeletal muscle proteins in fasted rats. Clin Exp Pharmacol Physiol 2006;33:952–7.

[36] Holloway KV, O'Gorman M, Woods P, Morton JP, Evans L, Cable NT, et al. Proteomic investigation of changes in human vastus lateralis muscle in response to interval-exercise training. Proteomics 2009;9:5155–74.

[37] Garabrant DH, Peters JM, Mack TM, Bernstein L. Job activity and colon cancer risk. Am J Epidemiol 1984;119:1005–14.

[38] Lee KJ, Inoue M, Otani T, Iwasaki M, Sasazuki S, Tsugane S, , et al.JPHC Study Group Physical activity and risk of colorectal cancer in Japanese men and women: the Japan public health center-based prospective study. Cancer Causes Control 2007;18:199–209.

[39] Mai PL, Sullivan-Halley J, Ursin G, Stram DO, Deapen D, Villaluna D, et al. Physical activity and colon cancer risk among women in the California teachers study. Cancer Epidemiol Biomarkers Prev 2007;16:517–25.

[40] Vena JE, Graham S, Zielezny M, Brasure J, Swanson MK. Occupational exercise and risk of cancer. Am J Clin Nutr 1987;45:318–27.

[41] Zheng W, Shu XO, McLaughlin JK, Chow WH, Gao YT, Blot WJ. Occupational physical activity and the incidence of cancer of the breast, corpus uteri, and ovary in Shanghai. Cancer 1993;71:3620–4.

[42] Michaud DS, Giovannucci E, Willett WC, Colditz GA, Stampfer MJ, Fuchs CS. Physical activity, obesity, height, and the risk of pancreatic cancer. J Am Med Assoc 2001;286:921–9.

[43] Wannamethee SG, Shaper AG, Walker M. Physical activity and risk of cancer in middle-aged men. Br J Cancer 2001;85:1311–6.

[44] World Cancer Research Fund, and American Institute for Cancer Research. Physical activity. In: Food, Nutrition, Physical Activity, and the Prevention of Cancer: a Global Perspective (World Cancer Research Fund, American Institute for Cancer Research editors.), Washington: 2007. p. 198–209.

[45] Hagio M, Matsumoto M, Yajima T, Hara H, Ishizuka S. Voluntary wheel running exercise and dietary lactose concomitantly reduce proportion of secondary bile acids in rat feces. J Appl Physiol 2010;109:663–8.

[46] Shephard RJ, Rhind S, Shek PN. The impact of exercise on the immune system: NK cells, interleukins 1 and 2, and related responses. Exerc Sport Sci Rev 1995;23:215–41.

[47] McTiernan A, Ulrich C, Slate S, Potter J. Physical activity and cancer etiology: associations and mechanisms. Cancer Causes Control 1998;9:487–509.

[48] Hoffman-Goetz L, Apter D, Demark-Wahnefried W, Goran MI, McTiernan A, Reichman ME. Possible mechanisms mediating an association between physical activity and breast cancer. Cancer 1998;83:621–8.

[49] Demarzo MM, Martins LV, Fernandes CR, Herrero FA, Perez SE, Turatti A, et al. Exercise reduces inflammation and cell proliferation in rat colon carcinogenesis. Med Sci Sports Exerc 2008;40:618–21.

[50] Song BK, Cho KO, Jo Y, Oh JW, Kim YS. Colon transit time according to physical activity level in adults. J Neurogastroenterol Motil 2012;18:64–9.

[51] Aoi W, Naito Y, Takagi T, Kokura S, Mizushima K, Takanami Y, et al. Regular exercise reduces colon tumorigenesis associated with suppression of iNOS. Biochem Biophys Res Commun 2010;399:14–9.

[52] Brekken RA, Sage EH. SPARC, a matricellular protein: at the crossroads of cell-matrix communication. Matrix Biol 2001;19:816–27.

[53] Bradshaw AD, Sage EH. SPARC, a matricellular protein that functions in cellular differentiation and tissue response to injury. J Clin Invest 2001;107:1049–54.

[54] Bornstein P. Diversity of function is inherent in matricellular proteins: an appraisal of thrombospondin 1. J Cell Biol 1995;130:503–6.

[55] Jendraschak E, Sage EH. Regulation of angiogenesis by SPARC and angiostatin: implications for tumor cell biology. Semin Cancer Biol 1996;7:139–46.

[56] Rentz TJ, Poobalarahi F, Bornstein P, Sage EH, Bradshaw AD. SPARC regulates processing of procollagen I and collagen fibrillogenesis in dermal fibroblasts. J Biol Chem 2007;282:22062–71.

[57] Chlenski A, Guerrero LJ, Salwen HR, Yang Q, Tian Y, Morales La Madrid A, et al. Secreted protein acidic and rich in cysteine is a matrix scavenger chaperone. PLoS One 2011;6:e23880.

[58] Nie J, Sage EH. SPARC functions as an inhibitor of adipogenesis. J Cell Commun Signal 2009;3:247–54.

[59] Nakamura K, Nakano S, Miyoshi T, Yamanouchi K, Matsuwaki T, Nishihara M. Age-related resistance of skeletal muscle-derived progenitor cells to SPARC may explain a shift from myogenesis to adipogenesis. Aging (Albany NY) 2012;4:40–8.

[60] Puolakkainen PA, Brekken RA, Muneer S, Sage EH. Enhanced growth of pancreatic tumors in SPARC-null mice is associated with decreased deposition of extracellular matrix and reduced tumor cell apoptosis. Mol Cancer Res 2004;2:215–24.

[61] Said N, Motamed K. Absence of host-secreted protein acidic and rich in cysteine (SPARC) augments peritoneal ovarian carcinomatosis. Am J Pathol 2005;167:1739–52.

[62] Yiu GK, Chan WY, Ng SW, Chan PS, Cheung KK, Berkowitz RS, et al. SPARC (secreted protein acidic and rich in cysteine) induces apoptosis in ovarian cancer cells. Am J Pathol 2001;159:609–22.

[63] Cheetham S, Tang MJ, Mesak F, Kennecke H, Owen D, Tai IT. SPARC promoter hypermethylation in colorectal cancers can be reversed by 5-Aza-2_deoxycytidine to increase SPARC expression and improve therapy response. Br J Cancer 2008;98:1810–9.

[64] Yang E, Kang HJ, Koh KH, Rhee H, Kim NK, Kim H. Frequent inactivation of SPARC by promoter hypermethylation in colon cancers. Int J Cancer 2007;121:567–755.

[65] Tai IT, Tang MJ. SPARC in cancer biology: its role in cancer progression and potential for therapy. Drug Resist Updat 2008;11:231–46.

[66] Tai IT, Dai M, Owen DA, Chen LB. Genome-wide expression analysis of therapy-resistant tumors reveals SPARC as a novel target for cancer therapy. J Clin Invest 2005;115: 1492–1402

[67] Taghizadeh F, Tang MJ, Tai IT. Synergism between vitamin D and secreted protein acidic and rich in cysteine-induced apoptosis and growth inhibition results in increased susceptibility of therapy-resistant colorectal cancer cells to chemotherapy. Mol Cancer Ther 2007;6:309–17.

[68] Takahashi M, Mutoh M, Kawamori T, Sugimura T, Wakabayashi K. Altered expression of β-catenin, inducible nitric oxide synthase and cyclooxygenase-2 in azoxymethane-induced rat colon carcinogenesis. Carcinogenesis 2000;21:1319–27.

[69] Hashimoto T, Hussien R, Oommen S, Gohil K, Brooks GA. Lactate sensitive transcription factor network in L6 cells: activation of MCT1 and mitochondrial biogenesis. FASEB J 2007;21:2602–12.

[70] van Hall G, Strømstad M, Rasmussen P, Jans O, Zaar M, Gam C, et al. Blood lactate is an important energy source for the human brain. J Cereb Blood Flow Metab 2009;29:1121–9.

[71] Valadi H, Ekström K, Bossios A, Sjöstrand M, Lee JJ, Lötvall JO. Exosome-mediated transfer of mRNAs and microRNAs is a novel mechanism of genetic exchange between cells. Nat Cell Biol 2007;9:654–9.

[72] McCarthy JJ. MicroRNA-206: the skeletal muscle-specific myomiR. Biochim Biophys Acta 2008;1779:682–91.

[73] McCarthy JJ. The MyomiR network in skeletal muscle plasticity. Exerc Sci Sport Rev 2011;39:150–4.

[74] Small EM, O'Rourke JR, Moresi V, Sutherland LB, McAnally J, Gerard RD, et al. Regulation of PI3-kinase/Akt signaling by muscle-enriched microRNA-486. Proc Natl Acad Sci USA 2010;107:4218–23.

[75] Sempere LF, Freemantle S, Pitha-Rowe I, Moss E, Dmitrovsky E, Ambros V. Expression profiling of mammalian microRNAs uncovers a subset of brain-expressed microRNAs with possible roles in murine and human neuronal differentiation. Genome Biol 2004;5:R13.

SECTION 5

MINERALS AND SUPPLEMENTS IN MUSCLE BUILDING

CHAPTER

31

The Role of Testosterone in Nutrition and Sports: An Overview

Jan Lingen, Hande Hofmann and Martin Schönfelder

Institute of Preventive Pediatrics, Technische Universität München, Munich, Germany

INTRODUCTION

Testosterone is a steroid hormone with powerful androgenic and anabolic effects. It is primarily released by the Leydig cells in the testes and to a small extent by the adrenal cortex [1]. It is generally known that testosterone can help to build muscle mass and change body composition in favor of fat-free mass [2]. Supraphysiologic doses of testosterone have been shown to increase fat-free mass massively within a short time (3.2 kg after 10 weeks) and to boost strength (19% higher one repetition maximum, 1RM, in squatting after 10 weeks) [2,3]. Additionally those effects can be enhanced if combined with strength training (6.1 kg and 38% higher 1RM in squatting after 10 weeks) [2].

Testosterone exerts its hypertrophic effects on muscles in an anabolic and anti-catabolic manner (Boxes 31.1 and 31.2), leading to increased protein synthesis and decreased protein breakdown [4,5]. This is possible by its stimulatory effects on the expression of muscle insulin-like growth factor (IGF-1) [6], as IGF-1 and its downstream targets are able to initiate protein synthesis and hypertrophy [7–9].

Furthermore, testosterone can activate resting myogenic stem cells, so-called satellite cells, which in turn can cause myonuclear accretion or fusion with existing muscle fibers to form or renew myotubes and thus support hypertrophy [1].

Naturally, its positive effects on body composition can be explained on the one hand by its properties described above and on the other by its potential ability to drive the development of mesenchymal pluripotent cells into cells of the myogenic lineage, while inhibiting differentiation to adipogenic cells [3]. Additionally it has been suggested that testosterone inhibits uptake of lipids in adipocytes and stimulates lipolysis [1].

It is known that testosterone has to bind to androgen receptors (AR) to initiate its effects. However, recent animal studies suggest that there may be an AR-independent, nongenomic mechanism of action for testosterone through increased intracellular Ca^{2+} concentration via activation of a G-protein-linked receptor of myoblasts from the rat neonatal hind limb. This signaling process could result in the phosphorylation of several transcription factors which are connected with cellular growth. The significance of this nongenomic effect of testosterone on human skeletal muscle is still unclear [1].

In sports the use of testosterone as an illegal drug to enhance physical performance is not limited to competitive athletes. The misuse has already infiltrated recreational sports, especially the fitness and bodybuilding area [1,2,4]. Abuse of testosterone has potential side effects which are known to occur in the hepatic, cardiovascular, reproductive, endocrine, dermatological and psychiatric systems. But there are also some adverse effects reported in the musculoskeletal system, e.g., bone fractures, tendon pathology and rhabdomyolysis [4].

It seems that those detrimental health-related impacts are either ignored, not only by professional athletes, or that there is a lack of knowledge leading to abuse of testosterone or anabolic steroids in general. This attitude becomes apparent in cases where those substances caused tremendous lesions such as ruptures of ligaments (e.g., the anterior cruciate ligament) and tendons (e.g., the patellar tendon). Despite warnings of impaired healing of the damaged structures, which

Nutrition and Enhanced Sports Performance.
DOI: http://dx.doi.org/10.1016/B978-0-12-396454-0.00031-X

BOX 31.1

DOCUMENTED TESTOSTERONE EFFECTS IN MAMMALIAN TISSUE

Increased Overall Fat-Free Mass

Increased appendicular fat-free mass
Increased fat-free mass of the trunk

Myogenesis

Increased type I and type II muscle fiber volume
Increased myonuclear number
Increased protein synthesis
Decreased protein degradation
Increased satellite cell number
Increased myogenesis of pluripotent stem cells

Cell Signaling

Increased androgen receptors in pluripotent stem cells

Decreased Overall Adipose Tissue

Decreased subcutaneous fat
Decreased deep intermuscular fat

Functionality

Increased grip strength
Increased leg press power

Motor Neurons

Increased androgen receptors
Increased motor neuron number and size

Decreased Overall Adipose Tissue

Decreased subcutaneous fat
Decreased deep intermuscular fat

Adapted from Herbst et al. [3]

BOX 31.2

SUGGESTIVE BUT UNCONFIRMED TESTOSTERONE EFFECTS IN MAMMALIAN TISSUE

Decreased intra-abdominal fat
Improved physical function
Outcomes for non-receptor androgen-mediated signaling pathways
Appetite stimulation

Adapted from Herbst et al. [3]

were obviously caused by those substances, abuse continued [4].

The reasons for testosterone abuse are mainly its mentioned effects on strength and muscle mass [1,2], and in many cases the desire to be attractive [1,4]. It is its anabolic potential that makes testosterone so popular among those who (ab)use it, and therefore for science too.

As the bioavailability of testosterone correlates negatively with fat mass and positively with muscle mass [3], and supraphysiologic exogenic testosterone administration leads to suppression of endogenous testosterone production as a result of blocked gonadotropin releasing hormone (GnRH) release in the hypothalamus pituitary axis [1], it is the authors' interest to show whether endogenous testosterone production may be regulated via nutrition and physical training in order to support muscle growth.

Of course, the benefits are expected to be inferior to those from an exogenous testosterone abuse; but the goal has to be to maximize muscle growth in a natural, legal and healthy way.

The aim of this book is to give recommendations on optimizing muscle growth. Thus, the authors will discuss the role of food on testosterone level(s) and the hormonal changes that occur in response to resistance training combined with immediate post-exercise food intake, while emphasizing testosterone.

MEANING OF NATURAL TESTOSTERONE LEVELS AND HYPERTROPHY/MUSCLE GROWTH

The power of exogenous testosterone administration in supraphysiologic doses on skeletal muscle has already been described [2]. However, the relevance of even physiological levels of this hormone on muscle size becomes clear if one observes the natural process of muscle growth and muscle loss. In boys, it is puberty, with its accrual in muscle mass concurrently with increase in testosterone, that highlights its meaning for muscle growth. On the other hand, testosterone declines by approximately 1–3% per year, from the age of 35–40 years on. Later in life this decline is associated with sarcopenia [5].

Considering that free testosterone levels correlate positively with muscle mass [3], hormonal and anthropometric comparisons of different ethnic groups give an indication of the potential relevance of different testosterone levels on muscle size. For superior muscle development, steroid hormone binding globulin (SHBG) should not increase in proportion with increase in testosterone, otherwise levels of free testosterone will remain unchanged.

Anthropometric measurements show that black men have a higher lean mass (absolute and relative) [10] and a higher fat-free mass index (FFMI) [11] than white controls. Those results could be explained by the data of Ross et al. [12]. They showed that, compared with whites, black college students have significantly higher total (19%) and free (21%) testosterone levels, while SHBG levels did not differ significantly between those two groups [12]. On the other hand, in a study where people of different ethnic groups (African Americans and Caucasians) took part in a resistance training of their dominant leg, significant changes in thigh muscle volumes were observed but no significant differences between the two ethnic groups were found [13]. This contrasts somewhat with the findings of Ross et al., as a higher free testosterone level would have been expected to result in a superior hypertrophic response.

It has to be mentioned that other authors found no significant differences in total and free testosterone between non-Hispanic black men (NHB) and non-Hispanic white men (NHW) (data were adjusted for age), although NHB had slightly higher total and free levels of testosterone compared with NHW. If adjusted for age, SHBG levels were slightly but not significantly higher in NHB than in NHW. When additionally adjusted for age, percent body fat, smoking, alcoholic consumption and physical activity, SHBG levels were significantly lower in NHW than in NHB. With those adjustments, differences in total and free testosterone remained but were not significant [14].

Thus, there can be significant differences between ethnic groups in total and free testosterone levels [12]. Those differences may have an effect on the natural development of muscle mass, which is indirectly shown by anthropometric measurements [10,11]. It remains unclear whether elevation of testosterone levels within the physiologic range confers a physiologic advantage.

ENERGY RESTRICTION, TESTOSTERONE LEVELS AND HYPERTROPHY

Several authors have examined the relationship between nutritional aspects and testosterone levels. Energy restriction and loss of body weight lead to diminished testosterone levels [15,16]. This decrease becomes very dramatic under rapid weight reduction [16]. For example, fasting in rats for 66 hours is accompanied by a 50% reduction in total testosterone [15]. In men, 8 days of vigorous physical activity combined with both energy restriction and sleep deprivation leads to a fall of 45% of free and total testosterone within the first 4 days [17].

A less extreme effect was described my Mäestu and colleagues [16]. In male bodybuilders testosterone levels decreased significantly (pre: 20.3 ± 6.0 nmol/L → post: 17.2 ± 6.5 nmol/L) during an 11-week competition preparation that included energy restriction and increased training volume. The hypocaloric diet started with an energy deficit of approximately 200 kcal/day, which was increased during the study to 950 kcal/day. The data were compared with a control group that did not vary those two variables and consequently did not show any significant variation in testosterone levels. As expected, in the hypocaloric group insulin fell significantly. In addition, insulin concentrations were significantly related to changes of IGF-1 ($r = 0.741$) and lean body mass ($r = 0.725$). Thus it can be concluded that energy restriction can result in a decrease of testosterone, insulin and IGF-1, each of which is an anabolic agent [16]. Furthermore, different studies also suggest that caloric intake and therefore a positive energy balance is fundamental for maximizing resistance training-induced hypertrophy [18,19]. Thus it seems that energy restriction creates a milieu that is not favorable for muscle building.

INFLUENCE OF DIET ON TESTOSTERONE LEVELS

The literature on effects of nutrition on testosterone levels shows inconsistent findings on this topic. With habituated diets, total testosterone levels in general do not seem to differ significantly between omnivores and vegetarians [20–22], although slight [22] or even significant decreases were observed [23]. To the authors' knowledge, only the study of Howie and colleagues has found significant decreases in testosterone levels in habituated vegetarians compared with habituated omnivores [23].

When it comes to free testosterone levels and SHBG, significant but opposite findings between vegetarians and omnivores were observed. Bélanger et al. discovered that vegetarian men had significantly higher SHBG levels and a significantly lower free androgen index (FAI: ratio of total testosterone to SHBG) compared with omnivores [21]. In women on the other hand, comparing vegetarians with omnivores, vegetarians displayed a significantly higher free testosterone level that was accompanied by a lower total testosterone (not significant) level and a lower (significant) SHBG level [22]. Reed et al. observed that omnivores and vegetarians consume the same amount of fat, but that intake of saturated fat is higher and intake of unsaturated fat is lower for omnivores (expressed by a higher polyunsaturated fat to saturated fat (P/S) ratio) [22]. Bélanger et al. on the other hand found differing results and showed that there are significant differences between omnivores and vegetarians. Energy intake was similar between omnivores and vegetarians but omnivores consumed significantly more fat, more protein, less dietary fiber, more saturated fat, less polyunsaturated fat, and more cholesterol [21]. There is some evidence that the influence of dietary fat on testosterone levels varies with the type of fat.

Volek et al. found that resting testosterone correlates positively ($r = 0.72$) with percent of fat and negatively with percent protein ($r = -0.71$) in the diet. Furthermore there are significant correlations between ingested mono-unsaturated fat and testosterone ($r = 0.79$), saturated fatty acids (SFA) and testosterone ($r = 0.77$), and last but not least between P/S ratio and testosterone ($r = -0.63$) [24]. Those results are based on habituated diets and are in agreement with a study that reported higher total and SHBG-bound testosterone with a lower P/S ratio in a dietary intervention [25].

The findings regarding changes in testosterone become more consistent when habituated diets get changed. Several studies examined the effect of the switch from a habituated diet to a low-fat diet, or from a high-fat diet to a low-fat diet, on testosterone and SHBG levels [26–29].

Switching from a habituated "high-fat" diet (around 40% of total energy) to a low-fat diet (around 25% of total energy) for 6 weeks resulted in significant decreases of total and calculated free testosterone [26] or a significant decrease in total testosterone and a nonsignificant increase in free testosterone [27]. After another 6 weeks on the previous habituated diet testosterone rose [26] or even reached baseline levels again [27]. These observations are in agreement with data that show a significant decrease in testosterone when switching from a habituated diet (40% fat) to a low-fat diet (20–25% fat) [28], or a significant decrease in free testosterone accompanied by a significant elevation in SHBG when going from a high-fat diet (>100 g/day) to a low-fat diet (<20 g/day) [30].

Another interesting study was carried out by Raben and colleagues [29]. They found that changing the diet of endurance athletes from their habituated diet to a mixed or a vegetarian diet for 6 weeks resulted in significant hormonal changes. The experimental diets did not differ in supplied energy, but the change of the habituated diet in terms of total nutrient composition was more extreme with the vegetarian diet (Table 31.1). Both intervention groups experienced significant decreases in both total and free testosterone after 3 weeks, but the decrease in the vegetarian group was more obvious than in the group with the mixed diet (total testosterone: mixed, 21.8 to 17.1 nmol/L; vegetarians, 21.1 to 14.7 nmol/L; free testosterone: mixed, 0.53 to 0.40 nmol/L; vegetarians, 0.53 to 0.39 nmol/L). After 6 weeks in the mixed-diet group, total testosterone rose again to 17.4 nmol/L, whereas it declined to 13.7 nmol/L in the vegetarian group. Free testosterone levels rose in the mixed-diet and vegetarian group between the 3rd and 6th week to 0.45 and 0.41 nmol/L, respectively [29]. To the authors' knowledge this is the only study that examined not only the effects of a change from a habituated high-fat to a low-fat diet but also the change to a mixed diet.

Despite the significantly higher P/S ratio, the mixed-diet group also displayed significant decreases in testosterone. These results lead to the suggestion that it is not only the composition of a diet intervention that may lead to hormonal changes but a diet change itself. Also Raben et al. state that a dietary change alone may be responsible for the observed hormonal changes [29].

Furthermore, the author suggests that the more extreme hormonal changes found in the vegetarian group may be due to the more extreme changes in the vegetarian intervention diet. Out of 17 dietary variables (percentage of macronutrients, cholesterol, essential

TABLE 31.1 Dietary Characteristics and Testosterone Course During Diet

	Habituated Diet	Mixed Diet	Vegetarian Diet
Testosterone at 0, 3, 6 weeks on diet			
Total testosterone (nmol/L)		21.8, 17.1, 17.4	21.1, 14.7, 13.7
Free testosterone (nmol/L)		0.53, 0.40, 0.45	0.53, 0.39, 0.41
Characteristics of diet			
Energy (MJ)	17.2 (13.2–20.5)	17.5 (15.4–19.1)	18.3 (15.7–20.4)
Protein			
Energy content (%)	14.0 (11.0–17.0)	13.9 (13.1–14.4)	14.7 (14.1–15.2)*
Absolute amount (g)	141 (119–156)	145 (124–151)	160 (133–172)##*
Intake (g/kg b.w.)	1.95 (1.70–2.20)	2.04 (1.71–2.27)	2.22 (1.88–5.54)##*
Animal origin (%)	64 (60–70)	69 (68–71)	16 (15–18)##*
Plant origin (%)	36 (30–40)	31 (29–33)	84 (82–85)##*
Carbohydrate			
Energy content (%)	51.5 (46.0–62.0)	57.2 (56.0–59.9)#	57.9 (56.9–58.8)#
Absolute amount (g)	523 (412–589)	591 (516–642)##	622 (539–686)##
Intake (g/kg b.w.)	7.2 (5.9.–7.9)	8.4 (6.9–9.6)#	8.9 (7.1–9.8)##
Simple sugars (%)	20 (16–30)	32 (30–34)##	20 (19–21)*
Dietary fibers (g)	47 (35–67)	47 (42–53)	98 (82–104)##*
Fat			
Energy content (%)	32.0 (21.0–42.0)	28.7 (26.9–29.3)	27.4 (26.8–27.7)*
Absolute amount (g)	134 (87–221)	131 (113–146)	136 (113–153)
Serum cholesterol (mg)	527 (296–770)	512 (444–574)	180 (150–210)##*
Essential fatty acids (%)	5 (3–6)	6 (5–10)#	8(4–7)##*
P/S ratio	0.43 (0.30–0.86)	0.49 (0.48–0.57)	1.14 (1.07–1.20)##*
Other			
Iron (mg)	21 (17–29)	23 (20–25)	34 (29–36)##*

**Difference between mixed and vegetarian diet, $P < 0.01$*
#Difference between indicated experimental diet and habitual diet, $P < 0.05$
##$P < 0.01$. [29]
Values are median for testosterone, median and range for dietary characteristics.
FA, fatty acids; P/S ratio, ratio between polyunsaturated and saturated fatty acids.
Adapted from Raben et al. [29]

fatty acids, P/S ratio, dietary fibers, ...) in the vegetarian group, 12 differed significantly from the habituated diet whereas in the mixed diet group only 5 differed significantly from the habituated diet (Table 31.1). This idea is supported by the fact that resting testosterone correlates inversely ($r = -0.63$, significant) with the P/S ratio [24], and despite the significantly lower P/S ratio in the mixed diet group compared with the vegetarian group there was a fall in testosterone [29].

To sum up, it seems that habituated diets usually have little or no influence on total testosterone levels. Changed SHBG levels and thus levels of free testosterone are more common, although findings vary [21,22,30,31]. On the other hand, changing a habituated diet can be expected to result in a transient decrease of testosterone [26]. Furthermore, dietary fat can increase testosterone concentrations, but the influence of different types of lipids is still unclear [24].

The question of the physiological relevance of elevated or diminished testosterone levels through nutrition in the context of muscle building unfortunately remains unanswered. Furthermore, the mechanisms whereby dietary factors such as different macronutrients and/or different dietary fats influence testosterone levels are not fully understood. One idea is that diet may have an influence on hepatic SHBG secretion and thus on free testosterone levels [21].

INFLUENCE OF PROTEIN AND CARBOHYDRATE INGESTION POST EXERCISE ON TESTOSTERONE LEVELS AND HYPERTROPHY/BODY COMPOSITION

Insulin plays a critical role in controlling blood sugar level and is secreted after ingestion of carbohydrates [32]. It is a factor capable of regulating both testosterone secretion and metabolism. Insulin inhibits SHBG synthesis in both males and females [33]. In addition, data of other authors have shown that it is negatively correlated with SHBG [34]. It has to be noted that there is a gender effect in the way insulin affects testosterone levels [33]. Since we suggest that most people interested in optimizing muscle building are men, the following studies on effects of macronutrients and insulin on testosterone are observed effects predominantly on men.

Some studies have shown inverse correlations of testosterone and insulin, whereas others observed concurrent rise of testosterone and insulin [34,35]. Pasquali et al. observed the testosterone response to acute hyperinsulinemia in obese and normal-weight men. Obese men expressed mainly a rise in testosterone with hyperinsulinemia, whereas normal-weight men mainly showed a fall in testosterone. In some, but fewer individuals, an opposite reaction took place [33]. This suggests different response mechanisms either due to interindividual differences or due to abnormal physiologic/hormonal characteristics of acute hyperinsulinemia during rest in obese men. The effect of macronutrient intake (carbohydrate, protein, carbohydrate + protein) and thus, indirectly at least to some extent, of insulin on testosterone levels after a session of resistance training was studied by several authors [34–36]. Ingestion of macronutrients (protein and carbohydrate) after a bout of resistance training can alter testosterone levels. It was found that food intake combined with resistance training lowers testosterone transiently [34–36] and those alterations may persist for up to 24 h post exercise/post feeding [35,36].

However, the most minimal falls of testosterone levels post training were observed with an almost non-caloric intake [34–36], but there was no relation between the degree of insulin elevation and the degree of testosterone decrease. It was observed that, as long as insulin was elevated, testosterone remained suppressed [36]. Thus it could be argued that insulin exerts more of an absolute effect in an "if–then manner" on testosterone and that different decreases could be due to individual response differences.

As protein ingestion alone is able to suppress testosterone after training compared with isocaloric protein/carbohydrate supplementation [36] and lowest testosterone suppressions took place after "non/minimum caloric" placebo supplements [34–36], it could be hypothesized that the amount of caloric intake is very important to testosterone suppression after resistance training. In addition, it could be a combination of the dietary composition, which exerts different insulin responses (inverse correlation with testosterone/known to suppress testosterone), and the caloric amount.

However, pre-exercise values of testosterone do not correlate well with energy intake ($r = -0.18$) [24], and composition of food seems to play a minor or a transient role concerning change in resting testosterone levels [29]. Volek et al. on the other hand found that resting testosterone correlates positively ($r = 0.72$) with percent of fat and negatively ($r = -0.71$) with percent protein contributing to the diet [24].

Furthermore Kraemer et al. found that despite a decrease in total testosterone after exercise + food or placebo, free testosterone is not significantly altered/diminished. This was measured indirectly via the free androgen index. Kraemer et al. even supposed that it was possible that free testosterone was elevated! [34]. This is in accordance with the findings that SHBG may be lowered after insulin rise [33] or negatively correlated with insulin, and that SHBG is lowered after exercise + supplement or exercise + placebo but it is diminished to a greater extent after exercise + supplement [34].

Concerning those findings, it seems that, in resting conditions, insulin may have different effects on total testosterone levels. Those effects may be due to factors including, but not limited to, gender, body weight, etc. On the other hand, after a resistance training and following carbohydrate or protein intake, results showed in general a transient decrease of at least total testosterone [35,36].

MECHANISMS WHEREBY INSULIN COULD ALTER TESTOSTERONE LEVELS

There are several possible mechanisms whereby insulin could alter testosterone levels. During rest the observed rises of testosterone during acute hyperinsulinemia may be due to stimulation of the Leydig cells. Specific insulin receptors in the Leydig cells of rats can stimulate testosterone synthesis in consequence of insulin binding. Furthermore, *in vitro* experiments have shown that testosterone production can be initiated by IGF-1. Insulin and IGF-1 receptors have similar biochemical and functional properties. It is possible that both IGF-1 and insulin exert their effects on testosterone production via both receptors. Additionally, a modulatory action of insulin on the aromatase system is possible. Studies on rats have already shown that chronic hyperinsulinemia may increase aromatase activity. Studies on hyperandrogenic women and obese men by contrast suggest exactly the opposite [33].

Furthermore the influence of insulin on SHBG is suggested to alter free testosterone levels, as insulin can suppress hepatic SHBG synthesis. Last but not least, an impaired luteinizing hormone (LH) secretion or a greater testosterone clearance rate may be responsible for altered testosterone levels after a bout of resistance training and a protein or carbohydrate intake [34,36]. As LH secretion was not altered by a carbohydrate/protein supplement after resistance training, Chandler et al. supposed that this is indirect evidence for elevated testosterone clearance rate [36].

The exact mechanisms by which testosterone and its binding proteins are regulated after training and supplement are still not fully understood. The same applies to the influence of quantity and composition of dietary nutrients on testosterone levels [34].

EFFECT OF RESISTANCE TRAINING PLUS DIETARY SUPPLEMENTATION ON TESTOSTERONE LEVELS

As described above, there are different findings on the impact of insulin or resistance training + supplement (and thus insulin) on testosterone levels. There is accordance that resistance training followed by a protein, carbohydrate, or protein–carbohydrate mixture ingestion leads to a transient decrease in at least total testosterone. The smallest fall in testosterone is seen in placebo supplements with almost noncaloric intake [33–36]. Thus, it can be concluded that these interventions led to a change in the hormonal milieu. As this book is about giving recommendations for muscle building, we examined the effects of different macronutrient supplements combined with resistance training (and thus indirectly the hormonal milieu) on hypertrophic responses. Intense literature research led to six publications dealing with this issue. Two of the studies compared a protein or branched-chain amino acid (BCAA)–carbohydrate mixture supplement against an isocaloric carbohydrate supplement, and four compared a protein supplement against an isocaloric carbohydrate supplement. The tested variables (mass gain, lean mass gain, muscle mass gain, cross sectional area, fiber cross sectional area) varied throughout the studies. No study elucidated any significant differences in hypertrophic outcomes between the intervention groups stated above [37–42].

Those results are consistent with findings that no significant changes in hormonal milieu (testosterone and insulin) are caused by carbohydrate, protein, or carbohydrate–protein mixture supplementation after resistance training [34–36].

The power/reliability of those results is underpinned by the fact that no study was able to show significant changes among intervention groups, even though the methods and the population and its size differed across the board. Participants ranged from untrained individuals to recreational bodybuilders; population size varied between 19 and 52 participants; intervention time ranged from 6 to 14 weeks with training frequencies between 2 and 4 times per week; training loads ranged from 70% to 90% of 1RM, and rest intervals of two consecutive sets varied from 1 to 3 minutes. Additionally, ingestion time of the supplements differed among the studies. Where some interventions administered the supplement before and after resistance training, others supplied it only post exercise. Furthermore, the tested variables for hypertrophy varied, as did the methods to determine them. Observed variables for hypertrophy were cross-sectional area, fiber cross-sectional area, lean mass gain, and muscle mass gain. Those variables were examined via ultrasound [37], muscle biopsy [38,42], and dual-energy x-ray absorptiometry (DEXA) [39–41].

Chandler et al. pointed out that, after resistance training plus macronutrient supplementation (post exercise), insulin was significantly more strongly elevated in 5 out of 8 time points during an 8 h follow up (after resistance training and supplementation) and was never significantly lower in the group supplied with a protein–carbohydrate mixture compared with the group supplied with protein only. Testosterone was lower (not significant) from the second hour of follow-up and was significantly lower from the 5th to the 7th hour of follow-up (6th and 7th time point) [36]. This indicates an overall diminished testosterone decrease in the protein + carbohydrate group compared with the protein-only group. Unfortunately, SHGB was not quantified in this study.

Insulin can lower SHBG, thus potentially increasing the FAI [34], and the bioavailability of testosterone correlates positively with muscle mass [3]. This circumstance leads to the idea that despite the decrease in total testosterone appearing concurrently with an increase in insulin, insulin supports or even enhances hypertrophy by creating "a more anabolic" milieu. Unfortunately, empirical studies were unable to find superiority in terms of muscle building between groups that participated in a resistance training supplemented by a pure protein or by a protein–carbohydrate mixture [39,40].

To the authors' knowledge there is no scientific (proven) evidence that habituated diets influence testosterone levels significantly in a way that could be linked to a more favorable anabolic milieu (and eventually leading to a higher hypertrophic response).

On the other hand, transient hormonal changes arise as acute effects of macronutrient supplements combined with resistance training. This was shown in the study described above where, after a bout of resistance training plus macronutrient supplementation, significant differences concerning anabolic agents were detected with different macronutrient supplements at some time points during an 8 hour follow-up period [36]. To date, to the authors' knowledge, there is no evidence that those differences in hormonal milieu are leading to significant advantages in terms of muscular hypertrophy initiated via resistance training. But for future studies the present overview suggests that there is a potential interplay between circulating testosterone levels, kind of resistance training, timing and composition of dietary intake, and extent of muscle hypertrophy. In addition, in these times of molecular biology, the influence of individual genetic and epigenetic aspects on training and nutritional responsiveness has to be taken into account. Possibly, the genetic background of each individual could explain the high variance and diverging results of the mentioned studies.

References

[1] Kadi F. Cellular and molecular mechanisms responsible for the action of testosterone on human skeletal muscle. A basis for illegal performance enhancement. Br J Pharmacol 2008;154(3): 522–8.

[2] Bhasin S, Store TW, Berman N, Callegari C, Clevenger B, Phillips J, et al. The effects of supraphysiologic doses of testosterone on muscle size and strength in normal men. N Engl J Med 1996;335(1):1–7.

[3] Herbst KL, Bhasin S. Testosterone action on skeletal muscle. Curr Opin Clin Nutr Metab Care 2004;7(3):271–7.

[4] Peters C, Sarikaya H, Schulz T. Congress manual: biomedical side effects of doping : international symposium, october 21st, 2006, Munich, Germany. 1st ed. München: Technische Univ; 2007.

[5] Vingren JL, Kraemer WJ, Ratamess NA, Anderson JM, Volek JS, Maresh CM. Testosterone physiology in resistance exercise and training: the up-stream regulatory elements. Sports Med 2010;40(12):1037–53.

[6] Wu Y, Zhao W, Zhao J, Pan J, Wu Q, Zhang Y, et al. Identification of androgen response elements in the insulin-like growth factor I upstream promoter. Endocrinology 2007;148 (6):2984–93.

[7] Dai Z, Wu F, Yeung EW, Li Y. IGF-IEc expression, regulation and biological function in different tissues. Growth Horm IGF Res 2010;20(4):275–81.

[8] Schoenfeld BJ. The mechanisms of muscle hypertrophy and their application to resistance training. J Strength Cond Res 2010;24(10):2857–72.

[9] Coffey VG, Hawley JA. The molecular bases of training adaptation. Sports Med 2007;37(9):737–63.

[10] Araujo AB, Chiu GR, Kupelian V, Hall SA, Williams RE, Clark RV, et al. Lean mass, muscle strength, and physical function in a diverse population of men: a population-based cross-sectional study. BMC Public Health 2010;10:508.

[11] Hull HR, Thornton J, Wang J, Pierson RN, Kaleem Z, Pi-Sunyer X, et al. Fat-free mass index: changes and race/ethnic differences in adulthood. Int J Obes (Lond) 2011;35(1):121–7.

[12] Ross R, Bernstein L, Judd H, Hanisch R, Pike M, Henderson B. Serum testosterone levels in healthy young black and white men. J Natl Cancer Inst 1986;76(1):45–8.

[13] Walts CT, Hanson ED, Delmonico MJ, Yao L, Wang MQ, Hurley BF. Do sex or race differences influence strength training effects on muscle or fat? Med Sci Sports Exerc 2008;40(4): 669–76.

[14] Rohrmann S, Nelson WG, Rifai N, Brown TR, Dobs A, Kanarek N, et al. Serum estrogen, but not testosterone, levels differ between black and white men in a nationally representative sample of Americans. J Clin Endocrinol Metab 2007; 92(7):2519–25.

[15] Steiner J, LaPaglia N, Kirsteins L, Emanuele M, Emanuele N. The response of the hypothalamic-pituitary-gonadal axis to fasting is modulated by leptin. Endocr Res 2003;29(2):107–17.

[16] Mäestu J, Eliakim A, Jürimäe J, Valter I, Jürimäe T. Anabolic and catabolic hormones and energy balance of the male bodybuilders during the preparation for the competition. J Strength Cond Res 2010;24(4):1074–81.

[17] Alemany JA, Nindl BC, Kellogg MD, Tharion WJ, Young AJ, Montain SJ. Effects of dietary protein content on IGF-I, testosterone, and body composition during 8 days of severe energy deficit and arduous physical activity. J Appl Physiol 2008;105(1):58–64.

[18] Asker Jeukendrup BW, editor. Kevin tipton: protein and gains in muscle Mass; 2009.

[19] Cardinale M, Newton R, Nosaka K. Strength and conditioning: biological principles and practical applications. Chichester, West Sussex, UK: John Wiley & Sons; 2011.

[20] Deslypere JP, Vermeulen A. Leydig cell function in normal men: effect of age, life-style, residence, diet, and activity. J Clin Endocrinol Metab 1984;59(5):955–62.

[21] Bélanger A, Locong A, Noel C, Cusan L, Dupont A, Prévost J, et al. Influence of diet on plasma steroids and sex hormone-binding globulin levels in adult men. J Steroid Biochem 1989;32(6):829–33.

[22] Reed MJ, Dunkley SA, Singh A, Thomas BS, Haines AP, Cruickshank JK. The role of free fatty acids in regulating the tissue availability and synthesis of sex steroids. Prostaglandins Leukot Essent Fatty Acids 1993;48(1):111–6.

[23] Howie BJ, Shultz TD. Dietary and hormonal interrelationships among vegetarian Seventh-Day Adventists and nonvegetarian men. Am J Clin Nutr 1985;42(1):127–34.

[24] Volek JS, Kraemer WJ, Bush JA, Incledon T, Boetes M. Testosterone and cortisol in relationship to dietary nutrients and resistance exercise. J Appl Physiol; 82(1):49–54.

[25] Dorgan JF, Judd JT, Longcope C, Brown C, Schatzkin A, Clevidence BA, et al. Effects of dietary fat and fiber on plasma and urine androgens and estrogens in men: a controlled feeding study. Am J Clin Nutr 1996;64(6):850–5.

[26] Hämäläinen EK, Adlercreutz H, Puska P, Pietinen P. Decrease of serum total and free testosterone during a low-fat high-fibre diet. J Steroid Biochem 1983;18(3):369–70.

[27] Hämäläinen E, Adlercreutz H, Puska P, Pietinen P. Diet and serum sex hormones in healthy men. J Steroid Biochem 1984;20 (1):459–64.

[28] Goldin BR, Woods MN, Spiegelman DL, Longcope C, Morrill-LaBrode A, Dwyer JT, et al. The effect of dietary fat and fiber on serum estrogen concentrations in premenopausal women under controlled dietary conditions. Cancer 1994;74(Suppl. 3): 1125–31.

[29] Raben A, Kiens B, Richter EA, Rasmussen LB, Svenstrup B, Micic S, et al. Serum sex hormones and endurance performance after a lacto-ovo vegetarian and a mixed diet. Med Sci Sports Exerc 1992;24(11):1290–7.

[30] Reed MJ, Cheng RW, Simmonds M, Richmond W, James VH. Dietary lipids: an additional regulator of plasma levels of sex hormone binding globulin. J Clin Endocrinol Metab 1987;64 (5):1083–5.

[31] McVey MJ, Cooke GM, Curran IHA, Chan HM, Kubow S, Lok E, et al. Effects of dietary fats and proteins on rat testicular steroidogenic enzymes and serum testosterone levels. Food Chem Toxicol 2008;46(1):259–69.

[32] Triplitt CL. Examining the mechanisms of glucose regulation. Am J Manag Care 2012;18(Suppl. 1):S4–10.

[33] Pasquali R, Macor C, Vicennati V, Novo F, de lasio R, Mesini P, et al. Effects of acute hyperinsulinemia on testosterone serum concentrations in adult obese and normal-weight men. Metab Clin Exp 1997;46(5):526–9.

[34] Kraemer WJ, Volek JS, Bush JA, Putukian M, Sebastianelli WJ. Hormonal responses to consecutive days of heavy-resistance exercise with or without nutritional supplementation. J Appl Physiol 1998;85(4):1544–55.

[35] Bloomer RJ, Sforzo GA, Keller BA. Effects of meal form and composition on plasma testosterone, cortisol, and insulin following resistance exercise. Int J Sport Nutr Exerc Metab 2000;10 (4):415–24.

[36] Chandler RM, Byrne HK, Patterson JG, Ivy JL. Dietary supplements affect the anabolic hormones after weight-training exercise. J Appl Physiol 1994;76(2):839–45.

[37] Vieillevoye S, Poortmans JR, Duchateau J, Carpentier A. Effects of a combined essential amino acids/carbohydrate supplementation on muscle mass, architecture and maximal strength following heavy-load training. Eur J Appl Physiol 2010;110 (3):479–88.

[38] Andersen LL, Tufekovic G, Zebis MK, Crameri RM, Verlaan G, Kjaer M, et al. The effect of resistance training combined with timed ingestion of protein on muscle fiber size and muscle strength. Metab Clin Exp 2005;54(2):151–6.

[39] Cribb PJ, Williams AD, Hayes A. A creatine-protein-carbohydrate supplement enhances responses to resistance training. Med Sci Sports Exerc 2007;39(11):1960–8.

[40] Cribb PJ, Williams AD, Stathis CG, Carey MF, Hayes A. Effects of whey isolate, creatine, and resistance training on muscle hypertrophy. Med Sci Sports Exerc 2007;39(2):298–307.

[41] Burke DG, Chilibeck PD, Davidson KS, Candow DG, Farthing J, Smith-Palmer T. The effect of whey protein supplementation with and without creatine monohydrate combined with resistance training on lean tissue mass and muscle strength. Int J Sport Nutr Exerc Metab 2001;11(3):349–64.

[42] Willoughby DS, Stout JR, Wilborn CD. Effects of resistance training and protein plus amino acid supplementation on muscle anabolism, mass, and strength. Amino Acids 2007;32(4):467–77.

CHAPTER

32

Nutritional Interventions to Reduce Immune Suppression Post Marathon

John C. Blocher[1], *Sonja E. Nodland*[1], *Don J. Cox*[1], *Brian K. McFarlin*[2], *Hiroyoshi Moriyama*[3] *and Yoshiaki Shiojima*[3]

[1]Biothera, Eagan, MN, USA [2]University of North Texas, Applied Physiology Laboratory, Denton, TX, USA
[3]Ryusendo Co. Ltd, Tokyo, Japan

There is a common perception, shared by athletes, coaches and many researchers, that there is a link between training for and participation in endurance events such as a marathon, and increased incidence of upper respiratory tract infections (URTI). This chapter will review the literature on the rate of URTI symptoms and changes in immune biomarkers associated with endurance running. We will also review the literature on the effect of nutritional intervention strategies on long-distance running-induced changes in immune biomarkers and URTI symptoms.

UPPER RESPIRATORY TRACT INFECTIONS AND RUNNERS

There is a large volume of research published over the past 35 years which has reported increased occurrence of URTI symptoms after completion of endurance running events [1–7]. The increased incidence of URTI after high-intensity exercise is in contrast to studies that have shown that moderate levels of exercise can improve markers of immune function and reduce the apparent incidence of URTI [8–13]. The effect of exercise intensity on URTI incidence has led to the J-curve hypothesis in which URTI incidence drops as exercise intensity increases until it reaches an inflection point and begins to increase as exercise intensity increases from moderate to intense (Figure 32.1) [14].

Recent reviews of the data on the link between exercise and the incidence of URTI [15–17] have highlighted a shortcoming in study design, since studies in this area have failed to identify an infectious agent responsible for the reported URTI. Thus, there is the lack of a direct causal relationship between the upper respiratory symptoms (URS) reported in these studies and an infectious agent. Walsh et al. [16] concluded that the current research data have not proven a direct link between URS and an infectious agent. This is due to the subjective nature of symptom data, even when collected by a physician, and a lack of identification of specific infectious agents in the majority of studies on URTI in athletes. The occurrence of URS after intensive training or an endurance running event appears to be common, but the underlying cause may not always be an infection. It is possible that the URS may be due to irritation of the upper respiratory tract during endurance running, allergies, temporary inflammation of the respiratory tract, or other events. The data behind the J-curve hypothesis describe the relationship between exercise intensity and the incidence of *symptoms* experienced by athletes, even if the underlying biological cause of the symptoms may not be fully understood. In light of this ambiguity, we will use the phrase upper respiratory symptoms (URS) for the rest of this review.

CHANGES IN IMMUNE BIOMARKERS AND RUNNERS

Changes in Circulating Immune Cell Numbers

There is a consistent pattern of changes in the number of circulating immune cells (neutrophils, NK cells,

Nutrition and Enhanced Sports Performance.
DOI: http://dx.doi.org/10.1016/B978-0-12-396454-0.00032-1

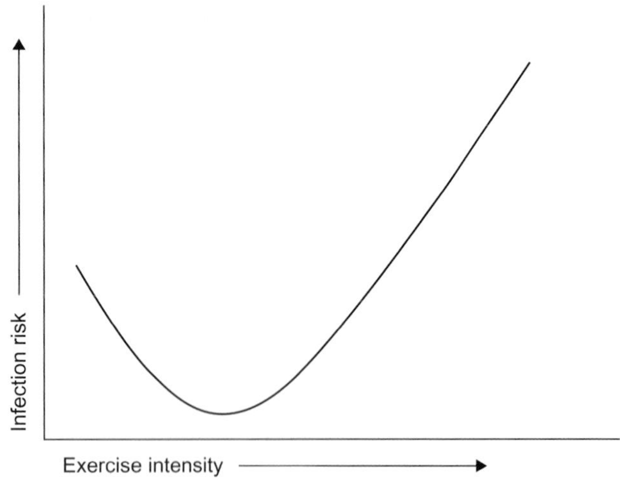

FIGURE 32.1 J-curve model of the effect of exercise intensity on the risk of upper respiratory tract infections/symptoms.

monocytes and lymphocytes) during and after exercise [16,18–21]. Circulating immune cell numbers increase above baseline levels during and immediately after exercise. Certain cell types have a greater flux in numbers than others, including neutrophils, NK cells and certain subtypes of T cells [18,22,23]. Monocytes have also been described to follow this pattern [24–27]. This increase is followed by a fairly rapid decline to below pre-exercise levels during early recovery (1 to 2 hours post exercise) and subsequent gradual recovery to baseline levels typically within 24 hours. Several other types of cells, including basophils and B cells, are relatively unaffected by exercise [18].

The changes in circulating cell numbers are thought to be due to recruitment from the marginated pool into the circulation during exercise and are likely the result of the increased levels of hormones such as cortisol, epinephrine and norepinephrine in the blood released during exercise [16]. The decline in circulating cell numbers after exercise are not due exclusively to destruction of any of the cell populations, rather it is thought to result from a temporary relocation of the cells to reservoir sites in tissues or the walls of peripheral veins [28].

Changes in Cellular Immune Function

Phagocytosis and oxidative burst have been shown to follow the same pattern described above for circulating immune cell numbers, with an increase during and immediately after exercise, followed by a drop and gradual return to baseline [19,20]. There are numerous reports of increased NK cell cytotoxicity during and after intense exercise, but this activity is a result of the increased number of NK cells in circulation, not an increase in per-cell cytotoxic potential [16]. T cell activation as measured by changes in expression of cell surface activation markers (CD69, CD25, CD45 and HLA-DR) increases in response to acute exercise [29,30]. Also, lymphocyte proliferation and immunoglobulin production from B cells have both been demonstrated to decrease in response to intense exercise [31].

Changes in Mucosal Immune Function

The largest part of the human immune system resides at mucosal surfaces in the airways and gut, and these are thought to be the main route of entry for most pathogens. Immunoglobulin-A (IgA), a type of antibody secreted into the mucosa by B cells, is thought to be one of the major ways in which the human body is able to resist infection. Since many viruses are transmitted through mechanisms that involve contact with the nose and mouth, studying the amount of IgA present in human saliva has been an active area of research for decades. Assessment of the effect of exercise on mucosal immune function has mainly focused on measurement of secretory (s) IgA in saliva [16]. The general pattern reported in the literature is a decrease in sIgA levels during and immediately after intense exercise; [16,32–36] however, sIgA response is highly variable both within individuals and between groups. Rates of salivary IgA secretion vary greatly between individuals and groups as well as with other biological variables such as hydration status. Methodology differences (sample collection, test methods) as well as biological variables contribute to the high degree of variability and make data comparison and interpretation between studies difficult [15]. Additionally, there is no agreement on what salivary IgA concentration constitutes "enough" protection and at what level an individual is functionally immunocompromised [33]. Despite these issues, Walsh et al. [16] concluded that there is reasonable evidence indicating that reduced sIgA concentration following exercise, relative to resting values, is associated with an increased risk of URS.

Changes in Humoral Immune Factors

Changes in the levels of plasma cytokines after long-distance running or other high-intensity exercise are well documented [16,37–39]. Increased interleukin-6 (IL-6) levels have been found in numerous studies and are thought to be due largely to release from muscle tissue [23,40]. Increases in a wide range of other cytokines and chemokines such as tumor necrosis factor alpha (TNFα), macrophage inflammatory protein-1 (MIP-1), interleukin-1 beta

(IL-1β), interleukin-10 (IL-10) and interleukin-1 receptor antagonist (IL-1ra) have also been observed, although it is unclear whether these increases originate from specific leukocytes or other cell types [23]. In addition to changes in cytokines, long-distance running and other high-intensity exercise have been shown to increase the concentration of a number of other stress-related hormones and immunomodulators, including C-reactive protein (CRP), epinephrine (adrenaline), cortisol, growth hormone and prolactin [16,41]. All of these factors are known to modulate the inflammatory immune response. The clinical impact of changes in plasma immune signaling compounds post acute exercise is uncertain, but it has been suggested that they may be linked to inflammation which may manifest itself as URS in some athletes [42].

IMMUNONUTRITION SUPPORT AND LONG-DISTANCE RUNNING

Over the past 10 years there have been numerous reviews and meta-analyses assessing the effects of nutritional interventions in countering exercise-induced immunosuppression [15,43–46]. The consensus recommendation from these publications is that endurance athletes, such as long-distance runners, need to consume a balanced healthy diet with adequate intakes of micronutrients (vitamins and minerals). The common conclusion in these reviews is that the research on efficacy of most specific nutritional countermeasures to immunosuppression associated with intense exercise is inconclusive, too preliminary to draw conclusions, or disappointing [46]. Rather than reiterating the work of the reviewers listed above, we will focus on research surrounding two specific nutritional countermeasures for immunosuppression and long-distance running that have promising research results: carbohydrate supplementation, and recently published studies on beta glucan derived from baker's yeast.

Carbohydrates

Carbohydrate (CHO) consumption during and immediately after intense exercise has been shown to blunt some changes in immune biomarkers post exercise [43,46]. Carbohydrate supplementation also attenuates the rise in a number of stress hormones associated with inflammation (cortisol, catecholamine, etc.) post exercise [15]. Consumption of 60 grams of CHO per hour (usually in the form of a beverage containing 6% CHO) is recommended, particularly if the endurance exercise session lasts longer than 90 minutes [44].

There is no convincing evidence that CHO consumption reduces the incidence of URS in endurance athletes [16,43]. Reduction in the post-exercise increase in plasma cortisol has been widely reported for both runners [24,47–51] and cyclists [52,53] who consumed CHO during intense exercise. Carbohydrate consumption during endurance exercise has been reported to attenuate the changes in immune biomarkers normally seen post exercise: a drop in salivary IgA [51], plasma cytokine increases [48,49], a drop in circulating neutrophil numbers [54], increased lymphocytosis [47] and IL-2-stimulated natural killer cell activity [55,56]. In general, CHO consumption during endurance exercise seems to blunt some of the exercise-induced inflammatory changes and immunosuppression, but it has not been shown to have any impact on actual health outcomes such as URS.

Baker's Yeast Glucan

Several recently published studies have investigated the effect on immune function after intense exercise of a well-characterized commercial yeast beta glucan derived from a proprietary strain of baker's yeast (Wellmune WGP®) [57–59]. There is the potential for confusion in the literature on "beta glucan" and exercise because of the many different types of long-chain polysaccharides that are grouped under this single term. While all beta glucans share a "common" β form of chemical bond between the individual glucose units, there are many subtle but important structural differences within the beta glucan family that result in significant differences in function and potential health benefits. Beta glucans from cereal grains have a linear form composed of sections of β 1,3 and β 1,4 linkages. This form of beta glucan has been shown to have little or no immune effect in exercise testing with either animals [60] or humans [61].

Baker's yeast beta glucan is a beta 1,3/1,6 linked glucan with a branching structure containing side chains several glucose subunits long (Figure 32.2). It is found in yeast and fungal cell walls and functions as a pathogen-associated molecular pattern (PAMP) recognized by specific receptors (Dectin-1 and Complement Receptor 3) found on many immune cells [62–65].

Animal models indicate that whole glucan particles (WGP) of baker's yeast beta glucan are directly transported from the gut lumen by cells of the Peyer's patch. The WGP particles are transported to peripheral immune tissues by macrophages, broken down into smaller fragments and presented via antigen presentation to a range of immune cells where they bind to specific receptors [66–73].

Two studies have been conducted to evaluate the effect of WGP consumption on the incidence of URS

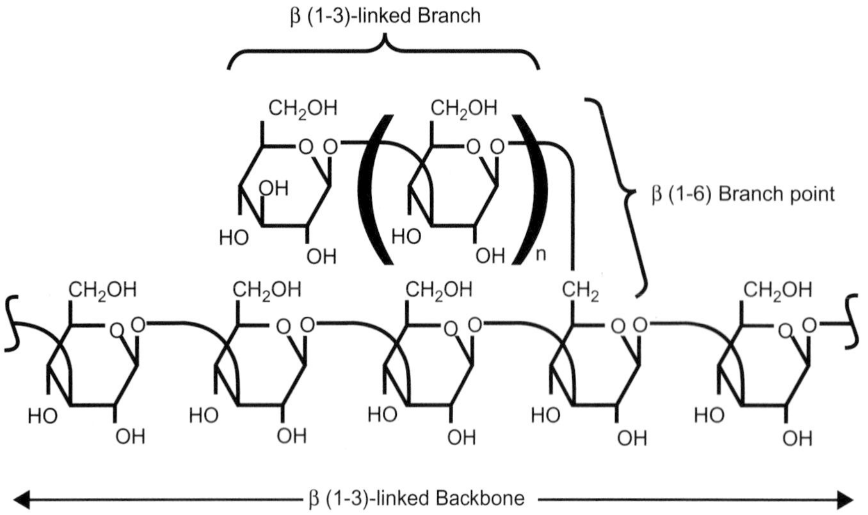

FIGURE 32.2 Baker's yeast beta glucan chemical structure.

after completing a marathon [57,59]. The first study [57] was a placebo-controlled, double blind study to evaluate the effect of WGP on URS in runners that had completed the 2007 Carlsbad Marathon. Seventy-five runners (35 men, 40 women) ranging in age from 18 to 53 years, mean age: 36 ± 9, participated in the study. Subjects were randomly distributed into three groups of 25. Treatments were placebo, 250 mg or 500 mg of WGP daily during the 4-week post-marathon trial period. Subjects completed a questionnaire-based health log measuring health status and URS after 2-and 4-week treatment administrations (Figure 32.3). During the course of the 4-week study, subjects in the treatment groups (250 mg and 500 mg Wellmune WGP per day) reported significantly fewer URS and better overall health scores compared with placebo. The authors concluded that Wellmune WGP may reduce URS and improve overall health following a competitive marathon.

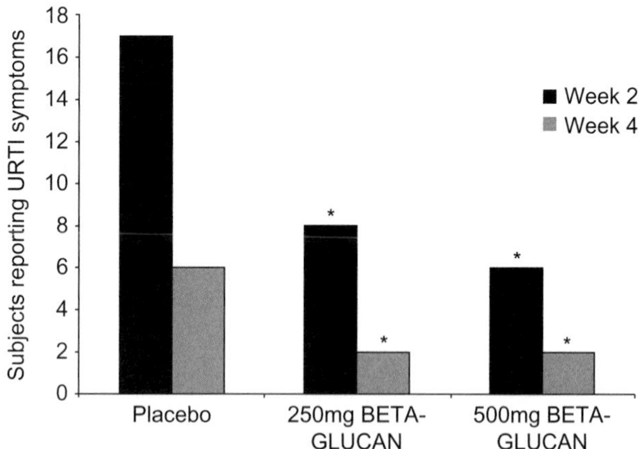

FIGURE 32.3 Total number of subjects reporting any of 11 pre-selected upper-respiratory tract infection symptoms. Subjects orally administered placebo, 250 mg or 500 mg baker's yeast beta glucan. Data analysis was by paired t-tests. *$p \leq 0.05$. *Reprinted from Talbott and Talbott [57] with permission from the* Journal of Sports Science and Medicine.

Another study with WGP was conducted with runners who completed the 2011 Austin Texas Live Strong marathon [59]. This study evaluated both insoluble and soluble forms of WGP. The study was similar in methodology to the marathon study discussed above. Subjects who had completed the marathon began daily supplementation with WGP or placebo immediately post race and continued for 28 days. Cold and flu symptoms were tracked via several survey tools daily [74,75]. The study was double blinded and placebo controlled, and the investigators did not know the identity of the variables until after all the data analysis was completed. There was a significant difference between both the insoluble and soluble WGP groups and the control ($p = 0.026$) for the following question: "Did you experience any health problems today (i.e., Cold, Flu, etc.)?" Subjects in the placebo group reported yes to this question on 5.8 ± 0.6 days compared with 3.5 ± 0.6 (insoluble BG) and 3.5 ± 0.8 (soluble BG). There was no significant difference in average number of symptom days between the insoluble and soluble groups. This study confirms that conversion from the insoluble form to the soluble form does not reduce the biological activity of WGP. In addition, this study represents the fifth independent research group that has found similar health benefits for this well-characterized commercial form of WGP [57–59,76–79].

The effect of WGP on immune system changes that are characteristic of the open-window response

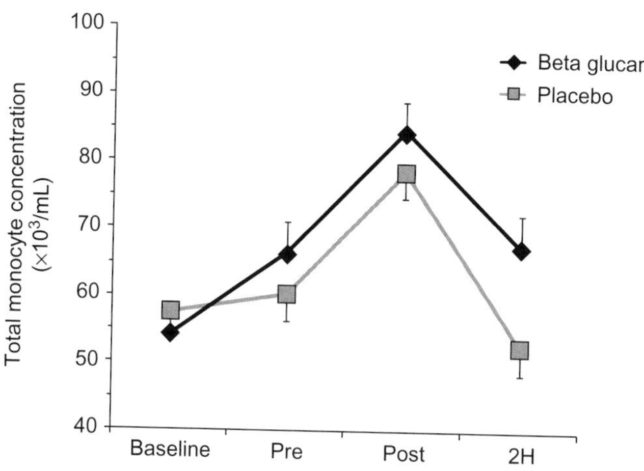

FIGURE 32.4 **Total CD14+ monocyte concentration.** Subjects were supplemented with 250 mg/day of either WGP (Wellmune) or placebo for 10 days, and total monocyte concentration was measured by flow cytometry. A significant increase in total circulating monocytes was observed in the WGP condition compared with placebo. *$p = 0.026$. *Reprinted with permission from Carpenter et al. [58].*

described above was also tested in a controlled laboratory environment. In a blinded, placebo-controlled crossover study participants consumed WGP or placebo supplements for 10 days and then completed an intense, timed bout of acute exercise. Blood and saliva were collected at baseline before either arm of the study began, after 10 days of supplementation before the exercise trial, immediately after exercise, and 2 hours after exercise. An improvement compared with placebo in several of the parameters measured in this study was observed.

One of the most pronounced changes was in cellular composition of the blood after exercise due to WGP supplementation, specifically the concentration of monocytes (Figure 32.4) [58]. Although the total monocyte concentration increased immediately post-exercise to a similar degree as in the placebo condition, the concentration returned to pre-exercise values within 2 hours when athletes were supplemented with WGP, compared with a drop in monocyte concentration below pre-exercise levels in the placebo condition.

Monocytes are the precursor cells for macrophages, a cell type that is integral in mediating a protective immune response. Therefore, given the importance of monocytes/macrophages for initiation of the immune response, maintaining pre-exercise monocyte concentration is likely to result in a net protective effect against onset and development of URTI.

The effect of WGP on salivary IgA production as well as on plasma and stimulated-cytokine production was measured in the same study. With WGP, suppression of salivary IgA concentration at 2 hours after exercise was prevented compared with the placebo arm of the study, where there was reduced salivary IgA at the same time point [59]. As discussed above, although no threshold of sufficient sIgA has been determined, maintaining sIgA at a pre-stress level is likely a beneficial effect of WGP.

When changes in plasma cytokines were measured, improved concentrations of interleukins -4 (IL-4), -5 (IL-5), -7 (IL-7), -8 (IL-8), -10 (IL-10) and interferon gamma (IFN-γ) were observed 2 hours after exercise compared with placebo [58]. Additionally, the capacity of blood cells supplemented with WGP to produce IL-2, IL-4, IL-5 and IFN-γ was increased both before and immediately after exercise when measured in an *ex vivo* stimulation assay. The observation of increased IL-4, IL-5 and IFN-γ in plasma 2 hours after exercise is particularly interesting in the context of this stimulated cytokine data given that these same cytokines were produced at an increased level from ex vivo stimulated cells at both pre and post exercise compared with placebo [58]. These data establish that WGP supplementation can moderate immune system changes observed in response to strenuous exercise.

CONCLUSIONS AND RECOMMENDATIONS

Upper respiratory symptoms are a common occurrence in endurance athletes. The cause of URS may be infectious agents, irritation of the upper respiratory tract, or other events. Immune biomarkers associated with URS include changes in circulating immune cell numbers, phagocytosis and oxidative burst, indicators of mucosal immune function, and changes in humoral immune factors such as cytokines. There are potential nutritional countermeasures for immunosuppression caused by long-distance running. Two nutritional countermeasures evaluated in this review were carbohydrate supplementation and baker's yeast beta glucan. There is no convincing evidence that carbohydrate consumption consistently reduces the incidence of URS in endurance athletes, but there is a growing body of clinical research that supports the efficacy of yeast beta glucan in reducing URS in endurance athletes. To further substantiate and support the role of nutritional interventions on URS, it is recommended that controlled randomized, well-designed clinical research continue to be conducted on these ingredients to provide a convincing argument in favor of specific nutritional intervention against URS in endurance runners. Clinical research that demonstrates a physical health benefit (reduction in URS) and correlating immune biomarkers will be well received by the scientific and medical community.

References

[1] Heath GW, Ford ES, Craven TE, Macera CA, Jackson KL, Pate RR. Exercise and the incidence of upper respiratory tract infections. Med Sci Sports Exerc 1991;23(2):152–7.

[2] Nieman DC, Johanssen LM, Lee JW. Infectious episodes in runners before and after a roadrace. J Sports Med Phys Fitness 1989;29(3):289–96.

[3] Nieman DC, Johanssen LM, Lee JW, Arabatzis K. Infectious episodes in runners before and after the Los Angeles Marathon. J Sports Med Phys Fitness 1990;30(3):316–28.

[4] Peters EM, Bateman ED. Ultramarathon running and upper respiratory tract infections. An epidemiological survey. S Afr Med J 1983;64(15):582–4.

[5] Peters EM, Goetzsche JM, Grobbelaar B, Noakes TD. Vitamin C supplementation reduces the incidence of postrace symptoms of upper-respiratory-tract infection in ultramarathon runners. Am J Clin Nutr 1993;57(2):170–4.

[6] Moreira A, Delgado L, Moreira P, Haahtela T. Does exercise increase the risk of upper respiratory tract infections? Br Med Bull 2009;90:111–31.

[7] Spence L, Brown WJ, Pyne DB, et al. Incidence, etiology, and symptomatology of upper respiratory illness in elite athletes. Med Sci Sports Exerc 2007;39(4):577–86.

[8] Nehlsen-Cannarella SL, Nieman DC, Balk-Lamberton AJ, et al. The effects of moderate exercise training on immune response. Med Sci Sports Exerc 1991;23(1):64–70.

[9] Nieman DC, Nehlsen-Cannarella SL, Markoff PA, et al. The effects of moderate exercise training on natural killer cells and acute upper respiratory tract infections. Int J Sports Med 1990;11(6):467–73.

[10] Chubak J, McTiernan A, Sorensen B, et al. Moderate-intensity exercise reduces the incidence of colds among postmenopausal women. Am J Med 2006;119(11):937–42.

[11] Ekblom B, Ekblom O, Malm C. Infectious episodes before and after a marathon race. Scand J Med Sci Sports 2006;16 (4):287–93.

[12] Nieman DC, Henson DA, Austin MD, Sha W. Upper respiratory tract infection is reduced in physically fit and active adults. Br J Sports Med 2011;45(12):987–92.

[13] Martin SA, Pence BD, Woods JA. Exercise and respiratory tract viral infections. Exerc Sport Sci Rev 2009;37(4):157–64.

[14] Nieman DC. Exercise, infection, and immunity. Int J Sports Med 1994;15(Suppl 3):S131–41.

[15] Gleeson M, Nieman DC, Pedersen BK. Exercise, nutrition and immune function. J Sports Sci 2004;22(1):115–25.

[16] Walsh NP, Gleeson M, Shephard RJ, et al. Position statement. Part one: immune function and exercise. Exerc Immunol Rev 2011;17:6–63.

[17] Shephard RJ. Special feature for the Olympics: effects of exercise on the immune system: overview of the epidemiology of exercise immunology. Immunol Cell Biol 2000;78 (5):485–95.

[18] Nieman DC, Berk LS, Simpson-Westerberg M, et al. Effects of long-endurance running on immune system parameters and lymphocyte function in experienced marathoners. Int J Sports Med 1989;10(5):317–23.

[19] Robson PJ, Blannin AK, Walsh NP, Castell LM, Gleeson M. Effects of exercise intensity, duration and recovery on in vitro neutrophil function in male athletes. Int J Sports Med 1999;20(2):128–35.

[20] Peake JM. Exercise-induced alterations in neutrophil degranulation and respiratory burst activity: possible mechanisms of action. Exerc Immunol Rev 2002;8:49–100.

[21] Kakanis MW, Peake J, Brenu EW, et al. The open window of susceptibility to infection after acute exercise in healthy young male elite athletes. Exerc Immunol Rev 2010;16:119–37.

[22] Shek PN, Sabiston BH, Buguet A, Radomski MW. Strenuous exercise and immunological changes: a multiple-time-point analysis of leukocyte subsets, CD4/CD8 ratio, immunoglobulin production and NK cell response. Int J Sports Med 1995;16 (7):466–74.

[23] Pedersen BK. Special feature for the Olympics: effects of exercise on the immune system: exercise and cytokines. Immunol Cell Biol 2000;78(5):532–5.

[24] Nieman DC, Fagoaga OR, Butterworth DE, et al. Carbohydrate supplementation affects blood granulocyte and monocyte trafficking but not function after 2.5 h of running. Am J Clin Nutr 1997;66(1):153–9.

[25] Field CJ, Gougeon R, Marliss EB. Circulating mononuclear cell numbers and function during intense exercise and recovery. J Appl Physiol 1991;71(3):1089–97.

[26] Nemet D, Mills PJ, Cooper DM. Effect of intense wrestling exercise on leucocytes and adhesion molecules in adolescent boys. Br J Sports Med 2004;38(2):154–8.

[27] Simpson RJ, McFarlin BK, McSporran C, Spielmann G, o Hartaigh B, Guy K. Toll-like receptor expression on classic and pro-inflammatory blood monocytes after acute exercise in humans. Brain Behav Immun 2009;23(2):232–9.

[28] Westermann J, Bode U. Distribution of activated T cells migrating through the body: a matter of life and death. Immunol Today 1999;20(7):302–6.

[29] Fry RW, Morton AR, Keast D. Acute intensive interval training and T-lymphocyte function. Med Sci Sports Exerc 1992;24 (3):339–45.

[30] Gabriel H, Schmitt B, Urhausen A, Kindermann W. Increased CD45RA+CD45RO+cells indicate activated T cells after endurance exercise. Med Sci Sports Exerc 1993;25(12):1352–7.

[31] Gleeson M. Immune function in sport and exercise. J Appl Physiol 2007;103(2):693–9.

[32] Gleeson M. Mucosal immune responses and risk of respiratory illness in elite athletes. Exerc Immunol Rev 2000;6:5–42.

[33] Gleeson M, Pyne DB, Callister R. The missing links in exercise effects on mucosal immunity. Exerc Immunol Rev 2004;10:107–28.

[34] Allgrove JE, Geneen L, Latif S, Gleeson M. Influence of a fed or fasted state on the s-IgA response to prolonged cycling in active men and women. Int J Sport Nutr Exerc Metab 2009;19 (3):209–21.

[35] Moreira A, Arsati F, de Oliveira Lima-Arsati YB, de Freitas CG, de Araujo VC. Salivary immunoglobulin A responses in professional top-level futsal players. J Strength Cond Res 2011;25 (7):1932–6.

[36] Usui T, Yoshikawa T, Orita K, et al. Changes in salivary antimicrobial peptides, immunoglobulin A and cortisol after prolonged strenuous exercise. Eur J Appl Physiol 2011;111 (9):2005–14.

[37] Ostrowski K, Rohde T, Asp S, Schjerling P, Pedersen BK. Chemokines are elevated in plasma after strenuous exercise in humans. Eur J Appl Physiol 2001;84(3):244–5.

[38] Ng QY, Lee KW, Byrne C, Ho TF, Lim CL. Plasma endotoxin and immune responses during a 21-km road race under a warm and humid environment. Ann Acad Med Singapore 2008;37(4):307–14.

[39] Waskiewicz Z, Klapcinska B, Sadowska-Krepa E, et al. Acute metabolic responses to a 24-h ultra-marathon race in male amateur runners. Eur J Appl Physiol 2012;112(5):1679–88.

[40] Pedersen BK, Akerstrom TC, Nielsen AR, Fischer CP. Role of myokines in exercise and metabolism. J Appl Physiol 2007;103 (3):1093–8.

[41] Murray DR, Irwin M, Rearden CA, Ziegler M, Motulsky H, Maisel AS. Sympathetic and immune interactions during dynamic exercise. Mediation via a beta 2-adrenergic-dependent mechanism. Circulation 1992;86(1):203–13.

[42] Cox AJ, Pyne DB, Saunders PU, Callister R, Gleeson M. Cytokine responses to treadmill running in healthy and illness-prone athletes. Med Sci Sports Exerc 2007;39(11):1918–26.

[43] Moreira A, Kekkonen RA, Delgado L, Fonseca J, Korpela R, Haahtela T. Nutritional modulation of exercise-induced immunodepression in athletes: a systematic review and meta-analysis. Eur J Clin Nutr 2007;61(4):443–60.

[44] Nieman DC. Immunonutrition support for athletes. Nutr Rev 2008;66(6):310–20.

[45] Senchina DS, Shah NB, Doty DM, Sanderson CR, Hallam JE. Herbal supplements and athlete immune function—what's proven, disproven, and unproven? Exerc Immunol Rev 2009;15:66–106.

[46] Walsh NP, Gleeson M, Pyne DB, et al. Position statement. Part two: maintaining immune health. Exerc Immunol Rev 2011;17:64–103.

[47] Henson DA, Nieman DC, Blodgett AD, et al. Influence of exercise mode and carbohydrate on the immune response to prolonged exercise. Int J Sport Nutr 1999;9(2):213–28.

[48] Nieman DC, Henson DA, Smith LL, et al. Cytokine changes after a marathon race. J Appl Physiol 2001;91(1):109–14.

[49] Nieman DC, Davis JM, Henson DA, et al. Carbohydrate ingestion influences skeletal muscle cytokine mRNA and plasma cytokine levels after a 3-h run. J Appl Physiol 2003;94 (5):1917–25.

[50] McAnulty SR, McAnulty LS, Nieman DC, et al. Influence of carbohydrate ingestion on oxidative stress and plasma antioxidant potential following a 3 h run. Free Radic Res 2003;37(8):835–40.

[51] Costa RJ, Jones GE, Lamb KL, Coleman R, Williams JH. The effects of a high carbohydrate diet on cortisol and salivary immunoglobulin A (s-IgA) during a period of increase exercise workload amongst Olympic and Ironman triathletes. Int J Sports Med 2005;26(10):880–5.

[52] Lancaster GI, Jentjens RL, Moseley L, Jeukendrup AE, Gleeson M. Effect of pre-exercise carbohydrate ingestion on plasma cytokine, stress hormone, and neutrophil degranulation responses to continuous, high-intensity exercise. Int J Sport Nutr Exerc Metab 2003;13(4):436–53.

[53] Bishop NC, Walsh NP, Haines DL, Richards EE, Gleeson M. Pre-exercise carbohydrate status and immune responses to prolonged cycling: II. Effect on plasma cytokine concentration. Int J Sport Nutr Exerc Metab 2001;11(4):503–12.

[54] Nieman DC. Risk of upper respiratory tract infection in athletes: an epidemiologic and immunologic perspective. J Athl Train 1997;32(4):344–9.

[55] McFarlin BK, Flynn MG, Stewart LK, Timmerman KL. Carbohydrate intake during endurance exercise increases natural killer cell responsiveness to IL-2. J Appl Physiol 2004;96(1):271–5.

[56] McFarlin BK, Flynn MG, Hampton T. Carbohydrate consumption during cycling increases in vitro NK cell responses to IL-2 and IFN-gamma. Brain Behav Immun 2007;21(2):202–8.

[57] Talbott ST, Talbott J. Effect of Beta 1,3/1,6 glucan on upper respiratory tract infection symptoms and mood state in marathon athletes. J Sports Sci Med 2009;8:509–15.

[58] Carpenter KC, Breslin WL, Davidson T, Adams A, McFarlin BK. Baker's yeast beta-glucan supplementation increases monocytes and cytokines post-exercise: implications for infection risk?. Br J Nutr 2012;1–9.

[59] McFarlin B, Carpenter K, Davidson T, McFarlin M. Baker's Yeast Beta Glucan Supplementation increases salivary IgA and decreases cold/flu symptomatic days after intense exercise. submitted 2012.

[60] Davis JM, Murphy EA, Brown AS, Carmichael MD, Ghaffar A, Mayer EP. Effects of moderate exercise and oat beta-glucan on innate immune function and susceptibility to respiratory infection. Am J Physiol Regul Integr Comp Physiol 2004;286(2): R366–72.

[61] Nieman DC, Henson DA, McMahon M, et al. Beta-glucan, immune function, and upper respiratory tract infections in athletes. Med Sci Sports Exerc 2008;40(8):1463–71.

[62] Mueller A, Raptis J, Rice PJ, et al. The influence of glucan polymer structure and solution conformation on binding to (1–>3)-beta-D-glucan receptors in a human monocyte-like cell line. Glycobiology 2000;10(4):339–46.

[63] Ross GD, Vetvicka V. CR3 (CD11b, CD18): a phagocyte and NK cell membrane receptor with multiple ligand specificities and functions. Clin Exp Immunol 1993;92(2):181–4.

[64] Taylor PR, Tsoni SV, Willment JA, et al. Dectin-1 is required for beta-glucan recognition and control of fungal infection. Nat Immunol 2007;8(1):31–8.

[65] Goodridge HS, Reyes CN, Becker CA, et al. Activation of the innate immune receptor Dectin-1 upon formation of a 'phagocytic synapse'. Nature 2011;472(7344):471–5.

[66] Brown GD, Taylor PR, Reid DM, et al. Dectin-1 is a major beta-glucan receptor on macrophages. J Exp Med 2002;196 (3):407–12.

[67] Hong F, Yan J, Baran JT, et al. Mechanism by which orally administered beta-1,3-glucans enhance the tumoricidal activity of antitumor monoclonal antibodies in murine tumor models. J Immunol 2004;173(2):797–806.

[68] Vetvicka V, Thornton BP, Ross GD. Soluble beta-glucan polysaccharide binding to the lectin site of neutrophil or natural killer cell complement receptor type 3 (CD11b/CD18) generates a primed state of the receptor capable of mediating cytotoxicity of iC3b-opsonized target cells. J Clin Invest 1996;98 (1):50–61.

[69] Xia Y, Vetvicka V, Yan J, Hanikyrova M, Mayadas T, Ross GD. The beta-glucan-binding lectin site of mouse CR3 (CD11b/ CD18) and its function in generating a primed state of the receptor that mediates cytotoxic activation in response to iC3b-opsonized target cells. J Immunol 1999;162(4):2281–90.

[70] Lavigne LM, Albina JE, Reichner JS. Beta-glucan is a fungal determinant for adhesion-dependent human neutrophil functions. J Immunol 2006;177(12):8667–75.

[71] Brown GD, Gordon S. Immune recognition. A new receptor for beta-glucans. Nature 2001;413(6851):36–7.

[72] Tsikitis VL, Albina JE, Reichner JS. Beta-glucan affects leukocyte navigation in a complex chemotactic gradient. Surgery 2004;136(2):384–9.

[73] Li B, Allendorf DJ, Hansen R, et al. Yeast beta-glucan amplifies phagocyte killing of iC3b-opsonized tumor cells via complement receptor 3-Syk-phosphatidylinositol 3-kinase pathway. J Immunol 2006;177(3):1661–9.

[74] Barrett B, Brown R, Mundt M, et al. The Wisconsin Upper Respiratory Symptom Survey is responsive, reliable, and valid. J Clin Epidemiol 2005;58(6):609–17.

[75] Nieman DC, Henson DA, Fagoaga OR, et al. Change in salivary IgA following a competitive marathon race. Int J Sports Med 2002;23(1):69–75.

[76] Domitrovich S, Domitrovich J, Ruby B. Effects of an immunomodulating supplement on upper respiratory tract infection symptoms in wildland firefighters. manuscript in preparation 2012.

[77] Feldman S, Schwartz HI, Kalman DS, Mayers A, Kohrman HM, Clemens R, Krieger DR. Randomized phase II clinical trials of Wellmune WGP for immune support during cold and flu season. J Appl Res 2009;9:30–42.

[78] Talbott S, Talbott J. Beta 1,3/1,6 glucan decrease upper respiratory tract infection symptoms and improves psychological well-being in moderate to highly-stressed subjects. AgroFood Industry Hi-Tech 2010;21(1):21–4.

[79] Fuller R, Butt H, Noakes PS, Kenyon J, Yam TS, Calder PC. Influence of yeast-derived 1,3/1,6 glucopolysaccharide on circulating cytokines and chemokines with respect to upper respiratory tract infections. Nutrition 2012;28(6):665–9.

CHAPTER

33

Carbohydrate and Muscle Glycogen Metabolism

Exercise Demands and Nutritional Influences

Anthony L. Almada

IMAGINutrition, Inc., Dana Point, CA, USA

"Science never gives up searching for truth, since it never claims to have achieved it. It is civilizing because it puts truth ahead of all else, including personal interests." ***John C. Polanyi, 1986 Nobel Laureate in Chemistry***

The performance of muscular work is reliant upon the delivery and utilization of fuel substrates to match the metabolic demands and oscillations of the chosen activity or sport. Alterations in fuel availability can thus influence the volume and duration of work performed, and alter the patterns of substrate selection, e.g., exercise in a prolonged fasting or postprandial state. Additionally, the rate and magnitude of repletion of fuel substrate can also influence subsequent or repeat performance, e.g., twice daily training, sequential/same day heats in competition, cycling stage races.

Skeletal muscle is "omnivorous" in its use of fuel substrates, which have both endogenous and exogenous origins. The metabolic plasticity of skeletal muscle enables the organism to adapt to its state of nutriture, and the varying metabolic demands it encounters. Carbohydrate is the only fuel substrate of exercising muscle that (i) is produced *de novo* (via gluconeogenesis), (ii) is capable of acutely influencing performance when delivered from exogenous sources, e.g., intravenous or oral, *and* (iii) can attain augmented intramuscular concentrations within a short duration of time (by glycogen supercompensation).

CARBOHYDRATE UTILIZATION DURING EXERCISE

Endurance Exercise

Arguably, most study of carbohydrate utilization during muscular work has been performed with moderate intensity endurance exercise, employing large muscle groups (whole body), with mixed diets being the macronutrient intake background. At light intensity work rates (25–30% of VO_{2max}) carbohydrate (muscle glycogen and plasma glucose) contributes 10–15% of the oxidized fuel substrate, increasing up to approximately 75% of the energy supply during work at 85% of VO_{2max}, progressing to 100% of the energy substrate during intensities of 100% VO_{2max} or greater [1,2]. More recent studies, employing direct measurements of free fatty acid (FFA) oxidation via use of [U-^{13}C] palmitate tracer and $^{13}CO_2$ production in expired air, coupled with serial measurements of whole muscle glycogen concentration and the rate of disappearance (Rd) of [6,6-^{2}H2] glucose, observed a significant decrement in total fat oxidation at higher work rates among trained cyclists [3]. van Loon and colleagues noted carbohydrate delivered 45% of oxidized energy substrates at 45% W_{max} (maximal power), 51% at 55% W_{max}, and 73% at 75% W_{max}. Total fat oxidation fell precipitously between 55% (49% of oxidized substrate) and 75% W_{max} (24%) [3], indicative of the markedly greater reliance upon carbohydrate at the higher exercise intensity and the compromised ability of working muscle to utilize lipid as a substrate.

Recent studies employing graded exercise intensities with *small* muscle groups (knee extensions, quadriceps) have also demonstrated a progressive increase in carbohydrate oxidation (from whole muscle glycogen and plasma glucose), yet unattended by a decline in total fat oxidation rates over exercise intensities of 25%, 65% and 85% W_{max} [4]. This maintenance of total fat oxidation rates during high intensity exercise with small muscle groups has been postulated to be due to

Nutrition and Enhanced Sports Performance.
DOI: http://dx.doi.org/10.1016/B978-0-12-396454-0.00033-3

the relative (to whole body exercise) blunting of catecholamine response and the relatively augmented blood flow, both of which can reduce glycolytic flux and thereby protect the availability of free carnitine, a permissive step for fatty acid oxidation [5].

Resistance Exercise

There do not appear to be any studies that have systematically examined, with quantitative measurements, e.g., stable isotopes, glucose oxidation *during* resistance or strength training. Nevertheless, several studies have examined muscle glycogen utilization during intensive resistance training bouts. One of the first studies enrolled male bodybuilders and had them perform a battery of resistance-training exercises to failure, targeting the upper legs [6]. After a total of 20 sets (five sets for each of four exercises; repetitions to self-determined muscle failure on each set) total muscle (left vastus lateralis) glycogen had fallen by 26% (the paper stated a 40% decrement yet the pre-exercise value was 160 mmol/kg wet weight (ww) muscle, and the post-exercise was 118 mmol/kg ww, or a change of 42 mmol/kg ww). A later paper from the same laboratory and with the same experimental data [7] reported a 28% decline (in the latter paper, muscle glycogen was expressed in mmol/kg *dry* weight muscle).

Robergs et al. evaluated eight resistance-trained males, subjecting them to two different, consecutive exercise regimens on consecutive days [8]. On the first day, all subjects performed six sets of six repetitions as single leg extensions at 70% of one repetition maximum (1RM). The second exercise was performed the following day, with subjects performing six sets of leg extensions with the contralateral leg, yet at 35% of 1RM. Muscle biopsies (vastus lateralis) were taken just prior to each exercise, after completion of the third and sixth sets, and 2 hours post exercise (with no nutrient intake during this period). The muscle glycogen decline was similar in both exercises (46.9 mmol/kg ww, or 39% for 70% 1RM, and 46.6 mmol/kg ww, or 38% for 35% 1RM). The rate of muscle glycogen utilization during the 70% 1RM bout was double that of the 35% 1RM, with the time to completion and repetitions/set during the 35% 1RM exercise being approximately twice that of 70% 1RM.

MacDougall and associates had eight experienced, bodybuilding-type resistance-training men perform one or three sets of single arm seated bicep curls at 80% 1RM [9]. Muscle biopsies of the biceps brachii were taken from the control arm prior to the exercise, and from the exercising arm after finishing either set. A statistically non-significant (12%) decline in muscle glycogen was observed after the single set, but after the completion of three sets a significant 24% decrement was measured.

In an elegant design, Churchley et al. had seven males, experienced in resistance training, undertake an acute diet and exercise regimen to achieve one leg exhibiting a glycogen adequate ("Normal") state, and the contralateral leg in a glycogen reduced ("Low") state [10]. Subjects performed eight sets of five repetitions of single leg presses at 80% 1RM, for each leg. The Low leg commenced the exercise with 60 seconds of rest before the Normal leg completed the same set. Weight was adjusted down by 5% in sets where the Low leg could not perform five repetitions. Muscle biopsies (vastus lateralis) were collected prior to the exercise, and upon completion of the last (eighth) set for each leg. Despite the resting glycogen concentration in the Low leg being significantly less than the Normal leg (193 ± 29 (SD) and 435 ± 87 mmol/kg dry muscle (dm), respectively), net glycogen utilization was similar between legs (Low 91 mmol/kg dm; Normal 123 mmol/kg dm). Relative glycogen utilization was 47% of baseline in the Low leg, and 28% in the Normal leg. The work capacity of the Low leg was typically less than that of the Normal leg (J. A. Hawley, personal communication).

Competitive Sport

Competitive sport is marked by modest to dramatic inter-individual differences in performance, which are likely reflected in muscle energy substrate preference and magnitude of use (dependent on training state, intra-competition nutrition/fueling strategies, and work rates). Work by Sherman et al. revealed near depletion of gastrocnemius muscle glycogen among experienced runners following a marathon [11]. These runners also engaged in a glycogen supercompensation program prior to the race and were able to maintain an average VO_{2max} (mL kg^{-1} min^{-1}) between 73.4 ± 8.3 (SD) and 75.6 ± 8.9 over the first two-thirds of the race. Average VO_{2max} over the latter third ranged between 66.2 ± 10.4 and 65.5 ± 12.4. Unfortunately, no information regarding intra-race nutrition was provided.

Bangsbo et al. reviewed seven studies describing muscle glycogen (vastus lateralis) decline after a soccer match, with four also providing muscle glycogen changes at half time [12]. Pre-game concentrations ranged from 112 ± 20 (SD) to 45 mmol/kg ww. In one study described in the review and conducted by the authors the pre-game range was 97–157 mmol/kg ww [13]. Within the four studies measuring muscle glycogen content at half time the percent range of decline from pre-game concentrations was 25–87%. Post-game percent decreases (from pre-game) ranged over 31–100%. The study by Krustrup and colleagues [13] examined muscle fiber-specific changes in glycogen utilization in a subset of ten Fourth Division male soccer players. Prior to the "friendly" match 73 ± 6% of all fibers were deemed glycogen full, yet post-match this fell significantly ($P < 0.05$) to 19 ± 4%. Additionally, almost half of the fibers analyzed post-game were classified as "almost empty" to "empty" (of glycogen).

A more recent study by Krustrup et al. measured the change in muscle glycogen in seven First and Second division Danish male soccer players after a competitive match [14]. A control muscle glycogen (vastus lateralis) biopsy was obtained from two of the players 64 hours following the last training session (biopsy timing relative to start of match not described). The other five players had a control muscle biopsy collected 5 days after the match. During the match the players were allowed to drink only water. Fifteen minutes after the match a muscle biopsy was collected from all seven players. Control muscle glycogen (449 ± 34 (SEM) mmol kg^{-1} dry weight (dw)) dropped 57% (193 ± 22 mmol kg^{-1} dw).

A revealing series of muscle glycogen utilization studies in an elite, age group triathlete were performed by Gillam et al. [15] and Cuddy et al. [16]. In the first study the subject competed in a half Ironman (1.2 mile swim + 56 mile bike + 13.1 mile run) [15]. Muscle biopsies (vastus lateralis) were obtained 90 minutes pre-race (just after breakfast) and immediately post-race. Total carbohydrate intake during the race was 308 grams (liquid and gel forms). Muscle glycogen fell precipitously (227.1 mmol kg^{-1} ww pre-race to 38.6 post-race), a decline of 83%.

In a follow-up study with the *same* triathlete, yet racing a full Ironman distance (World Ironman Championships, Kailua-Kona, Hawaii; 2.4 mile swim + 112 mile bike + 26.2 mile run), pre- and post-race muscle (vastus lateralis) biopsies were again obtained. The pre-race muscle glycogen value was 152 mmol kg^{-1} ww (33% less than pre-race muscle glycogen for the half Ironman [15]). Post-race glycogen, obtained 30 minutes post-race, fell 68% (to 48 mmol kg^{-1} ww). To complement muscle glycogen measures Cuddy et al. also implemented doubly labeled water ($^2H_2{}^{18}O$; DLW) ingestion to assess total energy expenditure (TEE) [17,18]. The dose of DLW was consumed approximately 64 hours prior to the race. Carbohydrate intake (reported by subject, from bars, liquids, and gels) intra-race was 632 grams (swim 0 g; bike 404 g; run 228 g). This was not accounted for in the calculated carbohydrate and fat oxidation measures. DLW-determined TEE was 8,926 kilocalories (kcal) over the race. Using a collection of laboratory and field metabolic cart measurements (conducted prior to the race), calculated energy expenditure during the swim [19], average hourly power output (watts) during the cycling leg of the race, and official split times during the run, the authors determined that 1370 g of carbohydrate and 348 g of fat were oxidized over the race. The TEE value comported with the indirect calorimetry (metabolic cart) calculated TEE of 9029 kcal.

McArdle Disease—A Human "Knockout" Model of Glycogen-Free/Alactic Exercise

The seminal work of Hermansen et al. [20] illuminated the role of muscle glycogen as a pivotal energy substrate during intense exercise. Additional insights can be gleaned from evaluating persons living with McArdle disease (MCD), first described in 1951 by British physician Brian McArdle. McArdle noted a patient unable to display an increase in venous lactate (or pyruvate) during an ischemic exercise test [21]. MCD is the most common muscle metabolic disease, defined by a lack of myophosporylase, the skeletal muscle-specific isoform of glycogen phosphorylase and the obligate enzyme involved in the degradation of glycogen in skeletal muscle [22]. Given the inability to activate glycogenolysis, persons with MCD commonly display "supraphysiological" muscle glycogen concentrations, up to twice that of non-MCD persons [23]. The conditional essentiality of muscle glycogen is revealed when a person with MCD attempts to undertake moderate to vigorous exercise. Within several minutes after the commencement of exercise they will demonstrate intolerance to further exercise, marked by fatigue, tachycardia, and/or severe muscle pain (in the absence of changes in muscle pH and lactate, or even in the presence of a modest increase and decrease, respectively).

The phenotypic hallmark of MCD is a "second wind" [24] that spontaneously manifests after reducing exercise intensity or a brief rest, following several minutes of sustained exercise. The attendant fall in heart rate and improved exercise tolerance is indicative of circulating extramuscular fuels (free fatty acids, hepatic glucose production, and possibly circulating lactate [25]) "catching" up with the bioenergetic demands of exercise. Although increased fat availability (through lipid infusion) does improve exercise capacity in MCD [26], and whole body fat oxidation is augmented in concert with the onset of the second wind [27], both are unable to compensate for the glycogenolytic impairment and reduced glycolytic flux.

Haller and Vissing elegantly demonstrated the role of glycogenolysis in persons with MCD, both those with frank deficiency of myophosphorylase activity and in an individual with a residual (3% of non-MCD control) amount of enzyme activity [28]. The spontaneous second wind was observed only in those lacking myophosporylase activity. In the single subject with partial myophosporylase activity, peak VO_2 during the first 6 to 8 minutes of exercise was 1.72 times greater than the average peak VO_2 of those devoid of myophosporylase activity. Moreover, glucose infusion in the subjects with frank myophosporylase deficiency produced a "glucose-induced second wind", and a 21% higher peak VO_2, yet was ineffectual in the subject with residual myophosporylase activity. Despite the additional, incremental second wind and resultant increase in oxidative capacity via glucose infusion, the oxidative capacity of all the subjects was about half that observed in healthy individuals.

A recent case report in a male adolescent with MCD, undertaking a 6-week resistance-training program (60–75% of 1RM; each training session was preceded by ingestion of a carbohydrate-containing drink and high carbohydrate meals), described notable improvements in strength and force [29]. No clinical or subjective measures of excessive muscle damage were observed, and his self-reported exercise tolerability improved to an "asymptomatic" level. Given the higher glycolytic/glycogenolytic demands of strength training, this isolated report temptingly suggests that progressive resistance training may alter the temporal and/or spatial kinetics of substrate availability in MCD during exercise.

Modulation of Muscle Glycogen I: Fasting

Submaximal exercise (approximately 75% of VO_{2max}) endurance capacity is compromised by reduced muscle glycogen concentrations present prior to exercise [30]. Numerous nutritional methods have been developed and tested, aiming to (i) enhance muscle glycogen stores, or (ii) shift substrate selection to favor greater fat oxidation, thus "sparing" or decreasing the utilization rate of muscle glycogen.

Ostensibly, the simplest dietary method to increase fat utilization is fasting, although the rigor of adhering to this "nutrition regimen" may render it the most challenging for an athlete. Paleolithic humans likely had to engage in "exercise-intensive" hunts in a starved/prolonged fasted state, and thus conferred, in an evolutionary sense, similar muscle metabolic plasticity to current day humans. A few investigations have explored glycogen substrate use during endurance exercise in a starved state.

Loy and coworkers assessed glycogen utilization rates among ten trained cyclists who did rides to exhaustion at either 79% ($n = 6$) or 86% ($n = 4$) of VO_{2max}[31]. Each subject performed the cycling bout 3 hours (postprandial; PPR) or 24 hours (FAST) after a 355 kcal liquid meal (Ensure® Plus; 52.8% carbohydrate, 32% fat, and 15.2% protein). The paper mentions that during the 24 hour period prior to the exercise tests the "...subjects were constantly supervised" yet no description of the supervisory method was provided. Pre-exercise muscle (vastus lateralis; immediately before exercise) glycogen values did not significantly differ between any treatments (likely due to the small sample sizes), yet values in the 79% VO_{2max}/FAST state were 30% lower than those for PPR. Despite dramatic differences in cycling times between PPR and FAST, pre- and post-exercise muscle glycogen values did not significantly differ between dietary treatments.

Knapik and colleagues persuaded eight untrained, lean male soldiers to undergo 3.5 days of starvation (STV) within a supervised metabolic ward (the only way to assure complete calorie deprivation; water and caffeine-free teas were allowed for consumption) and then perform cycling exercise at 50% of VO_{2max} [32]. This was compared, in the same subjects, to cycling in a post-absorptive (PA) state (14 hours without eating; likely longer than most athletes endure day to day). Expectedly, muscle glycogen (vastus lateralis) content prior to exercise did not differ between the treatments. However, muscle glycogen fell by $44\% \pm 6\%$ (SE) in the PA state and $28\% \pm 4\%$ in the STV state ($P < 0.01$). Although mean cycling time to volitional exhaustion in PA was 21 minutes (15%) longer than in STV, this difference was not statistically significant ($p < 0.09$; *athletically* significant?). In contradistinction, glycogen utilization rates did differ between treatments. When calculated as the quotient of the difference in glycogen (pre- minus post-exercise) and the time to fatigue, glycogen utilization (per minute) was $0.31 \pm 0.04\ \mu mol\ g^{-1}$ ww under PA conditions, and $0.19 \pm 0.02\ \mu mol\ g^{-1}$ ww under STV ($P < 0.01$), a 39% lower rate after prolonged fasting. The authors also employed an infusion of [6,6-D2]-glucose to assess endogenous glucose production (gluconeogenesis, hepatic glycogenolysis; Ra) and whole body oxidation (Rd). STV fostered a 40% relative drop in Rd and a 32% reduction in Ra, affirmed by the higher utilization (via Respiratory Quotient; RQ) of lipid substrate.

Modulation of Muscle Glycogen II: Low Carbohydrate/Ketogenic Diet

Restriction of carbohydrate intake in healthy individuals fosters the genesis of ketone bodies (β-hydroxybutyrate, acetoacetate) in metabolically significant quantities. Due to the protracted intervention period needed to achieve metabolic *adaptation* during a ketogenic diet [33], Phinney et al. had professional cyclists equilibrate to such a diet (<2% carbohydrate, 85% fat, ≈13% protein; KT) for 28 days after a baseline diet for 1 week (57% carbohydrate, 29% fat, 14% protein; CHO) [34]. Subjects were required to maintain their training volume and intensity over the 4-week KT diet. Cycling performance (continuous 62–64% of VO_{2max}, until volitional exhaustion) was assessed while on CHO and 4 weeks after a KT. Distinctively, time to exhaustion did not differ between treatments. Notably, muscle (vastus lateralis) glycogen content *after* the exercise bout during CHO (53 ± 5 (SEM) mmol glucose kg^{-1} ww) yielded a concentration only 30% less than the *pre*-exercise concentration (76 ± 4) after 4 weeks of KT. However, pre-exercise CHO muscle glycogen was 143 ± 10 mmol glucose kg^{-1} ww, 1.9 times that of pre-exercise KT glycogen. The finding of partial glycogen repletion suggests that gluconeogenesis was able to support modest muscle glycogenesis, despite the subjects being able to consume less than a surfeit of dietary protein (1.75 g kg^{-1} day^{-1}). Post-exercise (in both treatment stages) fiber-specific muscle glycogen concentrations revealed absolute depletion in type I (slow) fibers only. Glycogen utilization rate during the bout after KT was 21% of that during CHO.

A salient observation not reported in the publication [34] was that the cyclists experienced an impaired ability to perform hill climbs during training while on the KT diet [35], suggesting that the (presumably chronic) near-depleted glycogen state could not adequately support higher power outputs during training. The training program employed by the cyclists was also likely characterized by some variable intensity work (absent from the laboratory-based cycling bouts to exhaustion), which can modify fiber-specific glycogenolytic rates [36].

Lambert et al. [37] undertook a moderated approach to that of Phinney et al. [34]. They enrolled five trained cyclists with a relatively lower mean VO_{2max} (4.21 ± 0.3 (SEM) vs 5.1 ± 0.18) and an apparently lower weekly training volume. The subjects were randomly assigned to two diet treatments, each 2 weeks in duration, followed by a two-week "washout" *ad libitum*/"normal" diet and then crossing over to the other diet. The high carbohydrate (HC) diet phase allocated daily energy intake as 73.6% ± 5.0% carbohydrate, 12.0% ± 6.7% fat, and 13.5% ± 1.6% protein. The high fat (HF) diet provided 7.1% ± 2.2% carbohydrate, 67.3% ± 1.8% fat, and 25.5% ± 2.5% protein. Although the diets were isocaloric, total caloric intake was not described. Exercise tests were performed at the end of the HC and HF diet phases. Each subject performed a series of 5-second tests on a cycle ergometer at maximal cadence, with variable loads. Each subject then performed a 30-second Wingate test on the ergometer [38]. After a 30-minute rest period following the single Wingate test, subjects were biopsied (vastus lateralis) and then performed a cycling bout to exhaustion at approximately 90% of their VO_{2max} (measured prior to the diet period) with another biopsy collected from the same leg at exhaustion. A 20-minute rest interval followed the high intensity ride to exhaustion, which was then proceeded by another ride to exhaustion, yet at approximately 60% of each subject's pre-determined VO_{2max}. HC muscle glycogen prior to the high intensity cycling bout was 77% greater than at the same time point for HF, while after the end of the bout HC glycogen concentration was similar to the *pre*-exercise muscle glycogen with the HF treatment. Muscle glycogen utilization rates did not significantly differ between treatments. The time to exhaustion during the high intensity ride for HC was 12.5 ± 3.8 (SEM) minutes and 8.3 ± 2.3 for HF (a 51% greater duration) but did not achieve statistical significance (likely due to the sample size of five subjects). In contradistinction, the time to exhaustion during the moderate intensity bout was 42.5 ± 6.8 minutes for HC and 79.7 ± 7.6 for HF (88% greater for HF). Unfortunately muscle biopsies were not taken after the *moderate* intensity ride. The nearly twofold higher fat and 42% lower carbohydrate oxidation rates in HF may explain the dramatically greater endurance time to exhaustion for HF in the moderate intensity bout, contrasted to the HC diet treatment, which had a greater reliance on carbohydrate oxidation in the face of already compromised muscle glycogen concentrations.

These limited data suggest that submaximal/moderate intensity constant load exercise can be sustained in the face of reduced muscle glycogen concentrations (from fasting or carbohydrate restriction), with no apparent compromise in performance (compared with a "normal" mixed diet). However, near maximal/higher intensity, and perhaps variable load exercise capacity may be compromised [39], even after one day of glycogen depletion [10] (J. A. Hawley, personal communication).

Repletion of Muscle Glycogen I: Glycemic Index

As described above, muscle glycogen is a significant and, sometimes, predominant fuel substrate utilized during moderate to intense exercise, especially among a variety of elite to professional athletes in competition. Systemically available dietary carbohydrate sources remain the superior means of replacing muscle (and liver) glycogen. The systemic availability of dietary carbohydrates in relation to muscle glycogen repletion following exercise can be determined only by the quantitation of muscle glycogen concentrations, e.g., by muscle biopsy and ^{13}C magnetic resonance spectroscopy [40].

The introduction of the concept and tool of the glycemic index (GI) by Jenkins and collaborators in 1981 [41], as a means of assessing the physiological response to a dietary carbohydrate source, slowly ushered in a number of studies exploring the influence of foods and nutritional supplements with varying GI values upon muscle glycogen (reviewed in [42,43]). These few studies represent the majority of studies that have compared the glycogen replacement efficacy of various *foods.* An abundance of recommendations have been disseminated in both academic and lay publications, centered upon the assumption that a food or nutritional supplement (carbohydrate source) that has a higher GI value exhibits faster appearance in the blood than a low or lower GI food or supplement [44], and elicits a more robust insulin release reaction [45]. The glycemic index, which can be determined only *in vivo* in humans after oral ingestion of a candidate food or supplement, simply describes the increment in blood glucose concentration over a fixed period of time (usually 2 hours), relative to an "indexed" food (dextrose or white bread) [41]. *It does not describe the glucose kinetics derived from a food, beverage, or nutritional supplement.* This also suggests that the results of muscle glycogen repletion studies comparing foods with different GI values [42,43] may not be explained by the rate of glucose delivery to the circulation, as this has never been measured.

A pioneering study in 2003 by Schenk et al. evaluated the glycemic index of a classical low GI cereal (All Bran, [41]) and a classical high GI cereal (corn flakes, [41]) in fit males [46]. Additionally, they infused [6,6-^{2}H2] glucose into the subjects, to directly determine the Ra and Rd of glucose. Expectedly, the bran cereal yielded a low GI value (54.5 ± 7.2 (SEM)) and the corn flakes a GI value over twofold greater (131.5 ± 33.0). Distinctively, the Ra for each of the two cereal treatments was virtually identical over a 3-hour period. However, the bran cereal Rd over the 30–60 minute postprandial period was 31% greater than for corn flakes ($P < 0.05$), and was accompanied by a 54% greater glucose clearance rate ($P < 0.05$). The authors ascribed the greater glucose disposal rates observed with bran cereal to the significantly greater insulin concentration 20 minutes postprandial ($P < 0.05$) and the more than twofold greater insulin area under the curve 0–30 minutes postprandial ($P < 0.05$). Two additional studies by Eelderink et al., using various breads, fiber additives, and pastas, have confirmed the observations of Schenk et al. and underscored hepatic gluconeogenesis as another factor capable of modulating glucose kinetics from a dietary carbohydrate source [47,48].

Repletion of Muscle Glycogen I: Carbohydrate Plus Protein Combinations

Replacement and supercompensation of muscle glycogen through dietary means, although effective [30], may seem less than alluring to athletes assaulted by advertising, exhortations by teammates and consultants, and athlete/celebrity endorsements for nutritional supplements. The first nutritional supplement composition that showed superior muscle glycogen replacement was a combination of carbohydrates and protein. Zawadzki and coworkers [49] had nine male cyclists perform glycogen-depleting rides and then, at three different times, randomly assigned them to one of three post-exercise nutritional supplements, ingested immediately and 2 hours post-exercise. They hypothesized that the addition of protein to carbohydrate would magnify the insulinemic response and thus increase muscle glycogen resynthesis. The CARB treatment was 112 g of carbohydrate from a dextrose and chemically undefined maltodextrin mixture. The PRO treatment delivered 40.7 g of protein from a milk and whey protein isolate mixture. The CARB-PRO mixture provided 112 g carbohydrate and 40.7 g protein, of the same sources as the other two treatments. All treatments were consumed immediately 2 hours post-exercise. Muscle biopsies (vastus lateralis) were obtained immediately after exercise and 4 hours later. PRO elicited a modest, statistically non-significant increase in muscle glycogen, whereas CARB induced a significantly greater ($P < 0.05$) increase than PRO. CARB-PRO induced a significantly greater ($P < 0.05$) increase than CARB. Postprandial blood insulin response was significantly greater with CARB-PRO than CARB or PRO through most of the postprandial interval. In their discussion the authors (perhaps) presciently stated "...the addition of protein to a carbohydrate supplement may be beneficial only when an *inadequate* CHO source is used." [emphasis added].

A sequence of studies by the same group systematically addressed the caveat expressed by Zawadzki et al. van Loon and colleagues first demonstrated that post-exercise intakes of carbohydrate at 1.2 g kg^{-1} h^{-1} can attain greater muscle glycogen repletion than 0.8 g kg^{-1} h^{-1}[50]; thereafter, Jentjens et al. revealed that addition of protein to the higher carbohydrate intake rate (1.2 g kg^{-1} h^{-1}) confers no additional increment in muscle glycogen (over a 3-hour period), despite higher insulin concentrations [51]. Thus, for athletes *restricting* carbohydrate intakes, the addition of protein to a modest carbohydrate quantity will facilitate greater post-exercise muscle glycogen resynthesis. Moreover, ingestion of protein *alone* will promote only modest repletion of muscle glycogen [34,37,49].

Repletion of Muscle Glycogen I: Carbohydrate Comparisons

A variety of nutritional supplements with "exotic" carbohydrates or carbohydrate blends abound for the athlete to choose from, claiming faster and/or greater muscle glycogen restoration. Very few of these products have been the subject of controlled research investigations, let alone been assessed for their impact on muscle glycogen resynthesis after exercise. One of the first comparison studies was undertaken by Jozsi et al. in 1996 [52]. Eight trained cyclists completed a glycogen-depleting bout of cycling exercise, with prior dietary controls. Each subject was then immediately biopsied (vastus lateralis) and given one of four different carbohydrate drinks ($\approx$19% solution; aspartame sweetened): (a) dextrose, (b) maltodextrin (mean chain length of 5 glucose units), (c) waxy maize starch (Amioca™, National Starch, New York, USA; a slow digesting, low glycemic, low insulinemic starch; [53–55]), and (d) a resistant starch. Subjects then consumed a high carbohydrate diet with all of the carbohydrate coming from one of the four drinks, at 6.5 g kg^{-1} body weight for 12 hours prior to sleep. The subjects then returned 24 hours later for a second biopsy. No significant differences in 24-hour post-exercise muscle glycogen increment were noted, with the exception of the resistant starch, which was significantly less ($P < 0.05$) than the other three treatments. Thirty minutes after the biopsy the subjects performed a 30-minute time trial. No differences in work output between treatments were observed.

The findings of Jozsi et al. contrast with those of Piehl Aulin and colleagues, who compared a soluble, high molecular weight ($5–7 \times 10^5$) extract of potato starch (Vitargo®; Swecarb, Sweden) against a moderately hydrolyzed maltodextrin with an average molecular weight of 500 (Glucidex® IT 38, Roquette, Lille Cedex, France) [56]. Thirteen multi-sport trained males ran and cycled to exhaustion and were then given, at two separate times, 75 grams of a carbohydrate in an artificially sweetened and flavored (15%) solution at 0, 30, 60, and 90 minutes post-exercise. Muscle biopsies (vastus lateralis) were collected at 5 minutes and 2 and 4 hours post-exercise. The glycogen synthesis rate within 2 hours following 300 g of carbohydrate from the starch extract was 68% greater ($P < 0.06$), while the muscle glycogen increment was greater ($P < 0.05$), compared with the maltodextrin. No differences were seen between treatments after 4 hours.

Stephens and colleagues compared a cornstarch extract form of Vitargo to a commercial corn maltodextrin composition (Maxijul, Nutricia Ltd, Trowbridge, UK; $\approx$88% oligosaccharides with a mean chain length of 5 glucose units, 6.7% sugars) [57]. Eight recreationally active but untrained men performed a cycling bout to deplete muscle glycogen [58] and were then given 100 g of the starch extract, 100 g of the maltodextrin (both as a 10% solution), or an artificially sweetened, flavored, non caloric beverage. After 2 hours of rest the subjects then performed a maximal cycling endurance trial for 15 minutes. The starch ($P < 0.001$) and maltodextrin treatments ($P < 0.01$) were associated with significantly greater total work output than the placebo. Additionally, the starch extract treatment produced a significantly greater total work output than the maltodextrin (164.1 kJ $\pm$ 21.1 (SD) and 149.4 kJ $\pm$ 21.8, respectively; $P < 0.01$). The authors speculated that the observed greater recovery performance with the starch extract was due to more rapid muscle and, perhaps, hepatic glycogen repletion.

CONCLUSION

Skeletal muscle is metabolically promiscuous [59], exhibiting plasticity in its selection of fuel substrate. Uniquely, in healthy individuals, muscle glycogen is always utilized to a variable yet appreciable degree, independent of the work intensity. The metabolic inaccessibility of muscle glycogen in individuals with glycogen-related conditions such as McArdle disease illuminates the critical role muscle glycogen exerts in any form of exercise. Dietary strategies designed to restrict dietary carbohydrate in exercising/training individuals, and consequently produce dramatic drops in muscle glycogen, appear to be capable of sustaining constant load exercise but may be inadequate in variable load or high intensity training. Dietary and nutritional supplement strategies to replace muscle glycogen are best served by adequate carbohydrate intakes, and perhaps by the selection of carbohydrate supplements that have been demonstrated in comparison studies to produce greater muscle glycogen increments, rather than blood measures, e.g., glycemic index.

Disclosure

The author is an owner and executive of a company that markets and sells Vitargo-containing nutritional supplements.

References

[1] Holloszy JO, Kohrt WM. Regulation of carbohydrate and fat metabolism during and after exercise. Ann Rev Nutr 1996;16:121–38.

[2] Romijn JA, Coyle EF, Sidossis LS, Gastaldelli A, Horowitz JF, Endert E, et al. Regulation of endogenous fat and carbohydrate metabolism in relation to exercise intensity and duration. Am J Physiol Endocrinol Metab 1993;265:E380–91.

[3] van Loon LJC, Greenhaff PL, Constantin-Teodosiu D, Saris WHM, Wagenmakers. AJM. The effects of increasing exercise intensity on muscle fuel utilisation in humans. J Physiol 2001;536:295–304.

[4] Helge JW, Stallknecht B, Richter EA, Galbo H, Kiens B. Muscle metabolism during graded quadriceps exercise in man. J Physiol 2007;581:1247–58.

[5] Jeppesen J, Kiens B. Regulation and limitations to fatty acid oxidation during exercise. J Physiol 2012;1059–68 [590.5]

[6] Tesch PA, Colliander EB, Kaiser P. Muscle metabolism during intense, heavy-resistance exercise. Eur J Appl Physiol Occup Physiol 1986;55:262–366.

[7] Essén-Gustavsson B, Tesch PA. Glycogen and triglyceride utilization in relation to muscle metabolic characteristics in men performing heavy-resistance exercise. Eur J Appl Physiol Occup Physiol 1990;61:5–10.

[8] Robergs RA, Pearson DR, Costill DL, Fink WJ, Pascoe DD, Benedict MA, et al. Muscle glycogenolysis during differing intensities of weight-resistance exercise. J Appl Physiol 1991;70:1700–6.

[9] MacDougall JD, Ray S, Sale DG, McCartney N, Lee P, Garner S. Muscle substrate utilization and lactate production. Can J Appl Physiol 1999;24:209–15.

[10] Churchley EG, Coffey VG, Pedersen DG, Shield A, Carey KA, Cameron-Smith D, et al. Influence of preexercise muscle glycogen content on transcriptional activity of metabolic and myogenic genes in well-trained humans. J Appl Physiol 2007;102:1604–11.

[11] Sherman WM, Costill DL, Fink WJ, Hagerman FC, Armstrong LE, Murray TF. Effect of a 42.2-km footrace and subsequent rest or exercise on muscle glycogen and enzymes. J Appl Physiol 1983;55:1219–24.

[12] Bangsbo J, Iaia FM, Krustrup P. Metabolic response and fatigue in soccer. Int J Sports Physiol Perform 2007;2:111–27.

[13] Krustrup P, Mohr M, Steensberg A, Bencke J, Kjaer M, Bangsbo J. Muscle and blood metabolites during a soccer game: implications for sprint performance. Med Sci Sports Exerc 2006;38:1165–74.

[14] Krustrup P, Ortenblad N, Nielsen J, Nybo L, Gunnarsson TP, Iaia IA, et al. Maximal voluntary contraction force, SR function and glycogen resynthesis during the first 72 h after a high-level competitive soccer game. Eur J Appl Physiol 2011;111:2987–95.

[15] Gillum TL, Dumke CL, Ruby BC. Muscle glycogenolysis and resynthesis in response to a half Ironman triathlon: a case study. Int J Sports Physiol Perform 2006;1:408–13.

[16] Cuddy JS, Slivka DR, Hailes WS, Dumke CL, Ruby BC. Metabolic profile of the Ironman World Championships: a case study. Int J Sports Physiol Perform 2010;5:570–6.

[17] Schoeller DA. Measurement of energy expenditure in free-living humans by using doubly labeled water. J Nutr 1988;118:1278–89.

[18] Hoyt RW, Jones TE, Stein TP, McAninch GW, Lieberman HR, Askew EW, et al. Doubly labeled water measurement of human energy expenditure during strenuous exercise. J Appl Physiol 1991;71:16–22.

[19] Kimber NE, Ross JJ, Mason SL, Speedy DB. Energy balance during an ironman triathlon in male and female triathletes. Int J Sport Nutr Exerc Metab 2002;12:47–62.

[20] Hermansen L, Hultman E, Saltin B. Muscle glycogen during prolonged severe exercise. Acta Physiol Scand 1967;71:129–39.

[21] McArdle B. Myopathy due to a defect in muscle glycogen breakdown. Clin Sci 1951;10:13–33.

[22] Beynon RJ, Bartram C, Hopkins P, Toescu V, Gibson H, Phoenix J, et al. McArdle's disease: molecular genetics and metabolic consequences of the phenotype. Muscle Nerve 1995;3:S18–22.

[23] Nielsen JN, Vissing J, Wojtaszewski JFP, Haller RG, Begum N, Richter EA. Decreased insulin action in skeletal muscle from patients with McArdle's disease. Am J Physiol Endocrinol Metab 2002;282:E1267–75.

[24] Pearson CM, Rimer DG, Mommaerts WF. A metabolic myopathy due to absence of muscle phosphorylase. Am J Med 1961;30:502–17.

[25] Kitaoka Y, Ogborn DI, Mocellin NJ, Schlattner U, Tarnopolsky M. Monocarboxylate transporters and mitochondrial creatine kinase protein content in McArdle disease. Mol Genet Metab 2013;108:259–62.

[26] Andersen ST, Jeppesen TD, Taivassalo T, Sveen ML, Heinicke K, Haller RG, et al. Effect of changes in fat availability on exercise capacity in McArdle disease. Arch Neurol 2009;66:762–6.

[27] Ørngreen MC, Jeppesen TD, Tvede Andersen S, Taivassalo T, Hauerslev S, Preisler N, et al. Fat metabolism during exercise in patients with McArdle disease. Neurology 2009;72:718–24.

[28] Haller RG, Vissing J. Spontaneous "second wind" and glucose-induced second "second wind" in McArdle disease: oxidative mechanisms. Arch Neurol 2002;59:1395–402.

[29] García-Benítez S, Fleck SJ, Naclerio F, Martín MA, Lucia A. Resistance (weight lifting) training in an adolescent with McArdle disease. J Child Neurol 2012; doi:10.1177/0883073812451328.

[30] Bergström J, Hermansen L, Hultman E, Saltin B. Diet, muscle glycogen and physical performance. Acta Physiol Scand 1967;71:140–50.

[31] Loy SF, Conlee RK, Winder WW, Nelson AG, Arnall DA, Fisher AG. Effects of 24-hour fast on cycling endurance time at two different intensities. J Appl Physiol 1986;61:654–9.

[32] Knapik JJ, Meredith CN, Jones BH, Suek L, Young VR, Evans WJ. Influence of fasting on carbohydrate and fat metabolism during rest and exercise in men. J Appl Physiol 1988;64:1923–9.

[33] Phinney SD, Horton ES, Sims EA, Hanson JS, Danforth Jr E, LaGrange BM. Capacity for moderate exercise in obese subjects after adaptation to a hypocaloric, ketogenic diet. J Clin Invest 1980;66:1152–61.

[34] Phinney SD, Bistrian BR, Evans WJ, Gervino E, Blackburn GL. The human metabolic response to chronic ketosis without caloric restriction: preservation of submaximal exercise capability with reduced carbohydrate oxidation. Metabolism 1983;32:769–76.

[35] Brooks GA. Bioenergetics of exercising humans. Compr Physiol 2012;2:537–62.

[36] Palmer GS, Borghouts LB, Noakes TD, Hawley JA. Metabolic and performance responses to constant-load vs. variable-intensity exercise in trained cyclists. J Appl Physiol 1999;87:1186–96.

[37] Lambert EV, Speechly DP, Dennis SC, Noakes TD. Enhanced endurance in trained cyclists during moderate intensity exercise following 2 weeks adaptation to a high fat diet. Eur J Appl Physiol Occup Physiol 1994;69:287–93.

[38] Ayalon A, Inbar O, Bar-Or O. Relationships among measurements of explosive strength and anaerobic power. In: Nelson RC, Morehouse CA, editors. Biomechanic IV. International series on sport sciences, vol. 1. Baltimore: University Press; 1974. p. 572–7.

[39] Fleming J, Sharman MJ, Avery NG, Love DM, Gomez AL, Scheett TP, et al. Endurance capacity and high-intensity exercise performance responses to a high fat diet. Int J Sport Nutr Exerc Metab 2003;13:466–78.

[40] Taylor R, Price TB, Rothman DL, Shulman RG, Shulman GI. Validation of ^{13}C NMR measurement of human skeletal muscle glycogen by direct biochemical assay of needle biopsy samples. Magn Reson Med 1992;27:13–20.

[41] Jenkins DJ, Wolever TM, Taylor RH, Barker H, Fielden H, Baldwin JM, et al. Glycemic index of foods: a physiological basis for carbohydrate exchange. Am J Clin Nutr 1981;34:362–6.

[42] Mondazzi L, Arcelli E. Glycemic index in sport nutrition. J Am Coll Nutr 2009;28:455S–63S.

[43] O'Reilly J, Wong SHS, Chen Y. Glycaemic index, glycaemic load and exercise performance. Sports Med 2010;40:27–39.

[44] <http://www.livestrong.com/article/320361-glycemic-index-and-working-out/> [accessed 26.03.2013].

[45] Brown LJS, Midgley AW, Vince RB, Madden LA, McNaughton LR. High versus low glycemic index 3-h recovery diets following glycogen-depleting exercise has no effect on subsequent 5-km cycling time trial performance. J Sci Med Sport 2012; <http://dx.doi.org/10.1016/j.jsams.2012.10.006>.

[46] Schenk S, Davidson CJ, Zderic TW, Byerley LO, Coyle EF. Different glycemic indexes of breakfast cereals are not due to glucose entry into blood but to glucose removal by tissue. Am J Clin Nutr 2003;78:742–8.

[47] Eelderink C, Moerdijk-Poortvliet TCW, Wang H, Schepers M, Preston T, Boer T, et al. The glycemic response does not reflect the in vivo starch digestibility of fiber-rich wheat products in healthy men. J Nutr 2012;142:258–63.

[48] Eelderink C, Schepers M, Preston T, Vonk RJ, Oudhuis L, Priebe MG. Slowly and rapidly digestible starchy foods can elicit a similar glycemic response because of differential tissue glucose uptake in healthy men. Am J Clin Nutr 2012;96:1017–24.

[49] Zawadzki KM, Yaspelkis III BB, Ivy JL. Carbohydrate-protein complex increases the rate of muscle glycogen storage after exercise. J Appl Physiol 1992;72:1854–9.

[50] van Loon LJ, Saris WH, Kruijshoop M, Wagenmakers AJ. Maximizing postexercise muscle glycogen synthesis: carbohydrate supplementation and the application of amino acid or protein hydrolysate mixtures. Am J Clin Nutr 2000;72:106–11.

[51] Jentjens RLPG, van Loon LJC, Mann CH, Wagenmakers AJM, Jeukendrup AE. Addition of protein and amino acids to carbohydrates does not enhance postexercise muscle glycogen synthesis. J Appl Physiol 2001;91:839–46.

[52] Jozsi AC, Trappe TA, Starling RD, Goodpaster B, Trappe SW, Fink WJ, et al. The influence of starch structure on glycogen resynthesis and subsequent cycling performance. Int J Sports Med 1996;17:373–8.

[53] Goodpaster BH, Costill DL, Fink WJ, Trappe TA, Jozsi AC, Starling RD, et al. The effects of pre-exercise starch ingestion on endurance performance. Int J Sports Med 1996;17:366–72.

[54] Anderson GH, Catherine NLA, Woodend DM, Wolever TMS. Inverse association between the effect of carbohydrates on blood glucose and subsequent short-term food intake in young men. Am J Clin Nutr 2002;76:1023–30.

[55] Sands AL, Leidy HJ, Hamaker BR, Maguire P, Campbell WW. Consumption of the slow-digesting waxy maize starch leads to blunted plasma glucose and insulin response but does not influence energy expenditure or appetite in humans. Nutr Res 2009;29:383–90.

[56] Piehl Aulin K, Soderlund K, Hultman E. Muscle glycogen resynthesis rate in humans after supplementation of drinks containing carbohydrates with low and high molecular masses. Eur J Appl Physiol 2000;81:346–51.

[57] Stephens FB, Roig M, Armstrong G, Greenhaff PL. Post-exercise ingestion of a unique, high molecular weight glucose polymer solution improves performance during a subsequent bout of cycling exercise. J Sports Sci 2008;26:149–54.

[58] Casey A, Short AH, Hultman E, Greenhaff PL. Glycogen resynthesis in human muscle fibre types following exercise-induced glycogen depletion. J Physiol 1995;483:265–71.

[59] Yoshida Y, Jain SS, McFarlan JT, Snook LA, Chabowski A, Bonen A. Exercise-, and training-induced upregulation of skeletal muscle fatty acid oxidation are not solely dependent on mitochondrial machinery and biogenesis. J Physiol 2012; doi:10.1113/jphysiol.2012.238451.

CHAPTER

34

An Overview of Adaptogens with a Special Emphasis on *Withania* and *Rhodiola*

Pranay Wal and Ankita Wal

Pranveer Singh Institute of Technology, Kanpur, UP, India

INTRODUCTION

The term adaptogen refers to a substance that can effect a state of raised resistance [1] enabling an organism to cope with different kinds of stressful situations [2]. This concept is derived from the "general adaptation syndrome" [3,4] and proposes that an organism when facing a stressful situation goes through three physiologic phases: (a) alarm, (b) resistance, and (c) exhaustion. According to this syndrome, an organism has a limited capacity to cope with environmental aggression, and this capacity may decline with continuous exposure to such aggression, resulting in health disturbances and disease. An adaptogen is a phytonutrient that has the ability to help normalize or "adapt" the body's physiologic functions to a higher rate of functioning, regardless of the person's current health status (i.e., whether one is an athlete or chronically ill). As adjuvant to other specific treatments, adaptogens can also help in altering the course of the disease.

The adaptogenic herbs contain active plant constituents known as phytochemicals. Examples of phytochemicals that contribute to our wellbeing are triterpenes, phenylpropanes, oxylipins and polysaccharides. They help to stimulate the immune system and increase overall vital energy. Adaptogens are a pharmacotherapeutic group of herbal preparations used to increase attention and endurance during fatigue, and to prevent, mitigate or reduce stress-induced impairments and disorders related to neuroendocrine and immune systems [5].

Another definition of adaptogen is in terms of physiologic conditions: adaptogenic substances are stated to have the capacity to normalize body functions and strengthen systems compromised by stress. They are reported to have a protective effect on health against a wide variety of environmental assaults and emotional conditions. Adaptogens are compounds which could increase "the state of non-specific resistance during stress" [6,7]. Adaptogens are innocuous agents, non-specifically increasing resistance against physically, chemically, biologically and psychologically noxious factors ("stressors"), whose normalizing effect is independent of the nature of the pathologic state [8]. Adaptogens are substances which elicit in an organism a state of nonspecifically raised resistance allowing them to counteract stressor signals and to adapt to exceptional strain [9].

Adaptogens typically act upon the neuroendocrine immunologic system, which is an all-encompassing description of how the immune system and brain interact with hormones. The term adaptogen has not yet been accepted in medicine. This is probably due to the difficulties in discriminating adaptogenic drugs from immunostimulators, anabolic drugs, nootropic drugs, and tonics. There can be no doubt, however, that, at least in animal experiments, there are plant drugs capable of modulating distinct phases of the adaptation syndrome as defined by Seyle [2]. These drugs reduce stress reactions in the alarm phase, or retard/prevent the exhaustion phase, and thus provide a certain degree of protection against long-term stress. The small number of drugs whose anti-stress activity has been proven or reported includes, among others, the plant drugs *Withania*, *Rhodiola*, *Ginseng*, *Eleutherococcus*, *Ocimum* and *Codonopsis*. Though scientifically unproven, they are marketed as supplements to increase resistance to stress, trauma, anxiety and fatigue. The term is used mainly by herbalists who also refer to adaptogens as rejuvenating herbs, tonics,

Nutrition and Enhanced Sports Performance.
DOI: http://dx.doi.org/10.1016/B978-0-12-396454-0.00034-5

rasayanas, or restoratives. One specific characteristic of adaptogen action is that its effect is believed to help the body return to a balanced state. However, there is no strict definition of the adaptogenic characteristics of a plant product, leading to a generalized usage of the term for commercial or pseudoscientific reasons.

An adaptogen must have the following properties [6]:

1. Show a nonspecific activity, i.e., increase in power of resistance against physical, chemical or biological noxious agents;
2. Have a normalizing influence independent of the nature of the pathological state;
3. Be innocuous and not influence normal body functions more than required.

It is accepted that adaptogen plants, when chronically used, are able to increase the animal's capacity to endure physical, chemical, or environmental aggressions. As a consequence, there is a general improvement in health conditions, which can be manifested, among other things, through the betterment of cognitive functions (such as learning and memory capacities) and an increase in quality of sleep and sexual performances [6–8]. On the other hand, it is doubtful whether these beneficial effects are directly mediated through the CNS, it being very likely that the endocrine system plays a major role [2]. Adaptogens improve the response to stress [9]. They help the body to adapt by normalizing physiological processes in times of increased stress [10].

WITHANIA AS ADAPTOGEN

Withania somnifera is also known as ashwagandha. The Ayurveda, which is gaining in popularity in many western countries, states that, because the primary quality and flavor of ashwagandha is sharp and pungent, this indicates that it is warming, raises metabolism, stimulates digestion, clears mucus, and improves circulation. The Ayurveda also identifies a secondary post-digestive flavor, which for ashwagandha is sweet. It is this effect, which is not necessarily directly identified by one's sense of taste, that occurs when a substance is converted into a still purer nutritive extract [10]. Following this, the post-digestive sweet flavor of ashwagandha represents its deep nutritive, hormonal properties as well as its ability to strengthen and nourish the nervous system. The distinctive earthy odor and flavor of ashwagandha is due to the presence of certain steroidal lactones or withanolides [11]. It is from this characteristic odor that its Sanskrit name, "like a horse", derives. While most medicinal herbs are not particularly prized or known for their appealing flavor, ashwagandha for most may be promoted to the forefront of those herbs with the least taste-smell appeal. It is commonly referred to as Indian Ginseng. Ashwagandha is unique as a tonic herb in that it is exceptionally easy to cultivate and is ready for harvest in only one year of growth.

Withania has immunomodulatory, anti-inflammatory but most significantly adaptogenic effects, which may result from the complex of the many steroidal withanolides found in the root of the herb [12,13]. The withanolides are the major chemical constituents of *Withania*, and the plant has been the subject of considerable modern scientific attention [14]. Most of the studies on *Withania* have used rat experimental models, but in one human clinical trial the anxiolytic efficacy of an ethanolic extract of the herb was evaluated [15]. In that double-blind placebo-controlled study, 20 patients suffering from anxiety disorder received an extract of *Withania* in the form of a tablet whilst 19 people received a placebo, and patients were assessed at baseline, the end of week 2 and at the end of week 6 (the treatment endpoint). This study demonstrated a trend for the anxiolytic superiority of *Withania* over placebo and the authors concluded that *Withania* has useful anxiolytic potential and merits further investigation [13]. As anxiety can be one outcome of chronic or severe stress, *Withania* may well have a role in stress management. However, the study was not large enough with only 39 people involved and it was conducted over a very short period, only 6 weeks.

Another randomized human trial investigated the effectiveness of naturopathic care on anxiety symptoms, and *Withania* was used in the treatment group alongside other therapies rather than alone [16]. The trial was double-blinded, randomized, with 41 participants in the treatment group who received dietary counseling, deep breathing relaxation techniques, a standard multivitamin, and *Withania* root standardized to 1.5% withanolides, 300 mg twice daily [16]. The control group of 40 participants received psychotherapy, matched deep breathing relaxation techniques, and placebo. The primary outcome measure the Beck Anxiety Inventory (BAI), decreased by 56.5% in the treatment group compared with 30.5% in the control group. These scores were statistically significant, although it would be difficult to determine which treatment was of benefit: the counseling, relaxation, multivitamin, *Withania*, or a combination. A recent study was conducted on Swiss albino mice to see the effect of withanolide A (isolated from *Withania somnifera* root extract) on chronic stress-induced alterations on T lymphocyte subset distribution and corresponding cytokine secretion patterns [17].

The adaptogenic effects of *Withania* were studied in a clinical trial using a rat model of chronic stress produced by random regular foot shock. The chronic stress induced significant hyperglycemia, glucose intolerance, increased cortisol levels, immunosuppression and gastric ulceration. *Withania* administered 1 hour before foot shock was found to attenuate these effects and demonstrated significant anti-stress and adaptogenic activity similar to that of Panax Ginseng [18]. Withanolide A was orally administered once daily on the stressed experimental animals and was found to cause significant recovery of the stress-induced depleted T cell population, with an increase in the expression of IL-2 and IFN-gamma (a signature cytokine of Th1 helper cells) and a decrease in the concentration of corticosterone [17]. This study supports *Withania's* role in stress management including immune function regulation. The involvement of such immunoregulatory cells induced by *Withania somnifera* (WS) might have several functions: such as regulating antigen presentation and an immunosuppressive microenvironment along with a physiological cytokine milieu for an effector T cell function. It is intriguing that the treatment with *Withania* induces a Thl cell-mediated immune response and an elevation of IgG2a-mediated humoral immune responses [19,20]. In addition, aqueous suspension of WS shows anti-inflammatory and immunosuppressive effects by inhibiting the complement system, mitogen-induced lymphocyte proliferation and delayed type hypersensitivity (DTH) in rats [21]. Although in this investigation no effect on the humoral response was observed, others have reported elevated levels of IgG2a over lgGl in WS-treated BALBIc mice [20].

As previously stated, chronic stress can result in memory impairment. In another study, researchers investigated the effect of withanolide A on memory-deficient mice showing neuronal atrophy and synaptic loss in the brain [22]. Stress parameters were reduced in the treatment group compared with controls, and blood parameters revealed a decrease in CPK, lactase dehydrogenase (LDH) and lipid peroxidation (LPO) in the treatment group compared with controls. There was also a reduction in serum corticosterone in the treatment group compared with controls.

The anti-stress activity of WS root extract has been explained by the fact that it has antioxidant property. Chronic fatigue induced by forced swimming for 15 days induced a significant rise in brain malondialdehyde (MDA) levels as compared with naïve mice, indicating the oxidation of proteins, DNA and lipids. Administration of WS (100 mg/kg p.o.) significantly reversed the extent of lipid peroxidation [23]. *Withania* has demonstrated activity on the CNS and the HPA axis as well as having profound antioxidant properties. Identified actions on the CNS include neurotransmitter function (catecholamines), acetylcholine regulation, and a serotonergic effect which may be responsible for memory-enhancing and cognitive function.

WS demonstrated a GABA-mimetic effect which plays an important role in reducing psychological stress, a well-known factor in the evolution of disease states [24]. Activity on the HPA axis was postulated as a mechanism of action [25] in a study to test the adaptogenic effect of WS on chronic stress using a foot shock model. The symptoms of stress from foot shock manifest in a variety of nonspecific maladies including gastric ulcer, hyperglycemia, glucose intolerance, increased plasma cortisone, sexual dysfunction in males, cognitive deficits, immunosuppression and mental depression.

The anti-stressor effect of ashwagandha was investigated in rats using a cold water swimming stress test. The treated animals showed better stress tolerance [25]. A withanolide-free aqueous fraction isolated from the roots of WS exhibited anti-stress activity in a dose-dependent manner in mice [26].

RHODIOLA AS ADAPTOGEN

Rhodiola rosea L. is a valuable medicinal plant known mainly as an adaptogen, increasing resistance to harmful effects of various stressors [27–33]. Also known as golden root, rose root, or Arctic root, it belongs to the plant family Crassulaceae and genus *Rhodiola*. It is found at high altitudes in the Arctic and mountainous regions throughout Europe and Asia, and has been used medically in Russia, Scandinavia and many other countries for a range of conditions such as stress-induced depression and anxiety, fatigue, anemia, impotence, infections (including colds and influenza), cancer, nervous system disorders and headache. It is classified as an adaptogen. It has been widely studied in Russia and Scandanavia for over 35 years and is thought to stimulate the nervous system, decrease depression, enhance work performance, and eliminate or reduce fatigue [34]. It is also regarded as a tonic and stimulant and used to increase physical endurance, stress resistance, attention span, memory, work productivity, and resistance to high-altitude sickness.

Small doses of *R. rosea* increase the bio-electrical activity of the brain [35]. It prolongs the actions of neurotransmitters such as adrenaline, dopamine, serotonin and acetylcholine in the central nervous system and brain by inhibiting the activity of enzymes responsible for their degradation [36]. Consequently, the cognitive functions of the cerebral cortex, and the attention, memory and learning functions of the prefrontal and frontal cortex, are enhanced [37]. *R. rosea* prevents the rise in mediators of the stress response—phosphorylated stress-activated protein kinase, nitric oxide and cortisol—following immobilization stress [38]. *R. rosea* prevents exercise-induced ATP decrease in mitochondria after exhaustive swimming [39]. It can also be applied in cases of borderline nervous-mental diseases, neuroses, neurotic disorders and psychopathies [40]. In psychiatric practice, extracts of *R. rosea* are indicated for the correction of neurological side-effects associated with psychopharmacological therapy, and for the intensification and stabilization of remissions of asthenic and apathistical-aboulic type schizophrenia patients [35–41].

In human studies, this root has been shown to be effective for treating mild depression, neurasthenia [41], impaired cognitive function [42], erectile dysfunction, amenorrhea, and infertility in women. *Rhodiola* is useful for people with deficient (asthenic) depression, altitude sickness (use it with Cordyceps, Reishi, and Holy Basil), and to aid in recovery from head trauma injury. Traditionally, *Rhodiola* is used in Tibetan medicine for nourishing the lungs, to increase blood circulation, and for fatigue, altitude sickness, and weakness.

The compound salidroside (Also known as rhodioloside or rhodosin) is the most biologically active compound in *Rhodiola*. It shares many of its effects with its sister compound rosavin. Salidroside can also influence the uptake of glucose into muscle cells via activation of AMPK. The compound rosavin accounts for the highest percent of *R. rosea* active constituents per gram, and is the measure by which to standardize SHR-5 (a special extract from *R. Rosea*). The component rosidirin acts as a MAO-a/b inhibitor; monoamine oxidase (MAO) is an enzyme that degrades dopamine, serotonin and adrenaline (epinephrine). Supplementation with MAO inhibitors is related to temporary increased levels of certain neuropeptides (such as the three just named).

Rhodiola also contains proanthocyanidins (anthocyanidins are the beneficial compound in blueberries) and may be in part the reason why *Rhodiola rosea* exerts such a potent antioxidant effect *in vitro*. Proanthocyanidins exert a powerful antioxidant effect and can be measured via oxygen radical absorbance capacity (ORAC) but they are not causative of the increase in intrinsic antioxidant defenses typical of *R. rosea*. Proanthocyanidins are also quite neuroprotective in nature, but this is also mostly through antioxidant abilities.

It seems that *Rhodiola rosea* is capable of both inducing a stress response, and either the plant itself or a downstream effect of its ingestion protects cells from being overly damaged from this induced stress [43]. Whether piling one adaptogen onto another during exercise causes adaptation to a super-stressor, as each adaptogen has different mechanisms of action, is not known. The pharmacological effects of *R. rosea* are very enticing, as many seem to be able to induce and adapt to stresses (induce stress yet prevent it from becoming excessive and hurting the body, then allow the body to respond beneficially). This may lend some notions as to why the body 'adapts' with adaptogens. The ability of *R. rosea* to exert its own antioxidant protective effects in addition to helping produce some of the body's intrinsic antioxidant systems (primarily the H_2O_2 defense system via the stress response) is interesting, and *R. rosea*'s various protective effects on neurons and cardiac cells from stimulation-induced death via either substrates of *R. rosea* (salidroside) or via heat shock proteins (HSPs) should be of interest to anybody who takes the idea of 'excess' to the extremes.

Research into the adaptogenic effects of the herb has revealed that there are many different species of *Rhodiola*; however, *R. rosea* is the most extensively researched. *R. rosea* can also been found to affect memory abilities, have activity on neurotransmitters in neuronal pathways, suppress inhibition of acetylcholine with age-associated memory loss, and reduce oxidative damage, all of which are consequences of stress [44]. *R. rosea* has been seen to reduce symptoms of physically and psychiatrically induced asthenia and to increase intellectual capacity [44]. It has been shown to improve the effects of tricyclic antidepressants and to decrease their side effects [44]. In a study investigating the antioxidant potential of three adaptogen extracts—*Rhodiola rosea*, *Eleutherococcus senticosus* and *Emblica officinalis* (Indian gooseberry)—it was found that *Rhodiola* had the highest potential for singlet oxygen scavenging, hydrogen peroxide scavenging, ferric reduction, ferrous chelation, and protein thiol protection of the three extracts [44]. *Rhodiola* also exhibited the highest polyphenol content, which may not only have adaptogen properties but may decrease the risk of complications induced by oxidative stress [45].

Research has shown that the consumption of *Rhodiola rosea* for 20 days significantly improved physical fitness, and reduced mental fatigue [46]. The subjects in this study were students who were in the middle of an exam period and so were under a higher than usual level of stress. The *R. rosea* therefore helped their body to adapt and cope with the increased levels of stress.

Further research, looking at the effect of supplementation, with a low-dose of *Rhodiola rosea*, on the mental fatigue of 56 young and healthy medical physicians on night duty [43] found that mental fatigue was significantly reduced following supplementation. This indicates that the *R. rosea* supplement allowed them to better adapt to the increased stress of the situation, and reduced their level of physical and mental fatigue.

Rhodiola rosea appears to be a promising endurance supplement that may enhance fatigue resistance. Following 4 weeks of supplementation, subjects significantly increased their time to exhaustion, from 16.8 to 17.2 minutes, and significantly improved their aerobic capacity (VO_{2max}), from 50.9 to 52.9 mL kg^{-1} min^{-1}[32]. The swim time to exhaustion in rats increased by 139–159% following supplementation [47]. Other positive effects of *R. rosea* supplementation include: protection against stress-induced damage to the heart muscle [48], a protective effect against cancers [49,50], and possibly reduction in liver toxicity [50].

The stress-protective effect of *Rhodiola*, which increased survival of simple organisms and isolated cells in oxidative stress, is not purely associated with its antioxidant or pro-oxidant effects [50–52], because *Rhodiola* can enhance survival against oxidative stress at dose levels that do not elevate the major antioxidant defenses, activate the antioxidant response element, or degrade H_2O_2. The adaptogenic effect of *Rhodiola* root SHR-5 extract have been shown in several double-blind, randomized controlled clinical trials. Orally administrated for 2–6 weeks, dry SHR-5 extract prepared with ethanol–water (ethanol 70% v/v) in daily doses of 288–680 mg (1–4 tablets) has been shown to improve mood [53], cognitive performance and attention [54–56], and relieve fatigue [57] in stress-related conditions. A single-dose effect is achieved in 1–2 hours after administration of *Rhodiola* extracts [58–62]. The adaptogenic effect of *Rhodiola* root water–alcohol extracts have been confirmed in many preclinical studies [63–74].

Numerous *in vitro* and *in vivo* studies on animals have demonstrated CNS-stimulating [75–84] neuro-, cardio- and hepato-protective effects [84], increasing life-span [68], and MAO inhibitory [85], immunotropic [86] and antibacterial activity [78]. Using animal models, bioassay-guided fractionation of various extracts of plant adaptogens has shown that the active principles are mainly phenylpropane and phenyl ethane derivatives, including salidroside, rosavin, syringin, triandrin, tyrosol, etc. Of these, rhodioloside/salidroside and triandrin were reported to be the most active in a number of different test systems [77,78].

Other reviewed studies suggested that increased resistance to nonspecific stress may be due to a serotonergic action, an increase in [beta]-endorphins, and that it moderates opioid peptide, an excess of which may damage the brain and heart. *R. rosea* acts on the neuroendocrine system similarly to other adaptogens and has strong antioxidant properties that can reduce toxicity from drugs. Animal studies have shown that *R. rosea* decreases toxicity from cyclophosphamide, rubomycin and adriamycin (anti-cancer drugs), while it enhances their anticarcinogenic effects [88–90].

Further studies confirm that the anti-aging activity of herbal medicines such as those containing *Rhodiola* may be due to oxygen-scavenging molecules that reduce imbalanced redox reactions and restore defense against free radicals [90]. Induction of iNOS gene expression by *Rhodiola sachalinensis*, leading to NO synthesis, was another proposed mechanism of action [91]. In another study using rabbits, the objective was to ascertain which mediators of stress response are significantly involved in the mechanisms of action of adaptogens and to determine their relevance as biochemical markers for evaluating anti-stress effects. It was suggested that the inhibitory effects of *R. rosea* and *Schisandra chinensis* on phosphorylated kinase p-SAPK/p-JNK activation may be associated with their antidepressant activity as well as their positive effects on mental performance under stress [92]. Additionally, a study to test the effect of *Astragalus* and *Rhodiola* species on noise stress observed a reduction in hepatic glycogen, lactic acid and cholesterol which may be ultimately controlled by the HPA axis as an adaptive response [93]. Finally an anti-inflammatory action was seen as a mediator of adaptation, as levels of C-reactive protein and creatine kinase were reduced in untrained volunteers before exercise in a treatment group. In a recent study of 60 individuals (30 in treatment group, 30 in placebo group), *Rhodiola* was used to treat stress-related fatigue [56]. It was concluded that repeated administration of R. rosea extract SHR-5 exerts an anti-fatigue effect that increases mental performance, particularly the ability to concentrate, and decreases cortisol response to awakening stress in burnout patients with fatigue syndrome [56].

Based on the proposed mechanism of action and available experimental data, *Rhodiola* appears to offer an advantage over other adaptogens in circumstances of acute stress. A single dose of *Rhodiola rosea* (SHR-5) prior to acute stress produces favorable results and prevents stress-induced disruptions in function and performance. Since many stressful situations are acute in nature, and sometimes unexpected, an adaptation that can be taken acutely in these circumstances, rather than requiring chronic advance supplementation, could be potentially very useful. *Rhodiola* also offers some cardio-protective benefits not associated with other adaptogens. Its proposed ability to moderate stress-induced damage and dysfunction in cardiovascular tissue might make *Rhodiola* the adaptogen of choice among patients at higher risk for cardiovascular disease [94]. However, it is important to reproduce and confirm the nonclinical studies and plan for human trials conducted according to good clinical practice.

The clearest emerging indication for *R. rosea* preparation is as a drug as a tonic during convalescence to increase both mental and physical work capacity against a background of fatigue and/or stress. Some animal and preliminary clinical evidence suggests the need for a well-defined range of therapeutic dosage of *Rhodiola*. It may be concluded from the review of evidence presented in this chapter that encouraging support exists for *Rhodiola*'s beneficial effect on cognitive function and fatigue, as demonstrated by numerous nonclinical and several clinical studies. *Rhodiola*'s adaptogenic effect increases attention and endurance in situations of decreased performance caused by fatigue and sensation of weakness, and reduces stress-induced impairments and disorders related to the function of neuroendocrine and immune systems.

References

[1] Lazarev NV. 7th All-Union Congress of Physiology, Biochemistry and Pharmacology; Medgiz, Moscow. 1947;579.
[2] Wagner H, Nörr H, Winterhoff H. Plant adaptogens. Phytomedicine 1994;1:63–76.
[3] Selye H. Studies on adaptation. Endocrinology 1937;21:169–88.
[4] Selye H. Experimental evidence supporting the conception of "adaptation energy". Am J Physiol 1938;123:758–65.
[5] Panossian A, Wikman G, Wagner H. Plant adaptogens III, earlier and more recent aspects and concepts on their mode of action. Phytomedicine 1999;6(4):287–99.
[6] Breckhman II, Dardymov IV. New substances of plant origin which increase nonspecific resistance. Annu Rev Pharmacol 1969;9:419–30.
[7] Baranov AI. Medicinal use of ginseng and related plants in the Soviet Union: recent trends in the soviet literature. J Ethnopharmacol 1982;6:339–53.
[8] Carlini EA. Efeito adaptógeno ou resistógeno de algumas plantas. In: Buchillet D, editor. Medicinas tradicionais e medicina ocidental na Amazônia. Belém: Edições CEJUP; 1989. p. 45–59.
[9] Mills S, Bone K. Principles and practice of phytotherapy. Edinburgh: Churchill Livingstone; 2000.
[10] Sharma PV. Introduction to dravyaguna, i.e. Indian pharmacology. Varanasi, India: Chaukhambha Orientalia; 1976. p. 38–41.
[11] Bhatnagar SS, et al. The wealth of India: a dictionary of Indian raw material and industrial products, vol. 10. New Delhi: Publicity and Information Directorate Council of Social and Industrial Research; 1976582–585.
[12] Trickey R. Women, hormones & the menstrual cycle. 2nd ed. Sydney: Allen & Unwin; 2003.
[13] Braun L, Cohen M. Herbs & natural supplements: an evidence-based guide. 2nd ed. Sydney: Elsevier Australia; 2003.
[14] Mirjalili M, Moyano E, et al. Steroidal lactones from Withania somnifera, an ancient plant for novel medicine. Molecules 2009;14(7):2373–93.
[15] Andrade C, et al. A double-blind, placebo-controlled evaluation of the anxiolytic efficacy of an ethanolic extract of Withania Somnifera. Indian J Psych 2000;42(3):295–301.
[16] Cooley K, et al. Naturopathic care for anxiety: a randomized controlled trial ISRCTN78958974. PLoS ONE 2009;4(8)e6628. doi:10.1371/journal.pone.0006628.
[17] Kour K, et al. Restoration of stress-induced altered T cell function and corresponding cytokines patterns by Withanolide A. Int Immunopharmacol 2009;9(10):1137–44.
[18] Bhattacharya SK, Muruganandam AV. Adaptogenic activity of Withania somnifera: an experimental study using a rat model of chronic stress. Pharmacol Biochem Behav 2003;75(3):547–55.
[19] Malik F, Singh J, Khajuria A, Suri KA, Satti NK, Singh S, et al. A standardized root extract of Withania somnifera and its major constituent withanolide-A elicit humoral and cell-mediated immune responses by up regulation of Thl-dominant polarization in BALBlc mice. Life Sci 2007;80:1525–38.
[20] Bani S, Gautam M, Sheikh FA, Khan B, Satti NK, Suri KA, et al. Selective Thl up-regulating activity of Withania somnifera aqueous extract in an experimental system using flow cytometry. J Ethnopharmacol 2006;107:107–15.
[21] Rasool M, Varalakshmi P. Immunomodulatory role of Withania somnifera root powder on experimental induced inflammation: an in vivo and in vitro study. Vascul Pharmacol 2006;44:406–10.
[22] Kuboyama T, Tohda C, Komatsu K. Neuritic regeneration and synaptic reconstruction induced by Withanolide A. Brit J Pharmacol 2005;144(7):961–71.
[23] Dhuley JN. Adaptogenic and cardioprotective action of ashwaganda in rats and frogs. J Ethnopharmacol 2000;70:57–63.
[24] Sternberg EM, Gold PW. The mind-body interaction in disease. Distance learning module: symptomatology, diagnosis and pathology 2. Sydney: Nature Care College; 2000.
[25] Archana R, Namasivayam A. Anti-stressor effect of Withania somnifera. J Ethnopharmacol 1999;64(1):91–3.
[26] Khare CP. Indian medicinal plants—an illustrated dictionary. New Delhi: First Indian Reprint, Springer (India) Pvt. Ltd.; 2007. pp. 717–18.
[27] Boon-Niermeijer EK, Van den Berg A, Wikman G, Wiegant FAC. Phyto-adaptogens protect against environmental stress-induced death of embryos from the freshwater snail Lymnaea stagnalis. Phytomedicine 2000;7:389–99.
[28] Darbinyan V, Kteyan A, Panossian A, Gabrielian E, Wikman G, Wagner H. Rhaodiola rosea in stress induced fatigue – a double blind cross-over study of a standardized extract SHR-5 with a repeated low-dose regimen on the mental performance of healthy physicians during night duty. Phytomedicine 2000;7:365–71.
[29] Panossian A, Hambartsumyan M, Hovanissian A, Gabrielyan E, Wikman G. The adaptogens Rhodiola and Schizandra Modify the response to immobilization stress in rabbits by suppressing the increase of phosphorylated stress-activated protein kinase, nitric oxide and cortisol drug targets. Insights 1, 39–54, <http://la-press.com/cr_data/files/f_DTI-2-Panossian-et-al_290.pdfS>; 2007.
[30] Perfumi M, Mattioli L. Adaptogenic and central nervous system effects of single doses of 3% rosavin and 1% rhodioloside Rhodiola rosea L. extract in mice. Phytother Res 2007;21:37–43.
[31] Saratikov AS, Krasnov EA. Rhodiola rosea (Golden root). Tomsk: Tomsk University Publishing House; 2004. p. 292.
[32] Spasov AA, Wikman GK, Mandrikov VB, Mironova IA, Neumoin VV. A double-blind, placebo controlled pilot study of the stimulating and adaptogenic effect of Rhodiola rosea SHR-5 extract on the fatigue of students caused by stress during an examination period with a repeated low-dose regimen. Phytomedicine 2000;7:85–9.
[33] Shevtsov VA, Zholus BI, Shervarly VI, Vol'skij VB, Korovin YP, Khristich MP, et al. A randomized trial of two different doses of a SHR-5 Rhodiola rosea extract versus placebo and control of capacity for mental work. Phytomedicine 2003;10:95–105.
[34] Abidov M, Crendal F, Grachev S, et al. Effect of extracts from Rhodiola rosea and Rhodiola crenulata (Crassulaceae) roots on ATP content in mitochondria of skeletal muscles. Bull Exp Biol Med 2003;136:585–7.

[35] Saratikov AS, Krasnov EA. Rhodiola rosea (Golden root). 4th ed. Tomsk: Tomsk University Publishers; 1974. pp. 22–41.
[36] Krasik ED, Morozova ES, Petrova KP, Ragulina GA, Shemetova LA, Shuvaev VP. Therapy of asthenic conditions: clinical perspectives of application of Rhodiola rosea extract (goldenroot). In: Avrutskiy GY, editor. Proceedings modern problems in psycho-pharmacology. Kemerovo City: Siberian Branch of Russian Academy of Sciences; 1970. p. 298–330.
[37] Krasik ED, Petrova KP, Rogulina GA. About the adaptogenic and stimulating effect of Rhodiola rosea extract. In: Avrutskiy GY, editor. Proceedings of all-union and 5th Sverdlovsk area conference of neurobiologists, psychiatrists and neurosurgeons, 26–29 May 1970. Sverdlovsk: SverdlovskPress; 1970. p. 215–7.
[38] Komar VV, Kit SM, Sishchuk LV, et al. Effect of Rhodiola rosea on the human mental activity [Ukrainian]. Farmatsevtichnii Zhurnal 1981;36(4):62–4.
[39] Brichenko VS, Kupriyanova IE, Skorokhodova TF. The use of herbal adaptogens together with tricyclic antidepressants in patients with psychogenic depressions. In: Goldberg ED, editor. Modern problems of pharmacology and search for new medicines, vol. 2. Tomsk: Tomsk University Press; 1986. p. 58–60.
[40] Panossian A, Wikman G, Sarris J. Rosenroot (Rhodiolarosea): Traditional use, chemical composition, pharmacology and clinical efficacy Phytomedicine 2010;17:481–93.
[41] Mikhailova MN. Clinical and experimental substantiation of asthenic conditions therapy using Rhodiolarosea extract. In: Goldsberg ED, editor. Current problems of psychiatry. Tomsk: Tomsk State University Press; 1983. p. 126–7.
[42] Saratikov AS, Krasnow EA. Rhodiola rosea is a valuable medicinal plant. Tomsk: Medical Institute Tomsk; 1987. p. 252
[43] Klein R, Kindscher, K. Botanical medicines with stress adaptogen properties in the ethnobotanical literature: a review. Unpublished manuscript 6/21/03CFIDS, ADHD.
[44] Chen X, Liu J, et al. Salidroside attenuates glutamate induced apoptotic cell death in primary cultured hippocampal neurons of rats. Brain Res 2008;1238:189–98.
[45] Brown RP, Gerbarg PL, Ramazanov Z. Rhodiola rosea: a phytomedicinal overview. Herbal Gram 2002;56:40–52.
[46] Chen Q, et al. Effects of Rhodiola rosea on body weight and intake of sucrose and water in depressive rats induced by chronic mild stress. J Chinese Integr Med 2008;6(9):952–5.
[47] De Bock K, Eijnde BO, Ramaekers M, et al. Acute Rhodiola rosea intake can improve endurance exercise performance. International Journal of Sport Nutrition & Exercise Metabolism 2004;14(3):298–307.
[48] Azizov AP, Seifulla RD. The effect of Elton, leveton, fitoton and adapton on the work capacity of experimental animals. Eksp Klin Farmakol 1998;61:61–3.
[49] Afanas'ev SA, Alekseeva ED, Bardamova IB, et al. Cardiac contractile function following acute cooling of the body and the adaptogenic correction of its disorders. Biull Eksp Biol Med 1993;116:480–3.
[50] Udinstev SN, Schakov VP. The role of humoral factors of regenerating liver in the development of experimental tumors and the effect of Rhodiola rosea extract on this process. Neoplasma 1991;38:323–31.
[51] Udinstev SN, Schakov VP. decrease of cyclophosphamide haematotoxicity by Rhodiola rosea root extract in mice with Ehrlich and Lewis transplantable tumours. Eur J Cancer 1991;27:1182.
[52] Schriner SE, Avanesian A, Liu Y, Luesch H, Jafari M. Protection of human cultured cells against oxidative stress by Rhodiola rosea without activation of antioxidant defenses. Free Radic Biol Med 2009;47:577–84.
[53] Wiegant FA, Surinova S, Ytsma E, Langelaar-Makkinje M, Wikman G, Post JA. Plant adaptogens increase lifespan and stress resistance in C. elegans. Biogerontology 2009;10:27–42.
[54] Wiegant FAC, Limandjaja G, de Poot SAH, Bayda LA, Vorontsova ON, Zenina TA, et al. Plant adaptogens activate cellular adaptive mechanisms by causing mild damage. In: Lukyanova L, Takeda N, Singal PK, editors. Adaptation biology and medicine: health potentials, vol. 5. NewDelhi: Narosa Publishers; 2008. p. 319–32.
[55] Darbinyan V, Aslanyan G, Amroyan E, Gabrielyan E, Malmström C, Panossian A. Clinical trial of Rhodiola rosea L. extract SHR-5 in the treatment of mild to moderate depression. Nordic J Psychiatry 2007;61:2343–8.
[56] Olsson EM, von Scheele B, Panossian AG. A randomised, double-blind, placebo-controlled, parallel-group study of the standardised extract shr-5 of the roots of Rhodiola rosea in the treatment of subjects with stress-related fatigue. Planta Med 2009;75(2):105–12.
[57] Shevtsov VA, et al. A randomized trial of two different doses of a SHR-5 Rhodiola rosea extract versus placebo and control of capacity for mental work. Phytomed 2003;10:95–105.
[58] Schutgents FWG, Neogi P, can Wijk EPA, van Wijk R, Wikman G, Wiegant FAC. The influence of adaptogens on ultraweak biophoton emission: a pilot-experiment. Phytother Res 2009;23:1103–8.
[59] Schutgents FWG, Neogi P, van Wijk EPA, van Wijk R, Wikman G, Wiegant FAC. The influence of adaptogens on ultraweak biophoton emission: a pilot-experiment. Phytother Res 2009;23:1103–8.
[60] Perfumi M, Mattioli L. Adaptogenic and central nervous system effects of single doses of 3% rosavin and 1% salidroside Rhodiola rosea L. extract in mice. Phytother Res 2007;21:37–43.
[61] Panossian A, Wikman G. Evidence-based efficacy of adaptogens in fatigue, and molecular mechanisms related to their stress-protective activity. Curr Clin Pharmacol 2009;4:198–219.
[62] Mattioli L, Funari C, Perfumi M. Effects of Rhodiola rosea L. extract on behavioural and physiological alterations induced by chronic mild stress in female rats. J Psychopharmacol 2008;23:130–42.
[63] Panossian A, Wikman G. Evidence-based efficacy of adaptogens in fatigue and molecular mechanisms related to their stress-protective activity. In: Bonn K, editor. International evidence-based complementary medicine conference, 13–15 March. Armidale: University of New England; 2009. p. 10.
[64] Saratikov, AS. Adaptogenic action of eleutherococcus and golden root preparations. In: Brekhman, II, editors. Adaptation processes and biologically active compounds, 1976. pp. 54–62.
[65] Saratikov AS, Krasnov EA, Chnikina LA, Duvidson LM, Sotova MI, Marina TF, et al. Rhodiolosid, a new glycoside from Rhodiola rosea and its pharmacological properties. Pharmazie 1968;23:392–5.
[66] Aksenova RA, Zotova MI, Nekhoda MF, Cherdintsev SG. Comparative characteristics of the stimulating and adaptogenic effects of Rhodiola rosea preparations. In: Saratikov AS, editor. Stimulants of the central nervous system, vol. 2. Tomsk: Tomsk University Press; 1968. p. 3–12.
[67] Panossian A, Wagner H. Stimulating effect of adaptogens: an overview with particular reference to their efficacy following single dose administration. Phytother Res 2005;19:819–38.
[68] Jafari M, Felgner JS, Bussel II, Hutchili T, Khodayari B, Rose MR, et al. Rhodiola: apromising anti-aging Chinese herb. Rejuvenat Res 2007;10:587–602.
[69] Dieamant GC, Velazquez Pereda MC, Eberlin S, Nogueira C, Werka RM, Queiroz ML. Neuroimmunomodulatory compound for sensitive skin care: in vitro and clinical assessment. J Cosmet Dermatol 2008;7:112–9.

[70] Abidov M, Crendal F, Grachev S, Seifulla R, Ziegenfuss T. Effect of extracts from Rhodiola rosea and Rhodiola crenulata (Crassulaceae) roots on ATP content in mitochondria of skeletal muscles. Bull Exp Biol Med 2003;136:585–7.

[71] Iaremiĭ IN, Grigor'eva NF. Hepatoprotective properties of liquid extract of Rhodiola rosea. Eksp Klin Farmakol 2002;65:57–9.

[72] Qin YJ, Zeng YS, Zhou CC, Li Y, Zhong ZQ. Effects of Rhodiola rosea on level of 5-hydroxytryptamine, cell proliferation and differentiation, and number of neuron in cerebral hippocampus of rats with depression induced by chronic mild stress. Zhongguo Zhong Yao Za Zhi 2008;33:2842–8246.

[73] Wang H, Ding Y, Zhou J, Sun X, Wang S. The in vitro and in vivo antiviral effects of salidroside from Rhodiola rosea L. against coxsackie virus B3. Phytomedicine 2009;16:146–55.

[74] Pooja B, Khanum, F. AS. Anti-inflammatory activity of Rhodiola rosea—"a second-generation adaptogen. Phytother Res 2009;23:1099–102.

[75] Zdanowska D, Skopińska-Rŏżewska E, Sommer E, Siwiski AK, Wasiutynski A, Bany J. The effect of Rhodiola rosea extracts on the bacterial infection in mice. Centr Eur J Immunol 2009;34:35–7.

[76] Sokolov SY, Boyko VP, Kurkin VA, Zapesochnaya GG, Rvantsova NV, Grinenko HA. A comparative study of the stimulant property of certain phenylpropanoids. Khim Pharm Z 1990;24:66–8.

[77] Sokolov SY, Ivashin VM, Zapesochnaya GG, Kurkin VA, Shavlinskiy AN. Study of neurotropic activity of new substances isolated from Rhodiola rosea. Khim Pharm Z 1985;19:1367–71.

[78] Barnaulov OD, Limarenko AY, Kurkin VA, Zapesochnaya GG, Shchavlinskij AN. A comparative evaluation of the biological activity of compounds isolated from species of Rhodiola. Khim Pharm Z 1986;23:1107–12.

[79] Saratikov AS, Krasnov EA, Chnikina LA, Duvidson LM, Sotova MI, Marina TF, et al. Rhodiolosid, a new glycoside from Rhodiola rosea and its pharmacological properties. Pharmazie 1968;23:392–5.

[80] Saratikov A, Marina TF, Fisanova LL. Effect of golden root extract on processes of serotonin synthesis in CNS. J Biol Sci 1978;6:142.

[81] Saratikov AS, Marina TF, Fisanova LL. Mechanism of action of salidrozide on the metabolism of cerebral catecholamines. Vopr Med Khim 1978;24(5):624–8.

[82] Aksenova RA, Zotova MI, Nekhoda MF, Cherdintsev SG. Comparative characteristics of the stimulating and adaptogenic effects of Rhodiola rosea preparations. In: Saratikov AS, editor. Stimulants of the central nervous system, vol. 2. Tomsk: Tomsk University Press; 1968. p. 3–12.

[83] Kurkin VA, Dubishchev AV, Titova IN, Volotsueva AV, Petrova ES, Zhestkova NV, et al. Neurotropic properties of some phytopreparations containing phenylpropanoids. Rastit Resursi 2003;3:115–22.

[84] Wang SH, Wang WJ, Wang XF, Chen WH. Effects of salidroside on carbohydrate metabolism and differentiation of 3T3-L1 adipocytes. Zhong Xi Yi Jie He Xue Bao 2008;2:193–5.

[85] van Diermen D, Marston A, Bravo J, Reist M, Carrupt PA, Hostettmann. K. Monoamine oxidase inhibition by Rhodiola rosea L. roots. J Ethnopharmacol 2009;122:397–401.

[86] Siwicki AK, Skopińska-Rŏżewska E, Hartwich M. The influence of Rhodiola rosea extracts on non-specific and specific cellular immunity in pigs, rats and mice. Centr Eur J Immunol 2007;32:84–91.

[87] Udintsev SN, Schakhov VP. Decrease of cyclophosphamide haematotoxicity by Rhodiola rosea root extract in mice with Ehrlich and Lewis transplantable tumors. Eur J Cancer 1991;27(9):1182.

[88] Borovskaya TG, Fomina TI, Iaremenko KV. A decrease in the toxic action of rubomycin on the small intestine of mice with a transplantable tumor through the use of a Rhodiola extract. Antiobiot Khimioter 1988;33(8):615–7.

[89] Udintsev SN, Krylova SG, Fomina TI. The enhancement of the efficacy of adriamycin by using hepatoprotectors of plant origin in metastases of Ehrlich's adenocarcinoma to the liver in mice. Vopr Onkol 1992;38(10):1217–22.

[90] Ohsugi M, Fan W, Hase K, Xiong Q, Tezuka Y, Komatsu K. Active-oxygen scavenging activity of traditional nourishing-tonic herbal medicines and active constituents of Rhodiola sacra. J Ethnopharmacol 1999;67:111–9.

[91] Seo WG, Pae HO, Oh GS, Kim NY, Kwon TO, Shin MK. The aqueous extract of Rhodiola sachalinensis root enhances the expression of inducible nitric oxide synthase gene in RAW264.7 macrophages. J Ethnopharmacol 2001;76:119–23.

[92] Panossian A, et al. The adaptogens Rhodiola and Schizandra modify the response to immobilization stress in rabbits by suppressing the increase of phosphorylated stress activated protein kinase, nitric oxide and cortisol. Drug Target Insights 2007;1:39–54.

[93] Zhu B-W, Sun Y-M, Yun X, Han S, Piao M-L, Murata Y. Reduction of noise-stress-induced physiological damage by radices of Astragali and Rhodiolae: glycogen lactic acid and cholesterol contents in liver of the rat. Biosci Biotech Biochem 2003;67(9):1930–6.

[94] Maslov LN, Lishmanov IB, Naumova AV, Lasukova TV. Do endogenous ligands of peripheral mu- and delta-opiate receptors mediate anti-arrhythmic and cardioprotective effects of Rhodiola rosea extract? Biull Eksp Biol Med 1997;124:151–3.

CHAPTER

35

Anabolic Training Response and Clinical Implications

Jan Sundell

Department of Medicine, University of Turku, and Turku University Hospital, Turku, Finland

INTRODUCTION

An understanding of basic scientific principles related to resistance training is necessary in order to optimize training responses. Resistance training, adequate nutrition and appropriate rest period to recover from the exercise are needed for muscle and bone growth. Resistance training causes microscopic damage to the muscle cells (catabolism), which in turn are quickly repaired (anabolism). The muscle cells adapt to the extra workload by enlarging (hypertrophy) and recruiting greater numbers of nerve cells to aid contraction [1–2]. Optimal nutrition enhances the anabolic effect of resistance training. The present chapter reviews the anabolic effect of resistance training and nutrition. In addition, suggestions are provided for clinicians as to how to perform effective resistance training for patients with the frailty (sarcopenia and osteoporosis) and metabolic syndromes.

RESISTANCE TRAINING

The goal of resistance training is to progressively overload the musculoskeletal system. Regular progressive resistance training develops the strength and size of muscles [3] and increases bone mass [4] from young male athletes to older women. To receive lasting results, resistance training should be an uninterrupted part of lifestyle. Typically, each large muscle group (for a whole-body resistance training program) is trained for two to three times per week (on nonconsecutive days). One training session per week might be enough to maintain the gained results. However, more-advanced practitioners such as bodybuilders split the training program. A split resistance-training program involves working no more than two or three muscle groups or body parts per day. Muscles are trained once or twice per week and allowed roughly 72 hours to recover. It involves fully exhausting individual muscle groups during a workout, then allowing several days for the muscle to fully recover.

Higher training efficiency is superior to lower for improving muscle strength [5] and bone mineral density [4]. Therefore, each set (usually two to four) or at least the last set(s) of an exercise are performed to fatigue, the state where the subject cannot make one more repetition with good form. Therefore, to specifically increase muscle size (hypertrophy-specific training) medium to heavy loading is needed: 70–80% of one repetition maximum (1RM). When the training goal is muscular hypertrophy, sets with short rest intervals (about 1–2 minutes) are suitable, whereas longer resting periods (several minutes) are needed to increase especially absolute strength [6]. Optimal exercise time is below 45–60 minutes since, thereafter, training intensity reduces significantly. Resistance training under the supervision of a personal trainer usually leads to greater workout intensities [7]. By varying the number of repetitions and length of rest and to some extent the types of exercises over time (periodization), it may be possible to achieve greater benefits.

NUTRITION

In adults, daily protein supply, measured as intake per body weight, is 0.8–1.0 g/kg. However, active

Nutrition and Enhanced Sports Performance.
DOI: http://dx.doi.org/10.1016/B978-0-12-396454-0.00035-7

people and athletes require elevated protein intake. In intensive resistance training, adequate protein supply, from 1.5 to 2 g/kg per day [8], is required for maximal muscle growth. There is no significant evidence for a detrimental effect of this amount of protein intake on kidney function in healthy subjects [9]. However, dietary protein restriction might be necessary by an individual with existing renal disease. Daily protein supply should be divided across several meals (~20 g per meal) [10]. Therefore, it is reasonable to eat at regular intervals and split daily food intake into four to six protein rich meals instead of the traditional three meals a day. In addition, adequate amounts of vitamin D and calcium intake are important for bone growth. Also an adequate supply of carbohydrates is needed as a source of energy. A light, balanced meal prior to the workout ensures that adequate energy and amino acids are available during the intense bout of exercise. Water should be consumed throughout the course of the workout to prevent poor performance due to dehydration.

Whey protein and creatine seem to have positive effects on muscle size, strength and athletic performance without major adverse effects and high costs. However, creatine supplementation should not be used by an individual with existing renal disease. Most studies have shown that supplementation of whey (~15–30 g) alone or with carbohydrates immediately after and possibly before and during resistance exercise can enhance the muscle hypertrophy response to resistance training [11] because both protein uptake and protein usage are increased at this time. Some studies also suggest that whey protein may enhance recovery from heavy exercise and possibly decrease muscle damage and soreness [8]. Creatine has been studied in several clinical trials and has shown benefits including increased muscle strength, endurance and size [12]. In addition, creatine may increase bone mineral content and enhance muscle mass, with potentially greater tension on bone at sites of muscle attachment [13]. The most common creatine monohydrate loading program involves an initial loading phase of 20 g/day for 5–7 days, followed by a maintenance phase of 3–5 g/day for differing periods of time [12].

CLINICAL ASPECTS

The importance of resistance training is well recognized in physiatrics and rehabilitation medicine. In addition, resistance training is currently the most effective known strategy to combat sarcopenia, by stimulating hypertrophy and increasing muscle strength [14]. Based on a randomized controlled clinical trial Strasser et al. [15] concluded that to promote hypertrophy with resistance training in elderly adults, loading intensity should approach 60–80% of 1RM with an exercise volume ranging from three to six sets per muscle group per week of 10–15 repetitions per exercise. Progressive resistance training is an effective and safe tool against sarcopenia even in geriatric subjects [16]. In general, significant ameliorations (up to >50% strength gain) can be expected even with 6 weeks of resistance training at a rhythm of two to three sessions per week. Therefore, from a preventive viewpoint, all elderly subjects should be advised to begin such an exercise program and continue it as long as possible.

Osteoporosis can be thought of as a bone analog of sarcopenia. Many studies have shown that resistance training can maintain or even increase bone mineral density in postmenopausal women [17]. It seems that a combination of high-impact (i.e., jumping) and weight-lifting exercises might be superior for bone stimulation in adults [18]. Chilibeck et al. [13] demonstrated that resistance training, significantly increased whole-body and leg bone mineral density by approximately 0.5–1% in older men as early as 12 weeks. The BEST (Bone-Estrogen Strength Training) [19] project identified six specific resistance training exercises that yielded the largest improvements in bone mineral density. This project suggested squat, military press, lateral pull down, leg press, back extension, and seated row, with three weight-training sessions a week of two sets of each exercise, alternating between moderate (6–8 reps at 70% of 1RM) and heavy (4–6 reps at 80% of 1RM).

At rest, skeletal muscle consumes 54.4 kJ/kg (13.0 kcal/kg) energy per day, which is larger than adipose tissue at 18.8 kJ/kg (4.5 kcal/kg) [20]. Since resistance training increases muscle mass, it does not result in weight loss without caloric restriction. However, resistance training, even without caloric restriction, has a favorable effect on body composition since it decreases fat mass, including abdominal fat [21]. Skeletal muscle is an important determinant of insulin sensitivity. In many studies, resistance training has enhanced insulin sensitivity and improved glucose tolerance [21]. Improved glucose uptake appears not to be a mere consequence of the increased muscle mass associated with resistance training, but seems to be also a result of qualitative changes in resistance-trained muscle [22]. In a meta-analysis, progressive resistance training led to small but statistically significant absolute reductions in glycosylated hemoglobin A_{1c} (HbA_{1c}) of 0.3% in subjects with type 2 diabetes [23]. Bweir et al. [24] demonstrated that resistance training lowers HbA_{1c} by 18% with type 2 diabetes. In this study, the resistance-training group used seven exercises that encompassed knee and hip flexion/extension, shoulder flexion/extension, adduction/

abduction, elbow flexion/extension and a chest press. Three sets of 8–10 repetitions were performed for all exercises with a rest period of 2 minutes between sets in 30–35 minutes three times per week for 10 weeks.

More clinical studies are needed to further investigate the effect of different training protocols, long-term effects, and the role of optimal nutrition in frailty and metabolic syndromes. It is important to compare whole-body and split resistance training programs on these syndromes. Less is also known about the long-term (>6 months) hypertrophic response to resistance training. Therefore, studies with years of training are needed, as the beneficial effects would be even more pronounced. In addition, the effect of nutrition during resistance training has not been studied extensively in these syndromes. It is possible that optimal nutrition would improve the results in clinical studies. Malnutrition is a common feature associated with frailty syndrome. Esmarck et al. [25] demonstrated that immediate intake of a protein supplement following resistance training is more effective than delayed intake for the maximal development of muscle hypertrophy in elderly men. Therefore, nutritional evaluation is needed in future studies. Adequate protein supply (e.g., 1.5–2.0 g/kg per day) and a whey protein shake (15–30 g) immediately following the exercise might increase the beneficial effects of resistance training in frailty syndrome. Finally, subjects with frailty syndrome may benefit from creatine supplementation (3–5 g per day).

References

[1] Russell B, Motlagh D, Ashley WW. Form follows function: how muscle shape is regulated by work. J Appl Physiol 2000;88 (3):1127–32.

[2] Sale DG. Neural adaptation to resistance training. Med Sci Sports Exerc 1988;20(5 Suppl):S135–45.

[3] Charette SL, McEvoy L, Pyka G, Snow-Harter C, Guido D, Wiswell RA, et al. Muscle hypertrophy response to resistance training in older women. J Appl Physiol 1991;70(5):1912–6.

[4] Tsuzuku S, Ikegami Y, Yabe K. Effects of high-intensity resistance training on bone mineral density in young male powerlifters. Calcif Tissue Int 1998;63(4):283–6.

[5] Steib S, Schoene D, Pfeifer K. Dose-response relationship of resistance training in older adults: a meta-analysis. Med Sci Sports Exerc 2010;42:902–14.

[6] de Salles BF, Simão R, Miranda F, Novaes Jda S, Lemos A, Willardson JM. Rest interval between sets in strength training. Sports Med 2009;39(9):765–77.

[7] Ratamess NA, Faigenbaum AD, Hoffman JR, Kang J. Self-selected resistance training intensity in healthy women: the influence of a personal trainer. J Strength Cond Res 2008;22 (1):103–11.

[8] Kreider RB, Campbell B. Protein for exercise and recovery. Phys Sportsmed 2009;37(2):13–21.

[9] Martin WF, Armstrong LE, Rodriguez NR. Dietary protein intake and renal function. Nutr Metab 2005;2:25.

[10] Moore DR, Robinson MJ, Fry JL, Tang JE, Glover EI, Wilkinson SB, et al. Ingested protein dose response of muscle and albumin protein synthesis after resistance exercise in young men. Am J Clin Nutr 2009;89(1):161–8.

[11] Hulmi JJ, Kovanen V, Selänne H, et al. Acute and long-term effects of resistance exercise with or without protein ingestion on muscle hypertrophy and gene expression. Amino Acids 2009;37:297–308.

[12] Rawson ER, Volek JS. Effects of creatine supplementation and resistance training on muscle strength and weightlifting performance. J Strength Cond Res 2003;17:822–31.

[13] Chilibeck PD, Chrusch MJ, Chad KE, Shawn Davison K, Burke DG. Creatine monohydrate and resistance training increase bone mineral content and density in older men. J Nutr Health Aging 2005;9(5):352–3.

[14] Sundell J. Resistance training is an effective tool against metabolic and frailty syndromes. Adv Prev Med 2011; [Article ID 984683, 7 pages].

[15] Strasser B, Keinrad M, Haber P, Schobersberger W. Efficacy of systematic endurance and resistance training on muscle strength and endurance performance in elderly adults--a randomized controlled trial. Wien Klin Wochenschr 2009;121 (23–24):757–64.

[16] Binder EF, Yarasheski KE, Steger-May K, Sinacore DR, Brown M, Schechtman KB, et al. Effects of progressive resistance training on body composition in frail older adults: results of a randomized, controlled trial. J Gerontol A Biol Sci Med Sci 2005;60 (11):1425–31.

[17] Bonaiuti D, Shea B, Iovine R, Negrini S, Robinson V, Kemper HC, et al. Exercise for preventing and treating osteoporosis in postmenopausal women. Cochrane Database Syst Rev 2002;(3): CD000333.

[18] Guadalupe-Grau A, Fuentes T, Guerra B, Calbet JA. Exercise and bone mass in adults. Sports Med 2009;39(6):439–68.

[19] Houtkooper LB, Stanford VA, Metcalfe LL, Lohman TG, Going SB. Preventing osteoporosis the bone estrogen strength training way. ACSMs Health Fit J 2007;11:21–7.

[20] Heymsfield SB, Gallagher D, Kotler DP, Wang Z, Allison DB, Heshka S. Body-size dependence of resting energy expenditure can be attributed to nonenergetic homogeneity of fat-free mass. Am J Physiol Endocrinol Metab 2002;282(1):E132–8.

[21] Tresierras MA, Balady GJ. Resistance training in the treatment of diabetes and obesity: mechanisms and outcomes. J Cardiopulm Rehabil Prev 2009;29(2):67–75.

[22] Holten MK, Zacho M, Gaster M, Juel C, Wojtaszewski JF, Dela F. Strength training increases insulin-mediated glucose uptake, GLUT4 content, and insulin signaling in skeletal muscle in patients with type 2 diabetes. Diabetes 2004;53(2):294–305.

[23] Irvine C, Taylor NF. Progressive resistance exercise improves glycaemic control in people with type 2 diabetes mellitus: a systematic review. Aust J Physiother 2009;55(4):237–46.

[24] Bweir S, Al-Jarrah M, Almalty AM, Maayah M, Smirnova IV, Novikova L, et al. Resistance exercise training lowers HbA1c more than aerobic training in adults with type 2 diabetes. Diabetol Metab Syndr 2009;1:27.

[25] Esmarck B, Andersen JL, Olsen S, Richter EA, Mizuno M, Kjær M. Timing of postexercise protein intake is important for muscle hypertrophy with resistance training in elderly humans. J Physiol 2001;535(1):301–11.

CHAPTER

36

Requirements of Energy, Carbohydrates, Proteins and Fats for Athletes

Chad M. Kerksick[1] *and Michelle Kulovitz*[2]

[1]Health, Exercise and Sports Sciences Department, University of New Mexico, Albuquerque, NM, USA
[2]Department of Kinesiology, California State University - San Bernadino, San Bernadino, CA, USA

ENERGY REQUIREMENTS

Introduction to Energy Needs

The central component of success in sport begins with adequate energy intake to support caloric expenditure and promote the maintenance or improvement in strength, endurance, muscle mass, and health. Athletes consuming a well-designed diet that includes both adequate amounts and proportions of the macronutrients (carbohydrates, proteins, and fat) will promote peak performance [1,2]. Inadequate energy intake relative to energy expenditure will reduce athletic performance and even reverse the benefits of exercise training. The result of limited energy will cause the body to break down fat and lean tissue to be used as fuel for the body. Meanwhile, inadequate blood glucose levels will increase fatigue and perception of exercise effort and ultimately reduce performance. Over time this could significantly reduce strength and endurance performance, as well as compromise the immune system, endocrine, and musculoskeletal function [3]. Additionally, sport-specific energy requirements vary greatly between sports where sport-specific energy needs should be determined, but overall athletes and coaches are highly encouraged to focus upon daily energy intake before concerning themselves too much with optimal intakes of the macronutrients.

Estimating Energy Needs of Athletes and Active Individuals

Estimation of energy needs for active individuals as well as athletes can be done using several resources. Typically in the field, an accessible as well as practical way to estimate energy expenditure of an athlete or active individual is to use prediction equations that have been developed based on assessments of resting metabolic rate and energy cost of physical activity (see Table 36.1) [3]. It is important to keep in mind during assessment that height, weight, age, body composition, and gender will influence caloric expenditure and alter the quantification of daily caloric needs, thus the initial computed outcome from these predictive approaches should be viewed as a general guideline or simply a starting point and not a final and conclusive number. Athletes and coaches should always measure height and weight when utilizing a predictive equation. Ideally, those wanting to quantify their personal resting metabolic rate without the use of a prediction equation can have it assessed using indirect calorimetry. Measuring resting metabolic rate using this preferred approach, however, can be costly to athletes, and it may become difficult to find a credible laboratory or location for all athletes to be measured using standardized conditions (e.g., fasting state, no recent stressful bouts of exercise, refrain from caffeine, alcohol, nicotine, etc.).

Once resting energy expenditure has been estimated using an appropriate prediction equation or measured, the value is then multiplied by the daily total energy expenditure. For simplicity, a physical activity level (PAL) factor is applied in order to average the daily total energy expended (see Table 36.1) and are intended to adjust daily energy intake needs relative to the individual's activity level. Typically, individuals who participate in recreational exercise or an overall fitness program (30 to 45 min/day, 3 to 4

Nutrition and Enhanced Sports Performance.
DOI: http://dx.doi.org/10.1016/B978-0-12-396454-0.00036-9

TABLE 36.1 Mifflin-St Joer [4] and Harris-Benedict [5] Resting Metabolic Rate Prediction Equation and Physical Activity Level (PAL) Factors.

Mifflin-St Joer	
Men	RMR = (9.99 × weight in kg) + (6.25 × height in cm) − (4.92 × age) + 5
Women	RMR = (9.99 × weight in kg) + (6.25 × height in cm) − (4.92 × age) + 161
Harris-Benedict	
Men	RMR = 66.47 + (13.75 × weight in kg) + (5.0 × height in cm)−(6.75 × age)
Women	RMR = 665.09 + (9.56 × weight in kg) + 1.84 × height in cm)−(4.67 × age)
Physical Activity Level (PAL) factors[a]	
1.0−1.39	Sedentary, typical daily living activities (e.g., household tasks, walking to bus)
1.4−1.59	Low active, typical daily living activities plus 30−60 min of daily moderate activity (e.g., walking @ 5−7 km/hour)
1.6−1.89	Active, typical daily living activities plus 60 min of daily moderate activity
1.9−2.5	Very active, typical daily activities plus at least 60 min of daily moderate activity plus an additional 60 min of vigorous activity or 120 min of moderate activity.

[a]*Each factor is associated with a range that is intended to be viewed as a general starting point rather than a specific ending point. Manipulation within each range should be performed and should be performed on a largely individual basis.*

RMR, resting metabolic rate.

From Dietary Reference Intake (DRI) [6] and other sources [1,7].

times/week) do not typically need to alter their daily intake to meet nutritional needs. A typical diet of 25 to 35 kcal kg^{-1} day^{-1}, or approximately 1800 to 2400 calories per day, will likely be sufficient for a recreational athlete because caloric expenditure demands from exercise are not large (i.e., 200 to 400 kcal/session). However, athletes who are involved in moderate levels of exercise training for longer durations (i.e., 2 to 3 hours/day) multiple times per week (5 to 6 times/week), or high-intensity power or resistance training (3 to 6 hours/day) comprised of high-intensity or high-volume training multiple times per week (5 to 6 times/week) can expend 600 to 1200 or more kcal/hour of exercise [8,9].

Energy Needs of Endurance Athletes

Depending on the training schedule and exercise intensity of an endurance athlete, field research has documented hourly caloric expenditure in the range of 600 to 1200 kcal/hour. Consequently, estimated energy needs of such athletes are routinely in the range of 50−80 kcal kg^{-1} day^{-1}[8,9]. This means that depending on body size, a 50−100 kg endurance athlete will need to consume 2500 to 8000 calories per day in order to maintain energy balance to promote optimal endurance training and recovery. Extensive research has investigated the importance of ensuring adequate caloric intake for endurance athletes in order to maintain energy substrate during exercise, for mental function as well as muscular contraction. However, to delay the onset of fatigue from endurance activity, repletion of calories may be necessary during a training bout lasting longer than 60 to 90 minutes. Field research with ultra-endurance athletes recommends caloric intake to range from 100 to 430 calories per hour to maintain force output during exercise for endurance athletes [10].

Energy Needs of Strength and Power Athletes

Energy intake recommendations for strength and power athletes (i.e., sprinters, team sport athletes such as American football or rugby, weightlifters, throwing athletes, and bodybuilders) can vary greatly from those for endurance athletes. Unlike endurance athletes, quantification of caloric expenditure is much harder to determine for strength and power athletes, because of the variability in high-intensity bursts and power, varying lengths of recovery periods from training and competition, and a significant contribution of eccentric contractions which are known to instigate greater muscle damage and compromised recovery [11−13]. Similar to endurance athletes, caloric recommendations should be determined based on individual needs and goals as well as age, height, and weight (see Table 36.1). High-intensity activity requires a high level of energy production, typically followed by periods of rest intervals, which will create periods of high caloric expenditure to periods of recovery. For example, a sprint athlete during a 100-meter dash will perform for approximately 10 seconds or less supra-maximally followed by a recovery period. The ability of the athlete to recover between supra-maximal bouts can influence performance during training or competition. The variability in training volume, duration, and recovery periods adds to the complexity of energy needs and associated global recommendations of energy requirements for these athletic populations. Regardless, ensuring adequate energy balance will optimize force production per active bout, whether it is a sprint or weight workout, and will aid in optimal recovery.

Elite strength and power athletes utilize intermittent bouts of high-intensity force output or high-volume repetitive muscle contractions 3 to 6 hours/day up to 5 to 6 times/week. They can expend 600 to 1200 calories or more per hour of exercise [8,9]. The typical range of caloric expenditure per minute can be from 5.2 to 11.2 kcal/minute [9]. Variability occurs with body size, gender, age, amount of muscle mass activated during the lift, number of sets and repetitions completed, rest periods given, and time the contraction is held. Given the extreme muscularity of most strength athletes and the relationship between amount of muscle mass and total energy expenditure, it is not surprising that the current recommendations for energy intake range from 44 to 50 kcal kg^{-1} day^{-1} [9,14], particularly when one also considers that most of these athletes also seek to induce skeletal muscle hypertrophy, a process which demands even more energy.

Additional Considerations for Optimal Energy Intake

Regardless of athletic type, highly trained athletes who perform multiple bouts of high-volume, moderate- to high-intensity workouts each week have enhanced energy needs. Due to these increased energy demands in combination with other social or sport-specific factors, the athlete may be reticent about ingesting such large quantities of food for fear of the associated changes (perceived or real) to their bodies and physique. These concerns, in addition to the immense logistical planning which must be completed by the athlete and coaches to optimally meet energy needs can result in suboptimal energy intake. As mentioned previously in this section, inadequate energy intake puts the human body in a situation where it must unfavorably allocate various nutrient supplies to meet everyday cellular demand, which in the case of an exercising athlete can result in altered protein metabolism, poor recovery, and other associated outcomes linked to over-reaching/under-recovery. In this respect, the athlete and coach should be readily aware of this possibility and take great measures to ensure adequate energy as well as optimal amounts and ratios of the macronutrients are ingested, a point which will be developed in greater detail in the remaining sections.

CARBOHYDRATES

Structure and Function of Carbohydrates

Particularly in the context of increased energy demands from physical activity, carbohydrates are one of the most important nutrients for an exercising athlete. Carbohydrates serve as the primary fuel for working muscles during exercise, particularly as the intensity of exercise increases [15]. Moreover, carbohydrate in the form of glucose is often viewed as the exclusive fuel source for tissues such as the brain, spinal cord, and red blood cells. Generally speaking, the proportion of carbohydrates in the human diet is recommended to be around 55% of total calories, with an absolute daily requirement of 100–120 grams, but as will be explained in greater detail, the carbohydrate needs for endurance and resistance athletes surrounding workouts have much greater specificity.

Carbohydrate Types and Quality

Carbohydrates are found in the diet as grains, fruits, beans, legumes and dairy products and collectively are comprised of sugar units called saccharides. A common way of categorizing carbohydrates is based upon the number of saccharide units (e.g., mono-, di-, oligo- and polysaccharides) found within the overall carbohydrate molecule. The predominant forms of carbohydrate in the human diet are polysaccharides in the form of starch. This basis has also created a simple but easy to grasp concept of qualitatively assessing the complexity of a carbohydrate whereby mono- and disaccharides are commonly referred to as "simple" sugars and oligo- and polysaccharides are referred to as "complex" carbohydrates. While overly simplistic, this paradigm has meshed well with glycemic index and glycemic load, the most widely accepted means of objectively assessing carbohydrate quality.

Briefly, glycemic index refers to a rating or score assigned to a food that reflects the change in blood glucose which occurs after ingesting a standardized amount of carbohydrate of the food in question, relative to that for an identical amount of a standard test food such as white bread or pure glucose. Importantly, ratings have been established for a wide variety of carbohydrate-containing foods and, even though its application and utility have been met with much confusion and misuse, it remains as both the most recognized and accepted means of evaluating carbohydrate quality. Glycemic load refers to a number assigned to a food or meal that considers both the glycemic index of that particular food and the carbohydrate content of the food in question.

Carbohydrate Recommendations for Endurance Athletes

There is a great range of carbohydrate recommendations for an athlete, which depend largely upon intensity and duration of exercise. According to a recent position statement and other recent review articles, a recommended carbohydrate intake for athletes is 6–10 g kg^{-1} day^{-1} [1,3,8,16,17]. Importantly, as exercise intensity increases, so does the reliance on carbohydrates for energy—research has shown that approximately 50–60% of energy substrate utilization during 1–4 hours of continuous exercise at 70% VO_{2max} is derived from carbohydrates [15]. As endurance training proceeds, energy expenditure does not change, but the reliance on carbohydrate decreases in favor of lipids at any given exercise intensity [15]. Ensuring adequate carbohydrate intake is necessary to guarantee adequate glycogen concentration, and strategies exploiting both the composition and timing of carbohydrate intake can have an effect on glycogen stores within the muscle and liver. Specifically, increasing glycogen stores within the muscle can play an influential role on carbohydrate availability during exercise and subsequent exercise performance.

Utilization of a high-carbohydrate diet in endurance athletes will promote elevated glycogen stores. In endurance sports lasting >90 minutes, it is suggested that super-saturated glycogen stores within the muscle will improve performance for low- to moderate-intensity long-duration exercise. To maximize glycogen refueling in preparation for a race or to maximize recovery following an intense training session, endurance athletes should consume approximately 7–10 g kg^{-1} day^{-1}. Manipulating the timing of carbohydrate intake and type of carbohydrate in preparation for a race or intense training may provide advantages metabolically during the race as well as following the race for refueling. Carbohydrate recommendations for both endurance and strength and power athletes are summarized in Table 36.2, and subsequent sections will further detail strategies to meet carbohydrate requirements surrounding a workout or competitive bout.

Carbohydrate Recommendations for Strength and Power Athletes

Consuming adequate carbohydrates for strength and power athletes is vital for optimal power output and overall performance. Intense intermittent muscle contractions lasting 1–5 minutes in duration, using exercises that recruit large masses of muscle, combined with short rest intervals can decrease glycogen stores by 24–40% [19–22]. Certainly the magnitude of muscle glycogen depletion depends on the intensity, duration, and amount of muscle mass that is recruited during the training session. It is commonly recommended that strength and power athletes who utilize training regimens that include high repetitions with a moderate to high level of resistance to maximize both strength and power adaptations as well as muscle hypertrophy will deplete greater concentrations of glycogen. For these reasons, an intake of 5–10 g kg^{-1} day^{-1} is sufficient to maintain optimal glycogen stores in strength and power athletes [18].

Carbohydrate Intake for Pre-Training/Pre-Competition

The ideal pre-competition meal should contain 150 to 300 grams of carbohydrate (3 to 5 g kg^{-1} body weight) approximately 3 to 4 hours prior to exercise. This amount consumed prior to exercise will maximize muscle and liver glycogen stores and help to sustain blood glucose concentrations throughout prolonged bouts of moderate- to high-intensity exercise [23]. Additional considerations for the pre-exercise meal include food choices that contain little fat and fiber, to maximize gastric emptying and minimize gastric upset.

Carbohydrate Intake During Exercise

Moderate- to high-intensity exercise is characterized by high oxidation rates of carbohydrate whereby such values have commonly been reported to be in the

TABLE 36.2 Average Macronutrient Requirements for Athletes[a].

	Endurance Athletes	Strength Athletes
Carbohydrates[b]	6–10 g kg^{-1} day^{-1}	3.9–8.0 g kg^{-1} day^{-1}
Protein[b]	1.2–1.4 g kg^{-1} day^{-1}	1.2–1.7 g kg^{-1} day^{-1}
Fat	20–30% of Total Energy Intake (10% saturated, 10% polyunsaturated, 10% monounsaturated)	20–30% of Total Energy Intake (10% saturated, 10% polyunsaturated, 10% monounsaturated)

[a] *Variability depends on sport or mode, intensity, duration, and skill of the athlete.*
[b] *kg represents kilogram body weight.*
Adapted from Genton et al [18], The Institute of Medicine Guidelines 2005 [9], and The ADA/ACSM Position on Nutrition and Athletic Performance [1,17].

order of 1.0–1.2 grams of carbohydrate per minute (60–72 grams per hour) of exercise [24,25]. At these rates, high-intensity endurance exercise (e.g., >70% VO_{2max}) that lasts approximately 1 hour can exhaust liver glycogen stores and significantly deplete muscle glycogen stores in as little as 2 hours. For these reasons, optimal repletion of carbohydrates and energy is vital to continue exercise and/or maintain force output. According to research done with endurance athletes, it is recommended that 60 grams or 0.5–1.0 g kg^{-1} of liquid or solid carbohydrates be consumed each hour of moderate- to higher-intensity endurance exercise lasting longer than 1 hour [3,16]. Moreover, decades of sport nutrition research tells us that glucose-electrolyte solutions which deliver carbohydrate concentrations of 6–8% carbohydrate (6–8 grams of carbohydrate per 100 mL of fluid) offers the ideal balance between non-episodic gastric emptying and efficient energy delivery [3,24]. These solutions are recommended to be ingested every 15 to 30 minutes, which effectively provides a continual supply of carbohydrate to the working muscles. A host of positive effects arise from this strategy, including an optimal maintenance of blood glucose levels which aids in preventing common hypoglycemic symptoms such as headaches, lightheadedness, nausea, and muscular fatigue while also delivering a preferred fuel source which can be rapidly oxidized in favor of limited glycogen stores located in the liver and muscle. This feeding strategy has been shown in a number of studies and recent reviews to minimally maintain and likely have ergogenic benefits [1,3,8,26]. Finally, and while most of this research has used endurance modes of exercise, a number of studies are also available demonstrating that providing a glucose-only beverage or a combination of carbohydrate and protein or amino acids favorably impacts performance, muscle damage, and recovery [27–30].

Carbohydrate Intake into Recovery

The extent to which carbohydrate intake should be considered depends largely upon the duration and intensity of exercise, but an equally important factor is the time available for recovery to take place. A number of strategies including but not limited to the glycemic index of the carbohydrates being consumed, adding protein to carbohydrate, and adding caffeine have been examined for their ability to favorably influence both the rate and extent to which recovery of lost muscle glycogen occurs [3,31,32]. Collectively, these studies indicate that the single most important variable to optimize recovery of lost muscle glycogen is the absolute amount of carbohydrate intake [3,31]. Table 36.2 highlights specific recommendations regarding carbohydrate intake.

Briefly, carbohydrate intake following an exercise bout should begin immediately, to take advantage of favorable hormonal environments upon which timely nutrient administration can both facilitate recovery of lost glycogen and minimize muscle protein breakdown. As duration, intensity, or both increases, carbohydrate intake should also increase. For moderate-intensity exercise lasting 45 minutes to 1 hour, daily carbohydrate intake of 5–7 g kg^{-1} body weight day^{-1} is necessary. For moderate exercise lasting one to three hours, it is recommended that athletes consume 7–10 g kg^{-1} day^{-1}, while exercise lasting 4–5 hours or greater should consume 10–12 or more g kg^{-1} day^{-1} (see Table 36.2). The timing and amounts of carbohydrate ingested take on an even higher level of importance if time is short between the end of the exercise bout and commencement of subsequent bouts: e.g., for extremely long training sessions (4–8 hours) or multiple training sessions or competitions per day [3,26]. Generally speaking, if an exercise bout consists of moderate-intensity exercise spanning 30–45 minutes, carbohydrate replacement should not be a critical consideration.

Low-Carbohydrate High-Protein Diet: Is it a Good Idea for Athletes?

Popularity of higher-protein lower-carbohydrate diets has grown in our society, which can have potentially negative complications for some athletes. Athletes and coaches need to understand the appropriate energy intake for athletes because of the direct relationship it has with sport-specific energy substrate distribution that can help or hurt performance.

As previously mentioned, the current recommendation for carbohydrate intake for endurance and strength athletes is anywhere between 5 and 12 g kg^{-1} day^{-1} depending upon the intensity and duration of exercise. Because endurance athletes, especially, rely on glucose in the form of glycogen as a main energy source during endurance exercise, low blood glucose can cause symptoms such as mental fatigue or muscular fatigue where the athlete feels lethargic or tired, which will dramatically decrease force output as well as decrease the amount of time they can perform exercise. For strength and power athletes, the use of a low-carbohydrate diet will decrease the amount of force that can be exerted per muscle contraction, which can decrease strength performance. Other symptoms include changes in mood, constipation, headache, and dehydration. For a typical non-athlete a minimal intake of 150 grams of carbohydrate is recommended per day, while athletes require much more. If you are an athlete, or a coach with an athlete, who has these symptoms, increasing carbohydrate intake may be advantageous to performance.

PROTEIN

Structure and Function of Proteins

Proteins, carbohydrates, and fats are the three nutrients ingested in the human body that have the potential to produce energy for the body to perform various types of work. Proteins are distinguished from carbohydrates and fats by the presence of an amino or amine ($-NH_2$) group, which creates the framework for how dietary status and protein needs have evolved. While much of the focus for protein, particularly in the context of exercise and performance, seems to center upon muscle protein and its balance, nearly every one of the human body's 100 trillion cells is composed of various proteins. Thus, they are ubiquitous and function in numerous capacities within the physiology and biochemistry of the human body. The current recommended dietary allowance for protein is $0.8\,g\,kg^{-1}\,day^{-1}$ [17].

Essentiality of Amino Acids

Proteins are composed at the individual level of amino acids, and approximately 20 amino acids are used by the body to build proteins. Unlike carbohydrates or fat, no reservoirs of protein exist in the human body, but protein exists throughout the body as pools of amino acids. These pools are in a constant ebb and flow based largely upon physiological supply and demand [33,34]. This ongoing and dynamic state of amino acid movement highlights the importance of dietary intake of protein as well as the concepts of essentiality and protein completeness. Of the 20 amino acids used to build protein, the essential amino acids cannot be produced by the body, which creates an absolute requirement of their intake in the diet. The nonessential amino acids are subsequently considered as such because they can be made in vast amounts inside the human body. Finally, some amino acids are considered to be conditionally essential, which means that in a normal physiological setting, the body is able to produce adequate amounts, but if the body becomes stressed or physiologically challenged, the production rates become inadequate. In this respect, the indispensable or essential amino acids are histidine, isoleucine, leucine, lysine, methionine, phenylalanine, threonine, tryptophan, and valine). Optimization of human performance places a great deal of focus upon the balance of proteins found within skeletal muscle, and it is worth mentioning that studies indicate an absolute requirement exists for the essential amino acids to maximize muscle protein synthesis [35,36].

Protein Type and Quality

A number of protein types exist which are available, and a complete discussion of each is beyond the scope of this chapter. Within the framework of protein requirements, the notion of protein quality and therein protein completeness will be discussed. A number of ways exist for protein quality to be assessed, and the reader is referred to excellent summaries by Phillips [2] and Rodriguez [37]. Briefly, the protein digestibility corrected amino acid scores (PDCAAS) and protein efficiency ratios (PER) are commonly considered methods to assess protein quality. Using these approaches and all other approaches, the milk proteins (whey and casein) are typically rated as one of the highest qualities of proteins available. However, protein sources from egg, beef, poultry, fish, and other dairy sources should still be viewed as excellent sources of protein.

A complete protein is any protein source that provides all of the essential amino acids in both the correct amounts and proportion to stimulate and support the synthesis of new proteins [2]. In this light, incomplete protein sources fail to provide at least one (or more) of the essential amino acids in the correct amount and proportion. Moreover, even protein sources which lack only one amino acid in adequate amounts are viewed to be incomplete (e.g., most versions of soy protein lack adequate required amounts of methionine). For this reason, complete protein sources are considered to be of higher quality, and dietary protein sources of animal origin (e.g., egg, milk or dairy, and flesh proteins such as fish, poultry, beef, pork, bison, etc.) are broadly classified as complete protein sources. Protein sources derived from plants or vegetables are commonly void of one or more of the essential amino acids and must be combined with complementary incomplete protein sources to produce a complete protein.

Protein Requirements of Endurance Athletes

As with other nutrients, "blanket" or "cookie-cutter" recommendations are not appropriate for the dietary protein requirements of endurance athletes. Certainly, the provision of recommendations to multiple athletes lends itself to overgeneralization, but the diligent athlete, coach, or practitioner will closely evaluate and consider other important factors such as training status, exercise intensity, workout duration, gender of the athlete, and dietary energy and carbohydrate intake. In this regard, a prudent approach to recommendations was adopted by Tarnopolsky [38] where he classified athletes as either recreational athletes (those predominantly performing low- to moderate-intensity endurance exercise), modestly trained athletes, and top sport or elite endurance athletes.

For recreational athletes, a number of previously published reports have reached a consensus that this amount and intensity of endurance exercise does not appear to markedly alter the balance of protein or amino acids throughout the body, particularly when energy intake is adequate [6,38–42]. For example, El-Khoury and colleagues determined that a protein intake of 1.0 g kg^{-1} day^{-1} in young men performing two 90-minute bouts of exercise at 50% VO_{2max} yielded a neutral nitrogen balance [43]. Extensions of this work showed that, when additional protein was provided in the diet, increases in leucine oxidation (an indicator of excess protein intake) occurred [44,45].

The report by Tarnopolsky highlighted three studies which examined protein intake needs of modestly trained endurance athletes [38]. Meredith and investigators had younger (27 years, $VO_{2max} = 65$ mL kg^{-1} min^{-1}) and middle-aged men (52 years, $VO_{2max} = 55$ mL kg^{-1} min^{-1}) consume three different protein intakes (0.61, 0.92, and 1.21 g kg^{-1} day^{-1}) and reported that a protein intake of 0.94 g kg^{-1} day^{-1} resulted in a zero net balance of protein [46]. Additionally, Phillips determined nitrogen balance in a group of well-trained men and women who consumed a diet that contained 0.86 g kg^{-1} day^{-1} protein, which resulted in a net negative balance of protein [47]. Of particular interest, additional analyses revealed that these subjects were in energy balance. Finally, a diet which contained a protein intake of 1.0 g kg^{-1} day^{-1} was insufficient to prevent a net negative balance of protein [48]. Collectively, it can be concluded that in modestly trained endurance athletes, independent of gender, protein intake ranging from the current recommended daily allowance (RDA) of 0.86 g kg^{-1} day^{-1} up to 1.0 g kg^{-1} day^{-1} is inadequate to prevent a net loss of body protein [38].

A small collection of studies has examined the protein requirements and protein metabolism of top sport endurance athletes. Tarnopolsky and colleagues completed a nitrogen balance experiment in six elite male endurance athletes and determined a protein intake of 1.6 g kg^{-1} day^{-1} was needed [49]. Friedman and Lemon also determined nitrogen balance in a group of five elite endurance runners and concluded a protein intake of 1.49 g kg^{-1} day^{-1} was advised [50]. A Tour de France simulation by Brouns in well-trained cyclists yielded a protein requirement of 1.5–1.8 g kg^{-1} day^{-1} [51,52]. Furthermore, a randomized crossover approach using highly trained ($VO_{2max} = 70.6 \pm 0.1$ mL kg^{-1} min^{-1}) male endurance runners, where they consumed three diets providing varying amounts of dietary protein (low-protein 0.8, moderate-protein 1.8, high-protein 3.6 g kg^{-1} day^{-1}) determined that nitrogen balance was negative during the low intake of protein and was positive during both the moderate- and high-protein intakes. Additionally, markers of excessive protein intake and oxidation were evident for the high-protein intake. The authors concluded that a protein intake of 1.2 g kg^{-1} day^{-1} was needed to achieve a positive net protein balance [7].

In summary, the protein needs of endurance-training athletes depend upon many factors, including volume, intensity and duration of training, gender of the athlete, energy and carbohydrate intake, and the current fitness level or training status of the athlete. Of these factors, it is important to highlight the impact of gender on dietary protein needs. In this respect, Tarnopolsky used data from six published reports to clearly highlight that the protein requirements for men are approximately 25% greater than those for women.

It remains that recreational athletes performing low- to moderate-intensity endurance activity do not have increased requirements for dietary protein, whereas modestly trained athletes may have a 25% increase in protein needs (to ~1.1 g kg^{-1} day^{-1}) [38]. Only elite athletes or those with exemplary fitness status and who are performing extremely high volumes of training exhibit markedly increased protein requirements, which amount to approximately 1.6 g kg^{-1} day^{-1}.

Protein Requirements of Strength and Power Athletes

Acute resistance exercise increases rates of both muscle protein synthesis [53–56] and muscle protein breakdown [53–55]. If food intake is absent, net protein balance remains in a negative state [53,54]. While ingestion of carbohydrate helps to attenuate changes in muscle protein breakdown, which improves overall net protein balance [57–59], ingestion of protein and/or amino acids alone or in combination with carbohydrate is required to yield a net positive protein balance [58,60–63]. Increases in lean body mass are the result of chronic resistance training and provision of amino acids and/or protein which results in a robust increase in net protein balance. Additionally, regularly resistance training invokes additional sources of stress and trauma that in turn require greater amino acid/protein availability to repair any ultrastructural damage which may be occurring secondary to the resistance training. This theoretical framework suggests that resistance training athletes would have an increased requirement of dietary protein. Indeed, studies which have directly compared the protein requirements of individuals habitually performing resistance training with those of sedentary individuals indicate that protein needs are in fact greater [49,64,65].

As with endurance training, a number of factors impact protein turnover, but training status appears to greatly impact the efficiency with which the body processes protein [54,66]. For example, untrained or unaccustomed individuals who begin performing resistance training almost universally have been shown to have increased requirements of dietary protein. But as the resistance training becomes habitual, the efficiency with which the body handles or processes its protein stores goes up, and several published reports indicate that more-trained individuals have lesser dietary protein requirements [54,66]. As a result, no consensus exists on whether resistance exercise increases protein requirements, largely because of concerns over which is the best method to assess protein requirements [67].

Tarnopolsky and colleagues estimated the protein requirements for American football and rugby players by comparing protein turnover rates using a combination of nitrogen balance and kinetic measurements after the athletes consumed diets which contained a low, moderate, or high amount of protein: 0.86, 1.4, or 2.4 g kg^{-1} day^{-1}, respectively. The authors concluded that the lowest intake of protein compromised protein synthesis when compared with the diets which provided moderate and high amounts of protein [64]. Other studies which determined dietary protein needs in resistance-trained athletes to be in the range 1.4–1.7 g kg^{-1} day^{-1} [64,65] are commonly reported in the literature to justify increased protein intakes, but criticism regarding utilization of the nitrogen balance approach has precluded full acceptance of these recommendations [67]. While controversy abounds, a well-crafted position statement from the International Society of Sports Nutrition (http://www.sportsnutritionsociety.org), a professional organization devoted to the professional advancement of the field of sport nutrition, recommend protein intakes for resistance-training athletes to be in the range of 1.4–2.0 g kg^{-1} day^{-1} [68], and a recent consensus statement from the American Dietetic Association (http://www.eatright.org), American College of Sports Medicine (http://www.acsm.org), and Dietitians of Canada (http://www.dietitians.ca) recommended a general protein intake of 1.2–1.7 g kg^{-1} day^{-1} [1,17]. Moreover, a regression approach of nitrogen balance data from studies of people who were undergoing regular resistance training was employed by Phillips to determine an optimal daily intake of dietary protein [67]. Using this approach, he concluded that on average these athletes required approximately 49% more protein than the current RDA, or a value of 1.19 g kg^{-1} day^{-1}. When

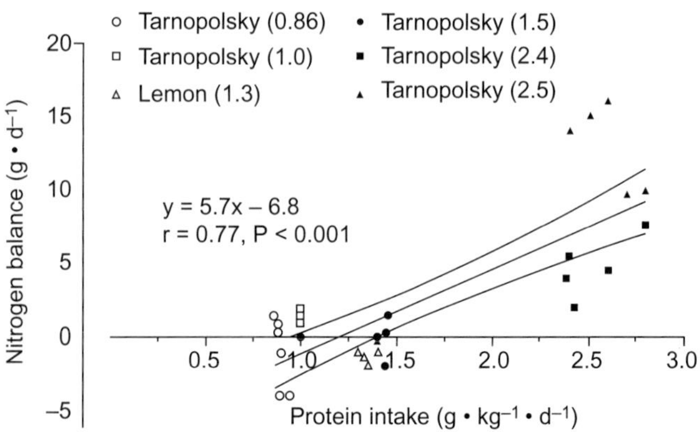

FIGURE 36.1 **A regression approach of nitrogen balance data indicating estimated protein requirements.** *Figure used with permission from Phillips [67].*

a 95% confidence interval was included, a protein intake of 1.33 g kg^{-1} day^{-1} was determined (see Figure 36.1).

In conclusion, optimal protein intake is a key factor to facilitate optimal adaptations and recovery from stressful exercise, regardless of whether it is endurance or resistive in nature. Studies clearly indicate that, when athletes are regularly performing exercise of appropriate intensity, volume and duration, protein needs are increased. The extent to which these values are increased invites controversy, but recent position statements and review papers reveal that a protein intake of 1.2–2.0 g kg^{-1} day^{-1} should more than capture the protein needs of exercising athletes. Outside of these general guidelines, specific considerations should be made after closely assessing key factors such as energy and carbohydrate intake, gender (particularly for endurance athletes), and training status. Finally, analyses of nutritional intakes of athletes routinely and almost unanimously indicate that athletes do an exceptional job of consuming even these greater recommended amounts of protein (see Figure 36.2) [67]. While these data provide sound evidence with which to advise athletes against protein supplementation, other factors such as optimal nutrient timing, and leucine and essential amino acid intake during acute or immediate recovery, will continue to fuel these recommendations.

FATS

Structure and Function of Fats

Dietary lipids are often long hydrocarbon chains, highly insoluble in water and, like carbohydrates, contain carbon, hydrogen and oxygen. Lipids are primarily incorporated into bodily tissues as triglycerides in adipose tissue. Triglycerides consist of two primary components: a glycerol backbone and three fatty acids.

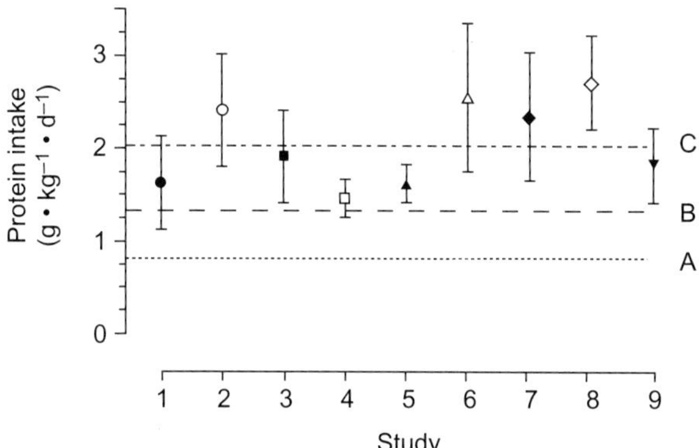

FIGURE 36.2 **Reported habitual protein intakes in resistance-trained athletes.** Line A is current RDA (0.8 g kg^{-1} day^{-1}). Line B is an extrapolated "safe" protein requirement (1.33 g kg^{-1} day^{-1}). Line C is mean reported protein intake (2.05 g kg^{-1} day^{-1}). Values are means ± standard deviation. *Figure adapted from Phillips [67] with permission.*

Glycerol production typically derives through its incorporation into the glycolytic process, where fatty acids can range from short-chains to the most common long-chains, typically 16–20 carbons in length. As nutrients, lipids or fats are the most energy dense and serve primary functions as insulator, energy supply, backbone of cholesterol and fat-soluble vitamins, and key components of cell membranes and nervous tissue. The American Dietetic Association recommends that 30% of total calories come from dietary fat.

Fat Types and Quality

A number of different types of fat exist. The primary means of differentiation is the degree of saturation whereby fully saturated fatty acids contain no double bonds, and mono- and polyunsaturated fatty acids contain one or more than one double bond, respectively. Nutritional epidemiology studies indicate that higher dietary intakes of saturated fatty acids and cholesterol exert negative impacts on cardiovascular health whereas mono- and polyunsaturated fatty acids exert more healthful impacts. Other classes of lipids include the phospholipids, which are incorporated into several different aspects of membrane structures, and the lipoproteins, which are a class of lipids with varying amounts of lipid, cholesterol and protein. Due to their insoluble nature, lipids must be incorporated into lipoproteins to allow for their transport in the highly aqueous medium of the blood. Finally, lipids incorporated into sterols, of which cholesterol and its esters are the most predominant, have central roles in sterol production.

Fat Requirements of Athletes

Increasing the proportion of dietary fat has been a dietary strategy employed by athletes. The rationale for this approach is based primarily upon enhancing endogenous stores of intramuscular triglycerides, which, as the theory goes, should improve prolonged exercise performance while preserving glycogen stores [18]. This theory and approach has been considered by endurance athletes to enhance performance of this type of exercise. In contrast, little consideration of modifying fat intake has been considered by strength and power athletes. While high-fat diets are employed by individuals interested in maximizing leanness as part of the sport of bodybuilding, these considerations lack the necessary relevance to enhancement of sporting performance and for this reason won't be discussed in this chapter. Burke and investigators appeared to offer exciting data after they demonstrated that 5 days of a high-fat (~65% of total calories from fat) and low carbohydrate (2.5 grams carbohydrate per kilogram body mass per day) enhanced fat utilization and still allowed the athletes to complete high-intensity/high-volume training [69]. Further, the increases in fat utilization even persisted after a carbohydrate loading protocol was followed and muscle glycogen levels were replenished. Collectively, this diet manipulation provided indications that high fats followed by a carbohydrate loading could create a favorable scenario where skeletal muscle was able to oxidize more fat while also having a plentiful supply of muscle glycogen. Subsequent research, however, failed to demonstrate increases in exercise performance [70], and in fact rates of muscle glycogen utilization were found to be reduced throughout the exercise bout [71]. When one considers that greater availability of carbohydrate should facilitate enhanced power production and exercise intensity, particularly towards latter periods of a prolonged exercise bout (i.e., the "kick" or "final push"), these findings were considered to be counterproductive.

While a good bit of research has been conducted to examine the efficacy of high-fat diets, a consensus appears to exist that increasing the proportion of dietary fat is not a recommended strategy for enhancement of sport performance. Johnson completed an excellent review of the literature and stated the following regarding the impact of a high-fat diet on physical activity performance: (i) no definitive conclusions can be drawn to indicate that depletion of intramuscular triglycerides negatively impacts performance (one of the underlying theories suggested to improve performance); (ii) high-fat diet consisting of >46% of total calories as fat and <21% of total calories as carbohydrate stimulates fat oxidation through mechanistic adaptations including increased fatty acid oxidation enzymes as well as enhancements of both fatty acid transport and beta-oxidation; and (iii) exercise performance was not improved and in some cases it was negatively impacted [18].

Increasing dietary fat intake has been suggested to favorably impact substrate utilization, but negative performance outcomes and reports of both reduced carbohydrate utilization and gastrointestinal upset have led to the consensus that high-fat diets are not recommended. Irrespective of whether the high intake of dietary fat or the likely concomitant reduction in dietary carbohydrate is responsible for untoward outcomes, the practice of high-fat diets is not advised. A consensus statement by the American College of Sports Medicine, American Dietetic Association, and Dietitians of Canada advise a fat intake of 20–35% of total calories from fat [1,17]. These groups further advised that diets with less than 20% of total calories from fat do not benefit performance, since fat is a source of energy and necessary for production of both fat-soluble vitamins and essential fatty acids.

CONCLUSIONS

Optimal exercise performance is first predicated upon adequate energy intake that will allow for effective delivery of required fuels not only during the energy-demanding exercise period, but also during needed recovery. Ensuring appropriate energy intake is a critical first step for any athlete who desires to achieve optimal performance. Carbohydrates are the preferred, but substantially limited, source of fuel used by nearly every type of athlete who performs at high levels of their ability. Aggressive strategies must be employed to ensure optimal carbohydrate is available. Equally important is meeting the required demands for dietary protein, which facilitates optimal recovery. Dietary needs and requirements for both carbohydrate and protein are increased in nearly every exercising population. Lastly, adequate fat intake is also important and, while fat loading or high-fat diets have proven to be ineffective, a diet which provides 20–30% of its total calories from fat is recommended to ensure optimal fat intake is achieved.

References

[1] American Dietetic Association, Dietitians of Canada, American College of Sports Medicine, Rodriguez NR, Di Marco NM, Langley S. American college of sports medicine position stand. Nutrition and athletic performance. Med Sci Sports Exerc 2009;41(3):709–31.

[2] Phillips SM, Tang JE, Moore DR. The role of milk- and soy-based protein in support of muscle protein synthesis and muscle protein accretion in young and elderly persons. J Am Coll Nutr 2009;28(4):343–54.

[3] Burke LM, Loucks AB, Broad N. Energy and carbohydrate for training and recovery. J Sports Sci 2006;24(7):675–85.

[4] Mifflin MD, St Jeor ST, Hill LA, Scott BJ, Daugherty SA, Koh YO. A new predictive equation for resting energy expenditure in healthy individuals. Am J Clin Nutr 1990;51(2):241–7.

[5] Harris JA, Benedict FG. A biometric study of basal metabolism in men. Publication no. 279. Washington, DC: Carnegie Institute of Washington; 1919.

[6] Young VR, Torun B. Physical activity: impact on protein and amino acid metabolism and implications for nutritional requirements. Prog Clin Biol Res 1981;77:57–85.

[7] Gaine PC, Pikosky MA, Martin WF, Bolster DR, Maresh CM, Rodriguez NR. Level of dietary protein impacts whole body protein turnover in trained males at rest. Metabolism 2006;55(4):501–7.

[8] Kreider RB, Wilborn CD, Taylor L, Campbell B, Almada AL, Collins R, et al. ISSN exercise & sport nutrition review: research & recommendations. J Int Soc Sports Nutr 2010;7:7.

[9] Otten J, Pitzi Hellwig J, Meyers L, editors. Dietary references intakes: the essential guide to nutrient requirements. Washington DC: The National Academies Press; 2006.

[10] Fallon KE, Broad E, Thompson MW, Reull PA. Nutritional and fluid intake in a 100-km ultramarathon. Int J Sport Nutr 1998;8(1):24–35.

[11] Widrick JJ, Costill DL, McConell GK, Anderson DE, Pearson DR, Zachwieja JJ. Time course of glycogen accumulation after eccentric exercise. J Appl Physiol 1992;72(5):1999–2004.

[12] Costill DL, Pascoe DD, Fink WJ, Robergs RA, Barr SI, Pearson D. Impaired muscle glycogen resynthesis after eccentric exercise. J Appl Physiol 1990;69(1):46–50.

[13] Kerksick CM, Kreider RB, Willoughby DS. Intramuscular adaptations to eccentric exercise and antioxidant supplementation. Amino Acids 2010;39(1):219–32.

[14] Slater G, Phillips SM. Nutrition guidelines for strength sports: sprinting, weightlifting, throwing events, and bodybuilding. J Sports Sci 2011;29(Suppl. 1):S67–77.

[15] Coyle EF, Jeukendrup AE, Wagenmakers AJ, Saris WH. Fatty acid oxidation is directly regulated by carbohydrate metabolism during exercise. Am J Physiol 1997;273(2 Pt 1):E268–75.

[16] Jeukendrup AE. Nutrition for endurance sports: marathon, triathlon, and road cycling. J Sports Sci 2011;29(Suppl. 1):S91–9.

[17] Rodriguez NR, Di Marco NM, Langley S. American college of sports medicine position stand. Nutrition and athletic performance. Med Sci Sports Exerc 2009;41(3):709–31.

[18] Genton L, Melzer K, Pichard C. Energy and macronutrient requirements for physical fitness in exercising subjects. Clin Nutr 2010;29(4):413–23.

[19] Haff GG, Koch AJ, Potteiger JA, Kuphal KE, Magee LM, Green SB, et al. Carbohydrate supplementation attenuates muscle glycogen loss during acute bouts of resistance exercise. Int J Sport Nutr Exerc Metab 2000;10(3):326–39.

[20] Pascoe DD, Costill DL, Fink WJ, Robergs RA, Zachwieja JJ. Glycogen resynthesis in skeletal muscle following resistive exercise. Med Sci Sports Exerc 1993;25(3):349–54.

[21] Pascoe DD, Gladden LB. Muscle glycogen resynthesis after short term, high intensity exercise and resistance exercise. Sports Med 1996;21(2):98–118.

[22] Robergs RA, Pearson DR, Costill DL, Fink WJ, Pascoe DD, Benedict MA, et al. Muscle glycogenolysis during differing intensities of weight-resistance exercise. J Appl Physiol 1991;70(4):1700–6.

[23] Coyle EF, Coggan AR, Hemmert MK, Lowe RC, Walters TJ. Substrate usage during prolonged exercise following a preexercise meal. J Appl Physiol 1985;59(2):429–33.

[24] Jeukendrup AE. Carbohydrate intake during exercise and performance. Nutrition 2004;20(7–8):669–77.

[25] Wallis GA, Dawson R, Achten J, Webber J, Jeukendrup AE. Metabolic response to carbohydrate ingestion during exercise in males and females. Am J Physiol Endocrinol Metab 2006;290 (4):E708–15.

[26] Kerksick C, Harvey T, Stout J, Campbell B, Wilborn C, Kreider R, et al. International society of sports nutrition position stand: nutrient timing. J Int Soc Sports Nutr 2008;5:17.

[27] Haff GG, Stone MH, Warren BJ, Keith R, Johnson RL, Nieman DC, et al. The effect of carbohydrate supplementation on multiple sessions and bouts of resistance exercise. J Strength Cond Res 1999;13:111–9.

[28] Bird SP, Tarpenning KM, Marino FE. Liquid carbohydrate/essential amino acid ingestion during a short-term bout of resistance exercise suppresses myofibrillar protein degradation. Metabolism 2006;55(5):570–7.

[29] Bird SP, Tarpenning KM, Marino FE. Independent and combined effects of liquid carbohydrate/essential amino acid ingestion on hormonal and muscular adaptations following resistance training in untrained men. Eur J Appl Physiol 2006;97(2):225–38.

[30] Beelen M, Koopman R, Gijsen AP, Vandereyt H, Kies AK, Kuipers H, et al. Protein coingestion stimulates muscle protein synthesis during resistance-type exercise. Am J Physiol Endocrinol Metab 2008;295(1):E70–7.

[31] Jentjens R, Jeukendrup A. Determinants of post-exercise glycogen synthesis during short-term recovery. Sports Med 2003;33 (2):117–44.

[32] Pedersen DJ, Lessard SJ, Coffey VG, Churchley EG, Wootton AM, Ng T, et al. High rates of muscle glycogen resynthesis after exhaustive exercise when carbohydrate is coingested with caffeine. J Appl Physiol 2008;105(1):7–13.

[33] Wolfe RR. Regulation of skeletal muscle protein metabolism in catabolic states. Curr Opin Clin Nutr Metab Care 2005;8 (1):61–5.

[34] Wolfe RR. The underappreciated role of muscle in health and disease. Am J Clin Nutr 2006;84(3):475–82.

[35] Tipton KD, Gurkin BE, Matin S, Wolfe RR. Nonessential amino acids are not necessary to stimulate net muscle protein synthesis in healthy volunteers. J Nutr Biochem 1999;10 (2):89–95.

[36] Volpi E, Kobayashi H, Sheffield-Moore M, Mittendorfer B, Wolfe RR. Essential amino acids are primarily responsible for the amino acid stimulation of muscle protein anabolism in healthy elderly adults. Am J Clin Nutr 2003;78(2):250–8.

[37] Rodriguez N, Lunn W. Proteins and amino acids the repair blocks their place in growth and recovery. In: Kerksick C, editor. Nutrient timing metabolic optimization for health, performance and recovery. Boca Raton, FL: CRC Press; 2011. p. 44–60.

[38] Tarnopolsky M. Protein requirements for endurance athletes. Nutrition 2004;20(7–8):662–8.

[39] Torun B, Scrimshaw NS, Young VR. Effect of isometric exercises on body potassium and dietary protein requirements of young men. Am J Clin Nutr 1977;30(12):1983–93.

[40] Butterfield GE, Calloway DH. Physical activity improves protein utilization in young men. Br J Nutr 1984;51(2):171–84.

[41] Todd KS, Butterfield GE, Calloway DH. Nitrogen balance in men with adequate and deficient energy intake at three levels of work. J Nutr 1984;114(11):2107–18.

[42] Millward DJ, Bowtell JL, Pacy P, Rennie MJ. Physical activity, protein metabolism and protein requirements. Proc Nutr Soc 1994;53(1):223–40.

[43] el-Khoury AE, Forslund A, Olsson R, Branth S, Sjodin A, Andersson A, et al. Moderate exercise at energy balance does not affect 24-h leucine oxidation or nitrogen retention in healthy men. Am J Physiol 1997;273(2 Pt 1):E394–407.

[44] Forslund AH, El-Khoury AE, Olsson RM, Sjodin AM, Hambraeus L, Young VR. Effect of protein intake and physical activity on 24-h pattern and rate of macronutrient utilization. Am J Physiol 1999;276(5 Pt 1):E964–76.

[45] Bowtell JL, Leese GP, Smith K, Watt PW, Nevill A, Rooyackers O, et al. Modulation of whole body protein metabolism, during and after exercise, by variation of dietary protein. J Appl Physiol 1998;85(5):1744–52.

[46] Meredith CN, Zackin MJ, Frontera WR, Evans WJ. Dietary protein requirements and body protein metabolism in endurance-trained men. J Appl Physiol 1989;66(6):2850–6.

[47] Phillips SM, Atkinson SA, Tarnopolsky MA, MacDougall JD. Gender differences in leucine kinetics and nitrogen balance in endurance athletes. J Appl Physiol 1993;75(5):2134–41.

[48] Lamont LS, Patel DG, Kalhan SC. Leucine kinetics in endurance-trained humans. J Appl Physiol 1990;69(1):1–6.

[49] Tarnopolsky MA, MacDougall JD, Atkinson SA. Influence of protein intake and training status on nitrogen balance and lean body mass. J Appl Physiol 1988;64(1):187–93.

[50] Friedman JE, Lemon PW. Effect of chronic endurance exercise on retention of dietary protein. Int J Sports Med 1989;10 (2):118–23.

[51] Brouns F, Saris WH, Stroecken J, Beckers E, Thijssen R, Rehrer NJ, et al. Eating, drinking, and cycling. A controlled Tour de France simulation study, Part II. Effect of diet manipulation. Int J Sports Med 1989;10(Suppl. 1):S41–8.

[52] Brouns F, Saris WH, Stroecken J, Beckers E, Thijssen R, Rehrer NJ, et al. Eating, drinking, and cycling. A controlled Tour de France simulation study, Part I. Int J Sports Med 1989;10(Suppl. 1): S32–40.

[53] Biolo G, Maggi SP, Williams BD, Tipton KD, Wolfe RR. Increased rates of muscle protein turnover and amino acid transport after resistance exercise in humans. Am J Physiol 1995;268(3 Pt 1):E514–20.

[54] Phillips SM, Tipton KD, Aarsland A, Wolf SE, Wolfe RR. Mixed muscle protein synthesis and breakdown after resistance exercise in humans. Am J Physiol 1997;273(1 Pt 1):E99–107.

[55] Sheffield-Moore M, Yeckel CW, Volpi E, Wolf SE, Morio B, Chinkes DL, et al. Postexercise protein metabolism in older and younger men following moderate-intensity aerobic exercise. Am J Physiol Endocrinol Metab 2004;287(3):E513–22.

[56] Tipton KD, Ferrando AA, Williams BD, Wolfe RR. Muscle protein metabolism in female swimmers after a combination of resistance and endurance exercise. J Appl Physiol 1996;81(5):2034–8.

[57] Borsheim E, Cree MG, Tipton KD, Elliott TA, Aarsland A, Wolfe RR. Effect of carbohydrate intake on net muscle protein synthesis during recovery from resistance exercise. J Appl Physiol 2004;96(2):674–8.

[58] Miller SL, Tipton KD, Chinkes DL, Wolf SE, Wolfe RR. Independent and combined effects of amino acids and glucose after resistance exercise. Med Sci Sports Exerc 2003;35(3):449–55.

[59] Roy BD, Tarnopolsky MA, MacDougall JD, Fowles J, Yarasheski KE. Effect of glucose supplement timing on protein metabolism after resistance training. J Appl Physiol 1997;82 (6):1882–8.

[60] Biolo G, Tipton KD, Klein S, Wolfe RR. An abundant supply of amino acids enhances the metabolic effect of exercise on muscle protein. Am J Physiol 1997;273(1 Pt 1):E122–9.

[61] Borsheim E, Tipton KD, Wolf SE, Wolfe RR. Essential amino acids and muscle protein recovery from resistance exercise. Am J Physiol Endocrinol Metab 2002;283(4):E648–57.

[62] Rasmussen BB, Tipton KD, Miller SL, Wolf SE, Wolfe RR. An oral essential amino acid-carbohydrate supplement enhances muscle protein anabolism after resistance exercise. J Appl Physiol 2000;88(2):386–92.

[63] Tipton KD, Ferrando AA, Phillips SM, Doyle Jr D, Wolfe RR. Postexercise net protein synthesis in human muscle from orally administered amino acids. Am J Physiol 1999;276(4 Pt 1): E628–34.

[64] Tarnopolsky MA, Atkinson SA, MacDougall JD, Chesley A, Phillips S, Schwarcz HP. Evaluation of protein requirements for trained strength athletes. J Appl Physiol 1992;73(5):1986–95.

[65] Lemon PW, Tarnopolsky MA, MacDougall JD, Atkinson SA. Protein requirements and muscle mass/strength changes during intensive training in novice bodybuilders. J Appl Physiol 1992;73(2):767–75.

[66] Phillips SM, Parise G, Roy BD, Tipton KD, Wolfe RR, Tamopolsky MA. Resistance-training-induced adaptations in skeletal muscle protein turnover in the fed state. Can J Physiol Pharmacol 2002;80(11):1045–53.

[67] Phillips SM. Protein requirements and supplementation in strength sports. Nutrition 2004;20(7–8):689–95.

[68] Campbell B, Kreider RB, Ziegenfuss T, La Bounty P, Roberts M, Burke D, et al. International society of sports nutrition position stand: protein and exercise. J Int Soc Sports Nutr 2007;4:8.

[69] Burke LM, Angus DJ, Cox GR, Cummings NK, Febbraio MA, Gawthorn K, et al. Effect of fat adaptation and carbohydrate restoration on metabolism and performance during prolonged cycling. J Appl Physiol 2000;89(6):2413–21.

[70] Burke LM, Kiens B. "Fat adaptation" for athletic performance: the nail in the coffin? J Appl Physiol 2006;100(1):7–8.

[71] Stellingwerff T, Spriet LL, Watt MJ, Kimber NE, Hargreaves M, Hawley JA, et al. Decreased PDH activation and glycogenolysis during exercise following fat adaptation with carbohydrate restoration. Am J Physiol Endocrinol Metab 2006;290(2):E380–8.

CHAPTER

37

An Overview of Branched-Chain Amino Acids in Exercise and Sports Nutrition

Humberto Nicastro, Daniela Fojo Seixas Chaves and Antonio Herbert Lancha Jr

Laboratory of Nutrition and Metabolism, School of Physical Education and Sports, University of São Paulo, São Paulo, Brazil

INTRODUCTION

Skeletal muscle is highly adaptable in response to exogenous stimuli such as mechanical overload (e.g., through resistance exercise) and endurance, hormonal and dietary (protein and amino acids) factors [1]. This responsiveness confers to muscle tissue the characteristic named "plasticity". Considering a healthy adult of 70 kg as reference, skeletal muscle comprises ~75% of lean body mass and ~40% of the peripheral tissue. Furthermore, 60–75% of all protein in the body is located in the skeletal muscle, and ~35% of the indispensable amino acids in muscle protein are branched-chain amino acids (BCAA): leucine, isoleucine, and valine [2]. Amino acids are the building blocks of muscle proteins. Twenty amino acids are required for skeletal muscle protein synthesis and, in healthy adults, nine of these amino acids (including BCAA) are considered essential because their carbon skeletons cannot be synthesized endogenously and must be acquired through diet or nutritional supplementation.

Some decades ago, skeletal muscle anabolism and catabolism were compared to a "brick wall": i.e., energy is spent in a specific manner to build it, but little energy is spent in its "demolition" (degradation), and that is spent in a totally unspecific way. However, molecular biology studies have identified transcription factors, proteins and molecular pathways that can modulate both synthetic and catabolic pathways. Additionally, it has been reported that myofibrillar protein degradation is not an unspecific process and requires energy [3]. Thus, protein synthesis and degradation are integrated processes. In this context, amino acids (especially the BCAA leucine) are well known to modulate some proteins at the post-translational (phosphorylation) level in order to both stimulate protein synthesis and indirectly attenuate muscle wasting and can also interact with the proteolytic machinery in order to attenuate muscle catabolism [4].

However, some studies have demonstrated that such proteins involved in molecular signaling differ in terms of time- and dose-response and are not always involved in muscle adaptation [5]. Therefore, it is important to emphasize that such signaling must be translated into a physiological response. Regarding skeletal muscle, muscle mass and strength are strong adaptive markers of muscle responsiveness and functionality and must be considered.

This chapter describes the metabolism and signaling pathways of essential amino acids and the physiological and functional effects of their supplementation on skeletal muscle mass and strength. The chapter highlights the importance and effects of the BCAA leucine in skeletal muscle remodeling and presents a great body of evidence describing its effects on muscle anabolism. The studies presented were conducted in humans in randomized, double-blind, and placebo-controlled design, thus establishing considerable external validity of the information described and promoting its practical application.

Nutrition and Enhanced Sports Performance.
DOI: http://dx.doi.org/10.1016/B978-0-12-396454-0.00037-0

AMINO ACIDS METABOLISM IN SKELETAL MUSCLE

Amino acids metabolism and function in skeletal muscle have been extensively investigated by many research groups focusing both on ergogenic and therapeutic effects. Some studies evaluating nutritional strategies capable of increasing sports performance (i.e., skeletal muscle strength) or to attenuate or even treat some atrophic muscular diseases (i.e., cancer, sarcopenia and muscle disuse) have revealed that such effects can be safely achieved by means other than pharmacological compounds. In view of this, it is understandable that the nutritional products industry has constantly invested and encouraged research into amino acids supplementation and its effects on skeletal muscle metabolism and function.

In this context, branched-chain amino acids (BCAA), especially leucine, have demonstrated the most promising effects regarding skeletal muscle mass. BCAA metabolism is unique in that it is finely regulated by enzyme compartmentalization. In general, BCAA metabolism can be summarized into two major steps: transamination and oxidative decarboxylation [6]. Transamination is a reversible reaction catalyzed by the enzyme branched-chain aminotransferase (BCAT) found both in cytosol (BCATc) and in mitochondria (BCATm). However, the proportion of the enzymes' reaction differs substantially by organ: ~75% and 22% cytosolic in brain and kidney of rats, respectively [6]. Ichihara et al. [7] proposed a classification for the BCAT enzyme into: isoform I, which accepts all three BCAA; isoform II, which is specific for leucine; and isoform III, which has not been totally characterized. BCAT is a pyridoxine phosphate-dependent enzyme, exhibiting the "ping pong" kinetic mechanism. Methionine aminotransferase is another enzyme that accepts leucine as substrate [6].

As described by Hutson et al. [8], the transamination reaction comprises the following steps. (a) Pyridoxine phosphate reacts with the amino group of the BCAA. (b) The pyridoxine monophosphate form of the enzyme releases a branched-chain keto acid (BCKA) which is specific to each BCAA: α-keto-isocaproate (α-KIC) from leucine; α-keto-β-methylvalerate (α-KMV) from isoleucine, and α-keto-isovalerate (α-KIV) from valine. (c) Pyridoxine monophosphate enzyme aminates the amino acceptor α-keto-glutarate (α-KG), reforming pyridoxine phosphate enzyme and releasing glutamate, or reaminates the BCKA to their respective amino acid. In the first reaction, BCAT presents preference for isoleucine > leucine > valine, and α-KG is the preferred substrate for the second half-reaction. Thus, the main products of BCAA transamination are BCKA and glutamate and, unless BCKA is rapidly removed (oxidized) or reaminated, nitrogen excretion is expected to occur between BCAA and glutamate. Figure 37.1 illustrates the BCAA metabolism in skeletal muscle.

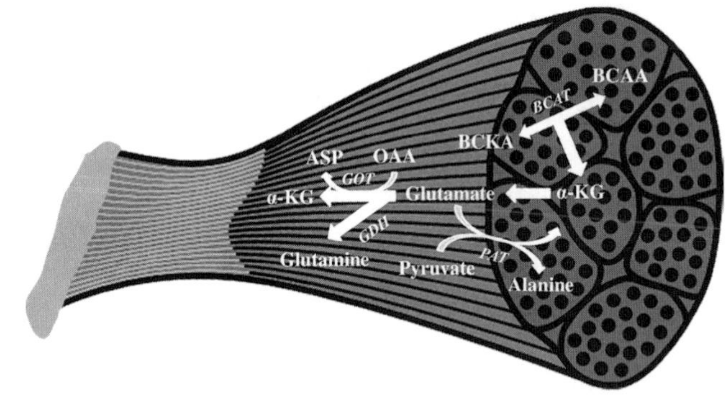

FIGURE 37.1 **BCAA metabolism in skeletal muscle.** BCAA are transaminated in skeletal muscle through the branched-chain amino transaminase (BCAT) enzyme, producing a branched-chain keto acid (BCKA). The amino group derived from BCAA transamination can be incorporated by α-keto glutarate (α-KG), producing glutamate. Accumulation of glutamate in skeletal muscle can lead to transamination of pyruvate to alanine through the pyruvate transaminase (PAT) enzyme. Glutamate can also favor the conversion of oxaloacetate (OAA) to aspartate (ASP) through the glutamate-oxaloacetate transaminase (GOT). Glutamate can also be converted to glutamine through glutamate dehydrogenase (GDH).

The major situ for BCKA oxidation are tissues that present elevated activity of branched-chain α-keto acid dehydrogenase complex (BCKDH). Human skeletal muscle presents a high BCAT:BCKDH activity ratio (~13% of whole body distribution of BCKDH capacity), which favors BCAA transamination by skeletal muscle and oxidation out of this tissue. The major tissue responsible for BCAA oxidation is the liver, which presents a high BCKDH:BCAT activity ratio [8]. In this complex, the BCKA is oxidatively decarboxylated (irreversible loss of BCAA carbon skeletons) in the mitochondria, forming acyl-CoA and NADH. Leucine is a ketogenic amino acid, and from its keto acid (α-KIC) derives aceto acetyl-CoA. The BCKA from isoleucine and valine (α-KMV and α-KIV) produce succinyl-CoA [6].

Three enzymes constitute the BCKDH complex: E1 (branched-chain α-keto acid decarboxylase), E2 (dihydrolipoyl transacylase), and E3 (dihydrolipoyl decarboxylase). The BCKDH complex is mainly controlled by post-translation modification through a kinase (BCKDH kinase, or BDK) and a phosphatase (BCKDP phosphatase, or BDP). Phosphorylation of this complex results in its inactivation, while the removal of the phosphate group results in its activation. Low-protein diets, protein synthesis, and/or enzymatic saving are situations characterized by attenuation of protein catabolism/degradation and promote activation of

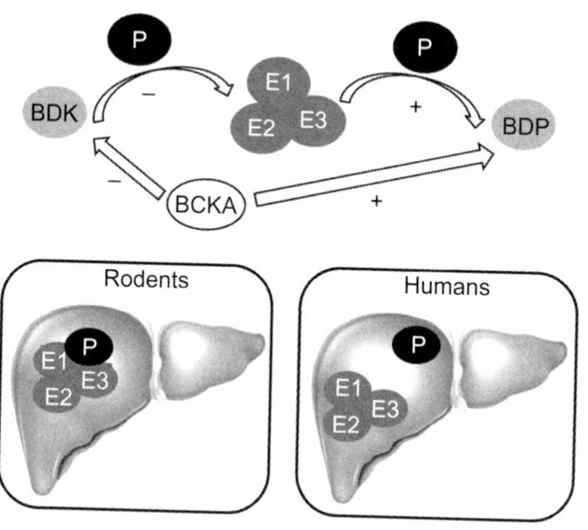

FIGURE 37.2 BCKDH complex regulation in liver of rodents and humans. The BCKDH complex is composed of three enzyme isoforms (E1, E2 and E3) and is regulated by post-translational modification (phosphorylation). The branched-chain dehydrogenase kinase (BDK) phosphorylates the BCKDH complex, leading to its inactivation. In contrast, the branched-chain phosphatase enzyme (BDP) removes the phosphate (P) from the BCKDH complex, leading to its activation. Branched-chain keto acids (BCKA) are indirect activators of the BCKDH complex through inhibition of BDK activity. In basal condition, rodents present BCKDH complex phosphorylated, while in humans it is dephosphorylated.

BDK and, consequently, inactivation of the BCKDH complex. In contrast, high-protein diets, fasting, diabetes, sepsis, cancer, and pro-inflammatory cytokines are well-characterized catabolic stimuli able to activate BDP and inhibit BDK activity, which results in BCKDH complex activation [9].

Some studies have reported that administration of an isolated BCAA, leucine, may result in the plasmatic imbalance of essential amino acids [10,11]. This observation was reported by studies investigating the potential ergogenic/therapeutic effects of leucine on skeletal muscle, showing reduced plasmatic concentrations of other essential amino acids, mainly isoleucine and valine. The proposed mechanisms for such a response were: (a) increase in net tissue protein synthesis, (b) transporters activity that could alter body amino acid distribution, and (c) increased BCAA oxidation. The first hypothesis was refuted by *ex vivo* and *in vivo* studies demonstrating no significant data to support the decrease in isoleucine and valine degradation through protein synthesis and degradation after increased plasma leucine. The second mechanism also is unlikely since studies in rats demonstrated that tissue and plasma pools change similarly during leucine infusion. The third mechanism appears to be more consistent since α-KIC is a potent inhibitor of BDK activity and, consequently, activates indirectly the BCKDH complex [6]. Thus, elevating leucine intake

TABLE 37.1 Distribution of BCAA Transamination and Oxidation Capacity in Human Tissues

Tissue	BCAT		BCKDH (actual/total)	
	Units/100 g BW	**% of Total**	**Units/100 g BW**	**% of Total**
Skeletal muscle	5.1	65.4	0.04/0.21	54/66
Kidney	0.3	3.8	0.01/0.03	13/9
Liver	0.6	7.7	0.006/0.04	8/13
Brain	1.2	15.4	0.015/0.03	20/9
Stomach + intestine	0.6	7.7	0.003/0.01	4/3
Total	7.8		0.074/0.32	

BCAA, branched-chain amino acids; BCAT, branched-chain amino transaminase; BCKDH, branched-chain α-keto acid dehydrogenase complex. BCKDH actual activity reflects the activity state of BCKDH in the tissue, and BCKDH complex activity is an estimate of enzyme amount measured after activation of the complex.
Adapted from Hutson et al. [8]; Data are from Suryawan et al. [12].

results in increased α-KIC production and enhances BCKDH complex activity which oxidizes isoleucine and valine. The result of such a "pathway" is the imbalance in plasmatic BCAA concentration that may impair protein balance in peripheral tissues, such as skeletal muscle.

It is important to emphasize that such a decrease in plasma amino acids may not impair physiological events such as skeletal muscle remodeling. Recently, some authors reported that isolated leucine administration to humans promoted a significant decrease in plasmatic concentration of isoleucine and valine. However, this reduction was still within physiological levels [10]. In order to know that the reduction was within physiological levels, one must know that there is a physiological range (for details see [11a]). Figure 37.2 illustrates the regulation of the BCKDH complex activity.

Table 37.1 describes the enzymatic distribution of BCAT and BCKDH in human tissues.

MOLECULAR PATHWAYS OF AMINO ACIDS IN SKELETAL MUSCLE

The major cellular pathways involved in skeletal muscle remodeling can be summarized as transcription and translation, which involves the generation of a messenger RNA (mRNA) and its subsequent translation into protein. These processes result in accumulation of contractile material (sarcomeric proteins), i.e., muscle hypertrophy [13]. In general, amino acids (especially the BCAA leucine) do not interact directly with the DNA in the nucleus, but can modulate transcription through post-translational modifications of transcription factors (i.e., phosphorylation) [14]. Regarding protein translation, amino acids can also promote post-translation modification in order to stimulate the first step of translation called initiation [15]. Therefore, amino acids can increase the efficiency and the effectiveness of the first step of protein translation in order to favor the binding of the ribosome to the mRNA. Thus, amino acids are signaling nutrients able to promote post-translational effects (mainly phosphorylation) that interact with both transcription and translation.

Some cellular pathways have been characterized to elucidate the stimulation of protein synthesis in response to mechanical stimuli combined with amino acids supplementation. The signaling pathway regulated by protein activity with serine/threonine kinase mTOR (mammalian target of rapamycin) is considered extremely important in the control of muscle protein synthesis [16,17]. Its activation has been linked to several conditions where protein synthesis is increased. For example, treatment of skeletal muscle cells with the insulin-like growth factor (IGF-1) can induce a significant increase in caliber of myofibers per muscle by stimulating the mTOR pathway and its effector, the 70 kDa ribosomal protein S6 kinase (p70S6K) and the eukaryotic translation initiation factor 4E-binding protein 1 (4E-BP1). Similarly, after ingestion of the amino acid leucine, protein synthesis suffers parallel increase in the phosphorylation of these proteins [14]. The protein 4E-BP1 appears to be more responsive to leucine. This protein activates the process of protein synthesis through the release of eukaryotic initiation factor 4E (eIF4E), favoring the formation of the eukaryotic initiation complex 4F (eIF4F) and subsequent interaction of the messenger RNA with the 40S ribosomal subunit, which is the beginning the protein translation process [18].

Importantly, some proteins were recently identified as candidates to mediate the nutritional signal of leucine to the biological effect of muscle protein synthesis. Among these, we highlight the hVps34 (human vacuolar protein sorting 34) and MAP4K3 (mitogen activated protein kinase kinase kinase-3). Therefore, activation of the mTOR pathway appears to be crucial in the control of muscle protein synthesis. Figure 37.3 summarizes the main mechanisms of muscular protein translation stimulated by leucine supplementation.

Recently, our research group described a hypothesis whereby leucine could modulate the innate immune response (Figure 37.4). Under pro-inflammatory and catabolic conditions, leucine could be transaminated in order to promote glutamine synthesis [19]. This, in turn, could be used as a substrate for macrophages and neutrophils and thus modulate cytokine expression. Consequently, conditions that are also characterized by increased oxidative stress have such a response attenuated. This modulation could reflect inflammatory benefits for muscle remodeling in order to reduce the blocking of the cellular pathways of protein synthesis and reduce the stimulation of the muscle degradation pathways.

EFFECTS OF AMINO ACIDS SUPPLEMENTATION ON MUSCLE MASS AND STRENGTH

There are no chronic studies in healthy adult humans evaluating the effects of BCAA or leucine supplementation combined with resistance exercise on muscle mass or strength, except for the studies of Verhoeven and Leenders which evaluated the effects of chronic leucine supplementation without resistance exercise. Thus, the current available evidence regarding amino acids supplementation and resistance exercise is derived from acute studies that measured the phosphorylation of proteins involved in the synthetic response of skeletal muscle after these interventions (supplementation + mechanical stimuli). It is expected that the chronic effect of supplementation and exercise in skeletal muscle is the result of the accumulation of repeated acute loads.

Karlsson et al. [20] conducted a randomized, double-blind, cross-over study in order to investigate the effects of resistance exercise, combined or not with BCAA supplementation, on the phosphorylation of proteins involved in muscle synthesis. The resistance exercise protocol consisted of leg press (4 sets of 10 repetitions at 80% 1RM) on two occasions. The supplementation protocol was composed of 150 mL placebo (flavored water) or BCAA (45% leucine, 30% valine, 25% isoleucine) totaling 100 mg/kg of BCAA. The subjects consumed the supplement prior to warming up, immediately before, during, immediately after, and 15, 30, 60 and 90 minutes after the exercise protocol. The exercise promoted a significant increase in the phosphorylation of p70S6k^{Ser424} and p70s6k^{Thr421} which persisted until 2 hours after the end of the exercise session. BCAA supplementation increased the phosphorylation 3.5-fold during the recovery period, and phosphorylation of p70S6k^{Thr389} increased only in the group supplemented with BCAA.

The same research group conducted a randomized, double-blind, cross-over trial in order to distinguish the influence of resistance exercise and BCAA supplementation on the expression of proteins involved in protein synthesis. The volunteers underwent two sessions of unilateral resistance exercise separated by an interval of 1 month. The exercise protocol performed consisted of leg press (4 sets of 10 repetitions at 80% 1RM, followed by 4 sets of 15 repetitions at

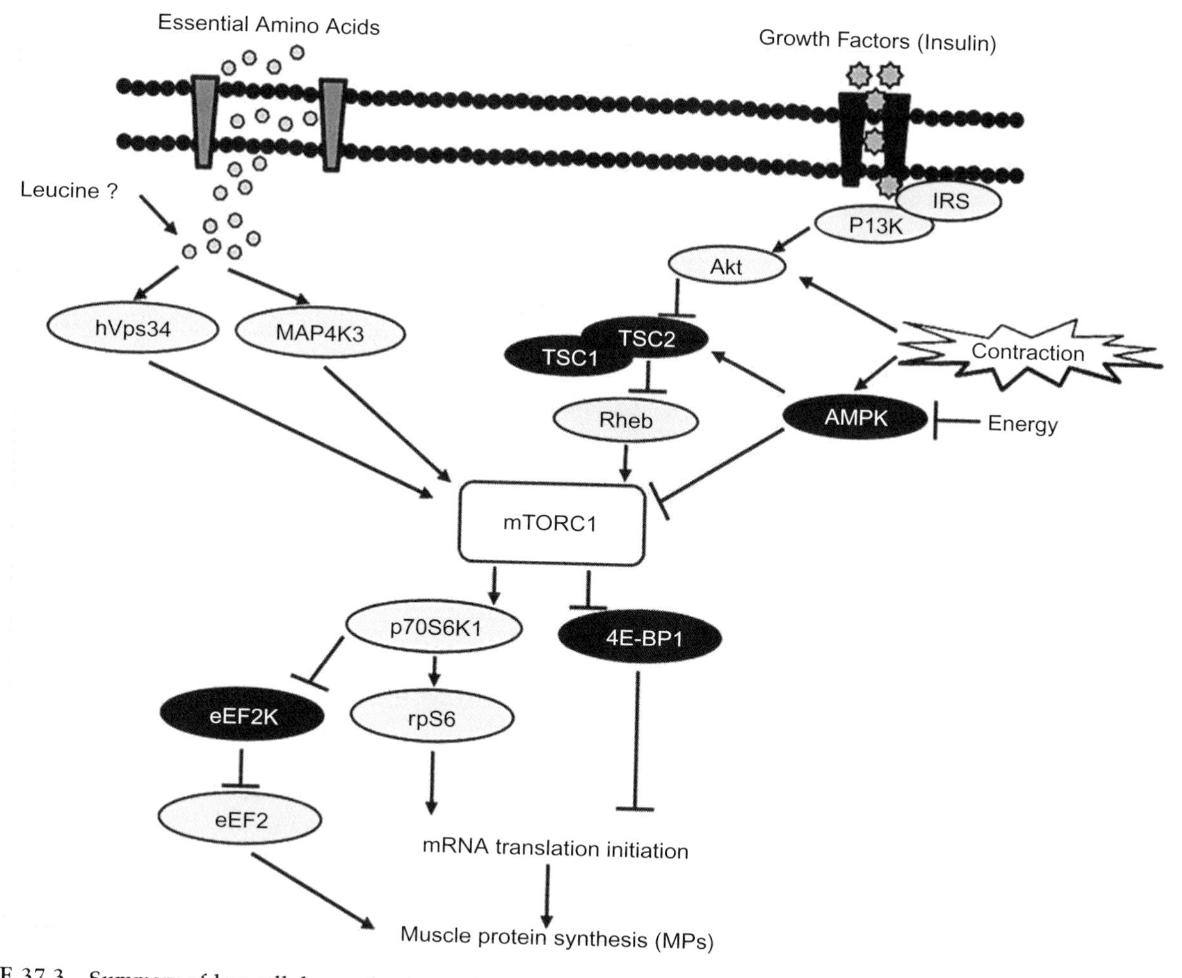

FIGURE 37.3 **Summary of key cellular mechanisms of action of leucine on protein synthesis in skeletal muscle.** 4E-BP1, eIF4E binding protein 1; AMPK, AMP-protein kinase; eEF2, eukaryotic elongation factor 2; eEF2k, eukaryotic elongation factor 2 kinase; hVps34, human vacuolar protein sorting 34; IRS, insulin receptor substrate; MAP4K3, mitogen activated protein kinase kinase kinase kinase-3; mTORC1, mammalian target of rapamycin complex 1; p70S6k1, 70 kDa ribosomal protein S6 kinase 1, PI3K, phosphoinositide 3-kinase; Rheb, Ras homolog enriched in brain; rpS6, ribosomal protein S6; TSC1/2, tuberous sclerosis complex 1/2. *Extracted from Pasiakos & McLung [18a].*

65% 1RM, with an interval of 5 minutes between sets). The subjects ingested 150 mL of a solution containing a mixture of BCAA (45% leucine, 30% valine, 25% isoleucine), totaling 85 mg/kg of BCAA or flavored water prior to warming up, immediately before protocol, after the fourth set of exercise, immediately after the protocol, and 15 and 45 minutes into the recovery period. The authors found that, regardless of supplementation, the phosphorylation of $mTOR^{Ser2448}$ increased significantly in both legs after exercise. Phosphorylation of $p70S6k^{Thr389}$ did not increase after the resistance exercise. However, with BCAA supplementation, phosphorylation of $p70S6k^{Thr389}$ increased 11- and 30-fold immediately and 1 hour after the exercise session, respectively, when compared with the non-exercised leg. Surprisingly, phosphorylation of $p70S6k^{Thr389}$ in the non-exercised leg increased 5- and 16-fold immediately and 1 hour after the exercise, respectively, when compared with baseline. Phosphorylation of eukaryotic elongation factor 2 (eEF2) was attenuated 1 hour after the exercise session in both control (10–40%) and exercised leg (30–50%) in both conditions [21]. The increase in phosphorylation in the control leg may indicate a possible neural effect (cross-education) that may have compromised the cross-over design of the study.

Dreyer et al. [22] randomized male and sedentary subjects into two groups (control and essential amino acids plus carbohydrate) to investigate the mechanisms of increased muscle protein synthesis observed with supplementation after exercise. The solution of carbohydrate and essential amino acids contained 35% of leucine, 8% of isoleucine and 10% of valine and 0.5 g/kg of lean body mass of sucrose. Subjects received supplementation 1 hour after the exercise bout. The exercise protocol consisted of 10 sets of 10 repetitions at 80% 1RM with 3 minutes interval between sets of knee extension. The authors found that muscle protein synthesis was elevated in both groups 1 hour after the

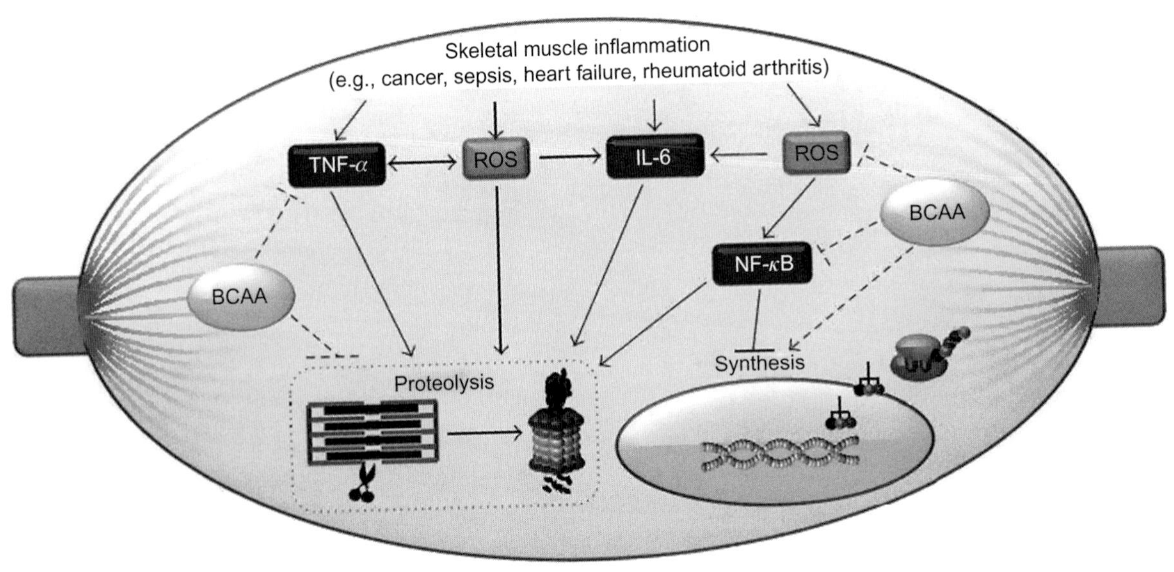

FIGURE 37.4 Cellular signaling pathways related to inflammation and muscle remodeling in pro-inflammatory conditions—possible role of leucine. BCAA, branched-chain amino acids; ROS, reactive oxygen species; IL-6, interleukin-6; NF-κB, nuclear factor kappa B; TNF-α, tumor necrosis factor alpha. *Extracted from Nicastro et al. [19].*

exercise session at the same magnitude but it was higher only in the supplemented group 2 hours after the exercise. This increase was associated with increased phosphorylation of S6K1 and $mTOR^{Ser2448}$. Phosphorylation of Akt^{Ser473} increased 1 hour after exercise and returned to baseline after 2 hours in the control group, but remained elevated in the supplemented group. Similarly, phosphorylation of $4E\text{-}BP1^{Thr37/46}$ remained high only in the supplemented group.

These acute studies demonstrate that isolated amino acids, particularly BCAA and leucine, are able to modulate phosphorylation of proteins involved in muscle protein synthesis. It is supposed that the accumulation of these effects can promote and elucidate the chronic effects on muscle mass and strength. However, chronic studies are necessary to confirm the acute evidences.

DIETARY SOURCES AND PRACTICAL APPLICATION OF PROTEIN SUPPLEMENTATION

Recently, studies have investigated the role of amino acids supplementation, from protein sources, post resistance exercise in order to optimize muscle mass and strength gains. These studies have demonstrated that protein supplementation post resistance exercise through bolus (1 dose) can promote significant increase in skeletal muscle protein synthesis. Tang et al. [23] investigated the supplementation of 21.4 g of whey, 21.9 g of casein, or 22.2 g of soy (each drink provided ~10 g of essential amino acids) at rest and after resistance exercise in healthy humans. The authors observed that whey protein supplementation promoted higher increments in muscle protein synthesis than casein and soy in both rest and after resistance exercise. This great response with whey protein supplementation can be attributed to two main factors: distinct concentrations of amino acids and/or digestion and absorption rate. The first argument does not appear to be determinant since the differences among the protein sources were quite small (i.e., 2.3 g of leucine in whey versus 1.8 g in both soy and casein). Thus, digestion and absorption rate appears to be determinant in modulating muscle protein synthesis since the plasmatic amino acid pattern observed after the supplementation was quite different. Whey protein supplementation promoted a high and transient plasmatic peak (30 minutes) of essential amino acids and leucine (~2.5 times the basal value), whereas casein and soy promoted a slow and prolonged increase in plasmatic amino acids.

However, one can argue that the compositions of these protein sources are different and that similar results would be observed if this variation were absent. In view of this, West et al. [24] investigated the effects of 1 dose of 25 g or 10 doses of 2.5 g of whey protein after resistance exercise. The authors intended to promote the same aminoacidemia pattern of casein and soy with whey protein and to compare the effects with the same protein consumed in 1 dose. Although 1 dose of whey protein promoted a rapid and transient

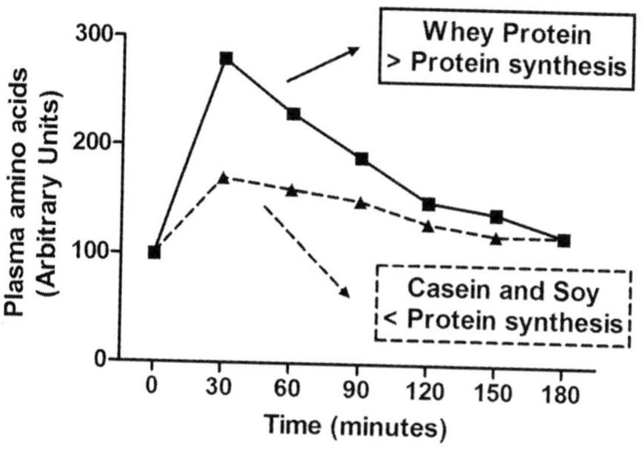

FIGURE 37.5 **Relation among dietary protein source, plasma amino acids, and muscle protein synthesis—the "muscle full effect".** Whey protein promotes a rapid and transient increase in plasma amino acids which reflects greater response of skeletal muscle protein synthesis after resistance exercise that remains elevated for up to 5 hours. In contrast, soy and casein, which are slowly digested and absorbed, promote a slow and prolonged increase in plasma amino acids resulting in a small increase in muscle protein synthesis that begins to increase ~3–5 hours after resistance exercise and is small compared with that for the whey protein. *Adapted from data of Tang et al. [23]. The values presented in* x *and* y *axes are didactic.*

increase in plasma amino acids, muscle protein synthesis was increased 1–3 hours post resistance exercise and remained elevated until 3–5 hours. The consumption of fractionated doses promoted a slow and prolonged increase in aminoacidemia. However, muscle protein synthesis was elevated only 3–5 hours post exercise but to a lesser extent than with bolus supplementation. Therefore, we can conclude that skeletal muscle response to protein and amino acids supplementation is greater when there is a rapid and transient increase in plasma amino acids concentration, especially leucine.

If skeletal muscle protein synthesis reaches a plateau after 25 g of a rapid digestive and absorptive protein source and this effect remains for 5 hours, how many doses of protein is it possible to supplement in 24 hours? In a dose-response design, Moore et al. [25] investigated the effects of 0, 5, 10, 20, or 40 g of whey protein after resistance exercise on muscle protein synthesis. The authors demonstrated that skeletal muscle is refractory to amino acids, i.e., presents a limited response. It was shown that 20 g of whey protein (the same amount used in the previous studies) promoted the maximal effect on protein synthesis. Considering that this dose easily reaches a plateau in protein synthesis and that this effect remains for 5 hours, even when plasma amino acid concentration returns to the baseline values, the authors suggested that 2 or 3 doses are the maximal tolerated in one day. Atherton et al. [5] named this phenomenon as "muscle full effect". Figure 37.5 illustrates the relation among dietary protein sources, plasma amino acids, and skeletal muscle protein synthesis.

TABLE 37.2 Content of BCAA and Leucine in 100 g of Protein Sources

	BCAA	Leucine
Whey protein isolate	26%	14%
Milk protein	21%	10%
Egg protein	20%	8.5%
Muscle protein	18%	8%
Soy protein isolate	18%	8%
Wheat protein	15%	7%

BCAA, branched-chain amino acids.
Adapted from Layman & Baum [26]; Source: *USDA Food Composition Tables.*

In conclusion, the consumption of 20–25 g of a rapidly digested and absorbed protein (composed of ~10 g of essential amino acids, including 2–3 g of leucine) can promote a rapid and transient increase in protein synthesis after resistance exercise, and supplementation with high doses does not promote further increase in protein synthesis.

AMINO ACIDS—PRACTICAL RECOMMENDATIONS

BCAA are mainly found in animal food sources, such as milk. Some vegetable foods also contain a smaller amount of amino acids (e.g., soy). In this context, the protein isolated from milk (whey protein) is rich in BCAA and leucine. Table 37.2 lists the amount of BCAA and leucine in some foods and dietary supplements.

Nitrogen balance studies are criticized because of the experimental limitation associated with higher energy intake and the exclusion of the miscellaneous losses. Rose [27] and Leverton and coworkers [28–32] demonstrated, through nitrogen balance, that leucine recommendation for humans is ~10 mg kg d^{-1}. The use of tracer methodology in more recent studies suggested that the daily recommendation for leucine is greater than that suggested by nitrogen balance studies.

Dietary reference intakes (DRI) recommend the daily ingestion of 34, 15 and 19 mg kg $d^{-1}$of leucine, isoleucine and valine, respectively. Therefore, the DRI for protein and amino acids is quite similar to the recommendations presented in the tracer and current practical studies. For example, considering an adult man of 80 kg of body weight, the recommendation of daily BCAA intake would be 2.7 g of leucine, 1.2 g of isoleucine and 1.5 g of valine based on DRI recommendation. Recently, Elango et al. [33] investigated the upper level for leucine intake in adult men. The authors provided 50, 150, 250, 500, 750, 1000 and 1250 mg kg d^{-1} acutely, which corresponds to 3–25 times the amount of Estimated Average Requirement (EAR), and identified the upper level through the measurement of plasma and urinary biochemical variables and changes in leucine oxidation (L-[1-^{13}C]leucine). It was observed that the upper level of leucine intake could be 500 mg kg d^{-1}, which is equivalent to 35 g/day for a 70 kg man. Therefore, at least acutely, the human body presents a high tolerability for amino acids excess even if this amino acid is provided alone. It is important to emphasize that long-term studies are needed in order to support this evidence.

However, it is important to point out that the physiological condition must be considered before consuming high levels of protein and amino acids. Some authors have identified that the association between high-fat diets and BCAA supplementation can promote insulin resistance. In this condition, BCAA promotes metabolic signatures related to the fatty acids metabolism that are related to insulin resistance [34]. This evidence suggests interplay between amino acids and fatty acids in the development of metabolic disease. Therefore, it is quite important to evaluate the dietary background of the individual before recommending the use of amino acid supplementation. Individuals consuming high-fat diets should adapt their dietary intake prior to the amino acid supplementation in order to avoid such metabolic deleterious effects.

Recently, Churchwarde-Venne et al. [35] investigated the effects of leucine supplementation on muscle protein synthesis in both rest and after resistance exercise in healthy humans. Subjects were supplemented with 25 g of whey protein, a well-characterized dose able to maximally stimulate muscle protein synthesis after resistance exercise (described above) with 3.0 g of leucine, 6.25 g of whey protein with 3.0 g of leucine, or 6.25 g of essential amino acids with 0.75 g of leucine. Plasma amino acids concentration demonstrated that whey protein promoted a higher peak of rapid and transient aminoacidemia (BCAA, essential amino acids, and total amino acids) when compared with the other interventions, and plasma leucine concentration was greater in the group supplemented with 6.25 g of whey protein and 3.0 g of leucine. Muscle protein synthesis at rest was stimulated after 1–3 hours and returned to baseline after 3–5 hours in both interventions. This evidence suggests that leucine alone can stimulate protein synthesis at rest at the same level as can whey protein and that a mixture of essential amino acids without leucine also can promote the same effect. Thus, the composition of the dietary protein and the pattern of plasma amino acids does not influence muscle protein synthesis at rest. The same results pattern was observed 1–3 hours after resistance exercise. However, only whey protein promoted a significant increase in muscle protein synthesis until 5 hours after resistance exercise, suggesting that leucine alone cannot maintain muscle protein synthesis elevation for more than 3 hours. Also, a mixture of essential amino acids without leucine can also promote the same effect as leucine alone, which suggests that other amino acids (e.g., isoleucine) can compensate for leucine deficiency.

Therefore, leucine alone can stimulate muscle protein synthesis after resistance exercise, but to a limited extent (for 1–3 hours). The absence of leucine in a mixture of essential amino acids can also promote the same molecular effect, indicating that other amino acids can play the same role on muscle signaling pathways. These results demonstrate that a high-quality protein with high digestibility and absorption is necessary to promote a significant increase in muscle protein synthesis.

Arginine has also been considered as an amino acid potentially able to modulate muscle mass because of its properties in stimulating nitric oxide (NO) metabolism and to possibly act, as does leucine, as a nutrient able to stimulate phosphorylation of proteins involved in translation initiation. Tang et al. [36] investigated such effects through supplementation of 10 g of arginine both at rest and after resistance exercise. Arginine supplementation was given with other essential amino acids in order to provide substrate for protein synthesis. No significant changes were observed in muscle blood flow after arginine supplementation at rest and after resistance exercise. Only the mechanical stimulus was able to increase muscle blood flow. Regarding muscle protein synthesis, arginine supplementation did not promote an increase in either mixed or myofibrillar synthetic rate at rest or after exercise. Therefore, arginine supplementation appears to be ineffective in stimulating muscle protein synthesis and increasing muscle blood flow even when provided with sufficient amounts of essential amino acids.

CONCLUSION AND PERSPECTIVES

Nutritional supplementation represents a wide field of investigation. Protein and amino acid supplementation to promote muscle mass and strength gains are constantly being debated in order to elucidate the best protein source, amount and timing of consumption. Recent evidence is helping professionals to develop more accurate supplementation protocols. Undoubtedly, amino acids are signaling nutrients able to modulate the remodeling processes (synthesis and degradation). However, it is important to emphasize that chronic studies must be conducted in order to verify whether the acute effects are reproduced chronically.

References

[1] Zanchi NE, Lancha Jr. AH. Mechanical stimuli of skeletal muscle: implications on mTOR/p70s6k and protein synthesis. Eur J Appl Physiol 2008;102:253–63.

[2] Smith AG, Muscat GE. Skeletal muscle and nuclear hormone receptors: implications for cardiovascular and metabolic disease. Int J Biochem Cell Biol 2005;37:2047–63.

[3] Lecker SH, Goldberg AL, Mitch WE. Protein degradation by the ubiquitin-proteasome pathway in normal and disease states. J Am Soc Nephrol 2006;17:1807–19.

[4] Nicastro H, Zanchi NE, da Luz CR, de Moraes WM, Ramona P, de Siqueira Filho MA, et al. Effects of leucine supplementation and resistance exercise on dexamethasone-induced muscle atrophy and insulin resistance in rats. Nutrition 2012;28:465–71.

[5] Atherton PJ, Etheridge T, Watt PW, Wilkinson D, Selby A, Rankin D, et al. Muscle full effect after oral protein: time-dependent concordance and discordance between human muscle protein synthesis and mTORC1 signaling. Am J Clin Nutr 2010;92:1080–8.

[6] Harper AE, Miller RH, Block KP. Branched-chain amino acid metabolism. Annu Rev Nutr 1984;4:409–54.

[7] Ichihara A, Noda C, Goto M. Transaminase of branched chain amino acids. X. High activity in stomach and pancreas. Biochem Biophys Res Commun 1975;67:1313–8.

[8] Hutson SM, Sweatt AJ, Lanoue KF. Branched-chain [corrected] amino acid metabolism: implications for establishing safe intakes. J Nutr 2005;135:1557S–64S.

[9] Shimomura Y, Honda T, Shiraki M, Murakami T, Sato J, Kobayashi H, et al. Branched-chain amino acid catabolism in exercise and liver disease. J Nutr 2006;136:250S–3S.

[10] Leenders M, Verdijk LB, van der Hoeven L, van Kranenburg J, Hartgens F, Wodzig WK, et al. Prolonged leucine supplementation does not augment muscle mass or affect glycemic control in elderly type 2 diabetic men. J Nutr 2011;141:1070–6.

[11] Verhoeven S, Vanschoonbeek K, Verdijk LB, Koopman R, Wodzig WK, Dendale P, et al. Long-term leucine supplementation does not increase muscle mass or strength in healthy elderly men. Am J Clin Nutr 2009;89:1468–75.

[11a] Wu G. Amino acids: metabolism, functions, and nutrition. Amino Acids 2009;37(1):1–17.

[12] Suryawan A, Hawes JW, Harris RA, Shimomura Y, Jenkins AE, Hutson SM. A molecular model of human branched-chain amino acid metabolism. Am J Clin Nutr 1998;68:72–81.

[13] Kimball SR, Jefferson LS. Control of translation initiation through integration of signals generated by hormones, nutrients, and exercise. J Biol Chem 2010;285:29027–32.

[14] Crozier SJ, Kimball SR, Emmert SW, Anthony JC, Jefferson LS. Oral leucine administration stimulates protein synthesis in rat skeletal muscle. J Nutr 2005;135:376–82.

[15] Dennis MD, Baum JI, Kimball SR, Jefferson LS. Mechanisms involved in the coordinate regulation of mTORC1 by insulin and amino acids. J Biol Chem 2011;286:8287–96.

[16] Glass DJ. Skeletal muscle hypertrophy and atrophy signaling pathways. Int J Biochem Cell Biol 2005;37:1974–84.

[17] Sandri M. Signaling in muscle atrophy and hypertrophy. Physiology (Bethesda) 2008;23:160–70.

[18] Layman DK. Role of leucine in protein metabolism during exercise and recovery. Can J Appl Physiol 2002;27:646–63.

[18a] Pasiakos SM, McClung JP. Supplemental dietary leucine and the skeletal muscle anabolic response to essential amino acids. Nutr Rev 2011;69(9):550–7.

[19] Nicastro H, da Luz CR, Chaves DF, Bechara LR, Voltarelli VA, Rogero MM, et al. Does branched-chain amino acids supplementation modulate skeletal muscle remodeling through inflammation modulation? possible mechanisms of action. J Nutr Metab 2012;2012:136937.

[20] Karlsson HK, Nilsson PA, Nilsson J, Chibalin AV, Zierath JR, Blomstrand E. Branched-chain amino acids increase p70S6k phosphorylation in human skeletal muscle after resistance exercise. Am J Physiol Endocrinol Metab 2004;287:E1–7.

[21] Apro W, Blomstrand E. Influence of supplementation with branched-chain amino acids in combination with resistance exercise on p70S6 kinase phosphorylation in resting and exercising human skeletal muscle. Acta Physiol (Oxf) 2010;200:237–48.

[22] Dreyer HC, Fujita S, Cadenas JG, Chinkes DL, Volpi E, Rasmussen BB. Resistance exercise increases AMPK activity and reduces 4E-BP1 phosphorylation and protein synthesis in human skeletal muscle. J Physiol 2006;576:613–24.

[23] Tang JE, Moore DR, Kujbida GW, Tarnopolsky MA, Phillips SM. Ingestion of whey hydrolysate, casein, or soy protein isolate: effects on mixed muscle protein synthesis at rest and following resistance exercise in young men. J Appl Physiol 2009;107:987–92.

[24] West DW, Burd NA, Coffey VG, Baker SK, Burke LM, Hawley JA, et al. Rapid aminoacidemia enhances myofibrillar protein synthesis and anabolic intramuscular signaling responses after resistance exercise. Am J Clin Nutr 2011;94:795–803.

[25] Moore DR, Robinson MJ, Fry JL, Tang JE, Glover EI, Wilkinson SB, et al. Ingested protein dose response of muscle and albumin protein synthesis after resistance exercise in young men. Am J Clin Nutr 2009;89:161–8.

[26] Layman DK, Baum JI. Dietary protein impact on glycemic control during weight loss. J Nutr 2004;134:968S–73S.

[27] Rose WC. The amino acid requirements of adult man. Nutr Abstr Rev Ser Hum Exp 1957;27:631–47.

[28] Leverton RM, Ellison J, Johnson N, Pazur J, Schmidt F, Geschwender D. The quantitative amino acid requirements of young women. V. Leucine. J Nutr 1956;58:355–65.

[29] Leverton RM, Gram MR, Brodovsky E, Chaloupka M, Mitchell A, Johnson N. The quantitative amino acid requirements of young women. II. Valine. J Nutr 1956;58:83–93.

[30] Leverton RM, Gram MR, Chaloupka M, Brodovsky E, Mitchell A. The quantitative amino acid requirements of young women. I. Threonine. J Nutr 1956;58:59–81.

[31] Leverton RM, Johnson N, Ellison J, Geschwender D, Schmidt F. The quantitative amino acid requirements of young women. IV. Phenylalanine, with and without tyrosine. J Nutr 1956;58:341–53.

[32] Leverton RM, Johnson N, Pazur J, Ellison J. The quantitative amino acid requirements of young women. III. Tryptophan. J Nutr 1956;58:219–29.

[33] Elango R, Chapman K, Rafii M, Ball RO, Pencharz PB. Determination of the tolerable upper intake level of leucine in acute dietary studies in young men. Am J Clin Nutr 2012;96:759–67.

[34] Newgard CB. Interplay between lipids and branched-chain amino acids in development of insulin resistance. Cell Metab 2012;15:606–14.

[35] Churchward-Venne TA, Burd NA, Mitchell CJ, West DW, Philp A, Marcotte GR, et al. Supplementation of a suboptimal protein dose with leucine or essential amino acids: effects on myofibrillar protein synthesis at rest and following resistance exercise in men. J Physiol 2012;590:2751–65.

[36] Tang JE, Lysecki PJ, Manolakos JJ, MacDonald MJ, Tarnopolsky MA, Phillips SM. Bolus arginine supplementation affects neither muscle blood flow nor muscle protein synthesis in young men at rest or after resistance exercise. J Nutr 2011;141:195–200.

CHAPTER

38

Water, Hydration and Sports Drink

Flavia Meyer[1], *Brian Weldon Timmons*[2] *and Boguslaw Wilk*[3]

[1]Universidade Federal do Rio Grande do Sul (UFRGS), Porto Alegre, Brazil [2]McMaster University, Hamilton, ON, Canada [3]Children's Exercise & Nutrition Centre, McMaster Children's Hospital, Hamilton, ON, Canada

INTRODUCTION

To avoid adverse effects of hypohydration due to exercise on performance and health, fluid ingestion is recommended in accordance with individual, environmental and exercise characteristics. Attention has been given not only to the timing and the volume of fluid intake, but also the type and the composition of the fluid to be ingested. Research has consistently demonstrated that under conditions of profuse sweating, electrolyte loss and glycogen depletion, ingestion of solutions containing a proper combination of electrolytes and carbohydrate (sports drink) may be advantageous compared with plain water [1–3]. Ingestion of a sports drink may therefore benefit those who practice prolonged exercises, especially in the heat, or high intensity and intermittent frequent efforts. Such drinks (e.g., Gatorade and Powerade) have the purpose of optimizing body hydration, replacement of electrolytes (mainly sodium), and maintaining high rates of carbohydrate oxidation as energy supply which may guarantee performance. Sports drinks are commercially available in different flavors and colors, with bottles ergonomically designed to stimulate voluntary drinking, although excessive fluid ingestion during exercise and hyperhydration are also undesirable. These drinks are commonly used by a wide age-range of athletes and active individuals during and after physical activities, being very popular among school-aged children [4]. "Energy drinks" containing other ingredients (caffeine, vitamins, or amino-acids such as taurine) have also been commercialized, but so far no convincing evidence exists to support their inclusion during exercise. In fact, energy drinks may cause adverse effects in the young population [5] and, according to a recent statement from the Committee on Nutrition and the Council on Sports Medicine and Fitness of the American Academy of Pediatrics [6], they should not be promoted for children, adolescents and young adults. The purpose of this chapter is to summarize the general impact of hypohydration in a variety of performance scenarios, highlight age-related concerns relative to children, adolescents and older individuals, and provide practical fluid intake recommendations.

EFFECTS OF HYPOHYDRATION

Aerobic Performance and Endurance Activities

There is sufficient evidence to indicate that hypohydration impairs aerobic and endurance performance and these effects are related to the degree of body water deficit [7,8]. The premature fatigue in sustained aerobic exercise due to hypohydration is explained by thermoregulatory, cardiovascular and metabolic factors. Compared with a euhydrated state, gradual dehydration increases core temperature and causes elevation in heart rate parallel to a decrease in blood flow, stroke volume, cardiac output, and skin blood flow. These responses, proportional to the hypohydration level, may be accentuated in a warm environment [7–9]. The reduced blood volume, combined with a greater blood flow demand to the periphery and exercising muscles, may impair sweating. As evaporation is the most effective manner of heat loss during exercise under warm ambient conditions, core body temperature rises at a greater extent as hypohydration levels increase. Hypohydration also affects muscle metabolism by accelerating the rate of glycogen depletion and affects central nervous system functioning by reducing motivation and effort [10].

Nutrition and Enhanced Sports Performance.
DOI: http://dx.doi.org/10.1016/B978-0-12-396454-0.00038-2

Intermittent High Intensity Activities and Sports

Hypohydration and its consequences may happen in sport modalities that require high intensity and intermittent bouts of effort over a period of hours. Another risky aspect is the carry-over effect of successive training and competition events performed in a single day [11]. Inadequate time between multiple events may not allow for optimal rehydration, as indicated in a recent review [12] that identified ten recent studies in which players arrived to their practices or competitions already in a state of hypohydration, according to urinary markers.

In soccer, for example, dribbling performance was impaired after a simulated game when the players reached levels of hypohydration of ~2.4% compared with when they were kept euhydrated [13]. Performance in an intermittent shuttle run test was also impaired when soccer players (~25 years old) reached hypohydration levels of ~2% [14].

In basketball, the consequences of hypohydration are inconsistent but can also be explained by limitations in attempts to simulate in-field conditions and by different study protocols and testing models. When ten basketball players (~17 years old) dehydrated by ~2% [15] no impairment was found in shooting performance, anaerobic power, vertical jumping height, or counter movement jump, compared with when they ingested water and were kept euhydrated. However, in players who were younger (12 to 15 years old) [16] or adolescent and young adults (17 to 28 years old) [17], hypohydration as mild as 2% impaired basketball skills. A recent study [18] showed no impairment in a set of basketball drills in adolescents (14 to 15 years old) that reached levels of hypohydration of 2.4% from a previous 90-min training session, although the rate of perceived exertion increased.

The inconsistent results from the above-mentioned studies demonstrate some limitations when simulating in-field conditions to research laboratory models. Different protocols, recovery times and/or age-maturational factors may influence a given performance test. It may also be difficult to isolate a given skill or ability from other factors that may also be affected by hypohydration, such as a cognitive function, which is discussed below.

Muscular Strength

It is still unclear how much hypohydration impairs muscular strength, and inconsistent outcomes may be due to the use of different protocols to achieve hypohydration and to the hypohydration level [19]. For example, if strength testing is preceded by hypohydration due to running or cycling in heat, increased fatigue is expected in the leg muscles [20,21]. An option could be to test arm strength; nonetheless, other factors such as the body temperature should also be taken into consideration. Increases in core temperature may affect the sequence of muscle strength production by reducing the motor cortex activation, the peripheral stimulus and power output [22–24]. Other confounding factors are the fitness and heat acclimatization levels, nutritional status and initial hydration status [19].

Another explanation for muscle strength impairment due to hypohydration is a reduction in the number of motor units recruited. Ftaiti et al. [25]. observed a reduction of ~40% and 25% in the respective isometric and isokinetic knee extensor contractions with a 2% hypohydration level induced by running as assessed by electromyography. However, Hayes and Morse [26] found no changes in the electromyography activity of the vastus lateralis muscle during isometric and isokinetic knee extensor strength tests, even at ~4% hypohydration level induced by running in the heat. Therefore, more research is needed to test the isolated effects of gradual hypohydration on muscle strength and to clarify the underlying mechanisms.

Cognitive Aspects

In addition to physical conditioning, many sport situations require cognitive and reaction abilities. Some studies indicate that a variety of cognitive functions such as vigilance, alertness, perceptual discrimination, arithmetic ability, visuomotor tracking, and psychomotor skills are affected by a moderate (~2%) level of hypohydration [27–29]. Short term memory was also shown to be altered across a wide age range [30]. However, cognitive parameters are not always impaired in a similar way and may be dependent on the methods employed to achieve hypohydration and to evaluate the cognitive parameter.

When subjects were dehydrated (~3%) by exercising on a treadmill in a warm environment, impairment was observed in tasks involving visual perception, short term memory, and psychomotor ability [28,29]. In another study [31], using exercise combined with water restriction to achieve hypohydration, less impairment was found in cognitive function of young adult athletes, but significant increases in fatigue, confusion and anger were observed. When only water restriction over a 24 h period was used to induce hypohydration by ~2.5%, no impairment in cognitive performance was observed [32]. It is therefore possible that heat stress alone may have an important influence on cognitive performance when it is used to induce hypohydration.

In the paediatric population, maintenance of euhydration seems to be important for cognitive performance [33]. A group of children (10 to 12 years old) who were restricted from drinking for a few hours had decreased task scores of short term memory and verbal analogy compared with a group that maintained the scores and were kept euhydrated [34]. Two studies [35,36] used a protocol in which children (6 to 9 years old) were tested before and after they ingested water, and some visual attention tasks were improved.

SPECIAL POPULATIONS

Children and Adolescents

Children may potentially dehydrate as much as adults [37] even though their sweat rate is lower when exercising at a similar intensity and in similar environmental conditions and when corrected for body surface area [38]. Both laboratory-based [39,40] and in-field (trainings and competitions) studies [12] have shown a trend towards body water deficit of ~2%, but it may reach higher levels. In a triathlon conducted in Costa Rica [41], it was observed that ~20% of the competitors from 9 to 13 years of age and ~25% of those from 14 to 17 years of age reached hypohydration levels >2%.

Insufficient fluid intake has been attributed to a thirst perception delay and lack of recognition of the need to replace the fluid lost from sweating. However, the type of fluid available (flavor and composition) seems to have a major impact on fluid intake in children and adolescents. After 29 children (9 to 13 years old) achieved mild hypohydration (~0.8%) by cycling in the heat, thirst perception level increased significantly and, during recovery, voluntary fluid intake was greater with flavored beverages (grape and orange) than with plain water [42].

Beverages containing electrolytes and carbohydrate in concentrations consistent with commercially available sports drinks seem to prevent dehydration by stimulating thirst and fluid retention. In a laboratory study [43], 12 heat-unacclimatized boys cycled intermittently for 3 hours in the heat (35°C, 43% relative humidity) in three separate sessions. In a randomized-order and double-blinded design, the boys in each of these sessions could drink as desired one of the following drinks: a grape-flavored sports drink, grape-flavored water, and plain water. Voluntary fluid intake was greater with the grape-flavored drinks, but further intake was observed in the sport drink session which was sufficient to avoid dehydration. Subsequent studies, one from the same laboratory [44] and another [45] testing heat-acclimatized athletic boys who cycled and rested intermittently for 180 min in the outdoor warm conditions of Puerto Rico, showed this consistency of increased volume intake with sports drinks at volumes that maintained euhydration. In another laboratory study [46] in which young adolescent athletes had access to a lemon-lime-flavored sport drink while running for 1 hour, no dehydration was observed, possibly because of the characteristic of the drink. However, in another study [47] neither drink flavor nor composition had significant effect on fluid intake of adolescent runners who exercised intermittently for 2 hours in the heat at higher intensity and with shorter rest periods.

The aforementioned studies where performed under a controlled design of endurance type of exercise and may not reflect the fluid ingestion pattern of sports that are characterized by stop-and-go intermittent bouts of high intensity exercise. When adolescent tennis players (15 years old) trained two sessions of 120 min of tennis in the heat, the volume of water intake (~1750 mL) was similar to that of a typical carbohydrate-electrolyte sport drink (~1900 mL) [48]. However, the resultant hypohydration level was lower in the sport drink training session (0.5%) compared with that of water (0.9%). This could be explained by a greater intestinal absorption of the fluid and lower urine output when, instead of plain water, sports drinks are chosen as a rehydration beverage [3].

The importance of hydration for children and adolescents is highlighted by the unique physical and physiological characteristics that may impair thermoregulation of children exercising in the heat [49]. It was observed that children, compared with adults, present a greater increase in core temperature as they become dehydrated [39] ,which may be related to their increased metabolic cost of locomotion [50]. Their higher surface area per body mass may cause a higher heat gain when air temperature is higher than skin temperature [51]. The heat acclimatization process, due to repeated heat exposure, may be delayed in children, as compared with adults [52]. Therefore, at the start of the warm season, it is possible that children are at greater risk for hypohydration and associated heat-related problems due to insufficient heat acclimatization parallel to an increase in physical activity levels [53,54]. Of concern is the elevation from 15 to 29 in the reported number of American high school and college deaths in football players due to heat stroke in the past 10 years [55]. A recent Policy Statement of the American Academy of Pediatrics [56] addresses the problem and provides preventive strategies to avoid heat-related illness among exercising children under climatic heat stress.

Older Individuals

There are some characteristics in older individuals that may interfere with their hydration status and thermoregulation during exercise. While some changes are expected, others are preventable and related to their trend for a lower level of physical activity.

Some of the expected aging changes are the decrease in: thirst perception, body water content, peripheral blood flow and possibly in sweat rate [57]. Lower thirst perception and fluid intake following water deprivation have been documented in older compared with younger individuals [58,59]. It was found that, for a given degree of hypohydration, older individuals, in comparison with younger adults, presented an inability to rehydrate, a decreased thirst perception and a greater core temperature [60]. This lower sensitivity for thirst perception may be explained by central regulatory mechanisms that affect both osmoreceptors and baroreceptors [61].

Because older individuals also have lower body water content and may have difficulty in identifying signs and symptoms of hypohydration, it is advisable to educate them about the importance of hydration in the context of exercising in the heat.

However, much of the heat intolerance of older individuals could be due to their less active or sedentary lifestyle, which impairs their aerobic fitness and acclimatization. Other aggravating factor can be the presence of chronic diseases such as hypertension and diabetes, kidney problems and the use of medications such as diuretics and beta-blockers. Old individuals who are physically active and on salt restriction (6 g of salt or 2.4 g of sodium a day) need not rule out sports drinks as a choice of rehydration when exercising. For example, the amount of sodium in 230 mL of a sports drink is about 0.110 g. It is unexpected that an individual ingests in one day such a large volume (5 L) of a sports drink to reach 2.4 g of sodium. Therefore, older individuals should be encouraged to rehydrate during or after exercise, but they should also consider the risks of excess water (i.e., hyponatremia) or sodium ingestion (i.e., hypertension) because they may be slower in excreting both the water and electrolytes.

HYDRATION FOR PHYSICAL ACTIVITIES

It is important to prepare effective hydration procedures to guarantee optimal performance and well-being in exercise activities that may potentially disturb body fluid homeostasis. Since sweating rate and fluid intake vary widely according to factors related to activity and environmental conditions, recommendations should, as far as possible, be individualized. Therefore, it is generally advisable to periodically evaluate the individual's tendency to body water changes during their usual physical activities. This can be simply determined by the change in body weight before and after exercise, having the individual wearing no or minimal clothes and their bladder emptied. On a regular basis, body hydration status of a resting condition and prior to an exercise event may be evaluated by blood and urine markers. Blood markers (e.g., osmolality and hematocrit) are more sensitive to acute changes of body hydration status, but they have been used mostly for research purpose since they are not as feasible to assess as urine markers (e.g., osmolality, specific gravity, and color). A urine osmolarity >700 mOsmol kg^{-1} is an indication of hypohydration [1]. The urine specific gravity can be determined using a handheld pocket refractometer in which values >1.020 indicate hypohydration, and >1.030 indicate significant hypohydration [62,63]. Even more feasible is the 8-level color chart which is handy for self-assessment and follow-up. This chart classifies as well hydrated (1–2), or minimal (3–4), significant (5–6) and serious (>6) hypohydration. Once an individual is aware of his/her hydration needs during exercise, other recommendations would include the type of beverage (composition) to be consumed and the timing of fluid ingestion before, during and after exercise.

Drink Composition and Choice of Rehydration Beverage

Plain water intake will likely be a proper rehydration choice for the majority of healthy individuals who follow adequate nutritional habits and perform physical activities lasting less than 60–80 min. However, under some circumstances, individuals may benefit from ingesting carbohydrate-electrolyte beverages which are commercially known as sports drinks and are available in various fruit-related flavors. The carbohydrate content will provide the energy required for prolonged or intermittent high intensity exercises. The electrolytes (sodium, chloride and potassium) help replace the losses from sweat, with sodium being the major one. Although sodium sweat concentration may vary widely (20 to 80 mmol L^{-1}), its concentration is lower than that of plasma (138–142 mmol L^{-1}). A risk of hyponatremia (serum sodium concentration <130 mmol L^{-1}) may exist during prolonged exercises in which heavy sweating is accompanied by excessive intake of sodium-free beverages [64]. The presence of sodium appears to stimulate thirst and retain body fluid, which may help a faster fluid replenishment after exercise (see below).

For comparative purposes, Table 38.1 shows the composition and osmolarity of common sports drinks and other common groups of beverages. The components of sports drinks are similar, as they have been developed on the basis of similar physiological principles to optimize rehydration under conditions of

TABLE 38.1 Carbohydrate-Electrolyte Composition and Osmolarity of Some Sports Drinks and Other Types of Beverages

	Carbohydrate	Sodium	Potassium	Osmolarity
		(g/100 mL)		(mOsm/kg.H_2O)
Sports Drinks				
Gatorade®	6	46	12	280
Powerade®	8	22.5	12	381
Lucozade®	6.4	50	12	285
Isostar®	7.7	70	18	322
Other Beverages				
Soda (Coca-Cola® Classic)	11	5	0	700
Skim milk (0.1%)	5.2	53	172	283
Orange Juice (Tropicana®)	10.8	0	190	663
Energy Drink (Red Bull®)	11.3	80.5	0	601
Oral Rehydration Solution (Pedialyte®)	2.5	104	79	250

considerable sweating. The carbohydrate portion of these beverages is generally composed of monosaccharides and disaccharides ranging from 6% to 9% weight/volume. The electrolyte portion of sports drinks may present some variation; however, concentrations are lower than (or at least equal to) those of the sweat. When combining these ingredients, further attention is given to obtaining an "isotonic" solution (i.e., beverage osmolality of ~300 mOsm.kg^{-1}). Such isotonic characteristics should facilitate gastric emptying and intestinal absorption, thus avoiding gastrointestinal discomfort and optimizing rehydration. Other beverages in Table 38.1 may not offer a suitable combination of carbohydrate and electrolyte content for an exercising condition. For example, a soda (or soft drink) may present an excess of carbohydrate, low in electrolytes, and still be hypertonic. "Energy drinks" usually add extra carbohydrate and substances that have no evidence of improvement in performance during exercise; these drinks may also contain stimulants and be potentially dangerous when ingested in excess [2]. Oral rehydration solutions (e.g., Pedialyte®) prescribed for the management of diarrhea-induced dehydration, have higher sodium and potassium concentrations and lower carbohydrate content.

Due to the presence of electrolytes, low fat milk has been shown to be effective for body fluid restoration in adults [65,66] and in children [67] after hypohydration induced by exercise. Skim milk has a similar amount of sodium compared with sports drinks and carbohydrate to replenish muscle glycogen; it also provides protein, which could enhance muscle recovery. Thus, as mentioned, skim milk could be a viable rehydration beverage for those individuals who are lactose tolerant.

Timing of Fluid Ingestion

Before Exercise

It is important to start physical activities in a euhydrated state, which can be achieved by daily adequate hydration. To ensure euhydration, especially in endurance events, fluid intake in the preceding 3 to 4 hours of an exercise event is recommended in amounts that may vary from 5 to 7 mL per kg of body mass (i.e., 350 to 500 mL for a 70 kg body weight subject) [1]. This gives time for the kidney to eliminate extra water. If no urine is eliminated, or if urine is dark (for example > #5 of the urine color chart previously mentioned), it is recommended to keep drinking gradually until a clearer urine is produced.

Hyperhydration prior to exercise does not appear to help performance or thermoregulation improvement [68]. Intake of substances that expand plasma volume may increase the risk of dilutional hyponatremia and associated symptoms, especially if drinking is excessive during exercise [1,64]. Another inconvenience of prior hyperhydration may be the need to void during competition [69]. In addition, as plasma expanders such as glycerol may work as a masking agent, their use is banned by the World Anti-Doping Agency [70].

During Exercise

During exercise, if necessary, individuals should drink periodically in amounts according to their sweating rate. Due to the great variability in body fluid losses, rather than stipulating fixed volume intakes, a reasonable approach is to educate the individual that body weight loss should not be greater than 2% during exercise [1]. As mentioned above, if the activity is prolonged (>1 h) or intense and intermittent, sports drinks may be advantageous compared with plain water.

After Exercise

It has been shown that, to achieve proper rehydration from prolonged exercise, the volume of fluid intake for every kg (or L) of body weight loss should be 1.5 L (or 1.5-fold) [71,72]. This may be due to the ongoing fluid losses from urine and sweat production that continues after prolonged exercise. Besides the volume, a beverage containing electrolytes (about 0.3–0.7 g of sodium per liter to replace losses), carbohydrate (to restore glycogen) and perhaps protein may be effective in promoting recovery [73]. This so-called "recovery" beverage has been used by endurance and by team-sport athletes between exercise sessions.

CONCLUSIONS

Body fluid may become disturbed under exercising conditions in which sweat losses are not matched by fluid intake. Hypohydration is more prevalent than hyperhydration or hyponatremia. Hypohydration in excess of 2% impairs endurance performance, as it "overloads" the cardiovascular and thermoregulatory systems, increases perceived exertion and decreases motivation, mainly when exercise is performed in the heat. Some studies have shown that hypohydration may also affect muscular strength and cognitive function; however, more studies are necessary to ascertain whether the performance of intermittent high intensity sports is impaired by hypohydration. Hypohydration is also an aggravating factor for exertional heat-illness with the young and older populations at particular risk.

To avoid hypohydration, fluid should be ingested during exercise so as to avoid a 2% loss of body weight. After exercise, body fluids should be restored rapidly, especially if recovery time is short between exercise events. Plain water intake is an option for most people who are involved in activities lasting <60–80 min. For longer or intense intermittent activities, sports drinks offer benefits due to their combination of carbohydrates and electrolytes that optimize body fluid restoration.

References

[1] Sawka MN, Burke LM, Eichner ER, Maughan RJ, Montain SJ, Stachenfeld NS. American college of sports medicine position stand. Exercise and fluid replacement. Med Sci Sports Exerc 2007;39:377–90.

[2] Rodriguez NR, Di Marco NM, Langley S. American college of sports medicine position stand. Nutrition and athletic performance. Med Sci Sports Exerc 2009;41:709–31.

[3] Murray R, Stofan J. Formulating carbohydrate-electrolyte drinks for optimal efficacy. In: Maughan RJ, Murray R, editors. Sports drinks: basic science and practical aspects. Boca Raton: CRC Press; 2001. p. 197–223.

[4] Lasater G, Piernas C, Popkin BM. Beverage patterns and trends among school-aged children in the US, 1989–2008. Nutr J 2011;10:103.

[5] Seifert SM, Schaechter JL, Hershorin ER, Lipshultz SE. Health effects of energy drinks on children, adolescents, and young adults. Pediatrics 2011;11(127):511–28.

[6] American Academy of Pediatrics. Clinical report–sports drinks and energy drinks for children and adolescents: are they appropriate? Pediatrics 2011;127(8):1182–9.

[7] Cheuvront SN, Kenefick RW, Montain SJ, Sawka MN. Mechanisms of aerobic performance impairment with heat stress and dehydration. J Appl Physiol 2010;109(6): 1989-695

[8] Sawka MN. Physiological consequences of hypohydration: exercise performance and thermoregulation. Med Sci Sports Exerc 1992;24:657–70.

[9] Nielsen B, Nybo L. Cerebral changes during exercise in the heat. Sports Med 2003;33(1):1–11.

[10] Jentjens RL, Wagenmakers AJ, Jeukendrup AE. Heat stress increases muscle glycogen use but reduces the oxidation of ingested carbohydrates during exercise. J Appl Physiol 2002;92:1562–72.

[11] Bergeron M. Youth sports in the heat, recovery and scheduling considerations for tournament play. Sports Med 2009;39:513–22.

[12] Meyer F, Volterman KA, Timmons BW, Wilk B. Fluid balance and dehydration in the young athlete: assessment considerations and effects on health and performance. AJLM 2012;6:489–501.

[13] McGregor S, Nicholas C, Lakomy H, Williams C. The influence of intermittent high intensity shuttle running and fluid ingestion on the performance of a soccer skill. J Sports Sci 1999;17:895–903.

[14] Edwards AM, Mann ME, Marfell-Jones MJ, Rankin DM, Noakes TD, Shillington DP. Influence of moderate dehydration on soccer performance: physiological responses to 45 min of outdoor match-play and the immediate subsequent performance of sport specific and mental concentration tests. Br J Sports Med 2007;41:385–91.

[15] Hoffman JR, Stavsky H, Falk B. The effect of water restriction on anaerobic power and vertical jumping height in basketball players. Int J Sports Med 1995;16:214–8.

[16] Dougherty KA, Baker LB, Chow M, Kenney WL. Two percent dehydration impairs and six percent carbohydrate drink improves boys basketball skills. Med Sci Sports Exerc 2006;38:1650–8.

[17] Baker LB, Dougherty KA, Chow M, Kenney WL. Progressive dehydration causes a progressive decline in basketball skill performance. Med Sci Sports Exerc 2007;39:1114–23.

[18] Carvalho P, Oliveira B, Barros R, Padrão P, Moreira P, Teixeira VH. Impact of fluid restriction and ad libitum water intake or an 8% carbohydrate-electrolyte beverage on skill performance of elite adolescent basketball players. Int J Sport Nutr Exerc Metab 2011;21:214–21.

[19] Judelson DA, Maresh CM, Anderson JM, Armstrong LE, Casa DJ, Kraemer WJ, et al. Hydration and muscular performance – Does fluid balance affect strength, power and high-intensity endurance. Sports Med 2007;37(10):907–21.

[20] Millet GY, Lepers R. Alterations of neuromuscular function after prolonged running, cycling and skiing exercises. Sports Med 2004;34(2):105–16.

[21] Taylor JL, Gandevia SCA. Comparison of central aspects of fatigue in submaximal and maximal voluntary contractions. J Appl Physiol 2008;104:542–50.

[22] Morrison S, Sleivert GG, Cheung SS. Passive hyperthermia reduces voluntary activation and isometric force production. Eur J Appl Physiol 2004;91:729–36.

[23] Nybo L, Nielsen B. Perceived exertion is associated with an altered electrical activity of the brain during exercise with progressive hyperthermia. J Appl Physiol 2001;91:2017–23.

[24] Nybo L, Nielsen B. Hyperthermia and central fatigue during prolonged exercise in humans. J Appl Physiol 2001;91:1055–60.

[25] Ftaiti F, Grélot L, Coudreuse JM, Nicol C. Combined effect of heat stress, dehydration and exercise on neuromuscular function in humans. Eur J Appl Physiol 2001;84:87–94.

[26] Hayes LD, Morse CI. The effect of progressive dehydration on strength and power: is there a dose-response? Eur J Appl Physiol 2009;108:701–7.

[27] Gopinathan PM, Pichan G, Sharma VM. Role of dehydration in heat stress-induced variations in mental performance. Arch Environ Health 1988;43:15–7.

[28] Cian C, Koulmann PA, Barraud PA, Raphel C, Jimenez C, Melin B. Influence of variations of body hydration on cognitive performance. J Psychophysiol 2000;14:29–36.

[29] Cian C, Barraud PA, Melin B, Raphel C. Effects of fluid ingestion on cognitive function after heat stress or exercise induced dehydration. Int J Psychophysiol 2001;42:243–51.

[30] Popkin BM, D'Anci KE, Rosenberg IH. Water, hydration, and health. Nutr Rev 2010;68(8):439–58.

[31] D'Anci KE, Vibhakar A, Kanter JH, Mahoney CR, Taylor HA. Voluntary dehydration and cognitive performance in trained college athletes. Percept Mot Skills 2009;109:251–69.

[32] Szinnai G, Schachinger H, Arnaud MJ, Linder L, Keller U. Effect of water deprivation on cognitive-motor performance in healthy men and women. Am J Physiol Regul Integr Comp Physiol 2005;289:R275–80.

[33] D'Anci KE, Constant F, Rosenberg IH. Hydration and cognitive function in children. Nutr Rev 2006;64:457–64.

[34] Bar-David Y, Urkin J, Kozminsky E. The effect of voluntary dehydration on cognitive functions of elementary school children. Acta Paediatr 2005;94:1667–73.

[35] Edmonds CJ, Burford D. Should children drink more water? The effects of drinking water on cognition in children. Appetite 2009;52:776–9.

[36] Edmonds CJ, Jeffes B. Does having a drink help you think? 6–7-year-old children show improvements in cognitive performance from baseline to test after having a drink of water. Appetite 2009;53:469–72.

[37] Meyer F, Bar-Or O. Fluid and electrolyte loss during exercise: the pediatric angle. Sports Med 1994;18:4–9.

[38] Meyer F, Bar-Or O, MacDougall D, Heigenhauser GJF. Sweat electrolyte loss during exercise in the heat: effects of gender and maturation. Med Sci Sports Exerc 1992;24:776–81.

[39] Bar-Or O, Dotan R, Inbar O, Rotshtein A, Zonder H. Voluntary hypohydration in 10–12-year-old boys. J Appl Physiol 1980;48:104–8.

[40] Bar-Or O, Blimkie CJR, Hay JA, MacDougall JD, Ward D, Wilson WM. Voluntary dehydration and heat intolerance in patients with cystic fibrosis. Lancet 1992;339:696–9.

[41] Aragón-Vargas LF, Wilk B, Brian W, Timmons BW, Bar-Or O. Body weight changes in child and adolescent athletes during a triathlon competition. Eur J Appl Physiol 2012; doi:10.1007/s00421-012-2431-8 June 7.

[42] Meyer F, Bar-Or O, Passe D, Salberg A. Hypohydration during exercise in children: effect on thirst, drink preferences and rehydration. Int J Sport Nutr 1994;4:22–35.

[43] Wilk B, Bar-Or O. Effect of drink flavor and NaCl on voluntary drinking and hydration in boys exercising in the heat. J Appl Physiol 1996;80:1112–7.

[44] Wilk B, Kriemler S, Keller H, Bar-Or O. Consistency in preventing voluntary dehydration in boys who drink a flavored carbohydrate-NaCl beverage during exercise in the heat. Int J Sport Nutr 1998;8:1–9.

[45] Rivera-Brown AM, Gutierrez R, Gutierrez JC, Frontera WR, Bar-Or O. Drink composition, voluntary drinking, and fluid balance in exercising trained, heat-acclimatized boys. J Appl Physiol 1999;86:78–84.

[46] Horswill CA, Passe DH, Stofan JR, Horn MK, Murray R. Adequacy of fluid ingestion in adolescents and adults during moderate-intensity exercise. Pediatr Exerc Sci 2005;17:41–50.

[47] Wilk B, Timmons BW, Bar-Or O. Voluntary fluid intake, hydration status, and aerobic performance of adolescent athletes in the heat. Appl Physiol Nutr Metab 2010;35:834–41.

[48] Bergeron M, Waller JL, Marinik EL. Voluntary fluid intake and core temperature responses in adolescent tennis players: sports beverage versus water. Br J Sports Med 2006;40:406–10.

[49] Falk B. Effects of thermal stress during rest and exercise in the pediatric population. Sports Med 1998;25:221–40.

[50] Krahenbuhl GS, Williams TJ. Running economy: changes with age during childhood and adolescence. Med Sci Sports Exerc 1992;24:462–6.

[51] Bar-Or O. Temperature regulation during exercise in children and adolescents. In: Gisolfi CV, Lamb DR, editors. Perspectives in exercise and sports medicine: youth and, exercise and sports. Indianapolis: Benchmark Press; 1989. p. 335–67.

[52] Inbar O, Dotan R, Bar-Or O, Gutin B. Conditioning versus exercise in heat as method for acclimatizing 8–10- year old boys to dry heat. J Appl Physio 1981;50:406–11.

[53] Bergeron MF, McKeag DB, Casa DJ, Clarkson PM, Dick RD, Eichner ER, et al. Youth football: heat stress and injury risk. Med Sci Sports Exerc 2005;37:1421–30.

[54] Godek SF, Godek JJ, Bartolozzi AR. Hydration status in college football players during consecutive days of twice-a-day preseason practices. Am J Sports Med 2005;33:843–51.

[55] Yard EE, Gilchrist J, Haileyesus T, Murphy M, Collins C, McIlvain N, et al. Heat illness among high school athletes—United States, 2005–2009. J Safety Res 2010;41:471–4.

[56] American Academy of Pediatrics. Policy statement—Climatic heat stress and exercising children and adolescents. Pediatrics 2011;128:1–7.

[57] Kenney WL. Thermoregulation at rest and during exercise in healthy older adults. Exerc Sport Sci Rev 1997;25:41–76.

[58] Phillips PA, Rolls BJ, Ledingham JG, Forsling ML, Morton JJ, Crowe MJ, et al. Reduced thirst after water deprivation in healthy elderly men. N Engl J Med 1984;311:753–9.

[59] Mack GW, Weseman CA, Langhans GW, Scherzer H, Gillen CM, Nadel ER. Body fluid balance in dehydrated healthy older men: thirst and renal osmoregulation. J Appl Physiol 1994;76:1615–23.

[60] Meisher E, Fortney SM. Responses to dehydration and rehydration during heat exposure in young and older men. Am J Physiol 1989;257:1050–6.

[61] Popkin BM, D'Anci KE, Rosenberg IH. Water, hydration, and health. Nutr Rev 2010;68(8):439–58.

[62] Armstrong LE, Pumerantz AC, Fiala KA, Roti MW, Kavouras SA, Casa DJ, et al. Human hydration indices: acute and longitudinal reference values. Int J Sport Nutr Exerc Metab 2010;20:145–53.

[63] Armstrong LE. Hydration assessment techniques. Nutr Rev 2005;63:40–54.

[64] Montain SJ, Sawka MN, Wenger CB. Hyponatremia associated with exercise: risk factors and pathogenesis. Exerc Sports Sc Rev 2001;29:113–7.

[65] Shirreffs SM, Watson P, Maughan RJ. Milk as an effective post-exercise rehydration drink. Br J Nutr 2007;91:173–80.

[66] Watson P, Love TD, Maughan RJ, Shirreffs SM. A comparison of the effects of milk and a carbohydrate-electrolyte drink on the restoration of fluid balance and exercise capacity in a hot, humid environment. Eur J Appl Physiol 2008;104(4):633–42.

[67] Volterman K, Obeid J, Wilk B, Timmons BW. Ability of milk to replace fluid losses in children after exercise in the heat. In: Williams CA, Armstrong N, editors. Children and exercise XXVII: the proceedings of the XXVII international symposium of the european group of pediatric work physiology. London and New York: Routledge; 2011. p. 101–5.

[68] van Rosendal SP, Osborne MA, Fassett RG, Coombes JS. Guidelines for glycerol use in hyperhydration and rehydration associated with exercise. Sports Med 2010;40:113–29.

[69] Latzka WA, Sawka MN, Montain SJ, Skrinar GS, Fielding RA, Matott RP, et al. Hyperhydration: tolerance and cardiovascular effects during uncompensable exercise-heat stress. J Appl Physiol 1998;84:1858–64.

[70] World Anti-Doping Agency. The 2012 prohibited list international standard, <http://www.ukad.org.uk/assets/uploads/Files/WADA_Prohibited_List_2012_EN.pdf>; 2012 [accessed 25.06.2012].

[71] Shirreffs SM, Maughan RJ. Volume repletion after exercise-induced volume depletion in humans: replacement of water and sodium losses. Am J Physiol 1998;274:F868–75.

[72] Shirreffs SM, Taylor AJ, Leiper JB, Maughan RJ. Post-exercise rehydration in man: effects of volume consumed and drink sodium content. Med Sci Sports Exerc 1996;28:1260–71.

[73] Spaccarotella KJ, Andzel WD. Building a beverage for recovery from endurance activity: a review. J Strength Cond Res 2011;25 (11):3198–204.

CHAPTER

39

Physiological Basis for Creatine Supplementation in Skeletal Muscle

William J. Kraemer[1], Hui-Ying Luk[1], Joel R. Lombard[2], Courtenay Dunn-Lewis[1] and Jeff S. Volek[1]

[1]Human Performance Laboratory, Department of Kinesiology, University of Connecticut, Storrs, CT, USA

[2]Department of Chemistry/Biology, Springfield College, Springfield, MA, USA

INTRODUCTION/OVERVIEW

Creatine (methylguanidino acetic acid) is a biochemical compound found mainly in skeletal muscle (~94–97%), with much smaller amounts in cardiac muscle [1,2]. Scholossmann and Tiegs first showed that diffused creatine (Cr) increased during muscular contraction, while Fiske and Subbarow [3] discovered phosphocreatine in the 1920s. Chevreul identified Cr in 1835 when investigating meat extract, and later, in 1847, Liebig discovered that Cr was in several different kinds of muscle and not in the other tissues that he examined [3]. The relationship between Cr and neuromuscular function has thus been of scientific interest for almost a century and its implications on muscular development are explored to this day.

Cr is an important physiological compound as part of the adenosine triphosphate (ATP)/phosphocreatine (PCr) phosphogen energy system. Cr and inorganic phosphate combine to form phosphocreatine, and a greater Cr pool allows for higher concentrations and/or rates of PCr biosynthesis in the muscle. The breakdown of PCr allows for increased and rapid biosynthesis of ATP. Thus, PCr is an immediate fuel reserve for the replenishment of ATP in the ATP/PCr energy system, and Cr is an important substrate supporting this system.

Skeletal muscle is the primary end point (or biological sink) for exogenous Cr taken in as a dietary supplement. Cr is not an essential dietary nutrient; the body can produce Cr through endogenous pathways, and the rates of biosynthesis maintain normal concentrations in the muscle. While not an essential nutrient, Cr is a safe, legal, and effective method to help enhance lean body mass by optimizing quality of the training as well as maximal strength and power performance. Cr has become one of the most popular nutritional supplements among athletes [4].

The purpose of this chapter is to overview the physiological basis for Cr supplementation. We will explore how this supplement is used as an ergogenic aid for athletes attempting to gain muscle and improve performance. We will further describe how these same attributes are important for other populations, including the elderly. Finally, the potential effects of Cr supplementation on medical disorders will be described.

CREATINE BIOSYNTHESIS, UPTAKE, AND DEGRADATION

The biosynthesis of Cr takes place primarily in the liver. It requires two amino acids: arginine and glycine. It also requires a methyl donor, typically S-adenosylmethionine (SAM) [2]. The initiation of Cr synthesis begins with the transfer of the amidino group of arginine to glycine through the enzyme L-Arginine : glycine amidinotransferase (AGAT). The products of this activity include L-ornithine and guanidinoacetic acid. Guanidinoacetic acid is then methylated by SAM through the enzyme S-adenosyl-L-methionine : N-guanidinoacetate methyltransferase (GAMT), resulting in the endogenous production of Cr (see Figure 39.1).

Nutrition and Enhanced Sports Performance.
DOI: http://dx.doi.org/10.1016/B978-0-12-396454-0.00039-4

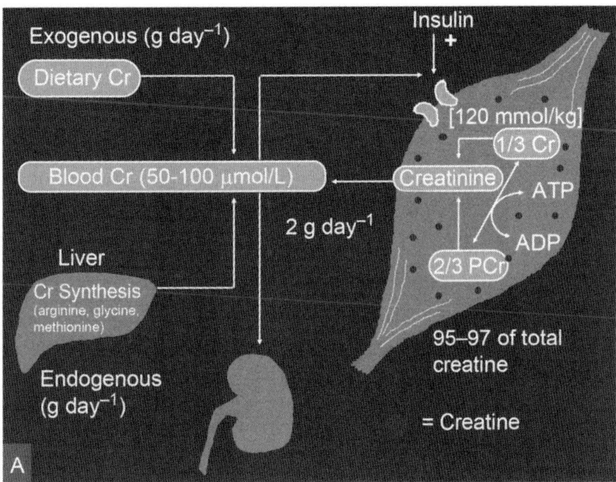

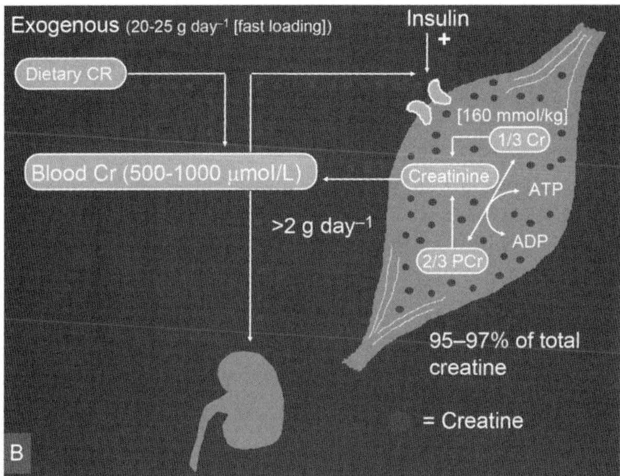

FIGURE 39.1 **Cybernetic interactions between creatine and muscle.** (A) Normal creatine dynamics without supplementation. (B) Creatine cybernetics with supplementation. Following supplementation, creatine from the liver plays a diminished role, but normal function returns after stopping supplementation. (This figure is reproduced in color plate section.)

TABLE 39.1 The Amount of Creatine in Common Foods

Food (1 g)	Cr Content (g)
Beef	0.0045
Pork	0.005
Cod	0.003
Herring	0.0065 to 0.01
Salmon	0.0045
Shrimp	trace
Tuna	0.004
Milk	0.0001
Cranberries	0.00002

In addition to endogenous synthesis, muscular Cr stores are obtained from exogenous dietary intake [2]. Foods highest in Cr come from the flesh of animals and fish. However, food preparation can have an impact on the total Cr obtained in a serving; well-cooked meats, for example, will have less Cr than red meat that is uncooked. Cr can be found in foods other than meat but typically in significantly smaller amounts. Table 39.1 lists some exogenous food sources of Cr and their respective amounts. Examination of this table reveals that the amount of Cr obtained from dietary sources is very low and that concentrations obtained through supplementation would be very hard to mirror through dietary intake alone. Cr loading (10 to 25 g per day) is very difficult to achieve with normal dietary intakes of regular food. Thus commercial products are needed for the higher concentrations of Cr that result in effective ergogenic effects (see Figure 39.2). It is not possible to perform a fast load with normal dietary food intakes.

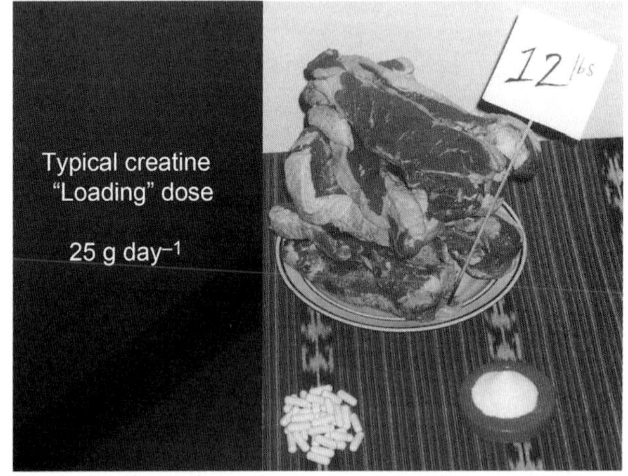

FIGURE 39.2 While many dieticians want to use natural food to produce the effects that supplements do (pills or powders) the amount of meat that would have to be consumed for 3 to 6 days for a fast-loading phase for creatine is not possible, as demonstrated in this comparison of steaks, pills and powder to equate to the 25 grams per day that would be ingested (e.g., 5 grams five times a day). (This figure is reproduced in color plate section.)

Creatine obtained through the diet has a high bioavailability, allowing it to pass through the digestive tract and directly into the bloodstream [2]. Plasma levels of Cr on average are $\sim 50\ \text{mmol L}^{-1}$. In one investigation, consuming 2 g of Cr in solution resulted in a peak plasma concentration of $400\ \text{mmol L}^{-1}$ at 30–60 min, while a similar dosing of Cr from steak resulted in a smaller peak but more extended plasma elevations [2]. Once in the bloodstream, Cr is either taken up by tissue (primarily skeletal muscle) or is excreted by the kidneys. Muscle uptake of Cr occurs via sodium-dependent

CreaT transporters within the muscle membrane [5]. Once Cr is in the myocyte (muscle cell), it becomes phosphorylated by the enzyme Cr kinase. The concentrations of phosphocreatine and Cr are based on the energy state of the cell, but typically ~40% is Cr and ~60% is in the phosphorylated form [6].

Cr degradation is on average approximately 2 g d^{-1}, which may be replenished exogenously to maintain muscular concentrations. Nonenzymatic breakdown of Cr occurs hepatically to form creatinine [2]. Creatinine, a membrane permeable molecule, is picked up by systemic blood flow and excreted to the urine by the kidneys.

BIOENERGETICS AND MECHANISMS OF ACTION

In addition to serving as a fuel reserve for ATP, Cr also plays a role in spatial energy and proton buffering as well as modulating glycolysis. Phosphocreatine (PCr) acts as an immediate fuel reserve for the replenishment of adenosine triphosphate (ATP) [5]. PCr works to rephosphorylate adenosine diphosphate (ADP) to ATP between rest and exercise via the enzymatic action of Cr kinase.

$$\text{PCr} + \text{ADP} \leftarrow \text{Cr kinase} \rightarrow \text{ATP} + \text{Cr} \qquad (39.1)$$

While ATP/PCr is a phosphogenic system, the resynthesis of PCr is an oxygen-dependent process that has both a fast and slow component to it [6]. The fast component lasts around 21–22 seconds, and the slow component tends to be over 170 seconds.

The amount of PCr in the sarcoplasm of the muscle determines both the rate and maintenance of the ATP turnover. The storage capacity for PCr is limited in the body and is depleted rapidly (with enough to last approximately 4–5 s) [7]. While Cr storage within skeletal muscle is low, its storage can grow with Cr supplementation to about 20%, with one third of Cr in the form of PCr [8,9]. This is particularly relevant to short, intense bouts of exercise, where PCr resynthesis affects the ATP turnover rate and is crucial to the success of exercise performance [7].

The primary energy system at the outset of short-burst, high-power, and high-force anaerobic activities is the ATP-PCr system. The ATP-PCr system will be central to the ergogenic effects of Cr as described further in our section on acute training effects.

LIMITS OF CREATINE STORAGE CAPACITY WITH SUPPLEMENTATION

Several studies have examined what is believed to be the maximal storage capacity for Cr within the muscle. In one investigation, pre- and post-exercise muscle biopsies of the vastus lateralis were examined after Cr supplementation of 5 g Cr monohydrate ingested 4–6 times a day for more than 2 days [9]. Mean total Cr content in the 17 subjects was found to be 126.8 mmol kg^{-1} dry muscle prior to the supplementation and 148.6 mmol kg^{-1} after. Both PCr and Cr content increased, but there was no subsequent increase in the ATP concentrations. The study concluded that approximately 155 mmol kg^{-1} of dry muscle may be the maximum amount of total Cr stored when supplementing dosages in the range 20–30 g day^{-1}. Many studies use rapid loading schemes to assure such increases.

Increases in Cr among the subjects may be anywhere between 20% and 40% of resting values. Increases in muscle Cr concentrations after supplementation appear to be inversely related to the initial concentration. For this reason, there is significant variability seen in the initial gain in mass; for example, the more Cr that enters the muscle, the more water will be retained. Individuals who have near maximal levels of Cr will not see great increases compared with those who start with minimal Cr stores. Thus, there appears to be a genetically determined dichotomy between "responders" and "non-responders", as identified by body mass gains after a loading phase is completed, and those who are non-responders do not see the benefits of Cr supplementation.

BODY MASS, BODY COMPOSITION, AND BODY WATER

During the initial loading phase of Cr supplementation, gains in body mass are typically attributed to an increase in water retention [10]. Water retention is caused by the osmotic activity of Cr; the increase in Cr concentration within the sarcoplasm of the muscle fiber draws water into the cell. It is because of this action that "cell swelling" or "Cr bloat" is initiated. While most mass gained during the loading phase of Cr supplementation is water, evidence indicates such a signal is part of the initial hypertrophic mechanisms that stimulate protein synthesis.

Increase in water weight influences hydrostatic weighing numbers such that gains obtained by the subjects appear with an increase in percent body fat [11]. If this occurs, a "false" increase in body fat should be observed. However, in one investigation with 7 days of Cr loading at 0.3 g kg^{-1}, older men had experienced a 7.7 to 11.0 lb (3.5 to 5.0 kg) increase in body mass despite a slight decrease in body fat; this indicated that the mass gained was not entirely due to increased water weight [12]. Even if all of the Cr ingested had been stored, it would have accounted for less than a pound of the weight gain observed. As previously described, the acute increases in body mass demonstrated after 1 week of supplementation may be due in part to hydrostatic stretching triggering a protein synthesis mechanism. While theoretical, no other mechanisms have been identified for how Cr may stimulate protein synthesis in this manner.

MUSCLE FIBER TYPE ADAPTATIONS

Muscle recruitment is related to the recruitment of Type I and Type II motor units containing Type I and Type II muscle fibers, respectively. Type I skeletal muscle fibers are characterized by thicker non-contractile proteins (e.g., Z lines and titan), slower myosin ATPase enzymes, and reliance upon reductions in degradation rates rather than increases in synthesis rates. Conversely, Type II skeletal muscle fibers have less dense non-contractile proteins, faster myosin ATPase enzyme, and rely more on increasing protein synthesis rates to increase cross-sectional fiber size. The electrical threshold of each motor unit dictates the recruitment of motor units and their associated muscle fibers, from low threshold (Type I muscle fibers) to high threshold (Type II muscle fibers) (this is known as the Size Principle). The recruitment of higher threshold motor units with heavy resistance also recruits the lower threshold motor units but stimulates overall greater muscle fiber hypertrophy.

The metabolic system used in obtaining ATP for the myosin motor of each muscle fiber is dependent upon the exercise stress. For aerobic activities, the repetitive use of lower-threshold motor units relies upon the oxidative metabolic system. Conversely, short-burst, heavy-load, and high-power exercises depend upon the ATP/PCr and anaerobic glycolysis metabolic systems. As previously described, this Type II ATP-PCr energy focus during exercise is where Cr loading has the opportunity to be beneficial to the bioenergetics of the tasks.

A characteristic of PCr within the sarcoplasm of skeletal muscle is that is appears to be fiber Type dependent. Type II fibers can generate a quick muscle action due to the fast myosin ATPase isoform that makes up most of its cross-bridge heads. The faster the ATPase is, the faster the hydrolysis of ATP, and the quicker the muscle actions. Type II fibers at rest have a 5–15% higher PCr concentration than Type I fibers because of their role in high force/power activities. The rate of PCr degradation is greater in Type II fibers than in Type I during sprint exercise lasting 10–30 seconds [13]. However, Type I fibers were shown to resynthesize PCr at a slightly faster rate than Type II fibers in recovery from sprint exercises, as they are recruited (and PCr depleted) whenever Type II fibers are recruited (according to the size principle) [14]. In one study, Cr supplementation increased the mRNA expression of myosin heavy chain contractile proteins when combined with resistance training [15]. Supplementation with Cr has been show to increase both fiber types' concentrations of Cr/PCr, with a trend towards larger increases in Type II fibers because of their larger size.

ACUTE ANAEROBIC BENEFITS

Cr monohydrate supplementation can improve the quality of any exercise that stresses the ATP-PCr system [7]. As previously described, the acute effect of Cr supplementation is to enlarge the pool of PCr for rapid ATP resynthesis, delaying the depletion of PCr and delaying fatigue. An increase in the rate of PCr resynthesis during rest can allow for a higher storage of PCr at the beginning of each subsequent set [8]. The increased quantity of free Cr and PCr also allows for improved buffering of the ADP that results from ATP hydrolysis. Cr can therefore enhance the capacity to perform short bouts of exercise, delay the onset of fatigue, promote recovery between sets, [16,17] and improve training quality. The overall mechanisms for Cr's ergogenic effects are shown in Figure 39.3.

In view of the metabolic pathways that mediate Cr action, it is evident that Cr loading works best for strength and power. Supplementation acutely increases muscular strength and increases the number of repetitions performed at a given resistance load after 5 to 7 days (with or without training) [6,16,18]. Within 5 to 7 days, Cr supplementation (20–30 g d^{-1}) also helps to maintain power and force output with repeated jumping [6,16,19], swimming [20,21], and exhaustive bouts of cycling [22]. (See Figure 39.4 for an example of the squat jump power output [6].) Arciero et al. [23] conducted an investigation on the effect of Cr without training and with resistance training. Groups that did not train had an 8% and 16% increase in maximal bench press and leg, respectively, while groups that trained showed an 18% and 42% increase, respectively. Acute Cr ingestion explained 40% of strength improvements, while the remaining 60% may be due to other mediating mechanisms. This, increases in intracellular Cr can improve strength and power exercises by increasing PCr storage.

CHRONIC ANAEROBIC ADAPTATIONS

Chronic adaptations to Cr supplementation may arise from the accumulation of acute improvements to

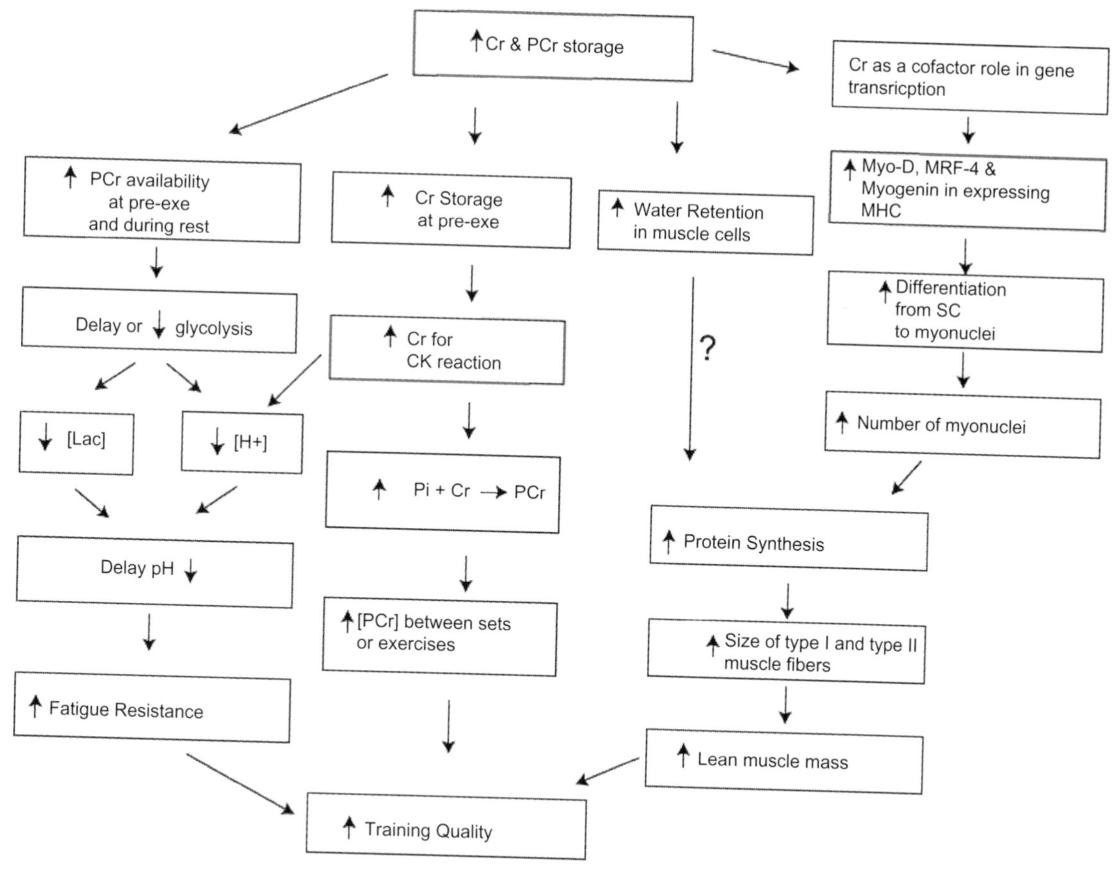

FIGURE 39.3 **A multitude of mechanisms exist to mediate creatine's effects in the body.**

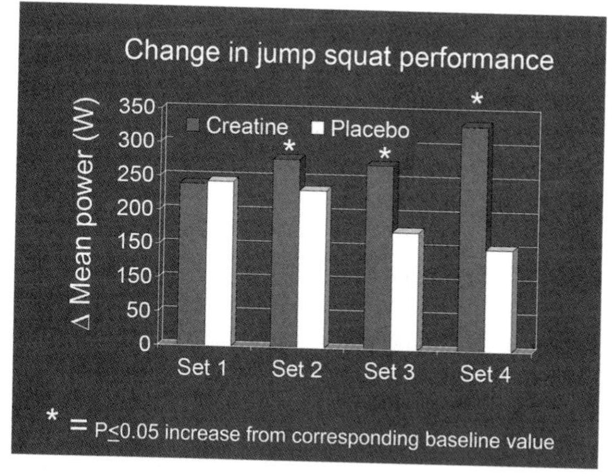

FIGURE 39.4 **Acute response of squat jump performances with and without short-term creatine loading.** *Modified from Volek and Kraemer [6].* (This figure is reproduced in color plate section.)

force/power production with each repetition and total exercise volume (repetitions × sets × load) [4]. In a training program with controlled workloads and repetitions, there was an increase in total work performed with Cr supplementation as compared with placebo groups over several weeks of training [16,24]. As previously mentioned, this increase in performance may be attributed to increases in the pool of Cr and PCr in the muscles, thereby increasing training intensity, training stimulus, and ultimately, physiological adaptations to training.

Muscular Hypertrophy and Activation of Satellite Cells

The increased work, force, or power production with Cr supplementation creates a higher-quality exercise stimulus and a faster rate of muscle fiber hypertrophy over time. Twelve weeks of heavy resistance training combined with either $6\ g\ d^{-1}$[15] or $25\ g\ d^{-1}$ for a 1 week loading phase with a $5\ g\ d^{-1}$ maintenance dose [25] increased fat free mass [15,25], thigh volume [15], cross-sectional area [25,26], and muscle strength [15,25,26]. Cr supplementation augments the training-related gains not only in the Type II muscle fibers but also the Type I muscle fibers, as shown in Figure 39.5 [25].

The various mechanisms that explain muscle hypertrophy (or an increase in cross-sectional area) are not completely understood, but our understanding is slowly evolving. Skeletal muscle is unique, with

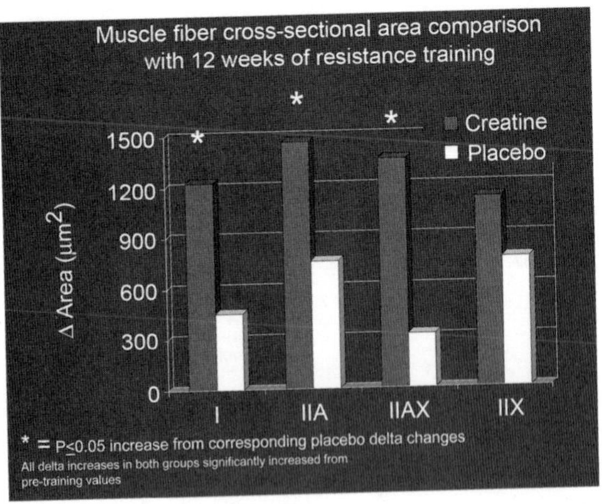

FIGURE 39.5 Changes in muscle fiber cross-sectional areas of Type I and Type II muscle fibers with resistance training with or without creatine supplementation. *Modified from Volek et al. [25].* (This figure is reproduced in color plate section.)

multinucleated cells and each myonucleus responsible for a given amount of fiber protein (i.e., myonuclear domain). An initial increase in cross-sectional area of muscle is mainly due to the increase in the myonucleus protein domain—in other words, an increase in the amount of protein in a fiber that is controlled by one myonucleus. However, increasing muscle fiber size greater than about 15–25% may require the addition of more myonuclei to manage the larger muscle fiber protein mass [27]. This is accomplished through satellite cells, an undifferentiated pool of stem cells found between the sarcolemma membranes of the skeletal muscle [28]. These cells contribute daughter myonuclei to address the demands for viable control of myonuclear domains with increased protein accretion. Satellite cells can also differentiate into myoblasts for muscle fiber repair and regeneration. In adulthood, the satellite cells are in a quiescent stage. Stimulation induced by loading (e.g., eccentric stress) and/or along with hormones and cytokines is required to activate the satellite cells, so that they can undergo differentiation and proliferation in part producing myoblasts or contribution to a new generation of myonuclei [28]. Interestingly, a group of investigators showed increases in the absolute satellite cell number per fiber and relative number of satellite cells in both a Cr and protein group during weeks 4, 8, and 16 of resistance training with $6\,g\,d^{-1}$ supplementation [26]. However, only the Cr group showed an increase in muscle myonuclei content per fiber and increase in mean muscle fiber cross-sectional area across the 16 weeks. These results may indicate that Cr plays an important role in differentiating the satellite cells into myonuclei, allowing an enhanced amount of protein synthesis when the number of myonuclei is a limiting factor in muscle fiber hypertrophy. As noted previously, a positive association has been shown between area of the muscle fibers and myonuclei content [26].

From stimulating the satellite cells to generating new myonuclei, the myogenic regulatory factors (MRF) play several roles in differentiating myonuclei. Myogenic regulatory factors include Myo-D, myogenin, MRF-4, and Myf5. These proteins act as transcription activators, initiating transcription and regulating gene expression for myosin heavy chains (MHC) and myosin light chains [29]. Cr supplementation has been shown to play a cofactor role in gene transcription together with Myo-D, MRF-4, and myogenin in regulating the expression of MHC [29]. One investigation used 2 weeks of lower-limb immobilization and followed it with 10 weeks of Cr supplementation [30]. The first week of Cr loading ($20\,g\,d^{-1}$) was followed by a 9-week maintenance dose using $5\,g\,d^{-1}$ during which resistance training was performed 3 days per week. The results demonstrated a positive relationship in muscle fiber size and the MRF-4 protein expression [30]. Based on these results, the authors suggested that the increase in the muscle fiber size was due to the increased number of myonuclei donated from the satellite cells, and Cr played a role in enhancing the myogenic transcription factor MRF-4 [30]. Thus, Cr supplementation appears to interface with the cellular mechanisms of muscle hypertrophy associated with resistance training.

ENDURANCE EXERCISE ACUTE EFFECTS AND ADAPTATION

Acute

While there is convincing evidence for the positive effects of supplementation with Cr on anaerobic exercise, there is less evidence for its benefit in endurance exercise. Cr supplementation does not appear to benefit performance with 120% maximal oxygen consumption for treadmill running [17], repeated 60 m dash [31], or the 700 m run [32]. Lack of an effect may be explained by confounding experimental factors; the bioenergetic ATP/PC mechanism may not have been the most significant mediating mechanism for the muscular force or power in the event, the subject populations in the study may have had non-responders as part of the subject pool, and/or more importantly, the test-retest reliability of the test may not have been high enough to pick up a 5–10% treatment effect. Many endurance athletes still use Cr, although not for their event; its purpose is to improve interval sprint training and strength training that enhance ground reaction forces, exercise efficiency, and exercise economy.

The enhanced muscle Cr and PCr storage with Cr supplementation can help reduce the dependence on the glycolytic pathway, cause shifts relative to the oxidative pathway, reduce blood lactate, maintain blood pH, and either prolong exercise duration or increase exercise intensity. Five days of Cr supplementation in rats lowered blood lactate in response to 6 bouts of high-intensity intermittent maximal exercise test when compared with a placebo group. Glycogen storage in the soleus was not significantly higher with Cr than with placebo supplementation; however, Cr supplementation showed significantly higher glycogen storage post exercise compared with placebo [33]. Another group observed a ~42% reduction in basal lactate production and a ~40% increase in basal CO_2 production in the Cr group when compared with the control [34]. Thus, intermittent exercise performance can be enhanced with Cr supplementation in part by maintaining pH.

Chronic

Maximum oxygen consumption (VO_{2max}), a measure of endurance performance, is measured via indirect spirometry using various incremental exercise-testing protocols (typically on a cycle ergometer or treadmill) to volitional fatigue. Following Cr supplementation, VO_{2max} either did not change [35,36] or slightly increased [17]. In terms of submaximal VO_2, studies have shown increases [37,38] and decreases at a given intensity (that is, improved performance at a given intensity) [39–42]. Authors have speculated that the significant reduction in submaximal VO_2 after Cr supplementation would be due to improved maintenance of the ATP:ADP ratio, postponing the activation on the aerobic metabolism pathway, and thus improving the exercise efficiency [39,43].

Time to exhaustion (enhanced resistance to fatigue) is another way to measure endurance capacity. Cr supplementation has been shown to be most effective for time to exhaustion with shorter-duration exercise [17,43–46] and the least effective in longer-duration exercise [37,39,40]. The anaerobic metabolic pathways of shorter-duration exercise may once again explain these results.

Substrate Use

In shorter-duration testing (3 to 12 minutes), Cr supplementation may play a role in preserving muscle glycogen during exercise by delaying the use of the glycolytic pathway in favor of the ATP-PCr pathway. Post-exercise muscle glycogen storage is enhanced by Cr supplementation with carbohydrate [47] and with regular diets, similar to a typical carbohydrate-loading diet [48]. Hyperglycemic mice experienced significant decrease in blood glucose with Cr supplementation, as well as improving the glucose responses after intravenous injection of glucose [49].

It is theorized that the glucose response in the muscle after Cr supplementation may be caused by direct alterations inside the cell membrane to glucose transporter 4 (GLUT4). GLUT4 is translocated to the surface of the cell (typically in response to a conformational signal from insulin) to take up glucose from the bloodstream. It was reported that GLUT 4 content increased by ~40% in vastus lateralis with Cr supplementation after an immobilization of the lower limb for 2 weeks followed by 10 weeks heavy resistance training [48]. Furthermore, an increase in GLUT 4 translocation has been shown in patients with Type 2 diabetes after a $5\,g\,d^{-1}$ supplement of Cr combined with $2\,d\,wk^{-1}$ exercise for 12 weeks [50]. Some *in vitro* studies of non-contracting cell lines have not seen the same effects. It is speculated the inconsistent results might be due to the different study protocol, since studies showed a positive alteration in exercising muscle glycogen storage [47,48] and GLUT 4 translocation [50] after Cr supplementation but not in the non-contracting cells [34,48,51]. Insulin is an important signal to trigger muscle glucose uptake from the circulation, but insulin does not increase after 5 days and 3 days of $20\,g\,d^{-1}$ Cr supplementation [52,53] or 5 days [51] at $300\,mg\,kg^{-1}\,d^{-1}$. As serum insulin concentrations are not altered with Cr ingestion, Cr may have a separate role in stimulating GLUT 4 activity.

Since an alteration in GLUT4 translocation and in muscle glucose storage has been shown to occur post-exercise, it is crucial to understand what causes it. AMP-activated protein kinase (AMPK) is one of the energy-sensing proteins that monitor the change in energy states in the muscle and trigger different metabolic pathways to maintain ATP concentrations [54]. AMPK is activated and inhibited by the increased concentration of adenosine monophosphate (AMP) and ATP, respectively. AMP concentrations increase and PCr concentrations decrease especially during and after exercise, which activates AMPK allosterically [54]. Activated AMPK causes an elevation in GLUT 4 concentration [55] with a positive association between changes in AMPK-α and GLUT 4 translocation [50]. The increases in either GLUT4 or muscle glycogen storage with Cr supplementation following exercise are believed to coincide with an alteration in energy state of the muscle. Contributors to this phenomenon may include a decrease in the ATP : AMP ratio, an increase in the AMPK activation, an increase in GLUT4 translocation, and enhanced muscle glycogen storage. However, elevations in AMPK can be counterproductive to protein synthesis, rendering the mechanisms that interact with this process unclear.

The ratio between PCr and TCr (total Cr) is also an indication of energy state in the muscle [48], and stimulation of AMPK can be altered by the ratio [56]. Copious studies have reported a fall in PCr : TCr ratio after Cr supplementation, altering the energy state in the cell [51–53]. The fall in the ratio may be mainly due to the disproportional increase in the total Cr and PCr after Cr supplementation [51–53]. For example, after 5 weeks of 300 mg g^{-1} d^{-1} Cr supplementation, rat plantaris muscles exhibited an increase of 23% and 12% in free Cr and TCr, respectively. The calculated PCr was not significantly different between Cr and control group. It follows that the PCr : TCr ratio decreased in the supplemented rats (0.26) when compared with the control rats (0.33) [51]. Ceddia [34] also observed a drop in the ratio between control (0.67) and Cr (0.37) cells with an ~1.8-fold increase in AMPK phosphorylation.

In summary, Cr and PCr storage in muscle increases, AMPK phosphorylation is enhanced, GLUT4 activation increases, and glycogen storage in muscle increases after Cr supplementation combined with exercise. How this constellation of events impacts protein synthesis under the circumstances of an increase in AMPK is counter to optimal anabolic conditions for muscle proteins. Some investigations suggest that Cr supplementation improves exercise economy and substrate utilization with exercise. While there appear to be some positive benefits of Cr supplementation on aerobic exercise performance, this area warrants further investigation.

ROLE OF CREATINE SUPPLEMENTATION FOR USE TO COMBAT AGING

Both muscular power and strength in men and women peak between the ages of 20 and 35 years [57]. While strength and power reductions are slow at first, at approximately the 60th year of life significant decreases start to be observed [58]. Decreased muscle mass (sarcopenia) is one of the primary reasons for the age-related reductions in strength and power [59]. Atrophy occurs primarily in Type II skeletal fibers and is responsible for the reduction in contractile strength and power of muscle [60,61]. As a result of decreased strength and power, common activities (such as getting up and down or climbing stairs) become progressively more challenging, leading to a diminished quality of life.

Cr supplementation in the elderly has been shown to increase dynamic muscular strength, isometric strength, lower-body explosive power, and lower-extremity functional capacity. Following 7 days of Cr supplementation in men aged 59–73, significant improvements were seen in knee extension/flexion maximal force, lower-body peak and mean power, sit and stand, and the tandem gait tests in the Cr supplemented group while not in the placebo group [12]. The Cr group also showed significant increases in body mass as well as fat free mass when compared with the placebo group [12]. In a similar study of elderly women, improvements were seen in the sit-stand and tandem gait test, leg press, maximal bench press, and mean upper- and lower-body in the Cr supplemented group. Neither group showed an increase in peak power. While no body fat changes were recorded, body mass and fat free mass were significantly increased in the Cr supplemented group. No significant changes were noted in the placebo group for any of the trials [57]. The previous two studies on the elderly indicate that Cr supplementation is beneficial for improving the quality of life through the increases in physical performances.

THERAPEUTIC USE OF CREATINE SUPPLEMENTATION

In addition to improving exercise performance, Cr supplementation has potential therapeutic roles. Others have described benefits for CNS function and connective tissue (for a more thorough description of Cr's therapeutic roles, please refer to Gualano et al. [62]). Here we describe the potential therapeutic role of Cr in muscle wasting and insulin resistance.

Muscle-wasting disorders (e.g., resulting from prolonged bed rest and a sedentary lifestyle) may benefit from Cr's muscle hypertrophic effects and increases to transcription factor post-exercise. As previously discussed, Cr supplementation influences MHC gene expression. Furthermore, Cr plays a role in differentiating satellite cells to myonuclei, meaning it may enhance the protein synthesis in muscle and therefore increase muscle mass and reduce fat mass. According to the previously mentioned studies, Cr supplementation combined with exercise attenuates muscle loss and facilitates muscle growth.

Another emerging role is the impact of Cr supplementation on insulin resistance. Exercise with Cr supplementation improves GLUT4 content and glucose tolerance in animals with Type II diabetes and Huntington's disease. The observations of enhanced AMPK signaling concurrent with supplementation may point to a possible mechanism. Stimulation of

AMPK signaling is associated with the energy state, ATP : AMP ratio, and PCr : TCr ratio after exercise; therefore, Cr might play a role in altering the ratio of PCr to TCr. This is theoretical, as there is no research directly showing an association between Cr and glucose uptake. Cr supplementation may enhance GLUT 4 by ~40% after 2 weeks immobilization followed by 10 weeks resistance training. Gualano et al. [62] showed improved glucose tolerance in sedentary men supplemented with $\sim 10\,\mathrm{g\,d^{-1}}$ Cr compared with placebo after 3 months of aerobic exercise.

SUMMARY

Cr is a safe and effective supplement. Although we have highlighted its potential role in muscle wasting and insulin resistance here, a number of possible therapeutic benefits may emerge through Cr supplementation. In addition to its beneficial role particularly for performance of anaerobic exercise in the context of aging, it may have wide-ranging implications for health and well-being in the future. A host of other potential applications, including impacting traumatic brain injury, ALS and other neurological pathologies, are opening fields of research in creatine. Its role in increasing the metabolic capability of the neuromuscular system is fundamental to the benefits realized by this unique and simple supplementation method.

References

[1] Greenhaff PL. Creatine and its application as an ergogenic aid. Int J Sport Nutr 1995;(Suppl: S100–10):5.

[2] Volek JS, Ballard KD, Forsythe CE. Overview of creatine metabolism. In: Stout JR, Antonio D, Kalman D, editors. Essentials of creatine in sports and health. Totowa: Humana Press; 2008.

[3] Needham DM. Cambridge Eng Machina carnis; the biochemistry of muscular contraction in its historical development. Cambridge University Press; 1971.

[4] Volek JS, Rawson ES. Scientific basis and practical aspects of creatine supplementation for athletes. Nutrition 2004;20(7–8):609–14.

[5] Terjung RL, Clarkson P, Eichner ER, Greenhaff PL, Hespel PJ, Israel RG, et al. American college of sports medicine roundtable. The physiological and health effects of oral creatine supplementation. Med Sci Sports Exerc 2000;32(3):706–17.

[6] Volek JS, Kraemer WJ. Creatine supplementation: its effect on human muscular performance and body composition. J Strength Cond Res 1996;10(3):200–10.

[7] Kraemer WJ, Volek JS. Creatine supplementation. Its role in human performance. Clin Sports Med 1999;18(3):651–66 ix.

[8] Greenhaff PL, Bodin K, Soderlund K, Hultman E. Effect of oral creatine supplementation on skeletal muscle phosphocreatine resynthesis. Am J Physiol 1994;266(5 Pt 1):E725–30.

[9] Harris RC, Soderlund K, Hultman E. Elevation of creatine in resting and exercised muscle of normal subjects by creatine supplementation. Clin Sci (Lond) 1992;83(3):367–74.

[10] Powers ME, Arnold BL, Weltman AL, Perrin DH, Mistry D, Kahler DM, et al. Creatine supplementation increases total body water without altering fluid distribution. J Athl Train 2003;38(1):44–50.

[11] Giandola RN, Wiswell RA, Romero G. Body composition changes resulting from fluid ingestion and dehydration. Res Q 1977;48:299–303.

[12] Gotshalk LA, Volek JS, Staron RS, Denegar CR, Hagerman FC, Kraemer WJ. Creatine supplementation improves muscular performance in older men. Med Sci Sports Exerc 2002;34 (3):537–43.

[13] Greenhaff PL, Nevill ME, Soderlund K, Bodin K, Boobis LH, Williams C, et al. The metabolic responses of human type I and II muscle fibres during maximal treadmill sprinting. J Physiol 1994;478(Pt 1):149–55.

[14] Soderlund K, Hultman E. ATP and phosphocreatine changes in single human muscle fibers after intense electrical stimulation. Am J Physiol 1991;261(6 Pt 1):E737–41.

[15] Willoughby DS, Rosene J. Effects of oral creatine and resistance training on myosin heavy chain expression. Med Sci Sports Exerc 2001;33(10):1674–81.

[16] Volek JS, Kraemer WJ, Bush JA, Boetes M, Incledon T, Clark KL, et al. Creatine supplementation enhances muscular performance during high-intensity resistance exercise. J Am Diet Assoc 1997;97(7):765–70.

[17] Balsom PD, Harridge SD, Soderlund K, Sjodin B, Ekblom B. Creatine supplementation per se does not enhance endurance exercise performance. Acta Physiol Scand 1993;149(4):521–3.

[18] Volek JS, Ratamess NA, Rubin MR, Gomez AL, French DN, McGuigan MM, et al. The effects of creatine supplementation on muscular performance and body composition responses to short-term resistance training overreaching. Eur J Appl Physiol 2004;91(5–6):628–37.

[19] Bosco C, Tihanyi J, Pucspk J, Kovacs I, Gabossy A, Colli R, et al. Effect of oral creatine supplementation on jumping and running performance. Int J Sports Med 1997;18(5):369–72.

[20] Grindstaff PD, Kreider R, Bishop R, Wilson M, Wood L, Alexander C, et al. Effects of creatine supplementation on repetitive sprint performance and body composition in competitive swimmers. Int J Sport Nutr 1997;7(4):330–46.

[21] Peyrebrune MC, Nevill ME, Donaldson FJ, Cosford DJ. The effects of oral creatine supplementation on performance in single and repeated sprint swimming. J Sports Sci 1998;16 (3):271–9.

[22] Dawson B, Cutler M, Moody A, Lawrence S, Goodman C, Randall N. Effects of oral creatine loading on single and repeated maximal short sprints. Aust J Sci Med Sport 1995;27 (3):56–61.

[23] Arciero PJ, Hannibal III NS, Nindl BC, Gentile CL, Hamed J, Vukovich MD. Comparison of creatine ingestion and resistance training on energy expenditure and limb blood flow. Metabolism 2001;50(12):1429–34.

[24] Burke DG, Chilibeck PD, Parise G, Candow DG, Mahoney D, Tarnopolsky M. Effect of creatine and weight training on muscle creatine and performance in vegetarians. Med Sci Sports Exerc 2003;35(11):1946–55.

[25] Volek JS, Duncan ND, Mazzetti SA, Staron RS, Putukian M, Gomez AL, et al. Performance and muscle fiber adaptations to creatine supplementation and heavy resistance training. Med Sci Sports Exerc 1999;31(8):1147–56.

[26] Olsen S, Aagaard P, Kadi F, Tufekovic G, Verney J, Olesen JL, et al. Creatine supplementation augments the increase in satellite cell and myonuclei number in human skeletal muscle induced by strength training. J Physiol 2006;573 (Pt 2):525–34.

[27] Kadi F, Thornell LE. Concomitant increases in myonuclear and satellite cell content in female trapezius muscle following strength training. Histochem Cell Biol 2000;113(2):99–103.

[28] Snijders T, Verdijk LB, van Loon LJ. The impact of sarcopenia and exercise training on skeletal muscle satellite cells. Ageing Res Rev 2009;8(4):328–38.

[29] Willoughby DS, Rosene JM. Effects of oral creatine and resistance training on myogenic regulatory factor expression. Med Sci Sports Exerc 2003;35(6):923–9.

[30] Hespel P, Op't Eijnde B, Van Leemputte M, Urso B, Greenhaff PL, Labarque V, et al. Oral creatine supplementation facilitates the rehabilitation of disuse atrophy and alters the expression of muscle myogenic factors in humans. J Physiol 2001;536(Pt 2):625–33.

[31] Redondo DR, Dowling EA, Graham BL, Almada AL, Williams MH. The effect of oral creatine monohydrate supplementation on running velocity. Int J Sport Nutr 1996;6(3):213–21.

[32] Terrillion KA, Kolkhorst FW, Dolgener FA, Joslyn SJ. The effect of creatine supplementation on two 700-m maximal running bouts. Int J Sport Nutr 1997;7(2):138–43.

[33] Roschel H, Gualano B, Marquezi M, Costa A, Lancha Jr. AH. Creatine supplementation spares muscle glycogen during high intensity intermittent exercise in rats. J Int Soc Sports Nutr 2010;7(1):6.

[34] Ceddia RB, Sweeney G. Creatine supplementation increases glucose oxidation and AMPK phosphorylation and reduces lactate production in L6 rat skeletal muscle cells. J Physiol 2004;555(Pt 2):409–21.

[35] Reardon TF, Ruell PA, Fiatarone Singh MA, Thompson CH, Rooney KB. Creatine supplementation does not enhance submaximal aerobic training adaptations in healthy young men and women. Eur J Appl Physiol 2006;98(3):234–41.

[36] Miura A, Kino F, Kajitani S, Sato H, Fukuba Y. The effect of oral creatine supplementation on the curvature constant parameter of the power-duration curve for cycle ergometry in humans. Jpn J Physiol 1999;49(2):169–74.

[37] Rico-Sanz J, Mendez Marco MT. Creatine enhances oxygen uptake and performance during alternating intensity exercise. Med Sci Sports Exerc 2000;32(2):379–85.

[38] Stroud MA, Holliman D, Bell D, Green AL, Macdonald IA, Greenhaff PL. Effect of oral creatine supplementation on respiratory gas exchange and blood lactate accumulation during steady-state incremental treadmill exercise and recovery in man. Clin Sci (Lond) 1994;87(6):707–10.

[39] Nelson AG, Day R, Glickman-Weiss EL, Hegsted M, Kokkonen J, Sampson B. Creatine supplementation alters the response to a graded cycle ergometer test. Eur J Appl Physiol 2000;83(1):89–94.

[40] Murphy AJ, Watsford ML, Coutts AJ, Richards DA. Effects of creatine supplementation on aerobic power and cardiovascular structure and function. J Sci Med Sport 2005;8(3):305–13.

[41] Jones AM, Carter H, Pringle JS, Campbell IT. Effect of creatine supplementation on oxygen uptake kinetics during submaximal cycle exercise. J Appl Physiol 2002;92(6):2571–7.

[42] Engelhardt M, Neumann G, Berbalk A, Reuter I. Creatine supplementation in endurance sports. Med Sci Sports Exerc 1998;30(7):1123–9.

[43] Thompson CH, Kemp GJ, Sanderson AL, Dixon RM, Styles P, Taylor DJ, et al. Effect of creatine on aerobic and anaerobic metabolism in skeletal muscle in swimmers. Br J Sports Med 1996;30(3):222–5.

[44] Prevost MC, Nelson AG, Morris GS. Creatine supplementation enhances intermittent work performance. Res Q Exerc Sport 1997;68(3):233–40.

[45] Chwalbinska-Moneta J. Effect of creatine supplementation on aerobic performance and anaerobic capacity in elite rowers in the course of endurance training. Int J Sport Nutr Exerc Metab 2003;13(2):173–83.

[46] Maganaris CN, Maughan RJ. Creatine supplementation enhances maximum voluntary isometric force and endurance capacity in resistance trained men. Acta Physiol Scand 1998;163(3):279–87.

[47] Robinson TM, Sewell DA, Hultman E, Greenhaff PL. Role of submaximal exercise in promoting creatine and glycogen accumulation in human skeletal muscle. J Appl Physiol 1999;87(2):598–604.

[48] Eijnde BO, Richter EA, Henquin JC, Kiens B, Hespel P. Effect of creatine supplementation on creatine and glycogen content in rat skeletal muscle. Acta Physiol Scand 2001;171(2):169–76.

[49] Ferrante RJ, Andreassen OA, Jenkins BG, Dedeoglu A, Kuemmerle S, Kubilus JK, et al. Neuroprotective effects of creatine in a transgenic mouse model of Huntington's disease. J Neurosci 2000;20(12):4389–97.

[50] Alves CR, Ferreira JC, de Siqueira-Filho MA, Carvalho CR, Lancha AHJ, Gualano B. Creatine-induced glucose uptake in type 2 diabetes: a role for AMPK-α? Amino Acids 2012;43:1803–7.

[51] Young JC, Young RE. The effect of creatine supplementation on glucose uptake in rat skeletal muscle. Life Sci 2002;71(15):1731–7.

[52] Green AL, Hultman E, Macdonald IA, Sewell DA, Greenhaff PL. Carbohydrate ingestion augments skeletal muscle creatine accumulation during creatine supplementation in humans. Am J Physiol 1996;271(5 Pt 1):E821–6.

[53] Green AL, Hultman E, Macdonald IA, Sewell DA, Greenhaff PL. Carbohydrate feeding augments skeletal muscle creatine accumulation during creatine supplementation in man. Am J Physiol 1996;271:E821–6.

[54] Winder WW. Energy-sensing and signaling by AMP-activated protein kinase in skeletal muscle. J Appl Physiol 2001;91(3):1017–28.

[55] Holmes BF, Kurth-Kraczek EJ, Winder WW. Chronic activation of 5′-AMP-activated protein kinase increases GLUT-4, hexokinase, and glycogen in muscle. J Appl Physiol 1999;87(5):1990–5.

[56] Ponticos M, Lu QL, Morgan JE, Hardie DG, Partridge TA, Carling D. Dual regulation of the AMP-activated protein kinase provides a novel mechanism for the control of creatine kinase in skeletal muscle. EMBO J 1998;17(6):1688–99.

[57] Gotshalk LA, Kraemer WJ, Mendonca MA, Vingren JL, Kenny AM, Spiering BA, et al. Creatine supplementation improves muscular performance in older women. Eur J Appl Physiol 2008;102(2):223–31.

[58] Aoyagi Y, Shephard RJ. Aging and muscle function. Sports Med 1992;14(6):376–96.

[59] Evans WJ, Campbell WW. Sarcopenia and age-related changes in body composition and functional capacity. J Nutr 1993;123(2 Suppl.):465–8.

[60] Brooks SV, Faulkner JA. Skeletal muscle weakness in old age: underlying mechanisms. Med Sci Sports Exerc 1994;26(4):432–9.

[61] Lexell J, Taylor CC, Sjostrom M. What is the cause of ageing atrophy? Total number, size and proportion of different fiber types studied in whole vastus lateralis muscles from 15- to 83-year old men. J Neurol Sci 1988;84:275–94.

[62] Gualano B, Artioli GG, Poortmans JR, Lancha Junior AH. Exploring the therapeutic role of creatine supplementation. Amino Acids 2010;38(1):31–44.

CHAPTER

40

Oral Bioavailability of Creatine Supplements

Insights into Mechanism and Implications for Improved Absorption

Donald W. Miller[1], *Samuel Augustine*[2], *Dennis H. Robinson*[3], *Jonathan L. Vennerstrom*[3] *and Jon C. Wagner*[4]

[1]Department of Pharmacology and Therapeutics, University of Manitoba, Winnipeg, MB, Canada [2]Department of Pharmacy Practice, Creighton University, Omaha, NE, USA [3]Department of Pharmaceutical Sciences, University of Nebraska Medical Center, Omaha, NE, USA [4]Vireo Resources LLC, Plattsmouth, NE, USA

INTRODUCTION

Creatine is found primarily in skeletal muscle as both free creatine and creatine phosphate. Creatine phosphate comprises 70% of the total creatine found in skeletal muscle. The cellular importance of creatine phosphate is as a readily available source of phosphate for regeneration of adenosine triphosphate (ATP) from adenosine diphosphate (ADP). In the absence of creatine phosphate, ATP cannot be regenerated, resulting in impaired muscle function due to lack of an available energy source in the muscle cells. Studies have demonstrated that dietary supplementation with creatine can increase total skeletal muscle creatine levels by approximately 20% [1–3]. Of the increased deposition of creatine in the muscle following dietary supplementation, approximately one third is in the form of creatine phosphate and available for immediate use [1,4,5]. The correlation between increased muscle stores of creatine and improved muscle performance is well established [6], and dietary supplementation with creatine is widely used and accepted by most governing sports bodies [7].

The performance benefits that result from creatine supplementation include (a) more cellular energy for short bursts of high intensity exercise, (b) improved energy transfer in muscle cells, and (c) greater buffering capacity resulting in less fatigue and shorter recovery time following intense exercise. [8–12]. While there is certainly individual variation regarding creatine response, the above-described effects are obtainable with daily dosing of creatine ranging from 3 to 10 grams. In addition to the well-known effects of creatine supplementation on muscle performance, more recent studies have reported potential use of creatine supplements in muscle repair following injury [13,14], as well as anti-inflammatory effects [15–17], that apply to both endurance and power athlete. The muscle repair and anti-inflammatory effects appear to require higher doses of creatine supplements, often in the 20–30 gram daily dose range [13–17].

There is little available scientific evidence to refute the effects of creatine supplementation on muscle performance. However, the relatively large amounts of creatine supplementation required to produce these desired effects suggests inefficiencies in either the bioavailability and/or tissue distribution of current creatine products. While there has been much research devoted to enhancing creatine uptake into muscle cells through co-administration of glucose [18], various fatty acids [19], and insulin-stimulating products [20], until recently, little effort was devoted to enhancing the oral bioavailability of creatine supplements. This is due in large part to the assumption that creatine monohydrate, the most widely used form of creatine, is completely absorbed from the gastrointestinal tract. However, there is sufficient evidence to suggest that the oral bioavailability of creatine monohydrate is far

Nutrition and Enhanced Sports Performance.
DOI: http://dx.doi.org/10.1016/B978-0-12-396454-0.00040-0

from complete. This chapter reviews the mechanisms governing creatine absorption in the epithelial cells of the gastrointestinal tract and the evidence supporting passive diffusion of creatine as the route for oral absorption. Based on this information, a critical re-examination of the bioavailability of creatine monohydrate, especially in comparison with newer salt forms of creatine, will be undertaken to provide potential insight into ways to improve the efficiency of the available dosage forms of creatine.

CELLULAR MECHANISM OF INTESTINAL ABSORPTION

The gastrointestinal tract (GIT) is well suited for the absorption of nutrients. Ingested material first enters the stomach where excreted enzymes and the low pH environment begin breaking down the material into more absorbable nutritional units such as glucose, amino acids and peptides. For any compound to be absorbed in the GIT, whether it is a nutrient, drug, or potential toxin, it must be in solution. The primary function of the digestive processes in the stomach is to solubilize the ingested material so that it can be absorbed in the small intestines. The basic anatomical features of the small intestine result in a large surface area for the absorption of nutrients. The small intestine is divided into three segments: the duodenum is the first part, the jejunum is the middle segment, and the ileum is the third and final segment adjacent to the large intestine.

There are three basic absorption pathways by which a nutrient can be taken up from the GIT and enter into the systemic circulation (Figure 40.1). These include transcellular diffusion, paracellular diffusion and transcellular transport. The transcellular diffusion route is a passive process governed by the concentration gradient that exists for the solute of interest (in this case creatine) in the lumen of the intestine and the epithelial cell interface and the permeability of the solute across a lipid bilayer. Permeability is determined by the physico-chemical properties of a solute, with parameters such as size/surface area, lipophilicity, charge and hydrogen bonding potential of the solute influencing passive diffusion across the cellular interface [21]. Indeed, based on solutes with high oral bioavailability through the transcellular diffusion route, ideal properties include a molecular weight less than 300 Daltons, a log P value (measure of lipophilicity) between 1 and 3, nonionized and with fewer than five hydrogen-bond acceptors [22].

While paracellular diffusion is also dependent on a concentration gradient, instead of moving through the epithelial cell, the solute travels in a bulk flow manner in the spaces that exist between the epithelial cells. For the absorptive epithelial cells that line the intestine, the junctions between the cells have a collection of membrane proteins that interact with each other to form what is referred to as a tight junction [23]. While the bulk flow movement of solutes through the tight junctions is restricted, the complexity and restrictiveness of the tight junctions vary depending on the location within the small intestine. Thus, tight junctions between epithelial cells in the duodenum and jejunum have a larger pore opening (approximately 8–13 Ångstroms in diameter) than in the ileum, where pore sizes of approximately 4 Ångstroms in diameter are observed [24]. Those solutes most likely to be absorbed via paracellular diffusion processes in the GIT are

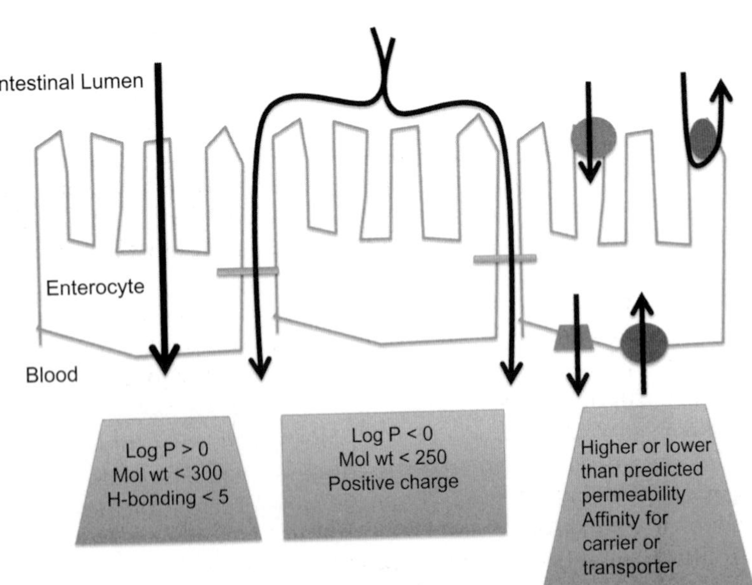

FIGURE 40.1 Schematic of different solute absorption pathways in the gastrointestinal tract and the general properties governing absorption through each specific pathway. (This figure is reproduced in color plate section.)

relatively small in size (<250 Daltons in molecular weight) with positive charged solutes having greater potential for movement through the paracellular pore than negative charged or zwitterionic solutes [21,23,25]. Another important feature of solutes that are absorbed through paracellular diffusion is that they typically display regionally specific and incomplete absorption characteristics due to the limited sites for paracellular diffusion within the GIT [23,25].

Transcellular transport of solutes requires specific membrane proteins that act as carriers to move the solute across the biological membrane. These carriers or transporters facilitate the movement of the solute into or out of the cell at a rate that is far greater than that possible with simple passive diffusion of the solute. Transcellular transport processes for solutes are often coupled with the movement of ions in either a co-transport or counter-transport fashion [26]. Examples in the small intestine include vitamin and peptide transporters that are driven by the co-transport of sodium and hydrogen ions, respectively [27,28]. There are also solute transporters that utilize adenosine triphosphate (ATP) as the cellular energy source to move solutes across the cell membrane [29]. Regardless of whether the solute transporter is driven by electrochemical gradients or hydrolysis of ATP, a feature of all solute transporters is selectivity and saturability of transport.

CREATINE ABSORPTION IN THE GIT

Creatine is a zwitterion with a positively charged guanidine functional group and a negatively charged carboxylic acid functional group. Within the acidic environment of the stomach and jejunum of the small intestine, the carboxylic acid functional group is likely to be protonated and the predominant form of ingested creatine is the positively charged species. As the ingested creatine progresses down the intestinal lumen the pH becomes neutral and the zwitterion and negative charged species will become more prevalent. Because creatine exists primarily as a charged molecule, its ability to partition into a lipophilic environment such as the plasma membrane of the intestinal epithelial cell is limited. Indeed, the logP value for creatine monohydrate, the most common form of creatine supplement, is approximately −1.0. By comparison, only solutes displaying LogP values between 1 and 3 are likely to be absorbed by transcellular diffusion in the gastrointestinal tract [22]. Thus, in considering the absorption of creatine from the gastrointestinal tract, passive diffusion via the transcellular route is likely to be minimal.

There are potentially multiple transporters for creatine within the GIT. The creatine transporter (CRT) is a solute transporter that selectively transports creatine and creatine analogs in a sodium dependent manner [30]. The CRT is expressed in high amounts in the brain, intestine and skeletal muscle, where it plays a crucial role in the distribution of creatine to target tissue [31,32]. Indeed, CRT genetic abnormalities are linked to reduced creatine distribution to the brain and the development of mental retardation [33]. Expression of CRT at both the messenger RNA and protein level has been reported in epithelial cells of the intestine [34]. Recent studies examining the expression and function of CRT during development in the rat, showed multiple forms of CRT within the colon during development [34]. However, the activity and expression level of CRT in the colon diminished during maturation, with little CRT activity observed in the adult rat [34]. If similar developmental patterns exist in humans, this would suggest that creatine absorption through CRT1 in the GIT is lowest in adulthood.

An additional consideration is the localization of CRT within the GIT. To aid in the oral absorption of creatine, CRT would need to be localized in the brush border membrane of the intestinal epithelial cells. Previous studies have reported brush border expression of CRT in the jejunal and ileal segments of the small intestine [35,36]. However, while brush border CRT is the first step in absorbing dietary sources of creatine, there would need to be transporters positioned on the basolateral membrane of the intestinal epithelial cells to move the absorbed creatine into the bloodstream. While an extensive search for CRT on the basolateral membrane did reveal a sodium-dependent transporter for creatine the directionality of transport was inward [37]. The directionality of this transporter means that it would only be able to transport creatine from the blood into the intestinal epithelial cell, and thus would not aid in the oral absorption of creatine.

A final consideration for transporter-mediated absorption of creatine in the GIT is the kinetics of the CRT. As absorption through CRT is saturable at high concentrations, the extent to which CRT can efficiently absorb creatine will be dependent on the amount of creatine in the gastrointestinal fluid. Given that creatine supplements are consumed at doses of 5–10 grams or more, the concentration of creatine in the gastrointestinal fluid is likely to be above those required for optimal transport function of CRT. Indeed, this may explain some of the dose-dependent observations where low doses of creatine administered more frequently appear to provide better results [38]. For this reason, and those discussed above, transcellular transport pathways for the absorption of creatine supplements are likely to be minimal.

Based on creatine permeability studies conducted in various intestinal models, the most likely route for creatine absorption in the GIT is through paracellular diffusion. Studies by Orsenigo and colleagues [37] examined creatine permeability across inverted jejunal segments of the rat intestine. While transporter-mediated uptake of creatine was observed in both brush border and basolateral intestinal membrane preparations, permeability across intact intestinal tissue was not transporter dependent, as demonstrated by the absence of concentration dependency and the inability to influence creatine permeability with various CRT inhibitors [37]. The paracellular diffusion pathway also fits from the perspective of what is known about the characteristics of solutes most likely to undergo paracellular diffusion. As creatine is below the 250 molecular weight cut-off and is primarily positively charged in the early portion of the GIT, where pH of the intestinal fluid is slightly acidic, solute diffusion through the paracellular route is ideal. Together, these studies provide compelling evidence for paracellular diffusion of creatine as the primary mechanism for oral absorption.

A paracellular diffusion pathway for creatine absorption in the GIT is also supported by Caco-2 cell permeability studies [39–41]. The Caco-2 cell line is a human transformed cell that is widely used to examine oral absorption within the pharmaceutical industry [42]. It expresses many of the transporters involved in absorption of nutrients in the GIT [42]. From the standpoint of creatine absorption, Caco-2 express CRT at the mRNA level [43], although potential changes in expression were observed during differentiation, consistent with the developmental expression analysis reported for CRT in rats [34]. Studies using radiolabeled creatine showed a very low permeability across Caco-2 monolayers, consistent with a solute with poor oral absorption profile [39]. Permeability of various creatine salt forms across Caco-2 monolayers was also consistent with minimal permeability [40]. As the Caco-2 have highly developed tight junctions, paracellular diffusion would be limited in this model and solutes undergoing paracellular diffusion would have low permeability [21]. Thus, the results with creatine monohydrate and various creatine salt forms in the Caco-2 reflect poorly permeable solutes. For solutes undergoing paracellular diffusion as their mechanism of permeability in the GIT, the Caco-2 system may underestimate actual intestinal permeability. Interestingly, recent studies with creatine ethyl ester, an esterified form of creatine with improved solubility and lipophilicity properties, demonstrated significantly greater permeability than did either creatine monohydrate or creatinine in Caco-2 monolayers [41].

HUMAN ORAL BIOAVAILABILITY OF CREATINE SUPPLEMENTS

Bioavailability is defined as the amount of an administered agent that is absorbed and present in the systemic bloodstream for distribution and use by the various tissues. It is typically expressed as a percentage or fraction of the amount administered. As creatine supplements are ingested, the oral bioavailability represents the fraction of the administered dose that is absorbed in the GIT and available in the system circulation for distribution to various tissue sites. The oral bioavailability of any compound is determined definitively by calculating the area under the curve (AUC) of the plasma concentration versus time profile following oral administration and comparing this to the resulting AUC plasma concentration versus time profile for the compound following intravenous injection. The direct intravenous administration of the compound results in a 100% bioavailability and the resulting plasma concentration profile provides the necessary data for comparison of absorption from other routes.

While there is abundant evidence in the literature for creatine supplementation and improved muscle performance using creatine monohydrate (CM), considerably less is known about the oral bioavailability of CM. Indeed, there are no published reports of the definitive oral bioavailability of any creatine supplement. Given the physico-chemical properties of CM, the relatively high doses required, and the previously discussed studies in the various intestinal absorption models that report low permeability of CM [37,39–41], the oral absorption of creatine supplements are likely to be incomplete. Despite this, CM based supplements are generally considered to have nearly complete absorption in the GIT. Such claims of complete oral absorption of CM are based on studies in which CM was administered orally and the increases in tissue levels of creatine combined with the increases in creatinine elimination in the urine were used to provide an index of the body burden of CM [44,45]. These methods used to obtain the estimates of oral bioavailability for CM have at least two major limitations. First, the accuracy of using urinary creatinine levels as an index of the amount of CM absorbed depends on the extent to which the creatinine excreted in the urine originated from the conversion of systemically absorbed creatine. The assumption is that increased levels of urinary creatinine following CM supplementation are the result of the conversion of creatine in the various tissues to creatinine which is readily excreted into the urine. However, both the intestinal epithelial cells and bacteria have the ability to take up and process creatine [31,34,46,47]. This ability to acquire and metabolically process creatine within the intestine provides a potential source of creatinine in the GIT. Thus the conversion of creatine to creatinine within the lumen of the GIT, and its subsequent absorption into the bloodstream would result in a potential overestimate of the amount of creatine that has been systemically absorbed.

A second issue with previous studies reporting complete oral absorption of CM is equating an absence of creatine or creatinine in fecal matter with the complete absorption of CM from the intestine. Such interpretations ignore the ability of bacteria to utilize creatine and its metabolic products [31,46,47]. The extent of creatine utilization in the GIT provides a potential pathway for the pre-systemic removal of ingested creatine and results in an overestimate of CM oral bioavailability.

While definitive oral bioavailability data are lacking for CM, there are studies examining the relative bioavailability of various CM formulations [45,48] that suggest absorption is less than complete. Studies by Harris and colleagues [48] examined the oral absorption and pharmacokinetics of single-dose CM supplements when given either in solution, as a suspension or as a solid dosage form lozenge. In these studies, CM delivered in solution in liquid formulation resulted in faster absorption of CM as well as more extensive absorption represented by the larger AUC of the plasma creatine concentration versus time profile compared with either lozenge or suspension formulations. Based on these data, the authors suggested that there was a decrease in oral bioavailability of CM when in suspension or solid dosage form. The importance of this is underappreciated. Given that the aqueous solubility of CM is approximately 12–15 mg/mL, athletes taking a standard dose of CM (ranging from 5 to 10 grams) would require 400–800 mL of fluid to ensure the dose is completely solubilized. As a result of this, most CM products are taken as suspensions and would be incompletely absorbed in the GIT. The authors also compared absorption of CM in solution to that of equivalent amounts of creatine contained within red meat. While the absorption of creatine was delayed, compared with CM in aqueous solution, the amount of creatine absorbed from the red meat was similar [48]. Although more studies are required, there may be advantages to a delayed but sustained absorption of creatine supplements in terms of more efficient loading into skeletal muscle [48,49].

This is in contrast to the recent studies of Deldicque et al. [45] examining the oral absorption and resulting plasma pharmacokinetics of CM when administered in aqueous solution compared with either protein or beta-glucan nutritional bars. The investigators reported a significant difference in the rate of absorption of CM based on the dosage formulation used, with the aqueous solution providing the most rapid intestinal absorption. However, when looking at the extent of intestinal absorption, despite the appearance of moderate alterations in bioavailability, there was no significant difference in the resulting AUC for plasma creatine with any of the formulations examined.

ABSORPTION CHARACTERISTICS OF ALTERNATIVE FORMS OF CREATINE

Perhaps the most convincing evidence that the oral absorption of CM is less than complete comes from the comparison of oral bioavailability of various creatine salt forms. As different salt forms have potentially different solubility parameters [49,50], identification of creatine salts with increased aqueous solubility is likely to result in improvements in oral absorption and more efficient dosage formulations. Unfortunately there is considerably less information on the performance of these newer forms of creatine supplements in terms of bioavailability and biological response. Of the many different creatine salt forms, there are three—creatine pyruvate (CPyr), creatine citrate (CCit) and creatine hydrochloride (CHCl)—for which human oral bioavailability studies have been reported [49,50]. The aqueous solubility and relative bioavailability of these three creatine salt forms in comparison with CM is summarized in Figure 40.2. Studies by Jagar and

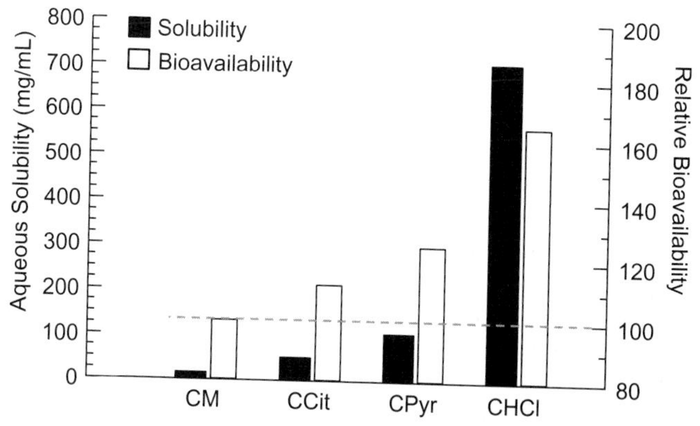

FIGURE 40.2 Comparison of aqueous solubility and human bioavailability of various creatine salt forms. The mean aqueous solubility values were obtained from [40]. Relative bioavailability was determined from the reported areas under the curve (AUC) for plasma creatine obtained with various creatine salt forms compared with that of creatine monohydrate (CM). The AUC values were taken from data in [49] for creatine citrate (CCit) and creatine pyruvate (CPyr) and [50] for creatine hydrochloride (CHCl). (This figure is reproduced in color plate section.)

colleagues compared the oral bioavailability of CCit and CPyr with that of CM. In these studies subjects were given 5-gram doses of CCit, CPyr or CM as an aqueous solution and the resulting plasma creatine levels were measured over time. For CCit, which has a slightly greater solubility than CM, there was a small increase (approximately 10%) in oral bioavailability compared with CM (Figure 40.2). For CPyr, which has an approximately 8-fold higher aqueous solubility, there was a modest, approximately 25%, increase in oral bioavailability compared with CM (Figure 40.2). Given the improved aqueous solubility of CPyr, an increase in oral bioavailability would be expected. Interestingly, the improved oral absorption observed with CPyr and CCit was statistically significant, though the differences were not considered important, primarily due to the perceived complete oral absorption of CM [49]. In this respect, regulatory standards indicate that different salt forms of a compound are considered bioequivalent when the relative bioavailability of one salt form is 75–120% of that observed with another salt form [51]. Under this criterion, CPyr, with an approximately 25% increase in oral bioavailability, would not be considered to have the same oral absorption properties as CM and would not be considered bioequivalent.

While the increases in oral bioavailability observed with CCit and CPyr could be classified as relatively modest, oral absorption studies with CHCl provide even more compelling evidence to suggest that improvements in bioavailability of creatine supplements are possible [50]. In these studies volunteers were given a 5-gram dose of either CM or CHCl in 8 ounces of water. A cross-over design was employed to allow direct comparisons of differences in oral absorption of the two creatine salts within the same subject. The results of these studies reported an approximately 60% increase in oral absorption of CHCl compared with CM [50]. With such increases in bioavailability observed with CHCl, it is difficult to claim complete oral absorption of CM. Together, these studies provide two important and fundamental findings. First, with increased oral absorption of the various creatine salts ranging from 10% to 60% over that of CM, it is clear that the bioavailability of CM is not nearly complete. Second, as CM bioavailability is not complete, there is potential for development of creatine supplements with improved oral absorption, which in turn could provide significant advancements in performance benefits, and allow for reduced dosages and more flexible dosing formulations (sports drinks and bars, fortified foods, etc.).

In addition to the multiple salt forms of creatine, there are modified forms of creatine such as creatine ethyl ester (CEE). In addition to having improved aqueous solubility compared with CM, the ester form of creatine also has improved octanol–water partitioning, an index of cell permeability (Table 40.1). As there are esterases throughout the blood and tissue, CEE was designed to be a pronutrient with enhanced oral absorption, which could then be hydrolyzed to creatine once absorbed into the body. There are few studies examining either the biological effects or oral absorption properties of CEE. The only direct comparison study of CEE and CM reported that CEE was "...not as effective at increasing serum and muscle creatine levels or in improving body composition, muscle mass, strength, and power" [52]. The study followed healthy volunteers over a 48-day period in which subjects were supplemented daily with 300 mg/kg of CM, CEE or placebo and underwent an exercise weight training regimen. Furthermore, the study reported high levels of creatinine in serum samples from the CEE-supplemented treatment group [52]. However, re-examination of the data looking at changes in muscle creatine, and peak and mean power measurements that occurred in the CEE, CM and placebo groups over the course of the 48-day study show CEE performance was as good or better than CM (Figure 40.3). This is due in part to the lower starting values for subjects in the CEE treatment group in terms of muscle creatine levels and power assessments [52]. As for the high serum creatinine levels in the CEE group, recent reports demonstrate that CEE is rapidly converted to creatinine in aqueous solutions at neutral pH [53]. Interestingly, CEE is very stable in aqueous solutions at low pH and appears to be more stable in lipophilic environments at neutral pH ranges [53], suggesting that CEE is intact during absorption in the GIT and stable within the membrane environment of cells. Thus, while initial findings suggested CEE is less effective, more studies are required to definitively address the issue.

TABLE 40.1 Comparison of Aqueous Solubility and Octanol–Water Partitioning of Creatine Ethyl Ester (CEE) and Creatine Monohydrate (CM)

Property	CEE HCl	CRT Monohydrate
Molecular weight g/mol	195.6	149.7
Percent by weight creatine	67	88
Aqueous solubility 25°C mg/mL	396	14.5
Ratio of solubility (relative to monohydrate)	27.4	1.00
Octanol–water partition coefficient	0.205	0.102
Ratio of Partition coefficient (relative to monohydrate)	2.01	1.00

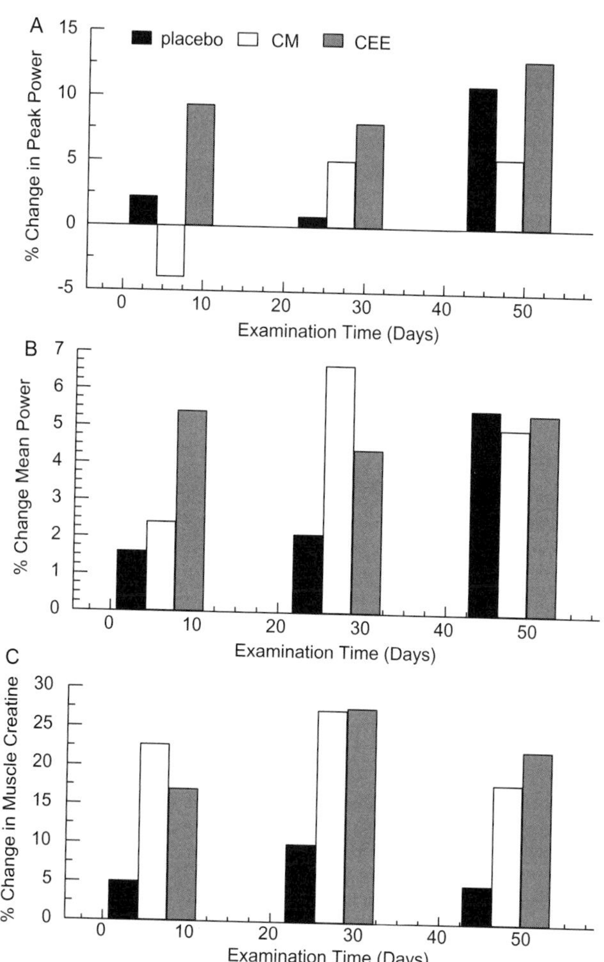

FIGURE 40.3 Comparison of changes in (A) peak power, (B) mean power and (C) muscle creatine observed during various days of supplementation with either placebo (black bars), creatine monohydrate (white bars) or creatine ethyl ester (grey bars). Values reported represent the changes from baseline measurements prior to initiation of the supplementation. The values reported were taken from data first reported in [52].

CONCLUSIONS

Creatine supplementation is an accepted and effective way to increase power output and speed muscle recovery. Because of these effects on skeletal muscle, creatine supplements are commonly used to improve athletic performance. While much effort has been given to improving the cellular processes governing accrual of creatine into the tissue, significantly less effort has been placed on increasing the oral absorption of creatine dietary supplements. Much of the reason for this is the perpetuated assumption that CM, the most widely used and studied form of creatine supplement, is completely absorbed from the GIT and is thus 100% bioavailable. Based on the mechanism of absorption of creatine in the GIT, there is ample evidence to suggest that creatine is not likely to have 100% bioavailability. Furthermore, based on the available human pharmacokinetic data, there is mounting evidence pointing to the less than complete absorption of creatine supplements.

Given the relatively large daily doses administered, improving the efficiency of creatine absorption could result in reductions in dosage as well as a more diverse range of formulations including sports drinks, bars and potentially fortified foods. Efforts to identify ways to increase creatine oral absorption, whether based on altering the absorption profile of CM through reduced doses given more frequently [38], or formulations with delayed release matrixes [45,48], or through identification of different creatine salt forms and compositions [49,50,53], have the potential to improve creatine supplementation options. Such efforts, in combination with appropriate safety and efficacy studies, will ultimately provide consumers and athletes with better options for creatine supplementation.

References

[1] Harris RC, Soderlund K, Hultman E. Elevation of creatine in resting and exercised muscle of normal subjects by creatine supplementation. Clin Sci 1992;83:367–74.

[2] Greenhaff PL, Bodin K, Soderlund K, Hultman E. Effect of oral creatine supplementation on skeletal muscle phosphocreatine resynthesis. Am J Physiol 1994;266:E725–30.

[3] Hultman E, Soderlund K, Timmons JA, Cederblad G, Greenhaff PL. Muscle creatine loading in men. J Appl Physiol 1996;81:232–7.

[4] Gordon A, Hultman E, Kaijser L, Kristjansson S, Rolf CJ, Nyquist O, et al. Creatine supplementation in chronic heart failure increases skeletal muscle creatine phosphate and muscle performance. Cardiovasc Res 1995;30:413–8.

[5] Casey A, Constantin-Teodosiu D, Howell S, Hultman E, Greenhaff PL. Creatine supplementation favorable: affects performance and muscle metabolism during maximal exercise in humans. Am J Physiol 1996;271:E31–7.

[6] Gualano B, Roschel H, Lancha-Jr AH, Brightbill CE, Rawson ES. In sickness and in health: the widespread application of creatine supplementation. Amino Acids 2012;43:519–29.

[7] Spriett LL, Perry CG, Talanian JL. Legal pre-event nutritional supplements to assist energy metabolism. Essays Biochem 2008;44:27–43.

[8] Becque MD, Lochmann JD, Melrose DR. Effect of creatine supplementation during strength training on 1RM and body composition. Med Sci Sports Exerc 1997;29:S146.

[9] Bosco C, Tihanyi J, Pucspk J, Kovacs I, Gabossy A, Colli R, et al. Effect of oral creatine supplementation on jumping and running performance. Int J Sports Med 1997;18:369–72.

[10] Ferriera M, Kreider R, Wilson M, Grinstaff P, Plisk S, Reinhardy J, et al. Effects of ingestion of a supplement designed to enhance creatine uptake on strength and sprint capacity. Med Sci Sports Exerc 1997;29:S146.

[11] Kreider RB, Ferriera M, Wilson M, Grindstaff P, Plisk S, Reinardy J, et al. Effects of creatine supplementation on body composition, strength, and sprint performance. Med Sci Sports Exerc 1998;30:73–82.

[12] Larson DE, Hunter GR, Trowbridge CA, Turk JC, Harbin PA, Torman SL. Creatine supplementation and performance during

off-season training in female soccer players. Med Sci Sports Exerc 1998;30:S264.

[13] Hespel P, Op't Eijnde B, Van Leemputte M, Urso B, Greenhaff PL, Labarque V, et al. Oral creatine supplementation facilitates the rehabilitation of disuse atrophy and alters the expression of muscle myogenic factors in humans. J Physiol 2001;536:625–33.

[14] Op't Eijnde B, Urso B, Richter EA, Greenhaff PL, Hespel P. Effect of oral creatine supplementation on human muscle GLUT4 protein content after immobilization. Diabetes 2001;50:18–23.

[15] Bassit BA, Curi R, Costa Rosa LF. Creatine supplementation reduces plasma levels of pro-inflammatory cytokines and PGE2 after a half-ironman competition. Amino Acids 2008;35:425–31.

[16] Santos RV, Bassit RA, Caperuto EC, Costa Rosa LF. The effect of creatine supplementation upon inflammatory and muscle soreness markers after a 30 km race. Life Sci 2004;75:1917–24.

[17] Rahimi R. Creatine supplementation decreases oxidative DNA damage and lipid peroxidation induced by a single bout of resistance exercise. J Strength Cond Res 2011;25:3448–55.

[18] Green AL, Hultman E, MacDonald IA, Sewell DA, Greenhaff PL. Carbohydrate ingestion augments skeletal muscle creatine accumulation during creatine supplementation in humans. Am J Physiol 1996;271:821–6.

[19] Burke DG, Chilibeck PD, Parise G, Tarnopolsky MA, Candow DG. Effect of alpha-lipoic acid combined with creatine monohydrate on human skeletal muscle creatine and phosphagen concentration. Int J Sport Nutr Exerc Metab 2003;13:294–302.

[20] S, teenge GR, Lambourne J, Casey A, Macdonald IA, Greenhaff PL. Stimulatory effect of insulin on creatine accumulation in human skeletal muscle. Am J Physiol 1998;275:E974–9.

[21] Travelin, S. New approaches to studies of paracellular drug transport in intestinal epithelial cell monolayers. Acta Universitatis Upsaliensis. In: Comprehensive Summaries of Uppsala Dissertations from the Faculty of Pharmacy vol. 285, 2003. p. 1–66.

[22] Andrews CW, Bennett L, Yu LX. Predicting human oral bioavailability of a compound: development of a novel quantitative structure-bioavailability relationship. Pharm Res 2000;17:639–44.

[23] Artursson P, Ungell AL, Lofroth JE. Selective paracellular permeability in two models of intestinal absorption: cultured monolayers of human intestinal epithelial cells and rat intestinal segments. Pharm Res 1993;10:1123–9.

[24] Fihn BM, Jodal M. Permeability of the proximal and distal rat colon crypt and surface epithelium to hydrophilic molecules. Pflugers Arch 2001;441:656–62.

[25] Tanaka Y, Taki Y, Sakane T, Nadai T, Sezaki H, Yamashita S. Characterizations of drug transport through tight-junctional pathway in Caco-2 monolayer: comparison with isolated rat jejunum and colon. Pharm Res 1995;12:523–8.

[26] Hediger MA, Kanai Y, You G, Nussberger S. Mammalian ion-coupled solute transporters. J Physiol 1995;482:7S–17S.

[27] Vadlapudi AD, Vadlapatla RK, Mitra AK. Sodium dependent multivitamin transporter (SMVT): a potential target for drug delivery. Curr Drug Targets 2012;13:994–1003.

[28] Broberg MI, Holm R, Tonsberg H, Frolund S, Ewon KB, Nielsen AI, et al. Function and expression of the proton-coupled amino acid transporter PAT1 along the gastrointestinal tract: implications for intestinal absorption of gaboxadol. Br J Pharmacol 2012;167:654–65.

[29] Li Y, Lu J, Paxton JW. The role of ABC and SLC transporters in the pharmacokinetics of dietary and herbal phytochemicals and their interactions with xenobiotics. Curr Drug Metab 2012;13:624–39.

[30] Hediger M, Romero MF, Peng JB, Rolfs A, Takanaga H, Bruford EA. The ABCs of solute carriers: physiological, pathological, and therapeutic implications of human membrane transport proteins. Pfluegers Arch 2004;447:465–8.

[31] Wyss M, Kaddurah-Daouk R. Creatine and creatinine metabolism. Pharmacol Rev 2000;80:1108–213.

[32] Brosnan JT, Brosnan ME. Creatine: endogenous metabolite, dietary, and therapeutic supplement. Annu Rev Nutr 2007;27:241–61.

[33] Garcia P, Rodrigues F, Valongo C, Salomons GS, Diogo L. Phenotypic variability in a portuguese family with x-linked creatine transporter deficiency. Pediatr Neurol 2012;46:39–41.

[34] Garcia-Miranda P, Garcia-Delgado M, Peral MJ, Calonge MI, Ilundain AA. Ontogeny regulates creatine metabolism in rat small and large intestine. J Physiol Pharmacol 2009;60:127–33.

[35] Tosco M, Faelli A, Sironi C, Gastaldi G, Orsenigo MN. A creatine transporter is operative at brush border level of rat jejunal enterocyte. J Membrane Biol 2004;202:85–95.

[36] Peral-Rubio MJ, Garcia-Delgado M, Calonge ML, Duran JM, De La Horra MC, Wallimann T, et al. Human, rat and chicken small intestinal Na+ −Cl—creatine transporter: functional, molecular characterization and localization. J Physiol 2002;545:133–44.

[37] Orsenigo MN, Faelli A, De Biasi S, Sironi C, Laforenza U, Paulmichl M, et al. Jejunal creatine absorption: what is the role of the basolateral membrane? J Membrane Biol 2005;207:183–95.

[38] Sale C, Harris RC, Florance J, Kumps A, Sanvura R, Poortmans JR. Urinary creatine and methylamine excretion following $4 \times 5\ \mathrm{g\ day^{-1}}$ or $20 \times 1\ \mathrm{g\ day^{-1}}$ of creatine monohydrate for 5 days. J Sports Sci 2009;27:759–66.

[39] Dash AK, Miller DW, Han H-Y, Carnazzo J, Stout JR. Evaluation of creatine transport using Caco-2 monolayers as an in vitro model for intestinal absorption. J Pharm Sci 2001;90:1593–8.

[40] Gufford BT, Sriraghavan K, Miller NJ, Gu X, Miller DW, Vennerstrom JL, et al. Physicochemical characterization of creatine *N*-methylguanidinium salts. J Dietary Suppl 2010;7:240–52.

[41] Gufford BT, Ezell EL, Robinson DH, Miller DW, Miller NJ, Gu X, Vennerstrom JL. pH-dependent stability of creatine ethyl ester: relevance to oral absorption. J Dietary Suppl 2013; (in press).

[42] Hidalgo IJ, Raub TJ, Borchardt RT. Characterization of the human colon carcinoma cell line (Caco-2) as a model system for intestinal epithelial permeability. Gastroenterology 1989;96:736–49.

[43] Landowski CP, Anderle P, Sun D, Sadee W, Amidon GL. Transporter and ion channel gene expression after caco-2 cell differentiation using 2 different microarray technologies. AAPS J 2004;6:1–10.

[44] Chanutin A. The fate of creatine when administered to man. J Biol Chem 1926;68:29–41.

[45] Deldicque L, Decombaz J, Zbinden Foncea H, Vuichoud J, Poortmans JR, Francaux M. Kinetics of creatine ingested as a food ingredient. Eur J Applied Physiol 2008;102:133–43.

[46] Ten Krooden E, Owens CWI. Creatinine metabolism by *Clostridium welchii* isolated from human faeces. Experintia 1975;31:1270–6.

[47] Twort FW, Mellanby E. On creatin-destroying *Bacilli* in the intestine, and their isolation. J Physiol (Lond) 1912;44:43–9.

[48] Harris RC, Nevill M, Beorn Harris D, Fallowfield JL, Bogdanis GC, Wise JA. Absorption of creatine supplied as a drink, in meat or in solid form. J Sports Sci 2002;20:147–51.

[49] Jager R, Harris RC, Purpura M, Francaux M. Comparison of new forms of creatine in raising plasma creatine levels. J Intl Soc Sports Nutr 2007;4:17.

[50] Miller DW, Vennerstrom JL, Faulkner MC. 2011. Creatine oral supplementation using creatine hydrochloride salt. US Patent Publication Patent No. 760864.

[51] Chereson R. Bioavailability, bioequivalence, and drug selection. In: Makoid MC, editor. Basic Pharmacokinetics, 1. 1996. p. 1–117. Chapter 8.

[52] Spillane M, Schoch R, Cooke M, Harvey T, Greenwood M, Kreider R, et al. The effects of creatine ethyl ester supplementation combined with heavy resistance training on body composition, muscle performance, and serum and muscle creatine levels. J Intl Sports Nutr 2009;6:6.

[53] Gufford BT, Ezell EL, Robinson DH, Miller DW, Miller NJ, Gu X, et al. pH-Dependent stability of creatine ethyl ester: relevance to oral absorption. J Dietary Suppl 2013; (in press).

CHAPTER

41

An Overview of Carnitine

Richard J. Bloomer, Tyler M. Farney and Matthew J. McAllister

Cardiorespiratory/Metabolic Laboratory, The University of Memphis, Memphis, TN, USA

INTRODUCTION

Carnitine is a naturally occurring substance in mammalian tissue [1], derived primarily from lysine and methionine (in addition to niacin, vitamin C, vitamin B_6, and iron; along with the precursor amino acids glycine, arginine, and taurine). It is manufactured in the liver and kidney at the rate of approximately $0.16-0.48\ mg\ kg^{-1}\ day^{-1}$. It is considered to be a *conditionally essential* nutrient and has a wealth of scientific data highlighting its importance in human physiology—inclusive of placebo-controlled trials documenting benefits for a wide variety of clinical conditions.

Carnitine was first discovered by two Russian scientists in 1905 within muscle extracts; however, the actual metabolic importance of carnitine was not known until 1955 [2]. There are two forms of carnitine: L-carnitine (biologically active form) and D-carnitine (biologically inactive form). Due to the water solubility of carnitine and the body's ability to synthesize it endogenously, it has been given status as a type of vitamin B, known as vitamin B_T. It is essential for a variety of physiological functions: in particular, fatty acid transport into the mitochondria for oxidation [3].

With regards to dietary sources, carnitine is more prevalent in animal products (in particular, red meat: ~80 mg per 3 oz serving) than in plant products (~0.08–0.36 mg per serving) [4]. Because of this discrepancy, vegetarians may present with lower carnitine concentrations than those consuming a mixed diet [5]; however, true carnitine deficiency is rare. Unfortunately, only 60–70% of readily available carnitine is absorbed from food sources, and the carnitine content in meat can be depleted if cooked over an open flame at high temperature. Because of the above factors, regular supplementation of carnitine is recommended by some, in an attempt to complement the carnitine content obtained in whole foods and through routine biosynthesis. Such supplementation makes sense in theory, in particular for individuals who perform acute physically stressful tasks, as the carnitine content in skeletal muscle can be depleted rapidly following strenuous physical work [6].

While L-carnitine is the active form of this nutrient, a variety of specific carnitine salts are available (e.g., base, fumarate, acetyl, tartrate, propionyl). Aside from assisting in fatty acid transport, carnitine has well-documented antioxidant activity [7], with human dosing ranging from as little as 500 mg/day to as much as 4–6 g/day. Certain forms have been studied with specific end goals in mind, such as improving cognitive function, aiding exercise recovery, or improving exercise performance and nitric oxide production. For example, acetyl-L-carnitine has the ability to cross the blood–brain barrier and has been used with success for purposes of enhancing cognitive function, memory, and mood [8]. L-carnitine L-tartrate (LCLT) has been reported to favorably impact selected markers of exercise recovery [9]. Propionyl-Carnitine (PLC) is a novel form of carnitine with multiple physiological roles [10]. It is created with the esterification of propionic acid and carnitine and has been used recently within dietary supplements as glycine propionyl-L-carnitine (GPLC). Both PLC [11–13] and GPLC [14] have been reported to improve physical function, with increased nitric oxide metabolites noted in studies using both PLC [15] and GPLC [16,17].

Carnitine is viewed as a safe nutrient [18–20], with adverse outcomes typically limited to mild gastrointestinal discomfort when individuals ingest high dosages (>5 g/day). A review of findings pertaining to antioxidant activity, nitric oxide production, and exercise performance and associated variables (e.g., muscle damage and soreness, androgen receptor alteration)

Nutrition and Enhanced Sports Performance.
DOI: http://dx.doi.org/10.1016/B978-0-12-396454-0.00041-2

will be discussed in more detail within the following sections. As this chapter is not meant to provide a comprehensive review of literature in relation to these areas of research, but rather to provide a brief overview of the topics indicated, the reader is referred to the following reviews [21–24].

ANTIOXIDANT ACTIVITY

Exercise of sufficient intensity and duration can lead to the formation of reactive oxygen and nitrogen species (RONS) [25], which when produced in amounts that overwhelm the antioxidant defense system, *may* lead to a condition of "oxidative stress." Carnitine has exhibited antioxidant activity in both animal [26–29] and man [30,31] and can limit the degree of oxidative stress. The antioxidant effects may be mediated by a number of different processes, including a reduction in ischemia-reperfusion-induced oxidative stress [32,33], a reduction in xanthine oxidase activity [27], and a free-radical scavenging activity [34]. Numerous studies involving carnitine supplementation of many different forms and dosages have noted a reduction in concentrations of oxidized biomolecules (e.g., malondialdehyde, glutathione). This is generally viewed as a positive finding and may have significant implications for those using carnitine for general health purposes, as increased oxidative stress is associated with human disease [35].

However, aside from potential improvements in general health, there exists little direct evidence that a decrease in oxidative stress leads to benefits with regards to exercise performance. This is particularly true in healthy, well-trained individuals, as the oxidative stress response to even very high-intensity exercise is minimal [36]. In such individuals, there are two important issues to consider: (i) most healthy, exercise-trained men and women have very low concentrations of oxidative stress biomarkers to begin with and (ii) most healthy, exercise-trained men and women experience a transient, low-grade oxidative stress in response to exercise. Considering the above, the questions might be: (i) Do such individuals really need an antioxidant (carnitine or otherwise) for purposes of attenuating oxidative stress? and (ii) If oxidative stress is attenuated in response to an acute exercise bout (or multiple exercise sessions over time), does this attenuation result in a favorable outcome in terms of improving exercise performance? These important questions should be considered before using supplemental carnitine for antioxidant purposes *alone*. Of course, carnitine has many other benefits which an individual may find attractive—in particular with regards to enhancing overall health (in which regard, receiving antioxidant protection may be of interest). With the above understanding, it should be noted that there exist some data describing an increase in oxidative stress in individuals exposed to extreme volumes of exercise [37,38]. In such cases, antioxidant supplementation (in the form of carnitine or otherwise) may prove beneficial.

NITRIC OXIDE

Nitric oxide, which was first referred to as endothelium-derived relaxing factor [39], is synthesized from the amino acid L-arginine, oxygen, and a variety of cofactors, by a family of enzymes known as nitric oxide synthases [40], including endothelial nitric oxide synthase (eNOS), inducible nitric oxide synthase (iNOS), and neuronal nitric oxide synthase (nNOS). Nitric oxide is a gaseous chemical that functions as a signaling molecule within the human body [41,42]. Very high concentrations of nitric oxide favor cell cycle arrest and apoptosis, while brief production at low (nanomolar) concentrations can promote beneficial physiological functions (e.g., enhanced blood flow and immune defense, regulation of neurotransmission and muscle atrophy/hypertrophy, and the stimulation of satellite cells—some of which may be related to muscle tissue hypertrophy) [42–44]. These effects are mediated through a cyclic guanosine monophosphate (cGMP)-dependent and -independent signaling cascade [41].

Within the sports supplement world, nitric oxide supplements have received a great deal of attention over the past several years. We have reported that oral intake of GPLC at a dosage of 4.5 g/day results in increased plasma nitric oxide, as measured by the metabolites nitrate/nitrite. These findings have been noted in previously sedentary men and women following an 8-week intervention of GPLC and aerobic exercise [16], as well as in resistance-trained men following a 4-week intervention of GPLC [17]. A decrease in NADPH oxidase activation [45] may be responsible for the increase in circulating nitric oxide, coupled with an increase in eNOS expression, which has been reported for PLC [46]. A single intake of GPLC has not been associated with the same increase in nitrate/nitrite [47], suggesting that chronic intake may be needed to exhibit these changes. Aside from PLC or GPLC, we are aware of one study using L-carnitine and demonstrating favorable effects for nitric oxide in a sample of hypertensive rodents [48], while one other study indicates suppression in nitric oxide production and iNOS protein expression (in cell culture) with L-carnitine treatment [49].

Despite the above-mentioned results of increased circulating nitric oxide metabolites following chronic carnitine supplementation in some studies, we are unaware of evidence indicating a cause and effect relationship between circulating nitric oxide and improved exercise performance [50]. Interestingly, recent work using either sodium nitrate [51] or beetroot juice [52] has demonstrated an increase in circulating nitrite following chronic use, coupled with an improvement in exercise performance. However, it is not clear as to the exact mechanism of action responsible for the noted performance gain. Certainly, more controlled laboratory studies are needed to elucidate the impact of circulating nitric oxide on exercise performance and related variables (e.g., recovery, hypertrophy).

EXERCISE PERFORMANCE-RELATED VARIABLES

Fatty Acid Oxidation and Muscle Metabolism

Carnitine has been investigated since the 1950s in relation to its primary role in the oxidation of long-chain fatty acids [3]. There are two primary roles that carnitine plays within skeletal muscle: the first being the translocation of long-chain fatty acids across the mitochondrial membrane into the matrix for β-oxidation, and the second involving the regulation of mitochondrial acetyl-CoA/CoASH ratio [22]. Within the body, over 95% of carnitine is stored in skeletal muscle as either free or acyl carnitine, which can be readily available for cellular processes [53].

Research focused on carnitine supplementation for purposes of improving exercise performance began to increase towards the end of the 20th century. The rationale for supplementation with carnitine is based on the idea that if an individual is capable of saturating the muscle with carnitine, then fatty acid transport may be enhanced, which may in turn increase fatty acid oxidation during exercise, spare glycogen stores, and thus delay the onset of fatigue [22]. Research within this area has since somewhat declined due to the lack of consistent evidence supporting alterations in fuel metabolism during exercise, as well as noted problems in increasing skeletal muscle carnitine content following regular supplementation.

The availability of free carnitine is rate limiting for skeletal muscle fat oxidation during exercise [54,55]. Considering that most carnitine is stored in skeletal muscle, it serves an essential role in β-oxidation with the translocation of long-chain fatty acids into the mitochondrial matrix [53]. Carnitine also plays a major role in the regulation of mitochondrial acetyl-CoA/CoASH ratio [56,57]. During high-intensity exercise, there is an accumulation of acetyl-CoA from the Pyruvate Dehydrogenase Complex (PDC). Carnitine's role with aceyl-CoA accumulation involves a buffering action by forming acylcarnitine via carnitine palmitoyltransferase (CPT)-1. However, as exercise intensity increases, this acetylation depletes the free carnitine pool. This in turn will have an effect on CPT-1, which is the rate-limiting step of long-chain acyl-CoAs (LCA-CoA) into the mitochondria. Roepstorff and colleagues reported that with high muscle glycogen, the availability of free carnitine may limit fat oxidation during exercise due to its increased use for acetylcarnitine formation [55].

In terms of increasing carnitine content within skeletal muscle (and improving subsequent exercise performance), most investigations have reported no significant effect for carnitine supplementation [58–61]. Most investigations have included a dosage of carnitine of 2–4 g/day. Vukovich et al. investigated the effects of 14 days of carnitine supplementation on muscle carnitine and glycogen content during exercise [59] and concluded that muscle glycogen content was not different from pre-exercise to post-exercise between carnitine and a no carnitine control condition. Lipid oxidation and muscle carnitine content was also not affected despite the increased availability of serum carnitine. Barnett and colleagues also investigated carnitine supplementation over 14 days [58]. These authors studied the effect of carnitine supplementation on muscle and blood carnitine fractions, and muscle and blood lactate concentrations during high-intensity sprint cycling exercise. They reported no increases in muscle carnitine content with carnitine supplementation, as well as no attenuation of lactate accumulation during exercise [58]. A longer-duration study was conducted by Wachter and coworkers [60], in which subjects consumed 4 g/day of carnitine over 3 months. Exercise was conducted at pre- and post-intervention, and was performed using a cycle ergometer until exhaustion. The authors observed no significant increases in muscle carnitine content, mitochondrial proliferation, or physical performance with long-term administration of carnitine [60].

One potential explanation for the lack of effect in terms of increasing skeletal muscle carnitine content may be the concentration gradient for bringing carnitine into the cell [60,62,63]. That is, carnitine transport into skeletal muscle is done via a high-affinity, saturable, Na^+-dependent, active transport process [64]. This process is performed against a considerable concentration gradient (>100-fold) [60,62,63]. As indicated above, many investigators have noted a lack of increase in muscle carnitine content or a lack of improvement in exercise performance after administering oral carnitine, even when provided for relatively long periods, such as 3 [60] or 6 [6] months.

Even though muscle carnitine levels often do not increase with conventional supplementation, methods have been studied in an attempt to achieve greater carnitine transport into the cell. For example, considering that sacrolemmal Na^+/K^+ ATPase pump activity is increased in the presence of insulin [65], Stephens and colleagues investigated whether or not insulin could augment carnitine levels within skeletal muscle [63]. This was done by infusing a supra-physiological concentration of carnitine (~500 μmol/L) intravenously for 5 hours along with inducing hyperinsulinemia (~150 mIU/L). The authors reported an increase in muscle total carnitine content and carnitine transporter protein mRNA expression via a combination of hypercarnitinemia and hyperinsulinemia [63]. In a follow-up study from the same laboratory, Wall and colleagues investigated the effects of chronic carnitine and carbohydrate ingestion on muscle total carnitine content, and exercise metabolism and performance in humans over 24 weeks [66]. Testing comprised of three visits, and included performing an exercise test of 30 minutes cycling at 50% VO_{2max}, 30 minutes at 80% VO_{2max}, and then a 30-minute work output performance trial. Supplementation involved subjects consuming daily either 80 g of carbohydrates or 80 g of carbohydrates and 2 g of LCLT. The authors concluded that muscle total carnitine content increased by 21% in the LCLT group, whereas the control group remained unchanged. Also noted was an increase in work output by 11% in the LCLT group.

These investigations may shed some light on the effectiveness of consuming carnitine along with carbohydrate with the intent of increasing muscle total carnitine content—and subsequent exercise performance. Of course, the consumer also needs to consider the fact that 80 g of carbohydrate may not be desirable, in particular if the individual is carbohydrate sensitive and/or carefully controlling dietary energy. An additional 300+ kcal in the form of carbohydrate may not be an option for many individuals—in particular if this is only being done in an attempt to increase muscle carnitine uptake.

Skeletal Muscle Function

Most carnitine is stored within skeletal muscle, with concentrations 50–200 times higher than in blood plasma [24]. However, as highlighted above, it is difficult to increase skeletal muscle carnitine concentrations with oral carnitine supplementation. While this is the case, it is possible to increase plasma carnitine levels by approximately 50% [67]. In fact, several investigations have confirmed that oral supplementation may increase plasma carnitine content, but fail to alter skeletal muscle carnitine content [58,68,69].

Aside from skeletal muscle carnitine content, increasing the carnitine pool in other areas may prove beneficial. As suggested by Kraemer et al. [70], it is possible that carnitine supplementation may help protect vascular endothelial cells from carnitine deficiency. This may also help improve muscle function and recovery by allowing for proper vascular function (i.e., improved blood flow, increase oxygen delivery to appropriate tissues, reduced tissue damage and muscle soreness, and a reduction in the accumulation of free radicals) [70]. The potential vasodilatory properties of carnitine may be one mechanism responsible for the increased muscle force output previously reported following carnitine supplementation [68]. This is partially supported by the findings of increased nitrate + nitrite in response to carnitine supplementation (in the form for GPLC) [16,17].

Muscle Injury and Soreness

Exercise-induced muscle injury is well described [71] and may be associated with impaired skeletal muscle function and increased muscle soreness. For example, Hunter et al. [72] reported that resistance exercise involving eccentric muscle actions can decrease maximal voluntary contraction torque and rate of torque development in the biceps brachii. Increased levels of muscle injury markers (e.g., creatine kinase activity, muscle soreness) suggest the presence of tissue (or cell membrane) damage. Excessive muscle soreness may be associated with impaired muscle force production—a key variable in overall sport performance. An early study found that 3 weeks of carnitine supplementation (3 g/day) may attenuate muscle soreness and creatine kinase activity during the recovery period following exercise [73]. Volek and coworkers also noted favorable effects of carnitine supplementation (delivered as LCLT) on the recovery following resistance exercise [32]. Specifically, these investigators hypothesized that LCLT supplementation would improve blood flow via decreased purine degradation, free radical accumulation, and muscle tissue damage. The markers of purine catabolism measured in this study included hypoxanthine, xanthine oxidase, and uric acid. Xanthine oxidase has been shown to be associated with increased production of superoxide [74], which has the potential to lead to lipid peroxidation and cellular damage [75]. In the Volek et al. study [32], purine catabolism was noted to be significantly lower in individuals who supplemented with LCLT. Also notable, muscle tissue disruption (assessed via MRI) and muscle soreness was significantly lower in the LCLT group, following the exercise protocol. Taken together with the findings of Giamberardino et al. [73], these reports suggest that carnitine supplementation may allow for improved recovery following exercise. However, more work is certainly needed in this area, in particular in resistance-trained individuals using carnitine on a *regular* basis and having assessments of muscle performance done repeatedly, rather than simply once or twice as part of a novel investigation. Such work would provide more-convincing evidence that supplemental carnitine is beneficial for purposes of improving muscle recovery following strenuous and damaging exercise bouts.

Androgens and Androgen Receptors

Although not a common theme for work related to carnitine, it has been reported that carnitine (in the form of LCLT) can impact androgen receptors [76,77]. Androgens are male sex hormones that are responsible for healthy reproductive function. Androgen hormones such as testosterone and 5α-dihydrotestosterone (DHT) function via interaction with the androgen receptor (AR), and help support healthy male reproduction, as well as bone and muscle development [78]. Testosterone is known to significantly improve muscle mass and strength [78]. Kadi et al. [79] found that resistance-trained individuals have an increased concentration of AR myonuclei per muscle fiber cross-section compared with untrained individuals. In addition, self-administration of androgenic-anabolic steroids (AAS) may further increase AR muscle fiber concentrations; however, these findings may be muscle specific [79]. It has also been shown that an androgen receptor antagonist known as oxendolone reduces hypertrophy in rats [80]. Taken together, these data support the role of ARs as moderators of exercise-induced hypertrophy. Furthermore, the evidence that AR concentrations increase in response to exercise [81,82] and that elevated testosterone levels via administration of AAS can significantly change AR-specific myonuclei in skeletal muscle [83], provides further support to ARs involvement in remodeling skeletal muscle in both physiological and supra-physiological conditions [78].

Kraemer and colleagues used LCLT to examine the effects of carnitine supplementation on anabolic hormones subsequent to resistance exercise [76]. The rationale stemmed from prior reports supporting carnitine supplementation as an aid to muscle recovery following exercise [30,32]. These authors sought to investigate the potential mechanism responsible for these positive effects, with the hypothesis that these findings may be related to increases in concentrations of anabolic hormones such as testosterone, growth hormone, and insulin-like growth factor [76]. Subjects in this investigation supplemented with 2 grams of LCLT/day for 3 weeks before a squat protocol, as well as during the recovery period. Subjects in the LCLT group displayed a significantly higher level of IGFBP-3 at 30, 120, and 180 minutes post exercise—with authors suggesting these findings as one mechanism by which increased protein metabolism and tissue repair could result. With regards to the analysis of muscle tissue disruption and damage, the LCLT group displayed significantly less tissue damage during the days subsequent to the squat protocol. The investigators suggested that the reduction in tissue disruption may be related to the accumulation of LCLT in capillary endothelial cells, which could allow for a vasodilatory effect on capillaries and result in improved oxygen delivery and reduced local hypoxia to working muscles [76].

A follow up study conducted by Kraemer and colleagues analyzed the effects of LCLT supplementation (equivalent to 2 g carnitine/day) and post resistance exercise feeding on hormonal and AR responses [77]. Subjects supplemented with either LCLT or a placebo for 3 weeks, and then performed two resistance training protocols (one followed by water, one followed by a feeding). The feeding consisted of a caloric beverage that comprised of 8 kcal/kg body mass with 1.1 g/kg carbohydrates, 0.3 g/kg protein, and 0.25 g/kg fat. The resistance exercise protocol consisted of four sets of 10 repetitions of the squat, bench press, bent row, and shoulder press (16 total sets). With regards to AR content, 3 weeks of LCLT supplementation significantly increased pre-exercise AR content in the vastus lateralis compared with the placebo group. Post-exercise feeding increased AR content in both LCLT- and placebo-treated groups. The findings for testosterone concentrations were such that the water and LCLT-supplemented group demonstrated higher testosterone concentrations up to 60 minutes post resistance exercise when compared with water only. While this was an acute study, further research is needed to determine whether or not regular use of carnitine (in the form of LCLT or other forms) could repeatedly influence AR and circulating hormones. More importantly, work is needed to determine what, if any, meaning these changes in AR or circulating hormones have in terms of the adaptive response to exercise training.

SUMMARY OF CARNITINE FOR ATHLETIC POPULATIONS

From a pure exercise performance perspective, the scientific literature does not strongly support the use of supplemental carnitine for purposes of improving exercise performance. That being said, there do exist some studies that indicate an ergogenic effect of this nutrient, in particular within selected individuals (i.e., responders). Despite the general lack of data supporting carnitine as an ergogenic aid, data do exist indicating a benefit in terms of exercise-related outcomes. These include both increased production of nitric oxide metabolites [16,17]—which may be associated with improved bloodflow and "muscle pumps"—as well as a reduction in muscle damage following strenuous exercise [32]. Unfortunately, the number of studies demonstrating a benefit in these areas is small, and additional work is certainly needed. Perhaps more importantly, the studies performed have been short-term (acute lab studies) with outcome measures (e.g., nitrate/nitrite, muscle soreness) often being used to suggest effects for other variables not measured within the research designs. For example, the increase in nitrate/nitrite may be suggestive of an increase in circulating nitric oxide, which may be suggestive of an increase in blood flow leading to increased exercise performance and recovery. While this is certainly possible, to date there are no direct data indicating this to be the case [50]. The same is true for findings of decreased muscle damage and soreness, as well as the alteration in AR content. While these results may suggest a positive effect that would then translate into improved recovery and favorable adaptations to *chronic* exercise training (e.g., hypertrophy, force output), longer-term supplementation studies are needed to generate such answers. At this point, such talk is mere speculation.

OVERVIEW OF CARNITINE FOR NON-ATHLETIC (HEALTHY AND DISEASED) POPULATIONS

While carnitine supplementation may be questionable specifically for purposes of improving physical performance, a wealth of evidence supports the use of carnitine for multiple purposes related to human health—in particular in those with known disease. Carnitine has been identified as having antioxidant properties [84], corrects metabolic and cardiovascular alterations induced by poor dietary intake [85], and may be capable of promoting healthy vascular function [86]—findings that may be of benefit to both healthy and diseased individuals. Carnitine has been shown to have favorable effects in a variety of pathological conditions, including Alzheimer's disease [87], AIDS [88], and thalassemia disease [89]. It has been reported that individuals suffering from congestive heart failure may be aided by carnitine supplementation [90], as are those with other forms of cardiovascular disease [91]. Treatment with carnitine may improve muscular strength [82] and walking performance [92] in individuals with peripheral arterial disease. Furthermore, improvements in muscular strength, walking tolerance, heart rate, blood pressure, oxygen saturation, and blood lactate clearance have been documented in association with carnitine supplementation in individuals with chronic obstructive pulmonary disease [93]. Carnitine has been shown to attenuate neurological damage observed following brain ischemia and reperfusion [94], as well as assist in cellular immune function, in particular with regard to the incidence of upper respiratory tract infections [24,95]. Recently, carnitine treatment has resulted in positive findings for those with autism spectrum disorders [96]. Collectively, the above findings indicate a clear role for carnitine supplementation in the treatment of a variety of disease states.

CONCLUSIONS

Carnitine is a conditionally essential nutrient, available in several forms, which possesses multiple physiological properties. In a manner similar to a multivitamin/mineral supplement, carnitine may be viewed as a supplement to support general health. For certain individuals, carnitine may prove beneficial for purposes of providing antioxidant protection, increasing nitric oxide and related blood flow, assisting with recovery from strenuous exercise, and/or improving exercise performance. That being said, several performance-specific studies have failed to note an ergogenic effect for carnitine. However, as with most nutritional supplements, results will vary considerably from person to person, with some individuals benefiting from carnitine supplementation for exercise-related purposes and others experiencing no observed benefit. While the possibility of a placebo effect always remains [97–99], one must at least entertain the possibility that carnitine supplementation provides a benefit to those who claim so enthusiastically that the supplement is "working". Only more well-controlled studies, ideally comparing different forms of carnitine within the research design and including a longer-term intervention approach, will provide additional information regarding the role of supplemental carnitine to improve exercise performance and related outcome measures—in particular within a sample of exercise-trained individuals.

References

[1] Vaz FM, Wanders RJ. Carnitine biosynthesis in mammals. Biochem J 2002;361(Pt 3):417–29.

[2] Kelly GS. L-Carnitine: therapeutic applications of a conditionally-essential amino acid. Altern Med Rev 1998;3 (5):345–60.

[3] Fritz I. The effect of muscle extracts on the oxidation of palmitic acid by liver slices and homogenates. Acta Physiol Scand 1955;34 (4):367–85.

[4] Rebouche CJ, Engel AG. Kinetic compartmental analysis of carnitine metabolism in the human carnitine deficiency syndromes. Evidence for alterations in tissue carnitine transport. J Clin Invest 1984;73(3):857–67.

[5] Lombard KA, Olson AL, Nelson SE, Rebouche CJ. Carnitine status of lactoovovegetarians and strict vegetarian adults and children. Am J Clin Nutr 1989;50(2):301–6.

[6] Arenas J, Ricoy JR, Encinas AR, Pola P, D'Iddio S, Zeviani M, et al. Carnitine in muscle, serum, and urine of nonprofessional athletes: effects of physical exercise, training, and L-carnitine administration. Muscle Nerve 1991;14(7):598–604.

[7] Calo LA, Pagnin E, Davis PA, Semplicini A, Nicolai R, Calvani M, et al. Antioxidant effect of L-carnitine and its short chain esters: relevance for the protection from oxidative stress related cardiovascular damage. Int J Cardiol 2006;107(1):54–60.

[8] Inano A, Sai Y, Nikaido H, Hasimoto N, Asano M, Tsuji A, et al. Acetyl-L-carnitine permeability across the blood-brain barrier and involvement of carnitine transporter OCTN2. Biopharm Drug Dispos 2003;24(8):357–65.

[9] Ho JY, Kraemer WJ, Volek JS, Fragala MS, Thomas GA, Dunn-Lewis C, et al. l-Carnitine l-tartrate supplementation favorably affects biochemical markers of recovery from physical exertion in middle-aged men and women. Metabolism 2010;59 (8):1190–9.

[10] Mingorance C, Rodriguez-Rodriguez R, Justo ML, Herrera MD, de Sotomayor MA. Pharmacological effects and clinical applications of propionyl-L-carnitine. Nutr Rev 2011;69(5):279–90.

[11] Barker GA, Green S, Askew CD, Green AA, Walker PJ. Effect of propionyl-L-carnitine on exercise performance in peripheral arterial disease. Med Sci Sports Exerc 2001;33(9):1415–22.

[12] Caponnetto S, Canale C, Masperone MA, Terracchini V, Valentini G, Brunelli C. Efficacy of L-propionylcarnitine treatment in patients with left ventricular dysfunction. Eur Heart J 1994;15(9):1267–73.

[13] Chiddo A, Gaglione A, Musci S, Troito G, Grimaldi N, Locuratolo N, et al. Hemodynamic study of intravenous propionyl-L-carnitine in patients with ischemic heart disease and normal left ventricular function. Cardiovasc Drugs Ther 1991;5(Suppl. 1):107–11.

[14] Jacobs PL, Goldstein ER, Blackburn W, Orem I, Hughes JJ. Glycine propionyl-L-carnitine produces enhanced anaerobic work capacity with reduced lactate accumulation in resistance trained males. J Int Soc Sports Nutr 2009;6:9.

[15] Loffredo L, Pignatelli P, Cangemi R, Andreozzi P, Panico MA, Meloni V, et al. Imbalance between nitric oxide generation and oxidative stress in patients with peripheral arterial disease: effect of an antioxidant treatment. J Vasc Surg 2006;44 (3):525–30.

[16] Bloomer RJ, Tschume LC, Smith WA. Glycine propionyl-L-carnitine modulates lipid peroxidation and nitric oxide in human subjects. Int J Vitam Nutr Res 2009;79(3):131–41.

[17] Bloomer RJ, Smith WA, Fisher-Wellman KH. Glycine propionyl-L-carnitine increases plasma nitrate/nitrite in resistance trained men. J Int Soc Sports Nutr 2007;4:22.

[18] Rubin MR, Volek JS, Gomez AL, Ratamess NA, French DN, Sharman MJ, et al. Safety measures of L-carnitine L-tartrate supplementation in healthy men. J Strength Cond Res 2001;15 (4):486–90.

[19] Cruciani RA, Dvorkin E, Homel P, Malamud S, Culliney B, Lapin J, et al. Safety, tolerability and symptom outcomes associated with L-carnitine supplementation in patients with cancer, fatigue, and carnitine deficiency: a phase I/II study. J Pain Symptom Manage 2006;32(6):551–9.

[20] Hathcock JN, Shao A. Risk assessment for carnitine. Regul Toxicol Pharmacol 2006;46(1):23–8.

[21] Cha YS. Effects of L-carnitine on obesity, diabetes, and as an ergogenic aid. Asia Pac J Clin Nutr 2008;17(Suppl. 1):306–8.

[22] Stephens FB, Constantin-Teodosiu D, Greenhaff PL. New insights concerning the role of carnitine in the regulation of fuel metabolism in skeletal muscle. J Physiol 2007;581 (Pt 2):431–44.

[23] Brass EP. Carnitine and sports medicine: use or abuse? Ann N Y Acad Sci 2004;1033:67–78.

[24] Karlic H, Lohninger A. Supplementation of L-carnitine in athletes: does it make sense? Nutrition 2004;20(7–8):709–15.

[25] Fisher-Wellman K, Bloomer RJ. Acute exercise and oxidative stress: a 30 year history. Dyn Med 2009;8:1.

[26] Loster H, Bohm U. L-carnitine reduces malondialdehyde concentrations in isolated rat hearts in dependence on perfusion conditions. Mol Cell Biochem 2001;217(1–2):83–90.

[27] Di Giacomo C, Latteri F, Fichera C, Sorrenti V, Campisi A, Castorina C, et al. Effect of acetyl-L-carnitine on lipid peroxidation and xanthine oxidase activity in rat skeletal muscle. Neurochem Res 1993;18(11):1157–62.

[28] Bucioli SA, De Abreu LC, Valenti VE, Vannucchi H. Carnitine supplementation effects on nonenzymatic antioxidants in young rats submitted to exhaustive exercise stress. J Strength Cond Res 2012;26(6):1695–700.

[29] Siktar E, Ekinci D, Siktar E, Beydemir S, Gulcin I, Gunay M. Protective role of L-carnitine supplementation against exhaustive exercise induced oxidative stress in rats. Eur J Pharmacol 2011;668(3):407–13.

[30] Corbucci GG, Montanari G, Mancinelli G, D'Iddio S. Metabolic effects induced by L-carnitine and propionyl-L-carnitine in human hypoxic muscle tissue during exercise. Int J Clin Pharmacol Res 1990;10(3):197–202.

[31] Sachan DS, Hongu N, Johnsen M. Decreasing oxidative stress with choline and carnitine in women. J Am Coll Nutr 2005;24 (3):172–6.

[32] Volek JS, Kraemer WJ, Rubin MR, Gomez AL, Ratamess NA, Gaynor P. L-Carnitine L-tartrate supplementation favorably affects markers of recovery from exercise stress. Am J Physiol Endocrinol Metab 2002;282(2):E474–82.

[33] Loffredo L, Marcoccia A, Pignatelli P, Andreozzi P, Borgia MC, Cangemi R, et al. Oxidative-stress-mediated arterial dysfunction in patients with peripheral arterial disease. Eur Heart J 2007;28(5):608–12.

[34] Vanella A, Russo A, Acquaviva R, Campisi A, Di Giacomo C, Sorrenti V, et al. L-propionyl-carnitine as superoxide scavenger, antioxidant, and DNA cleavage protector. Cell Biol Toxicol 2000;16(2):99–104.

[35] Dalle-Donne I, Rossi R, Colombo R, Giustarini D, Milzani A. Biomarkers of oxidative damage in human disease. Clin Chem 2006;52(4):601–23.

[36] Farney TM, McCarthy CG, Canale RE, Schilling BK, Whitehead PN, Bloomer RJ. Absence of blood oxidative stress in trained men following strenuous exercise. Med Sci Sports Exerc 2012;44 (10):1855–63.

[37] Tanskanen MM, Uusitalo AL, Kinnunen H, Hakkinen K, Kyrolainen H, Atalay M. Association of military training with oxidative stress and overreaching. Med Sci Sports Exerc 2011;43 (8):1552–60.

[38] Zoppi CC, Macedo DV. Overreaching-induced oxidative stress, enhanced HSP72 expression, antioxidant and oxidative enzymes downregulation. Scand J Med Sci Sports 2008;18 (1):67–76.

[39] Furchgott RF, Zawadzki JV. The obligatory role of endothelial cells in the relaxation of arterial smooth muscle by acetylcholine. Nature 1980;288(5789):373–6.

[40] Collier J, Vallance P. Physiological importance of nitric oxide. BMJ 1991;302(6788):1289–90.
[41] Bian K, Doursout MF, Murad F. Vascular system: role of nitric oxide in cardiovascular diseases. J Clin Hypertens (Greenwich) 2008;10(4):304–10.
[42] Thomas DD, Ridnour LA, Isenberg JS, Flores-Santana W, Switzer CH, Donzelli S, et al. The chemical biology of nitric oxide: implications in cellular signaling. Free Radic Biol Med 2008;45(1):18–31.
[43] Anderson JE. A role for nitric oxide in muscle repair: nitric oxide-mediated activation of muscle satellite cells. Mol Biol Cell 2000;11(5):1859–74.
[44] Salanova M, Schiffl G, Puttmann B, Schoser BG, Blottner D. Molecular biomarkers monitoring human skeletal muscle fibres and microvasculature following long-term bed rest with and without countermeasures. J Anat 2008;212(3):306–18.
[45] Pignatelli P, Lenti L, Sanguigni V, Frati G, Simeoni I, Gazzaniga PP, et al. Carnitine inhibits arachidonic acid turnover, platelet function, and oxidative stress. Am J Physiol Heart Circ Physiol 2003;284(1):H41–8.
[46] de Sotomayor MA, Mingorance C, Rodriguez-Rodriguez R, Marhuenda E, Herrera MD. l-carnitine and its propionate: improvement of endothelial function in SHR through superoxide dismutase-dependent mechanisms. Free Radic Res 2007;41 (8):884–91.
[47] Bloomer RJ, Farney TM, Trepanowski JF, McCarthy CG, Canale RE, Schilling BK. Comparison of pre-workout nitric oxide stimulating dietary supplements on skeletal muscle oxygen saturation, blood nitrate/nitrite, lipid peroxidation, and upper body exercise performance in resistance trained men. J Int Soc Sports Nutr 2010;7:16.
[48] Mate A, Miguel-Carrasco JL, Monserrat MT, Vazquez CM. Systemic antioxidant properties of L-carnitine in two different models of arterial hypertension. J Physiol Biochem 2010;66 (2):127–36.
[49] Koc A, Ozkan T, Karabay AZ, Sunguroglu A, Aktan F. Effect of L-carnitine on the synthesis of nitric oxide in RAW 264.7 murine macrophage cell line. Cell Biochem Funct 2011;29 (8):679–85.
[50] Bloomer RJ. Nitric oxide supplements for sports. Strength Cond J 2010;32(2):14–20.
[51] Larsen FJ, Weitzberg E, Lundberg JO, Ekblom B. Effects of dietary nitrate on oxygen cost during exercise. Acta Physiol (Oxf) 2007;191(1):59–66.
[52] Cermak NM, Gibala MJ, van Loon LJ. Nitrate supplementation's improvement of 10-km time-trial performance in trained cyclists. Int J Sport Nutr Exerc Metab 2012;22(1):64–71.
[53] Brass EP. Pharmacokinetic considerations for the therapeutic use of carnitine in hemodialysis patients. Clin Ther 1995;17 (2):176–85 [discussion 175].
[54] van Loon LJ, Greenhaff PL, Constantin-Teodosiu D, Saris WH, Wagenmakers AJ. The effects of increasing exercise intensity on muscle fuel utilisation in humans. J Physiol 2001;536 (Pt 1):295–304.
[55] Roepstorff C, Halberg N, Hillig T, Saha AK, Ruderman NB, Wojtaszewski JF, et al. Malonyl-CoA and carnitine in regulation of fat oxidation in human skeletal muscle during exercise. Am J Physiol Endocrinol Metab 2005;288(1):E133–42.
[56] Constantin-Teodosiu D, Carlin JI, Cederblad G, Harris RC, Hultman E. Acetyl group accumulation and pyruvate dehydrogenase activity in human muscle during incremental exercise. Acta Physiol Scand 1991;143(4):367–72.
[57] Harris RC, Foster CV, Hultman E. Acetylcarnitine formation during intense muscular contraction in humans. J Appl Physiol 1987;63(1):440–2.
[58] Barnett C, Costill DL, Vukovich MD, Cole KJ, Goodpaster BH, Trappe SW, et al. Effect of L-carnitine supplementation on muscle and blood carnitine content and lactate accumulation during high-intensity sprint cycling. Int J Sport Nutr 1994;4(3):280–8.
[59] Vukovich MD, Costill DL, Fink WJ. Carnitine supplementation: effect on muscle carnitine and glycogen content during exercise. Med Sci Sports Exerc 1994;26(9):1122–9.
[60] Wachter S, Vogt M, Kreis R, Boesch C, Bigler P, Hoppeler H, et al. Long-term administration of L-carnitine to humans: effect on skeletal muscle carnitine content and physical performance. Clin Chim Acta 2002;318(1–2):51–61.
[61] Smith WA, Fry AC, Tschume LC, Bloomer RJ. Effect of glycine propionyl-L-carnitine on aerobic and anaerobic exercise performance. Int J Sport Nutr Exerc Metab 2008;18(1):19–36.
[62] Brass EP, Hoppel CL, Hiatt WR. Effect of intravenous L-carnitine on carnitine homeostasis and fuel metabolism during exercise in humans. Clin Pharmacol Ther 1994;55 (6):681–92.
[63] Stephens FB, Constantin-Teodosiu D, Laithwaite D, Simpson EJ, Greenhaff PL. Insulin stimulates L-carnitine accumulation in human skeletal muscle. FASEB J 2006;20(2):377–9.
[64] Rebouche CJ. Carnitine movement across muscle cell membranes. Studies in isolated rat muscle. Biochim Biophys Acta 1977;471(1):145–55.
[65] Clausen T. Na+ -K+ pump regulation and skeletal muscle contractility. Physiol Rev 2003;83(4):1269–324.
[66] Wall BT, Stephens FB, Constantin-Teodosiu D, Marimuthu K, Macdonald IA, Greenhaff PL. Chronic oral ingestion of L-carnitine and carbohydrate increases muscle carnitine content and alters muscle fuel metabolism during exercise in humans. J Physiol 2011;589(Pt 4):963–73.
[67] Kraemer W, Volek J. L-carnitine supplementation for the athlete. A new perspective. Ann Nutr Metab 2000;44:88–9.
[68] Dubelaar ML, Lucas CM, Hulsmann WC. Acute effect of L-carnitine on skeletal muscle force tests in dogs. Am J Physiol 1991;260(2 Pt 1):E189–93.
[69] Lennon DL, Shrago ER, Madden M, Nagle FJ, Hanson P. Dietary carnitine intake related to skeletal muscle and plasma carnitine concentrations in adult men and women. Am J Clin Nutr 1986;43(2):234–8.
[70] Kraemer WJ, Volek JS, Dunn-Lewis C. L-carnitine supplementation: influence upon physiological function. Curr Sports Med Rep 2008;7(4):218–23.
[71] Brentano MA, Martins Kruel LF. A review on strength exercise-induced muscle damage: applications, adaptation mechanisms and limitations. J Sports Med Phys Fitness 2011;51(1):1–10.
[72] Hunter AM, Galloway SD, Smith IJ, Tallent J, Ditroilo M, Fairweather MM, et al. Assessment of eccentric exercise-induced muscle damage of the elbow flexors by tensiomyography. J Electromyogr Kinesiol 2012.
[73] Giamberardino MA, Dragani L, Valente R, Di Lisa F, Saggini R, Vecchiet L. Effects of prolonged L-carnitine administration on delayed muscle pain and CK release after eccentric effort. Int J Sports Med 1996;17(5):320–4.
[74] Heunks LM, Vina J, van Herwaarden CL, Folgering HT, Gimeno A, Dekhuijzen PN. Xanthine oxidase is involved in exercise-induced oxidative stress in chronic obstructive pulmonary disease. Am J Physiol 1999;277(6 Pt 2):R1697–704.
[75] Duarte JA, Appell HJ, Carvalho F, Bastos ML, Soares JM. Endothelium-derived oxidative stress may contribute to exercise-induced muscle damage. Int J Sports Med 1993;14 (8):440–3.
[76] Kraemer WJ, Volek JS, French DN, Rubin MR, Sharman MJ, Gomez AL, et al. The effects of L-carnitine L-tartrate

supplementation on hormonal responses to resistance exercise and recovery. J Strength Cond Res 2003;17(3):455–62.

[77] Kraemer WJ, Spiering BA, Volek JS, Ratamess NA, Sharman MJ, Rubin MR, et al. Androgenic responses to resistance exercise: effects of feeding and L-carnitine. Med Sci Sports Exerc 2006;38(7):1288–96.

[78] Li J, Al-Azzawi F. Mechanism of androgen receptor action. Maturitas 2009;63(2):142–8.

[79] Kadi F, Bonnerud P, Eriksson A, Thornell LE. The expression of androgen receptors in human neck and limb muscles: effects of training and self-administration of androgenic-anabolic steroids. Histochem Cell Biol 2000;113(1):25–9.

[80] Inoue K, Yamasaki S, Fushiki T, Okada Y, Sugimoto E. Androgen receptor antagonist suppresses exercise-induced hypertrophy of skeletal muscle. Eur J Appl Physiol Occup Physiol 1994;69(1):88–91.

[81] Deschenes MR, Maresh CM, Armstrong LE, Covault J, Kraemer WJ, Crivello JF. Endurance and resistance exercise induce muscle fiber type specific responses in androgen binding capacity. J Steroid Biochem Mol Biol 1994;50(3–4):175–9.

[82] Bamman MM, Shipp JR, Jiang J, Gower BA, Hunter GR, Goodman A, et al. Mechanical load increases muscle IGF-I and androgen receptor mRNA concentrations in humans. Am J Physiol Endocrinol Metab 2001;280(3):E383–90.

[83] Kadi F. Cellular and molecular mechanisms responsible for the action of testosterone on human skeletal muscle. A basis for illegal performance enhancement. Br J Pharmacol 2008;154 (3):522–8.

[84] Li JL, Wang QY, Luan HY, Kang ZC, Wang CB. Effects of L-carnitine against oxidative stress in human hepatocytes: involvement of peroxisome proliferator-activated receptor alpha. J Biomed Sci 2012;19(1):32.

[85] Mingorance C, Duluc L, Chalopin M, Simard G, Ducluzeau PH, Herrera MD, et al. Propionyl-L-carnitine corrects metabolic and cardiovascular alterations in diet-induced obese mice and improves liver respiratory chain activity. PLoS One 2012;7(3): e34268.

[86] Volek JS, Judelson DA, Silvestre R, Yamamoto LM, Spiering BA, Hatfield DL, et al. Effects of carnitine supplementation on flow-mediated dilation and vascular inflammatory responses to a high-fat meal in healthy young adults. Am J Cardiol 2008;102 (10):1413–7.

[87] Pettegrew JW, Levine J, McClure RJ. Acetyl-L-carnitine physical-chemical, metabolic, and therapeutic properties: relevance for its mode of action in Alzheimer's disease and geriatric depression. Mol Psychiatry 2000;5(6):616–32.

[88] De Simone C, Tzantzoglou S, Famularo G, Moretti S, Paoletti F, Vullo V, et al. High dose L-carnitine improves immunologic and metabolic parameters in AIDS patients. Immunopharmacol Immunotoxicol 1993;15(1):1–12.

[89] El-Beshlawy A, El Accaoui R, Abd El-Sattar M, Gamal El-Deen MH, Youssry I, Shaheen N, et al. Effect of L-carnitine on the physical fitness of thalassemic patients. Ann Hematol 2007;86 (1):31–4.

[90] Kobayashi A, Masumura Y, Yamazaki N. L-carnitine treatment for congestive heart failure—experimental and clinical study. Jpn Circ J 1992;56(1):86–94.

[91] Ferrari R, Merli E, Cicchitelli G, Mele D, Fucili A, Ceconi C. Therapeutic effects of L-carnitine and propionyl-L-carnitine on cardiovascular diseases: a review. Ann N Y Acad Sci 2004;1033:79–91.

[92] Hiatt WR. Carnitine and peripheral arterial disease. Ann N Y Acad Sci 2004;1033:92–8.

[93] Borghi-Silva A, Baldissera V, Sampaio LM, Pires-DiLorenzo VA, Jamami M, Demonte A, et al. L-carnitine as an ergogenic aid for patients with chronic obstructive pulmonary disease submitted to whole-body and respiratory muscle training programs. Braz J Med Biol Res 2006;39(4):465–74.

[94] Calvani M, Arrigoni-Martelli E. Attenuation by acetyl-L-carnitine of neurological damage and biochemical derangement following brain ischemia and reperfusion. Int J Tissue React 1999;21(1):1–6.

[95] Gleeson M, McDonald WA, Pyne DB, Clancy RL, Cripps AW, Francis JL, et al. Immune status and respiratory illness for elite swimmers during a 12-week training cycle. Int J Sports Med 2000;21(4):302–7.

[96] Geier DA, Kern JK, Davis G, King PG, Adams JB, Young JL, et al. A prospective double-blind, randomized clinical trial of levocarnitine to treat autism spectrum disorders. Med Sci Monit 2011;17(6):PI15–23.

[97] Trojian TH, Beedie CJ. Placebo effect and athletes. Curr Sports Med Rep 2008;7(4):214–7.

[98] Beedie CJ, Foad AJ. The placebo effect in sports performance: a brief review. Sports Med 2009;39(4):313–29.

[99] McClung M, Collins D. "Because I know it will!": placebo effects of an ergogenic aid on athletic performance. J Sport Exerc Psychol 2007;29(3):382–94.

CHAPTER

42

An Overview of the Influence of Protein Ingestion on Muscle Hypertrophy

Koji Okamura

Graduate School of Sport and Exercise Sciences, Osaka University of Health and Sport Sciences, Kumatori, Japan

THE IMPORTANCE OF DIETARY ENERGY

Protein is the nutrient that athletes are most interested in because of its role in building muscle. Protein is the most abundant component in muscle besides water. However, there has to be sufficient dietary energy to promote muscle protein synthesis. When the dietary energy is not sufficient, dietary protein is consumed as an energy source, and is not utilized for muscle synthesis. For example, when glycogen stores are low, exercise-induced proteolysis increases [1]. Therefore, not only ingesting a proper amount of protein, but also consuming sufficient energy, is important for building muscle. It should be noted that the protein intake described in the following section is the amount needed when there is sufficient dietary energy intake.

It is known that skeletal muscle strength increases prior to skeletal muscle hypertrophy. However, it is also known that muscle hypertrophy is associated with an increase in muscle strength. A recent study showed that significant skeletal muscle hypertrophy appears to occur around 3–4 weeks after the commencement of a training program, suggesting that training-induced skeletal muscle hypertrophy may occur earlier than was previously thought [2]. However, skeletal muscle hypertrophy is a relatively slow process. Therefore, the effect of nutrients on skeletal muscle hypertrophy requires some time to emerge.

PROTEIN REQUIREMENTS

Daily Requirements

Protein is the most popular supplement among athletes and is thought to enhance muscle hypertrophy. However, no scientific evidence has demonstrated that protein supplementation facilitates exercise-induced skeletal muscle hypertrophy [3]. Therefore, it seems unnecessary for athletes to increase their protein intake above that required for normal nutrition. There is no justification that intakes in excess of about 1.7 gram per kilogram body weight per day ($g\,kg^{-1}\,d^{-1}$) enhance training adaptation or facilitate the gains in muscle mass and strength [3,4]. Dietary surveys have shown that most athletes consume diets containing more than 1.2–1.6 $g\,kg^{-1}\,d^{-1}$ of protein, even without the use of protein supplements. Some resistance-trained athletes and body builders consume more than 2–3 $g\,kg^{-1}\,d^{-1}$ of protein [3,4].

Resistance exercise training increases dietary nitrogen retention, and a recent study indicated that muscle mass increases without any increase in protein intake during a 12-week training period in untrained young males, as shown in Table 42.1 [5]. The authors of that study concluded that an increase in nitrogen balance after training demonstrates a more efficient utilization of dietary protein, and that the protein requirements for novice weight lifters are not elevated. However, the authors noted that their study cannot be directly extrapolated to more experienced weightlifters who may have maximized their lean mass gains. However, considering that changes in muscle strength and muscle size occur more rapidly in untrained compared with trained individuals, it is likely that the protein requirements would be greatest during the early stages of a resistance-training program. Therefore, the authors speculated that, because a protein intake of 1.4 $g\,kg^{-1}\,d^{-1}$ was adequate to maintain a positive nitrogen balance in novice weightlifters, the dietary protein requirements would be below the habitual

Nutrition and Enhanced Sports Performance.
DOI: http://dx.doi.org/10.1016/B978-0-12-396454-0.00042-4

TABLE 42.1 Daily Nitrogen Balance in Young Males Before (Pre) and After (Post) a 12-Week Whole Body Resistance Training Program [a]

	Pre		Post	
	mg/(kg LBM day)			
	Day 1	Day 2	Day 1[b]	Day 2
Modified NBAL[c]	55.9 ± 13.8	55.3 ± 11.2	100.3 ± 19.0	90.1 ± 19.6
NBAL[d]	14.0 ± 13.5	13.3 ± 8.9	38.5 ± 13.9	28.3 ± 17.0

[a]*Values are the means ± SD. Time (pre vs post) was significant for both variables, P < 0.01. The days within each period did not differ.*
[b]*Final bout of resistance exercise performed on Day 1.*
[c]*Modified NBAL based on urine only.*
[d]*NBAL estimated using fecal, sweat, and miscellaneous nitrogen losses for sedentary and strength- training athletes based on the previous literature (Ref. [6]).*
NBAL, nitrogen balance.
Reproduced from Moore et al. [5].

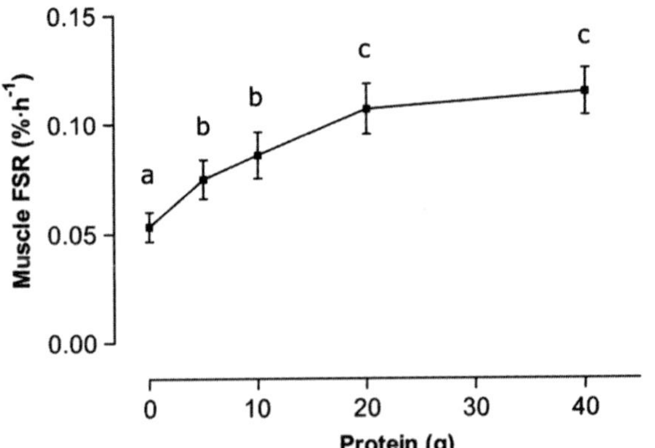

FIGURE 42.1 **The muscle fractional synthesis rate (FSR) after resistance exercise in response to increasing amounts of dietary protein.** The means with different letters indicate values that are significantly different from each other (P < 0.01; n = 6). *Reproduced from Moore et al. [7].*

intake (typically $\geq 2\ g\ kg^{-1}\ d^{-1}$) of many resistance-trained individuals.

Requirements for a Single Meal

Studies on the protein requirement for one meal are lacking in comparison with the daily requirement. Figure 42.1 shows that the skeletal muscle protein synthesis rate reaches a plateau upon ingesting 20 g of protein after resistance exercise [7]. The same study also demonstrated that albumin synthesis in the liver is maximally stimulated by consumption of 20 g of protein, and that leucine oxidation increases significantly after 20 and 40 g protein are ingested. Ingesting a large amount of protein at a time increases the amino acid concentration rapidly, resulting in a large amount of amino acids being delivered to skeletal muscle. However, skeletal muscle synthesis can become saturated, and excess amino acids which are not utilized for muscle protein synthesis remain. It appears that oxidation is the major metabolic pathway for dealing with excess amino acids. A typical meal contains 20–30 g of protein. Therefore, skeletal muscle protein synthesis is maximally stimulated after typical meals.

It has been shown that the protein retention in the body is greater when about 80% of the daily protein is provided in one meal compared with when the daily protein is spread over four meals in elderly females [8]. In young females, however, the protein retention in the body did not significantly differ when 80% of the daily protein was given in one meal compared with when the daily protein was given over four meals [9].

In a study of post-exercise healthy adult males, both an increase in the amino acid concentration in the blood and the skeletal muscle protein synthesis were greater after consuming 25 g of protein at a time than after consuming 2.5 g of protein every 20 min for 10 times [10]. On the other hand, in a study that examined the pre-exercise protein ingestion pattern, no differences in post-exercise muscle protein synthesis were observed between subjects who ingested 25 g protein in a single dose and those who ingested the same amount of protein in 13 small doses [11].

The Consensus Statement on Sports Nutrition prepared by the International Olympic Committee (IOC) states that ingesting foods or drinks providing 15–25 g of protein after each training session will maximize the synthesis of muscle protein.

HABITUAL HIGH PROTEIN INTAKE

The mass of body protein is determined based on the balance between the synthesis and breakdown of body protein. Skeletal muscle hypertrophy is achieved when synthesis surpasses degradation for a certain duration of time.

As shown in Figure 42.2, a habitual high protein intake increases both the postprandial synthesis (positive N balance) and post-absorptive breakdown (negative N balance) [12]. Therefore, the protein requirement is likely to reflect the demand for repletion of post-absorptive losses, which increases with increasing habitual protein intake.

It has been reported that the postprandial whole-body retention of protein nitrogen after meal ingestion decreased when the subjects switched from a normal protein diet to a high-protein diet [13]. This decrease is considered to be due to an increased splanchnic utilization of dietary nitrogen for urea production, whereas its incorporation into splanchnic proteins is

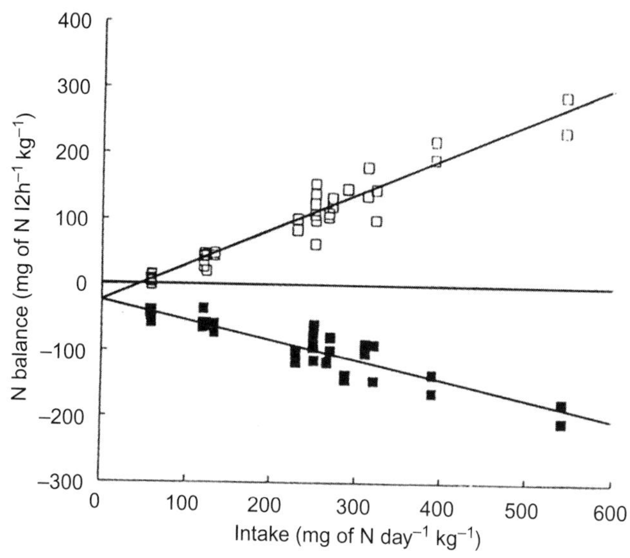

FIGURE 42.2 **Diurnal changes in the nitrogen balance in subjects habituated to increased dietary protein intake.** □ postprandial; ■ post-absorptive. *Reproduced from Price et al. [12].*

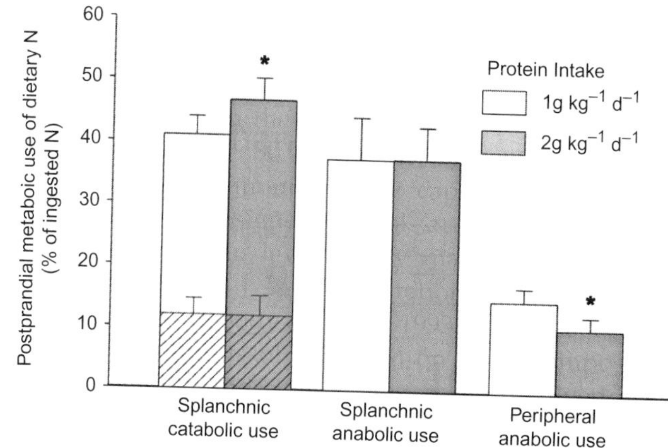

FIGURE 42.3 **The model-predicted 8 h balance for the postprandial utilization of dietary nitrogen in the splanchnic and peripheral areas in response to a single, solid mixed meal containing [^{15}N]-labeled wheat protein after adaptation for 7 days.** Hatched bars: dietary N recycling into the metabolic N pool through urea hydrolysis at 8 h. *Reproduced from Juillet et al. [13].*

unchanged, resulting in a decrease in peripheral availability and anabolic use in subjects adapted to high protein consumption compared with subjects adapted to normal protein intake (Figure 42.3). The authors conclude that increasing the protein intake reduces the postprandial retention of protein nitrogen, mainly by diminishing the efficiency of its peripheral availability and anabolic use. These results suggest that the habitual consumption of a high-protein diet may not be effective for inducing skeletal muscle hypertrophy and it may even be disadvantageous for muscle building.

AMINO ACIDS

Amino acids are a substrate of, as well as a stimulatory factor for, muscle protein synthesis. In particular, leucine, a branched-chain amino acid (BCAA), stimulates muscle protein synthesis [14]. The mechanisms underlying the stimulating effect of leucine on muscle protein synthesis have been extensively investigated. For example, the effects of leucine on muscle protein metabolism were examined in young subjects who ingested 10 g of essential amino acids (EAA), which contained either 1.8 g or 3.5 g of leucine [15]. The muscle protein synthesis increased similarly after both treatments. This result indicates that the leucine content of 10 g of typical high-quality protein (~1.8 g) is sufficient to induce a maximal skeletal muscle protein anabolic response in young adults [15]. On the other hand, ingesting l0 g EAA containing 3.5 g leucine during moderate exercise led to a 33% greater muscle protein synthesis compared with ingesting the same amount of EAA containing 1.8 g leucine [16]. It is important to note that an increase in muscle protein synthesis does not necessarily lead to muscle hypertrophy, because an increase in muscle mass requires a positive balance between muscle synthesis and breakdown. However, studies investigating the long-term effect of leucine supplementation on muscle mass are lacking. Therefore, whether an increase in muscle protein synthesis induced by amino acid intake leads to muscle hypertrophy is currently unclear.

An increase in the amino acid concentration is one of the factors known to enhance muscle protein synthesis. Figure 42.4 shows that increasing the plasma amino acid concentration by infusing amino acids increases muscle protein synthesis, but that the synthesis decreases after 120 min, even if the amino acid infusion is continued and the plasma amino acid concentration remains high [17]. Therefore, maintaining the increased plasma amino acid concentration is not enough to maintain the increased level of muscle protein synthesis.

Figure 42.5 shows that muscle protein synthesis in rats increases 90 min after a meal [18]. The synthesis decreases at 180 min without nutrient supplementation at 135 min, whereas the synthesis remained high when the meal was supplemented with either carbohydrate, leucine, or both leucine and carbohydrate at 135 min [18]. The authors of that study stated that supplementation with leucine or carbohydrate ~2 h after a meal maintains the cellular energy status and extends the postprandial duration of muscle protein synthesis.

In aged rats, the sensitivity of muscle protein synthesis to leucine was lower than that in young rats [19]. However, because aged rats are still able to

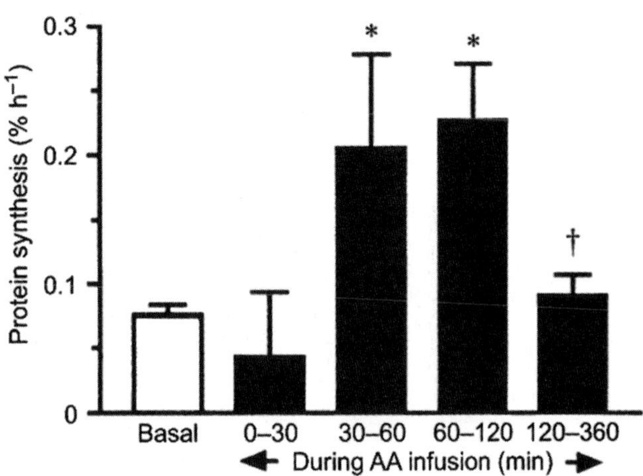

FIGURE 42.4 **The time course of the rate of synthesis of mixed muscle proteins.** AA, amino acid. *$P < 0.05$ vs basal; †$P < 0.01$ vs peak value. *Reproduced from Bohé et al. [17].*

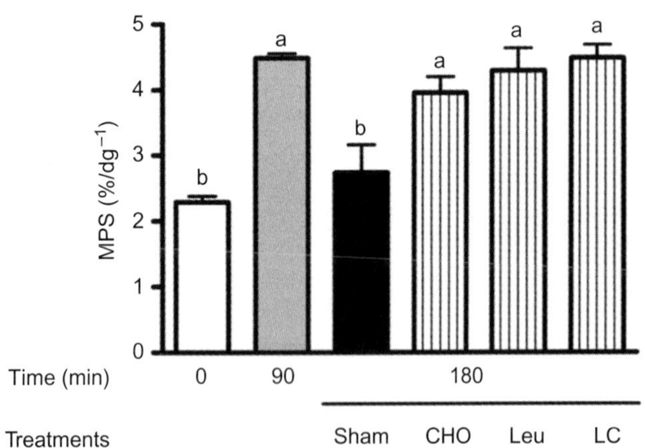

FIGURE 42.5 **Postprandial changes in muscle protein synthesis (MPS) in rats.** The treatment groups were intubated 135 min after the meal with water (sham), CHO (carbohydrate), Leu (leucine) or LC (leucine and carbohydrate). Means with different lower case letters are significantly different ($P < 0.05$). *Reproduced from Wilson et al. [18].*

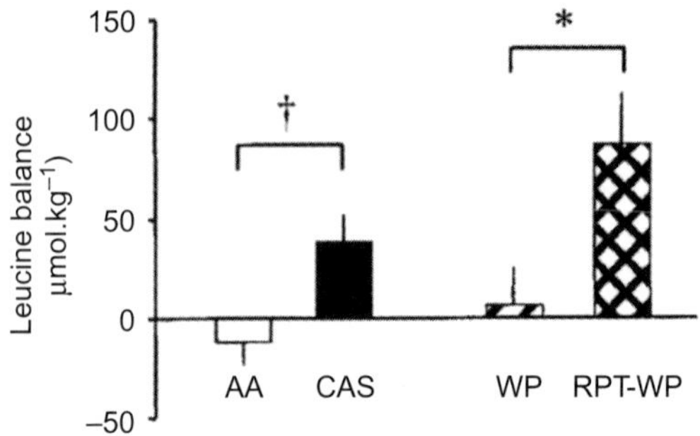

FIGURE 42.6 **Postprandial leucine balance over 420 min after meal ingestion in young men.** AA, amino acid; CAS, casein, WP, whey protein consumed at one time; RPT-WP, whey protein consumed in small portions. *$P < 0.05$, †$P < 0.01$. *Reproduced from Dangin et al. [22].*

respond normally to high leucine concentrations, it was hypothesized that nutritional manipulation increasing the availability of leucine to muscle could be beneficial to stimulate muscle protein synthesis [19]. Amino acid supplementation has also been reported to stimulate muscle protein synthesis in elderly humans, as well as young humans [20].

ABSORPTION RATE

The rate of absorption is another factor that influences body protein metabolism. The increase in whole-body protein synthesis is greater, but only temporarily, after consuming whey protein, which is absorbed rapidly, in comparison with consuming the same amount of casein, which is absorbed slowly [21]. This study also shows that the increase in protein synthesis is smaller, but the decrease in proteolysis lasts longer, after casein consumption than after whey consumption. In addition, the extent of amino acid oxidation is higher after consuming whey than casein. As a result, the leucine balance, which indicates the net whole-body protein balance for 420 min after consumption, is positive for casein ($141 \pm 96\ \mu mmol\ kg^{-1}$) and not significantly different from zero for whey ($11 \pm 36\ \mu mmol\ kg^{-1}$; $P < 0.05$, casein vs whey) [21]. In line with this finding, the leucine balance is larger after casein ingestion than after amino acid ingestion (Figure 42.6), because amino acids are absorbed more quickly [22]. However, as shown in Figure 42.6, the leucine balance is larger when whey is consumed in small portions than when it is consumed at one time [22]. These results suggest that factors other than the absorption rate, such as the amount of ingestion, may affect the body protein balance.

The above-mentioned results were observed under sedentary conditions. Similar effects of the absorption rate of protein sources on muscle protein synthesis have been reported after exercise. Figure 42.7 shows that the plasma amino acid concentration increases more rapidly and to a higher level after ingesting soy protein, which is absorbed quickly, than after ingesting skim milk, which is absorbed slowly, while the uptake of amino acids by muscle is greater after consumption of skim milk than after soy protein [23]. This difference is associated with the fact that the uptake of amino acids by muscle lasts longer for skim milk than for soy (Figure 42.7). This short-term effect after ingestion was confirmed by a 12-week study that examined the effects of post-exercise ingestion of milk, soy, or carbohydrate, which demonstrated a greater increase in the lean body mass and muscle fiber hypertrophy for milk than for soy or carbohydrate [24].

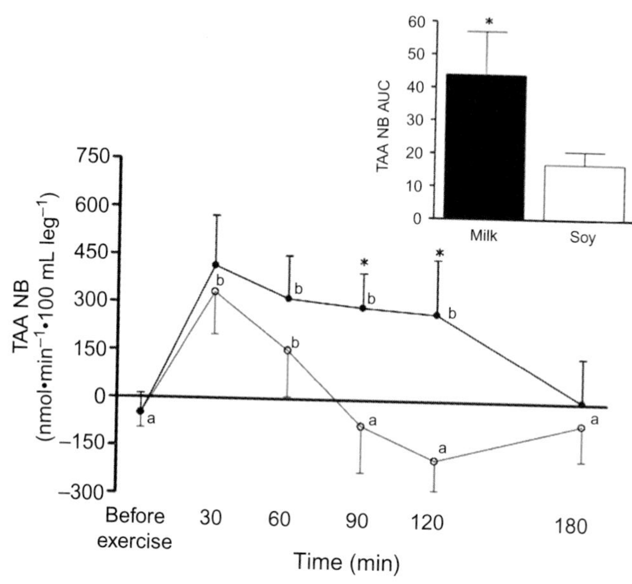

FIGURE 42.7 **The total amino acid (TAA) chemical net balance (NB) after consumption of a nonfat milk-protein beverage (●) or an isonitrogenous, isoenergetic, macronutrient-matched soy-protein beverage (○).** Inset: the positive area under the curve (AUC) for the TAA NB after consumption of the milk or soy beverage. *Significantly different from the soy group, $P < 0.05$. Significant differences across time are represented by lowercase letters; means with different lowercase letters are significantly different, $P < 0.05$. *Reproduced from Wilkinson et al. [23].*

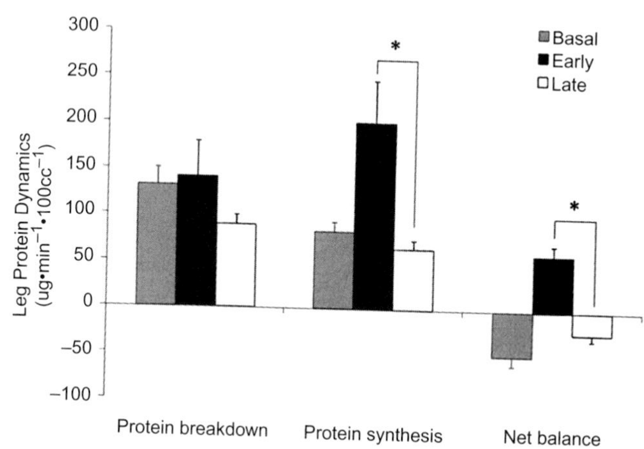

FIGURE 42.8 **Rates of leg protein dynamics for 10 subjects given an oral nutrient supplement either immediately after exercise (Early) or 3 h post-exercise (Late).** *Significantly different ($P < 0.05$), Early vs Late. *Reproduced from Levenhagen et al. [30].*

These observations suggest that protein sources inducing a rapid and large increase in the plasma amino acid concentration appear to be less effective for increasing muscle mass than are protein sources inducing a less rapid and moderate increase in the plasma amino acid concentration.

CO-INGESTION OF PROTEIN WITH CARBOHYDRATE

The ingestion of carbohydrate increases the plasma insulin concentration. Insulin stimulates muscle protein synthesis and inhibits muscle protein breakdown. Therefore, insulin can be used as a muscle hypertrophic agent. However, since administering insulin is associated with a risk of inducing hypoglycemia, insulin is banned as a doping agent.

However, utilizing the action of endogenous insulin secreted by carbohydrate ingestion is not regarded as doping. It has been reported that nitrogen retention in the body is greater when protein is consumed together with sugar, compared with when protein is consumed alone or with fat [25]. Since fat does not stimulate insulin secretion, the carbohydrate consumed with protein appears to enhance nitrogen retention through its effects on insulin. In addition, post-exercise urinary urea excretion has been shown to decrease when amino acids are administered with glucose in comparison with when amino acids are administered alone [26].

While the above-mentioned studies suggest that muscle building appears to be facilitated by co-ingesting carbohydrate with protein, it has been reported that neither muscle protein synthesis nor degradation differ after ingesting 25 g whey with 50 g maltodextrin versus ingesting only 25 g whey [27]. Further studies are needed to elucidate the effect of carbohydrate ingestion.

INGESTION TIMING

Ingesting protein as soon as possible after exercise has been reported to be effective in enhancing muscle building [28–31]. Figure 42.8 shows that the net balance of human leg protein is greater when protein is consumed immediately after exercise than when protein is consumed 3 hours after exercise, and this is due to the higher protein synthesis after the early consumption after exercise [30]. In addition, the skeletal muscle hypertrophy induced by 12-week resistance training was greater in elderly subjects who consumed protein right after each training session than in elderly subjects who consumed it 2 hours after exercising [31]. Increasing the amino acid supply to the muscle, as a result of increased blood flow to the muscles after exercise, is considered to be responsible for this effect. An increase in the sensitivity of the muscles to insulin after exercise is also a plausible reason for the observed effect.

On the other hand, there was a report that the consumption of a supplement containing amino acids and carbohydrate immediately after exercise did not

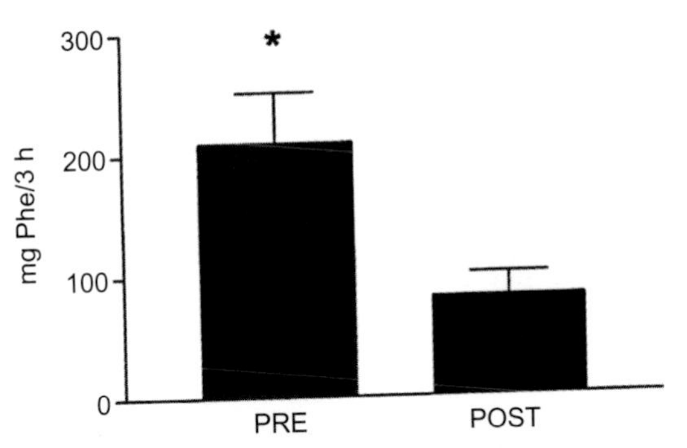

FIGURE 42.9 **Net muscle protein synthesis represented as the net phenylalanine uptake across the leg over 3 h (mg Phe/3 h) for pre-exercise ingestion (PRE) and post-exercise ingestion (POST) of a mixture of essential amino acids and carbohydrate.** *Significantly different from the POST value ($P < 0.013$). *Reproduced from Tipton et al. [33].*

enhance muscle strength [32]. Consuming the supplement immediately after exercise was associated with a 33% greater maximal muscle strength; however, this increase was not statistically significant [32].

It takes some time to digest and absorb dietary protein. This suggests that ingesting protein before and/or during exercise can be effective for increasing the plasma amino acid concentration, leading to facilitated post-exercise skeletal muscle synthesis. As shown in Figure 42.9, the net muscle protein synthesis was greater in subjects who had pre-exercise ingestion of a mixture of essential amino acids and carbohydrate than in those who had post-exercise ingestion [33]. A supplement containing protein ingested both pre- and post-exercise was also reported to be more effective for inducing muscle hypertrophy compared with the same supplement ingested early in the morning and late in the evening [34]. However, whether pre-exercise protein consumption alone facilitates exercise-induced muscle hypertrophy has not been fully investigated. It should be noted that ingesting a large amount of food or beverage before exercise can result in deterioration of the exercise performance.

Protein ingested immediately before sleep has been shown to stimulate muscle protein synthesis and improve the whole-body protein balance during post-exercise overnight recovery [35]. Further studies including an investigation of the long-term effects of such supplementation are needed to determine whether this dietary regimen is effective for inducing skeletal muscle hypertrophy.

References

[1] Lemon PW, Mullin JP. Effect of initial muscle glycogen levels on protein catabolism during exercise. J Appl Physiol 1980;48:624–9.

[2] DeFreitas JM, Beck TW, Stock MS, Dillon MA, Kasishke II PR. An examination of the time course of training-induced skeletal muscle hypertrophy. Eur J Appl Physiol 2011;111:2785–90.

[3] Nutrition for Athletes. A practical guide to eating for health and performance. Nutrition Working Group of the International Olympic Committee. <http://www.olympic.org/Documents/Reports/EN/en_report_833.pdf>.

[4] Nutrition for athletics. A practical guide to eating and drinking for health and performance in track and field. <http://www.iaaf.org/mm/Document/imported/42817.pdf>.

[5] Moore DR, Del Bel NC, Nizi KI, Hartman JW, Tang JE, Armstrong D, et al. Resistance training reduces fasted- and fed-state leucine turnover and increases dietary nitrogen retention in previously untrained young men. J Nutr 2007;137:985–91.

[6] Tarnopolsky MA, Atkinson SA, MacDougall JD, Chesley A, Phillips S, Schwarcz HP. Evaluation of protein requirements for trained strength athletes. J Appl Physiol 1992;73:1986–95.

[7] Moore DR, Robinson MJ, Fry JL, Tang JE, Glover EI, Wilkinson SB, et al. Ingested protein dose response of muscle and albumin protein synthesis after resistance exercise in young men. Am J Clin Nutr 2009;89:161–8.

[8] Arnal MA, Mosoni L, Boirie Y, Houlier ML, Morin L, Verdier E, et al. Protein pulse feeding improves protein retention in elderly women. Am J Clin Nutr 1999;69:1202–8.

[9] Arnal MA, Mosoni L, Boirie Y, Houlier ML, Morin L, Verdier E, et al. Protein feeding pattern does not affect protein retention in young women. J Nutr 2000;130:1700–4.

[10] West DW, Burd NA, Coffey VG, Baker SK, Burke LM, Hawley JA, et al. Rapid aminoacidemia enhances myofibrillar protein synthesis and anabolic intramuscular signaling responses after resistance exercise. Am J Clin Nutr 2011;94:795–803.

[11] Burke LM, Hawley JA, Ross ML, Moore DR, Phillips SM, Slater GR, et al. Preexercise aminoacidemia and muscle protein synthesis after resistance exercise. Med Sci Sports Exerc 2012;44:1968–77.

[12] Price GM, Halliday D, Pacy PJ, Quevedo MR, Millward DJ. Nitrogen homeostasis in man: influence of protein intake on the amplitude of diurnal cycling of body nitrogen. Clin Sci (Lond) 1994;86:91–102.

[13] Juillet B, Fouillet H, Bos C, Mariotti F, Gaussères N, Benamouzig R, et al. Increasing habitual protein intake results in reduced postprandial efficiency of peripheral, anabolic wheat protein nitrogen use in humans. Am J Clin Nutr 2008;87:666–78.

[14] Anthony JC, Anthony TG, Layman DK. Leucine supplementation enhances skeletal muscle recovery in rats following exercise. J Nutr 1999;129:1102–6.

[15] Glynn EL, Fry CS, Drummond MJ, Timmerman KL, Dhanani S, Volpi E, et al. Excess leucine intake enhances muscle anabolic signaling but not net protein anabolism in young men and women. J Nutr 2010;140:1970–6.

[16] Pasiakos SM, McClung HL, McClung JP, Margolis LM, Andersen NE, Cloutier GJ, et al. Leucine-enriched essential amino acid supplementation during moderate steady state exercise enhances postexercise muscle protein synthesis. Am J Clin Nutr 2011;94:809–18.

[17] Bohé J, Low JFA, Wolfe RR, Rennie MJ. Latency and duration of stimulation of human muscle protein synthesis during continuous infusion of amino acids. J Physiol 2001;532(Pt 2):575–9.

[18] Wilson GJ, Layman DK, Moulton CJ, Norton LE, Anthony TG, Proud CG, et al. Leucine or carbohydrate supplementation reduces AMPK and eEF2 phosphorylation and extends postprandial muscle protein synthesis in rats. Am J Physiol Endocrinol Metab 2011;301:E1236–42.

[19] Dardevet D, Sornet C, Balage M, Grizard J. Stimulation of in vitro rat muscle protein synthesis by leucine decreases with age. J Nutr 2000;130:2630–5.

[20] Paddon-Jones D, Sheffield-Moore M, Zhang XJ, Volpi E, Wolf SE, Aarsland A, et al. Amino acid ingestion improves muscle protein synthesis in the young and elderly. Am J Physiol Endocrinol Metab 2004;286:E321–8.

[21] Boirie Y, Dangin M, Gachon P, Vasson MP, Maubois JL, Beaufrère B. Slow and fast dietary proteins differently modulate postprandial protein accretion. Proc Natl Acad Sci USA 1997;94:14930–5.

[22] Dangin M, Boirie Y, Garcia-Rodenas C, Gachon P, Fauquant J, Callier P, et al. The digestion rate of protein is an independent regulating factor of postprandial protein retention. Am J Physiol Endocrinol Metab 2001;280:E340–8.

[23] Wilkinson SB, Tarnopolsky MA, Macdonald MJ, Macdonald JR, Armstrong D, Phillips SM. Consumption of fluid skim milk promotes greater muscle protein accretion after resistance exercise than does consumption of an isonitrogenous and isoenergetic soy-protein beverage. Am J Clin Nutr 2007;85:1031–40.

[24] Hartman JW, Tang JE, Wilkinson SB, Tarnopolsky MA, Lawrence RL, Fullerton AV, et al. Consumption of fat-free fluid milk after resistance exercise promotes greater lean mass accretion than does consumption of soy or carbohydrate in young, novice, male weightlifters. Am J Clin Nutr 2007;86:373–81.

[25] Gaudichon C, Mahé S, Benamouzig R, Luengo C, Fouillet H, Daré S, et al. Net postprandial utilization of [15N]-labeled milk protein nitrogen is influenced by diet composition in humans. J Nutr 1999;129:890–5.

[26] Hamada K, Matsumoto K, Minehira K, Doi T, Okamura K, Shimizu S. Effect of glucose on ureagenesis during exercise in amino acid-infused dogs. Metabolism 1998;47:1303–7.

[27] Staples AW, Burd NA, West DW, Currie KD, Atherton PJ, Moore DR, et al. Carbohydrate does not augment exercise-induced protein accretion versus protein alone. Med Sci Sports Exerc 2011;43:1154–61.

[28] Okamura K, Doi T, Hamada K, Sakurai M, Matsumoto K, Imaizumi K, et al. Effect of amino acid and glucose administration during postexercise recovery on protein kinetics in dogs. Am J Physiol 1997;272:E1023–30.

[29] Suzuki M, Doi T, Lee SJ, Okamura K, Shimizu S, Okano G, et al. Effect of meal timing after resistance exercise on hindlimb muscle mass and fat accumulation in trained rats. J Nutr Sci Vitaminol (Tokyo) 1999;45:401–9.

[30] Levenhagen DK, Gresham JD, Carlson MG, Maron DJ, Borel MJ, Flakoll PJ. Postexercise nutrient intake timing in humans is critical to recovery of leg glucose and protein homeostasis. Am J Physiol Endocrinol Metab 2001;280:E982–93.

[31] Esmarck B, Andersen JL, Olsen S, Richter EA, Mizuno M, Kjær M. Timing of postexercise protein intake is important for muscle hypertrophy with resistance training in elderly humans. J Physiol 2001;535(Pt 1):301–11.

[32] Williams A, Van den Oord M, Sharma A, Jones D. Is glucose/amino acid supplementation after exercise an aid to strength training? Br J Sports Med 2001;35:109–13.

[33] Tipton KD, Rasmussen BB, Miller SL, Wolf SE, Owens-Stovall SK, Petrini BE, et al. Timing of amino acid-carbohydrate ingestion alters anabolic response of muscle to resistance exercise. Am J Physiol Endocrinol Metab 2001;281:E197–206.

[34] Cribb PJ, Hayes A. Effects of supplement timing and resistance exercise on skeletal muscle hypertrophy. Med Sci Sports Exerc 2006;38:1918–25.

[35] Res PT, Groen B, Pennings B, Beelen M, Wallis GA, Gijsen AP, et al. Protein ingestion before sleep improves postexercise overnight recovery. Med Sci Sports Exerc 2012;44:1560–9.

CHAPTER

43

An Overview of Ornithine, Arginine and Citrulline in Exercise and Sports Nutrition

Kohei Takeda and Tohru Takemasa

Exercise Physiology, Health and Sport Sciences, University of Tsukuba, Japan

INTRODUCTION

Many nutritional supplements are designed to enhance sports performance for athletes, and the intended effects of such supplements cover a broad range. For example, protein and amino acid supplements are designed to enhance muscle protein synthesis. Recently, ornithine, arginine and citrulline (components of the urea cycle) have been highlighted in the context of sports nutrition, and are thought to exert an ergogenic effect as nitric oxide (NO) donors. NO induces vasodilatation by relaxing endothelial cells and improving blood flow. During exercise, blood flow increases to supply more oxygen and nutrients to muscles and thus enhance exercise performance. Ornithine, arginine and citrulline are also related to ammonia detoxification in the liver. During physical exercise, ammonia is produced through resynthesis of ATP, and accumulates in the body, causing both peripheral and central fatigue. Hence, ammonia detoxification is important for enhancement of exercise performance. These amino acids have other physiological effects in the body. For instance, ornithine and arginine promote the secretion of growth hormone, which enhances protein synthesis, and therefore these amino acids supplements are expected to aid muscle hypertrophy.

This chapter focuses on the effects of ornithine, arginine and citrulline supplementation on skeletal muscle structure and function.

ORNITHINE, ARGININE AND CITRULLINE: BIOGENESIS AND METABOLISM

Ornithine is a free amino acid that is not encoded by DNA, but synthesized through the urea cycle from arginine. Ornithine is present in fish and cheese, but also in *Corbicula* (a fresh water clam), which contains considerably more ornithine than other foods. Ornithine is absorbed via the intestinal tract and incorporated into liver, kidney and skeletal muscle [1].

Arginine is a conditional essential amino acid for mammals, including humans. It is present in all foods that contain protein, particularly meat, seafood, nuts and soy [2]. Orally ingested arginine is absorbed via the intestine and incorporated into blood vessels [3–5].

Citrulline is the one of the non-essential amino acids and is present in most foods, especially watermelon, *Citrullus vulgaris*, from which it derives its scientific name. Orally administrated citrulline is absorbed via the intestinal tract and transported to the kidney, where it is metabolized to arginine and transported to the whole body [3].

Acute toxicity tests in rodents have shown that the 50% lethal dose (LD_{50}) of ornithine, arginine and citrulline is over 10, 12, and 5 g/kg (body weight), respectively. No adverse effects of long-term administration of any of the three have been observed.

Ornithine, arginine and citrulline are components of the urea cycle in the liver, and are related to the

Nutrition and Enhanced Sports Performance.
DOI: http://dx.doi.org/10.1016/B978-0-12-396454-0.00043-6

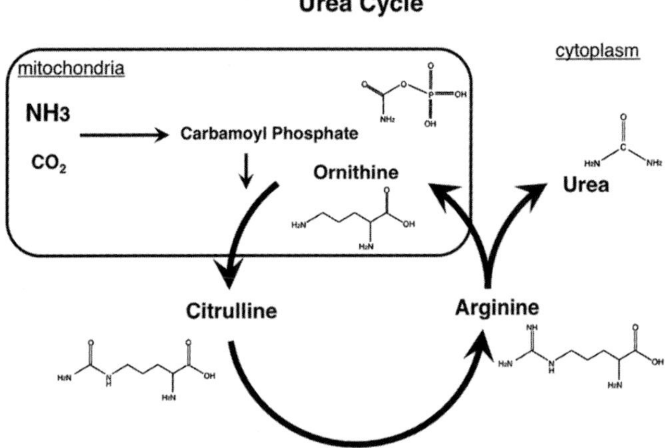

FIGURE 43.1 **Ammonia detoxification pathway in the urea cycle.** The urea cycle in the liver mainly participates in ammonia detoxification.

production of urea. Details of the metabolic pathways involving ornithine, arginine and citrulline are illustrated in Figure 43.1.

Arginine is the main physiological precursor of NO. Arginine and citrulline participate in the NOS-dependent pathway in a reaction catalyzed by specific NOS enzymes.

NITRIC OXIDE SYNTHESIS AND ITS FUNCTION IN THE BODY

Nitric oxide (NO), an unstable gas, with a half-life of only a few seconds, is synthesized in several parts in the body. NO is formed during the reaction of arginine with oxygen to produce citrulline through the action of nitric oxide synthases (NOS), which exist in three isoforms [6]: neuronal NOS (nNOS), inducible NOS (iNOS) and endothelial NOS (eNOS).

In skeletal muscle, eNOS and nNOS are localized in the cytoplasmic membrane of endothelial cells and in skeletal muscle cells, respectively [7,8]. nNOS is upregulated as a result of crush injury, chronic stimulation, and aging [9–11]. eNOS expression is increased by mechanical loading, exercise and shear stress [9,12]. iNOS in skeletal muscle is upregulated by inflammatory cytokines and lipopolysaccharides [13,14].

NO is known to play important roles in physiological processes [15,16]. One of the functions of NO is blood vessel vasodilatation, which was first clarified in the 1980s [17]. NO production is increased by exercise, and it has been suggested that NO may contribute to skeletal muscle vasodilatation through relaxation of vascular smooth muscle cells [18,19]. Vasodilatation causes an increase in blood flow, thus enabling more oxygen and nutrients to be supplied to skeletal muscle. This physiological response can enhance muscle work capacity and exercise performance. One study of the effect of NO inhibition (by the NOS inhibitor L-NAME) on human exercise performance revealed a reduction of maximal oxygen uptake during incremental cycle exercise in humans [20].

EFFECTS OF ORNITHINE, ARGININE AND CITRULLINE SUPPLEMENTATION ON NITRIC OXIDE SYNTHESIS AND PERFORMANCE

It has been shown that skeletal muscle blood flow increases in proportion to the metabolic demands of the tissue during exercise. There is a direct relationship between the increase in muscle blood flow and increased contraction frequency, exercise intensity and muscle oxygen consumption [21–25]. Arginine is the main source of NO as a result of the action of NOS enzymes. It has been proposed that arginine supplementation promotes vasodilatation through increased production of NO in working muscle [26]. Studies of NO have evaluated exercise performance in terms of arginine used (Table 43.1). Olek et al. studied the effect of a single acute dose of arginine supplement (2 g) 60 min before a 30-second Wingate test. They showed that this amount of arginine had no effect on any component of exercise performance (total work, mean power, and VO_2). Additionally, plasma NOx was also unchanged after arginine supplementation compared with placebo [27]. Acute arginine supplementation (6 g) did not stimulate any increase in NO synthesis in healthy men in response to exercise, despite an increase in muscle blood volume. In addition, arginine supplementation did not enhance isokinetic concentric elbow extension strength performance [26]. Liu et al. investigated the effect of acute arginine supplementation on blood NOx and intermittent anaerobic exercise capacity in judo athletes. The subjects had ingested 6 g/day arginine for 3 days and carried out 13 sets of a 20-second all-out ergometer test with 15-second rests. There were no differences in peak and mean power during exercise, and no difference in NOx between arginine supplementation and a control group [28]. Furthermore, arginine supplementation alone for athletes did not affect exercise performance parameters such as VO_2 during treadmill running [29,30]. However, arginine supplementation has been reported to have beneficial effects in mice. Maxwell et al. fed arginine-containing water (6%) to mice for 4–8 weeks and found that chronic arginine supplementation increased aerobic running capacity (approximately 8% increase in VO_{2max} and 5% increase in running

TABLE 43.1 Studies with Ornithine, Arginine and Citrulline Supplementation in Relation to NO Production and Exercise Performance

Supplements	Dose/Day	Duration	Animal	Effects	Ref.
Arginine	2 g	1 day	Human	No change in NO synthesis and performance	[27]
Arginine	6 g	1 day	Human	Increased muscle blood volume. No change in NO and strength	[26]
Arginine	6 g	3 days	Human	No change in NOx and ergometer performance	[28]
Arginine	15 g	3 days	Human	NO change in NOx and oxygen uptake	[29]
Arginine	12 g	28 days	Human	No change VO_{2max} and VT	[30]
Arginine	unknown	4–8 weeks	Mouse	Increased NOx concentration. Increased VO_{2max} and running distance	[31]
Citrulline	3 or 9 g	1 day	Human	Decreased NO marker and running time to exhaustion	[32]
Arginine, vitamins and amino acids	6 g	1 day	Human	Increased NO synthesis and time to task failure	[33]
Arginine	500 mg/kg body weight	14 days	Rat	Prevention of loss of soleus wet weight. Increased total/phospho p70S6K protein. Decreased Atrogin-1 and MuRF-1 mRNA expression	[34]

distance), which was linked to increases in endothelial NO function [31].

One study investigated the acute effect of citrulline supplementation on NO and exercise performance. Hickner et al. assessed the dose effect of citrulline administered as a supplement 3 h (3 g) or 24 h (9 g) before an incremental treadmill test until exhaustion in young subjects. The results showed that citrulline supplementation impaired exercise performance measured as treadmill time to exhaustion compared with placebo. Lower levels of plasma NO markers (nitrates/nitrites) were also noted following citrulline supplementation compared with placebo [32]. The authors considered that citrulline supplementation reduced the level of NO by inhibiting the arginine/NO reaction.

Another study analyzed the effect of arginine in combination with other nutrients on NO synthesis and exercise performance in healthy men. Acute dietary supplementation (total 20 g), including 6 g arginine with vitamins (E, C, B_6 and B_{12}), other amino acids (citrulline, glutamine, leucine, valine carnithine, cystaine and isoleucine) and 11 g fructose, increased NO synthesis and prolonged the time until task failure during high-intensity exercise (Figure 43.2) [33].

Reduced muscle activity, such as that resulting from bed rest or unloading, leads to muscle atrophy [34]. Lomonosova et al. examined the effects of arginine administration on hindlimb suspension-induced muscle atrophy in Wistar rats, with or without oral administration of arginine (500 mg/kg) for 14 days. Hindlimb suspension led to significant loss of soleus mass, relative NO content and total/phosphorylated p-70S6K (factors of protein synthesis), increased atrogin-1 and MuRF-1 (both being ubiquitin ligases related to protein degradation). The arginine-administered group showed higher soleus weight, a decrease of NO content and total/phosphorylated p70S6K protein, and decreased expression of atrogin-1 and MuRF-1 mRNA compared with rats subjected to hindlimb suspension alone [35]. The authors considered that NO synthesis following arginine administration inhibited hindlimb suspension-induced protein degradation. However, the precise mechanism responsible was not clarified.

Recently, citrulline has attracted attention as a precursor of arginine. It has been indicated that citrulline administration could be a more effective way of increasing extracellular arginine levels [36]. Orally ingested citrulline is delivered directly to the kidney, and not taken up by the liver. On the other hand, about 30% of arginine is metabolized, and the remainder is taken up into the urea cycle and metabolized to urea [3]. In a study using human volunteers, intake of 3.3 kg of watermelon increased blood citrulline level from 22 to 593 μmol/L. At the same time, the blood arginine level was also increased from 65 to 199 μmol/L. This result suggests that citrulline ingestion enhances the blood level not only of citrulline but also of arginine [37]. So far, few studies have investigated the effects of citrulline supplementation on NO production and exercise performance, and therefore there is a need to evaluate the effects of supplemented citrulline as a NO donor.

EXERCISE AND AMMONIA

Although it is well known that exercise-induced fatigue affects performance, fatigue is attributable to many factors, including accumulation of metabolites, depletion of muscle glycogen, and so on [38–40]. Ammonia (NH_3 and NH_4^+) is a metabolite and intermediate of several biochemical pathways in the body, and produced in the gut, brain and kidney [41].

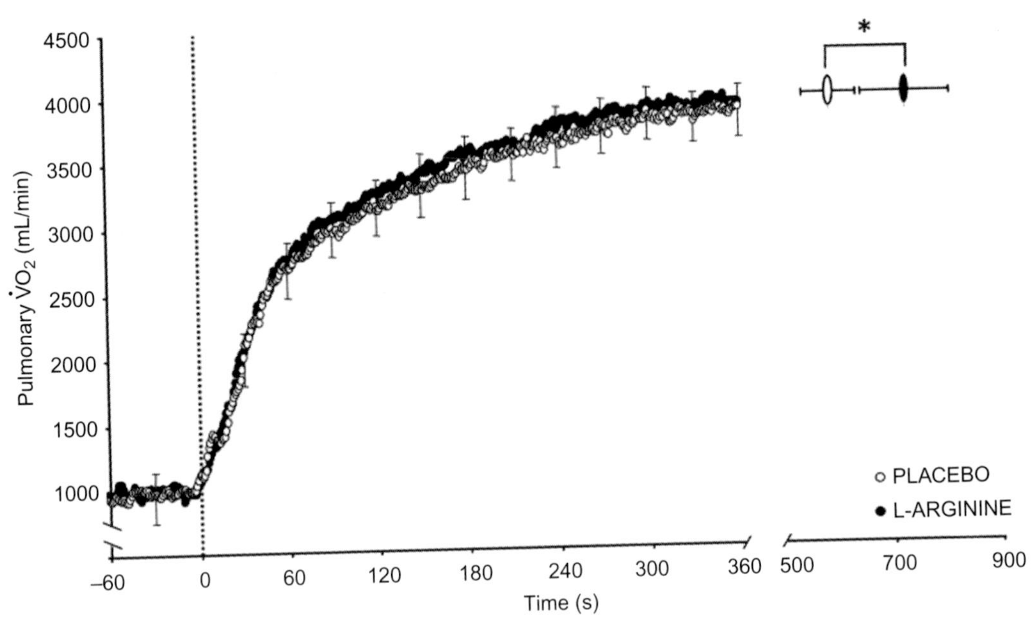

FIGURE 43.2 Group mean VO_2 response to 6 min of severe-intensity exercise and mean VO_2 at task failure. Subjects were randomly assigned to receive arginine-based supplement or placebo. They performed one 6-min bout of severe-intensity cycling and continued until task failure as a measure of exercise tolerance. The time until task failure was extended by 20% following arginine administration. *Time to exhaustion is significantly different from placebo ($P < 0.05$). *Reproduced from Bailey et al. [33].*

Parnas and colleagues first reported the discovery of ammonia production by muscle in the 1920s [42]. During exercise, ammonia is a product that accumulates in skeletal muscle when AMP is deaminated to IMP during the resynthesis of ATP ($AMP + H_2O \rightarrow IMP + NH_3$). Some studies using animals and humans have reported that blood ammonia is increased after exercise. Exhaustive treadmill running has been reported to increase the blood level of ammonia from 37.8 to 68.1 μM [43]. Barnes et al. have reported that rats that were forced to swim to exhaustion showed a significantly elevated level of ammonia [44]. Other studies of high-intensity exercise in rat and humans have obtained similar findings [45,46]. On the other hand, low-intensity exercise (below 50% VO_{2max}) is reportedly unassociated with significant ammonia accumulation [47–49]. The source of ammonia during exercise is deamination of not only AMP but also branched chain amino acids (BCAAs). During exercise, oxidation of amino acids, including BCAAs, reaches about 10% of total energy production [50].

Ammonia is very toxic and has a harmful influence on the body, activating phosphofructokinase (PFK), which is the rate-limiting enzyme in glycolysis and inhibits the oxidation of pyruvate to acetyl CoA [51,52]. Activated PFK facilitates the production of lactate, causing a decline of intercellular pH, decreased release of Ca^{2+} from the sarcoplasmic reticulum, and consequently a decrease of contractility [53]. Inhibition of pyruvate oxidation hinders the supply of ATP to skeletal muscle, thus causing exhaustion.

Ammonia has also been suggested to play a role in fatigue in the central nervous system. During exercise, ammonia is produced in peripheral tissues and its level increases in the brain [54]. Ammonia has a negative effect on energy metabolism and induces dysfunction of neurotransmission through depletion of glutamate and GABA, which are essential neurotransmitters in the brain [55–57].

To avoid accumulation of ammonia, the urea cycle in the liver is responsible for ammonia detoxification. Ammonia produced in skeletal muscle is metabolized into harmless urea via the urea cycle in hepatocytes after transportation to the liver via blood [58].

EFFECTS OF ORNITHINE, ARGININE AND CITRULLINE SUPPLEMENTATION ON AMMONIA AND PERFORMANCE

Ornithine, arginine and citrulline play a role in ammonia detoxification via the urea cycle in the liver. Some studies have tested the effects of supplementation of these amino acids on ammonia detoxification during exercise in humans and other mammals. Meneguello et al. investigated the effects of 8-day supplementation with a mixture of ornithine, arginine and citrulline (0.2, 0.4, 0.026 g/kg body weight, respectively) on plasma ammonia and period in rats forced to swim to exhaustion with loading (5% body weight). Administration of these amino acids prolonged the

TABLE 43.2 Studies with Ornithine, Arginine and Citrulline Supplementation in Relation to Ammonia Levels and Exercise Performance

Supplements	Dose/Day	Duration	Animal	Effects	Ref.
Ornithine/Arginine/ Citrulline	0.2, 0.4, 0.026 g/kg body weight	7 days	Rat	Lower ammonia level. Increased swimming time to exhaustion	[59]
Ornithine	0.1 g/kg body mass	1 day	Human	Lower ammonia level. No change in ergomerter performance and maximal O_2 consumption	[60]
Ornithine	2 g/day for 7 days 6 g/day for 1 day	8 days	Human	Lower ammonia level and higher urea level. No change in bicycle performance	[61]
Arginine asparate	5 g	10 days	Human	Lower ammonia and lactate. Performance was not measured	[63]
Arginine asparate	15 g/day (total)	2 weeks	Human	No change in ammonia level during or after running. Performance was not measured	[64]
Arginine glutamate	20 g (total)	1 day	Human	Lower blood ammonia level	[65]
Citrulline	250 mg/kg body weight	7 days	Mouse	Increased time to exhaustion. Lower blood ammonia and lactate	[66]

period until exhaustion by about 50% (control: about 12 min, supplement: about 19 min). Also, they found that accumulation of plasma ammonia after exercise was suppressed (control: 12.2, supplement: 8.5 μg/mL) [59]. Other studies have also analyzed the effects of supplementation with either ornithine, arginine, or citrulline alone, rather than as a mixture. Demura et al. reported that acute ingestion of ornithine (0.1 g/kg body mass) before incremental exhaustive ergometer bicycle exercise was unable to improve exercise time or maximal oxygen consumption. However, exercise-induced blood ammonia elevation was suppressed [60]. Sugino et al. administered ornithine for 8 days (2 g/day for 7 days and 6 g/day for 1 day) to human subjects, who then performed bicycle exercise for 4 hours at 80% of the anaerobic threshold. At the end of the exercise, the ornithine-supplemented group showed a lower level of blood ammonia and a higher level of blood urea (ornithine 4662 nmol/mL, placebo 3884.8 nmol/mL). However, no clear improvement of 10-second maximal ergometer performance was observed [61]. Arginine is also a component of the urea cycle, and arginine supplementation has been reported to increase the level of blood urea significantly [62], suggesting that it enhances ammonia detoxification in the urea cycle. It has been reported that 10 days of arginine aspartate supplementation (5 g per day) resulted in a reduction of both plasma ammonia and lactate levels after 80% VO_{2max} cycling [63]. However, 15 g/day arginine aspartate supplementation for 2 weeks had no effect on the plasma ammonia concentration during or after long-distance running [64]. One report has indicated that dose supplementation with arginine and glutamate (total 20 g) was effective at reducing blood ammonia levels [65], although exercise performance was not measured.

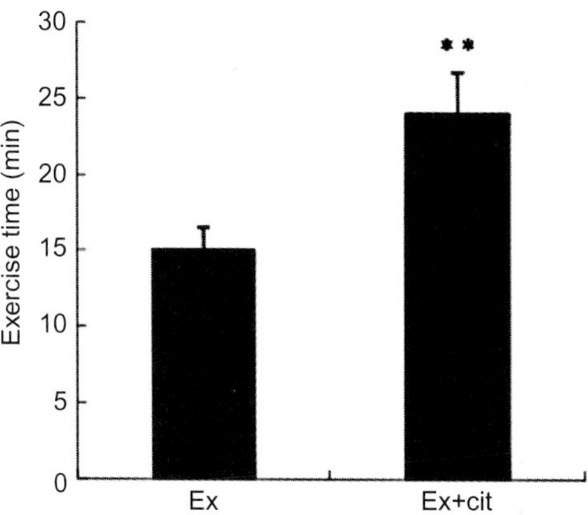

FIGURE 43.3 Effect of citrulline supplementation on exercise time to exhaustion. The citrulline-supplemented group (right bar) swim significantly longer with a weight burden of 5% body weight than the non-supplemented group (left bar). Values represent the means ± SE. $^{**}P < 0.01$ vs Ex group. *Reproduced from Takeda et al. [66].*

Takeda et al. studied the effect of citrulline supplementation on exercise performance and blood ammonia in mice. The mice were given citrulline at a dose of 250 mg/kg/day for one week, and then subjected to exhaustive swimming with a load of 5% body weight. Citrulline supplementation significantly increased the swimming time by about 67% (control 15 min, citrulline 24 min). Blood ammonia in the control group after exercise was significantly increased to about 450 μg/dL, but no change was observed in the citrulline group. The blood lactate level was significantly increased by swimming exercise in both the control and supplemented groups, being about 6 mM and 4.5 mM, respectively, although the increase was significantly lower in the latter (Figure 43.3) [66].

Fewer studies of ornithine, arginine and citrulline supplementation have focused on ammonia detoxification than has been the case for studies of NO (Table 43.2).

There is a need for further investigation of the effects of these amino acids under various conditions according to subjects, acute or chronic supplementation, and training status.

OTHER EFFECTS OF ORNITHINE, ARGININE AND CITRULLINE SUPPLEMENTATION

Other studies have evaluated the effects of ornithine, arginine and citrulline, but did not focus on NO synthesis or ammonia detoxification. In one such study, postomenopausal women were given a large amount (about 15 g/day) of arginine supplement for 6 months, and this resulted in a significant increase in peak jumping power and force [67].

It is known that arginine administration facilitates the secretion of growth hormone (GH) [68,69]. An *in vitro* study using anterior pituitary gland cells showed that arginine acts directly on the somatotrophs [70]. GH stimulates protein synthesis and enhances amino acid uptake into muscle [71]. Yao et al. fed neonatal pigs a diet containing 0.6% arginine for 7 days, and found that this led to activation of the mTOR signaling pathway and protein synthesis in neonatal skeletal muscle, relative to pigs receiving a normal diet [72]. In a human study, Collier et al. noted a significant GH response compared with placebo after ingestion of 5 or 9 g of arginine. The rise in GH began at 30 min after ingestion, and peaked at 60 min (Figure 43.4) [73]. Other studies have also observed a rise in the GH level resulting from infusion of arginine in postmenopausal women [74] and elderly men [75]. These results suggest that arginine may facilitate protein synthesis through secretion of GH, thus inducing muscle hypertrophy. Recently, Tang et al. investigated the ergogenic potential of arginine on NO and skeletal muscle protein synthesis [76]. They hypothesized that NO synthesis induced an increase of amino acid delivery through an increase in blood flow and skeletal muscle anabolism. The study subjects drank a mixture of 10 g arginine and essential amino acids after unilateral leg resistance training. Arginine ingestion did not stimulate any increase of NO synthesis or muscle blood flow at rest, or after resistance exercise. Arginine supplementation resulted in a greater GH response than was seen in the controls, but did not enhance muscle protein synthesis immediately after resistance training. Ornithine also promotes GH secretion by stimulating the pituitary gland. In a study of bodybuilders, acute ornithine administration (170 mg/kg body weight) significantly increased the serum GH level about fourfold in comparison with that before ingestion (before ingestion 2.2 ng/mL, 90 min after ingestion 9.2 ng/mL)

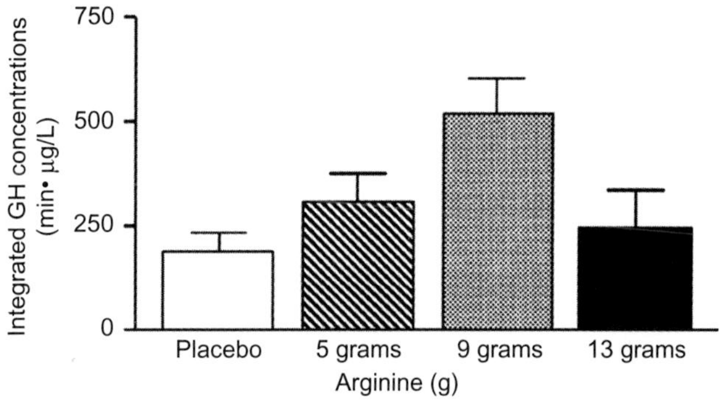

FIGURE 43.4 Mean integrated growth hormone concentrations in response to arginine ingestion. Subjects were given intentioned placebo, or 5, 9 or 13 g arginine on different days. A significant increase in the integrated AUC was observed on days when the 5 and 9 g were administered compared with the placebo day. Values represent the means ± SE. *Reproduced from Collier et al. [73].*

[77]. Evain-Brion et al. performed an ornithine infusion test (12 g/m^2 body area) in 54 children with constitutionally short stature. They confirmed a significant elevation of the serum GH level of about eight-fold over the basal level after ingestion of ornithine [78].

Elam et al. surveyed the effect of an arginine and ornithine combination supplement with 5 weeks of resistance training. The study participants were given a supplement containing 1 g arginine and 1 g ornithine or placebo each day during a 5-week strength-training program. The results suggested that the supplement in combination with the training program significantly decreased body weight and percentage body fat, while muscle strength and lean body mass were significantly increased [79,80]. Zajac et al. investigated the effects of arginine and ornithine supplementation during 3 weeks of resistance training on serum GH and insulin-like growth factor-1 (IGF-1). After the training, arginine and ornithine supplementation showed significant increases in the serum levels of both GH and IGF-1 [81].

Citrulline supplementation in old malnourished rats increased the muscle protein content by stimulating protein synthesis [82]. Rats aged 19 months were subjected to 50% dietary restriction for 12 weeks, after which they were fed a standard diet or a citrulline-containing diet (5 g/kg/day) for 1 week. The protein content of the tibialis muscle (163 mg/organ) was significantly higher than in the normal diet group (117 mg/organ). Protein synthesis rates were also increased in the citrulline-fed group (citrulline 0.56 mg/h, normal 0.30 mg/h). These findings suggested that citrulline supplementation might enhance muscle hypertrophy.

Ornithine, arginine and citrulline supplementation is probably related to muscle hypertrophy, as mentioned above (Table 43.3). However, precise details are still unclear, and more studies will be needed to clarify the muscle-hypertrophic effects of these amino acids.

TABLE 43.3 Studies with Ornithine, Arginine and Citrulline Supplementation in Relation to GH Secretion and Muscle Hypertrophy

Supplements	Dose/day	Duration	Animal	Effects	Ref.
Arginine	0.6% Arginine containing diet	7 days	Neonatal pig	Increased protein synthesis and expression of mTOR signaling pathway	[72]
Arginine	5 or 9 g	1 day	Human	Increased blood GH	[73]
Arginine	9 g	4 weeks	Human	Increased blood GH	[74]
Arginine	8 g	1 day	Human	Increased blood GH	[75]
Arginine and other amino acids	10 g	1 day	Human	Did not change NO synthesis, blood flow and protein synthesis. Increased blood GH	[76]
Arginine/ornithine	1 g (each)	5 weeks	Human	Decreased body weight, % body fat. Increased strength, lean body mass	[80]
Arginine/ornithine	3, 2.2 g twice a day	3 weeks	Human	Increased serum GH and IGF-1	[81]
Citrulline	5 g/kg body weight	7 days	Malnourished rats	Increased protein content and synthesis	[82]

CONCLUSION

We have reviewed the effects of ornithine, arginine and citrulline supplementation from the viewpoint of molecular exercise physiology. First, arginine and citrulline supplements are suggested to increase NO synthesis and to enhance exercise performance in humans and other mammals. Some studies have suggested that arginine or citrulline supplementation alone can increase NO synthesis. However, there are few data to suggest that exercise performance is improved after such supplementation. Combination supplementation with other amino acids and vitamins is suggested to be effective for NO synthesis and improving exercise performance.

Comparatively few studies have focused on the effects of amino acid supplementation and ammonia detoxification on exercise performance. Although such supplementation may reduce the level of ammonia or lactate, any improvement of exercise performance is equivocal in humans. In studies using mice, on the other hand, enhancement of exercise capacity has been suggested. Therefore, further investigations are needed to clarify the effects of these amino acids supplements on ammonia levels.

Ornithine, arginine and citrulline reportedly exert some effects on GH secretion. Although acute or chronic supplementation with these amino acids increases the blood GH level, few studies have demonstrated any muscle-building effect of chronic supplementation with ornithine, arginine or citrulline. Since it takes a long time for muscle to gain weight by resistance training, there would be a need to study the effects of long-term supplementation of these amino acids in power athletes.

Ornithine, arginine and citrulline have recently been promoted as commercial sports supplements. However, the number of studies that have demonstrated the benefits of such supplementation is still limited. Therefore, it will be necessary to carry out further studies under various conditions, for example by varying the amount taken, the duration of supplementation, and its timing in combination with exercise/training.

References

[1] Vaubourdolle M, Jardel A, Coudray-Lucas C, Ekindjian OG, Agneray J, Cynober L. Fate of enterally administered ornithine in healthy animals: interactions with alpha-ketoglutarate. Nutrition 1989;5:183–7.

[2] Visek WJ. Arginine needs, physiological state and usual diets. A reevaluation. J Nutr 1986;116:36–46.

[3] Curis E, Nicolis I, Moinard C, Osowska S, Zerrouk N, Benazeth S, et al. Almost all about citrulline in mammals. Amino Acids 2005;29:177–205.

[4] Tangphao O, Grossmann M, Chalon S, Hoffman BB, Blaschke TF. Pharmacokinetics of intravenous and oral L-arginine in normal volunteers. Br J Clin Pharmacol 1999;47:261–6.

[5] Bode-Boger SM, Boger RH, Galland A, Tsikas D, Frolich JC. L-arginine-induced vasodilation in healthy humans: pharmacokinetic-pharmacodynamic relationship. Br J Clin Pharmacol 1998;46:489–97.

[6] Schmidt HH, Nau H, Wittfoht W, Gerlach J, Prescher KE, Klein MM, et al. Arginine is a physiological precursor of endothelium-derived nitric oxide. Eur J Pharmacol 1988;154:213–6.

[7] Frandsen U, Lopez-Figueroa M, Hellsten Y. Localization of nitric oxide synthase in human skeletal muscle. Biochem Biophys Res Commun 1996;227:88–93.

[8] Grozdanovic Z, Gosztonyi G, Gossrau R. Nitric oxide synthase I (NOS-I) is deficient in the sarcolemma of striated muscle fibers in patients with Duchenne muscular dystrophy, suggesting an association with dystrophin. Acta Histochem 1996;98:61–9.

[9] Balon TW, Nadler JL. Evidence that nitric oxide increases glucose transport in skeletal muscle. J Appl Physiol 1997;82:359–63.

[10] Capanni C, Squarzoni S, Petrini S, Villanova M, Muscari C, Maraldi NM, et al. Increase of neuronal nitric oxide synthase in

rat skeletal muscle during ageing. Biochem Biophys Res Commun 1998;245:216–9.

[11] Rubinstein I, Abassi Z, Coleman R, Milman F, Winaver J, Better OS. Involvement of nitric oxide system in experimental muscle crush injury. J Clin Invest 1998;101:1325–33.

[12] Tidball JG, Lavergne E, Lau KS, Spencer MJ, Stull JT, Wehling M. Mechanical loading regulates NOS expression and activity in developing and adult skeletal muscle. Am J Physiol 1998;275: C260–6.

[13] El Dwairi Q, Guo Y, Comtois A, Zhu E, Greenwood MT, Bredt DS, et al. Ontogenesis of nitric oxide synthases in the ventilatory muscles. Am J Respir Cell Mol Biol 1998;18:844–52.

[14] Boczkowski J, Lanone S, Ungureanu-Longrois D, Danialou G, Fournier T, Aubier M. Induction of diaphragmatic nitric oxide synthase after endotoxin administration in rats: role on diaphragmatic contractile dysfunction. J Clin Invest 1996;98:1550–9.

[15] Moncada S, Palmer RM, Higgs EA. Nitric oxide: physiology, pathophysiology, and pharmacology. Pharmacol Rev 1991;43:109–42.

[16] Stamler JS, Meissner G. Physiology of nitric oxide in skeletal muscle. Physiol Rev 2001;81:209–37.

[17] Palmer RM, Ferrige AG, Moncada S. Nitric oxide release accounts for the biological activity of endothelium-derived relaxing factor. Nature 1987;327:524–6.

[18] Luiking YC, Engelen MP, Deutz NE. Regulation of nitric oxide production in health and disease. Curr Opin Clin Nutr Metab Care 2010;13:97–104.

[19] Balon TW, Nadler JL. Nitric oxide release is present from incubated skeletal muscle preparations. J Appl Physiol 1994;77:2519–21.

[20] Jones AM, Wilkerson DP, Campbell IT. Nitric oxide synthase inhibition with L-NAME reduces maximal oxygen uptake but not gas exchange threshold during incremental cycle exercise in man. J Physiol 2004;560:329–38.

[21] Armstrong RB, Laughlin MH. Rat muscle blood flows during high-speed locomotion. J Appl Physiol 1985;59:1322–8.

[22] Bockman EL. Blood flow and oxygen consumption in active soleus and gracilis muscles in cats. Am J Physiol 1983;244: H546–51.

[23] Laughlin MH, Armstrong RB. Muscular blood flow distribution patterns as a function of running speed in rats. Am J Physiol 1982;243:H296–306.

[24] Mackie BG, Terjung RL. Blood flow to different skeletal muscle fiber types during contraction. Am J Physiol 1983;245:H265–75.

[25] Mohrman DE, Regal RR. Relation of blood flow to VO_2, PO_2, and PCO_2 in dog gastrocnemius muscle. Am J Physiol 1988;255: H1004–10.

[26] Alvares TS, Conte CA, Paschoalin VM, Silva JT, Meirelles Cde M, Bhambhani YN, et al. Acute l-arginine supplementation increases muscle blood volume but not strength performance. Appl Physiol Nutr Metab 2012;37:115–26.

[27] Olek RA, Ziemann E, Grzywacz T, Kujach S, Luszczyk M, Antosiewicz J, et al. A single oral intake of arginine does not affect performance during repeated Wingate anaerobic test. J Sports Med Phys Fitness 2010;50:52–6.

[28] Liu TH, Wu CL, Chiang CW, Lo YW, Tseng HF, Chang CK. No effect of short-term arginine supplementation on nitric oxide production, metabolism and performance in intermittent exercise in athletes. J Nutr Biochem 2009;20:462–8.

[29] Bescos R, Gonzalez-Haro C, Pujol P, Drobnic F, Alonso E, Santolaria ML, et al. Effects of dietary L-arginine intake on cardiorespiratory and metabolic adaptation in athletes. Int J Sport Nutr Exerc Metab 2009;19:355–65.

[30] Sunderland KL, Greer F, Morales J. VO_{2max} and ventilatory threshold of trained cyclists are not affected by 28-day L-arginine supplementation. J Strength Cond Res 2011;25:833–7.

[31] Maxwell AJ, Ho HV, Le CQ, Lin PS, Bernstein D, Cooke JP. L-arginine enhances aerobic exercise capacity in association with augmented nitric oxide production. J Appl Physiol 2001;90:933–8.

[32] Hickner RC, Tanner CJ, Evans CA, Clark PD, Haddock A, Fortune C, et al. L-citrulline reduces time to exhaustion and insulin response to a graded exercise test. Med Sci Sports Exerc 2006;38:660–6.

[33] Bailey SJ, Winyard PG, Vanhatalo A, Blackwell JR, DiMenna FJ, Wilkerson DP, et al. Acute L-arginine supplementation reduces the O_2 cost of moderate-intensity exercise and enhances high-intensity exercise tolerance. J Appl Physiol 2010;109:1394–403.

[34] Tidball JG. Mechanical signal transduction in skeletal muscle growth and adaptation. J Appl Physiol 2005;98:1900–8.

[35] Lomonosova YN, Kalamkarov GR, Bugrova AE, Shevchenko TF, Kartashkina NL, Lysenko EA, et al. Protective effect of L-Arginine administration on proteins of unloaded m. soleus. Biochemistry (Mosc) 2011;76:571–80.

[36] Hartman WJ, Torre PM, Prior RL. Dietary citrulline but not ornithine counteracts dietary arginine deficiency in rats by increasing splanchnic release of citrulline. J Nutr 1994;124:1950–60.

[37] Mandel H, Levy N, Izkovitch S, Korman SH. Elevated plasma citrulline and arginine due to consumption of Citrullus vulgaris (watermelon). J Inherit Metab Dis 2005;28:467–72.

[38] Bergstrom J, Hermansen L, Hultman E, Saltin B. Diet, muscle glycogen and physical performance. Acta Physiol Scand 1967;71:140–50.

[39] Hermansen L, Hultman E, Saltin B. Muscle glycogen during prolonged severe exercise. Acta Physiol Scand 1967;71:129–39.

[40] Sahlin K. Metabolic factors in fatigue. Sports Med 1992;13:99–107.

[41] Olde Damink SW, Deutz NE, Dejong CH, Soeters PB, Jalan R. Interorgan ammonia metabolism in liver failure. Neurochem Int 2002;41:177–88.

[42] Parnas JK, Mozolowski W. Über die Ammoniakbildung im Muskel und Ihren Zusammenhang mit Tätigkeit und Zustandsänderung. J Mol Med 1927;6:998–9.

[43] Wilkerson JE, Batterton DL, Horvath SM. Exercise-induced changes in blood ammonia levels in humans. Eur J Appl Physiol Occup Physiol 1977;37:255–63.

[44] Barnes RH, Labadan BA, Siyamoglu B, Bradfield RB. Effects of exercise and administration of aspartic acid on blood ammonia in the rat. Am J Physiol 1964;207:1242–6.

[45] Broberg S, Sahlin K. Adenine nucleotide degradation in human skeletal muscle during prolonged exercise. J Appl Physiol 1989;67:116–22.

[46] Dudley GA, Staron RS, Murray TF, Hagerman FC, Luginbuhl A. Muscle fiber composition and blood ammonia levels after intense exercise in humans. J Appl Physiol 1983;54:582–6.

[47] Babij P, Matthews SM, Rennie MJ. Changes in blood ammonia, lactate and amino acids in relation to workload during bicycle ergometer exercise in man. Eur J Appl Physiol Occup Physiol 1983;50:405–11.

[48] Buono MJ, Clancy TR, Cook JR. Blood lactate and ammonium ion accumulation during graded exercise in humans. J Appl Physiol 1984;57:135–9.

[49] Katz A, Sahlin K, Henriksson J. Muscle ammonia metabolism during isometric contraction in humans. Am J Physiol 1986;250: C834–40.

[50] Brooks GA. Amino acid and protein metabolism during exercise and recovery. Med Sci Sports Exerc 1987;19:S150–6.

[51] Lowenstein JM. Ammonia production in muscle and other tissues: the purine nucleotide cycle. Physiol Rev 1972;52:382–414.
[52] Katunuma N, Okada M, Nishii Y. Regulation of the urea cycle and TCA cycle by ammonia. Adv Enzyme Regul 1966;4:317–36.
[53] Fitts RH, Balog EM. Effect of intracellular and extracellular ion changes on E-C coupling and skeletal muscle fatigue. Acta Physiol Scand 1996;156:169–81.
[54] Banister EW, Cameron BJ. Exercise-induced hyperammonemia: peripheral and central effects. Int J Sports Med 1990;11 (Suppl 2):S129–42.
[55] Mutch BJ, Banister EW. Ammonia metabolism in exercise and fatigue: a review. Med Sci Sports Exerc 1983;15:41–50.
[56] McCandless DW, Schenker S. Effect of acute ammonia intoxication on energy stores in the cerebral reticular activating system. Exp Brain Res 1981;44:325–30.
[57] Hindfelt B, Plum F, Duffy TE. Effect of acute ammonia intoxication on cerebral metabolism in rats with portacaval shunts. J Clin Invest 1977;59:386–96.
[58] Hirai T, Minatogawa Y, Hassan AM, Kido R. Metabolic interorgan relations by exercise of fed rat: carbohydrates, ketone body, and nitrogen compounds in splanchnic vessels. Physiol Behav 1995;57:515–22.
[59] Meneguello MO, Mendonca JR, Lancha Jr. AH, Costa Rosa LF. Effect of arginine, ornithine and citrulline supplementation upon performance and metabolism of trained rats. Cell Biochem Funct 2003;21:85–91.
[60] Demura S, Yamada T, Yamaji S, Komatsu M, Morishita K. The effect of L-ornithine hydrochloride ingestion on performance during incremental exhaustive ergometer bicycle exercise and ammonia metabolism during and after exercise. Eur J Clin Nutr 2010;64:1166–71.
[61] Sugino T, Shirai T, Kajimoto Y, Kajimoto O. L-ornithine supplementation attenuates physical fatigue in healthy volunteers by modulating lipid and amino acid metabolism. Nutr Res 2008;28:738–43.
[62] Barbul A. Arginine: biochemistry, physiology, and therapeutic implications. JPEN J Parenter Enteral Nutr 1986;10:227–38.
[63] Denis C, Dormois D, Linossier MT, Eychenne JL, Hauseux P, Lacour JR. Effect of arginine aspartate on the exercise-induced hyperammoniemia in humans: a two periods cross-over trial. Arch Int Physiol Biochim Biophys 1991;99:123–7.
[64] Colombani PC, Bitzi R, Frey-Rindova P, Frey W, Arnold M, Langhans W, et al. Chronic arginine aspartate supplementation in runners reduces total plasma amino acid level at rest and during a marathon run. Eur J Nutr 1999;38:263–70.
[65] Eto B, Peres G, Le Moel G. Effects of an ingested glutamate arginine salt on ammonemia during and after long lasting cycling. Arch Int Physiol Biochim Biophys 1994;102:161–2.
[66] Takeda K, Machida M, Kohara A, Omi N, Takemasa T. Effects of citrulline supplementation on fatigue and exercise performance in mice. J Nutr Sci Vitaminol (Tokyo) 2011;57:246–50.
[67] Fricke O, Baecker N, Heer M, Tutlewski B, Schoenau E. The effect of L-arginine administration on muscle force and power in postmenopausal women. Clin Physiol Funct Imaging 2008;28:307–11.
[68] Besset A, Bonardet A, Rondouin G, Descomps B, Passouant P. Increase in sleep related GH and Prl secretion after chronic arginine aspartate administration in man. Acta Endocrinol (Copenh) 1982;99:18–23.
[69] Knopf RF, Conn JW, Floyd Jr. JC, Fajans SS, Rull JA, Guntsche EM, et al. The normal endocrine response to ingestion of protein and infusions of amino acids. Sequential secretion of insulin and growth hormone. Trans Assoc Am Physicians 1966;79:312–21.
[70] Villalobos C, Nunez L, Garcia-Sancho J. Mechanisms for stimulation of rat anterior pituitary cells by arginine and other amino acids. J Physiol 1997;502:421–31.
[71] DK. G. Chapter 45: Pituitary and Hypothalamic Hormones. NY: McGraw-Hill/Appleton&Lange; 1999.
[72] Yao K, Yin YL, Chu W, Liu Z, Deng D, Li T, et al. Dietary arginine supplementation increases mTOR signaling activity in skeletal muscle of neonatal pigs. J Nutr 2008;138:867–72.
[73] Collier SR, Casey DP, Kanaley JA. Growth hormone responses to varying doses of oral arginine. Growth Horm IGF Res 2005;15:136–9.
[74] Blum A, Cannon III RO, Costello R, Schenke WH, Csako G. Endocrine and lipid effects of oral L-arginine treatment in healthy postmenopausal women. J Lab Clin Med 2000;135:231–7.
[75] Ghigo E, Ceda GP, Valcavi R, Goffi S, Zini M, Mucci M, et al. Low doses of either intravenously or orally administered arginine are able to enhance growth hormone response to growth hormone releasing hormone in elderly subjects. J Endocrinol Invest 1994;17:113–7.
[76] Tang JE, Lysecki PJ, Manolakos JJ, MacDonald MJ, Tarnopolsky MA, Phillips SM. Bolus arginine supplementation affects neither muscle blood flow nor muscle protein synthesis in young men at rest or after resistance exercise. J Nutr 2011;141:195–200.
[77] Bucci LR, Hickson Jr. JF, Pivarnik JM, Wolinsky I, McMahon TJ, Tuner SD. Ornithine ingestion and growth hormone release in bodybuilders. Nutr Res 1990;10:239–45.
[78] Evain-Brion D, Donnadieu M, Roger M, Job JC. Simultaneous study of somatotrophic and corticotrophic pituitary secretions during ornithine infusion test. Clin Endocrinol (Oxf) 1982;17:119–22.
[79] Elam RP, Hardin DH, Sutton RA, Hagen L. Effects of arginine and ornithine on strength, lean body mass and urinary hydroxyproline in adult males. J Sports Med Phys Fitness 1989;29:52–6.
[80] Elam RP. Morphological changes in adult males from resistance exercise and amino acid supplementation. J Sports Med Phys Fitness 1988;28:35–9.
[81] Zajac A, Poprzecki S, Zebrowska A, Chalimoniuk M, Langfort J. Arginine and ornithine supplementation increases growth hormone and insulin-like growth factor-1 serum levels after heavy-resistance exercise in strength-trained athletes. J Strength Cond Res 2010;24:1082–90.
[82] Osowska S, Duchemann T, Walrand S, Paillard A, Boirie Y, Cynober L, et al. Citrulline modulates muscle protein metabolism in old malnourished rats. Am J Physiol Endocrinol Metab 2006;291:E582–6.

CHAPTER

44

An Overview of Glycine-Arginine-Alpha-Ketoisocaproic Acid (GAKIC) in Sports Nutrition

Bruce R. Stevens

Department of Physiology and Functional Genomics, University of Florida College of Medicine, Gainesville, FL, USA

INTRODUCTION

Muscle physiology and performance is impacted during acute and chronic phases of intense exercise that includes an anaerobic component. Dynamic high-intensity use of skeletal muscle rapidly leads to fatigue and reductions in muscle force and work, and therefore diminished athletic performance, especially when initiated during the anaerobic phase. The damaging effects which occur during this phase translate into poor long-term training results in athletes, and also impair body-building efforts. The ergogenic aid GAKIC fills the gap between the short (seconds) anaerobic energetic benefits of creatine and long-term aerobic benefits imparted by sodium-dependent carbohydrate/rehydration technology.

ISOKINETIC DYNAMOMETER STUDIES

Studies by Stevens and coworkers [1,2] demonstrated that GAKIC, a glycine and L-arginine amino acid salt of α-ketoisocaproic acid, significantly enhanced human muscle dynamic performance in an isolated quadriceps model of intense, exhaustive, anaerobic exercise involving simultaneously applied concentric plus eccentric fatigue to the point of complete exhaustion in each bout. Figure 44.1 demonstrates the time course of the GAKIC test protocol which employed isokinetic dynamometry of isolated quadriceps; in essence, the testing employed three anaerobic bouts over a 15-minute period during which isolated quadriceps were completely fatigued by eccentric plus concentric exercise. The enhancements attributable [2] to GAKIC included increased isokinetic torque and gain in overall muscle work by about 10% as compared with isocaloric control treatment. An additional benefit of the treatment was the delay in onset of fatigue during the early phases of exhaustive dynamic exercise as assessed using a Fatigue Resistance Index (FRI). In this study [2], the FRI was increased up to 28% over isocaloric control treatment (Figure 44.2). FRI was obtained by dividing the mean peri-exhaustion peak torque by the maximal torque generated during a baseline testing session. The peri-exhaustion values were obtained by measuring the power output values for both the concentric and eccentric phases of the last five repetitions of the 35-rep fatigue set [1] with GAKIC or with isocaloric control.

Notably, the Stevens' isokinetic dynamometry results were obtained under conditions of concentric plus eccentric fatigue of the isolated quadriceps group, unlike cycle ergometry studies which tax various muscle groups and fiber types, and do not emphasize eccentric exhaustion in parallel with concentric utilization in the manner of an isokinetic dynamometer. Advances in quantifying the effects of training or metabolic treatments on acute intense dynamic skeletal muscle performance have been limited in the literature by the difficulty of objectively measuring reproducible performance parameters during this phase of exercise. Stevens and colleagues [1,2] and others (reviewed in [1]) approached this problem by employing objective techniques in the isolated quadriceps muscle group

Nutrition and Enhanced Sports Performance.
DOI: http://dx.doi.org/10.1016/B978-0-12-396454-0.00044-8

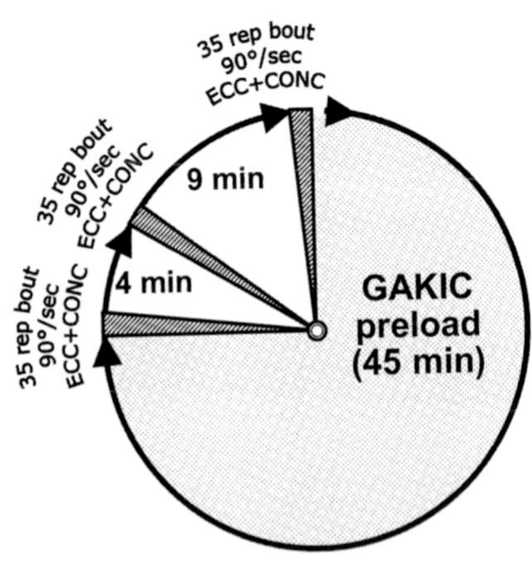

FIGURE 44.1 **Protocol clock showing isokinetic dynamometry of repeated bouts of maximal anaerobic exhaustive work output by quadriceps with GAKIC.** Individual subjects provided concentric work output as measured from three maximal exertion bouts executed within a 15-minute period beginning 45 minutes after consuming GAKIC treatment. Each bout was a sequence of 35 continuous maximal isokinetic concentric plus eccentric reps at a rate of 90°/sec. Force and work were measured during the concentric portion. The entire period lasted 60 min from start of oral treatment. Experimental design was randomized same subject double-blind crossover repeated measures, with 1 week washout. *Adapted from Stevens et al. [2].*

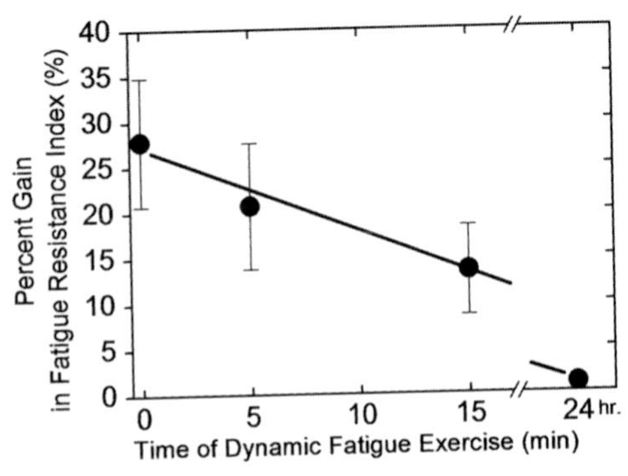

FIGURE 44.2 **Gain in concentric isolated quadriceps fatigue resistance with GAKIC compared with isocaloric control, assessed by isokinetic dynamometry.** The maximal torque value was 241 ± 9 N m. *Reproduced from Stevens et al. [2].*

undergoing contractions in both concentric and eccentric phases leading to fatigue. Others [3,4] have utilized cycle ergometry, or assessed body performance based on jumping or leg/arm press [5].

CYCLE ERGOMETRY STUDIES

Two studies [3,4] examined the effects of GAKIC on Wingate cycle ergometry performance parameters. The Koch group [4] demonstrated that GAKIC gave a beneficial ergogenic effect by attenuating the decline in mean power output during repeated exhaustive cycling sprints. They used untrained subjects who developed initial peak powers of roughly 800 W which dropped to 700 W by sprint #5; the GAKIC group demonstrated a significant reduction in FRI. The initial fatigue FRI of 5% was enhanced to 10% after sprint #5. In these experiments FRI was obtained as the ratio between peak power value and the minimum value, as determined from:

$$\text{fatigue index\%} = [(\text{peak power} - \text{minimum power})/\text{peak power}] \times 100.$$

The minimum power value used was the lowest power output after the subject achieved peak, rather than an overall minimum.

On the other hand, Beis et al. [3] utilized highly trained athletes, which gave initial peak powers of 1353 W that dropped to approximately 1200 W at sprint #5, and ending at 1061 W at sprint #10. The Beis group measured an initial FRI of roughly 38%, which remained fairly constant through sprint #10. Their data [3] indicated that GAKIC provided no significant differences in FRI as compared with isocaloric control treatment.

It is notable that there are systematic differences between the Buford & Koch study [4] and the Beis et al. study [3]; this prevents direct comparisons and global interpretations between the two cycle studies, and further prevents direct comparison of Beis with the previous isolated quadriceps isokinetic dynamometer study of Stevens et al. [1,2]. The Beis group states [3] that all possible confounding factors were minimized in their study, including speculation that the differences between studies may involve more consistent performances by the trained subjects. Nonetheless, care was taken by Buford & Koch [4] to ensure reliability of subjects' efforts. An interpretation of the differences further extends beyond experimental regimens, and must account for the physiologic differences between highly trained and untrained subjects utilizing GAKIC. The isokinetic dynamometer study of Stevens et al. [1,2] employed both eccentric and concentric fatigue in each bout, making disparate comparisons between cycle ergometry and isolated quadriceps fatigue. Using Stevens' [1,2] isokinetic dynamometry data in untrained athletes, it can be computed that in control subjects roughly 150 watts was exerted by isolated quadriceps in the concentric phase following both eccentric plus concentric fatigue, and that with GAKIC on board this work increased significantly to 167 watts.

In the light of these three studies [1–4], one conclusion that may be reached is that GAKIC is more effective in untrained subjects than in highly conditioned and trained athletes. Additional studies are warranted to expand comparisons among these modalities of testing.

KIC MONOTHERAPY

Yarrow et al. [5] explored the efficacy of short-term monotherapy supplementation of KIC given orally immediately prior to moderate- and high-intensity single-bout exercise performance measurements. In this study, resistance-trained men completed a trial with either 1.5 g or 9.0 g of either KIC or isocaloric placebo control. The other components of GAKIC, glycine and L-arginine, were not included. Subjects completed leg and chest press repetitions to failure and 30 s of repeated maximal vertical jumping (VJ) on a force plate. It was found that acute KIC ingestion alone with no other supplement immediately prior to exercise did not alter single-bout moderate- or high-intensity exercise. Therefore, it may be interpreted that KIC together with glycine and L-arginine are necessary to incur performance enhancement. The Yarrow study addressed single-dose single-bout performance events, and therefore the effect of KIC monotherapy on repeated high-intensity exercise bouts and long-term exercise training has not yet been studied.

METABOLISM

In vitro experiments and hospital studies have shown [6,7] that metabolic manipulation of nitrogen metabolism using selected ketoacids in conjunction with certain amino acids, or alone, can reduce various acute pathophysiological effects of trauma, including clinically dysfunctional skeletal muscle. A single bout of exhaustive exercise—indeed any acute muscle trauma—perturbs inter-organ and intra-organ metabolic and energetic events, resulting in attenuated performance and recovery [2,8].

GAKIC is a glycine and L-arginine salt of α-ketoisocaproic acid. The physiological effects of GAKIC treatment likely affect metabolic pathways individually or synergistically associated with L-arginine, glycine, and α-ketoisocaproate (KIC) as the ketoacid parent of the branched-chain amino acid L-leucine [9]. The exact mechanism by which the components of GAKIC enhanced acute exhaustive muscle performance is currently not known. However, the mechanism likely included effects on energetics, modifications of muscle bed hemodynamics, metabolic acidosis, or ammonia levels, and/or enzymatic pathways concerning branched-chain ketoacid and nitrogen metabolism [6,8–14].

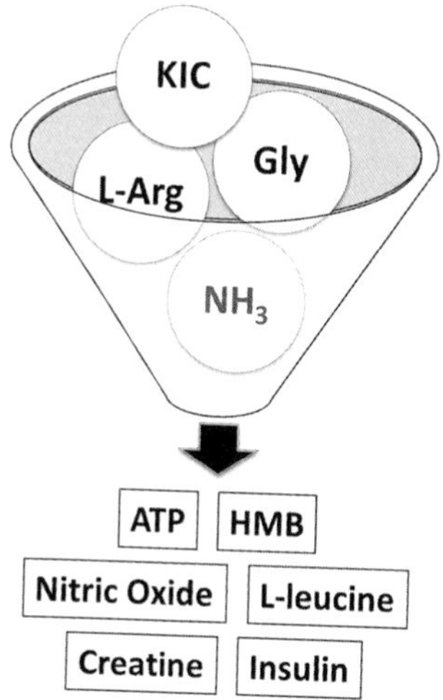

FIGURE 44.3 Components of GAKIC—glycine, L-arginine, α-ketoisocaproate—yield key metabolic products relevant to training enhancement physiology. Details are shown in Figure 44.4. KIC, α-ketoisocaproate; L-Arg, L-arginine; Gly, glycine; HMB, into β-hydroxy-β-methylbutyrate.

GAKIC components interact synergistically to enhance overall muscle metabolic efficiency, providing measurable enhancement of overall muscle performance. GAKIC is effective when on board during the beginning of maximal exhaustive anaerobic muscle work output, and then again during the critical early phase of post-bout recovery [2]. GAKIC components can synergistically accomplish the following benefits: (i) increase muscle work and force output during workouts or performance; (ii) reduce exhaustive anaerobic fatigue during workouts or performance; (iii) prolong useful muscle function during workouts or performance; and (iv) reduce protein catabolism and increase protein anabolism. Enhancement of metabolic events in the early phase of exhaustive exercise initiates immediate rebound repair of the tissue, prevents or reduces further muscle breakdown, and permits long-term muscle growth and strength gains.

The observed biochemical and physiological effects of oral KIC on muscle recovery likely involve several control points, including key enzymes. Figure 44.3 summarizes the primary metabolic events associated with GAKIC-enhanced muscle function. The details are amplified in Figure 44.4.

As summarized in Figure 44.4, GAKIC benefits occur due to several mechanisms. First, the mixture removes the deaminated nitrogen waste created by intense, exhaustive bouts (e.g., ammonia and trauma-released amino acid transamination products are removed, thereby reducing toxic effects and acidosis pH shifts in local muscle beds). Secondly, it can regenerate L-leucine released by overworked/damaged muscle, via KIC covalently binding to exhaustive

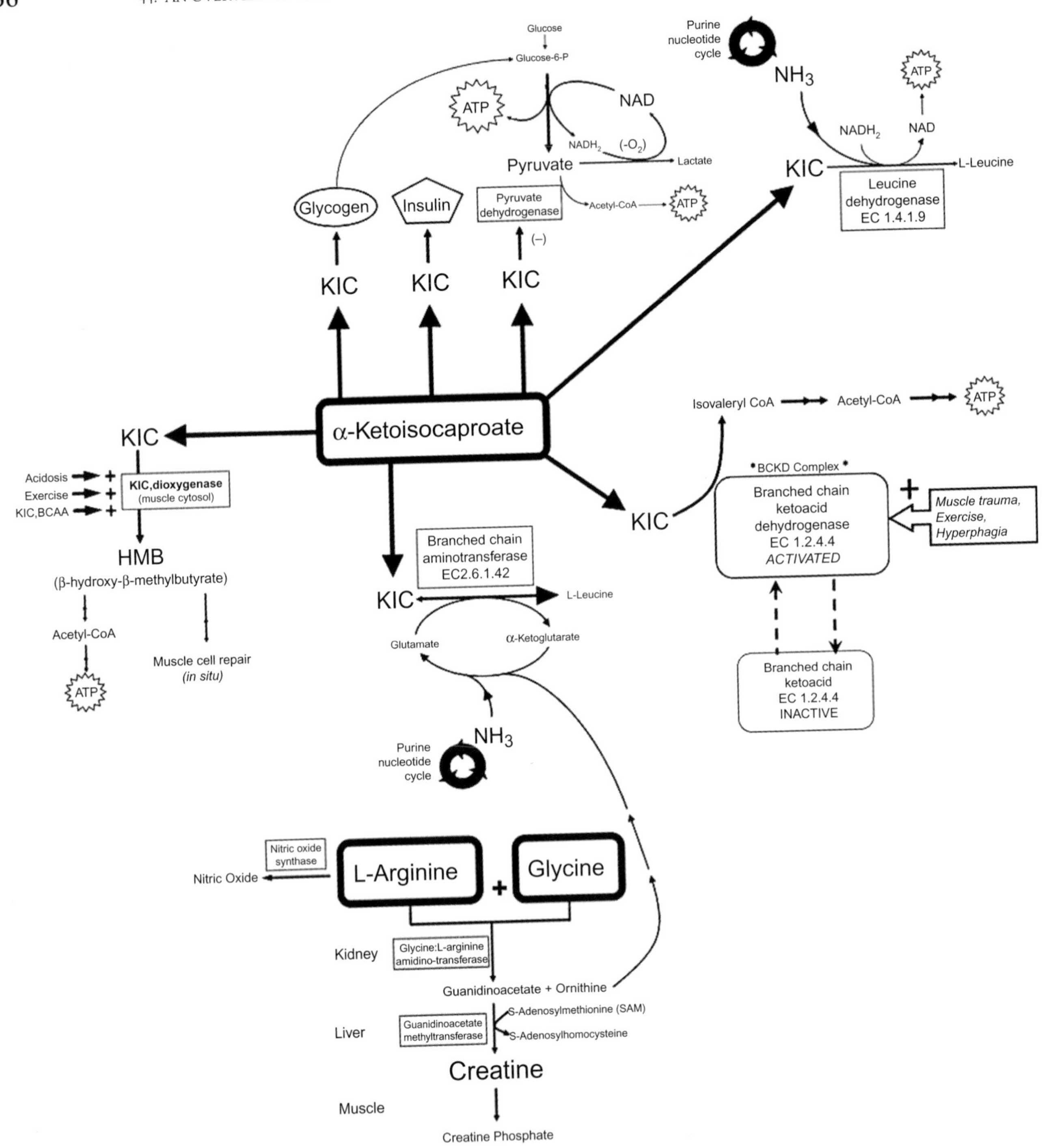

FIGURE 44.4 **Intermediary metabolism pathways relevant to the components of GAKIC: α-ketoisocaproate, L-arginine and glycine.**

exercise-induced nitrogen waste. The formed leucine, a branched-chain amino acid, is then immediately available for reincorporating into the damaged muscle. Thirdly, components of GAKIC metabolically shunt into HMB synthesis locally *in situ* in tissues when demanded by the muscle's metabolic status. Fourth, ATP levels are raised anaerobically. Fifth, GAKIC permits athlete-controlled transient insulin release, yet its glycemic index is virtually zero. Sixth, intermediary metabolism of L-arginine plus glycine generates creatine which provides an immediate energy product for muscle performance. Finally, a byproduct of L-arginine metabolism is nitric oxide, the beneficial vasodilator.

KIC helps to reduce the endogenous ammonia that hinders sarcomere contractions associated with intense exercise. In the presence of the excessive free NH_3 liberated from the purine nucleotide cycle activated during exercise, or from glutamate, L-leucine dehydrogenase (EC 1.4.1.9) provides a beneficial pathway that catalyzes the NH_3 amination of KIC to yield L-leucine, a branched-chain essential amino acid. Matthews and coworkers [15,16] have demonstrated that orally delivered KIC in humans is covalently aminated to form leucine in the splanchnic bed. Further, KIC can be shunted into β-hydroxy-β-methylbutyrate (HMB) via cytosolic KIC dioxygenase *in situ* in muscle tissues, and this is a process favored under physiological conditions accompanying intense exercise [17–19]. In laboratory studies, orally delivered KIC may contribute to survival of septic rats by conversion to various keto energy alternatives [20], and notably the survival benefits are specific to KIC rather than to ketoacids in general.

KIC rapidly activates pancreatic beta-cells to transiently release insulin [21,22]. KIC is as insulinogenic as glucose [23], yet the glycemic index of KIC is essentially zero. KIC-induced insulin release from the pancreas is initiated by activating beta-cell K^+ ion channels in direct response to Ca^{+2} sensitive KIC-dependent ATP synthesis, and this is potentiated by L-arginine [22,24].

Other key enzymes (Figure 44.4) include branched-chain aminoacid-aminotransferase (EC 2.6.1.42), and branched-chain ketoacid dehydrogenase (EC 1.2.4.4) contained within a branched-chain ketoacid dehydrogenase (BCKD) complex that contains several redox enzymes [25–27]. Enzyme levels of branched-chain aminoacid-aminotransferase EC 2.6.1.42 are relatively unregulated at fairly steady-state levels in muscle. Therefore, transamination is reversibly catalyzed by branched-chain aminoacid-aminotransferase activity through mass action of available concentrations of KIC, L-leucine, L-glutamate, α-ketoglutarate, and their metabolites. In contrast to unregulated branched-chain aminoacid-aminotransferase, the activity of branched chain ketoacid dehydrogenase (EC 1.2.4.4) is highly regulated during exercise in muscle and liver [25–27]. This is part of a multienzyme complex [28] (BCKD complex in Figure 44.4) that catalyzes the irreversible oxidative decarboxylation of branched-chain ketoacids, as it reduces NAD to NADH. The activity of EC 1.2.4.4 greatly increases immediately after strenuous exercise [29,30], with subsequent return to resting baseline levels by 10 minutes post-exercise. Within the BCKD complex, branched-chain ketoacid-dehydrogenase EC 1.2.4.4 enzymatic activity is regulated by an ATP phosphorylation (inactivation)–dephosphorylation (activation) mechanism. It is notable that KIC is a key stimulator of this enzyme complex, whereby it inhibits the ATP-mediated kinase allosteric inactivation of branched-chain ketoacid-dehydrogenase [25,26]; the potency of KIC is several orders of magnitude greater than that of the other branched-chain amino acids. Branched-chain ketoacid dehydrogenase activity is therefore mediated by exercise and nutritional factors at the levels of allosteric and substrate mass action.

L-arginine and glycine are other components of GAKIC. The dibasic amino acid L-arginine is a "conditionally essential" amino acid that becomes essential under certain metabolic conditions including muscle trauma and injury. L-arginine can rapidly induce vasodilation in skeletal muscle via vascular smooth muscle nitric oxide biosynthesis [31] (Figure 44.4). L-arginine aids in conversion of ammonia to urea, with the metabolic effects of removal of ammonia by arginine enhanced by simultaneously administrating glycine (the third component of GAKIC) (Figure 44.4).

The time frame during which GAKIC affects high-intensity exercise energetics likely initially overlaps with that of creatine energetics [32], and then extends beyond the short time frame of creatine. In the light of known metabolic pathways converting arginine plus glycine to creatine [33,34] (Figure 44.4), and given the supporting role of KIC in stabilizing purine nucleotide cycle waste, components of GAKIC likely also work through muscle energetic pathways in common with creatine. Additional *in vitro* and *in vivo* research is required to pursue this.

The time frame of onset and duration of the effects of orally delivered GAKIC are consistent with previously reported natural kinetics of endogenous KIC *in situ* in humans. Fielding and coworkers [35] demonstrated in humans that endogenous KIC naturally occurring levels in skeletal muscle rose by about 50% during acute high-intensity cycle exercise, while natural levels of plasma KIC peaked at 15 min following the exercise period and returned to pre-exercise levels by 60 min. These data suggested that KIC diffuses into the circulating blood from exercising muscle following a lag period [35]. Hospital and *in vitro* studies of sepsis, muscle trauma, liver disease and attendant central portal encephalopathy, or renal disease, have shown that administered combinations of ketoacids and amino acids improve muscle trauma recovery time, reduce serum ammonia, and enhance acute and long-term injury repair [20].

CONCLUSION

In concert, exercise studies and metabolic studies indicate that employing all the components of GAKIC may be necessary to demonstrate full benefits in the muscle fiber types observed in isolated quadriceps group of skeletal muscle tested by isokinetic dynamometry. Benefits of the GAKIC mixture are extended to cycle ergometry training, but trained elite cyclists may reach a point of diminishing returns in whole-body utilization of GAKIC. A variety of metabolic pathways are involved in the effects of GAKIC.

References

[1] Kaminski TW, Godfrey MD, Braith RW, Stevens BR. Measurement of acute dynamic anaerobic muscle fatigue using a novel fatigue resistance index. Isokinet Exerc Sci 2000;8:95–101.

[2] Stevens BR, Godfrey MD, Kaminski TW, Braith RW. High-intensity dynamic human muscle performance enhanced by a metabolic intervention. Med Sci Sports Exerc 2000;32 (12):2102–8.

[3] Beis L, Mohammad Y, Easton C, Pitsiladis YP. Failure of glycine-arginine-alpha-ketoisocaproic acid to improve high-intensity exercise performance in trained cyclists. Int J Sport Nutr Exerc Metab 2011;21(1):33–9.

[4] Buford BN, Koch AJ. Glycine-arginine-alpha-ketoisocaproic acid improves performance of repeated cycling sprints. Med Sci Sports Exerc 2004;36(4):583–7.

[5] Yarrow JF, Parr JJ, White LJ, Borsa PA, Stevens BR. The effects of short-term alpha-ketoisocaproic acid supplementation on exercise performance: a randomized controlled trial. J Int Soc Sports Nutr 2007;4:2.

[6] Buckspan R, Hoxworth B, Cersosimo E, Devlin J, Horton E, Abumrad N. alpha-Ketoisocaproate is superior to leucine in sparing glucose utilization in humans. Am J Physiol 1986;251 (6 Pt 1):E648–53.

[7] Cynober LA. The use of alpha-ketoglutarate salts in clinical nutrition and metabolic care. Curr Opin Clin Nutr Metab Care 1999;2(1):33–7.

[8] Blomstrand E, Saltin B. Effect of muscle glycogen on glucose, lactate and amino acid metabolism during exercise and recovery in human subjects. J Physiol 1999;514(Pt 1):293–302.

[9] Tessari P, Garibotto G. Interorgan amino acid exchange. Curr Opin Clin Nutr Metab Care 2000;3(1):51–7.

[10] van Hall G, van der Vusse GJ, Soderlund K, Wagenmakers AJ. Deamination of amino acids as a source for ammonia production in human skeletal muscle during prolonged exercise. J Physiol 1995;489(Pt 1):251–61.

[11] Fitts RH. Muscle fatigue: the cellular aspects. Am J Sports Med 1996;24(6 Suppl.):S9–13.

[12] Lewis SF, Fulco CS. A new approach to studying muscle fatigue and factors affecting performance during dynamic exercise in humans. Exerc Sport Sci Rev 1998;26:91–116.

[13] Wagenmakers AJ. Fuel utilization by skeletal muscle during rest and during exercise. In: Stipanuk MH, editor. Biochemical and physiological aspects of human nutrition. W.B. Saunders; 1999. p. 882–900.

[14] van Someren KA, Edwards AJ, Howatson G. Supplementation with beta-hydroxy-beta-methylbutyrate (HMB) and alpha-ketoisocaproic acid (KIC) reduces signs and symptoms of exercise-induced muscle damage in man. Int J Sport Nutr Exerc Metab 2005;15(4):413–24.

[15] Matthews DE, Harkin R, Battezzati A, Brillon DJ. Splanchnic bed utilization of enteral alpha-ketoisocaproate in humans. Metabolism 1999;48(12):1555–63.

[16] Toth MJ, MacCoss MJ, Poehlman ET, Matthews DE. Recovery of (13)CO(2) from infused [1-(13)C]leucine and [1,2-(13)C(2)] leucine in healthy humans. Am J Physiol Endocrinol Metab 2001;281(2):E233–41.

[17] Holecek M, Muthny T, Kovarik M, Sispera L. Effect of beta-hydroxy-beta-methylbutyrate (HMB) on protein metabolism in whole body and in selected tissues. Food Chem Toxicol 2009;47 (1):255–9.

[18] Kovarik M, Muthny T, Sispera L, Holecek M. Effects of beta-hydroxy-beta-methylbutyrate treatment in different types of skeletal muscle of intact and septic rats. J Physiol Biochem 2010;66(4):311–9.

[19] Zanchi NE, Gerlinger-Romero F, Guimaraes-Ferreira L, et al. HMB supplementation: clinical and athletic performance-related effects and mechanisms of action. Amino acids 2011;40 (4):1015–25.

[20] Hirokawa M, Walser M. Enteral infusion of sodium 2-ketoisocaproate in endotoxic rats. Crit Care Med 1999;27 (2):373–9.

[21] Escobar J, Frank JW, Suryawan A, et al. Leucine and alpha-ketoisocaproic acid, but not norleucine, stimulate skeletal muscle protein synthesis in neonatal pigs. J Nutr 2010;140(8):1418–24.

[22] Zhou Y, Jetton TL, Goshorn S, Lynch CJ, She P. Transamination is required for {alpha}-ketoisocaproate but not leucine to stimulate insulin secretion. J Biol Chem 2010;285(44):33718–26.

[23] Heissig H, Urban KA, Hastedt K, Zunkler BJ, Panten U. Mechanism of the insulin-releasing action of alpha-ketoisocaproate and related alpha-keto acid anions. Mol Pharmacol 2005;68(4):1097–105.

[24] McClenaghan NH, Flatt PR. Metabolic and K(ATP) channel-independent actions of keto acid initiators of insulin secretion. Pancreas 2000;20(1):38–46.

[25] Brosnan JT, Brosnan ME. Branched-chain amino acids: enzyme and substrate regulation. J Nutr 2006;136(1 Suppl.):207S–11S.

[26] Shimomura Y, Honda T, Shiraki M, et al. Branched-chain amino acid catabolism in exercise and liver disease. J Nutr 2006;136 (1 Suppl.):250S–3S.

[27] Islam MM, Nautiyal M, Wynn RM, Mobley JA, Chuang DT, Hutson SM. Branched-chain amino acid metabolon: interaction of glutamate dehydrogenase with the mitochondrial branched-chain aminotransferase (BCATm). J Biol Chem 2010;285(1):265–76.

[28] Lynch CJ, Halle B, Fujii H, et al. Potential role of leucine metabolism in the leucine-signaling pathway involving mTOR. Am J Physiol Endocrinol Metab 2003;285(4):E854–63.

[29] Shimomura Y, Yamamoto Y, Bajotto G, et al. Nutraceutical effects of branched-chain amino acids on skeletal muscle. J Nutr 2006;136(2):529S–32S.

[30] Tarnopolsky M. Protein requirements for endurance athletes. Nutrition 2004;20(7–8):662–8.

[31] Maxwell AJ, Ho HV, Le CQ, Lin PS, Bernstein D, Cooke JP. L-arginine enhances aerobic exercise capacity in association with augmented nitric oxide production. J Appl Physiol 2001;90 (3):933–8.

[32] Nakagawa Y. Energy metabolism and characteristics of muscle during anaerobic exercise. Hokkaido Igaku Zasshi 1998;73 (4):379–88.

[33] Humm A, Fritsche E, Steinbacher S. Structure and reaction mechanism of L-arginine:glycine amidinotransferase. Biol Chem 1997;378(3–4):193–7.

[34] Valayannopoulos V, Boddaert N, Chabli A, et al. Treatment by oral creatine, L-arginine and L-glycine in six severely affected patients with creatine transporter defect. J Inherit Metab Dis 2012;35(1):151–7.

[35] Fielding RA, Evans WJ, Hughes VA, Moldawer LL, Bistrian BR. The effects of high intensity exercise on muscle and plasma levels of alpha-ketoisocaproic acid. Eur J Appl Physiol Occup Physiol 1986;55(5):482–5.

CHAPTER

45

L-Arginine and L-Citrulline in Sports Nutrition and Health

Rachel Botchlett[1], *John M. Lawler*[1,2] *and Guoyao Wu*[1,3]

[1]Faculty of Nutrition [2]Department of Health and Kinesiology [3]Department of Animal Science, Texas A&M University, College Station, TX, USA

INTRODUCTION

Diet has a profound effect on human health. Nutrients can have both positive and negative effects, depending on the quantity we ingest, the quality of the diet, and what other foods they react with in digestion. Over- or under-eating can also severely affect health. Maintaining a balanced diet is therefore extremely important in order to maintain optimal metabolism and overall health. In fact, a diet that closely matches the USDA's Healthy Eating Index, which includes adequate portion sizes and a variety of grains, fruits and vegetables, has been related to a lower incidence of developing major chronic diseases [1].

In addition to maintaining a healthy lifestyle, contents of the diet before and after exercise are of paramount importance in sports nutrition. Olympic athletes, body builders, or exercise enthusiasts alike all must ingest foods and nutrients that complement their sports training in order to optimize their performance. Depending on the sport and athletic ability, athletes may have a higher requirement for energy and protein [2,3] and must therefore provide their body with a higher amount of carbohydrates, fats, and protein, which are the major constituents of foods we eat. Carbohydrates are broken down to simple sugars and glucose and are the predominant form of energy intake in a normal diet. Dietary fats are also used as metabolic fuel, but there is a danger in ingesting too much fat as compared with carbohydrates because fats are more easily stored as triglycerides in white adipose tissue, thereby contributing to overweight and obesity.

Dietary proteins are broken down to amino acids, which are vital to health and metabolism. Amino acids are used extensively in body tissues and cell types primarily to synthesize body proteins [4,5]. However, amino acids are also highly involved in cell signaling [6], the synthesis of other amino acids [7], and as metabolic intermediates [8]. Many metabolic pathways (e.g., synthesis of creatine and carnitine) utilize multiple amino acids, and their function is dependent on amino acid availability. Maintaining adequate concentrations of amino acids is therefore extremely important, especially in sports nutrition, to beneficially promote positive changes in the structure and function of skeletal muscle.

There are 20 different amino acids generally required for the synthesis of proteins and peptides. Not all amino acids are used for all proteins; however, some amino acids do play multiple roles and are highly involved in metabolic pathways. As such, some amino acids have higher rates of endogenous synthesis. For example, because glutamine is the most abundant amino acid in many species, its rate of synthesis is high in many tissues and cell types. Of the common physiological amino acids, arginine (Arg) and citrulline (Cit) especially play significant roles throughout the body. Both are necessary to meet the body's requirement for maintenance, optimal growth, and health and can also positively influence sports nutrition requirements. The major objective of this chapter is to highlight important roles of Arg and Cit in overall health and metabolism and their benefits in enhancing sports nutrition.

ARGININE

Arg is generally considered a non-essential amino acid for adult humans because, in addition to being

Nutrition and Enhanced Sports Performance.
DOI: http://dx.doi.org/10.1016/B978-0-12-396454-0.00045-X

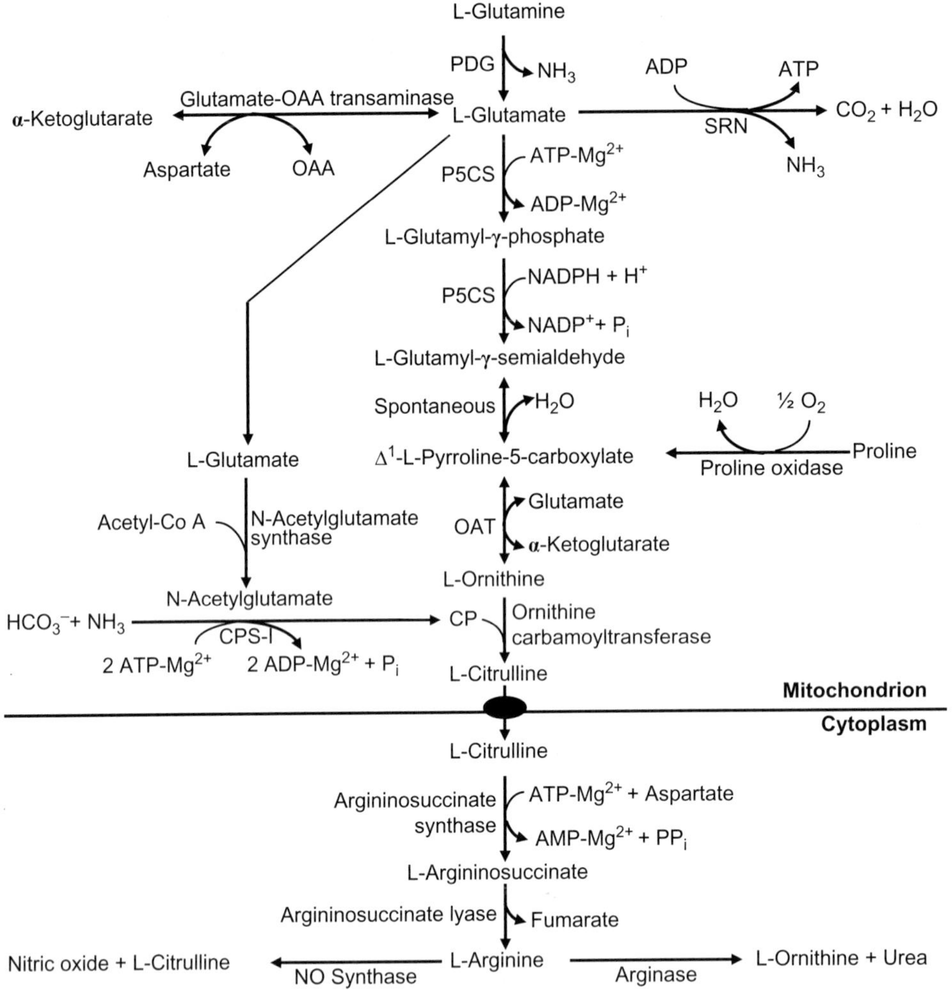

FIGURE 45.1 **Pathways for arginine and nitric oxide synthesis.** The small intestine of most mammals (including humans, pigs and rats) can synthesize citrulline from glutamine, glutamate, and proline. Citrulline is either converted into arginine in the gut or released into the circulation for arginine production by the kidneys as well as other tissues and cell types. Arginine is hydrolyzed to ornithine plus urea by arginase, and NO is synthesized from arginine by NO synthase, virtually in all cell types. CP, carbamoylphosphate; CPS-I, carbamoyl-phosphate synthase I; NO, nitric oxide; OAT, ornithine aminotransferase; PDG, phosphate dependent glutaminase; P5CS, pyrroline-5-carboxylate synthetase; SRN, a series of reactions.

absorbed from the diet, it can be synthesized within the body. In most mammals (including humans), Arg is synthesized in the small intestine from proline, glutamate, and glutamine [9] (Figure 45.1). In fact, the small intestine is the only mammalian organ with this ability. Arg plays an extremely important role in metabolism that occurs in multiple tissues and cell types, and is involved in several pathways including ammonia detoxification, nitric oxide (NO) production, and polyamine and creatine synthesis [10–13]. Here the beneficial roles of Arg in sports nutrition are discussed, with specific emphasis on the role of nitric oxide in skeletal muscle and heart health, the significance of Arg in stimulating the release of growth hormone and in creatine synthesis, and how Arg aids fat loss in obese patients.

Skeletal Muscle Health

One of the most important benefits of Arg is its role in NO production. NO is a cell-signaling molecule involved in many physiological systems, including the immune system, to kill pathogens, viruses, and parasites, and most significantly in vasodilation to reduce vascular resistance and high blood pressure. NO is synthesized from Arg via NO synthase (NOS) which is highly expressed in skeletal muscle. In fact, there are three types of NOS present in muscle: endothelial NOS (eNOS), neuronal NOS (nNOS), and inducible NOS (iNOS), with nNOS generally being the most prevalent. eNOS and nNOS are Ca^{2+}-dependent, while iNOS is Ca^{2+}-independent. The main functions of NO in muscle are to help regulate force generation, mechanotransduction, myocyte differentiation, protein turnover, satellite cell activation, mitochondrial biogenesis, fatty acid oxidation, and glucose homeostasis [14,15]. Further, NO modulates muscle respiration in that it helps control the delivery and uptake of oxygen and energy substrates for contraction [16]. NOS has been shown to be important in eliciting skeletal muscle hypertrophy in response to increased loading [17]. For all of these reasons maintaining physiological concentrations of Arg is of utmost importance for skeletal muscle health in order to sustain the production of NO.

In addition to maintaining normal levels of Arg, some research has shown that ingesting an above average amount of Arg within physiological ranges via supplementation may be beneficial for increasing muscle mass and strength. A study in 2006 found that a daily combination of 12 g L-Arg + alpha-ketoglutarate (α-KG) for 8 weeks increased males' 1 repetition maximum bench press strength [18]. It is important to note, however, that, because NO is a free radical species, it can be toxic to cells at high concentrations. Too much NO (from over-production) can negatively affect health by inducing oxidative damage to macromolecules (e.g., proteins and lipids) and cells [19], and can contribute to disease states such as diabetes and muscle injury [20,21]. Nonetheless, interest is growing regarding the benefits of Arg supplementation on increasing muscle mass, but more research is needed to fully understand accurate dosages and potential positive effects.

Arg has also been shown to stimulate the release of growth hormone (GH) [22]. GH is a hormone secreted by the pituitary gland that helps promote cell growth and regulate the mobilization of fuels in the body. Specifically, GH increases both DNA/RNA and protein synthesis and the amount of glucose taken up by muscle. As such, GH is considered an anabolic hormone because it positively contributes to skeletal muscle growth. The proposed mechanism of action of Arg on increasing GH secretion is that Arg enhances the pituitary somatotroph responsiveness to growth hormone releasing hormone (GHRH), while at the same time suppressing the endogenous growth hormone inhibiting hormone (GHIH). Both of these actions culminate in the increased release of GH [23]. Several studies have therefore investigated the potential role of Arg supplementation in increasing overall body mass, or muscle mass when combined with exercise or sports training. For example, pigs that were fed 5–9 g Arg showed a 6.5% body weight gain and a 5.5% improvement in muscle content [24]. In humans, however, mixed results have been reported. For instance, a study in 1997 found that males and females who participated in resistance training 2–4 times per week had a 2.7-fold increase in GH 60 minutes after oral supplementation with 1.5 g Arg and 1.5 g lysine [25]. Further, an intravenous infusion of 550 mg Arg per kilogram body weight resulted in an increase of GH response in both healthy males and females, with females displaying a larger increase than males [26]. On the other hand, seven male body builders who orally ingested 2.4 g Arg and 2.4 g lysine after an 8 hour fast showed no difference in GH response, as compared with placebo controls [27]. However, one caveat to this study may be that the male participants were only given the supplement on four separate occasions and did not ingest it every day. Although the mechanisms by which Arg influences GH release are better known, more research is needed to fully elucidate the accurate dosage and conditions in which Arg supplementation would be most effective in increasing skeletal muscle or overall body mass.

Creatine, a product of Arg catabolism, has been shown to play a significant role in increasing muscle strength [28,29] and fiber size [30]. It is synthesized in the liver and kidney from Arg, methionine (Met), and glycine (Gly) and, when phosphorylated, is an important source of energy for both the brain and muscle. In fact, skeletal muscle accounts for about 95% of all creatine stores in the body. Approximately 15–20% of Arg is used for creatine synthesis in adults. Creatine contributes to muscle mass and strength because an increase in intramuscular creatine increases the amount of phosphocreatine (a storage form of energy). This sustains the provision of energy and contributes to a greater exercise-training intensity and thus an increase in muscle mass [31]. Therefore, Arg is very important in not only maintaining energetic requirements for muscle, but also in potentially improving muscle mass and strength.

Arg is also important in both muscle and overall health because of its role in maintaining physical function, especially with aging. Aging is a naturally occurring process; however, the loss of muscle and strength can substantially contribute to more severe health issues, especially as the incidence for falls and injuries increases with age. Recent research is finding promising support for the use of Arg in combating sarcopenia. For example, low plasma Arg levels have been associated with a weak overall response to exercise performance in aged individuals [32]. Therefore, Arg supplementation could potentially prevent the decline of exercise performance, which is normally associated with aging. When combined with antioxidant therapies, Arg supplementation has also been reported to be beneficial in improving exercise performance in adult humans [33]. Further, supplementation with a blend of nutritionally essential amino acids and Arg has been reported to improve lean body mass, strength and physical function in insulin-resistant elderly individuals [34]. Research continues to focus on studying the full beneficial effects of Arg supplementation on aging, but it is clear thus far that Arg is a promising nutrient to prevent sarcopenia and other age-related health problems, particularly when combined with exercise.

Heart Health

Arg plays a key role in heart health and blood flow because it is a substrate for NOS in the production of NO. Although NO has many uses as a cell-signaling molecule, its main physiological function is its action as a vasodilator. In fact, endothelial NO secretion modulates blood flow and thus the amount of oxygen supplied throughout the entire body [35]. NO is, therefore, extremely important in both the resting and active states as the balance between vasodilation and vasoconstriction plays a major role in the degree of aerobic exercise capacity.

Exercise capacity may be limited by endothelial dysfunction [36], which is a pathological state of the inner lining of blood vessels. Endothelial dysfunction is a major component of vascular disease and is related to a variety of health concerns, including hypertension, hypercholesterolemia, and diabetes. Impaired cardiovascular function can be caused by a variety of factors including a decrease in the bioavailability of NO and the uncoupling of eNOS caused by inadequate Arg or tetrahydrobiopterin (BH4). As such, Arg supplementation may play a preventative role against vascular disease in its ability to maintain physiological concentrations of NO [37]. In fact, recent research has shown evidence that Arg supplementation can restore NO in patients who have experienced a heart event or have recently undergone a transplant. For example, in heart transplant recipients, oral L-Arg supplementation stimulated the NO pathway and restored endothelial function, and thus exercise capacity [38], which can both be severely reduced after such an event [39]. Further, an intravenous infusion of L-Arg improved endothelial dysfunction of both the heart microvasculature and epicardial coronary arteries in cardiac transplant recipients [40]. Arg is therefore highly beneficial for maintaining exercise capacity after cardiac events.

Arg may also play a role in reducing the risk of coronary heart disease. In 2008, researchers found that L-Arg supplementation is especially effective in increasing the number of endothelial progenitor cells (EPC) in mice after moderate exercise [41]. EPCs are important in heart health because they stimulate angiogenesis and are responsible for repairing the lining of blood vessels after vascular damage or other heart events. Therefore, Arg may further play a significant part in preventing the risk of vascular disease since this amino acid also positively contributes to the production and secretion of circulating EPCs. Vascular disease is the leading cause of death among men and women in the United States, and any potential treatments or preventative therapies, such as Arg supplementation, are of paramount importance.

Fat Loss

Obesity is a major health concern within the United States. In fact, the Center for Disease Control estimates that over one-third of American adults (~35%) are obese. This is of utmost importance as obesity increases the risk of developing other severe health problems including hypertension and diabetes [42]. Researchers are therefore focusing on potential treatments to reduce adiposity and prevent the onset of such metabolic diseases. Because Arg plays a positive role in several metabolic pathways, it has recently gained significant interest for its use in combating and potentially preventing obesity. Indeed, Arg has been linked to several mechanisms that help reduce obesity, including lipolysis, the oxidation of glucose and long-chain fatty acids (LCFAs) and, more recently, the development of brown adipose tissue.

Several studies have supported the involvement of Arg in both increasing lipolysis (the breakdown of lipids) and decreasing lipogenesis (the formation of long-chain fatty acids). Thus, Arg plays a significant role in the balance of both processes, which is important in preventing the accumulation of excess fat. Arg directly contributes to lipolysis via increasing cAMP [43], which is a known activator of hormone-sensitive lipase [44], the enzyme used to break down fat. Therefore, an increase in cAMP contributes to enhanced lipolysis. Further, physiological concentrations of both NO and carbon monoxide (whose synthesis is modulated by Arg) stimulate the oxidation of LCFA and glucose [45]. In 2005 a similar study reported that dietary Arg supplementation reduced the weight of abdominal adipose tissue, and both serum glucose and triglyceride concentrations in Zucker diabetic fatty rats [46]. As such, Arg supplementation seems to also indirectly enhance lipolysis by enhancing the production of NO within physiological ranges. NO is also involved in decreasing lipogenesis in that it reacts with and inhibits coenzyme A (CoA) [47], which is a central factor for fatty acid synthesis. Taken together, Arg can reduce lipogenesis, while increasing fatty acid oxidation.

Recent evidence shows that Arg may also play a role in preventing obesity because of its involvement in brown adipose tissue (BAT) development [48]. In fact, Arg supplementation increases BAT in fetal sheep [49], Zucker diabetic rats [50,51], and cold-acclimated rats [52]. BAT is responsible for non-shivering thermogenesis in mammals (Figure 45.2). This finding is especially significant because BAT contains uncoupling protein-1 (UCP1) and has a high capacity for oxidizing energy substrates [53]. Arg, which increases NO synthesis, contributes to BAT activation since NO helps control BAT cell differentiation and mitochondrial biogenesis [54,55]. This is significant for the prevention of excess fat accumulation because, when BAT is activated, high amounts of lipids and glucose are broken down within the tissue. Further, increased levels of NO may contribute to the conversion of white adipose tissue (WAT) into BAT [56].

Findings such as these are important to the field of nutritional biochemistry and in clinical settings for the potential treatment and prevention of obesity. Arg supplementation is quickly gaining interest as a potential treatment, especially because of the beneficial role that NO plays in several pathways associated with obesity. Research on the full benefits of Arg supplementation in obesity is ongoing, but nonetheless it seems thus far to be a crucial factor in potentially combating the development of this and other metabolic disorders.

CITRULLINE

Cit is an immediate and effective precursor for Arg synthesis in almost all cell types. Cit is considered as a nutritionally non-essential amino acid and is primarily synthesized from glutamine (Gln), glutamate and proline in enterocytes. Enterocytes are actually the only cell type containing the necessary enzymes required for this conversion, making the small intestine an integral part in the endogenous synthesis of Cit. Cit can also be synthesized from Arg and ornithine, with plasma glutamine and Arg serving as the main precursors during food deprivation [57]. Cit is used in several metabolic pathways throughout the body, the most notable of which being Arg synthesis and the urea cycle. The specific role of Cit in each pathway will be discussed, as well as its importance in muscle health and endurance.

Arginine Synthesis

As discussed previously, Arg plays an extremely important role in metabolism. Cit is therefore another

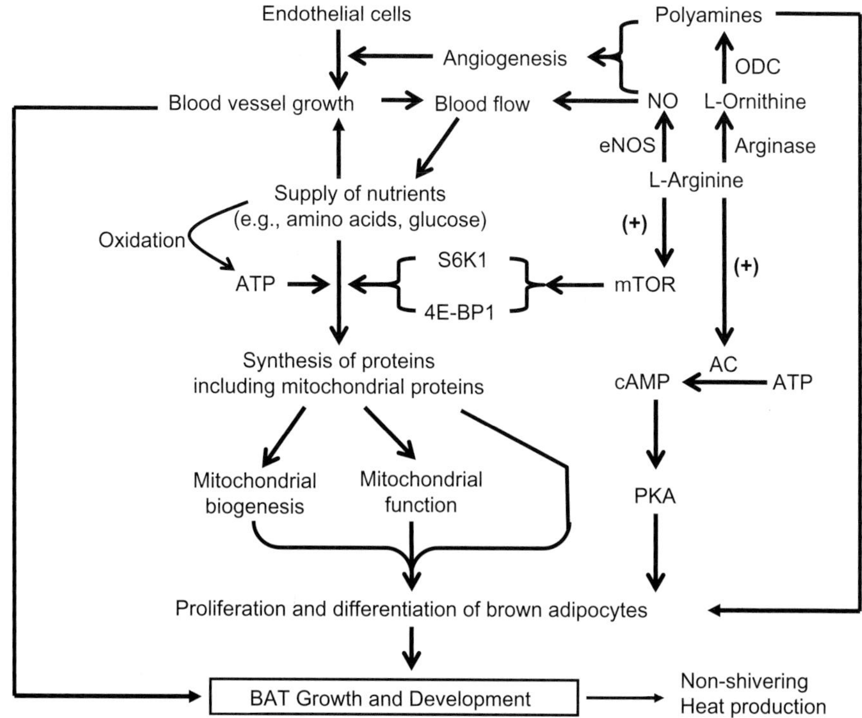

FIGURE 45.2 **L-Arginine enhances growth of brown adipose tissue (BAT) via NO-, polyamine- and cAMP-dependent mechanisms.** Both NO and polyamines, products of arginine catabolism, are essential for angiogenesis (the formation of new blood vessels from existing vessels). Growth of blood vessels increases the supply of nutrients (including amino acids, glucose, and fatty acids) to BAT and other tissues (including skeletal muscle, heart, and liver) via the circulation. The macronutrients are oxidized to provide ATP, and arginine may activate mTOR, resulting in S6 and 4E-BP1 phosphorylation, leading to increases in protein synthesis and mitochondrial biogenesis in BAT. BAT produces a large amount of heat via non-shivering mechanisms. AC, adenylate cyclase; BAT, brown adipose tissue; eNOS, endothelial NO synthase; 4E-BP1, eukaryotic initiation factor 4E-binding protein-1; mTOR, mechanistic target of rapamycin; ODC, ornithine decarboxylase; PKA, cAMP-dependent protein kinase A; S6K1, ribosomal protein S6K1; UCP1, uncoupling protein-1. The symbol (+) denotes activation.

integral amino acid throughout the body as it is extensively used for Arg synthesis in many cell and tissue types including hepatocytes, macrophages, endothelial cells, enterocytes, and in the kidney [58–61]. Specifically, argininosuccinate (AS) synthetase converts Cit to AS, which is then converted to Arg via AS lyase (Figure 45.1). This synthetic pathway consumes ammonia, a product of amino acid catabolism in skeletal muscle and other tissues. Supplementation with Cit can therefore be used to potentially treat and prevent arginine deficiencies [62], or restore impaired or reduced NO production [63,64]. Because both Arg and, more specifically, NO play vital roles in overall cardiovascular health, maintaining physiological concentrations of Cit is of utmost importance. Cit supplementation has even been suggested to more efficiently increase plasma Arg levels than the ingestion of Arg itself [65], because of a longer half-life of Cit than Arg in the circulation.

Cit is further involved in Arg synthesis through the Cit–Arg cycle, which combines Cit and aspartate to produce Arg and fumarate. This cycle is physiologically significant because it constantly recycles Cit in order to maintain a sufficient concentration of Arg within cells [66]. For example, male cyclists who received 6 g Cit supplementation prior to cycling events showed increased plasma levels of Cit, Arg and NO concentrations [67]. Cit is therefore of paramount importance in the body. Research is continually studying the effects of treatments combining Arg and Cit supplementation, as it seems the combination of these amino acids is vital to muscle health and sports metabolism.

Skeletal Muscle Health

One of the ways in which Cit is involved in maintaining muscle health and endurance is through its role in the urea cycle. The urea cycle is a metabolic pathway which occurs primarily in the liver, and to a lesser extent in the small intestine [8], and that serves to produce urea from ammonia. This pathway is significant because it removes ammonia from multiple cell types throughout the body and thus prevents ammonia toxicity. Cit is an amino acid intermediate within the urea cycle and, as such, has been shown to play an extremely important role in maintaining muscle health through its involvement in reducing ammonia build-up in muscle. For example, Cit supplementation actively decreased both exercise-induced blood ammonia levels and the blood-lactate increment in mice subjected to high-intensity exercise [68]. Cit therefore aids in the detoxification of ammonia and may inhibit additional glycolysis during exercise. The use of Cit in this manner is especially significant in sports nutrition as it is a neutral amino acid that serves to reduce fatigue and improve exercise performance. The full effects of Cit supplementation on improving exercise capacity in humans are now gaining interest as Cit seems to be a major factor in muscle health.

Cit may also be beneficial to muscle in that it potentially improves force production. Cit-malate (CM) supplementation has been a popular treatment choice for these types of studies as CM has been reported to both promote aerobic energy production [69] and normalize energy metabolism [70]. It seems that, when paired with malate, an intermediate of the tricarboxylic acid (TCA) cycle, Cit further serves to remove muscle metabolism by-products (i.e., ammonia and lactate) and plays a role in improving muscle function [71]. In fact, a recent study reported that supplementation with CM led to a 23% increase in specific force production in rat gastrocnemius muscle [72].

Although many of the full benefits of Cit supplementation on improving muscle mass and exercise performance and capacity are still unknown, Cit remains a vital amino acid for muscle and overall body health mainly through its use as a precursor to Arg. Several metabolic diseases and health conditions can be linked to Arg or NO deficiencies, which makes maintaining adequate levels of Cit so important. In subjects with impaired absorption and transport of Arg, Cit can be effectively used to provide Arg in virtually all cell types, including myocytes.

CONCLUSION

In summary, both Arg and Cit play extremely important roles in health and sports nutrition. Arg is a major precursor for the synthesis of NO, which helps regulate force production, myocyte differentiation, mitochondrial biogenesis, fatty acid oxidation, glucose homeostasis, and muscle respiration in skeletal muscle. This is significant in not only maintaining muscle health but in overall body metabolism because skeletal muscle constitutes 40–45% of the human body. Arg supplementation has therefore been extensively studied for its use in improving muscle function and overall body health. Arg also contributes to the release of growth hormone and in creatine synthesis, both of which have been shown to improve muscle mass and strength. Vascular disease, a major factor which can hinder metabolism and negatively influence nutrition, seems to be prevented and even potentially restored with Arg. Maintaining physiological concentrations of Arg is therefore of utmost importance in order to sustain heart health and exercise capacity. Further, Arg contributes to fat loss in obesity as it both directly stimulates lipolysis and decreases the accumulation of abdominal and sub-cutaneous fat. Another significant action of Arg is its role in the development of brown adipose tissue, which when activated oxidizes high amounts of lipids and glucose.

The importance of Cit in sports nutrition and overall metabolism is its role in Arg synthesis. Cit is converted to Arg in almost all cell types, including myocytes, hepatocytes, macrophages, endothelial cells and enterocytes. The Cit–Arg cycle is also a highly significant metabolic process because it constantly recycles Cit for Arg synthesis. Cit further helps maintain muscle health through its use as an intermediate in the hepatic urea cycle. The cycle functions to remove ammonia from the body and as such reduces the risk of ammonia toxicity. Cit therefore prevents the buildup of ammonia which would otherwise negatively impact skeletal muscle and overall health. The benefits of both Arg and Cit supplementation in muscle health are continually being studied; however, thus far both amino acids play vital roles in maintaining the health and integrity of muscle and in improving overall body nutrition.

References

[1] Chiuve SE, Fung TT, Rimm EB, Hu FB, McCullough ML, Wang M, et al. Alternative dietary indices both strongly predict risk of chronic disease. J Nutr 2012;142:1009–18.

[2] Tarnopolsky MA, MacDougall JD, Atkinson SA. Influence of protein intake and training status on nitrogen balance and lean body mass. J Appl Physiol 1988;64:187–93.

[3] Tarnopolsky MA, Atkinson SA, MacDougall JD, Chelsey A, Phillips S, Schwarcz HP. Evaluation of protein requirements for trained strength athletes. J Appl Physiol 1992;73(5):1986–95.

[4] Coeffier M, Claeyssens S, Hecketsweiler B, Lavoinne A, Ducrotte P, Dechelotte P. Enteral glutamine stimulates protein synthesis and decreases ubiquitin mRNA level in human gut mucosa. Am J Physiol 2003;285:G266–73.

[5] Kim SW, Wu G. Regulatory role for amino acids in mammary gland growth and milk synthesis. Amino Acids 2009;37:89–95.

[6] Rhoads JM, Wu G. Glutamine, arginine, and leucine signaling in the intestine. Amino Acids 2009;37:111–22.

[7] Wu G. Synthesis of citrulline and arginine from proline in enterocytes of postnatal pigs. Am J Physiol 1997;272:G1382–90.

[8] Wu G. Urea synthesis in enterocytes of developing pigs. Biochem J 1995;312:717–23.

[9] Wu G. Amino acid metabolism in the small intestine. Trends Comp Biochem Physiol 1998;4:39–74.

[10] Wu G, Morris Jr SM. Arginine metabolism: nitric oxide and beyond. Biochem J 1998;336:1–17.

[11] Wu G, Bazer FW, Davis TA, Jaeger LA, Johnson GA, Kim SW, et al. Important roles for the arginine family of amino acids in swine nutrition and production. Livest Sci 2007;112:8–22.

[12] Wu G, Bazer FW, Davis TA, Kim SW, Li P, Marc Rhoads J, et al. Arginine metabolism and nutrition in growth, health and disease. Amino Acids 2009;37:153–68.

[13] Blachier F, Davila AM, Benamouzig R, Tome D. Channelling of arginine in NO and polyamine pathways in colonocytes and consequences. Front Biosci 2011;16:1331–43.

[14] Stamler JS, Meissner G. Physiology of nitric oxide in skeletal muscle. Physiol Rev 2001;81(1):209–37.

[15] Anderson J, Pilipowicz O. Activation of muscle satellite cells in single-fiber cultures. Nitric Oxide 2002;7(1):36–41.

[16] Wolin MS, Hintze TH, Shen W, Hohazzabh KM, Xie Y-W. Involvement of reactive oxygen and nitrogen species in signaling mechanisms that control tissue respiration in muscle. Biochem Sos Trans 1997;25(3):934–9.

[17] Smith LW, Smith JD, Criswell DS. Involvement of nitric oxide synthase in skeletal muscle adaptation to chronic overload. J Appl Physiol 2002;92:2005–11.

[18] Campbell B, Roberts M, Kerksick C, Wilborn C, Marcello B, Taylor L, et al. Pharmacokinetics, safety, and effects on exercise performance of L-arginine α-ketoglutarate in trained adult men. Nutrition 2006;22:872–81.

[19] Inoue S, Kawanishi S. Oxidative DNA damage induced by simultaneous generation of nitric oxide and superoxide. FEBS Lett 1995;371(1):86–8.

[20] Cosentino F, Hishikawa K, Katusic ZS, Luscher TF. High glucose increases nitric oxide synthase expression and superoxide anion generation in human aortic endothelial cells. Circulation 1997;96:25–8.

[21] Stangel M, Zettl UK, Mix E, Zielasek J, Toyka KV, Hartung HP, et al. H_2O_2 and nitric oxide-mediated oxidative stress induce apoptosis in rat skeletal muscle myoblasts. J Neuropath Exp Neur 1996;55(1):36–43.

[22] Kanaley JA. Growth hormone, arginine and exercise. Curr Opin Clin Nutr Metab Care 2008;11:50–4.

[23] Ghigo E, Arvat E, Valente F, Nicolosi M, Boffano GM, Procopio M, et al. Arginine reinstates the somatotrope responsiveness to intermittent growth hormone releasing hormone administration in normal adults. Neuroendocrinology 1991;54(3):291–4.

[24] Tan B, Yin Y, Liu Z, Li X, Xu H, Kong X, et al. Dietary L-arginine supplementation increases muscle gain and reduces body fat mass in growing finishing pigs. Amino Acids 2008;37:169–75.

[25] Suminski RR, Robertson RJ, Goss FL, Arslanian S, Kang J, DaSilva S, et al. Acute effects of amino acid ingestion and resistance exercise on plasma growth hormone concentration in young men. Int J Sport Nutr 1997;7(1):48–60.

[26] Merimee TJ, Rabinowitz D, Fineberg SE. Arginine-initiated release of human growth hormone. N Engl J Med 1969;280:1434–8.

[27] Lambert MI, Hefer JA, Millar RP, Macfarlane PW. Failure of commercial oral amino acid supplements to increase serum growth hormone concentrations in male bodybuilders. Int J Sport Nutr 1993;3:298–305.

[28] Vandenberghe K, Goris M, Van Hecke P, Van Leemputte M, Vangerven L, Hespel P. Long-term creatine intake is beneficial to muscle performance and pharmacokinetic considerations. J Appl Physiol 1997;83:2055–63.

[29] Candow DG, Chilibeck PD, Burke DG, Mueller KD, Lewis JD. Effect of different frequencies of creatine supplementation on muscle size and strength in young adults. J Strength Cond Res 2011;25(7):1831–8.

[30] Santos RS, Pacheco MTT, Martins RABL, Villaverde AB, Giana HE, Baptista F, et al. Study of the effect of oral administration of L-arginine on muscular performance in healthy volunteers: an isokinetic study. Isokinet Exer Sci 2002;10(3):153–8.

[31] Chrusch MJ, Chilibeck PD, Chad KE, Davison KS, Burke DG. Creatine supplementation combined with resistance training in older men. Med Sci Sports Exerc 2001;33:2111–7.

[32] Hauer K, Hildebrandt W, Sehl Y, Edler L, Oster P, Droge W. Improvement in muscle performance and decrease in tumor necrosis factor level in old age after antioxidant treatment. J Mol Med 2003;81:118–25.

[33] Chen S, Kim W, Henning SM, Carpenter CL, Zhaoping L. Arginine and antioxidant supplement on performance in elderly male cyclists: a randomized controlled study. J Int Soc Sports Nutr 2010;7:13.

[34] Børsheim E, Bui Q-UT, Tissier S, Kobayashi H, Ferrando AA, Wolfe RR. Effect of amino acid supplementation on muscle mass, strength and physical function in elderly. Clin Nutr 2008;27:189–95.

[35] Böger RH. L-Arginine therapy in cardiovascular pathologies: beneficial or dangerous? Curr Opin Clin Nutr Metab 2008;11:55–61.

[36] Patel AR, Kuvin JT, DeNofrio D, Kinan D, Sliney KA, Eranki KP, et al. Peripheral vascular endothelial function correlates with exercise capacity in cardiac transplant recipients. Am J Cardiol 2003;91:897–9.

[37] Wu G, Meininger CJ. Arginine nutrition and cardiovascular function. J Nutr 2000;130:2626–9.

[38] Doutreleau S, Rouyer O, Di Marco P, Lonsdorfer E, Richard R, Piquard F, et al. L-Arginine supplementation improves exercise capacity after a heart transplant. Am J Clin Nutr 2010;91:1261–7.

[39] Marconi C, Marzorati M. Exercise after heart transplantation. Eur J Appl Physiol 2003;90:250–9.

[40] Drexler H, Fischell TA, Pinto FJ, Chenzbraun A, Botas J, Cooke JP, et al. Effect of L-arginine on coronary endothelial function in cardiac transplant recipients. Relation to vessel wall morphology. Circulation 1994;89:1615–23.

[41] Fiorito C, Balestrieri ML, Crimi E, Giovane A, Grimaldi V, Minucci PB, et al. Effect of L-Arg on circulating endothelial progenitor cells and VEGF after moderate physical training in mice. Int J Cardiol 2008;126:421–3.

[42] National Institutes of Health. Clinical guidelines on the identification, evaluation, and treatment of overweight and obesity in adults—The evidence report. Obes Res 1998;6(2):51S–209S.

[43] McKnight JR, Satterfield MC, Jobgen WS, Smith SB, Spencer TE, Meininger CJ, et al. Beneficial effects of L-arginine on reducing obesity: potential mechanisms and important implications for human health. Amino Acids 2010;39:349–57.

[44] Kersten S. Mechanisms of nutritional and hormonal regulation of lipogenesis. EMBO Rep 2001;2:282–6.

[45] Jobgen WS, Fried SK, Fu WJ, Meininger CJ, Wu G. Regulatory role for the arginine-nitric oxide pathway in metabolism of energy substrates. J Nutr Biochem 2006;17:571–88.

[46] Fu WJ, Haynes TE, Kohli R, Hu J, Shi W, Spencer TE, et al. Dietary L-arginine supplementation reduces fat mass in Zucker diabetic fatty rats. J Nutr 2005;135(4):714–21.

[47] Roediger WE. Nitric oxide-dependent nitrosation of cellular CoA: a proposal for tissue responses. Nitric Oxide Biol Chem 2001;5:83–7.

[48] Wu Z, Satterfield MC, Bazer FW, Wu G. Regulation of brown adipose tissue development and white fat reduction by L-arginine. Curr Opin Clin Nutr Metab Care 2012;15:529–38.

[49] Satterfield MC, Dunlap KA, Keisler DH, Bazer FW, Wu G. Arginine nutrition and fetal brown adipose tissue development in nutrient-restricted sheep. Amino acids 2011; doi:10.1007/s00726-011-1168-8.

[50] Wu G, Bazer FW, Cudd TA, Jobgen WS, Kim SW, Lassala A, et al. Pharmacokinetics and safety of arginine supplementation in animals. J Nutr 2007;137:1673S–80S.

[51] Wu G, Collins JK, Perkins-Veazie P, Siddig M, Dolan KD, Kelley KA, et al. Dietary supplementation with watermelon pomace juice enhances arginine availability and ameliorates the metabolic syndrome in Zucker diabetic fatty rats. J Nutr 2007;137:2680–5.

[52] Petrović V, Buzadžić B, Korać A, Korać B. Antioxidative defense and mitochondrial thermogenic response in brown adipose tissue. Genes Nutr 2010;. doi:10.1007/s12263-009- 0162-1.

[53] Nisoli E, Clementi E, Paolucci C, Cozzi V, Tonello C, Sciorati C, et al. Mitochondrial biogenesis in mammals: the role of endogenous nitric oxide. Science 2003;299:896–9.

[54] Cannon B, Nedergaard J. Brown adipose tissue: function and physiological significance. Physiol Rev 2004;84(1):277–359.

[55] Cannon B, Nedergaard J. Nonshivering thermogenesis and its adequate measurement in metabolic states. J Exp Biol 2011;214:242–53.

[56] Nisoli E, Carruba MO. Emerging aspects of pharmacotherapy for obesity and metabolic syndrome. Pharmacol Res 2004;50:453–4.

[57] Marini JC. Arginine and ornithine are the main precursors for citrulline synthesis in mice. J Nutr 2012;142(3):572–80.

[58] Dhanakoti SN, Brosnan JT, Herzberg JR, Brosnan ME. Renal arginine synthesis: studies in vitro and in vivo. Am J Physiol 1990;259:E437–42.

[59] Wakabayashi Y, Yamada E, Hasegawa T, Yamada RH. Enzymological evidence for the indispensability of small intestine in the synthesis of arginine from glutamate. I. Pyrroline-5-carboxylate synthase. Arch Biochem Biophys 1991;291:1–8.

[60] Wu G, Brosnan JT. Macrophages can convert citrulline into arginine. Biochem J 1992;281:45–8.

[61] Wu G, Knabe DA, Flynn NE. Synthesis of citrulline from glutamine in pig enterocytes. Biochem J 1994;311:115–21.

[62] Moinard C, Nicolis I, Neveux N, Darguy S, Benazath S, Cynober L. Dose-ranging effects of citrulline administration on plasma amino acids and hormonal patterns in healthy subjects: the Citrudose pharmacokinetic study. Br J Nutr 2008;99:855–62.

[63] Koeners MP, van Faassen EE, Wesseling S, de Sain-van der Velden M, Koomans HA, Braam B, et al. Maternal supplementation with citrulline increases renal nitric oxide in young spontaneously hypertensive rats and has long-term antihypertensive effects. Hypertension 2007;50(6):1077–84.

[64] El-Hattab AW, Hsu JW, Emrick LT, Wong L-JC, Craigen WJ, Jahoor F, et al. Restoration of impaired nitric oxide production in MELAS syndrome with citrulline and arginine supplementation. Mol Genet Metab 2012;105(4):607–14.

[65] Hartman WJ, Torre PM, Prior PL. Dietary citrulline but not ornithine counteracts dietary arginine deficiency in rats by increasing splanchnic release of citrulline. J Nutr 1994;124:1950–60.

[66] Wu G, Meininger CJ. Regulation of L-arginine synthesis from L-citrulline by L- glutamine in endothelial cells. Am J Physiol 1993;265(6 pt 2):H1965–71.

[67] Sureda A, Cordova A, Ferrer MD, Tauler P, Perez G, Tur JA, et al. Effects of L-citrulline oral supplementation on polynuclear neutrophils oxidative burst and nitric oxide production after exercise. Free Radical Res 2009;43(9):828–35.

[68] Takeda K, Machida M, Kohara A, Omi N, Takemasa T. Effects of citrulline supplementation on fatigue and exercise performance in mice. J Nutr Sci Vitaminol 2011;57(3):246–50.

[69] Bendahan D, Mattei JP, Ghattas B, Confort-Gouny S, Le Guern ME, Cozzone PJ. Citrulline/malate promotes aerobic energy production in human exercising muscle. Br J Sports Med 2002;36:282–9.

[70] Giannesini B, Izquierdo M, Le Fur Y, Cozzone PJ, Verleye M, Le Guern ME, et al. Beneficial effects of citrulline malate on skeletal muscle function in endotoxemic rat. Eur J Pharmacol 2009;602:143–7.

[71] Verleye M, Heulard I, Stephens JR, Levy RH, Gillardin JM. Effects of citrulline malate on bacterial lipopolysaccharide induced endotoxemia in rats. Arzneimittel-forsch 1995;45:712–5.

[72] Giannesini B, Fur YL, Cozzone PJ, Verleye M, Le Guern M-E, Bendahan D. Citrulline malate supplementation increases muscle efficiency in rat skeletal muscle. Eur J Pharmacol 2011;667:100–4.

CHAPTER

46

Roles of Chromium(III), Vanadium, and Zinc in Sports Nutrition

John B. Vincent[1] *and Yasmin Neggers*[2]

[1]Department of Chemistry [2]Department of Human Nutrition and Hospitality Management, The University of Alabama, Tuscaloosa, AL, USA

INTRODUCTION

While deficiencies in some minerals can have detrimental effects on health and on athletic performance, the use of mineral supplements (including multivitamin/mineral supplements) by athletes or physically active individuals who consume well-balanced diets with appropriate caloric intake appears to be unnecessary. The only well-documented exception is for iron deficiency (particularly in female athletes), where the anemia is readily reversed by iron supplementation. This review examines the use of three transition metals—vanadium, chromium, and zinc—that have been previously touted as mineral supplements for athletes, particularly in regard to their potential use to build lean muscle mass.

VANADIUM

Nutritional Background

No nutritional requirement has been set for vanadium (V), and no biological role for V in mammals has been established [1]. A few isolated reports of vanadium deficiency in rats and goats have been reported (e.g., [2,3]); however, distinguishing between nutritional and pharmacological effects is difficult [4], as V has a distinct pharmacological effect on glucose metabolism. Human dietary intake of V is normally in the range of 5–20 μg/day [5,6]. An upper tolerable level (UL) for V has been established at 1.8 mg/day, about 100 times the average intake [1]. The amount of V used in clinical studies as a potential treatment for diabetes is far in excess of the UL (allowable for clinical studies with careful safety monitoring) [1]. While acute V poisoning has not been noted for humans, adverse effects in humans are primarily gastrointestinal, including cramping, diarrhea, and loose stools [1].

Pharmacological Effects

In terms of pharmacological effects, the primary mode of action of V appears reasonably clear, although definitive *in vivo* experiments are still lacking. As vanadate, VO_4^{3-}, V inhibits the active sites of phosphatases and related enzymes involved in the hydrolysis of phosphate esters. VO_4^{3-} is a structural analog of one product of these enzymes, PO_4^{3-} (although this ion is protonated to varying degrees at physiological pH values). When acting as an inhibitor, the V in the active site has a trigonal bipyramidal geometry [7]. The three equatorial sites are filled by oxide oxygens, while one apical site is filled by an oxide oxygen and the other apical site is filled by the hydroxyl oxygen from a serine residue.

In rat adipocytes pretreated with VO_4^{3-}, insulin-stimulated insulin receptor kinase activity is augmented as is the autophosphorylation of insulin receptor [8]. In high glucose-treated rat adipocytes (that are insulin resistant), VO_4^{3-} stimulation of insulin receptor autophosphorylation and insulin receptor kinase activity are enhanced, as is Akt serine phosphorylation [9]. Thus, V appears to serve as an insulin mimetic. While V is able to inhibit PTP1B, the phosphatase responsible for dephosphorylation of insulin receptor, it also basically inhibits all phosphatases, although the K_i values can differ considerably. Given

Nutrition and Enhanced Sports Performance.
DOI: http://dx.doi.org/10.1016/B978-0-12-396454-0.00046-1

the range of functions regulated by phosphatase enzymes and that VO_4^{3-} administration could potentially affect them all to some degree, the potential for side effects from V treatments is easily understood.

V has never been shown to have any effects on body composition or muscle mass development. Clinical studies with V are limited in number [10] and have generally been limited to diabetic subjects. Studies using vanadyl sulfate (75 or 150 mg daily) have not been double blind in design, but most reported statistically significant improvement in fasting blood glucose and glycated hemoglobin. All the studies reported a high incidence of adverse gastrointestinal side effects [10]. The first results of a phase I and phase IIa clinical trial using a V chelate complex, bis(ethylmaltolato)oxovanadium(IV), have recently been reported [11,12]. The compound was "consistently well tolerated" [11] and had mixed results in the studies, which were not double-blinded or large enough for meaningful statistical treatment. Use of vanadium as a supplement for sports nutrition cannot be recommended.

CHROMIUM

Nutritional Background

Chromium was first proposed to be an essential trace element for mammals just over five decades ago. Rats fed a diet with Torula yeast as the sole protein source developed an apparent inability to properly dispose of glucose after a glucose challenge [13]. Adding high doses of a variety of Cr(III) complexes (200 μg Cr/kg diet), but not over 50 different elements, to the diet restored this ability. Similarly, the addition of Brewer's yeast or porcine kidney powder, both rich in chromium, also apparently reversed the condition. Unfortunately, the study was flawed. For example, the rats were kept in metal cages, and the Cr content of diet was never measured. Additionally, large doses of chromium(III) have subsequently been shown to have pharmacological effects on rodents [14]. Thus, these results and those of other earlier studies using similar diets and treatments [15] could readily be interpreted as suggesting a pharmacological effect from high doses of chromium.

In 1980, the Food and Nutritional Board of the National Academy of Science (USA) determined that chromium was an essential element and set an Estimated Safe and Adequate Daily Dietary Intake (ESADDI) of 50–200 μg [16]. In 2002, this was changed to an Adequate Intake (AI) of 30 μg per day [1]. This value was chosen as it reflects the Cr content of a nutritionist-designed diet [17]. By definition, the adequate intake means that >98 % of Americans show no signs of deficiency at this intake.

Recently, providing rats a diet with as little chromium as reasonably possible has been shown not to have any deleterious effects [18]. Rats were provided the AIN-93 G purified diet with added chromium in the mineral mix (<20 μg Cr/kg diet) and were kept in cages with no access to metal for 6 months. This Cr content is similar to that of a human consuming 30 μg Cr daily (the AI value). No differences in body mass, insulin sensitivity, or response to a glucose challenge were observed compared with rats on the complete AIN-93 G diet (with 1000 μg Cr/kg). Adding additional Cr(III) to the diet (200 μg/kg or 1000 μg/kg, clearly supra-nutritional or pharmacological doses) also had no effects on body mass but resulted in increased insulin sensitivity [18].

Other data have been proposed to support chromium being an essential trace element in mammals. The studies most often cited include those on the effects of chromium supplementation of rats on high-sugar [19] or high-fat diets [20] or humans on total parenteral nutrition (TPN) that developed diabetes-type symptoms [21]. In these cases, the animals had altered carbohydrate and lipid metabolism that was at least partially restored by supplementation with high, supra-nutritional doses of chromium. These actually only provide evidence for a pharmacological role for chromium, not one as an essential trace element [14], as the animals' glucose and lipid metabolism was compromised independently of their Cr intake.

Anderson and coworkers have reported that, for humans, absorption varies inversely with intake; low intakes (~15 μg/day) lead to "high" rates of absorption (~2%) while high intake (~35 μg/day) reduces absorption to ~0.4% [22]. However, the effects were only observed for female subjects; no dependence of absorption on dose was observed for male subjects. The greater number of female subjects resulted in an effect being observed when all subjects were pooled. Close scrutiny of this work suggests that propagation of error analysis needs to be carefully performed with these results. Studies of this type are not trivial to perform, and the error associated with the results is probably easily as large as the apparently observed effects. The study also needs to be reproduced. Recently, Kottwitz et al. [23] demonstrated in rats that absorption of chromium from oral $CrCl_3$ was independent of dose over seven different doses in the range 0.01–20 μg Cr. These results are consistent with perfusion studies that demonstrate that dietary Cr is absorbed by passive diffusion [24]. Cr would be a unique trace metal as its intake would not be controlled. Consequently, unless additional evidence should appear, chromium can no longer be considered an essential trace element [14,18].

Body Mass and Composition

However, as Cr supplementation has beneficial pharmacological effects at large doses, the question arises as to whether Cr can be used in a beneficial fashion related to sports nutrition. The use of Cr, particularly as [Cr(pic)$_3$], as a nutritional supplementation or nutraceutical for weight loss and muscle mass development was initiated by the publication of a paper in the *International Journal of Biosocial and Medical Research* by Gary W. Evans, the patent inventor for the use of the compound, in 1989 [25]. In the first study described, ten males between 18 and 21 years of age were involved. Half the students received a supplement containing 200 μg Cr as [Cr(pic)$_3$] for 40 days; the other half received a placebo. The subjects engaged in 40-minute exercise periods twice a week. By measuring thickness of skin folds and bicep and calf circumferences, body composition was estimated. Subjects on the supplement on average gained 2.2 kilograms (kg) body mass, had no significant change in % body fat, and gained 1.6 kg of lean body mass. In contrast, subjects on the placebo on average gained 1.25 kg body mass, had an increase in body fat of 1.1%, and increased their lean body mass by 0.04 kg. The increase in lean body mass for the subjects receiving chromium was said to be statistically greater than that for the control (or placebo) group ($P = 0.019$) [25]. In the second study [25], 31 college football players completed a 42-day program. Half the players were given 200 μg Cr as [Cr(pic)$_3$], while the other half received a placebo. The subjects exercised 1 hour per day for 4 days per week. Body composition was estimated by measuring thigh, abdomen, and chest skin folds and thigh, bicep, and calf circumference. After 14 days, subjects receiving Cr on average lost 2.7% of their body fat and had an increase in lean body mass of 1.8 kg, while no changes were observed in the control group. After 6 weeks, the chromium group on average lost 1.2 kg, lost 3.6% (or 3.4 kg) of their body fat, and had an increase in lean body mass of 2.6 kg, while the control group had a loss of 1 kg of body fat and a 1.8 kg increase in lean body mass. Both the loss of body fat and the increase in lean body mass were said to be significantly greater for the chromium group ($P = 0.001$ and $P = 0.031$, respectively). These results were rapidly challenged [26–29]. Body composition by skin fold measurements and circumference measurements is only an indirect estimation, especially in young males [27]; more accurate techniques such as underwater weighing were available in 1989. No method to determine compliance of subjects was indicated. The standard deviation of the data was not presented. Evans and coworkers followed this with a study reported in 1990 [30]. The study was double-blind, crossover in design. Groups received supplements or placebos for 42 days, then received neither for 14 days, and finally received for 42 days the opposite (placebo or supplement) from that in the original 42 days. Subjects varying from 25 to 80 years in age were utilized. Compliance was monitored by capsule count. While the study was designed to look at serum cholesterol and apolipoprotein levels, body mass data were also collected. [Cr(pic)$_3$] supplementation, 200 μg per day, had no effect on body mass. In 1993, Evans reported a study of the effects of [Cr(pic)$_3$] on body composition [31]. 12 males and 12 females were involved in a weekly aerobics class. Subjects were 25 to 36 years of age. Males received 400 μg Cr per day as [Cr(pic)$_3$] or 400 μg of Cr as chromium nicotinate. Females received half as much of either Cr source. Lean body mass was measured by resistivity. Data were presented with standard errors. For males receiving [Cr(pic)$_3$], the lean body mass increased by 2.1 kg and was statistically equivalent to the initial value and to the values of the group receiving chromium nicotinate. For females, the lean body mass increased by 1.8 kg and was statistically equivalent to the initial value and to those of the group receiving chromium nicotinate. Yet, Evans claimed, despite this, that the change in lean body mass for both males and females on [Cr(pic)$_3$] was significant ($P < 0.01$). The statistical analysis that indicated that, while final and initial values were equivalent, the difference between them was significant failed to incorporate the error in both the initial and final values in the calculation of the error of the difference. This study has been criticized by Lefavi [32], who states

> "It is likely that reviewers well-read in exercise physiology would find the notion of a 4.6-lb *lean* body mass (LBM) increase in males and a 4.0-lb LBM increase in females resulting from 12 weeks of a weekly aerobics class preposterous. A LBM increase that dramatic is not typically seen in subjects who are weight-training three times per week for 12 weeks, no matter what they are taking. ... Investigators familiar with this type of research would suggest either (a) that was one great aerobics class, or (b) people in Bemidji, MN, respond in a highly unusual manner to aerobic exercise and/or are extremely chromium deficient, or (c) Dr. Evan's group is consistently having difficulty accurately measuring LBM." (p. 121).

Curiously, none of the studies by Evans and coworkers described above reported a source of funding.

Subsequent studies of body composition have generally used underwater weighing, dual X-ray absorptiometry, or magnetic resonance imaging to measure the fat and lean body content and consequently are more accurate than these initial studies. These studies have failed to confirm the early results of the patent inventor [33], even when using athletes, with their greater nutritional requirements, as subjects. Thus, the overwhelming evidence from human studies is that chromium has no effect on body mass or body composition, including the enhancement of muscle mass. The results of rodent studies are equally compelling. Most notable is a study commissioned by NIH in which male and female rats and mice were provided with up to 5% of their diet as chromium picolinate for up to 2 years [34]; no effects were found on body mass and composition.

Largely on account of these early studies on chromium picolinate and subsequent work funded by the licensee of the patents, Nutrition21 (formerly AMBI), products containing chromium picolinate grew in sales to about one half billion dollars annually [35]. Consequently, when many people think of the element chromium from a nutritional point of view, the first thing to come to mind is probably one or more of the following:

- reduces body fat,
- causes weight loss,
- causes weight loss without exercise,
- causes long-term or permanent weight loss, and
- increases lean body mass or builds muscle.

The Federal Trade Commission (FTC) of the USA ordered entities associated with the nutritional supplement chromium picolinate to stop making each of these representations in 1997 because of the lack of "competent and reliable scientific evidence" [36]. Nutrition21 filed for bankruptcy in 2011 and sold its assets at auction in 2012.

Thus, as individuals in the USA and other developed nations should not be Cr deficient, and as Cr does not promote changes in body mass or composition, *healthy individuals have no need for Cr supplementation.* Studies continue to examine the effects of chromium supplementation of subjects with type 2 diabetes and related conditions. To date, studies with rodent models of diabetes demonstrate beneficial effects from chromium supplementation while human studies have failed to do so convincingly. Human studies have failed to use doses as high in comparison to those used in rodent studies, such that studies with higher doses are necessary in humans (with, of course, appropriate safety monitoring) [14].

Related Animal Studies

Several studies have appeared on the use of chromium supplements to prevent the effects of stress of livestock and farm animals [37]. While some studies indicate that meat quality and other variables may be improved, the results are exceedingly contradictory, so that no firm conclusions can be made.

Toxicity

The potential toxicity of chromium picolinate has recently been an area of intense debate, but consensus about its toxicity and the toxicity of Cr(III) compounds generally has probably recently been reached [14]. Only Cr(III) complexes with ligands such as imines, e.g., [Cr(III)(1,10−phenanthroline)$_3$], where the redox potential of the Cr(III) ion is altered so that the Cr center can enter into redox chemistry at the potentials in cells [38], are likely to have any toxic effects. Chromium picolinate with three imine nitrogens coordinated could potentially be toxic. In mammalian cell culture studies and studies in which the complex is given intravenously, chromium picolinate is clearly toxic and mutagenic, unlike other commercial forms of Cr(III) [14,39]. (Recently, studies funded by the company selling the supplement observed no effects in cell culture studies [40,41]; however, subsequent research has shown that DMSO (used as a solvent for the supplement and that can serve as a radical trap) quenched the deleterious effects [42].) However, when given orally to mammals, the complex does not appear to be toxic nor appear to be a mutagen or carcinogen. (Curiously, the complex appears to be a potent mutagen in fruit flies while other forms tested are not, as it apparently is absorbed intact [43,44].) The NIH study of the effects of up to 5% of the diet (by mass) of rats and mice for up to 2 years found no harmful effects on female rats or mice and at most ambiguous data for one type of carcinogenicity in male rats (along with no changes in body mass in either sex of rats or mice) [34]. This leads to a No Observed Adverse Effect Level (NOAEL) of 5 mg Cr/kg body mass/day or 300 mg Cr/day for an average 60 kg adult. These apparently contradictory results can readily be explained. The complex undergoes hydrolysis in the stomach, releasing the chromium before the intact complex can be absorbed to an appreciable level ($\sim$1%) [45], in contrast to the cell studies and fruit fly studies where the very stable, neutral complex could be absorbed intact. The World Health Organization and the European Food Safety Authority consider chromium supplementation of 250 μg Cr/day to be without safety concerns, while the Expert Group on Vitamins and Minerals (United Kingdom) has indicated that a daily dose of 10 mg Cr/day would be anticipated to be without adverse effects [46]. Thus, health concerns over the use of commercial chromium supplements is minimal; but given the lack of beneficial effect, any risk is unwarranted from chromium supplementation in healthy individuals.

ZINC

Nutritional Background

Zinc is a versatile trace element and is present in all organs, tissues and fluids of the body. It is needed as a cofactor for more than 300 enzymes [47]. Zn is involved in macronutrient metabolism and is required for nucleic acid and protein metabolism, thereby regulating cell differentiation and replication [48]. In addition, some Zn-dependent enzymes, such as lactate dehydrogenase and carbonic anhydrase, regulate glycolysis and are crucial for exercise metabolism [49]. Zn is also needed for the antioxidant enzyme superoxide dismutase. The need for zinc increases during periods of rapid growth such as pregnancy, infancy, and puberty [48].

In the USA, the 2001 recommended dietary allowance (RDA) for Zn was set at 8 mg/day for adult women and 11 mg/day for adult men [50]. These recommendations are lower than those set in 1989, which were 12 mg/day for adult women and 15 mg/day for adult men. Clinical Zn deficiency is rare in the general population when a balanced diet is consumed. The median zinc intake in the USA was reported to be approximately 9 mg/day for adult women and 13 mg/day for adult men in 2000 [50]. The highest concentrations of Zn are found in foods of animal origin such as oysters, lobster, and red meats like beef, lamb and liver [51]. Cooked dried beans, sea vegetables, fortified cereals, nuts, peas, and seeds are the major plant sources of Zn [51]. Although some concern exists that some vegetarian diets may not provide enough bioavailable Zn [52], studies indicate that in Western countries vegetarians and vegans do not suffer from overt zinc deficiencies any more than people who consume meat [53]. Some evidence suggests that more than 15 mg of zinc daily may be needed in those whose diet is high in phytates, such as with some vegetarians [50]. Excessive Zn intakes can be toxic and can produce nausea, vomiting and bloody diarrhea. A very large dose of Zn supplement (160 mg/day for 16 weeks) has been reported to reduce high-density lipoprotein concentrations [54]. The tolerable upper intake level for Zn has been set at 40 mg/day by the Food and Nutrition Board of the National Research Council.

Nutritional Status Assessment

A variety of indices have been used to asses Zn status, including measurement of Zn in plasma/serum, red blood cells, leukocytes and neutrophils. Evaluation of Zn status is difficult because of homeostatic control of body Zn. Due to ease of measurement, the most common method of assessment is serum/plasma Zn, with a fasting concentration of less than 70 μg/dL (10.7 μmol/L) suggesting deficiency [55]. Plasma Zn concentration should be interpreted with caution because it can be affected by many factors other than Zn depletion, including meals, time of the day, stress, infection, steroid therapy, and use of oral contraceptives [56].

Zinc Supplements

Zinc supplements are available in many forms, including oral forms such as tablets, lozenges and spray. The compounds used as Zn supplements include zinc oxide, zinc acetate, zinc sulfate, zinc chloride and zinc gluconate. The amount of Zn and bioavailability of Zn differs in various supplement forms. The amount of elemental Zn is as low as 14.3% in the gluconate form, whereas zinc sulfate and zinc chloride contain 23% and 28% Zn, respectively. Zn lozenges are supposed to assist in the treatment of common cold, but a meta-analysis has reported no significant benefit associated with use of Zn lozenges in treatment of the common cold [57].

Zinc and Sports Performance

Many athletes use dietary supplements commonly containing vitamins, minerals, protein, creatine, and some ergogenic compounds as a part of their regular or completion regimen [58]. Widespread belief exists among athletes that mineral supplements can enhance performance. This idea is popular because of the assumption that athletes have higher than usual requirements for minerals and that even a mild deficiency may have a detrimental effect on performance [59]. Striegel et al. [60] have reported that 30% of master athletes participating in the World Masters Athletics Championships in 2004 used mineral supplements. Research involving zinc supplements and exercise performance is very limited. Also, several researchers have noted that flawed study designs and small sample size further limit the ability to provide recommendations regarding Zn supplementation for athletes [61,62].

Early interest in a relationship between physical activity and zinc status was generated by a report of a decrease in serum Zn levels in some endurance runners as compared with sedentary subjects [63]. Dietary habits, particularly the avoidance of animal products, a diet high in carbohydrates, and increased Zn losses, were postulated to result in low serum Zn concentrations in some of the runners. Since then other studies have also reported low circulating Zn levels in physically active adults, including a group of female marathon runners [64,65]. An explanation given for these finding was that dietary Zn may have been inadequate in athletes with the low circulating Zn. However, based on self-reported dietary intakes, Zn consumption for athletes generally exceeds the estimated average requirement (EAR) of 9.4 and 6.8 mg/day for adult males and females, respectively [54]. Actually, the range of Zn intake/day for adult male and female athletes of 13.7–17.9 and 10.3–15.8 mg/day, respectively, as reported by several investigators, is close to or higher than the recommended dietary allowance (RDA) in 2001 for adult males and females [66–68].

Some data support the hypothesis that exercise induces Zn redistribution. Exercise is a significant stressor and affects circulating Zn Levels [54]. Longer-duration activities, such as distance running or skiing, have been observed to decrease plasma/serum Zn hours after the activity, while short-duration, high-intensity activity has been reported to increase plasma/serum concentration immediately [69]. Interpretation of these changes in circulating Zn is difficult without accounting for dietary Zn intake and other confounders, but they are usually considered evidence of redistribution of Zn due to exercise.

The effect of Zn supplementation on muscle function and strength has been evaluated by several investigators, but the evidence that Zn supplementation is needed for optimal muscle function and performance is equivocal [59,70–72]. Since various Zn-dependent enzymes that regulate glycolysis (lactate dehydrogenase) are needed for elimination of carbon dioxide from cells (carbonic anhydrase) and Zn is a cofactor for the antioxidant enzyme superoxide dismutase, Zn supplementation has been hypothesized to improve muscle strength and enhance athletic performance. Preliminary studies indicated that Zn enhances *in vitro* muscle contraction [73]. Dynamic isokinetic strength and isometric endurance significantly improved in 16 middle aged women supplemented with 30 mg Zn/day for a 14-day period in a double-blind placebo-controlled, cross-over designed study [70]. Muscle functions had been hypothesized to depend on recruitment of fast-twitch glycolytic muscle fibers as Zn supplementation enhances the activity of Zn-dependent enzyme lactate dehydrogenase. Since neither dietary Zn nor Zn status was measured, the large dose of Zn supplement makes it difficult to determine if Zn supplementation had physiological or pharmacological effects on muscle strength and endurance. In a recent study, Ali et al. [74] investigated the effect of Zn supplementation on the upper and lower trunk strength in athletic women with at least one year prior resistance training at a frequency of three times per week. Twenty-four women aged 18–35 years were randomly assigned to a control group and test group (25 mg Zn/day). The two groups participated in resistance training for 8 weeks. Testing sessions at week 1 and 8 included performing 1-RM and 80% of 1-RM tests on the bench press and the leg press. Fasting serum Zn concentrations were evaluated before and after resistance training. Dietary zinc intake was not measured. No significant difference existed in serum Zn concentrations before and after the 8 weeks of resistance training in the control or the supplemented group. The results indicated that Zn supplementation (25 mg/day) for 8 weeks had no significant effect on the upper trunk (triceps) or the lower trunk (quadriceps) strength of the subjects. However, the authors reported improvement in muscle function in the upper and lower trunk in the Zn supplemented group. The procedure for measurement of muscle function was not discussed in this study, making interpretation of the results difficult.

Studies evaluating the effect of graded dietary Zn or low Zn intake on physical performance have produced contradictory results. In untrained men fed diets containing variable Zn content (3.6–33.6 mg/day), Zn status measures were not affected [75]. On the other hand, the activity of the Zn-dependent enzyme carbonic anhydrase decreased significantly along with oxygen use and carbon dioxide elimination when dietary Zn was reduced to 4 mg/day as compared with 18 mg/day in physically active men [64]. In another study, men fed a formula-based diet severely low in Zn (1 mg/day), compared with men on a diet with an adequate amount of Zn (12 mg/day), had significantly decreased serum Zn associated with decreases in knee and shoulder extensor and flexor muscle strength [76]. Thus, evidence indicates that severe Zn deficiency, as evaluated by dietary and biochemical measures of Zn status, negatively affects muscle strength.

Research regarding the role of Zn supplementation for improvement of athletic performance is limited. The benefits of Zn supplementation to physical performance have not been established. Most studies have indicated that, for athletes who consume adequate amounts of dietary Zn, with Zn status indices within normal range, Zn supplementation does not enhance muscle strength or athletic performance. However, athletes who restrict energy intake, use severe weight loss regimens, or consume diets severely restricted in Zn may experience decreased muscle strength and endurance.

Future Research

Long-term, double-blind, placebo-controlled clinical trials with graded zinc supplements are needed to definitively clarify whether zinc supplementation has any positive effect on athletic performance.

CONCLUSION

Vanadium has not been used previously in sports nutritional supplements and is extremely unlikely to be used in the future. While chromium supplements have been used extensively in the past, current research does not support any benefits from their use; thus, the use of chromium as a sports supplement cannot be recommended at the current time. Zinc supplements may have beneficial effects, particularly in terms of maintaining muscle strength. The evidence, however, is inconclusive so that the use of zinc supplements cannot currently be recommended.

References

[1] National Research Council. Dietary Reference Intakes for Vitamin A, Arsenic, Boron, Chromium, Copper, Iodine, Iron, Manganese, Molybdenum, Nickel, Silicon, Vanadium, and Zinc. A report of the Panel on Micronutrients, Subcommittee on Upper Reference Levels of Nutrients and of Interpretations and Uses of Dietary Reference Intakes, and the Standing Committee on the Scientific Evaluation of Dietary Reference Intakes. Washington, D. C: National Academy of Sciences; 2002.

[2] Schwarz K, Milne DB. Growth effects of vanadium in the rat. Science 1971;4007:426–8.

[3] Uthus EO, Nielsen FH. Effect of vanadium, iodine and their interaction on growth, blood variables, liver trace elements and thyroid status indices in rats. Magnes Trace Elem 1990;9:219–26.

[4] Nielsen FH. Nutritional requirements for boron, silicon, vanadium, nickel, and arsenic: current knowledge and speculation. FASEB J 1991;5:2661–7.

[5] Myron DR, Zimmerman TJ, Shuler TR, Klevay LM, Lee DM, Nielsen FH. Intake of nickel and vanadium by humans. A survey of selected diets. Am J Clin Nutr 1978;31:527–31.

[6] Pennington JA, Jones JW. Molybdenum, nickel, cobalt, vanadium, and strontium in total diets. J Am Diet Assoc 1987;87:1644–50.

[7] Peters KG, Davis MG, Howard BW, Pokross M, Rastogi V, Diven C, et al. Mechanism of insulin sensitization by BMOV (bis maltolato oxo vanadium); unliganded vanadium (VO_4) as the active component. J Inorg Biochem 2003;96:321–30.

[8] Fantus IG, Ahmad F, Deragon G. Vanadate augments insulin-stimulated insulin receptor kinase activity and prolongs insulin action in rat adipocytes: evidence for transduction of amplitude of signaling into duration of response. Diabetes 1994;43:375–83.

[9] Lu B, Ennis D, Lai R, Bogdanovic E, Nikolov R, Salamon L, et al. Enhanced sensitivity of insulin-resistant adipocytes to vanadate is associated with oxidative stress and decreased reduction of vanadate (+5) to vanadyl (+4). J Biol Chem 2001;276:35589–98.

[10] Smith DM, Pickering RM, Lewith GT. A systematic review of vanadium oral supplements for glycaemic control in type 2 diabetes mellitus. Q J Med 2008;101:351–8.

[11] Thompson KH, Lichter J, LeBel C, Scaife MC, McNeill JH, Orvig C. Vanadium treatment of type 2 diabetes: A view to the future. J Inorg Biochem 2009;103:554–8.

[12] Thompson KH, Orvig C. Vanadium in diabetes: 100 years from Phase 0 to Phase I. J Inorg Biochem 2006;100:1925–35.

[13] Schwarz K, Mertz W. Chromium(III) and the glucose tolerance factor. Arch Biochem Biophys 1959;85:292–5.

[14] Vincent JB. Chromium: celebrating 50 Years as an Essential Element? Dalton Trans 2010;39: [3878–3794]

[15] Vincent JB, Bennent R. Potential and purported roles for chromium in insulin signaling: the search for the holy grail. In: Vincent JB, editor. The Nutritional Biochemistry of Chromium. Amsterdam: Elsevier; 2007. p. 139–60.

[16] National Research Council. Recommended Dietary Allowances, 9th Ed. Report of the Committee on Dietary Allowances, Division of Biological Sciences, Assembly of Life Science, Food and Nutrition Board, Commission on Life Science, National Research. Washington, DC: National Academy Press; 1980.

[17] Anderson RA, Polansky MM. Dietary and metabolite effects on trivalent chromium retention and distribution in rats. Biol Trace Elem Res 1995;50:97–108.

[18] Di Bona KR, Love S, Rhodes NR, McAdory D, Sinha SH, Kern N, et al. Chromium is not an essential trace element for mammals: Effects of a "low-chromium" diet. J Biol Inorg Chem 2011;16:381–90.

[19] Striffler JS, Law JS, Polansky MM, Bhathena SJ, Anderson RA. Chromium improves insulin response to glucose in rats. Metabolism 1995;44:1314–20.

[20] Striffler JS, Polansky MM, Anderson RA. Dietary chromium decreases insulin resistance in rats fed a high fat mineral imbalanced diet. Metabolism 1998;47:396–400.

[21] Jeejeebhoy KN. Chromium and parenteral nutrition. J Trace Elem Exp Med 1999;12:85–9.

[22] Anderson RA, Kozlovsky AS. Chromium intake, absorption and excretion of subjects consuming self-selected diets. Am J Clin Nutr 1985;41:1177–83.

[23] Kottwitz K, Laschinsky N, Fischer R, Nielsen P. Absorption, excretion and retention of ^{51}Cr from labeled Cr(III)picolinate in rats. Biometals 2009;22:289–95.

[24] Dowling HJ, Offenbacher EG, Pi-Sunyer FX. Absorption of inorganic, trivalent chromium from the vascularly perfused rat small intestine. J Nutr 1989;119:1138–45.

[25] Evans GW. The effect of chromium picolinate on insulin controlled parameters in humans. Int J Biosocial Med Res 1989;11:163–80.

[26] Clarkson PM. Nutritional erogogenic aids: chromium, exercise, and muscle mass. Int J Sport Nutr 1991;1:289–93.

[27] Moore RJ, Friedl KE. Ergogenic aids: physiology of nutritional supplements: chromium picolinate and vanadyl sulfate. Nat Strength Conditioning Assoc J 1992;14:47–51.

[28] Lefavi RG, Anderson RA, Keith RE, Wilson GD, McMillan JL, Stone MH. Efficacy of chromium supplementation in athletes: emphasis on anabolism. Int J Sport Nutr 1992;2:111–22.

[29] Whitmire D. Vitamins and Minerals: A Perspective in Physical Performance. In: Berning JR, Steen SN, editors. Sports Nutrition for the 90s. Gaithersburg: Aspen Publishers, Inc.; 1991. p. 129–51.

[30] Press RI, Geller J, Evans GW. The effect of chromium picolinate on serum cholesterol and apolipoprotein fractions in human subjects. West J Med 1990;152:41–5.

[31] Evans GW, Pouchnik DJ. Composition and biological activity of chromium-pyridine carboxylate complexes. J Inorg Biochem 1993;49:177–87.

[32] Lefavi RG. Response [letter]. Int J Sport Nutr 1993;3:120–2.

[33] Vincent JB. The potential value and toxicity of chromium picolinate as a nutritional supplement, weight loss agent, and muscle development agent. Sports Med 2003;33:213–30.

[34] Stout MD, Nyska A, Collins BJ, Witt KL, Kissling GE, Malarkey DE, et al. NTP toxicology and carcinogenesis studies of chromium picolinate monohydrate (CAS No. 27882-76-4) in F344/N rats and B6C3F1 mice (feed studies). Food Chem Toxicol 2009;47:729–33.

[35] Mirasol F. Chromium picolinate market sees robust growth and high demand. Chem Market Rep 2000;257:26.

[36] Federal Trade Commission. Docket No. C-3758 Decision and Order, <http://www.ftc.gov/os/1997/07/nutritid.pdf>; 1997 [accessed 05.03.12].

[37] Lindemann MD. Use of chromium as an animal feed supplement. In: Vincent JB, editor. The Nutritional Biochemistry of Chromium(III). Amsterdam: Elsevier; 2007. p. 85–118.

[38] Sugden KD, Geer RD, Rogers SJ. Oxygen radical-mediated DNA damage by redox-active Cr(III) complexes. Biochemistry 1992;31:11626–31.

[39] Stearns DM. Evaluation of chromium(III) genotoxicity with cell culture and in vitro assays". In: Vincent JB, editor. The Nutritional Biochemistry of Chromium(III). Amsterdam: Elsevier; 2007. p. 209–24.

[40] Slesinski RS, Clarke JJ, San RHC, Gudi R. Lack of mutagenicity of chromium picolinate in the hypoxanthine phosphoribosyltransferase gene mutation assay in Chinese hamster ovary cells. Mutat Res 2005;585:86–95.

[41] Gudi R, Slesinski RS, Clarke JJ, San RH. Chromium picolinate does not produce chromosome damage in CHO cells. Mutat Res 2005;587:140–6.

[42] Coryell VH, Stearns DM. Molecular analysis of hprt mutations induced by chromium picolinate in CHO AA8 cells. Mutat Res 2006;610:114–23.

[43] Hepburn DDS, Xiao J, Bindom S, Vincent JB, O'Donnell J. Nutritional supplement chromium picolinate causes sterility and lethal mutations in Drosophila melanogaster. Proc Natl Acad Sci USA 2003;100:3766–71.

[44] Stallings DM, Hepburn DDD, Hannah M, Vincent JB, O'Donnell J. Nutritional supplement chromium picolinate generates chromosomal aberrations and impedes progeny development in Drosophila melanogaster. Mutat Res 2006;610:101–13.

[45] Research Triangle Institute, Project Report, [^{14}C]Chromium Picolinate Monohydrate: Disposition and Metabolism in Rats and Mice, submitted to National Institutes of Environmental Health Sciences, 2002.

[46] European Food Safety Authority. Scientific opinion on the safety of trivalent chromium as a nutrient added for nutritional purposes to foodstuffs for particular nutritional uses and foods intended for the general population (including food supplement. EFSA J 2010;8:1–46.

[47] Cousins RJ. Zinc. In: 9th ed. Bowman BA, Russell RM, editors. Present Knowledge in Nutrition, I. Washington, DC: ILSI Press; 2006. p. 445–57.

[48] Chaney SG. Vitamins and Minerals: Requirements and Functions. In: Devlin TM, editor. Textbook of Biochemistry with Clinical Correlation. 7th ed. Hoboken, New Jersey: John Wiley Publisher; 2011. p. 1063–98.

[49] Lukaski HC. Micronutrients (magnesium, zinc, and copper): are mineral supplements needed for athletes? Int J Sport Nutr 1995;5:S74–83.

[50] Dietary reference Intakes. The essential guide to nutrient requirements, Part III: Vitamins and Minerals, Institute of Medicine, Washington, D.C: The National Academies Press; 2006. p 402–413.

[51] Berdanier CD, Dwyer JT, Feldman EB. Handbook of Nutrition and Food. Boca Raton, Florida: CRC Press; 2007.

[52] Position of the American Dietetic Association: Vegetarian Diets. J Am Diet Assoc 2009;109:1266–1282.

[53] Freeland-Graves JH, Bodzy PW, Epright MA. Zinc status of vegetarians. J Am Diet Assoc 1980;77: [665–661].

[54] Lukaski HC. Zinc. In: Driskell JA, Wolinsky I, editors. Sports Nutrition. 2nd ed. Boca Raton, Florida: CRC Press; 2006. p. 217–34.

[55] King J. Assessment of zinc status. J Nutr 1990;120:1474–779.

[56] Cousins RJ. Systematic transport of zinc. In: Mills CF, editor. Zinc in Human Biology. London: Springer-Verlag; 1989.

[57] Singh M, Das RR. Zinc for the common cold. Cochrane Data base of Systematic Reviews 2011;2: [CD001364].

[58] Maughan RJ, Frederic D, Geyer H. The use of dietary supplements by athletes. J Sports Sciences 2007;25(S1):S103–13.

[59] Micheletti A, Rossi R, Rufini S. Zinc status in athletes. Sports Med 2001;31:577–82.

[60] Striegel H, Simpn P, Wurster C, Niess AM, Ulrich R. The use of nutritional supplements among master athletes. Int J Sports Med 2006;27:236–41.

[61] Williams MH. Dietary supplements and sports performance: minerals. J Inter Soc Sports Nutr 2005;2:43–9.

[62] Lukaski H. Magnesium, zinc and chromium nutrition and athletic performance. Can J Appl Physiol 2001;26:S13–22.

[63] Dressendorfer RH, Sockolov R. Hypozincemia in runners. Phys Sports Med 1980;8:97–100.

[64] Lukaski HC. Vitamin and mineral status: effects on physical performance. Nutrition 2004;20:632–64.

[65] Singh A, Deuster PA, Moser PB. Zinc and Copper status of women by physical activity and menstrual status. J Sports Med Phys Fitness 1990;30:29–36.

[66] Lukaski HC, Hoverson B, Gallagher SK, Bolonchuk WW. Physical training and copper, iron, and zinc status of swimmers. Am J Clin Nutr 1990;51:1093–9.

[67] Fogelhom M, Laakso J, Lehto J, Ruokonene I. Dietary intake and indicators of magnesium and zinc status in male athletes. Nutr Res 1991;11:1111–8.

[68] Fogelhom M, REhunen S, Gref CG, Laakso JT, Ruokonene I, Himberg JJ. Dietary intake and thiamin, iron and zinc status in elite Nordic skiers during different training periods. Int J Sport Nutr 1992;2:351–65.

[69] van Rij AM, Hall M,T, Dohm GL, Bray J, Pories WJ. Changes in zinc metabolism following exercise in human subjects. Biol Trace Elem Res 1986;10:99–105.

[70] Krotkiewski M, Gudmundsson M, Backstrom P, Mandroukas K. Zinc and muscle strength and endurance. Acta Physiol Scand 1982;116:309–11.

[71] Weight LM, Noakes TM, Labadarios D, Graves J, Jacobs P, Berman PA. Vitamin and mineral status of trained athletes including the effects of supplementation. Am J Clin Nutr 1988;47:186–91.

[72] Lukaski HC. Magnesium, zinc, and chromium nutriture and physical activity. Am J Clin Nutr 2000;72(suppl):585S–93S.

[73] Richardson JH, Drake PD. The effects of zinc on fatigue of striated muscle. J Sports Med Phys Fitness 1979;19:133–4.

[74] Ali PA, Hanachi P, Golkhoo P. Effect of zinc supplements on the upper and lower trunk strength on athletic women. Res J Int Studies 2009;9:59–64.

[75] Lukaski HC, Bolonchuk WW, Klevay LM, Milne DB, Sandstead HH. Am J Physiol 1984;247:E88–93.

[76] Van Loan MD, Sutherland B, Lowe NM, Turnlund JR, King JC. The effects of zinc depletion on peak force and total work of knee and shoulder extensor and flexor muscles. Int J Sport Nutr 1999;9:125–35.

CHAPTER

47

An Overview on Beta-hydroxy-beta-methylbutyrate (HMB) Supplementation in Skeletal Muscle Function and Sports Performance

Carlos Hermano J. Pinheiro[1], Lucas Guimarães-Ferreira[2], Frederico Gerlinger-Romero[1] and Rui Curi[1]

[1]Department of Physiology and Biophysics, Institute of Biomedical Sciences, University of São Paulo. São Paulo, Brazil
[2]Exercise Metabolism Research Group, Center of Physical Education and Sports, Federal University of Espírito Santo. Vitória, Brazil

INTRODUCTION

Recently, amino acids such as leucine and its metabolite α-ketoisocaproate (KIC) have returned to the focus of several studies. It was observed that these compounds, or even a sub-product of their metabolism, could inhibit proteolysis and promote muscle hypertrophy in mice, leading to a reduction of urinary nitrogen loss and protein catabolism [1,2]. *In vitro*, KIC or another metabolite of leucine inhibits protein degradation, since the use of inhibitors of leucine transamination to KIC prevented this effect [3]. It is noteworthy that other branched-chain amino acids (BCAAs) isoleucine or valine, as well as their metabolites, did not promote these effects, supporting the hypothesis that some metabolite of leucine may act as a key element in triggering this anticatabolic effect [4]. Nissen et al. [5] suggested that beta-hydroxy-beta-methylbutyrate (HMB) is the metabolite responsible for these effects. In fact, direct effects of HMB were reported, with decreased proteolysis (~80%) and increased protein synthesis (~20%) in skeletal muscle of rats and chickens incubated with various concentrations of HMB [2].

Based on these observations, HMB is claimed to increase strength and fat-free mass (FFM), and its supplementation has been used as a potential strategy in the treatment of patients with muscular atrophy conditions, such as cachexia [6] and sarcopenia [7], or even to maximize gains in muscle mass during resistance training [8,9].

AN OVERVIEW ON HMB METABOLISM

HMB is a metabolite of the essential BCAA leucine [10]. The first step in HMB formation is the reversible transamination of leucine to form α-KIC, a process that occurs mainly extrahepatically [11]. Following this enzymatic reaction, α-KIC can follow two different pathways. In the first, HMB is produced from α-KIC by the cytosolic enzyme KIC dioxygenase in liver [12]. Following this pathway, HMB is first converted to cytosolic beta-hydroxy-beta-methylglutaryl-CoA (HMG-CoA), which can then be directed for cholesterol synthesis [13]. In the second pathway, α-KIC generates isovaleryl-CoA in the liver through the enzymatic action of branched-chain ketoacid dehydrogenase (BCKD), and after several steps, HMG-CoA is produced through the enzyme HMG-CoA synthase (Figure 47.1). Nissen and Abumrad [2] provided evidence that the primary fate of HMB is the conversion to HMG-CoA in the liver, for cholesterol biosynthesis.

Nutrition and Enhanced Sports Performance.
DOI: http://dx.doi.org/10.1016/B978-0-12-396454-0.00047-3

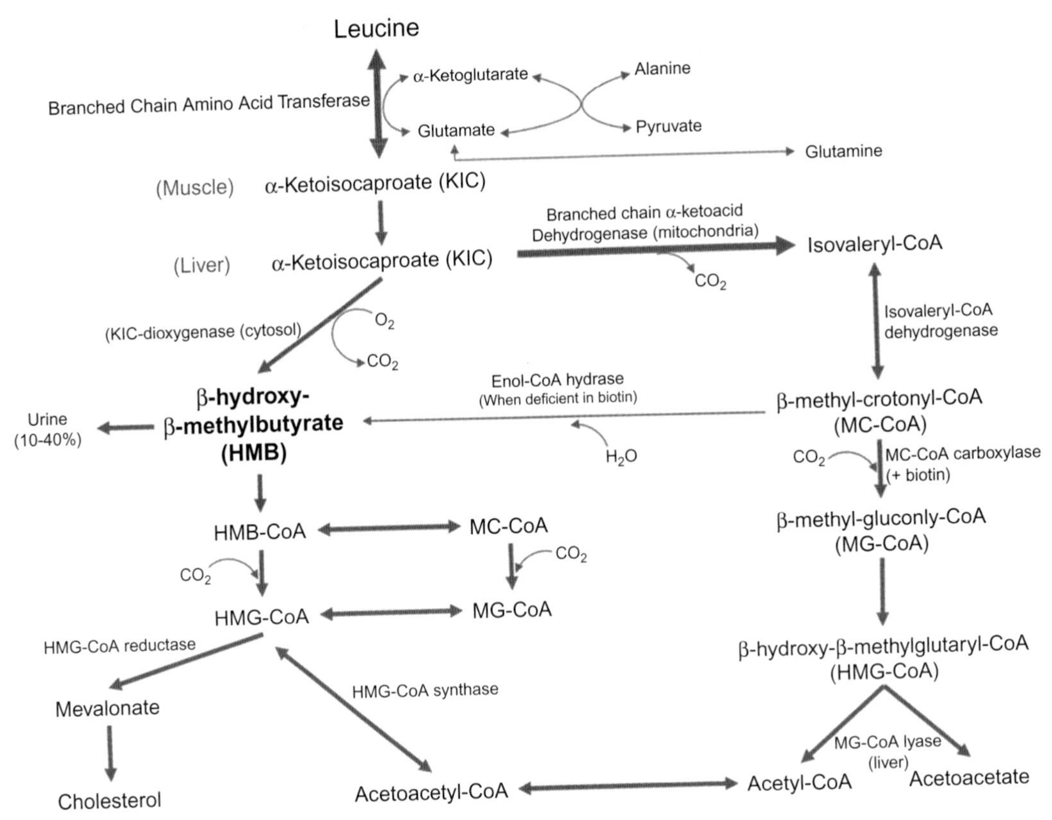

FIGURE 47.1 **HMB metabolism.** *Modified from Nissen and Abumrad [2].* (See color plate at the back of the book)

TABLE 47.1 HMB Pharmacokinetics

Parameter	1 g HMB	3 g HMB	3 g HMB + 75 g Glucose
Peak plasma concentration	115 nmol/L	~480 nmol/L	~350 nmol/L
Time to peak plasma concentration	2.0 h	1.0 h	1.9 h
Plasma half-life	2.37 h	2.38 h	2.69 h
Accumulation in urine	14%	29%	27%

Adapted from Vukovich et al. [14].

Van Koevering and Nissen [10] estimated that approximately 2–10% of leucine is converted to HMB.

The endogenous production of HMB is small. An individual weighing 70 kg produces about 0.2–0.4 g of HMB per day, from dietary leucine. Although the liver is the major site of production of HMB, by presenting high dioxygenase activity, studies with homogenates and perfusates also indicate that muscle and other tissues produce HMB, although in low amounts [2]. Plasma HMB half-life is approximately 2.5 h (Table 47.1). After approximately 9 h following ingestion, plasma HMB reaches baseline levels. Some HMB accumulates in urine, but 70–85% is retained in the body for further metabolism [14]. Leucine oxidation is affected by exercise, increasing up to 7-fold during exercise [15,16]. However, plasma HMB levels do not appear to be affected by acute exercise, since they did not change during a maximal oxygen consumption test in a cycle ergometer [17].

HMB SUPPLEMENTATION

HMB is present in foods such as citrus fruits, some fish and breast milk [5]. However, it is very difficult and impractical to provide by diet the quantities of HMB used in studies that demonstrate inhibition of proteolysis and muscle mass gains (~3 g/day). For this reason, HMB supplementation has been used as an alternative by the practitioners of weight training, for individuals subject to extreme muscular stress, the elderly, or in patients with diseases associated with muscle wasting syndromes, such as cancer.

Most studies with human subjects have employed 3 g/day of HMB. Moreover, Gallagher et al. [18] demonstrated that 3.0 g/day produced better results on FFM gain than did 1.5 and 6.0 g/day. Compared with HMB as calcium salt, HMB as free acid gel resulted in quicker and greater plasma concentrations (+185%) and improved clearance (+25%) of HMB from plasma. Therefore, HMB free acid gel could improve the availability and efficacy to tissues in health and disease [19]. Up to 76 mg/kg/day (equivalent to ~5.3 g/day in a 70 kg individual) for 8 weeks appears to be safe and does not adversely affect hepatic and renal function, or hematological parameters, in young male adults [20]. In fact, Nissen et al. [21] demonstrated that 3.0 g HMB/day promotes a decrease in total cholesterol (5.8%), low-density lipoprotein (LDL) cholesterol (7.3%), and systolic blood pressure (4.4 mmHg) when compared with placebo group.

EFFECTS OF HMB SUPPLEMENTATION ON STRENGTH AND BODY COMPOSITION

HMB Supplementation in Untrained and Healthy Individuals

The study of Nissen et al. [5] was the first to address the effect of different doses of HMB on skeletal muscle mass in humans. Three doses of HMB were used in this study (0, 1.5 and 3.0 g/day) with two dietary protein intakes (117 and 175 g/day) in untrained individuals subjected to a resistance-training regimen for 3 weeks. According the authors, supplementation prevented exercise-induced proteolysis and muscle damage, as indicated by a 20% decrease in urinary 3-methyl-histidine and 20–60% decreases in serum creatine kinase (CK) and lactate dehydrogenase (LDH) activity. However, utilization of these blood markers may not provide a reliable indication of muscle protein breakdown and damage, respectively. Despite no alteration in body composition, total strength increased in a dose-responsive manner: 8, 13 and 18.4% for 0, 1.5 and 3 g/day, respectively. In a later study with non-trained individuals, 3.0 g/day of HMB for 3 weeks results in a 37.5% cumulative increase in strength. Also, CK levels were lower in the HMB-supplemented group than in the placebo group, but fat-free mass (FFM) was not different between groups [22].

Doses higher than 3 g/day do not appear to further increase FFM or strength in untrained individuals. Gallagher et al. [18] found that doses of 3 g of HMB per day promoted an increase in FFM of 1.96 kg, but 6 g/day did not elicit the same effect. The 3 g/day dose also promoted greater increase in peak isometric torque. Increase in plasma CK levels following the initial training was attenuated by HMB supplementation, with no effect of dosage. The effects of HMB supplementation on strength and muscle damage did not differ by gender [23].

Vukovich et al. [24] randomized 31 70-year-old men and women into two groups, placebo and HMB-supplemented (3 g/day), in conjunction with a 5 day/week exercise program. HMB supplementation promoted an increased percentage of body fat loss assessed by skin fold estimation and computerized tomography. The increase in FFM tended to be greater in the HMB group (placebo -0.2 ± 0.3 kg; HMB 0.8 ± 0.4 kg), but the differences were not statistically different. In sedentary females, HMB (3 g/day) or placebo supplementation for 4 weeks did not promote any change in body composition [25]. When a resistance-training program (3 sessions per week) was associated with the supplementation protocol, greater gains in FFM (placebo 0.37 kg; HMB 0.85 kg) and strength (placebo 9%; HMB 16%) in the HMB-supplemented group were reported [25]. These results indicate that HMB supplementation enhanced gains in FFM and strength only in the presence of the stimulus provided by resistance training.

Short-term HMB supplementation in untrained subjects had no effect on swelling, muscle soreness, or torque following a bout of maximal isokinetic eccentric contractions of the elbow flexors [26]. Acute ingestion of 3 g HMB before 55 maximal eccentric knee extension/flexion contractions also had no effect on markers of skeletal muscle damage, while preventing increase in LDH [27]. Based on these results, for HMB-induced attenuation of muscle soreness and damage, a longer supplementation period may be necessary.

HMB Supplementation in Trained Individuals and Athletic Population

If in untrained individuals there are some data indicating positive effects of HMB supplementation, in trained individuals the results are less clear. While some demonstrated a possible role of HMB supplementation in strength and FFM gains in athletes, most studies do not confirm these findings.

Thomson et al. [28] supplemented 22 resistance-trained men with 3 g/day of HMB for 9 weeks with resistance training. They report an increase in lower-body (9.1% in leg press) but no alterations in upper-body strength (−1.9% in bench press and −1.7% in biceps curl). In distance runners, reductions in muscle damage markers represented by serum CK and LDH were found [29].

In trained collegiate football players, 4 weeks of HMB supplementation resulted in no alterations in body fat, body weight, or muscular strength [30], nor in markers of muscle catabolism (CK and LDH) and repetitive sprint performance [31]. Also in resistance-trained water polo and rowing athletes, 6 weeks of 3 g/day HMB supplementation did not result in increases in FFM and strength compared with a placebo group [32].

Vukovich and Dreifort [17] supplemented endurance-trained cyclists with 3 g/day of HMB for 14 days and evaluated VO_{2peak} and the onset of blood lactate accumulation (OBLA). Although VO_{2peak} (control $-2.6 \pm 2.6\%$; HMB $4.0 \pm 1.4\%$) and lactate accumulation peak (control 7.5 ± 1.3 mM; HMB 8.1 ± 1.1 mM) did not differ between groups, OBLA increased after HMB supplementation (control $0.75 \pm 2.1\%$; HMB $9.1 \pm 2.4\%$), suggesting that HMB promotes a decrease in lactate production, an increase in lactate removal, or a combination of both. In elite adolescent volleyball players, HMB supplementation promoted gains in anaerobic performance, as demonstrated by Portal et al. [33]. In this study, the effects of 7 weeks of HMB supplementation upon body composition, muscle strength, anaerobic and aerobic capacity, as well hormonal profile (growth hormone, IGF-I and testosterone) and inflammatory mediators (IL-6 and IL-1) were evaluated. HMB supplementation led to greater increase in FFM and knee flexion isokinetic force. Knee extension and upper-body isokinetic force were not affected by supplementation. Peak and mean anaerobic power, evaluated by lower-body Wingate test, was significantly greater in the HMB group compared with placebo. Aerobic capacity and hormonal and inflammatory profile were not affected. The positive effects on body composition contrast with most studies with well-trained athletes, in which HMB showed no effects. This may be due to the low age of players evaluated by Portal et al. [33] (13.5 to 18 years, Tanner stage 4 to 5) compared with studies using adult athletes.

It is possible that HMB supplementation exerts positive effects only in conditions in which muscle proteolysis is more pronounced, such as in untrained individuals acutely exposed to exercise training. Athletes and well-trained individuals may not respond to HMB supplementation, because of training-induced suppression of muscle protein breakdown.

Table 47.2 summarizes studies that evaluated the effects of HMB supplementation on muscle strength, muscle mass and muscle damage.

MECHANISMS OF ACTION OF HMB

Molecular and Metabolic Actions

The possible role of HMB in protection against contractile activity-induced muscle damage may be associated with increased stability of muscle plasma membrane and increased metabolic efficiency. HMB is converted to β-methylglutaryl-CoA (HMG-CoA) for cholesterol synthesis, and drugs that inhibit HMG-CoA reductase affect the electrical properties of skeletal muscle fiber membrane [43]. Besides, Nissen et al. [5] have suggested that HMB may act as a precursor of structural components of cell membranes, enhancing the repair of sarcolemma after contractile activity, but more research is needed to confirm this hypothesis. The effect of HMB supplementation upon acute muscle fatigue may also involve an increase in acetyl-CoA content, possibly through the conversion of HMG-CoA into acetoacetyl-CoA by HMG-CoA synthase inside the mitochondria [2,10]. We have also demonstrated that HMB supplementation results in increased ATP and glycogen content in gastrocnemius muscle [40]. It is possible that HMB could accelerate the tricarboxylic acid (TCA) cycle, increasing ATP content, malate-aspartate shuttle, and providing a carbon skeleton for glycogen synthesis.

In vitro data also demonstrate that HMB can promote an increase in fatty free acids (FFA) oxidation in skeletal muscle [44,45]. HMB supplementation could also promote an increase in lipid availability due to increased lipolysis. Our group has demonstrated that, in rats, HMB supplementation for 28 days resulted in decreased content of epididymal adipose tissue [42], as well as an increased plasma FFA concentration and increased hormone-sensitive lipase (HSL) gene and protein expression in white adipose tissue (unpublished data). This effect may be due to an increase of GH in response to HMB supplementation [42], since GH stimulates lipolytic enzymes as HSL [46].

EFFECTS OF HMB SUPPLEMENTATION ON PROTEIN HOMEOSTASIS IN SKELETAL MUSCLE

HMB on Protein Synthesis and Muscle Growth

In skeletal muscle, some evidence indicates that HMB supplementation affects protein metabolism by distinct pathways. For example, Kornasio et al. [47] demonstrated in chicken and human myoblasts an increase in thymidine incorporation, indicating DNA synthesis stimulation, which was 2.5-fold greater in HMB-treated than in control cells. Also, gene expression of some myogenic regulatory factors—MyoD, myogenin and MEF2—was stimulated by addition of HMB to the culture medium for 24 h in a dose-dependent fashion (25 to 100 g/mL). As result, HMB treatment promotes an increase in the number of cells, indicating an effect upon proliferation and differentiation of myoblasts. These authors also demonstrated an *in vitro* effect of HMB on

TABLE 47.2 Summary of Studies that Evaluated the Effects of HMB Supplementation on Muscle Strength, Muscle Mass and Muscle Damage

Reference	Samples	Supplementation Protocol		Results
Nissen et al. [5]	Human	1.5 or 3.0 g HMB day^{-1}	3 and 7 weeks with RT	Significant decrease in muscle proteolysis induced by exercise
Panton et al. [23]	Human	3.0 g HMB day^{-1}	4 weeks with RT	Lower levels of muscle damage markers
Gallagher et al. [18]	Human	3.0 g HMB day^{-1}	8 weeks with RT	Induced increases in muscle mass
Gallagher et al. [18]	Human	6.0 g HMB day^{-1}	8 weeks with RT	No response to supplementation
Gallagher et al. [20]	Human	6.0 g HMB day^{-1}	8 weeks with RT	No risk regarding renal and liver function or hematological parameters
Clark et al. [6]	Human	3 g of HMB + 14 g arginine + 14 g glutamine day^{-1}	8 weeks	In individuals with HIV, increased lean mass and improved immune system function
Slater et al. [32]	Human	3.0 g HMB day^{-1}	6 weeks with RT	No changes in body composition, muscular strength or biochemical markers
Jówko et al. [22]	Human	3.0 HMB day^{-1}	3 weeks with RT	No changes in muscle mass
Vukovich et al. [24]	Human	3.0 g HMB day^{-1}	8 weeks with RT	Increased body fat loss in elderly subjects
Vukovich and Dreifort [17]	Human (endurance-trained cyclists)	3.0 g HMB day^{-1}	2 weeks	Increased onset of blood lactate accumulation (OBLA) with no changes in lactate accumulation peak or VO_{2peak}
Ransone et al. [30]	Human (football players)	3.0 g HMB day^{-1}	4 weeks[a]	No gain in strength or muscle mass
Hoffman et al. [34]	Human (football players)	3.0 g HMB day^{-1}	10 days[a]	No gain in strength or muscle mass
Flakoll et al. [35]	Human	2.0 g HMB + 5.0 g arginine + 1.5 g lysine day^{-1}	12 weeks	Gain in maximum strength and muscle mass
Van Someren et al. [36]	Human	3.0 g HMB + 3.0 g KIC day^{-1}	2 weeks with RT	Gain in maximum strength and increased skeletal muscle repair
Kreider et al. [37]	Human	3.0 or 6.0 g HMB day^{-1}	4 weeks with RT	No gain in strength or muscle mass
Kreider et al. [31]	Human (football players)	3.0 g HMB day^{-1}	28 days[a]	No differences in catabolism markers, body composition, or repetitive sprint performance
Thomson et al. [28]	Human	3.0 g HMB day^{-1}	9 weeks with RT	Increased lower-body strength with no effects on body composition
O'Connor & Crowe [38]	Human (rugby players)	3.0 g HMB day^{-1}	6 weeks[b]	No gain in strength, nor in anthropometrical parameters
Lamboley et al. [39]	Human	3.0 g HMB day^{-1}	5 weeks[c]	Aerobic capacity enhancement with no alterations in body composition
Portal et al. [33]	Human (adolescent volleyball players)	3.0 g HMB day^{-1}	7 weeks[d]	Greater increase in FFM; increase in knee extension isokinetic force and in lower-body anaerobic power. No alterations in knee extension or upper-body isokinetic force

(*Continued*)

TABLE 47.2 (*Continued*)

Reference	Samples	Supplementation Protocol		Results
Pinheiro et al. [40]	Rat	0.320 g HMB day^{-1}	30 days	Increase in maximal tetanic force and resistance to acute muscle fatigue
Soares et al. [41]	Rat	0.002 g HMB day^{-1}	7 days	Decreased muscle damage and increased muscle fiber diameter under hindlimb immobilization procedures
Gerlinger-Romero et al. [42]	Rat	0.320 g HMB day^{-1}	28 days	No alterations in muscle weight; increases in GH and IGF-1

[a]*associated with specific exercise program for football players*
[b]*associated with specific exercise program for rugby players*
[c]*associated with an aerobic exercise training*
[d]*associated with specific exercise program for adolescent volleyball players.*
FFM, fat free mass; HMB, beta-hydroxy-beta-methylbutyrate; KIC, α-keto isocaproate; RT, resistance training.

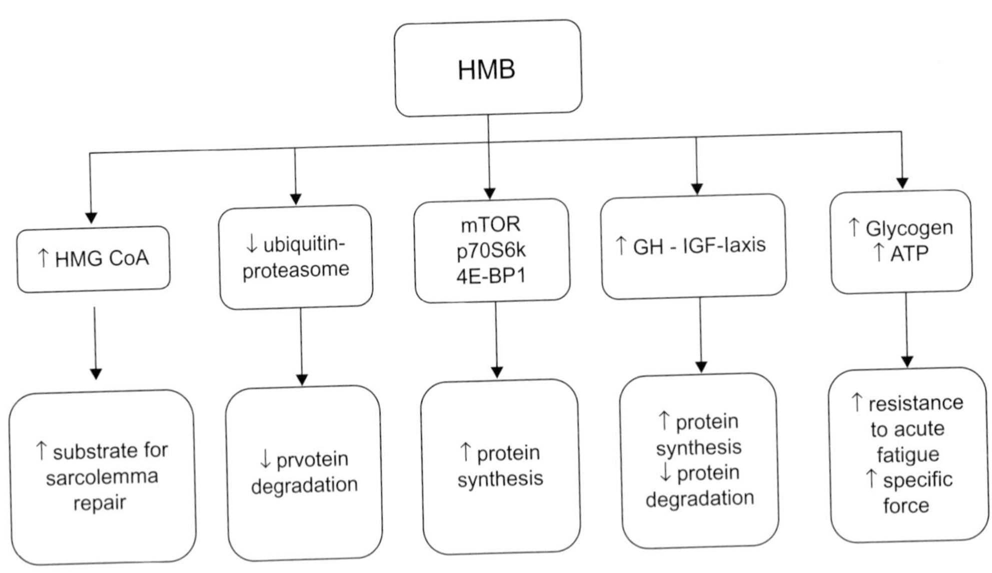

FIGURE 47.2 **Summary of the molecular and metabolic actions described for HMB.**

IGF-I gene expression, which was ~2-fold higher than in control untreated cells. In contrast, no alteration in IGF-I gene expression was verified after 30 days of HMB supplementation in soleus and EDL (extensor digitorum longus) muscles of rats. Liver mRNA and serum IGF-I, however, was stimulated [42]. In murine myoblasts, incubation with 50 mM of HMB significantly stimulated phosphorylation of mTOR and two important substrates, 4EBP-1 and p70S6k. Protein synthesis rate was increased in response to HMB, but this stimulatory effect was completely abolished in the presence of mTOR inhibitor rapamycin [47]. Is well known that mTOR can stimulate the translation initiation step of protein synthesis by several downstream targets, including p70S6k, eIF4E and eIF2B (Figure 47.2) [48].

HMB on Protein Degradation

Protein degradation is exacerbated in conditions such as fasting, hypogravity, immobilization, bed rest and cancer cachexia. The ubiquitin–proteasome system is a proteolytic system dependent on ATP, whereas the proteasome complex degrades intracellular proteins marked with a polyubiquitin chain [49]. Smith et al. [50] supplemented mice implanted with the MAC16 tumor with HMB and reported an attenuation of muscle mass loss and a reduced muscle proteolysis, reflected by a decrease in the catalytic activity of the proteasome. The data are scarce, but HMB exerts its anticatabolic action via a modulation on components of the ubiquitin–proteasome system. In addition to the activation of mTOR pathway [48] and increases in satellite cell content, myonuclei number and total DNA content [51], IGF-I promotes inhibition of ubiquitin-ligases in skeletal muscle, which could result in less protein directed to degradation [52]. However, more studies are needed to determine whether IGF-I produced by skeletal muscle and/or liver mediates some effects attributed to HMB supplementation upon skeletal muscle protein synthesis and degradation signaling pathways.

Impact on Skeletal Muscle Contractile Function

HMB supplementation can alter skeletal muscle strength in catabolic conditions because of its already known protective action on protein degradation and atrophy. We will discuss herein the possible actions in healthy subjects and in athletes.

Skeletal muscle strength production is determined by the amount of muscle mass (specifically muscle cross-sectional area, CSA) and intrinsic factors regulating metabolism and calcium handling within myofiber. Results from a meta-analysis of nine studies [53] indicate that HMB supplementation promotes gain in muscle mass and strength generation with resistance training. On the other hand, Rowlands and Thomson [54], in a more recent meta-analysis of eleven studies, show that HMB supplementation induces trivial gain in muscle mass and small gain in strength generation in untrained and trivial gain in trained lifters. Despite, the lack of effect of HMB supplementation on muscle mass within resistance training in healthy untrained subjects and athletes, which can be explained by many factors unconsidered in meta-analysis [55] regarding the study design—such as use of similar groups for comparison, small samples, non-periodized training protocols, lack of specificity between training and testing conditions and indirect measurement of muscle size—the gain in muscle strength generation in untrained lifters could be due to improvement of intrinsic capacity of tension production within myofibers. In a recent publication [40] of our group, HMB supplementation increased ATP and glycogen content and citrate synthase activity in rat skeletal muscle. Increase in specific force generation (obtained when the tension is normalized by muscle mass or muscle CSA), resistance to acute fatigue without changes in contraction and relaxation velocities in electrostimulated rat skeletal muscle were also observed. It has been demonstrated that leucine stimulates mitochondrial biogenesis genes SIRT-1, PGC-1α and NRF-1 as well mitochondrial mass (by 30%) and oxygen consumption in C2C12 myotubes. HMB is endogenously produced in consequence of leucine oxidation. Approximately 5% of KIC generated by leucine transamination is converted to HMB by α-keto-acid dioxygenase. In hypothesis, the increased HMB content in myofibers can inhibits α-keto-acid dioxygenase potentializing the effect of leucine on oxidative metabolism through conversion of KIC in isovaleryl-coenzyme A, accelerating the citric acid cycle, sparing glycogen, and improving specific muscle force. Thus, in an untrained lifter the effect of HMB supplementation in strength generation can be observed, although in trained lifters (who already have metabolic improvement due to the resistance training) there is a lack of this effect when no additional gain in muscle mass is observed.

CONCLUSION

According to the evidence discussed above, HMB supplementation induces metabolic and molecular effects which are associated with improvement in skeletal muscle contractile function. These effects can be useful in catabolic conditions and in untrained subjects subjected to resistance training.

References

[1] Hider RC, Fern EB, London DR. Relationship between intracellular amino acids and protein synthesis in the extensor digitorum longus muscle of rats. Biochem J 1969;114:171–8.

[2] Nissen SL, Abumrad NN. Nutritional role of the leucine metabolite β-hydroxy β-methylbutyrate (HMβ). J Nutr Biochem 1997;8:300–11.

[3] Tischler ME, Desautels M, Goldberg AL. Does leucine, leucyl-tRNA, or some metabolite of leucine regulate protein synthesis and degradation in skeletal and cardiac muscle? J Biol Chem 1982;257:1613–21.

[4] Holecek M, Muthny T, Kovarik M, Sispera L. Effect of beta-hydroxy-beta-methylbutyrate (HMB) on protein metabolism in whole body and in selected tissues. Food Chem Toxicol 2009;47:255–9.

[5] Nissen S, Sharp R, Rathmacher JA, Rice J, Fuller Jr JC, Connely AS, et al. The effect of leucine metabolite β-hydroxy β-methylbutyrate (HMβ) on muscle metabolism during resistance-exercise training. J Appl Physiol 1996;81:2095–104.

[6] Clark RH, Feleke G, Din M, Yasmin T, Singh G, Khan FA, et al. Nutritional treatment for acquired immunodeficiency virus-associated wasting using beta-hydroxy beta-methylbutyrate, glutamine, and arginine: a randomized, double-blind, placebo-controlled study. J Parenter Enteral Nutr 2000;24:133–9.

[7] Fuller Jr JC, Baier S, Flakoll P, Nissen SL, Abumrad NN, Rathmacher JA. Vitamin D status affects strength gains in older adults supplemented with a combination of β-hydroxy-β-methylbutyrate, arginine, and lysine: a cohort study. J Parenter Enteral Nutr 2011;35:757–62.

[8] Kraemer WJ, Hatfield DL, Volek JS, Fragala MS, Vingren JL, Anderson JM, et al. Effects of amino acids supplement on physiological adaptations to resistance training. Med Sci Sports Exerc 2009;41:1111–21.

[9] Wilson JM, Kim JS, Lee SR, Rathmacher JA, Dalmau B, Kingsley JD, et al. Acute and timing effects of beta-hydroxy-beta-methylbutyrate (HMB) on indirect markers of skeletal muscle damage. Nutr Metab (Lond) 2009;4:6.

[10] Van Koevering M, Nissen S. Oxidation of leucine and alpha-ketoisocaproate to beta-hydroxy-beta-methylbutyrate in vivo. Am J Physiol 1992;262:27–31.

[11] Block KP, Buse MG. Glucocorticoid regulation of muscle branched-chain amino acid metabolism. Med Sci Sports Exerc 1990;22:316–24.

[12] Sabourin PJ, Bieber LL. Formation beta-hydroxyisovalerate by an a-ketoisocaproate oxygenase in human liver. Metabolism 1983;32:160–4.

[13] Rudney H. The biosynthesis of beta-hydroxy-beta-methylglutaric acid. J Chem Biol 1957;227:363–77.

[14] Vukovich MD, Slater G, Macchi MB, Turner MJ, Fallon K, Boston T, et al. Beta-hydroxy-beta-methylbutyrate (HMB) kinetics and the influence of glucose ingestion in humans. J Nutr Biochem 2001;12:631–9.

[15] Hagg SA, Morse EL, Adibi SA. Effect of exercise on rates of oxidation turnover, and plasma clearance of leucine in human subjects. Am J Physiol 1982;242:407–10.

[16] Wolfe RR, Goodenough RD, Wolfe MH, Royle GT, Nadel ER. Isotopic analysis of leucine and urea metabolism in exercising humans. J Appl Physiol 1982;52:458–66.

[17] Vukovich MD, Dreifort GD. Effect of beta-hydroxy beta-methylbutyrate on the onset of blood lactate accumulation and V(O)(2) peak in endurance-trained cyclists. J Strength Cond Res 2001;15:491–7.

[18] Gallagher PM, Carrithers JA, Godard MP, Schulze KE, Trappe SW. Beta-hydroxy-beta-methylbutyrate ingestion, part I: effects on strength and fat free mass. Med Sci Sports Exerc 2000;32:2109–15.

[19] Fuller Jr JC, Sharp RL, Angus HF, Baier SM, Rathmacher JA. Free acid gel form of β-hydroxy-β-methylbutyrate (HMB) improves HMB clearance from plasma in human subjects compared with the calcium HMB salt. Br J Nutr 2011;105:367–72.

[20] Gallagher PM, Carrithers JA, Godard MP, Schulze KE, Trappe SW. Beta-hydroxy-beta-methylbutyrate ingestion, part II: effects on hematology, hepatic and renal function. Med Sci Sports Exerc 2000;32:2116–9.

[21] Nissen S, Sharp RL, Panton L, Vukovich M, Trappe S, Fuller Jr JC. beta-hydroxy-beta-methylbutyrate (HMB) supplementation in humans is safe and may decrease cardiovascular risk factors. J Nutr 2000;130:1937–45.

[22] Jówko E, Ostaszewski P, Jank M, Sacharuk J, Zieniewicz A, Wilczak J, et al. Creatine and beta-hydroxy-beta-methylbutyrate (HMB) additively increase lean body mass and muscle strength during a weight-training program. Nutrition 2001;17:558–66.

[23] Panton LB, Rathmacher JA, Baier S, Nissen S. Nutritional supplementation of the leucine metabolite beta-hydroxy-beta-methylbutyrate (hmb) during resistance training. Nutrition 2000;16:734–9.

[24] Vukovich MD, Stubbs NB, Bohlken RM. Body composition in 70-year-old adults responds to dietary beta-hydroxy-beta-methylbutyrate similarly to that of young adults. J Nutr 2001;131:2049–52.

[25] Nissen S, Panton L, Fuller J, Rice D, Ray M, Sharp R. Effect of feeding β-hydroxy-β-methylbutyrate (HMB) on body composition and strength of women [abstract]. FASEB J 1997;11:A150.

[26] Paddon-Jones D, Keech A, Jenkins D. Short-term beta-hydroxy-beta-methylbutyrate supplementation does not reduce symptoms of eccentric muscle damage. Int J Sport Nutr Exerc Metab 2001;11:442–50.

[27] Wilson JM, Kim JS, Lee SR, Rathmacher JA, Dalmau B, Kingsley JD, et al. Acute and timing effects of beta-hydroxy-beta-methylbutyrate (HMB) on indirect markers of skeletal muscle damage. Nutr Metab (Lond) 2009;6:1–8.

[28] Thomson JS, Watson PE, Rowlands DS. Effects of nine weeks of beta-hydroxy-beta-methylbutyrate supplementation on strength and body composition in resistance trained men. J Strength Cond Res 2009;23:827–35.

[29] Knitter AE, Panton L, Rathmacher JA, Petersen A, Sharp R. Effects of β-hydroxy-β-methylbutyrate on muscle damage after a prolonged run. J Appl Physiol 2000;89:1340–4.

[30] Ransone J, Neighbors K, Lefavi R, Chromiak J. The effect of beta-hydroxy beta-methylbutyrate on muscular strength and body composition in collegiate football players. J Strength Cond Res 2003;17:34–9.

[31] Kreider RB, Ferreira M, Greenwod M, Wilson M, Grindstaff P, Plisk S, et al. Effects of calcium b-HMB supplementation during training on markers of catabolism, body composition, strength and sprint performance. J Exerc Physiol Online 2000;3:48–59.

[32] Slater G, Jenkins D, Logan P, Lee H, Vukovich M, Rathmacher JA, et al. Beta-hydroxy-beta-methylbutyrate (HMB) supplementation does not affect changes in strength or body composition during resistance training in trained men. Int J Sport Nutr Exerc Metab 2001;11:384–96.

[33] Portal S, Zadik Z, Rabinowitz J, Pilz-Burstein R, Adler-Portal D, Meckel Y, et al. The effect of HMB supplementation on body composition, fitness, hormonal and inflammatory mediators in elite adolescent volleyball players: a prospective randomized, double-blind, placebo-controlled study. Eur J Appl Physiol 2011;111:2261–9.

[34] Hoffman JR, Cooper J, Wendell M, Im J, Kang J. Effects of beta-hydroxy beta-methylbutyrate on power performance and indices of muscle damage and stress during high-intensity training. J Strength Cond Res 2004;18:747–52.

[35] Flakoll P, Sharp R, Baier S, Levenhagen D, Carr C, Nissen S. Effect of betahydroxy betamethylbutyrate, arginine, and lysine supplementation on strength, functionality, body composition, and protein metabolism in elderly women. Nutrition 2004;20:445–51.

[36] Van Someren KA, Edwards AJ, Howatson G. Supplementation with beta-hydroxy-beta-methylbutyrate (HMB) and alpha-ketoisocaproic acid (KIC) reduce signs and symptoms of exercise induced muscle damage in man. Int J Sport Nutr Exerc Metab 2005;15:413–24.

[37] Kreider RB, Ferreira M, Wilson M, Almada L. Effects of calcium beta-hydroxy-beta-methylbutyrate (HMB) supplementation during resistance-training on markers of catabolism, body composition and strength. Int J Sports Med 1999;20:503–9.

[38] O'Connor DM, Crowe MJ. Effects of six weeks of betahydroxy-beta-methylbutyrate (HMB) and HMB/creatine supplementation on strength, power, and anthropometry of highly trained athletes. J Strength Cond Res 2007;21:19–423.

[39] Lamboley CR, Royer D, Dionne IJ. Effects of beta-hydroxybeta-methylbutyrate on aerobic-performance components and body composition in college students. Int J Sport Nutr Exerc Metab 2007;17:56–69.

[40] Pinheiro CH, Gerlinger-Romero F, Guimarães-Ferreira L, de Souza Jr AL, Vitzel KF, Nachbar RT, et al. Metabolic and functional effects of beta-hydroxy-beta-methylbutyrate (HMB) supplementation in skeletal muscle. Eur J Appl Physiol 2012;112:2531–7.

[41] Soares JMC, Póvoas S, Neuparth MJ, Duarte JA. The effects of beta-hydroxy-beta-methylbutyrate (HMB) on muscle atrophy induced by immobilization. Med Sci Sports Exerc 2001;33:S140.

[42] Gerlinger-Romero F, Guimarães-Ferreira L, Giannocco G, Nunes MT. Chronic supplementation of beta-hydroxy-beta methylbutyrate (HMβ) increases the activity of the GH/IGF-I axis and induces hyperinsulinemia in rats. Growth Horm IGF Res 2011;21:57–62.

[43] Pierno S, De Luca A, Tricarico D, Roselli A, Natuzzi F, Ferrannini E, et al. Potential risk of myopathy by HMG-CoA reductase inhibitors: a comparison of pravastatin and simvastatin effects on membrane electrical properties of rat skeletal muscle fibers. J Pharmacol Exp Ther 1995;275:1490–6.

[44] Cheng W, Phillips B, Abumrad N. Beta-hydroxy-beta-methylbutyrate increases fatty acid oxidation by muscle cells. FASEB J 1997;11:381.

[45] Cheng W, Phillips B, Abumrad N. Effect of HMB on fuel utilization, membrane stability and creatine kinase content of cultured muscle cells. FASEB J 1998;12:950.

[46] Carrel AL, Allen DB. Effects of growth hormone on adipose tissue. J Pediatr Endocrinol Metab 2000;13(Suppl. 2):1003–9.

[47] Kornasio R, Riederer I, Butler-Browne G, Mouly V, Uni Z, Halevy O. beta-hydroxy-beta-methylbutyrate (HMB) stimulates

myogenic cell proliferation, differentiation and survival via the MAPK/ERK and PI3K/Akt pathways. Biochim Biophys Acta 2009;23:836–46.
[48] Glass DJ. Skeletal muscle hypertrophy and atrophy signaling pathways. Int J Biochem Cell Biol 2005;37:1974–84.
[49] Ciechanover A. Intracellular protein degradation: from a vague Idea thru the lysosome and the ubiquitin-proteasome system and onto human diseases and drug targeting. Cell Death Differ 2005;12:1178–11790.
[50] Smith HJ, Mukerji P, Tisdale MJ. Attenuation of proteasome-induced proteolysis in skeletal muscle by {beta}-hydroxy-{beta}-methylbutyrate in cancer-induced muscle loss. Cancer Res 2005;65:277–83.
[51] Fiorotto ML, Schwartz RJ, Delaughter MC. Persistent IGF-1 overexpression in skeletal muscle transiently enhances DNA accretion and growth. FASEB J 2003;17:59–60.
[52] Dehoux M, Deneden RV, Pasko N, Lause P, Verniers J, Underwood L, et al. Role of the insulin-like growth factor I decline in the induction of atrogin-1/MAFbx during fasting and diabetes. Endocrinology 2004;145:4806–12.
[53] Nissen SL, Sharp RL. Effect of dietary supplements on lean mass and strength gains with resistance exercise: a meta-analysis. J Appl Physiol 2003;94:651–9.
[54] Rowlands DS, Thomson JS. Effects of beta-hydroxy-beta-methylbutyrate supplementation during resistance training on strength, body composition, and muscle damage in trained and untrained young men: a meta-analysis. J Strength Cond Res 2009;23:836–46.
[55] Wilson GJ, Wilson JM, Manninen AH. Effects of beta-hydroxy-beta-methylbutyrate (HMB) on exercise performance and body composition across varying levels of age, sex, and training experience: a review. Nutr Metab (Lond) 2008;5:1–17.

CHAPTER

48

Role of Astaxanthin in Sports Nutrition

Bob Capelli, Usha Jenkins and Gerald R. Cysewski
Cyanotech Corporation, Kailua-Kona, HI, USA

INTRODUCTION

Astaxanthin is one of over 700 known natural carotenoids and is the most abundant carotenoid in the marine environment. It is what colors salmon flesh, lobster shells and shrimp red. Only natural astaxanthin derived from the microalgae *Haematococcus* and the yeast *Xanthophyllomyces* (formerly *Phaffia*) is allowed for sale for direct human consumption as a supplement. Natural astaxanthin was released in the human nutrition market in the late 1990s after the US Food & Drug Administration (FDA) reviewed safety parameters and allowed it use for human consumption. A new dietary ingredient submitted to the FDA in 2011 allows for consumption of natural astaxanthin at 12 mg/day. Based on research of its strong antioxidant capacity, astaxanthin was originally marketed as "The World's Strongest Natural Antioxidant" [1]. Further research has also shown astaxanthin never becomes a pro-oxidant [2] and has anti-inflammatory properties and suppresses a number of different inflammatory pathways [3–5]. Additional research has validated astaxanthin's ability to increase strength and endurance in human and animal populations. New *in vitro* studies demonstrate the ability of astaxanthin to combat oxidation, which is of particular interest to athletes who generate excessive levels of free radicals and wish to lengthen workouts and improve recovery time. Furthermore, astaxanthin is emerging as a safe and natural alternative to over-the-counter and prescription anti-inflammatories to help with overuse injuries as well as joint, tendon, and muscle pain [1]. In this paper, we will first analyze chronologically the published literature on astaxanthin's ability to enhance sports performance and then examine its antioxidant and anti-inflammatory properties and discuss how these properties can benefit athletes.

BENEFITS OF ASTAXANTHIN FOR ATHLETES

Part 1: Sports Performance

It has long been known that strenuous exercise causes oxidative stress, and to combat exercise-induced oxidative stress, consumption of antioxidant supplements have been recommended [6–8]. Antioxidant supplements that have been recommended include vitamins E and C, coenzyme Q10, beta-carotene, and isoflavonoids. As discussed in detail below, astaxanthin has been shown to be the strongest natural antioxidant known. We can therefore expect that astaxanthin would be an important supplement for sports nutrition.

In the animal kingdom, the animal that has the greatest amount of astaxanthin in its body is the salmon. Salmon cannot synthesize astaxanthin, but rather obtain astaxanthin through their diet with the primary source of astaxanthin in the marine environment being microalgae. Astaxanthin concentrates to the highest levels in the muscles of the salmon. It has been theorized that it is astaxanthin that gives salmon the strength and endurance to swim upstream for weeks. In salmon, as in other species (including humans), high levels of free radicals are generated when exerting force [9]. The presence of a strong antioxidant such as natural astaxanthin in the fish's muscle tissue is hypothesized to mitigate or eliminate the excess free radicals generated by this extreme exertion of swimming upstream for weeks at a time. It appears that this same effect of astaxanthin concentrating in the muscles may occur in humans. Indeed, studies demonstrating astaxanthin's value in sports nutrition include both human and animal studies as discussed below.

Nutrition and Enhanced Sports Performance.
DOI: http://dx.doi.org/10.1016/B978-0-12-396454-0.00048-5

The first indication of astaxanthin's ability to enhance muscle function was in US Patent 6,245,818 entitled "Medicament for Improvement of Duration of Muscle Function or Treatment of Muscle Disorders of Diseases" granted in 1998 [10]. The patent describes a study on trotting-horses suffering from exertional rhabdomyolysis, a severe tightening of muscles, which necessitates a halt in training. Horses receiving 100 mg (approximately 0.2 mg/kg) per day of natural astaxanthin in the form of *Haematococcus* algal meal for 2 weeks were cured of exertional rhabdomyolysis and remained cured so long as consumption of astaxanthin continued. The patent also describes a double-blind, placebo-controlled human study with 40 male college students. The students were equally divided between the placebo group and the treatment group, with the treatment group receiving 4 mg/day of astaxanthin. Strength/endurance was estimated by the maximum number of knee bends a subject could perform in a Smith machine with a 40 kg load under standardized conditions and was measured at the beginning of the study and at the end of the 6-month treatment period. Strength/endurance of the treatment group consuming astaxanthin increased by 61.74% compared to an increase of only 23.78% for the placebo group. In other words, the strength/endurance of those taking astaxanthin increased almost three times faster than the placebo group.

The first published study on astaxanthin's ability to affect muscle function was in 2002. Research in Japan examined sports performance benefits of oral consumption of astaxanthin [11]. This double-blind, placebo-controlled experiment examined muscle fatigue in 16 adult males. An equal number of subjects were in the placebo group and the treatment group with the treatment group consuming 6 mg/day of natural astaxanthin per day for 4 weeks. The muscle fatigue study involved measuring the level of serum lactic acid and creatine kinase 2 min after subjects ran 1200 m. The level of creatine kinase showed a decreasing tendency and level of lactic acid decreased by a significant 28.6% in the treatment group compared to the placebo group. Creatine kinase and lactic acid are recognized as indicators of fatigue, and the study suggests that supplementation with astaxanthin can reduce muscle fatigue and improve sports performance.

An animal study also conducted in Japan and published in 2003 supports the results of the above study [12]. In this study, mice ran on a treadmill until they were exhausted. The mice were separated into three different groups: Group A was the control group that was not exercised at all and was not given astaxanthin. Group B was exercised until exhaustion, but was not given astaxanthin either. Group C was exercised similar to Group B, but their diets were supplemented with natural astaxanthin. After the exhaustive exercise, the mice were sacrificed and examined. Their heart and calf muscles were checked for oxidative damage. The researchers found that various markers of oxidative damage were reduced in both the heart muscles and the calf muscles of Group C. They found a corresponding reduction of oxidation in the plasma as well. The cell membranes in the treatment group's calf and heart muscles suffered significantly less peroxidation damage. Also, damage to DNA and proteins was significantly reduced in the mice supplemented with astaxanthin. In addition, a significant decrease in inflammation damage indicators and serum creatine kinase resulted. Muscle inflammation was found to decrease by more than 50% in the mice fed with astaxanthin. Results indicated that astaxanthin was absorbed and transported into the heart and skeletal muscle, although most carotenoids concentrate mainly in the liver and do not normally concentrate to peripheral tissues. In conclusion, astaxanthin attenuated exercise-induced damage by scavenging reactive oxygen species and by decreasing inflammation. This study demonstrates that natural astaxanthin is available in the two very different areas in rodents' bodies following oral ingestion unlike other carotenoids. This is a unique and important difference between natural astaxanthin and other antioxidants and carotenoids which are generally poorly distributed in the body.

A second human study in Japan examined the effect of astaxanthin consumption on exercise-induced physiological changes [13]. Eighteen healthy male subjects with an average age of 35.8 years (±4.51) were involved in this double-blind, placebo-controlled, cross-over study. Subjects took either 5 mg/day of astaxanthin or a placebo for 2 weeks, and exercise stress tests were conducted before and after the ingestion period. The exercise stress tests involved running on a treadmill at intensities of 30%, 50%, and 70% of maximum heart rate. Expired gas analysis and blood biochemical parameters were measured as well as the activity of the sympathetic and parasympathetic nervous systems. The study found that consumption of astaxanthin lead to a decrease of sympathetic nervous activity and an increase of parasympathetic nervous activity, a significant decrease in LDL cholesterol after exercise, and a decrease in respiratory quotient. The authors concluded that astaxanthin consumption may enhance lipid metabolism and improve the efficiency of energy metabolism during exercise.

A 2006 study done with mice was designed to measure the effects of astaxanthin on endurance [14]. This study took course over a 5-week period. Mice were divided into two groups and their endurance was tested by seeing how long they could swim until exhaustion. The mice fed with astaxanthin showed a significant increase in swimming time before exhaustion. Blood lactose levels were measured in both groups, and, as expected, the levels of the astaxanthin group were significantly lower than the control group. Another effect measured was that astaxanthin supplementation significantly reduced fat accumulation. This is the first mention of such an effect and further proof is needed before putting any credence into this potential benefit. The study's authors suggested that astaxanthin enabled the mitochondria to burn more fat: "These results suggest that improvement in swimming endurance by the administration of astaxanthin is caused by an increase in utilization of fatty acids as an energy source."

A subsequent mouse study backed up the results found in the human clinical trials discussed above as well as earlier mouse trials [15]. This study was set up to investigate the effects of astaxanthin supplementation on muscle lipid metabolism in mice that exercised heavily. The outcome was that mice that were fed astaxanthin for 4 weeks and then exercised (i) had better fat utilization, (ii) had longer running time until exhaustion, (iii) had reduced fat tissue, and (iv) had better muscle lipid metabolism. The researchers concluded that astaxanthin supplementation led to improvement in endurance.

The first human study on astaxanthin and endurance/strength was published in 2008 [16]. The study was done with healthy male students between the ages of 17 and 19. The researcher used 40 men with an equal number (twenty) in the treatment group and in the placebo group. Each subject took one 4-mg capsule per day with a meal for 6 months. The subjects' strength was measured at the beginning of the experiment, halfway through (after 3 months), and again at the end of the experiment (after 6 months) by counting the maximum number of knee bends to a 90° angle that each subject could do. This was controlled by an adjustable stool in a "Smith machine." (The Smith machine is specifically designed for measuring strength and endurance in clinical trials.) The subjects were properly warmed up for a set time and in a similar manner before each strength measurement. There was a significant difference in strength/endurance between the two experimental groups even though the number of knee bends increased in both groups. The placebo group on average could do 9.0 more knee bends and the group consuming natural astaxanthin on average could do 27.0 more knee bends.

The results showed that, in 6 months, the students taking natural astaxanthin improved their strength and endurance by 62%. This was achieved at the relatively low dose of only 4 mg/day. The students taking a placebo increased their strength by 22%, which is normal

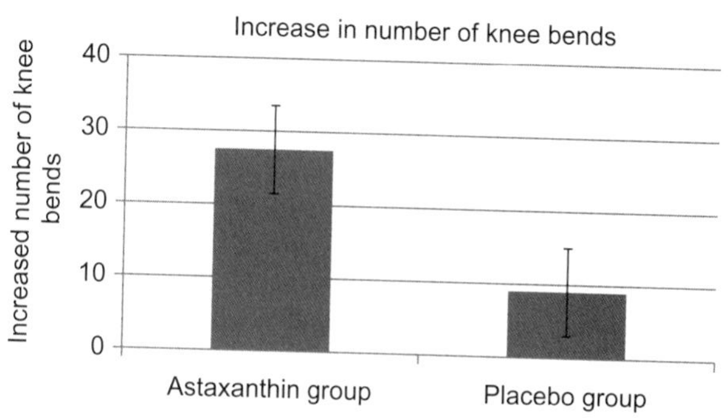

FIGURE 48.1 **Increase in number of knee bends after 6 months with (astaxanthin group) or without (placebo group) supplementation with 4.0 mg/day of astaxanthin per day [16].**

for people in this age group over a 6-month period, as they were generally involved in sports and physical activity. Results indicated that the treatment group's endurance increased almost three times more than the placebo group (Figure 48.1).

The authors concluded that the significant improvement in endurance cannot be explained by improved fitness (step-up test) or improved lactic acid tolerance (Wingate test). There was no significant increase in body weight, hence increased muscle mass does not account for the endurance improvement. It appears that increase in strength and endurance is caused by astaxanthin supplementation. This study demonstrated that astaxanthin supplementation has a positive effect on physical performance, and is supported by earlier animal trials in mice showing increased swimming time before exhaustion, and also human research indicating that biomarkers of muscle fatigue decrease after exercise due to astaxanthin supplementation.

It is hypothesized that astaxanthin helps with endurance via mitochondrial organelles, which are numerous in muscle tissue and produce up to 95% of our body's energy by burning fatty acids and other substances. But this energy that is produced also generates highly reactive free radicals. The free radicals in turn can damage cell membranes and lead to mitochondrial dysfunction. An *in vitro* study published in 2009 demonstrated that astaxanthin protects mitochondrial redox state and the functional integrity of mitochondria against oxidative stress [17].

In a recent clinical study funded by Gatorade, the leading sports drink company, competitive cyclists were supplemented with a placebo or 4 mg of natural astaxanthin each day for 4 weeks [18]. In a 20-km (approximately 12.5 miles) cycling time trial, the performance of the subjects taking astaxanthin significantly improved, while the subjects taking placebo showed no improvement. Natural astaxanthin made these competitive cyclists on average 5% faster in only 28 days. Also, the cyclists taking astaxanthin

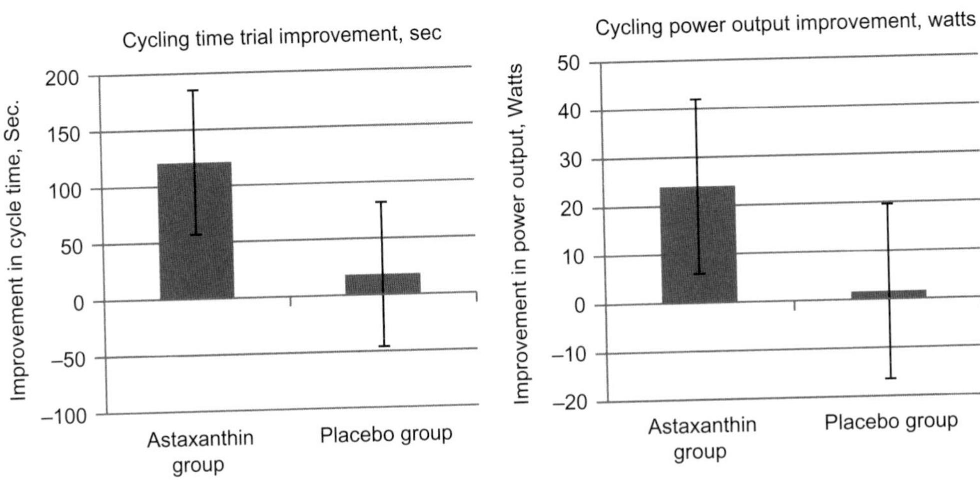

FIGURE 48.2 **Improved cycling performance after 28 days with (astaxanthin group) or without (placebo group) supplementation with 4.0 mg/day of astaxanthin per day [18].**

demonstrated significant improvement in their power output, which increased by 15% on average over the same 28-day period (Figure 48.2). Although this study did not establish a mechanism of action for astaxanthin's improvement in performance, strength, and endurance for these cyclists, the fact that it is the second human clinical trial showing that natural astaxanthin can improve strength and endurance is of particular significance. Of particular interest to competitive athletes is the 5% improvement in speed in less than a month. In many sports where fractions of a second separate gold medal winners from silver, bronze, and being off the podium, a potential 5% improvement is significant.

A paper presented at the Seventh EFSMA—European Congress of Sports Medicine in 2011 documented the benefits of astaxanthin supplementation for young soccer players [19]. A double-blind, placebo-controlled study was conducted with 60 healthy young (mean age 17.7 years) split into equal placebo and treatment groups. The treatment group received 4 mg astaxanthin per day and the study ran for 3 months. Performance parameters were measured before and at the end of the treatment period. A significant improvement in maximal running speed and running time to exhaustion was found in the treatment group consuming astaxanthin. Also, the placebo group had a significant increase in the blood level of creatine kinase (an indicator of muscle fatigue) after a training session which was not observed in the treatment group. This study supports previous studies on the potential benefits of astaxanthin as the authors concluded: "Astaxanthin supplementation could improve endurance that may lead to better sports performance."

In contrast to the numerous studies cited above demonstrating astaxanthin's potential value in sports nutrition, a study reported in 2012 did not yield any positive results [20]. This double-blind, parallel-design study used 32 well-trained cyclists or triathletes supplemented with 20 mg astaxanthin per day or with a placebo for 4 weeks. Before and after the supplementation period subjects performed 60 min of submaximal exercise followed by a cycling time trial lasting approximately 1 h. Whole-body fat oxidation rates during the submaximal exercise did not differ between the treatment and the placebo group and no improvement in time trial performance was observed in either group. The lack of any positive effect of astaxanthin being demonstrated in this study cannot currently be explained, especially in light of the preponderance of positive results cited in all previous studies. One major difference in this study is that subjects consumed rather high amounts of astaxanthin (20 mg/day) compared to other studies in which 4–6 mg astaxanthin per day was consumed.

In summary, four out of five human clinical trials validate and three pre-clinical animal trials corroborate strength and endurance benefits for athletes who use natural astaxanthin. Highlights of these benefits include:

- 5% time improvement for competitive cyclists using 4 mg natural astaxanthin per day for 28 days.
- 15% power output improvement for the same competitive cyclists using 4 mg natural astaxanthin per day for 28 days.
- 62% more deep knee bends by 17- to 19-year-old men taking 4 mg natural astaxanthin per day for 6 months.
- 28.6% decrease in serum lactic acid in 20-year-old men taking 6 mg natural astaxanthin per day for 6 weeks when measured after a 1200-m run.

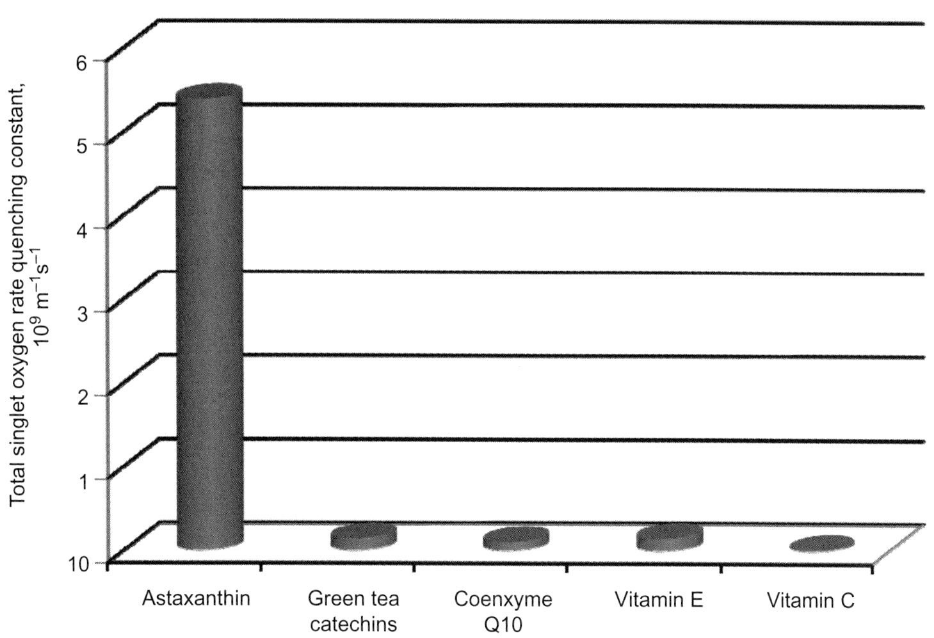

FIGURE 48.3 Total singlet oxygen-quenching activities of some antioxidants [21].

- Human clinical results above are substantiated in animal trials.
- Additional potential benefits in animal trials include:
  - reduced fat accumulation,
  - improved muscle lipid metabolism,
  - reduced markers for oxidative damage in heart and calf muscles after heavy exercise,
  - reduced oxidation in the plasma,
  - reduced peroxidation damage to DNA and proteins,
  - improved modulation of serum creatine kinase.

Part 2: A Diverse and Unique Antioxidant

It is well documented that athletes, particularly endurance athletes and anyone doing heavy workouts, generate high levels of free radicals. Some potential benefits of neutralizing free radicals are:

- faster recovery,
- increased length of workouts,
- reduced muscle soreness.
- as cited above, increased endurance [1].

It is well documented that astaxanthin is an extremely powerful antioxidant. A number of *in vitro* studies comparing it to many other antioxidants such as Vitamin C, Vitamin E, Pycnogenol®, green tea catechins, CoQ10, alpha-lipoic acid, as well as other carotenoids (beta-carotene, lutein, and lycopene) generally show astaxanthin to be, at minimum, an order of magnitude more powerful as an antioxidant. This is true regardless of the type of antioxidant test—whether it be free radical elimination or singlet oxygen quenching. For example, astaxanthin proved to be 800× stronger than CoQ10; 550× stronger than both green tea catechins and Vitamin E; and 6000× stronger than Vitamin C in eliminating singlet oxygen [21,22] (Figure 48.3).

Astaxanthin's strength as an antioxidant is only one factor in why it is different from other antioxidants. In addition astaxanthin can:

- span the cell membrane and bring antioxidant protection to both the fat-soluble and the water-soluble parts of the cell [23],
- target both cardiac and skeletal muscles [12],
- cross the blood–brain and blood–retinal barriers and bring antioxidant protection to the eyes and brain [22],
- *never* become a pro-oxidant [2].

The combination of strong antioxidant activity; the ability to distribute throughout the body and protect organs vital to athletic performance like the heart, eyes, and brain [12,22]; the ability to bond with muscle tissue; and the fact that it can never turn into a pro-oxidant and potentially *cause* oxidation—as is the case with Vitamins C and E, zinc, and carotenoids like beta-carotene, lycopene, and zeaxanthin [2,24]—demonstrate astaxanthin's potential in sports nutrition.

Part 3: Anti-Inflammatory Activity of Astaxanthin

In addition to being a strong and unique antioxidant, astaxanthin also possesses anti-inflammatory activity to help athletes with joint, tendon, and muscle soreness. There is a distinguishing factor between astaxanthin and other anti-inflammatories—natural astaxanthin has a long history of safe use without side effects or contraindications.

When one considers current anti-inflammatory products on the market, most have serious side effects. Aspirin has anti-inflammatory effects, but prolonged use can cause stomach bleeding and ulcers. Nonsteroidal anti-inflammatory drugs (NSAIDs) such as acetaminophen (Tylenol®) can cause liver damage. Prescription anti-inflammatory drugs such as Vioxx® and Celebrex® can cause heart problems. After many years of consumer use and extensive safety studies, natural astaxanthin has never been documented to have any side effect or contraindication. It is a safe alternative in the high-risk category of anti-inflammatories.

One possible reason why astaxanthin is safe compared to alternative anti-inflammatories is that it works on several different inflammatory pathways, but at lower activity levels. Astaxanthin is documented to suppress the production of inflammation-causing agents in our bodies including tumor necrosis factor-alpha, prostaglandin E-2, nitric oxide, interleukin 1-B, cox-1 enzyme, and cox-2 enzyme [3–5,25,26].

The most recent of these five studies citing astaxanthin's multiple pathways in combating inflammation found activity on all six of the pathways listed above, and noted that astaxanthin's anti-inflammatory activity was very efficient [26]. Working in a gentler manner on six different causes of inflammation is much safer than working intensely on one cause (as is the case with Vioxx and Celebrex which work intensely on the cox-2 enzyme). Professor of Medicine and Neurology, Greg Cole, PhD, from the University of California at Los Angeles explains, "While anti-inflammatory drugs usually block a single target molecule and reduce its activity dramatically, natural anti-inflammatories gently tweak a broader range of inflammatory compounds. You'll get greater safety and efficacy reducing five inflammatory mediators by 30% than by reducing one by 100%" [27].

CONCLUSIONS

Natural astaxanthin is documented to increase strength and endurance in five out of six human clinical studies as well as four supporting animal trials. Competitive cyclists taking 4 mg/day of natural astaxanthin for 28 days improved cycling times by 5% and power output by 15% on average. In earlier research, young men taking 4 mg/day of natural astaxanthin for 6 months performed 62% more deep knee bends. In addition to improving strength and endurance, astaxanthin has other beneficial properties for athletes due to its strong and unique antioxidant activity and its safe and natural anti-inflammatory effects. As an antioxidant, its reach extends throughout the body to all organs and muscle tissues, combating excessive free radical production by athletes, and is stronger than other common antioxidants. As an anti-inflammatory, it reduces six different inflammatory pathways and provides a safe alternative to over-the-counter and prescription anti-inflammatories, most of which have serious side effects. In conclusion, natural astaxanthin has a strong potential in sports nutrition.

References

[1] Capelli B, Cysewski G. The World's Best Kept Health Secret: Natural Astaxanthin. 2012; [ISBN-13: 978-0-9792353-0-6].
[2] Martin H, Jager C, Ruck C, Schmidt M. Anti- and pro-oxidant properties of carotenoids. J Pract Chem 1999;341 (3):302–8.
[3] Lee S, Bai S, Lee K, Namkoong S, Na H, Ha K, et al. Astaxanthin inhibits nitric oxide production and inflammatory gene expression by suppressing IkB kinase-dependent NFR-kB activation. Mol and Cells 2003;16(1):97–105.
[4] Ohgami K, Shiratori K, Kotake S, Nishida T, Mizuki N, Yazawa K, et al. Effects of Astaxanthin on lipopolysaccharide-induced inflammation in vitro and in vivo. Investig Ophthalmol Vis Sci 2003;44(6):2694–701.
[5] Choi SK, Park YS, Choi DK, Chang HI. Effects of Astaxanthin on the production of NO and the expression of COX-2 and iNOS in LPS-stimulated BV2 microglial cells. J Microbiol Biotech 2008;18(12):1990–6.
[6] Dekkers J, van Doornen L, Kemper H. The role of antioxidant vitamins and enzymes in the prevention of exercise-induced muscle damage. Sports Med 1996;21(3):213–38.
[7] Witt E, Reznick C, Viguie P, Starke-Reed P, Packer L. Exercise, oxidative damage and effects on antioxidant manipulation. J Nutr 1992;122:766–73.
[8] Goldfarb A. Nutritional antioxidants as therapeutic and preventative modalities in exercise-induced muscle damage. Can J Appl Physiol 1999;24(3):249–66.
[9] Hammouda O, Chtourou H, Chahed H, Ferchichi S, Kallel C, Chamari K, et al. Effect of short-term maximal exercise on biochemical markers of muscle damage, total antioxidant status, and homocysteine levels in football players. Asian J Sports Med 2012;3(4):239–46.
[10] Lignell A. Medicament for improvement of duration of muscle function or treatment of muscle disorders or diseases. 1998; [US Patent # 6,245,818].
[11] Sawaki K, Yoshigi H, Aoki K, Koikawa N, Azumane A, Keneko K, et al. Sports performance benefits from taking natural astaxanthin characterized by visual acuity and muscle fatigue improvements in humans. J Clin Ther Med 2002;18 (9):73–88.
[12] Aoi W, Naito Y, Sakuma K, Kuchide M, Tokuda H, Maoka T, et al. Astaxanthin limits exercise-induced skeletal and cardiac muscle damage in mice. Antioxid Redox Signal 2003;5 (1):139–44.
[13] Taeko T, Akira N. Effects of astaxanthin ingestion on exercise-induced physiological changes. Health Behav Sci 2004;3 (1):5–10.

[14] Ikeuchi M, Koyama T, Takahashi J, Yazawa K. Effects of astaxanthin supplementation on exercise-induced fatigue in mice. Biol Pharmaceut Bull 2006;29(10):2106–10.

[15] Aoi W, Naito Y, Takanami Y, Ishii T, Kawai Y, Akagiri S, et al. Astaxanthin improves muscle lipid metabolism in exercise via inhibitory effect of oxidative CPT I modification. Biochem Biophys Res Commun 2008;366(4):892–7.

[16] Malmsten C, Lignell A. Dietary supplementation with Astaxanthin-rich algal meal improves strength endurance- A double placebo controlled study on male students. Carotenoid Sci 2008;13:20–2.

[17] Wolf AM, Asoh S, Hiranuma H, Ohsawa I, Iio K, Satou A, et al. Astaxanthin protects mitochondrial redox state and functional integrity against oxidative stress. J Nutr Biochem 2009;21 (5):882–8.

[18] Earnest CP, Lupo M, White KM, Church TS. Effects of astaxanthin on cycling time trial performance. Int J Sports Med 2011;32(11):882–8.

[19] Radivojevic N, Dikic N, Baralic I, Djordjevic B, Vujic S, Andjelkovic M. Novel antioxidant supplementation in young soccer players. Seventh EFSMA-European Congress of Sports Medicine. October 26–29, Salzburg, Austria; 2011.

[20] Res PT, Cermak NM, Stinkens R, Tollakson TJ, Haenen GR, Bast A, et al. Astaxanthin supplementation does not augment fat use or improve endurance performance. Med Sci Sports Exerc 2012; [Epub ahead of print].

[21] Nishida Y, Yamashita E, Miki W. Comparison of Astaxanthin's singlet oxygen quenching activity with common fat and water soluble antioxidants. Results presented at the 21st Annual Meeting on Carotenoid Research held at Osaka, Japan on September 6 and 7, 2007.

[22] Shimidzu N, Goto M, Miki W. Carotenoids as singlet oxygen quenchers in marine organisms. Fish Sci 1996;62(1):134–7.

[23] Goto S, Kogure K, Abe K, Kimata Y, Kitahama K, Yamashita E, et al. Efficient radical trapping at the surface and inside the phospholipid membrane is responsible for the highly potent antiperoxidative activity of the carotenoid astaxanthin. Biochim Biophys Acta 2001;1512:251–8.

[24] Beutner S, Bloedron B, Frixel S, Blanco I, Hoffman T, Martin H, et al. Quantitative assessment of antioxidant properties of natural colorants and phytochemicals: carotenoids, flavonoids, phenols and indigoids. The role of B-carotene in antioxidant functions. J Sci Food Agric 2000;81:559–68.

[25] Saki S, Sugawara T, Matsubara K, Hirata T. Inhibitory effect of carotenoids on the degranulation of mast cells via suppression of antigen-induced aggregation of high affinity IgE receptors. J Biol Chem 2009;284(41):28172–9.

[26] Kishimoto Y, Tani M, Uto-Kondo H, Iizuak M, Saita E, Sone H, et al. Astaxanthin suppresses scavenger receptor expression and matrix metalloproteinase activity in macrophages. Eur J Nutr 2010;49(2):119–26.

[27] Cole G. Professor of Medicine and Neurology at UCLA, as reported to Anne Underwood, Newsweek Magazine, Special Summer Issue, August 2005. p. 26–28.

[28] Tso M, Lam T. Method of retarding and ameliorating central nervous system and eye damage. 1996; [US Patent #5527533].

CHAPTER

49

Ursolic Acid and Maslinic Acid

Novel Candidates for Counteracting Muscle Loss and Enhancing Muscle Growth

Raza Bashir, MSc.

Iovate Health Sciences International Inc., Oakville, ON, Canada

INTRODUCTION

The quest to increase lean body mass is widely pursued by professional and amateur athletes, bodybuilders and recreational weightlifters. Skeletal muscle hypertrophy is defined as an increase in muscle mass due to an increase in the size of pre-existing skeletal muscle fibers from the accumulation of new muscle proteins. Skeletal muscle atrophy on the other hand is associated with decreased contractile proteins. Thus, skeletal muscle mass is directly proportional to its protein content. However, skeletal muscle protein undergoes rapid turnover, and maintaining homeostasis is a fine balance between the rates of protein synthesis and protein degradation.

Many factors mediate the hypertrophic process, such as an increase in functional demand, as seen with resistance training. Diet and nutrition also plays a major role in exercise-induced muscle growth. Given the strong correlation between muscle cross-sectional area and muscular strength [1], maximizing muscle mass has important implications for professional and amateur athletes alike. Accordingly, the use of performance-enhancing supplements is increasing in popularity among athletes for increasing lean body mass and optimizing performance. Two novel, naturally occurring triterpenoid compounds: ursolic acid or maslinic acid are emerging as promising therapeutic agents with the potential to counteract muscle loss and enhance muscle growth.

URSOLIC ACID

Ursolic Acid Inhibits Muscle Atrophy

Ursolic acid is a water-insoluble pentacyclic triterpenoid (Figure 49.1) that is the major waxy component naturally occurring in apple peels [2]. It is also found in other edible plants such as *Ilea parguariensis* [3], *Urtica dioica* roots [4] and *Isodon excisus* [5]. Interestingly, it has been previously proposed to have therapeutic use in various conditions such as cancer [4–6] and diabetes [7,8]. More recently ursolic acid has been investigated for preventing muscle atrophy [9,10]. Research has demonstrated that skeletal muscle atrophy is driven by changes in skeletal muscle gene expression [11,12]. In order to identify ursolic acid as a natural compound to treat skeletal muscle atrophy, Kunkel et al. [9] determined the effects of two distinct atrophy-inducing stresses (fasting and spinal cord injury) on skeletal muscle mRNA levels in humans. This information was then used to generate mRNA expression signatures of muscle atrophy. These signatures were used to query the connectivity map for compounds with expression signatures negatively correlated with muscle atrophy. Applying these techniques, Kunkel et al. [9] singled out ursolic acid as a compound with a signature opposite to those of atrophy-inducing stresses and the most likely inhibitor of muscle atrophy among more than 1300 compounds. Muscle Ring Finger 1 (MuRF1) and Muscle Atrophy

Nutrition and Enhanced Sports Performance.
DOI: http://dx.doi.org/10.1016/B978-0-12-396454-0.00049-7

F-box (MAFbx), also known as atrogin-1, are two genes whose expression has been shown to be significantly elevated in multiple models of skeletal muscle atrophy [11,13]. The ubiquitin–proteasome pathway of skeletal muscle atrophy is well established, and both atrogin-1 and MuRF1 act on this by encoding E3 ubiquitin ligases [11]. As predicted by the connectivity map, acute ursolic acid treatment in fasted mice or those who had undergone surgical muscle denervation, did in fact reduce catabolic gene expression of atrogin-1 and MuRF1 mRNA levels in association with reduced muscle atrophy. Similarly, chronic ursolic acid treatment for 5 weeks in unstressed mice in the absence of any atrophy stimulus also reduced atrogin-1 and MuRF1 expression [9].

Ursolic Acid Enhances Muscle Hypertrophy and Exercise Capacity

Five weeks of ursolic acid treatment in unstressed mice reduced atrogin-1 and MuRF1 expression but also induced muscle hypertrophy [9]. Consistent with the hypertrophy findings, IGF-1 was found to be one of the most up-regulated muscle mRNA, which is known to be transcriptionally induced in hypertrophic muscle [14–16]. The IGF-1/IRS1/PI3K/Akt signaling pathway is fundamental in governing hypertrophy by increasing protein synthesis [17]. Furthermore, this pathway is also important for blocking protein degradation, and IGF-1 is known to repress atrogin-1 and MuRF1 mRNAs [18,19] as well as DDIT4L mRNA [18,19], which, after atrogin-1 mRNA, was the most repressed mRNA in ursolic acid-treated mice muscle. Therefore, 5 weeks of dietary ursolic acid altered skeletal muscle gene expression in a manner known to reduce atrophy and promote hypertrophy.

The anabolic effect of ursolic acid has also been observed in high-fat-fed mice [10]. In this mouse model of diet-induced obesity, ursolic acid treatment for 6 weeks was shown to increase Akt activity as well as downstream mRNAs that promote glucose utilization (hexokinase-II), blood vessel recruitment (Vegfa) and autocrine/paracrine IGF-1 signaling. Akt is downstream of the IGF-1 pathway and an important mediator of skeletal muscle hypertrophy [17]. As expected on the basis of previous studies on transgenic mice expressing elevated Akt specifically in skeletal muscle [20], ursolic acid increased grip strength, skeletal muscle weight, and the size of both fast and slow skeletal muscle fibers without altering the ratio of fast to slow fibers [10]. As anticipated with increased size of both slow (oxidative) and fast (glycolytic) muscle fibers, ursolic acid-treated mice ran significantly farther than control mice on an exercise treadmill. Furthermore, ursolic acid did not alter blood pressure and it induced a slight but significant reduction in resting heart rate [10]. Thus, in addition to stimulating skeletal muscle Akt activity and muscle hypertrophy, ursolic acid improved exercise capacity and lowered resting heart rate.

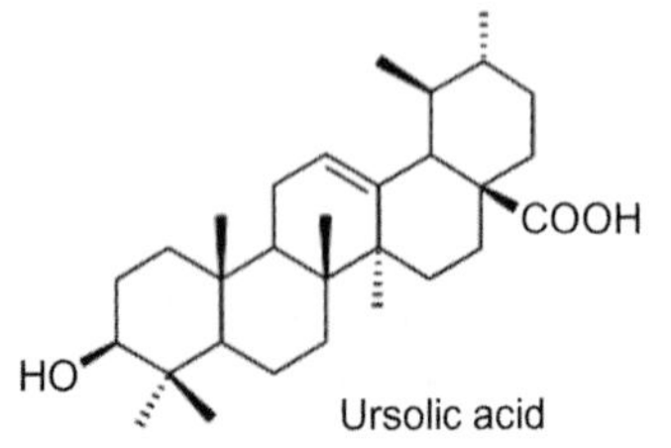

COOH
HO
Ursolic acid

FIGURE 49.1 **Chemical structure of ursolic acid.**

Ursolic Acid Decreases Body Fat and Improves Body Composition

Since ursolic acid increased Akt activity, and because muscle-specific increases in Akt activity are associated with reduced adiposity as a secondary consequence of muscle hypertrophy [20,21], it was speculated that ursolic acid Akt activation could increase energy expenditure, reduce adiposity and impart resistance to diet-induced obesity [20,21]. Indeed, ursolic acid has been shown to reduce total body weight, white fat, glucose intolerance and hepatic steatosis in high-fat-fed mice [22,23]. In non-obese mice, 7 weeks of dietary ursolic acid was shown to reduce weight of epididymal and retroperitoneal fat depots [9]. Ursolic acid reduced adipose weight by reducing adipocyte size. Correspondingly, muscle and fat weights were inversely related. In addition, plasma triglyceride and cholesterol were both significantly reduced without increasing plasma markers of hepatotoxicity or nephrotoxicity [9]. Consistent with its effects on epididymal and retroperitoneal white fat, ursolic acid reduced interscapular white fat. Interestingly, ursolic acid increased brown fat [10], which shares its developmental origins with skeletal muscle [24] and protects against obesity [25]. Since skeletal muscle and brown fat have relatively high rates of energy expenditure, treatment with ursolic acid was able to significantly increase energy expenditure [9,10]. Thus, ursolic acid not only increases skeletal muscle but also another tissue that can improve body composition: brown fat. This could have important implications for improving the overall fat to lean mass ratio and aid in sports such as wrestling and boxing that require acute reductions in body weight.

Ursolic Acid Mechanism of Action

The anabolic effects of ursolic acid seen *in vivo* could be mediated by a number of factors such as increased hormone/growth factors, neural input or satellite cell proliferation. To determine whether ursolic acid can act directly in the muscle cell or if it is mediated by other neural or systemic factors such as increased hormone/growth factors, or satellite cells, Figueriedo and Nader [45] measured protein content in pure differentiated myotube cultures in response to ursolic acid administration. Ursolic acid directly promoted protein accretion in cultured myotubes but did not modify myoblast proliferation, therefore demonstrating a direct effect on the muscle cell [10]. Within myotubes, ursolic acid has been shown to increase Akt activity at least in part by enhancing ligand-dependent activation of the insulin receptor and insulin-like growth factor I (IGF-1) receptor (Figure 49.2). The protein growth factor IGF-1 has been demonstrated to be sufficient to induce skeletal muscle hypertrophy [26]. Over the past few years, signaling pathways which are activated by IGF-1, and which are responsible for regulating protein synthesis pathways, have been defined. More recently, it has been shown that IGF-1 can also block the transcriptional up-regulation of key mediators of skeletal muscle atrophy, the ubiquitin-ligases MuRF1 and atrogin-1 [26]. Thus, muscle-specific IGF-1 induction is likely the contributing mechanism in ursolic acid-induced muscle hypertrophy. In C2C12 skeletal myotubes, a well-established *in vitro* model of skeletal muscle [12,27], ursolic acid increased Akt phosphorylation in the presence of IGF-1 and insulin. Thus, ursolic acid enhanced IGF-1-mediated and insulin-mediated Akt phosphorylation. Additionally, ursolic acid increased IGF-1 receptor phosphorylation and insulin receptor phosphorylation in the presence of insulin and IGF-1. Thus these data show that ursolic acid promotes muscle hypertrophy by increasing activity of the IGF-1 and insulin receptors.

Brown fat shares developmental origins with skeletal muscle; however, how ursolic acid increases brown fat remains uncertain. One possibility is that increased brown fat is a secondary effect of reduced insulating white fat, while another is that ursolic acid increases sympathetic activity, which is known to expand brown fat [28]. However, Kunkel et al. [10] witnessed slightly decreased resting heart rate alongside no effect on blood pressure. This may rule out a systemic increase in sympathetic activity. Nevertheless, without directly measuring sympathetic outflow to brown fat, there is still a possibility that

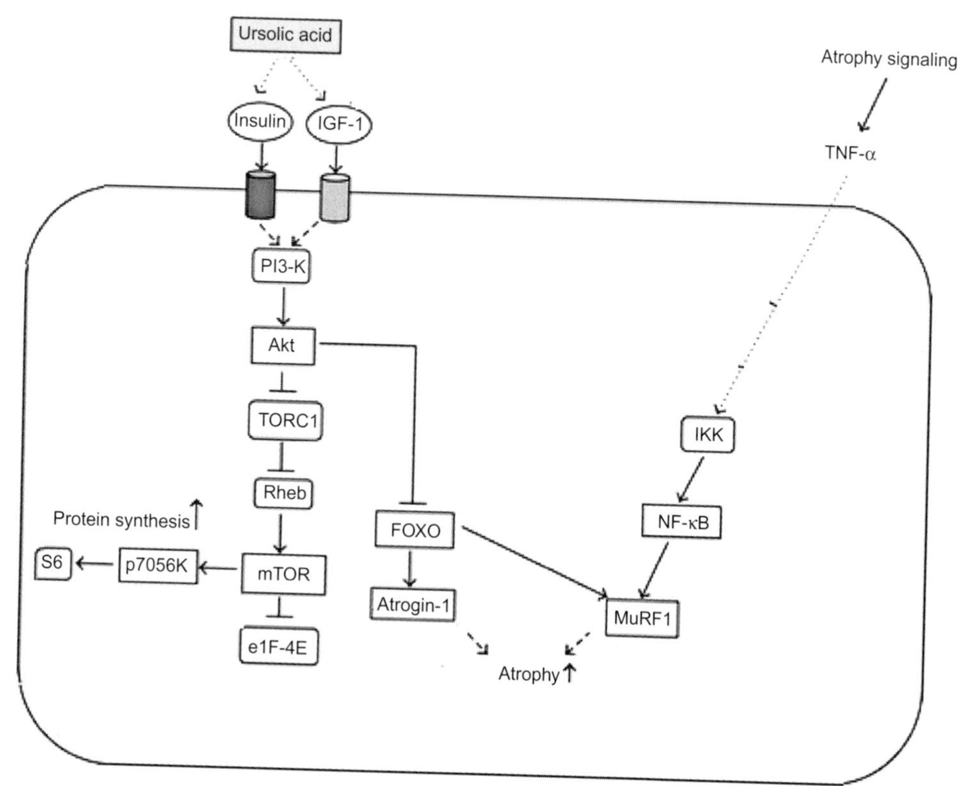

FIGURE 49.2 **Anabolic and anti-catabolic pathways relevant to ursolic acid.** Supplementation with ursolic acid up-regulates the amount of IGF-1 and insulin, stimulating protein synthesis and inhibiting atrophy via the Akt pathway. (See color plate at the back of the book).

sympathetic outflow to brown fat can be dissociated from sympathetic outflow to other tissues [29,30]. Finally, since brown fat and skeletal muscle arise from the same precursor cells [24], and since insulin/ IGF-1 signaling promotes brown fat growth [31], one possibility is that ursolic acid increases brown fat and skeletal muscle through a common molecular mechanism. This is an important area for future investigation.

Ursolic acid has also been shown to efficiently inhibit aromatase *in vitro* in a dose-dependent fashion, which was comparable to that of the potent aromatase inhibitor apigenin [2]. This is another potential underlying mechanism for ursolic acid and muscle hypertrophy. The irreversible conversion of androgens to estrogens by the enzymatic complex called aromatase has been an area of interest in indirect anabolic strategies such as aromatase inhibitors [32]. This has led to the use of aromatase inhibitors in attempts to increase blood testosterone levels. However, the magnitude of the rise in blood testosterone concentration is particularly important, and this direct effect with ursolic acid treatment has not been studied.

MASLINIC ACID

Maslinic acid, a triterpenoid compound derived from oleanolic acid (3-β-hydroxyolean-12-en-28-oic acid) (Figure 49.3), is widely distributed in plants, and is particularly abundant in the surface wax on the fruits and leaves of *Olea europaea* [33]. Maslinic acid, obtained from many plant species, is also present in considerable proportion in the solid waste from olive oil production [34]. The protease-inhibiting properties of Maslinic acid have been studied by multiple research groups in the treatment of several pathologies, including those caused by human immunodeficiency viruses [35,36]. More recently it has emerged as a growth-stimulating factor, and researchers have investigated the way in which the addition of maslinic acid to a standard fish diet can influence growth, protein turnover rates and nucleic acid concentrations in the liver of rainbow trout (*Oncorhynchus mykiss*) [37]. For instance, feeding rainbow trout diets containing maslinic acid ranging from 1 to 250 mg per kg of diet resulted in whole body and liver weight as well as growth rates that were higher than in controls. The highest weight increase observed was in the group fed 250 mg/kg, which yielded a 29% increase over controls. Furthermore the total hepatic DNA or liver cell hyperplasia levels in this group of trout were 68% higher than in controls. The liver plays an important role in regulating the energy metabolism of fish and is a major site for fatty acid synthesis and gluconeogenesis. Thus, based on these functions, it plays a role in regulating fish growth [37]. Additionally, fractional and absolute hepatic protein synthesis rates were significantly higher than in control, and significant increments in hepatic protein synthesis efficiency and protein synthesis capacity were reported. Microscopy studies also confirmed that trout fed on 25 and 250 mg/kg maslinic acid exhibited hepatocytes that were more compact, with a larger rough-endoplasmic reticulum and larger glycogen stores, than those of controls.

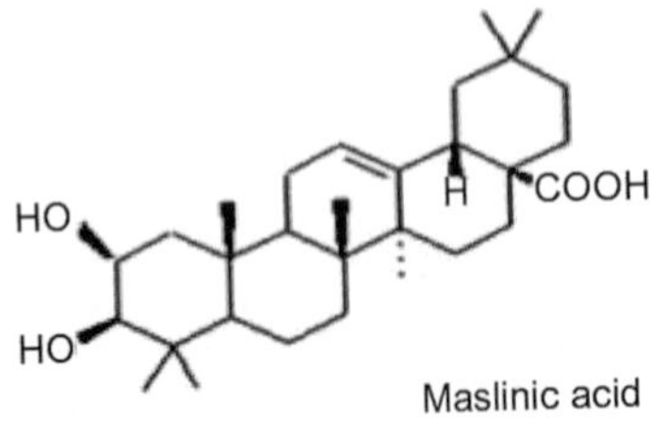

HO
HO
COOH
H
Maslinic acid

FIGURE 49.3 **Chemical structure of maslinic acid.**

The effect of maslinic acid in trout white muscle complements the previous work in rainbow trout liver [38]. For example, white muscle weight and protein accumulation rate of trout fed with maslinic acid were higher than in controls. Whole body growth of trout proved to be similar to that of white muscle weight. The major differences in comparison with control were detected in the 25 mg/kg and 250 mg/kg groups after 225 days, when the whole body growth was 19.3% and 29.2% higher than in control, respectively. The total content of DNA, RNA and protein in trout fed with 25 and 250 mg/kg of maslinic acid were significantly higher than in control. The protein : DNA ratio was also slightly higher than that of control. Furthermore, fractional (KS) and absolute (AS) protein synthesis rates increased to more than 80% over the control values while no differences were found in the fractional protein degradation rate (KD) [38]. These results were similar to previous findings in liver [37], showing that maslinic acid can act as a growth factor when added to a standard trout diet. Under the current experimental conditions, the increase in cell number was greater than in the cell size for all groups of trout [38]. This agrees with previous reports stating that differences in cell number usually make a large contribution to tissue growth [39,40]. Together these results suggest that maslinic acid can significantly boost protein synthesis rates and act as a growth factor when added to trout diet. These enhanced growth results, if translated to fish farm conditions, could represent an important boost in trout production and even justify its use as a potential feed additive. Furthermore, it would be useful to apply this study to other species and humans to assess its effect on total nitrogen balance and muscle growth.

Maslinic Acid Mechanism of Action

Maslinic acid has been shown to exhibit antioxidant activity against oxygen and nitrogen reactive species [41,42]. It has also showed a suppressive effect on pro-inflammatory cytokines such as TNF-α and IL-6 in murine macrophages [42]. Both these mechanisms may play a role in the greater protein synthesis and growth rates found in white muscle after maslinic acid feeding [38]. Furthermore, maslinic acid can act as a glycogen phosphorylase inhibitor in mouse liver [43,44], which was supported by the higher accumulation of glycogen in rainbow trout liver [37]. Glycogen phosphorylases, which catalyze the first step of glycogen breakdown, play an important role in glucose metabolism, especially in the glycogenolytic pathway. The anabolic effect of this molecule can therefore result in glycogen accumulation that could provide sustained energy supply during exercise and aid in performance. Overall, the higher number of white muscle cells, mediated by increases in the DNA, RNA and protein content of maslinic acid-fed trout, seem to result from a stimulation of the biosynthesis pathways similar to those produced by a growth factor [38]. Researchers also theorize that, due to its chemical structure that resembles steroid hormones, it could act in a cell-signaling pathway in a manner similar to that of hormones [37]—for example, crossing the plasma membrane and binding to specific cytoplasm receptors or nucleus receptors in order to activate specific genes related to growth and protein synthesis [37]. However, more research is needed in mammals to discover the underlying mechanisms involved in the effects attributed to maslinic acid.

CONCLUSIONS

A considerable amount of progress has been made recently in our understanding of the distinct set of genes and signaling pathways that mediate skeletal muscle hypertrophy and atrophy. Given the strong correlation between muscle cross-sectional area and muscular strength, professional and amateur athletes are turning to performance-enhancing supplements as a means to enhance lean body mass and performance. Two novel, naturally occurring triterpenoid compounds: ursolic acid and maslinic acid are emerging as promising therapeutic agents with the potential to counteract muscle loss and enhance muscle growth. Ursolic acid has been shown to enhance skeletal muscle insulin/IGF-1 signaling, leading to Akt activation, muscle hypertrophy and reduced atrophy and adiposity in animal models. Maslinic acid can significantly boost protein synthesis rates and act as a growth factor and glycogen phosphorylase inhibitor in trout. Thus, the initial research is compelling and both compounds may prove to be an important strategy to favor muscle hypertrophy and reduce atrophy, however placebo-controlled human studies are needed to confirm efficacy and determine optimal dosing before any strong recommendations can be made to athletes and bodybuilders.

References

[1] Maughan RJ, Watson JS, Weir J. Strength and cross-sectional area of human skeletal muscle. J Physiol 1983;338:37–49.

[2] Frighetto RTS, Welendorf RM, Nigro EN, Frighetto N, Siani. AC. Isolation of ursolic acid from apple peels by high speed counter-current chromatography. Food Chem 2008;106 (2):767–71.

[3] Gnoatto SCB, Dassonville-Klimpt A, Da Nascimento S, Galéra P, Boumediene K, Gosmann G, et al. Evaluation of ursolic acid isolated from Ilex paraguariensis and derivatives on aromatase inhibition. Eur J Med Chem 2008;43:1865–77.

[4] Pinon A, Limami Y, Micallef L. A novel form of melanoma apoptosis resistance: melanogenesis up-regulation in apoptotic B16-F0 cells delays ursolic acid-triggered cell death. Exp Cell Res 2011;3:1669–76.

[5] De Angel RE, Smith SM, Glickman RD. Antitumor effects of ursolic acid in a mouse model of postmenopausal breast cancer. Nutr Cancer 2010;62:1074–86.

[6] Kim DK, Baek JH, Kang CM. Apoptotic activity of ursolic acid may correlate with the inhibition of initiation of DNA replication. Int J Cancer 2000;87:629–36.

[7] Zhang W, Hong D, Zhou Y. Ursolic acid and its derivative inhibit protein tyrosine phosphatase 1B, enhancing insulin receptor phosphorylation and stimulating glucose uptake. Biochim Biophys Acta 2006;1760:1505–12.

[8] Jayaprakasam B, Olson LK, Schutzki RE. Amelioration of obesity and glucose intolerance in high-fat-fed C57BL/6 mice by anthocyanins and ursolic acid in Cornelian cherry (Cornus mas). J Agric Food Chem 2006;54:243–8.

[9] Kunkel SD, Suneja M, Ebert SM. mRNA expression signatures of human skeletal muscle atrophy identify a natural compound that increases muscle mass. Cell Metab 2011;13:627–38.

[10] Kunkel SD, Elmore CJ, Bongers KS, Ebert SM, Fox DK, Dyle MC, et al. Ursolic acid increases skeletal muscle and brown fat and decreases diet-induced obesity, glucose intolerance and fatty liver disease. PLoS One 2012;7(6).

[11] Bodine SC, Latres E, Baumhueter S, Lai VK, Nunez L, Clarke BA, et al. Identification of ubiquitin ligases required for skeletal muscle atrophy. Science 2001;294:1704–8.

[12] Sandri M, Sandri C, Gilbert A, Skurk C, Calabria E, Picard A, et al. Foxo transcription factors induce the atrophy-related ubiquitin ligase atrogin-1 and cause skeletal muscle atrophy. Cell 2004;117:399–412.

[13] Gomes MD, Lecker SH, Jagoe RT, Navon A, Goldberg AL. Atrogin-1, a muscle-specific F-box protein highly expressed during muscle atrophy. Proc Natl Acad Sci USA 2001;98:14440–5.

[14] Adams GR, Haddad F. The relationships among IGF-1, DNA content, and protein accumulation during skeletal muscle hypertrophy. J Appl Physiol 1996;81:2509–16.

[15] Gentile MA, Nantermet PV, Vogel RL, Phillips R, Holder D, Hodor P, et al. Androgen-mediated improvement of body composition and muscle function involves a novel early

transcriptional program including IGF-1, mechano growth factor, and induction of {beta}-catenin. J Mol Endocrinol 2010;44:55–73.

[16] Hameed M, Lange KH, Andersen JL, Schjerling P, Kjaer M, Harridge SD, et al. The effect of recombinant human growth hormone and resistance training on IGF-1 mRNA expression in the muscles of elderly men. J Physiol 2004;555:231–40.

[17] Banerjee A, Guttridge DC. Mechanisms for maintaining muscle. Curr Opin Support Palliat Care 2012;6(4):451–6.

[18] Frost RA, Huber D, Pruznak A, Lang CH. Regulation of REDD1 by insulin-like growth factor-I in skeletal muscle and myotubes. J Cell Biochem 2009;108:1192–202.

[19] Sacheck JM, Ohtsuka A, McLary SC, Goldberg AL. IGF-1 stimulates muscle growth by suppressing protein breakdown and expression of atrophy-related ubiquitin ligases, atrogin-1 and MuRF1. AJP Endocrinol Metab 2004;287:E591–601.

[20] Izumiya Y, Hopkins T, Morris C, Sato K, Zeng L, et al. Fast/Glycolytic muscle fiber growth reduces fat mass and improves metabolic parameters in obese mice. Cell Metab 2008;7:159–72.

[21] Lai KM, Gonzalez M, Poueymirou WT, Kline WO, Na E, Zlotchenko E, et al. Conditional activation of akt in adult skeletal muscle induces rapid hypertrophy. Mol Cell Biol 2004;24:9295–304.

[22] Aayaprakasam B, Olson LK, Schutzki RE, Tai MH, Nair MG. Amelioration of obesity and glucose intolerance in high-fat-fed C57BL/6 mice by anthocyanins and ursolic acid in Cornelian cherry (Cornus mas). J Agric Food Chem 2006;54:243–8.

[23] Rao VS, Melo CL, Queiroz MG, Lemos TL, Menezes DB, et al. Ursolic acid, a pentacyclic triterpene from Sambucus Australis, prevents abdominal adiposity in mice fed a high-fat diet. J Med Food 2011;14:1375–82.

[24] Kajimura S, Seale P, Spiegelman BM. Transcriptional control of brown fat development. Cell Metab 2010;11:257–62.

[25] Cypess AM, Kahn CR. Brown fat as a therapy for obesity and diabetes. Curr Opin Endocrinol Diabetes Obes 2010;17:143–9.

[26] Glass DJ. Skeletal muscle hypertrophy and atrophy signaling pathways. Int J Biochem Cell Biol 2005;37(10):1974–84.

[27] Stitt TN, Drujan D, Clarke BA, Panaro FJ, Timofeyva Y, Kline WO, et al. The IGF-1/PI3K/Akt pathway prevents expression of muscle atrophy-induced ubiquitin ligases by inhibiting FOXO transcription factors. Mol Cell 2004;14:395–403.

[28] Cannon B, Nedergaard J. Brown adipose tissue: function and physiological significance. Physiol Rev 2004;84:277–359.

[29] Morrison SF. Differential control of sympathetic outflow. Am J Physiol Regul Integr Comp Physiol 2001;281:R683–98.

[30] Mark AL, Agassandian K, Morgan DA, Liu X, Cassell MD, et al. Leptin signaling in the nucleus tractus solitarii increases sympathetic nerve activity to the kidney. Hypertension 2009;53:375–80.

[31] Tseng YH, Butte AJ, Kokkotou E, Yechoor VK, Taniguchi CM, et al. Prediction of preadipocyte differentiation by gene expression reveals role of insulin receptor substrates and necdin. Nat Cell Biol 2005;7:601–11.

[32] Handelsman DJ. Indirect androgen doping by oestrogen blockade in sports. Br J Pharmacol 2008;154(3):598–605.

[33] Bianchi G, Vlahov G, Pozzi N. The lipids of Olea europaea. Pentacyclic triterpene acids in olives. Phytochemistry 1994;37:205–7.

[34] García-Granados A, Dueñas J, Moliz JN, Parra A, Pérez FL, Dobado JA, et al. Semi-synthesis of triterpene A-ring derivatives from oleanolic and maslinic acids. Part II. Theoretical and experimental C-13 chemical shifts. J Chem Res, Synop 2000;2:0326–39.

[35] Xu HX, Zeng PQ, Wan M, Sim KY. Anti-HIV triterpene acids from Geum japonicum. J Nat Prod 1996;59:643–5.

[36] Vlietinck AJ, De Bruyne T, Apers S, Pieters LA. Plant-derived leading compounds for chemotherapy of human immunodeficiency virus (HIV) infection. Planta Med 1998;64:97–109.

[37] Fernández-Navarro M, Peragón J, Esteban FJ, de la Higuera M, Lupiáñez JA. Maslinic acid as a feed additive to stimulate growth and hepatic protein-turnover rates in rainbow trout (Onchorhynchus mykiss). Comp Biochem Physiol C Toxicol Pharmacol 2006;144(2):130–40.

[38] Fernández-Navarro M, Peragón J, Amores V, De La Higuera M, Lupiáñez JA. Maslinic acid added to the diet increases growth and protein-turnover rates in the white muscle of rainbow trout (Oncorhynchus mykiss). Comp Biochem Physiol C Toxicol Pharmacol 2008;147(2):158–67.

[39] Peragón J, Barroso JB, de la Higuera M, Lupiáñez JA. Relationship between growth and protein turnover rates and nucleic acids in the liver of rainbow trout (Oncorhynchus mykiss) during development. Can J Fish Aquat Sci 1998;55:649–57.

[40] Conlon I, Raff M. Size control in animal development. Cell 1999;96:235–44.

[41] Montilla MP, Agil A, Navarro MC, Jimenez MI, Garcia-Granados A, Parra A, et al. Antioxidant activity of maslinic acid, a triterpene derivative obtained from Olea europaea. Planta Med 2003;69:472–4.

[42] Márquez-Martín A, De la Puerta-Vázquez R, Fernández-Arche A, Ruíz-Gutiérrez V. Suppressive effect of maslinic acid from pomace olive oil on oxidative stress and cytokine production in stimulated murine macrophages. Free Rad Res 2006;40:205–302.

[43] Wen X, Sun H, Liu J, Wu G, Zhang L, Wu X, et al. Pentacyclic triterpenes. Part 1: the first examples of naturally occurring pentacyclic triterpenes as a new class of inhibitors of glycogen phosphorylases. Bioorg Med Chem Lett 2005; 15(22):4944–8.

[44] Wen X, Zhang P, Liu J, Zhang L, Wu X, Ni P, et al. Pentacyclic triterpenes. Part 2: synthesis and biological evaluation of maslinic acid derivatives as glycogen phosphorylase inhibitors. Bioorg Med Chem Lett 2006;16(3):722–6.

[45] Figueiredo VC, Nader GA. Ursolic acid directly promotes protein accretion in myotubes but does not affect myoblast proliferation. Cell Biochemistry & Function 2012;30(5):432–7.

CHAPTER

50

Plant Borates and Potential Uses to Promote Post-training Recovery and to Mitigate Overtraining Syndrome

Zbigniew Pietrzkowski[1], *John Hunter*[2], *Brad Evers*[2] *and Hartley Pond*[2]

[1]Applied BioClinical, Inc., Irvine, CA, USA [2]VDF FutureCeuticals, Inc., Momence, IL, USA

INTRODUCTION

Overtraining syndrome (OTS) is a condition that occurs when the body is subjected to chronic daily intense physical activity without adequate time for rest and recovery. Damage from chronic intense physical activity can be cumulative and can be marked by a number of symptoms. These symptoms include (but are not limited to) mood changes, reduced immune response, vulnerability to infections, loss of coordination, decreased exercise performance, irregular sleep patterns, loss of muscle and cardiovascular endurance, loss of muscle strength, chronic fatigue, and loss of appetite. It is important to mention that overtraining should not be confused with moderate overreaching or occasional exercise exhaustion, both of which are more short-term or acute consequences of physical overload. In other words, OTS symptoms generally manifest only after a number of weeks of "over training". Currently, there is no universally accepted hypothesis explaining the molecular etiology of OTS. Based on new scientific findings, the role of myokines and inflammatory factors involved in the process of OTS will be reviewed here as potentially important components of this syndrome. Additionally, the use of anti-inflammatory plant boron-carbohydrate conjugates for the promotion of faster recovery after training, and as potential mediators of OTS and muscle inflammation, will be discussed. Results showing the anti-inflammatory potency of these conjugates as clinically measured by CRP, calcitriol, YKL-40, TNF-alpha, and IL-1b will be presented.

Although a number of boron-containing materials such as boron citrate, boron aspartate, boron glycinate chelates, boron ascorbate and sodium borate are currently available in the nutraceutical industry today [1] the *borates*, in the sub-class of boro-carbohydrates (such as calcium fructoborate), are arguably the most interesting from a physiological vantage. Certain of these plant-based boro-carbohydrates have a very high association constant (e.g., >6000 daltons) (US Patent #5,962,049) and therefore can at least partially enter the bloodstream intact rather than being converted to free boric acid in the acidic environment of the gut. Due to Patrick Brown's work we know that the boro-carbohydrates, such as calcium fructoborate, are the only boron-containing compounds found in plants [2]. Consequently, it can reasonably be inferred, especially since mammals typically do not eat soil or ore, that the only *natural* sources of borates for humans and other mammals are borate-rich fruits and vegetables incorporated into healthy diets. This argument is compelling, especially since the recent discovery of a borate receptor in mammals [1,3,4] and recent published reports regarding the antioxidant activity of calcium fructoborate (CFB).

Unfortunately, due to boron-depleted soils, dietary shifts towards nutrient-deficient, fat-, sugar- and salt-laden fast foods, and a general Western trend towards aversion to vegetable and fruit consumption, many individuals may not be getting sufficient levels of these healthy plant-based borates. Of the boron-based nutraceuticals, calcium fructoborate, a natural sugar-borate-ester [5], is the most clinically studied.

Nutrition and Enhanced Sports Performance.
DOI: http://dx.doi.org/10.1016/B978-0-12-396454-0.00050-3

Currently CFB is synthesized as a "nature–identical" entity (Miljkovic US patent #5,962,049) and is commercially marketed as an active component of dietary supplemental products in the USA for bone health, and for modulation of the symptoms of arthritis and joint degeneration based upon its clinical potency to reduce inflammation, and to improve WOMAC Index and McGill Index [6]. Of primary interest to this discussion is CFB's anti-inflammatory potency and consequent potential applications for muscle recovery, endurance and overall sports health.

THE TRACE ELEMENT BORON, BORATES AND BORO-CARBOHYDRATES

The element boron, while not yet considered to be an "essential" mineral, has been widely studied and has been suggested to be pivotal in many physiological processes, including embryonic development, bone health, and steroid hormone homeostasis. Recent investigations have even postulated that boron may have been essential to the origin of life by contributing to the formation of, and stabilizing, ribose (the "R" in RNA) [7,8,9]. Forrest Nielsen and Curtis Hunt from USDA Grand Forks have contributed greatly to current understanding of boron [10–14,1,15]. Borates are the most common forms of boron-containing compounds occurring in nature, mainly as carbohydrate-boron conjugates and borate minerals (salts derived from simple boric acid conjugated with other mineral compounds) such as boro-silicate. As reported by Hu and Brown et al., boron is vital in the transport of nutrients throughout the plant and to the maintenance of plant cell wall integrity. Plant borates generally present as salts conjugated with a sugar, an alcohol or a poly-ol [2]. Natural plant boron compounds discovered to date include sugar alcohol borate complexes (e.g., fructo borates, including CFB), pectic polysaccharide borate complexes, organic acid borate esters, and amino acid borate esters.

MYOKINES, INFLAMMATION, RECOVERY AND OVERTRAINING SYNDROME

The term "Myokines" has been proposed as a generic descriptor for cytokines or other peptides that are produced and released by skeletal muscle fibers *per se*, and that subsequently exert their effects in muscle tissue as well as other organs of the body [16,17]. Interleukin-6 (IL-6) has been identified as a factor that is produced by muscle fibers and released into the circulation during exercise. Several other factors have also been identified. Some of them, such as IL-6 and interleukin-15 (IL-15), are released to the bloodstream, while others, such as leukemia inhibitory factor (LIF), remain active locally in the muscles. Generally, myokines represent a possible link between working skeletal muscles, adipose tissues, liver, brain, and vascular compartments. In view of accumulating evidence suggesting that regular exercise can protect against chronic disorders such as cardiovascular diseases, type 2 diabetes, dementia, and depression, the activities of myokines may at least partially explain how regular muscle activity can influence mood, performance, fatigue and cognitive functions.

Based upon recent scientific findings, exercise and muscle contractions results in the release of myokines such as IL-6, IL-15, LIF, or brain-derived neurotrophic factor (BDNF). Each of these has specific roles in stimulation of muscle fibers to repair damage caused by exercise, especially extensive and frequent exercise.

An acute increase in myokines resulting from muscle contractions is important to initiate the process of muscle repair and regeneration by activating satellite cells within inflamed muscles. In other words, these myokines, in response to stress, serve as triggers to initiate repair processes within damaged muscles. Satellite cells function as precursors of myoblasts and myofibers and are crucial for skeletal muscle adaptation to extensive exercise and regeneration [18]. Satellite cells play a key role in the process of hypertrophy by providing new myoblasts capable of regenerating muscle myofibers [19–21]. It is of interest to mention that routinely exercised muscles may contain up to twice as many satellite cells than untrained muscles [22,23]. Importantly, the use of non-steroidal anti-inflammatory drugs (NSAIDs) can decrease satellite cell response to exercise [24] and can reduce exercise-induced protein synthesis [25]. These effects may result in reduced muscle regeneration. This indicates that the NSAID effect is likely due to other mechanism(s) since chronic inflammation as such causes muscle loss [26,27]. Furthermore, an increased level of myostatin, a protein that inhibits muscle differentiation and growth [28], was found in people with chronic low-grade inflammation [29]. All of these findings suggest that muscles can secrete several different active myokines, each with defined function [29]. Myokines such as IL-6, interleukin-7 (IL-7) and myostatin are involved in the process of muscle hypertrophy and myogenesis. On the other hand, IL-6 and BDNF are involved in adenosine monophosphate-activated protein kinase (AMPK)-mediated fat oxidation. Moreover, IL-6 may affect liver function, pancreatic islets, adipose tissue and the immune system [29]. This broad spectrum of physiological activity suggests that these myokines, if secreted by muscle, play more metabolic and regulatory roles than inflammation regulatory roles. Indeed, more peptides secreted by contracting muscles, and associated physiological functions specific to these peptides, are expected to be identified [30]. However, myokines still remain under-investigated in regards to exercise [31–34]. More research on this line may find that OTS is mediated by improperly functioning myokines or by disturbances in the production of specific myokines. If so, production of myokines would be more of an adaptive response resulting in muscle regeneration and growth rather than a pro-inflammatory event. Any reduction of such an adaptive mechanism may result in severe inflammation due to extended muscle damage and impaired muscle regeneration. According to [35], glucose ingestion during endurance training attenuates expression of myokine receptors such as IL-6. This suggests that fast recovery of muscle fibers could be limited. Consequently, repetitive and frequent training combined with ingestion of glucose may create increased damage to muscles and may induce more inflammatory reactions. Based upon this premise, it is suggested that OTS could be the result of improper metabolic background during exercise and post-exercise that subsequently results in disturbances in the secretion and proper functioning of myokines by contracting muscles. For example, Blank recently reported that exercising under sodium-depleted conditions causes symptoms strikingly similar to OTS [36]. This further supports the importance of metabolic conditions in the development of symptoms associated with overtraining. These facts are important since they provide evidence that disturbing the natural path of exercise-induced production of myokines could be important for developing OTS.

It could also be hypothesized that overtraining causes excessive damage to muscle fibers and, because of this damage, myokines that are produced cannot efficiently repair and regenerate the tissue. Following this scenario, muscles damaged by micro-injury may induce more inflammatory reactions. In such a case, paradigms that can limit inflammatory reactions may become important in order to prevent or diminish OTS. Finally, overtraining may also result in stimulation of production of interleukin-6 soluble receptor (IL-6sR) and/or glycoprotein 130 (gp130) resulting in the triggering of pro-inflammatory action of IL-6 rather than a myokine-type action as discussed in the section of this chapter on exercise and inflammation.

INFLAMMATION

In general, inflammation is a complex biological response to harmful stimuli such as pathogens, damage, injury (microinjury), or irritants [37]. Inflammation is classified as either acute or chronic. Acute inflammation is the initial response of the body to insult and is achieved by increased movement of plasma and leukocytes from the blood to the injured and affected sites within the body. Acute inflammation appears within minutes or hours and diminishes upon the removal of the injurious insult [38]. Cardinal signs such as an increased blood flow to the damaged tissue, causing increased temperature, redness, swelling and pain, are characteristic of an acute response. Whereas under chronic inflammatory conditions the healing process is impaired, acute inflammation is a protective response that helps remove the consequences of pro-inflammatory insults and that initiates the healing process. Persistent and repetitive acute inflammation due to non-removable pathogens, infection, foreign bodies, autoimmune reactions, or any other pro-inflammatory insults results in chronic inflammation. Chronic inflammation leads to a progressive shift in the type of cells present in the site of inflammation, affecting various processes: for example, the healing process [39–42]. It may also lead to over-activation of monocytes and lymphocytes, resulting in systemic inflammation manifested by a two- to three-fold increase in the systemic concentrations of cytokines such as TNF-alpha, IL-6, IL-8, or the acute-phase protein C-reactive protein (CRP) [43–45]. The amount of such cytokines, chemokines and proteins could be measured in blood as a means of estimating levels of chronic inflammation. A number of health conditions are associated with increased blood levels of these biomarkers, including diabetes, cardiovascular disease and conditions, osteoarthritis, osteoporosis and associated increased risk of fractures [46–51]. Other blood molecules besides cytokines and chemokines (such as histamine, leukotrienes and prostaglandins) may also serve as biomarkers and indicators of inflammatory conditions.

EXERCISE AND INFLAMMATION

Acute inflammation may result from extensive exercise. More precisely, acute inflammation of muscle cells and tissue may result after eccentric or concentric muscle training [52]. Interestingly, in response to muscular contractions, the acute inflammatory response initiates the breakdown and removal of damaged muscle tissue [53]. Furthermore, muscle can synthesize certain cytokines such as interleukin-1 beta (IL-1b), tumor necrosis factor-alpha (TNF-α) and IL-6 [54–56] in response to contraction. IL-6 blood levels can increase up to 100-fold over resting levels during the first 4 hours and remain high for up to 24 hours [57,58].

IL-6 was discussed above as a myokine playing an important role in the process of muscle repair. It could also cause an anorexigenic effect [59]. This mode of action is mediated by activation of IL-6 receptors present in muscle tissue. This way, IL-6 acts more as an anti-inflammatory molecule. However, interleukin-6 (IL-6) is a cytokine largely induced during infection, inflammation, and cancer. In the liver, IL-6 induces the synthesis of acute-phase proteins (including CRP), which are believed to support the response of the body during infection and inflammation. As mentioned, IL-6 binds to an IL-6 receptor (IL-6R) on the surface of cells expressing this receptor to form an IL-6/IL-6R complex, which associates with a homodimer of a second receptor subunit, gp130, in order to initiate downstream intracellular signaling. Gp130 is present on all cells of the body, whereas IL-6R is only expressed on hepatocytes, some leukocytes, and some epithelial and muscle cells. Since gp130 has no measurable affinity for IL-6, cells which do not express IL-6R are unresponsive to the cytokine. A soluble form of IL-6R, which is found in the blood, can still bind IL-6, and the complex of IL-6/sIL-6sR can also bind to cellular gp130 on cells without IL-6R expression. This signaling mechanism has been called "trans-signaling". Interestingly, a soluble form of gp130 (sgp130) blocks IL-6 trans-signaling without affecting classic IL-6 signaling via the membrane-bound IL-6R. Finally, pro-inflammatory activities of IL-6 are mediated via trans-signaling, whereas anti-inflammatory or regenerative activities are mediated via classic signaling, via IL-6R [60]. This new finding is absolutely critical to understanding the functioning of IL-6 during and after physical exercise. It points to the importance of circulating IL-6sR and gp130 levels in the blood and also their ratio(s) with circulating blood IL-6. This is a good example of how important it is to analyze the profiles of various pro-inflammatory cytokines and proteins (IL-1b, IL-8, TNF-α, CRP) and receptors (IL-6sR and gp130) in response to exercise, rather than only measuring IL-6.

Another important cytokine released after exercise is interleukin-10 (IL-10). IL-10 is known to play an important role in regulating changes in the phenotype of macrophages during muscle growth, and also in regeneration during injury [61]. Micro-injury occurs during extensive and frequent exercise. IL-10 is also known as anti-inflammatory cytokine. Measuring IL-10 in blood before and after exercise, and comparing same with changes in blood levels of IL-6, IL-6sR, gp30, TNF-α, IL-8, IL-15, BDNF, and CRP may improve our understanding of the balance between muscle regeneration and inflammation.

Altogether, the above information indicates that the "cytokine hypothesis" could actually be re-named as a "myokine hypothesis", denoting the importance of cytokines and other peptides secreted *only* by contracting muscles. The function of these myokines is to repair damaged muscles as quickly as possible. This process could be disrupted if contracting muscles produce insufficient amounts of IL-10 and insufficient levels of cyclooxygenase-2 (COX-2) expression (target of NSAIDs mentioned earlier) [62]. This suggests that overtraining may induce OTS-specific symptoms if muscles are unable to recover in a timely manner after each extensive exercise. In such cases, damaged muscle tissue may trigger inflammatory responses and would be mediated by blood cell derived cytokines and chemokines, or by IL-6/IL-6sR complex and gp130, TNF-α, IL-1b and likely by other factors as well.

Since myokines are relatively new in sports science, it is important to more deeply investigate their role in the process of muscle recovery under various metabolic conditions and various levels of exercise. Therefore, the prevention of inflammation in muscles subjected to extensive contractions prior to adequate recovery could be a main target of investigation (measured by creatine kinase (CK)) to develop means of preventing OTS and promoting faster muscle recovery following vigorous training. In such a case, the use of anti-inflammatory intervention would be justified to prevent inflammation induced by damaged muscles. As reviewed, high levels of CK in blood would be a sign of extensive muscle damage. It would be important to prevent development of inflammatory conditions induced by injured muscles and characteristic of increased blood level of IL-6 in relation to IL-6sR and the IL-6/IL-6sR complex, IL-15, IL-8, IL-1b, TNF-α, myostatin, BDNF, IL-10 and CRP. Here we would like to discuss the use of calcium fructoborate, a "boro-carbohydrate", that was clinically investigated and was found to modulate blood levels of TNF-α, IL-1b, calcitriol and CRP.

BIOLOGICAL POTENCY OF CALCIUM FRUCTOBORATE

In comparison with all known organic forms of boron listed above, calcium fructoborate (CFB) is the most investigated, with over a dozen published studies on its unique chemical and clinical properties. Clinical publications describing its anti-inflammatory activity and potency to acutely increase blood levels of endogenous calcitriol are the most relevant to OTS. As reported in [6], CFB at a dose of 108 mg may reduce CRP by 37% (Study #1 in Figure 50.1). Scorei and Rotaru [5], published results indicating CFB at a dose of 56 mg/day may reduce CRP by 60.25% (Study #2 in figure above). Militaru, et al., 2012, found a reduction of CRP by 39.7% at a serving of 112 mg/day (Study #3 in figure above) [63]. In [6] the CRP results were

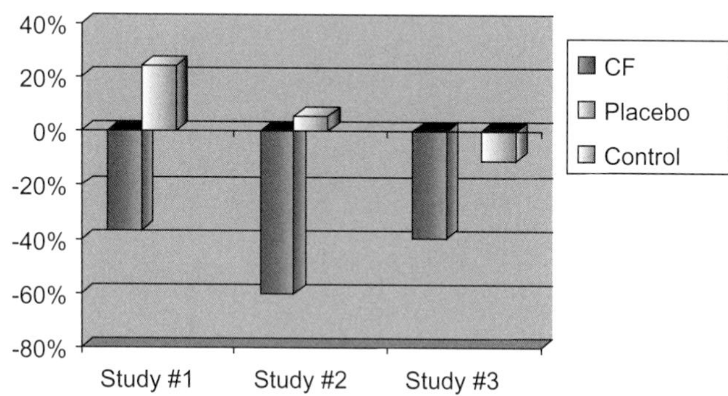

FIGURE 50.1 **Reduction of serum levels of C-reactive protein by calcium fructoborate (CF).** *Data from: Study #1 [6]; Study #2 [5]; Study #3 [63].* (See color plate at the back of the book.)

associated with a statistically significant improvement of WOMAC and McGill Indexes, reflecting potency of CFB to improve performance and pain in osteoarthritic subjects. Interestingly, in the same study, it was found that treatment with CFB increased blood levels of endogenous calcitriol, an active form of vitamin D. An additional clinical study has been completed (manuscript in preparation) showing that indeed CFB can acutely increase the blood levels of endogenous calcitriol in healthy subjects in a statistically significant manner.

As published, calcitriol is known to reduce inflammation [64–66]; however, it is a rather new finding that this active form of vitamin D exerts a strong effect on muscles. As published recently, calcitriol plays an important role in muscle physiology, and may affect the proliferation of myoblasts during the proliferating and differentiating phases, and the expression of the fast myosin heavy chain isoform in the differentiated phase at doses as low as 1nM. In addition, calcitriol may have an anabolic effect on differentiated skeletal muscle [67]. 1α,25-dihydroxyvitamin D_3 enhances fast-myosin heavy chain expression in differentiated C2C12 myoblasts [68]. This observation is supported further by the demonstration that myogenesis can be activated by calcitriol [69] by very rapid activation of the AKT pathway (also known as the protein kinase B (PKB)) in muscle cells. 1α,25$(OH)_2$D3-dependent modulation of Akt in proliferating and differentiating C2C12 skeletal muscle cells has also been reported [69]. An earlier study by Garcia has suggested that calcitriol may increase myogenesis as well but by a different mechanism involving inhibition of myostatin [70]. 1,25$(OH)_2$ vitamin D3 stimulates myogenic differentiation by inhibiting cell proliferation and modulating the expression of promyogenic growth factors and myostatin in C2C12 skeletal muscle cells [70]. Altogether, calcitriol as an active form of vitamin D shows an interesting potency to increase myogenesis of muscles. This activity could be beneficial for muscles damaged due to overtraining and remaining under inflammatory conditions.

Additional research was conducted in order to better understand CFB's mechanism of action and the effect of CFB on inflammation, as previously measured by improvements in blood levels of CRP. It was found that a 108 mg single dose of CFB may reduce blood levels of YKL-40, TNF-α, and IL-1b (internal results as yet unpublished). These are primary observations which justify further research to obtain more results suitable for peer-review publication.

SUMMARY

Scientific research shows that overtraining syndrome (OTS) is based on a multifactorial mechanism involving regulation of metabolism, muscle micro-injuries, follow-up recovery via myokines and finally, inflammation leading to symptoms such as depression, loss of appetite, muscle strength, predisposition to infections and other characteristics of overtraining. This complex scheme requires additional clinical investigation in order to better understand the delicate balance between muscle response to overtraining under various metabolic conditions, and the muscle repair system mediated by myokines under normal or inflammatory conditions. Overtraining may prevent all healthy benefits of exercise mediated by myokines, particularly under inflammatory conditions. Therefore, prevention of inflammation may be important in promoting optimal muscle repair and growth, to maintain myokine-mediated health benefits of training, rapid recovery, better endurance and to prevent development of OTS. Based upon published results thus far, the use of boron-containing calcium fructoborate is proposed, since this material shows acute anti-inflammatory activities as measured by CRP, TNF-α, IL-1b, YKL-40, and calcitirol, criteria that have been associated with positive effects on muscle function, growth and differentiation and thus may contribute to protection of, and recovery from, muscle damage from overtraining.

References

[1] Hunt CD. Boron. Encyclopedia of dietary supplements. 2nd ed. New York, London: Informa Healthcare; 2010. pp. 82–90.

[2] Hu H, Penn SG, Lebrilla CB, Brown PH. Isolation and characterization of soluble boron complexes in higher plants. The mechanism of phloem mobility of boron. Plant Physiol 1997;113:649–55.

[3] Dembitsky VM, Al-Quntar AAA, Srebnik M. Natural and synthetic small boron-containing molecules as potential inhibitors of bacterial and fungal quorum sensing. Chem Rev 2011;111:209–37.

[4] Nieves JW. Skeletal effects of nutrients and nutraceuticals, beyond calcium and vitamin D. Osteoporos Int 2013;24771–86. doi:10.1007/s00198-012-2214-2214.

[5] Scorei RI, Rotaru P. Calcium fructoborate—potential anti-inflammatory agent. Biol Trace Elem Res 2011;1431223–38. doi:10.1007/s12011-011-8972-6.

[6] Reyes-Izquierdo T, Nemzer B, Gonzalez AE, Zhou Q, Argumedo R, Shu C, et al. Short-term intake of calcium fructoborate improves WOMAC and McGill scores and beneficially modulates biomarkers associated with knee osteoarthritis: a pilot clinical double-blinded placebo-controlled study. Am J Biomed Sci 2012;4111–22. doi:10.5099/aj120200111.

[7] Scorei R, Cimpoiaşu VM. Boron enhances the thermostability of carbohydrates. Orig Life Evol Biosph 2006;36:1–11.

[8] Scorei R. Is Boron a prebiotic element? a mini-review of the essentiality of boron for the appearance of life on earth. Orig Life Evol Biosph 2012;42:3–17.

[9] Benner SR, Ricardo A, Illangkoon H, Kim MJ, Carrigan M, Frye F, Benner DS. Elements and the origin of life. Boron and Molybdenum. American Geophysical Union, Fall Meeting, 2008.

[10] Nielsen FH. The saga of boron in food: from a banished food preservative to a beneficial nutrient for humans. Curr Topics Plant Biochem Physiol 1991;10:274–86.

[11] Nielsen FH. Facts and fallacies about boron. Nutr Today 1992;27:6–12.

[12] Nielsen FH. Biochemical and physiologic consequences of boron deprivation in humans. Environ Health Perspect 1994;102:59–63.

[13] Hunt CD. The biochemical effects of physiologic amounts of dietary boron in animal nutrition models. Environ Health Perspect 1994;102:35–43.

[14] Hunt CD. Biochemical effects of physiological amounts of dietary boron. J Trace Elem Exp Med 1996;9:185–213.

[15] Hunt CD. Dietary boron: Progress in establishing essential roles in human physiology. J Trace Elem Exp Med 2012;26:157–60.

[16] Henriksen T, Green C, Pedersen BK. 2012. Myokines in Myogenesis and Health. Recent Pat Biotechnol Oct 22 [Epub ahead of print].

[17] Mathers JL, Farnfield MM, Garnham AP, Caldow MK, Cameron-Smith D, Peake JM. Early inflammatory and myogenic responses to resistance exercise in the elderly. Muscle Nerve 2012;46:407–12.

[18] Serrano AL, Baeza-Raja B, Perdiguero E, Jardí M, Muñoz-Cánoves P. Interleukin-6 is an essential regulator of satellite cell-mediated skeletal muscle hypertrophy. Cell Metab 2008;7:33–44.

[19] Grounds MD, White JD, Rosenthal N, Bogoyevitch MA. The role of stem cells in skeletal and cardiac muscle repair. J Histochem Cytochem 2002;50:589–610.

[20] Hawke TJ, Garry DJ. Myogenic satellite cells: physiology to molecular biology. J Appl Physiol 2001;91:534–51.

[21] Hawke TJ. Muscle stem cells and exercise training. Exerc Sport Sci Rev 2005;33:63–8.

[22] Kadi F, Charifi N, Denis C, Lexell J, Andersen JL, Schjerling P, et al. The behaviour of satellite cells in response to exercise: what have we learned from human studies? Pflugers Arch 2005;451(2):319–27.

[23] Eriksson A, Kadi F, Malm C, Thornell LE. Skeletal muscle morphology in power-lifters with and without anabolic steroids. Histochem Cell Biol 2005;124:167–75.

[24] Mikkelsen UR, Langberg H, Helmark IC, Skovgaard D, Andersen LL, Kjaer M, et al. Local NSAID infusion inhibits satellite cell proliferation in human skeletal muscle after eccentric exercise. J Appl Physiol 2009;107:1600–11.

[25] Trappe TA, White F, Lambert CP, Cesar D, Hellerstein M, Evans WJ. Effect of ibuprofen and acetaminophen on postexercise muscle protein synthesis. Am J Physiol Endocrinol Metab2 2002;82:E551–6.

[26] Visser M, Pahor M, Taaffe DR, Goodpaster BH, Simonsick EM, Newman AB, et al. Relationship of interleukin-6 and tumor necrosis factor-alpha with muscle mass and muscle strength in elderly men and women: the Health ABC Study. J Gerontol A Biol Sci Med Sci 2002;57:M326–32.

[27] Reardon KA, Davis J, Kapsa RM, Choong P, Byrne E. Myostatin, insulin-like growth factor-1, and leukemia inhibitory factor mRNAs are upregulated in chronic human disuse muscle atrophy. Muscle Nerve 2001;24:893–9.

[28] Louis E, Raue U, Yang Y, Jemiolo B, Trappe S. Time course of proteolytic, cytokine, and myostatin gene expression after acute exercise in human skeletal muscle. J Appl Physiol 2007;103:1744–51.

[29] Pedersen BK, Febbraio MA. Muscles, exercise and obesity: skeletal muscle as a secretory organ. Nat Rev Endocrinol 2012;8:457–65.

[30] Roca-Rivada A, Al-Massadi O, Castelao C, Senin L, Alonso J, Seoane L, et al. Muscle tissues as en endocrine organ: comparative secretome profiling of slow-oxidative and fast-glycolytic rat muscle explants and its variation with exercise. J Proteomics 2012;75:5414–25.

[31] Brandt C, Pedersen BK. The role of exercise-induced myokines in muscle homeostasis and the defense against chronic diseases. J Biomed Biotechnol 2010;2010520258. doi:10.1155/2010/520258 Epub 2010 Mar 9. Review.

[32] Pedersen BK. Exercise-induced myokines and their role in chronic diseases. Brain Behav Immun 2011;25:811–6.

[33] Tamura Y, Watanabe K, Kantani T, Hayashi J, Ishida N, Kaneki M. Upregulation of circulating IL-15 by treadmill running in healthy individuals: is IL-15 an endocrine mediator of the beneficial effects of endurance exercise?. Endocr J 2011;58:211–5.

[34] Yeo N, Woo J, Shin K, Park J, Kang S. The effect of different exercise intensity on myokine and angiogenesis factors. J Sport Med Phys Fitness 2012;52:448–54.

[35] Akerstrom TC, Fischer CP, Plomgaard P, Thomsen C, van Hall G, Pedersen BK. Glucose ingestion during endurance training does not alter adaptation. J Appl Physiol 2009;106:1771–9.

[36] Blank MC, Bedarf JR, Russ M, Grosch-Ott S, Thiele S, Unger JK. Total body Na(+)-depletion without hyponatraemia can trigger overtraining-like symptoms with sleeping disorders and increasing blood pressure: explorative case and literature study. Med Hypotheses 2012;79:799–804.

[37] Ferrero-Miliani L, Nielsen OH, Andersen PS, Girardin SE. Chronic inflammation: importance of NOD2 and NALP3 in interleukin-1beta generation. Clin Exp Immunol 2007;147:227–35.

[38] Cotran RS, Kumar V, Collins T. Robbins pathologic basis of disease. Philadelphia: W.B Saunders Company; 1998.

[39] Tidball JG. Inflammatory processes in muscle injury and repair. Am J Physiol Integr Physiol 2005;288:R345–53.

[40] Mountziaris P, Spicer P, Kasper F, Mikos A. Harnessing and modulating inflammation in strategies for bone regeneration. Tissue Eng Part B Rev 2011;17:339–402.

[41] Leaper D, Schultz G, Carville K, Fletcher J, Swanson T, Drake R. Extending the TIME concept: what have we learned in the past 10 years? Int Wound J 2012;9(Suppl 2):1–19.

[42] Khattak M, Ahmad T, Rehman R, Umer M, Hasan SH, Ahmed M. Muscle healing and nerve regeneration in muscle contusion model in the rat. J Bone Joint Surg Br 2010;92:894–9.

[43] Loffreda S, Lin HZ, Karp CL, Brengman ML, Wang DJ, Klein AS, et al. Leptin regulates proinflammatory immune responses. FASEB J 1998;12:57–65.

[44] Esposito K, Marfella R, Giugliano G, Giugliano F, Ciotola M, Quagliaro L, et al. Inflammatory cytokine concentrations are acutely increased by hyperglycemia in humans: role of oxidative stress. Circulation 2002;106:2067–72.

[45] Petersen AM, Pedersen BK. The anti-inflammatory effect of exercise. J Appl Physiol 2005;98:1154–62.

[46] Pradhan AD, Manson JE, Rifai N, Buring JE, Ridker PM. C-reactive protein, interleukin 6, and risk of developing type 2 diabetes mellitus. JAMA 2001;286327–34. doi:10.1001/jama.286.3.327.

[47] Dehghan A, Kardys I, de Maat MP, Uitterlinden AG, Sijbrands EJ, Bootsma AH, et al. Genetic variation, C-reactive protein levels, and incidence of diabetes. Diabetes 2007;56:872–8.

[48] Dehghan A, van Hoek M, Sijbrands EJ, Stijnen T, Hofman A, Witteman JC. Risk of type 2 diabetes attributable to C-reactive protein and other risk factors. Diabetes Care 2007;30:2695–9.

[49] Cesari M, Penninx BW, Newman AB, Kritchevsky SB, Nicklas BJ, Sutton-Tyrell K, et al. Inflammatory Markers and onset of cardiovascular events: results from the Health ABC study. Circulation 2003;108:2317–22.

[50] Yeh ET, Willerson JT. Coming of age of C-reactive protein: using inflammation markers in cardiology. Circulation 2003;107:370–1.

[51] Spector TD, Hart DJ, Nandra D, Doyle DV, Mackillop N, Gallimore JR, et al. Low-level increases in serum C-reactive protein are present in early osteoarthritis of the knee and predict progressive disease. Arthritis Rheum 1997;40:723–7.

[52] Wilmore J. Physiology of sport and exercise. Champaign, IL: Human Kinetics; 2008. pp. 26–36, 98–120, 186–250, 213–218.

[53] Cannon JG, St Pierre BA. Cytokines in exertion-induced skeletal muscle injury. Mol Cell Biochem 1998;179:159–67.

[54] Lang CH, Hong-Brown L, Frost RA. Cytokine inhibition of JAK-STAT signaling: a new mechanism of growth hormone resistance. Pediatr Nephrol 2005;20:306–12.

[55] Pedersen BK, Toft AD. Effects of exercise on lymphocytes and cytokines. Br J Sports Med 2000;34:246–51.

[56] Bruunsgaard H, Galbo H, Halkjaer-Kristensen J, Johansen TL, MacLean DA, Pedersen BK. Exercise-induced increase in serum interleukin-6 in humans is related to muscle damage. J Physiol 1997;499(Pt 3):833–41.

[57] McKay BR, De Lisio M, Johnston AP, O'Reilly CE, Phillips SM, Tarnopolsky MA, et al. Association of interleukin-6 signalling with the muscle stem cell response following muscle-lengthening contractions in humans. PLoS One 2009;4:e6027.

[58] MacIntyre DL, Sorichter S, Mair J, Berg A, McKenzie DC. Markers of inflammation and myofibrillar proteins following eccentric exercise in humans. Eur J Appl Physiol 2001;84:180–6.

[59] Almada C, Cataldo LR, Smalley SV, Diaz E, Serrano A, Hodgson MI, et al. Plasma levels of interleukin-6 and interleukin-18 after an acute physical exercise: relation with post-exercise energy intake in twins. J Physiol Biochem 2013;69:85–95.

[60] Chalaris A, Schmidt-Arras D, Yamamoto K, Rose-John S. Interleukin-6 trans-signaling and colonic cancer associated with inflammatory bowel disease. Dig Dis 2012;30:492–9.

[61] Deng B, Wehling-Henricks M, Villalta SA, Wang Y, Tidball JG. IL-10 triggers changes in macrophage phenotype that promote muscle growth and regeneration. J Immunol 2012;189:3669–80.

[62] Xiao W, Chen P, Dong J. Effects of overtraining on skeletal muscle growth and gene expression. Int J Sports Med 2012;33:846–53.

[63] Militaru C, Donoiu I, Craciun A, Scorei ID, Bulearca AM, Scorei RI. Oral resveratrol and calcium fructoborate supplementation in subjects with stable angina pectoris: effects on lipid profiles, inflammation markers, and quality of life. Nutrition 2013;29(1):178–83.

[64] Alkharfy KM, Al-Daghri NM, Yakout SM, Ahmed M. N-Calcitriol attenuates weight-related systemic inflammation and ultrastructural changes in the liver in a rodent model. Basic Clin Pharmacol Toxicol 2013;112:42–9.

[65] Kavandi L, Collier MA, Nguyen H, Syed V. Progesterone and calcitriol attenuate inflammatory cytokines CXCL1 and CXCL2 in ovarian and endometrial cancer cells. J Cell Biochem 2012;113:3143–52.

[66] Vanoirbeek E, Krishnan A, Eelen G, Verlinden L, Bouillon R, Feldman D, et al. The anti-cancer and anti-inflammatory actions of 1,25(OH)$_2$D$_3$. Pract Res Clin Endocrinol Metab 2011;25:593–604.

[67] Okuno H, Kishimoto KN, Hatori M, Itoi E. 1α,25-dihydroxyvitamin D$_3$ enhances fast-myosin heavy chain expression in differentiated C2C12 myoblasts. Cell Biol Int 2012;36:441–7.

[68] Buitrago CG, Arango NS, Boland RL. 1α,25(OH)2D3-dependent modulation of Akt in proliferating and differentiating C2C12 skeletal muscle cells. J Cell Biochem 2012;113:1170–81.

[69] Buitrago C, Costabel M, Boland R. PKC and PTPα participate in Src activation by 1α,25OH2 vitamin D3 in C2C12 skeletal muscle cells. Mol Cell Endocrinol 2011;339:81–9.

[70] Garcia LA, King KK, Ferrini MG, Norris KC, Artaza JN. 1,25(OH)2vitamin D3 stimulates myogenic differentiation by inhibiting cell proliferation and modulating the expression of promyogenic growth factors and myostatin in C2C12 skeletal muscle cells. Endocrinology 2011;152:2976–86.

CHAPTER

51

An Overview on Caffeine

Brittanie M. Volk and Brent C. Creighton

Human Performance Laboratory, Department of Kinesiology, University of Connecticut, Storrs, CT, USA

INTRODUCTION

Caffeine, a naturally occurring stimulant, is the most widely used psychoactive drug in the world and a common constituent in the diet of many athletes [1]. First discovered around 2737 BC [2], caffeine was later popularized in Egypt in 850 AD [3]. Today, nearly 90% of adults in Europe and North America consume caffeine daily [4]. In its truest form, it is a bitter, white powder originating in over 60 plant sources, including coffee beans, cocoa beans, and tea leaves [5]. Caffeine is naturally found in a wide variety of foods and beverages and is artificially added to products such as colas, energy beverages, dietary supplements, chewing gum, bottled water, and medications [6].

Physiologically, caffeine has divergent effects on the central nervous system (CNS) and cardiovascular, metabolic, muscular, respiratory, and renal functions both during exercise and at rest. Increased alertness and cognition, enhanced mood, and improvements in performance-based tasks are well known responses of caffeine ingestion. These effects are considered safe for the average moderate consumer but vary significantly based on dose, frequency, and individual sensitivity. This chapter will explore the global effects of caffeine on performance outcomes, proposed mechanisms for modes of action, as well as the practical use and considerations of caffeine consumption.

CAFFEINE PHARMACOLOGY

Caffeine (1,3,7-trimethylxanthine) is a methylated purine base (xanthine alkaloid) formed as a product of purine degradation [7]. Purines are most commonly associated with the nucleotide bases adenine, guanine, and uracil, building blocks for DNA and RNA, respectively. The primary method of caffeine ingestion occurs through oral consumption, after which it is absorbed through the intestinal track and distributed to varying tissues and organs throughout the body. Appearance in the bloodstream following ingestion usually occurs within 30 minutes. The form of caffeine: coffee, chocolate, pills, etc., does not affect its rate of absorption [8] via the gastrointestinal route. One exception is caffeinated gum, which is absorbed orally through the buccal (cheek) mucosa more rapidly than through ingestion [9]. Caffeine pills, when taken independently (in laboratory settings or by the general consumer), are absorbed at a rapid rate. However, co-ingestion of caffeine with other substances may [10] or may not [11] slow the appearance of caffeine in the blood. Once absorbed, caffeine maintains a half-life that is largely dependent on dose and individual variation. For a dose less than 10 mg kg^{-1}, the half-life can range from 2.5 to 10 hours [12] with peak plasma concentrations seen 15–120 minutes after ingestion (variation is due in part to varying rates of gastric emptying) [13].

Caffeine metabolism occurs primarily in the liver and is carried out via the enzyme cytochrome P450 1A2, a member of the larger cytochrome P450 superfamily. Both the brain and kidney also produce P450 1A2, suggesting that these tissues may also readily metabolize caffeine [14,15]. Figure 51.1 displays the metabolic fate of caffeine and its first stage metabolites (paraxanthine, theobromine and theophylline), which are ultimately excreted in the urine. Like caffeine, its metabolites exhibit their own ergogenic effects and will be discussed later in the chapter.

The metabolism of caffeine is affected by a number of factors including gender, exercise, diet, genetics, regularity/frequency of caffeine consumption, as well as the use of specific drugs such as acetaminophen [16]. Exercise can up-regulate P450 1A2 [17], which

Nutrition and Enhanced Sports Performance.
DOI: http://dx.doi.org/10.1016/B978-0-12-396454-0.00051-5

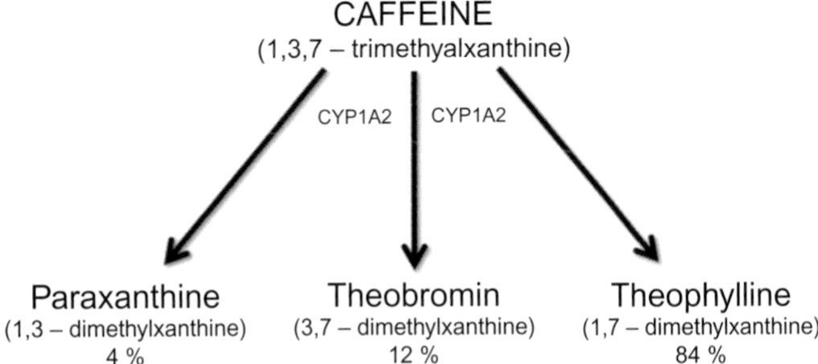

FIGURE 51.1 Caffeine's primary metabolites and representative proportions via metabolism.

increases the rate of caffeine metabolism, as well as decreases peak plasma levels and half-life [18], likely due to increased circulation. Flavenoids, smoking [19], and regular consumption of cruciferous vegetables [17] are additional factors that can increase the rate of caffeine metabolism. Conversely, alcohol has been shown to inhibit caffeine metabolism [16]. Disease states, most notably liver disease, can lead to the accumulation of caffeine, thus increasing its half-life [20].

As with many cellular enzymes, the chronic ingestion of caffeine leads to the up-regulation and increased affinity of cytochrome P450 1A2 for caffeine [15]. This may help explain why habitual caffeine users do not observe the same stimulant effects as novel consumers and why athletes choose to refrain from caffeine consumption leading up to an athletic event. This cessation lowers caffeine tolerance, enhancing its effects once it is reintroduced.

Caffeine is most well known for its ability to induce wakefulness and restore alertness by acting as a central nervous system (CNS) stimulant. The interaction of caffeine with the CNS and brain is the primary mode of caffeine's temporary cognitive arousal. Caffeine, both water and lipid soluble, readily crosses the blood–brain barrier and in doing so increases several essential neurotransmitters that are critical to its physiological actions [21]. Additional effects are discussed in the following section, including mechanisms of catecholamine release, potassium release, lipolysis, levels of intracellular calcium, respiratory ventilation, and inhibition of cAMP phosphodiesterase. These stimulatory effects are thought to occur primarily through adenosine receptor antagonism (the binding of the caffeine molecule to the adenosine receptor) [22].

MECHANISMS OF CAFFEINE AND EFFECTS ON PERFORMANCE

Both caffeine and methylxanthines have been shown to improve athletic performance via central and peripheral mechanisms, although results can be equivocal. Caffeine's ability to cross membranes of various tissues throughout the body makes it difficult to distinguish between cognitive CNS effects and peripheral effects such as muscular performance. It is likely that, rather than being centered on one independent mechanism, the paradigm for improved performance from caffeine use is multifactorial [23]. Countless publications with strong scientific standards have investigated caffeine and its global effects, but many of the conclusions still require additional research. Numerous factors influence these discrepancies: methodology, population of interest (gender/ethnicity), age, dose of caffeine given, prior caffeine use (habituation), and cessation. Commonly consumed in a variety of forms by athletes of many sport disciplines, caffeine is a well-known ergogenic aid in a range of doses of 3–6 mg kg^{-1} and even as low as 1–3 mg kg^{-1}[24]. These ergogenic effects and their proposed mechanisms are explored below.

Performance Effects

The ergogenic effects of caffeine on endurance exercise are well established [24,25], whereas influences on sprint and power activities provide equivocal results [26]. Increased time-to-exhaustion and greater work output are commonly observed when caffeine is consumed before and/or during endurance exercise, primarily in running and cycling [27]. Other observed benefits of caffeine include decreased time to complete a fixed workout and increased amount of work performed in a set amount of time.

Research looking at anaerobic performance has focused mostly on bouts of 4–180 seconds [26]. Improved peak power, speed and isokinetic strength are seen in events lasting 10 seconds or less [28]. This is likely due to improved muscular force production resulting from increased intracellular Ca^{2+} concentration and improved Na^+/K^+ ATPase pump activity [29,30]. Additionally, reaction time improves with caffeine ingestion. During bouts of 15 seconds to 3 minutes, with a primary reliance on anaerobic glycolysis, caffeine shows no impact on sprint and power performance and a negative effect with repeated bouts. This may be due to an increase in ammonia levels and decrease in intracellular pH [31].

Little research has been carried out to examine isokinetic peak torque, isometric maximal force, muscular endurance for upper body musculature, and maximal strength, all of which seem to be minimally affected by caffeine [26]. Thirty-second Wingate protocols are a simple model to use for anaerobic assessment, but caffeine shows minimal effect [26]. Conversely, caffeine can have moderately to highly favorable effects on speed endurance (60–180 seconds in duration), high-intensity exercise such as sprinting or sprint cycling power, and studies attempting to mimic sporting activities with quick movements (i.e., 4–6-second bursts of soccer, rugby, lacrosse, football activities) [26,32].

Perceived Exertion

Ratings of perceived exertion (RPE) is a way to measure the central effects of caffeine on performance and has been studied under numerous endurance exercise conditions [33]. Caffeine lowers RPE during and after exercise by 5.6% compared with placebo, according to a meta-analysis [34]. This may explain nearly 30% of the performance increases seen with caffeine ingestion.

It has also been demonstrated that caffeine can improve performance and power output in subjects with similar RPE, heart rates and leg pain ratings, compared with placebo trials [35]. In endurance-trained individuals, a moderate dose of caffeine taken 1 hour prior to exercise enhances the athlete's perception of the exercise and helps maintain a more positive experience during a prolonged workout [33]. Interestingly, caffeine has been shown to help lower pain perception in the heat but not in cold conditions [36].

Extensive research surrounding the effects of caffeine on perceived exertion in endurance exercise is well documented; however, less research has been conducted examining caffeine's effects on RPE in anaerobic exercise. Research conducted using high intensity exercise has shown no change [37,38], decreased [27], or even increased RPE [39] with caffeine use. However, when RPE remains the same during anaerobic activity, several studies show increases in total work and peak power, suggesting that caffeine does reduce perceived exertion but is not necessarily detected by RPE scales [26]. During resistance exercise is well documented; however, RPE remains similar with and without caffeine consumption. Interestingly, an increase in repetitions (more work) has been shown, which suggests a reduction in perceived exertion [38,40].

Overall, caffeine reduces RPE with and without caffeine during most prolonged exercise. Absence of differences in RPE can often be explained by greater total work, because the athletes are able to work at a greater intensity or extend the duration of exercise with the use of caffeine. Examination of RPE following caffeine ingestion and resistance exercise shows equivocal results. When no change or an increase in RPE occurs, performance is often improved with caffeine consumption. As with all caffeine use, the dose required to exert these effects without accompanying negative effects vary by individual.

Mechanisms

Adenosine Receptor Antagonism

Though caffeine's ergogenic effects are likely due to a combination of actions, adenosine receptor antagonism is the most favored theory [26,41]. This mechanism plays a role in central fatigue and thus alters RPE, pain perception, and delays fatigue, all of which can alter exercise performance.

Adenosine, a purine nucleoside, functions both as an inhibitory neurotransmitter and neuromodulator [42]. It is well known for its critical roles in energy transfer as a component of adenosine tri- and diphosphate (ATP/ADP) and as a signal transducer in the form of cyclic adenosine monophosphate (cAMP). The molecular similarity of caffeine (a methylated purine) to adenosine (a purine nucleoside) illustrates why caffeine competitively binds to the adenosine receptors [43] (Figure 51.2).

Adenosine's primary role is thought to be neuroprotective: promoting sleep, suppressing arousal, and reducing motor activity through accumulation of adenosine following prolonged mental activity [44]. It does so by inhibiting the major neurotransmitters: serotonin, dopamine, acetylcholine, epinephrine, norepinephrine, and glutamate [45,46]. Thus, caffeine functions to oppose adenosine and fight off "fatigue" observed with normal adenosine binding. In athletic performance when energy demands are challenged, caffeine can clearly provide a stimulus to override or delay these feelings of fatigue, making it a practical approach for improving response to physical and mental demands [45].

The downstream effects of caffeine consumption are modulated by the binding of caffeine to three of the four adenosine receptors: A_1, A_{2A}, and A_{2B} [47,48].

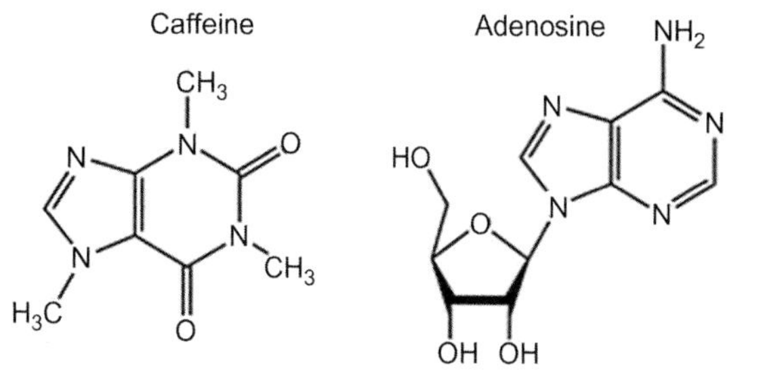

FIGURE 51.2 Structural comparison of caffeine and adenosine.

Caffeine's metabolites, theophylline and theobromine, also act as antagonists and bind to adenosine A_1 and A_2 receptors. Adenosine receptors are present throughout the nervous system and can also be found in muscle, adipose tissue, heart and vascular endothelium [49,50]. Thus, caffeine's effects depend on which adenosine receptor it binds to and where the receptor is located. Adenosine receptor binding is widely thought to occur presynaptically [50].

The direct cognitive stimulatory effects of caffeine appear to be linked to the blockade of presynaptic A_1 receptors and postsynaptic A_1 and A_{2A} receptors. Once activated, A_1 receptors have been shown to respond by opening potassium and chloride channels [47]. This conversely causes calcium channel inhibition, which consequently decreases voltage-dependent neurotransmitter release or adenosine release [47]. By preventing the release of adenosine (or simply blocking it) and subsequently removing its inhibitory influence, caffeine increases the release of excitatory neurotransmitters, increasing neuronal excitability [51]. The neurotransmitters identified as largely responsible for caffeine's excitatory effects include dopamine, serotonin, and gamma-aminobutyric acid (GABA) [52].

Metabolic and Hormonal Effects

Among the mechanisms supporting the ergogenic effects of caffeine, increased fat oxidation and glycogen sparing may be the most contentious. For many years, improved endurance performance via caffeine was thought to be the result of increased fat oxidation (fat breakdown) from adipose tissue and intramuscular triglycerides, which increases the circulation of fatty acids [53,54] for enhanced energy availability. The mobilization of fatty acids into the blood can help delay carbohydrate utilization and spare glycogen stores, a useful strategy for endurance athletes looking to increase exercise time to exhaustion [53]. Many studies, however, have been cited as having weak methodology and a small sample size [55]. If caffeine does play a role in glycogen sparing, it is likely that these effects occur during the initial 15–30 minutes of exercise [56,57]. As described previously, the central nervous system effects seen with caffeine ingestion are now believed more likely to be the largest contributor to performance improvements [26].

Still, increased blood epinephrine levels seen in some studies [58,59] following caffeine consumption help support these metabolic effects, as epinephrine is a potent stimulator of lipolysis. Contrastingly, recent scientific literature suggests that caffeine may not directly favor the sparing of glycogen or increase fat oxidation; rather they are independent of epinephrine release [60]. Possible individual differences and genetic variation may be at the root of this controversy. Ultimately, more research is needed to fully determine caffeine's role in fat oxidation and carbohydrate sparing.

Caffeine has also been shown to directly affect glycogen metabolism independently of any impact on fat oxidation. Essential to the metabolism of glycogen, glycogen phosphorylase (GP) is responsible for cleaving glucose units from glycogen during glycolysis. Caffeine binds to GP at the purine inhibitor site [61], inhibiting glycogen breakdown [12]. Conversion of the enzyme while in the active form prevents the cleavage of glucose from glycogen, promoting glycogen storage. Endurance athletes may benefit from this delayed response in glycogen breakdown, as glycogen perseveration in the initial stages of exercise can help prolong energy stores (glycogen) for use later in the event.

Additional metabolic effects occur through caffeine's competition with phosphodiesterase inhibitors. Methylated xanthines, such as caffeine, function as competitive nonselective inhibitors of phosphodiesterases [62]. Molecules that inhibit phosphodiesterase enzymes (PDEs) prevent the cleavage of a phosphodiester bond and inhibit the inactivation of two important intracellular second messengers: cyclic adenosine monophosphate (cAMP) and cyclic guanosine monophosphate (cGMP). Both second messengers are critically important for intracellular signaling (relaying signals from the cell surface to the cell interior and amplifying the signal). Interestingly, cAMP is important in the regulation of fat oxidation, and caffeine may promote the breakdown of fat [12]. Caffeine may also inhibit PDEs by up to 40% [52], increasing intracellular concentrations of both cAMP and cGMP [63]. Increased levels of these second messengers may lead to further amplification (specific to the ligand, receptor, and intracellular message being triggered); however, this concept is speculative. It should be noted that many studies investigating this proposed mechanism have administered doses well above normal caffeine intakes; it is unlikely that such levels would be reached under normal physiological conditions.

Ion Flux in Skeletal Muscle

Caffeine along with its three primary metabolites are thought to promote Ca^{2+} release from the sarcoplasmic reticulum (SR) [64] while simultaneously reducing the rate of Ca^{2+} reuptake by the SR [65]. These compounds interfere with the excitation-contraction coupling mechanism. In short, caffeine causes the activation of ryanodine-sensitive Ca^{2+} channels in the SR and may render the ryanodine receptors more sensitive to activators such as Ca^{2+} and ATP [12,52]. A great deal of human research is still needed to identify the exact mechanism whereby methylxanthines initiate Ca^{2+} efflux, as much existing literature supporting these theories has been observed using animal models [66].

Increased Na^+/K^+-ATPase (Na^+/K^+ pump) activity is a second ion-related mechanism thought to occur following caffeine ingestion, with paraxanthine acting more potently than caffeine [12]. The Na^+/K^+-ATPase functions primarily to maintain cellular resting membrane potential and cell volume. Stimulation of the pump in inactive skeletal muscle can result in increased K^+ uptake and Na^+ efflux. Exercising muscle causes a rise in plasma K^+, and an increased discharge of Na^+ could help to maintain the membrane potential [67]. This mechanism is thought to occur indirectly on the pump, possibly involving Ca^{2+} release or cAMP as a result of caffeine binding [67]. Either of these two proposed mechanisms may play a role in increased skeletal muscle force production [46]. Overall, it is difficult to assess exactly how caffeine affects such ion-handling mechanisms and to what extent they aid in exercise performance. Like PDE inhibition, ion flux changes are unlikely to occur with normal physiological doses of caffeine.

Respiratory Effects

Caffeine's ability to effect breathing capacity is an area of interest to many athletes, particularly since research has shown improvements in CO_2 sensitivity [68], VO_{2max} [69,70], and minute ventilation. Theophylline, a metabolite of caffeine, is used to treat asthma (the therapeutic dose used is higher than physiological levels seen with caffeine metabolism) as it has been shown to relax the smooth muscles of the bronchi. Thus far, two proposed mechanisms have been postulated to explain caffeine's effect: (i) increased epinephrine is believed to be responsible for decreasing bronchodilation [71] while (ii) adenosine receptor blockade may increase respiration [72].

SOURCES OF CAFFEINE

Caffeine is found both naturally and artificially in many foods, beverages, and medications. Coffee, however, is the world's leading source of caffeine [73], with over 50% of adults in America consuming an average of three cups each day [74]. Soft drinks, tea [73], and energy drinks comprise the other more common dietary sources. Caffeine levels present in these sources can be highly variable. For example, one cup of brewed coffee can contain an amount of caffeine ranging from 70 to 280 mg [75], depending on the brand and method of preparation. See Table 51.1 for the caffeine content of selected sources.

The use of caffeine is popular in sports, with nearly 70% of athletes regularly consuming caffeine [25]. Once prohibited by the World Anti-doping Agency (WADA) and other governing organizations, caffeine

TABLE 51.1 Caffeine Content of Popular Foods, Beverages, and Medications

Beverage	**Caffeine (mg)**
Coffee (8 fl oz unless otherwise noted)	
Coffee, Brewed	108
Coffee, Decaf, Brewed	6
Coffee, Decaf, Instant	3
Coffee, Drip	145
Coffee (Instant)	57
Espresso, 1.5 fl oz	77
Dunkin Donuts Coffee, Brewed, medium	178
Dunkin Donuts Coffee, Latte, medium	97
Dunkin Donuts, Espresso, single shot (small)	75
Starbucks Coffee, grande	330
Starbucks Coffee, Decaf, grande	13
Tea (8 fl oz unless otherwise noted)	
Tea, Black	47
Tea, Green	25
Tea, White	15
Tea, Iced	47
Nesta, Iced Tea, 16 fl oz	34
Tea, Instant	26
Soft Drinks (12 fl oz)	
Coca-Cola Classic	34
Diet Coke	45
Pepsi Cola	38
Diet Pepsi	50
Mountain Dew	54
A&W Cream Soda	29
Energy Drinks (16 fl oz unless otherwise noted)	
Amp	142
Full throttle	144
NOS	260
Monster	160
Monster, Absolutely Zero	135
Redbull, 8.4 fl oz	80
Rockstar Zerb Carb	240
5 Hour Energy, 2 fl oz	138
Vitamin Water Energy Citrus, 20 fl oz	42

* Caffeine data was obtained from the USDA National Nutrient Database (http://ndb.nal.usda.gov/) and from manufacturers via product labels and websites.

is now legally permitted for use in sports by WADA, the International Olympic Committee (IOC), and the National Collegiate Athletic Association (NCAA). One year following caffeine's removal from the list of banned substances, detection among elite athletes for caffeine via urinary analysis [76] showed that its frequency of use hadn't changed. The IOC and NCAA continue to monitor the use of caffeine but allow doses that appear in the urine as $\leq 15\,\mu g\,mL^{-1}$, which would result from the consumption of approximately 800 mg of caffeine by the average adult [77], twice the amount of the suggested moderate intake and far above what can be seen as ergogenic.

Endurance athletes, especially marathoners, triathletes and those competing in other ultra-endurance events [78], commonly supplement with caffeine prior to and during events (gels, fruit chews, cola, sports drinks, energy bars, etc.). The use of such products varies depending on the individual athlete and sport demands. Furthermore, stimulatory effects can be optimized when habitual consumers abstain from caffeine for at least 7 days prior to an athletic event [24].

Energy Drinks

Well known to most, the primary and most physiologically active ingredient in the majority of energy drinks is caffeine. With increased use, these beverages can be a significant source of caffeine in the diet. Since the introduction of Red Bull to the American market in 1997, the sales and consumption of energy beverages has grown remarkably [4], with the United States being one of the top ten consuming countries. This 5.7 billion dollar industry markets more than 500 energy drinks worldwide [79]. Caffeine content among energy drinks varies but typically ranges from 80 to 140 mg per 8 ounces [80], compared with 100 mg in an average cup of brewed coffee or 23 mg in a cup of cola. Additionally, many cans and bottles consist of two or three 8-ounce servings and can contain nearly 400 mg of caffeine (energyfiend.com).

Both safety and frequency of consumption remain a common concern with energy drinks due to the lack of regulatory control over the ingredients. There is significant discrepancy in the regulation of energy products between countries; in the USA, there is little to no control of the sale, promotion, or use of caffeinated beverages. Although the amounts and details of specific ingredients are not required to be printed on the label, there have been no identified negative effects associated with common ingredients (B vitamins, ginseng, guarana, and taurine) in the amounts found within most energy drinks [81]. Many side effects reported are likely due to the high levels of caffeine.

Additionally concerning is the amount of sugar contained within a majority of energy drinks. Sugar-free alternatives are available; however, most energy drinks contain 25–35 grams (6–7 teaspoons) of sugar per 8 ounces. When there are two or more servings per container, as much as 100 grams of sugar can be consumed. As a source of calories and "energy," these beverages are often placed near sports drinks and are thus commonly chosen by athletes. Research surrounding energy beverages and their effect on performance is mixed, but the most likely source of stimulatory effects is caffeine. When consumed in moderation, the use of energy beverages for a caffeine "kick" can be safe for most individuals. Nevertheless, just as with other caffeine sources, it is recommended that moderate amounts be consumed, especially due to variation in individual sensitivity.

SAFETY AND SIDE EFFECTS

The US Food and Drug Administration has classified caffeine under the category of substances "Generally Recognized as Safe" (GRAS). A moderate caffeine intake of ≤ 400 mg day^{-1} for a healthy adult is considered safe without adverse side effects [5,82]. While just a recommendation, some individuals may experience adverse effects with less than this dose and others may tolerate more. As mentioned previously, ingestion of caffeine is known to have effects on the cardiorespiratory, endocrine, and neurological systems, but effects vary among individuals. The quality needed to produce adverse effects, such as restlessness, nervousness, insomnia, diarrhea, and headaches [83], varies based on individual sensitivity and the dose consumed per kg bodyweight [84]. Other common side effects seen from caffeine intake include rapid heartbeat, tremors, abdominal pain, nausea, vomiting, and diuresis [84]. Additionally, there are potential effects on performance, including sleep disturbance, gastric upset, and inability to focus. Redosing of caffeine regularly throughout the day negatively impacts sleep habits [85], so timing of consumption is an important consideration.

Tolerance and Withdrawal

Regular consumers of caffeine can develop a physical dependence, and over time the stimulatory effects are reduced as a tolerance develops. Caffeine has addictive properties; when habitual consumption abruptly stops, symptoms of withdrawal include headaches, fatigue, irritability, muscle pain, difficulty focusing, mood changes, nausea or vomiting [84]. Significant withdrawal symptoms can be noticed with intakes as low as 100 mg day^{-1} (one cup of coffee) but are more prevalent for greater amounts [86].

Symptoms experienced with caffeine overdose include irregular heartbeat, vomiting, agitation, and delirium. Again, the amount of caffeine needed to experience these symptoms varies based on individual tolerance. An even greater caffeine ingestion of 15–30 mg kg^{-1} body weight, or 1000–2000 mg of caffeine, can result in severe toxicity [84]. In cases of extreme intake, it is even possible to consume a life-threatening dose. The LD-50, or lethal oral dose required to kill 50% of the population for caffeine, is 150–200 mg kg^{-1} body weight [87]. Though unrealistic, it is not impossible to consume this amount, which equates to about 80–100 cups of coffee.

Caffeine Effects on Fluid and Electrolyte Status (a Common Misconception)

According the American College of Sports Medicine, "Caffeine ingestion has a modest diuretic effect in some individuals but does not affect water replacement on habitual caffeine users, so caffeinated beverages can be ingested during the day by athletes who are not caffeine naïve". Therefore, for individuals who commonly consume caffeine, moderate amounts will not increase urine output more than would a comparable amount of water. A review of thirteen studies examined the effect of caffeine on hydration status and concluded that fluid balance resulting from the intake of caffeinated beverages was not statistically different from control fluids of water or a placebo [88]. Still, it is often recommended by nutrition and medical professionals to refrain from or limit the use of caffeine around athletic events, because of the diuretic effects. While caffeine does acutely promote the excretion of urine by increasing blood flow to the kidneys, it does not affect 24-hour fluid balance [89] and is not the cause of dehydration as many still believe. Furthermore, during exercise, blood flow is shunted away from the kidneys towards exercising muscle and skin for heat dissipation. Therefore, the acute diuretic effects of caffeine are likely blunted during exercise in non-naïve individuals. Still, the safest and most appropriate approach is to consume caffeine only in amounts that can be tolerated based on individual responses in order to avoid potential adverse performance effects.

Dehydration can result in an electrolyte imbalance. Therefore, the acute fluid loss that accompanies caffeine consumption raises a concern of electrolyte imbalance. Sodium (Na^+) and potassium (K^+) are two of the primary electrolytes responsible for homeostatic maintenance of fluid balance (controlling water loss at the kidneys). Acutely, caffeine acts on the renal nephrons to inhibit the reabsorption of sodium for a period of increased losses. However, similar to 24-hour fluid balance, chronic sodium balance is not affected. Likewise, research shows that urinary potassium levels over a 24-hour period do not change with caffeine consumption. Since electrolyte and fluid indices are not chronically affected by moderate caffeine consumption, the notion of having to restrict caffeine for these reasons is not supported [88].

SUMMARY

Caffeine is consumed globally through a variety of sources. It is both a popular stimulant known for its effects on mood and cognitive enhancement, as well as its role as an ergogenic aid in athletics. Primarily for running and cycling endurance events, 3–6 mg kg^{-1} caffeine helps to increase time-to-exhaustion, promote a greater work capacity, and reduce perceived exertion. Caffeine peaks in the blood within 30–75 minutes after consumption, so timing is an important component to its effects on performance. A person's caffeine intake, including dosing and timing, should be based on individual tolerance. The effect of caffeine weakens over time in habitual users, so to maximize athletic performance, one should abstain from caffeine for at least 7 days prior to an athletic event. Likewise, if using caffeine to enhance performance, an individual should experiment with caffeine intake prior to the event, as it can produce negative side effects in the caffeine naïve individual. An average intake of <400 mg/day is shown to be safe for a healthy person. Therefore, the notion of having to limit moderate caffeine consumption due to its chronic effects on hydration and electrolyte indices is not supported. Provided that one remembers the importance of dosing, timing, and frequency of caffeine intake, combined with a person's habitual intake, caffeine can be a successful tool to enhance cognition, alertness, and athletic performance.

References

[1] Chester N, Wojek N. Caffeine consumption amongst British athletes following changes to the 2004 WADA prohibited list. Int J Sports Med 2008;29(6):524–8.

[2] Arab L, Blumberg JB. Introduction to the proceedings of the fourth international scientific symposium on tea and human health. J Nutr 2008;138(8):1526S–8S.

[3] Chou T. Wake up and smell the coffee. Caffeine, coffee, and the medical consequences. West J Med 1992;157(5):544–53.

[4] Reissig CJ, Strain EC, Griffiths RR. Caffeinated energy drinks—a growing problem. Drug Alcohol Depend 2009;99(1–3):1–10.

[5] Heckman MA, Weil J, Gonzalez de Mejia E. Caffeine (1, 3, 7-trimethylxanthine) in foods: a comprehensive review on consumption, functionality, safety, and regulatory matters. J Food Sci 2010;75(3):R77–87.

[6] Mednick SC, Cai DJ, Kanady J, Drummond SP. Comparing the benefits of caffeine, naps and placebo on verbal, motor and perceptual memory. Behav Brain Res 2008;193(1):79–86.

[7] Spiller GA. Caffeine. Boca Raton: CRC Press; 1998.
[8] Mumford GK, Benowitz NL, Evans SM, et al. Absorption rate of methylxanthines following capsules, cola and chocolate. Eur J Clin Pharmacol 1996;51(3–4):319–25.
[9] Kamimori GH, Karyekar CS, Otterstetter R, et al. The rate of absorption and relative bioavailability of caffeine administered in chewing gum versus capsules to normal healthy volunteers. Int J Pharm 2002;234(1–2):159–67.
[10] Cox GR, Desbrow B, Montgomery PG, et al. Effect of different protocols of caffeine intake on metabolism and endurance performance. J Appl Physiol 2002;93(3):990–9.
[11] Graham TE, Hibbert E, Sathasivam P. Metabolic and exercise endurance effects of coffee and caffeine ingestion. J Appl Physiol 1998;85(3):883–9.
[12] Jones G. Caffeine and other sympathomimetic stimulants: modes of action and effects on sports performance. Essays Biochem 2008;44:109–23.
[13] Grab FL, Reinstein JA. Determination of caffeine in plasma by gas chromatography. J Pharm Sci 1968;57(10):1703–6.
[14] Agundez JA, Gallardo L, Martinez C, Gervasini G, Benitez J. Modulation of CYP1A2 enzyme activity by indoleamines: inhibition by serotonin and tryptamine. Pharmacogenetics 1998;8 (3):251–8.
[15] Goasduff T, Dreano Y, Guillois B, Menez JF, Berthou F. Induction of liver and kidney CYP1A1/1A2 by caffeine in rat. Biochem Pharmacol 1996;52(12):1915–9.
[16] Le Marchand L, Franke AA, Custer L, Wilkens LR, Cooney RV. Lifestyle and nutritional correlates of cytochrome CYP1A2 activity: inverse associations with plasma lutein and alpha-tocopherol. Pharmacogenetics 1997;7(1):11–9.
[17] Vistisen K, Poulsen HE, Loft S. Foreign compound metabolism capacity in man measured from metabolites of dietary caffeine. Carcinogenesis 1992;13(9):1561–8.
[18] Collomp K, Caillaud C, Audran M, Chanal JL, Prefaut C. Effect of acute or chronic administration of caffeine on performance and on catecholamines during maximal cycle ergometer exercise. C R Seances Soc Biol Fil 1990;184(1):87–92.
[19] Swanson JA, Lee JW, Hopp JW, Berk LS. The impact of caffeine use on tobacco cessation and withdrawal. Addict Behav 1997;22 (1):55–68.
[20] Verbeeck RK. Pharmacokinetics and dosage adjustment in patients with hepatic dysfunction. Eur J Clin Pharmacol 2008;64 (12):1147–61.
[21] Nehlig A, Daval JL, Debry G. Caffeine and the central nervous system: mechanisms of action, biochemical, metabolic and psychostimulant effects. Brain Res Brain Res Rev 1992;17 (2):139–70.
[22] Fredholm BB. Astra award lecture. Adenosine, adenosine receptors and the actions of caffeine. Pharmacol Toxicol 1995;76 (2):93–101.
[23] Sokmen B, Armstrong LE, Kraemer WJ, et al. Caffeine use in sports: considerations for the athlete. J Strength Cond Res 2008;22(3):978–86.
[24] Ganio MS, Klau JF, Casa DJ, Armstrong LE, Maresh CM. Effect of caffeine on sport-specific endurance performance: a systematic review. J Strength Cond Res 2009;23(1):315–24.
[25] Keisler BD, Armsey II TD. Caffeine as an ergogenic aid. Curr Sports Med Rep 2006;5(4):215–9.
[26] Davis JK, Green JM. Caffeine and anaerobic performance: ergogenic value and mechanisms of action. Sports Med 2009;39 (10):813–32.
[27] Doherty M, Smith P, Hughes M, Davison R. Caffeine lowers perceptual response and increases power output during high-intensity cycling. J Sports Sci 2004;22(7):637–43.
[28] Schneiker KT, Bishop D, Dawson B, Hackett LP. Effects of caffeine on prolonged intermittent-sprint ability in team-sport athletes. Med Sci Sports Exerc 2006;38(3):578–85.
[29] Clausen T. Na+ -K+ pump regulation and skeletal muscle contractility. Physiol Rev 2003;83(4):1269–324.
[30] Herrmann-Frank A, Luttgau HC, Stephenson DG. Caffeine and excitation-contraction coupling in skeletal muscle: a stimulating story. J Muscle Res Cell Motil 1999;20(2):223–37.
[31] Williams JH, Signorile JF, Barnes WS, Henrich TW. Caffeine, maximal power output and fatigue. Br J Sports Med 1988;22 (4):132–4.
[32] Burke LM. Caffeine and sports performance. Appl Physiol Nutr Metab 2008;33(6):1319–34.
[33] Backhouse SH, Biddle SJ, Bishop NC, Williams C. Caffeine ingestion, affect and perceived exertion during prolonged cycling. Appetite 2011;57(1):247–52.
[34] Doherty M, Smith PM. Effects of caffeine ingestion on exercise testing: a meta-analysis. Int J Sport Nutr Exerc Metab 2004;14 (6):626–46.
[35] Astorino TA, Roupoli LR, Valdivieso BR. Caffeine does not alter RPE or pain perception during intense exercise in active women. Appetite 2012;59(2):585–90.
[36] Ganio MS, Johnson EC, Klau JF, et al. Effect of ambient temperature on caffeine ergogenicity during endurance exercise. Eur J Appl Physiol 2011;111(6):1135–46.
[37] Astorino TA, Terzi MN, Roberson DW, Burnett TR. Effect of caffeine intake on pain perception during high-intensity exercise. Int J Sport Nutr Exerc Metab 2011;21(1):27–32.
[38] Green JM, Wickwire PJ, McLester JR, et al. Effects of caffeine on repetitions to failure and ratings of perceived exertion during resistance training. Int J Sports Physiol Perform 2007;2 (3):250–9.
[39] Crowe MJ, Leicht AS, Spinks WL. Physiological and cognitive responses to caffeine during repeated, high-intensity exercise. Int J Sport Nutr Exerc Metab 2006;16(5):528–44.
[40] Hudson GM, Green JM, Bishop PA, Richardson MT. Effects of caffeine and aspirin on light resistance training performance, perceived exertion, and pain perception. J Strength Cond Res 2008;22(6):1950–7.
[41] Kalmar JM, Cafarelli E. Caffeine: a valuable tool to study central fatigue in humans? Exerc Sport Sci Rev 2004;32(4):143–7.
[42] Fernstrom MH, Bazil CW, Fernstrom JD. Caffeine injection raises brain tryptophan level, but does not stimulate the rate of serotonin synthesis in rat brain. Life Sci 1984;35(12):1241–7.
[43] Fisone G, Borgkvist A, Usiello A. Caffeine as a psychomotor stimulant: mechanism of action. Cell Mol Life Sci 2004;61 (7–8):857–72.
[44] Basheer R, Strecker RE, Thakkar MM, McCarley RW. Adenosine and sleep-wake regulation. Prog Neurobiol 2004;73 (6):379–96.
[45] Sinclair CJ, Geiger JD. Caffeine use in sports. A pharmacological review. J Sports Med Phys Fitness 2000;40(1):71–9.
[46] Spriet L. Caffeine. In: Bahrke MaY CE, editor. Performance enhancing supplements in sport and exercise. Champaign, IL: Human Kinetics; 2002.
[47] Olah ME, Stiles GL. Adenosine receptor subtypes: characterization and therapeutic regulation. Annu Rev Pharmacol Toxicol 1995;35:581–606.
[48] Daly JaF BB. Mechanism of action of caffeine on the nervous system. In: Nehlig A, editor. Coffee, tea, chocolate, and the brain. Boca Raton, FL: CRC Press, LLC; 2004. p. 1–12.
[49] Dixon AK, Gubitz AK, Sirinathsinghji DJ, Richardson PJ, Freeman TC. Tissue distribution of adenosine receptor mRNAs in the rat. Br J Pharmacol 1996;118(6):1461–8.

[50] Fredholm BB, Battig K, Holmen J, Nehlig A, Zvartau EE. Actions of caffeine in the brain with special reference to factors that contribute to its widespread use. Pharmacol Rev 1999;51 (1):83–133.
[51] Meeusen R, De Meirleir K. Exercise and brain neurotransmission. Sports Med 1995;20(3):160–88.
[52] Daly JW. Caffeine analogs: biomedical impact. Cell Mol Life Sci 2007;64(16):2153–69.
[53] Costill DL, Dalsky GP, Fink WJ. Effects of caffeine ingestion on metabolism and exercise performance. Med Sci Sports 1978;10 (3):155–8.
[54] Ivy JL, Costill DL, Fink WJ, Lower RW. Influence of caffeine and carbohydrate feedings on endurance performance. Med Sci Sports 1979;11(1):6–11.
[55] Graham TE, Battram DS, Dela F, El-Sohemy A, Thong FS. Does caffeine alter muscle carbohydrate and fat metabolism during exercise? Appl Physiol Nutr Metab 2008;33(6):1311–8.
[56] Spriet LL, MacLean DA, Dyck DJ, Hultman E, Cederblad G, Graham TE. Caffeine ingestion and muscle metabolism during prolonged exercise in humans. Am J Physiol 1992;262(6 Pt 1): E891–8.
[57] Graham TE, Rush JW, van Soeren MH. Caffeine and exercise: metabolism and performance. Can J Appl Physiol 1994;19 (2):111–38.
[58] Chesley A, Hultman E, Spriet LL. Effects of epinephrine infusion on muscle glycogenolysis during intense aerobic exercise. Am J Physiol 1995;268(1 Pt 1):E127–34.
[59] Chesley A, Howlett RA, Heigenhauser GJ, Hultman E, Spriet LL. Regulation of muscle glycogenolytic flux during intense aerobic exercise after caffeine ingestion. Am J Physiol 1998;275 (2 Pt 2):R596–603.
[60] Van Soeren M, Mohr T, Kjaer M, Graham TE. Acute effects of caffeine ingestion at rest in humans with impaired epinephrine responses. J Appl Physiol 1996;80(3):999–1005.
[61] Tsitsanou KE, Skamnaki VT, Oikonomakos NG. Structural basis of the synergistic inhibition of glycogen phosphorylase a by caffeine and a potential antidiabetic drug. Arch Biochem Biophys 2000;384(2):245–54.
[62] Essayan DM. Cyclic nucleotide phosphodiesterases. J Allergy Clin Immunol 2001;108(5):671–80.
[63] Ribeiro JA, Sebastiao AM. Caffeine and adenosine. J Alzheimers Dis 2010;20(Suppl. 1):S3–15.
[64] Hawke TJ, Allen DG, Lindinger MI. Paraxanthine, a caffeine metabolite, dose dependently increases [Ca(2+)](i) in skeletal muscle. J Appl Physiol 2000;89(6):2312–7.
[65] Tarnopolsky MA. Caffeine and endurance performance. Sports Med 1994;18(2):109–25.
[66] Kumbaraci NM, Nastuk WL. Action of caffeine in excitation-contraction coupling of frog skeletal muscle fibres. J Physiol 1982;325:195–211.
[67] Lindinger MI, Willmets RG, Hawke TJ. Stimulation of Na+, K (+)-pump activity in skeletal muscle by methylxanthines: evidence and proposed mechanisms. Acta Physiol Scand 1996;156 (3):347–53.
[68] D'Urzo AD, Jhirad R, Jenne H, et al. Effect of caffeine on ventilatory responses to hypercapnia, hypoxia, and exercise in humans. J Appl Physiol 1990;68(1):322–8.
[69] Wiles JD, Bird SR, Hopkins J, Riley M. Effect of caffeinated coffee on running speed, respiratory factors, blood lactate and perceived exertion during 1500-m treadmill running. Br J Sports Med 1992;26(2):116–20.
[70] Kaminsky LA, Martin CA, Whaley MH. Caffeine consumption habits do not influence the exercise blood pressure response following caffeine ingestion. J Sports Med Phys Fitness 1998;38 (1):53–8.
[71] Brown DD, Knowlton RG, Sullivan JJ, Sanjabi PB. Effect of caffeine ingestion on alveolar ventilation during moderate exercise. Aviat Space Environ Med 1991;62(9 Pt 1):860–4.
[72] Schmidt C, Bellingham MC, Richter DW. Adenosinergic modulation of respiratory neurones and hypoxic responses in the anaesthetized cat. J Physiol 1995;483(Pt 3):769–81.
[73] Frary CD, Johnson RK, Wang MQ. Food sources and intakes of caffeine in the diets of persons in the United States. J Am Diet Assoc 2005;105(1):110–3.
[74] Chen Y, Parrish TB. Caffeine's effects on cerebrovascular reactivity and coupling between cerebral blood flow and oxygen metabolism. NeuroImage 2009;44(3):647–52.
[75] McCusker RR, Goldberger BA, Cone EJ. Caffeine content of specialty coffees. J Anal Toxicol 2003;27(7):520–2.
[76] Del Coso J, Munoz G, Munoz-Guerra J. Prevalence of caffeine use in elite athletes following its removal from the world anti-doping agency list of banned substances. Appl Physiol Nutr Metab 2011;36(4):555–61.
[77] Ellender L, Linder MM. Sports pharmacology and ergogenic aids. Prim Care 2005;32(1):277–92.
[78] Getzin AR, Milner C, LaFace KM. Nutrition update for the ultraendurance athlete. Curr Sports Med Rep 2011;10 (6):330–9.
[79] Boyle M, Castillo V. Monster on the loose. Fortune 2006;:116–22.
[80] Malinauskas BM, Aeby VG, Overton RF, Carpenter-Aeby T, Barber-Heidal K. A survey of energy drink consumption patterns among college students. Nutr J 2007;6:35.
[81] Clauson KA, Shields KM, McQueen CE, Persad N. Safety issues associated with commercially available energy drinks. J Am Pharm Assoc (2003) 2008;48(3):e55–63 [quiz e4–e7].
[82] Nawrot P, Jordan S, Eastwood J, Rotstein J, Hugenholtz A, Feeley M. Effects of caffeine on human health. Food Addit Contam 2003;20(1):1–30.
[83] Higdon TA, Heidenreich JW, Kern KB, et al. Single rescuer cardiopulmonary resuscitation: can anyone perform to the guidelines 2000 recommendations? Resuscitation 2006;71 (1):34–9.
[84] Higdon JV, Frei B. Coffee and health: a review of recent human research. Crit Rev Food Sci Nutr 2006;46(2):101–23.
[85] Hindmarch I, Rigney U, Stanley N, Quinlan P, Rycroft J, Lane J. A naturalistic investigation of the effects of day-long consumption of tea, coffee and water on alertness, sleep onset and sleep quality. Psychopharmacology (Berl) 2000;149(3):203–16.
[86] Dews PB, Curtis GL, Hanford KJ, O'Brien CP. The frequency of caffeine withdrawal in a population-based survey and in a controlled, blinded pilot experiment. J Clin Pharmacol 1999;39 (12):1221–32.
[87] Peters JM. Factors affecting caffeine toxicity: a review of the literature. J Clin Pharmacol 1967;7(3):131–41.
[88] Armstrong LE, Casa DJ, Maresh CM, Ganio MS. Caffeine, fluid-electrolyte balance, temperature regulation, and exercise-heat tolerance. Exerc Sport Sci Rev 2007;35(3):135–40.
[89] Armstrong LE, Pumerantz AC, Roti MW, et al. Fluid, electrolyte, and renal indices of hydration during 11 days of controlled caffeine consumption. Int J Sport Nutr Exerc Metab 2005;15 (3):252–65.

CHAPTER

52

Role of Quercetin in Sports Nutrition

John Seifert

Montana State University, Bozeman, MT, USA

Quercetin is one of the most commonly occurring polyphenolic flavonoids. It is a plant-derived flavonoid found in various fruits, berries, grains, and certain vegetables [1]. Many physiological benefits have been associated with quercetin ingestion. These benefits include ergogenic enhancements (primarily through mitochondrial biogenesis), improved antioxidant status, enhanced immune response, cardiovascular protection, and anti-inflammatory activity.

CHEMISTRY

The quercetin molecule consists of a three-ring structure with two aromatic centers and a central oxygenated heterocyclic ring (see Figure 52.1). Its structure lends itself to being a natural antioxidant of oxygen free radicals and lipid peroxides. Quercetin's potent antioxidant properties occur through a variety of reactive sites, but for the most part are associated with the two hydroxyl groups found on the catachol-type B ring [2].

Flavonoids, and quercetin specifically, appear in two basic forms: a glycoside and an aglycone [3]. A quercetin glycoside is when quercetin is conjugated with a sugar moiety. If quercetin is found without an attachment to a sugar molecule, it is known as aglycone. The sugar moiety of a quercetin glycoside is an important determinant of absorption and bioavailability. This conjugation with glucose appears to enhance absorption of quercetin from the small intestine.

Intestinal absorption and subsequent metabolism are the keys to establishing a cause-and-effect relationship between quercetin and the desired response. In clinical patients, 52% of the quercetin was absorbed when it was orally ingested in the form of the quercetin glycoside. In comparison, 24% of the quercetin was absorbed when it was ingested in the quercetin aglycone form [4]. With the apparent rapid absorption of the quercetin glycoside, peak plasma concentrations typically occur from 1 to 3 hours after ingestion, and the half life is 6 to 12 hours after ingestion [5]. Little is known about how the glycoside form is absorbed faster than the aglycone form, especially since the glycoside has to be hydrolyzed in the small intestine before absorption. Perhaps the sodium/glucose active transport system plays a role in dragging quercetin across the membrane or there are specialized transporters of quercetin. Nonetheless, because of its long half-life of elimination, repeated consumption of quercetin-containing foods will cause accumulation of quercetin in blood and may, theoretically, enhance functional capacity.

PERFORMANCE

Performance outcomes of numerous studies show equivocal results pertaining to performance and quercetin ingestion. When investigating the role of quercetin ingestion on exercise performance, one needs to pay attention to dosing of quercetin, timing of the ingestion, duration of dosing (days or weeks), the type of exercise involved, and the fitness level of the subjects. Even though various studies may differ in protocols, they can be compared statistically by a meta-analysis to find if an intervention does indeed have an effect.

Kressler et al. [6] performed a meta-analysis on studies that investigated endurance exercise performance and quercetin ingestion. These authors reported that, in eight studies that met the inclusionary criteria, the quercetin-ingesting group outperformed the placebo group. In three other studies that met inclusionary criteria, the placebo group performed better than the quercetin group. When statistically analyzed, the quercetin groups improved by 3% over the placebo groups in the eight studies that demonstrated improved endurance performance.

Nutrition and Enhanced Sports Performance.
DOI: http://dx.doi.org/10.1016/B978-0-12-396454-0.00052-7

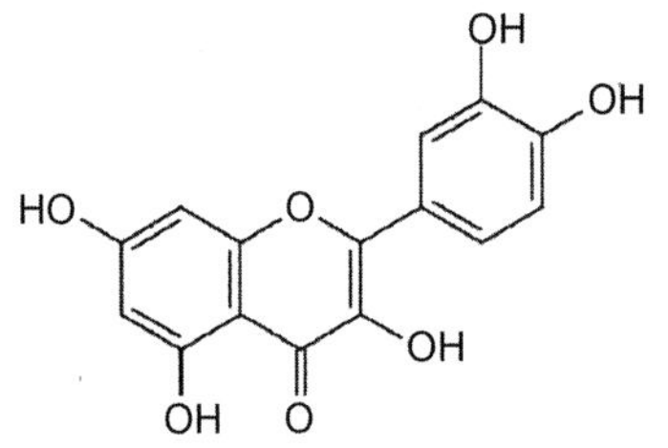

FIGURE 52.1 Quercetin flavonoid.

There are nuances with every study that sets it apart from others in that cohort. Studies that have used untrained human participants have typically shown improved performance in time to exhaustion or time trial type tests. Case in point, Davis et al. [7] fed 1 g/day of quercetin for 7 days. A small, but statistically significant, increase of 3.9% in maximal oxygen uptake (VO_{2max}) was observed with quercetin treatment. More importantly, these authors noted a 13.2% improvement in ride time to exhaustion when quercetin was ingested compared with the placebo. In another human study, Nieman et al. [8] reported that untrained males improved their 12-minute run for distance by 2.9% over the placebo after ingesting quercetin. As with the Davis et al. study, quercetin dosing in the Nieman study was 1 g/day over 2 weeks.

Not all studies that used untrained subjects reported positive effects from quercetin ingestion, however. Cureton et al. [9] reported their findings on quercetin ingestion and untrained subjects. The subjects were active, but not endurance trained. Subjects received 1 g/day of quercetin. No significant differences between quercetin and placebo were observed for VO_{2max}, perception of effort, substrate utilization, or total work completed during a 10-minute maximal effort cycling trial. However, quercetin feeding took place over 7–16 days. Some subjects received quercetin for only 7 days, whereas other subjects received the treatment for up to 16 days. This issue alone may have impacted the results. In another study, Ganio et al. [10] observed similar findings to Cureton's results. Ganio et al. fed sedentary subjects 1 g/day quercetin, or a placebo, for 5 days. No differences between treatments were observed for VO_{2max} or any other cardiorespiratory measures.

Other quercetin-feeding studies have used well trained athletes to test the treatment effects. In a study by Dumke et al. [11], trained cyclists were fed 1 g/day of quercetin or a placebo for 3 weeks. No differences between treatments were found for indices of performance, such as cycling efficiency, power output, or substrate utilization. An improvement in efficiency implies that a subject can produce a given amount of work at a lower metabolic cost. Using trained subjects may actually mask the effects of quercetin, because of the limited ability to increase mitochondria density within the muscle, possessing an already high antioxidant capacity, or improved anti-inflammatory environment within the muscle. Casuso et al. [12] published a report of a study where there were no significant differences between quercetin and placebo for VO_{2peak} or run time to exhaustion in trained rats.

It is unclear why performance is improved in endurance exercises rather than in VO_{2max} tests. Davis et al. [7] noted that a time to fatigue test may be a better choice to test endurance performance, due to the increased mitochondrial capacity. Apparently, a greater percentage of the endurance work is dependent on mitochondrial development, whereas VO_{2max} tests are dependent not only on mitochondrial development, which would influence arterial–venous oxygen difference, but is also dependent upon cardiac output. To date, there is no evidence that quercetin has a significant effect on cardiovascular dynamics during exercise.

MITOCHONDRIAL BIOGENESIS

Notwithstanding, there have been numerous endurance performance studies that have shown a benefit with quercetin ingestion. This raises the question of how this result occurs. Quercetin ingestion may improve endurance performance by promoting mitochondrial biogenesis. Commonly used markers of mitochondrial biogenesis are PGC-1α and SIRT-1. Their increased expression is responsible for mitochondrial biogenesis. This fact, by itself, provides supporting evidence for the improvement of aerobic work as a result of quercetin ingestion.

Davis et al. [13] demonstrated that running time to fatigue was improved, by about 37%, and significant increases in oxidative enzymes were observed through the stimulation of mitochondrial biogenesis in mice. Lagouge et al. [14] also reported improved mitochondrial biogenesis in mice with quercetin feeding. Although not statistically significant, there were trends towards increased expression of mitochondrial biogenesis markers in untrained human subjects [15].

In contrast, however, mitochondrial biogenesis has not been observed in other human studies [8,9]. There may be greater benefits of quercetin supplementation in untrained muscle, where there is greater potential for mitochondrial development since the density is lower than in trained muscle. It may be that quercetin supplementation increases skeletal muscle oxidative capacity through mitochondrial biogenesis and enhances endurance performance in untrained to moderately trained individuals, but not in the well trained, as a physiological ceiling for mitochondrial content may already exist in these individuals [16,17].

Theoretically, improved mitochondrial proliferation should lead to an increased preference for oxidation of fat, over carbohydrates, as a fuel source during endurance exercise. However, altered metabolism from quercetin ingestion is equivocal. Wu et al. [18] noted a significant shift in energy production from glycolysis to lipolysis in swim-trained mice when fed quercetin. In contrast to Wu et al.'s finding, Dumke et al. [11] reported that substrate utilization was not affected by quercetin ingestion in well-trained subjects who cycled for 3 hours. Subjects in that study were fed 1 g of quercetin per day for 3 weeks. The authors reported that there was no difference between quercetin and the placebo treatment for substrate utilization, although there was a trend ($p = 0.11$) for fat utilization to be higher in the quercetin treatment. It is difficult to come to a conclusion about quercetin's influence on metabolism on the basis of these studies, because they are so different with respect to subject base, modality of exercise, and the fact that one is an animal study while the other is a human study.

IMMUNE & INFLAMMATORY RESPONSES

Although there have been few studies of the relationship between quercetin ingestion and exercise-induced inflammation and oxidative stress, it has been reported that quercetin has antioxidant and anti-inflammatory potential [19,20]. The antioxidant properties are primarily associated with two hydroxyl groups on the catechol-type B-ring [2]. It is widely accepted that high intensity or prolonged exercise increases oxidative stress. To test this supposition, Quindry et al. [21] fed subjects 1 g of quercetin per day or a placebo for 3 weeks before a 160 km running race. No differences were found between the quercetin and placebo treatments for oxidative damage or antioxidant capacity.

Quercetin may not only affect parameters of endurance performance, it may also influence the immune system by minimizing the inflammatory response. Maintaining health status, by minimizing respiratory infections, is important for the competitive athlete who is trying to improve his/her training status. It has been reported that quercetin ingestion decreased the occurrence of upper respiratory traction infection (URTI) in athletes [22]. Nieman et al. found that ingesting 1 g of quercetin per day in a multi-ingredient concoction for 3 weeks reduced the incidence of URTI during the 2-week period which followed 3 days of exhaustive exercise. These authors noted significant decreases in C-reactive protein, plasma interleukin 6 (IL-6), and interleukin 10 (IL-10) [22]. Findings by Davis et al. [23] support those of Nieman et al. Davis et al. reported that quercetin reduced susceptibility to infection following exercise in mice.

Not all authors have reported positive effects of quercetin feeding on markers of inflammation. O'Fallon [24] found no significant differences between quercetin and a placebo for markers of muscle damage and inflammation, creatine kinase and IL-6, when healthy subjects underwent a series of eccentric muscle contractions. Subjects ingested 1 g of quercetin per day for 7 days before and 5 days after the exercise. The type or mode of exercise may have played a role in the contrast between these results and those of the previously cited studies.

Another potential ergogenic effect of quercetin supplementation is the psychostimulant role it may play during exercise. Numerous studies have reported on quercetin's ability to block adenosine receptors in the brain. This may, in turn, reduce the perception of effort and pain during exercise. Alexander [25] noted that quercetin has a high affinity for the adenosine-A1 antagonist receptor, resulting in effects similar to those of caffeine. This suggestion has been supported by results in mice from Davis et al. However, this finding has not been supported by human studies. Cheuvront et al. [26] reported that quercetin did not influence rating of perceived exertion (RPE) on a short performance task. Subjects in that study were given a dose of 2 g of quercetin prior to exercise that was conducted in 40°C heat. It is not known whether heat influenced results or whether enough quercetin was in the single dose to produce an effect.

Quercetin may also exhibit another beneficial effect through the regulation of blood pressure in the hypertensive patient. In a number of studies performed on hypertensive humans and animals, quercetin ingestion led to a decrease in blood pressure [27]. Blood pressure is controlled by both neural mechanisms through the autonomic nervous system and humoral mechanisms involving nitric oxide (NO) and endothelin-1 (ET-1). Nitric oxide is a potent vasodilator while ET-1 is a vasoconstrictor. Both NO and ET-1 are released by different cell types at the endothelial level. It is known that polyphenolic flavonoids improve endothelial function [28]. Perhaps quercetin is an influencing factor in blood flow during exercise, but this is not yet known. Additionally, these results are generalizable only to hypertensive patients, as the influence of quercetin on endothelial function in normotensives is not yet known.

FUTURE RESEARCH AREAS

Some areas in which further research should be conducted include elucidating effective quercetin dosing. Most human studies have used 1 g of quercetin per day for a given period of time. Positive effects from quercetin ingestion have been found in as little as 7 days of feeding, while other studies have fed quercetin for 3 weeks and did not report differences between quercetin and a placebo. Another area of investigation is whether there are gender effects in quercetin ingestion. While some studies have included females, there has not been a definitive focus on studying responses to quercetin feeding in female subjects. Little is known about how, or if, females respond similarly, or differently, than males to quercetin ingestion. Another research area involves subject ages. Most published studies use younger (college age) subjects. In the aging population, mitochondrial density typically decreases due to inactivity. This provides potential for quercetin to have a significant effect. There are also changes in the cells as we age, leading to a decrease in the antioxidant capacity of the cells. This by itself could have repercussions, especially in diseased states. Another promising avenue is the combining of quercetin with other compounds. There is the potential of a synergistic effect when multiple compounds are used, as noted by Nieman et al. [15].

CONCLUSIONS

In conclusion, quercetin appears to have the potential to provide ergogenic effects in very specific arenas. Fitness level plays a key role in the effectiveness of quercetin supplementation. Numerous authors have reported that quercetin exhibits little to no effect in highly trained subjects. However, quercetin supplementation does hold promise in those subjects who are sedentary or have a low fitness level or in those with low tissue quercetin concentration. There is a need for further research into quercetin usage and physiological responses.

References

[1] Chun OK, Chung SJ, Song WO. Estimated dietary flavonoid intake and major food sources of U.S. adults. J Nutr 2007;137:1244–52.

[2] Rice-Evans CA, Miller NJ. Antioxidant activities of flavonoids as bioactive components of food. Biochem Soc Transactions 1996;24:790–5.

[3] Hollman PCH, Hertog MGL, Katan MB. Analysis and health effects of flavonoids. Food Chem 1996;57:43–6.

[4] Hollman PCH, de Vries JHM, van Leeuwen SD, Mengelers MJB, Katan MB. Absorption of dietary quercetin glycosides and quercetin in healthy ileostomy volunteers. Am J Clin Nutr 1995;62:1276–82.

[5] Egert S, Wolffram S, Bosy-Westphal A, Boesch-Saadatmandi C, Wagner AE, Frank J. Daily quercetin supplementation dose-dependently increases plasma quercetin concentrations in healthy humans. J Nutr 2008;138:1615–21.

[6] Kressler J, Millard-Stafford M, Warren GL. Quercetin and endurance exercise capacity: a systematic review and meta-analysis. Med Sci Sports Exerc 2011;43:2396–404.

[7] Davis JM, Carlstedt CJ, Chen S, Carmichael MD, Murphy EA. The dietary flavonoid quercetin increases VO_2max and endurance capacity. Int J Sport Nutr and Exer Metab 2010;20:56–62.

[8] Nieman DC, Williams AS, Shanely RA, Jin F, McAnulty SR, Triplett NT, et al. Quercetin's influence on exercise performance and muscle mitochondrial biogenesis. Med Sci Sports Exer 2010;42:338–45.

[9] Cureton KJ, Tomporowski PD, Singhal A, Pasley JD, Bigelman KA, Lambourne K, et al. Dietary quercetin supplementation is not ergogenic in untrained men. J Appl Physiol 2009;107:1095–104.

[10] Ganio MS, Armstrong LE, Johnson EC, Klau JF, Ballard KD, Michniak-Kohn B, et al. Effect of quercetin supplementation on maximal oxygen uptake in men and women. J Sports Sci 2010;28:201–8.

[11] Dumke CL, Nieman DC, Utter AC, Rigby MD, Quindry JC, Triplett NT, et al. Quercetin's effect on cycling efficiency and substrate utilization. J Appl Physiol Nutr Metab 2009;34:993–1000.

[12] Casuso RA, Martinez-Amat A, Martinez-Lopez EJ, Camiletti-Moiron D, Porres JM, Aranda P. Ergogenic effects of quercetin supplementation in trained rats. J Int Soc Sports Nutr 2013;10:3.

[13] Davis JM, Murphy EA, Carmichael MD, Davis B. Quercetin increases brain and muscle mitochondrial biogenesis and exercise tolerance. Am J Physiol Regul Integr Comp Physiol 2009;296:R1071–7.

[14] Lagouge M, Argmann C, Gerhart-Hines Z, Meziane H, Lerin C, Daussin F, et al. Resveratrol improves mitochondrial function and protects against metabolic disease by activating SIRT1 and PGC-1alpha. Cell 2006;127:1109–22.

[15] Nieman DC, Henson DA, Maxwell KR, Williams AS, McAnulty SR, Jin F, et al. Effects of quercetin and EGCG on mitochondrial biogenesis and immunity. Med Sci Sports Exerc 2009;41:1467–75.

[16] Brooks SV, Vasilaki A, Larkin LM, McArdle A, Jackson MJ. Repeated bouts of aerobic exercise lead to reductions in skeletal muscle free radical generation and nuclear factor kappaB activation. J Physiol 2008;586:3979–90.

[17] Hood DA, Irrcher I, Ljubicic V, Joseph AM. Coordination of metabolic plasticity in skeletal muscle. J Exp Biol 2006;209:2265–75.

[18] Wu J, Gao W, Wei J, Yang J, Pu L, Guo C. Quercetin alters energy metabolism in swimming mice. Appl Physiol Nutr Metab 2012;37:912–22.

[19] Filipe P, Lanca V, Silva JN, Morliere P, Santus R, Fernandes A. Flavonoids and urate antioxidant interplay in plasma oxidative stress. Mol Cell Biochem 2001;221:79–87.

[20] Silva MM, Santos MR, Caroco G, Rocha R, Justino G, Mira L. Structure antioxidant activity relationships of flavonoids: a re-examination. Free Radic Res 2002;36:1219–27.

[21] Quindry JC, McAnulty SR, Hudson MB, Hosick P, Dumke C, McAnulty LS, et al. Oral quercetin supplementation and blood oxidative capacity in response to uUltramarathon competition. Int J of Sport Nutr Exer Metab 2008;18:601–16.

[22] Nieman DC, Henson DA, Gross SJ, Jenkins DP, Davis JM, Murphy EA, et al. Quercetin reduces illness but not immune perturbations after intensive exercise. Med Sci Sports Exerc 2007;39:1561–9.

[23] Davis JM, Murphy EA, McClellan JL, Carmichael MD, Gangemi JD. Quercetin reduces susceptibility to influenza infection following stressful exercise. Am J Physiol Regul Integr Comp Physiol 2008;295:R505–9.

[24] O'Fallon KS, Kaushik D, Michniak-Kohn B, Dunne CP, Zambraski EJ, Clarkson PM. Effects of quercetin supplementation on markers of muscle damage and inflammation after eccentric exercise. Int J Sport Nutr Exer Metab 2012;22:430–7.

[25] Alexander SP. Flavonoids as antagonists at A1 adenosine receptors. Phytother Res 2006;20:1009–12.

[26] Cheuvront SN, Ely BR, Kenefick RW, Michniak-Kohn BB, Rood JC, Sawka MN. No effect of nutritional adenosine receptor antagonists on exercise performance in the heat. Am J Physiol Regul Integr Comp Physiol 2009;296:R394–401.

[27] Larson AJ, Symons JD, Jalili T. Therapeutic potential of quercetin to decrease blood pressure: review of efficacy and mechanisms. Am Soc Nutr Adv Nutr 2012;3:39–46.

[28] Edwards RL, Lyon T, Litwin SE, Rabovsky A, Symons JD, Jalili T. Quercetin reduces blood pressure in hypertensive patients. J Nutr 2007;137:2405–11.

CHAPTER

53

Human Performance and Sports Applications of Tongkat Ali (*Eurycoma longifolia*)

Shawn M. Talbott

GLH Nutrition, LLC, Draper, UT, USA

Eurycoma longifolia is an herbal medicinal plant found in Southeast Asia (Malaysia, Vietnam, Java, Sumatra, Thailand). In Malaysia, it is commonly called tongkat ali and has a range of medicinal properties as a general health tonic, including improvement in physical and mental energy levels and overall quality of life [1,2]. The roots of tongkat ali, often called "Malaysian ginseng," are used as an adaptogen and as a traditional "anti-aging" remedy to help older individuals adapt to the reduced energy, mood, and libido that often comes with age [3–7]. In modern dietary supplements, tongkat ali can be found in a variety of products intended to improve libido and energy, restore hormonal balance (cortisol/testosterone levels), and enhance both sports performance and weight loss.

In both men and women, testosterone levels peak between 25 and 30 years of age and thereafter drop approximately 1–2% annually [8,9]. At the age of 60, testosterone levels are typically only 40–50% of youthful levels and may be lower due to stress and related lifestyle issues such as diet, exercise, and sleep patterns [10,11]. The benefits of maintaining a youthful testosterone level are many, including increased muscle mass and reduced body fat, high psychological vigor (mental/physical energy), and improved general well-being [12,13].

Eurycoma contains a group of small peptides referred to as "eurypeptides" and known to have effects in improving energy status and sex drive in rodents [14–16]. The effect of tongkat ali in restoring normal testosterone levels appears to arise less from actually "stimulating" testosterone synthesis than from increasing the release rate of "free" testosterone from its binding hormone, sex-hormone-binding globulin (SHBG) [17,18]. In this way, eurycoma may be considered not so much a testosterone "booster" (such as an anabolic steroid), but rather a "maintainer" of normal testosterone levels and a "restorer" of normal testosterone levels (from "low" back "up" to normal ranges) [19]. This would make eurycoma particularly beneficial for individuals with subnormal testosterone levels, including those who are dieting for weight loss, middle-aged individuals suffering with fatigue or depression, and intensely training athletes who may be at risk for overtraining [20,21].

TRADITIONAL USE

Decoctions of tongkat ali roots have been used for centuries in Malaysia and elsewhere in Southeast Asia as an aphrodisiac for loss of sexual desire and impotence, as well as to treat a range of ailments including post-partum depression, malaria, high blood pressure, and fatigue [22].

Tongkat ali has been referred to as Malaysia's "home-grown Viagra" in respected research journals [4], with the Malaysian government investing considerable effort to license, develop, and sustain research into the potential health benefits of *Eurycoma longifolia* through a variety of governmental organizations, including the Forest Research Institute of Malaysia (FRIM) [22].

MODERN EXTRACTS

Numerous commercial tongkat ali supplements claim "extract ratios" from 1:20 to 1:200 without any information about bioactive constituents, extraction

Nutrition and Enhanced Sports Performance.
DOI: http://dx.doi.org/10.1016/B978-0-12-396454-0.00053-9

methodology (e.g., ethanol versus water), or extract purity. Alcohol extracts of eurycoma have been studied in mice for antimalarial effects of concentrated eurycomalactone [23], but also exhibit toxic effects at high doses (LD50 at 2.6 g/kg), which would preclude safe use in humans as a long-term dietary supplement [24,25]. In contrast, hot-water root extracts standardized for known bioactive components (1% eurycomanone, 22% protein, 30% polysaccharides, 35% glycosaponin) have been demonstrated to be extremely safe at high doses and for long-term consumption [26–28].

Properly standardized hot-water extracts [2,26,29] have a distinctly bitter taste due to the presence of quassinoids, which are recognized as some of the bitterest compounds in nature [30,31]. Tongkat ali extracts that do not taste bitter are either not true *Eurycoma longifolia* root (there are many commercial examples of "fake" tongkat ali extracts) or are subpotent in terms of bioactive constituents, and thus would also be expected to have low efficacy. Because of tongkat ali's reputation for libido benefits, there are several examples of dietary supplements labeled as *Eurycoma longifolia* but containing none of the actual root, and instead being "spiked" with prescription erectile dysfunction drugs (tadalafil/Cialis, sildenafil/Viagra, vardenafil/Levitra).

LABORATORY AND ANIMAL RESEARCH

Bhat and Karim [1] conducted an ethnobotanical and pharmacological review on tongkat ali, noting that laboratory research such as cell assay studies offers possible mechanistic support for the myriad traditional uses of tongkat ali, including aphrodisiac [32], antimalarial [33], antimicrobial [34], anticancer [35] and antidiabetic effects [36].

Numerous rodent studies exist demonstrating reduced anxiety and improved sexual performance following tongkat ali feeding [37–40], with such effects thought to be due to a restoration of normal testosterone levels. Eurycoma's anxiolytic effects have been demonstrated in a variety of behavioral tests, including elevated plus-maze, open field, and anti-fighting, suggesting an equivalent anti-anxiety effect to diazepam as a positive control [37].

Animal studies have shown that many of the effects of the extract are mediated by its glycoprotein components [14]. The mechanism of action of the bioactive complex polypeptides ("eurypeptides" with 36 amino acids) has been shown to be activation of the CYP17 enzyme (17 alpha-hydroxylase and 17,20 lyase) to enhance the metabolism of pregnenolone and progesterone to yield more DHEA (dehydroepiandrosterone) and androstenedione, respectively [29]. This glycoprotein water-soluble extract of *Eurycoma longifolia* (tradename: Physta™) has been shown to deliver anti-aging and anti-stress benefits subsequent to its testosterone-balancing effects [41,42].

HUMAN-FEEDING TRIALS

Based on a long history of traditional use and confirmation of biological activity via cell culture and animal-feeding studies, several human supplementation studies have been conducted to evaluate the potential benefits of tongkat ali for sexual function, exercise performance, weight loss, and vigor (mental/physical energy).

Importantly, all of the human trials have used the same water-extracted and standardized eurycoma root (Physta™, Biotropics Malaysia) for which a patent has been issued jointly to the Government of Malaysia and the Massachusetts Institute of Technology (United States Patent #7,132,117) [29]. The patent discloses a process whereby *Eurycoma longifolia* roots undergo an aqueous extraction combined with HPLC and size-exclusion chromatography to yield a bioactive peptide fraction (a 4300-dalton glycopeptide with 36 amino acids) that is responsible for its effects in maintaining testosterone levels. Physta™ is a freeze-dried standardized extract of the root of *Eurycoma longifolia* which contains numerous active compounds including phenolic components, tannins, high molecular weight polysaccharides, glycoproteins, and mucopolysaccharides. The bioactive fraction of *Eurycoma longifolia* root delivers a demonstrated ability to improve testosterone levels [41], increase muscle size and strength [43,44], improve overall well-being [45,46], accelerate recovery from exercise [47], enhance weight loss [48,49], reduce stress [50], and reduce symptoms of fatigue [51–53].

In two recent studies of young men undergoing a weight-training regimen [43,44] tongkat ali supplementation (100 mg/day of Physta™) improved lean body mass, 1 RM strength, and arm circumference to a significantly greater degree compared with a placebo group.

In a recent 12-week trial [46] of *Eurycoma longifolia* supplementation (300 mg/day of Physta™ in men aged 30–55 years), subjects showed significant improvement compared with placebo in the Physical Functioning domain of the SF-36 survey. In addition, sexual libido was increased by 11% (week 6) and 14% (week 12) and abdominal fat mass was significantly reduced in subjects with BMI > 25 kg/m^2.

In men with low testosterone levels (average age 51 years), 1 month of daily supplementation with tongkat ali extract (Physta™, 200 mg/day) resulted in a significant improvement in serum testosterone levels and quality-of-life parameters [41], suggesting a role for tongkat ali as an "adaptogen" against aging-related stress. In another study of healthy adult males (average age 25 years), 100 mg/day of tongkat ali extract (Physta™) added to an intensive strength-training program (every other day for 8 weeks) resulted in significant improvements in fat-free mass, fat mass, maximal strength (1 RM) and arm circumference compared with a placebo group [43]. These results indicate that tongkat ali extract can enhance muscle mass and strength gains, while accelerating fat loss, in healthy exercisers, and thus may be considered a natural ergogenic aid for athletes and dieters alike.

In a recent study from our group (submitted), we supplemented 63 subjects (32 Men and 31 women) daily with tongkat ali root extract (Physta, 200 mg/day) or a look-alike placebo for 4 weeks. Significant ($p < 0.05$) mood state improvements were found in the tongkat ali group for Tension (−11%), Anger (−12%), and Confusion (−15%). Hormone profile (salivary cortisol and testosterone) was significantly improved by tongkat ali supplementation, with reduced cortisol exposure (−16%), increased testosterone status (+37%), and overall improved cortisol:testosterone ratio (−36%). These results indicate that daily supplementation with tongkat ali (Physta) improves stress hormone profile and certain mood state parameters, suggesting an effective natural approach to shielding the body from the detrimental effects of chronic stress, which may include the "stress" of intense exercise training.

One study of middle-aged women (aged 45–59 years) found that twice-weekly strength training plus 100 mg/day of *Eurycoma longifolia* extract for 12 weeks enhanced fat-free mass to a greater degree than for women adhering to the same strength-training program and taking a placebo [44]. Additional studies in dieters [48–50] and athletes [47] have shown 50–100 mg/day of Tongkat ali extract to help restore normal testosterone levels in supplemented dieters (compared with a typical drop in testosterone among non-supplemented dieters) and supplemented athletes (compared with a typical drop in non-supplemented athletes). In one trial of endurance cyclists [47], cortisol levels were 32% lower and testosterone levels were 16% higher in supplemented subjects compared with placebo, indicating a more favorable biochemical profile for promoting an "anabolic" hormone state.

For a dieter, it would be expected for cortisol to rise and testosterone to fall following several weeks of dieting [54]. This change in hormone balance (elevated cortisol and suppressed testosterone) is an important factor leading to the familiar "plateau" that many dieters hit (when weight loss slows/stops) after 6–8 weeks on a weight loss regimen. By maintaining normal testosterone levels, a dieter could expect to also maintain their muscle mass and metabolic rate (versus a drop in both, subsequent to lower testosterone levels) and thus to continue to lose weight without plateauing.

For an athlete, the same rise in cortisol and drop in testosterone is an early signal of "overtraining"—a syndrome characterized by reduced performance, increased injury rates, suppressed immune system activity, increased appetite, moodiness, and weight gain [55]. Maintenance of normal cortisol/testosterone levels in eurycoma-supplemented subjects may be able to prevent or reduce some of these overtraining symptoms as well as help the athlete to recover more quickly and more completely from daily training bouts.

SAFETY

Oral toxicity studies (Wistar rats) have determined the LD50 of Physta™ as 2000 mg/kg body weight (acute) and the NOAEL (no observed adverse effect level) as greater than 1000 mg/kg body weight (28-day sub-acute feeding), resulting in a classification as Category 5 (extremely safe) according to the United Nations Globally Harmonized System of Classification and Labeling of Chemicals (GHS).

In addition to the very high safety profile demonstrated in the rodent toxicity studies, there are no reported adverse side effects in human studies of tongkat ali supplementation. For example, one 2-month human supplementation trial [27] of 20 healthy males (age range 38–58), found high doses of *Eurycoma longifolia* extract (600 mg/day) to have no influence on blood profiles (hemoglobin, RBC, WBC, etc.) or any deleterious effects on measures of liver or renal function. In our own recent supplementation trial (200 mg/day for 4 weeks), there were no changes in measures of liver enzymes (ALT/AST). Typical dosage recommendations, based on traditional use and on the available scientific evidence in humans, including dieters and athletes, call for 50–200 mg/day of a water-extracted tongkat ali root standardized to 22% eurypeptides.

SUMMARY

A wide range of investigations, from laboratory research, to animal feeding studies, to human supplementation trials, have confirmed the health benefits and traditional use of tongkat ali root extract. Laboratory evidence shows that eurycoma peptides stimulate release of free testosterone from its binding proteins and improve overall hormone profiles. More than a dozen rodent-feeding studies have demonstrated improved sex drive, balanced hormonal profiles, and enhanced physical function. Human supplementation trials show a clear indication of reduced fatigue, heightened energy and mood, and greater sense of well-being in subjects consuming tongkat ali root extracts. It is important to note that the majority of these studies, and all of the human supplementation trials, have been conducted on specific hot-water-extracts of *Eurycoma longifolia* (which is the traditional Malaysian preparation) produced using a patented extraction process to isolate and concentrate the bioactive compounds (Physta™). Some of the tongkat ali extracts currently on the US market are alcohol-extracts, which provide a substantially different chemical profile, and may not be as effective or as safe as the more extensively studied hot-water-extracts.

In conclusion, tongkat ali, used for centuries in traditional medicine systems of Southeast Asia for treating lethargy, low libido, depression, and fatigue, appears to have significant potential for restoring hormone balance (cortisol/testosterone) and overall well-being in humans exposed to various modern stressors, including aging, dieting, and exercise stress.

Disclosure

The author (S. Talbott) has served as principal investigator on several studies of tongkat ali, including a recent study of Physta, a brand of tongkat ali extract that was funded by the manufacturer, Biotropics Malaysia. Dr. Talbott has no direct financial relationship with Biotropics or with Physta.

References

[1] Bhat R, Karim AA. Tongkat ali (Eurycoma longifolia Jack): a review on its ethnobotany and pharmacological importance. Fitoterapia 2010;10:1–11.

[2] Ali JM, Saad JM. Biochemical effect of Eurycoma longifolia jack on the sexual behavior, fertility, sex hormone, and glycolysis. (Dissertation). Department of Biochemistry, University of Malaysia 1993.

[3] Adimoelja A. Phytochemicals and the breakthrough of traditional herbs in the management of sexual dysfunctions. Int J Androl 2000;23(Suppl. 2):82–4.

[4] Cyranoski D. Malaysian researchers bet big on home-grown Viagra. Nat Med 2005;11(9):912.

[5] Joseph S, Sugumaran M, Lee KLW. Herbs of Malaysia. An introduction to the medicinal, culinary, aromatic and cosmetic use of herbs. Federal Publications Sdn Berhad; 2005.

[6] Wan Hassan, WE. Healing herbs of Malaysia. Federal Land Development Authority (FELDA) 2007.

[7] Zhari I, Norhayati I, Jaafar L. Malaysian herbal monograph, 1. Malaysian Monograph Committee; 1999. pp. 67–70.

[8] Araujo AB, Wittert GA. Endocrinology of the aging male. Best Pract Res Clin Endocrinol Metab 2011;25(2):303–19.

[9] Traish AM, Miner MM, Morgentaler A, Zitzmann M. Testosterone deficiency. Am J Med 2011;124(7):578–87.

[10] Henning PC, Park BS, Kim JS. Physiological decrements during sustained military operational stress. Mil Med 2011;176 (9):991–7.

[11] Gatti R, De Palo EF. An update: salivary hormones and physical exercise. Scand J Med Sci Sports 2011;21(2):157–69.

[12] Miller KK. Androgen deficiency: effects on body composition. Pituitary 2009;12(2):116–24.

[13] Grossmann M. Low testosterone in men with type 2 diabetes: significance and treatment. J Clin Endocrinol Metab 2011;96 (8):2341–53.

[14] Asiah O, Nurhanan MY, Ilham MA. Determination of bioactive peptide (4.3 kDa) as an aphrodisiac marker in six Malaysia plants. J Trop For Sci 2007;19(1):61–3.

[15] Zanoli P, Zavatti M, Montanari C, Baraldi M. Influence of Eurycoma longifolia on the copulatory activity of sexually sluggish and impotent male rats. J Ethnopharmacol 2009;126:308–13.

[16] Ang HH, Ikeda S, Gan EK. Evaluation of the potency activity of aphrodisiac in Eurycoma longifolia Jack. Phytother Res 2001;15 (5):435–6.

[17] Chaing HS, Merino-chavez G, Yang LL, Wang FN, Hafez ES. Medicinal plants: conception/contraception. Adv Contracept Deliv Syst 1994;10(3–4):355–63.

[18] Tambi, MI. Water soluble extract of Eurycoma longifolia in enhancing testosterone in males. In: Proceedings of the International Trade Show and Conference, Supply Side West, Las Vegas, NV, USA; 2003. October 1–3.

[19] Talbott S, Talbott J, Christopulos AM, Ekberg C, Larsen W, Jackson. V. Ancient wisdom meets modern ailment – traditional asian medicine improves psychological vigor in stressed subjects. Prog Nutr 2010;12(1): April.

[20] Tambi MI. Eurycoma longifolia jack: a potent adaptogen in the form of water-soluble extract with the effects of maintaining men's health. Asian J Androl 2006;8(Suppl. 1):49–50.

[21] Talbott S, Christopulos AM, Richards. E. A lifestyle approach to controlling holiday stress and weight gain. Med Sci Sports Exerc 2006;38(5): Supplement, Abstract #2478, May.

[22] Azmi MMI, Fauzi A, Norini H. Economic analysis of E. longifolia (Tongkat Ali) harvesting in peninsular Malaysia. New Dimensions in Complementary Health Care 2004;:91–9.

[23] Satayavivad J, Soonthornchareonnon N, Somanabandhu A, Thebtaranonth Y. Toxicological and antimalarial activity of eurycomalactone and Eurycoma lingifolia Jack extracts in mice. Thai J Phytopharm 1998;5(2):14–27.

[24] Chan KL, Choo CY. The toxicity of some quassinoids from Eurycoma longifolia. Planta Med 2002;68(7):662–4.

[25] Le-Van-Thoi, Nguyen-Ngoc-Suong. Constituents of Eurycoma longifolia Jack. J Org Chem 1970;35(4):1104–9.

[26] Athimulam A, Kumaresan S, Foo DCY, Sarmidi MR, Aziz RA. Modelling and optimization of E. longifolia water extract production. Food Bioprod Process 2006;84(C2):139–49.

[27] Mohd MAR, Tambi I, Kadir AA. Human toxicology and clinical observations of Eurycoma longofiola on men's health. Int J Androl 2005;28(Suppl. 1): June.

[28] Lin LC, Peng CY, Wang HS, Lee KWW. Reinvestigation of the chemical constituents of Eurycoma longifolia. Clin Pharm J 2001;53:97–106.

[29] Sambandan, TG, Rha CK, Kadir AA, Aminudim N, Saad J-M. Bioactive fraction of Eurycoma longifolia. U.S. Patent No. 7132117, November 2007.

[30] Bedir E, Abou-Gazar H, Ngwendson JN, Khan IA. Eurycomaoside: a new quassinoid-type glycoside from the roots of Eurycoma longifolia. Chem Pharm Bull (Tokyo) 2003;51(11):1301–3.

[31] Darise M, Kohda H, Mizutani K, Tanaka O. Eurycomanone and eurycomanol, quassinoids from the roots of Eurycoma longifolia. Phytochemistry 1982;21:2091–3.

[32] Ang HH, Cheang HS, Yusof AP. Effects of Eurycoma longifolia Jack (Tongkat Ali) on the initiation of sexual performance of inexperienced castrated male rats. Exp Anim 2000;49(1):35–8.

[33] Kuo PC, Shi LS, Damu AG, Su CR, Huang CH, Ke CH, et al. Cytotoxic and antimalarial beta-carboline alkaloids from the roots of Eurycoma longifolia. J Nat Prod 2003;66(10):1324–7.

[34] Farouk AE, Benafri A. Antibacterial activity of Eurycoma longifolia Jack. A Malaysian medicinal plant. Saudi Med J 2007;28 (9):1422–4.

[35] Nurhanan MY, Azimahtol HLP, Ilham MA, Shukri MMA. Cytotoxic effects of the root extracts of Eurycoma longifolia Jack. Phytother Res 2005;19(11):994–6.

[36] Husen R, Pihie AH, Nallappan M. Screening for antihyperglycaemic activity in several local herbs of Malaysia. J Ethnopharmacol 2004;95:205–8.

[37] Ang HH, Cheang HS. Studies on the anxiolytic activity of Eurycoma longifolia Jack roots in mice. Jpn J Pharmacol 1999;79 (4):497–500.

[38] Ang HH, Ngai TH, Tan TH. Effects of Eurycoma longifolia Jack on sexual qualities in middle aged male rats. Phytomedicine 2003;10(6–7):590–3.

[39] Ang HH, Lee KL. Effect of Eurycoma longifolia Jack on libido in middle-aged male rats. J Basic Clin Physiol Pharmacol 2002;13(3):249–54.

[40] Ang HH, Ngai TH. Aphrodisiac evaluation in non-copulator male rats after chronic administration of Eurycoma longifolia Jack. Fundam Clin Pharmacol 2001;15(4):265–8.

[41] Tambi MI, Imran MK, Henkel RR. Standardised water-soluble extract of Eurycoma longifolia, Tongkat ali, as testosterone booster for managing men with late-onset hypogonadism? Andrologia 2011;15.

[42] Tambi MI, Imran MK. Eurycoma longifolia Jack in managing idiopathic male infertility. Asian J Androl 2010;12(3):376–80.

[43] Hamzah S, Yusof A. The ergogenic effects of Tongkat ali (Eurycoma longifolia): A pilot study. Br J Sports Med 2003;37:464–70.

[44] Sarina MY, Zaiton Z, Aminudin AHK, Nor AK, Azizol AK. Effects of resistance training and Eurycoma longifolia on muscle strength, lipid profile, blood glucose, and hormone level in middle-aged women. Abstract from fourth Asia-Pacific Conference on Exercise and Sport Science & eighth International Sports Science Conference; 2009.

[45] Udani JK, George A, Mufiza M, Abas A. Gruenwald J, Miller M. Effects of a proprietary freeze-dried water extract of Eurycoma longifolia on sexual performance and well-being in men with reduced sexual potency: a randomized, double-blind, placebo-controlled study. Presented at Scripps Natural Supplements Conference, January 2011.

[46] George A, Shaiful Bahari I, Zahiruddin WM, Abas A. Randomised Clinical Trial on the Use of PHYSTA Freeze-dried Water Extract of Tongkat Ali for the Improvement in Sexual Well-Being and Quality of Life in Men. Evid Based Complement Alternat Med 2012; doi:10.1155/2012/429268.

[47] Talbott S, Talbott J, Negrete J, Jones M, Nichols M, Roza J. Effect of eurycoma longifolia extract on anabolic balance during endurance exercise. J Int Soc Sports Nutr 2006;3(1):S32.

[48] Talbott S, Christopulos AM, Ekberg. C. Effect of a 12-week lifestyle program on mood state and metabolic parameters in overweight subjects. Med Sci Sports Exerc 2007;39(5): Supplement (227), Abstract #1503, May.

[49] Talbott S, Christopulos AM, Richards E. Effect of a lifestyle program on holiday stress, cortisol, and body weight. J Am Coll Nutr 2005;24(5): Abstract #31, October.

[50] Talbott S, Talbott J, Larsen W, Jackson V. Significant improvements in mood state and hormone profile associated with a "low-attrition" weight loss program. J Am Coll Nutr 2007;26(5): Abstract 24.

[51] Tambi MI. Glycoprotein water-soluble extract of Eurycoma longifolia Jack as a health supplement in management of healthy aging in aged men. In: Lunnenfeld B, editor. Abstracts of the third World Congress on the Aging Male, February 7–10, Berlin, Germany. Aging Male; 2002.

[52] Tambi MI. Standardized water soluble extract of Eurycoma longifolia maintains healthy aging in man. In: Lunenfeld B., editor. Abstracts of the fifth World Congress on the Aging Male. February 9–12, Salzburg, Austria; 2006.

[53] Tambi MI. Standardized water soluble extract of Eurycoma longifolia on men's health. In: Abstracts of the eighth International Congress of Andrology, 12–16 June, Seoul, Korea. Int. J. Androl 2005; 28 (Suppl. 1): 27.

[54] Foss B, Dyrstad SM. Stress in obesity: cause or consequence? Med Hypotheses 2011;77(1):7–10.

[55] Kraemer WJ, Ratamess NA. Hormonal responses and adaptations to resistance exercise and training. Sports Med 2005;35 (4):339–61.

SECTION 6

DIETARY RECOMMENDATIONS

CHAPTER

54

Nutrition and Dietary Recommendations for Bodybuilders

Philip E. Apong

Iovate Health Sciences International, Inc., Oakville, ON, Canada

INTRODUCTION

"Tell me what you eat and I'll tell you what you are," also paraphrased as "You are what you eat," is an expression that has become the seminal central dogma of nutrition, describing one of the most fundamental principles implicating dietary intake in physiological health status. This formative concept of nutrition should serve as a starting point for health and wellbeing. Moreover, it ought to be a notion that is continually disseminated globally and appreciated by the masses. Unfortunately, not everyone ascribes to the tenets of proper nutrition, as is evidenced by the mobs frequenting popular fast food establishments across North America. As the hedonist would agree, we all need to "live it up" once in a while, with the endangered Twinkie in hand, and enjoy life. Nonetheless, the easy accessibility to decadent food is facilitating an undesirable expansion of waistlines from coast to coast. Despite the escalating body of the corpulent populace, there exists, on the opposite end of the spectrum, the dedicated few who are extremely body-image concerned and usually more health-conscious individuals. These people work fervently towards physical perfection, and as such, consider themselves unique physique athletes or bodybuilders. Unlike their weaker-hearted, sedentary counterparts, bodybuilders are amenable to spending long arduous hours, seemingly ascetically punishing their muscles with high intensity heavy weights. All of their forceful efforts are put forth in the hopes of promoting herculean-proportioned musculature.

While it has probably been known since ancient times that intense exercise facilitates an increase in strength and muscle, it has only been in recent years that meticulous researchers have begun to truly elucidate the biological mechanisms behind exercise-induced muscle remodeling, and the critical interplay between dietary components in the process. This has resulted in an explosion of available information regarding nutritional strategies to support extreme increases in muscle size and strength. Consequently, a general grasp of some basic nutritional approaches grounded upon contemporary scientific findings may allow the dedicated bodybuilder to fully realize his or her physical potential. Therefore, this chapter will focus on some of the evidence available to help individuals tailor their nutritional requirements for extreme muscle growth.

According to many nutritionists, the foundations of any dietary regimen for the general population should include a variety of nutrients in order to ensure a well-balanced diet without insufficiencies. An example of attempts to guarantee dietary adequacy is the food guide pyramid which was first developed by the US Department of Agriculture in 1992 [1], and later updated in subsequent years [2]. For the most part, guidelines like these were originally conceived for the average individual, and so may not be sufficient for hard-training bodybuilders who are seeking to maximize their genetic muscle-building potential. Moreover, the perfect diet for an athlete or bodybuilder is contingent upon several factors, such as genetics, age, body size, and gender [1–3]. Even training conditions, including exercise frequency and intensity, can affect dietary recommendations.

It was only within the last couple of decades that there emerged a "nutritional renaissance" of sorts, where inquisitive scientists began to reveal the mechanistic processes of muscle remodeling while unraveling

Nutrition and Enhanced Sports Performance.
DOI: http://dx.doi.org/10.1016/B978-0-12-396454-0.00054-0

BOX 54.1

ESSENTIAL AND NONESSENTIAL AMINO ACIDS FOUND IN COMPLETE PROTEINS

Essential Amino Acids	Nonessential Amino Acids
Histidine[a,b]	Alanine
Isoleucine	Arginine[b]
Leucine	Asparagine
Lysine	Aspartic acid
Methionine	Cysteine[b]
Phenylalanine	Glutamic acid
Threonine	Glutamine[b]
Tryptophan	Glycine
Valine	Proline[b]
	Serine
	Tyrosine[b]

[a]*Literature states that some adults can synthesize histidine. However, for other adults and for infants, histidine is considered an essential amino acid [1]. Therefore, under certain circumstances histidine may be grouped among the conditionally essential amino acids [86].*

[b]*Sometimes these amino acids are considered conditionally essential under certain conditions since they are rate limiting for protein synthesis, especially under extreme conditions such as in the absence of certain nonessential amino acids from the diet and the presence of limited amounts of essential amino acids or in times of metabolic stress [86].*

the subtleties between nutrition and muscle protein accretion. For example, Biolo et al. [4] hypothesized that the net muscle protein synthesis after exercise involves an interaction between exercise and nutritional factors. They reported that, during recovery from resistance exercise in the fasted state, muscle amino acid transport and protein turnover (both synthesis and degradation) were accelerated [4]. Results from elaborate amino acid infusion experiments began to show that increased availability of free amino acids to skeletal muscle can modulate or promote anabolic activity [4–6]. Interestingly, research was also corroborating what bodybuilders had already known through experience—that resistance exercise can independently stimulate muscle protein synthesis [7–11]. Moreover, the explorative science indicated that feeding and resistance exercise has a synergistic additive effect and can maximize anabolic activity [4,12–15]. This type of work raised the question of whether protein requirements for those seeking to build more muscle are higher than those for the sedentary person.

PROTEIN REQUIREMENTS FOR THE BODYBUILDER

In the most rudimentary sense, proteins comprise long chains of amino acids which are bound by peptide bonds. Protein is essential to the structural integrity and functionality of muscles, bones, tissues and organs. Moreover, besides playing a structural role, protein can be used to produce hormones, enzymes, hemoglobin, and plasma proteins such as albumin. Proteins can even be used as a source of energy when other energy substrates are consumed in insufficient amounts. With respect to human biological requirements, there are 20 requisite amino acids identified as needed to support adult human growth and metabolism (Box 54.1). Of these, 11 amino acids are called nonessential, meaning that they can be synthesized in the body and do not need to be consumed in the diet. The remaining 9 amino acids cannot be synthesized in the body and are termed essential because they need to be consumed in the diet. A shortage or absence of any of these amino acids can hinder proper tissue maintenance and repair, and stymie maximal muscle growth.

The recommended daily protein requirement for the bodybuilder and athlete in general has been a topic of debate over the years and has not been without controversy [3,16,17]. The daily protein requirement for athletic individuals is influenced by a wide array of conditions running the gamut from exercise type, age, body size, to training status (e.g., gym neophyte or seasoned bodybuilder), etc. Indeed, the science as we understand it today, by and large, supports an increased requirement for daily protein consumption in resistance-training individuals when compared with their sedentary counterparts. However, the body is a fantastic biological machine, and some research even suggests that protein requirements may actually decrease during resistance or endurance training, due to biological adaptations that improve net protein retention [3,18]. Nevertheless, many experts today would agree that the bodybuilder requires more protein than the sedentary person in order to sustain an awe-inspiring level of musculature. In addition, ultimately meeting or slightly exceeding daily requirements for skeletal muscle adaptation to training will support optimal muscle mass accretion. It is worthy to note that any protein that is consumed in excess of what is required to stimulate total body and muscle protein synthesis can be oxidized for energy production [19–21]. This is significant since protein may become an important energy substrate under certain circumstances, such as when a bodybuilder undergoes a purposeful calorie deficit (usually limiting either carbohydrates or fat intake). Many contestants do this during their pre-contest diets in order to shed superficial body fat. Besides the need for a compensatory energy substrate during reduced fat or carbohydrate diets, bodybuilders have good reason to monitor their protein during their dieting or "cutting" phase of their pre-contest preparation. That is because research shows that consuming dietary protein at levels above the recommended daily allowance (RDA) during energy deprivation may actually attenuate skeletal muscle loss by affecting the intracellular regulation of muscle anabolism and proteolysis [22].

The current recommended daily allowance for protein in healthy adults is 0.8 g/kg bodyweight per day [3,23,24]. However, as demonstrated by research such as a study done by Tarnopolsky et al. [19], typical RDA protein amounts may not be sufficient to meet the true needs of resistance-training individuals. In their study, leucine kinetic and nitrogen balance (NBAL) methods were used to determine the dietary protein requirements of strength athletes compared with sedentary subjects. The results from Tarnopolsky et al. [19] show that protein requirements for athletes performing strength training are greater than for sedentary individuals and are above contemporary US and Canadian recommended daily protein intake levels. In fact, the protein intake to achieve a steady-state zero nitrogen balance (indicative of neither a net gain nor a net loss of bodily amino acids) was 1.41 g/kg per day for male strength athletes and 0.69 g/kg per day for sedentary male subjects (104% greater for strength athletes) [19]. Based on these numbers, a 90 kg male bodybuilder would need to consume a minimum of approximately 127 g of high quality protein daily to maintain steady-state nitrogen balance. It is interesting to note that the results from Tarnopolsky et al. [19] suggest a potential upper limit for daily protein intake, beyond which point, protein synthesis in response to training becomes maximally saturated and seemingly immutable. Therefore, excessive protein intake (e.g., some anecdotal reports of athletes consuming as much as 4 g/kg exist [25]) for the purpose of attempting to build muscle is not warranted. This is good to know in light of the fact that protein-based foods are typically on the costly side. Tarnopolsky et al. [19] found that when dietary protein intake was increased from 0.86 to 1.41 g/kg per day, whole body protein synthesis was increased in men who resistance trained; however, when intakes were increased to 2.4 g/kg per day, there was no increase in protein synthesis. Results such as demonstrated by this experiment support the current position stand of the International Society of Sports Nutrition (ISSN), which states that dietary protein intake RDA of 0.8 g/kg may be sufficient to meet the needs of the general population, but is most likely insufficient for athletes, especially when taking into account caloric requirements of athletes and increased potential for amino acid oxidation and substrate required for muscle tissue growth. The ISSN currently recommends 1.4–2.0 g/kg protein per day of bodyweight [16,23]. Thus, a 90 kg male bodybuilder would need approximately 180 grams of protein on a daily basis. Obtaining this daily quota is certainly feasible with whole foods, such as meat and dairy products. However, it can be quite a labor intensive and costly venture to prepare copious amounts of chicken breasts or lean cuts of steak on a regular basis. In addition, high dietary intake of red meat may be undesirable for those concerned with excessive dietary fat intake. This is why many bodybuilders choose to use supplemental protein sources, including whey and casein, which have become staples in many nutritional regimens of amateur and professional bodybuilders alike.

PROTEIN TYPE AND DIGESTIBLY

Whole-Foods Protein Sources

There is a wide array of protein sources available to the bodybuilder, including everyday dietary sources from common animal-derived whole foods (e.g., chicken, beef, fish, eggs, cheese, milk) and from legumes and vegetables (e.g., soy, bean, rice). For individuals unable to obtain their daily quota from food or for those who are looking for a more convenient source of protein, there is an extensive choice of protein supplements on the market (e.g., soy, whey, casein, and milk protein comprising a combination of whey and casein). With the notable exception of collagen, protein derived from animal sources are considered whole or complete proteins since they contain the full spectrum of the 20 amino acids including all essential amino acids. Whereas in general, protein derived from plant sources (an exception being soy protein) are considered incomplete sources of dietary protein because they typically lack essential amino acids [26]. Unless limited by lifestyle restrictions such as vegetarianism or religious beliefs, it should be the goal of every bodybuilder to attempt to consume complete proteins, since research supports the ingestion of complete protein (especially dairy) after resistance exercise in promoting positive muscle protein balance [27,28]. Research has explored the effect of fluid milk [27] or its constituent protein fractions, whey and casein [28], on muscle protein balance, and the results emphasize the importance of complete protein sources such as dairy for supporting muscle-building effects. Milk comprises both casein and whey (80% and 20%, respectively), and each of these proteins has distinct physicochemical properties. Because of their inherently different digestibility characteristics, the effects of administration of these discrete components on the kinetics of blood amino acids and anabolic/catabolic activity has been the subject of a variety of studies. Indeed the varying results from some of these trials are somewhat difficult to reconcile and provide some conflicting evidence regarding which of these protein sources may be superior for eliciting muscle anabolic response with acute and chronic use. Despite the differences in the results from various studies, bodybuilders should probably not fret too much over the minutiae and ensure that they include various complete protein sources in their diets. This pragmatic approach will allow for the provision of a full spectrum of amino acids throughout the day and add some variety to the nutritional regimen of the bodybuilder, which tends to be inherently repetitive and mundane.

Whey

Whey protein has received much attention and has a large fanfare among bodybuilders since research shows that it is among the top quality proteins in terms of various markers of protein digestibility and assimilation, such as the biological value score and protein digestibility corrected amino acid (PDCAA) score [26]. Furthermore, in comparison with other protein sources, whey protein generally contains a higher concentration of essential amino acids [29], and has rapid absorption kinetics [30–32]. Unlike casein, the ingestion of whey protein has been reported to lead to a rapid, yet transient spike in plasma amino acid levels [30]. This fast digestibility and ability to quickly increase plasma amino acid availability to muscles can make whey protein a good source of protein to influence the muscle-building (mTOR) machinery, which is sensitive to amino acid (namely leucine) concentration or availability [33–36]. Moreover, whey protein may have great implications for the aged bodybuilder or even, for that matter, for muscle preservation in an elderly population in general. This is because some research supports the phenomenon of an age-related increased susceptibility to impaired anabolic activity in response to protein or leucine ingestion [37–39], sometimes referred to in the literature as "anabolic resistance". It seems that whey protein may be an effective tool to combat this phenomenon. In fact, whey protein was shown to be better than casein at promoting protein anabolism in old men with lean mass atrophy [31]. Even with the popularity of whey protein and the general perception that it is the best source for muscle growth, it is interesting to note that in physically active healthy young men at rest [30] or after resistance exercise [28] the superiority of whey over casein may actually be less definitive than once thought and is likely contingent upon various factors including the level of physical activity.

In the prolonged absence of exercise, such as during a period of recovery where a bodybuilder may purposefully abstain from the gym if he is feeling overtrained, whey protein may be an ideal mediator to stimulate and maintain anabolic activity. This concept could be supported by the results of Antonione et al.'s [40] prolonged bed rest study in which whey protein was observed to be better than casein protein in young inactive men at promoting anabolic effects. Antonione et al. [40] proposed that the relative ability of whey and casein to stimulate net-whole body protein synthesis is contingent upon physical activity level. Other research highlights the ambiguity of the scientific finding between acute and chronic effects of administration of whey or casein, and it is somewhat difficult to resolve the findings. For example, Cribb et al. [29], using two groups of matched, resistance-trained males, showed that whey isolate provided significantly greater gains in strength, lean body mass, and a decrease in fat mass, compared with supplementation with casein during an intense 10-week resistance-training program. Another trial, by Tang et al. [41], showed that whey protein appears to promote a larger muscle protein synthesis response than either casein or soy, at least during the first 3 hours after ingestion, both at rest and after resistance exercise in young, healthy males. Thus with results like these, some may contend that whey is better than casein for bodybuilders. However, other research investigating acute effects of casein and whey after exercise shows that casein is capable of stimulating muscle anabolism after resistance exercise as effectively as whey protein [28]. It has been noted in the literature that the synthesis rate of skeletal muscle protein is up-regulated for several hours following exercise [8,11], leading to enhanced anabolic efficiency of dietary proteins [4,42], an effect which may "normalize" or provide a more level playing field, so to speak, regarding any immediate beneficial effects driven by varying digestibility rates of complete proteins. More research is warranted on this topic, but it seems that hard training is the key to providing the anabolic impetus.

Casein

Whereas whey protein is considered a fast protein, casein is noted in the literature as a slowly digesting protein [30]. This is because upon ingestion the casein protein coagulates in the acidic environment of the stomach, possibly delaying gastric emptying and/or hindering easy accessibility of the amino acid residues to hydrolytic digestion [30,43]. In their archetypical study, Boirie et al. [30] investigated the effects of speed of protein digestion on postprandial protein accretion in resting healthy, physically active male subjects, and showed that in young healthy men at rest, casein administration induced a prolonged plateau of moderate hyper-aminoacidemia. Moreover, whole body protein breakdown was inhibited after casein ingestion, but not after whey protein ingestion. Interestingly, even though whey protein administration caused a greater increase in postprandial protein synthesis than did casein administration, the net leucine balance over the 7 hours after the test meal (i.e., casein protein or whey protein) was more positive with casein than with whey. So it seems that slowly digesting protein such as casein may have an anti-catabolic effect which can lead to better overall net whole body leucine balance, at rest, in healthy, physically active young men [30]. Therefore, quickly digestible protein such as whey can stimulate rapid aminoacidemia and acute activation of protein synthetic machinery, whereas more slowly digesting protein promotes anti-catabolic effects and better net leucine balance at the whole body level [30–32]. Casein's inhibitory effect on protein breakdown can allow for muscle preservation as well as preservation of net protein the splanchnic region [44,45] at least in young individuals. This paradigm seems true for the younger generation of subjects studied, but contradictory results in experiments in elderly subjects with respect to whole body protein metabolism suggest that the ingestion of a quickly digesting protein is associated with a greater whole body leucine balance [31,39]. Ultimately, for young bodybuilders, supplementing with casein-based protein may be a good way to stave off catabolism, whereas whey protein can promote acute increases in anabolic activity. Since milk comprises both fast and slow protein components, it is worthwhile asking if milk protein can deliver benefits from each of its constituent parts.

Milk Proteins

To date, the evidence is that complete proteins and especially dairy sources are able to elicit anabolic activity and lean mass accretion. However, as noted by the ISSN, the superiority of whey protein or any complete protein source over another for chronic muscle-building effects remains to be conclusively shown and is not exactly clear cut [23]. Therefore, it's probably a good idea for bodybuilders to add variety to their diets and consume various proteins at different times in order to strategically capitalize on the potential physicochemical properties of each of the individual protein sources. It has become common practice among many bodybuilders to strategically consume faster-digesting protein sources close to the time of physical exertion in order to promote anabolic activity, while consuming slowly digesting protein such as casein throughout the late evening or between meals, to promote an anti-catabolic milieu, especially prior to retiring to bed for the night. There is also an ostensible rationale and an increasing trend in the marketplace for combined fast and slow protein matrices encompassing various sources of protein, including casein and whey. That being said, the evolution of mammalian species has given rise to examples such as bovine milk which contains both fast and slowly digesting proteins. Milk is presumably designed to be the ideal sustenance for mammalian offspring during their most critical formative stages in the life cycle, when lean mass accrual is vital as a matter of survival. So it stands to reason that this natural evolutionary outcome over the millennia underscores the importance of consuming a mixture of slow and fast proteins. This type of strategy can influence plasma aminoacidemia to promote anabolic activity as well as support anti-catabolic effects on muscle and whole body protein. The net effect could quite possibly lead to synergism, especially when combined with resistance training, which in and of itself is the impetus for adaptive anabolic response. This contention of milk protein synergy is supported by Lacroix et al. [46], who compared the postprandial utilization of dietary nitrogen from three [^{15}N]-labeled milk products—micellar casein, milk soluble protein isolate (whey protein), and total milk protein—in healthy volunteers. Total milk protein and casein sustained prolonged plasma aminoacademia in comparison with the whey protein fraction, which was rapidly deaminated. Lacroix et al. [46] contend that the rate of amino acid delivery for the milk soluble protein isolate (whey) is too rapid to sustain the anabolic requirement during the postprandial period and that total milk protein had the best nutritional quality. Therefore there is good reasoning behind the ever-growing popularity of multi-phase fast/slow milk protein products among nutrition savvy bodybuilders.

Soy

With regards to soy protein, even though it is considered a complete protein source, some research seems to suggest that it may not be the preferential source of protein when attempting to facilitate a prolonged anabolic environment within the body. For example, Wilkinson et al. [47] conducted a study which the authors claim to be the first to show that the source of intact dietary protein (i.e., milk vs soy) is important for determining the degree of post-exercise anabolism. They observed that milk protein promoted a more sustained net positive protein balance after resistance exercise than did soy protein. Wilkinson et al. [47] proposed that a difference in digestion rate of milk and soy protein affects the pattern of amino acid appearance, which leads to differences in net amino acid uptake and muscle protein synthesis after resistance exercise. This work was corroborated by Tang et al. [41], who showed that soy appears to be less effective at stimulating muscle protein synthesis than whey protein, despite inducing a similar rise in circulating essential amino acids. Moreover, Hartman et al. [48] observed greater gains in fat-free mass and muscle hypertrophy in response to 12 weeks of resistance training with fat-free milk vs soy. Although soy protein may not be the ideal choice for bodybuilders, there are some positive data pertaining to it. For example, Candow et al. [49] found that both whey and soy protein increased lean tissue mass more than an isocaloric carbohydrate placebo in healthy young men. Candow et al. [49] and Brown et al. [50] reported no significant differences in changes in body composition in response to combined resistance training in either whey or soy protein ingestion. Therefore, although soy protein may not be the ideal protein source for bodybuilders, it is still suitable for those who are seeking to avoid dairy-based products.

PROTEIN TIMING FOR BODYBUILDERS

Research suggests that strategic nutritional timing may allow bodybuilders to maximize muscle protein synthetic machinery. With regards to sustained amino acid delivery (at least during prolonged resting periods) in which amino acids have been administered to subjects in a lab setting, an interesting phenomenon has been reported, known as refractoriness of protein synthesis [25]. This phenomenon manifests itself as a limited temporal window of maximum protein synthesis lasting approximately 2 hours before returning to baseline [6,51], despite the continuous availability of sustained elevated plasma amino acid concentrations via intravenous infusion. This means that in the absence of exercise, maximum muscle protein synthetic response to hyperaminoacidemia may occur within a limited temporal window. Norton et al. [52] demonstrated that this so-called anabolic window of maximum protein synthetic rate in response to nutritional stimulus may be prolonged for up to 3 hours by consuming a complete meal. In their review paper, Norton and Wilson [25] propose a nutritional strategy to help combat this refractoriness of protein synthesis in which the authors recommend several bolus doses of protein that contain sufficient leucine (e.g., 3 grams per meal, as research seems to suggest that anything significantly higher than 3 grams is not any more effective at driving the anabolic signaling cascade) throughout the day to maximize mTOR signaling and muscle protein synthesis, while permitting enough time for postprandial amino acid levels to fall sufficiently between meals to re-sensitize the innate biological system.

Adding exercise to the equation seems to change the situation somewhat regarding protein timing, as research shows that administration of amino acids in combination with resistance exercise augments acute protein synthesis [4,15], whereas, if subjects remain fasted after a bout of resistance training, muscle protein balance can be pushed into a net catabolic state [8,11,53]. Indeed, increased amino acid availability immediately after a training bout has been shown to improve acute net protein balance [4,8,11,14]. Moreover, post-exercise carbohydrate ingestion may also be beneficial because of a decreased rate of muscle protein breakdown [54,55] which is associated with the anti-catabolic properties of insulin [56]. Thus, research emphasizes the benefits of consuming carbohydrate and protein macronutrients post-workout. It is noteworthy that although carbohydrate consumption post-exercise is associated with increased insulin output and anti-catabolic effects, it is the provision of protein that seems to be the key to promoting adaptive response to training. This was demonstrated by Andersen et al. [57] who showed that 14 weeks of resistance training in healthy young males combined with protein versus carbohydrate supplementation induced similar gains in mechanical muscle performance, but only protein supplementation induced muscle hypertrophy. In their study, subjects ingested a protein supplement (containing whey, casein, egg, and glutamine) immediately before and immediately after training on training days, and in the mornings on non-training days, in order to maximize potential anabolic activity. Andersen et al. [57] chose this "before and after" bracket timing paradigm in light of the fact that some evidence suggests that the acute effect of protein supplementation on muscle anabolism may even be greater if protein is ingested just before the training bout [58]. Andersen et al. [57] assert that their work is the first to suggest the long term importance of timed protein ingestion as compared with isoenergetic carbohydrate intake on muscle fiber hypertrophy in healthy young men.

The notion of a period of time within which the body responds best to post-workout nutrition, often coined the "post-workout anabolic window", is one put forth by many trainers, and evidence does exist to support this, especially in an aged population. For example, in a training study Esmarck et al. [59], the ingestion of protein immediately after resistance training by elderly subjects resulted in muscle hypertrophy, whereas postponed protein intake for 2 hours did not. Interestingly, there seems to be more of a post-workout grace period for younger individuals. For the non-elderly it has been noted that the contractile activity associated with intense resistance exercise results in increased rates of muscle protein synthesis that are sustained for approximately 48 hours in the fasted state [8]. Even though the apparent window of opportunity for nutrient administration seems to be extended in young individuals, it is highly unlikely that a hard-training bodybuilder would purposefully refrain from consuming a meal proximal to training. So unless extenuating circumstances prohibit eating, it is recommended that the bodybuilder consume a meal as close to the training session as conveniently possible, either before or after training. A moderate meal up to 1 hour or more before exercise may provide energy and support aminoacidemia during training without promotion of stomach discomfort or cramps. Some bodybuilders also choose to use commercially available intra-workout amino acid and carbohydrate based drinks to keep their plasma amino acids stores replete while training. This strategy is also supported by research which shows that the provision of essential amino acids and carbohydrates while training suppresses cortisol levels and protein breakdown while promoting muscular gains [60–62].

PROTEIN DOSE PER MEAL

Research suggests that muscle protein synthesis rate can be maximized by an acute bolus dose of protein in a given meal. For example, work by Moore et al. [63] showed that there appears to be a maximal rate at which dietary amino acids can be assimilated into muscle tissue after training and that, with increasingly higher concentrations of amino acid provision post-workout, there is no apparent further stimulation of protein synthesis. In their study, Moore et al. [63] used amino acid tracer techniques to assess muscle protein synthesis to various doses of high quality protein (egg protein was used) after a bout of resistance training. They showed that a dose of 20 grams of a high quality complete protein post workout was enough to maximally stimulate muscle protein synthesis. Moreover, leucine oxidation was increased at 20 grams, and even at 40 grams of post-workout protein, which indicates that a dose of protein as high as about 20 grams can not only stimulate muscle protein synthesis, but also stimulate oxidation of amino acids for fuel. The research has ramifications not only for the acute dose of protein, but also for meal frequency. In fact, Moore et al. [63] had speculated that ingestion of about 20 grams of high quality protein 5 to 6 times daily could be one strategy to maximally stimulate protein synthesis. This strategy could reconcile well with the aforementioned strategy proposed by Norton and Wilson [25] and could help ensure proper temporal "staccato-type" maximum peaking of plasma amino concentrations (especially leucine) to counter any potential refractory effects of protein synthesis noted in the literature. Of course, many protein supplements on the market today contain higher than 20 grams of protein per recommended serving. The rationale for exceeding 20 grams of protein per serving could be that the bodybuilder is on a lower carb/fat diet and is looking for compensatory caloric substrate. Nevertheless, unless on a carbohydrate or fat restricted diet, it is probably a good idea for bodybuilders to first assess their lean mass gains and recuperative ability with approximately 20–25 gram dose of protein per serving for a few weeks before contemplating increasing their dose per serving.

ENERGY REQUIREMENTS

Research shows that maintaining dietary protein levels above the recommended daily allowance of 0.8 g/kg during energy deprivation may attenuate skeletal muscle loss by affecting the intracellular regulation of muscle anabolism and proteolysis [22]. The data from related research shows that whole body protein turnover is endergonic (i.e., requiring energy consumption) and is down-regulated in response to sustained energy deficit, probably to conserve endogenous protein stores when insufficient protein is consumed [22]. As a natural corollary, since protein deposition and muscle growth requires energy, bodybuilders interested in increasing lean body mass must ingest extra calories to facilitate the adaptive response to training. In a recent review, Stark et al. [64] noted that, for maximal hypertrophy to occur, weightlifters should consume greater than 44–50 kilocalories per kilogram of body weight daily. In another study, Rozenek et al. [65] showed that extra calories added to a normal diet of healthy male subjects, in the form of either a high calorie protein/carbohydrate supplement or an isocaloric carbohydrate supplement, facilitated greater increases in fat-free mass and total body mass in response to an 8-week moderate intensity resistance-training program in comparison with subjects in a control group who received no supplement. Both the carbohydrate group and the protein/carbohydrate group experienced similar significant increases in body mass and fat-free mass, indicating that in their study the extra protein did not enhance the efficacy of the supplement [65]. Rozenek et al. [65] surmised that the results from their study corroborate other studies and support the concept that once individual protein requirements are met, the total energy intake may be the most important dietary factor related to body composition changes, rather than specific ingredients used to provide additional energy. In their study, the subjects' supplement yielded an extra 2010 kilocalories above their normal diet [65]. This is quite a considerable amount of extra calories and may not be easy to ingest for every bodybuilder. Therefore, one recommendation when developing a nutritional regimen is to concentrate on obtaining the daily quota of protein and then to experiment by adding daily calories in increments, on a week-by-week basis using healthy carbohydrates (e.g., brown rice, yams, oatmeal) and healthy unsaturated fats, while assessing the rate of lean mass gains. This may allow the bodybuilder to build muscle and help minimize excessive body fat accumulation. Extra calories in the form of a mass-gaining protein/carbohydrate dietary supplement could also be another strategy as this could add a level of convenience, since these types of commercially available products are often easier to prepare than whole foods. Moreover, a properly engineered product would have organoleptic properties and a taste profile to ensure user compliance, since it might be otherwise considered a chore to guzzle down these voluminous protein/carbohydrate shakes.

CARBOHYDRATES

With all the emphasis placed on protein consumption by many bodybuilders, the importance of carbohydrates can easily become overshadowed. Since evidence suggests that there is a level of protein intake which can saturate the body's own protein synthetic machinery, combined with the fact that bodybuilders need to intake a surplus of calories to grow, it makes neither financial nor nutritional sense to intake daily calories strictly from protein alone, in the absence of proper daily carbohydrate sources. Moreover, carbohydrates are the body's principal source of energy substrate, so they should definitely receive proper scrutiny. Because carbohydrates are the chief fuel source in the body, total carbohydrate stores in the body can be readily depleted by a single bout of exercise [66]. This is an important consideration since in athletes the depletion of stored carbohydrates in the form of glycogen within the muscles has been correlated with fatigue in prolonged exercise [67,68]. Therefore, keeping muscles replete with stored glycogen is a good idea to help prolong exercise capacity. Additionally, it is worthy to note that research indicates carbohydrate consumption immediately after a glycogen-depleting bout of exercise can enhance subsequent muscle glycogen re-synthesis when compared with the same intake several hours later [69]. So it is evident that post-workout carbohydrate supplementation definitely has its potential benefits.

The consumption of protein and carbohydrate macronutrients has been associated with insulin production. Insulin is known for its role in promoting nutrient uptake into muscle, which is important for facilitating an environment conducive to anabolism. In fact, there is a body of research investigating hyperinsulinemia for the purposes of increasing the uptake of various ergogenic aids such as creatine and L-carnitine [70–72]. Many bodybuilders have had great success with high glycemic index carbohydrate post-workout drinks in combination with various ergogenic aids such as creatine to help "shuttle" them into the muscle tissue. With regards to insulin and anabolic response, the effects of insulin on muscle protein accumulation seem to be attributable not only to protein synthesis *per se*, but also to its anti-catabolic effects. In fact, insulin has been shown to have a strong inhibitory effect on muscle protein breakdown [55,56,73], which can improve net protein balance [13,54,74–76]. However, even though large doses of carbohydrates can drive insulin levels in the body, the intake of carbohydrates in the absence of protein or amino acid availability does not result in a net positive muscle protein balance [13,76]. It seems that insulin plays more of a permissive role with regards facilitating muscle anabolism and rather seems to partake in a more anti-catabolic role as was demonstrated by Greenhaff et al. [56]. In their elaborate experimental design, Greenhaff et al. [56] examined the rates of muscle protein synthesis and breakdown in response to graded doses of insulin and amino acid infusion. This research group showed that there is low threshold plasma concentration of insulin (5 mU/L) that appears to be required to promote maximal amino acid induced stimulation of leg protein synthesis, and further increasing plasma insulin up to 30 mU/L caused a further reduction of leg protein breakdown but no concomitant increase in anabolic activity. Additionally, a further increase in insulin concentration did not have any effect on decreasing leg protein breakdown. These findings support the contention that carbohydrates post workout along with a protein source could promote anabolism via availability of amino acids and facilitate better positive nitrogen balance via anti-catabolic effects of insulin. So bodybuilders can experiment with a combination of protein and high glycemic carbohydrates post workout to support amino acid deposition and positive nitrogen balance. In the review paper by Kerksick et al. [77], the ISSN concluded that, irrespective of timing, regular ingestion of snacks or meals providing both carbohydrate and protein in a 3:1 ratio helps to promote recovery and replenishment of muscle glycogen.

FATS

In general, dietary fat has received an undesirable stigma in recent years, especially with the more progressive health-promotional media spokespeople berating diets excessively high in fat. Of course, the negative viewpoint on fat stems from valid research implicating excessive dietary fat intake in chronic diseases such as cardiovascular disease, cancer, diabetes, and obesity. So it is quite understandable that health conscious individuals would wish to limit their fat intake. However, overly reducing or eliminating all fat from the diet can be counterproductive to the bodybuilder's goals. That is because dietary fat can play a variety of roles beyond simply providing an energy dense fuel substrate. For example, diets comprising very low fat intake have been associated with reduced sex hormone concentrations and compromised intake or absorption of fat-soluble vitamins and essential fatty acids [78]. On the other hand, higher fat diets may help to maintain circulating testosterone better than low fat diets [79–81]. Experts generally recommend that the acceptable intake of dietary fat represent 20% to 35% of a person's daily energy intake [82]. For athletes attempting to lose weight, diets containing 0.5 to 1 g/kg per day of fat have been recommended [16]. There currently does not appear to be any strong scientific validation to justify excessively surpassing common recommendations for the general public in terms of daily fat intake for the bodybuilder [3,16,78].

Besides being concerned with the daily amount of fat consumption, more research indicates that the bodybuilder should be concerned with the type of fat consumed. For example, evidence is accumulating to support the potential of omega-3 fatty acids, namely eicosapentaenoic acid (EPA) and docosahexaenoic acid (DHA), in facilitating an internal physiological milieu conducive to muscle growth. In a study by Noreen et al. [83], the effects of supplemental fish oil on resting metabolic rate (RMR), body composition, and cortisol production in healthy adults was examined. Six weeks of supplementation with fish oil supplying 1600 mg EPA and 800 mg DHA daily, significantly increased lean mass and decreased fat mass [83]. In their study, Noreen et al. [83] added the fish oil on top of an *ad libitum* diet with no regimented exercise program. So it would indeed be interesting to see more research using the same supplement protocol combined with a structured resistance-training program. Other research by Smith et al. [84,85] shows that 8 weeks of supplementation with EPA and DHA facilitated anabolism by sensitizing muscles to insulin and amino acid availability in young, middle-aged, and elderly subjects. That is to say, the EPA/DHA *per se* did not elicit changes in basal levels of protein synthesis, but rather enhanced the anabolic response to nutritional stimuli, such as amino acids in the presence of insulin. The results of this research may not only have implications for bodybuilders, but also hold significance in the treatment of age-related muscle loss due to sarcopenia and from senescence-induced anabolic resistance. Although further research is warranted on a bodybuilding demographic, it is conceivable that bodybuilders who supplement with a high quality omega-3 fish oil product yielding sufficient fatty acid content may experience the benefits associated with potentially improved anabolism in response to nutritional stimuli. Many nutrition savvy bodybuilders already include fish oil in their dietary regimens and may already be experiencing metabolic benefits.

CONCLUSION

The dedicated bodybuilder is a different breed of athlete who is interested primarily in sustaining a level of musculature that may be deemed excessive or undesirable to other athletes. To achieve this goal, bodybuilders must endure the rigors of repetitive heavy resistance training to elicit extreme muscle growth stimulus. It is because they subject their bodies to such physical hardships that bodybuilders need nutritional strategies that go beyond the mere provision of adequate calories and macronutrients to sustain daily life support functions. The bodybuilder should seek to

BOX 54.2

SAMPLE MUSCLE-BUILDING MEAL PLAN

Meal 1: Breakfast

Egg white omelet made with chicken and vegetables
Oatmeal (plain) made with cinnamon and low calorie natural or artificial sweetener
1 grapefruit

Meal 2: Snack

1 scoop Whey Protein Supplement
1 cup skim milk or almond milk
1 palmful of almonds

Meal 3: Lunch

1 to 2 grilled BBQ chicken breasts
2 palmfuls of brown rice or quinoa
1 serving of any green vegetable (e.g., broccoli, green beans, asparagus)
1 piece of fruit

Meal 4: Snack

1 cup Greek-style yogurt or low-fat cottage cheese
Mixed berries

Meal 5: Dinner

1 lean steak grilled (e.g., top sirloin) or baked salmon fillet
1 baked potato or sweet potato
1 large salad with mixed raw vegetables (e.g., peppers, cucumbers)
1 serving pineapple

Meal 6: Late night/before bed

1 scoop Casein Protein Supplement
1 cup skim milk or almond milk
1 to 2 plain rice cakes topped with natural peanut butter or almond butter

continually learn about nutrition and begin with the tenets of basic nutrition in order to ensure the prevention of micronutrient inadequacies. Consuming a variety of high quality whole foods, including complete proteins, wholesome carbohydrates, and fruits and vegetables, can help underpin a solid nutritional foundation. A sample muscle-building meal plan is shown in Box 54.2. The bodybuilder can then fine tune his or her diet, paying close attention to the principles discovered over the past couple of decades regarding amino acid/protein, carbohydrate, and energy provision, along with the temporal aspects of nutrition timing. With that being said, however, it is ultimately hard work in the gym which serves as the key anabolic impetus for muscle and strength gains. Nonetheless, in the end, proper nutrition will serve to reinforce the results of a strong gym work ethic.

References

[1] Reimers K, Ruud J. Nutritional factors in health and performance. In: Baechle T, Earle R, editors. Essentials of strength training and conditioning. 2nd ed. United States: Human Kinetics; 2000. p. 229–57.

[2] Reimers K. Nutritional factors in health and performance. In: Baechle T, Earle R, editors. Essentials of strength training and conditioning. 3rd ed. United States: Human Kinetics; 2008. p. 201–34.

[3] American Dietetic Association, Dietitians of Canada, America College of Sports Medicine, Rodriguez NR, Di Marco NM, Langley S. American college of sports medicine position stand. Nutrition and athletic performance. Med Sci Sports Exerc 2009;41(3):709–31.

[4] Biolo G, Tipton KD, Klein S, Wolfe RR. An abundant supply of amino acids enhances the metabolic effect of exercise on muscle protein. Am J Physiol 1997;273:E122–9.

[5] Bohe J, Low A, Wolfe RR, Rennie MJ. Human muscle protein synthesis is modulated by extracellular, not intramuscular amino acid availability: a dose-response study. J Physiol 2003;552:315–24.

[6] Bohe J, Low JF, Wolfe RR, Rennie MJ. Latency and duration of stimulation of human muscle protein synthesis during continuous infusion of amino acids. J Physiol 2001;532:575–9.

[7] Chesley A, MacDougall JD, Tarnopolsky MA, Atkinson SA, Smith K. Changes in human muscle protein synthesis after resistance exercise. J Appl Physiol 1992;73:1383–8.

[8] Phillips SM, Tiption KD, Aarsland A, Wolf SE, Wolfe RR. Mixed muscle protein synthesis and breakdown after resistance exercise in humans. Am J Physiol 1997;273:E99–107.

[9] Phillips SM, Tipton KD, Ferrando AA, Wolfe RR. Resistance training reduces the acute exercise-induced increase in muscle protein turnover. Am J Physiol 1999;276:E118–24.

[10] Yarasheski KE, Zachwieja JJ, Bier DM. Acute effects of resistance exercise on muscle protein synthesis rate in young and elderly men and women. Am J Physiol 1993;265:E210–4.

[11] Biolo G, Maggi SP, Williams BD, Tipton KD, Wolfe RR. Increased rates of muscle protein turnover and amino acid transport after resistance exercise in humans. Am J Physiol 1995;268:E514–20.

[12] Borsheim E, Tipton KD, Wolf SE, Wolfe RR. Essential amino acids and muscle protein recovery from resistance exercise. Am J Physiol Endocrinol Metab 2002;283:E648–57.

[13] Miller SL, Tipton KD, Chinkes DL, Wolf SE, Wolfe RR. Independent and combined effects of amino acids and glucose after resistance exercise. Med Sci Sports Exerc 2003;35:449–55.

[14] Rasmussen BB, Tipton KD, Mille SL, Wolf SE, Wolfe RR. An oral essential amino acid-carbohydrate supplement enhances muscle protein anabolism after resistance exercise. J Appl Physiol 2000;88(2):386–92.

[15] Tipton KD, Ferrando AA, Philiips SM, Doyle Jr D, Wolfe RR. Post-exercise net protein synthesis in human muscle from orally administered amino acids. Am J Physiol 1999;276:E628–34.

[16] Kreider RB, Wilborn CD, Taylor L, Campbell B, Almada AL, Collins R, et al. ISSN exercise & sport nutrition review: research and recommendations. J Int Soc Sports Nutr 2010;7:7.

[17] Bilsborough S, Mann N. A review of issues of dietary protein intake in humans. Int J Sport Nutr Exerc Metab 2006;16 (2):129–52.

[18] Rennie MJ, Tipton KD. Protein and amino acid metabolism during and after exercise and the effects of nutrition. Annu Rev Nutr 2000;20:457–83.

[19] Tarnopolsky MA, Atkinson SA, MacDougall JD, Chesley A, Phillips S, Schwarcz HP. Evaluation of protein requirements for trained strength athletes. J Appl Physiol 1992;73 (5):1986–95.

[20] Young VR, Bier DM. A kinetic approach to the determination of human amino acid requirements. Nutr Rev 1987;45:289–98.

[21] Young VR, Bier DM, Pellett PL. A theoretical basis for increasing current estimates of amino acid requirements in adult man, with experimental support. Am J Clin Nutr 1989;50:80–92.

[22] Carbone JW, McClung JP, Pasiakos SM. Skeletal muscle responses to negative energy balance: effects of dietary protein. Adv Nutr 2012;3(2):119–26.

[23] Campbell B, Kreider RB, Ziegenfuss T, La Bounty P, Roberts M, Burke D, et al. International society of sports nutrition position stand: protein and exercise. J Int Soc Sports Nutr 2007;4:8.

[24] U.S. Department of Agriculture, U.S. Department of Health and Human Services. Dietary guidelines for americans. 7th ed. Washington, DC: U.S. Government Printing Office; December 2010.

[25] Norton LE, Wilson GJ. Optimal protein intake and frequency for athletes. Agrofood Ind Hi-Tech 2009;20(2):54–7.

[26] Hoffman JR, Falvo MJ. Protein – which is best? J Sports Sci Med 2004;3:118–30.

[27] Elliott TA, Cree MG, Sanford AP, Wolfe RR, Tipton KD. Milk ingestion stimulates net muscle protein synthesis following resistance exercise. Med Sci Sports Exerc 2006;38:1–8.

[28] Tipton KD, Elliott TA, Cree MG, Wolf SE, Sanford AP, Wolfe RR. Ingestion of casein and whey proteins result in muscle anabolism after resistance exercise. Med Sci Sports Exerc 2004;36:2073–81.

[29] Cribb PJ, Williams AD, Carey MF, Hayes A. The effect of whey isolate and resistance training on strength, body composition, and plasma glutamine. Int J Sport Nutr Exerc Metab 2006;16 (5):494–509.

[30] Boirie Y, Dangin M, Gachon P, Vasson MP, Maubois JL, Beaufrère B. Slow and fast dietary proteins differently modulate postprandial protein accretion. Proc Natl Acad Sci USA 1997;94 (26):14930–5.

[31] Dangin M, Guillet C, Garcia-Rodenas C, Gachon P, Bouteloup-Demange C, Reiffers-Magnani K, et al. The rate of protein digestion affects protein gain differently during aging in humans. J Physiol 2003;549(Pt2):635–44.

[32] Dangin M, Boirie Y, Garcia-Rodenas C, Gachon P, Fauquant J, Callier P, et al. The digestion rate of protein is an independent regulating factor of postprandial protein retention. Am J Physiol Endocrinol Metab 2001;280:E340–8.

[33] Karlsson HK, Nilsson PA, Nilsson J, Chibalin AV, Zierath JR, Blomstrand E. Branched-chain amino acids increase p70S6k phosphorylation in human skeletal muscle after resistance exercise. Am J Physiol Endocrinol Metab 2004;287:E1–7.

[34] Kimball SR, Jefferson LS. Signaling pathways and molecular mechanisms through which branched-chain amino acids mediate translational control of protein synthesis. J Nutr 2006;136:227S–31S.

[35] Liu Z, Jahn LA, Long W, Fryburg DA, Wei L, Barrett EJ. Branched chain amino acids activate messenger ribonucleic acid translation regulatory proteins in human skeletal muscle, and glucocorticoids blunt this action. J Clin Endocrinol Metab 2001;86:2136–43.

[36] Drummond MJ, Rasmussen BB. Leucine-enriched nutrients and the regulation of mammalian target of rapamycin signalling and human skeletal muscle protein synthesis. Curr Opin Clin Nutr Metab Care 2008;11:222–6.

[37] Guillet C, Prod'homme M, Cachon P, Giraudet C, Morin L, Girzard J, et al. Impaired anabolic response of muscle protein synthesis is associated with S6K1 dysregulation in elderly humans. FASEB J 2004;18:1586–7.

[38] Guillet C, Zangarelli A, Mishellany A, Rousset P, Sornet C, Dardevet D, et al. Mitochondrial and sarcoplasmic proteins, but not myosin heavy chain, are sensitive to leucine supplementation in older rat skeletal muscle. Exp Gerontol 2004;39:745–51.

[39] Cuthbertson D, Smith K, Babraj J, Leese G, Waddell T, Atherton P, et al. Anabolic signaling deficits underlie amino acid resistance of wasting, aging muscle. FASEB J 2005;19:422–4.

[40] Antonione R, Caliandro E, Zorat F, Guarnieri G, Heer M, Biolo G. Whey protein ingestion enhances postprandial anabolism during short term bedrest in young men. J Nutr 2008;138 (11):2212–6.

[41] Tang JE, Moore DR, Kujbida GW, Tarnopolsky MA, Phillips SM. Ingestion of whey hydrolysate, casein, or soy protein isolate: effects on mixed muscle protein synthesis at rest and following resistance exercise in young men. J Appl Physiol 2009;107:987–92.

[42] Tipton KD, Elliot TA, Cree MG, Aarsland AA, Sanford AP, Wolfe RR. Stimulation of net muscle protein synthesis by whey protein ingestion before and after exercise. Am J Physiol Endocrinol Metab 2007;292:E71–6.

[43] Mahé S, Roos N, Benamouzig R, Davin L, Luengo C, Gagnon L, et al. Gastrojejunal kinetics and the digestion of [15N]beta-lactoglobulin and casein in humans: the influence of the nature and quantity of the protein. Am J Clin Nutr 1996;63 (4):546–52.

[44] Nakshabendi IM, McKee R, Downie S, Russell RI, Rennie MJ. Rates of small intestinal mucosal protein synthesis in human jejunum and ileum. Am J Physiol Endocrinol Metab 1999;277: E1028–31.

[45] Nakshabendi IM, Obeidat W, Russell RI, Downie S, Smith K, Rennie MJ. Gut mucosal protein synthesis measured using intravenous and intragastric delivery of stable tracer amino acids. Am J Physiol Endocrinol Metab 1995;269:E996–9.

[46] Lacroix M, Bos C, Léonil J, Airinei G, Luengo C, Daré S, et al. Compared with casein or total milk protein, digestion of milk soluble proteins is too rapid to sustain the anabolic postprandial amino acid requirement. Am J Clin Nutr 2006;84 (5):1070–9.

[47] Wilkinson SB, Tarnopolsky MA, Macdonald MJ, Macdonald JR, Armstrong D, Phillips SM. Consumption of fluid milk promotes greater muscle protein accretion after resistance exercise than does consumption of iso-nitrogenous and iso-energetic soy-protein beverage. Am J Clin Nutr 2007;85:1031–40.

[48] Hartman JW, Tang JE, Wilkinson SB, Tarnopolsky MA, Lawrence RL, Fullerton AV, et al. Consumption of fat-free fluid milk after resistance exercise promotes greater lean mass accretion than does consumption of soy or carbohydrate in young, novice, male weightlifters. Am J Clin Nutr 2007;86:373–81.
[49] Candow DG, Burke NC, Smith-Palmer T, Burke DG. Effect of whey and soy protein supplementation combined with resistance training in young adults. Int J Sport Nutr Exerc Metab 2006;16:233–44.
[50] Brown EC, DiSilvestro RA, Babaknia A, Devor ST. Soy versus whey protein bars: effects on exercise training impact on lean body mass and antioxidant status. Nutr J 2004;3:22.
[51] Anthony JC, Lang CH, Crozier SJ, Anthony TG, MacLean DA, Kimball SR, et al. Contribution of insulin to the translational control of protein synthesis in skeletal muscle by leucine. Am J Physiol Endocrinol Metab 2002;282:E1092–101.
[52] Norton LE, Layman DK, Bunpo P, Anthony TG, Brana DV, Garlick PJ. The leucine content of a complete meal directs peak activation but not duration of skeletal muscle protein synthesis and mammalian target of rapamycin signaling in rats. J Nutr 2009;139(6):1103–9.
[53] Pitkanen HT, Nykanen T, Knuutinen J, Lahti K, Keinanen O, Alen M, et al. Free amino acid pool and muscle protein balance after resistance exercise. Med Sci Sports Exerc 2003;35:784–92.
[54] Roy BD, Tarnopolsky MA, MacDougall JD, Fowles J, Yarasheski KE. Effect of glucose supplement timing on protein metabolism after resistance training. J Appl Physiol 1997;82:1882–8.
[55] Biolo G, Williams BD, Fleming RY, Wolfe RR. Insulin action on muscle protein kinetics and amino acid transport during recovery after resistance exercise. Diabetes 1999;48:949–57.
[56] Greenhaff PL, Karagounis LG, Peirce N, Simpson EJ, Hazell M, Layfield R, et al. Disassociation between the effects of amino acids and insulin on signaling, ubiquitin ligases, and protein turnover in human muscle. Am J Physiol Endocrinol Metab 2008;295(3):E595–604.
[57] Andersen LL, Tufekovic G, Zebis MK, Crameri RM, Verlaan G, Kjaer M, et al. The effects of resistance training combined with timed ingestion of protein on muscle fiber size and muscle strength. Metabolism 2005;54(2):151–6.
[58] Tipton KD, Rasmussen BB, Miller SL, Wolf SE, Owens-Stovall SK, Petrini BE, et al. Timing of amino acid carbohydrate ingestion alters anabolic response of muscle to resistance exercise. Am J Physiol Endocrinol Metab 2001;281:E197–206.
[59] Esmarck B, Andersen JL, Olsen S, Richter EA, Mizuno M, Kjaer M. Timing of postexercise protein intake is important for muscle hypertrophy with resistance training in elderly humans. J Physiol 2001;535:301–11.
[60] Bird SP, Tarpenning KM, Marino FE. Independent and combined effects of liquid carbohydrate/essential amino acid ingestion on hormonal and muscular adaptations following resistance training in untrained men. Eur J Appl Physiol 2006;97:225–38.
[61] Bird SP, Tarpenning KM, Marino FE. Effects of liquid carbohydrate/essential amino acid ingestion on acute hormonal response during a single bout of resistance exercise in untrained men. Nutrition 2006;22(4):367–75.
[62] Bird SP, Tarpenning KM, Marino FE. Liquid carbohydrate/essential amino acid ingestion during a short-term bout of resistance exercise suppresses myofibrillar protein degradation. Metabolism 2006;55(5):570–7.
[63] Moore DR, Robinson MJ, Fry JL, Tang JE, Glover EI, Wilkinson SB, et al. Ingested protein dose response of muscle and albumin protein synthesis after resistance exercise in young men. Am J Clin Nutr. 2009;89(1):161–8.
[64] Stark M, Lukaszuk J, Prawitz A, Salacinki A. Protein timing and its effects on muscular hypertrophy and strength in individuals engaged in weight-training. J Int Soc Sports Nutr 2012;9:54.
[65] Rozenek R, Ward P, Long S, Garhammer J. Effects of high-calorie supplements on body composition and muscular strength following resistance training. J Sports Med Phys Fitness 2002;42(3):340–7.
[66] Lemon PW. Beyond the zone. Protein needs of active individuals. J Am Coll Nutr 2000;19(5 Suppl.):513S–21S.
[67] Costhill DL, Hargreaves M. Carbohydrate nutrition and fatigue. Sports Med 1992;13:86–92.
[68] Hawley JA, Schabort EJ, Noakes TD, Dennis SC. Carbohydrate-loading and exercise performance an update. Sports Med 1997;24(2):73–81.
[69] Ivy JL, Katz AL, Culter CL, Sherman WM, Coyle EF. Muscle glycogen synthesis after exercise: effect of time of carbohydrate ingestion. J Appl Physiol 1988;64:1480–5.
[70] Green AL, Hultman E, Macdonald IA, Sewell DA, Greenhaff PL. Carbohydrate ingestion augments skeletal muscle creatine accumulation during creatine supplementation in humans. Am J Physiol 1996;271(5 Pt 1):E821–6.
[71] Steenge GR, Lambourne J, Casey A, Macdonald IA, Greenhaff PL. Stimulatory effect of insulin on creatine accumulation in human skeletal muscle. Am J Physiol 1998;275 (6 Pt 1):E974–9.
[72] Stephens FB, Evans CE, Constantin-Teodosiu D, Greenhaff PL. Carbohydrate ingestion augments L-carnitine retention in humans. J Appl Physiol 2007;102(3):1065–70.
[73] Gelfand RA, Barrett EJ. Effect of physiologic hyperinsulinemia on skeletal muscle protein synthesis and breakdown in man. J Clin Invest 1987;80:1–6.
[74] Glynn EL, Fry CS, Drummond MJ, Dreyer HC, Dhanani S, Volpi E, et al. Muscle protein breakdown has a minor role in the protein anabolic response to essential amino acid and carbohydrate intake following resistance exercise. Am J Physiol Regul Integr Comp Physiol 2010;299(2):R533–40.
[75] Staples AW, Burd NA, West DW, Currie KD, Atherton PJ, Moore DR, et al. Carbohydrate does not augment exercise-induced protein accretion versus protein alone. Med Sci Sports Exerc 2011;43:1154–61.
[76] Borsheim E, Cree MG, Tipton KD, Elliott TA, Aarsland A, Wolfe RR. Effect of carbohydrate intake on net muscle protein synthesis during recovery from resistance exercise. J Appl Physiol 2004;96:674–8.
[77] Kerksick C, Harvey T, Stout J, Campbell B, Wilborn C, Kreider R, et al. International society of sports nutrition position stand: nutrient timing. J Int Soc Sports Nutr 2008;5:17.
[78] Lowery LM. Dietary fat and sport nutrition: a primer. J Sports Sci Med 2004;3:106–17.
[79] Dorgan JF, Judd JT, Longcope C, Brown C, Schatzkin A, Clevidence BA, et al. Effects of dietary fat and fiber on plasma and urine androgens and estrogens in men: a controlled feeding study. Am J Clin Nutr 1996;64(6):850–5.
[80] Hamalainen EK, Adlercreutz H, Puska P, Pietinen P. Decrease of serum total and free testosterone during a low-fat high-fiber diet. J Steroid Biochem 1983;18(3):369–70.
[81] Reed MJ, Cheng RW, Simmonds M, Richmond W, James VH. Dietary lipids: an additional regulator of plasma levels of sex hormone binding globulin. J Clin Endocrinol Metab 1987;64 (5):1083–5.
[82] Institute of Medicine (IOM). Dietary reference intakes for energy, carbohydrate, fiber, fat, fatty acids, cholesterol, protein and amino acids. Washington DC: National Academies Press; 2002335–432.

[83] Noreen EE, Sass MJ, Crowe ML, Pabon VA, Brandauer J, Averill LK. Effects of supplemental fish oil on resting metabolic rate, body composition, and salivary cortisol in healthy adults. J Int Soc Sports Nutr 2010;7:31.

[84] Smith GI, Atherton P, Reeds DN, Mohammed BS, Rankin D, Rennie MJ, et al. Omega-3 polyunsaturated fatty acids augment the muscle protein anabolic response to hyperinsulinaemia-hyperaminoacidaemia in healthy young and middle-aged men and women. Clin Sci (Lond) 2011;121 (6):267–78.

[85] Smith GI, Atherton P, Reeds DN, Mohammed BS, Rankin D, Rennie MJ, et al. Dietary omega-3 fatty acid supplementation increases the rate of muscle protein synthesis in older adults: a randomized controlled trial. Am J Clin Nutr 2011;93(2):402–12.

[86] Di Pasquale MG. Amino acids and protein for the athlete: the anabolic edge. 2nd ed. Boca Raton: CRC Press; 2008.

CHAPTER

55

Performance Nutrition for Young Athletes

JohnEric W. Smith[1] *and Asker Jeukendrup*[2,3]

[1]Gatorade Sports Science Institute, Bradenton, FL, USA [2]Gatorade Sports Science Institute, Barrington, IL, USA [3]Loughborough University, Loughborough, UK

INTRODUCTION

Nutrition is an integral part of all athletes' lives; however, the importance of proper nutrition is even greater for the adolescent athlete. Proper nutrition provides all athletes with fuel for energy and the foundation necessary to recover from exercise stress. With the rising rate of childhood and adolescent obesity, a great deal of attention is given to overall general youth nutrition; however, little focus is given to the nutritional needs of the young athlete. In youth athletes, nutrition still plays the same vital role observed in adult athletes but, more importantly, nutrition in the young athlete is primarily required to maintain development associated with growth and maturation [1]. Additionally, young athletes and their influencers may not fully realize/appreciate the impact nutrition has on health and performance, and therefore they make food choices that do not provide needed nutrients, they skip meals, or they fail to plan their nutrition accordingly for upcoming practices or competitions.

The role of nutrition on growth and development has been extensively studied [2]. Nutrition not meeting the minimum intake requirements can result in delayed growth and development [3]. Even in western countries with adequate access to nutrition, energy restrictions by children concerned with obesity have been shown to result in slowed growth and delayed puberty [4]. The impact of training on growth and maturation in young athletes is a commonly heard concern. Research has not found evidence to support such concerns for athletes involved in recreational and club level sports with activity consisting of more than 15 hours per week [5,6], nor in athletes participating in light to moderate resistance training twice a week [7].

It is well established that training for sport and competition increases energy, carbohydrate and protein requirements of adults. Because of the limited research examining whether similar differences exist as a result of activity level in children and adolescents, many use nutrition findings in adults to make recommendations for active youth. Young athletes have been shown to require greater caloric intakes than their non-exercising counterparts because of the increased energy expenditures related to training and competition [8]. This increase in energy needs compared with their non-athletic peers includes an increased need for protein and, likely, carbohydrate [9]. While not meeting the energy intake needs may lead to slowed growth and development, excess energy intake also leads to adverse effects. Because of the importance of nutrition for growth and development, participation in sport and physical activity is usually recommended to treat childhood obesity, rather than energy restriction [1].

Due to factors including the underdeveloped glycolytic metabolism and likely motor recruitment, young athletes have reduced exercise efficiency compared with adults [10]. Research has also shown that the ratio of fat to carbohydrate oxidation at a given relative intensity is greater in children than in adults [11]. Research has demonstrated that energy expenditure is often underestimated in youth (i.e., the energy cost of a given activity may be elevated in youth compared with adults). This higher rate of energy expenditure at similar relative workloads exists until near the end of puberty [12,13]. The underestimation of energy expenditures in different settings makes it difficult to gauge without accurate measurement the increased energy needs of young athletes during exercise [14,15].

Nutrition and Enhanced Sports Performance.
DOI: http://dx.doi.org/10.1016/B978-0-12-396454-0.00055-2

All sports require specific physiological capabilities and skills. Most of these capabilities are influenced in some form by nutritional choices. Sports nutrition is often the subject of attention during competition, but its role encompasses not only competition but also preparation for competition during training and general daily function. This chapter will highlight the current understanding of sports nutrition for young athletes and identify some of the gaps that exist in our knowledge and understanding.

CARBOHYDRATE

As in adults, fat and carbohydrate are the primary fuel sources during exercise in youth. As intensity increases, a shift is made from fat to carbohydrate metabolism. The recommendations for adults suggest athletes consume 5–12 g/kg per day of carbohydrate dependent on sport, intensity, sex, and environmental conditions [16]. This is a wide range of intake, and the exact recommended intake will depend on the activity level and the specific training goals of the athlete. There is a lack of research to determine specific daily carbohydrate recommendations for young athletes, but it is commonly suggested that at least 50% of the young athlete's total energy intake come in the form of carbohydrate [17–19]. However, it has also been pointed out that carbohydrate intake guidelines as percentages are relatively meaningless, as energy expenditures may vary significantly and carbohydrate requirements are better expressed in g/kg per day.

The ergogenic effects and importance of carbohydrate around the active occasion have been repeatedly demonstrated in adult athletes. Carbohydrate ingestion aids the athlete by maintaining blood glucose during exercise and replenishing glycogen stores during non-active periods. Reductions in both blood glucose and muscle glycogen have been shown to reduce endurance performance [20], while ingestion of carbohydrate during exercise has demonstrated the ability to maintain blood glucose and improve prolonged exercise performance. The specific mechanism by which performance is improved with carbohydrate likely varies based on type, duration, and intensity of exercise. Carbohydrate's ability to maintain performance has been reported in both endurance [21] and skill-based team sports [22,23]. To support the potential for carbohydrate's providing positive impacts through multiple means, recent research demonstrates improvements in shorter bouts of exercise performance when muscle glycogen and blood glucose levels are not compromised to an extent that will affect performance. The current theory for this beneficial effect is the potential for stimulation of an oral-glucose receptor to enhance exercise performance [24].

While quite a bit is known about the role of carbohydrate ingestion for adults during exercise, less is known about the role of carbohydrate in youth. Like their adult counterparts the exact amounts of carbohydrate needed will likely vary greatly based on the specific activity being preformed [25], as well as the intensity, and duration of that activity [8,16]. Young athletes have been shown to have lower glycogen stores than their adult counterparts. Endogenous stores of glycogen provide a great deal of carbohydrate for energy during moderate to intense exercise [26,27]. The benefits of carbohydrate loading in adults are well known in both the scientific and lay communities and can be seen through the actions of many endurance athletes. Smaller glycogen stores in children and adolescents [28] require that they rely more heavily on exogenous carbohydrate and fat for energy during exercise. Carbohydrate ingestion in the diet is important for the youth athlete, but a need for glycogen loading has not been shown and is not well understood in this population [29].

In addition to lower glycogen stores, incomplete development of metabolic pathways associated with carbohydrate metabolism results in the adolescent athlete's inability to utilize carbohydrate as effectively as adults during exercise [30]. After puberty, development of these metabolic pathways allows young athletes to metabolize carbohydrate in a manner more in line with their adult counterparts [12]. Young athletes have been shown to utilize carbohydrate for energy during heavy exercise at a rate of 1.0–1.5 g/kg/h [31]. Even with their inability to utilize carbohydrate as well as adults, simple carbohydrates will aid young athletes in supporting the energy demands of exercise [29]. The ingestion of glucose alone before [32] or during [33] endurance exercise has not been shown to improve performance in young athletes. However, 10–14-year-old boys who performed an endurance cycling test with glucose combined with fructose ingestion during prolonged exercise demonstrated a performance benefit [33]. Beyond the impact of carbohydrate on prolonged exercise performance, intermittent team sport skills have also been shown to benefit from carbohydrate ingestion. The ingestion of a carbohydrate beverage during basketball-related skills exercise resulted in improved performance in 10–15-year-old boys [22].

When carbohydrate concentration ingestion increases above 6%, there can be reductions in gastric emptying, slowing of intestinal absorption, and increases in gastrointestinal complaints [34]. In another study, ingestion of a 6% carbohydrate solution improved intermittent endurance performance. In this study, performance was enhanced to a greater extent with a 6% solution than with a 10% solution [35].

Following exercise, carbohydrate ingestion will promote the replenishment of muscle glycogen used for energy during exercise [18]. Adult recommendations suggest carbohydrate be ingested at 1–1.5 g/kg of body weight during the first 30 min following exercise and again every 2 hours afterward for 4–6 hours [8]. Due to the reduced glycogen stores in children and adolescents at the beginning of exercise, post-exercise carbohydrate replenishment is likely beneficial to aid in restoration of their glycogen stores. Even though increased levels of carbohydrate intake (resulting in excess energy intake) are a legitimate concern for inactive youth, the young athlete will benefit from increasing carbohydrate intake in line with their increased energy expenditure related to exercise [9].

PROTEIN

While carbohydrate is relied on heavily for energy production during exercise, protein intake is crucial in providing metabolic building blocks for muscle, enzymes, and other body tissues. Current recommendations for protein intake in adults are 12–15% of energy intake or 0.8–1.2 g/kg/day [17,36]. Compared with their sedentary counterparts, adult athletes have greater protein needs to maintain positive protein balance and support growth [37]. Protein is important for adult athletes to maintain normal physiological function and to support recovery from exercise but has a greater importance in youth to facilitate proper development and maturation. The World Health Organization recommends protein intakes of 0.85–0.92 g/kg of body mass from 3 to 18 years old for non-active to moderately active individuals [2]. This level of protein intake has been shown to be inadequate for adolescent athletes [27]. The American College of Sports Medicine in a joint position stand with the American Dietetics Association recommended a protein intake of 1.2–1.8 g/kg per day in adults [8,38]. This is in line with the recommendations of other authors for adolescent athletes [27,30]. One of the most common misconceptions by athletes and their coaches is that protein supplements are necessary on account of their strenuous training. However, the standard western diet has a protein intake that typically exceeds the RDA by 2–3-fold [39], providing more than the additional needs associated with exercise [27,30,40]. Because of this elevated protein intake, additional supplementation does not likely provide benefit.

Protein intake shortly before exercise is not recommended, due to its slow gastric emptying times. Recent research in adults has demonstrated mixed results for ergogenic benefits with protein ingestion during exercise [41,42]. Therefore the role of protein ingestion during exercise not only needs to be studied in children and adolescents, but more research is also required in adults before clear conclusions can be made in regard to its potential benefits.

Unlike its role during exercise, the role of protein after exercise is better understood. Protein ingestion following exercise serves to maintain positive protein balance and support recovery from exercise. Research suggests ~20 g of intact protein and/or 9 g of essential amino acids should be ingested shortly after exercise to maximize muscle protein synthesis and inhibit protein breakdown, consequently leaving the body in a positive protein balance [43]. Specific needs for the young athlete are not available but, with the similarities found in protein needs for adult and youth athletes, it can be hypothesized that this would likely be sufficient for young athletes as well.

FAT

Fat has a crucial role as fuel for energy production, as a facilitator of fat-soluble vitamin uptake, in the maintenance of cell membranes, and as an insulator to provide aid in the protection of vital organs as well as thermoregulation. Less is known in regard to specific fat requirements of young athletes as compared with carbohydrate and protein. Research has shown young athletes to be more reliant on fat metabolism during exercise than adults [1]. It is recommended that 25–30% of the total energy intake of children or adolescents come from fat [15]. The need to focus on consuming enough fat is not as great a concern as it is for other macronutrients. Increases in energy intake, to meet energy needs and support carbohydrate and protein requirements, will usually add fat to the total intake. When insufficient fat is present in the diet, there can also be a deficit in energy intake, resulting in decrements in growth and development [40]. Fat intakes greater than 30% may result in an excess energy intake, promoting weight gain or poorly balanced nutrient intake; however, reducing energy intake to less than 20% fat does not provide additional performance benefits [8]. Fat ingestion prior to exercise has been shown to reduce carbohydrate metabolism but does not provide a benefit during exercise [44]. Even with young athletes' increased reliance on fat metabolism during exercise, fat ingestion before exercise is not recommended, due to its lack of ergogenic benefit and potential negative effects on gastric emptying and absorption of fluid and carbohydrate. It also has the potential to reduce growth hormone secretion during exercise, resulting in negative effects on adaptation [45].

MICRONUTRIENTS

Inadequate energy intake is often associated with an inadequate intake of micronutrients. Micronutrients are important for energy production, bone health, and maintenance of other body functions [18]. Any increased need for certain micronutrients by a young athlete is often met through increased energy intake. However, a few micronutrients may still be inadequately consumed. Commonly inadequately consumed micronutrients are calcium and iron [1]. Calcium is crucial in its role in the maintenance of bone mass. Calcium intake during childhood and adolescence is important for building the foundation of future bone mass, and an inadequate calcium intake has been associated with stress injuries [36]. A calcium intake of 800–1200 mg/day is recommended for individuals aged 6–10 years and an intake of 1200–1500 mg/day for individuals aged 11–24 years [46]. Iron is needed for the formation of hemoglobin to maintain the oxygen-carrying capabilities of the blood. It is recommended that adolescents consume 18 mg of iron per day [18]. Iron deficiencies have also been reported in young athletes, with higher incidences occurring in males between 11 and 14 years old and females 15–19 years old [47]. These iron deficiencies are likely the result of low iron intake combined with muscle growth in the males and menstrual losses in the females.

Specific to the young athlete, as opposed to their sedentary counterpart, sodium and potassium intake is important to offset losses in sweat. This applies even though young athletes typically lose less salt in their sweat due to their activity levels than do adults. The cumulative sweat loss from these small losses can result in overall losses equivalent to daily intakes [36]. This deficit can be partially offset by the ingestion of sports drinks during the active occasion and with additional salting of food and dietary inclusion of salty foods.

HYDRATION

As with other aspects of nutrition, proper hydration is often underemphasized in youth populations. While adults are usually in situations that allow personal choice in accessing fluid, it is not uncommon for young people to face situations of fluid restriction based on school/activity rules and/or the inability to have a direct influence on situational decisions. As awareness of the value of hydration on cognitive performance increases [48], the prevalence of these restrictions is declining.

Maintaining an appropriate fluid intake is important for performance and safety purposes. Nonathletic children have been reported to have daily fluid turnover rate of ~1.6 L/day, about 50–75% of the turnover rate seen in adults [49]. In both adult and young athletes an increased fluid intake is needed to account for fluid losses associated with sweating. Prolonged exercise performance and changes in core body temperature have been positively influenced by fluid ingestion as compared with dehydrated situations. However, overhydrating has been linked to life-threatening cases of hyponatremia [8]. For these reasons, appropriate fluid intake requires that an athlete ingest enough fluid to offset some of the fluid losses associated with sweating, whilst not ingesting more fluid than the body is losing.

The difficulty in applying lessons learnt from adult athletes to youth athletes is increased by youth's increased participation in team sports and lower participation in endurance sport, which are the most commonly studied activities in adults. That being said, the limited amount of research that has been done in team sport athletes has also shown decrements in performance with dehydration [50]. Research on young basketball players has demonstrated decrements in skill performance with 2% reductions in body mass [22].

Large variability in sweat rates exists between athletes as a result both of genetics and of acclimatization. Sweat rates of 0.3– ~5.0 L/h have been reported in adults [8,36]. Due to youth and adolescents' incompletely developed sweating mechanism, young athletes do not typically sweat as much as their adult counterparts [51], with sweat rates of 0.5 L/h being reported during exercise [52]. As mentioned previously, children are not as efficient as adults working at the same relative intensity, thus adding to their thermal load during activity [12]. Along with the increased thermal load related to increased relative workload, children's increased surface area to body mass ratio increases the risk of heat-injury and illness when environmental conditions reduce the athlete's ability to dissipate heat [53,54].

Proper fluid intake is important but, due to large variations in genetics, acclimatization, and changes in sweat rate as a result of maturation, it is ill advised to give an overarching quantifiable recommendation regarding proper fluid intake. It is recommended that both adult and young athletes consistently monitor the amount of body weight change that occurs during practice and conditioning to ensure adequate hydration. If large weight reductions occur over an exercise/training session, athletes should increase exercise fluid intakes. If weight is gained during exercise/training sessions, athletes should reduce fluid intake accordingly.

SUPPLEMENTS

Supplement use in young athletes is becoming increasingly common as athletes look to gain performance advantages. The lack of a benefit from supplementing protein above the recommended daily levels has already been discussed. Other supplements commonly sought by young athletes include creatine, caffeine, and multi-vitamins.

Creatine has been estimated to be used by 25–75% of collegiate athletes [19] and has been shown to provide ergogenic effects to training adult athletes [55]. Little is known about the long-term effects of creatine supplementation in young athletes; therefore the American College of Sports Medicine Roundtable on creatine supplementation states: "creatine monohydrate is not recommended for children under 18 years of age" [55].

Also commonly used in sport, caffeine has been studied extensively in adults related to its impact on endurance performance and vigilance. While many of these studies have demonstrated an ergogenic effect for caffeine [56], high levels have also been associated with adverse effects. Due to differences in maturation, children may be more vulnerable to the adverse effects of caffeine than are adults and should likely avoid its use for performance enhancement [39].

The use of multi-vitamins to support an athlete in meeting their daily need for nutrient intake is common. If the athlete is consuming a well-balanced diet which meets their energy needs, a multi-vitamin will likely not provide additional benefit. In summary, supplements are not recommended for young athletes unless there is a clinical need [30].

GAPS IN OUR KNOWLEDGE

What we know about sports nutrition specifically for child and adolescent athletes is actually quite limited. Our current knowledge is based on a small number of studies and on the combination of general nutritional recommendations for children and adolescents, on the one hand, with differences between the nutritional needs of adult athletes and those of the general adult population. In attempting to further our understanding, it would likely be ill-advised to conduct certain nutrition studies requiring energy restriction on the developing bodies of children and adolescents [29]. However, other types of nutritional interventions and observational studies could provide greater insight into the differences that exist, not only between the active and inactive adolescent and child, but also the differences that exist between active younger and older athletes.

PUTTING IT ALL TOGETHER

As with adults, child and adolescent athletes require alterations to their nutrition to account for the demands of their physical activity. These alterations include an increase in energy intake, which if done properly will adequately provide the increases in protein and carbohydrate intakes that are needed to support adaptations associated with exercise and provide the nutrients necessary to maintain growth and development. Sports nutrition needs vary based on the demands of specific sports and activities. Based on our current knowledge, the needs of a young athlete do not appear to differ greatly from those of their adult counterparts within similar sporting environments. Therefore, in general, recommendations for adult athletes are not too different from those for young athletes with similar sporting demands, until future research provides more detailed insights.

The benefits of carbohydrate ingestion before exercise have not been studied thoroughly in young athletes. However, some carbohydrate ingestion prior to exercise would likely be beneficial for athletes participating in high-intensity training, because of young athletes' lower stores of endogenous carbohydrate. During prolonged exercise, carbohydrate ingestion may provide ergogenic effects. While consistent benefits are observed in adults, more research needs to be done in children and adolescents. Following exercise, carbohydrate ingestion should be provided to help promote muscle glycogen replenishment. Refer to Table 55.1 as a quick reference guide to the recommended nutritional intakes.

In addition to carbohydrate, proper protein ingestion is critical for young athletes as their bodies recover from exercise stress in addition to the needs associated with growth and development as part of normal maturation. While in the process of meeting carbohydrate and protein needs, fat and micronutrient intake should be sufficiently met by normal dietary food intake from a well-balanced diet. In addition to energy intake, young athletes should give special attention to hydration. Fluid losses associated with exercise must be accounted for, whilst also learning to minimize the adverse effects associated with dehydration. Finally, the use of supplements is not recommended for young athletes unless clinically advised.

Disclaimer

JohnEric W. Smith and Asker Jeukendrup are employees of PepsiCo Inc. The views expressed in this article are those of the authors and do not necessarily reflect the position or policy of PepsiCo, Inc.

TABLE 55.1 Nutrition Reference Guide

	Protein	Fat	Carbohydrate	Micronutrients
Daily intake	1.2–1.8 g/kg per day in adults [8,38]	Balance of daily energy needs	5–12 g/kg per day of carbohydrate dependent on sport, intensity, sex, and environmental conditions [16]	Calcium: 800–1200 mg/day for individuals 6–10 yr and 1200–1500 mg/day for individuals 11–24 yr [46]. Iron: 18 mg/day [18] Sodium and potassium: Liberal salting of food and salty snacks
Within 90–120 min of beginning exercise	Not recommended shortly before exercise	Not recommended shortly before exercise	No need to avoid	No specific recommendation
During exercise	Mixed findings; likely no benefit	No benefit	30–60 g/h for exercise lasting longer than 60 min [8]; for endurance exercise lasting more than 2.5 h higher carbohydrate intakes may be advantageous [16]	Sodium to offset sodium lost in sweat
Post exercise	~20 g of intact protein or 9 g of essential amino acids shortly after exercise [43]	No specific recommendation	1–1.5 g/kg body weight during the first 30 minutes and again every 2 h afterward for 4–6 h [8]	Sodium to offset sodium lost in sweat

References

[1] Thompson JL. Energy balance in young athletes. Int J Sport Nutr 1998;8(2):160–74.

[2] Consultation FWUE. Protein and amino acid requirements in human nutrition. Geneva, Switzerland; 2002.

[3] Borer KT. The effects of exercise on growth. Sports Med 1995;20 (6):375–97.

[4] Pugliese MT, Lifshitz F, Grad G, Fort P, Marks-Katz M. Fear of obesity. A cause of short stature and delayed puberty. N Engl J Med 1983;309(9):513–8.

[5] Bonen A. Recreational exercise does not impair menstrual cycles: a prospective study. Int J Sports Med 1992;13(2):110–20.

[6] Fogelholm M, Rankinen T, Isokaanta M, Kujala U, Uusitupa M. Growth, dietary intake, and trace element status in pubescent athletes and schoolchildren. Med Sci Sports Exerc 2000;32(4):738–46.

[7] Sandres E, Eliakim A, Constantini N, Lidor R, Falk B. The effect of long-term resistance training on anthropometric measures, muscle strength and self-concept in pre-pubertal boys. Pediatr Exerc Sci 2001;13:357–72.

[8] Rodriguez NR, Di Marco NM, Langley S. American college of sports medicine position stand. Nutrition and athletic performance. Med Sci Sports Exerc 2009;41(3):709–31.

[9] Mundt CA, Baxter-Jones AD, Whiting SJ, Bailey DA, Faulkner RA, Mirwald RL. Relationships of activity and sugar drink intake on fat mass development in youths. Med Sci Sports Exerc 2006;38(7):1245–54.

[10] Bar-Or O. Nutritional considerations for the child athlete. Can J Appl Physiol 2001;26(Suppl.):S186–91.

[11] Duncan GE, Howley ET. Substrate metabolism during exercise in children and the crossover concept. Pediatr Exerc Sci 1999;11:12–21.

[12] Harrell JS, McMurray RG, Baggett CD, Pennell ML, Pearce PF, Bangdiwala SI. Energy costs of physical activities in children and adolescents. Med Sci Sports Exerc 2005;37(2):329–36.

[13] Haymes EM, Buskirk ER, Hodgson JL, Lundegren HM, Nicholas WC. Heat tolerance of exercising lean and heavy prepubertal girls. J Appl Physiol 1974;36(5):566–71.

[14] Blaak EE, Westerterp KR, Bar-Or O, Wouters LJ, Saris WH. Total energy expenditure and spontaneous activity in relation to training in obese boys. Am J Clin Nutr 1992;55(4):777–82.

[15] Bolster DR, Pikosky MA, McCarthy LM, Rodriguez NR. Exercise affects protein utilization in healthy children. J Nutr 2001;131(10):2659–63.

[16] Burke LM, Hawley JA, Wong SH, Jeukendrup AE. Carbohydrates for training and competition. J Sports Sci 2011;29(Suppl. 1):S17–27.

[17] Bonci L. Sports nutrition for young athletes. Pediatr Ann 2010;39(5):300–6.

[18] Cotunga N, Vickery CE, McBee S. Sports nutrition for young athletes. J Sch Nurs 2005;21(6):323–8.

[19] Nemet D, Eliakim A. Pediatric sports nutrition: an update. Curr Opin Clin Nutr Metab Care 2009;12(3):304–9.

[20] Jeukendrup AE. Carbohydrate feeding during exercise. Eur J Sport Sci 2008;8(2):77–86.

[21] Jeukendrup AE. Carbohydrate intake during exercise and performance. Nutrition 2004;20(7–8):669–77.

[22] Dougherty KA, Baker LB, Chow M, Kenney WL. Two percent dehydration impairs and six percent carbohydrate drink improves boys basketball skills. Med Sci Sports Exerc 2006;38 (9):1650–8.

[23] Ali A, Williams C, Nicholas CW, Foskett A. The influence of carbohydrate-electrolyte ingestion on soccer skill performance. Med Sci Sports Exerc 2007;39(11):1969–76.

[24] Carter JM, Jeukendrup AE, Jones DA. The effect of carbohydrate mouth rinse on 1-h cycle time trial performance. Med Sci Sports Exerc 2004;36(12):2107–11.

[25] Christmass MA, Dawson B, Passeretto P, Arthur PG. A comparison of skeletal muscle oxygenation and fuel use in sustained continuous and intermittent exercise. Eur J Appl Physiol Occup Physiol 1999;80(5):423–35.

[26] Aucouturier J, Baker JS, Duche P. Fat and carbohydrate metabolism during submaximal exercise in children. Sports Med 2008;38(3):213–38.

[27] Boisseau N, Vermorel M, Rance M, Duche P, Patureau-Mirand P. Protein requirements in male adolescent soccer players. Eur J Appl Physiol 2007;100(1):27–33.

[28] Boisseau N, Delamarche P. Metabolic and hormonal responses to exercise in children and adolescents. Sports Med 2000;30 (6):405–22.
[29] Burke LM, Millet G, Tarnopolsky MA. Nutrition for distance events. J Sports Sci 2007;25(Suppl. 1):S29–38.
[30] Meyer F, O'Connor H, Shirreffs SM. Nutrition for the young athlete. J Sports Sci 2007;25(Suppl. 1):S73–82.
[31] Riddell MC, Bar-Or O, Schwarcz HP, Heigenhauser GJ. Substrate utilization in boys during exercise with [13C]-glucose ingestion. Eur J Appl Physiol 2000;83(4–5):441–8.
[32] Hendelman DL, Ornstein K, Debold EP, Volpe SL, Freedson PS. Preexercise feeding in untrained adolescent boys does not affect responses to endurance exercise or performance. Int J Sports Nutr 1997;7:207–18.
[33] Riddell MC, Bar-Or O, Wilk B, Parolin ML, Heigenhauser GJ. Substrate utilization during exercise with glucose and glucose plus fructose ingestion in boys ages 10–14 yr. J Appl Physiol 2001;90(3):903–11.
[34] Shi X, Horn MK, Osterberg KL, Stofan JR, Zachwieja JJ, Horswill CA, et al. Gastrointestinal discomfort during intermittent high-intensity exercise: effect of carbohydrate-electrolyte beverage. Int J Sport Nutr Exerc Metab 2004;14(6):673–83.
[35] Phillips SM, Turner AP, Sanderson MF, Sproule J. Beverage carbohydrate concentration influences the intermittent endurance capacity of adolescent team games players during prolonged intermittent running. Eur J Appl Physiol 2012;112 (3):1107–16.
[36] Petrie HJ, Stover EA, Horswill CA. Nutritional concerns for the child and adolescent competitor. Nutrition 2004;20(7–8):620–31.
[37] Tipton KD, Jeukendrup AE, Hespel P. Nutrition for the sprinter. J Sports Sci 2007;25(Suppl. 1):S5–15.
[38] Phillips SM, Van Loon LJ. Dietary protein for athletes: from requirements to optimum adaptation. J Sports Sci 2011;29 (Suppl. 1):S29–38.
[39] Jeukendrup A, Cronin L. Nutrition and elite young athletes. Med Sport Sci 2011;56:47–58.
[40] Bresson JL. Protein and energy requirements in healthy and ill paediatric patients. Baillières Clin Gastroenterol 1998;12(4):631–45.
[41] Osterberg KL, Zachwieja JJ, Smith JW. Carbohydrate and carbohydrate + protein for cycling time-trial performance. J Sports Sci 2008;26(3):227–33.
[42] Saunders MJ, Moore RW, Kies AK, Luden ND, Pratt CA. Carbohydrate and protein hydrolysate coingestions improvement of late-exercise time-trial performance. Int J Sport Nutr Exerc Metab 2009;19(2):136–49.
[43] Beelen M, Burke LM, Gibala MJ, van Loon LJ. Nutritional strategies to promote postexercise recovery. Int J Sport Nutr Exerc Metab 2010;20(6):515–32.
[44] Hargreaves M, Hawley JA, Jeukendrup A. Pre-exercise carbohydrate and fat ingestion: effects on metabolism and performance. J Sports Sci 2004;22(1):31–8.
[45] Galassetti P, Larson J, Iwanaga K, Salsberg SL, Eliakim A, Pontello A. Effect of a high-fat meal on the growth hormone response to exercise in children. J Pediatr Endocrinol Metab 2006;19(6):777–86.
[46] NIH Consensus conference. Optimal calcium intake. NIH consensus development panel on optimal calcium intake. JAMA 1994;272(24):1942–48.
[47] Wahl R. Nutrition in the adolescent. Pediatr Ann 1999;28 (2):107–11.
[48] D'Anci KE, Constant F, Rosenberg IH. Hydration and cognitive function in children. Nutr Rev 2006;64(10 Pt 1):457–64.
[49] Balluff A, Kersting M, Manz F. Do children have adequate fluid intake? Water balance studies carried our at home. Ann Nutr Metab 1988;68:17–22.
[50] Baker LB, Conroy DE, Kenney WL. Dehydration impairs vigilance-related attention in male basketball players. Med Sci Sports Exerc 2007;39(6):976–83.
[51] Bergeron MF, McKeag DB, Casa DJ, Clarkson PM, Dick RW, Eichner ER, et al. Youth football: heat stress and injury risk. Med Sci Sports Exerc 2005;37(8):1421–30.
[52] Rivera-Brown AM, Gutierrez R, Gutierrez JC, Frontera WR, Bar-Or O. Drink composition, voluntary drinking, and fluid balance in exercising, trained, heat-acclimatized boys. J Appl Physiol 1999;86(1):78–84.
[53] Bar-Or O. Temperature regulation during exercise in children and adolescents. In: Gisolfi CV, Lamb DR, editors. Youth, exercise and sport. Indianapolis, IN: Benchmark Press; 1989. p. 335–67.
[54] Godek SF, Godek JJ, Bartolozzi AR. Hydration status in college football players during consecutive days of twice-a-day preseason practices. Am J Sports Med 2005;33(6):843–51.
[55] Terjung RL, Clarkson P, Eichner ER, Greenhaff PL, Hespel PJ, Israel RG, et al. American college of sports medicine roundtable. The physiological and health effects of oral creatine supplementation. Med Sci Sports Exerc 2000;32(3):706–17.
[56] Graham TE. Caffeine and exercise: metabolism, endurance and performance. Sports Med 2001;31(11):785–807.

SECTION 7

CONCLUDING REMARKS

CHAPTER

56

Commentary

Debasis Bagchi[1], *Sreejayan Nair*[2] *and Chandan K. Sen*[3]

[1]University of Houston College of Pharmacy, Houston, TX, USA [2]University of Wyoming School of Pharmacy, Laramie, WY, USA [3]The Ohio State University Wexner Medical Center, Columbus, OH, USA

The urge to make the most of the human body was nested in the cradle of the most ancient human civilizations. In the vicinity of the noted Sphinx of Giza, the Stela of Amenophis II is a landmark that reminds us about sports in ancient Egypt. Archery, running, rowing and love for horses were the pride of Amenophis II. Sculptures and paintings show that ancient Egyptians were fit, with strong muscular men and slender erect women. The mural of the Zoser (5000–3000 BC) depicts the Pharaoh participating in the running program of the Heb Sed festival. On a wall of her sanctuary in the Karnak Temple, Queen Hatshepsut of the 18th dynasty is represented in a similar attitude in the Heb Sed. Heb Sed was one of the oldest feasts of ancient Egypt, celebrated by the king after 30 years of rule and repeated every 3 years thereafter. In India, the practice of sports dates back to the Vedic era. The Vedic era (1200–500 BC) was a period during which the Vedas, the oldest scriptures of Hinduism, were composed. Martial arts such as *Karate* and *Kung Fu* are rooted in India. In the 5th century a Buddhist monk from India named Bodhidharma (the Chinese called him Po-ti-tama) introduced Kalari into China and Japan. The temple in which he taught his art is now known as the Shaolin temple. The *Rig Veda* (1200–900 BC) places emphasis on proper food and diet for good health. In 1889, Ralph T.H. Griffith published in London his translation as The Hymns of the Rig Veda. In praise of food, Griffith translated the hymn "*In thee, O Food, is set the spirit of great Gods. Under thy flag brave deeds were done*". The vedas placed equal emphasis on eating right as well as on fasting or caloric restriction.

In the history of the Olympics, alcohol was commonly taken as an ergogenic aid through the early 1900s. The legendary Milo of Croton, the wrestler who won five successive Olympic Games from 532 to 516 BC, ate 20 pounds of meat, 20 pounds of bread and 8.5 liters of wine a day. The Olympic gladiators' meal plan inspired the discipline of sports nutrition. The Roman gladiator meal was rich in protein from different sources, cereals and vegetables. Carbohydrate loading was achieved by eating fermented bread made of farro (a Roman cereal) and a soup made of farro and orzo. Barley was commonly regarded as a source of strength and stamina. Roasted meat, dry fruits, fresh cheese, goat milk and eggs were the primary sources of protein. Onions, garlic, wild lettuce and dill dominated among the vegetables. In particular, onions had a special place in ancient Greek sports nutrition as they were thought to "lighten the balance of the blood". After Rome conquered Greece, onion became a staple in the Roman diet. Roman gladiators were rubbed down with onion juice to "firm up the muscles". Olive oil was used frequently with meals. Fried cakes, boiled meat and cold drinks were prohibited. A great snack for energy in the Roman gladiator's diet was goats milk with honey and walnuts. At the public banquet *Coena Libera*, or the last dinner before the fight, athletes "stuffed themselves" so much that the dinner lasted many hours. They were advised to chew their food well in order to extract the maximum energy from it! Much of what was practiced in those days can be justified by today's science! Compared with the average inhabitant of Ephesus, gladiators ate more plants and very little animal protein, consistent with the advocacy for meat-limited diet in the vedas. Those were the days when nutritional choices were driven by desirable outcomes and not by commercial interests.

As the global sports nutrition industry rapidly approaches the $100 billion dollar mark, most products

Nutrition and Enhanced Sports Performance.
DOI: http://dx.doi.org/10.1016/B978-0-12-396454-0.00060-6

seem to be driven more by marketing strategies than by rigorous research and development. Sports nutrition supplements (powders, pills and "hardcore" bodybuilding ready-to-drink products), nutrition bars and gels, and sports and energy drinks and shots currently flood the market. Most solutions seem to be better marketed than developed through proper basic science research and clinical trials. Picking the most fitting solutions from an overcrowded marketplace represents a serious challenge and will depend on the scientific awareness of the consumer. It is our responsibility to verify the validity of claims made in advertisements. Of highest priority is to minimize potential risk to health in response to long-term use of any product. Next, potential benefits need to be critically evaluated. Both safety and efficacy can be established only through well controlled studies. Original research articles in established high impact peer-reviewed journals provide a reliable source of information. Look for registered clinical trial data in clinicaltrials.gov, and do not be misled by attractively delivered marketing gimmicks unsupported by peer-reviewed research publications. We are pleased to have been able to put together this digest aimed at providing general guidance to the physically active. Each chapter is supported by references to the source articles which readers may want to consult in developing their own opinion about a nutritional solution. We hope the readers enjoy this volume as much as we have enjoyed putting it together!

Index

Note: Page numbers followed by "*f*", "*t*" and "*b*" refers to figures, tables and boxes respectively.

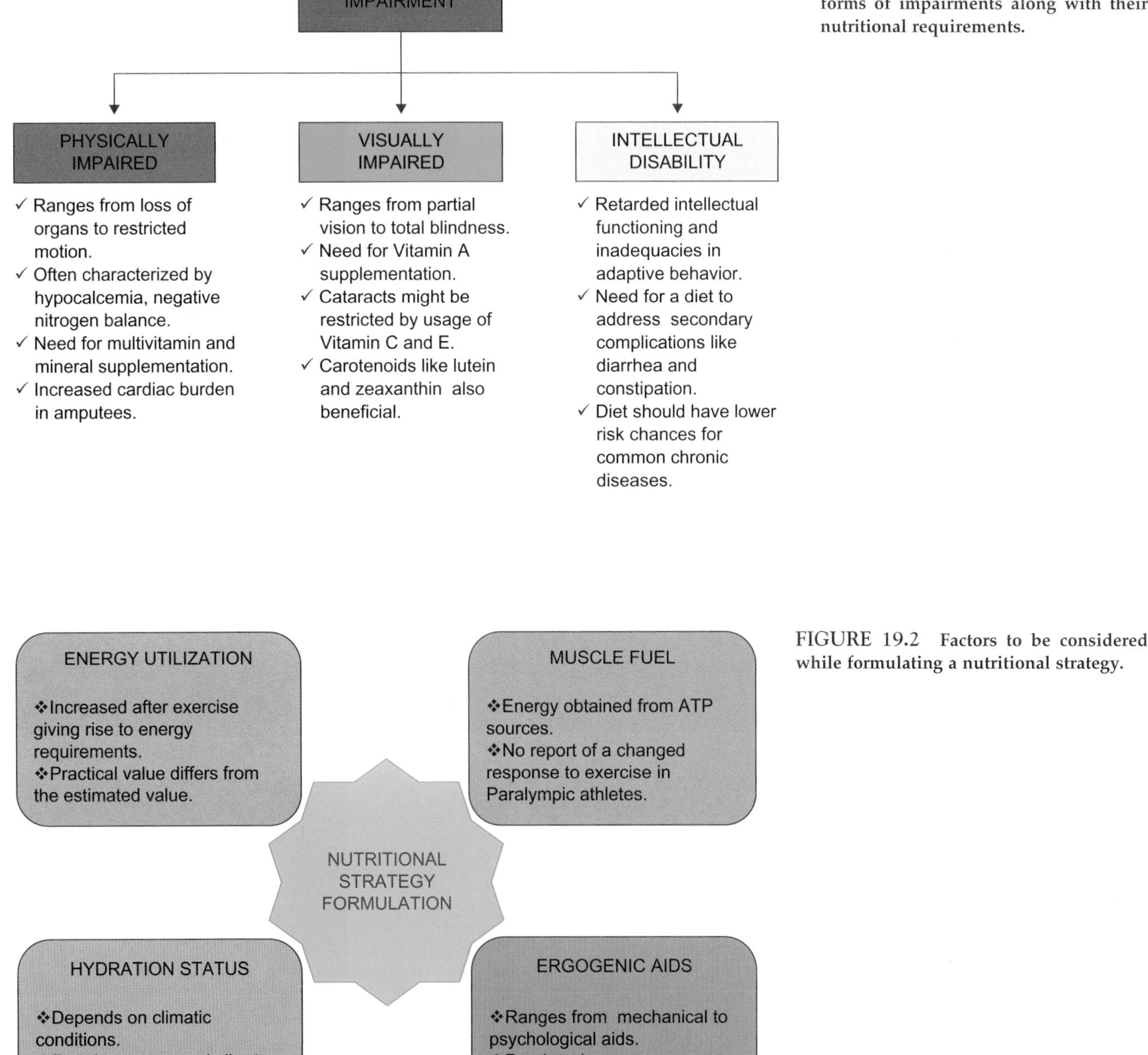

FIGURE 19.1 **Summary of different forms of impairments along with their nutritional requirements.**

FIGURE 19.2 **Factors to be considered while formulating a nutritional strategy.**

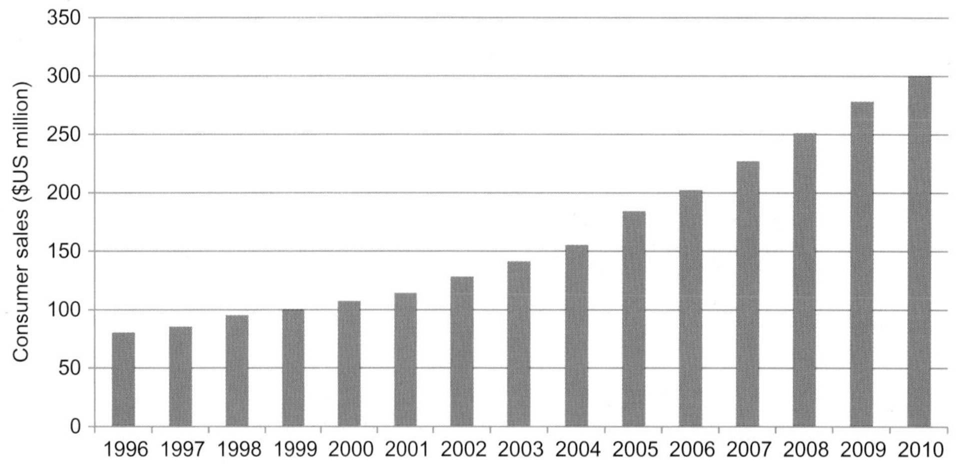

FIGURE 20.1 **Sales of Hardcore Sports Drinks in the USA 1996–2010 [20].**

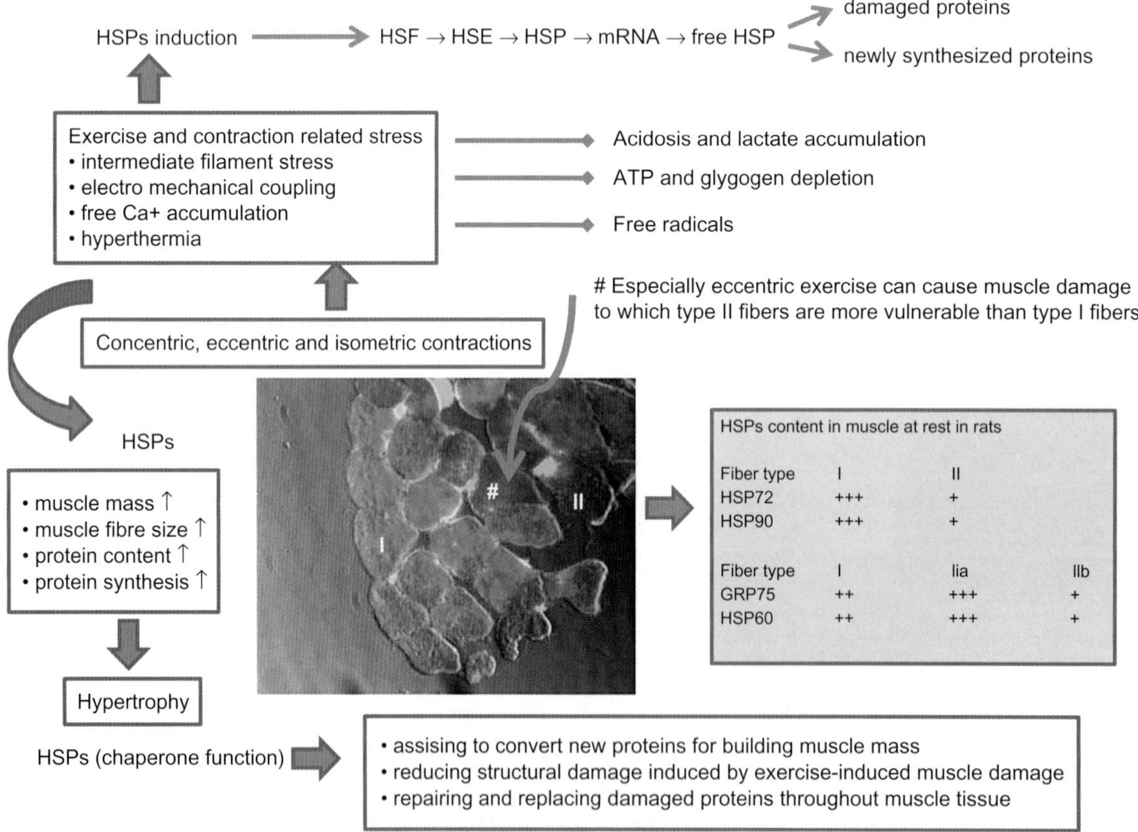

FIGURE 23.1 **Exercise-related factors that can induce HSPs in the skeletal muscle and HSP protein contents at rest.** A differential interference contrast microscopy (DIC) image of frozen section from vastus lateralis muscle of human subjects was double-stained with slow myosin heavy chain antibody (type I fiber) with red fluorochrome and anti-nitrotyrosine antibody (green) using Zeiss Axiovert 200 M (20× magnification). Type II (fast myosin heavy chain) fibers are unstained. Ca +, calcium; GRP75, glucose regulated protein; HSE, heat shock elements; HSF, heat shock factor; HSP, heat shock protein.

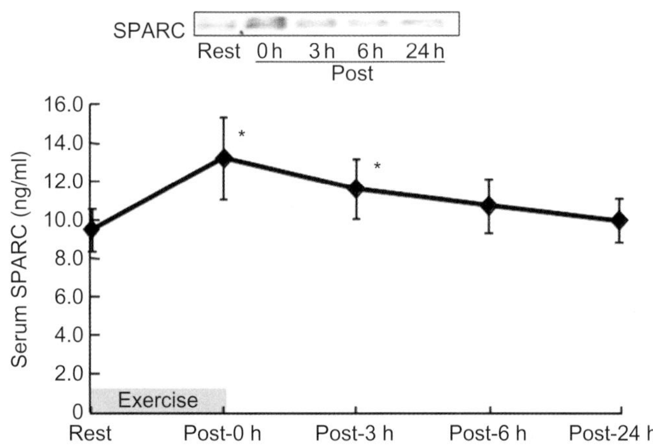

FIGURE 30.1 A single bout of exercise increases circulating levels of SPARC in humans. Time course of serum SPARC level after steady-state cycling at 70% maximal oxygen uptake (VO_{2max}) for 30 min (n = 10). *$P < 0.05$ versus resting state (Rest). Results are shown as mean ± standard error. *Data from Aoi et al. [22].*

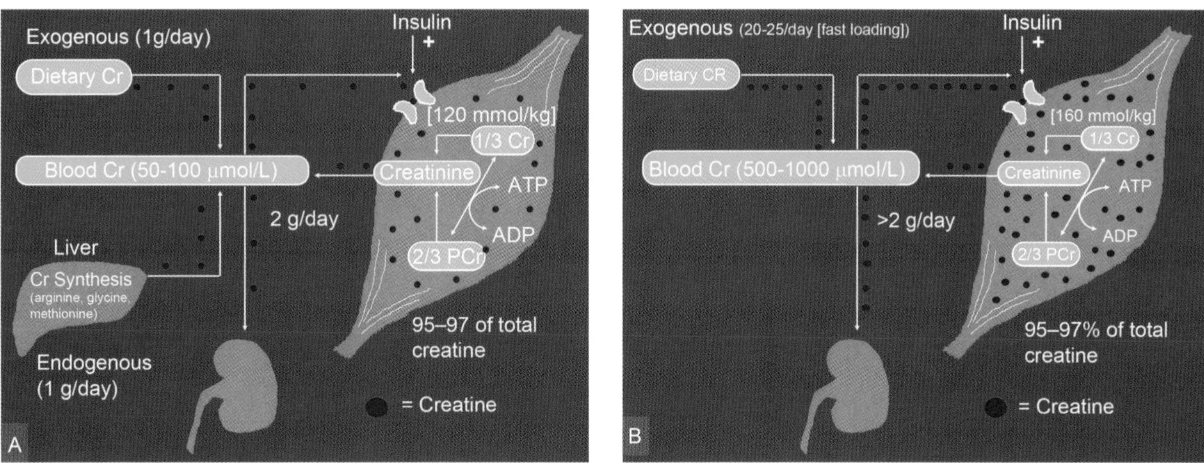

FIGURE 39.1 Cybernetic interactions between creatine and muscle. (A) Normal creatine dynamics without supplementation. (B) Creatine cybernetics with supplementation. Following supplementation, creatine from the liver plays a diminished role, but normal function returns after stopping supplementation.

FIGURE 39.2 While many dieticians want to use natural food to produce the effects that supplements do (pills or powders) the amount of meat that would have to be consumed for 3 to 6 days for a fast-loading phase for creatine is not possible, as demonstrated in this comparison of steaks, pills and powder to equate to the 25 grams per day that would be ingested (e.g., 5 grams five times a day).

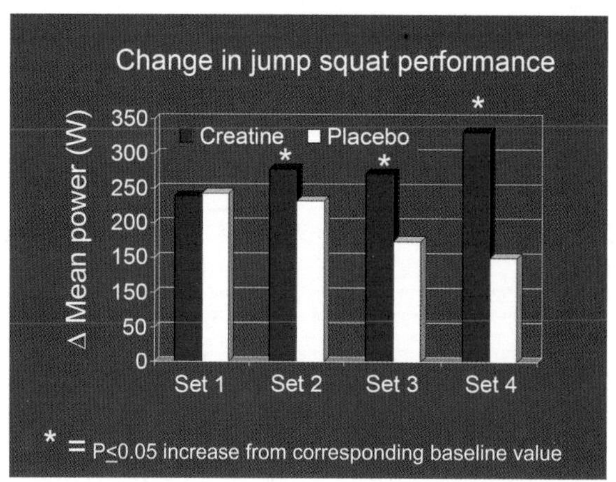

FIGURE 39.4 **Acute response of squat jump performances with and without short-term creatine loading.** *Modified from Volek and Kraemer [6].*

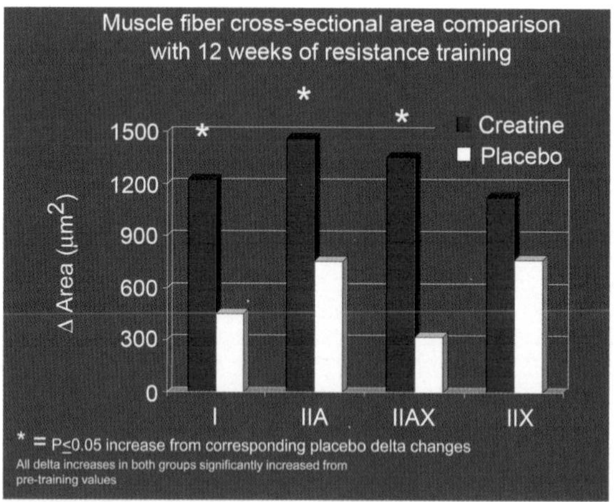

FIGURE 39.5 **Changes in muscle fiber cross-sectional areas of Type I and Type II muscle fibers with resistance training with or without creatine supplementation.** *Modified from Volek et al. [25].*

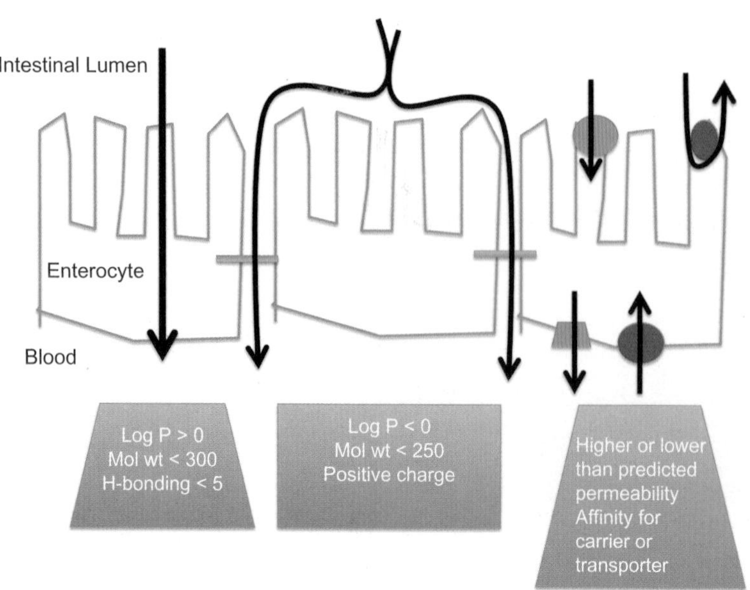

FIGURE 40.1 **Schematic of different solute absorption pathways in the gastrointestinal tract and the general properties governing absorption through each specific pathway.**

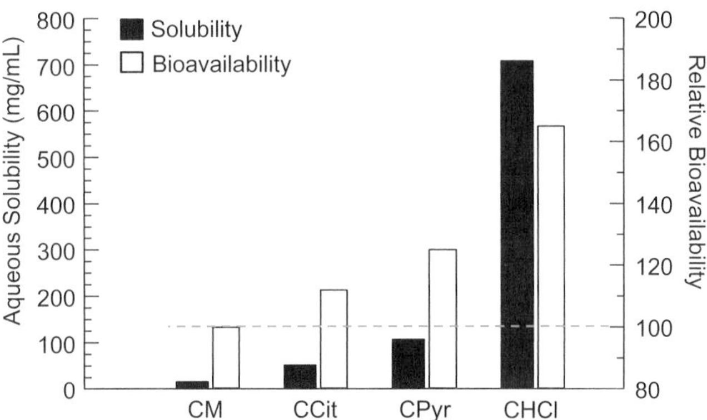

FIGURE 40.2 **Comparison of aqueous solubility and human bioavailability of various creatine salt forms.** The mean aqueous solubility values were obtained from [40]. Relative bioavailability was determined from the reported areas under the curve (AUC) for plasma creatine obtained with various creatine salt forms compared with that of creatine monohydrate (CM). The AUC values were taken from data in [49] for creatine citrate (CCit) and creatine pyruvate (CPyr) and [50] for creatine hydrochloride (CHCl).

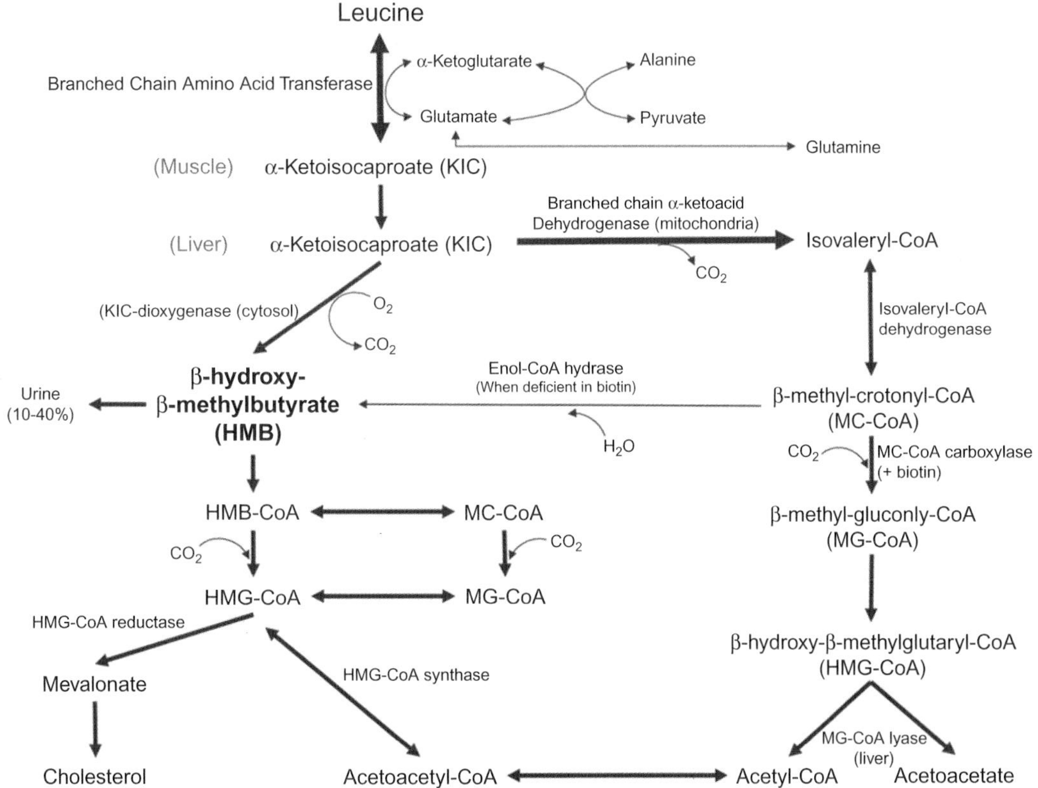

FIGURE 47.1 **HMB metabolism.** *Modified from Nissen and Abumrad [2].*

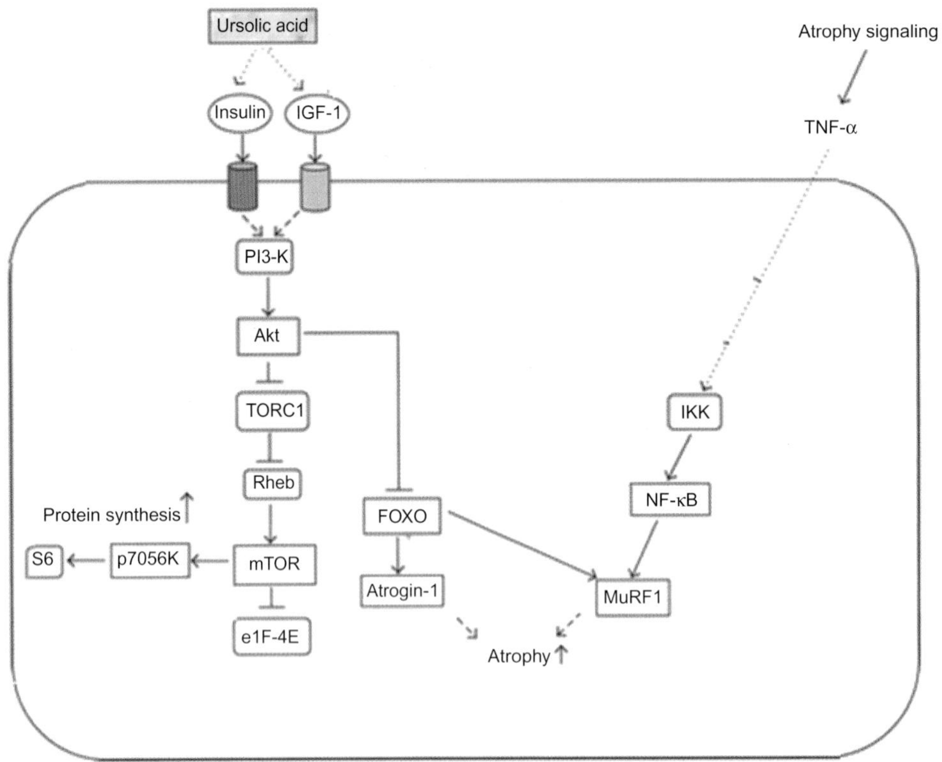

FIGURE 49.2 **Anabolic and anti-catabolic pathways relevant to ursolic acid.** Supplementation with ursolic acid up-regulates the amount of IGF-1 and insulin, stimulating protein synthesis and inhibiting atrophy via the Akt pathway.

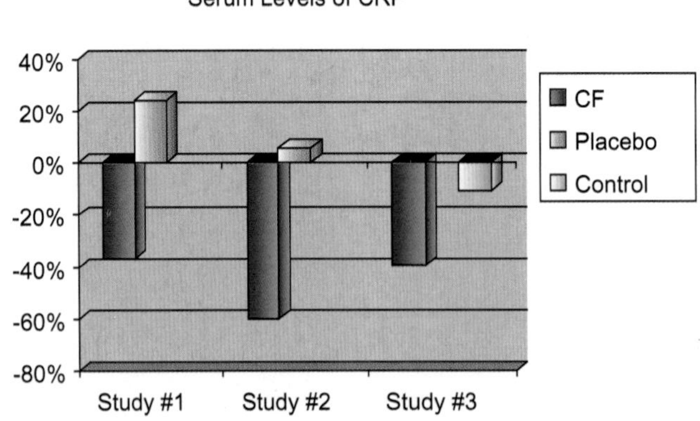

FIGURE 50.1 **Reduction of serum levels of C-reactive protein by calcium fructoborate (CF).** *Data from: Study #1 [6]; Study #2 [5]; Study #3 [63].*